Magnesium Technology 2015

TMS2015

144th Annual Meeting & Exhibition

March 15-19, 2015 • Walt Disney World • Orlando, Florida, USA

Magnesium Technology 2015

Proceedings of a symposium sponsored by

Magnesium Committee of the Light Metals Division of

The Minerals, Metals & Materials Society (TMS)

held during

TMS2015

144th Annual Meeting & Exhibition

March 15-19, 2015

Walt Disney World • Orlando, Florida, USA

Edited by:

Michele V. Manuel • Alok Singh

Martyn Alderman • Neale R. Neelameggham

Editors
Michele V. Manuel
Alok Singh
Martyn Alderman
Neale R. Neelameggham

ISBN 978-3-319-48611-6
ISBN 978-3-319-48185-2 (eBook)
DOI 10.1007/978-3-319-48185-2

Chemistry and Materials Science: Professional

Published by Springer International Publishers, Switzerland, 2016
Reprint of the original edition published by John Wiley & Sons, Inc., 2015, 978-1-119-08243-9

Printed on acid-free paper

This Springer imprint is published by Springer Nature
The registered company is Springer International Publishing AG
The registered company address is: Gewerbestrasse 11, 6330 Cham, Switzerland

TABLE OF CONTENTS
Magnesium Technology 2015

Keynote Session

Elevated Temperature and Creep

Primary, Sustainability, Recycling and Processing

Deformation I

Deformation II

Wrought

Forming and Alloy Design

Corrosion, Coatings, Fatigue, and Fracture

Casting and Metal Matrix Composites

Biomedical Applications

Functional and Emerging Alloys

Strip Casting of Light Metals

Modeling and Properties

Process Technology

PREFACE

This is the 16th annual volume of papers from the Magnesium Technology symposium, which contains all aspects of magnesium technology, ranging from primary production to applications and recycling, as well as basic research to industrialization. Continuing with the longstanding tradition of presenting the most recent and highest quality work on magnesium, all presenters have submitted papers to this edited proceedings volume or other TMS-sponsored peer-reviewed journals. All papers in this compendium were thoroughly reviewed by experts, leading researchers, and industrialists in the field, and their contribution continues to be an important aspect to the success of this symposium. This proceedings volume is highly regarded as a centralized repository of magnesium research carried out globally and includes the finest and latest breaking articles in this area. Papers in this collection originated from 19 countries, highlighting the global competitiveness and interest in magnesium technology. This collection of papers represents the spectrum of research that will be presented during the 2015 Magnesium Technology Symposium at the 144th TMS Annual Meeting & Exhibition held in Orlando, Florida from March 15-19, 2015. This is one of the largest yearly gatherings of magnesium specialists in the world.

The symposium will start with a keynote session that consists of several distinguished invited speakers who were carefully selected as world-class experts who could provide their perspective on the future possibilities and opportunities on the topic of magnesium alloy research and development. The keynote speaker, Alan Taub, professor at the University of Michigan and the Chief Technical Officer of the American Lightweight Materials Manufacturing Innovation Institute (ALMMII), will provide us an overview of the activities of the institute, which constitutes one arm of the federal National Network for Manufacturing Innovation (NNMI) initiative, proposed by President Obama, and serves as a hub to accelerate the development and adoption of cutting-edge advanced lightweight materials and manufacturing technologies. His talk will be followed by a presentation by Bruce Davis from Magnesium Elektron, who will discuss the ongoing and recently standardized efforts by the Federal Aviation Authority to integrate magnesium alloys into structural applications in commercial aircraft. Tresa Pollock, professor at the University of California Santa Barbara, will summarize the outcomes of a recently held National Science Foundation–sponsored workshop to identify emerging and transformative research areas in metals and metallic nanostructures, where magnesium technology was highlighted as a topical area of interest. These broad-ranging, overview keynote session talks will be followed by a set of distinguished scientists, Jian-Feng Nie (Monash University), Matthew Barnett (Deakin University), and Sean Agnew (University of Virginia) who will discuss their latest technical research.

This year, integrated into the plenary are two new activities: (1) poster authors will be presenting elevator-pitch style presentations of their posters which will be later displayed during the poster session following the completion of the keynote session, and (2) the keynote speakers will participate in a panel for an interactive discussion with the audience on the future of magnesium research and applications. This discussion will be dictated, summarized, and then transcribed into a manuscript that will be subsequently released to the community.

Due to the large number of manuscripts submitted, each of the four days of the symposium are organized in parallel sessions of technical presentations, with two parallel sessions dedicated to the specialty symposium on Strip Casting of Light Metals organized by Kai Karhausen (Hydro Aluminium Rolled Products GmbH), Wim Sillekens (European Space Agency), Murat Dündar (Assan Aluminium), Jan Bohlen (Helmholtz-Zentrum Geesthacht), and Dietmar Letzig (Helmholtz-Zentrum Geesthacht). This symposium is co-sponsored by the Aluminum and Magnesium Committees and is dedicated to continuous casting/rolling methods to directly transform aluminum or magnesium melts into (semi-finished) strip, thereby emphasizing scientific technological communalities in the processing of both light-metal classes. Implemented as twin-roll casting, the manufacturing of magnesium-alloy strip is currently one of the highlights in magnesium research and development with particular relevance to automotive applications.

We would like to gratefully acknowledge and thank all the authors for their contributions and, in addition, the reviewers, session chairs, and poster judges for their willingness to volunteer their time in order to ensure a successful, high quality symposium and proceedings volume.

2015 Magnesium Technology Conference Organizers

Michele V. Manuel, Chair
Alok Singh, Vice Chair
Martyn Alderman, Past Chair
Neale R. Neelameggham, Advisor

ABOUT THE LEAD EDITOR

MICHELE V. MANUEL
MAGNESIUM TECHNOLOGY 2015 LEAD EDITOR

Michele Manuel is an associate professor in the Department of Materials Science and Engineering at the University of Florida. She received her Ph.D. in materials science and engineering at Northwestern University in 2007 and B.S. in materials science and engineering at the University of Florida. Prior to arriving at the University of Florida, she performed her post-doctoral work at the General Motors Technical Center in Warren, Michigan. She is the recipient of the NSF CAREER award, NASA Early Career Faculty award, ASM Bradley Stoughton Award for Young Teachers, AVS Recognition for Excellence in Leadership, TMS Early Career Faculty award, TMS Young Leaders Professional Development award, and TMS/JIM International Scholar award. Her research lies in the basic understanding of the relationship between processing, structure, properties, and performance. She uses a systems-based materials design approach that couples experimental research with theory and mechanistic modeling for the accelerated development of materials. Her current research is focused on the use of systems-level design methods to advance the development of new materials through microstructure optimization. In the TMS Magnesium Committee, she has previously served in the roles of Vice Chair, Secretary, and *JOM* Advisor, and she currently serves as Chair. Additionally, she serves as the Vice Chair of the Content Development and Dissemination Committee, as well as a member of the Education, Young Leaders, Integrated Computational Materials Engineering Committee (ICME), and the Women in Science Committees.

MAGNESIUM TECHNOLOGY 2015
Session Chairs

Keynote Session

Michele V. Manuel,
University of Florida

Alok Singh,
National Institute for Materials Science (NIMS)

Elevated Temperature and Creep

Amit Pandey,
LG Fuel Cell Systems

Karl Kainer,
Helmholtz-Zentrum Geesthacht

Primary, Sustainability, Recycling, and Processing

Neale R. Neelameggham,
IND LLC

James Saal,
QuestTek Innovations, LLC

Deformation I

Bilal Mansoor,
Texas A&M University at Qatar

Haitham El Kadiri,
Mississippi State University

Deformation II

Suveen Mathaudhu,
University of California, Riverside

Kiran Solanki,
Arizona State University

Wrought

Bin Li,
University of Nevada, Reno

Eric Nyberg,
Pacific Northwest National Laboratory

Forming and Alloy Design

James Saal,
QuestTek Innovations, LLC

Mesut Varlioglu,
John Deere Moline Technology Innovation Center

Corrosion, Coatings, Fatigue, and Fracture

Brian Jordon,
University of Alabama

Syamala Pulugurtha,
Medtronic

Casting and Metal Matrix Composites

Lei Zhang,
University of Alaska, Fairbanks

Wim Sillekens,
European Space Agency

Biomedical Applications

Raymond Decker,
Nanomag

Norbert Hort,
Helmholtz-Zentrum Geesthacht

Functional and Emerging Alloys

Vineet Joshi,
Pacific Northwest National Laboratory

STRIP CASTING OF LIGHT METALS
Session Chairs

Modeling and Properties

Dietmar Letzig,
Helmholtz-Zentrum Geesthacht

Murat Dündar,
Assan Aluminium

Process Technology

Kai F. Karhausen,
Hydro Aluminium Rolled Products

Jan Bohlen,
Helmholtz-Zentrum Geesthacht

POWELL, Adam	Infinium Metals
PRENTICE, Leon	CSIRO
PULUGURTHA, Syamala Rani	Medtronic
RAVINDRA, N.M	New Jersey Institute of Technology
ROHATGI, Aashish	Pacific Northwest National Laboratory
ROSALIE, Julian	National Institute for Materials Science
SAAL, James	QuesTek Innovations, LLC
SEVER, James	Alpha/Omega Engineering
SILLEKENS, Wim	European Space Agency
SOLANKI, Kiran	Arizona State University
ST JOHN, David	University of Queensland
TALEFF, Eric	University of Texas
TARTAGLIA, John	Element Materials Technology
TOLNAI, Domonkos	MagIC Magnesium Innovation Center
VERMA, Ravi	General Motors
WILKS, Tim	Magnesium Elektron

Magnesium Technology 2015

Keynote Session

Magnesium Technology 2015
Edited by: Michele V. Manuel, Alok Singh, Martyn Alderman, and Neale R. Neelameggham
TMS (The Minerals, Metals & Materials Society), 2015

REDUCING WEIGHT FOR TRANSPORTATION APPLICATIONS: TECHNOLOGY CHALLENGES AND OPPORTUNITIES

Alan I. Taub
University of Michigan, Ann Arbor, 2300 Hayward Street, Ann Arbor, MI, 48109, USA

Keywords: lightweight materials, manufacturing processes, systems engineering principles

Today's land, sea and air transportation industries – as a business necessity – are focused on technology solutions that will make vehicles more sustainable in terms of energy, the environment, safety and affordability. Reducing vehicle weight is a key enabler for meeting these challenges as well as increasing payload and improving performance. The potential weight reductions from substituting lightweight metals (advanced high-strength steels, aluminum, magnesium and titanium alloys) are well established. For magnesium castings, weight savings of 60% have been reported [1]. The value of weight reduction depends on the transportation sector and ranges from about \$5/kg saved for automobiles to over \$500/kg saved for aircraft [2]. The challenge is to optimize the material properties and develop robust, high volume, manufacturing technologies and the associated supply chain to fabricate components and subsystems at the appropriate cost for each application.

The American Lightweight Materials Manufacturing Innovation Institute (ALMMII) was established as a public-private partnership to accelerate the adoption of these materials. ALMMII is one of four new institutes that are part of the growing National Network of Manufacturing Institutes and serves as the nation's essential bridge between basic research and final product commercialization for lightweight metals [3].

The institute will take advantage of opportunities to drive synergies across the various industries serving both commercial and defense transportation needs. Advanced manufacturing processes will be developed through projects executed by our industry partners in collaboration with an extensive network of universities and the national and federal laboratories. Commercialization of the resulting technologies is enabled by the early integration of the capabilities of the full production supply chain from material producers and metal formers to OEM's.

Improved manufacturing processes for magnesium in both cast and wrought forms are required to expand the alloy utilization. Our industry and government partners have identified a number of focus areas including higher integrity and larger thin-walled castings, overcasting of magnesium on aluminum and steel substrates, and joining of mixed-material components with particular emphasis on mitigation of galvanic corrosion.

The projects will employ systems engineering principles as shown in Figure 1. The industry members will select the particular alloy and an associated component and platform. In developing the new manufacturing process, the project team will employ ICME enabled models that can support both the design and certification/validation phases of the work.

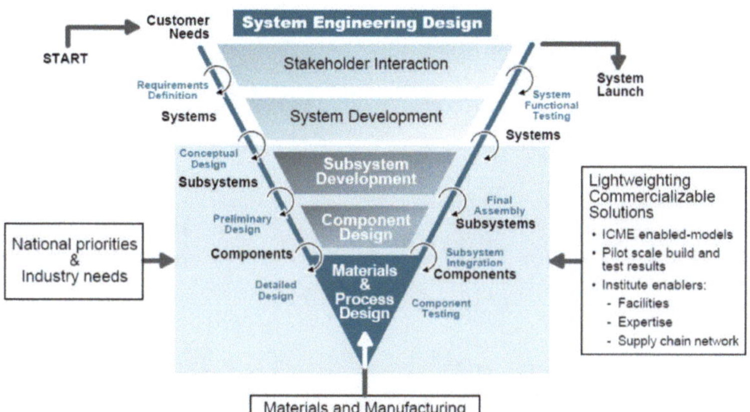

Figure 1: Each ALMMII project will utilize systems engineering principles to deliver innovative manufacturing processes integrated with product design modeling

References

[1] Alan I. Taub, Paul E. Krajewski, Alan A. Luo, and John N. Owens, The Evolution of Technology for Materials Processing over the Last 50 Years: The Automotive Example, JOM v 59, n 2, p 48-57, February 2007.

[2] Materials Research to Meet 21st Century Defense Needs, NRC Committee on Materials Research for Defense After Next, National Academy Press, 2003.

[3] National Network for Manufacturing Innovation: A Preliminary Design, Executive Office of the President, National Science and Technology Council, Advanced Manufacturing National Program Office, January, 2013

Magnesium Technology 2015
Edited by: Michele V. Manuel, Alok Singh, Martyn Alderman, and Neale R. Neelameggham
TMS (The Minerals, Metals & Materials Society), 2015

THE APPLICATION OF MAGNESIUM ALLOYS IN AIRCRAFT INTERIORS – CHANGING THE RULES

Bruce Davis
Magnesium Elektron North America; 1001 College St. PO Box 258; Madison, IL, 62060, USA

Extended Abstract

The commercial aircraft market is forecast to steadily grow over the next two decades. Part of this growth is driven by the desire of airlines to replace older models in their fleet with newer, more fuel efficient designs, to realize lower operating costs and to address the rising cost of aviation fuel. As such the aircraft OEMs are beginning to set more and more demanding mass targets on their new platforms.

The cabin interior, including furnishings and operational items, account for 19% of the overall aircraft weight. This establishes it as a significant area for weight savings. Modern magnesium alloys can offer up to 30% mass reduction when directly replacing aluminum components, which makes them very attractive to designers and engineers. However, until recently, there have been restrictions on the use of magnesium in aircraft interiors that have been enforced through the SAE and the FAA.

The FAA has been working for the last seven years to assess the safety issues associated with the introduction of magnesium into the cabin interior. This work began with simplistic flammability testing of a wide range of alloys that identified different ignition behaviors of magnesium alloy types. This lead the FAA to undertake full scale fuselage staged burn tests of seat structures. The typical structural seat components are shown in Figure 1. The seats tested in the full-scale fuselage assessment contained five key parts fabricated from magnesium: spreader bars, seat leg assembly, cross tubes, baggage bars and the seat back frame.

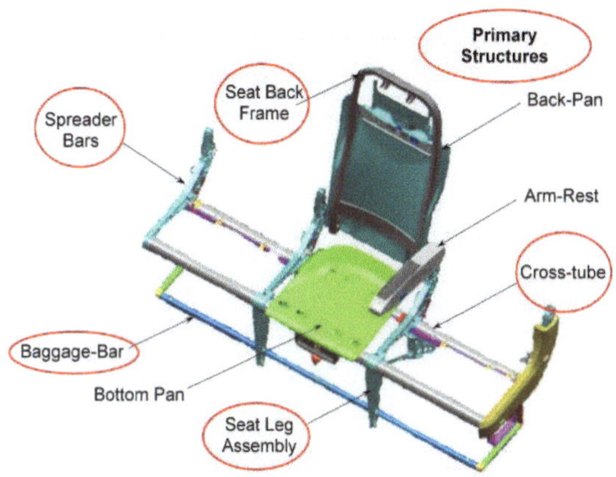

Figure 1. Schematic of major structural seat components

The testing arrangement is shown in Figure 2, which indicates the monitoring systems that were in place. The test assessed standard seat ignition properties compared to two magnesium containing seats containing either AZ31B alloy or Elektron®43.

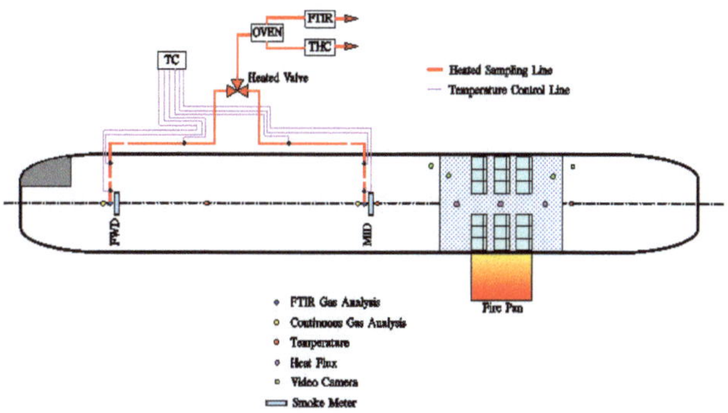

Figure 2. Overview of full-scale fuselage test.

The results of this testing showed that the presence of magnesium did not compromise passenger safety. The FAA has now issued guidelines on a laboratory scale flammability test. The test setup is shown in Figure 3. Magnesium alloys that meet the criteria of this test will be considered allowable for use in aircraft seat applications. The test method is available in the FAA's Aircraft Materials Fire Test Handbook Chapter 25 Oil Burner Flammability Test for Magnesium Alloy Seat Structure

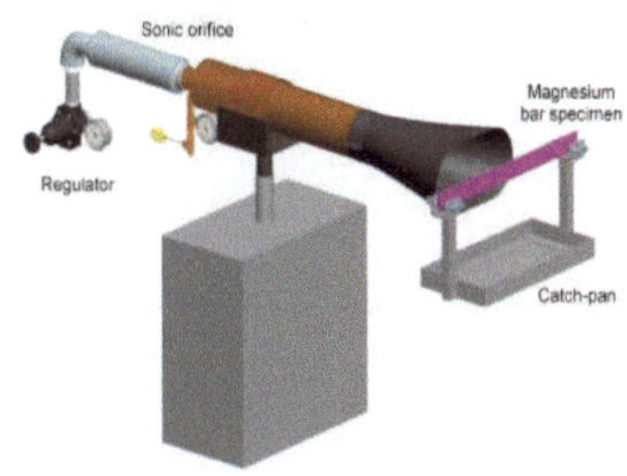

Figure 3. Lab scale burner test setup

The work by the FAA has culminated with a lifting of the ban of magnesium's application in aircraft seat applications. The FAA will now accept application for special conditions that will allow magnesium containing aircraft seat certification in commercial transport aircraft. Future work by the FAA will investigate methods of compliance (MOC) to allow magnesium in other aircraft interior applications, thereby extending the weight saving potential and benefit of today's advanced magnesium alloys.

Furthermore, the SAE standard AS8049 Rev B, which defines the engineering requirements of a commercial aircraft seat, restricts the use of magnesium in paragraph 3.3.3. AS8049 is currently undergoing its scheduled five year review. Revision C is scheduled to be issued in 2015 and paragraph 3.3.3 has been altered to allow the use of magnesium. The document now states that magnesium alloys maybe used in aircraft seats so long as the conform to the requirements set out in the Aircraft Materials Fire Test Handbook.

The application of the next generation magnesium alloys in aircraft interiors is now positioned to grow.

Magnesium Technology 2015
Edited by: Michele V. Manuel, Alok Singh, Martyn Alderman, and Neale R. Neelameggham
TMS (The Minerals, Metals & Materials Society), 2015

EMERGING SCIENCE AND RESEARCH OPPORTUNITIES FOR METALS AND METALLIC NANOSTRUCTURES: A REPORT ON THE NSF MMN WORKSHOP

Tresa Pollock[1], Carol Handwerker[2]
[1]University of California Santa Barbara; Materials Department; Santa Barbara, CA; 93106-5050; USA
[2]Purdue University; 701 W. Stadium Avenue; West Lafayette, IN; 47907; USA

Extended Abstract

In the next decade, fundamental research in metals and metallic nanostructures (MMN) has the potential to continue to transform science into innovative materials, devices, and systems. This talk summarizes the findings of a workshop to identify emerging and potentially transformative research areas in MMN. The metals and metallic nanostructures (MMNs) workshop aimed to identify significant research trends, scientific fundamentals, and recent breakthroughs that can enable new or enhanced MMN performance, either alone or in a more complex materials system, for a wide range of applications. Additionally, the role that MMN research can play in high-priority research and development (R&D) areas such as the U.S. Materials Genome Initiative, the National Nanotechnology Initiative, the Advanced Manufacturing Initiative, and other similar initiatives that exist internationally was assessed. The workshop also addressed critical issues related to materials research instrumentation and the cyberinfrastructure for materials science research and education, as well as science, technology, engineering, and mathematics (STEM) workforce development, with emphasis on the United States but with an appreciation that similar challenges and opportunities for the materials community exist internationally.

The societal and economic impacts of MMN research will continue to be significant. Metals and metallic nanostructures underlie systems and technologies critical to solving national and global grand challenges. Innovations in metals, metal alloys, and metallic nanostructures are needed to advance systems for:

- energy generation, harvesting and storage – thermoelectrics, batteries, fuel cells, advanced turbines, magnetic induction, motors, hydrogen production and storage, nuclear reactor materials, separation membranes
- propulsion and transportation – high strength, high performance lightweight structures, materials designed for improved creep, impact, fatigue or corrosion resistance, high temperature metals, alloys and coatings for extreme environments, non-destructive evaluation
- electronics, telecommunication, and information technology – electronic packaging, process modeling, device nanostructures to overcome the end to Moore's law, thin film and nanostructured electrical interconnects, magnetic devices and motors, and sensors
- sustainability – nanostructured metals for catalytic convertors, polymer catalysis, hydrogen and ammonia production, low pollution smelting and refining processes
- health care – imaging systems, integrated in-vivo sensors, electronics, and communication systems for health monitoring and drug delivery, prosthetics and implant materials and systems, cancer treatment technologies
- manufacturing – additive and nano-manufacturing technologies, non-equilibrium processing approaches, recycling and reuse, rapid deployment of alternatives to "critical" materials.

A central theme of the Workshop was the view that research in metals and metallic nanostructures has provided and will continue to provide benefit to the nation through the coupling of experiment, theory, and simulation to link atomistic, nanoscale, microscale, and mesoscale phenomena across time scales for an ever widening range of applications. Within this overarching theme, the workshop participants identified emerging research opportunities that are categorized in terms of:

- 3-D and 4-D Materials Science
- Structure Evolution and the Challenge of Heterogeneous and Multicomponent Systems
- The Science Base for Property Prediction Across the Lengthscales
- Nanoscale Phenomena at Surfaces – Experiment, Theory, Simulation
- Prediction and Control of the Morphology, Microstructure, and Properties of "Bulk" Nanostructured Metals
- Functionality and Control of Materials Far from Equilibrium
- Hybrid and Multifunctional Materials Assemblies
- Materials Discovery and Design: Enhancing the Theory – Simulation - Experiment Loop

This talk will summarize the findings of the workshop and will provide a brief overview of these emerging research opportunities, along with challenges and opportunities for the materials community in the areas of instrumentation, cyberinfrastructure, education, and workforce development.

Magnesium Technology 2015
Edited by: Michele V. Manuel, Alok Singh, Martyn Alderman, and Neale R. Neelameggham
TMS (The Minerals, Metals & Materials Society), 2015

SOLUTE SEGREGATION AND AGGREGATION IN Mg ALLOYS

J.F. Nie[1], Y.M. Zhu[1], N.C. Wilson[2]
[1] Department of Materials Engineering, Monash University; Victoria 3800, Australia
[2] CSIRO Manufacturing Flagship; Clayton, Victoria 3168, Australia

Keywords: Magnesium alloys, Solute segregation, Solute aggregation, Microstructures

Extended Abstract

The addition of adequate amounts of rare-earth (RE) alloying elements can produce appreciable effects on microstructures, formability and mechanical properties of magnesium alloys. At low concentrations, it can weaken the basal texture in the as-deformed and the recrystallized conditions in wrought alloys and improve the creep resistance of casting alloys at elevated temperatures; at high concentrations, it can bring significant age hardening effects and thus improve the alloy strength. While this knowledge has been known for some years, the precise roles of RE solutes in texture weakening and the early stage of precipitation remain unclear.

It has been demonstrated in recent years that the application of high-angle annular dark-field scanning transmission electron microscopy (HAADF-STEM) can reveal the distribution of RE solutes and the structure of precipitate phases at the atomic scale. The use of this Z-contrast imaging technique, combined with atom probe tomography or other characterization techniques such as energy dispersive X-ray spectroscopy and electron energy loss spectroscopy, has revealed the occurrence of solute segregation in high-angle grain boundaries even in solid solution single phase alloys based on the Mg-Gd system [1–4]. The segregation of RE atoms in the grain boundaries will no doubt impede the boundary migration [5] and thus change the recrystallization behaviours and perhaps even the recrystallization texture. In addition to the well-established solute segregation in the partially-coherent high-angle grain boundaries, a very recent study using HAADF-STEM [6] has revealed the occurrence of a periodic segregation of solute atoms in fully-coherent twin boundaries of $\{10\bar{1}1\}_\alpha$, $\{10\bar{1}2\}_\alpha$ and $\{10\bar{1}3\}_\alpha$. This segregation results from the minimization of elastic strain energies associated with twin boundaries and the over-sized individual solute atoms, and it provides a strong pinning of the twin boundaries and a concomitant annealing hardening effect when the alloy is deformed under certain circumstances. Subsequent studies using first-principles density functional theory computations indicate that solute atoms of most commonly used alloying elements have a strong tendency to segregate to $\{10\bar{1}2\}_\alpha$ twin boundaries [7,8], and that solute segregation can also occur in a symmetrical tilt boundary [9].

In concentrated Mg-RE alloys, the intermediate precipitate phases such as β' and β_1 have been relatively well characterized in recent years [10]. However, the structures and compositions of precipitate phases, GP zones and clusters that form in the early stage of decomposition are still poorly understood [11,12]. It remains to be firmly established whether solute clusters form and, if so, how they interact with lattice defects such as dislocations and twins. Such information is critically needed if some of the beneficial and yet mysterious effects of RE additions are to be rationalized. This presentation will also provide a review of our recent findings in the study of solute aggregates in a group of magnesium alloys using HAADF-STEM and first-principles density functional theory.

References

[1] M. Bugnet, A. Kula, M. Niewczas, and G.A. Botton, "Segregation and Clustering of Solutes at Grain Boundaries in Mg-Rare Earth Solid Solutions," *Acta Mater.* 79 (2014), 66-73.

[2] J.P. Hadorn, T.T. Sasaki, T. Nakata, T. Ohkubo, S. Kamado, and K. Hono, "Solute Slustering and Grain Boundary Segregation in Extruded Dilute Mg-Gd Alloys," *Scripta Mater.* (2014) in press.

[3] N. Stanford, G. Sha, J.H. Xia, S.P. Ringer, and M.R. Barnett, "Solute Segregation and Texture Modification in An Extruded Magnesium Alloy Containing Gadolinium," *Scripta Mater.* 65 (2011), 919-921.

[4] J.P. Hadorn, K. Hantzsche, S.B. Yi, J. Bohlen, D. Letzig, J.A. Wollmershauser, and S.R. Agnew, "Role of Solute in the Texture Modification during Hot Deformation of Mg-Rare Earth Alloys," *Metall. Mater. Trans. A* 43A (2012), 1347-1362.

[5] J.D. Robson, "Effect of Rare-Earth Additions on the Texture of Wrought Magnesium Alloys: The Role of Grain Boundary Segregation," *Metall. Mater. Trans. A* 45A (2014), 3205-3212.

[6] J.F. Nie, Y.M. Zhu, J.Z. Liu, and X.Y. Fang, "Periodic Segregation of Solute Atoms in Fully Coherent Twin Boundaries," *Science*, 340 (2013), 957-960.

[7] J. Zhang, Y. Dou, and Y. Zheng, "Twin-Boundary Segregation Energies and Solute-Diffusion Activation Enthalpies in Mg-Based Binary Systems: A First-Principles Study," *Scripta Mater.* 80 (2014), 17-20.

[8] M. Ghazisaeidi, L.G. Hector Jr, and W.A. Curtin, "Solute Strengthening of Twinning Dislocations in Mg Alloys," *Acta Mater.* 80 (2014), 278-287.

[9] L. Huber, J. Rottler, and M. Militzer, "Atomistic Simulations of the Interaction of Alloying Elements with Grain Boundaries in Mg," *Acta Mater.* 80 (2014), 194-204.

[10] J.F. Nie, "Precipitation and Hardening in Magnesium Alloys," *Metall. Mater. Trans. A* 43A (2012), 3891-3939.

[11] W. Lefebvre, V. Kopp, and C. Pareige, "Nano-Precipitates Made of Atomic Pillars Revealed by Single Atom Detection in a Mg-Nd Alloy," *Appl. Phys. Lett.* 100 (2012), 141906.

[12] Y. Matsuoka, K. Matsuda, K. Watanabe, J. Nakamura, W. Lefebvre, D. Nakagawa, S. Saikawa, and S. Ikeno, "Precipitation Sequence in the Mg-Gd-Y System Investigated by HRTEM and HAADF-STEM," *Mater. Trans.* 55 (2014), 1051-1057.

Magnesium Technology 2015

Elevated Temperature and Creep

Magnesium Technology 2015
Edited by: Michele V. Manuel, Alok Singh, Martyn Alderman, and Neale R. Neelameggham
TMS (The Minerals, Metals & Materials Society), 2015

THREE DECADES OF ELECTRON BACKSCATTER DIFFRACTION OF MAGNESIUM: WHAT HAS IT TAUGHT US?

Matthew R. Barnett
Institute for Frontier Materials, Deakin University; Pigdons Rd; Geelong; 3217; Australia

Abstract

Electron backscatter diffraction has transformed microstructural analysis. Its advent enabled the microstructure to be correlated with crystallographic orientation in an unprecedented manner. This correlation has taught us many things about the physical metallurgy of magnesium and its alloys. Some highlights from three decades of research are presented in this talk. Chief among these is the insight the technique has provided into mechanical twinning, which plays a major role in the deformation of magnesium. The technique has also shed important light on deformation mechanisms, heterogeneity of material flow and, of course, the processes of texture development during wrought processing. It also promises more. In coming years, careful analysis of electron backscatter diffraction patterns and maps will in all likelihood continue to highlight the dominating mechanisms and phenomena that must be captured in microstructurally honest models of material response.

Some Examples

Electron backscatter diffraction (EBSD) capability became commercially available in 1986 [1]. Pettersen and Ryum [2] applied backscattered pattern analysis to a magnesium alloy in 1989. Since then many studies have applied the technique to this material, thereby shedding important new light on its physical metallurgy.

The technique involves automated indexing of Kikuchi diffraction patterns captured using a camera mounted behind a phosphor screen. Combined with beam rastering, this permits microstructure maps to be re-constructed from crystallographic orientation data. The technique has proven ideal for studying the occurrence of deformation twinning in magnesium and its alloys. Figure 1 provides an example of an image generated using the technique along with the secondary twins it was used to identify [3].

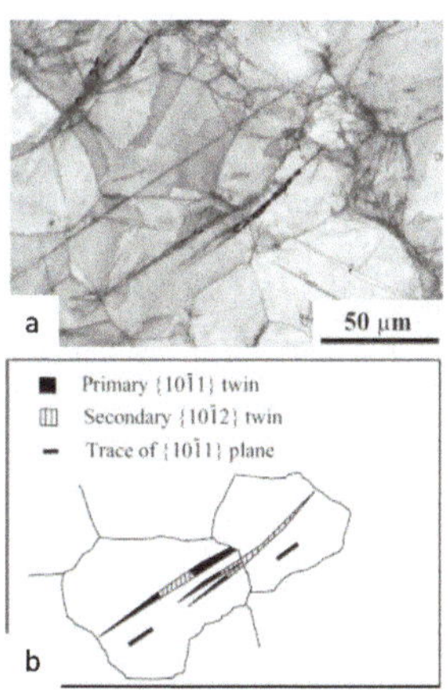

Figure 1. (a) EBSD map of pure Mg generated using Kikuchi pattern contrast, (b) identification of the double twins present [3].

The technique has been instrumental in verifying the prevalence of this secondary twinning reaction in commercial magnesium. EBSD also provided indication that such twinning events may be involved with void formation and ductile failure [4]. An example is provided in Figure 2. Although twin identification is difficult in this image, twins appear to be associated with the voids and fragments of double twin boundaries are able to be identified in twins parallel to the voids. Other notable studies that have exploited EBSD to shed important light on twinning in magnesium include those by the McGill [5] and Los Alamos [6] groups, for example.

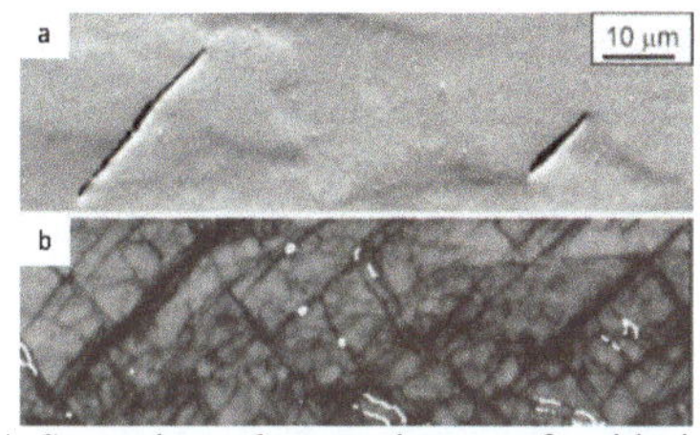

Figure 2. (a) Secondary electron image of voids in a sample of extruded ZK60 deformed to failure in tension, (b) fragments of double-twin boundaries highlighted in white [4].

The technique has also revealed the microstructural origin of the deformation texture of magnesium alloys containing rare-earth elements. One example is shown in Figure 3. Here, the inverse pole figure of the extrusion direction is shown for horizontally aligned grain clusters and for grain clusters formed in shear bands. It is seen that the two microstructural locations produce different orientations. Unique "rare-earth" textures can, in certain circumstances, be due to a predominance of the nucleation of recrystallization in shear bands [7].

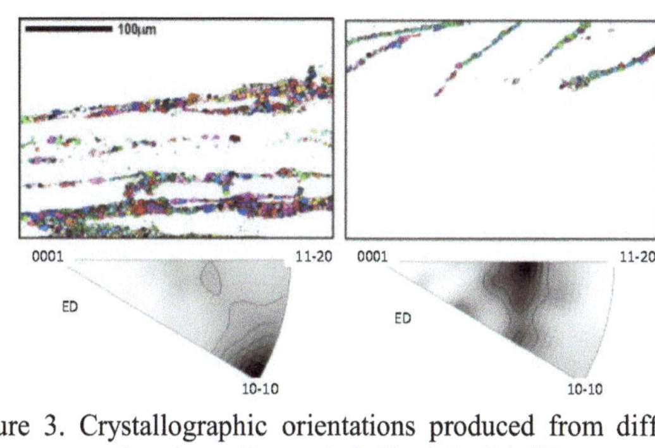

Figure 3. Crystallographic orientations produced from different microstructural regions of extruded alloy Mg-Gd. [7]

[1] Maitland, T and Sitzmanin, S. in Scanning Microscopy for Nanotechnology, Zhou, W, Wang Z.L. eds, Springer (2007), 41.
[2] Pettersen, K., Ryum, N. Metall. Trans. A, 20(1989) 847.
[3] Barnett, M.R., Nave, M.D., Bettles, C.J. Mat. Sci. Eng. A, (2004) 386, 205.
[4] Barnett, M.R., Mat. Sci. Eng. A (2007) 464, 8.
[5] Jiang, L., Jonas, et al. Scripta Mat. (2006) 54, 771.
[6] Beyerlein, I.J., Capolungo, et al. Phil. Mag. (2010) 90, 2161.
[7] Stanford, N., Barnett, M.R. Mat. Sci. Eng. A, (2008) 496, 399.

Magnesium Technology 2015
Edited by: Michele V. Manuel, Alok Singh, Martyn Alderman, and Neale R. Neelameggham
TMS (The Minerals, Metals & Materials Society), 2015

MEASURING AND MODELING THE EFFECTS OF MECHANICAL TWINNING ON THE BEHAVIOR OF MAGNESIUM ALLOYS

Sean Agnew[1], Peidong Wu[2], Kaan Inal[3], Haitham El Kadiri[5], Jian Wang[6], Carlos Tome[6]

[1]University of Virginia; 395 McCormick Rd; Charlottesville, VA; 22904-4745; USA
[2]McMaster University; 1280 Main Street West; Hamilton; L8S4L7; Canada
[3]University of Waterloo; 200 University Avenue West; Waterloo; N2L3G1; Canada
[5]Mississippi State University; 200 Research Blvd.; Starkville, MS; 39759; USA
[6]Los Alamos National Laboratory; P.O. Box 1663; Los Alamos; 87545; USA

Keywords: twinning, detwinning, crystal plasticity, texture, asymmetry, plateau

Abstract

Our community has learned a great deal about twinning in Mg alloys over the past decade. Some of these things were known qualitatively in prior decades, but we have now developed a proficiency in characterization and computational modeling which permits a quantitative description of these twinning-induced effects over a wide range of strain rates, temperatures, loading conditions, and in a variety of alloy families. These capabilities could only be dreamed of by prior generations. This lecture will review the effects of the main twinning mode, {10.2} extension twinning (e.g., tension-compression yield asymmetry, yield plateau, anisotropy, rapid strain hardening, detwinning, etc.); characterization, primarily by diffraction-based techniques (electron, X-ray, and neutron); and modeling by crystal plasticity-based methods. Strategies to control these effects through microstructure, texture, and alloy design will be proposed. Finally, outstanding questions which merit further research will be highlighted.

Extended Abstract

One of the most important tools for improving our understanding of the effects of twinning on deformation has been crystal plasticity modeling. The effect of deformation twinning has now been incorporated within polycrystal plasticity models for over three decades. Initially, the interest was in explaining the different texture evolution observed within rolled face centered cubic (fcc) metals and alloys with high and low stacking fault energies, respectively. A noteworthy demonstration was made by Van Houtte [1], who developed a generalized Taylor model, which incorporated twinning as a polar, pseudo-slip mechanism that would generate new grains with volume fractions in proportion to the amount of shear strain accommodated. In part due to computational limitations of the date, he introduced a strategy designed to minimize the number of new grain orientations, which would have to be introduced into the polycrystal plasticity calculation over the course of incremental, large strain deformation.

A few years later, Tomé, Lebensohn, and Kocks [2] introduced two more strategies to overcome the problem of grain proliferation. One was termed the volume fraction transfer scheme and the other was termed the predominant twin reorientation (PTR) scheme. The former eliminates the proliferation of grain orientations by simply populating orientation space at the beginning of the calculation and then transforming volume fraction among the available orientations which describe the polycrystal. However, it suffers from failing to adequately account for the various sources of strain hardening within each of the discrete "grains," since they are actually nothing more than regions of orientation space which may contain contributions from a variety of different initial grains, due to slip and twinning reorientation. This strategy has been largely abandoned.

The latter strategy, PTR, involves reorienting entire grains when a threshold amount of twinning has occurred within them. This strategy has been largely successful and very popular among modelers who employ the various self-consistent polycrystal modeling strategies. It does make the assumption that the crystal reorientation associated with twinning can be accounted for, on average, by tracking on a single twin variant within each grain. The increasing threshold strategy for reorienting entire grains employed in this model has been successful in preventing unnatural discontinuities in the predicted polycrystalline stress-strain behavior. The applications of the model to hexagonal close packed (hcp) metals and alloys is too long to list here. That said, if one wishes to accurately account for effects related to the amount of twin reorientation at small plastic strain levels or cases where multiple variants of twins are known to occur within a given grain, the model has limitations.

Kalidindi [3] proposed quite a different scheme, which has been popular within Taylor (full constraints) and finite element implementations of crystal plasticity. In this approach, the grains (nodal descriptions) are subdivided into virtual matrix and twin domains upon the activation of twinning. The constitutive response at that physical location is the volume weighted average response of the matrix and twin domains, all subject to the Taylor (upper bound) approximation. This approach has been employed with some success for a variety of material types, including the aforementioned fcc metals [e.g. 3], as well as hcp metals such as Ti [4] and Mg [5]. Notably, the calculations are performed in a relaxed (referential) configuration in which the misorientation between the twinned and the untwinned parent domains are fixed. In this way, the author avoids unnecessary twin proliferation and, yet, permits the possibility that the parent and twin domains may not maintain their specific twin misorientation relationship with subsequent deformation.

None of the above models seek to deal with the spatial details of twinning within the parent grain. Rather, the twin-induced strain within the parent is linked to the volume fraction of material which has twinned. A so-called composite grain model was introduced to overcome this limitation [e.g., 6]. A primary benefit of this model was to introduce latent hardening effects of twinning on other slip systems. The concept in view here is that the twinning interface provides a barrier (like a general grain boundary) to slip. In addition, the model employs a geometric concept to harden those slip systems whose glide planes are nearly parallel to the predominant twin interfaces more than the those which are nearly perpendicular to the twin interfaces. An estimate of the mean free glide path is computed and this is used in a Hall-Petch type hardening model. One critique of this and other Hall-Petch arguments, for describing hardening due to twinning in Mg alloys is that one would expect the most hardening effect early in the deformation (when twin thicknesses are small) where as the hardening effects of twinning are typically observed later, when twinning is nearing the point of exhaustion and twin thicknesses are large. Another argument is that twin boundaries do not appear to be simple barriers to dislocation deformation, since there is a significant probability of transmission or absorption [c.f., 7]. Numerous recent works aiming to clarify these details employing modern interatomic potentials have been performed by co-authors J. Wang et al. and El Kadiri et al.

Recently, H. Wang et al. [8] introduced a strategy to account not only for twinning during monotonic deformation but also detwinning, which occurs during strain path changes. They dubbed their model the twinning-detwinning (TDT) model. Like all prior models, the TDT model acknowledges that twinning is a polar, one-way deformation mechanism, but only during the nucleation step. Once a twin is nucleated, the twin may either continue accommodating strain in the forward direction or it may detwin, hence accommodating strain in the reverse sense. This is not a violation of the polarity of twinning.

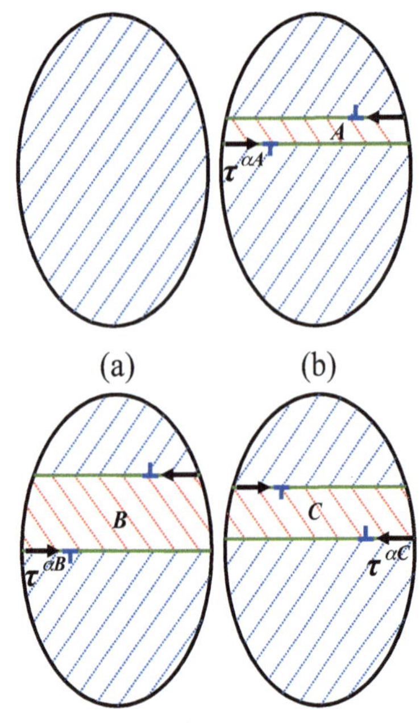

Figure 1. Schematic representation of twinning and detwinning in a grain, after H. Wang et al. [8]. (a) The grain is twin-free and is called the matrix in the text. (b) The grain undergoes twin initiation (denoted Operation A). The solid green lines represent twin boundaries (TBs). (c) Operation B is growth of the twin, thereby consuming the matrix. (d) Operation C is the reverse of operation B, involving propagation of the TBs back into the twin, causing growth of the matrix.

There is another key feature of this model, which was missing from all prior models, and it is critical, even in monotonic deformation without detwinning. That is recognition of the fact that twinning is a process involving interfacial motion. As such, it is the stress on the twin interface, which determines whether or not the interface will move (during twinning (Fig. 1(c)) or detwinning (Fig. 1(d)). As a first approximation, the TDT model interrogates stresses on both sides of the interface (i.e. within the matrix and within the twin itself) to determine if the stress condition required for interfacial propagation is met. If so, a degree of twinning occurs whether that condition is met in the parent grain or in the twin (or both). Some readers may wonder if this strategy may lead to "double counting" of the possible twinning strain, since twinning may be activated from the parent grain and/or from within the twinned domain. However, the volume fraction of twinning permitted in any domain (parent or twin) is scaled by the volume fraction of that domain (parent or twin, respectively). And the total volume fraction of the parent and twin(s) is always held constant (at the original volume fraction of the parent). Of course, it requires some amount of bookkeeping to insure this manifestation of the conservation of mass is respected.

Throughout the lecture, specific attention will be drawn to the most common deformation twinning mode in hcp metals, including Mg, the {10.2} twinning mode. The TDT twinning model will be introduced by way of example, demonstrating that it can describe many features of the mechanical response which are common to textured Mg alloys: 1) tension/compression yield strength asymmetry, 2) strain hardening plateau or yield point elongation [e.g. 9], 3) rapid strain hardening [e.g. 10], and 4) temperature [e.g. 11] and strain rate [e.g. 12] insensitivity of the twinning stress.

In-situ neutron diffraction data will be used as the experimental method of choice for validating the modeling, especially as pertains to characterizing the twin volume fraction evolution and the internal strains within twins, as originally demonstrated by Brown et al. [13] and Clausen et al. [14]. The latter set of experiments draw particular attention to the notion of a "relaxation" or "back-stress" within the twins upon their formation. Experimental data from commercial Mg alloys: AZ31B [e.g. 11,13,14], AM30 [10], ZK60A [15], and WE43 [12], will be used to illustrate certain material behaviors and to validate certain aspects of the crystal plasticity modeling approach.

Acknowledgements

SRA would like to thank the National Science Foundation, grant number CMMI 1235259 (Mary Toney) and the U.S. Army Research Office for their support under contract number W911NF-12-1-0455 (Suveen Mathaudhu and David Stepp as contract monitors), for their support of research targeting a better understanding of the role of twinning in plasticity and fracture. Gratitude is also extended to McMaster University and the Initiative for Automotive Manufacturing Innovation (IAMI) for sponsoring a visiting faculty appointment during which many of the thoughts contained herein were first conceived or understood.

References

[1] P. Van Houtte, "Simulation of the rolling and shear texture of brass by the Taylor theory adapted for mechanical twinning." *Acta Metall.*, vol. 26, pp. 591-604, 1978.

[2] C.N. Tomé, R.A. Lebensohn, U.F. Kocks. "A model for texture development dominated by deformation twinning: application to zirconium alloys." *Acta Metall. Mater.,* vol. 39, pp. 2667-2680, 1991. "Incorporation of deformation twinning in crystal plasticity models," *J. Mech. Phys. Solids*, 46, pp. 267-290, 1998.

[4] A.A. Salem, S.R. Kalidindi, R.D. Doherty. "Strain hardening of titanium: role of deformation twinning." *Acta Mater,* vol. 51, pp. 4225-4237, 2003.

[5] J. Lévesque et al. "Numerical modeling of formability of extruded magnesium alloy tubes." *Int. J. Plasticity*, vol. 26, pp. 65-83, 2010.

[6] G. Proust et al. "Modeling the effect of twinning and detwinning during strain-path changes of magnesium alloy AZ31." *Int. J. Plasticity*, vol. 25, pp. 861-880. 2009.

[7] A. Serra, D.J. Bacon, R.C. Pond. "Dislocations in interfaces in the hcp metals—I. Defects formed by absorption of crystal dislocations." *Acta Mater.*, vol. 47, pp. 1425-1439, 1999.

[8] H. Wang et al., "A constitutive model of twinning and detwinning for HCP polycrystals," *Mater. Sci. Eng. A,* vol. 555, pp. 93-98, 2012.

[9] O. Muránsky et al "Investigation of deformation twinning in a fine-grained and coarse-grained ZM20 Mg alloy: Combined *in situ* neutron diffraction and acoustic emission." *Acta Mater.*, vol. 58, pp. 1503-1517, 2010.

[10] A.L. Oppedal et al., "Anisotropy in hexagonal close-packed structures: improvements to crystal plasticity approaches applied to magnesium alloy." *Phil. Mag.*, vol. 93, pp. 4311-4330 (2013).

[11] A. Jain, S.R. Agnew. "Modeling the temperature dependent effect of twinning on the behavior of magnesium alloy AZ31B sheet." *Mater. Sci. Eng. A*, vol. 462, pp. 29-36, 2007.

[12] J. Bhattacharyya et al. "Crystal plasticity modeling of the dynamic behavior of magnesium alloys, WE43-T5, plate," *Magnesium Technology 2015*, present volume.

[13] D.W. Brown et al., "Internal strain and texture evolution during deformation twinning in magnesium." *Mater. Sci. Eng. A*, vol. 399, pp. 1-12, 2005.

[14] B. Clausen et al., "Reorientation and stress relaxation due to twinning: modeling and experimental characterization for Mg." *Acta Mater*. vol. 56, pp. 2456-2468, 2008.

[15] H. Qiao, S.R. Agnew, P.D. Wu. "Modeling twinning and detwinning behavior of Mg alloy ZK60A during monotonic and cyclic loading." *Int. J. Plasticity*, vol. 65, pg. 61-84, 2015.

MICROSTRUCTURE AND PROPERTIES OF AGED VS. CREPT Mg-Al-Zn-Sn ALLOYS WITH ADDITIONS OF Nd AND Ce

Uri Vainberg[1] Shaul Avraham[2], Alexander Katsman[1], Menachem Bamberger[1]
[1]Department of Materials Engineering, Technion - Israel institute of technology, Haifa, Israel
[2]Department of Material Engineering, NRCN – Nuclear Research Center Negev, Beer-Sheva, Israel.

ABSTRACT

Mg-alloys are highly applicative at the automobile and aerospace industries due to their high specific mechanical properties. Mg-Al-RE alloy design was motivated by the formation of Al-RE intermetallic at the grain boundaries (GB) and Mg_2Sn precipitates inside the α-Mg matrix. The presence of precipitates at the GB can inhibit GB sliding, while the precipitation inside the α-Mg matrix grains inhibits dislocation climb responsible for a bulk creep. Thermodynamic simulations were used for optimal composition and process parameters derivation. Mg-Al-Zn-Sn alloys with additions of Nd and Ce were examined in terms of creep resistance and microstructure development during aging and creep. The evolution of the microstructure was monitored by scanning electron microscopy and X-ray diffraction. Mechanical properties were examined by creep testing and Vickers microhardness measurements. A directional growth of the $MgZn_2$ precipitates was found in the crept specimens in contrast to the aged counterparts.

1. INTRODUCTION

Mg alloys are highly attractive for engineering applications. However, in comparison to Al alloys, Mg alloys are featured with poor creep resistance at elevated temperatures (T>200°C). Alloying of Mg-Al alloys with rare earth (RE) elements results in formation of intermetallic phases at the GB, thus hindering GB sliding (1). Alloying the Mg-Al alloy with Sn causes precipitation of Mg_2Sn inside the α-Mg matrix, thus promoting precipitation hardening (2). The alloying of the Mg-Al-Sn alloy with Zn may cause formation of $MgZn_2$ phase with elongated morphology (3). In the present work the Mg-Al-Sn-Zn-RE (Ce, Nd) alloys were designed in order to yield thermally stable intermetallic phases in the grain boundaries, and precipitates inside the α-Mg matrix.

2. EXPERIMANTAL

2.1. Casting

Thermodynamic simulations (Thermo-Calc®) were used for optimal composition and process parameters derivation. The composition of the base alloy (NB) is presented in Table 1. Two different NB based alloys with additions of Ce or Nd mischmetal (MM) (Magnesium Research Institute, DSM, Beer Sheva, Israel) were melt (730°C) in a steel (DIN ST37) crucible under CO_2+Freon134 protective atmosphere. The Ce MM composition was 50%Ce-25%La-20%Nd-5%Pr, and the Nd MM composition was Mg-40%Nd (4). Tensile samples (E 8M-96, Ø9 mm) (5) were permanent mold cast in the AISI H13 steel mould. The compositions of the alloys were confirmed by induced coupled plasma optical emission (BAIRD, DV-5). The NB, NB+Nd and NB+Ce alloys compositions are presented in Table 1, as well the compositions of reference alloys (AM50 & AE42).

Table 1: Compositions of the NB+Nd\Ce alloys.

Alloy	Al Wt.%	Zn Wt.%	Mn Wt.%	Sn Wt.%	Ce Wt.%	La Wt.%	Nd Wt.%	Mg Wt.%
NB	4.61	0.97	0.33	2.21	---	---	---	balance
NB+1%Nd	4.71	0.9	0.41	1.66	<.0.01	<.0.01	1.23	balance
NB+1%Ce2	4.96	0.98	0.36	2.3	0.5	0.28	0.13	balance
AM50	4.9	0.2	0.45	---	---	---	---	balance
AE42	3.65	0.01	---	---	1.3875	0.6858	0.4301	balance

Heat treatments were performed under Ar protective atmosphere with a rapid elevation of temperature from room temperature up to 400°C, following an hour of dwelling before rising the temperature up to 440°C in 4 hours, dwelling at that temperature for 16 hours, following by water quenching.

The as-cast tensile specimens were sliced at the neck region into 5mm length sections. These sections were cold-mounted with Buehler EpoThin™ Epoxy. The metallurgical samples were polished by SiC papers and 50 nm Gamma Alumina (Buehler). The samples were analyzed by Vickers micro hardness (Matsuzawa Seiki co. LTD.), HRSEM (Zeiss Ultra-

Plus) and XRD analysis (Rigaku Smart Lab X-Ray Diffractometer) for microstructure evolution.

2.2 Microstructure Evolution of NB+Nd During Creep-Aging

The NB+Nd tensile samples were solute treated (T4) followed by creep under load of 65 MPa at 200°C. The samples were tested until failure and analyzed by HRSEM in the parallel and in the perpendicular directions to the load direction. The results were compared to the microstructure of solute treated and aged at 200°C (T6) unloaded samples.

2.3 Tensile Stress Tests of NB+Nd and NB+Ce Alloys at Elevated Temperatures

Tensile tests were performed at temperature 160±5°C on NB+Nd and NB+Ce tensile samples by Instron 3369, using Instron SFL heatwave, in order to measure the alloys' yield stress.

2.4 Creep Tests

Creep tests were conducted by a lever-arm SATEC-like creep machine equipped with a circulated air furnace, under controlled load and temperature conditions. The sample displacement was measured using a Singer Instruments & Control LVDT ST-25. The data was collected and analyzed using MATLAB®, by 4th degree polynomial fitting and finding the minimal strain rate $\dot{\varepsilon}_s$, which is equivalent to secondary, steady-state creep rate.

Two sets of data were acquired: the first was a set of four loads at the same temperature - (185°C) under loads of 60, 65, 70 and 85 MPa; the second was a set of four different temperatures under the same load: 155°C, 165°C, 175°C and 185°C under the load of 65 MPa. The first set was used for extracting the strength factor, n, and the threshold stress, σ_{th}, using the following algorithm:

Steady state creep is given by equations [2]:

$$(1)\ \dot{\varepsilon}_s = \overbrace{\frac{ADb}{k}}^{A'}\frac{G(T)}{T}\left(\frac{\sigma-\sigma_{th}}{G(T)}\right)^n$$

$$(2)\ \frac{\sigma_{th}(T)}{G(T)} = \sigma_{th}^{0}e^{\left(\frac{Q_{th}}{RT}\right)}$$

where D is the diffusion coefficient, G is the shear modulus, k is the Boltzmann constant, σ is the applied stress, Q_{th} is the threshold stress activation energy, R is the gas constant, b is the Burgers vector, A is a material constant.

Following to eq.(1), a function of $\dot{\varepsilon}_s^{\frac{1}{n}}$ $vs\ \sigma$ was plotted for several possible values of n (3, 5, 7) (Fig. 3). A linear extrapolation was made to zero to determine the threshold stress [6]. The best fit of linear trend line was selected according to the R^2 values. Another way to find n was employed by using the graph $\ln\left(\dot{\varepsilon}_s\right)$ vs $\ln\left(\sigma-\sigma_{th}\right)$. The slope was derived as the strength factor, n.

In order to find the creep activation energy, an assumption was made that the threshold stress activation energy, Q_{th}=6 kJ/mol, that was based on the results for the similar designed NB+Ce alloy [2], where the Mg_2Sn precipitates were supposed to have a similar role interacting with dislocations and providing a threshold stress. Thus, using eq.(2), σ(T) were derived for 155, 165, 175°C. Temperature dependent shear modulus G(T) was derived by linear extrapolation from previous work [7].

The values of σ(T) and G(T) were incorporated into eq.(1) and a function of ln(A') vs. 1/T was plotted. The creep activation energy Q_{cr} was derived by multiplying the slope to the gas constant, $R_{gas.}$

3. RESULTS

3.1 Microstructure Evolution During Aging

XRD diffraction patterns were acquired from aged samples after 48 hours and 768 hours in the parallel and perpendicular direction, in respect to the loading direction (AG # PARA/PERP) (Fig.1).

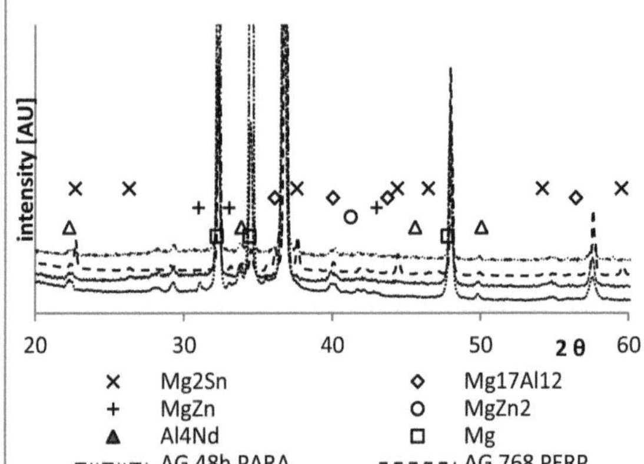

Figure 1: XRD diffraction patterns of NB+Nd (T6) samples for various orientations and different aging times (200°C, 48h in the parallel direction, 48h perpendicular, 0h perpendicular, 768h perpendicular).

The XRD spectra of the cross section perpendicular and parallel to the creep loading direction after aging for 48h (labeled: AG 48h perpendicular and AG 48h parallel, respectively) are similar, there is no preferred orientation at aged (unloaded) samples. Al_4Nd and $Mg_{17}Al_{12}$ were detected in all samples. Mg_2Sn is discernible for the 48h aged and the over-aged for 768h specimens. MgZn phase was detected in the AG 0h and the AG 768h specimens.

3.2 Creep Analysis

3.2.1 Derivation of n and σ_{th}

Following the steps that are described in section 2.4, the $\dot{\varepsilon}_s$ was derived from the fittings of creep curves at 185°C (Figs.2 - 4).

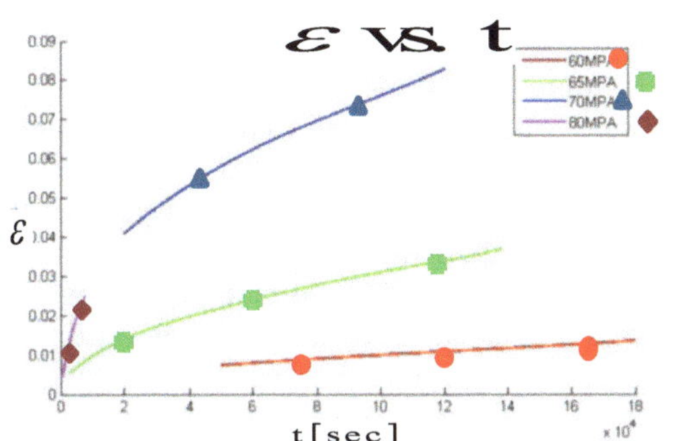

Figure 2: 4th degree polynomial fittings for the creep tests of solution treated (T6) NB+Nd alloy under 60, 65, 70, 80 MPa loads at 185°C.

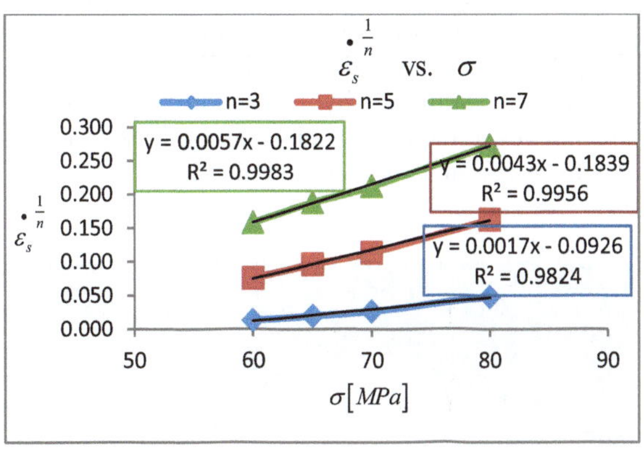

Figure 3: Plot of $\dot{\varepsilon}_s^{\frac{1}{n}}$ vs. $\sigma[MPa]$ for NB+Nd (T6) alloy at 185°C, comparing 3 different values of n.

The different *n* values are related to the different creep mechanisms. For example, n≈2 indicates GB sliding mechanism, n≈3-7 indicates dislocation climb [8]. Deriving σ_{th} for n=3, 5, 7 yields σ_{th}=54.5, 42.8, 32.0 [MPa], respectively. As it can be seen the best fitting (according to the R^2 values), is for n=7. As it can be seen the slope in the logarithmic coordinates (Fig.4) derived for n=7 has the most close value to the initial guess (6.988), thus the strength factor was determined to be n=7, that is indicative of dislocation climb as a governing mechanism.

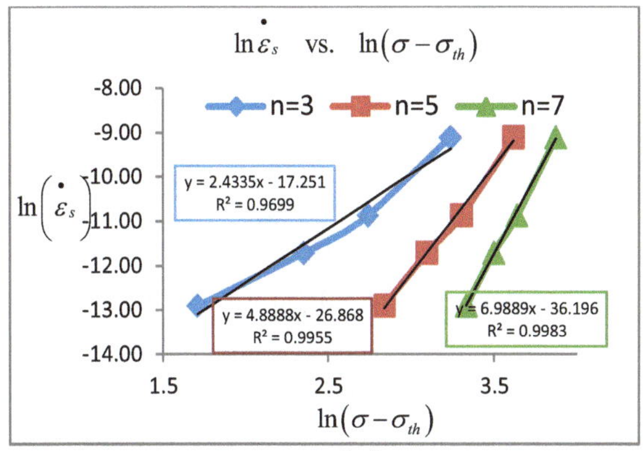

Figure 4: Isothermal plot of $\ln\left(\dot{\varepsilon}_s\right)$ vs $\ln(\sigma-\sigma_{th})$

3.2.2 Creep Activation Energy

The value of creep activation energy, Q_{cr}, is derived from the slope of ln(*A*') vs (1/T) (eq.(1), Fig.5), Q_{cr}=134.84 kJ/mol≈135 kJ/mol, that coincides with the α-Mg self-diffusion activation energy, and this is in agreement with dislocation climb activation energy [9].

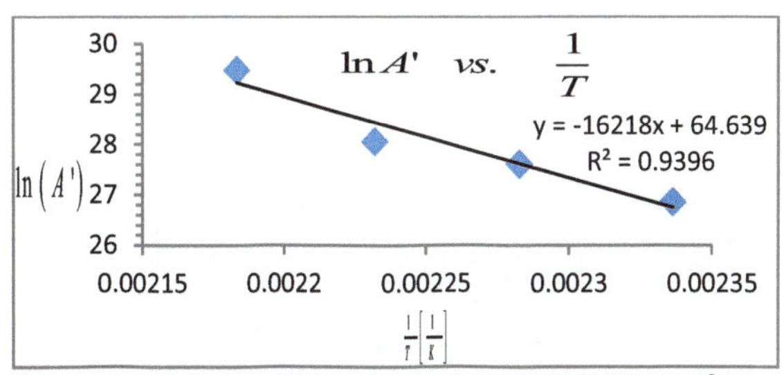

Figure 5: ln(A') (eq. (1))vs. 1/T, at T=155,165,175,185°C.

3.3 Comparison of Creep in the NB+Nd and NB+Ce Alloys

The NB+Nd alloy has a similar creep resistance to AM50 alloy. It has inferior creep resistance in comparison with NB+Ce despite that both alloys have similar microstructure design and similar precipitates structure (Fig. 6).

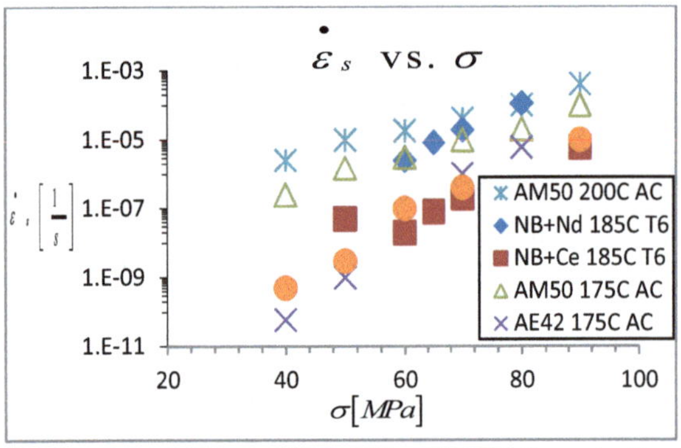

Figure 6: comparison of NB+Nd with NB+Ce and other Mg-Al alloys [(2), (10), (11)].

Strength factor of NB+Nd was derived as n=7 while for NB+Ce the best fitting corresponds to n=6.3 [(2)]
The creep activation energy, Q_{cr}≈135 kJ/mol for both alloys, is in agreement with dislocation climb mechanism [(9)].
The yield stresses that were derived from the tensile tests at 160±5°C were 29.5±0.9 MPa for the NB+Nd alloy and 40.8±3.5 MPa for the NB+Ce alloy. The difference in the yield stress indicates the superior mechanical properties of the NB+Ce alloy.

3.4 Microstructure Evolution of NB+Nd Alloy During Aging Vs. Creep-Aging Conditions

Micrographs of T6 aged for 48 hours sample were compared with T4 sample that underwent creep under 200°C & 65 MPa for 48 hours (Fig. 7).

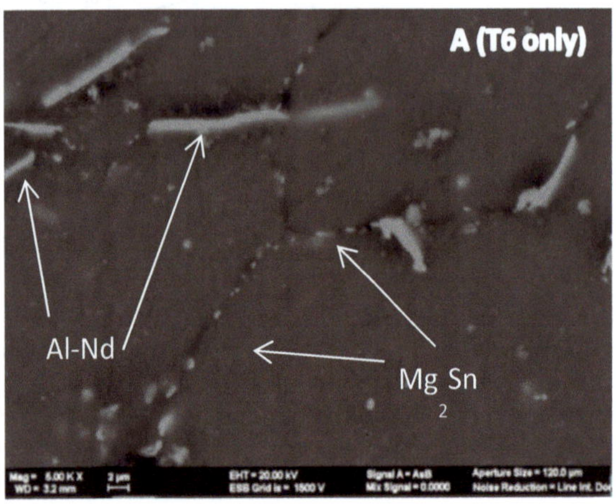

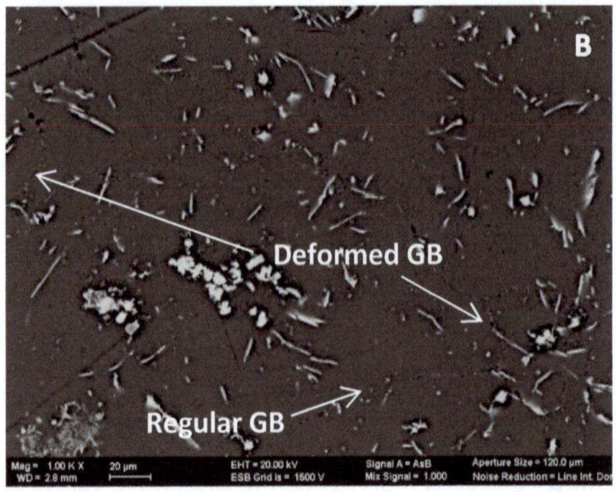

Figure 7: A - regular GBs that have been exposed only to aging. B - Coexistence of regular GB and deformed GB. The load direction is horizontal.

There were found many GBs deformed by the exposure to stress with embedded elongated precipitates (Fig. 7B), unlike to the non-stressed sample, where GBs have no such precipitates (Fig. 7A). As can be seen in Figs. 8(C) and 8(C1), there are precipitates inside the bulk of grain that are arranged in bands with about 45° orientation with respect to the applied stress direction.

The formation of elongated precipitates at GBs with about 45° orientation to the loading direction was observed only in the crept samples, and was not found in the aged only samples. Formation of such precipitates was accompanied by formation of near-GB precipitate depleated zones (PDZ), which, in turn, are farther followed by formation of elongated precipitates in the regions close to the PDZ, forming the precipitation "crust" around the PDZ.

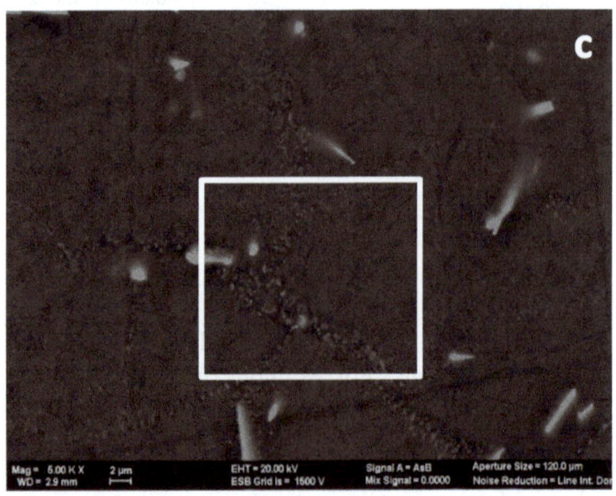

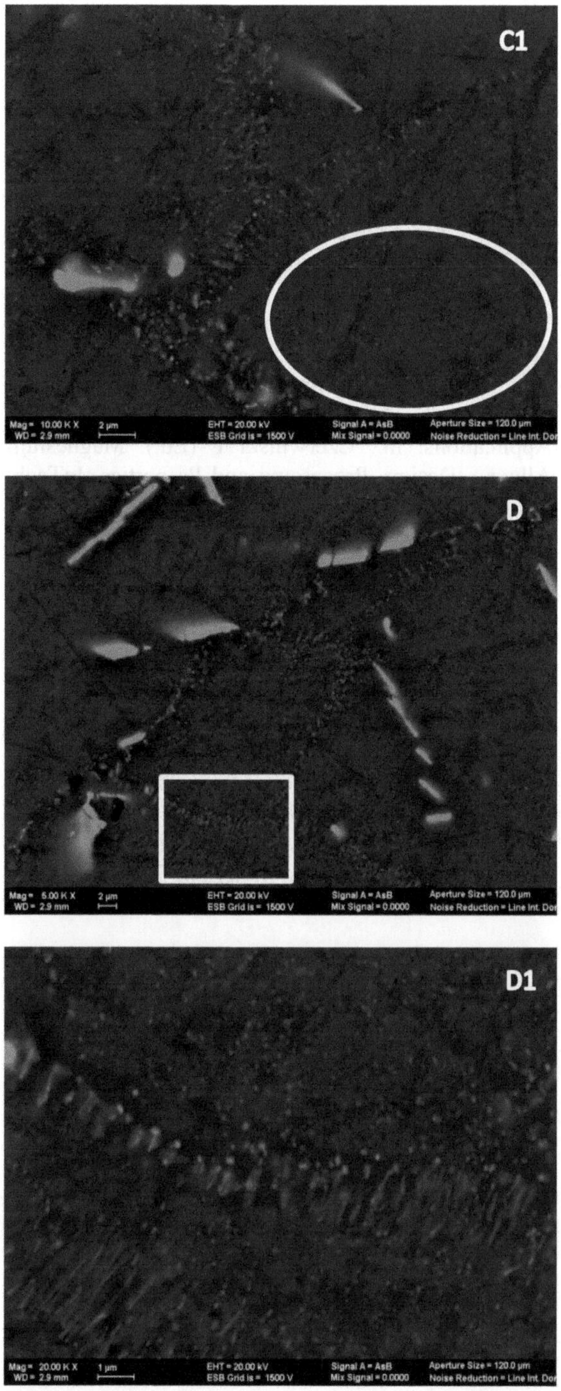

Figure 8: C, C1 - slip bands with about 45° orientation to the loading direction. D, D1 - elongated, needle-like, $MgZn_2$ precipitates at the GB and in the near-GB zone (precipitation “crust”) with about 45° with respect to the loading direction. The load direction is horizontal.

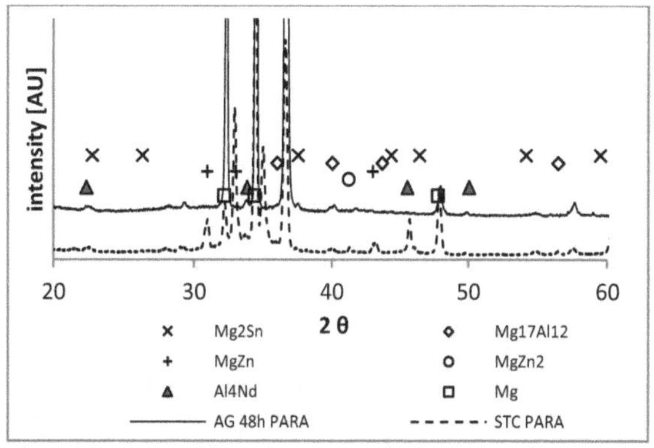

Figure 9: XRD diffraction patterns of T6 sample aged for 48h at 200°C (AG48) in the parallel direction to the load and of T4 sample in the parallel direction that was crept under load of 65MPa until failure (about 48h) at 200°C (STC).

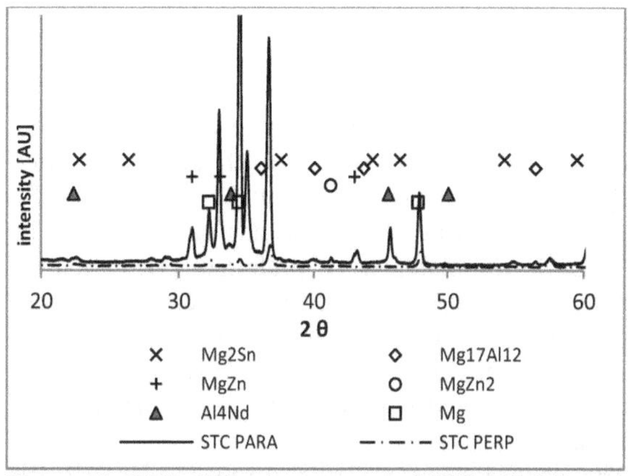

Figure 10: XRD diffraction patterns of STC (T4 followed by creep at 200°C under 65 MPa) sample in the parallel and the perpendicular directions to the load.

As can be seen from Figs. 9 & 10 the $MgZn_2$ phase was detected only in the crept samples and only in the direction parallel to the loading direction.

4. DISCUSSION

Analysis of creep in the NB+Nd and NB+Ce alloys at elevated temperatures revealed that the deformation is controlled by dislocation climb mechanism with strength factor n=7 and a threshold stress σ_{th}=32.0 MPa (Figs.3, 4)

It was found that precipitation during aging in tensile crept samples of the NB+RE alloys differed from that in unloaded sample. As seen in Fig. 8 (C1), bands of precipitates were formed under loading conditions at elevetaed temperature. By observing an orientation of about 45° to the loading direction, it is sugested that those bands are the dislocation slip lines. As dislocations provide a quick diffusion lane, they become preferred precipitation zones, thus the prcipitates nucleate in these patterns of bands.

The elongated precipitates that are formed at the GBs and in the near-GB zones in the creep loaded specimens, end up with a small round precipitates (Fig. 8 D1). Considering Figs. 9 and 10, and the about 45 degrees bands orientation in respect to the loading direction, one can conclude that the elongated precipitates are $MgZn_2$ phase that nucleates with preferred orientation, the Mg_2Sn precipitates being the preferred nucleation sites [3].

The creep resistance of the NB+Ce alloy is superior to that of the NB+Nd alloy. This can be attributed to the yield stress of the NB+Ce alloy at elevated temperatures. It can be seen that after the solution treatment, $Mg_{17}Al_{12}$ phase was presented in the NB+Nd alloy, which contributes to the poor creep properties of the alloy [(12), (13)]. It seems that the chosen heat treatment process parameters for the NB+Nd alloy were not the optimal ones.

5. SUMMARY AND CONCLUSIONS

It was found that during aging under tensile loading condition in the Mg-Al-Sn-Zn-RE alloy, elongated $MgZn_2$ precipitates are formed, with Mg_2Sn precipitates being the likely nucleation sites. The $MgZn_2$ precipitates have preferred orientation of about 45° in respective to the loading direction. These precipitates have been detected only in the loaded samples, thus one can conclude that the creep deformation under tensile stress was helping the formation of $MgZn_2$.

The controlling creep mechanism in the NB+RE alloys at elevated temperature was determined to be dislocation climb, by finding the strength factor n=7 and the creep activation energy $Q_{cr} \approx 135$ kJ/mol, that is the α-Mg self-diffusion activation energy.

By comparing creep resistance of the NB+Ce and NB+Nd alloys, it can be concluded that Nd and Ce have different effect on creep resistance. Different heat treatments are required for preventing unwanted brittle eutectic phase.

6. REFERENCES

[1] Avraham S, Katsman A, Bamberger M., Strengthning Mechanisms in Mg-Al-Sn-Zn Based Alloys, in:. S. Agnew, Neale R. Neelameggham, Eric A. Nyberg, W. Sillekens (Eds.). Magnesium Technology 2010. Washington State Convention Center, Seattle, WA: 2010.

[2] Avraham S, Strengthening Mechanisms In Mg-Al-Sn-Zn Based Alloys. Reserch thesis, Haifa, Israel : Department of Materials Engineering, Technion - Israel institute of technology. 2010.

[3] Gorny A, Katsman A., Precipitation and stress-influenced coarsening in Mg-based Mg–Zn–Sn–Y and Mg–Zn–Sn–Sb alloys, Journal of Materials Research, 23 (2008)1228.

[4] The Magnesium research institute, DMS, Israel. http://www.dsmag.co.il/.

[5] ASTM E 8M-96. Test Methods for Tension Testing of Metallic Materials. ASTM International; 2013.

[6] Li Y, Langdon T.G., Scripta Materialia 1997;36:1457.

[7] Somekawa H, Hirai K, Watanabe H, Takigawa Y, Higashi K., Materials Science and Engineering: A 2005;407:53.

[8] Meshinchi K., Improving the Properties of Magnesium Alloys for High Temperature Applications, in:. Czerwinski F (Ed.). Magnesium Alloys - Design, Processing and Properties. InTech; 2011.

[9] Pekguleryuz M.O., Kaya A.A., Advanced Engineering Materials 2003;5:866.

[10] Dieringa H, Hort N, Kainer K.U., Materials Science and Engineering A 2009;510-511:382.

[11] Ishimatsu N, Terada Y, Sato T, Ohori K., Metallurgical and Materials Transactions A: Physical Metallurgy and Materials Science 2006;37:243.

[12] Bryła K, Dutkiewicz J, Faryna M, Dobatkina T, Rokhlin L., Archives of Materials Science and Engineering 2008;Vol. 29:40.

[13] Moreno I.P., Nandy T.K., Jones J.W., Allison J.E., Pollock T.M., Scripta Materialia 2003;48:1029.

Magnesium Technology 2015
Edited by: Michele V. Manuel, Alok Singh, Martyn Alderman, and Neale R. Neelameggham
TMS (The Minerals, Metals & Materials Society), 2015

HOT COMPRESSION BEHAVIOR OF MAGNESIUM ALLOYS ZE20 AND AM30

Scott Sutton[1], Alan A. Luo[1, 2]
[1]Department of Materials Science and Engineering, The Ohio State University, Columbus, OH 43210, USA
[2]Department of Integrated Systems Engineering, The Ohio State University, Columbus, OH 43210, USA

Keywords: Magnesium Alloys, Extrusion, Deformation Behavior, Hot Compression, Microstructure

Abstract

Magnesium alloys have recently received much attention for their weight-saving potential in transportation systems. Their implementation is hindered, however, by poor formability. The present work compares the mechanical response to warm isothermal compression of two recently developed magnesium extrusion alloys, AM30 (Mg-3Al-0.4Mn)[1] and ZE20 (Mg-2Zn-0.2Ce). It was found that AM30 exhibits higher initial hardening hardening followed by softening, while ZE20 exhibited only slight softening and ended at a higher steady-state stress than AM30.

Introduction

Magnesium has the lowest density of all structural metals, making it an attractive choice to the automotive and aerospace industries for lightweighting applications. Extruded magnesium components have received much attention from these industries, but their application is limited by low room temperature ductility and high anisotropy in mechanical properties.

Magnesium's hexagonal structure and crystalline anisotropy are the primary reasons for these disadvantages compared to steels and aluminum alloys. At low temperatures, magnesium only has two independent slip systems along the basal planes and does not satisfy the Von Mises criterion for ductility. In such temperature regimes, magnesium will initiate twinning in order to accommodate strain along its c-axis [1, 2]. Heavy twinning can lead to crack formation and, eventually, failure. This is especially important in wrought structures, as magnesium alloys often form sharp basal textures in forming operations.

The conventional approach to increase the formability of magnesium is to use elevated temperatures, thereby thermally activating enough slip systems to make the material more ductile. This, however, increases processing costs to the point where magnesium is no longer cost-competitive with other materials [3]. Recent research has focused on alloying in order to improve magnesium's formability at lower temperatures. Alloying to modify grain size or texture are popular strategies. Regardless the mechanism, several alloying elements, including Al, Ca, Zn, Li, Y, Ce, La, Nd, Sm, and Gd have been identified as effective at improving low temperature ductility [4, 5].

The present work compares the hot compression behavior of two relatively new magnesium extrusion alloys, ZE20 (Mg-2Zn-0.2Ce) and AM30 (Mg-3Al-0.4Mn), both considered to have extrudability superior to the commercial wrought alloy, AZ31 (Mg-3Al-1Zn) [4, 6, 7]. While AM30 is relatively high-strength alloy (similar to AZ31 but with slightly improved ductility) [4], ZE20 alloy has shown remarkable ductility (about 27% elongation) with reasonable strength (135 MPa tensile yield strength) [6]. There is very limited data available in literature on the thermo-mechanical behaviors of these newly developed alloys. In this paper, mechanical data are presented for these alloys at various temperatures and strain rates. The ultimate purpose of this experiment is to provide data which will be used to build integrated computational materials engineering (ICME) models in order to accelerate the development of structural applications of these new alloys.

Experimental Methods

Materials

ZE20 and AM30 alloy billets (diameter 182 mm) were supplied by the United States Automotive Materials Partnership (USAMP). Both ZE20 and AM30 billets were cast using direct-chill casting process. However, the AM30 billet from a previous production batch was slightly pre-extruded (about 20% reduction of billet diameter by extrusion for improved workability). Since both materials have been used by USAMP for component manufacturing and testing, the two alloys are evaluated in this study, despite the minor difference in processing history.

Mechanical (Gleeble) Testing

Cylindrical compression specimens of 10mm diameter and 15mm length were machined from as-received AM30 and ZE20 billets. Specimens were machined such that the compression direction was parallel to the billet processing directions.

Isothermal compression tests were performed on a Gleeble 3800 thermo-mechanical testing system at three strain rates and three temperatures. The complete test matrix for this experiment is shown in Table I. Compression testing was controlled by a three-segment program consisting of a heating period, a thermal "soak" period, and compression. Specimens were first heated at 5°C/s to the desired temperature, then held for a three minutes to ensure thermal equilibrium, and compressed to 70% nominal strain at the desired nominal strain rate. Compression testing was quickly followed by a water quench. Temperature was monitored and controlled during each test by a single K-type thermocouple percussion welded to the midline of the specimen. Multiple tests were performed at each temperature and strain rate to ensure repeatability.

Table I. Testing Conditions

Alloy	Temperature	Strain Rate (Targeted)	Strain Rate (Actual)
AM30 ZE20	200°C 375°C 425°C	0.01 s^{-1} 5.0 s^{-1} 10 s^{-1}	0.01 s^{-1} 4.74 +/- 0.05 s^{-1} 9.21 +/- 0.29 s^{-1}

[1] All compositions in wt.% except otherwise stated.

Optical Microscopy

Specimens were sectioned with a low-speed saw and mounted in epoxy, so as not to introduce additional deformation or alter the specimens' thermal history. Specimens were ground then polished down to 1μm diamond abrasive; final polish was obtained using colloidal silica for 60 seconds. Specimens were etched for five seconds by immersion in a solution consisting of 5% nitric acid, 15% acetic acid, 20% water, and 60% ethanol. Microstructures were obtained via optical microscopy in an IM-3000 microscope.

Results

Mechanical Response

Figures 1 and 2 show typical results of compression tests for ZE20 and AM30, respectively. In general, both materials exhibit typical response to variation in temperature and strain rate. As the temperature increases, the materials' flow stresses decrease; and, as the strain rate increases, the flow stresses also increase. Similar studies have attributed this relationship between flow stress and strain rate to increased dynamic recrystallization (DRX) at low strain rates (longer time scales) [8, 9].

AM30 specimens exhibited more initial hardening and had a higher peak stress than ZE20 specimens tested under the same conditions. AM30 specimens also often showed significant softening after the peak stress, eventually ending at a steady-state stress at high strains. ZE20 specimens exhibited softening to a lesser degree, and ended at a higher steady-state stress.

In cases where the specimens exhibited fracture or failure, the above-mentioned trends were not observed. Figures 2b and 2c show failure of AM30 specimens tested at 200°C and 5/s and 10/s. All AM30 specimens tested at these conditions fractured catastrophically at 45 degrees to the loading axis. Some ZE20 specimens also fractured at 200°C and 10/s; but failure was inconsistent, only occurring in two of six specimens. ZE20 specimens failed more consistently at 0.01/s and 200°C, with specimens developing internal cracks at 45 degrees to the loading axis, though there is no discernable drop in stress in Figure 1a.

Optical Microstructure

Optical microscopy revealed that the deformation microstructures of the two materials are very different, which is also complicated by the considerably different starting microstructures. Figures 3 and 4 show the microstructure of each material before and after compression testing.

Figure 3a shows that the initial ZE20 microstructure consisted mainly of large grains containing some twins with a second phase distributed along most grain boundaries, likely Mg-Ce-Si intermetallic particles [6]. Figure 3b shows that the AM30 specimens retained some microstructural artifacts due to pre-extrusion. The microstructure consists of long, columnar grains which have mostly recrystallized. There are also strings of second phase particles aligned with the processing direction; these were reported to be Al_8Mn_5 in previous work [7].

The microstructures after deformation were also considerably different. ZE20 specimens generally formed many intersecting shear bands, with relatively unaffected areas between these shear bands. There was also heavy recrystallization apparent in the shear bands, giving the appearance of the early stages of a necklace-type microstructure in those areas. Figure 4a shows a tested ZE20 microstructure with these characteristics. The micrograph in Figure 4a focuses on a shear band running from lower left to upper right with much recrystallization visible, but a relatively undeformed region can also be seen at the upper left.

AM30 specimens displayed much more uniform material flow, with material symmetrically flowing outward from the center of the specimen, causing some barreling. Figure 4b shows a deformed AM30 microstructure. Most grains visible in Figure 4b are small and equiaxed, evidence that recrystallization progressed through testing. Though not easily visible in Figure 4b, AM30 specimens do retain microstructural artifacts of the pre-extrusion process, which complicates comparison of these materials.

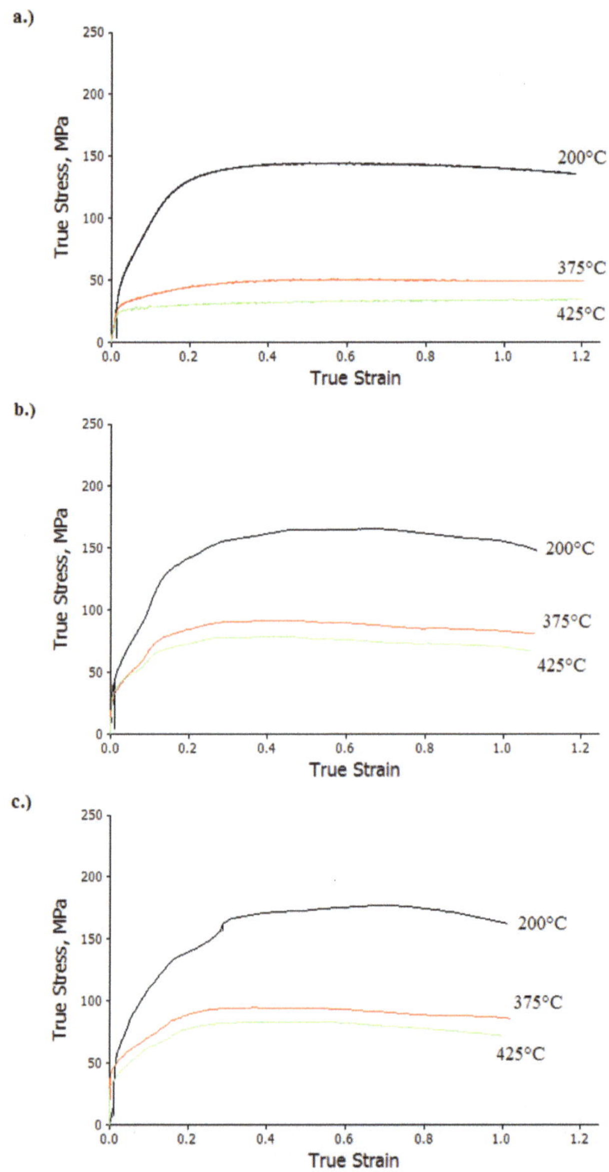

Figure 1. True stress-true strain curves obtained for hot compression of ZE20 at a) $0.01s^{-1}$, b) $5.0s^{-1}$, and c) $10s^{-1}$.

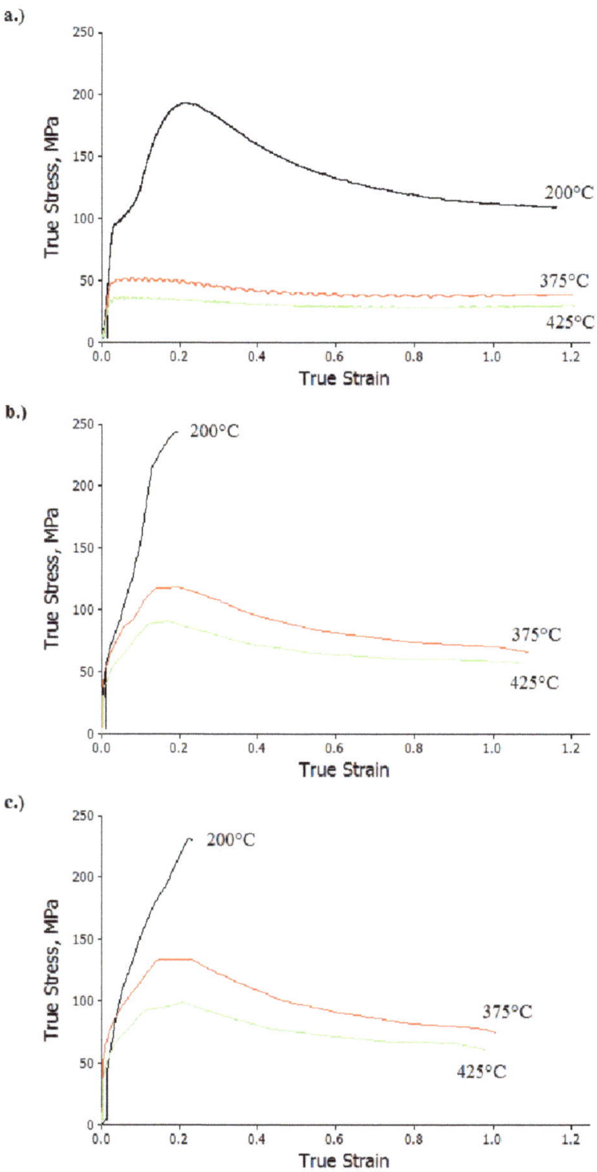

Figure 2. True stress-true strain curves obtained for hot compression of AM30 at a) $0.01s^{-1}$, b) $5.0s^{-1}$, and c) $10s^{-1}$.

Summary and Future Work

The Gleeble test data in this paper provide a good description of thermo-mechanical response of two relatively new magnesium extrusion alloys in the temperatures and strain rates of interest to industry. The data are currently being used to develop material models for extrusion process simulation.

Additional data are required to make a full assessment of the deformation mechanisms of these alloys. Electron backscatter diffraction (EBSD) maps will be collected and used to characterize the texture evolution during hot deformation. Compression tests halting at various final strains (interrupted tests) paired with EBSD maps are currently underway to provide further understanding in deformation mechanisms of these alloys.

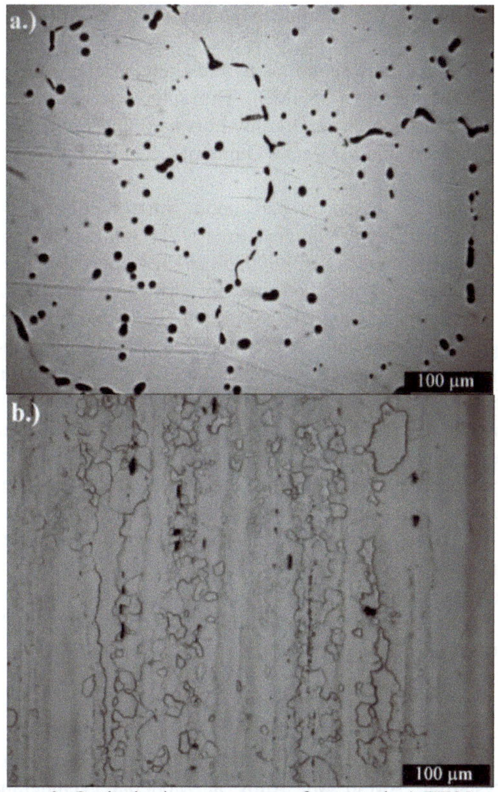

Figure 3. Optical microstructures of untested a.) ZE20 and b.) AM30. Billet processing direction is vertical.

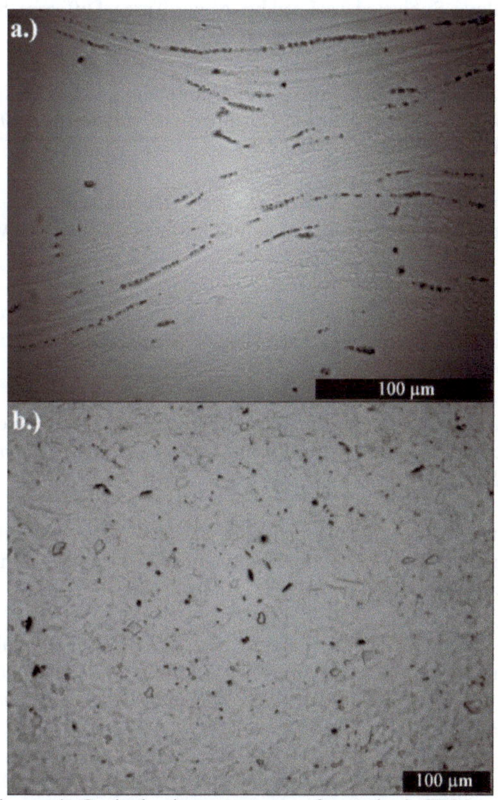

Figure 4. Optical microstructures of tested a.) ZE20 and b.) AM30. Compression direction is vertical.

Acknowledgements

This work is partially supported by the United States Automotive Materials Partnership (USAMP). The authors acknowledge the technical discussions with Drs. Joy Forsmark and Mei Li of Ford Motor Company and Mr. James Quinn of General Motors.

This material is based in part upon work supported by the Department of Energy National Energy Technology Laboratory under Award Number DE-EE0003583. This report was prepared as an account of work sponsored by an agency of the United States Government. Neither the United States Government nor any agency thereof, nor any of their employees, makes any warranty, express or implied, or assumes any legal liability or responsibility for the accuracy, completeness, or usefulness of any information, apparatus, product, or process disclosed, or represents that its use would not infringe privately owned rights. Reference herein to any specific commercial product, process, or service by trade name, trademark, manufacturer, or otherwise does not necessarily constitute or imply its endorsement, recommendation, or favoring by the United States Government or any agency thereof. The views and opinions of authors expressed herein do not necessarily state or reflect those of the United States Government or any agency thereof.

References

1. L. Jiang, et al. "Twinning-Induced Softening in Polycrystalline AM30 Mg Alloy at Moderate Temperatures." *Scripta Materialia* 54.5 (2006), 771-75.

2. L. Jiang, et al. "Twinning and Texture Development in Two Mg Alloys Subjected to Loading Along Three Different Strain Paths." *Acta Materialia* 55.11 (2007), 3899-910.

3. L. Gao, and A.A. Luo. "Hot Deformation Behavior of as-Cast Mg-Zn-Mn-Ce Alloy in Compression." *Mater. Sci. Eng. A* 560 (2013), 492-99.

4. Luo, A. A., and A. K. Sachdev. "Development of a New Wrought Magnesium-Aluminum-Manganese Alloy AM30." *Metall. Mater. Trans. A* 38A.6 (2007), 1184-92.

5. R.K. Mishra, et al. "Influence of Cerium on the Texture and Ductility of Magnesium Extrusions." *Scripta Materialia* 59.5 (2008), 562-65.

6. A.A. Luo, R.K. Mishra, and A.K. Sachdev. "High-Ductility Magnesium-Zinc-Cerium Extrusion Alloys." *Scripta Materialia* 64.5 (2011), 410-13.

7. Luo, A. A., C. Zhang, and A. K. Sachdev. "Effect of Eutectic Temperature on the Extrudability of Magnesium-Aluminum Alloys." *Scripta Materialia* 66.7 (2012), 491-94.

8. A.J. Thome, W.Z. Misiolek, and A.A. Luo. "Workability Evaluation of New Mg-Zn-Ce Alloys." *9th International Conference on Magnesium Alloys and Their Applications, Proceedings* (2012), 703-8

9. L. Jiang, et al. "Microstructure and Texture Evolution During the Uniaxial Tensile Testing of AM30 Magnesium Alloy." *Magnesium Technology 2006, Proceedings* (2006), 233-38.

Magnesium Technology 2015
Edited by: Michele V. Manuel, Alok Singh, Martyn Alderman, and Neale R. Neelameggham
TMS (The Minerals, Metals & Materials Society), 2015

CREEP DEFORMATION MECHANISMS AND RELATED MICROSTUCTURE DEVELOPMENT OF AZ31 MAGNESIUM ALLOY

Peiman Shahbeigi Roodposhti[1], Apu Sarkar[2], Korukonda Linga Murty[2]
[1]North Carolina State University Mater. Sci. Eng. Department; 911 Partner Way, 3035 EB1, Raleigh, NC 27606-7907 USA
[2]North Carolina State University; 2500 Stinson Drive, 3143 Burlington Engineering Labs, Raleigh, NC 27606-7909 USA

Keywords: Creep, Magnesium, AZ31, Dislocation density, Intermetallic phases

Abstract

Because of ever increasing demanded of Magnesium alloys in various industries, high temperature deformation of Mg-Al-Zn alloys (AZ31) at constant stress (i.e. creep) were studied at a wide range of stresses and temperatures to characterize underlying deformation mechanism and dynamic recrystallization (DRX) Various microstructures (e.g. grain growth & DRX) are noted during steady-state creep mechanisms such as grain boundary sliding (GBS), dislocation glide creep (DGC) and dislocation climb creep (DCC). Although a combination of DRX and grain growth is characteristic of low stacking fault energy materials like Mg alloys at elevated temperatures, observation reveals grain growth at low strain-rates (GBS region) along with dynamic recovery (DRV) mechanism. Scanning Electron Microscopic (SEM) characterization of the fracture surface reveals more inter-granular fracture for large grains (i.e. GBS region with DRV process) and more dimple shape fracture for small grains (i.e. DGC & DCC region with DRX).

Introduction

Researchers tried to investigate Magnesium alloys behavior at high temperature to characterize their creep resistance and fracture mechanisms[1–13] to be able to improve their mechanical properties at sever condition of stress and temperature . It is well known that their poor creep resistance at higher temperatures is due to intermetallic phase dissolving in the matrix[14, 15] which leads to clamping load reduction in bolted joints in power-train components of automobiles[10]. A well-known classification of the creep behavior of materials is based on Born, Mukherjee and Dorn power-low equation in which the minimum strain rate is defined as a function of stress, temperature and grain size:

$$\dot{\varepsilon} = \frac{ADGb}{kT}\left(\frac{b}{d}\right)^p\left(\frac{\sigma}{G}\right)^n \quad (1)$$

where n is the stress exponent, p is the inverse grain size exponent, A is a material constant, D is diffusion coefficient, G is the shear modulus, k is the Boltzmann's constant, b is the burgers vector, d is the grain size and T is the absolute testing temperature. The diffusion coefficient can be related to activation energy via $D=D_0 \exp(-Q/RT)$ where the Q is activation energy for specific creep mechanism, D_0 is a pre-exponential term and R is the gas constant. The stress exponent depends on the underlying creep mechanism: n~2 is related to the grain boundary sliding, n=3~3.5 dislocation glide and n=4.5~7.5 is related to dislocation climb.[1–3, 10, 14–19] DRX is one of the most significant mechanisms for grain refinement and improved mechanical properties during hot working[20], which reduces the flow stress of deformation[21]. Unlike high stacking fault energy metals, DRX is a common phenomenon during high temperature deformation of low or medium stacking fault energy metals, such as magnesium[22–27]. In Mg-Al alloys, limited number of slip systems and higher grain boundary diffusion rate can also stimulate DRX[21]. Various investigations showed that during DRX new equiaxed grains appear on the preexistent grain boundaries when the dislocation density reaches a critical value[27–29]. Although there are some reports on the influence of primary grain size on DRX[21, 30], new grain size is generally independent of the initial grain size and mostly related to the deformation conditions[21]. It is well understood that by increasing the temperature and strain-rate more nucleation sites appear which leads to easier DRX and more grain refinement[31] while, stress, another essential parameter for DRX, does not influence the nucleation behavior and may strongly affect the kinetics of plane strain deformation[32]. At higher temperatures the microstructure is more homogeneous albeit there is more grain growth due to higher grain boundary migration and diffusion[33]. In this study, the creep deformation of AZ31 magnesium alloy was investigated at high temperatures ranging from 503 to 623 K (i.e 0.53 ~ 0.68 T_m) to evaluate underlying deformation mechanisms, grain size distribution and the existence of DRX at different stages.

Experimental procedure

The experimental material was a hot rolled, rod shape AZ31 magnesium alloy with a dimension of 1" × 1" × 12" and chemical composition of Mg-2.7Al-1.2Zn (in wt. %). Dog-bone type tensile specimens of 25 mm gage length were machined out of the rod. The samples were annealed for 2 hrs in argon atmosphere at 450°C and furnace cooled to room temperature to attain homogeneous equiaxed grains. High temperature creep tests at constant load were performed at 230°C, 270°C and 350°C with the tensile axis parallel to the rolling direction. The experimental setup used to perform direct load creep tests consists of an extensometer and a Linear Variable Differential Transducer (LVDT) to measure the axial strain. Creep tests were performed under stresses ranging from 1 MPa to 70 MPa for dwell times of 3 minutes up to few weeks. By the end of the tests, the specimens were cooled under load to room temperature to preserve final stage crept microstructure. The microstructure and chemical composition of the sample were examined by a Hittachi-3200 Scanning Electron Microscope (SEM) equipped with X-ray energy dispersive spectroscopy (EDS). In situ high temperature XRD study was performed on the starting Mg alloy sample to investigate the thermal stability of the precipitates. Sample was placed inside a furnace at a vacuum of 5×10^{-5} torr and heated from ambient temperature to 450 °C at a rate of 5 °C/min. Diffraction profiles were recorded at 50 °C interval. To prepare Transmission

Electron Microscopy (TEM) foils, samples were thinned by standard twin jet electro-polishing technique with a solution of 5% $HClO_4$ and 95% ethanol.

Results and discussion

Figure 1 shows the TEM microstructure of the annealed sample and corresponding diffraction pattern for matrix and precipitates. Diffraction pattern of the matrix with zone axis of $[10\bar{1}0]$ is indicative of the hcp crystal structure, while the precipitates have different crystal structures.

A large number of creep tests have been performed at stresses ranging from 1 to 70 MPa. Depending on the creep conditions (stress & temperature) a wide range of creep curves with very low to large primary creep (stage I) regimes was noted. During this stage creep rate continuously decreased as a function of time followed by steady-state creep regime (stage II) in which the strain-rate remained constant. With increasing stress and/or temperature the secondary stage becomes shorter while the sample experiences a large primary creep[10].

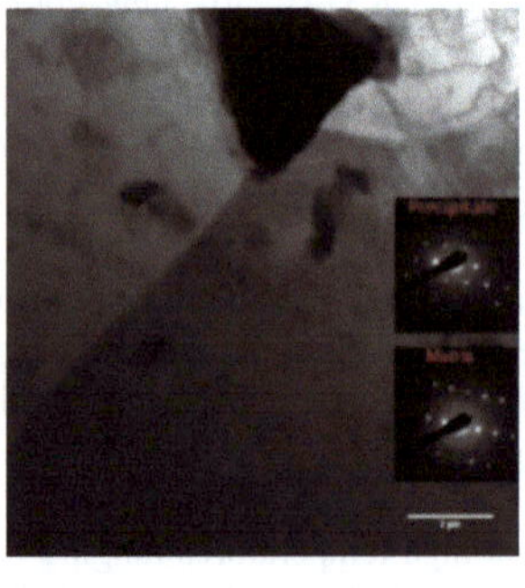

Figure 1 TEM micrograph of annealed sample along with diffraction pattern of matrix and precipitates.

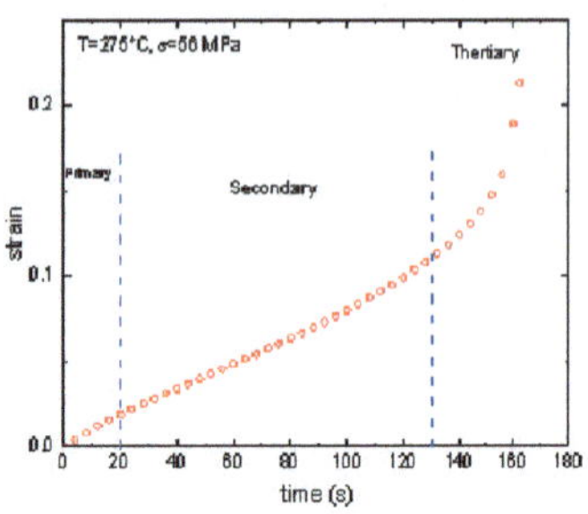

(a)

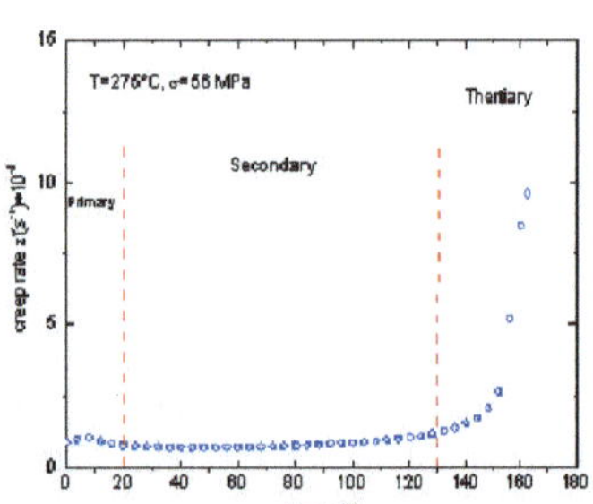

(b)

Figure 2 (a) Typical creep curve of AZ31 at high stress (56 MPa) T= 275 °C and (b) corresponding strain rate vs. time.

Figure 2 includes creep curves at high stresses along with the corresponding strain-rate vs. time plots. Figure 3 shows a double log plot of steady-state strain-rate vs. normalized stress (σ/E, E being temperature dependent elastic modulus) of several creep tests conducted at different temperatures and stresses. The temperature dependence of elastic modulus was obtained from eq. (2)[13].

$$E = 2.6 \times (1.92 \times 10^4 - 8.6T) \text{ in MPa (T in K)} \qquad (2)$$

It is evident from the plot (Figure 3a) that the data for steady-state or minimum strain-rate vs normalized stress follows power law behavior for a particular temperature. However, the slope changes with stress indicating transitions in creep mechanism(s). Data points in Figure 3 ware fitted using equation (1) to determine the stress exponent (n). Three different values of n were obtained: n~2 related to the grain boundary sliding at low stresses, n~3 dislocation glide at intermediate stresses and n~7 dislocation climb in high stress regime. Similar results were obtained for other Mg alloys[1, 10]. A semi-log plot of steady-state strain-rate vs. inverse of temperature (T-1, T in K) is used to determine the activation energy for creep at different regions. An activation energy value of 87 kJ/mol was obtained for region I, 119 kJ/mole for region II and 122 kJ/mole for region III. Activation energy value of 87 kJ/mole is identified with that for grain boundary diffusion (Q_{GB}) in Mg and its alloys while the higher value of 120 kJ/mole in regions II and III is close to that for self-diffusion[1]. The well-known Zener-Hollomon parameter (Z) can relate the strain-rate and temperature$(Z = \dot{\varepsilon}(\exp Q/RT))$. Figure 3 reveal the Zener-Hollomon parameter vs. dimension-less stress with $Q=Q_{GB}$ and $Q=Q_L$ respectively. Calculation of Z parameter with Q_{GB} leads to coalescence of data in region I while Q_L coalesce data in region II and III into a single curve. This parameter plays a significant role in microstructure of the crept samples and the appearance of DRX as discussed in the following.

A stress exponent of n~2 along with activation energy of 87 kJ/mole (close to grain boundary diffusion) is related to the GBS creep mechanism. It is well known that Q_{GB}~0.5-0.8Q_L and in present study it is Q_{GB}~0.7Q_L. The stress exponent of n~3 combined with activation energy of 119 kJ/mole suggests the operation of dislocation glide creep mechanism in region II[2, 34] as noted in many class-I alloys, and finally a stress exponent of n~7 and activation energy of 122 kJ/mole in region III is the characteristics of edge dislocation climb[2, 34] noted in pure metals and class-II alloys. Because dislocation glide and climb are sequential processes, slower one can be rate controlling. In region II (n~3) solute atoms lock moving dislocations and make their movements slower than annihilation of dislocation by climb. Thus in class A alloys the rate of dislocation glide can control creep process and this is commonly known as micro-creep. A uniform distribution of dislocations in the grain with no sub-grain boundaries (described by Murty[35, 36]) is characteristic of this region. In pure metals (or class II alloys), however, there are no obstacles against glide movement (except Peierls hills) leading to dislocation climb to be the dominant creep mechanism. Tilt and twist sub-grain formation is the predominant structure in the grains in this region[36]. It is interesting to note that such a transition from class-I to class-II is predicted by Murty[37] to occur at around 10^{-3}E given by the critical stress needed to unlock the dislocations from solute atmospheres. After this point, higher stress exponent is expected as noted in the current region III.

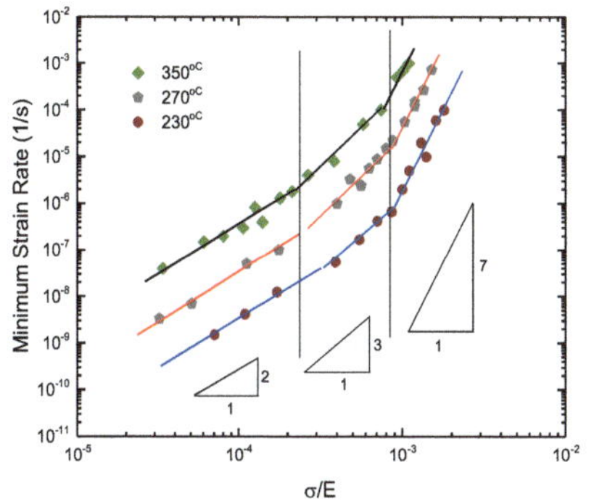

(a)

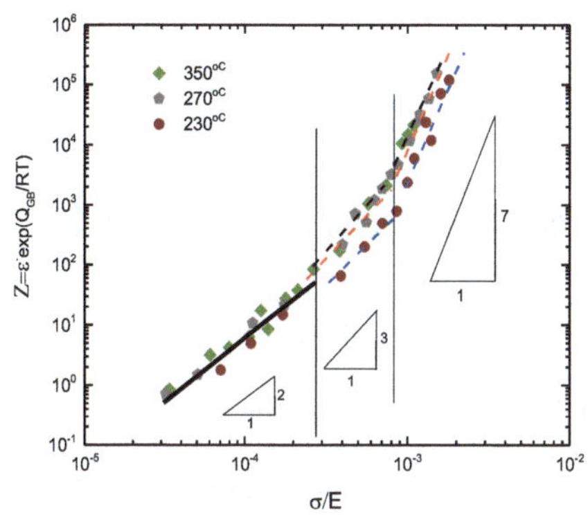

(b)

Figure 3 (a) Double log plot of steady-state strain-rate vs. stress at three different temperatures of 230, 270 and 350 °C and (b) Zener-Holman parameter calculated with grain boundary diffusion activation energy (Q_{GB}~87 kJ/mole)

Microstructural evolution during creep

There is clear evidence of DRX (i.e. the creation of new grains nucleating at previous grain boundaries[38]) at higher stresses at 350°C. In region II (DGC) and III (DCC) there is a combination of DRX and grain growth. Figure 4a show the optical micrographs of the sample tested at 350°C and 13 MPa after 24 hours (i.e. the early stages of steady-state creep deformation). Two different grain size distributions are clear in the micrograph. The small grains are new born grains which appear close to the pre-existing grain boundaries. Another test has been done at same conditions of temperature and stress to reveal the microstructure after 72 hours which is close to the end of steady-state deformation and a homogeneous microstructure is the result of long time DRX (Figure 4b) includes the optical micrograph of sample crept at 350°C , 36 MPa. Appearance of a large number of small grains on the preexisting grain boundaries indicates that severe DRX occurs at high stresses. Figure 4(c) shows the microstructure of the crept sample at 350°C and 36 MPa. Large number of dynamically recrystallized grains consumed most of the preexisting grains. Investigation of the microstructure of the crept samples at lower temperatures (220°C & 270°C), by contrast, reveals higher grain growth at GBS regions (region I), however at higher stresses (45 MPa, DCC region) very small number of dynamically recrystallized grains along with serrated boundaries appear (Figure 4d). Elevated temperatures can help the structure to reorganize under the influence of strain and stress, and provide an opportunity for new born grains to grow.

Fractography

Fracture surfaces of the AZ31 Mg alloy at different stages reveal inter-granular fracture at low strain-rates (corresponding to GBS Figure 5a&b) while dimple shape fracture with blade type edges between them (Figure 5c&d) at higher strain-rates (corresponding to DGC & DCC with DRX). Blade type edges appear due to void coalescence between dimples whenever they reach to a certain size with a large stress concentration in the walls between them[20].

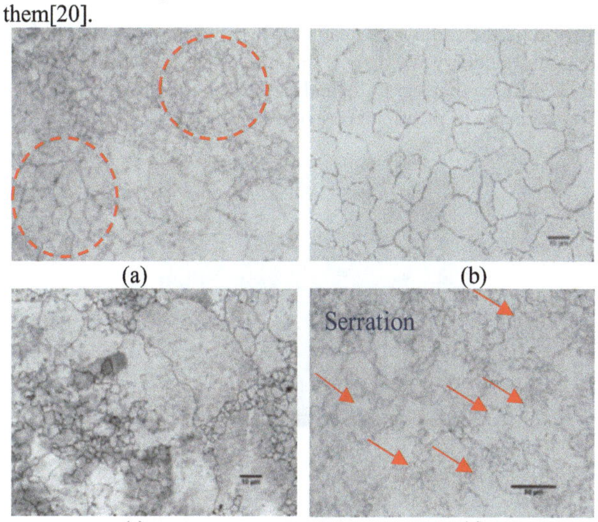

(a) (b) (c) (d)

Figure 4 (a) Microstructure distribution of crept AZ31 at 350°C, 13 MPa after 24 hours, (b) 350°C, 13 MPa after 72 hours, (c) 350°C , 36 MPa and (d) 270°C , 45 MPa.

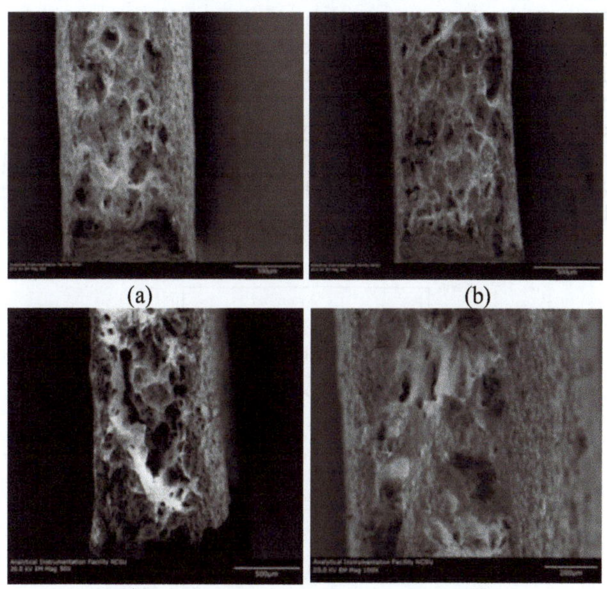

(a) (b) (c) (d)

Figure 5 Fracture surfaces of crept samples of AZ31 at 350°C: (a) 6 MPa, (b) 9 MPa, (c) 13MPa and (d) 36 MPa.

$Mg_{17}Al_{12}$, due to their lower shape factor and stress concentration[17, 39] can act as nucleation site for cavity formation (Figure **6**) and make the magnesium alloys susceptible to premature failure, however cavities are not be formed in all of them because precipitates can relieves stress concentrations via possible paths: along interphase boundaries, through the particle and lattice[39]. Chokshi [39] demonstrated the critical particle size above which stress concentration cannot be relieved via lattice or grain boundary diffusion:

$$d_c = \left(\frac{28D_l b^3\sigma}{\dot{\varepsilon}kT}\right)^{1/2} \quad (3)$$

$$d_c = \left(\frac{33\delta D_{gb} b^3\sigma}{\dot{\varepsilon}kT}\right)^{1/3} \quad (4)$$

where d_c is the critical particle size, b is the burgers vector, δ is the grain boundary width, $\dot{\varepsilon}$ is the strain rate, D_l and D_{gb} are the lattice diffusion and grain boundary diffusion respectively, other parameters are as mentioned previously.

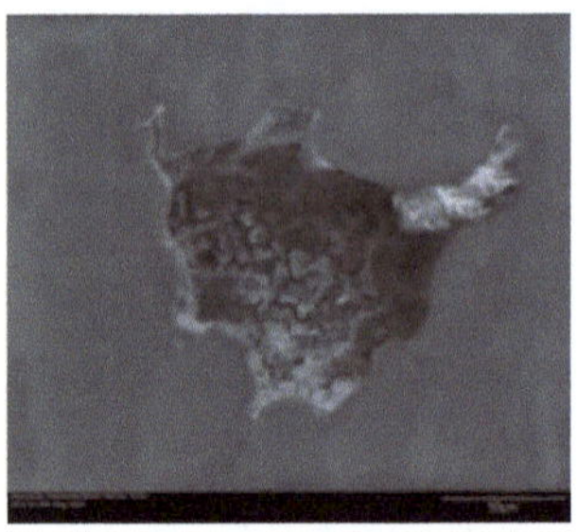

Figure 6 $Mg_{17}Al_{12}$ can act as nucleation site for cavity formation. Crept sample at 350°C 10 MPa.

Figure **7** shows the critical particle size for cavity formation for stress concentration relieving via grain boundary and lattice path. It is clear that any particle with dimension of more than $\sim 3\mu m$ can act as a stress concentration source for cavity formation. Figure 8 reveals some cavities appear close to fracture surface of crept samples at 4 and 13 Mpa. Because most of the particles have dimensions more than 3μm it is predictable to have lots of cavities in the material. By increasing the stress and strain rate the critical size of the particles for cavity nucleation increase lead to reduction in the number of cavities per unit of area. Figure 8 a & b compare the number of cavities at two different stresses of 4 and 13 MPa.

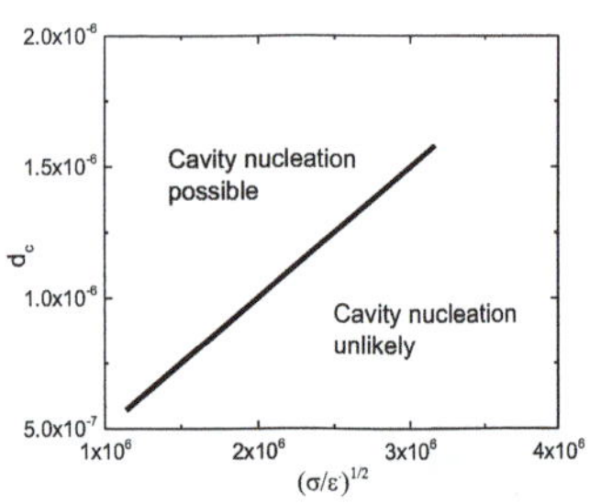

(a)

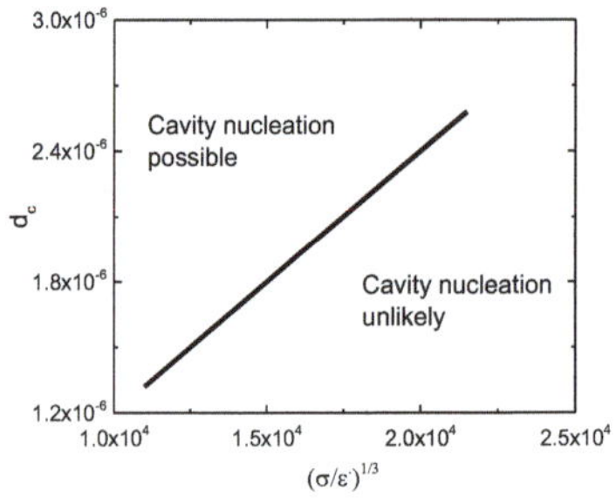

(b)

Figure 7 critical particle size for stress concentration relieving via (a) grain boundary path (b) lattice path.

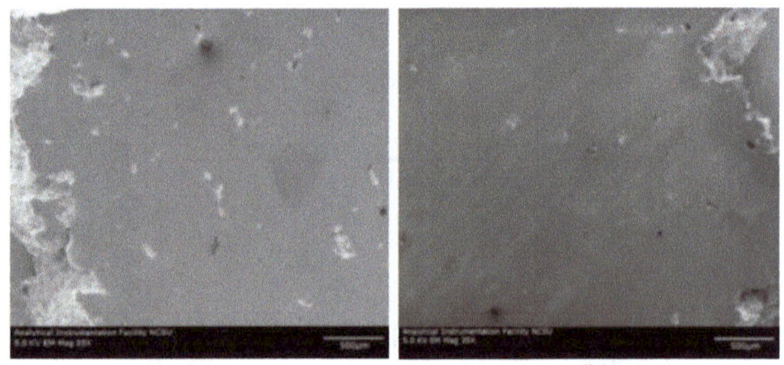

(a) (b)

Figure 8 Number of cavities per unit of area of crept sample at 350°C (a) 4 MPa (b) 13 MPa.

Thermal stability of $Mg_{17}Al_{12}$ precipitates

Figure 9a show XRD profiles of annealed AZ31. Along with the peaks for Mg, a peak for $Mg_{17}Al_{12}$ around 2θ of 37° is indexed in the plot. To investigate the thermal stability of the precipitate, high temperature XRD scans up to 450 °C were carried out at 50 °C intervals. Figure 9b shows the segments of the XRD profiles at different temperatures. It can be seen that the intensity of the $Mg_{17}Al_{12}$ peak decreased with increasing temperature and disappeared after 350 °C. This indicates that the $Mg_{17}Al_{12}$ precipitates in the AZ31 alloy dissolve in the matrix at around 350 °C. This can facilitate GBS at elevated temperature and decreases the possibility of cavity formation.

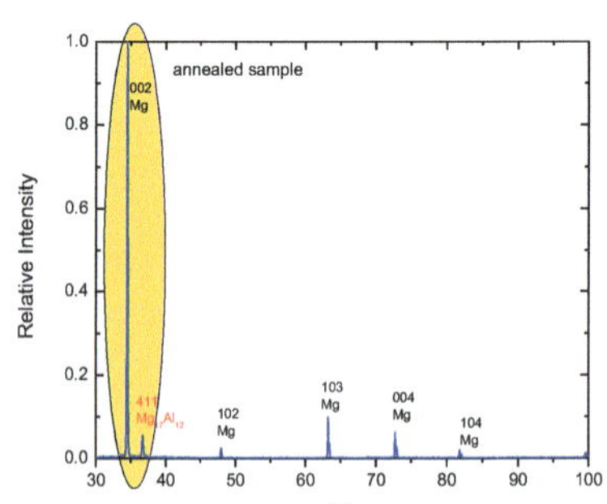

(a)

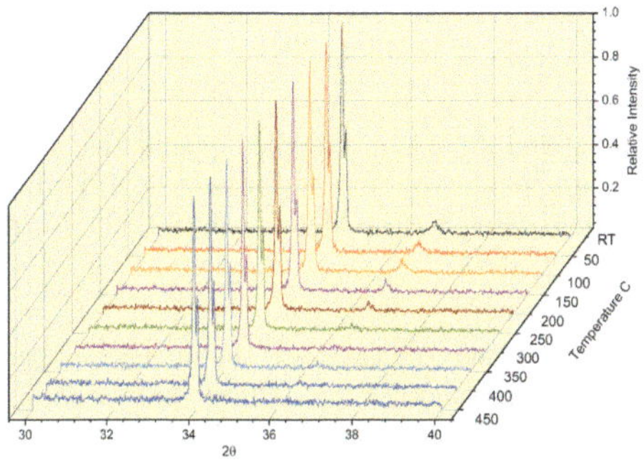

(b)

Figure 9 XRD profiles of annealed Mg alloy AZ31 (a) at room temperature, $30° < 2\theta < 100°$ and (b) elevated temperatures $30° < 2\theta < 40°$.

Conclusions

The high temperature deformation of AZ31 Mg alloy was carried out to characterize the transitions in creep mechanisms and corresponding microstructural evolution and fracture behaviors. Following conclusions are drawn from this study:

1- Three different creep regimes were noted. At low stresses in region I, the stress exponent of n~2 and activation energy of Q~87kJ/mole are related to the GBS mechanism. Grain growth, DRV and a high proportion of inter granular fracture surfaces are characteristics of this stage.

2- A stress exponent of n~3 and an activation energy of Q~119kJ/mole in region II are indicative of dislocation glide mechanism usually noted in class-I alloys. A moderate DRX, grain growth and more dimple shape fracture surfaces are characteristics of this stage.

3- A stress exponent of n~7 and activation energy of Q~122 kJ/mole in region III at high stresses reveal creep due to climb of edge dislocations as expected in alloy class due to unlocking of the dislocations from solute atmospheres. An advanced DRX, large grain growth and totally dimple shape fracture are characteristics of this stage.

Acknowledgement

This research is supported by the National Science Foundation grant 0968825.

References

1. S. W. Chung, H. Watanabe, W.-J. Kim, and K. Higashi, "Creep Deformation Mechanisms in Coarse-Grained Solid Solution Mg Alloys," Mater. Trans., vol. 45, no. 4, pp. 1266–1271, 2004.

2. R. Korla and A. H. Chokshi, "A Constitutive Equation for Grain Boundary Sliding: An Experimental Approach," Metall. Mater. Trans. A, vol. 45, no. 2, pp. 698–708, Oct. 2013.

3. R. B. Figueiredo and T. G. Langdon, "Developing superplasticity in a magnesium AZ31 alloy by ECAP," J. Mater. Sci., vol. 43, no. 23–24, pp. 7366–7371, Jul. 2008.

4. S. Spigarelli, M. El Mehtedi, D. Ciccarelli, and M. Regev, "Effect of grain size on high temperature deformation of AZ31 alloy," Mater. Sci. Eng. A, vol. 528, no. 22–23, pp. 6919–6926, Aug. 2011.

5. S. Spigarelli, M. El Mehtedi, M. Cabibbo, E. Evangelista, J. Kaneko, A. Jäger, and V. Gartnerova, "Analysis of high-temperature deformation and microstructure of an AZ31 magnesium alloy," Mater. Sci. Eng. A, vol. 462, no. 1–2, pp. 197–201, Jul. 2007.

6. H. Somekawa, K. Hirai, H. Watanabe, Y. Takigawa, and K. Higashi, "Dislocation creep behavior in Mg–Al–Zn alloys," Mater. Sci. Eng. A, vol. 407, no. 1–2, pp. 53–61, Oct. 2005.

7. K. Ishikawa, H. Watanabe, and T. Mukai, "High temperature compressive properties over a wide range of strain rates in an AZ31 magnesium alloy," J. Mater. Sci., vol. 40, no. 7, pp. 1577–1582, Apr. 2005.

8. A. G. Beer and M. R. Barnett, "Influence of initial microstructure on the hot working flow stress of Mg–3Al–1Zn," Mater. Sci. Eng. A, vol. 423, no. 1–2, pp. 292–299, May 2006.

9. S.-H. Choi, J. K. Kim, B. J. Kim, and Y. B. Park, "The effect of grain size distribution on the shape of flow stress curves of Mg–3Al–1Zn under uniaxial compression," Mater. Sci. Eng. A, vol. 488, no. 1–2, pp. 458–467, Aug. 2008.

10. H.-K. Kim and W.-J. Kim, "Creep behavior of AZ31 magnesium alloy in low temperature range between 423 K and 473 K," J. Mater. Sci., vol. 42, no. 15, pp. 6171–6176, Apr. 2007.

11. J. A. D. E. L. Valle and O. A. Ruano, "Deformation Mechanisms Responsible for the High Ductility in a Mg AZ31 Alloy Analyzed by Electron Backscattered Diffraction," Metallurgical and Material Trans A., vol. 36, no. June, pp. 1427–1438, 2005.

12. S. W. Chung, C. S. Chung, and D. Kum, "Super plasticity in thin Magnesium alloy sheets and deformation mechanism maps for magnesium.," Acta Mater., vol. 49, pp. 3337–3345, 2001.

13. K. Kitazono, E. Sato, and K. Kuribayashi, "Internal stress superplasticity in polycrystalline AZ31 magnesium alloy," Scr. Mater., vol. 44, no. 12, pp. 2695–2702, Jun. 2001.

14. P. Shahbeigi Roodposhti, A. Sarkar, and K. L. Murty, "A review of the influence of production methods and

intermetallic phase on the creep properties of AZ91," Magnes. Technol., pp. 59–64, 2014.

15. K. L. Murty, "Grain boundary sliding and viscous glide mechanisms of high temperature creep in Pb-6% Sn," Scr. Metall., vol. 7, pp. 1083–1088, 1973.

16. T. G. Langdon, "A unified approach to grain boundary sliding in creep and superplasticity," Acta Metall., vol. 42, no. 7, pp. 2437 – 2443, 1994.

17. H. Somekawa, H. Watanabe, and T. Mukai, "Effect of solute atoms on grain boundary sliding in magnesium alloys," Philos. Mag., vol. 94, no. 12, pp. 1345–1360, Apr. 2014.

18. S. Ansary, R. Mahmudi, and M. J. Esfandyarpour, "Creep of AZ31 Mg alloy: A comparison of impression and tensile behavior," Mater. Sci. Eng. A, vol. 556, pp. 9–14, Oct. 2012.

19. J. Koike, R. Ohyama, T. Kobayashi, M. Suzuki, and K. Maruyama, "Grain-Boundary Sliding in AZ31 Magnesium Alloys at Room Temperature to 523 K," Mater. Trans., vol. 44, no. 4, pp. 445–451, 2003.

20. J. Deng, Y. C. Lin, S. Li, J. Chen, and Y. Ding, "Hot tensile deformation and fracture behaviors of AZ31 magnesium alloy," Materials & Design, vol. 49, pp. 209–219, 2013.

21. a. G. Beer and M. R. Barnett, "Microstructural Development during Hot Working of Mg-3Al-1Zn," Metall. Mater. Trans. A, vol. 38, no. 8, pp. 1856–1867, Jul. 2007.

22. S. Gourdet and F. Montheillet, "A model of continuous dynamic recrystallization," Acta Mater., vol. 51, no. 9, pp. 2685–2699, May 2003.

23. S. Gourdet and F. Montheillet, "Effects of dynamic grain boundary migration during the hot compression of high stacking fault energy metals," Acta Mater., vol. 50, no. 11, pp. 2801–2812, Jun. 2002.

24. N. Xiao, C. Zheng, D. Li, and Y. Li, "A simulation of dynamic recrystallization by coupling a cellular automaton method with a topology deformation technique," Comput. Mater. Sci., vol. 41, no. 3, pp. 366–374, Jan. 2008.

25. G. Kugler and R. Turk, "Modeling the dynamic recrystallization under multi-stage hot deformation," Acta Mater., vol. 52, no. 15, pp. 4659–4668, Sep. 2004.

26. X. LIU, L. LI, F. HE, J. ZHOU, B. ZHU, and L. ZHANG, "Simulation on dynamic recrystallization behavior of AZ31 magnesium alloy using cellular automaton method coupling Laasraoui–Jonas model," Trans. Nonferrous Met. Soc. China, vol. 23, no. 9, pp. 2692–2699, Sep. 2013.

27. P. Shahbeigi Roodposhti, "A Review on the Equal Channel Angular Process of Commercially pure Titanium," Proc. MS&T14, pp. 1559–1566, 2014.

28. R. Ding and Z. . Guo, "Coupled quantitative simulation of microstructural evolution and plastic flow during dynamic recrystallization," Acta Mater., vol. 49, no. 16, pp. 3163–3175, Sep. 2001.

29. W. Roberts and B. Ahlblom, "A nucleation criterion for dynamic recrystallization during hot working," Acta Metall., vol. 26, no. 5, pp. 801–813, May 1978.

30. H. Watanabe, et al. "Grain size control of commercial wrought Mg-Al-Zn alloys utilizing dynamic recrystallization.pdf." Mater. Trans., Vol. 42, pp. 1200–1205, 2001.

31. Y. C. Lin, M.-S. Chen, and J. Zhong, "Microstructural evolution in 42CrMo steel during compression at elevated temperatures," Mater. Lett., vol. 62, no. 14, pp. 2132–2135, May 2008.

32. S. Mandal, a. K. Bhaduri, and V. Subramanya Sarma, "Influence of State of Stress on Dynamic Recrystallization in a Titanium-Modified Austenitic Stainless Steel," Metall. Mater. Trans. A, vol. 43, no. 2, pp. 410–414, Dec. 2011.

33. Y. C. Lin, M.-S. Chen, and J. Zhong, "Effect of temperature and strain rate on the compressive deformation behavior of 42CrMo steel," J. Mater. Process. Technol., vol. 205, no. 1–3, pp. 308–315, Aug. 2008.

34. F. A. Mohamed and T. G. Langdon, "The transition from dislocation climb to viscous glide in creep of solid solution alloys," Acta Metall., vol. 22, no. 6, pp. 779–788, Jun. 1974.

35. K. L. Murty and J. Ravi, "Transitions in creep mechanisms and creep anisotropy in Zr-1Nb-1Sn-0.2Fe sheet," Nucl. Eng. Des., vol. 156, no. 3, pp. 359–371, Jun. 1995.

36. Y. Zhou, B. Devarajan, and K. . Murty, "Short-term rupture studies of Zircaloy-4 and Nb-modified Zircaloy-4 tubing using closed-end internal pressurization," Nucl. Eng. Des., vol. 228, no. 1–3, pp. 3–13, Mar. 2004.

37. K. Linga Murty, "Transitional creep mechanisms in Al-5Mg at high stresses," Scr. Metall., vol. 7, no. 9, pp. 899–903, Sep. 1973.

38. S. Spigarelli and M. El Mehtedi, "Creep as an extension of hot working: A unified approach to high temperature deformation of AZ31 alloy," Mater. Sci. Eng. A, vol. 527, no. 21–22, pp. 5708–5714, Aug. 2010.

39. A. H. Chokshi, "Cavity nucleation and growth in superplasticity," Mater. Sci. Eng. A, vol. 410–411, no. May, pp. 95–99, Nov. 2005.

Magnesium Technology 2015
Edited by: Michele V. Manuel, Alok Singh, Martyn Alderman, and Neale R. Neelameggham
TMS (The Minerals, Metals & Materials Society), 2015

Creep Response of a Zn Containing Mg-Nd-La Alloy

Deep Choudhuri[1], David L Jaeger[1], Srinivasan Srivilliputhur[1], Mark A Gibson[2] and Rajarshi Banerjee[1]

[1] Center for Advanced Research and Technology, Department of Materials Science and Engineering, University of North Texas, Denton, TX-76207, USA
[2] CSIRO Process Science & Engineering, Private Bag 33, Clayton South MDC, Clayton, 3169 Vic., Australia

Keywords: Mg-RE, creep, precipitation, deformation

Abstract

A significant improvement in creep resistance was achieved by adding Zn to a Mg-Nd-La alloy. Such an improvement is indicated by an order of magnitude reduction in minimum creep rate by and maintaining remarkably low creep-strain for prolonged duration compared to the non-Zn containing alloy. Addition of Zn resulted in the formation of high number density of fine scale γ'' precipitates which presumably enhanced the load bearing capacity of Mg-Nd-La-Zn alloy. Observation of dislocation substructures further revealed that both intragranular precipitates as well as the interdendritic phase retarded dislocation motion.

Introduction

Precipitation hardenable lightweight magnesium alloys containing rare earth (RE) elements are seen as potential candidates for automobile powertrain applications [1-3]. During service such automobile parts are subjected to a combination of high operating temperature and stresses; resulting in creep-like conditions [3-7]. Therefore, favorable creep resistance of Mg-RE based alloys within 150-300°C allows their application over a wide operating temperature range [3-7].

Among the possible candidates, this paper focuses on role Zn addition on the creep behavior of Mg-Nd based alloys. In a recent work on creep tested Mg-Nd alloy (without Zn) it was found that the dynamic nucleation of β_1 (*fcc*, Fm$\bar{3}$m, a = 0.74 nm) precipitate on dislocation lines and stress-assisted formation of β'(*orthorhombic*) phase contributed to the creep resistance of Mg-Nd [8]. Zn was latter added to Mg-Nd based alloys (specifically Mg-Nd-La [9]) using high pressure die casting, and was found to significantly improve the resistance of Mg-Nd-La-based alloy [9]. Zn addition resulted in the formation of very different precipitate phases: fine scale hexagonal plates of γ'' (*hcp*, P$\bar{6}$2m) and plate-like γ (*fcc*, Fm$\bar{3}$m, a = 0.70nm) [9,10]. Furthermore, 3D atom probe tomography performed on the Mg-Nd-La-Zn alloy indicated preferential segregation of Zn at defect structures [9].

Here we examine the microstructure of Mg-Nd-La-Zn alloy at different stages of creep testing (till tertiary creep regime), as well as the dislocation structure present at the vicinity of minimum creep rate. These results allowed us to further investigate precipitate evolution during creep deformation and, importantly, the enhanced creep response of Mg-Nd-La-Zn compared to Mg-Nd-La.

Experimental Details

Two alloys containing Zn, Mg-0.6Nd-0.4La-0.3Zn (in at%), and without Zn, Mg-0.6Nd-0.4La (in at%) was prepared via high pressure high-pressure die cast (HPDC) process. La was added to facilitate the casting process, and in the microstructure La was primarily present within the interdendritic phase [9]. Cylindrical specimens, machined from the two cast alloys, were subjected to constant stress of 90 MPa at 177°C. Details of the testing process is described elsewhere [7,8]. Microstructure was examined by interrupting creep tests at various stages. A FEI Tecnai-TF20 transmission electron microscope (TEM), operating at 200kV, was utilized to perform microstructural characterization.

Results and Discussion

The bright field (BF) TEM presented in Fig.1(a) shows that the initial cast microstructure consisted of α-Mg dendrites surrounded by larger interdendritic phases. Presence of precipitation within α-Mg matrix was examined by recording selected area diffraction patterns (SADPs) along $[0001]_\alpha$ (Fig. 1(c)) and $[2\bar{1}\bar{1}0]_\alpha$ (Fig. 1(d)) zone axes (ZAs). These SADPs indicated only the primary Mg reflections and extra precipitate reflections were not detected – indicating either the absence of precipitation or very low volume fraction of precipitate formation within α-Mg matrix.

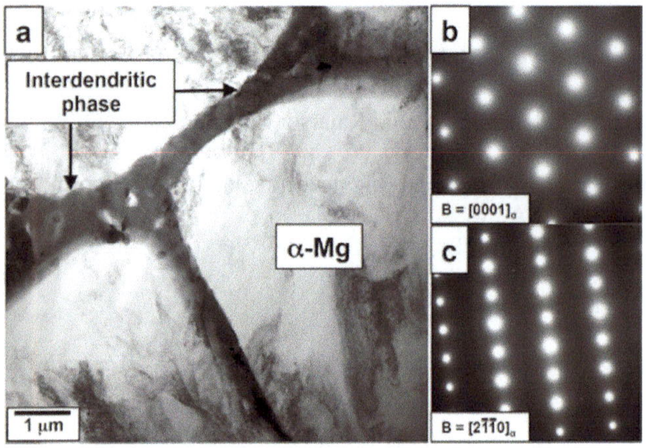

Figure 1. (a) BF TEM image showing the initial microstructure obtained after HPDC. SADPs from the α -Mg matrix recorded along (b) $[0001]_\alpha$ and (c) $[2\bar{1}\bar{1}0]_\alpha$ ZAs.

Comparison of the creep response of the two cast alloys (Fig.2(a)) showed a clear drop in the creep strain in the Zn containing alloy. Minimum creep rates (inset Fig.2(a)) for the two alloys were noted at ~100 Hrs (Mg-Nd-La, $\dot{\varepsilon}_{min} = \sim 10^{-3}$ Hr^{-1}) and ~1500 (Mg-Nd-La-Zn, $\dot{\varepsilon}_{min} = \sim 4 \times 10^{-4}$ Hr^{-1}) - an order of magnitude reduction in $\dot{\varepsilon}_{min}$ due to Zn addition. The Mg-Nd-La specimen failed within ~1000 Hrs, while creep testing of Mg-Nd-La-Zn was stopped after ~4800 Hrs.

Precipitate phases present in the microstructure of Mg-Nd-La and Mg-Nd-La-Zn (after interrupting the creep test near their respective $\dot{\varepsilon}_{min}$) are depicted with BF TEM images in Fig.2(b) and 3(a) respectively. BF image in Fig.2(b) was recorded close to [0001]α zone axis (ZA), while in Fig.3(a) the BF image was captured with the beam approximately parallel to either $[2110]_\alpha$. Our previous work suggested that fine scale precipitates in Mg-Nd-La-Zn are best viewed in convectional mode with the beam near $[2110]_\alpha$ [9]. After ~100 Hrs Mg-Nd-La microstructure consisted of GP zones and lenticular β' lying on $\{10\bar{1}0\}_\alpha$ and $\{11\bar{2}0\}_\alpha$ planes respectively, and β_1 phases decorating dislocation lines. In contrast, Mg-Nd-La-Zn after 1500 Hrs consisted of very fine scale (length ~ 8-10 nm) γ'' precipitates lying on $\{0002\}_\alpha$ planes, as well as relatively larger plate-like γ precipitates (Fig.3(a)). The γ' precipitates, like β_1 in Mg-Nd-La, were also observed to decorate dislocation lines [9]. Following ~4800 Hrs (Fig.3(b)) of creep γ'' phases in Mg-Nd-La-Zn had grown in length significantly (~20-25 nm). Interestingly, γ'' were distinctly elongated along the $[01\bar{1}0]_\alpha$ instead of $[0002]_\alpha$. Comparison of Figs.3(a) and 3(b) further suggests that the larger plate-like γ precipitate exhibit anisotropic growth along $[0002]_\alpha$ (unlike γ''). Moreover, since both γ'' and γ retain their respective morphologies after prolonged exposure (~4800 Hrs) to 90MPa/177°C, it appears that precipitate / α-Mg interface may have a crucial role towards their evolution during creep deformation.

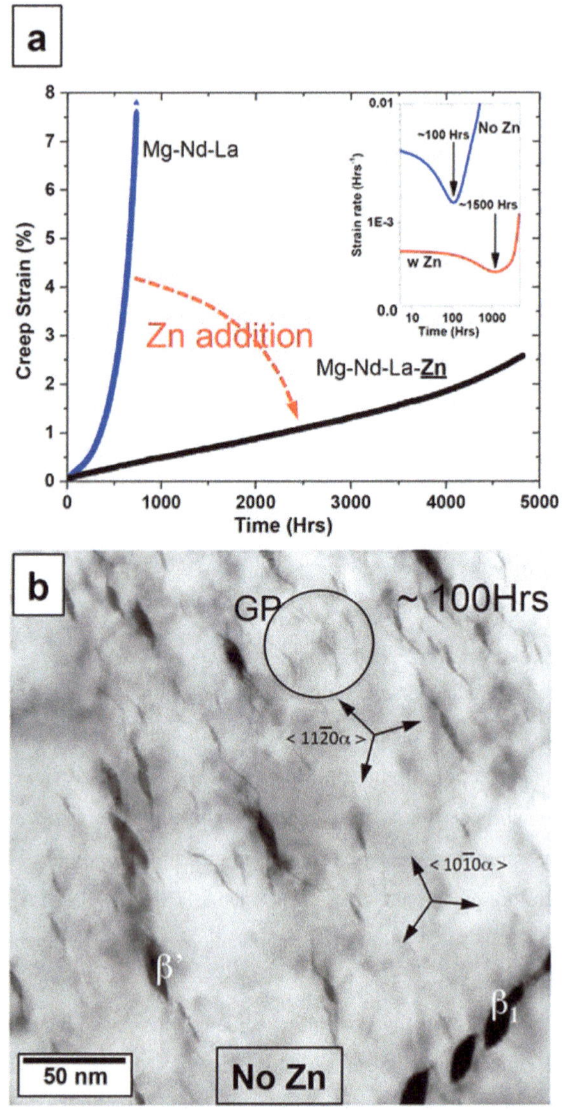

Figure 2. (a) Creep strain vs. time plots in the main figure , while the inset show strain-rate vs. time plot in log-log scale. (b) BF TEM image near $[0001]_\alpha$ showing precipitates in Mg-Nd-La. BF

Distribution of the fine scale microstructural entities within the parent α-Mg matrix (GP zones and β' in Mg-Nd-La and γ'' in Mg-Nd-La-Zn) was further quantified by conservatively evaluating their number densities (No. of particles / m^2) from several TEM images. Results of such an evaluation are plotted in Fig. 4. At $\dot{\varepsilon}_{min}$ conditions (~100 Hrs in Mg-Nd-La and ~1500 Hrs in Mg-Nd-La-Zn), number density of γ'' was significantly larger (by an order of magnitude) than the GP zones and β'. Following ~4800 Hrs of creep γ'' number density showed a significant drop.

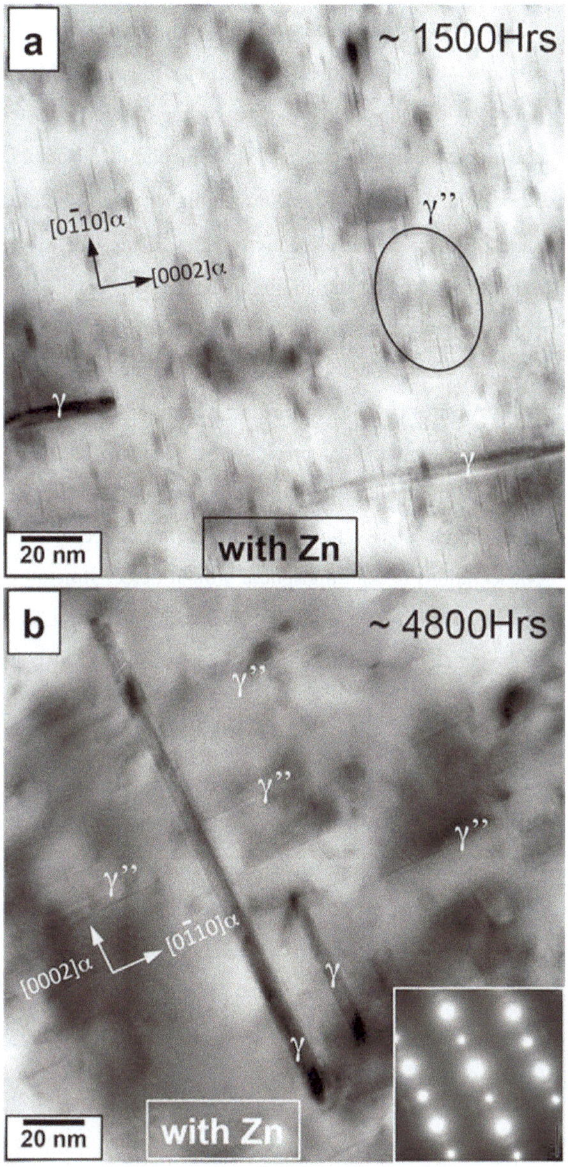

Figure.3 TEM images near [2110]$_\alpha$ ZA depicting microstructural evolution in Mg-Nd-La-Zn after (a) ~ 1500 Hrs and (b) ~4800 Hrs creep testing. Inset in (b) shows the corresponding [$2\bar{1}\bar{1}0$]$_\alpha$ ZA selected area diffraction pattern.

However, it is interesting to note that γ'' number density at ~4800 Hrs was discernably more compared to the number densities of GP zones and β' (after ~100 Hrs). Thus the relatively higher γ'' number densities (compared to the microstructural entities in Mg-Nd-La) enhance the load bearing capacity of Mg-Nd-La-Zn , and may facilitate in retarding dislocation motion.

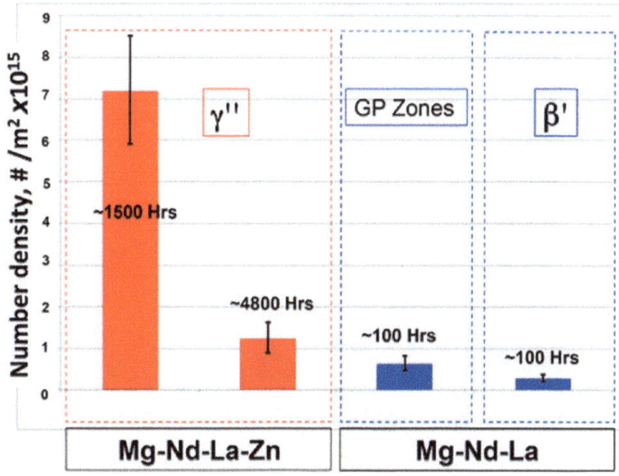

Figure 4 Plot showing the number densities of different fine scale microstructural entities in Mg-Nd-La-Zn and Mg-Nd-La alloys. Number densities of γ'' clearly exceed the number densities of GP zones and β' in the microstructures extracted after minimum creep rate.

Dislocation structures near $\dot{\varepsilon}_{\min}$ was further examined using **g.b** analysis. Fig.5(a), imaged with **g** = [$10\bar{1}\bar{1}$]$_\alpha$ and **B** = [$01\bar{1}1$]$_\alpha$, shows dislocation pile-up near the larger interdendritic phase. Such a dislocation arrangement suggest that the larger interdendritic phases (Fig.1(a) and Fig.5(a)) may retard the further dislocation motion due the back-stress generated by the dislocation pile-up. Imaging with **g** = [$10\bar{1}\bar{1}$]$_\alpha$ and **B** = [$01\bar{1}1$]$_\alpha$ excites <a>, <c+a> and <c> dislocations i.e. they become visible, thus identification of Burgers vector is practically impossible in this condition. Therefore, Burgers vector analysis was carried out using the strategy outlined by Zhu et.al. [11], where invisibility criteria and trace analysis was used to identify the dislocation type [11]. Most dislocation become invisible when imaged with **g** = [0002]$_\alpha$ and **B** = [$2\bar{1}\bar{1}0$]$_\alpha$ (Fig.5(b)), while the dislocations became visible for **g** = [$01\bar{1}\bar{1}$]$_\alpha$ and **B** = [$2\bar{1}\bar{1}0$]$_\alpha$ (Fig.5(c)). An arrow indicates point of reference in both Fig 5(b) and 5(c). Trace analysis and examination of **g.b** criterion indicated that the dislocations visible in Fig.5(c) are primarily <a> type dislocations. Additionally, dislocation loops (marked with a white arrow in Fig.5(c)) around precipitates were also observed. These dislocation loops or Orowan loops suggest that the intragranular precipitate phases in Mg-Nd-La-Zn are effectively retarding dislocation motion. Currently, studies are in progress to fully characterize and compare the dislocation structures present in creep tested Mg-Nd-La-Zn and Mg-Nd-La alloys.

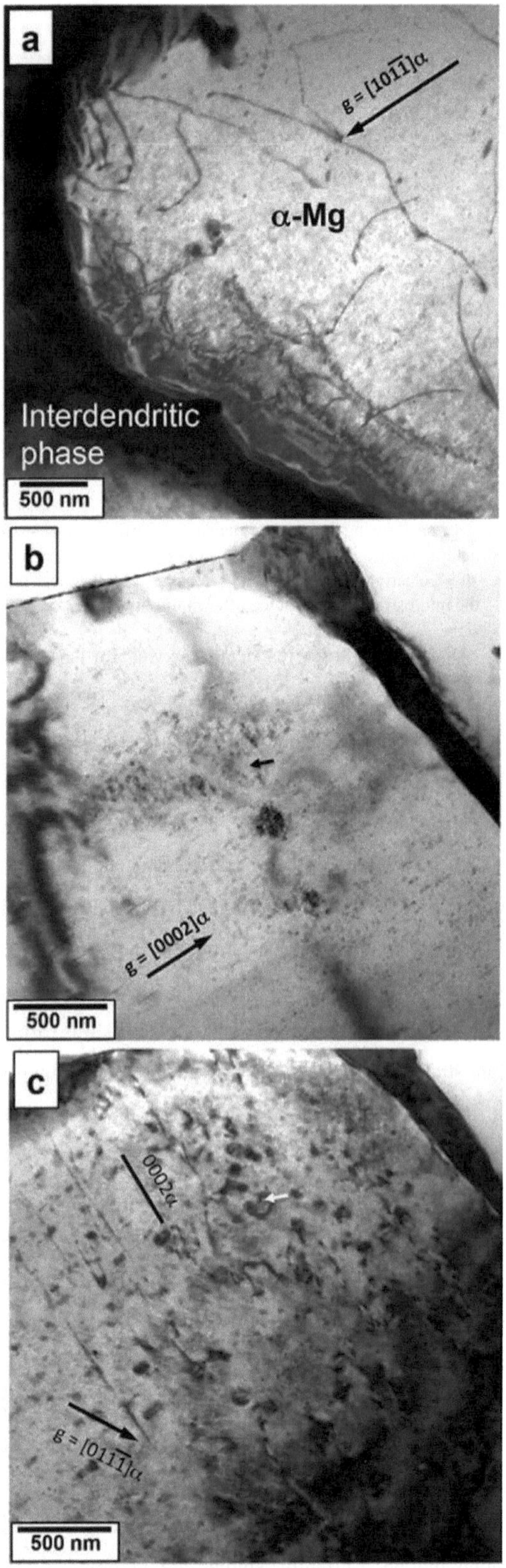

Figure 5. TEM observations of dislocation substructures at 90MPa / 177°C : (a) dislocation pile-up near the interdendritic phase, BF TEM images from the same area with (b) **g** = $[0002]_\alpha$ and **B** = $[2\bar{1}\bar{1}0]_\alpha$ and (c) **g** = $[01\bar{1}\bar{1}]_\alpha$ and **B** = $[2\bar{1}\bar{1}0]_\alpha$. Point of reference in (b) and (c) is indicated by an arrow

Summary

Zn addition significantly improved the creep resistance of Mg-Nd-La alloy by reducing the minimum creep rate by an order of magnitude and maintaining very low creep-strain for prolonged duration of testing. Fine scale γ'' and relatively larger plate-like γ phases were observed in intragranular regions of Mg-Nd-La-Zn alloy, while the microstructure of Mg-Nd-La consisted of GP zones, lenticular β' and plate-like β_1 phases. Notably, number density of fine scale γ'' was much higher than those of GP zones and β' phase in the respective microstructures of Mg-Nd-La-Zn and Mg-Nd-La at the minimum creep rate conditions. Such higher number density is expected to enhance the load bearing capacity of the Zn containing alloy. Examination of dislocation substructures (of Mg-Nd-La-Zn) at minimum creep rate indicated that both intragranular precipitates and interdendritic phases are contributing the creep resistance by regarding dislocation motion. The current observations, together with our recent experimental observations of Zn atmosphere around defect structures in creep tested Mg-Nd-La-Zn alloy [9], highlight Zn as an effective alloying element for developing creep resistant Mg-RE alloys.

Acknowledgements

The authors gratefully acknowledge the Center for Advanced Research and Technology (CART) at the University of North Texas. CAST Cooperative Research Centre (CAST CRC) was established under, and is supported in part by, the Australian Government's Cooperative Research Centres scheme. MAG would like to acknowledge the technical assistance of Mrs. M. Gershenzon and Mr. A. Yob in the HPDC of the samples used in this work.

References

1. T.L. Chia, M.A. Easton, Z.M. Zhu, M.A. Gibson, N. Birbilis, J.F. Nie Intermetallics 17 (2009) 481
2. M.A. Easton, M.A. Gibson, D. Qiu, S.M. Zhu, J. Grobner, R. Schmid-Fetzer, J.F. Nie, M.X. Zhang, Acta. Mater. 60 (2012) 4420
3. S.M. Zhu, M.A. Gibson, M.A. Easton, J.F. Nie, Scr. Mater. 63 (2010) 698
4. J. Yan, Y. Sun, F. Xue, S. Xue, W. Tao, Mater. Sci. Eng. A 476 (2008) 366

5. V. Janik, D.D. Yin, Q.D. Wang, S.M. He, C.J. Chen, C.J. Boehlert, Mater. Sci. Eng. A 528 (2011) 3105
6. Z.L. Ning, H.H. Liu, F.Y. Cao, S.T. Wang, J.F. Sun, M. Qian, Mater. Sci. Eng. A 560 (2013) 163
7. D. Choudhuri, N. Dendge, S. Nag, M.A. Gibson, R. Banerjee, Metall. Mater. Trans. A Comm. 44 (2013) 2905
8. D. Choudhuri, N. Dendge, S. Nag, M.A. Gibson, R. Banerjee, Mater. Sci. Eng. A 612 (2014) 140
9. D. Choudhuri, D. Jaeger, M.A. Gibson, R. Banerjee, Scripta Mater. 86 (2014) 32
10. J.F. Nie, Metall. Mater. Trans. A 43A (2012) 3891
11. S. M. Zhu, J.F. Nie, M.A. Gibson, M.A. Easton, P. Bakke 43A (2012) 4137

Magnesium Technology 2015

Primary, Sustainability, Recycling, and Processing

Magnesium Technology 2015
Edited by: Michele V. Manuel, Alok Singh, Martyn Alderman, and Neale R. Neelameggham
TMS (The Minerals, Metals & Materials Society), 2015

Thermal Electrolytic Production of Mg from MgO: Reflections on Commercial Viability

R Palumbo[1], M. Korenko[1, 3], C. Larson[1], L. J. Venstrom[1], S. Duncan[1], S. Nudehi[1], J. Schoer[1], J. Toberman[1], W. Prusinksi[1], D. Johnson[1], B. Robbinson[1], S. Barkely[1], and K. Warren[1]
R. Diver[2],
F. Šimko[3], M. Boča[3]

[1]Valparaiso University 1900 Chapel Drive Valparaiso IN 46383 USA
[2]Diver Solar LLC 1112 Monte Largo NE Albuquerque, NM 87123
[3]Institute of Inorganic Chemistry, Slovak Academy of Sciences, Dubravska cesta 9, Bratislava, 84536 Slovakia
Keywords: MgO, Mg, thermal electrolysis process

Abstract

We are exploring the commercial viability for producing Mg from MgO for which thermal energy is supplied to the cell as a substitute for some electric energy. The thermal input source may be concentrated sunlight or natural gas. Laboratory-scale electrochemical studies near 1250 K for two cell concepts show that we reached current densities above 0.5 $A\text{-}cm^{-2}$at an overvoltage of 1.0 V. Current efficiency values exceeded 80%. The discussion of the relationship between these bench-top experimental results and the industrial potential of the process has been initiated.

Introduction

We are developing an electro-thermal process for the efficient production of Mg from MgO that operates from 1200 – 1300 K. This process has the potential of being substantially simpler than current electrolytic processes using $MgCl_2$ [1]. Furthermore, an MgO electrolysis process theoretically utilizes less energy and produces less CO_2 per kg of Mg than both the $MgCl_2$ electrolysis process and the thermal reduction Pidgeon process [1]. Although some of the energy required to effect the reaction must be electrical, our goal is to use as much thermal energy as nature will allow. This change is anticipated to improve the economics and reduce CO_2 emissions for Mg production.

Specifically, process heat will be used to meet four energy requirements: (1) a portion of the enthalpy of reaction; (2) the sensible energy required to bring the reactants to temperature; (3) energy required to recover liquid Mg outside of the cell; and (4) the energy required to balance heat transfer losses for isothermal operation. The goal of substituting thermal energy for electrical energy is to produce Mg with an electrical energy input less than 3.5–5 $kWh\text{-}kg_{Mg}^{-1}$ and with an environmental penalty less than 2 $kg\text{-}CO_2\text{-}kg_{Mg}^{-1}$ at the cell, all at a cost that is commercially competitive. Doing so makes sense: Mg is already a valuable industrial commodity and its importance is likely to increase as the auto industry decreases vehicle weight to meet future CAFÉ standards.

The electrolysis is carried out in a fluoride-based molten salt electrolyte. The anode is carbon and the cathode is steel or possibly molybdenum. Mg forms as a liquid at the cathode and is channeled to a septum where it is isolated from the anode. When the Mg rises to the top of the electrolyte, it is then vaporized in an inert gas stream and finally condensed as a refined metal in a condenser downstream of the cell. CO_2/CO mixtures form at the anode where they rise and exit the cell. When it is economically and strategically imperative to do so, the thermal energy will be supplied with concentrated sunlight. In today's economic environment, natural gas is likely the thermal input to light metals production processes.

In developing this process, we face substantial technical and economic challenges. First, industrially relevant current densities must be attained concurrent with high current efficiency. High current density leads to small electrolytic cells for a specified Mg production rate, and high current efficiency reduces the electric energy required to produce the Mg. Secondly, for the case of solar input with tower-based optics, the receiver must match the concentrated solar flux delivered to the electrochemical reaction rate.

In this paper, we describe our progress toward meeting the technical challenges of attaining high current density and current efficiency. We present electrochemical data from bench-top experiments that describe our cell performance and our understanding of some of the physical chemistry controlling the performance. The experiments include voltammetry studies, solubility studies, and electrochemical impedance spectroscopy. To help interpret our findings within the context of a commercial process, we begin with the thermodynamic case for an ideal thermal-electrochemical process. We end by describing the quantitative significance of our current laboratory results on large scale performance, and we qualitatively describe the consequence of the current results on economics. The conclusions, in effect, create the red line connecting where we are technically to where we need to be for industrial viability.

The Thermodynamic Argument

The potential for substituting thermal energy for electric energy in the production of Mg from MgO is highlighted in Fig. 1. The figure shows the thermal and electric energy loads of a cell operating at 1200 K producing Mg according to the overall reaction

$$MgO_{298K} + ½C_{298K} = Mg(l)_{1200K} + ½CO_{2\,1200K}$$

in which thermal energy supplies a portion of the 463 kJ/mol Mg required to heat the reactants and drive the reaction, a portion of the 130 kJ/mol Mg required to flash the Mg(l) to Mg(g), and a portion of the 60 kJ/mol Mg required to overcome heat transfer losses for isothermal operation. The heat transfer loss is assumed to be 10% of the total energy required, the overvoltage-current density relationship is assumed to be linear with 0.75 A cm^{-2} at 1 V overvoltage, and the current efficiency is assumed to be 100%.

At 1200 K, the thermodynamic minimum potential for Mg production is 1.41 V. At 1.41 V, however, there is no current because the overvoltage is 0 V, and the thermal input heats the cell rather than driving Mg production. As the overvoltage increases, the capacity for thermal energy to drive Mg production increases, but the fraction of energy utilized for Mg production that is thermal decreases. At the thermo-neutral overvoltage of 2 V, all of the energy required in the cell is provided electrically. As a result of these competing factors, the applied potential that maximizes the magnitude of the thermal input is 2.4 V with a corresponding current density of 0.75 A-cm^{-2}. The fraction of energy supplied thermally at this operating point is 28%. At applied potentials below 2 V, the fraction of energy supplied thermally increases to greater than 40%, but the current density is less than 0.5 A-cm^{-2}. Given the high cost of electricity relative to the cost of thermal energy, the opportunity to supply 28% or more of the required energy at the cell thermally promises to substantially lower energy costs. The best operating point balances the capital and energy costs.

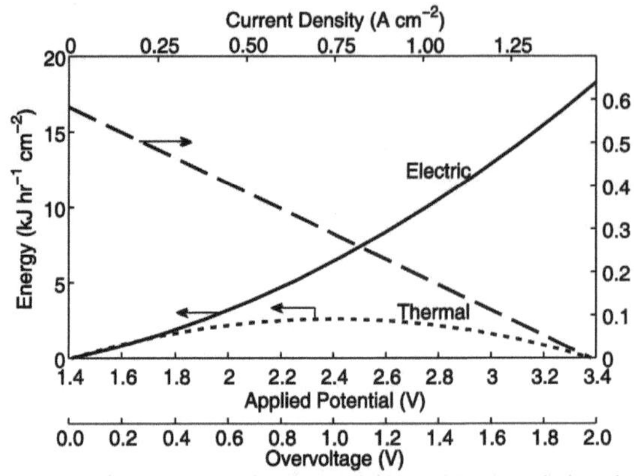

Figure 1: The area-specific thermal (dotted line) and electric (solid line) energy requirements and the fraction of total energy that is thermal (dashed line) as a function of the cell potential and assumed current density at 1200 K. The current efficiency is assumed to be 100%.

The Chemistry Experiments

In the analysis that lead to the results presented in Fig. 1, we made assumptions about the cell's current efficiency, current density, and although not obvious, the solubility of MgO in the electrolyte. In this section we describe what values we have actually obtained for these performance variables.

We define the current efficiency (CE) as the ratio of the amount of Mg recovered after electrolysis to the amount of magnesium expected when all of the charge transfer produces Mg. A CE of less than 100% means an increase in the amount of energy expended to produce Mg. But there are a number of physical pathways for current inefficiency and they impact the energy required to produce Mg in different ways.

One pathway to current inefficiency is the oxidation of Mg with the gas product evolving at the anode. Magnesium can oxidize via a gas phase reaction, a heterogeneous gas-liquid reaction, or a reaction that takes place in solution between dissolved Mg and dissolved gas product. When oxidation occurs, more electrical energy and less thermal energy is required to produce the Mg because the oxidation is exothermic, adding thermal energy to the cell. The fraction of thermal energy required in the cell thus decreases.

A second pathway to low CE is an electronic pathway for current flow as a result of the dissolution of Mg [2]. In this case, the CE drops because some of the charge transfer is non-faradic and by consequence the amount of electrical energy required to produce Mg increases.

Current efficiency is also compromised when some of the faradic current produces something other than Mg. For example, it is possible to electrolyze an impurity in the electrolyte, such as water. In this case, only a fraction of the current produces Mg at a given applied potential and thus more electrical energy is required to produce Mg.

Experimental Set-up

We are exploring two laboratory-scale cell concepts that have the potential for operating at high CE and are scalable to an industrial size. In this section, we present one of them. It is a concept first described in ref [3]. We call it the chimney concept (CC). We refer to our second cell as the baffle-channel concept (BC). The details are presented in ref [4].

Our experiments were conducted in the reaction vessel shown in Fig. 2. A typical anode is a Pt wire or foil or a graphite rod. The cathode is Mo-wire connected to Mo foil. For the CC cell the foil is coiled in a spiral form with circa 1 mm spacing between coils. Our solvent is currently a circa 50-50 weight % mixture of MgF_2 and CaF_2. We have been adding up to 1% by mass MgO to the mixture.

During electrolysis, we expect O_2 at a Pt anode or CO_2/CO mixtures at a graphite anode to evolve. Mg(l) forms at the cathode. The Mg(l) is less dense than the electrolyte and is expected to rise to the top of the Boron Nitride septum. The electrolyte temperature is measured with an S-type thermocouple. An electric furnace holds the temperature near 1000° C. A ~250 ccm flow of 4.8 grade Ar is maintained through the cell with an Omega mass flow meter. The reaction vessel is kept a few mm Hg above atmospheric pressure with a valve located downstream of the cell. At start-up, during a run, and during cool-down we use a Varion 490 Micro-gas chromatograph to monitor the

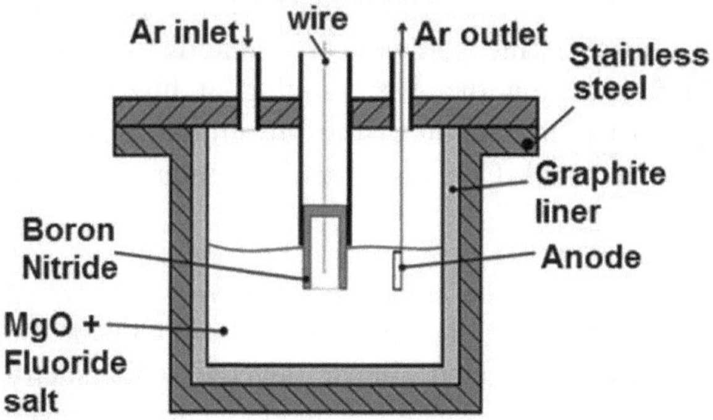

Figure 2: Schematic of the chimney electrolytic cell

O_2 concentration in the exit stream. By controlling the Ar flow rate and the pressure in the vessel, we keep the O_2 concentration in the range of 0.01-0.03 %. The Ar flowing through the vessel passes through a Meissner particle filter before exiting the laboratory through a fume-hood ventilation system.

The cell's anode and cathode are connected to a Gamry 3000 potentiostat. Once the electrolyte is melted, we obtain a linear sweep voltammogram to ensure the cell is behaving as expected and we conduct an electrochemical impedance spectroscopy analysis (EIS). With the linear sweep voltammetry we look to see that the start of electrolysis begins near the thermodynamically expected potential. We establish the electrolyte's resistance with EIS. In order to reduce the chance that we are electrolyzing an impurity, we conduct constant potential electrolysis for more than an hour at potentials below that expected for the production of Mg. Once we are satisfied that we have minimal current from a side reaction, we then set the cell potential to a given value and allow electrolysis to take place for one hour. After an hour, we obtain another linear sweep voltammogram and another EIS. The EIS scans from this point forward detect changes in the cell's resistance. This pattern is followed for up to 4 hours. At the end of an experiment, with an X-ray diffraction analysis, we confirm the metal is Mg. We quantify the mass with an HCl acid test [4] We produce between 0.25 and 0.9 g of Mg.

Discussion of Current Efficiency Results

The CEs are shown in Fig 3. The uncertainty interval at a 95% confidence interval is ±1% [4]. Three graphs are chosen to separate important differences between experiments: (1) CC cell with Pt anode; (2) CC cell with graphite anode, and (3) BC with a graphite anode. This grouping suggests how the current efficiency depends on cell design.

Comparison between the CC with Pt and graphite anodes shows for similar current levels better CE when O_2 is evolving rather than CO_2. This apparent paradox suggests the pathway to the oxidation of Mg is not a gas phase reaction between the metal and the oxidant, but rather a reaction between Mg and the dissolved gas product. The solubility of CO_2 in cryolite is reported to be 1000 times higher than that of O_2 [5]. Thus it is reasonable to suspect CO_2 is more soluble than O_2 in our fluoride-based electrolyte. The data supports the hypothesis that the gas threatening

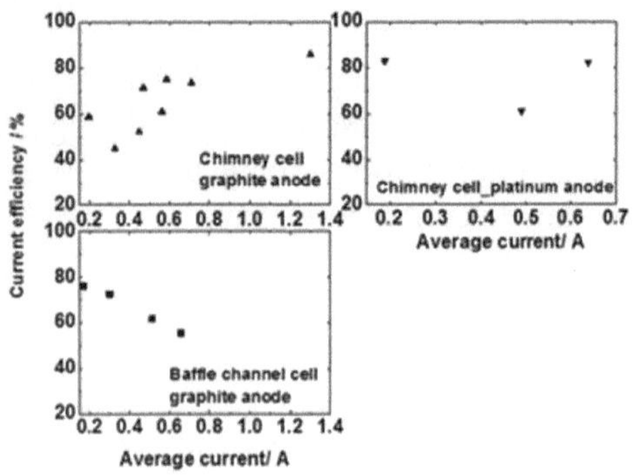

Figure 3: Current efficiency vs. average cell current for three cell arrangements. The cell temperature was between 1250 K and 1275 K.

our CE is CO_2, and the path to oxidation thus occurs via a dissolved anode product.

At this juncture we can only speculate as to whether or not our CE is being compromised by such an oxidation reaction. We do, however, have strong evidence that part of the inefficiency can be associated with an electronic conduction path due to the solubility of Mg in our electrolyte. This pathway to low CE has been reported in the literature where it lead to CE as low as 40% in a cell operating at temperatures in excess of 1600 K [2].

The strongest evidence for electronic conduction in the cell is in our EIS data. Figure 4 is a Nyquist plot that summarizes the frequency response of our BC cell to a 0.10 mV rms 10 Hz – 100 kHz sinusoidal signal. The electrodes were 1 cm apart. Where the imaginary part of the cell's impedance is zero, we have the ohmic resistance of the cell; it is the value corresponding to the real part of the impedance

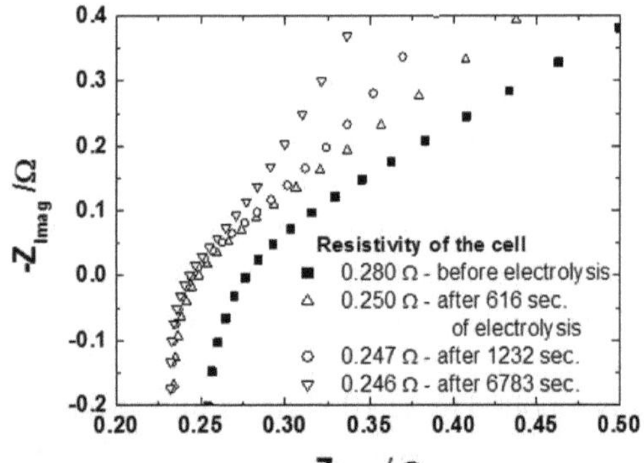

Figure 4: Nyquist plot of the frequency response of the Baffle-Channel cell at 1250 K. Carbon anode, Mo Cathode.

shown on the abscissa. The data showing a cell resistance of 0.250 Ω was taken before Mg was produced. The resistance drop to 0.240–0.243 Ω occurred after bulk electrolysis. By assuming this drop in resistance is associated with a current path that is parallel to that of ionic current, we can ascribe an estimate of the resistance of this path and the electronic transfer number. The techniques for doing so with a system like ours are described in [2]. For the case shown in the figure, the electronic resistance was just over 2 Ω giving an electronic transfer number between 0.1and 0.15. We thus cannot have a current efficiency better than 85–90%.

We envision two approaches to mitigate the drop in CE due to the electronic conduction path. First, we anticipate that the best control is temperature. The CC and BC cells operate below 1300 K thereby reducing the solubility of Mg leading to the electronic conduction that reduces CE. We are also exploring the possibility that stopping the electrolysis and restarting it from time to time interrupts the electronic transfer path. An on-off control algorithm may be a technique for maintaining a higher current efficiency.

In Fig. 3 one sees a difference in CE vs. average current between the BC and CC cells. We currently believe the oxidation path is the reason for the difference. In the BC cell, the cathode-anode gap is 1 cm. Furthermore, there is no shroud surrounding the cathode. High efficiency at low current suggests that for the BC cell, CO_2 gas evolving at the anode reaches the cathode; the cloud of this gas would be larger at higher currents. Adding support for this pathway is the fact that we find more Mg on the back side of the cathode as opposed to its front side at the end of an experiment. By contrast, the anode-cathode gap is nearly 5 cm for the CC concept and there is a physical barrier making it difficult for CO_2 bubbles to reach the anode. The data suggest that for the CC cell a higher current is better than a lower one. Such a conclusion makes sense if the pathway to oxidation is CO_2 in solution attacking the Mg: running at high current implies a larger fraction of the Mg produced outruns the rate at which CO_2 dissolves, and migrates to the cathode.

Current Density: Initial Results and Discussion

To begin assessing the current density (CD) at the anode, we conducted linear sweep voltammetry experiments. We estimate the anode's current density from the cell current and the geometric area of the

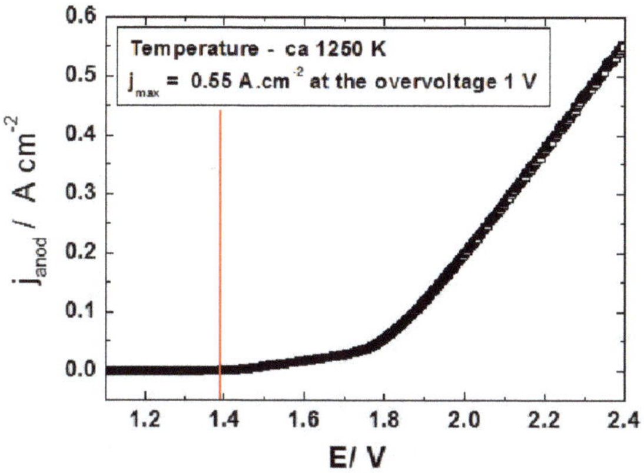

Figure 5: Linear sweep voltammogram for the Baffle-Channel Cell. The scan rate was 3 mv-s^{-1}.

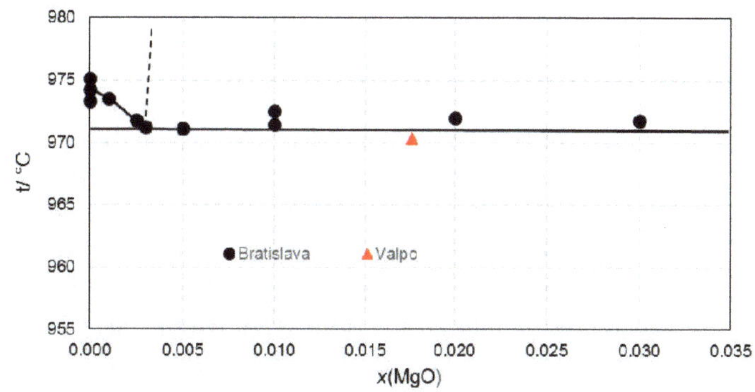

Figure 6: Eutectic cross-section of the ternary phase diagram of the $(CaF_2\text{-}MgF_2)_{EUT}$-MgO system. Black circles represent measurements of the Slovak Academy of Science in Bratislava.

electrode in contact with the electrolyte. In the case of graphite, this area is well defined: at the end of an experiment, we see that the anode diameter is uniformly reduced around its circumference. With caliper measurements, we estimate the uncertainty in the area participating in the electrolysis as +/- 0.3 cm^2. Because the uncertainty in the current is insignificant, the uncertainty in the CD is circa +/- 0.05 A-cm^{-2} for our current levels.

To help maximize the current density at a given overvoltage, we worked to develop electrodes with minimal resistance. The anode was a 0.635 cm diameter rod into which we threaded a 0.317 diameter 101 copper rod. The active graphite anode length was 1.27 cm long. The copper rod extended out of the cell and the electric furnace so its length was circa 50 cm long. We used 3 Mo wires 0.5 mm in diameter also about 50 cm long to connect to a 0.127 mm thick Mo foil 1 cm x 1.5 cm serving as the cathode.

With this cell arrangement for our CaF_2-MgF_2 electrolyte, the BC cell had a total cell resistance of circa 0.250 Ω at 1250 K. Figure 5 is a typical linear sweep voltammogram for the BC cell. The experimental decomposition potential, assuming CO_2 is evolving at the anode is very near the theoretical value of 1.38 V. For a 1 V overvoltage we have a current density just above 0.5 A-cm^{-2}.

Increasing the current density at 1 V overvoltage is certainly possible. We will explore reducing the resistance of the electrolyte by adding LiF to the electrolyte and closing the anode-cathode gap from 1 cm to 0.5 cm. The cell temperature can also be increased by 25–50 °C to improve kinetics.

Initial MgO Solubility Study

The solubility of MgO in the eutectic mixture of CaF_2 – MgF_2 was measured by thermal analysis under an argon atmosphere. A platinum crucible with the salt mixture (35 g) was placed in a resistance furnace. The temperature was measured with a Pt/Pt10Rh thermocouple located directly in the mixture. One measurement was carried out in a large graphite crucible (triangle in Figure 6) 250 g of salt.

Figure 6 shows the data defining the liquidus temperature of the eutectic mixture of CaF_2 – MgF_2 plotted as a function of the mole fraction of MgO up to 3 mole %. Based on the measured data, we predict the coordinates of the ternary eutectic of the system $(CaF_2 – MgF_2)_{EUT}$ – MgO as 0.3 mole % at 1245 K. Further addition of MgO beyond this concentration has no influence on the temperature of the primary crystallization of the binary eutectic mixture CaF_2 – MgF_2. Thus the aforementioned coordinates of the ternary eutectic represent the maximum solubility of MgO at this temperature in the system CaF_2 – MgF_2.

Conclusions

The proposed electrothermal process to produce Mg from MgO in a molten fluoride electrolyte shows promise. A substantial fraction of the energy required to produce Mg can be supplied thermally rather than electrically, between 28% and 50% depending on the current density. Low current densities increase the fraction of the thermal input, but also increase the capital cost to reach a desired rate of production. In order to determine the most appropriate operating point, capital cost will ultimately be compared against the energy savings realized by substituting electric energy with thermal energy.

We demonstrated a current density of 0.55 A-cm^{-2} for a 1 V overvoltage. We argued that it is reasonable to expect this value to increase. But if it does not, the consequence on the commercial viability of the process will be an increased capital cost. We are currently developing an economic model that can be used to quantify this cost. With 0.55 A-cm^{-2} at 1 V overvoltage, Mg is produced at 5.4 kWh_e-kg_{Mg}^{-1}, a fossil fuel energy intensity for a process at the cell well below that of any reported Mg production process.

Adding further support to the notion that we have a promising process is that we achieved a CE of 85.9% with CO_2 evolving at the anode. Our data suggest that a CE of 90% is possible. The CE is compromised in our cell primarily by an electronic short circuit associated with dissolved Mg. When we account for the current efficiency, our electric energy intensity per kg of Mg is 5.5 kWh_e-kg_{Mg}^{-1},

These energy intensity values that are based on our bench-top experimental results correspond to under 1 kg of CO_2 emissions per kg of Mg for a solar process. The value presumes an electric power plant operating on coal with a thermal efficiency of 40%.

The solubility of MgO in our electrolyte is low, but we have not experienced any negative consequences such as an undesirable interruption in the current. This fact implies that the rate of MgO dissolution kept up with our current level or the electrochemical reaction mechanism does not require MgO to be in solution. Consider the following possible mechanism:

$2MgF_2 \rightarrow MgF_4^{2-} + Mg^{2+}$ (dissolution)

Anode:

$MgF_4^{2-} + MgO(cr) + 0.5\ C(gr) \rightarrow 2MgF_2 + 0.5CO_2 + 2e$

Cathode: $Mg^{2+} + 2e \rightarrow Mg(l)$

We will conduct more fundamental voltammetry experiments in order to gain insight on the reaction mechanism. We must learn to operate the cell at both high CE and CD simultaneously. We need to develop the economic model so that the economics of the process can be directly linked to the performance of the technology. This step requires us to better define the technology, specifically the details for how the thermal input is brought into the cell. We invite the reader to stay tuned to future publications in which we will report our progress on these issues.

Acknowledgements

We are grateful to the US Department of Energy ARPA-E program for financial support for this work under the cooperative agreement DE-AR0000421.

We thank S. Lyle at Saint-Gobain for supplying Boron Nitride samples used as components in our electrochemical cells.

References

1. U. B. Pal, D.E. Wooley, A. Krishman, T. Keenan, C.P. Manning, and G.B. Kenney "Solid-Oxide-Oxygen-Ion-Conducting Membrane (SOM) Technology for Green Synthesis of Magnesium from its Oxides" *Magnesium Technology TMS* (2002).
2. Xiaofei Guan, Uday B. Pal, and Adam C. Powel, "Energy-Efficient and Environmentally Friendly Solid Oxide Membrane Electrolysis Process for Magnesium Oxide Reduction: Experiment and Modeling"; *Metallurgical and Materials Transactions E,* 1A (2014) pp. 132-144.
3. Qui Zhuxian, Zhang, Mingjie, and W. Jing, "Preparation of Aluminum-Magnesium Master Alloys by electrolysis of Magnesium Oxide in Fluoride Melts" *Light Metals* (1991)
4. R. Palumbo, L. Venstrom, S. Dunan, S. Nudhehi, M. Korenko, and J. Schoer, "The solar thermal electrolytic production of Mg from MgO with Concentrated Solar Energy" (ARPA-E DE-AR0000421 Quarter 2 Report (July, 2014).
5. Jomar Thonstad, Pavel Fellner, Geir Martin Haarberg, Jan Hives, Halvor Kvande, Asmund Sterten, *Aluminum Electrolysis: Fundamentals of the Hall-Heroult Process.* Dusseldorf, Aluminum–Verlag Marketing & Communication (2001).

Magnesium Technology 2015
Edited by: Michele V. Manuel, Alok Singh, Martyn Alderman, and Neale R. Neelameggham
TMS (The Minerals, Metals & Materials Society), 2015

STUDY ON COMPRESSIVE STRENGTH OF PELLETS FOR THE NOVEL SILICOTHERMIC PROCESS

FU Daxue, GUAN Lukui, WEN Ming, DOU Zhi-he, ZHANG Rui, ZHANG Ting-an

School of Materials and Metallurgy of Northeastern University, Key Laboratory of Ecological Untilization of Multi-metal Intergrown Ores of Education Ministry, Shenyang, 110819, China

Keywords: Dolomite; Compressive; Magnesium; Silicothermic

Abstract

A novel process of magnesium production has been developed by changing the preparation method of pellets of silicothermic process. For the method, the pellets consist of dolomite, ferrosilicon, fluorite and binder, which need to be roasted before reduction. However, the discharge of carbon dioxide reduces compressive strength of the pellets during the calcinations. Therefore, the present paper reports the effects of calcination atmosphere, temperature and insulation post-processing conditions on compressive strength of pellets. The results indicated that compressive strength was increased with increasing calcination temperature and post-processing temperature. The value of compressive strength obtained in air was larger than that in other atmospheres. The compressive strength value reached 80N in the experimental condition. The pellet was reduced at 1473K for 60min and the reduction ratio reached 86.75%.

Introduction

Magnesium is the lightest metal among the commonly used structured metals and its wide application in many fields, such as metallurgy, chemistry, automobile industry,[1,2] implies the huge demands for magnesium in the world. Magnesium is produced by two principal routes: electrolysis of molten magnesium chloride and thermal reduction of magnesia. The proportion of electrolysis is getting smaller due to the high cost [3]. As an alternative route to electrolytic process, the Solid Oxide Membrane (SOM) process[4] was proposed by Woolley and Pal in recent years. However, there are still several scientific challenges in the SOM process.[5] The thermal reduction processes are divided into carbothermal, aluminothermic and silicothermic processes. Several workers[6,7] studied the carbothermal process of magnesia. However, the well-known primary issue with the carbothermal process is the easy reverse reaction. CSIRO's MagSonic™ process [8,9], under development in Australia since 2003, achieves minimal reversion by cooling the reaction products (gases) at up to 10^{6}°C/s by supersonic acceleration through a Laval nozzle [10]. Aluminothermic routes offer attractive choice over other production methods since aluminum is a stronger reducing agent compared to ferrosilicon and carbon. However, this process can be only possible by finding a route to cut down on the cost. Silicothermic processes include continuous and semi-continuous processes. Several continuous processes for magnesium production were proposed, including Mintek process[11] and Zuliani process[12]. They are being studied in a pilot scale. The Pidgeon process is well-known as a semi-continuous one. In addition, some researchers [13,14] tried to use other alloys as reductant, such as Al-Si-Fe and Si-Cu, in thermal reduction processes which were still studied in the lab scale.

In all the processes, the Pidgeon process is widely used in thermal reduction of magnesia, which has been in the mature stage[15,16]. Magnesium produced by Pidgeon process in China accounts for 80% of the yields in the world. Although the Pidgeon process was modified and there was a big progressive on energy efficient and environmental protection, there are still the following unsolved problems. (1) Five percent fine powder produced during the calcination of dolomite is unavailability. (2) The temperature of calcined dolomite discharged from the cooler is still in 423-493K and the energy carried by the calcined dolomite is lost to the environment in the following processes. (3) The temperaute of calcined dolomite preparing for briquetting is decreased from 1473K to environment temperature However, calcined dolomite in pellets is heated to 1473K again in the stage of reduction. This will reduce the production efficiency

In order to solve these problems, Zhang et al[17] proposed a novel method of feed preparation for silicothermic process. In this method, dolomite, ferrosilicon, fluorite and binder are mixed to produce pellets and then calcined. The pellets are charged into the retorts at once after calcinations. Pelletization is helpful for solid-solid reaction and diffusion of magnesium vapor. Pellet compressive strength is very important for avoiding pellet fragmentation resulted from the extrusion between pellets in the retorts. The discharge of carbon dioxide reduces compressive strength of the pellets during the calcinations. Therefore, the effects of calcination atmosphere, temperature and insulation post-processing conditions on compressive strength of pellets were studied in the present paper.

Thermodynamic considerations

The reaction (1) occurs during the silicothermic process of magnesia.

$$2CaO(s)+2MgO(s)+Si(s)=2Mg(g)+2CaO\cdot SiO_2(s) \quad (1)$$

The relationships between equilibrium magnesium partial pressure and temperature of the reaction (1) calculated from thermodynamic data are plotted in Figure 1. The equilibrium magnesium partial pressure is exponentially increased with temperature. The lower calcined temperature can inhibit the reaction (1) in the stage of calcinations since the reaction is easy to reach equilibrium. If the calcined temperature is larger than boiling point of magnesium which is about 1373K, magnesium vapor is produced which will promote the reaction (1). The value of the equilibrium magnesium partial pressure is 1261Pa at 1373K, which is increased to 4661Pa at 1473K. Therefore, the calcined temperature should be lower than 1373K.

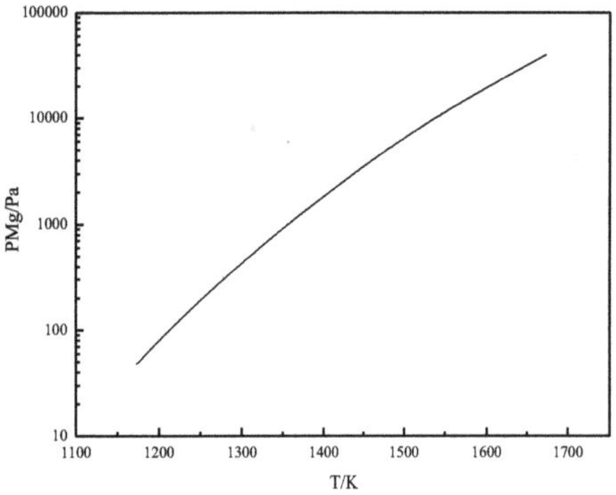

Figure 1 Relationship between Mg equilibrium partial pressure and temperature

Experiment

Material

The chemical compositions of dolomite are shown in Table1.

Figure 2 shows XRD pattern of dolomite, which includes the main dolomite phase and a small amount of calcium carbonate phase. The content of silicon in reduction agent ferrosilicon is 75.6% in mass fraction.

Procedures

The raw ore including dolomite, reduction agent and fluorite were crushed and pulverized. The finely powder were mixed with binder according to a certain ratio. Then, the mixture was made into pellets by disk balling machine.

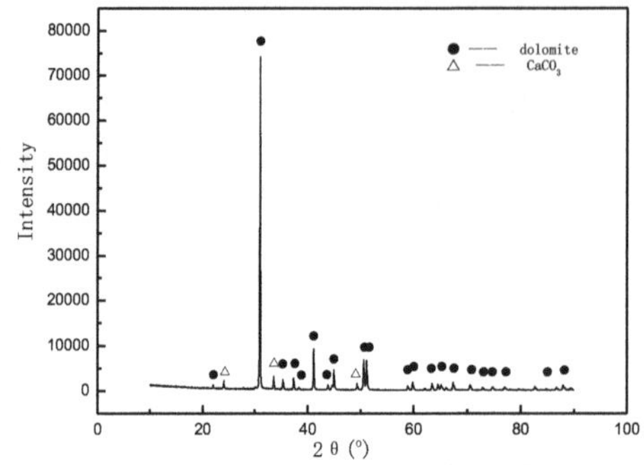

Figure 2 XRD pattern of dolomite

Since the fast discharge of carbon dioxide may result in cracks on the pellets during the calcinations, the pellets were firstly calcined at 800 °C for 30min and then at a higher calcined temperature (900°C, 1000 °C, 1100°C) for 60min by means of a tubular resistance furnace. The different calcined atmospheres were investigated, including CO_2, stagnant air, flowing air, argon, combination gas containing 20%CO_2 and 80% argon. The compressive strength of pellets after calcinations was investigated by a press machine. Some pellets were used for the measurement of compressive strength and the mean value was chosen.

Results and discussion

Figure 3 shows compressive strength of pellets after calcinations at different atmospheres, including CO_2, stagnant air, flowing air, argon, gas mixture containing 20%CO_2 and 80% argon.

Table 1 Chemical composition of dolomite

Compound	MgO	CaO	Al_2O_3	SiO_2	Na	K	Fe
Content / %	21.73	31.05	0.16	0.44	0.02	0.005	0.07

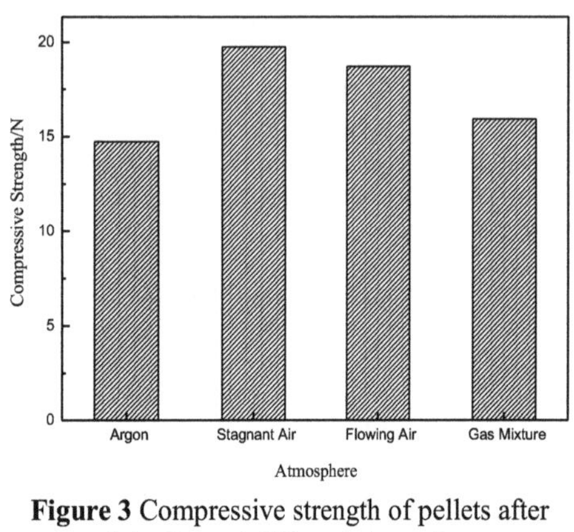

Figure 3 Compressive strength of pellets after calcinations at different atmospheres

The phases in the pellets after calcinations were analyzed by XRD. The main phases in the samples obtained from different atmospheres were almost the same. The phase $2CaO{\cdot}SiO_2$ was only found in the samples obtained from oxidizing atmosphere. However, the peak intensity of $2CaO{\cdot}SiO_2$ was very weak due to the low content. In order to conclusively prove the existence of $2CaO{\cdot}SiO_2$, the pellets including calcium oxide and ferrosilicon were calcined at the same conditions as a comparative experiment. Figure 4 shows XRD patterns of the comparative experiment samples obtained from stagnant air and argon. It is clearly seen that the phase $2CaO{\cdot}SiO_2$ was produced in stagnant air.

Based on the above analysis, it is deduced that silicon dioxide was produced on the surface of ferrosilicon. The phase $2CaO{\cdot}SiO_2$ was produced from the reaction between SiO_2 and CaO which strengthened the connection between the particles and increased the compressive strength of the pellets. The schematic diagram of the effect of calcined atmospheres on compressive strength of pellets is shown in Figure 5.

Ferrosilicon was oxidized when the pellets were calcined in oxidizing atmosphere according to XRD. The oxidation amount of ferrosilicon has a big effect on the reduction ratio. In order to calculate the oxidation amount of ferrosilicon, two samples, only ferrosilicon and a mixture of calcined dolomite and ferrosilicon, were calcined in stagnant air. The mixture was made into pellets by means of a press machine. According to Guo et al[18], Silicon is oxidized to SiO_2 when ferrosilicon is heated under the temperature ranging from 25 °C to 1200 °C in air. Therefore, the change in the pellet mass is taken as the oxidation amount of silicon.

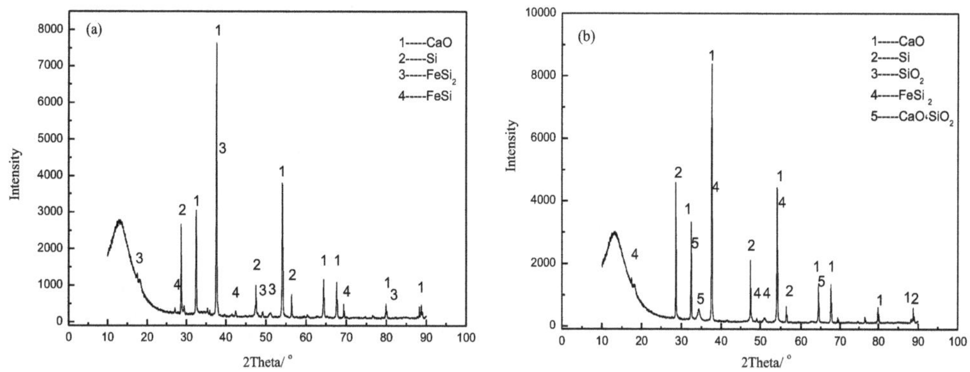

Figure 4 XRD patterns of the comparative experiment samples obtained from (a) argon; (b) stagnant air

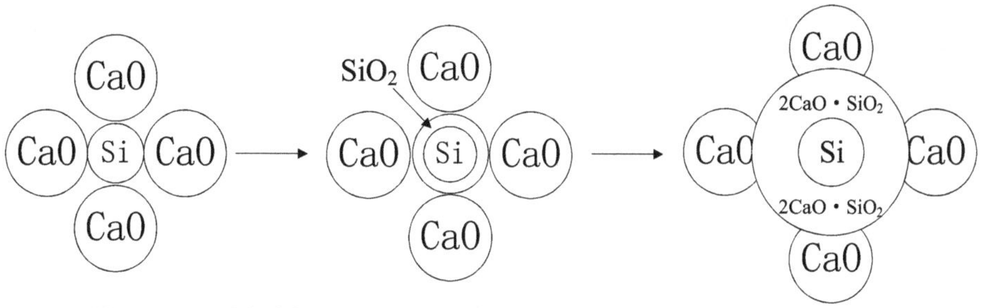

Figure 5 Model of the increase of compressive strength in oxidizing atmosphere

The results are shown in Table 2. The oxidation amount of silicon was 1.54% when only ferrosilicon was calcined in air. The value of the oxidation amount of silicon was increased to 2.76% when the pellets include both ferrosilicon and calcined dolomite. From these data, the effects of oxidation of silicon on the reduction ratio can be avoided by adding 3% excess ferrosilicon.

The effects of temperature on compressive strength and hydration activity of pellets were studied. The experimental conditions are shown in Table 3 and the results are shown in Figure 6.

Table 2 Oxidation amount of silicon

	Ferrosilicon	Ferrosilicon and calcined dolomite
Before calcinations	5.000g	36.079g
Silicon conten	5.000g	5.830g
After calcinations	5.077g	36.234g
Mass gain	0.077g	0.155g
Oxygenation ratio of silicon	1.54%	2.76%

Table 3 Experimental conditions

Atmosphere	heating rate	First stage		Second stage	
		Temperature/°C	Soaking time /min	Temperature/°C	Soaking time /min
Stagnant air	4℃/min	800	30	900	60
		800	30	1000	60
		800	30	1100	60

The compressive strength of pellets is the worst at 900 °C and the value is only 3.52N. The compressive strength of pellets is the best at 1100 °C and the value reaches 55.32N. However, hydration activity was the worst at 1100 °C and the value is only 7.23%. Hydration activity markedly decreased when the calcined temperature increased from 1000 °C to 1100 °C. Hydration activity at 1000 °C was appropriate.

In order to further increase compressive strength of pellets, the insulation post-processing was carried out after calcinations. The pellets were calcined in argon and the calcined temperature was the same as be shown in Table 3. After calcinations, the tubular resistance was sealed and the temperature was increased to 1125 °C, 1150 °C, 1175 °C, 1200 °C for a soaking time of 20min. The effects of the post-processing temperature on compressive strength of pellets were shown in Figure 7.

The compressive strength was increased with increasing post-processing temperature and added amount of ferrosilicon. The pellets with adding 20% excess ferrosilicon can obtain the largest compressive strength at the post-processing temperature of 1200 °C and the value reaches 80N. The pellet was reduced at 1473K for 60min and the reduction ratio reached 86.75%.

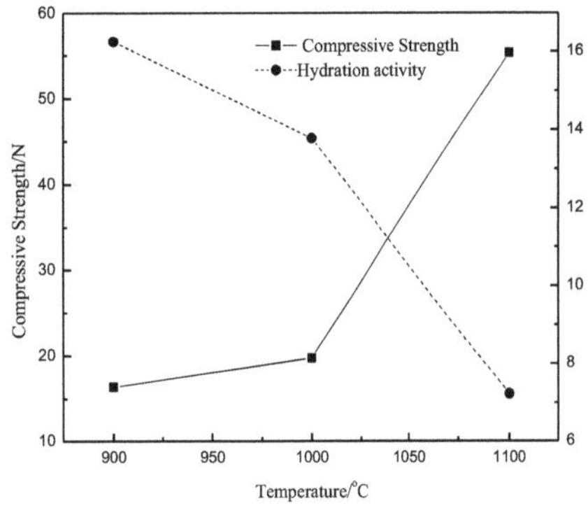

Figure 6 Effects of temperature on compressive strength and hydration activity of pellets

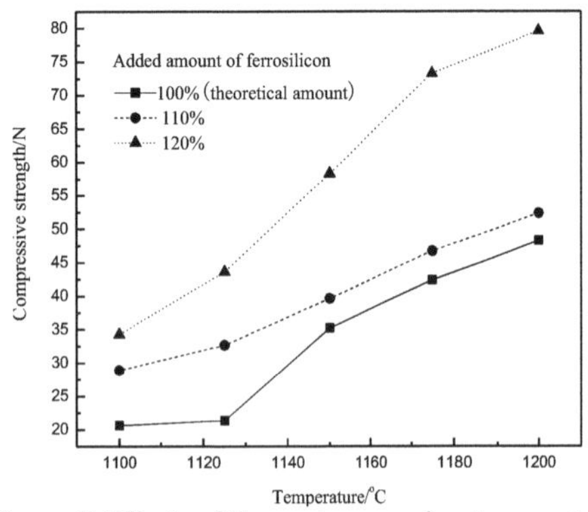

Figure 7 Effects of the post-processing temperature on compressive strength of pellets

Conclusions

1. The value of compressive strength obtained in air was larger than that in other atmospheres. The small amount of silica, which was produced in oxidizing atmosphere increased compressive strength of pellets and had no effects on the reduction of calcined dolomite.
2. The optimal calcined temperature was determined as 800 °C for 30min at the first stage and 1000 °C for 60min at the second stage.
3. The compressive strength of pellets was increased by insulation post-processing. The largest value of compressive strength reached 80N.

Acknowledgement

This research was supported by the Fundamental Research Funds for the Central Universities of China (N130302006, N130702001).

References

[1] MAO Ping-li, LIU Zheng, WANG Chang-yi, GUO Quan-ying, SUN Jin, WANG Feng, LIN Li. Fatigue behavior of magnesium alloy and application in auto steering wheel frame [J]. Transactions of Nonferrous Metals Society of China, 2008, 18(1): s218-s222.

[2] LIU Bin, TANG Ai-tao, PAN Fu-sheng, ZHANG Jing, PENG Jian, WANG Jing-feng. Improvement and application of neural network models in development of wrought magnesium alloys [J]. Transactions of Nonferrous Metals Society of China, 2011, 21(4): 885-891.

[3] Lebedev O A, Brusakov Y, Shkuryakov N P. Express monitoring of $MgCl_2$ concentration in the electrolyte of magnesium electrolyzers [J]. Russian Journal of Applied Chemistry, 2005, 78(8): 1276-1279.

[4] Krishnan A, Pal U B, Lu X G. Solid oxide membrane process for magnesium production directly from magnesium oxide[J]. Metallurgical and Materials Transactions B, 2005, 36(4): 463-473.

[5] Winny Wulandari, Geoffrey A.Brooks, Muhammad A.Rhamdhani. Magnesium: Current and alternative production routes[C], Chemeca 2010: Engineering at the Edge, 2010: 347-357.

[6] Winand R, Gysel M V, Fontana A, Segers L, Carlier J C. Production of magnesium by vacuum carbothermic reduction of calcined dolomite [J]. Transactions of the Institution of Mining and Metallurgy Section C, 1990, 99: C105-C112.

[7] Li Rong-ti., Pan Wei Sano M. Kinetics and mechanism of carbothermic reduction of magnesia [J]. Metallurgical and Materials Transactions B, 2003, 34B: 433-437.

[8] Brooks G, Trang S, Witt P, Khan M N H, Nagle M. The carbothermic route to magnesium[J]. JOM, 2006, 58(5): 51-55.

[9] Tassios S, Barton T R D, Constanti-Carey K K, Nagle M W, Prentice L H. Manufacture of metal E.G. magnesium, involves performing carbothermal reduction of metal oxide, preventing reformation of metal oxide, and cooling stream using Nozzle heated with unit other than gas under specific condition[P]. WO2010012042-A1, 2010.

[10] Donaldson A, Cordes R A. Rapid plasma quenching for the production of ultrafine metal and ceramic powders[J]. JOM, 2005, 57(4): 58-63.

[11] Abdellatif M. Review of the development work on the Mintek thermal magnesium process(MTMP)[J], The Journal of The South African Insitiute of Mining and Metallurgy, 2011, 111: 393-399.

[12] Zuliani D J, Reeson D. Making magnesium a more cost and environmentally competitive option[C], Conference of Magnesium Alloy in Vancouver, 2012. (http://www.gossan.ca/pdfs/Conference9MagALLOY-Paper-Vancouver-July2012.pdf)

[13] HU Wen-xin, Liu Jian, Feng Nai-xiang, PENG Jian-ping. Vacuum thermal reduction kinetics of calcined dolomite with Al-Si-Fe alloy[J]. The Chinese Journal of Process Engineering, 2010, 10(1): 127-132. (in Chinese)

[14] XIE Wei-dong, DANG Chun-mei, LI Zhao-nan, PENG Xiao-dong, WANG Hao. Preparation of Mg using Si-Cu reduction and its thermodynamics[J]. Chinese Journal of Rare Metals, 2012, 36(2): 213-217. (in Chinese)

[15] Morsi I M, Barawy K A E, Morsi M B, Abdel-Gawad S R. Silicothermic reduction of dolomite ore under inert atmosphere[J]. Canadian Metallurgical Quarterly, 2002, 41:15-28.

[16] Holywell G C. Magnesium: the first quarter millennium[J]. JOM, 2005, 57: 26-33.

[17] WEN Ming, ZHANG Ting-an, DOU Zhi-he, REN Xiao-dong, ZHANG Rui, ZHOU Lian. Research on new type materials preparation for magnesium production by silicothermic process[J]. Magnesium Technology, 2013:75-79.

[18] GUO Yunchun, YOU Guoqiang, LIU Yong. Research on oxidation characteristics of FeSi75 in silicothermic process [J]. Material & Heat Treatment, 2012, 41(6): 18-20.

Magnesium Technology 2015
Edited by: Michele V. Manuel, Alok Singh, Martyn Alderman, and Neale R. Neelameggham
TMS (The Minerals, Metals & Materials Society), 2015

Carbothermal Production of Magnesium in Vacuum

Tao Qu[1,2], Bin Yang[1,2], Yang Tian[1,2], Yongnian Dai[1,2]
(1. National Engineering Laboratory for Vacuum Metallurgy, Kunming University of Science and Technology, Kunming 650093, China; 2. State Key Laboratory of Complex Nonferrous Metal Resources Clear Utilization, Kunming University of Science and Technology, Kunming 650093, China)

Abstract:

Carbothermal production has been recognized as conceptually the cleanest and energy-efficient route to magnesium metal, but has suffered from technical challenges of development and scale-up. Work by National Engineering Laboratory for Vacuum Metallurgy of China has overcome some barriers of carbothermal production. By changing the condition of condensation, the magnesium vapor is condensed into bulk, so that the risk of magnesium powder explosion can be avoided, and it can reduce the oxygen content, enhance the recovered of magnesium. The mechanism of the carbothermic reduction process has be investigated, the reversion reactions are favored below 1373K at 30~100Pa, and reversion below 9% at bulk in condensation. Purification of magnesium by distillation is known technology, but difficulties with continuous operation are also known. In terms of process control, distillation temperature and pressure are controlled, the magnesium vapor is condensed into liquid, and then solidified into a solid, the solid magnesium is purified (>99.9%).

Keywords: Magnesium; Carbothermal production; Reversion; Purification

Introduction

Magnesium is light metal, it's density is 1.74 g/cm^3 (aluminum is 2.7 g/cm^3, iron is 7.8 g/cm^3) , and it become an important material for lightweight equipped. Magnesium alloy has lots of good properties, such as machining property, corrosion resistance, shock resistance, magnetic shielding. Therefore, magnesium alloys are widely used in aviation, automobile, computer, communication and other fields [1].

Currently, manufacturing auto parts by magnesium alloy is an important research direction for automobile companies. It can reduce the weight of the vehicle and energy consumption, and it also can improve the vehicle acceleration, braking performance. Car which is used magnesium alloy wheel hub can reduce 15% fuel-consumption than aluminum alloy wheel hub [2, 3]. The main methods of magnesium production are pigeon process and electrolytic process, but high energy consumption were needed [4, 5]. Carbothermic reduction is an alternative processes for the production of magnesium. National Engineering Laboratory for Vacuum Metallurgy of China has done some works of carbothermic reduction processes since 1992. We have extracted magnesium from dolomite [6], magnesite [7] and garnierite [8, 9]. And we have studied the mechanism of reversion reaction [10, 11] and catalyst of CaF_2 [12].

Magnesium vapor condensation

Carbothermic reduction process: magnesia and carbon with a certain molar ratio (MgO/C=1/1.4) were mixed, and cast into pellets of Φ30mm×10mm under 2-10MPa as shown in the Fig. 1. And the flow chart is shown as Fig. 2.

Fig. 1 Magnesia-carbon briquette

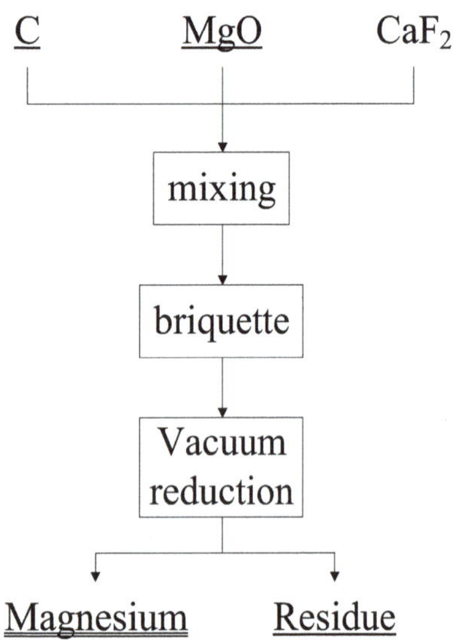

Fig. 2 flow chart of carbothermic Reduction experimental [13]

We control the temperature field of reduction furnace by changing the structure of the furnace, for getting the magnesium bulk. Through experiments, we designed condenser on the side of reactor crucible, condenser pipe is cooled by water and connected the vacuum pump. The structure of furnace is shown as Fig. 3. We can get some magnesium ribbon in the condenser (shown as Fig. 4), but the recovery is low, and can produce magnesium powder at the top of the crucible, and when we open the furnace, the powder will explode.

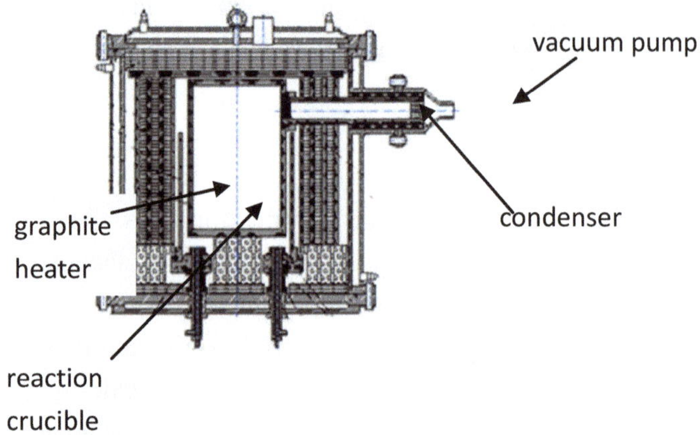

Fig. 3 The structure of furnace 1

Fig. 4 Magnesium ribbon in the condenser

In order to enhance recovery and avoid the production of magnesium powder, we redesigned the reduction furnace, the condenser is on the top of the crucible, no water cooling, is shown as Fig. 5.

Preliminary experiments we can't get magnesium in the graphite condensing pipe, some magnesium vapor condensing on the top of the furnace. After that, we get magnesium bulk through the extension of graphite condensing pipe which is well-crystallized, as shown in Fig. 6.

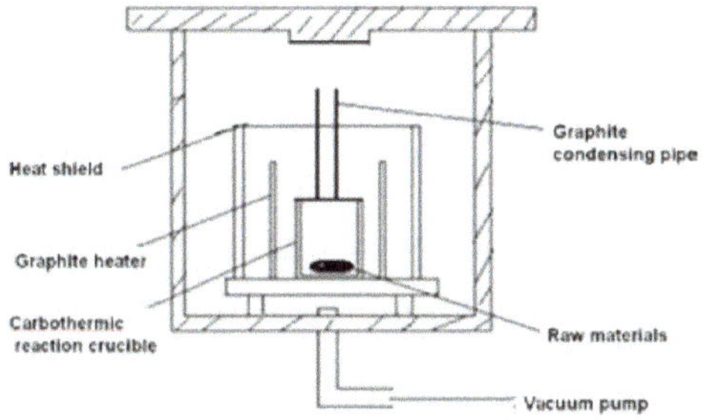

Fig. 5 The structure of furnace 2

Fig. 6 Magnesium bulk in the condenser

Reverse reaction

As is known, the reverse reaction will happen through Carbothermic reduction process, and the reaction temperature is 1100℃ in vacuum. Condensation product of Carbothermic reduction contains Mg, MgO, Mg_2C_3 and C. We can separate magnesium from impurities by distillation.[14]

The reaction temperature of MgO and C is 1250℃ in vacuum, and the vapor pressure of magnesium is 4000Pa at 800℃, so we chose 900℃-1000℃ for the distillation temperature, and the distillation time is 1 hour.

We can calculate the reversion by the following formula via detecting the content of MgO and Mg_2C_3 in condensate. [14]

$$\gamma = \frac{W_1}{W_0} \times 100\%$$

γ - reversion /%;

W_0 - mass of condensate /g;
W_1 - mass of MgO and Mg_2C_3/g。

We choose the condensates of four different reaction time which the mole ratio of C/MgO is 1.4 for vacuum distillation experiments. The results are shown as Tab. 1.
the whole process adverse reaction rate is below 8% when the condensate is bulk.

Purification

Purification of magnesium by distillation is known technology, but difficulties with continuous operation are also known. In terms of process control, distillation temperature and pressure are controlled, the magnesium vapor is condensed into liquid, and then solidified into a solid, the solid magnesium is purified (>99.9%)[15].

We prepare high-purity magnesium metal from raw magnesium (>96%) by vacuum distillation method. For continuous operating, we need the magnesium vapor is condensed into liquid. But as is known, magnesium is no liquid under high vacuum, so we use argon gas to protect magnesium vapor avoid oxidation, we keep the distillation pressure of the whole process between 0.4-0.6atm, and we get a purified magnesium bulk. The composition of raw magnesium and pure magnesium listed in Tab. 2 and Tab. 3 respectively, and the picture of pure magnesium bulk is shown as Fig. 7.

Fig. 7 the photo of pure magnesium by distillation under argon gas protection

Conclusion

National Engineering Laboratory for Vacuum Metallurgy of China has researched vacuum carbothermal magnesium technology nearly 20 years. Recent advances have demonstrated that magnesium bulk can be achieved, and the purified process can be sustained operation. Combining with previous research, the major barrier is sustained operation of carbothermal reduction; maybe the low vacuum can solve the barrier.

[1] Years 2010 Primary Magnesium Production. International Magnesium Association. http://www.intlmag.org.

[2] Mordike B L, Eben T. Magnesium properties application-potential. Materials Science Engineering A, 2001, 302: 37-45.

[3] Brooks G, Cooksey M, Wellwood G, Goodes C. Challenges in light metals production [J]. Transactions of the Institutions of Mining and Metallurgy, Section C, 2007,116 (1): 25-33.

[4] Prentice.L. Psuedo-Steady-State Control of High Temperature Gas-Solid Reaction. in Chemeca2011[C]. 2011. Sydney, Australia: Engineers Australia.

[3] Sujit Das, Primary magnesium production costs for automotive applications [J]. JOM, 60 (11) (2008), pp. 58–63.

[4] Hu Wen-xin, Liu Jian, Feng Nai-xiang, et al. Theoretical analysis and experimental study on thermal reduction of calcined dolomite with Al-Si-Fe ternary alloy[C]. Proceedings of the 9th Vacuum Metallurgy and Surface Engineering Conference, 2009:27～31.

[5] Leon H. Prentice; Michael W. Nagle; Timothy R. D. Barton, et al, Carbothermal production of magnesium: CSIRO'S MagSonic™ process, TMS, Magnesium Technology 2012:31-36;

[6] Wang Y.C, Tian Y, Yang C.B, et al, Thermodynamic Analysis of Mg Extraction from Calcined Dolomite by Carbothermic Reduction in Vacuum, Chinese Journal of Vacuum Science and Technology, 2013(11):88-94;

[7] Li Y.F, Dai Y.N, Yang B, et al, Study of preparing magnesium metal by carbothermal reduction of magnesite in Vacuum, Chinese Journal of Light Metal, 2011(9):52-57;

[8] Qu Tao, Tian Yang, Yang Bin, et al. Study on the Thermodynamic and Experimental Carbothermic Reduction of Garnierite[C].TMS2012. Magnesium Technology, 2012:505-510;

[9] Luo Qi, Qu Tao, Liu Dachun, et al, Behavior of magnesium in saprolite nickel laterite by carbothermic reduction in vacuum, Chinese Journal of Central South University, 2012, Vol43(11): 4191-4198;

[10] Tian Yang, Qu Tao, Yang Bin, et al. Mechanism of Carbothermic Reduction of Magnesia and Reversion Reaction[C].TMS2012. Magnesium Technology, 2012:511-516.

[11]Yang Cheng-bo, Tian Yang, Qu tao, et al. Analysis of the behavior of magnesium and CO vapor in the

carbothermic reduction of magnesia in a vacuum [J], Journal of Magnesium and Alloys 2 (2014): 50-58
[12] Tian Yang, Qu Tao, Yang Bin, Dai Yongnian, Xu Baoqiang, Geng Sen. Behavior analysis of CaF_2 in magnesia Carbothermic reduction Process in Vacuum [J].Metallurgical and Materials Transaction B, 2012(2).
[13] Zhong sheng. The study on magnesia carbothermic reduction in vacuum. Kunming University of Science and Technology. Dissertation for master's degree 1999.
[14]Tian Yang. Basic theory and experimental research on magnesia carbothermal reduction in vacuum. Kunming University of Science and Technology. Dissertation for doctor's degree 2012.
[15]Wang Yuchao, Study of preparing pure magnesium by vacuum distillation, Kunming University of Science and Technology. Dissertation for master's degree 2014.

Tab.1 The reversion of Carbothermic Reduction

C/MgO (mole ratio)	Reaction time/h	Condensate		Residue		W_0/g	W_1/g	Reversion /γ%
		mass/g	C%	mass/g	MgO%			
1.4	2	7.2	1.4	0.3	80.87	7.099	0.2426	3.42
	3	7.2	2.1	0.4	81.81	7.049	0.3272	4.64
	4	8.4	1.08	0.5	82.65	8.309	0.4132	4.97
	5	9.3	1.52	0.8	84.87	9.159	0.6789	7.41

Tab. 2 The composition of the raw magnesium(wt%)

element	Mg	Ca	Zn	Mn	Pb	Fe	Al	Ni
content	96.54	0.56	0.0035	0.0097	0.0011	0.066	0.014	0.019

Tab. 3 The composition of the pure magnesium(wt%)

element	Mg	Ca	Zn	Mn	Pb	Fe	Al	Ni
content	99.97	0.017	0.0045	0.0027	0.0007	-	0.002	-

Magnesium Technology 2015
Edited by: Michele V. Manuel, Alok Singh, Martyn Alderman, and Neale R. Neelameggham
TMS (The Minerals, Metals & Materials Society), 2015

EFFECT OF ARGON FLOW RATE ON THE CONDENSATION OF MAGNESIUM VAPOR FROM CARBOTHERMIC REDUCTION OF MAGNESIA

GuangYong Bin, Yu Wang*, SiYa Wang, XiaoPing Liang

College of Materials Science and Engineering, Chongqing University, Chongqing 400044, China

Keywords: Carbothermic reduction, magnesium vapor, argon, dilution of condensation stream

*Corresponding author: wangyu@cqu.edu.cn

Abstract

To increase both the recovery rate and purity of magnesium in carbothermic reduction process, magnesium vapor was condensed in a certain concentration of argon to avoid the occurrence of reverse reaction of CO and magnesium vapor. A cylinder by pressing a mixture of graphite and MgO with a molar ratio (2:1) was reacted at 1723K for 2.5h in an vacuum atmosphere of 1(±0.05) KPa. Then the mixture was cooled in different flow rates of argon atmosphere at a constant cooling rate. The condensation products were collected and characterized by XRD and SEM/EDS. The results show that increasing the argon flow rate enhanced diffusion of magnesium vapor and CO, meanwhile, diluted the CO concentration resulting in restraining the adverse reaction. The recovery rate of the condensation product reached a maximum of 95.6% with magnesium (strip-like) purity of 94.8% when the argon flow rate was 3.5 L/min.

Introduction

Magnesium is well known for its good machinability and formability[1], at the same time, it is the lightest structural metal and apply in various fields[2]. At the present, magnesium is produced by two principal processes: electrolytic process and silicothermic processes. Molten salt electrolysis is the mainly methods of electrolytic processes, the raw materials($MgCl_2$) is very extensive; such as seawater, natural brines, magnesium-rich brines from potash production and carnallite contain a lot of $MgCl_2$[3-4]. But this process is characterized by high energy consumption[5]. Magnesium production by Pidgeon process is the most widely used in silicothermic process in China[6-8]. However, the reduction of Pidgeon Process cycle is too long and the production efficiency is low(2Kg/h). The Pidgeon process need to consume energy is 354.5MJ and release of greenhouse gases CO_2 is 42Kg for product Per 1kg of magnesium ingot [9]. Scientists pay more attention to study the Carbothermic reduction process, and solve the problem about high-energy consumption and environmental pollution.

Carbothermic reduction method is an alternative to electrolytic process and silicothermic processes for the production of magnesium[10]. The reaction temperature as high as 1800 ℃ under the condition of atmospheric pressure, with the decline of the system pressure, the reaction temperature is reduced. Researchers focused on study carbothermic reduction mechanism in recent years. LI[11]studied the reduction kinetics and two stage reaction mechanism. Nusheh[12]studied effect of mechanical milling on the reaction.etc. So the researchers study on reduction mechanism of more mature. The great majority of work at the present time concentrated on Condensation and collection of magnesium, particularly focus on controlling the reversion reactions, it is magnesium vapor and CO recombining during cooling to Solid metal severe contaminated by MgO. To increase the purity of magnesium in carbothermic reduction process, the reversion reaction must be restrained or reduced the reaction.

This study was conducted to investigate the effect of argon flow rate on the condensation of magnesium vapor and avoid the occurrence of reverse reaction of CO and magnesium vapor at different flow rate were studied, respectively. For this purpose, fundamental characteristics of the product were examined using X-ray diffraction (XRD), scanning electron microscopy (SEM) and energy-dispersive X-ray spectroscopy (EDS).

Experimental

The raw materials used analytically pure magnesia(≥ 98.0%), graphite(≥99.8%) and CaF2(≥98.5%) in the experiments. Magnesia and graphite were mixed in a molar ratio 1:2 with the addition of total quality 5 pct CaF2 and made into a cylinder under the sample preparation pressure 15 MPa. Ac contactor and digital vacuum gauge could effectively steady pressure of the system. The experimental installation picture of vacuum furnace is shown in Figure 1. Firstly, the sample put in the Al_2O_3 crucible of vacuum furnace. Secondly, the furnace body was filled with argon and the gas was released by vacuum pump, the processes were be recycled five times, and ensure air would been discharged the furnace. Thirdly, the samples were heated from room temperature to 1723K for 2.5h, and condensation products were collected in the condenser of the vacuum furnace, at the same time, the furnace filled 0 to 4.0L/min argon gas. The phase of the condensation products were identified by XRD. The surface structural and chemical composition were characterized by SEM and EDS, respectively.

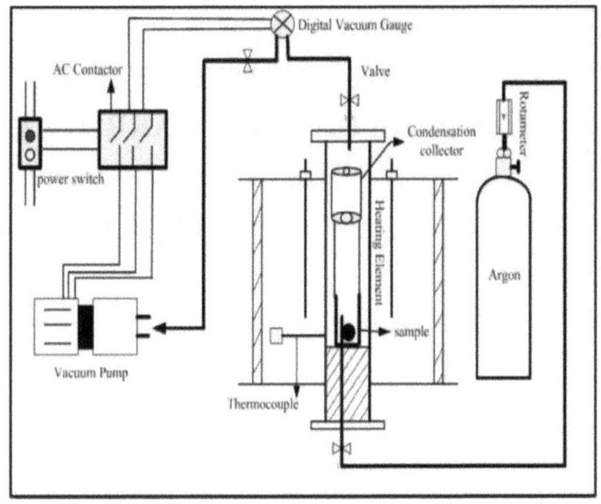

Figure1: The experimental installation picture of vacuum furnace

Results and Discussions

Theoretical analysis of magnesium vapor condensation

The theoretical analysis aims at calculate the magnesium vapor condensing temperature, in order to determine the products collect location in the furnace. There are two approaches to magnesium vapor condensation, From gas with liquid to solid(a) and directly from gas to solid(b) were shown in Figure 2. (a) is a gas-liquid-solid three phase transition that is easier to make magnesium vapor condenses into pieces, and (b) is direct from gas to solid that is easier to obtain magnesium powder[13].

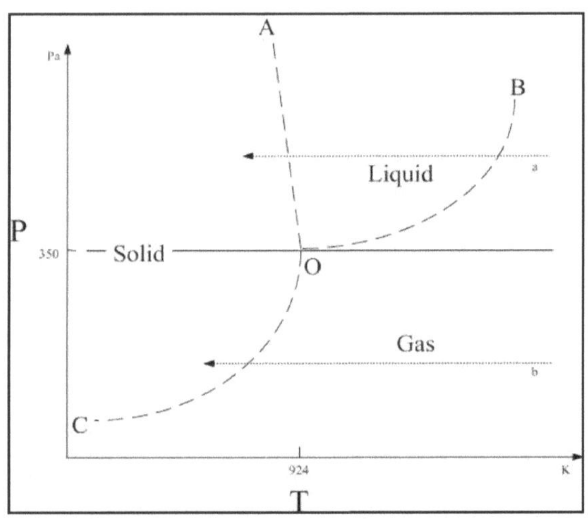

Figure2: Magnesium vapor solidification curve

Magnesium vapor condensation must make the magnesium vapor pressure in the system(P_{Mg}) more than saturated vapor pressure of magnesium(P^*_{Mg}). The P^*_{Mg} can calculate following equation(1) and (2) at different temperatures[14], it is shown in Figure 3. With the temperature increasing, the magnesium vapor of saturated vapor pressure increased, when the temperature reaches to 1000K, the rapid increase of P^*_{Mg}.

$$\lg p(pa) = 13.54 - 7780/T - 0.86\lg T$$

(298k≤T≤924K) (1)

$$\lg p(pa) = 14.92 - 7550/T - 1.41\lg T$$

(924k≤T≤1373k) (2)

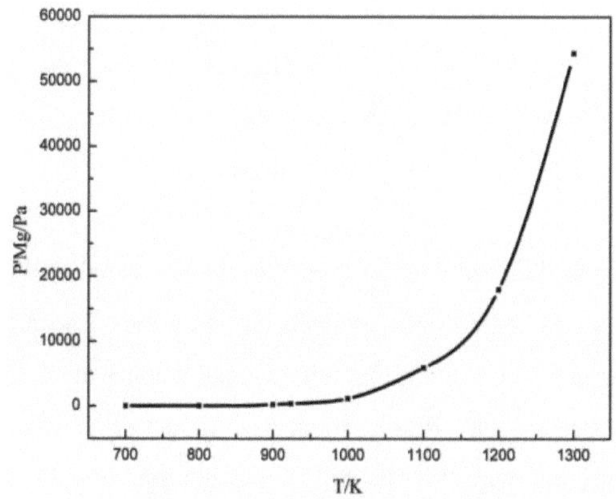

Figure3: Relationship between P^*_{Mg} and T for equation (1) and (2).

The Carbothermic reduction main reaction was described as equation (3).

$$MgO_{(s)} + C_{(s)} = Mg_{(g)} + CO_{(g)} \quad (3)$$

The reaction products Mg vapor and CO are equimolar, and hypothesize the system pressure P_{system} is equal to the sum of P_{Mg} and P_{CO}[15]. The Gibbs free energy ΔG_T was written as follows:

$$\Delta G_T = \Delta G^{\theta}{}_T + 19.144T\lg\left(\frac{\frac{1}{2}P_{system}}{P^{\theta}}\right)^2 \quad (4)$$

the ΔG_T was calculated at 1 KPa, and the ΔG_T equal to -3.62×10^3 when the reaction temperature was 1723 k. So the Carbothermic reduction can be happened. With the reaction progressing, the P_{Mg} of theoretical value reach to 500 pa, it is slightly smaller than the practical value. Condensation temperature range was under 1000K based on the comparison of P_{Mg} and P^*_{Mg}. Thermocouple measured the temperature of furnace body that determine the appropriate installation position of condenser.

Carbothermic reduction occurs and produce magnesium vapor and CO mixed gas, through the mass diffusion of magnesium vapor arrival in condenser, meanwhile cooling condensation. The product yield (Y) was shown in the following formula (5)

$$Y = \frac{6(M_0 - M_1)}{13M} \times 100\% \quad (5)$$

Where M_0 is the quality of the raw material before reaction, M_l is the quality of the materials after reaction, and M is the quality of the condensation product.

Effect of argon flow rate

The study was conducted with experimental runs of 2.5 hours in length at the temperature 1723K. The effect of argon flow rate on the magnesium metal yield was given in Table 1. It shows the product yield increased with increasing argon flow rate. The magnesium metal yield without argon(Ar) was only 67.6 pct. However, the percentage was up to 95.6 pct with the argon flow rate was 3.5 L/min. The product yield increased with Ar(3.5 L/min) more than without Ar, when the argon gas flow more than 3.5 L/min, the product yield began to descend. So 3.5 L/min argon flow rate was the optimal choice.

The increase of argon flow reduces the partial pressure of CO, and makes the carbothermic reduction reaction to the positive. Meanwhile, accelerate the diffusion of the CO and the magnesium vapor. Condensation product of magnesium steam output increases. But, as the argon flow further increases, it would take away a portion of magnesium vapor, so the argon flow rate is 4 L/min that reduce the magnesium vapor condensation product yield.

Table 1: Relation between argon flow rate and product yield.

Number	Argon flow rate (L/min)	Product yield (percent)
1	0	67.6
2	2	79.7
3	2.5	85.8
4	3	90.9
5	3.5	95.6
6	4	92.5

Condensation products were collected in the condensation collector of the vacuum furnace between 900 K and 1000 K. the products of with 3.5 L/min Ar and without Ar were examined by XRD as shown in figures 4 and 5, respectively. The XRD pattern of the condensation products demonstrates that all peaks were sharp and well defined, which explained that the product was well crystallized. Therefore, the final products were Mg and MgO in Fig. 4. Magnesia was found from the following reverse reaction:

$$Mg_{(g)} + CO_{(g)} = MgO_{(s)} + C_{(s)} \quad (6)$$

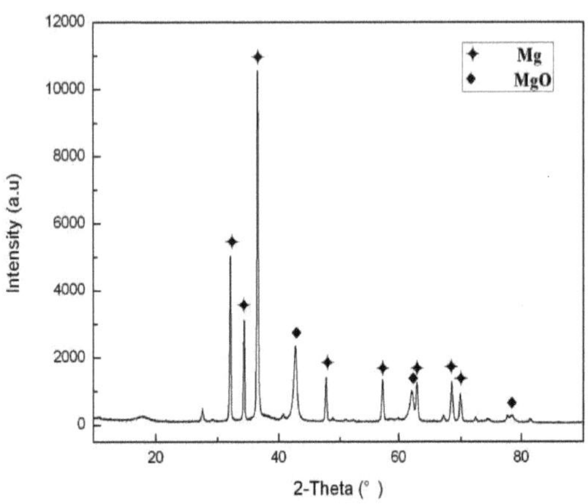

Figure 4: X-Ray diffraction patterns of condensation products of with Ar(3.5 L/min).

Condensation products of without Ar in Fig.5 and the final products were Mg, MgO and a few Mg_2C_3. As time progresses, a lot of magnesium vapor condensation in the cooling area, due to the CO was no timely discharge, and in the magnesium metal surface reaction to generate small amounts of Mg_2C_3 and MgO. Following reaction:

$$5Mg_{(s)} + 3CO_{(g)} = 3MgO_{(s)} + Mg_2C_{3(s)} \quad (7)$$

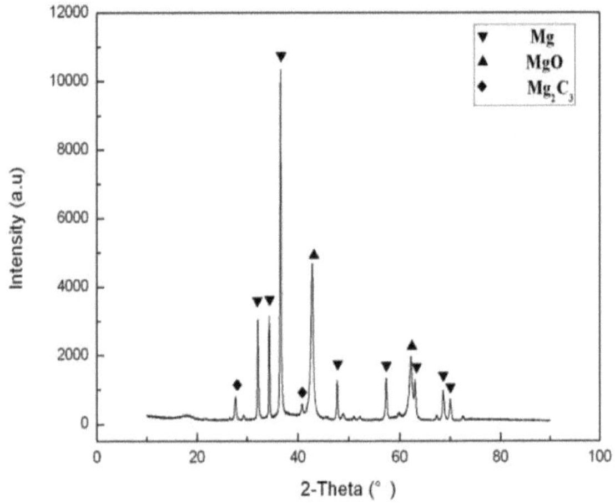

Figure 5: X-Ray diffraction patterns of condensation products of without Ar.

From the Figure 4 and Figure 5 could be concluded that the argon gas reduced produce the impurities. The argon flow rate is 3.5 L/min that only a small amount of MgO in the product, and without argon gas product of Mg_2C_3 and MgO in Figure 5. The argon gas dilution of CO, and reduced the CO in the condenser by the collision of between the Ar molecules and the CO molecules. So reduce and restrain the reaction of (6) and (7), respectively.

Figure 6 showed the SEM and EDS images of condensates that were obtained from the condensation collector. The product of magnesium metal had a strip structure and the metal surface had some small particles with different shapes. As shown in the EDS images, A was obvious that the content of magnesium in smooth condensates was 100.00 wt pct, and B was show that the content of magnesium in raised condensates was 89.64 wt pct, and oxygen was 10.36 wt pct. The average purity of magnesium metal reached 94.8 wt pct. The result was obtained when the argon flow rate was 3.5 L/min.

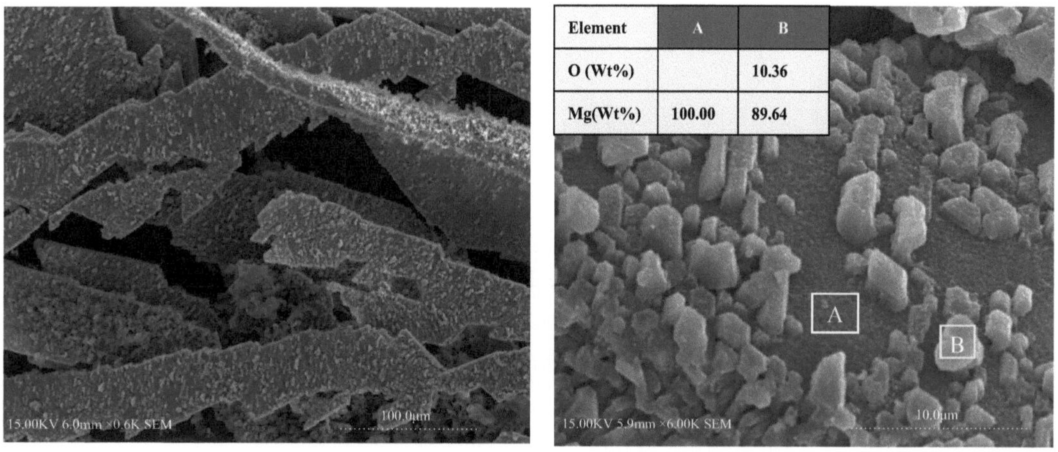

Element	A	B
O (Wt%)		10.36
Mg(Wt%)	100.00	89.64

Fig. 6: SEM and EDS images of condensation products.

Conclusions

1. The theoretical analysis of magnesium vapor condensation temperature range is obtained in 1000K under the reaction temperature 1723K at the system pressure is 1Kpa. It can determine the installation position of the condenser.
2. The experimental results indicated that ventilation with argon gas can improve the product yield of magnesium metal, argon flow accelerated the carbothermic reduction rate and condensation products. Metal yield was just only 67.6% without Ar. It increased with increasing argon flow rate. The percentage was up to 95.6% with 3.5 L/min Ar gas.
3. The experimental results showed that argon flow diluted the CO concentration resulting in restraining the adverse reaction, to improve the purity of magnesium metal. The average purity of magnesium metal reached 94.8 wt pct, when the argon flow rate was 3.5 L/min. The microcosmic structure is strip of magnesium metal.

References

1. Lan Hong, Hong Yong Sohn, and Masamichi Sano, *Scandinavian Journal of Metallurgy* 2003;32:171-176.
2. R.E.Brown, *Light metal age*, 1996.VOL.8.
3. Geoffrey Brooks, Simon Trang, and Michael Nagle, *JOM*,2006 May.
4. R.Winand, M. van Gysel, A.Fontana, L. Segers, and J-C. Carlier. *Recycling of Metalliferous Materials.* 99(1990):C105.
5. Rao YK, *Metall Trans*,1971:2:1439.
6. Chengbo YANG, et al, *MEMS* (2012):475-477.
7. G.J.Kipouras, and D.R. Sadoway. *Advances in Molten Salt Chemistry 6.* 1987,pp.127-209.
8. A. Froats, et al, *Light metal*, Metallurgical Society of AIME, (NY,1980),969-79.
9. S.Ramakrishnan, et al, *Resources Conservation and Recycling* 2004, 42,pp.49-64.
10. G.Brooks, et al *Mineral Processing and Extractive Metallurgy*,2007,vol 116, NO 1.
11. Li RongTi, and Pan Wei, et al, *Metallurgical and Materials Transactions B*.(Volume 34B,August 2003)-433.
12. M.NUSHEH, and H.YOOZBASIADEH, et al *ISIJ International*, Vol.50(2010),NO.5,pp.668-672.
13. Benny T, and Kuan n, et al, *Chemical Engineering Science* 87 (2013) 23–39.
14. DAI Yong-nian, and YANG Bin et al, *The vacuum metallurgy of nonferrous metals* [M]. Beijing: Metallurgical Industry Press, 2000: 65−72. (in Chinese).
15. YANG TIAN, and TAO QU, et al, *Metallurgical and Materials Transactions B*. (Volume 43b,JUNE 2012) 652-657.

Magnesium Technology 2015
Edited by: Michele V. Manuel, Alok Singh, Martyn Alderman, and Neale R. Neelameggham
TMS (The Minerals, Metals & Materials Society), 2015

ENVIRONMENTAL IMPACT OF MAGNESIUM PRODUCTION BY CARBOTHERMIC REDUCTION IN VACUUM

LIU Hong-xiang[1,2,3,4], TIAN Yang[1,2,3,4], YANG Bin[1,2,3,4], XU Bao-qiang[1,2,3,4], LIU Da-chun[1,2,3,4], DAI Yong-nian[1,2,3,4]

1. State Key Laboratory of Complex Nonferrous Metal Resources Clear Utilization in Yunnan Province, Kunming 650093, China;
2. National Engineering Laboratory for Vacuum Metallurgy, Kunming 650093, China;
3. Key Laboratory for Nonferrous Vacuum Metallurgy of Yunnan Province, China;
4. Faculty of Metallurgical and Energy Engineering, Kunming University of Science and Technology, Kunming 650093, China;

Keywords: Magnesium; Vacuum; Carbothermic reduction; Environmental impact

Abstract

Based on the practice of magnesium production via carbothermic reduction in vacuum, the pollutants were calculated for exhaust gas, waste water, waste residue, noise. And a quantitative evaluation of the environmental impact was carried out.

Introduction

Magnesium was named the most promising of lightweight materials and green metallic engineering materials in the 21st century [1]. By 2013, the production of primary magnesium was 910000t [2]. To extract metallic magnesium, two different technologies are commercially used: Pidgeon method in calcined dolomite and molten-salt electrolysis, but these processes suffer from high energy usage and low productivity [3]. Now Pidgeon process dominates the production of magnesium in the world [4, 5]. Carbothermic reduction is an alternative to both Pidgeon and electrolytic processes for the production of magnesium [6, 7].

Process Description

Magnesite(MgO:45%) used as the raw materials, calcined 2-3h in furnace(rotary kiln, shaft furnace), calcination temperature controlled at 1273K by soft coal(C:80%, S:0.8%, A:20%) burning, then 15-40mm regular MgO was obtained.

Magnesia, coking coal(C:73.5%, S:0.35%, A:9.7%) and fluorite were main reaction mass. The preparation process included crush, fine grinding, mixing and briquetting. Magnesia, fluorite and coking coke as the starting materials were mixed in a molar ratio (MgO/C=1/2) with addition of 3% CaF2, and the mixed mass were made into pieces of Φ30mm×10mm under 8-10MPa. The heating rate of system temperature was about 15K•min-1. In the magnesia carbothermic reduction process, the prepared pellets were heated by electricity from room temperature to 1273K, and held at this temperature for 20 minutes to remove the adsorbed gases and water when the vacuum degree of the reduction system is 1-13Pa. Then, the temperature was raised to 1473K-1773K. After the reaction time (2h), the reaction product was cooled until room temperature and took out.

C and MgO were also obtained by vacuum carbothermic reduction in rough magnesium. In order to get the high purity magnesium, rough magnesium refining was needed. Electrical heating was the heat source in this process; impurities removed from magnesium meltwater through added 15% 2# (Calcium solvent) refining flux in the finery. In order to avoid magnesium combustion, SF_6 or SO_2 used as protective gas in magnesium solution casting. Anticorrosive passivation was needed after casting of magnesium ingot condensing, due to the lower corrosion resistance of magnesium. Then the product was Packaging warehousing or on sales.

The flowchart of carbothermic reduction in vacuum and environmental pollution was shown in Fig1. The inventory of main resources and energy consumption of the carbothermic reduction process was shown in Table1.

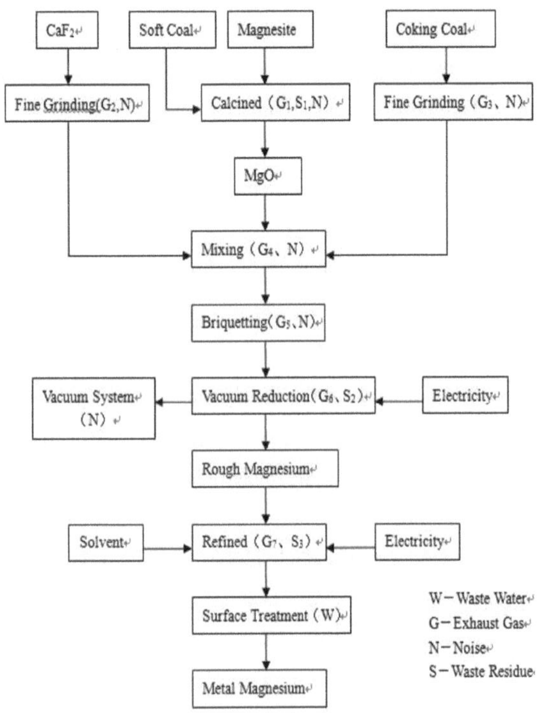

Fig1 Flowchart of carbothermic reduction in vacuum and environmental pollution

Table1 The inventory of main resources and energy consumption for 1t Mg of the carbothermic reduction process

Input	Data
Magnesite/(10^3kg·t^{-1})	3.70
Fluorite/(10^2kg·t^{-1})	1.98
Soft Coal/(10^3kg·t^{-1})	3.60
Coking Coal/(10^3kg·t^{-1})	1.36
Electric power/(10^4kW·h·t^{-1})	4.27

Pollutants Inventory

It can be seen from Fig1, the pollutants of carbothermic reduction process included exhaust gas, waste water, waste residue and noise. The different pollutants were calculated by material balance method, and were consulted by the Handbook of environmental protection calculation [8] and the first national census of pollution sources of industrial pollution sewage coefficient manual [9].

Exhaust gas (G)

The exhaust gas of carbothermic reduction process included: calcined magnesite (G_1), reaction mass preparation (G_2, G_3, G_4, and G_5), vacuum carbothermic reduction (G_6), rough magnesium refining (G_7).

Exhaust gas from calcined magnesite(G_1)

The exhaust gas resulted from two main resources: one was from fuel burning; the other was from magnesite decomposition. Soft coal was generally used as fuel. From soft coal burning, it would produce 10.56t CO_2, 0.06t SO_2, and 0.01t NOx and 0.374t dust for 1t Mg. From magnesite decomposition, it would produce 1.83t CO_2, 0.045t dust for 1t Mg.

Exhaust gas from reaction mass preparation (G_2, G_3, G_4, G_5)

The exhaust gas was dust during the reaction mass preparation. It resulted from four resources: 1 dust(G_2) produced by fluorite fine grinding, the main composition were CaF_2, then SiO_2, Fe_2O_3, Al_2O_3. 2 dust (G_3) produced by coking coal fine grinding, the main composition was C. 3 dust (G_4) produced by materials (magnesia, fluorite, coking coal) mixing, the main composition were MgO and C, then CaF_2, SiO_2, Fe_2O_3, Al_2O_3. 4 dust (G_5) produced by materials briquetting, the main composition were MgO and C, then CaF_2, SiO_2, Fe_2O_3, Al_2O_3. It would produce 0.028t dust for 1t Mg.

Exhaust gas from vacuum carbothermic reduction (G_6)

The exhaust gas resulted from magnesia carbothermic reduction; the key pollutant was CO, then CO_2, SO_2, dust. The raw materials in vacuum carbothermic reduction process were MgO, coking coal and fluorite, the weight ratio was accorded 54:43:3. It would consume 1.67t MgO and 1.36t coking coal (Only 1/2 to act) for 1t Mg and produces 0.93t CO, 0.37t CO_2, 0.004t SO_2, 0.034 t dust.

Rough magnesium refining (G_7)

In order to avoid magnesium combustion, SF_6 or SO_2 used as protective gas in magnesium solution casting. The SO_2 which produced by dropped sulfur on the liquid magnesium surface was used to prevent magnesium combustion before, but now the more and more applications by SF_6 used as protective gas for preventing magnesium combustion, it would consume 0.8kg SF_6 for 1t Mg [10].

The inventory of exhaust gas of magnesium production via carbothermic reduction in vacuum was shown in Table2.

Table 2

The inventory of exhaust gas [tons] for 1t Mg of the carbothermic reduction proces

Pollutants	Calcined magnesite	Mass preparation	Carbo thermic reduction	Mg Metal refining	Total
CO2	12.39	-	0.37	-	12.76
CO	-	-	0.93	-	0.93
SO2	0.06	-	0.004	-	0.064
NOx	0.01	-	-	-	0.01
Dust	0.419	0.028	0.034	-	0.481
SF6	-	-	-	0.0008	0.0008

Waste water (W)

Magnesium was one of the worst corrosion resistance metals, in order to ease for saving, it is necessary to be antiseptic treatment of magnesium ingot surface, the method was Chromate passivation, it would produce 1t waste water with chromium ion(Cr^{3+}, Cr^{6+}) for 1t Mg. The toxicity of Cr^{3+} was smaller, but Cr^{6+} was worse, the waste water was treated by the combined processes of neutralization and chemical sedimentation [11].

Waste Residue(S)

The waste residue of carbothermic reduction process included: calcined magnesite (S_1), carbothermic reduction (S_2) and rough magnesium refining (S_3).

Waste residue from calcined magnesite (S_1)

The waste residue included $MgCO_3$, MgO, SiO_2 and CaO. It would produce 2.03t waste residue for 1t Mg.

Waste residue from vacuum carbothermic reduction (S_2)

The raw materials in vacuum carbothermic reduction were magnesia, fluorite and coking coal (Only 1/2 to act); the weight ratio was accorded 54:43:3. It would produce 0.88t waste residue for 1t Mg, the main composition were C, then CaF_2, few MgO.

Waste residue from rough magnesium refining (S_3)

The rough magnesium refining used 2# solvent (calcium solvent), it would produce 0.24t waste residue for 1t Mg, and the main composition were $MgCl_2$ and KCl.

The calcined and carbothermic reduction residues(S_1, S_2) belonged to general industrial solid waste, its environmental effect was smaller, in order to develop circular economy, the residues could send to cement plant as agent, the application of residues was more and more widely in brick material, new-type wall material, desulfurizer and road building material [12]. The main composition of rough magnesium refining residues were MgCl2 and KCl, but few BaCl2 was also contained, it belonged to hazardous wastes (HW47) [13], so the refining residues should be properly handled, according to the Standard for pollution control on hazardous waste storage

GB18597 — 2001 and the Standard for Pollution Control on the Security Landfill Site for Hazardous Wastes GB18598—2001.

Noise

The noise in vacuum carbothermic reduction process was produced by mechanical equipment operation, the main resources included: air blower, crusher, ball mill, briquette machine, vacuum pump etc., the noise source strength was shown in Table3 [14].

Table3 The noise of equipment in vacuum carbothermic reduction (dB (A))

Device name	air blower	crusher	ball mill	briquette machine	vacuum pump
Limits	80~126	85~114	87~128	80~95	80~95

The noise influence can be reduced by using damping and sound insulation.

Environmental Impact

Environmental impact category

Through the analysis of link of producing pollution in the whole vacuum carbothermic reduction and calculation of producing pollution coefficient, the key pollutants were exhaust gas; secondary pollutants were waste water, waste residue and noise. The main exhaust gas was CO_2, then CO, SO_2, NOx, dust and SF_6. Environmental impact category of different pollutants was shown in Table4.

Table4 Environmental impact category of different pollutants

Environmental impact category	Pollutants
Global warming potential(GWP)	CO_2, SF_6
Acidification potential(AP)	SO_2, NO_x
Photochemical Ozone Creation Potential(POCP)	CO, NO_x
Human toxicity potential(HTP)	SO_2, NO_x, Dust

The environmental impact characterization

The aim of characterization was made every different materials of impact category transformed and gathered into unified unit [15]. The characterization result of environmental impact category of vacuum carbothermic reduction was calculated by the formula multiplied by characterization factor [16]. Pollutants characterization factor was shown in Table5 [17] and calculation of environmental impact characterization was shown in Table6.

Table5 Pollutants characterization factor（kg/kg）

Pollutants	GWP	AP	POCP	HTP
CO_2	1	-	-	-
SF_6	23900	-	-	-
SO_2	-	1	-	0.096
NOx		0.7	0.028	1.2
CO			0.027	
Dust				0.82
ETH			1	
1,5- DCB				1

Table6 Environmental impact characterization for 1t Mg by vacuum carbothermic reduction

Environmental impact category	Unit	Characterization
GWP	kg/CO_2	3.19E+01
AP	kg/SO_2	7.10E-02
POCP	kg/ ETH	2.54E-02
HTP	kg/1,5-DCB	4.13E-01

It can be seen from Table6, GWP was the most important factor in the four environmental impact category of magnesium production by carbothermic reduction in vacuum, and then HTP and AP were the secondary factor, the least important factor was POCP.

Conclusions

Through the assessing environmental impact of magnesium production by carbothermic reduction in vacuum, the key pollutants were exhaust gas; secondary pollutants were waste water, waste residue and noise. The main exhaust gas was CO_2, then CO, particulate matter, SO_2, NOx and SF_6.

GWP was the most important factor in the four environmental impact category of magnesium production by carbothermic reduction in vacuum, then HTP and AP were the secondary factor， the least important factor was POCP.

Reference

[1] Xu Riyao. Metal magnesium production technology [M].Changsha: Central South University Press, 2003.12:21.

[2] USGS Mineral Commodity Summaries, 2014.

[3] Feng Naixiang, Peng Jianping, Wu Xiuming. Improve Chinese magnesium smelting technology and increase competitive power of magnesium industry [J].Light Metals, 2002, 9:51.

[4] Das Sujit, Primary magnesium production costs for automotive applications[J].JOM, 2008,60:63-69.

[5] Chen Lini. Light metal smelting and environmental protection [M]. Shenyang: Northeastern University Press, 1991:177-190.

[6] Tian Yang, Qu Tao, Yang Bin, et al. Behavior analysis of CaF_2 in magnesia Carbothermic reduction Process in Vacuum[J]. Metallurgical and Materials Transaction B, 2012,43(3):657-661.

[7] Cherubini Francesco, Raugei Marco, Ulgiati Sergio. LCA of magnesium production technological overview and worldwide estimation of environmental burdens[J]. Resources Conservation and Recycling, 2008,52:1093-1100.

[8] Xi Yuanfu. Environmental protection reckoner [M].Chengdu: Sichuan Science and Technology Press,1991.6: 97-166.

[9] The first National Pollution Source Census Leading Group Office. Nonferrous metal smelting and rolling processing industry [Z].2008.2:158-161.

[10] Scott Bartos. U.S. EPA's SF6 Emmissions Reduction Partnership for the Magnesium Industry: An Update on Early Success[C]. Magnesium Technology 2001.New Orleans Louisiana:Pergamon Press,2001:46.

[11] Guangdong Environmental Protection Engineering Research and Design Institute. Two phase expansion project environmental impact report for Foshan aluminum industry limited company [Z].2008.11:26-31.

[12] Xu Xiangbin, Luo Xuyan, Li Changyong. Situation of magnesium slag applications[J].Light Metals,2009,7:50.

[13] Ministry of Environmental Protection of the People's Republic of China, National hazardous waste list.2008.8:25.

[14] Gao Hongwu. noise control engineering [M].Wuhan: Wuhan University of Technology Press, 2003.7:4-101.

[15] Deng Shengnan, Wang Xiaobing. Life cycle assessment[M].Beijing: Chemical Industry Press,2003.

[16] Yang Jianxin, Wang Rusong, Liu Jingru. Methodology of life cycle impact assessment for Chinese products[J]. Acta Scientiae Circumstantiae, 2001, (21), 234～237.

[17] Gao Feng. Study on life cycle assessment and apply on commercial magnesium[D].Beijing University of Technology.

Magnesium Technology 2015
Edited by: Michele V. Manuel, Alok Singh, Martyn Alderman, and Neale R. Neelameggham
TMS (The Minerals, Metals & Materials Society), 2015

COMPARATIVE ENVIRONMENTAL BENEFITS OF LIGHTWEIGHT DESIGN IN THE AUTOMOTIVE SECTOR: THE CASE STUDY OF RECYCLED MAGNESIUM AGAINST CFRP AND STEEL

Fabrizio D'Errico[1], Luigi Ranza[2]
[1]Politecnico di Milano, Department of Mechanical Engineering, Via La Masa 34, 20156 Milan, Italy
[2]CiaoTech-PNO Consultant Group, Via G. Pacini 11, 20131 Milan, Italy

Keywords: Magnesium alloys, Life Cycle Assessment, Eco-Magnesium, Lightweight design, CO_2 reduction

Abstract

A LCA feasibility study was undertaken to determine the environmental impact of an Eco-magnesium process route by recycled chips to manufacture panel for the automotive sector to be compared with comparative scenarios, a non-recycled carbon fiber reinforced polymer (CFRP) and a baseline steel-made component scenario. The objective of this LCA study was to assess the actual benefits of a lightweight solution considering the whole life cycle, including the dirty-phase (i.e. the "cradle-to-exit gate" stage) that impacts differently for the different materials. For this reason the analysis has regarded the net "cradle-to-grave" scenario. Different automotive floor pans were then compared considering the rate of fuel consumption during vehicle operation - i.e. the fuel-mass correlation factor - and the different material substitution factors allowed by the different materials selected.

The background: lightweight strategies in the automotive sector in brief

Lightweighting in advanced technology vehicles— such as plug-in hybrids, all-electric vehicles, and hydrogen fuel cell-powered vehicles—is important in order to reduce the effect of the heavier and more expensive powertrain. In addition, composites technology, better suited to low annual production volumes in the range of 20,000–40,000 vehicles, matches well with the market for niche hybrid vehicles today. Recent changes to the Corporate Average Fuel Economy (CAFE) are driving automakers to seek more aggressive methods for fuel consumption reductions. Light weighting of vehicles will be a factor in meeting these requirements due to the inherent relationship between mass and fuel consumption. In addition, light weighting may benefit other advanced fuel saving but load-constrained technologies, such as battery-powered vehicles. The ability to introduce new lightweight materials into vehicles is not a trivial matter. Many see a new concept, or limited production, vehicle introduced to the market with lightweight "space-aged" materials and feel that its adoption by mass produced vehicles is a simple matter of "remove and replace." However, this is not the case; factors such as existing infrastructure, material cost, and high volume capacity become of great importance for mass production vehicles. In addition, many of the low production vehicles incorporate these lightweight materials as a method for gaining experience on their performance. Without significant data to support durability, the risk-averse automotive culture will not adopt new materials. Therefore, it often takes many years to implement lightweight technology in mainstream production vehicles. There is a high emphasis on greenhouse gas reductions and improving fuel efficiency in the transportation sector, all car manufacturers, suppliers, assemblers, and component producers are investing significantly in lightweight materials research and development and commercialization. All are moving towards the objective of increasing the use of lightweight materials and to obtain more market penetration by manufacturing components and vehicle structures made from these materials. Because the single main obstacle in the application of lightweight materials is their high cost, priority is given to activities to reduce costs through the development of new materials, forming technologies, and manufacturing processes. Yet weight reduction is still the most cost-effective means to reduce fuel consumption and greenhouse gases from the transportation sector. The reasoning behind this is because it takes less work to accelerate and move a lighter object. It has been estimated in simulations that for every 10% of weight eliminated from a vehicle's total weight, fuel economy improves by 7%[1]. Lighter weight materials have the advantage of providing sustained greenhouse gas emission reductions over the use cycle of the vehicle. Another strategy used to reduce vehicle mass is through a complete vehicle redesign. Examples of redesign may be a switch from body-on-frame to unibody construction or reducing non-structural elements of vehicles. In this case, the recent example of *BMW-i* is worth noting; it has an architecture that was actually custom-built for electric cars: it is made up of two separate units, the passenger cell (made of carbon fiber composite) and the drive module with suspension and drive components and the high-voltage battery. There is no tunnel running through the middle of the car, thus leaving more room for passengers and the desired compact design. However, in many instances this is not possible. For example, changing the body construction affects the overall volume of vehicles produced and may increase costs due to complex assembly techniques. For this reason, the best preferred strategy of automakers that intend to buy and sell large vehicles at comparable cost and increased functionality, consists in the wide spreading use of advanced lightweight materials capable of mitigating increased weight despite larger volumes, meet increased fuel economy standards (or travel distance and battery costs for e-vehicles) and keep high safety stringent requirements on crash performance.

Life cycle modelling to production of a low environmental impact automotive part

Looking to major systems for guidance on what to expect, impacts associated with automobiles in literature are dominated by the consumption of fuel, indicating that improvements that reduce fuel consumption are environmentally preferable and will have the reatest impact. The Life Cycle Assessment (LCA) is the process

[1] Recent literature on lightweight design in automotive has been using data provided by ADVISOR (Advanced Vehicle Simulator) simulator developed by U.S. Department of Energy (DOE) and the National Renewable Energy Laboratory (NREL) with industry partners that simulates and analyze conventional, advanced, light and heavy vehicles, including hybrid electric and fuel cell vehicles.

of quantifying material, energy and potential environmental impact of technological systems. It involves the evaluation of various environmental aspects of a product in order to assess the impact of the product throughout its life. Generally it starts with the compilation of relevant environmental exchanges during the life cycle of a product, with the evaluation of the potential environmental impact expressed by global warming potential (GWP) measured in $kgCO_{2eq}$, that can be estimated and calculated for all the exchanges. A full product life cycle is usually divided into two stages:

- the *production phase* that goes from raw material extraction and refining, to final product;
- the *usage phase* that covers the estimated carbon footprint emissions from product use, recycling and disposal.

As the production of manufactured products results from the overall supply of raw materials (e.g. mineral resource extraction for metals), some intermediate phases for mixing/processing raw materials, and following secondary manufacturing processes, it can be critical for the potential release of pollutant emissions and the consumption of a significant number of various sources of non-renewable energy.To sum up all the stages capable of producing direct and indirect emissions, a Life Cycle Assessment (LCA) that addresses the life cycle of a manufactured product is divided into three stages [2]:

- *cradle-to-entry gate* (raw material extraction and refining);
- *entry gate-to-exit gate* (product manufacture);
- *exit gate-to-grave* (product use, recycling and disposal).

Particularly for the automotive sector, the rationale of assessing the emissions and energy consumed in the life cycle of a component onboard vehicles is schematically shown in Figure 1, while the LCA assessment for fossil-fuel vehicles therefore is based on a general scheme for calculation as show in Figure 2.

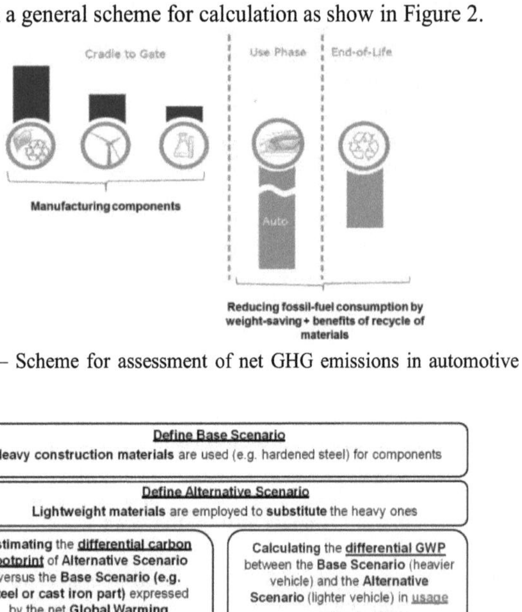

Figure 1 – Scheme for assessment of net GHG emissions in automotive sector.

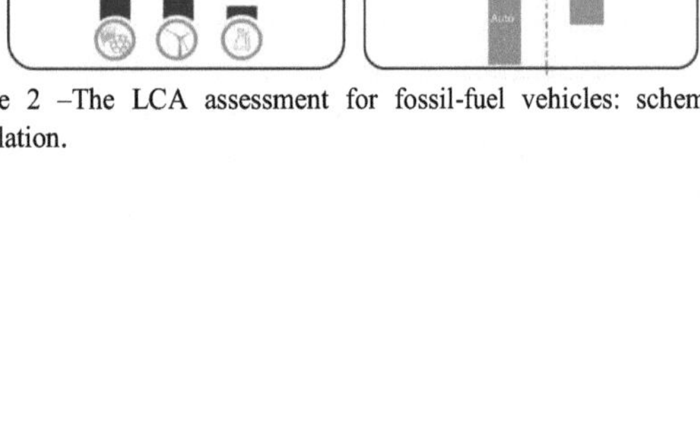

Figure 2 –The LCA assessment for fossil-fuel vehicles: scheme for calculation.

LCA modeling for specific production process

The model of the product system is typically a static simulation model that is composed of unit processes, each one representing one or several activities - such as production processes, transport, or retail. For each unit process, data is recorded on the inputs of natural resources, the emissions, waste flows, and other environmental exchanges. The environmental exchanges are typically assumed to be linearly related to one of the product flows of the unit process throughout its life. Each step of process activity is simplified with a black box into which materials flow to be transformed (or delivered) from an input to an output with the eventual production of some waste material and some direct emissions produced by chemical reactions inside the black box. The transformation is realized by consuming some energy (i.e. MJ) that can be provided by various sources (electricity, steam, fossil fuels, etc.). The majority of the modeling and analysis is performed using the CCaLC software, a specific free-tool developed by The University of Manchester reviewed with the collaboration of several industries. The tool is based on two databases for material and energy inventories, the specifically developed CCaLC and Ecoinvent databases. It uses a transparent framework for constructing LCA models of systems of interest.

The fuel-mass correlation factor

The material substitution factor for specific material vs baseline material represents the mass ratio to calculate how much new material is required to substitute baseline material to achieve similar mechanical response (i.e. satisfy design requirements)[2].

The fuel-mass correlation factor

The *fuel-mass correlation factor* describes the rate of fuel consumption during vehicle operation that can be presented as a simple linear function of vehicle mass, *M* (Keoleian and Sullivan, 2012) as follows:

(eq.1) $FCv = A \times Mv + B$

where *FCv* is the fuel consumption rate in L/km, constant A characterizes the fuel consumption associated with rolling, gradient, and acceleration resistance in L/(km·kg), Mv is vehicle mass in kg, and constant *B* represents parasitic loss in L/km mostly related to aerodynamic drag. The constant A here is often called *fuel reduction value (FRV)* or *fuel consumption reduction coefficient (FRC)* and is used as a measure of fuel-mass correlation. It varies depending on driving cycle, vehicle design, mass, powertrain type, and whether the powertrain is rematched for performance equivalence of the lightweight vehicle. According to a literature survey and simulations by Whohlecker et al. [3] the *FRV* of internal combustion engines lie in the range 0.15–0.7 L/(100 km·100 kg) depending on the factors discussed above, while the LCA studies reviewed here used values in the range 0.3÷0.6 L/(100 km·100 kg). Thus, considering 2.31 kgCO2eq are CO2 emissions resulting from each liter of motor fuel consumed, the above range can be represented in terms of reduction of carbon dioxide emission per each km when 1 kg is saved on board with respect to baseline scenario, namely $3.47 \times 10^{-5} \div 1.62 \times 10^{-4}$ $KgCO_{2eq}$/ (km·kg). Within this range is therefore 1.083×10^{-4} $kgCO2_{eq}$/(Km·kg), namely one major literature reference that we refer to in the following impact analysis [4].

[2] The material substitution factors for this case study are explained in the next section and are given in the Table I.

The scope of work: comparative case-studies scenarios

Case study addresses LCA of a floor pan in the automotive sector as produced by three different manufacturing process cycles, namely:

- *Cycle 1* - Conventional stamped steel floor pan;
- *Cycle 2* - Conventional Polyacrylonitrile (PAN) carbon fiber blended with epoxy resin;
- *Cycle 3* - Alternative low-impacting manufacturing technology process addressed by recycled Eco-Magnesium that uses a high recycling rate for magnesium machined chips as precursor materials [1]. A double scenario is assumed: i) 30% in situ recycled mixed with 70% of fresh material[3]; ii) 30% in situ recycled mixed with 70% of recycled chips from (future) Eco-Mg recycling market.

In Figure 3 are shown the 3 process cycles and relative GWP per kg of material produced.

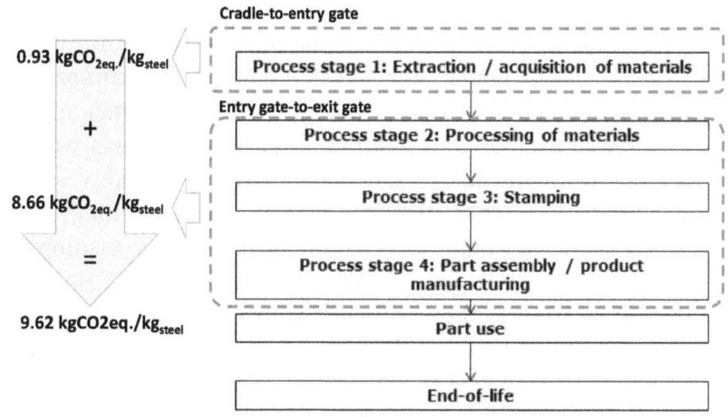

(a)

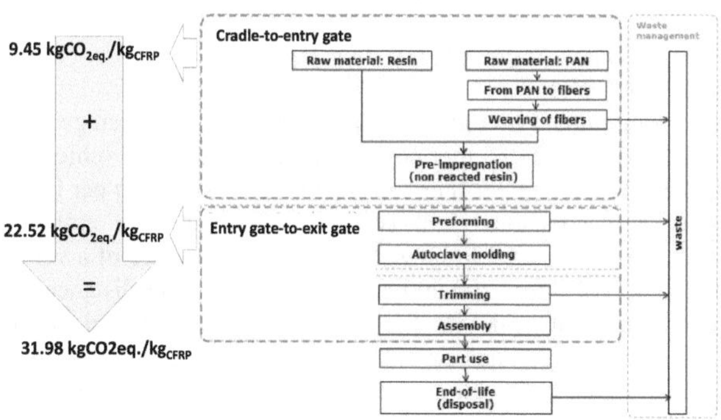

(b)

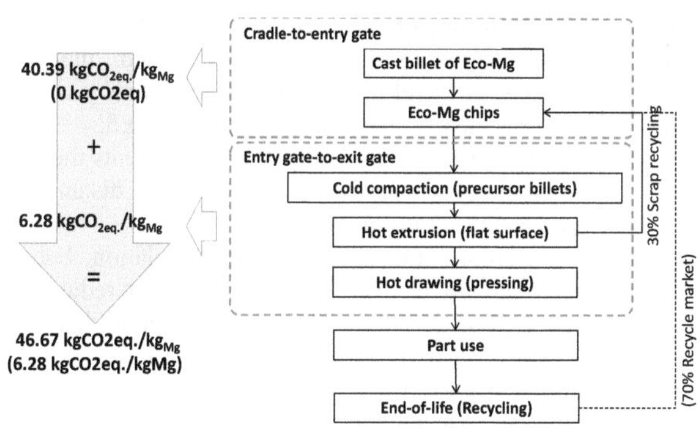

(c)

Figure 3 – Process cycles and GWP results by CCalC software calculation: a) LCA of conventional steel floor pan; b) LCA of CFRP floor pan; c) LCA for novel process route of extrusion of in-situ recycled Eco-chips mixed with: i) 70% of fresh material provided (cast billet and Eco-chips pathway); ii) (in parenthesis) 70% of raw material entirely provided by recycling market [1].

The material substitution factors for the pan case study

The materials substitution factor is defined as the mass ratio between the lightweight- and baseline-component. The materials substitution factor is determined by the physical properties of the material, design constraints, i.e., geometry and economic considerations. Material properties such as density, strength, and stiffness are often used to estimate the mass of lightweight materials for the same functionality. Fig.4 represents a simply case study here analyzed. A panel subjected to bend momentum *Mf* where design constraints fix the width *a* and the length *L* (namely the size and geometry) while the thickness *b* of the cross-section is free; the problem of reducing the mass would be solved by simply reducing the cross-section, but there is a constraint: the section-area *A* must be sufficient to carry the bending moment *Mf*.

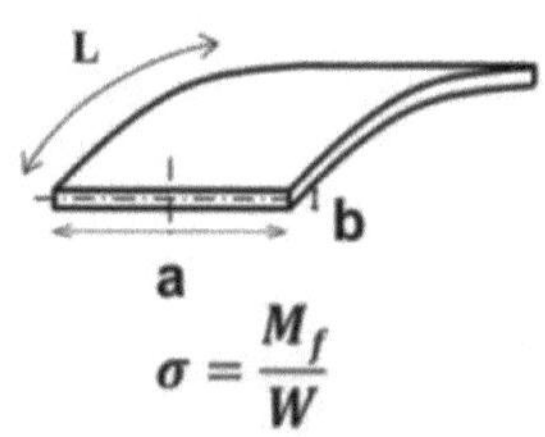

Figure 4 - Critical stress to calculate in order to compare the resistance of a material in case of bending load.

Such an optimization problem is therefore managed by material change; it is usually solved this way. The state of stress σ induced by the external momentum Mf is easily calculated by the formula in Fig.4 as:

(eq.2) $$\sigma_{appl} = M_f / W$$

where W is the moment of resistance to bending. For the study case it is calculated as:

(eq.3) $$W = \frac{a \cdot b^2}{6} = \frac{a}{a} \cdot \frac{ab^2}{6} = \frac{A^2}{6a}$$

In the eq.3 the W parameter is expressed in a more convenient way, as the function of the area A and the geometry constraint we have for the design, as the width of the panel *a*. The design load constraint that requires the pan to resist safely to the bending moment Mf can therefore be expressed by a relationship that states that the strength of material σ_f, the stress to failure, shall be higher than the maximum stress applied to panel σ_{appl} when it is supporting the external load bending momentum Mf:

(eq.4) $$\sigma_f \geq \sigma_{appl}$$

Combining the design failure constraint in (eq.4) with (eq.2) and (eq.3), we can rewrite (eq.4) as the following:

(eq.5) $$\sigma_f \geq \frac{6a \cdot M_f}{A^2}$$

[3] In the short term a more realistic scenario requires considerable primary production at very high GWP to get that much magnesium metal into the vehicle materials stream and develop the recycling market.

This last relationship means that the panel should be designed by a cross-area A capable of keeping the second member below the threshold σ_f, which is the strength limit of the chosen material. Thus, considering that we are addressing a minimizing weight problem, the larger the area A is, the heavier the component will be. Thus (eq.5) will actually be considered with its lower bound value, namely:

$$\text{(eq.6)}\ \sigma_f = \frac{6a \cdot M_f}{A^2}$$

As our first step consists in calculating the substitution factor for alternative materials, we need to write the ratio between the mass m (obtained multiplying density ρ, by section area A to be expressed by inverting the eq.6, by the length of panel L, in case of constant thickness) of the alternative panel against the mass of the baseline panel, that is:

$$\text{(eq.7)}\ \frac{m_{alternative}}{m_{baseline}} = \frac{\varrho_{alt}}{\varrho_{base}} \cdot \frac{\sigma_{base}{}^{1/2}}{\sigma_{alt}{}^{1/2}}$$

Table I gathers the results of the calculation of the mass of alternative panels made in two new materials against baseline steel, considering the constraint of thickness for the alternative light panel that will be limited to twice the baseline steel panel.

Table I. Comparison of weight saved onboard when either Eco-Magnesium considered in this case study are used to substitute a steel large pan loaded in bending mode to external momentum M_f.

Features	CFRP	Magnesium alloy (type Eco-Mg AZ31B) [ref]	Steel (type AISI 4140)
Density (kg/dm^3)	1.60	1.81	7.87
Resistance limit (MPa)	120	125	450
Substitution factor ($kg_{alternative}/kg_{baseline}$)	0.39	0.44	1.00
a (dm)	15.30	15.30	15.30
b (dm)	0.014	0.013	0.007
L (dm)	15.30	15.30	15.30
Volume (dm^3)	3.17	3.11	1.64
Total weight onboard (kg)	5.08	5.63	12.90
Weight saved (kg)	7.82	7.27	-
	60.6%	56.4%	-

Discussion: the net "cradle-to-grave" GWP of the pan

The first impact scenario addresses what happens when substituting the baseline case, the 7.7 kg steel metal floor pan, with about 5 kg of CFRP-made floor pan or about 5.7 kg of Mg pan in a vehicle, thus satisfying same design specifications (refer again to Table I). As stated, a LCA analysis in the automotive sector is based in fact on constructing a differential scenario for assessing the net global warming potential of a lighter component vs a heavier component, considering in differential analysis between the two scenarios, the new and baseline one, what is the net value of $kgCO_{2eq}$ for the new component over its life cycle (from "cradle" to "grave"). This method is therefore required to assess the net reduction of CO_2 emissions that will be achieved during the usage phase, since lighter components means lower fuel consumption, thus lower emissions per km travelled.

The NET "cradle-to-exit gate" GWP of material substitution

The first step consists in calculating the net "cradle-to-exit gate" GWP of the alternative material component (namely the functional unit of this LCA analysis) expressed as $kgCO_{2eq}$ per kg of component to be put on board to substitute the steel made component, i.e. the baseline case.

The net "cradle-to-exit gate" GWP for CFRP floor pans put on board

By the GWP value 31.98 $kgCO_{2eq}$ / kg_{CFRP} calculated by use of CCalC software over the cradle-to-exit gate process stages (refer to Fig.3b), it is possible to determine that 159.89 $kgCO_{2eq}$ is the GWP of 5kg of CFRP pans. Starting from this value, we are interested in calculating the Net Value of GWP over the "cradle-to-exit gate" phase, namely the adjusted value as it is compared to the baseline scenario consisting in a 13 kg steel pan, due to substitution factor 0.39 kg_{CFRP}/kg_{steel} in Table I. It means we have to subtract from the above calculated CO_{2eq} emissions for the manufacturing of the CRFP-made floor pan, the total CO_{2eq} emissions of the baseline steel component. By this assumption, the net "cradle-to-exit gate" GWP for the 5kg CFRP floor pan calculated referring to the baseline scenario 13 kg steel made pan accounts for each 5 kg of CFRP pan:

$$\text{(eq.8)}\ [31.98 kgCO_{2eq}/kg_{CFRP} \times 5\ kg_{CFRP}/pan] - [2.56\ kg_{steel}/kg_{CFRP} \times 5kg_{CFRP} \times 9.59\ kgCO_{2eq}/kg_{steel}] = 34.97\ kgCO_2eq/pan_{(CFRP)}$$

The net "cradle-to-grave" GWP for total of floor pans on board made of CFRP

By combining literature data referring to average fuel consumption – i.e. 8.5 liter per km - for a medium size vehicle and the kgCO2 emitted per liter consumed – 2.85 $kgCO_2eq$ per liter of motor fuel consumed - it is easy to estimate the emissions produced during the life-long standard travel distance of a vehicle (literature considers 200,000 km the average travel distance of a vehicle until it ends its life and being disposal and several construction materials recycled after dismantling). This value is around 48,500 $kgCO_2eq$ emitted over total 200,000 km. Finally it is considered in the GWP model the potentialities of CO_2 emission reduction per km once weight is reduced from the above baseline scenario by use of the fuel-mass correlation factor; by using $1.08 \times 10^{-4} kgCO_2eq/(Km \cdot kg)$ as above stated, the linear function emissions vs travel distance is scaled down by constant value. This result is clearly shown in the graph of Fig.5:

- the "baseline case scenario", the A-line represents the linear correlation between emissions and travel distance for baseline scenario (baseline vehicle);
- the B-line represents the shift of correlation between emissions and travel distance when a weight reduction is achieved on-board, but realized with material that has a certain impact in its production phase (see the "net GWP" interception to y-axis);

In this case, we can state actually that the substitution of materials has produced an increase in terms of net CO_2 emitted at the manufacturing stage which is then counter-balanced by the reduction in fuel consumption during vehicle life; it is obviously relevant to consider the "dirty" phase. As it clearly shown by dotted lines, the more is the NET "cradle-to-exit gate" GWP of the material substitution, the higher is the break-even point. It could even happen that no interception exists: in this worst case, the lighter solution against baseline solution is not actually convenient in terms of net CO_2.

In the Table II are summarized final results that have been calculated considering a total number of 6 CFRP pans put onboard to substitute 6 steel made pans. Starting from the net GWP of CFRP pan, the "break-even point", total mass saved, the net reduced emissions at the end of life (according to literature, 200,000 km).

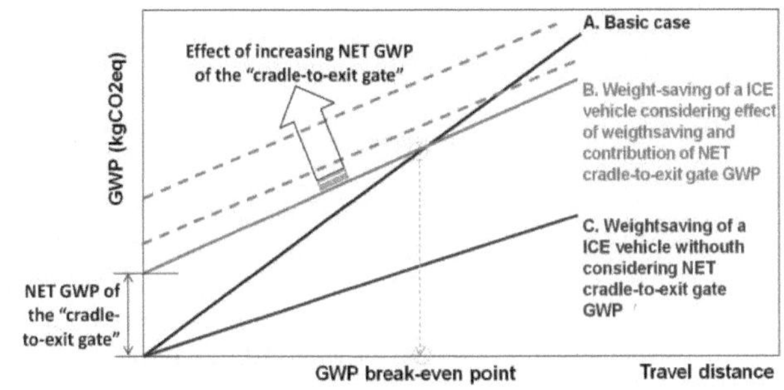

Figure 5 – Scheme for assessment of positive or negative GWP of automotive lighter component: seeking the GWP break-even point.

Table II. Scenario of n.6 CRFP pans versus n.6 steel made pans onboard; relevant results by model (refer to Figure 5 scheme).

Relevant features	Data calculated by model
Mass of CFRP pans (n.6) onboard (kg)	30.48
Mass of substituted steel pans (n.6) onboard (kg)	78.15
Mass saved onboard by substitution (kg)	47.67
NET GWP Break-Even Point (km)	41,500
Net percentage of emissions cut (at end of life, 200,000 km)[1]:	-2.10%

(1) kg saved without "cradle-to-exit gate" contribute.

The net "cradle-to-grave" GWP of the Mg made pan against steel made pan

The second impact scenario addressed analyses what happen if 7.7 kg steel metal floor pan are substituted with 5.63 kg Mg-made floor pan fabricated with cycle 3 process route (refer to Fig.3). The first step consists again in calculating the net "cradle-to-exit gate" GWP of Mg component (namely the functional unit of this LCA analysis) expressed as kgCO2eq per kg of component to put on board to substitute the steel made component. This value has been calculated by considering the average CO_2 equivalent emitted for producing 1 kg of component to put on board the vehicle; it has been estimated 46.50 $kgCO_{2eq}$ per kg of Eco-Mg processed by direct extrusion of recycled chips (30% over virgin chips) [1].

Otherwise, we are interested in the net value of GWP over the "cradle-to-exit gate" phase, hence we have to subtract from above calculated CO2 emission for the manufacturing of the CRFP-made floor pan the CO2 emissions of the baseline steel component. Following the calculation scheme above illustrated, we obtain for for each 5.63 kg Mg-floor pan:

(eq.9) $[46.50 kgCO_{2eq}/kg_{Mg} \times 5.63 kg_{Mg}/pan] - [2.27\ kg_{steel}/kg_{Mg} \times 33.78 kg_{Mg} \times 9.59\ kgCO_{2eq}/kg_{\ steel}] = 140.04\ kgCO_{2eq}\ /\ pan_{(Eco\text{-}Mg\text{-}30\%\ rec.)}$

In the Table III are summarized the final results that have been calculated considering a total number of 6 Eco-Mg pans produced by cycle 3 with 30% of in-situ recycling put onboard to substitute 6 steel made pans. Starting from the net GWP of CFRP pan, the "break-even point", total mass saved, the net reduced emissions at the end of life (according to literature, 200,000 km).

Table III. Scenario Eco-Mg pans versus steel made pans: relevant results by model calculated with 30% of in-situ recycled scenario and 70% of fresh material (refer to Figure 6 scheme).

Relevant features	Data calculated by model
Mass of Mg pans (n.6) onboard (kg)	33.78
Mass of substituted steel pans (n.6) onboard (kg)	76.77
Mass saved onboard by substitution (kg)	42.99
NET GWP Break-Even Point (km)	~176,500
Net percentage of emissions cut (at end of life, 200,000 km)[1]:	-0.20%

(1) kg saved without "cradle-to-exit gate" contribute

In the best case 100% recycled chips are used in the process as proposed in the cycle 3 as alternative process path (30% in-situ recycling, 70% from recycling market - not yet developed) [1], the resulting data is aligned with the case of CFRP scenario, as shown in Table IV. Finally a snapshot of variation of net "cradle-to-grave" GWP curves for the 4 scenarios of n.6 pans made of steel (cycle 1), CFRP (cycle 2), Eco-Mg chips with 30% recycling (cycle 3) and Eco-Mg chips with 100% recycling (cycle 3) versus travel distance curve accordingly with Fig.5 are shown limited for facilitating visualization to quarter of lifespan (50,000 km).

Table IV. Scenario Eco-Mg pans versus steel made pans: relevant results by 100% recycling model (refer to Figure 6 scheme, parenthesis values).

Relevant features	Data calculated by model
Mass of Mg pans (n.6) onboard (kg)	33.78
Mass of substituted steel pans (n.6) onboard (kg)	76.77
Mass saved onboard by substitution (kg)	42.99
NET GWP Break-Even Point (km)	0[2]
Net percentage of emissions cut (at end of life, 200,000 km)[1]:	-3.01%

(1) kg saved without "cradle-to-exit gate" contribute; (2) A Net GWP break-even point at 0 km means that component at the exit gate accounts lower CO2eq emissions than steel-made baseline scenario.

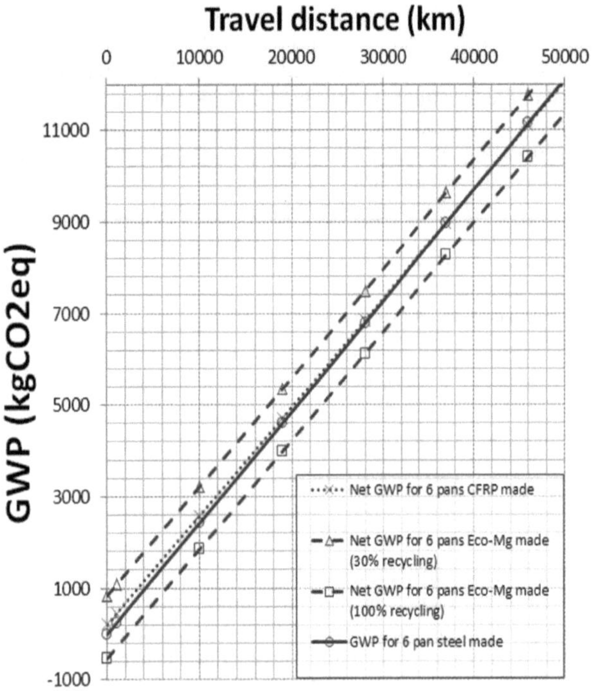

Figure 6 – Scheme for assessment of positive or negative GWP of automotive lighter component: seeking the GWP break-even point

Conclusions

The objective of this LCA case study was to determine the environmental impact of Eco-Mg based process used in the experimental pre-industrial scale manufacturing process developed by a EU project concluded in August 2014. A novel prototype machinery, project demonstrator, is estimated to produce 6.2 kgCO_2eq/kg semi-finished flat sheet produced by flat shaped extrusion profiles made of recycled chips in the (almost) theoretical case of 100% of recycled material is used in the process (i.e. 30% from the in-situ recycling, 70% supplied by a recycling market). On the other hand, today and for forthcoming years, 100% recycled material is not realistic. The 30% recycling would account for 176,500 break-even point. The 62% is accounted for the recycled Eco-Mg fraction for mixing fresh material to make cradle-to-gate GWP of Mg based pan comparable to the CRFP part.. Once the pre-industrialized process has demonstrated the suitability of employing recycling chips of Eco-Mg type as feedstock material, the case study conducted for analysing life cycle environmental impact in the automotive sector shows that pans in CRFP are usually less impacting over the total lifespan of vehicle, 200,000 km according to literature. On the other hand, novel Mg process has not any technology limit in using 100% of recycled feedstock materials. This option allows to align final GW values of low-impacting Eco-Mg produced by chips to those low values typical of a CRFP pan

Acknowledgements

This work makes use of results produced by the Green Metallurgy project (http://www.green-metallurgy.eu), co-funded by the European Commission within the 2009 LIFE+ Programme. Authors thank support provided by European Commission for supporting project dissemination activities. This work reflects only the authors' views. The Community is not liable for any use that may be made of the information contained therein.

References

1. F.D'Errico, G.Garces Plaza, F. Giger and Shae K. Kim, *"Final Assessment of Preindustrial Solid-State Route for High-Performance Mg-System Alloys Production: Concluding the EU Green Metallurgy Project"*, *JOM*, October 2013, Volume 65, Issue 10, 1293-1302.

2. ISO 14040 (1997) Environmental Management – Life Cycle Assessment – Principles and Framework. ISO/FDIS.

3. R.Wohlecker, M.Johannaber, M.Espig, Determination of Weight Elasticity of Fuel Economy for ICE, Hybrid and Fuel Cell Vehicles. SAE World Congress; April 16–19, Detroit, MI, SAE International, 2007

4. L. Ridge, EUCAR - Automotive LCA Guidelines - Phase 2. Total Life Cycle Conference and Exposition, Graz, Austria, December 1–3; Society of Automotive Engineers, 1998

Magnesium Technology 2015
Edited by: Michele V. Manuel, Alok Singh, Martyn Alderman, and Neale R. Neelameggham
TMS (The Minerals, Metals & Materials Society), 2015

In Situ Synchrotron Radiation Diffraction during Solidification of Mg15Gd: Effect of Cooling Rate

G. Szakács[a]*, C.L. Mendis[a], D. Tolnai[a], A. Stark[a], N. Schell[a], H. Ovri[a], M. Wolff[a], K.U. Kainer[a], J. Gröbner[b], R. Schmid-Fetzer[b] and N. Hort[a]

[a]Institute of Materials Research, Helmholtz-Zentrum Geesthacht, Max-Planck-Straße 1, D-21502 Geesthacht, Germany
[b]Institute of Metallurgy, Clausthal University of Technology, Robert-Koch-Str. 42, D-38678 Clausthal-Zellerfeld, Germany

*Corresponding author. Tel: +49 4152 871963, Fax: +49 4152 871909, e-mail address: gabor.szakacs@hzg.de

Keywords: Mg-Gd binary alloy, phase formation, *in situ*, solidification, synchrotron radiation diffraction

Abstract

In situ synchrotron radiation diffraction experiments were performed during the solidification of Mg15Gd at the P07 (HEMS) Beamline of PETRA III at DESY. The measurements were carried out in the chamber of a modified DIL 805A/D dilatometer with a beam energy of 100 keV. The temperature was controlled by type S thermocouples welded on the steel lid of the graphite crucibles containing the samples. The two dimensional diffraction patterns were recorded with a Perkin Elmer 1621 Flatpanel. The phase evolution observed during cooling at rates of 5, 20 and 100 K/min show formation of $GdMg_3$ at elevated temperatures, which transforms into $GdMg_5$ during continued cooling. Phases were identified with the information from the Pearson´s database for crystalline structures. This is different from that predicted with thermodynamic databases. Although the equilibrium phase diagram suggests a simple eutectic solidification, the experiments show a metastable phase formation and its transformation. The formation of $GdMg_3$ becomes more pronounced at higher cooling rates.

Introduction

Due to their light-weight and high specific strength, Mg-RE (RE - Rare Earth) alloys are promising for structural applications **[1]**. RE elements can be divided into two groups; light- (La, Ce, Pr, Nd, etc.) and heavy-RE (Gd, Dy). Recently the heavy RE elements have become attractive as degradable biomaterials **[2, 3]**. The addition of Y, Gd, Dy or combination of these elements, due to their high solubility in hcp Mg, allows tailoring the mechanical properties and corrosion behavior of these alloys in a wide range **[4]**.

Among these elements Gd shows good biocompatibility **[5]**, and it is currently used safely in medicine as a contrast agent in magnetic resonance imaging **[2]**. Phase equilibria and the thermodynamics of the binary Mg–Gd system have been extensively studied experimentally. The first experimental study of the binary Mg–Gd phase diagram was by Savitskii *et al.* **[6]** in 1961 using differential thermal analysis (DTA) and X-ray diffraction (XRD) coupled with hardness and micro hardness measurements. Four intermetallic phases; $GdMg_9$, $GdMg_3$, $GdMg_2$ and GdMg were reported. These results were revised by Manfrinetti and Gschneidner **[7]** and they determined the entire phase diagram, utilizing thermal, metallographic and X-ray analysis. Four intermetallic compounds $GdMg_5$, $GdMg_3$, $GdMg_2$ and GdMg were found and all of them form via a series of peritectic reactions each with lower formation temperature than the previous when described from the Gd rich side.

Gd has a maximum solubility of 23.49 wt.% (4.53 at.%) at 548 °C in Mg **[8]** and offers the possibility of both solid solution strengthening and precipitation hardening. The solvus line on the Mg rich side was experimentally determined by Manfrinetti and Gschneidner **[7]**. Thermodynamic descriptions of the Mg-Gd system were determined by Cacciamani *et al.* **[9,10]** and most recently by M. Hampl *et al.* **[11]**. New thermodynamic description of Mg-Gd are available **[12]**.

Majority of the Mg alloys are produced in the form of castings, thus their microstructure and consequently macroscopic mechanical properties are determined during the solidification and the subsequent thermo-mechanical treatment. The solidification process and the phase evolution can be modeled based on the thermodynamic databases **[13]**. Although they provide information on the presumed solidification path, the commercial castings normally deviate from the equilibrium descriptions. Therefore, an understanding of the sequence of the intermetallic phases evolution during solidification with different cooling rates (CR) is necessary.

The evolution of the intermetallic phases during solidification, previously could be investigated only with *ex situ* methods, i.e. interrupting the solidification at different temperatures (different stages of microstructure development) by quenching the sample to perform the characterization **[14]**. The major disadvantage of this method is the lack of time-temperature resolution. The continuous development of the X-ray acquisition systems at synchrotron sources provides a unique tool to characterize the phase evolution during the solidification *in situ*. Recording the diffraction patterns while cooling an alloy allows identification of the intermetallic phases to determine their solidification sequence **[15]** and to use these results as experimental validation of the existing thermodynamic databases and to contribute to their completion.

The present study reports the results of *in situ* synchrotron diffraction studies performed during solidification of a Mg15Gd alloy with different cooling rates (5, 20 and 100 K/min), their comparison with results from other techniques and with existing thermodynamic data of this system.

Experimental procedures

The Mg15Gd (wt.%) was cast by permanent mould direct chill casting **[16]**. For producing the alloys pure Mg (99.99%) and Gd (99.98%, Grirem Metal) were used. After melting pure Mg, the melt was kept at 720 °C for 15 min under a mixture of SF_6 (2%) and Ar protective atmosphere. After the addition of the Gd the melt was stirred continuously with 200 rpm by Eurostar P-CV type agitator to ensure melt homogeneity for 15 min. The melt was poured into a steel mould preheated to 650 °C. The mould with the melt was placed into a holding furnace at 680 °C for further 15 minutes. After the holding time, the alloy was solidified by immersing the mould slowly (10 mm/s) into running water.

Cylindrical samples were machined from the as-cast material in order to fit into the graphite crucibles designed for the synchrotron diffraction experiments (ø = 3 mm, h = 2 mm) to study the solidification. The measurements were conducted at the P07 High Energy Materials Science (HEMS) Beamline of Helmholtz-Zentrum Geesthacht (HZG) at the Petra III, Deutsches Elektronen-Synchroton (DESY). The measurements were performed in the chamber of a DIL 805A/D dilatometer in Ar flow. The dilatometer has been modified for *in situ* synchrotron measurements, there are two windows on the sides covered by Kapton foil, which is transparent for the X-ray beam, and the induction coil is opened in the middle so the beam passes through only the sample and the crucible **[17]**. During the test the sample was contained in a graphite crucible closed with steel caps in order to hold the melt and isolate it from the surrounding. Type S thermocouples welded on the caps were used to control the temperature. The samples were heated 750 °C and held for 5 min, to ensure melt homogeneity, then cooled to 200 °C with controlled CR of 5, 20 and 100 K/min. The *in situ* diffraction experiments were performed using a beam with a cross section of 1×1 mm^2. The beam energy was set to 100 keV, corresponding to a wavelength of $\lambda = 0.0124$ nm. During the experiments the 2D diffraction patterns were recorded with an acquisition time of 1, and 0.1 s depending on the cooling rate to ensure the temperature resolution of ~ 1 K in all the 3 cases with the Perkin Elmer XRD 1621 Flat Panel detector with a pixel size of 200 μm x 200 μm **[18]**. The 2048×2048 pixel diffraction patterns were integrated to azimuthal line profiles by using the fit2d software **[19]**.

The crystallographic information on the intermetallic phases was obtained from the Pearson's Crystal Structure Database **[20]**, the d-spacings and 2θ angles for the phases were calculated using CaRIn Crystallography 3.1™ software.

The as-cast alloy and the 3 as-solidified samples (after the *in situ* synchrotron solidification experiments) were ground with SiC papers to a grit size of 2500 and then polished with a mixture of water free OP-S and 1 μm diamond suspensions for microscopy examinations. Scanning Electron Microscopy (SEM) was performed on the as-solidified samples with a VEGA3 Tescan microscope, operating at 20 kV using back scattered electron (BSE) imaging mode.

In order to verify the results from the *in situ* synchrotron radiation diffraction experiment number of samples were quenched from various temperatures of interest to retain the high temperature microstructures. Small specimens were heated in stainless steel crucibles to 750 °C, cooled at a rate of 100 K/min near to the eutectic temperature and then quenched by rapidly flowing N_2.The specimens for transmission electron microscopy (TEM) investigations were prepared from regions containing intermetallic particles using focused ion beam (FIB) milling with a FEI Helios NanoLab 600 DualBeam FIB/SEM. The specimen surface was protected with a 300 nm Pt layer prior to FIB processing with Ga^+ ions to protect the specimen from damage. The TEM samples were mounted on Cu grids for TEM investigations Transmission electron microscopy was performed on a FEI CM200 Transmission electron microscope equipped with EDAX energy dispersive X ray (EDX) spectrometer operating at 200 kV.

The phase transformation temperatures during solidification was measured with differential scanning calorimetry (DSC) and differential thermal analysis (DTA) to correlate the transformation temperatures with those recorded during the *in situ* synchrotron radiation diffraction during solidification.

The experimentally recorded temperatures for the solidification of various phases were compared with the results from the Scheil solidification simulations performed with the Pandat 8.1™ thermodynamic software using PanMg8 database.

Results and discussion

In situ synchrotron radiation diffraction

The line profiles from the azimuthal integration of the 2D diffraction patterns acquired during the *in situ* synchrotron radiation diffraction experiments during solidification at a cooling rate of 20 K/min are shown in Figure 1. At 750 °C, a diffuse background originating from the melt and diffraction peaks of the graphite crucible can be observed (red). The solidification sequence starts with the formation of the α-Mg phase at a measured temperature of 640 °C (blue). As the cooling advances the Mg peaks become more pronounced and at a measured temperature of 561 °C $GdMg_3$ intermetallic phase forms from the remaining melt which transforms into $GdMg_5$ at 551 °C after the alloy has fully solidified.

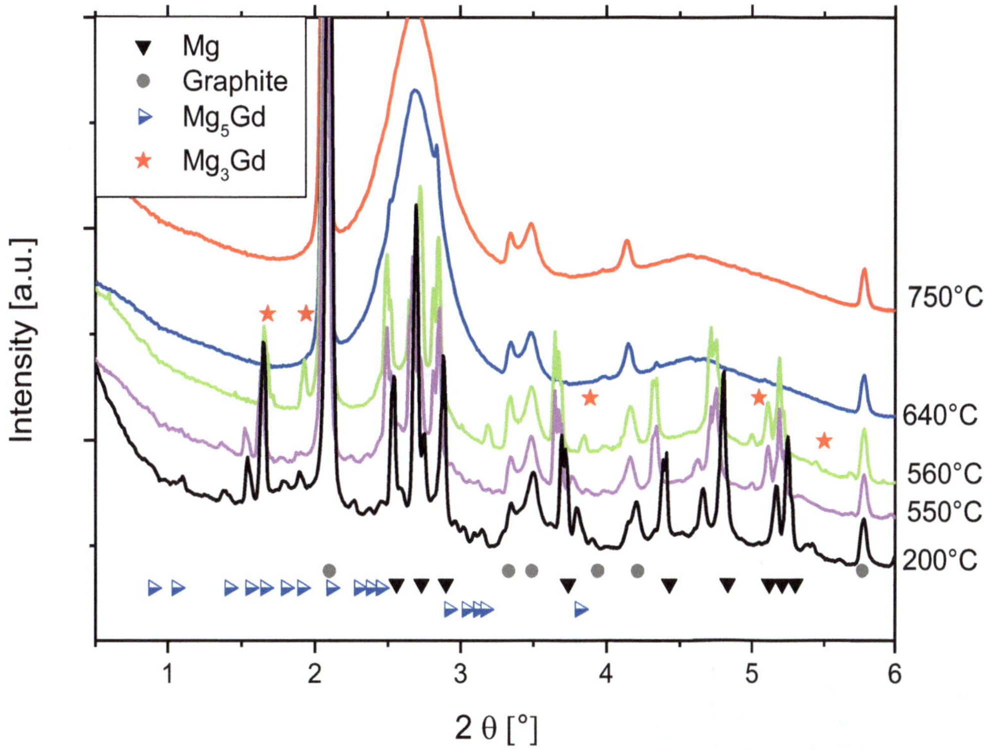

Figure 1 Sequence of XRD line profiles obtained by the integration of the diffraction patterns recorded during solidification of Mg15Gd at a cooling rate of 20 K/min. (wavelength λ = 0.0124 nm)

The *in situ* experiments show the same solidification path for all three cooling rates (CR), but the onset temperatures for the formation of various phases are different. The α-Mg phase first appears at measured temperatures of 613, 640 and 593 °C for CR of 5, 20 and 100 K/min respectively (Table1). The measured temperatures for the formation of $GdMg_3$ phase are 536, 561, 513 °C for a CR of 5, 20 and 100 K/min. However the measured temperatures of phase formations (Table 1) show a deviation from the expected. However, the real liquidus temperature differences between the experiments cannot be more than 5 °C based on the diffraction data for the graphite crucible at the measured temperatures.

Table 1. The measured temperatures for the formation of phases during the solidification of Mg15Gd alloy at a CR of 5, 10 and 100 K/min.

Phase	Measured Formation Temperature [°C]		
	CR 5 K/min.	CR 20 K/min.	CR 100 K/min.
α-Mg	613	640	593
$GdMg_3$	536	561	513
$GdMg_5$	535	551	510

All three experiments show that with the metastable phase formation of the diffuse background of the molten phase has disappeared, i.e. the sample has solidified. At measured temperatures of 535, 551 510 °C (for CR of 5, 20, 100 K/min. respectively) the transformation of the metastable $GdMg_3$ phase to stable $GdMg_5$ phase was observed. The formation of the stable phase occurs in within a temperature range of 2-3 K. , At measured temperatures below 532 °C, 548 °C, 505 °C (in CR 5, 20, 100 K/min) the metastable phase Mg_3Gd were no longer detected on the 2D diffraction patterns. The observed peaks can be indexed according those calculated for the $GdMg_5$, the phase expected from the Mg-Gd binary phase diagram for a composition of 15 wt.% Gd. The formation of the metastable phase from the liquid phase in hypo-eutectic Mg-Gd alloys has not been reported previously. Since the formation of the $GdMg_3$ phase is more pronounced at higher cooling rates, it suggests that the kinetics of the solidification plays a major role.

Microstructure

SEM micrograph showing the microstructure of the Mg-15Gd alloy after solidification at a cooling rate of 100 K/min is shown in Figure 2. The change in cooling rate did not affect the phases observed in the microstructures or the segregation of Gd along the interdendritic boundaries but the volume fraction of the intermetallic phase is slightly increased by the CR.

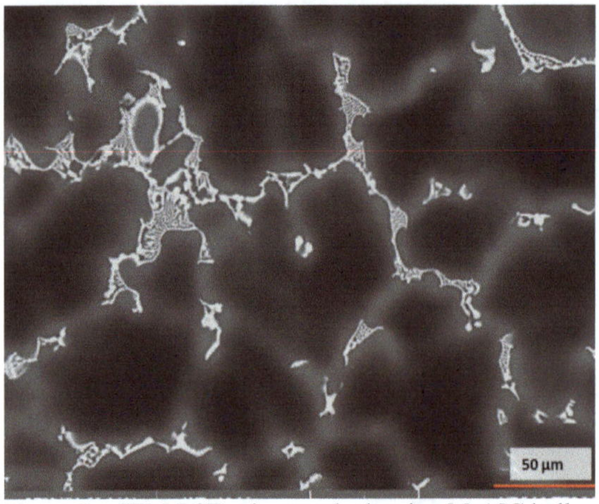

Figure 2: SEM-BSE micrograph showing Mg15Gd alloy after solidification at a cooling rate of 100 K/min.

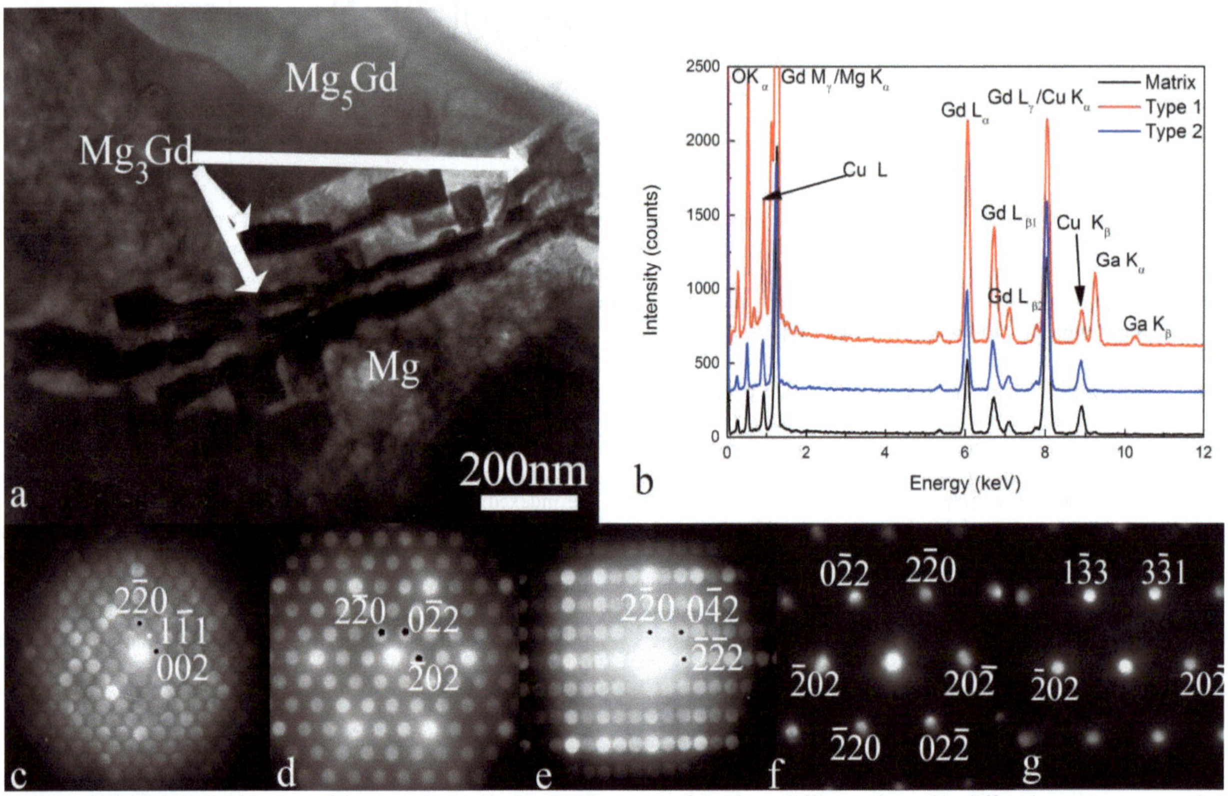

Figure 3 TEM-bright field image of the quench sample (a), EDX spectra of the three different phase (b), diffraction patterns recorded from phase type 1 identified as $GdMg_5$(c-e), diffraction patterns recorded from phase type 2 identified as $GdMg_3$ (f,g).

The interrupted solidification samples quenched from 556 °C was used for the TEM investigation. Figure 3 (a) shows a TEM-bright field image showing α-Mg and two intermetallic phases identified as $GdMg_5$ and $GdMg_3$. The diffraction patterns recorded from phase identified as $GdMg_5$, Figure 3 (c-e) corresponds to <110>, <111> and <112> zones of $GdMg_5$, respectively. The EDX spectra recorded from this phase, intermetallic phase type 1, was quantified with a composition of Mg 81±3 at.% and Gd 19±3 at. %, Figure 3 (b). This composition is close to that of $GdMg_5$. The diffraction patterns recorded from $GdMg_3$ phase, Figure 3 (f,g) corresponds to <111> and <343> zones, while the quantified EDX spectra shows a composition of Mg 70±5 at. % and Gd 30 ±5 at%, which provides a Mg:Gd ratio of 2.4. Cu and Ga peaks observed in the EDX spectra are unavoidable artefacts due to the Cu grid and the remnants Ga from milling process.

Thermal analysis and solidification calculations

As shown above the phase formation temperatures oscillate with the cooling rate. Thus the measured temperatures were correlated with experimental DTA and DSC results. The DTA measurements were conducted at a CR of 5 and 20 K/min. and the DSC measurement result at a CR of 20 K/min. The DTA results show two peaks in the signal first at 624 and 623 °C and the second at 545 and 544 °C at a CR 5 and 20 K/min. respectively. The first peak corresponds to the formation of the α-Mg and the second to the eutectic structure. The DSC result, Figure 4, shows that the first phase transformation starts at 623 °C and the second at 546 °C which correlate well with the DTA results (Table 2.). Although the measured temperatures from the synchrotron measurements do not correspond to the DSC and DTA results, the measurements show a systematic shift. The temperature differences between the start and the end of the solidification are 77 °C, 79 °C, 80 °C in the CR of 5, 20 and 100 K/min, respectively, from the temperature measurement of synchrotron experiment.. And this temperature differences correlate with the results from the DTA and DSC measurements.

Table 2. Measured temperatures for the formation of various phases during the solidification of Mg15Gd alloy with DTA (CR 5 and 20 K/min.) and DSC (CR 20 K/min).

Phase	Measured Formation Temperature [°C]		
	DTA		DSC
	5 K/min.	20 K/min	20 K/min
α-Mg	624	623	623
Eutectic str	545	544	546

The formation temperatures were calculated with thermodynamic simulations of the solidification using Scheil model with Pandat 8.1 software. The Scheil calculation predicts 623 °C for the start of the solidification and 547 °C for the end of the solidification with a formation of $GdMg_5$ intermetallic phase from the melt. If the formation of the $GdMg_5$ phase is suppressed in the Scheil calculation predicts the formation of $GdMg_3$ phase but at a lower temperature of 526 °C. The predicted temperatures from the Scheil calculation are in good correlation with the DTA and DSC results. However, the resolution of both DSC and DTA is not sufficient to observe the phase transformation from $GdMg_3$ to $GdMg_5$ phase.

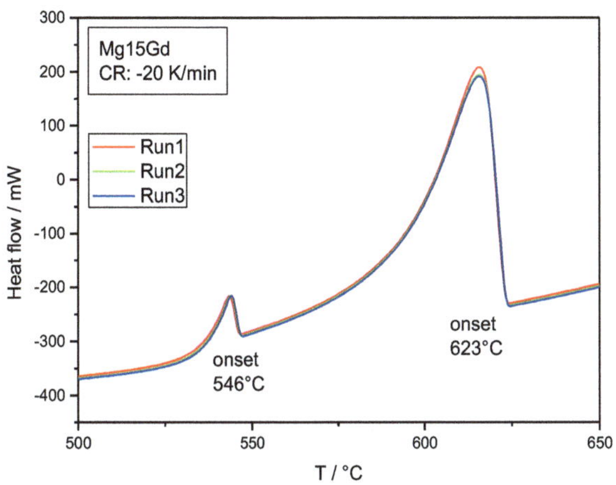

Figure 4. DSC cooling curves of the Mg15Gd alloy with a CR of 20 K/min.

Table 3. Predicted formation temperatures for various phases during the solidification of Mg15Gd calculated with Pandat 8.1 software and Scheil solidification model.

Phase	Temperatures of Scheil Simulation [°C]	
	Equilibrium Condition	Suppressed Mg5Gd
α-Mg	623	623
GdMg3	-	526
GdMg5	547	-

Conclusions

In situ experiments during solidification were performed on Mg15Gd alloy with synchrotron radiation diffraction, DSC and DTA. The experiment results were correlated with Scheil calculation using Pandat software. From these results the following conclusions can be drawn:

- The specimens investigated with *in situ* synchrotron radiation diffraction only show α-Mg and $GdMg_5$ phases.
- At temperatures close to the eutectic temperatures measured at three different cooling rates a metastable phase forms from the melt and subsequently transforms into $GdMg_5$ phase.
- The diffraction peaks from this metastable phase are found to correlate with the $GdMg_3$.
- The formation of $GdMg_3$ phase at elevated temperatures prior to the formation of $GdMg_5$ has been verified with interrupted solidification experiments where samples were quenched from temperatures close to the transformation temperatures measured with *in situ* synchrotron radiation diffraction experiments.
- This experimental work shows that the use of synchrotron radiation diffraction during solidification gives extra information on the phases that evolve during solidification which cannot be observed with other methods.

Acknowledgements

The authors acknowledge the Deutsches Elektronen-Synchrotron for the provision of synchrotron radiation facilities in the framework of the proposal I-20130434. G.S. acknowledges the European Union FP7-PEOPLE-ITN-2011 program under REA Grant Agreement No 289163: MagnIM project for the financial support.

References

[1] L. L. Rokhlin, *Magnesium Alloys Containing Rare Earth Metals, Structure and Properties*, (London: Taylor & Francis, 2003).

[2] N. Hort *et al.*; "Magnesium alloys as implant materials-Principles of property design for Mg-RE alloys" *Acta Biomater.*, 6 (2010), 1714-1725.

[3] F. Witte, "The history of biodegradable magnesium implants: A review", *Acta Biomater.*, 6 (5) (2010), 1680-1692.

[4] P.J. Apps *et al.*, "Precipitation reactions in magnesium-rare earth alloys containing yttrium, gadolinium or dysprosium", *Scripta Mater.*, 48 (2003), 475-481.

[5] F. Feyerabend *et al.*, "Evaluation of short-term effects of rare earth and other elements used in magnesium alloys on primary cells and cell lines", *Acta Biomater.*, 6 (5) (2010), 1834-1842.

[6] E. M. Savitskii *et al.*, "Diagramma Sostoyaniya Splavov System Mg-Gd", *Khim* 6 (7) (1961), 1734-1739.

[7] P. Manfrinetti and K. A. Gshneidner Jr., "Phase equilibrium in the La-Mg (0 – 65 at.% Mg) and Gd-Mg systems", *J. Less Common Metals,* 123 (1-2) (1986), 267-275.

[8] T. B. Massalski, *Binary Alloy Phase Diagrams*, (Materials Park Ohio: ASM International 1990).

[9] G. Cacciamiani *et al.*, "Computer coupling of thermodynamics and phase diagrams: the gadolinium-magnesium system as an example" *Thermochim. Acta,* 199 (1992) 17-24.

[10] G. Cacciamani *et al.*, "The Al–R–Mg (R=Gd, Dy, Ho) systems. Part II: Thermodynamic modelling of the binary and ternary systems" *Intermetallics,* 11 (11-12) (2003) 1135-1151.

[11] M. Hampl *et al.*, "Thermodynamic assessment and experimental study of Mg–Gd alloys" *J. Alloy Compd.,* 581 (2013) 166-177.

[12] S. Das *et al.*, "Thermodynamic Modeling and Diffusion Kinetic Experiments of Binary Mg-Gd and Mg-Y Systems", Acta Materialia, 71, 2014, 164-175

[13] W. J. Boettinger *et al.*, "Phase-field simulation of solidification", *Annual Review of Materials Scie*, 32 (2002) 164-194.
[14] Y. L. Liu, S. B. Kang; "The solidification process of AI-Mg-Si alloys" *Journal of Materials Science* 32 (6) (1997) 1443-1447
[15] D. Tolnai *et al.,* „In situ synchrotron diffraction of the solidification of Mg4Y3Nd" Mater. Lett. 102-103 (2013) 62-64.
[16] F. R. Elsayed *et al.,* "Magnesium Permanent Mold Castings Optimization" *Mater. Sci. Forum* 690 (2011) 65-68.
[17] P. Staron *et al.,"* In situ experiments with synchrotron high-energy X-rays and neutrons" *Adv. Eng. Mater.* 13 (2011) 658-663
[18] http://www.perkinelmer.com/pdfs/downloads/009299B_01%20PRD.pdf (downloded on 24th September 2014)
[19] http://www.esrf.eu/computing/scientific/FIT2D/index.html
[20] P. Villars, and K. Cenzual, Pearson's *Crystal Data - Crystal Structure Database for Inorganic Compounds* (on CD-ROM), (Ohio: ASM International, Materials Park, USA, Release 2012).

Magnesium Technology 2015
Edited by: Michele V. Manuel, Alok Singh, Martyn Alderman, and Neale R. Neelameggham
TMS (The Minerals, Metals & Materials Society), 2015

MICROSTRUCTURES AND TENSILE PROPERTIES OF Mg–4Al–4La–0.4Mn–XB (X = 0, 0.01, 0.02, 0.03) ALLOY

Jian Meng; Qiang Yang; Zheng Tian; Xin Qiu; Fanzhi Meng
State Key Laboratory of Rare Earth Resource Utilization, Changchun Institute of Applied Chemistry, Chinese Academy of Sciences; 5625 Renmin Street; Changchun 130022, P.R. of China

Keywords: Trace boron, Eutectic morphology, Tensile property, Mg–4Al–4La-based alloy

Abstract

Influences of trace boron on microstructures and tensile properties of Mg–4Al–4La-based alloys prepared by cold-chamber high-pressure die-casting method were thoroughly investigated. The results indicate that adding trace boron to Mg–4Al–4La-based alloy can refine the dendrite arm spacing of primary alpha-Mg phases, which are mainly due to the little inoculating AlB_2 particles. In addition, we found that adding 0.01–0.02 wt.% boron can drastically changes the eutectic morphology, with secondary particle dispersion becoming irregular and eutectic volume fraction being reduced. These phenomena can be attributed to the competitive nucleation between alpha-Mg and AlB_2 particles for $Al_{11}La_3$ phases, and to the fact that more Al and La atoms saturate into the α-Mg matrix. Considering the tensile properties, although adding 0.01–0.02 wt.% boron decreased the strength of Mg–4Al–4La-based alloy, adding 0.03 wt.% boron significantly improved the tensile properties due to dispersion strengthening and, to a certain extent, solid-solution strengthening.

Introduction

Magnesium alloys have recently become the focus of numerous studies due to their high specific strength, low density, good yield strength and machine ability which are important in automobile, aerospace and communication industries [1–5]. However, the high temperature mechanical properties of conventional magnesium alloys (e.g. AZ91) are poor due to the existence of β-$Mg_{17}Al_{12}$ phase (455 °C melting point), limiting their application to temperatures below 120 °C [5,6]. Recently, efforts to further improve the mechanical properties of magnesium alloys, adding some alloying elements has been an effective method [7–18]. For instance, Suresh et al. reported that adding boron in the form of Al–4B master alloy could significantly refine the grain size of AZ91 alloy due to the presence of AlB_2 particles which act as potential nucleation sites for alpha-Mg [18].

As is well known, AE42 and AE44 have excellent high temperature mechanical properties and creep resistance, which are mainly attributed to the formation of thermally stable $Al_{11}RE_3$ and Al_2RE phases, as well as to the suppression of the $Mg_{17}Al_{12}$ phase [19,20]. $Mg_{17}Al_{12}$ precipitates from the supersaturated alpha-Mg matrix in AE42 when the creep temperature is higher than 150 °C, which leads to deterioration in creep resistance. However, because most Al is consumed by rare earth metals having higher content, no discernible $Mg_{17}Al_{12}$ particles are observed in AE44; thus it has better high temperature properties and creep resistance [20–22]. Nevertheless, their mechanical properties are still lower than the requirements for power train parts. Therefore, to facilitate the application of magnesium alloys in automotive power trains, further improvements on the properties of Mg–Al–RE-based alloys should be studied.

Recently, numerous researchers have paid attention to the microstructures, mechanical properties and creep resistance of Mg–Al–RE alloys [19–23]. However, studies on Mg–Al–RE-based alloys [20,24] that deal with improving mechanical properties by the addition of other alloying elements are scarce. Xiao et al. investigated the influence of Zn additions on the casting defects, microstructures, mechanical properties, and creep properties of high-pressure die-casting (HPDC) Mg–xZn–4Al–4RE (x=1–10) alloys [25–27]. Such studies highlight the importance of investigating the influence of adding micro-alloying elements (e.g. boron) into HPDC AE44 alloys.

Although all RE elements have similar physical structures, the influence of the elements on AE44 alloy can be varied. For instance, substituting the Ce-rich misch metal of AE44 alloy with La results in the improvement of tensile and creep properties, which can be attributed to the higher thermal stability of $Al_{11}La_3$ phase, compared with $Al_{11}Ce_3$, $Al_{11}Pr_3$, and $Al_{11}Nd_3$ phases [1,22]. Moreover, adding boron could significantly refine the grain sizes of as-cast AZ91 alloy [16] and pure aluminum [28], because of the nucleation potency of AlB_2 in Al–B master alloys. However, researchers rarely investigate to the influence of boron addition on the microstructures and mechanical properties of Mg–Al–RE-based alloys, particularly the HPDC AE44 alloy. In this study, we selected HPDC Mg–4Al–4La-based (ALa44) alloy and boron, in the form of Al–4B master alloy, as the parent and microalloying elements, respectively. Primary focus was given to determine the influence of trace boron on the microstructure and mechanical properties of ALa44 alloy.

Experimental procedures

Pure magnesium, aluminum, and Mg–20 wt.% La master alloy were used to prepare Mg–4Al–4La (ALa44) alloy. Manganese was subsequently added in the form of Mg–2 wt.% Mn master alloy. Boron, as the microalloying element, was added in the form of Al–4 wt.% B master alloy. All the alloys were melted in an electric resistance furnace, and continuously flushed with CO_2 + 2 vol.% SF_6 mixed gas during the whole melting process. Tensile samples with gauge lengths (60 mm) and gauge diameters (6 mm) were produced using a cold chamber die-cast machine and a die equipped with an oil heating/cooling system. The temperature of the oil heater was set at 175 °C. The main chemical compositions of the specimens were determined by inductively coupled plasma atomic emission spectroscopy (ICP-AES). The results are shown in Table 1.

Table 1 Main chemical compositions (wt.%) of the ALa44 + xB (x = 0, 0.01, 0.02, 0.03 wt.%) alloys determined by ICP-AES.

Alloy	Al	La	Mn	B	Fe	Mg
ALa44	3.826	4.074	0.426	0	0.012	Balance
ALa44+0.01B	3.713	3.803	0.394	0.008	0.008	Balance
ALa44+0.02B	3.806	4.233	0.410	0.018	0.010	Balance
ALa44+0.03B	3.995	4.076	0.413	0.027	0.012	Balance

Samples for metallography were cut from the middle segments of the tensile bars, and ground using different grades of polishing paper, after which they were polished with Al_2O_3 paste. The cross-sectional microstructures in the core region are different from those near the surface of the samples produced by HPDC method [6,25,29]. Therefore, the microstructures of the middle portions, particularly those between the core and the skin, were present in this work. The microstructures and secondary particles were characterized using optical microscopy (OM), scanning electron microscopy equipped with an energy dispersive spectrometer (SEMEDS), and transmission electron microscopy equipped with an energy dispersive X-ray spectroscope (TEM-EDS). The foils for TEM observations were prepared by low-energy ion beam thinning. In addition, selected area electron diffraction (SAD), scanning transmission electron microscopy (STE) and high-resolution transmission electron microscopy (HRTEM) were used for phase identifications. X-ray diffraction (XRD) investigations were conducted using a Siemens diffractometer operated at 40 kV and 40 mA with Cu Kα radiation (λ = 0.15418 nm). Measurements were obtained by continuous scanning of 2θ from 10° to 80°. The tensile tests were performed from ambient temperature to 225 °C, with a strain rate of $5.56 \times 10^{-4}\ s^{-1}$. The values presented in this work were the average of at least six measurements from every test temperature.

Results

Microstructures Fig.1a–d shows the optical micrographs of HPDC ALa44 + *x*B (*x* = 0, 0.01, 0.02, 0.03 wt.%) alloys. Trace boron refines the dendrite arm spacing of primary α-Mg phases in the alloys [29,30]. Measurements obtained using linear intercept method show that the dendrite arm spacing average sizes of (a), (b), (c), and (d) alloys are 15.6 ± 1.2, 13.7 ± 0.9, 13.2 ± 1.1, and 10.7 ± 0.8 μm, respectively. Furthermore, adding trace boron into ALa44 alloy significantly influences the morphologies of the microstructures. Fig. 2 shows the XRD patterns, demonstrating that the main phases present in the alloys are alpha-Mg, $Al_{11}La_3$, and Al_2La, with potentially lesser amounts of Al_3La and Al–Mn(–La) phases [31,32]. After the addition of trace boron (0.01–0.03 wt.%), no new discernible peaks emerge and no preexisting peaks disappear indicating that the phase constitution remains unchanged. However, variations in the relative intensities of certain peaks are observed, which indicates that the number of secondary particles accordingly changes. Although peaks of the AlB_2 phase are not observed in the XRD patterns and no AlB_2 particles are observed in SEM or TEM, the existence of the said compound in the alloys have been confirmed in previous studies [17,33,34]. Therefore, we know that AlB_2 (hexagonal structure, a = 0.30054 nm, c = 0.32528 nm) particles form hexagonal platelets or blocky patterns, and their melting point is 892 °C.

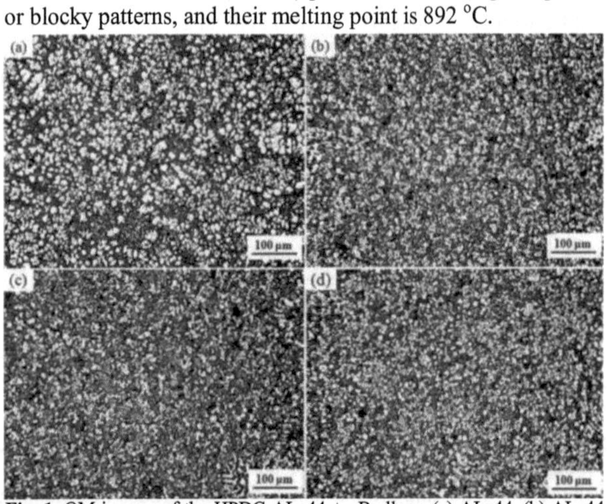

Fig. 1. OM images of the HPDC ALa44 + *x*B alloys: (a) ALa44, (b) ALa44 + 0.01B, (c) ALa44 + 0.02B, and (d) ALa44 + 0.03B.

SEM images of the alloy microstructures are shown in Fig. 3. The structures observed are similar to those reported in Ref. [35], wherein all the alloys consisted of continuous or semi-continuous reticular cell boundaries and alpha-Mg, with most of the secondary particles distributed in the cell boundaries. The morphologies of the eutectic also vastly differ with the secondary particle distribution, becoming irregular with the decrease in eutectic volume fraction after the addition of trace boron (0.01 and 0.02 wt.%) into the ALa44 alloy (Fig. 3b and c). On the other hand, the addition of 0.03 wt.% boron into the ALa44 alloy, results in a more regular secondary particle dispersion, and the eutectic volume fraction appropriately increases to the level of ALa44 (Fig. 3d). To estimate the changes in the eutectic volume fraction, we assumed that the eutectic area fraction on SEM images is proportional to the volume fraction in the investigated alloys, thus enabling the estimation of the eutectic volume fraction from the SEM images (Fig. 4). The addition of 0.01 wt.% boron into ALa44 alloy, visibly decreases the volume of secondary particles.

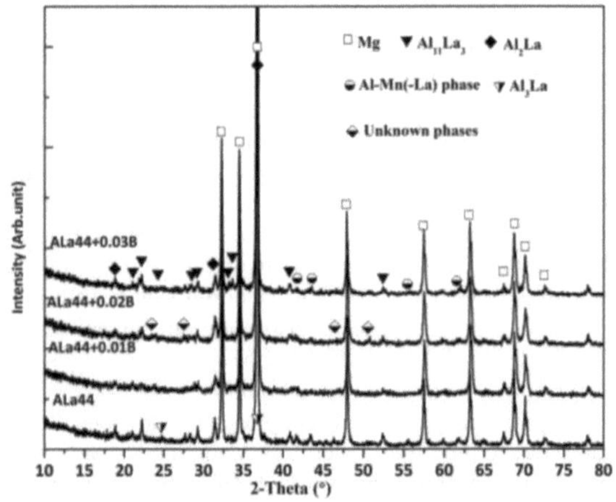

Fig. 2. XRD patterns of the HPDC ALa44 + *x*B (*x* = 0, 0.01, 0.02, 0.03 wt.%) alloys.

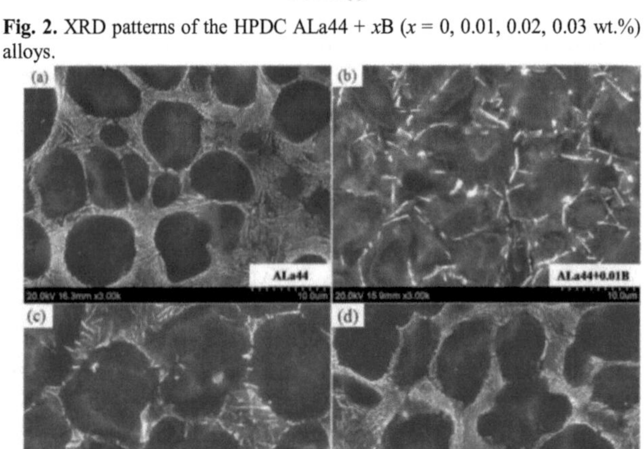

Fig. 3. SEM images of the HPDC ALa44 + *x*B alloys: (a) ALa44, (b) ALa44 + 0.01B, (c) ALa44 + 0.02B, and (d) ALa44 + 0.03B.

Figs. 5 and 6 show the SEM and TEM bright-field images of Al_{11}-La_3 particles and Al_2La particles, respectively. Fig. 5g and h shows the spot EDS results of branched particles (A) and lamellar particles (B), respectively, whose analysis showed that the atomic ratio of Al:La is in agreement with the 11:3 ratio of $Al_{11}La_3$. The EDS results together with the SAD patterns of the branched particles and the lamellar particles (Fig. 5c and f), confirmed that the phase with branched structure or lamellar structure is $Al_{11}La_3$ (body-centered orthorhombic structure, a = 0.44 nm, b = 1.30 nm, and c = 1.01 nm) [36]. Fig. 6c and d shows the spot EDS result and the SAD pattern of the blocky particles, respectively, indicating that the blocky particles are Al_2La (diamond structure, a = 0.81 nm) [32,36], with a size of approximately 0.3–2.5 μm.

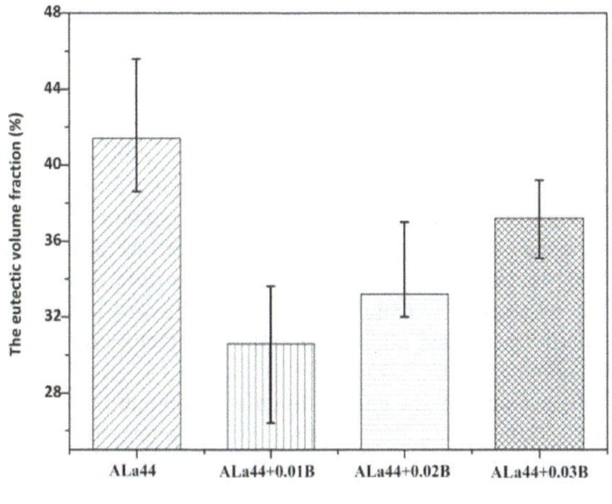

Fig. 4. The eutectic volume fraction of HPDC ALa44 + xB (x = 0, 0.01, 0.02, 0.03 wt.%) alloys.

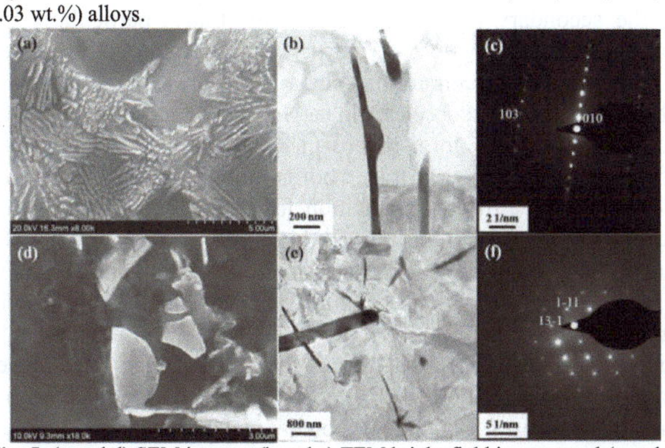

Fig. 5. (a and d) SEM images, (b and e) TEM bright-field images, and (c and f) SAD diffraction patterns of branched and lamellar $Al_{11}La_3$ particles.

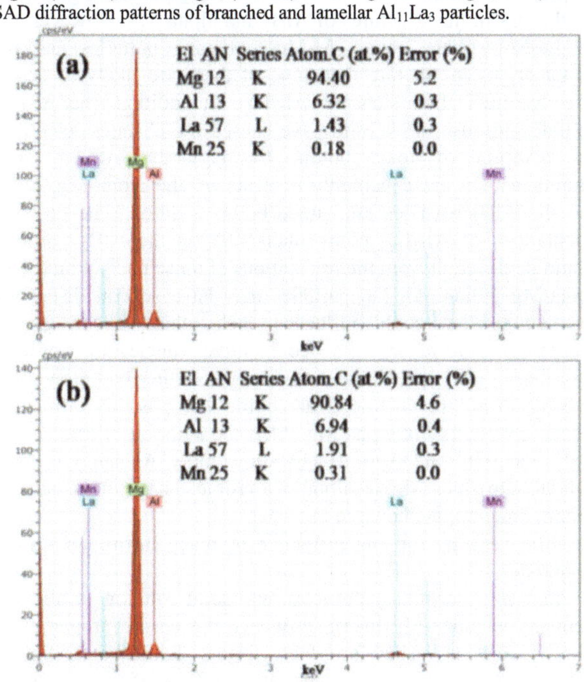

Fig.6. The representative EDS spectra with the analysis results of the branched (a) and lamellar (b) phases.

Some reports have demonstrated that one of the typical Al–Mn–RE phases in the Mn-containing Mg–Al–RE-based alloys is $Al_{10}RE_2Mn_7$ (hexagonal, a = 0.90 nm, c = 1.30 nm), with hexagonal shape and size in the range of 0.2–10 μm [21,32]. Fig. 7 shows the SEM images of the $Al_{10}La_2Mn_7$ particles. The EDS results demonstrated that the atomic ratio of Al:La:Mn is consistent with the 10:2:7 ratios, indicating that the particles are indeed $Al_{10}La_2Mn_7$. However, the $Al_{10}La_2Mn_7$ particles also have blocky (Fig. 7a) or rectangular shapes (Fig. 7c), and their sizes range from 0.3 to 3.6 μm. Furthermore, another Al-Mn-containing phase is distributed on the edge of the secondary particles. The STEM image with EDS elemental mappings, TEM bright-field image, and HRTEM image are shown in Fig. 8a–c, respectively. The micro-particles, as indicated by the yellow tetragon in Fig. 8a and by the white arrowheads in Fig. 8b, contain Al and Mn, with sizes ranging from 8 nm to 45 nm. Moreover, they are not identified to be Al_8Mn_5 phase, as depicted in Fig. 8c. The structure and lattice parameters of these micro Al-Mn-containing particles are still unknown. Particles of Al_3La phase are observed in the tested alloys, which often exist together with the $Al_{11}La_3$ phase (Fig. 9a) in small numbers [36]. Additional Al-rich phases are also observed in the alloys, with various morphologies: irregular lamellar, branched, blocky, and short-rod shapes (Fig. 9b). These observed phases possibly correspond to the peaks of the unknown phases, as shown in Fig. 2.

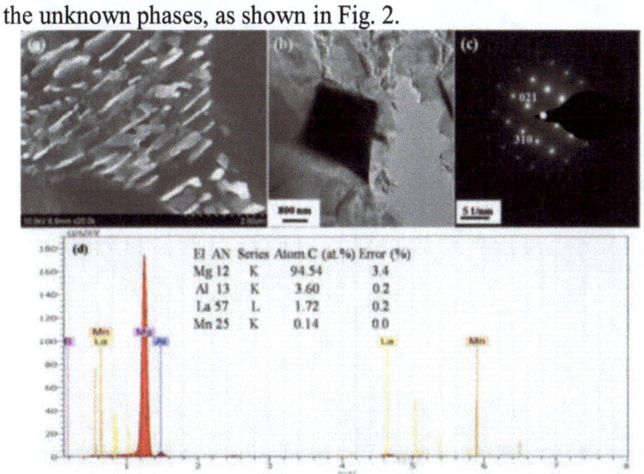

Fig.7. (a) SEM image, (b) TEM bright-field image, (c) SAD diffraction pattern and (d) EDS spectrum of Al_2La particles.

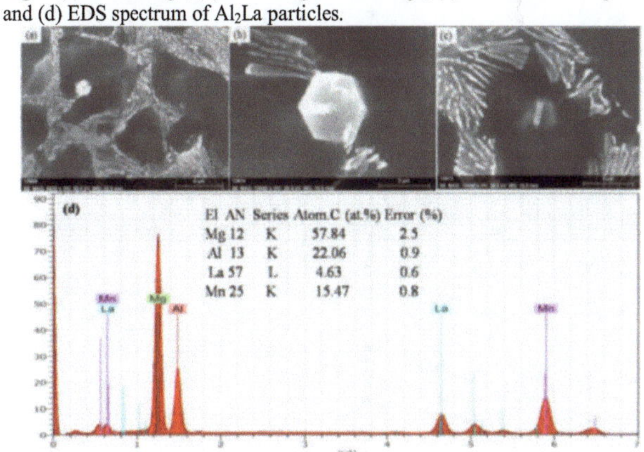

Fig. 8. (a-c) SEM images and (d) EDS pattern of the $Al_{10}La_2Mn_7$ particles.

Tensile properties

The typical tensile curves of alloys at ambient temperature are shown in Fig. 10a. The results of ultimate tensile strength (UTS), tensile yield strength (TYS), and elongation to failure (E) tests are shown in Fig. 10b–d, respectively. Clearly, 0.03 wt.% boron addition can significantly improve the mechanical properties of the ALa44 alloy at all tested temperature s, whereas the addition of 0.01–0.02 wt.% boron can decrease its strength, particularly when the test temperature is above 150 °C. For instance, the addition of 0.03 wt.% boron to ALa44 alloy, improves the UTS, TYS, and E at ambient temperature by 14.5%, 6.2%, and 84.6%, respectively. The addition of trace boron into ALa44 alloy is beneficial for E at all tested temperatures.

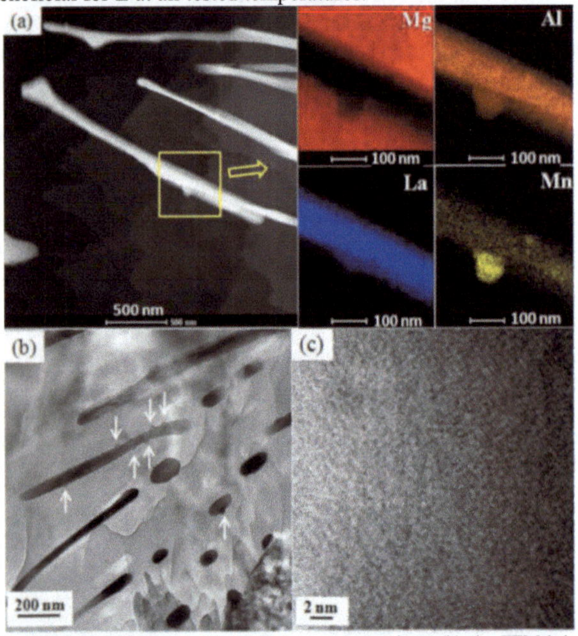

Fig. 9. (a) STEM image with EDS elemental mappings, (b) the TEM bright field image and (c) the HRTEM image of Al-Mn containing phase.

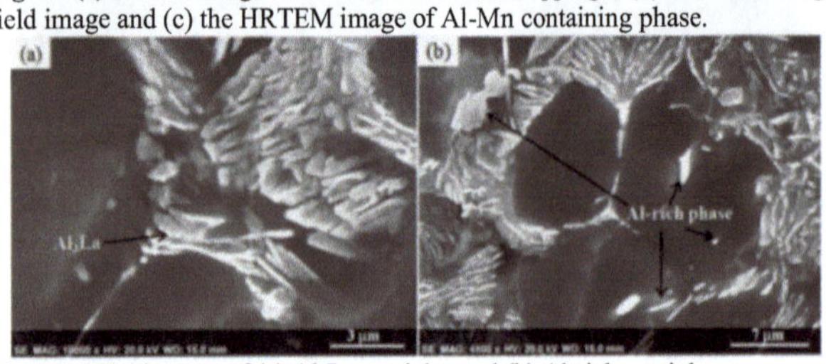

Fig. 10. SEM images of (a) Al_3La particles and (b) Al-rich particles.

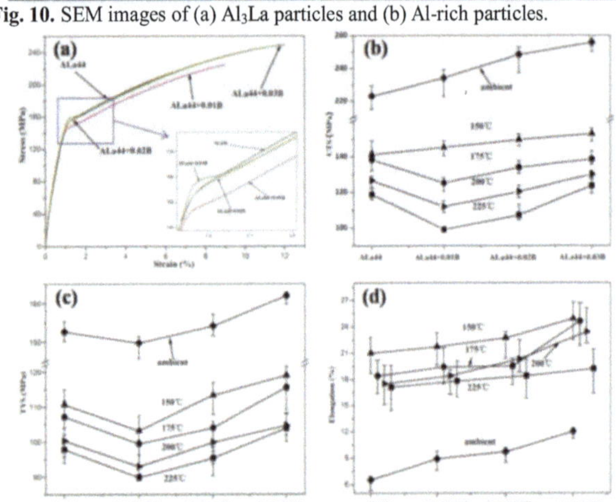

Fig. 11. (a) The alloys' typical tensile curves at ambient temperature and (b-d) the tensile properties of HPDC ALa44+xB (x = 0, 0.01, 0.02, 0.03 wt.%) alloys.

Discussion

Possible reasons for dendrite arm spacing refinement after the addition of trace boron

Adding trace boron into ALa44 alloy can refine the dendrite arm spacing of primary alpha-Mg phases. AlB_2 particles act as effective heterogeneous nucleation sites for alpha-Mg, which can lead to good grain refinement [17,33,34]. However, the refinement is mild because of the limited amount of AlB_2 particles. Al–B compounds can readily combine with soluble manganese [33]. Therefore, we have to consider the refinement of Mn-containing particles because of the existence of Al-Mn-containing particles in the alloys (Fig. 8). However, the potential nucleating particles have to satisfy the following factors: sufficient number, favorable particle size, and a relative spread of particle size distribution [37,38]. Thus, the micro Al-Mn-containing particles cannot act as effective nucleation sites due to rare numbers and local distribution in the alloys. Therefore, the mild dendrite arm spacing refinement is mainly attributed to the small inoculating AlB_2 particles formed after the addition of trace boron into the ALa44 alloy.

Possible reasons for the change in eutectic morphology

After adding trace boron into ALa44 alloy, the eutectic morphology significantly changes a lot with the secondary particle dispersion, becoming irregular with decrease in the eutectic volume fraction. This observation can be attributed to the changes in the secondary particle nucleation substrates, particularly the nucleation substrates for $Al_{11}La_3$ particles, and the fact that more Al and La atoms saturate into alpha-Mg. Considering the above observations, AlB_2 particles, alpha-Mg grains, and possibly some $Al_{10}La_2Mn_7$ phases form before the $Al_{11}La_3$ phase is nucleated after the addition of trace boron into the ALa44 alloy. Consequently, a direct competition for heterogeneous nucleation occurs among the co-existing solid particles. Solid particles that satisfy the nucleation criteria (small crystallographic misfit, suitable sizes, favorable distribution, and sufficient numbers) will prevail in the competition [37,38]. Using HRTEM, Zhang et al. [6] found that the $Al_{11}La_3$ phase tends to grow with its (020) planes parallel to that of the (0002) planes of alpha-Mg indicating that the a Mg phases can act as potent nucleation substrate s for $Al_{11}La_3$. The same can be inferred for AlB_2 particles, given the similarities in the structures of AlB_2 and alpha-Mg. However, the $Al_{10}La_2Mn_7$ phase, micro Al-Mn-containing particles and the Al_3La phase are so rare that none of them are likely to act as effective nucleation sites for $Al_{11}La_3$. In addition, the Al-rich particles are also unlikely to have caused the changes seen after the addition of trace boron. Consequently, $Al_{11}La_3$ may simultaneously or sequentially nucleate on the alpha-Mg grains and the AlB_2 particles after trace boron addition. The growing orientations of $Al_{11}La_3$ phase nucleating on the AlB_2 particles would be different from the orientations of those nucleating on the alpha-Mg grains. $Al_{11}La_3$ particles may have several dispersion orientations that could influence each other as the growth progresses. However, the competitive nucleating process described is still speculative and requires further comprehensive research to be proven true. Trace boron also possibly influences the solid solubility in alpha-Mg. After the addition of trace boron (0.01 and 0.02 wt.%) to ALa44 alloy, more Al and La atoms saturate into alpha-Mg, resulting in a visible change in the eutectic volume fraction, as shown in Fig. 4.

Possible reasons for the influence of trace boron on tensile properties

Grain size, secondary particles, and solid solution of alloying elements in alpha-Mg matrix influence the tensile properties of alloys [36,39–41]. Therefore, these factors should be considered in analyzing the possible reasons for the effects of trace boron on the tensile properties [6,36,42]. First, alloys with mild differences in the microstructure have negligible grain refinement on the mechanical properties. Second, for the supersaturated solutes in the alpha-Mg matrix, solid-solution strengthening can improve UTS and TYS to some extent [36]. However, solid-solution strengthening can also lead to a decrease in the amount of secondary particles, which lowers the UTS and TYS to some extent. Finally, secondary particles with large volumes in the grain boundaries have negative effects on the mechanical properties. Previous studies have reported that cracks easily initiate at the interfaces between the coarse secondary phases and the matrix in tensile tests [42,43]. Conversely, if the secondary particles are quite small, to the extent that they would not cause any significant damage to the matrix, they could effectively strengthen the alloy. Uniform distribution of secondary particles, coupled with good matrix-reinforcement interfacial integrity, can increase the internal stress between the secondary phases and the matrix, resulting in the improvement of elastic modulus [44,45]. Based on the above analysis, adding trace boron can significantly influence tensile properties. The improvement in tensile properties by adding 0.03 wt.% boron to ALa44 alloy can be attributed to the dispersion and possible solid-solution strengthening. On the other hand, the decrease in UTS and TYS by adding 0.01–0.02 wt.% boron into ALa44 alloy can be mainly attributed to the eutectic volume fraction being reduced and the presence of some large secondary particles.

Conclusions

The microstructures and the tensile properties of HPDC ALa44 + xB (x = 0, 0.01, 0.02, 0.03 wt.%) alloys have been thoroughly investigated. The following conclusions can be drawn:

(1) Although the inoculating AlB_2 particles can act as effective heterogeneous nucleation sites for alpha-Mg, the refinement on DAS is mild because only little AlB_2 particles form in the alloys with trace boron.

(2) In the alloys with trace boron, the main intermetallics are still $Al_{11}La_3$ and Al_2La, with low amounts of Al_3La, Al–Mn(–La), and AlB_2. However, adding 0.01–0.02 wt.% boron significantly changes the eutectic morphology, with the secondary particle dispersion becoming irregular and the eutectic volume fraction decreasing. These finding may be ascribed to the competitive nucleation between alpha-Mg and AlB_2 particles for $Al_{11}La_3$, and that more Al and La atoms saturate into the alpha-Mg matrix.

(3) Considering the tensile properties, although addition of 0.01–0.02 wt.% boron mildly decreases the alloys' strength, addition of 0.03 wt.% boron significantly improves the tensile properties due to dispersion strengthening and solid-solution strengthening to a certain extend. In conclusion, investigating the influence of microalloying elements on the properties of magnesium alloys is important, in discovering an efficient and more economical method to further improve the properties of existing magnesium alloys and in exploring new ones.

Acknowledgements

This work is supported by the National Key Technologies R&D Program 2012BAE01B04, 2011BAE22B01 and Guangdong Innovative Team Program 201001C0104669453, 2010A090100033.

References

[1] J. Zhang, K. Liu, D. Fang, X. Qiu, D. Tang, J. Meng, Microstructure, tensile properties, and creep behavior of high-pressure dis-cast Ng-4Al-4RE-0.4Mn (RE = La, Ce) alloys, J. Mater. Sci. 44 (2009)2046–2054.

[2] Z.-Z. Shi, The crystallography of lath-shaped Mg_2Sn precipitates in a Mg-Sn-Zn-Mn alloy, J. Alloys Comp. 559 (2013) 158–161.

[3] S. Zhang, W. Liu, X. Gu, C. Lu, G. Yuan, W. Ding, Effect of solid solution and aging treatments on the microstructures evolution and mechanical properties of Mg-14Gd-3Y-1.8Zn-0.5Zr alloy, J. Alloys Comp. 557 (2013) 91–97.

[4] K.M. Asl, A. Tari, F. Khomamizadeh, The effect of different content of Al, RE and Si element on the microstructure, mechanical and creep properties of Mg-Al alloys, Mater. Sci. Eng. A 523 (2009) 1–6.

[5] X. Tian, L.M. Wang, J.L. Wang, Y.B. Liu, J. An, Z.Y. Cao, The microstructure and mechanical properties of Mg-3AL-3RE alloys, J. Alloys Comp. 465 (2008)412–416.

[6] J. Zhang, M. Zhang, J. Meng, R. Wu, D. Tang, Microstructures and mechanical properties of heat-resistant high-pressure die-cast Mg-4Al-xLa-0.3Mn (x = 1, 2, 4, 6) alloys, Mater. Sci. Eng. A 527 (2010)2527–2537.

[7] J. Zhang, Y. Sun, W. Cheng, Z. Que, Y. Li, L. Liushan, The effect of Ca addition on microstructures and mechanical properties of Mg-RE based alloy, J. Alloys Comp. 554 (2013)110–114.

[8] R. Mahmudi, S. Moeendarbari, Effects of Sn additions on the microstructure and impression creep behavior of AZ91 magnesium alloy, Mater. Sci. Eng. A 566 (2013) 30–39.

[9] N. Stanford, The effect of rare earth elements on the behaviour of magnesium-base alloys: Part 2 – recrystallisation and texture development, Mater. Sci. Eng. A 565 (2013) 469–475.

[10] J. Wei, J. Chen, H. Yan, B. Su, X. Pan, Effects of minor Ca addition on microstructure and mechanical properties of the Mg–4.5Zn–4.5Sn–2Al-based alloy system, J. Alloys Comp. 548 (2013) 52–59.

[11] J. Zhou, J. Wang, J. Wang, Y. Yang, Effects of RE and Sr additions on dendrite growth and phase precipitation in AZ91D magnesium alloy, Trans. Nonferr. Metal Soc. China 20 (2010) s331–s335.

[12] S.S. Li, B. Tang, D.B. Zeng, Effects and mechanism of Ca on refinement of AZ91D alloy, J. Alloys Comp. 437 (2007) 317–321.

[13] A. Srinivasan, J. Swaminathan, U.T.S. Pillai, Krishna Guguloth, B.C. Pai, Effect of combined addition of Si and Sb on the microstructure and creep properties of AZ91 magnesium alloy, Mater.Sci. Eng. A 485 (2008) 86–91.

[14] M. Suresh, A. Srinivasan, U.T.S. Pillai, B.C. Pai, The effect of charcoal addition on the grain refinement and ageing response of magnesium alloy AZ91, Mater. Sci. Eng. A 528 (2011)8573–8578.

[15] Y.M. Kim, C.D. Yim, B.S. You, Grain refining mechanism in Mg–Al base alloys with carbon addition, Scripta Mater. 57 (2007) 691–694.

[16] B. Cicek, Y. Sun, A study on the mechanical and corrosion properties of lead added magnesium alloys, Mater. Des. 37 (2012) 369–372.

[17] M. Suresh, A. Srinivasan, K.R. Ravi, U.T.S. Pillai, B.C. Pai, Influence of boron addition on the grain refinement and mechanical properties of AZ91 Mg alloy, Mater. Sci. Eng. A 525(2009) 207–210.

[18] S. Candan, M. Unal, M. Turkmen, E. Koc, Y. Turen, E. Candan, Improvement of mechanical and corrosion properties of magnesium alloy by lead addition, Mater. Sci. Eng. A501 (2009) 115–118.

[19] B.R. Powell, V. Rezhets, M.P. Balogh, R.A. Waldo, Microstructure and creep behavior in AE42 magnesium die-casting alloy, JOM 54 (2002) 34–38.

[20] M.S. Dargusch, S.M. Zhu, J.F. Nie, G.L. Dunlop, Microstructural analysis of the improved creep resistance of a die-cast magnesium–aluminium–rare earth alloy by strontium

additions, Scripta Mater. 60 (2009) 116–119.
[21] S.M. Zhu, M.A. Gibson, J.F. Nie, M.A. Easton, T.B. Abbott, Microstructural analysis of the creep resistance of die-cast Mg-4Al-2RE alloy, Scripta Mater. 58(2008) 477–480.
[22] J. Zhang, P. Yu, k. Liu, D. Fang, D. Tang, J. Meng, Effect of substituting cerium-rich mischmetal with lanthanum on microstructure and mechanical properties of die-cast Mg-Al-RE alloys, Mater. Des. 30 (2009) 2372–2378.
[23] J. Bai, Y. Sun, F. Xue, J. Qiang, Microstructures and creep properties of Mg–4Al–(1–4) La alloys produced by different casting techniques, Mater. Sci. Eng. A 552 (2012) 472–480.
[24] J. Zhang, S. Liu, Z. Leng, X. Liu, Z. Niu, M. Zhang, R. Wu, Structure stability and mechanical properties of high-pressure die-cast Mg-AL-La-Y-based alloy, Mater. Sci. Eng. A 531(2012) 70–75.
[25] W.J. Xiao, M.A. Easton, M.S. Dargusch, S.M. Zhu, M.A. Gibson, The influence of Zn additions on the microstructure and creep resistance of high pressure die cast magnesium alloy AE44, Mater. Sci. Eng. A539 (2012) 177–184.
[26] W.J. Xiao, M.A. Easton, S.M. Zhu, M.S. Daegusch, M.A. Gibson, S.S. Jia, J.F. Nie, Casting defects and mechanical properties of high pressure die cast Mg-Zn-Al-RE alloys, Adv. Eng. Mater. 14 (2012) 68–76.
[27] W.J. Xiao, S.M. Zhu, M.A. Easton, M.S. Dargusch, M.A. Gibson, J.F. Nie, Microstructural characterization of high pressure die cast Mg-Zn-Al-RE alloys, Mater.Charact. 65 (2012) 28–36.
[28] T. Wang, Z. Chen, H. Fu, J. Xu, Y. Fu, T. Li, Grain refining potency of Al–B master alloy on pure aluminum, Scripta Mater. 64 (2011) 1121–1124.
[29] J. Zhang, Z. Leng, S. Liu, M. Zhang, J. Meng, R. Wu, Structure stability and mechanical properties of Mg-Al-based alloy modified with Y-rich and Ce-rich misch metals, J. Alloys Comp. 509 (2011)L187–L193.
[30] M. Yang, F. Pan, R. Cheng, A. Tang, Effect of Mg-10Sr master alloy on grain refinement of AZ31 magnesium alloy, Mater. Sci. Eng. A 491 (2008) 440–445.
[31] Y.D. Huang, H. Dieringa, N. Hort, P. Maier, K.U. Kainer, Y.L. Liu, Evolution of microstructure and hardness of AE42 alloy after heat treatment, J. Alloys Comp.463 (2008) 238–245.
[32] K. Wei, L.-Y. Wei, R. Warren, Influence of annealing on microstructures, mechanical and creep properties of Mg–4Al–2Sr alloy, Mater. Sci. Technol. 28 (2012) 311–320.
[33] N. Nishino, H. Kawahara, Y. Shimizu, H. Iwahori, Grain refinement of magnesium casting alloys by boron addition, in: K.U. Kainer (Ed.), Magnesium Alloys and their Applications, WILEY-VCH Verlag Gmbh, Weinheim, 2000, p59–p64.
[34] J.C. Viala, J. Bouix, G. Gonzalez, C. Esnouf, Chemical reactivity of aluminum with boron carbide, J. Mater. Sci. 32 (1997) 4559–4573.
[35] J. Zhang, D. Zhang, Z. Tian, J. Wang, K. Liu, H. Lu, D. Tang, J. Meng, Microstructures, tensile properties and corrosion behavior of die-cast Mg-4Al-based alloys containing La and/or Ce, Mater. Sci.Eng. A 489 (2008) 113–119.
[36] S.M. Zhu, J.F. Nie, M.A. Easton, P. Bakke, Microstructure and creep behavior of high-pressure die-cast magnesium alloy AE44, Metall. Mater. Trans. A. 43 (2012) 4137–4144.
[37] Z. Fan, Y. Wang, M. Xia, S. Arumuganathar, Enhanced heterogeneous nucleation in AZ91D alloy by intensive melt shearing, Acta Mater. 57 (2009) 4891–4901.
[38] T.E. Quested, A.L. Greer, The effect of the size distribution of innoculant particles on as-cast grain size in aluminium alloys, Acta Mater. 52 (2004) 3859–3868.
[39] C.D. Lee, Effect of grain size on the tensile properties of magnesium alloy, Mater. Sci. Eng. A 459 (2007) 355–360.
[40] J.D. Robson, N. Stanford, M.R. Barnett, Effect of precipitate shape on slip and twinning in magnetism alloys, Acta Mater. 59 (2011) 1945–1956.
[41] N. Stanford, J. Geng, Y.B. Chun, C.H.J. Davies, J.F. Nie, M.R. Barnett, Effect of plate-shaped particle distributions on the deformation behaviour of magnesium alloy AZ91 in tension and compression, Acta Mater.60 (2012) 218–228.
[42] B.H. Kim, K.C. Park, Y.H. Park, I.M. Park, Investigations of the properties of Mg–4Al–2Sn–1Ca–*x*Ce alloys, Mater. Sci. Eng. A 527 (2010) 6372–6377.
[43] S.G. Lee, J.J. Jeon, K.C. Park, Y.H. Park, I.M. Park, The effects of alloying elements on microstructure and properties of gravity casting Mg–Sn–Al–Si alloy, Mater. Chem. Phys. 128 (2011) 208–213.
[44] X. Zhang, H. Wang, L. Liao, X. Teng, N. Ma, The mechanical properties of magnesium matrix composites reinforced with (TiB_2 + TiC) ceramic particulates, Mater. Lett. 59 (2005) 2105–2109.
[45] A. Luo, Processing, microstructure, and mechanical behavior of cast magnesium metal matrix composites, Metall. Mater. Trans. A (1995) 2445–2455.

Magnesium Technology 2015
Edited by: Michele V. Manuel, Alok Singh, Martyn Alderman, and Neale R. Neelameggham
TMS (The Minerals, Metals & Materials Society), 2015

THE ROLE OF BISMUTH IN GRAIN REFINEMENT OF MAGNESIUM AND ITS ALLOYS

Utsavi Joshi[1], Nadendla Hari Babu[2]
[1,2]BCAST (Brunel Centre for Advanced Solidification technology), Brunel University, Uxbridge, Middlesex UB8 3PH, UK

Keywords: Bismuth, Grain refinement, AZ31, Direct chill (DC) casting

Abstract

The effect of Bismuth addition on the microstructure evolution of Mg and AZ31 Mg-alloy is studied at different cooling rates. A wedge-shaped copper mould was used to achieve continuous variation in cooling rates. Mg containing a small amount of bismuth produced a large decrease in the α-Mg grain size and suggested that that the presence of bismuth makes Mg less sensitive to the cooling rates employed. The macrostructure study of the static-mould direct-chill (DC) cast billets of AZ31 also supports the solute effect of bismuth on grain refinement of Mg and its alloys. The efficiency of Bi as a grain refiner solute can be seen with 50% average grain size reduction of AZ31 alloy at 0.2wt% Bi addition.

Introduction

The wrought Mg alloy billets or slabs are commonly produced through direct-chill (DC) casting process. The semi-finished products are obtained by the intensive thermo-mechanical processing which includes forging, extrusion or rolling. However, one of the problems encountered in conventional DC casting of wrought Mg alloys is the formation of coarse and non-uniform microstructures accompanied by severe chemical segregation, porosity and hot tearing, which affects the mechanical properties of the final products. A fine grain size billet is expected to reduce the size of the defects such as microporosity and second-phase particles. Therefore, grain refinement is considered effective for improving the quality of the wrought components. For effective refinement, the presence of some solute in addition to the grain refining particles is required to enhance the efficiency of the particles by restricting the growth of the solid. The solute could be present either at the columnar growth front competing with equiaxed solidification or from the particles which have already nucleated [1, 2]. For the spherical growth restricted by the partitioning of a single solute, the crystal growth rate for a given undercooling is given by an inverse proportion to the parameter Q [3], where:

$$Q = C_0 m(k-1) \quad (1)$$

m_l is the slope of the liquidus, c_o is the solute concentration in a binary alloy and k is the partition coefficient. The parameter Q is referred to as the growth restricting factor and is used as a measure of the effects of a solute on the grain refinement in the absence of solute interactions [4]. The average grain size $\bar{l}$ is described as the linear function of $1/Q$ for various Mg alloys [5]:

$$\bar{l} = a + \frac{b}{Q} \quad (2)$$

where the slope b is related to the efficiency of the nucleating particles and a corresponds to the number density of active nucleating particles.

This theory is verified for a large number of Al alloys however, it is not so widely applied on Mg alloys due to the difficulty of assessing the extent of solute particle dissolution being dependent on its particle size and processing conditions [5].

Numerous potential additions based on C, Sr, RE, Th, Si, Ca, B and Sb have been tried to refine the grain size of Mg-Al alloys but none have yet been made commercially available [6, 7, 8, 9, 10]. Although previous literature provides invaluable information on the role of solute addition to magnesium alloys, the study on individual Bi additions is limited [11, 12] and poorly understood. Also, these prior researches has focussed towards the effect of Bi additions only on the high aluminium containing Mg alloys (AZ91 and AZ80) while, this paper discusses the grain refinement effect on pure Mg and AZ31 wrought alloy.

Experimental Procedure:

The chemical composition of AZ31 alloy considered for this study is 3.1% Al, 0.92%Zn, 0.53%Mn, <0.01%Si, 0.002%Fe, <0.001%Cu, 0.0006%Ni, <0.0001 Be and remaining Mg. For the casting experiments, pure Mg (commercial purity) and AZ31 alloy each are separately melted in a conventional furnace lined with a steel mould. When the temperature of the melt reaches 720°C, the melt is held at this temperature for at least an hour before casting the reference sample or addition of grain refiner in form of Bi (size -100 mesh) solute. All the samples are obtained at 710±2°C. Pure Mg as-cast and grain refined samples were cast in a preheated (250°C) copper wedge mould, which has a cooling rate range from approximately 20°C/s at the narrow edge to 150°C/s at the broad edge of the wedge. AZ31 alloy samples were cast in a static mould DC simulator as shown in Figure 1.

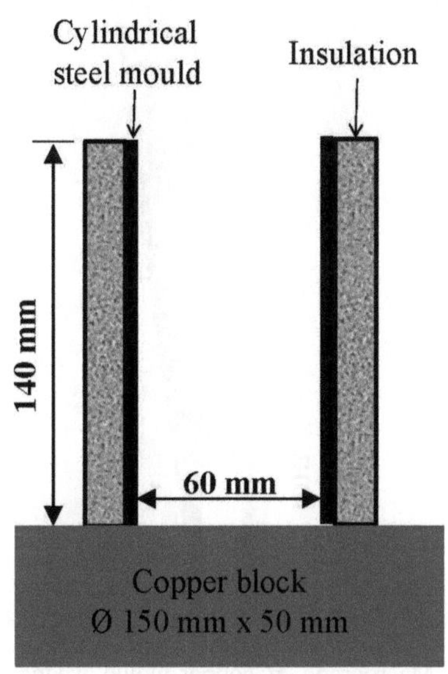

Figure 1.Schematic of the static mould direct-chill (DC) simulator

AZ31 samples were also taken in a cone mould with inner dimensions of 55mm diameter and 80mm height to study the grain refinement effect of Bi solute on Mg alloys. In all the grain refined samples, the refiner Bi is left for 20 minutes before casting at 710±2°C. Cover gas of a mixture of sulphurhexafluoride and nitrogen is used during melting and holding of the melt. No flux is used during this experiment. For microstructural analysis, the cone samples are vertically sectioned and polished using standard procedures for optical microscopy. Polarised light microscopy was applied on samples etched with a mixture of 5 ml acetic acid; 10 ml distilled water; 100 ml ethanol and 6 gm picric acid known as the acetic- picral reagent. The linear intercept method was used to measure the grain size using a Zeiss Axioskop2 MAT optical microscope.

Results and Discussion

1. Grain size as a function of cooling rate:

Copper wedge-shaped mould samples were prepared to study the effect of Bi addition on the average grain size of Mg at different cooling rates. The macrographs of pure Mg wedge samples with and without Bi addition are as shown in Figure 2. Addition of 0.4wt% Bi results in transition of columnar grains to equiaxed grains. In the low cooling rate region of the wedge sample, the average grain size of Mg is 1 to 2 mm, which reduces to 300 to 400 μm on addition of Bi powder.

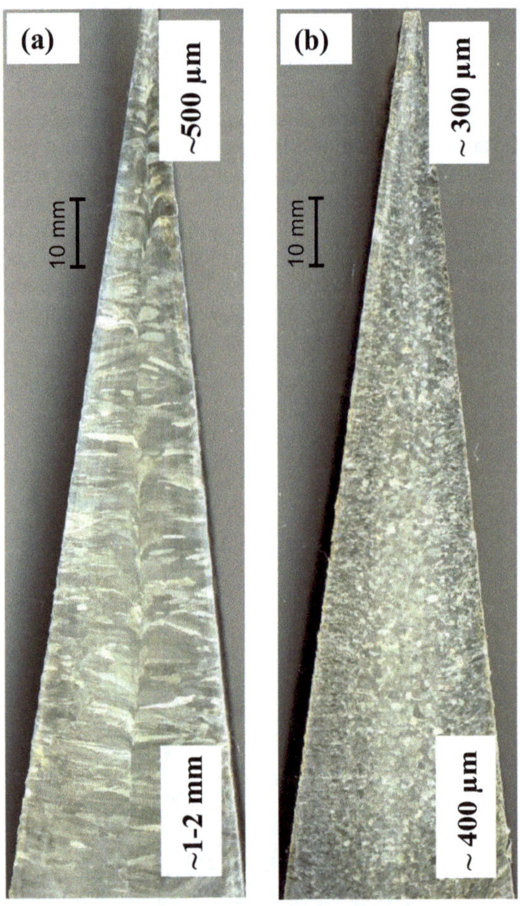

Figure 2. Macrograph of copper-mould wedge samples of Mg (a) without Bi addition and (b) with 0.4wt.%Bi addition

The grain size of Mg as-cast sample is observed to be highly sensitive to the cooling rate as there is an important spatial variation from the tip to the top of the wedge-shaped samples. Conversely, after the addition of Bi powder, the spatial variation is significantly reduced indicating that the final grain size is less dependent on the cooling rates employed. Based on these results, sand casting and direct-chill (DC) casting experiments on Mg and Mg alloys are also expected to show similar grain refinement effect on Bi addition.

2. Effect of Bi addition on direct chill (DC) casting of AZ31 alloy:

Specially designed experiments using a static-mould DC simulator were performed to observe the effect of Bi addition on the macrostructure of AZ31 cast billets. The static-mould DC simulator consists of a tapered cylindrical steel mould (pre-heated to 800°C) on top of a copper block pre-heated to 250°C). The steel mould is pre-heated and insulated to ensure that the heat extraction is mainly through the copper block. The macrostructures of vertically sectioned AZ31 billets is as shown in Figure 3. The macrograph of the as-cast AZ31 static-mould DC cast billet shows a distinctive 2-zone structure: columnar grain structure at the bottom and an equiaxed coarse grain structure towards the top of the billet.

Figure 3. Macrograph of vertically sectioned AZ31 Mg-alloy billet processed through the static-mould DC simulator with (a) without Bi and (b) columnar grains at the bottom chill zone with 0.1wt.% Bi addition

The columnar dendritic structure at the bottom section is formed due to the fast heat extraction from the chilling effect of copper block, which is at much lower temperature compared to the vertical length of the cylindrical mould. The addition of 0.1wt.% Bi leads to an increase in the number of equiaxed grains, with only a small columnar region at the bottom of the billet (Figure 3b). Also, the grain size of AZ31 alloy is observed to decrease on addition of Bi solute.

Figure 4 shows the micrographs of the AZ31-0.1Bi static-mould DC billet. The equiaxed grain as observed on the top of the billet are of the size ~500μm. The transition zone from columnar to equiaxed grains as seen at the bottom of the billet consists of small equiaxed grains inbetween the long columnar grains.

Figure 4. Micrograph of vertically sectioned AZ31-0.1Bi Mg-alloy billet processed through the static-mould DC simulator with (a) equiaxed grains at the top and (b) columnar grains at the bottom chill zone

3. Grain size sensitivity to Bi addition:

Figure 5 shows the variation in the average grain size of AZ31 alloy with the change in the amount of Bi addition. The average grain size of AZ31 alloy without Bi addition is 405 μm , which reduces to 185 μm at 0.2wt% Bi addition and remains unchanged between 0.02 to 0.1wt.% Bi additions.

This effect on grain refinement may be attributed to the growth restriction effect of Bi solute. The role of solutes in grain refinement has been extensively discussed in the literature [5, 13, 14] and it has been accepted that the segregation of elements play a major role in grain refinement. The solute Bi addition is expected to generate a constitutional supercooling ΔT_c due to its segregation ahead of the growing dendrite/grains. This constitutional supercooling (CS) suppresses the growth of the Mg grains through promoting consequent nucleation when ΔT_c reaches the undercooling required for nucleation ΔT_n in the CS zone.

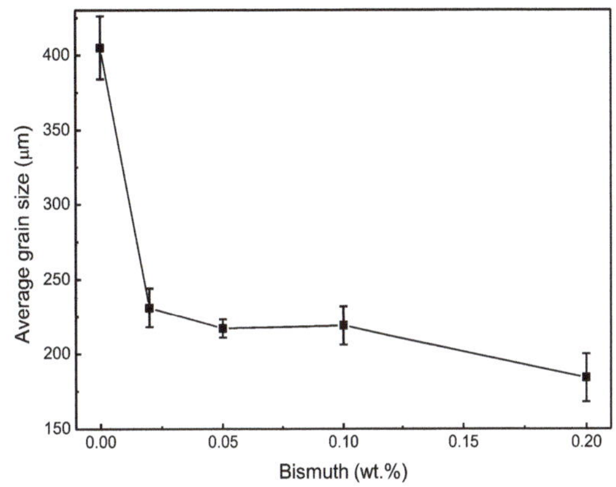

Figure 5. Graph of variation in the average grain size with the amount of Bi addition

A solute's contribution to the rate of development of the constitutionally undercooled zone can be expressed by a term known as the growth restriction factor Q [4, 5], which allows approximate prediction of the effect of solute on the grain size. Larger Q value will have a higher possibility to produce finer grain size.

The Q value for Bi in Mg at C_o = 1.0wt% is reported as 1.551 [15]. Despite a very low Q factor of Bi compared to Zr (Q = 38.29) [5], with Mg binary alloy, a significant grain refinement is observed through Bi addition. The actual Q value of Bi in Mg is expected to be very high owing to the high solid solubility of Bi in Mg reported as 8.9wt% [16]. The dissolved Bi could rapidly build up a CS zone ahead of the growing Mg crystals and hence, the solute effect of Bi on Mg could be substantial.

Conclusions

1. The solute effect of Bi results in a significant grain refinement of AZ31 alloy. At 0.2wt% additions, 50% reduction in average grain size of AZ31 is obtained.

2. The equiaxed grain structure obtained through 0.1wt%Bi additions in static-mould DC casting experiments indicates that better quality of AZ31 billets might be produced on mixing of small amount of Bi with the heterogeneous nucleating particles, which are currently being used as grain refiner compounds.

3. The final grain size of the Mg sample with Bi solute additions is less sensitive to the change in the cooling rates. This indicates the efficiency of Bi solute towards grain refinement.

References

1. A.M. Bunn. et al., "Grain refinement by Al–Ti–B alloys in aluminium melts: a study of the mechanisms of poisoning by zirconium," *Materials Science and Technology*, 15 (1999), 1115-1122

2. M. Gaumann, R. Trivedi and W. Kurz, "Nucleation ahead of the advancing interface in directional solidification," *Material Science and Engineering A*, 226–228 (1997), 763-769

3. I. Maxwell and A. Hellawell, "A simple model for grain refinement during solidification," *Acta Metallurgica*, 23 (1975), 229-237

4. M.A Easton and D.H StJohn, "A model of grain refinement incorporating alloy constitution and potency of heterogeneous nucleant particles," *Acta Materialia*, 49 (2001), 1867- 1878

5. D.H StJohn et al., "Grain Refinement of Magnesium Alloys," *Metallurgical and Materials Transactions A*, 36A (2005), 1669-1679

6. Suresh et al., "Influence of boron addition on the grain refinement and mechanical properties of AZ91 Mg alloy", *Materials Science and Engineering A*, 525 (2009) 207-210

7. L Piejie, T. Bin and E.G. Kandalova, "Microstructure and properties of AZ91D alloy with Ca additions", *Materials Letters*, 59 (6) (2005), 671-675

8. Kim,Y., Yim,C., and You,B., "Grain refining mechanism in Mg-Al base alloys with carbon addition", *Scripta Materialia*, 57 (8) (2007), 691-694

9. Nishino et al., "Grain refinement of magnesium casting alloys by boron addition", In: *Proceedings of the International Conference on Magnesium Alloys and their Applications*, September 2000. Germany: Wiley, 59-64

10. R. Cheng et al., "Effect of Sr addition on the grain refinement of AZ31 magnesium alloys," *Progress in Natural Science: Materials International*, 23:1 (2013), 7-12

11. W. Ya-xiao, et al., "Effect of Bi addition on microstructures and mechanical properties of AZ80 magnesium alloy," *Trans. Nonferrous Met. Soc. China*, 21 (2011), 711-716

12. Y. Guangyin, S. Yangshan and D. Wenjiang, "Effects of bismuth and antimony additions on the microstructure and mechanical properties of AZ91 magnesium alloy," *Materials Science and Engineering A*, 308 (2001), 38–44

13. Y.C. Lee., A.K. Dahle and D.H. StJohn, "The Role of Solute in Grain Refinement of Magnesium," *Metallurgical and Materials Transactions A*, 31 (2001), 2895-2906.

14. H. Men and Z. Fan, "Effects of solute content on grain refinement in an isothermal melt," *Acta Materialia*, 59(7), (2011), 2704-2712.

15. Y. Ali et. al., "Current research progress in grain refinement of cast magnesium alloys: a review article," *Journal of Alloys and Compounds* (2014), doi: http://dx.doi.org/10.1016/j.jallcom.2014.09.061

16. M.M. Avedesian. and Hugh Baker, *ASM Speciality Handbook: Magnesium and Magnesium Alloys*, 1999: ASM International.

Magnesium Technology 2015

Deformation I

Magnesium Technology 2015
Edited by: Michele V. Manuel, Alok Singh, Martyn Alderman, and Neale R. Neelameggham
TMS (The Minerals, Metals & Materials Society), 2015

In-Situ Neutron Diffraction Study of the Deformation Mechanisms in Solutionized Mg-Zn Alloys

R. P. Mulay[1], S.R. Agnew[2] and C. H. Caceres[3]

[1] Lawrence Livermore National Laboratories, Livermore, California, 94550, USA

[2] Department of Materials Science and Engineering, University of Virginia, Charlottesville, VA 22904-4745, USA

[3] Materials Engineering, School of Engineering, University of Queensland Brisbane, Queensland 4072, Australia

Keywords: neutron diffraction, solid solution strengthening, magnesium, elastic strains

Abstract

In-situ neutron diffraction experiments were carried out on solutionized and randomly textured Mg-Zn alloy castings with similar grain sizes but variation in Zn content from 1.7 to 6.6 wt %. The evolution in internal elastic strains and diffraction peak intensities with increasing load were analyzed. The macroscopic stress strain curve shows an increase in yield strength with an increase in zinc content. Neutron diffraction results indicate that the strength of basal slip, tension twinning and <c+a> slip/compression twinning modes increases with increase in zinc content. However, the strength of prismatic slip appears to be unaffected by zinc content at lower concentrations and increases with zinc content only at higher concentrations. These results are discussed in light of prior work on the Mg-Zn system.

Introduction

There is a great interest in the use of magnesium alloys as structural materials due to their low density. Efforts are being made to increase the strength of magnesium alloys through solid solution strengthening and age hardening [1,2,3]. Several studies have examined the deformation mechanisms in magnesium. The primary deformation mechanisms active at room temperature are basal slip, prismatic slip and tension twinning [4]. Single crystal studies have shown that basal slip is the softest deformation mechanism [5]. However, studies on polycrystalline samples have shown that the activation of tension twinning or prismatic slip are necessary for macroscopic yielding to occur [6,7].

Single crystal studies have shown that the prismatic slip mode softens with the addition of common alloying elements, such as Al and Zn [8]. However, basal slip shows the conventional solid solution strengthening effect [9]. Prior work on polycrystalline Mg-Zn alloys with concentrations ranging from 0 to 6.6 wt % Zn suggest that solid solution softening of the prism planes can reduce the yield strength for concentrations up to 1.7 wt% Zn. Beyond this concentration the softening effects are offset by the overall solid solution hardening effects [10]. However, effects of twinning were not considered in that study. Recently, Stanford and Barnett [11] examined textured Mg-Zn alloys in the concentration range 0 to 2.8 wt % Zn under deformation conditions designed to favor a certain deformation mode. They found that the critical resolved shear stress (CRSS) for twinning was not affected by zinc content while that of basal slip increased with increase in zinc content. In case of prismatic slip they found solid solution softening to be present only above grain sizes of ~ 50 um.

The present work examines the effect of Zn additions in the range of 1.7 to 6.6 wt % on the deformation mechanisms in solutionized Mg-Zn alloys using in-situ neutron diffraction. The experiments are done on cast Mg-Zn alloys with a weak (essentially random) initial texture.

Experimental Procedure

Material

Mg-Zn castings with nominal compositions of around 1.5 to 6.5 wt % Zn were made by melting commercially pure Mg (99.7 wt%) and Zn using an electrical resistance furnace in small boron nitride coated steel crucibles. Grain refinement was carried out by using a Mg–Zr master alloy (77 wt % Mg and 23 wt % Zr). An atmosphere of 0.5% SF_6 in dry air was used to protect the molten alloy from oxidation. Alloying additions were made at 700 °C. The melt was then mixed with a mechanical stirrer for 15 minutes and subsequently poured into split steel moulds preheated to 200 °C under protective atmosphere. The cast cylinders were 90 mm in height, 50 mm in diameter.

The cast cylinders were then sectioned in half, solution heat treated under nitrogen atmosphere and quenched into water. The actual compositions were determined using inductively coupled plasma-atomic emission spectroscopy (ICP-AES) [10, 12] and are listed in Table 1. In this paper, the alloys are referred to by their determined composition in wt% Zn. For each composition a longitudinal section was mechanically polished down to 1 μm and then etched with acetic-picral acid. The grain size was subsequently determined using the linear intercept method according to ASTM E112-88 and at least 300 intercepts were counted. The grain sizes are listed in Table 1. More details of the casting, heat treatment and grain size measurement procedures can be found elsewhere [10,12]. The castings used in this study are from the same batch used in references 10 and 12. All three compositions have similar grain sizes (within a factor of 2). Metallographic examination of the samples showed that they were single phase [12]. Cylinders 20 mm in length and 10 mm in diameter were machined from the castings for in-situ testing.

Table 1: Compositions (determined by ICP-AES) and grain sizes of the Mg-Zn alloys examined

Zn (wt %)	Zn (at.%)	Zr (wt %)	Zr (at.%)	Grain size (μm)
1.7	0.64	0.22	0.06	61
4.1	1.56	0.51	0.14	38
6.6	2.59	0.54	0.15	41

Initial texture measurements

Prior to compression testing, the initial texture was measured using the HIgh-Pressure Preferred Orientation diffractometer (HIPPO), at the Manuel Lujan Jr. Neutron Scattering Center (LANSCE, Los Alamos, NM). HIPPO allows for rapid characterization of full orientation distributions of bulk samples and has been designed to optimize the collection of texture data through the use of high neutron intensities and a large number of time of flight detectors. Since this is a neutron diffraction based method, it is not surface limited like X-rays or electron based techniques. A full description of HIPPO can be found elsewhere [13].

Compression testing

The compression tests were performed on a custom built Instron (Norwood, MA) load frame in the flight path of the time-of-flight neutron beam line, Spectrometer for MAterials Research at Temperature and Stress (SMARTS), at the Manuel Lujan Jr. Neutron Scattering Center (LANSCE, Los Alamos, NM). The tests were carried out at room temperature in position control at a constant velocity 0.01 mm/s which corresponds to an initial strain rate of 5×10^{-4}/s. An extensometer was attached to the sample using rubber bands to measure the strain. The compression axis was aligned with the cylindrical axis of the samples. The samples were lubricated using molybdenum disulfide (MoS_2) grease. One sample of each composition was tested due to the time limitations on the neutron beam line.

In-situ neutron diffraction

A detailed description of SMARTS is given elsewhere [14]. A short description is provided here. The loading axis is oriented at 45° relative to the incident neutron beam. Two detector banks are situated at ±90°, making it possible to simultaneously collect two complete diffraction patterns, one parallel and one perpendicular to the applied load. Since the in-situ testing setup is horizontal, the compression sample was preloaded to a stress of ~5 MPa. The initial d_0^{hkl} spacing measured at ~5 MPa is considered to be the ''stress-free'' spacing. The neutron diffraction measurements were carried out at several points along the stress-strain curve. The load frame was held at a constant position during diffraction pattern collection, and the hold time was varied to obtain a constant neutron count at the detectors (governed by the source current and the material neutron absorption and scattering cross sections). For this material, the collection time is approximately 10 to 12 minutes per pattern.

Internal strains within grains that satisfy the Bragg condition for a given {hkl} reflection are calculated from the difference between the measured and stress-free lattice spacing,

$$\varepsilon^{hkl} = \frac{d^{hkl} - d_0^{hkl}}{d_0^{hkl}}$$

where d_0^{hkl} is the stress free lattice spacing and d^{hkl} is the measured lattice spacing. The peak shifts and corresponding internal elastic strains were then calculated using single peak fits with an instrument specific code developed at LANSCE [15] which uses the General Structural Analysis System (GSAS) [16].

Results

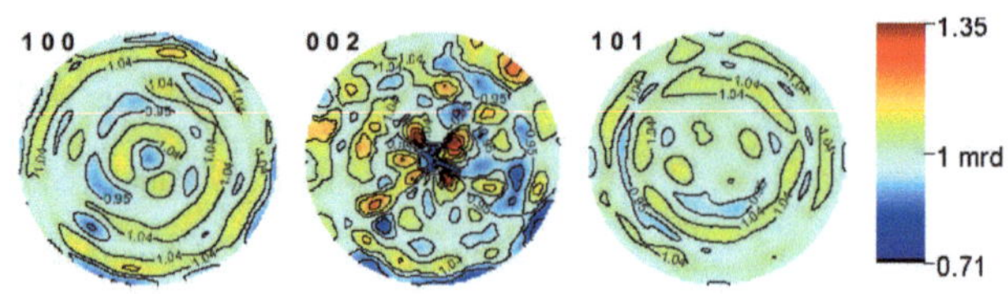

Figure 1: Pole figures showing the initial texture measured for the Mg-1.7 wt % Zn sample. The cylindrical axis of the sample is pointing out of the paper. The texture is essentially random.

The initial texture measured in the Mg-1.7 wt % Zn sample is shown in Figure 1. Similar textures were observed in case of the other two compositions examined in this study. The texture was very weak (essentially random) and axisymmetric. The cylindrical axis of the sample is pointing out of the paper.

Figure 2 shows the stress-strain curves obtained for the three zinc concentrations tested. The rubber bands holding the extensometer broke in the middle of the test in case of the 6.6 wt % Zn sample and hence extensometer strain data is not available for this sample. Thus, the strain plotted in Figure 2 has been calculated from the crosshead displacement to enable equitable comparison of all three samples. Strain obtained from extensometer measurements for the 1.7 and 4.1 wt % Zn samples has been shown in Figure 4a and 4b respectively. The macroscopic yield point (0.2 % strain offset) increases with increase in zinc concentration as seen in Figure 2. The yield point is ~55 MPa for the 1.7 wt % Zn sample, ~80 MPa for the 4.1 wt % Zn sample and ~92.5 MPa for the 6.6 wt % Zn sample. The strain hardening rate appears to be similar in all three samples. None of the samples were strained to failure. The load drops in the stress-strain curves correspond to the points where the test was stopped and held and a constant position to collect neutron diffraction patterns.

Figure 3 shows a diffraction pattern obtained from the 6.6 wt % Zn sample in the stress free condition parallel to the loading axis. No second phase peaks were present in any of the diffraction patterns suggesting that the samples were in the solutionized condition, though it must be admitted that this diffraction technique is not sensitive to very small volume fractions of second phase particles. The diffraction patterns obtained from all three concentrations examined were similar.

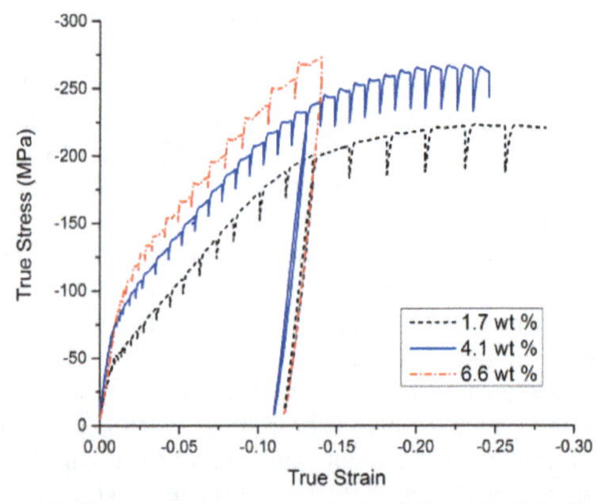

Figure 2: Stress- strain curves obtained from the Mg-Zn alloys. The load drops correspond to the points where diffraction data was collected by holding the test at a constant position.

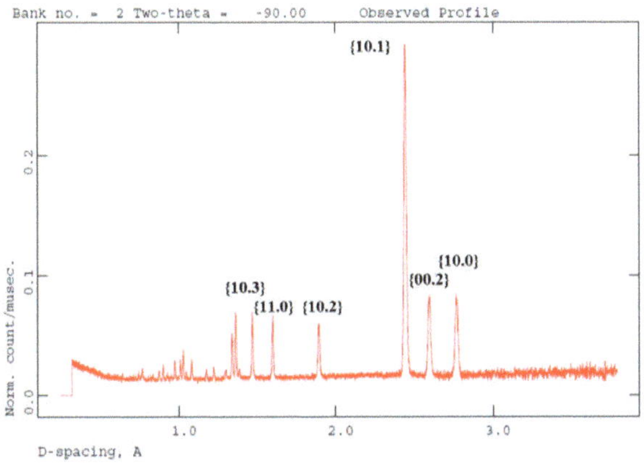

Figure 3: Diffraction pattern obtained from the 6.6 wt % Zn sample in the stress free condition from the bank parallel to the loading axis. No second phase peaks are present.

Figure 4 shows the stress-strain curves, the internal strains and peak intensity evolution from the 1.7, 4.1 and 6.6 wt % Zn samples respectively. As mentioned above, in the case of the 6.6 wt % Zn sample strain has been calculated from crosshead displacement. In Figure 4 c the apparent elastic strain has been subtracted and only the true plastic strain has been plotted. Consider the 1.7 wt % Zn sample first. Initially the internal strains in all the sets of grains increase at the same rate, which is reflective of the elastic isotropy in Mg. Past the micro-yield point (20-25 MPa) (Fig. 4a) there is a bifurcation in the internal strains. The {10.2} (ideally oriented for basal slip) and {10.1} (also well oriented for basal slip) diverge to the left, indicating these sets of grains have yielded and reduce their share of elastic strains. This suggests that the critical resolved shear stress (CRSS) is approximately 10 MPa in this dilute Mg-Zn alloy. On the other hand, the {11.0} and {00.2} sets of grains that diverge to the right are still accommodating most of the strain elastically. Hence, they must start sharing a greater amount of the elastic strain. If one could average the lattice strains in all grains, the result would correspond to Hooke's law (i.e. a linear relation between the applied stress and the average lattice strain with slope equal to Young's modulus).

The next inflection occurs at an applied stress of approximately 40 MPa, which is close to the macroscopic yield point (~55 MPa). At this point, the {11.0} grains and {00.2} grains diverge. A simple Schmid analysis (which must be applied with caution, given the fact that stresses are non-uniform among the grains at this stage) shows that this could be due to the activation of non – basal (prismatic) slip or tensile twinning [17], provided a CRSS of about 20 MPa. Distinction between the two modes can be made by observing peak intensity changes. From figure 4a (right hand side) it can be seen that, beyond this stress level, the intensity of the {00.2} reflection starts increasing rapidly and that of the {11.0} starts decreasing. The {10.1} and {10.2} intensities exhibit less rapid change. This type of texture evolution is a signature of tension twinning (whereby grains most favorably oriented for twinning, with c-axes perpendicular to the compression axis, are reoriented by nearly 90° to closely align with the compression axis).

At an applied stress of 135-140 MPa, the lattice strain evolution in the {11.0} grains slows. This indicates that the prismatic slip system has become active in this material. With such widely varying internal strain levels in the grains, it would be unwise to speculate too confidently about the CRSS value for prismatic slip without employing a polycrystal plasticity model, however, it is certainly higher than that of twinning and would appear to be in the vicinity of 60 MPa, given the fact that the lattice strain level in the {11.0} grains does not depart far from a linear extrapolation of the internal strain evolution during the elastic region and the Schmid factor for prismatic slip in {11.0} grains is m = 0.43.

At higher stress levels, the only significant inflection of note is the gradual bending of the {00.2} lattice strain curve at an applied stress which corresponds with the point where the linear hardening behavior gives way to a decreasing strain hardening rate and, finally, saturation of the flow curve (beginning at about 150 MPa). This could be because the grains have essentially exhausted the twinning mechanism reflected by the fact that the intensity of the {00.2} reflection is only increasing slowly. This twinning has placed a significant volume fraction of grains in a hard orientation (c-axes parallel to the compression axis), which can only be relaxed by <c+a> slip or compression twinning [6, 7]. Saturation of the flow curve at 225 MPa is probably reflective of the activity of one or more of these hard modes.

Similar observations can be made in the sample with 4.1 wt % Zn. The difference is that in this sample these inflections occur at higher stress (Fig. 4 b). Microyielding of the {10.2} grains occurs at approximately 45 MPa and the macroscopic yield point beyond which twinning is active occurs around 70 MPa. This suggests that there is solid solution hardening of both the basal slip (CRSS of about 20 MPa) and tension twinning deformation modes. It is admitted that the effect on twinning could be secondary, given the fact that nucleation of twinning is generally associated with stress concentrations related to basal slip [e.g.18]. The prismatic slip mode again appears to have been activated at an applied stress level of 140 MPa. Again, this suggests a CRSS value in the vicinity of 60 MPa, which indicates that prismatic slip is neither radically hardened nor softened in this range of solute addition. Finally, the linear strain hardening behavior terminates at around 175 MPa, and saturation occurs at 270 MPa. This suggests that the <c+a> and/or compression twinning modes are solution strengthened.

For the highest concentration sample examined in this study (6.6. wt % Zn, shown in Fig. 4 c), the {10.2} grains, favorably oriented for basal slip, appear to yield at 60 MPa, suggesting a CRSS of 30 MPa. The second inflection occurs quickly thereafter, ~75 MPa, at which point the tensile twinning mode appears to have been activated, given the onset of rapid intensity changes in the {00.2} and {11.0} peaks, in particular. The next inflection in the internal strain development of {11.0} grains again suggests activation of prismatic slip. This time it occurs at around 170 MPa. This suggests that either solid solution strengthening has finally set in, or precipitation at a level insufficient to be observed by neutron diffraction has begun to influence the activity of prismatic slip. In this final test, the flow curve never saturates and the internal strain evolution in the {00.2} grains has not slowed, even at 275 MPa. This suggests that the <c+a> and/or compression twinning modes continue to be hardened by the additional alloy content.

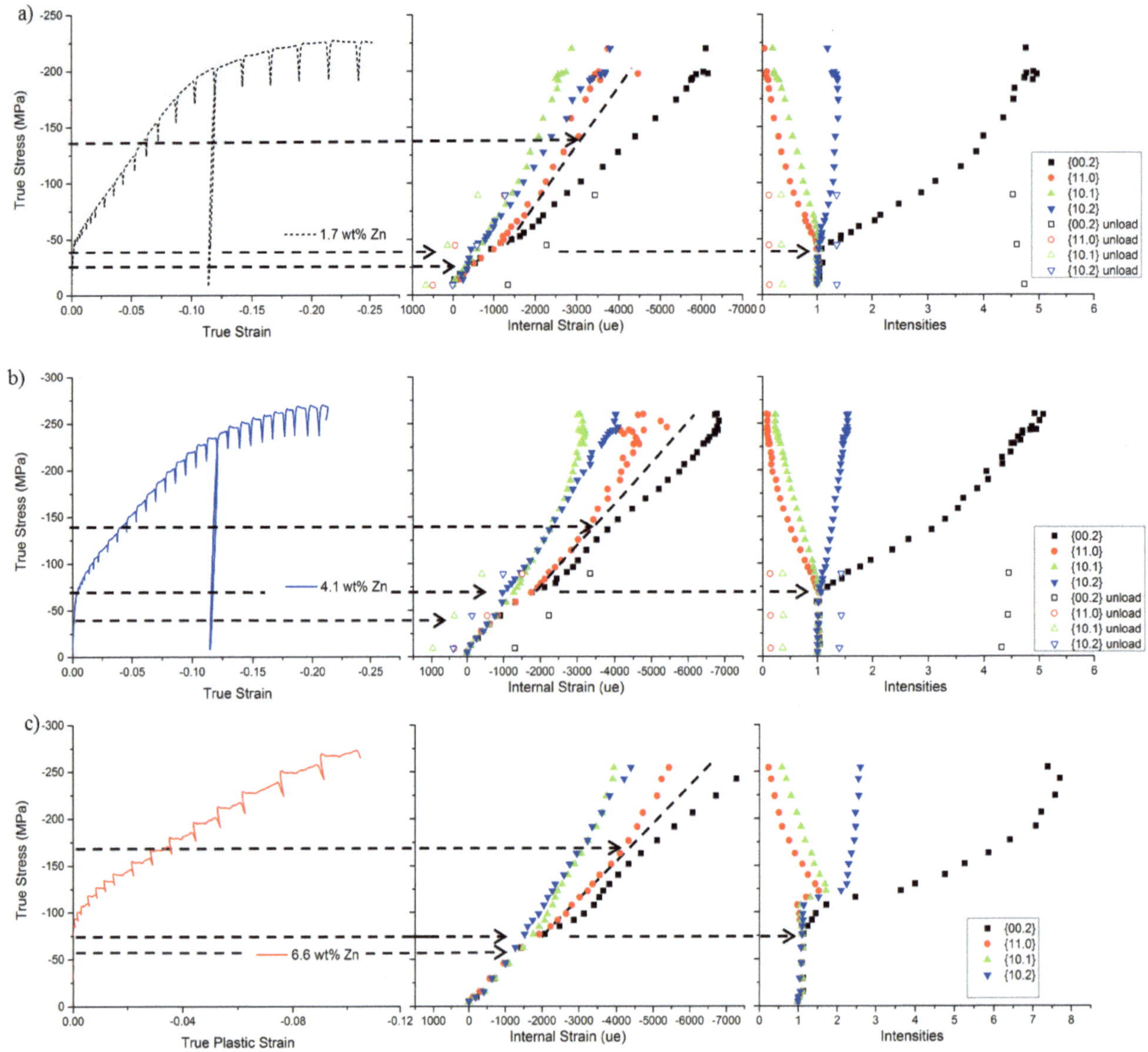

Figure 4: Stress-strain, internal elastic strains and diffraction peak intensity evolution for a) 1.7 b) 4.1 and c) 6.6 wt % Zn samples. The arrows are indicative of points where transitions are taking place.

The initial texture is essentially axisymmetric (since it is near random for all three samples) and since the compression process is inherently axisymmetric, it is possible to reconstruct the full texture at every point along the stress-strain curve where diffraction data is collected from the two diffraction patterns measured using SMARTS. Figure 5 shows the evolution of the axial distribution of the {00.2} poles as multiples of random distribution (MRD) with increasing stress in the 1.7 wt % Zn sample. Similar results were obtained for the 4.1 and 6.6 wt % samples. The intensity of the {00.2} peak increases significantly near the compression axis and reduces perpendicular to the compression axis. Assuming that the evolution of the {00.2} peak intensity is primarily due to deformation twinning (i.e., ignoring the rotations due to slip), it is possible to estimate the volume fraction of twinning as a function of deformation. The following relationship was employed,

$$f_{tw} = \int_{0}^{48} \{I(\chi) - I_0(\chi)\} \sin \chi d\chi$$

where χ is the angle of tilt between the {00.2} poles and the compressions axis and $I(\chi)$ is the intensity of the distribution at that tilt angle. If $I(\chi)\sin\chi$ is integrated from 0-90° the result will always be 1. A similar method was employed in previous work [17,19]. In the present case, we subtract the initial angular distribution $I_0(\chi)\sin\chi$, since we are interested in the change in intensity and not the absolute values. In the present work, the intensity of the axial distributions was equal around 48°, i.e., the volume fraction of grains with this orientation neither increase nor decrease. Hence, this angle was selected as the angle dividing regions associated with parents and twins. Figure 6 shows the twin volume fractions plotted as a function of stress for the three samples examined in this study. It can be seen that as the concentration of zinc increases, the stress necessary to produce the same volume fraction of twins increases. This observation is consistent with the observations made from the internal strains where a higher stress was found necessary to activate twinning with increase in zinc concentration.

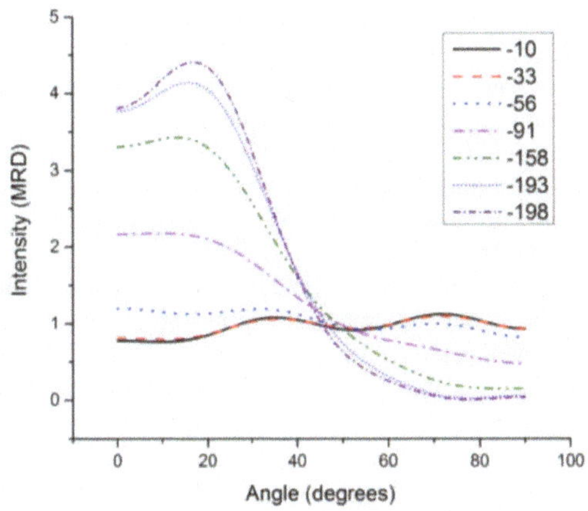

Figure 5: Axial distibution of the {00.2} poles in the 1.7 wt % Zn sample with increasing stress (MPa). The angles are measured with respect to the compression axis. There is a large evolution in the intensity of the {00.2} peak which is used to determine the amount of deformation twinning.

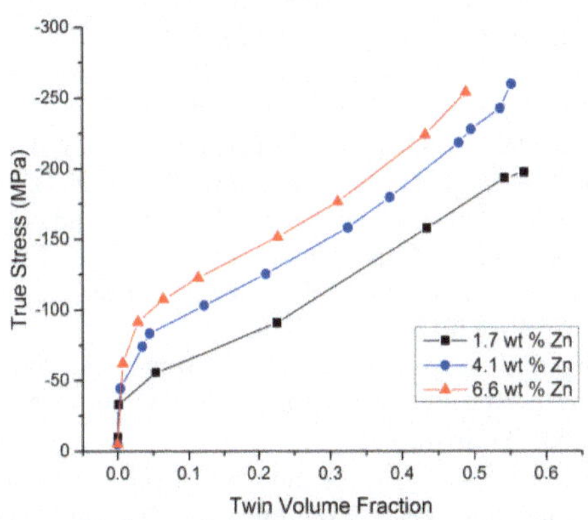

Figure 6: Twin volume fraction as a function of stress in 1.7, 4.1 and 6.6 wt % Zn samples.

A final point about the intensity evolution for the highest Zn content sample, shown in Fig. 4 c, is in order. Note that the relative intensity of all peaks suddenly increase at around 100 MPa. This is not indicative of a change in texture, since that would require some peaks to loose intensity. Because texture evolution is used as a fingerprint of twin activation, understanding the fundamental cause of this intensity change is of interest. However, no satisfying explanation has been forthcoming, as of yet. The aforementioned approach to estimating twin volume fraction is based upon a direct assessment of the texture evolution itself. The fact that the texture evolution is continuous (see Fig. 6) through the region where the discontinuity in the intensity is observed gives some confidence that there some other experimental artifact is responsible.

Discussion

The CRSS value of the basal slip mode is found to increase with an increase in zinc concentration from about 10 MPa for 1.7 wt % Zn to 30 MPa for 6.6 wt % Zn (Table 2). Without employing crystal plasticity modeling, the uncertainty in these estimates is nearly 5 MPa, so further work is planned. That said, prior work on Mg-Zn single crystals [9] and highly textured polycrystalline alloys [11] have shown a similar trend for concentrations up to 2.8 wt % Zn. While the estimated values of the CRSS are higher than those measured in the case of the single crystals (~ 1 MPa) [8,9], the current values show a continued solid solution strengthening trend even at higher Zn concentrations than those characterized in those original studies. Further, the CRSS values are in the same range as those obtained in a recent study of textured polycrystals favoring basal slip for samples with similar grain sizes [11]. Recent polycrystal plasticity modeling work [20, 21] on alloys similar to the present study (i.e., randomly textured cast alloys with a composition range of 0 to 6.6 wt % Zn) predicted a similar trend in the CRSS values.

Table 2: CRSS values (MPa) estimated for the different deformation modes based upon internal strain measurements.

	Basal slip	Tension twinning	Prismatic Slip
1.7 wt % Zn	10	20	60
4.1 wt % Zn	20	35	60
6.6 wt % Zn	30	38	75

The CRSS for twinning was also found to increase with increase in Zn content, from 20 to 38 MPa (Table 2). Also, the stress necessary to produce the same volume fraction of twins was found to increase with increase in zinc content (Fig. 6). Prior work on Mg single crystals found the CRSS for twinning to be independent of zinc concentration in the range examined (up to 2.8 wt % Zn) [5]. Stanford and Barnett [11] also found that the CRSS for twinning was ~15 MPa for a grain size of ~40 um irrespective of the Zn concentration. However, these samples were strongly textured. As multiple deformation mechanisms are necessary for yielding in randomly textured polycrystals, it is possible that the increase seen in the current work is a result of the increase in CRSS of basal slip. However, polycrystal plasticity modeling work carried out by Raeisinia and Agnew on randomly textured cast alloys [21] also showed an increase in twinning stress with zinc concentration up to around 2.1 wt % Zn followed by a saturation or slight decrease in CRSS. However, there was sufficiently large scatter in the data which precludes drawing firm conclusions. Again, it is admitted that the effect of Zn solute on twinning in these randomly textured castings is distinct from the insensitivity reported recently, based upon studies of textured samples [11]. Further work is planned involving crystal plasticity modeling of the results in order to lower the uncertainly levels in the estimates of the CRSS values of all the individual deformation mechanisms.

The CRSS of the prismatic slip mode seems to be unaffected by zinc content in the lower concentration range examined in this study and finally appears to increase with an increase in zinc content. Work carried out on single crystal Mg-Zn alloys [8] with zinc concentrations up to 1.1 wt % zinc showed a decrease in CRSS of the prismatic slip mode, especially at low temperatures. Stanford and Barnett observed a decrease in the CRSS value only in samples with grain sizes larger than 50um and in the concentration range of 0-1.1 wt % Zn [11]. In the range of 1.1 to 2.8 wt % Zn the CRSS values were independent of concentration for all grain sizes examined. Raeisinia et al. [20], in their reanalysis of early single crystal data [8,9], also showed that the CRSS of the prismatic slip mode is largely unaffected by zinc concentrations above ~1.1 wt % Zn. In the present work, all samples have concentrations above 1.1 wt % Zn, which explains the relative independence of the CRSS on zinc concentration. The increase in CRSS observed in the 6.6 wt % sample is indicative of either solid solution hardening setting in or precipitation beginning to occur which cannot be detected by neutron diffraction.

Finally, the present work provides some insight into the effect of zinc content on the stress necessary to activate the hard modes of <c+a> slip and / compression twinning. The stress necessary to activate these modes appears to increase with an increase in the zinc content.

Conclusions

Solutionized Mg-Zn alloys with zinc content from 1.7 to 6.6 wt % were examined. The macroscopic yield point was found to increase with increase in zinc concentration. In-situ neutron diffraction results revealed that the strength necessary to activate basal slip, tension twinning and <c+a> slip/compression twinning modes all increased with an increase in zinc content. On the other hand, the stress necessary to activate prismatic slip is unaffected at lower concentrations of zinc but increases at higher concentrations. Similar to recent findings obtained from studies of textured polycrystals [11], no solute softening of the prismatic slip mode was observed.

Acknowledgements

A portion of this work was performed under the auspices of the U.S. Department of Energy by Lawrence Livermore National Laboratory under Contract DE-AC52-07NA27344 and has been assigned the document release ID #LLNL-PROC-661274.This work has benefited from the use of the Lujan Neutron Scattering Center at LANSCE, funded by the Department of Energy's Office of Basic Energy Sciences. Los Alamos National Laboratory is operated by Los Alamos National Security LLC under DOE Contract DE-AC52-06NA25396. The assistance of Lujan Center researchers D.W. Brown and B. Clausen is gratefully acknowledged.

References

1. J. B. Clark, "Age hardening in a Mg-9 wt.% Al alloy," *Acta Metallurgica*, 16(2) (1968), 141-152.

2. L. Gao, R. S. Chen, and E. H. Han, "Solid solution strengthening behaviors in binary Mg–Y single phase alloys," *Journal of Alloys and Compounds*, 472(1) (2009), 234-240.

3. J. B. Clark, "Transmission electron microscopy study of age hardening in a Mg-5 wt.% Zn alloy," *Acta Metallurgica*, 13(12) (1965), 1281-1289.

4. A. V. Beck, ed., *The technology of magnesium and its alloys* (London: FA Hughes & Co Ltd, 1943)

5. E. W. Kelley, and W. F. Hosford JR, "Plane-strain compression of magnesium and magnesium alloy crystals," *Trans Met Soc AIME*, 242(1) (1968), 5-13.

6. S. R. Agnew, D. W. Brown, and C. N. Tomé, "Validating a polycrystal model for the elastoplastic response of magnesium alloy AZ31 using in situ neutron diffraction," *Acta materialia*, 54(18) (2006), 4841-4852.

7. S. R. Agnew, et al., "Study of slip mechanisms in a magnesium alloy by neutron diffraction and modeling," *Scripta Materialia*, 48(8) (2003), 1003-1008.

8. A. Akhtar, and E. Teghtsoonian, "Solid solution strengthening of magnesium single crystals—II the effect of solute on the ease of prismatic slip," *Acta Metallurgica*, 17(11) (1969), 1351-1356.

9. A. Akhtar, and E. Teghtsoonian, "Solid solution strengthening of magnesium single crystals—I alloying behaviour in basal slip," *Acta Metallurgica*, 17(11) (1969),1339-1349.

10. A. H. Blake, and C. H. Cáceres, "Solid-solution hardening and softening in Mg–Zn alloys," *Materials Science and Engineering: A*, 483 (2008), 161-163.

11. N. Stanford, and M. R. Barnett, "Solute strengthening of prismatic slip, basal slip and twinning in Mg and Mg–Zn binary alloys," *International Journal of Plasticity*, 47 (2013), 165-181.

12. C. H. Cáceres, and A. Blake, "The strength of concentrated Mg–Zn solid solutions," *Physica Status Solidi A*, 194(1) (2002), 147-158.

13. H. R. Wenk, L. Lutterotti, and S. Vogel, "Texture analysis with the new HIPPO TOF diffractometer," *Nuclear Instruments and Methods in Physics Research Section A: Accelerators, Spectrometers, Detectors and Associated Equipment*, 515(3) (2003), 575-588.

14. M. A. M. Bourke, D. C. Dunand, and E. Ustundag, "SMARTS–a spectrometer for strain measurement in engineering materials," *Applied Physics A*, 74 (2002), S1707-S1709.

15. B. Clausen, *SMARTSware Manual* (Los Alamos, NM: The Regents of the University of California, 2003), 1–22.

16. R. B Von Dreele, "Quantitative Texture Analysis by Rietveld Refinement, " *J. Applied Crystallography*, 30 (4) (1997), 517–25.

17. S. R. Agnew, et al., "In situ neutron diffraction and polycrystal plasticity modeling of a Mg–Y–Nd–Zr alloy: Effects of precipitation on individual deformation mechanisms," *Acta Materialia*, 61(10) (2013), 3769-3780.

18. J. Wang, I. J. Beyerlein, and C. N. Tomé, "An atomic and probabilistic perspective on twin nucleation in Mg," *Scripta Materialia*, 63(7) (2010), 741-746.

19. B. Clausen, et al., "Reorientation and stress relaxation due to twinning: modeling and experimental characterization for Mg," *Acta Materialia*, 56(11) (2008), 2456-2468.

20. B. Raeisinia, S. R. Agnew, and A. Akhtar, "Incorporation of solid solution alloying effects into polycrystal modeling of Mg alloys," *Metallurgical and Materials Transactions A*, 42A (2011), 1418-1430.

21. B. Raeisinia and S. R. Agnew, "Using polycrystal plasticity modeling to determine the effects of grain size and solid solution additions on individual deformation mechanisms in cast Mg alloys," *Scripta Materialia*, 63(7) (2010), 731-736

Magnesium Technology 2015
Edited by: Michele V. Manuel, Alok Singh, Martyn Alderman, and Neale R. Neelameggham
TMS (The Minerals, Metals & Materials Society), 2015

Investigation of Compression Behavior of Mg-4Zn-2(Nd,Gd)-0.5Zr at 350°C by *In Situ* Synchrotron Radiation Diffraction

*R. H. Buzolin[1,2], D. Tolnai[1], C. L. Mendis[1], A. Stark[1], N. Schell[1], H. Pinto[2], K. U. Kainer[1], N. Hort[1]

[1]Institute of Materials Research, Helmholtz-Zentrum Geesthacht, Germany
Max-Planck-Strasse 1,
D 21502, Geesthacht, Germany
(*Corresponding author: ricardo.buzolin@usp.br)

[2]Department of Material Engineering, University of Sao Paulo, Brazil
Av. Joao Dagnone, 1100 Jd. Sta Angelina,
13563-120, Sao Carlos, Brazil

Keywords: ZK40, RE additions, deformation behaviour, *in situ*, synchrotron radiation

Abstract

As-cast Mg-4Zn-0.5Zr, Mg-4Zn-2Gd-0.5Zr and Mg-4Zn-2Nd-0.5Zr alloys were investigated by *in situ* synchrotron radiation diffraction during hot compression at 350 °C using the facilities of P07 beamline of Petra III at Deutsches Elektronen Synchrotron (DESY), Hamburg, Germany. The specimens were heated at a rate of 100 K/min and compressed with an initial strain rate of 1.1 x 10^{-3} s^{-1} up to 30 % strain. The addition of rare earth elements improved the yield strength from 23 MPa in the Mg-4Zn-0.5Zr alloy up to 40 MPa in the alloy with Nd and Gd. Continuous dynamic recrystallization played an important role in the Mg-4Zn-0.5Zr alloy during deformation and twinning was not dominant. Discontinuous dynamic recrystallization was observed in the in the Mg-4Zn-2Gd-0.5Zr along the grain boundary regions while the grains remained largely without any recrystallization. In the Mg-4Zn-2Gd-0.5Zr and Mg-4Zn-2Nd-0.5Zr alloys the contribution of twinning to deformation was observed at 350 °C. Reasons for these differences will be discussed with respect to microstructures of the alloys.

Introduction

The increased demand from the automotive industry for weight reduction and more environmentally-friendly materials, with a view to reduce exhaust gases, has initiated research in the field of lightweight materials, especially in Mg alloys [1]. Significant research is still needed on processing, alloy development, joining, surface treatment, corrosion resistance and mechanical properties improvement of Mg alloys.

Conventional magnesium alloys exhibit poor formability at ambient temperatures [2], due limited number of independent active slip systems available at room temperatures to allow an arbitrary shape change. Even though Mg alloys display some plasticity at room temperature, especially with a finer grain size and through solute additions, [2], wrought processing, such as extrusion, rolling or forming are problematic at room temperatures. Therefore, higher temperatures (>225 °C) have been used to deform Mg alloys, at which non-basal slip and deformation twinning can be activated [3].

Wu et al. [4] investigated the compression behaviour of ZK60 Mg alloy at different temperatures and found out that 250-350 °C was the most suitable range of temperature for the ZK60 alloy. The microstructure was more homogeneous at this range of temperature and exhibited almost no twins with hot compression temperature greater than 350 °C. 350 °C is a typical extrusion temperature for ZK alloys [2], thus understanding the deformation behavior of the ZK based alloys at 350°C is of importance for future alloy development.

ZK alloys are one of the high strength wrought magnesium alloys that are commercially available and are also used in the cast form. The addition of Rare Earth (RE) elements to these alloys [5] results in improved castability and elevated temperature strength [6]. RE additions refine grain size mainly due to constitutional supercooling [7]. Furthermore they weaken the anisotropy thus improve the ductility of Mg alloys [8]. Therefore, Mg-Zn-RE-Zr system is of great interest for research.

In situ high energy X-ray diffraction (HEXRD) is a unique method to investigate the dynamic microstructural processes during mechanical loading in the bulk of the materials [9]. Azimuthal-angle vs. Time plots (AT-plots) hold information about grain size evolution, grain imperfection, grain rotation, and grain orientation relationships, texture evolution. These can be correlated with crystallographic slip due to dislocation motion, sub-grain formation, twinning, recrystallization, recovery and grain growth; which are the responsible physical processes for the microstructural changes associated with deformation [9].

The present work investigates the influence of Gd and Nd addition on the compression behaviour of the ZK40 alloy combining *in-situ* radiation synchrotron study, which provides new insights into the real-time evolution of the microstructure, and electron backscattered diffraction (EBSD) in order to study the deformed structure.

Experimental Procedures

The alloys were prepared by permanent mould indirect chill casting [10] with nominal compositions Mg4Zn-0.5Zr (Mg2.2Zn-0.14Zr) Mg-4Zn-0.5Zr-2Gd (Mg2.2Zn-0.14Zr-.65Gd) and Mg-4Zn-0.5Zr-2Nd (Mg2.2Zn-0.14Zr-0.70Nd) in wt.% (at.% compositions are shown in brackets). The actual chemical compositions in wt.% are listed in Table 1. The Gd was added as an Mg-4 wt.%Gd alloy. The Nd was added as pure element. The melt was held at 750 °C for 10 min and then poured into a mould preheated to 660 °C and held for another 15 min in another furnace at 660 °C for final cooling with water. The permanent mould direct chill casting procedure was used for a homogeneous

Figure 1: Optical micrographs typical of as-cast (a) ZK40; (b) ZK40-2Gd; (c) ZK40-2Nd.

distribution of alloying additions. The specimens for the *in situ* compression were machined with 5 mm diameter and 10 mm length.

For metallographic characterization of the as-cast and the deformed samples, they were mounted in epoxy and ground using SiC paper down to 2500 grit and then polished using 3 μm diamond solution. OPS solution with 1 μm diamond suspension was used as the last step of polishing. The specimens for optical microscopy (OM) were etched with acetic-picral solution and the specimens for EBSD were washed with 0.5 vol% nitric acid in ethanol solution for 5 seconds.

The optical analyses were performed using the light optical microscopy Leica DMI 5000. The SEM used was a Zeiss FEG-SEM Ultra 55 attached with a Hikari detector and a TSL-OIM software package for EBSD analysis. The measurements were performed at a voltage of 15 kV using a working distance of 13 mm and an area of 80 μm x 80 μm using a step size of 0.1μm.. The maps were measured in the centre of the deformed specimen to ensure a similar area to that measured during diffraction analysis

In-situ synchrotron high energy X-ray diffraction has been performed using the facilities of DESY (Deutsches Elektronen-Synchrotron) at P07 – Petra III with a beam having an energy of 100 keV (λ=0.124 nm) and a cross section of 1.0 mm x 1.0 mm. Diffraction patterns were recorded at a frame rate of 1/s by a PerkinElmer 1621 flat panel detector (200 μm x 200 μm) which was placed at a distance of 1535 mm from the specimen position. The acquisition time was 1.0 s. the detector to sample distance was calibrated with a LaB_6 standard powder sample and the software Fit2D® was used for evaluating the data.

The compression specimens were mounted into a chamber of a dilatometer DIL 805A/D (TA Instruments, Hüllhorst, Germany), combined with an induction coil modified so the beam can pass through the sample [11]. The specimens were heated to the intended temperature (350 °C) at a rate of 30 $°Cs^{-1}$ and held at this temperature for 3 min before the compression started. The specimens were compressed with a constant deformation speed of 0.01 mms^{-1} and the tests were carried out up to 0.3 of strain. The Debye-Scherrer rings were then analysed using Fit2D software. Selected reflections throughout the whole experiment were converted to azimuthal angle – time (A-t) plots whereas the grey scale represents the intensity from white (0) to black (high intensity).

Results

The optical micrographs of the as-cast ZK40, ZK40-2Gd and ZK40-2Nd alloys are presented in Figure 1. The ZK40 and the ZK40-2Gd alloys exhibited cellular structures whereas the ZK40-2Nd showed a more pronounced dendritic structure. The ZK40-2Gd alloy exhibited a heterogeneous microstructure with coarse grains surrounded by fine grains. For the ZK40-2Gd and ZK40-2Nd alloys the intermetallic phases were semi-continuously distributed along the grain boundaries, whereas the ZK40 had small eutectic compounds along the grain boundaries.

Table 1: The chemical compositions of the alloys investigated.

Chemical Composition				
Alloys	Gd %wt	Nd %wt	Zn %wt	Zr %wt
ZK40	-	-	5.00	0.53
ZK40-2Gd	1.70	-	4.50	0.55
ZK40-2Nd	-	2.35	4.60	0.50

The true stress-strain curves for the *in situ* compression tests are shown in Figure 2. ZK40 exhibited a 0.2% proof stress of 23 MPa and a maximum stress of 28 MPa. For the ZK40-2Gd alloy and ZK40-2Nd alloy, an increase in both the 0.2% proof stress and the maximum stress was observed. For the ZK40-2Gd alloy maximum stress and 0.2% proof stress were 56 and 40 MPa respectively and for ZK40-2Nd alloy they were 47 MPa and 39 MPa, respectively. In addition, the alloys modified with Gd or Nd show reduction in applied stress after reaching the maximum stress which was not evidenced in the ZK40 alloy.

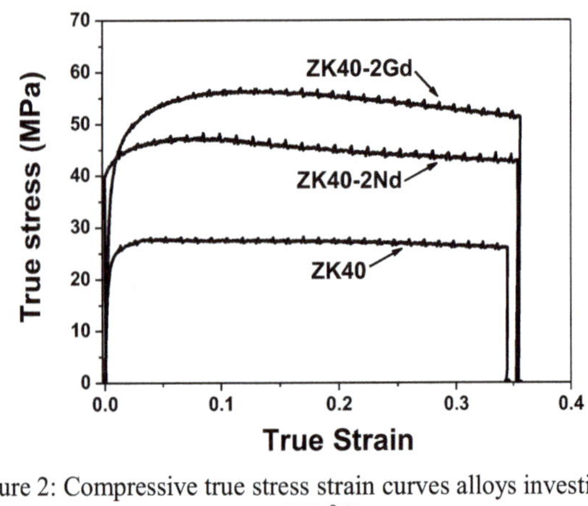

Figure 2: Compressive true stress strain curves alloys investigated at 350 °C

The Debye-Scherrer rings recorded during the *in situ* synchrotron radiation compression tests were converted into azimuthal-time plots to investigate the phenomena that are ongoing during the deformation and are shown in Figure 3. The planes of interest were the $\{11\bar{2}0\}$ and the $\{10\bar{1}3\}$ as twinning can be observed comparing the timelines evolution for these planes. Figures 3 (a), 3 (b) and 3 (c) display the AT-plots for the ZK40, ZK40-2Gd and ZK40-2Nd, respectively. Blurring of the timelines after the initial stages of compression is observed for the ZK40 alloy. The ZK40-2Nd alloy differs at the initial stages, where thicker timelines can be observed and the blurring of the timelines occurred at a higher strain than for the ZK40. The ZK40-2Gd does not show pronounced blurring of the timelines. The appearance of new timelines and dots in the AT-plots and the thickening of these spots after the initial stages can be observed for the ZK40-2Gd. The change in the intensity of the coupled pairs of diffraction lines $\{11\bar{2}0\}$-$\{10\bar{1}3\}$ indicate tensile twinning in Mg. The maximum normalised values of these peaks are plotted against the deformation in Figure 4. It is noted that all alloys show increase in intensity of one reflection with a corresponding decrease in the other, which is representative of tensile twinning.

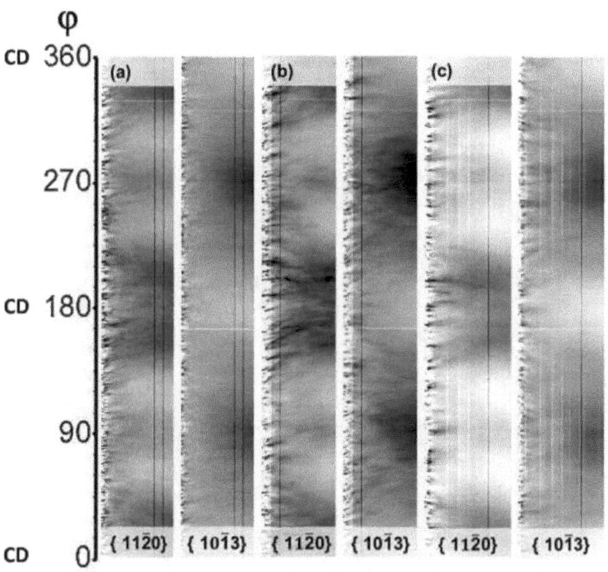

Figure 3: AT-plots derived from the *in situ* synchrotron radiation diffraction during compressive deformation of (a) ZK40, (b) ZK40-2Gd and (c) ZK40-2Nd alloys. CD represents the compression direction.

The SEM micrographs of the deformed microstructures are shown in Figure 5. The red circles highlight the areas where the fracture of the intermetallic particles was observed for the ZK40-2Gd alloy (Figure 5 (b)) and ZK40-2Nd alloy (Figure 5 (c)). This was not observed for the ZK40 alloy (Figure 5 (a)).

EBSD inverse pole figure maps are shown in Figure 6. The ZK40 contain recrystallized grains along the grain boundaries, region A in Figure 6 (a). Sub-grain formation is observed in region B while no tensile twins were observed. The ZK40-2Gd contains a significant number of recrystallized grains at the grain boundary as indicated by C, Figure 6 (b). In addition, tensile twins are observed in the final microstructure, region D. Tensile twins were also observed in the ZK40-Nd, Figure 6 (c). However, almost no recrystallized grains at the grain boundaries were observed for ZK40-Nd. In order to investigate in more details the sub-grain formation, grain reorientation average deviation (GROD) maps was used to study the effect of recrystallization on the microstructure of the alloys and are shown in Figure 6 (d-f). GROD maps show the deviation in orientation of a measurement point from the average orientation of the grain to which the point belongs [12]. The highest misorientation values were observed for the ZK40-2Gd alloy, whereas the lowest was for the ZK40 alloy. The highest values are at the grain boundary and the large misorientation is observed for the alloys modified with Gd or Nd addition.

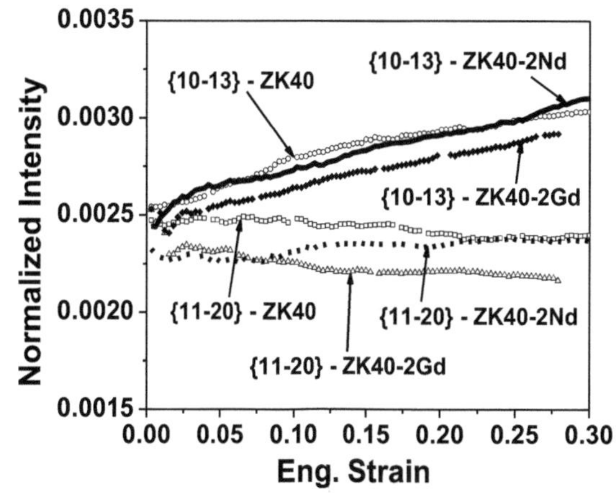

Figure 4: Change in the intensity for the $\{11\bar{2}0\}$ and $\{10\bar{1}3\}$ reflections during compressive deformation.

Discussion

In an *in situ* experiment it is possible to investigate the evolution of various deformation, recovery and recrystallization mechanisms and the manner they act during compression. There are essentially four phenomena that are observed on the AT-plots: recrystallization; recovery, which was not clearly observed for the alloys in study; grain rotation, which is shown by the bending of the timelines, and grain growth, which is very unlikely to happen during deformation. In addition, the discrete timelines correspond to individual grains. Investigating the AT-plots (Figure 3) an intensive recrystallization is noticeable at the initial stages, which is showed by the appearance of new timelines. Subsequently, the timelines bend due to the rotation of the grains by the reason of the plastic deformation process. Afterwards, these timelines blur as a result of the great amount of sub-grain formation. Eventually, the reflections broaden, merging into the final texture of the material. Such sequence is common for the three alloys investigated.

The impact on the mechanical behaviour during compression at 350 °C of the addition of Gd or Nd was significant as indicated by the increase in the 0.2% proof stress and maximum stress measured for ZK40-2Gd ZK40-2Nd alloys, Figure 2. The intermetallic particles were partially broken by cleavage for the ZK40-2Gd and ZK40-2Nd during compression, indicating that and these particles hinder the stress transfer between grains at the grain boundaries.

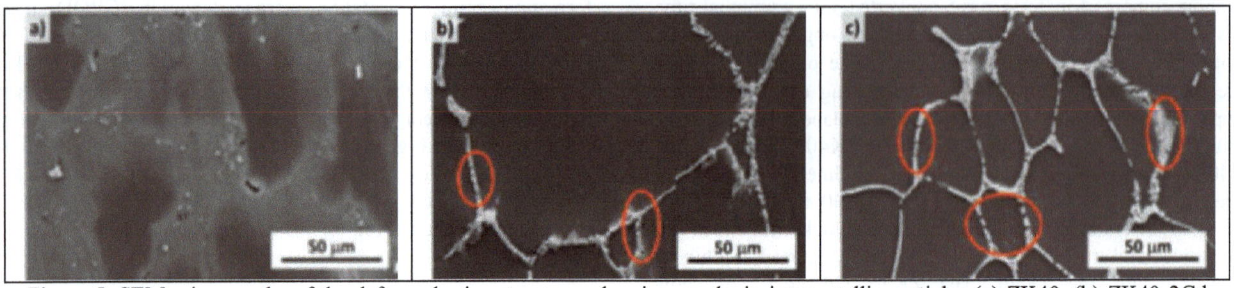

Figure 5: SEM micrographs of the deformed microstructures showing cracks in intermetallic particles (a) ZK40, (b) ZK40-2Gd and (c) ZK40-2Nd.

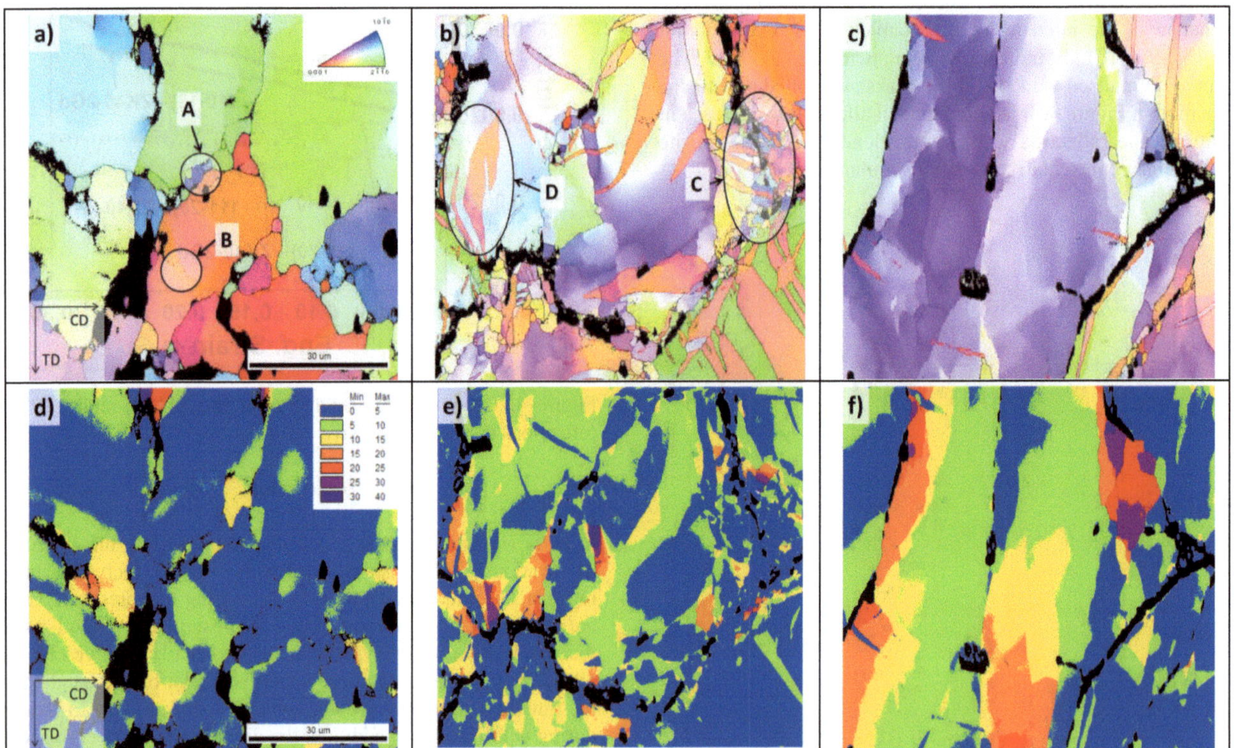

Figure 6: (a-c) EBSD inverse pole figure maps of samples after a total compressive strain of 30% at 350 °C and (d-f) Grain reorientation deviation maps for the same areas as in (a-c) (a, d) ZK40, (b, e) ZK40-2Gd and (c, f) ZK40-2Nd.

In the ZK40 alloy recrystallization is more pronounced from the early stages of deformation. It is possible that some grains had twinned because theirs orientation was energetically favourable. However, such event was not pronounced for this alloy and residual twins were not observed in the inverse pole figure map, Figure 6 (a). The highly deformed and elongated grains of the ZK40 alloy, Figure 6 (a), exhibited a large fraction of sub-grains highlighted in region B.

In the case of the ZK40 alloy no intermetallic phases hinder the grain boundary which allows grain sliding, thus lowering the energy stored at the grain boundary during deformation, which was not favourable for twinning. However, the changes of the maximum value of selected diffraction lines of the integrated line profiles during deformation show presence of some twinning. The decrease of the intensity of the $\{11\bar{2}0\}$ and increase of the intensity $\{10\bar{1}3\}$ indicates increase in the twin volume fraction. It is noted in Figure 6 (d) that the smallest GROD values was found for the ZK40, highlighting the impact of the particle pinning the grain boundary of ZK40-2Gd and ZK40-2Nd alloys. This is likely to be one of the reasons, twinning was more favourable for alloys with GD and Nd; and responsible for the higher strength observed.

AT plots of ZK40-2Gd alloy, Figure 3 (b), shows that the timelines are thicker which suggest that the CDRX was not so pronounced for this alloy. In addition, the number of new timelines that appear during the intermediate stages of deformation is much higher than for the other alloys, which explain the higher volume of recrystallized grains found along the grain boundaries. On the other hand, the ZK40-2Nd alloy shows AT plot similar to that of ZK40, Figure 3. For the ZK40-2Gd alloy the less pronounced activity of the CDRX caused a higher strength compared with the ZK40-2Nd, which exhibit a higher activity as observed in Figure 3 (b).

Conclusion

In situ synchrotron radiation diffraction was performed for as-cast specimens of ZK40, ZK40-2Gd and ZK40-2Nd at 350°C and results reveal that the deformation mechanisms differ between the ZK40 and the alloys modified with Nd or Gd addition. CDRX is observed for all alloys to some extent, evidenced by the sub-grain formation and the blurring of the timelines in the AT-plots. CDRX was more pronounced for the ZK40 alloy compared with the other alloys. The broadening of the reflections for the ZK40 alloy suggests the grains had more freedom to slide than for the modified alloys. The intermetallic phases of the modified alloys and the presence of twinning and lack of recrystallization and recovery observed in ZK40-2Nd and ZK40-2Gd alloys are responsible for the enhanced maximum strength.

Acknowledgements

The authors acknowledge the Deutsches Elektronen-Synchrotron (DESY) for the provision of facilities within the framework of proposal I20130434. RB acknowledges University of Sao Paulo for granting the fellowship *'Bolsa Empreendedorismo'* and Helmholtz-Zentrum Geesthacht (HZG) for financial support during his stay in Germany.

References

1 M. O. Pekguleryuz, K. U. Kainer, A. A. Kaya, eds. *Fundamentals of magnesium alloy metallurgy* (
Philadelphia, PA, USA: Woodhead Publishing, 2013).
2 M. M. Avedesian, H. Baker, *Magnesium and magnesium alloys* Metal Park, OH, USA: *ASM Speciality Handbook*, (ASM International, 1999).
3 M. H. Yoo, *et al.*, *''Non basal deformation modes of HCP metals and alloys: Role of dislocation source and mobility,'' Metall. Mater. Trans A*, 33 (2002), 813-822.
4 Y. Wu, *et al.*, *''Flow Behavior and microstructure of ZK60 magnesium alloy compressed at high strain rate''. Trans. Nonferrous Metals Soc China,* 24 (2014), 930-939.
5 T. E. Leontis, *"Magnesium and Magnesium Alloys - The Room and Elevated Temperature Properties of Some Sand Cast Magnesium-base Alloys Containing Zinc," Transactions of the Metallurgical Society of AIME*, 185 (1949), 327.
6 Q. Li, Q. Wang, Y. Wang, X. Zeng, W. Ding, *"Effect of Nd and Y addition on microstructure and mechanical properties of as-cast Mg–Zn–Zr alloy," J Alloys and Compd,* 427.1 (2007), 115-123.
7 M. B. Yang, *et al.*, *"Microstructure, tensile and creep properties of as-cast Mg-3.8Zn-2.2Ca-xCe (x=0, 0.5, 1 and 2 wt.%) magnesium alloys," J Rare Earths*, 30 (2012), 181-188.
8 N. Stanford *et al.*, *"Effect of microalloying with rare-earth elements on the texture of extruded magnesium-based alloys," Scripta Mater.,* 59.7 (2008), 772-775.
9 K.-D. Liss, and K. Yan., *"Thermo-mechanical processing in a synchrotron beam," Mater. Sci. Eng.: A,* 528.1 (2010), 11-27.
10 F. R. Elsayed, *et al., "Magnesium Permanent Mold Castings Optimization"* Mater. Sci. Forum 690 (2011), 65-68.
11 D. Tolnai, *et al. "Study of the solidification of AS alloys combining in situ synchrotron diffraction and differential scanning calorimetry" Mater. Sci. Forum,* 765 (2013), 286-290.
12 S .I. Wright, M. M. Nowell, and D. P. Field, *"A Review of Strain Analysis Using Electron Bakcscatter Diffrection," Microscopy and Microanalysis*, 17 (2011), 316-329.

Magnesium Technology 2015
Edited by: Michele V. Manuel, Alok Singh, Martyn Alderman, and Neale R. Neelameggham
TMS (The Minerals, Metals & Materials Society), 2015

THE DEFORMATION BEHAVIOR, MICROSTRUCTURE AND MECHANICAL PROPERTIES OF CAST AND EXTRUDED Mg-1Mn-xNd (wt%) AT TEMPERATURES BETWEEN 50°C AND 250°C

A. Chakkedath[1], J. Bohlen[2], S. Yi[2], D. Letzig[2], Z. Chen[1,3], and C.J. Boehlert[1,4]

[1]Michigan State University, East Lansing, Michigan 48824-1226, USA
[2]Magnesium Innovation Centre MagIC, Helmholtz-Zentrum Geesthacht Zentrum für Material- und Küstenforschung GmbH, Max-Planck-Str. 1, D-21502 Geesthacht, Germany
[3] University of Michigan, Ann Arbor, Michigan 48109, USA.
[4]IMDEA Materials Institute, C/Eric Kandel 2, 28906-Getafe, Madrid, Spain

Keywords: Microstructure, Texture, Tensile, Slip, Twinning, Cast, Extruded, Magnesium, Lightweight Alloys

Abstract

The tensile deformation behavior of as-cast and cast-then-extruded Mg-1Mn-1Nd(wt%) and Mg-1Mn-0.3Nd(wt%) alloys was studied by performing in-situ tests inside a SEM. A slip/twin trace analysis technique was used to identify the distribution of the deformation systems. Cast-then-extruded Mg-1Mn-1Nd(wt%) exhibited superior elevated-temperature strength retention compared to cast-then-extruded Mg-1Mn-0.3Nd(wt%). Basal slip and extension twinning were observed in the as-cast Mg-1Mn-1Nd(wt%) and Mg-1Mn-0.4Nd(wt%) alloys deformed at 50°C. In cast-then-extruded Mg-1Mn-1Nd(wt%), basal slip, prismatic slip, and pyramidal <c+a> slip were active at all temperatures. In cast-then-extruded Mg-1Mn-0.3Nd(wt%), at lower temperatures, twinning dominated the deformation and no non-basal slip activity was observed. The extent of twinning decreased with increasing temperature and basal slip was the major deformation mode at 150-250°C in both cast-then-extruded materials. The estimated CRSS ratio of extension twinning with respect to basal slip in Mg-1Mn-1Nd(wt%) was close to unity, suggesting that the addition of Nd results in an increase in the CRSS of basal slip.

Introduction

Magnesium (Mg) alloys are desirable to use in applications where light-weighting is at premium [1]. Conventional Mg alloys tend to exhibit low formability partially due to their large mechanical anisotropy [2]. This anisotropy can be reduced by alloying Mg with rare earth (RE) elements, and weaker wrought textures result [3, 4]. However, the origin of such weak textures is still not clear. The effect of RE elements, such as lanthanum (La), yttrium (Y), gadolinium (Gd), cerium (Ce), and neodymium (Nd), on the texture of wrought Mg alloys has been reported in a number of studies [3, 4]. It was found that for extruded binary Mg-RE alloys, dilute RE elements additions resulted in a texture similar to that of pure Mg (extrusion axis tilted 14° from the arc [10-10]-[11-20]) [3]. With increasing RE addition, the texture initially shifts toward a "conventional texture" (extrusion axis perpendicular to the c-axes), while the so-called "RE texture" (extrusion axis aligned with the [10-10] and the [-12-11] directions) was only observed with sufficiently high RE concentration (specifically, 0.03 at% for La, 0.06 at% for Gd, and 0.04 at% for Ce) [3]. It was found in another study that Nd is a much stronger texture modifier than Ce or Y in extruded Mg-Mn alloys [4].

Although Nd has been shown to be a stronger texture modifier compared with other RE elements, a systematic study has yet to be performed to understand the effect of Nd content on the texture. It would be expected that as the Nd content decreases, a more conventional wrought texture would be exhibited. However, it is not clear how much Nd is required for texture modification. In addition, an understanding of the effect of Nd on the deformation behavior and deformation mechanisms is lacking. It has been reported that the activity of non-basal slip at low temperatures in Mg-RE alloys is enhanced, compared with conventional Mg alloys, because the critical resolved shear stress (CRSS) of prismatic and pyramidal systems in Mg-RE alloys becomes closer to that of extension twinning and basal slip [5]. It has been hypothesized that the changes in the CRSS of the various slip systems might be attributed to the influence of RE solute elements on crystal structure features such as the c/a ratio, the stacking fault energy, and/or the width of the dislocation cores. However, no clear experimental evidence of the distribution of the deformation systems that are active under different testing conditions exists to date. Thus, there is a need to both quantify the relative activities of the different deformation modes present in Mg-RE alloys as a function of temperature and determine the CRSS values of the active slip systems.

Experimental

The Mg-1Mn-1Nd(wt%) and Mg-1Mn-0.3Nd(wt%) alloys were gravity cast to produce a billet of 125 mm diameter. The billet was then machined to a diameter of 93 mm and then homogenized at 350°C for 15 hours prior to extrusion. Indirect extrusion was carried out at 300°C and/or 275°C to produce round bars of 17 mm diameter, which corresponds to an extrusion ratio of 1:30. The extrusion rate was 5.6 mm/s which corresponded to a profile exit speed of 10 m/min. For simplicity, the Mg-1Mn-1Nd(wt%) alloy extruded at 300°C and 275°C are referred as MN11-300 and MN11-275, respectively. The Mg-1Mn-0.3Nd(wt%) alloy (referred as MN10-300) was extruded at 300°C. The as-cast Mg-1Mn-1Nd(wt%) and Mg-1Mn-0.4Nd(wt%) are referred as MN11-cast and MN10-cast, respectively. The bulk chemical composition of the as-cast materials and cast-then-extruded rods, measured using spark emission spectroscopy, is provided in Table 1.

Flat dog-bone samples with gage dimensions of 3x2x10 mm were electro-discharge machined from the as-cast billet and cast-then-extruded rod. In-situ tensile tests were performed at 50°C, 150°C, and 250°C, at a strain rate of $\sim 10^{-3} s^{-1}$, using a screw-driven tensile stage placed inside either a Tescan MIRA III FEG SEM or Zeiss EVO MA15 SEM. In the case of the extrusions, the tensile axis was parallel to the extrusion direction. One surface of the sample was mechanically polished to a mirror finish with the final step using a 0.06 μm colloidal silica solution followed by chemical

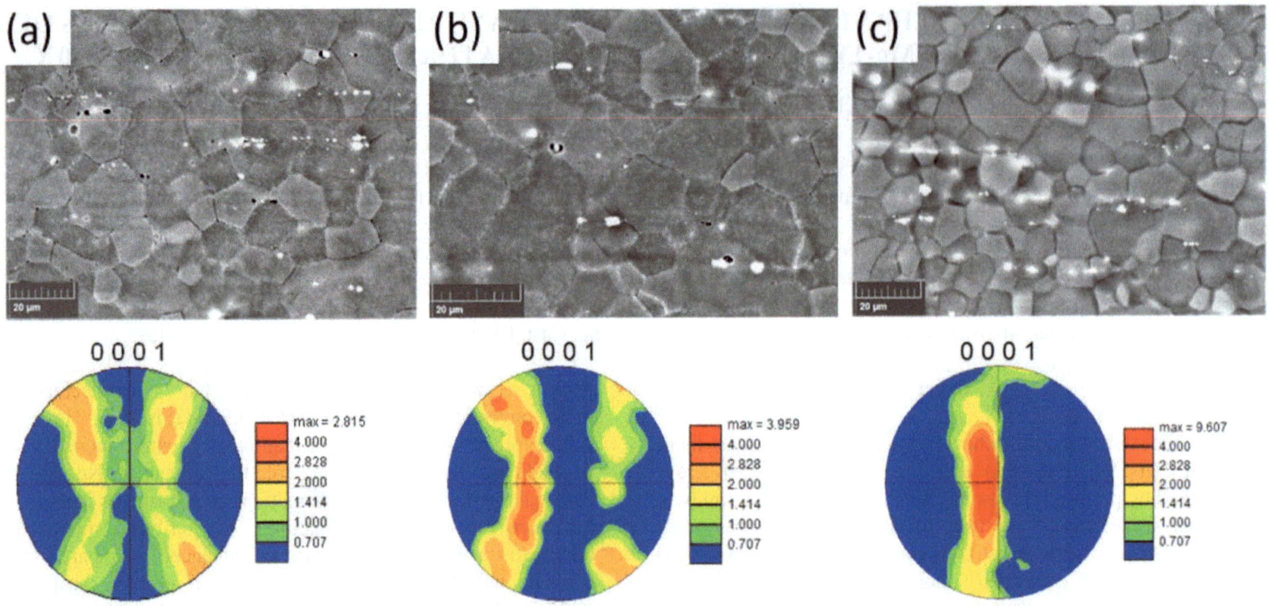

Figure 1 Secondary electron SEM images of the microstructure of (a) MN11-300 (b) MN11-275 and (c) MN10-300 and the corresponding EBSD (0001) pole figures. The extrusion direction is horizontal.

polishing (using a solution containing 100 ml methanol + 12 ml HCl + 8 ml nitric acid) and etching (using an etchant containing 60 ml ethanol + 20 ml H_2O + 15 ml acetic acid + 5 ml nitric acid). Secondary electron (SE) SEM images were taken before loading and during interruptions at different displacements throughout the tensile tests. Details of the testing apparatus and technique can be found elsewhere [6]. The microstructure in the center of the gage section was analyzed before and after each tensile test by electron backscatter diffraction (EBSD) performed using a 6500 F JEOL FEG SEM or Tescan MIRA III FEG SEM equipped with an EDAX-TSL (Mahwah, NJ, USA) EBSD system. Local strains, indicated in the stress versus displacement curves depicted in this work, were estimated by measuring the relative displacements of obvious microstructural features on SE SEM images acquired during the test. In some cases the local strains calculated were different from the global strains calculated using the displacement data recorded by the tensile stage and this was expected to be due to inhomogeneous deformation. The average grain size was determined by the TSL OIM Analysis software based on ASTM standards for grain size measurement, where only grains containing neighboring grain misorientations greater than 15° were considered.

The active slip and twinning systems were identified using the EBSD data and the SE SEM images acquired during the tensile experiments. Slip/twin trace analysis was performed by inputting the Euler angles obtained from the EBSD orientation measurements from a grain and all possible slip/twin systems in to a Matlab™ code that provided a visual representation of all the possible plane traces of each system on the sample surface as output. The Schmid factor values for the different deformation systems were calculated based on the EBSD-determined orientation and the assumption of a uniaxial tension stress state. A more detailed description of this deformation analysis technique is provided elsewhere [6].

From the 50°C experiments, the CRSS ratios of the different deformation modes in MN11-300 and MN11-275 were estimated based on the distribution of identified deformations modes and initial texture, using a methodology described in Ref. [7].

The chemical composition of the precipitates observed in the as-cast and cast-then-extruded alloys was examined using energy dispersive spectroscopy (EDS) inside JEOL JEM 3000F or JEOL JEM 2100 transmission electron microscope (TEM).

Results and Discussion

Microstructure and Texture

The cast-then-extruded MN11-300 and MN11-275 rods exhibited weak textures while the cast-then-extruded MN10-300 rod exhibited a stronger texture where the basal plane normals tended to be oriented perpendicular to the extrusion axis. The microstructure and texture (in the form of pole figures) of the MN11-300, MN11-275 and MN10-300 are shown in Figures 1a-c. A minimum of ~1000 grains were present in the microstructure from where the texture measurement was obtained. It has been reported that very small additions of RE elements can inhibit texture formation during extrusion in Mg-1Mn (wt.%) alloys [4]. However, the relatively strong texture in MN10-300 compared to the weak texture in MN11-300 and MN11-275 (See Figure 1) suggests that more than 0.3 wt.% Nd (RE) content is required to effectively weaken the wrought texture in Mg-1Mn (wt.%) alloys during extrusion. The texture in MN10-300 was similar to the texture observed in conventional alloys [6]. The average equiaxed grain diameter was between 10-15 μm in both alloys. Second-phase particles of ~1 μm diameter were observed in both alloys. EDS analysis using a TEM showed that some of these particles were enriched in Nd while others were enriched in Mn. TEM analysis also showed the presence of precipitates of ~200 nm size in the extruded microstructures of both alloys. These precipitates were found to be enriched in Mn. Similar precipitates were observed in extruded Mg-1Mn(wt%), Mg-3Al-1Zn(wt%), Mg-

1Mn-0.3Ce(wt%), and rolled Mg-1Zn-0.2Nd-0.3Zr(wt%) [8, 9]. However, the influence of these precipitates on the deformation behavior is not yet understood.

Alloy	Chemical composition
MN11-300	Mg-0.95Mn-0.94Nd (wt.%)
MN11-275	Mg-1Mn-1Nd (wt.%)
MN10-300	Mg-1.12Mn-0.26Nd (wt.%)
MN11-cast	Mg-0.89Mn-0.77Nd (wt%)
MN10-cast	Mg-0.97Mn-0.40Nd (wt%)

Table 1 Bulk chemical analysis of all of the alloys studied.

Figure 2a-b shows the SE SEM images of the microstructures of MN11-cast and MN10-cast. The grain size was ~200-1000 µm in both alloys. Stringers of second phase particles of ~200 nm size were observed within the grains in both alloys. From the EBSD orientation analysis, it appears that these stringers were aligned parallel to the basal planes of the hcp matrix grains. EDS analysis using TEM showed that these particles were enriched in Mn in both alloys. Nd-rich precipitates decorated the grain boundaries in MN11-cast. Compared to MN11-cast, the precipitates observed in MN10-cast were less in volume fraction and most of these were enriched in Mn. Particles of both Nd-rich and Mn-rich precipitates of ~1 µm were observed within grains in both alloys. The Nd-rich precipitates along grain boundaries in MN11-cast might have influenced the weak texture formation in MN11-300 and MN11-275 during extrusion. An in-depth study of the structure and morphology of the precipitates in the as-cast and extruded material is required to fully understand their influence on texture evolution. This will be the focus of a subsequent study.

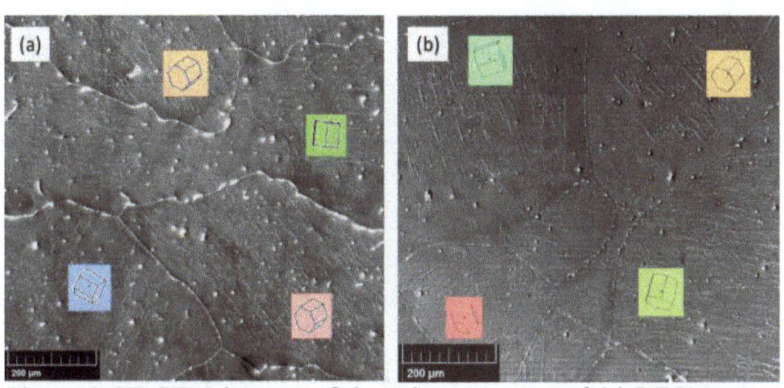

Figure 2 SE SEM images of the microstructure of (a) MN11-cast and (b) MN10-cast. The crystallographic orientation of the grains is represented by unit cells.

Tensile properties

Representative stress versus displacement curves for MN11-300, MN11-275, and MN10-300 samples tested at 50°C, 150°C and 250°C along the extrusion direction are plotted in Figure 3a-b. Some samples were not taken to failure in order to facilitate EBSD orientation mapping after the deformation without further sample preparation. At 50°C the MN10-300 yield stress (YS) was approximately 1.4 and 1.8 times that at 150°C and 250°C, respectively. This reduction in the tensile YS at elevated temperatures is similar, but less dramatic, to that reported in conventional Mg alloys [6]. For MN11, the 50°C YS was approximately 1.3 and 1.4 times that at 150°C and 250°C, respectively. Thus, the MN11-300 and MN11-275 alloys retained their strength more at elevated temperatures than MN10-300. This is expected to be a result of the greater Nd content, which is expected to result in both solid solution strengthening and precipitation strengthening. An insufficient content of Nd for effective (solid-solution and/or precipitation) strengthening is expected to be the reason behind the greater reduction in the elevated-temperature strength exhibited by MN10-300 compared with MN11-300 and MN11-275. This suggests that the amount of Nd content plays an important role in retaining tensile strength at elevated temperatures in Mg-1Mn-xNd(wt.%) alloys, and this effect is noticeable even with ~1 wt% RE addition. It is noted however that ultimate tensile strength of MN10-300 was slightly higher than that for MN11-300 and MN11-275 at 50°C. This is expected to be a result of the stronger texture exhibited by MN10-300 due to the lower Nd content.

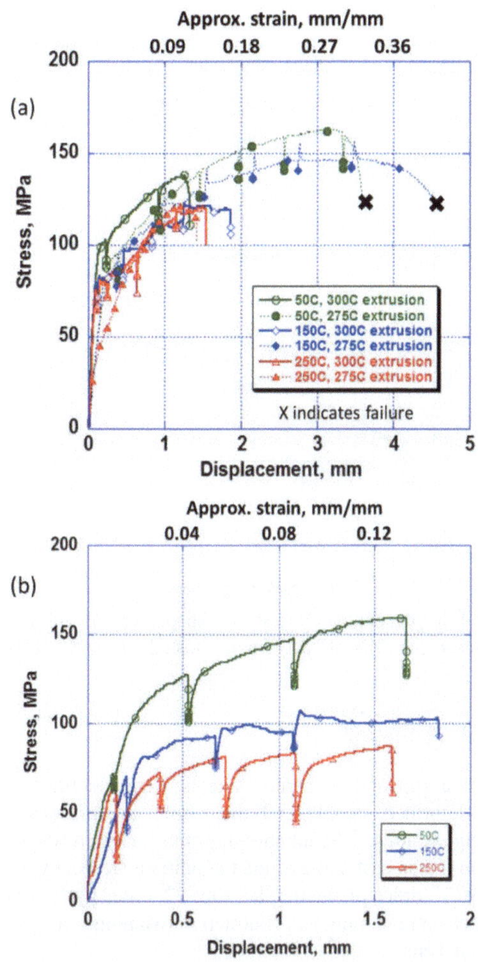

Figure 3 Representative stress vs. displacement curves for (a) MN11-300 and MN11-275, and (b) MN10-300 tensile tested at 50°C, 150°C and 250°C.

Figure 4a-b shows the representative stress versus displacement curves for MN11-cast and MN10-cast. The cast material exhibited lower tensile strength compared to the extruded counterpart. This is expected to be due to the larger grain size observed in the cast alloys. Both MN11-cast and MN10-cast exhibited similar elevated-temperature strengths. The precipitates observed in the cast alloys might have an influence in the elevated-temperature strengths, however, a more detailed study of the influence of precipitates on the deformation behavior at elevated temperatures is required.

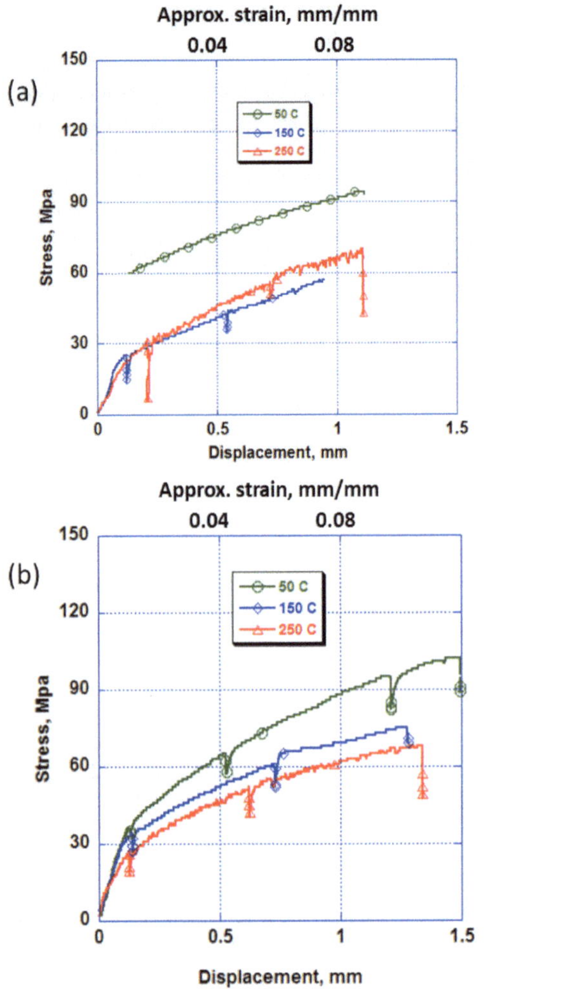

Figure 4 Representative stress vs. displacement curves for (a) MN11-cast (b) MN10-cast tensile tested at 50°C, 150°C and 250°C.

<u>Deformation behavior</u>

The active slip and twinning systems were identified for each tensile deformed sample and the results are presented in the following sections. The number of grains analyzed was ~100 and ~2000 in the as-cast and extruded conditions, respectively. Basal <a> slip, 1st order prismatic <a> slip, 2nd order pyramidal <c+a> slip, extension twinning and contraction twinning were considered for the analysis.

Figure 5a-c summarizes the Schmid factor distribution for the deformation modes observed in MN11-300 and MN11-275 at 50°C, 150°C and 250°C. Both alloys exhibited a similar deformation mode distribution at all tested temperatures. At 50°C, basal <a> slip, first order prismatic <a> slip, second order pyramidal <c+a> slip and extension twinning were observed. At 150°C and 250°C, basal <a> slip, first order prismatic <a> slip, and second order pyramidal <c+a> slip were observed. The extent of extension twinning decreased with increasing temperature. Compared to 50°C deformation, less extension twinning was observed at 150°C in both alloys. At 250°C, no twinning was observed in MN11-300, and significantly less twinning activity was observed in MN11-275. On the other hand, the amount of slip, especially basal slip, increased with increasing temperature. This suggests that the CRSS of basal slip could be higher than that of extension twinning at lower temperatures and the opposite would be true at elevated temperature. However, this suggestion is contrary to the belief that the CRSS values for both twinning and basal slip are relatively insensitive to temperature, where the CRSS for twinning is expected to be slightly higher than that for basal slip [10, 11]. This issue is discussed in the following section.

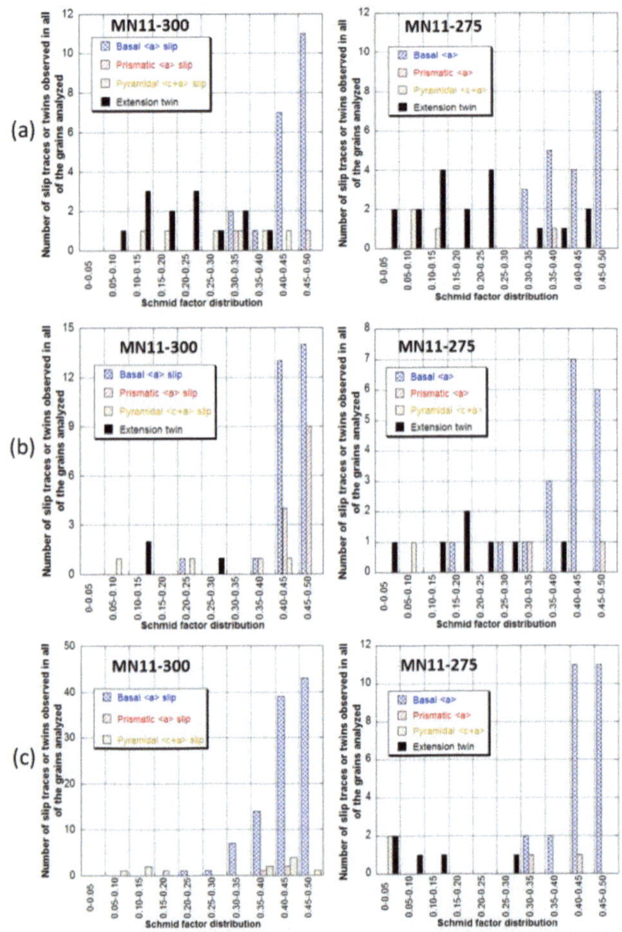

Figure 5 The Schmid factor distribution of the deformation modes observed in MN11-300 and MN11-275 at (a) 50°C, (b) 150°C, and (c) 250°C. The deformation modes in MN11-300 were quantified at ~29%, ~25%, and ~24% strain, and the deformation modes in MN11-275 were quantified at ~14%, ~17%, and ~14% strain at 50°C, 150°C, and 250°C, respectively.

Figure 6a-c summarizes the Schmid factor distribution for the deformation modes observed in MN10-300 at 50°C, 150°C and 250°C. Significantly less slip and/or twin traces were observed for MN10-300 compared to MN11-300 and MN11-275 under tension at similar strain values for all the test temperatures. At 50°C, no obvious slip traces were observed in MN10-300. This might be due to the strong basal texture observed in this alloy. The alignment of basal planes parallel to the tensile axis (extrusion axis) makes basal slip difficult. However, both extension and contraction twinning were observed. At 150°C, basal <a> slip, extension twinning and contraction twinning were observed. At 250°C, basal <a> slip, first order prismatic <a> slip

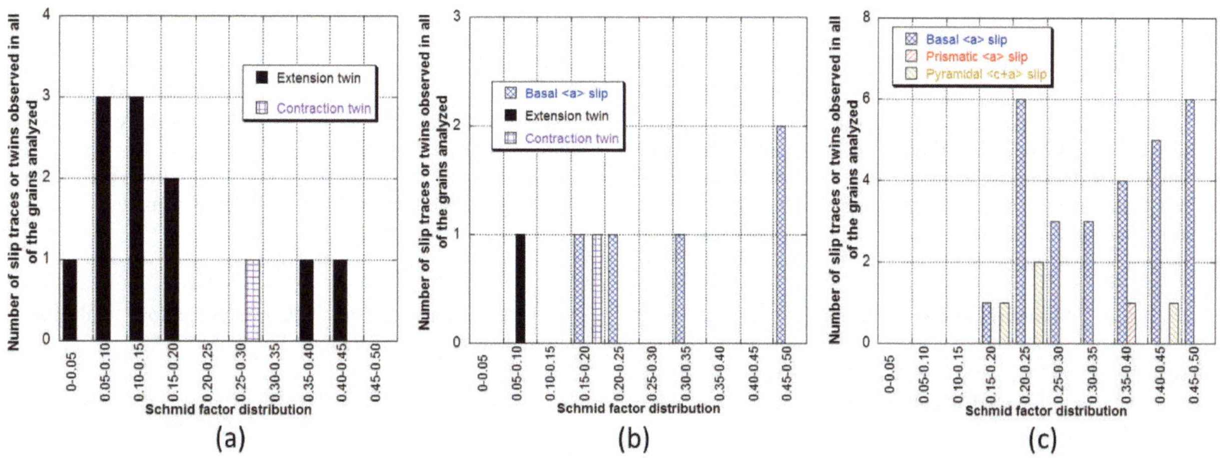

Figure 6 The Schmid factor distribution of the deformation modes observed in MN10-300 at (a) 50°C, (b) 150°C, and (c) 250°C. The slip traces were quantified at ~35%, ~40%, and ~27% strain and twin traces were quantified at ~6%, ~18%, and ~15% strain at 50°C, 150°C, and 250°C, respectively. Note that no slip traces were observed at 50°C and no twinning was observed at 250°C.

and second order pyramidal <c+a> slip were observed. No extension or contraction twinning was observed. At elevated temperatures, the deformation mode distribution of MN10-300 was similar to that for MN11-300 and MN11-275. Similar to MN11-300 and MN11-275, the twinning activity decreased and the slip activity increased with increasing temperature in MN10-300. It is interesting to note that even with a relatively strong basal texture, prismatic and pyramidal slip were observed only at 250°C in MN10 under tension, though the frequency of these observations was very low.

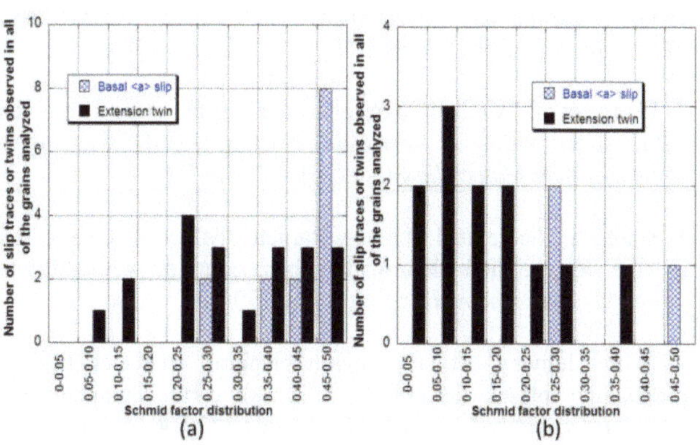

Figure 7 The Schmid factor distribution of the deformation modes observed in (a) MN11-cast at ~8% strain and (b) MN10-cast at ~10% strain at 50°C.

Figure 7 summarizes the Schmid factor distribution for the deformation modes observed in MN11-cast and MN10-cast at 50°C. Basal <a> slip and extension twinning were observed in both cases, however, extension twinning dominated the deformation. The basal slip traces observed in both alloys were parallel to the stringers of precipitates observed within the hcp grains. However, more studies are required to understand the interaction of these stringers with dislocations.

The validity of Schmid law for twinning systems in Mg has often been a subject of debate and for modeling purposes it is often assumed that Schmid law is valid for twinning [11]. However, in this study, the extension twinning observed in all the alloys studied was distributed over the entire Schmid factor range 0-0.50, especially at 50°C (See Figures 5a, 6a and 7). Hence, a strict dependence of extension twinning on orientation (Schmid factor) was not observed in this study.

Estimated CRSS ratios

The relative CRSS values of basal <a> slip, first order prismatic <a> slip, second order pyramidal <c+a> slip, and extension twinning were estimated based on the relative activity of these deformation modes in MN11-300 and MN11-275 at 50°C. The estimated CRSS ratios are listed in Table 2. The relative CRSS value for extension twinning with respect to basal <a> slip was estimated to be less than 1 at 50°C in both cases. The relative CRSS value for prismatic <a> slip and pyramidal <c+a> slip with respect to basal <a> slip was greater than 1.

Material/ Testing condition	Prismatic<a> / Basal <a>	Pyramidal<c+a> / Basal <a>	Extension twinning / Basal <a>
MN11-300	12.6 (3.1, 30.4)	12.2 (0.6, 40.0)	0.6 (0.2, 1.4)
MN11-275	11.0 (3.4, 18.7)	875.9 (29.8, 4033.5)	0.8 (0.3, 1.5)

Table 2 The mean CRSS ratios and the corresponding 90% confidence intervals (listed in parenthesis) for prismatic <a> slip, pyramidal <c+a> slip, and extension twinning in MN11-300 and MN11-275 estimated using the tensile deformation results observed at 50°C.

It is generally reported in literature that the CRSS of basal slip is lower than that of extension twinning. According to the Taylor-based model by Barnett [10], the CRSS of extension twinning is approximately 6 times than that of basal slip in AZ31 polycrystals. Other authors have reported that the CRSS of extension twinning is approximately two times that of basal slip in a Mg single crystal at temperatures close to 21°C [11]. A recent study showed that the addition of non-RE elements increases the CRSS of basal slip in Mg binary alloy single crystals [12]. However, the effect of RE elements on the relative CRSS values of the different deformation modes in Mg is currently unknown.

A recent study using in-situ TEM uniaxial nanocompression experiments on single crystal Mg-0.2Ce (wt%) reported similar CRSS values for basal slip and extension twinning [13]. The CRSS ratios estimated using our in-situ studies points to similar behavior in Nd-containing Mg alloys. The estimated CRSS ratio less than 1 for extension twinning relative to basal slip suggests that the CRSS value of extension twinning can be very close or even less than that of basal slip in RE-containing Mg alloys at temperatures close to room temperature.

Summary and Conclusions

Nominally Mg-1Mn-1Nd(wt%) and Mg-1Mn-0.3Nd(wt%) alloys in the as-cast condition and cast-then-extruded condition were tested in tension inside a SEM in the temperature range of 50-250°C. Using the SE SEM images acquired during the deformation and the EBSD orientation maps acquired before and after deformation on the same microstructural patch, the distribution of the deformation modes was acquired. A methodology was used to estimate the CRSS ratios of the different deformation modes. The following conclusions can be drawn from this study:

1, Due to reduced Nd content, the extruded MN10-300 alloy exhibited a stronger basal texture than the extruded MN11-300 and MN11-275 alloys. Thus, Nd contents greater than 0.3 wt% are needed for weakening the wrought texture commonly found in conventional Mg alloys.

2, The tensile strength decreased with increased temperature for both MN11-300, MN11-275, and MN10-300. However, with greater Nd content, MN11-300 and MN11-275 retained their strength more at elevated temperatures compared with MN10-300. Thus, the greater Nd content led to greater strength retention.

3, In MN11-300 and MN11-275, the deformation behavior at all temperatures was dominated by basal slip. The extent of extension twinning during the tensile deformation decreased with increasing temperature in MN11-300, MN11-275 and MN10-300. With the reduction in Nd content, much less slip deformation was observed during tensile deformation.

4, The activation of extension twinning did not strictly follow Schmid's law in the Nd-containing Mg-1Mn (wt%) alloys studied. However, basal slip generally followed Schmid's law at all of the test temperatures evaluated.

5, The current results suggest that the CRSS for extension twinning is close to that for basal slip at 50°C in Nd-containing Mg-1Mn (wt%) alloys and the CRSS value for basal slip may decrease with increasing temperature. The CRSS values for prismatic <a> and pyramidal <c+a> slip were always greater than those for basal slip and extension twinning.

Acknowledgements

This work was supported by the National Science Foundation Division of Material Research (Grant No. DMR1107117) and by the Spanish Ministry of Science and Innovation (grants PRI-PIBUS-2011-990 and PRI-PIBUS-2011-917) through the Materials World Network program. The authors acknowledge Professors Thomas Bieler and Martin Crimp of Michigan State University and Darren Mason of Albion College for their prior work in developing the slip trace analysis and CRSS ratio determination techniques. The authors acknowledge Drs. Maria Teresa Perez Prado and Javier Llorca of the IMDEA Materials Institute (Madrid, Spain) for useful discussions and Dr. Ivan. Gutiérrez-Urrutia of Max Planck Institute for Iron Research (Düsseldorf, Germany) for performing the sample preparation and EBSD analysis on some of the MN11 alloy samples.

References

1, Easton, Mark, et al. "Magnesium alloy applications in automotive structures."*JOM* 60.11 (2008): 57-62.

2, Ulacia, I., et al. "Mechanical behavior and microstructural evolution of a Mg AZ31 sheet at dynamic strain rates." *Acta materialia* 58.8 (2010): 2988-2998.

3, Stanford, N. "Micro-alloying Mg with Y, Ce, Gd and La for texture modification—A comparative study." *Materials Science and Engineering: A* 527.10 (2010): 2669-2677.

4, Bohlen, Jan, et al. "Effect of rare earth elements on the microstructure and texture development in magnesium–manganese alloys during extrusion."*Materials Science and Engineering: A* 527.26 (2010): 7092-7098.

5, Mishra, Raja K., et al. "Influence of cerium on the texture and ductility of magnesium extrusions." *Scripta Materialia* 59.5 (2008): 562-565.

6, Boehlert, C. J., et al. "In situ analysis of the tensile and tensile-creep deformation mechanisms in rolled AZ31." *Acta Materialia* 60.4 (2012): 1889-1904.

7, Li, H., et al. "Methodology for estimating the critical resolved shear stress ratios of α-phase Ti using EBSD-based trace analysis." *Acta Materialia* 61.20 (2013): 7555-7567.

8, Stanford, Nicole, and Matthew Barnett. "Effect of composition on the texture and deformation behaviour of wrought Mg alloys." *Scripta materialia* 58.3 (2008): 179-182.

9, Min, Junying, et al. "Spatio-temporal characteristics of propagative plastic instabilities in a rare earth containing Mg alloy." *International Journal of Plasticity* 57 (2014): 52-76.

10, Barnett, M. R. "A Taylor model based description of the proof stress of magnesium AZ31 during hot working." *Metallurgical and materials transactions A* 34.9 (2003): 1799-1806.

11, Chapuis, Adrien, and Julian H. Driver. "Temperature dependency of slip and twinning in plane strain compressed Mg single crystals." *Acta Materialia* 59.5 (2011): 1986-1994.

12, Kim, Hyung Lae, Jae Sin Park, and Young Won Chang. "Effects of lattice parameter changes on critical resolved shear stress and mechanical properties of Mg binary single crystals." *Materials Science and Engineering: A* 540 (2012): 198-206.

13, Ye, Jia, et al. "In situ TEM compression testing of Mg and Mg–0.2 wt.% Ce single crystals." *Scripta Materialia* 64.3 (2011): 292-295.

Magnesium Technology 2015
Edited by: Michele V. Manuel, Alok Singh, Martyn Alderman, and Neale R. Neelameggham
TMS (The Minerals, Metals & Materials Society), 2015

Effect of Dynamic Strain Aging on the Strain Rate Sensitivity of a Mg-2Zn-2Nd Alloy

Tong Wang, Stephen Yue, John J. Jonas
Department of Materials Engineering, McGill University, Montreal, Q.C., H3A 2B2, Canada

Keywords: Mg alloys, Rare earth addition, Dynamic strain aging, Strain rate sensitivity, Micro shear bands.

Abstract

Constant strain rate compression tests were performed on Mg-2Zn-2Nd samples over the temperature range 25 to 400 °C and strain rate range 0.001 to 0.1/s. The deformation conditions under which dynamic strain aging (DSA) takes place were determined. In this range, a peak is observed in the temperature/flow stress relationship and the strain rate sensitivity (SRS) is negative. At temperatures above the negative rate sensitivity region, abnormally high positive SRS's were detected. Interrupted compression tests were also carried out to follow the microstructural evolution. Metallographic examination indicated that shear bands form when samples are deformed under negative SRS conditions and are absent when the rate sensitivity is highly positive.

Introduction

When the coefficient of strain rate hardening is positive, the material being deformed has the ability to transfer deformation from potential high strain rate regions to parts being subjected to lower strain rates. Conversely, when it is negative, post-uniform deformation has a tendency to become unstable [1]. The strain rate sensitivity (SRS) exponent m is a measure of these tendencies. The lower its value, the less stable the flow and marked localization can be expected to take place when it is negative.

When dynamic strain aging (DSA) is taking place, the flow curves become serrated, the normally positive rate sensitivity changes sign and becomes negative, and the flow stress/temperature curves display peaks [2-4]. The occurrence of DSA has been reported in several rare earth (RE) containing Mg alloys. For example, serrated flow was detected in the WE54 alloy over the temperature range 150 to 225 °C at a testing strain rate of 5.6×10^{-4}/s [5]. Stanford et al. [6] observed that the SRS was negative when a Mg-Gd alloy was deformed over the temperature range from 200 to 250 °C. Jiang et al. [7] studied the DSA behavior of extruded Mg-Ce and Mg-Gd alloys. They reported that RE texture components are produced when these alloys are extruded under DSA conditions. A further study of Mg-Ca alloys [8] confirmed that deformation under DSA conditions leads to the formation of the RE type <11-21> texture component. In the latter work, texture weakening was attributed to the initiation of dynamic recrystallization, which was linked in turn to negative SRS and the promotion of flow localization.

In the present study, the conditions under which DSA takes place were determined in a Mg-2Zn-2Nd alloy. The domains within which the rate sensitivity is negative, normal positive and high positive were also established. Furthermore, compression tests were interrupted within the negative, normal positive, and high positive regions to assess the influence of rate sensitivity on the characteristics of the microstructure.

Experimental

An as-cast Mg-2Zn-2Nd alloy was selected for investigation. Its composition is specified in Table 1.

Table 1 Composition of the Mg-2Zn-2Nd alloy.

Zn	Nd	Ni	Fe	Cu	Si	Mg
2.04	2.04	<0.005	<0.005	<0.005	<0.005	balance

Uniaxial compression tests were conducted on a servo-controlled MTS machine at McGill University. Cylindrical compression samples with heights of 9 mm and diameters of 6 mm were machined from the as-received material. Constant strain rate compression tests were performed over the temperature range 25 to 400 °C and strain rate range 0.001 to 0.1/s to a true strain of 0.5 or until the sample fractured.

Strain rate jump compression tests were carried out at 150 and 200 °C to determine the instantaneous strain rate sensitivity (SRS) at those temperatures. The initial strain rate was 0.01/s; this was decreased to 0.001/s at a strain of 0.08, increased back to 0.01/s at a strain of 0.12, and finally held constant at this rate to a total strain of 0.16.

Interrupted compression tests were also performed for three different conditions: 300 °C - 0.01/s, 250°C - 0.01/s, and 250 °C - 0.001/s. In these cases, testing was terminated at a strain of 0.15, after which the samples were water quenched to preserve the deformation microstructure.

Electron backscatter diffraction (EBSD) analysis was employed on these samples to determine the texture changes brought about by deformation under these conditions. The EBSD samples were electropolished after grinding with 800 and 1200 grit SiC paper, polishing with 3 μm and 1 μm diamond paste, and finishing with a colloidal silica suspension.

Results and Discussion

Mechanical tests

Some true stress-true strain constant strain rate compression curves determined at various temperatures and strain rates are presented in Fig. 1. It can be seen that, when the strain rate is held constant, the flow stress generally decreases with increasing temperature. The flow stress also decreases with decreasing strain rate in the tests carried at 100, 250, 300 and 400 °C. However, this behavior is not followed at 150 and 200 °C; i.e. the 0.001/s flow stress is higher than the stresses determined at 0.01 and 0.1/s. This indicates that the strain rate sensitivity (SRS) is negative under these conditions.

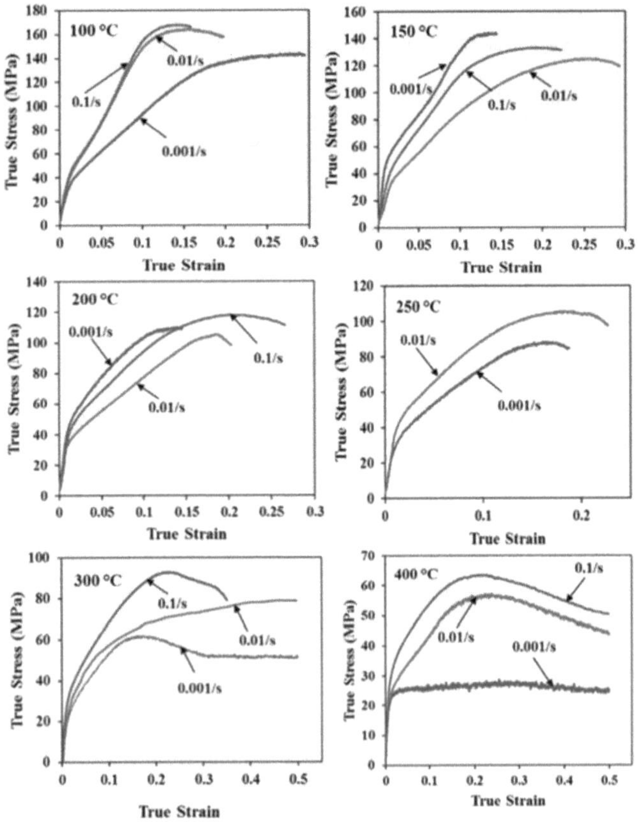

Fig. 1 True stress-true strain curves determined in compression under various conditions.

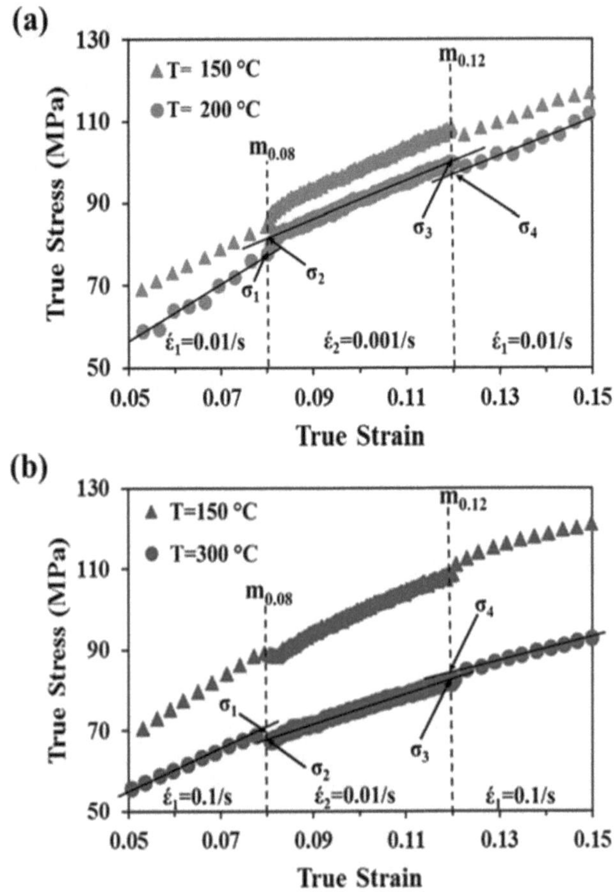

Fig. 2 Sections of true stress-true strain curves determined when the strain rate was decreased from (a) 0.01/s to 0.001/s (b) 0.1/s to 0.01/s and then increased again. Note that the 150 °C flow stress levels are higher than those that correspond to 200 and 300 °C.

Strain rate jump tests were then performed at 150, 200 and 300°C to measure the instantaneous SRS at these temperatures. Portions of the relevant flow curves are displayed in Fig. 2, where examples of the negative SRS measurements are displayed in Fig. 2 (a) and of the positive SRS results in Fig. 2 (b). The magnitudes and signs of the flow stress changes were determined using back- and forward- extrapolation, as shown in this figure and specified by equation (1) and (2). The average rate sensitivity m was then calculated using equation (3). Values of -0.017 and -0.019 were measured at 150 °C - 0.001/s and 200 °C - 0.001/s, respectively. The strain rate jump test results can be seen to support the earlier conclusions regarding the negativity of the SRS that was detected from the constant strain rate compression tests.

$$m_{0.08} = \frac{\log\sigma_2 - \log\sigma_1}{\log\dot{\varepsilon}_2 - \log\dot{\varepsilon}_1} \tag{1}$$

$$m_{0.12} = \frac{\log\sigma_4 - \log\sigma_3}{\log\dot{\varepsilon}_1 - \log\dot{\varepsilon}_2} \tag{2}$$

$$m = \frac{m_{0.08} + m_{0.12}}{2} \tag{3}$$

The flow stresses determined at 0.1 strain in the constant strain rate compression tests are plotted against temperature in Fig. 3. Peaks are observed on the 0.001, 0.01 and 0.1/s temperature/flow stress curves at around 170, 250 and 300 °C, respectively. Such flow stress peaks are generally produced when DSA is taking place. As indicated in Fig. 3, the appearance of such peaks implies that when

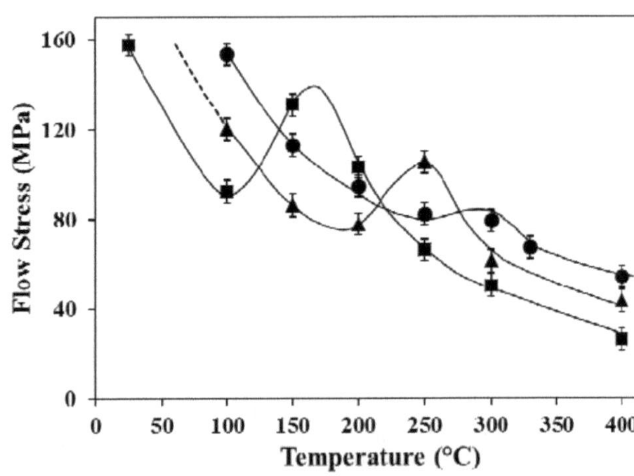

Fig. 3 Flow stress/temperature plots illustrating the presence of DSA peaks on the curves that correspond to strain rates of 0.001, 0.01 and 0.1/s.

the rate sensitivity is negative at a particular temperature, it must adopt high positive values at somewhat higher temperatures [9, 10].

In this work, 'high positive SRS' refers to m values that are higher than the 'normal' values that can be assessed from the portions of the flow stress/temperature curves that are well-removed from the peaks.

The negative SRSs and the peaks present on the flow stress/temperature curves indicate that dynamic strain aging (DSA) is taking place in this alloy under the conditions employed in these tests. The temperature and strain rate ranges over which the rate sensitivity can be expected to be negative, 'normal' positive and 'high positive' are displayed in Fig. 4. Here the experimental data employed in the construction of this figure are identified using three kinds of symbols. The data illustrated in Fig. 3 were also used in the construction of this diagram.

Deformation microstructures

In order to examine the effect of DSA on the microstructural characteristics of this alloy, interrupted compression tests were conducted in the negative, normal positive and high positive SRS regions. These conditions correspond to 250 °C - 0.01/s, 250 °C - 0.001/s and 300 °C - 0.01/s. They are identified by the labels **N**, **H** and **P**, respectively.

An example of the deformation microstructure of an **N** sample is displayed in Fig. 5. The misorientation angle distribution extracted from the EBSD data of Fig. 5 is presented here as Fig. 6. The misorientation relationships for the most frequently observed twin types in Mg alloys are listed in Table 2.

The presence of extension twins is clearly evident in sample **N**. These twins formed in the interiors of shear bands, within which the local misorientation differs from that of the matrix by about 2.5 °. These orientation differences appear to have been responsible for the resolved shear stresses on the habit planes for extension twinning being higher within the bands than in the matrix.

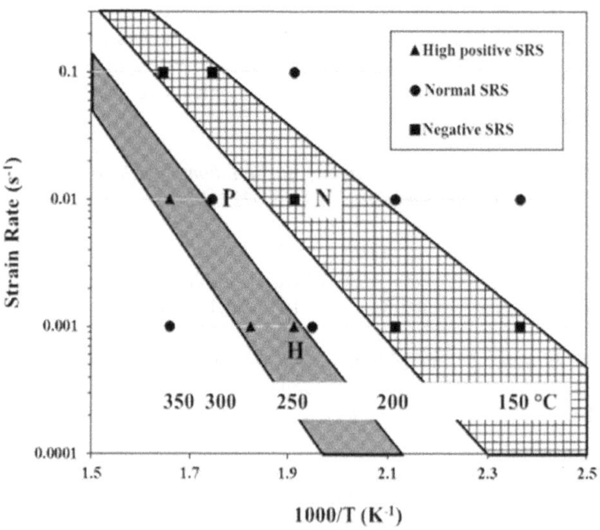

Fig. 4 Temperature and strain rate domains in which the rate sensitivity is negative (cross-hatched), normal (white) and high positive (grey).

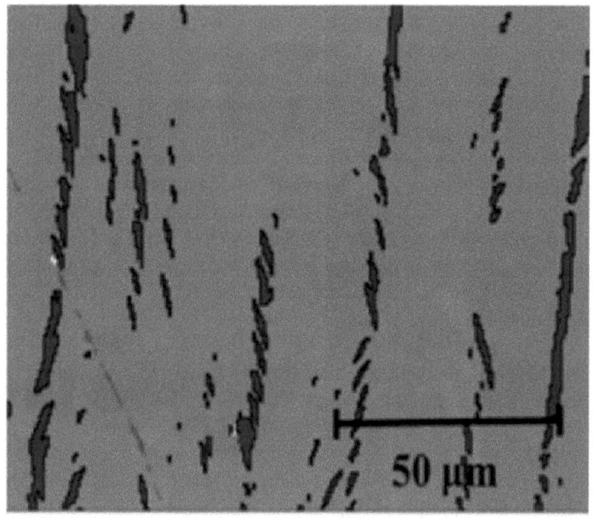

Fig. 5 Deformation microstructure of the interrupted compression sample N at 250 °C 0.01/s. Bold black lines: 86 ° <11-20> extension twins.

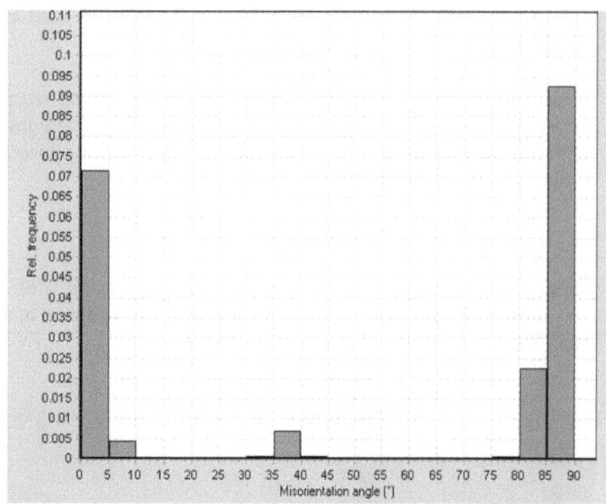

Fig. 6 Misorientation angle distribution from the EBSD data of sample N.

Table 2 Misorientation relationships for the most frequently observed twin types in Mg [13]. (ET: extension twin. CT: contraction twin.)

Type of twin	Misorientation angle/ axis
{10-12} **ET**	86 ° <1-210>
{10-11} **CT**	56 ° <1-210>
{10-11}-{10-12} **CT-ET**	38 ° <1-210>
(10-12)-(10-12) **ET-ET**	7.4 ° <1-210>
(10-12)-(01-12) **ET-ET**	60.0 ° <10-10>
(10-12)-(0-1-12) **ET-ET**	60.4 ° <8-1-70>

This interpretation is supported by the misorientation distributions of Fig. 6, in which the peaks at 85~90 ° and 0~5 ° correspond to the 86 ° <11-20 > extension twin boundaries and microshear band boundaries, respectively. The microstructure of sample **N** is displayed at higher magnification in Fig. 7. Here the extension twins can be seen more clearly and the alternating nature of the twinned and twin-free bands is more readily apparent. The widths of the bands in this figure range from 3 to 6 μm, and those of the extension twins within the bands from 0.5 to 2 μm. The formation of these alternating structures is considered here to have resulted from the flow localization brought about by the negative SRS attributable to DSA.

Some typical microstructures of samples **P** and **H** are displayed in Fig. 8. These were deformed at 300 ° 0.01/s and 250 °C 0.001/s, respectively. It can be seen that sample **P** is characterized by the presence of numerous extension twins and the complete absence of shear bands (Fig. 8a). Finally, there are few twins and no shear bands at all in sample **H** (Fig. 8b).

Conclusions

1. Dynamic strain aging takes place under well-defined temperature and strain rate conditions in the present Mg-2Zn-2Nd alloy.

2. Dynamic strain aging produces regions of negative as well as high positive strain rate sensitivity.

3. Microshear bands containing extension twins are formed when the sample is deformed under negative rate sensitivity conditions. The latter are thus responsible for the flow localization that appears in the form of microshear bands.

Acknowledgements

Tong Wang is grateful to the Chinese Scholarship Council (CSC) for scholarship support and to McGill University for the award of a McGill Engineering Doctoral Award (MEDA).

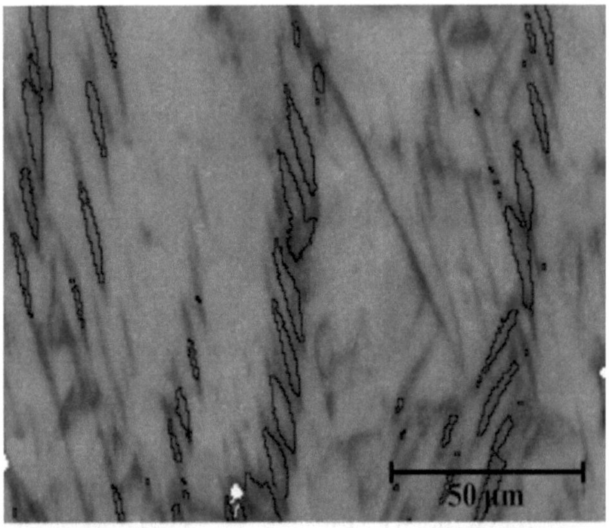

Fig. 7 Band contrast maps for N-250°C 0.01/s. Bold black lines identify the boundaries of 86 ° <11-20> extension twins.

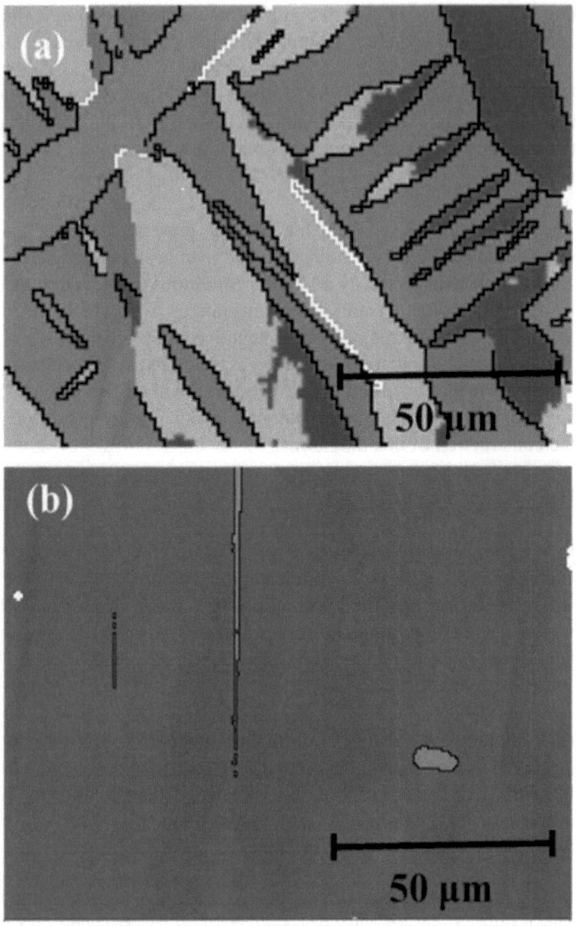

Fig. 8 Deformation microstructures of (a) sample **P** deformed at 300 °C 0.01/s and (b) sample **H** deformed at 250 °C 0.001/s. The bold lines shown here are extension twin boundaries.

References

1. Semiatin, S.L. and J.J. Jonas, *Formability & Workability of Metals: Plastic Instability and Flow Localization.* ASM Series in Metal Processing, ed. H.L. Gegel. 1984, Ohio: American Society for Metals.

2. Baird, J. *Dynamic strain aging.* in *The Inhomogeneity of Plastic Deformation, ASM Seminar. 1973, 191-222.* 1973.

3. Vandenbeukel, A., *THEORY OF EFFECT OF DYNAMIC STRAIN AGING ON MECHANICAL-PROPERTIES.* Physica Status Solidi a-Applied Research, 1975. **30**(1): p. 197-206.

4. Robinson, J.M. and M.P. Shaw, *Microstructural and mechanical influences on dynamic strain aging phenomena.* International Materials Reviews, 1994. **39**(3): p. 113-122.

5. Zhu, S.M. and J.F. Nie, *Serrated flow and tensile properties of a Mg–Y–Nd alloy.* Scripta Materialia, 2004. **50**(1): p. 51-55.

6. Stanford, N., Sabirov, I., Sha, G., Fontaine, A. La, Ringer, S.P., and Barnett, M. R., *Effect of Al and Gd Solutes on the Strain Rate Sensitivity of Magnesium Alloys.* Metallurgical and Materials Transactions A, 2010. **41**(3): p. 734-743.

7. Jiang, L., J.J. Jonas, and R. Mishra, *Effect of dynamic strain aging on the appearance of the rare earth texture component in magnesium alloys.* Materials Science and Engineering: A, 2011. **528**(21): p. 6596-6605.

8. Wang, T., Jiang, L., Mishra, R. J., and Jonas, J. J., *Effect of Ca Addition on the Intensity of the Rare Earth Texture Component in Extruded Magnesium Alloys.* Metallurgical and Materials Transactions A, 2014. **45**(10): p. 4698-4709.

9. Jonas, J.J., *Effects of shear band formation on texture development in warm-rolled IF steels.* Journal of Materials Processing Technology, 2001. **117**(3): p. 293-299.

10. Barnett, M.R. and J.J. Jonas, *Influence of Ferrite Rolling Temperature on Microstructure and Texture in Deformed Low C and IF Steels.* ISIJ International, 1997. **37**(7): p. 697-705.

11. Nave, M.D. and M.R. Barnett, *Microstructures and textures of pure magnesium deformed in plane-strain compression.* Scripta Materialia, 2004. **51**(9): p. 881-885.

12. Jiang, L., Jonas, J. J., Luo, A. A, Sachdev, A. K, and Godet, S., *Influence of {10-12} extension twinning on the flow behavior of AZ31 Mg alloy.* Materials Science and Engineering: A, 2007. **445–446**(0): p. 302-309.

13. Mu, S., J.J. Jonas, and G. Gottstein, *Variant selection of primary, secondary and tertiary twins in a deformed Mg alloy.* Acta Materialia, 2012. **60**(5): p. 2043-2053.

Magnesium Technology 2015
Edited by: Michele V. Manuel, Alok Singh, Martyn Alderman, and Neale R. Neelameggham
TMS (The Minerals, Metals & Materials Society), 2015

The Deformation Gradient of Interfacial Defects on Twin-like Interfaces

Christopher Barrett[1,2], Haitham El Kadiri[1,2]

[1]Department of Mechanical Engineering, Carpenter Building, Mississippi State University, Mississippi State, MS, 39762, USA
[2]Center for Advanced Vehicular Systems, 200 Research Blvd., Mississippi State University, Mississippi State, MS, 39762, USA

Keywords: Deformation gradient, Grain boundary, Interfacial defects

Abstract

Twinning dislocations (disconnections) are examples of a unique type of defect which migrate an interface as a result of glissile propagation. These defects produce a well-defined plastic deformation due to their motion occurring exclusively in the interface. In most cases, their deformation gradient is more complex than the simple shear of slip and twin systems. Twins and relatable asymmetric tilt boundaries have deformation gradients which are associated by rotation. For example, the $\{10\bar{1}2\}$ twin interface and the basalprismatic interface have misorientations which differ by only four degrees and are joined on a faceted boundary by an interfacial disclination line. The deformation due to glissile disconnections on the basal–prismatic boundary may be multiplicatively decomposed into the simple shear deformation of the $\{10\bar{1}2\}$ twin and a four degree rotation. A comprehensive understanding of the deformation on faceted and asymmetric boundaries enables more accurate analysis and modeling of twinning.

Introduction

The effects of grain boundaries at the mesoscale are at present poorly quantified. Grain boundaries are known as dislocation sources and sinks, as well as contributing directly to plasticity via sliding and migration. However, these effects are at present only roughly correlated to the grain boundary misorientation, with little or no consideration of the core structure and geometry at the interface.

The core structure of the grain boundary interface plays a crucial role in the quantitative description of grain boundary behavior. For example, the hexagonal close-packed twin, $\{10\bar{1}2\}$, can rapidly consume entire grains because the $\{10\bar{1}2\}$ twin disconnection has a wide core and low shear, and moves on a relatively low-energy twin boundary (Serra and Bacon, 1996). Moreover, the twin boundary is able to facet onto asymmetric tilt interfaces and absorb lattice dislocations all without impeding its mobility (El Kadiri et al., 2014). This extreme case of grain boundary migration potential is understood based on the core structure and energy of the boundary and its constituent defects. Without understanding of the core properties of this boundary, there would be no way to identify its potential. An understanding based only on misorientation fails to capture the necessary data to predict boundary behavior at higher scales.

On the other hand, raw geometric and energetic information about the interface, provided from molecular dynamics and quantum level studies of grain boundaries, is alone insufficient to predict higher scale behavior. For example, the $\{10\bar{1}1\}$ twin boundary in hexagonal close-packed (HCP) metals has a much lower energy than the $\{10\bar{1}2\}$ twin, yet it is far less mobile than the $\{10\bar{1}2\}$ twin. The means to analyze raw atomistic data and extract higher scale properties is crucial to bridging the length scales and formulating accurate mesoscale models.

The tools needed to effectively analyze this information are being developed. Recently, Stukowski et al. (2012) developed an algorithm to automatically extract dislocation lines from atomistic data, with their Burgers vectors. This algorithm is also capable of identifying grain boundary disconnections (Barrett and El Kadiri, 2014a). By using this information in conjunction with interfacial defect theory the analytical disconnection Burgers vector and step height can be obtained. Grain boundary energy and disconnection core width are easily extracted from atomistic data. These provide quantitative information on favorability and mobility of interfaces. Wang and Beyerlein (2012) and Wang et al. (2012) have characterized tilt boundary energies in HCP materials along the $\langle\bar{1}2\bar{1}0\rangle$ and $\langle10\bar{1}0\rangle$ axes, as well as describing their core structure in terms of intrinsic dislocation content on reference twin boundaries. Khater et al. (2012) have likewise studied $\langle0001\rangle$ axis boundaries.

Much work still remains, however. Of primary importance is the question, how do properties observed at the atomistic scale determine higher scale effects? In this work, we will address this question specifically for grain boundary mobility. At the atomistic scale, grain boundary motion is described by the action of disconnections, which are defined by a Burgers vector and step height. However, at the mesoscale, the deformation gradient is required to characterize the evolution of the grains due to grain boundary migration. A bridge from one method of description to the other is needed.

The problem is exacerbated considering that grain boundaries are not flat. For a given misorientation relationship between two grains, there are an infinite variety of theoretical boundaries with various interface angles. To completely enclose one grain in another, at least four unique interfaces with the same misorientation are needed. Pile-up of disconnection lines sometimes provides grain boundaries with the flexibility to adjust their interface angle. However, in many cases, a completely unique interface with a similar misorientation, but a different core structure and energy is instead joined along a disclination line to the original interface. A boundary which is formed by several flat segments at different angles is termed faceted, with each segment being a facet (Pond and Hirth, 1994; Barrett and El Kadiri, 2014b; Ostapovets and Gröger, 2014).

Faceting of tilt boundaries fundamentally impacts their mobility. A tilt boundary that is able to conservatively facet may adopt a wide variety of average interface angles and curve around defects which would normally restrict its motion (Barrett and El Kadiri, 2014b). Faceting may even allow the boundary to convert incoming lattice dislocations into multiple disconnections which encourage mobility (Serra and Bacon, 1996; Serra et al., 1999; Pond et al., 1999; Wang et al., 2013; El Kadiri et al., 2014). Adjoining facets generally do not share exactly the same misorientation and thus are separated by a disclination line. During interface migration due to disconnection motion, the disclination and the disconnection will interact, as described by Barrett and El Kadiri (2014c) and Barrett and El Kadiri (2014a). This is depicted in Figure 2.

Characterization of the deformation due to grain boundary migration on a faceted boundary is more complex than a single interface because a different set of disconnections exists on each unique facet. Under a special set of circumstances, the disconnections on each facet interact favorably with the disclination junctions to enable conservative mobility of the entire boundary with disconnections transforming from one facet onto another. Herein, we will examine the deformation gradient in such a process.

Deformation due to disconnection glide

The deformation gradient due to disconnection glide on a flat interface may be derived from expressions for the engineering strain matrix in Howe et al. (2009) and Hirth and Pond (2011). It is given by:

$$\mathbf{F} = \mathbf{I} + \frac{\mathbf{b} \otimes \mathbf{i}}{h} \tag{1}$$

Here $\mathbf{F}$ is the deformation gradient, $\mathbf{I}$ is the identity matrix, $\mathbf{b}$ is the Burgers vector, $\mathbf{i}$ is the unit normal to the interface, and h is the step height. The Burgers vector of a disconnection may be written as:

$$\mathbf{b} = \mathbf{n}^{\lambda} - \mathbf{n}^{\mu} \tag{2}$$

where $\mathbf{n}^{\lambda}$ and $\mathbf{n}^{\mu}$ are lattice vectors in the final and initial grain respectively. In this language, the deformation is given by:

$$\mathbf{F} = \mathbf{I} + \frac{(\mathbf{n}^{\lambda} - \mathbf{n}^{\mu}) \otimes \mathbf{i}}{\mathbf{n}^{\mu} \cdot \mathbf{i}} \tag{3}$$

The deformation gradient for a disconnection is similar to that of a dislocation, with the step height being used rather than the interplanar spacing, and relaxation of the condition that the Burgers vector must lie in the interface. In fact, there are many examples of disconnections with Burgers vectors lying outside the plane of glide.

Such disconnections accomplish a dilational deformation in addition to the simple shear component. It may be demonstrated that this deformation may be decomposed analytically into a simple shear followed by a dilation given as:

$$\mathbf{F_{shear}} = \mathbf{I} + \frac{\mathbf{b_s} \otimes \mathbf{i}}{h}, \quad \mathbf{F_{dilation}} = \mathbf{I} + \frac{\mathbf{b_d} \otimes \mathbf{i}}{h} \tag{4}$$

where $\mathbf{b} = \mathbf{b_s} + \mathbf{b_d}$, $\mathbf{b_s} \cdot \mathbf{i} = 0$, and $\mathbf{b_s} \times \mathbf{i} = 0$. This decomposition is illustrated in Figure 1.

Whether or not a disconnection is glissile is given by the following equation:

$$|\mathbf{F}| = \frac{V_{new}}{V_{old}} \tag{5}$$

The determinant of the deformation gradient indicates the volumetric change due to the passage of a disconnection. If the volume ratio of the unit cells of the two grains matches this determinant, then the disconnection is glissile.

Effect of coherency strain

Some interfaces contain coherency strains which may be relieved by occasional dislocations. When these interfaces

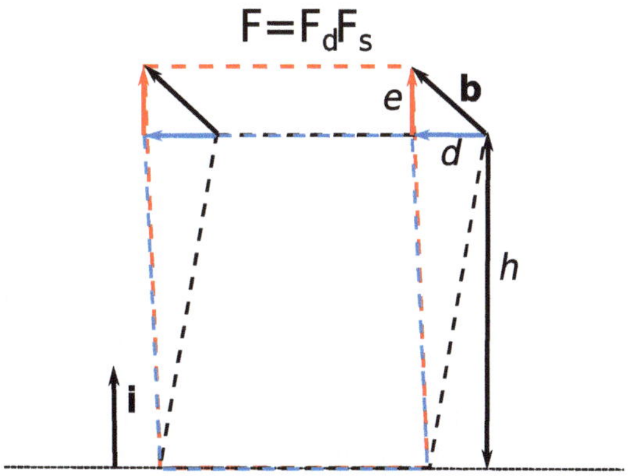

Figure 1: Schematic depicting the deformation due to a disconnection moving along the interface. The disconnection may be separated into a component lying in the interface, and a component normal to the interface. The component lying in the interface produces a simple shear deformation with a characteristic shear $s = d/h$. The component normal to the interface produces a dilational deformation with an engineering strain $\epsilon = e/h$. Thus, the general disconnection produces a deformation which a product of these two deformations.

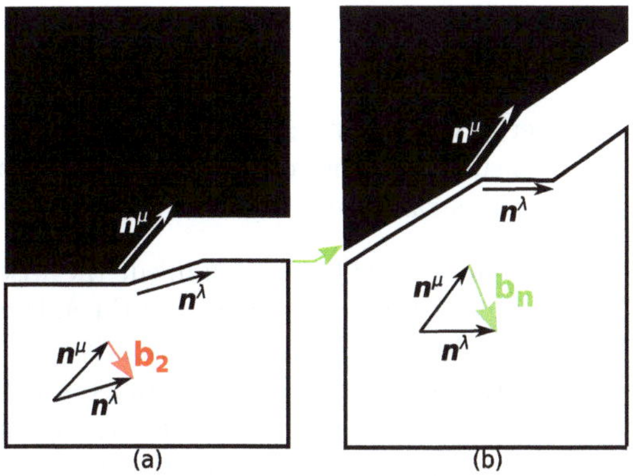

Figure 2: Transformation of a mobile disconnection from one facet to another one. For this to take place, $\mathbf{b_2}$ must be mobile on the initial facet, and $\mathbf{b_n}$ must be mobile on the new facet. The Burgers vector is transformed by the rotation of the two lattice vectors it is derived from by the disclination at the facet junction.

form as facets, they may be strained coherent. For a disconnection to move through such an interface, it must not only accomplish its own intrinsic deformation, but also migrate the coherency strains to remain in the interface (Howe et al., 2009). The deformation due to coherency strain may be described as follows:

$$\mathbf{F_c} = \mathbf{I} + \lambda_1 \mathbf{v_1} \otimes \mathbf{v_1} + \lambda_2 \mathbf{v_2} \otimes \mathbf{v_2} \tag{6}$$

The coherency strain lies in a flat interface, thus it must be two dimensional. Here λ_1 and λ_2 are the principle strains in the interface, and $\mathbf{v_1}$ and $\mathbf{v_2}$ are the corresponding strain vectors. Since the strain tensor is symmetric, $\mathbf{v_1}$ and $\mathbf{v_2}$ are orthogonal.

Coherency strains deform both grains near a grain boundary in opposite directions, thus the whole deformation due to migration of a coherent grain boundary is usually expressed as:

$$\mathbf{F} = \mathbf{F_{c2}F_dF_{c1}}, \quad \mathbf{F_{c2}F_{c1}} = \mathbf{F_c} \tag{7}$$

so that first a coherency strain is applied, then the disconnection deformation, then finally the second coherency strain. The partitioning of strains is generally assumed to be equal, but depends on the isotropy of the metal considered (Hirth et al., 2012).

The coherency strain affects the determinant of the deformation gradient. If there is no volumetric distortion from one grain to the other such that the volume ratio is unity, and there is no coherency strain, then the disconnection Burgers vector must lie in the interface to be glissile. However, if there is coherency strain, the Burgers vector must not lie in the interface, but instead have a normal component which induces a volumetric change opposite to that caused by the movement of the coherency strain.

Relating deformations on interfaces joined by a disclination line

The processes of facet formation, facet elimination, and disconnection transformation have been discussed briefly in Barrett and El Kadiri (2014c) and in detail in Barrett and El Kadiri (2014a). The effect of the disclination on the disconnections is also discussed there in more detail. Under the action of a disclination, the vector $\mathbf{n}^\lambda$ becomes $\mathbf{Rn}^\lambda$ while $\mathbf{n}^\mu$ is unchanged, as depicted in Figure 2. Thus the deformation gradient on one side of a facet is given by Equation (3) while on the other side it is given by:

$$\mathbf{F_f} = \mathbf{I} + \frac{(\mathbf{Rn}^\lambda - \mathbf{n}^\mu) \otimes \mathbf{f}}{\mathbf{n}^\mu \cdot \mathbf{f}} \tag{8}$$

Here $\mathbf{f}$ is the unit normal to the new facet plane.

We have determined that in many cases the disconnection associated with this new deformation gradient is actually glissile on the new facet. The following conditions are sufficient to guarantee conservative mobility of the disconnection on the new facet:

1. The facets are both tilt boundaries with the same axis.
2. The coherency strain which would be produced if one facet nucleated the other one is equal to the amount of coherency strain which the facet actually has.

Iff the first of these conditions is fulfilled, the disclination will accomplish a rotation about the tilt axis which is equal to the difference in misorientation between the two facets. If the second of these conditions is fulfilled, the change in the coherency strain and the change in the Burgers vector component normal to the interface will exactly offset one another.

Deformation gradient fields along a faceted boundary

In the isotropic approximation, the linear rotational distortion outside of a disclination core is distributed uniformly about its axis. This distortion is superimposed upon the coherency strain on one or both sides of the disclination.

The deformation from some initial boundary state to a final boundary state may then be considered by first expressing the deformation from some imaginary reference, in which the rotational distortion and coherency strain are removed, to the initial state. Secondly express the deformation from the same reference to the final state, and thirdly, combine the two to get the actual deformation gradient field from the initial real state to the final real state.

Summary

In summary, disconnections have a well defined deformation gradient which may be used to characterize grain boundary mobility at higher scales. A couple of considerations in particular have a dramatic effect on mobility. The first is the question of whether a particular interface has glissile disconnections. This question may be resolved directly by examining the determinant of the deformation gradient due to the disconnections motion, including the movement of any coherency strains which it drags along.

Secondly, the question of whether any favorable faceting relationships exist that a particular interface may use is very important for mobility. At present, this question is considered only for tilt boundaries, but nevertheless it remains critical because all twins are in this category and are predominant factors in grain boundary mobility. This question cannot be resolved without resorting to interfacial defect theory, and in some cases atomistic modeling or transmission electron microscopy. Once candidate facets are known from these methods, the possibility of glissile motion along these facets may be resolved using the deformation gradient method demonstrated above.

An adequate knowledge of the action of glissile disconnections and faceting for a given boundary between two grains of known orientation enables a quantitative description of the plastic deformation of material due to grain boundary migration, and the evolution of the grain boundary shape. It also enables more accurate prediction of the response of the grain boundary to applied stress or an increase in matrix dislocation densities.

References

Serra, A., Bacon, D.J.. Philosophical Magazine A 1996;73(2):333–343.

El Kadiri, H., Barrett, C., Wang, J., Tome, C.. Acta Materialia 2014;In review.

Stukowski, A., Bulatov, V.V., Arsenlis, A.. Modelling and Simulation in Materials Science and Engineering 2012;20(8):085007.

Barrett, C.D., El Kadiri, H.. Acta Materialia 2014a;70:137–161.

Wang, J., Beyerlein, I.. Modelling and Simulation in Materials Science and Engineering 2012;20(2):024002.

Wang, J., Beyerlein, I., Hirth, J.. Modelling and Simulation in Materials Science and Engineering 2012;20(2):024001.

Khater, H., Serra, A., Pond, R., Hirth, J.. Acta Materialia 2012;60(5):2007–2020.

Pond, R., Hirth, J.. Solid State Physics 1994;47:287–365.

Barrett, C.D., El Kadiri, H.. Acta Materialia 2014b;63(0):1 – 15.

Ostapovets, A., Gröger, R.. Modelling and Simulation in Materials Science and Engineering 2014;22(2):025015.

Serra, A., Bacon, D., Pond, R.. Acta materialia 1999;47(5):1425–1439.

Pond, R., Serra, A., Bacon, D.. Acta materialia 1999;47(5):1441–1453.

Wang, J., Liu, L., Tomé, C., Mao, S., Gong, S.. Materials Research Letters 2013;1(2):81–88.

Barrett, C.D., El Kadiri, H.. Scripta Materialia 2014c;84:15–18.

Howe, J., Pond, R., Hirth, J.. Progress in Materials Science 2009;54(6):792–838.

Hirth, J., Pond, R.. Progress in Materials Science 2011;56(6):586–636.

Hirth, J., Pond, R., Hoagland, R., Liu, X.Y., Wang, J.. Progress in Materials Science 2012;.

Magnesium Technology 2015
Edited by: Michele V. Manuel, Alok Singh, Martyn Alderman, and Neale R. Neelameggham
TMS (The Minerals, Metals & Materials Society), 2015

GEOMETRICALLY NECESSARY TWINS IN BENDING OF AN AZ31 SHEET

Bin Li[1], Z. McClelland[2], S.J. Horstemeyer[2]
[1]Department of Chemical and Materials Engineering, University of Nevada, Reno, NV 89557, USA
[2]Center for Advanced Vehicular Systems, Mississippi State University, Starkville, MS 39795, USA

Keywords: Magnesium; Twinning.

Abstract

Discrete arcs of twins were observed in an AZ31 rolled sheet Mg alloy after three point bending at room temperature. The arcs geometrically followed the contour of the bent specimens and were present in the tension zone where the stress state strongly disfavors the extension twinning. Electron backscatter diffraction analysis indicated that the twins in the arcs are $\{10\bar{1}2\}<10\bar{1}\bar{1}>$ extension twins, instead of $\{10\bar{1}1\}<10\bar{1}\bar{2}>$ contraction twins. Calculations showed that the twinned grains in the arcs have very low values of Schmid factor, even close to zero. The main tensile stress component in the tension zone was nearly perpendicular to the *c*-axis of the parent grains. Thus $\{10\bar{1}2\}<10\bar{1}\bar{1}>$ twinning in the tension zone was activated in a most unlikely scenario. The mechanism for such unusual twinning behavior was analyzed from the perspective of strain components that are generated by the $\{10\bar{1}2\}<10\bar{1}\bar{1}>$ twinning. It was demonstrated that the $\{10\bar{1}2\}<10\bar{1}\bar{1}>$ twins in the arcs in the tension zone of the bent specimen can be treated as "geometrically necessary twins", similar to geometrically necessary dislocations and boundaries.

Introduction

In dislocation-controlled plastic deformation, an important class of dislocations, i.e. geometrically necessary dislocations (GNDs) was first proposed by Nye [1]. GNDs are generated when a strain gradient is present across a deformed material. For instance, in bending tests of crystalline metals, the slip planes for dislocations are curved by the deformation. Thus the curvature of the slip plane gives rise to a slip gradient on successive slip planes that can be compensated by GNDs [2]. Additionally, GNDs are distinguished from statistically stored dislocations (SSDs) in the sense that GNDs are associated with organization of dislocations into low energy configurations such as dislocation cells or walls such that locally non-uniform strain field is developed [3]. Closely related to GNDs, geometrically necessary boundaries (GNBs) were suggested to account for the flow stress anisotropy in deformed metals with medium to high stacking fault energies [4]. These GNBs include dense dislocation walls (DDWs) and microbands (MBs) which are necessitated by the accommodation of the lattice misorientation between the dislocation cells [5].

In light of the concept of GND in dislocation dominated plastic deformations, Sevillano [6] first proposed the concept of "geometrically necessary twins" (GNTs). A GNT was treated as a "pseudo-dislocation line" at the twin edge with an effective Burgers vector in the direction of the twinning shear. The resultant hardening effect and strain accommodation by GNTs was also discussed [6]. Despite this proposition of GNTs, very little experimental evidence of GNTs has been reported since. In this work, we present arcs of twins in an AZ31 Mg alloy after three-point bending, which have the characteristic of GNTs.

Experiments

A twin-roll cast sheet of AZ31 Mg alloy [7] with thickness of 1.0 mm was cut into strips with dimensions $76\times7\times1$ mm^3. Three point bending was performed using an Instron 5882 testing system and an in-house three point bend fixture. The span distance is 28 mm. The specimens were bent at a rate of 10 mm/min. The bend was parallel to the rolling direction (RD). One of the specimens was immediately cold mounted and the microstructure of the cross-section through the thickness was studied using optical microscopy. The specimen was mechanically polished and then etched using an acetic picral solution containing 3 gram picric acid, 100 ml ethanol, 10 ml distilled water, and 5 ml acetic acid. The specimen was then examined using an optical microscope (OM). To better reveal the deformation twins, differential interference contrast (DIC) illumination was used in imaging. DIC improves the contrast of the twins [8]. After optical microscopy the specimen was then re-polished mechanically followed by electrochemical polishing with an electrolyte composed of 160 mg sodium thiocyanate, 800 ml ethanol, 80 ml ethylene glycol monobutyl ether, and 20 ml distilled water. Electron backscatter diffraction (EBSD) was then performed on the through-thickness cross-section using a field emission gun scanning electron microscope (SEM) equipped with an EBSD detection system. When the scans were completed the data was rotated at 90 degrees about the transverse direction using the data analysis software so that the data were aligned with the normal direction.

Results

Figure 1 shows optical microscopy of the microstructure of the specimen after three point bending. The image was taken on the through-thickness cross-section area. With the DIC illumination, two zones can be clearly distinguished. In the compression zone, very high density deformation twins can be observed. Away from the surface of the compression zone, twins are localized in bands. Such localized twin bands were first reported by Baird et al. [8]. The twin bands are aligned in two directions that are nearly perpendicular to each other. EBSD analysis showed that the twins are $\{10\bar{1}2\}<10\bar{1}\bar{1}>$ extension twins. At relatively low bending angles, separate twin bands formed and the size of the twin bands is much larger than the grain size. But between the twin bands, very few twins were observed. As the bending angle increases, the twin bands coalesce and the region close to the surface of the compression zone is almost totally twinned. In stark contrast, at this magnification, no twins can be observed in the tension zone. But as shown in the following, a special pattern of deformation twins was observed in the tension zone at higher magnifications.

With the DIC illumination, the twin bands can be clearly resolved as shown in Figure 1. However, after removing the DIC illumination, we observed another pattern of twins in the region at the vicinity of the tips of the twin bands, as shown in Figure 2a, the twin bands present a different contrast than that in Figure 1. As indicated by the arrows, the tips of the twin bands can still be seen. An arc of twins, which have a different contrast than the twin bands, can be observed between the red lines. Following the tips of the twin bands, lenticular twins are localized in an arc separating the twin bands from the region that is not twinned yet. The geometry of the arc follows exactly the contour of the bent specimen. Away from the compression zone that contains high density twins in the bands, in the tension zone, a small arc of twins can be observed in the upper right region.

Further away from the compression zone, a long arc of twins can be observed (Figure 2b) in the tension zone. This arc extends diagonally in the field of view, following the shape of the bent specimen. As shown below, multiple arcs of twins were captured. More surprisingly, high density twins were even observed near the surface of the tension zone, where the tensile strain is maximum in three point bending, as outlined in Figure 2c.

The arcs of twins can be better resolved at higher magnifications, as displayed in Figure 3. Because the grain size of the rolled sheet is rather small (~5 μm), the thickness of the twin laths is only a couple of microns. Figure 3a shows that an arc of twins splits into two separate arcs. Figure 3b shows another arc of twins in which more deformation twins are localized inside. All the arcs of twins follow the contour of the bent specimen.

The optical microscopy reveals arcs of twins with different image contrast than those twin bands in the compression zone, suggesting that the twins in the tension zone may be a different twin variant than those in the compression zone. But these images do not provide definitive crystallographic orientations of the twins in the tension zone in which the stress state strongly disfavors $\{10\bar{1}2\}<10\bar{1}\bar{1}>$ extension twins. To identify the type of the twins in the tension zone, we performed EBSD on the bent specimen. The results are shown in Figure 4.

Figure 4a shows the inverse pole figure (IPF) map of a portion of the tension zone. A few twin variants can be observed. The parent grains and the twins inside the rectangular box were cropped and magnified in Figure 4b. From the pole figure (PF), the AZ31 sheet presents a strong basal texture, which is typical of rolled sheets of Mg alloys.

Analysis and discussion

The capture of the twins in the tension zone in EBSD makes it possible to analyze the type, orientation relationship and the Schmid factor of the twins. In Figure 4b, three parent grains are marked as P1, P2 and P3, whereas their corresponding twins are marked as T1, T2 and T3. The grain boundaries that satisfy $86.3\pm5°<11\bar{2}0>$ are highlighted with black lines. Clearly, the twins T1, T2 and T3 are indeed $\{10\bar{1}2\}<10\bar{1}\bar{1}>$ extension twins, instead of $\{10\bar{1}1\}<10\bar{1}\bar{2}>$ compression twins. To help identify the rotation axis of these specific twins, hexagonal lattices of the parent grains and the twins are also plotted in Figure 4b.

Since the direction of the tensile stress is roughly parallel to the TD, we can estimate that the Schmid factor of the twinning planes should be very low. We quantitatively calculated the Schmid factor of the observed twinning mode in Figure 4, using the Euler angles of the parent grains, and the results are shown in Table 1.

As shown in Table 1, the Schmid factors of the twinning systems in the three parent grains have very low values. Especially for grain P2, the Schmid factor of the twinning systems is close to zero. On the other hand, it can be seen that, for grain P2 and P3, the direction of the tensile stress is almost perpendicular to the *c*-axis [0001]. This stress state strongly disfavors $\{10\bar{1}2\}<10\bar{1}\bar{1}>$ twinning but favors $\{10\bar{1}1\}<10\bar{1}\bar{2}>$ compression twinning [9]. These facts raise an interesting question as to how and why $\{10\bar{1}2\}<10\bar{1}\bar{1}>$ extension twinning still occurred in the tension zone of the three point bending specimen.

Table 1 Schmid factor of the possible twin variants in Figure 4

Euler angles (deg.)	Schmid factor
P1 (24.9, 155.3, 335.8)	0.008 – 0.025
P2 (336.3, 173.2, 314.5)	0.005 – 0.007
P3 (318.4, 7.3, 171.1)	0.049 – 0.055

Most commonly, $\{10\bar{1}2\}<10\bar{1}\bar{1}>$ twinning is deemed as a shear deformation with a theoretical twinning shear 12.9%. However, if the twinning shear should equal 12.9%, the reversible twinning widely observed in the experiments [10-11] could not be rationalized. As stated by Christian and Mahajan [12], "twinning is unidirectional" and "reverse twinning is *not* possible". There is no physics to de-shear a shear deformation in twinning. This statement seemingly contradicts the experimental observations of reverse twinning. However, this conflict can be reconciled inside the classical framework laid out by Christian and Mahajan [12], by realizing that the twinning shear of $\{10\bar{1}2\}<10\bar{1}\bar{1}>$ is actually zero, as revealed in the most recent in-situ transmission electron microscopy (TEM) [13]. Thus this twinning can only be shuffling dominated and no twinning dislocations should be involved [14]. The Christian-Mahajan framework allows different choices of shear and shuffle to best fit experimental observations. Because this twinning mode involves no shear deformation, it can be reversed, as shown by Li and Zhang [15].

Li and Zhang [15] also showed that after twinning, the parent lattice is reoriented by ~90° around the $<11\bar{2}0>$ axis, and two misfit strain components are created: a negative, compressive strain in the direction perpendicular to the *c*-axis of the parent, and a positive tensile strain in the direction parallel to the *c*-axis of the parent. The compressive strain equals $(\gamma-\sqrt{3})/\sqrt{3}\approx-6.24\%$. The tensile strain equals $(\sqrt{3}-\gamma)/\gamma\approx6.65\%$. No shear strain is created.

The arcs of twins in the tension zone of the three point bending specimen can be accounted for by the misfit strains generated by the twinning. For those twins in the tension zone, the major component of the stress is a tensile stress. This stress component introduces a compressive strain component in the direction along the normal of the sheet specimen. Without the extension twins, the strain would have to be accommodated mainly by non-basal dislocations with limited number of independent slip systems and high critical stresses. $\{10\bar{1}1\}<10\bar{1}\bar{2}>$ compression twinning might be activated, but experimentally this type of twins has very low volume fraction due to the high critical stress as well. The presence of the $\{10\bar{1}2\}<10\bar{1}\bar{1}>$ extension twins, however, provides an extra mechanism for strain accommodation. Although the stress state strongly disfavors the extension twinning, the high local strain generated in three point bending at large bending angles is able to activate the extension twinning, such that the compressive strain component can be effectively accommodated by the compressive strain generated by the twinning in the direction perpendicular to the *c*-axis of the parent grains. From this perspective, these stress-unfavorable twins present the characteristic of "geometrically necessary twins".

Conclusions

We performed three point bending tests on an AZ31 Mg rolled sheet. In the tension zone where the stress state strongly disfavors $\{10\bar{1}2\} < 10\bar{1}\bar{1} >$ twinning, arcs of $\{10\bar{1}2\} < 10\bar{1}\bar{1} >$ twins were observed. These twins are localized in the arcs that follow the contour of the bent specimen.

The extension twins in the most unlikely stress state can be understood from the misfit strain components produced by this twinning mode. In the tension zone of the bent specimen, the compressive strain along the normal direction of the sheet specimen activates the $\{10\bar{1}2\} < 10\bar{1}\bar{1} >$ twins, such that the compressive strain is accommodated by the contraction strain component generated by the twinning. Hence, the stress-unfavorable $\{10\bar{1}2\} < 10\bar{1}\bar{1} >$ twins present the characteristic of "geometrically necessary twins".

References

[1] J.F. Nye, Some geometrical relations in dislocated crystals. Acta Met 1 (1953) 153–162.

[2] M.F. Ashby, The deformation of plastically non-homogeneous materials. Phil Mag 21 (1970) 399–424.

[3] A. Needleman, J.G. Sevillano, Preface to the viewpoint set on: geometrically necessary dislocations and size dependent plasticity. Scripta Mater 48 (2003) 109–111.

[4] N. Hansen, D.J. Jensen, Flow stress anisotropy caused by geometrically necessary boundaries. Acta Met 40 (1992) 3265–3275.

[5] G. Winther, D.J. Jensen, N. Hansen, Modelling flow stress anisotropy caused by deformation induced dislocation boundaries. Acta Mater 45 (1997) 2455–2465.

[6] J.G. Sevillano, Geometrically necessary twins and their associated size effects. Scripta Mater 59 (2008) 135–138.

[7] C. Dreyer, F. Polesak, T. Shultz, S. Agnew, in: E.A. Nyberg, S.R. Agnew, N.R. Neelameggham, M.O. Pekguleryuz (Eds.), Magnesium Technology 2009, Proceedings of the 2009 TMS Annual Meeting and Exhibition, San Francisco, CA, 2009, pp. 491–496.

[8] J.C. Baird, et al., Localized twin bands in sheet bending of a magnesium alloy. Scripta Mater 67 (2012) 471–474.

[9] M.H. Yoo, Slip, twinning, and fracture in hexagonal close-packed metals. Met Trans A 12 (1981) 409–418.

[10] C.H. Cáceres, T. Sumitomo, M. Veidt. Pseudoelastic behaviour of cast magnesium AZ91 alloy under cyclic loading–unloading. Acta Mater 51 (2003) 6211–6218.

[11] L. Wu et al., Internal stress relaxation and load redistribution during the twinning–detwinning-dominated cyclic deformation of a wrought magnesium alloy, ZK60A. Acta Mater 56 (2008) 3699–3707.

[12] J.W. Christian, S. Mahajan, Deformation twinning. Prog Mater Sci 39 (1995) 1–157.

[13] B.Y. Liu et al., Twinning-like lattice reorientation without a crystallographic twinning plane. Nat Commun 5 (2014).

[14] B. Li, E. Ma, Shuffling dominated twinning mechanism in magnesium, Phys Rev Lett 103 (2009) 035503.

[15] B. Li, X.Y. Zhang, Global strain generated by shuffling-dominated twinning. Scripta Mater 71 (2014) 45–48.

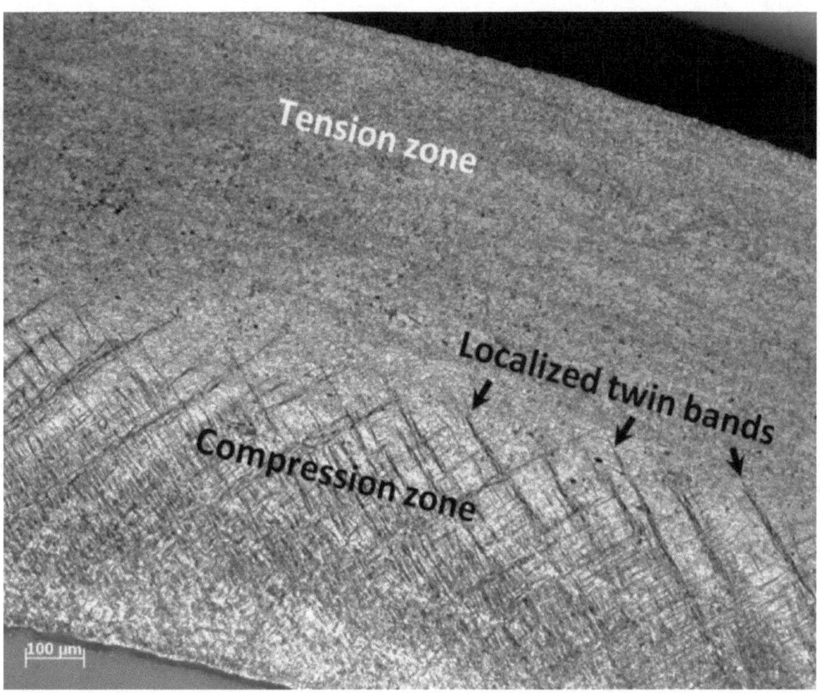

Figure 1. Optical microscopy of localized twin bands in the compression zone of the sample after three point bending. Differential interference contrast (DIC) illumination was used in imaging. Each band comprises a high density of extension twins. In the tension zone where the stress disfavors extension twinning, no twins can be observed at the low magnification.

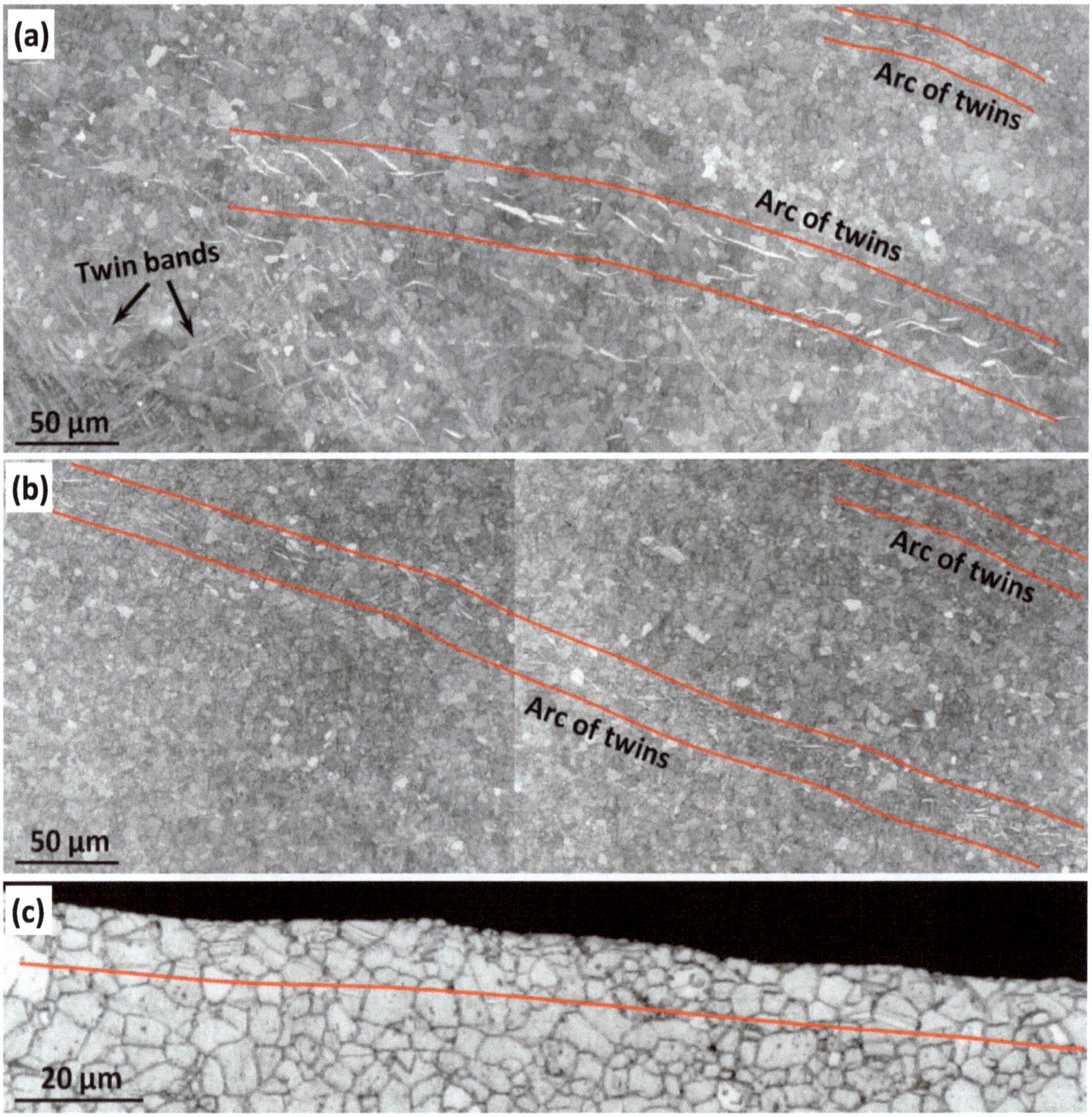

Figure 2. (a) Near the tips of the localized twin bands (indicated by the arrows), arcs of twins were observed when the differential interference contrast (DIC) illumination was removed. Note that the twins in the arcs and those twins in the bands present different contrast, indicating that they are different twin variants. (b) In the tension zone, long arcs of twins were observed, following the contour of the bent specimen. Outside the arcs, twin density is much lower. (c) Deformation twins even appear near the surface of the tension zone with maximum tensile strain.

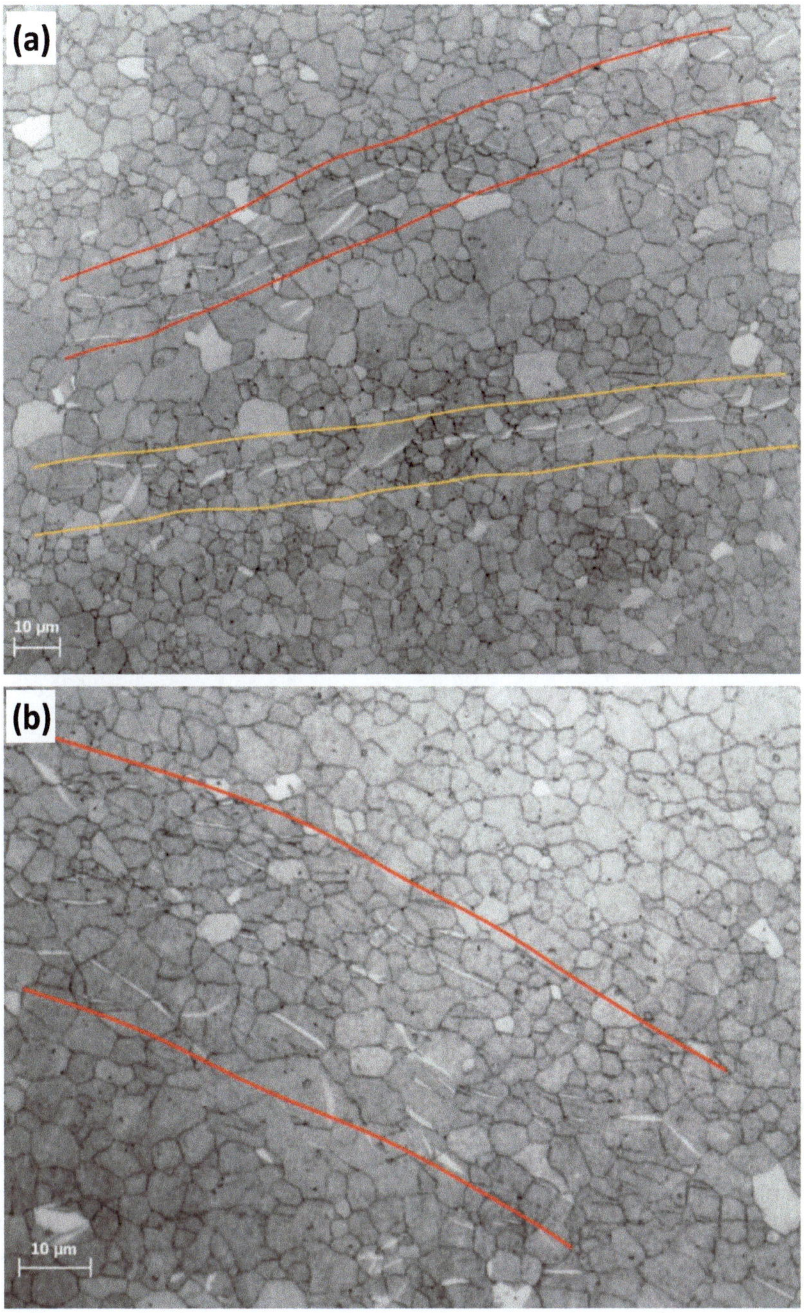

Figure 3. Arcs of twins at higher magnifications. (a) Split of an arc into two sub-arcs of twins. (b) Another arc of twins. The arcs follow the contour of the bent specimen.

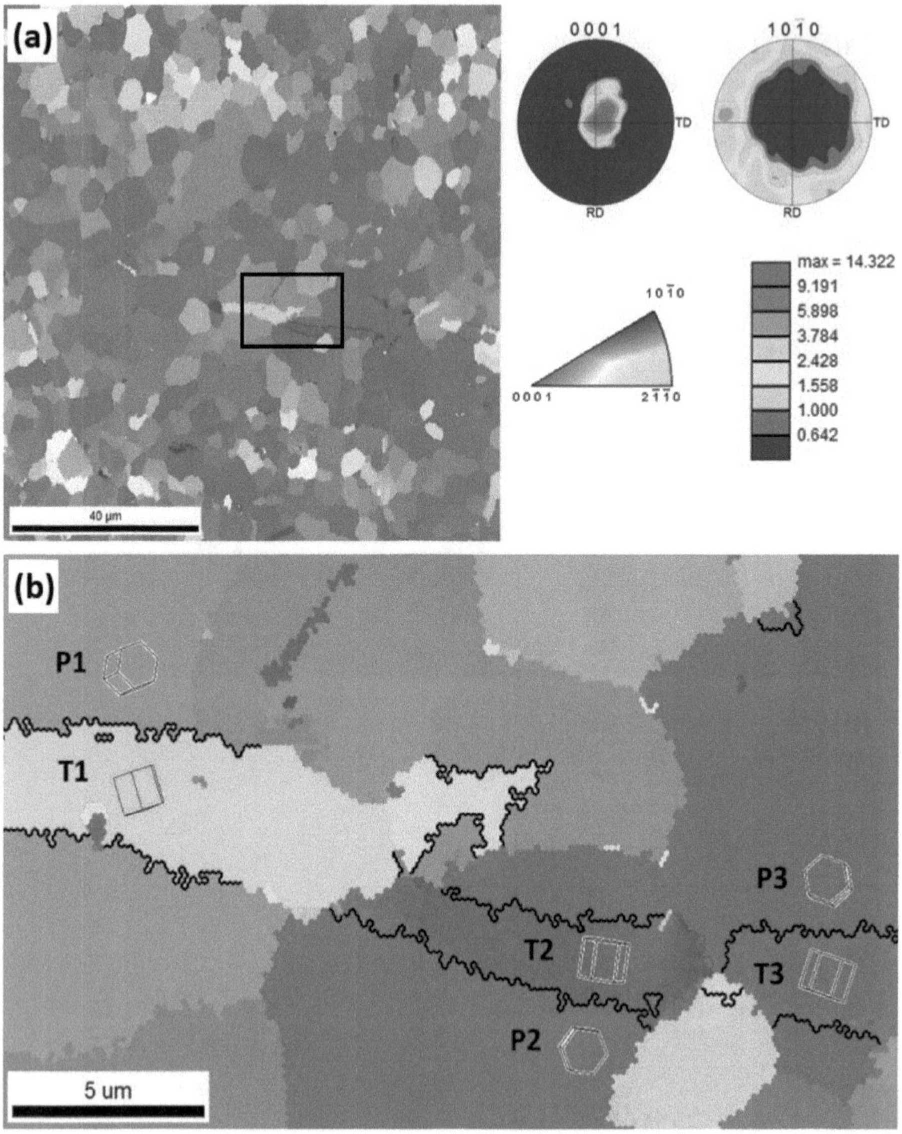

Figure 4. EBSD inverse pole figure (IPF) maps of the tension zone. As expected, a strong basal texture can be seen, typical of rolled Mg sheets. (a) Twins were captured after repeated scanning efforts. (b) Magnified view of the region enclosed by the rectangle in (a). Lattices of the parent grains (P1, P2 and P3) and the corresponding twins (T1, T2 and T3) were plotted to show the orientation relationship. Twin boundaries that satisfy 86.3±5°$< 11\bar{2}0 >$ are highlighted with black lines.

Magnesium Technology 2015
Edited by: Michele V. Manuel, Alok Singh, Martyn Alderman, and Neale R. Neelameggham
TMS (The Minerals, Metals & Materials Society), 2015

DEFORMATION TWINNING EFFECTS ON TEXTURE AND MICROSTRUCTURE OF AZ31B MAGNESIUM ROLLED SAMPLES

Litzy L Choquechambi Catorceno[1], Luis Flavio Gaspar Herculano[1], Hamilton Ferreira Gomes de Abreu[1]
[1]UFC (Federal University of Ceará- Department of Metallurgical Engineering and Materials); Bloco 729, Campus do Pici; Fortaleza, CE 60440-554, Brazil.

Keywords: Magnesium, Texture, Twinning

Abstract

Different twinning types can be formed on rolling deformation, even at high rolling temperatures. In this work, mechanical twins were studied on cold (at 25°C) and warm (at 250°C) rolled samples of AZ31B magnesium alloy. The influence of the mechanical twins on microstructure and texture evolution was examined by electron back scatter diffraction (EBSD), comparing the twin morphologies and determining their orientation relationships with matrices around them. The results showed that $\{10\bar{1}2\}\langle 12\bar{1}0\rangle$ and $\{10\bar{1}1\}\langle 12\bar{1}0\rangle$ twins and double twinning affect grain orientations during rolling processes. In addition, on rolled samples were evidenced that the shear bands were developed from doubly-twinned regions, but most of them quickly lost their morphology when applying high rolling temperature due to dynamic recrystallization.

Introduction

The low density of magnesium (Mg) alloys makes them a good candidate to produce lightweight components in automotive industry and portable electronic devices [1]. However, the use of rolled sheets is still limited because of their poor ductility and high propensity to corrosion. Their limited formability at ambient temperature is related to the insufficient number of active slip systems in the HCP lattice. According to the Von-Mises criterion, five independent slip systems are required for plastic deformation of polycrystalline materials.

Since the critical resolved shear stress (CRSS) for basal slips and mechanical twinning are significantly lower than for non- basal slip at low temperatures, they are dominant deformation mechanisms on cold deformation processes [2-3]. The activation of different deformation mechanisms depends mainly on rolling temperature, as well as, strain rate and texture [3].

The transition from twinning to slip dominated mode can occur increasing rolling temperature or decreasing grain sizes. Thus, the formability of magnesium alloys greatly increases at rolling temperatures up to 200ºC owing to the decrease of critical resolved shear stress (CRSS) of non-basal slip systems [4,7]. However, twinning is highly sensitive to the grain size. Samples with fine grains exhibit low volume fraction of twins because of their formation required more energy (higher average stress) in fine grained polycrystals [5]. Other factors, such as temperature, strain rate, stacking fault energy, texture and stress state can also influence twin development [5, 7].

It has long been recognized that an intense twin formation on cold rolled surfaces deteriorates considerably magnesium alloy ductility due to twin boundaries act as effective barriers for slips [8]. Nevertheless, it was not considered the importance of twin types during rolling processes. Twins can effectively strengthen a material under some circumstances or weaken under others.

In the case of AZ31B magnesium alloys, three types of twins are commonly observed during rolling processes: extension, contraction and secondary or double twins. The designations "extension" and "contraction" refer to the sense of the normal stress required to induce a shear in the appropriate direction on the relevant twin plane. The orientations of the crystal axes with respect to the stress and strain axes affect the types and volume fractions of the twins that result from the deformation significantly. Thus, it is of great importance to point out that the twinning type in Mg depends on the loading modes or the stress field.

The extension twinning occurs commonly along the $\{10\bar{1}2\}$ planes and is activated when there is an extension strain component parallel to the c-axis or when contraction is being applied perpendicular to the c-axis of the principal texture component. Their formation increases the flow stress as a result of the introduction of additional barriers to dislocation glide [8].These twins are easy to be activated in large grains at ambient or lower temperatures due to their low nucleation stress and are associated with hardening behavior. This type of twins is rarely observed in Mg alloys at temperatures higher than 350°C and strain rates less than $0.01 s^{-1}$ [9–11].

Extension twins cause misorientations of 86º between twinning planes and non-twinning planes. They can be represented by $\{10\bar{1}2\}\langle 10\bar{1}10\rangle$ or 86° $\langle 1\bar{2}10\rangle$ according to misorientation relationship.

In the case of contraction twins, they can accommodate compression or contraction along the c-axis. It is activated when there is a contraction strain component parallel to the c-axis or when macroscopic extension is being applied perpendicular to the c-axis. Thus, a contraction along the c-axis is generated by $\{10\bar{1}1\}$ planes, and basal planes are reoriented about 56º around $\langle 12\bar{1}0\rangle$ axis [10]. Their formation has the effect of rotating the basal planes towards more favorable orientations. Fractures are mainly originated from compression twins or shear bands developed from compression twins. As both basal slip and tension twinning lead to basal texture which possesses a high Schmid factor for compression twinning during rolling, the formation of compression twins is inevitable. In addition, the (dynamic) recrystallization nucleation at compression twins or shear bands was observed to be more effective than that at tension twins or grain boundaries.

Secondary twinning occurs when a twin twins again on contraction twins, this type of twins usually occurs within coarse lamellar twins, subdividing them. Generally $\{10\bar{1}1\}$ or $\{10\bar{1}3\}$ contraction twins form first, after which $\{10\bar{1}2\}$ extension twins are propagated within the original contraction twins [9,11].

The intensive occurrence of secondary twinning on the $\{10\bar{1}2\}$ planes creates localized flows followed by failures [12]. The misorientation relationships commonly produced between the matrix and twins during deformation of hexagonal structure alloys are summarized in Table 1.

Table 1 Twinning misorientations commonly observed in magnesium alloys [9].

Twin type	**Misorientation angle/axis**	
$\{10\bar{1}1\}$	$56^\circ\langle 12\bar{1}0\rangle$	Contraction twins or compression twins
$\{10\bar{1}2\}$	$86^\circ\langle 12\bar{1}0\rangle$	Extension twins or tension twins
$\{10\bar{1}3\}$	$64^\circ\langle 12\bar{1}0\rangle$	
$\{10\bar{1}1\}$-$\{10\bar{1}2\}$	$38^\circ\langle 12\bar{1}0\rangle$	Double twins
$\{10\bar{1}3\}$-$\{10\bar{1}2\}$	$22^\circ\langle 12\bar{1}0\rangle$	

Since the reorientations resulting from twinning may lead to favorable slip orientations, the local strain concentrations caused by twinning can induce non-basal slips or cracks. Therefore, if twinning can be controlled appropriately, it could provide an alternative way to improve the formability of magnesium alloys.

In this sense, this work aims to examine the different types of twins produced on rolling samples at room temperature and 250ºC. Firstly comparing their morphologies; secondly determining their orientation relationships with the matrix around them.

Experimental

The material considered in the current study was a commercial-grade AZ31 magnesium alloy (Mg–3%Al–1%Zn– 0.2%Mn) provided by Magnesium Elektron in the form of twin roll caster plate with 2 mm thickness. Two small strips were cold rolling with 10 and 15% of thickness reduction in one single pass. Further reduction was not possible due to the extensive lateral cracking. However, increasing the rolling temperature to 250ºC, it was possible to deform with 20% and 30% of thickness reduction.

The microstructure and texture of rolled samples were analyzed using scanning electron microscopy (SEM) and electron back-scattered diffraction (EBSD). All samples were ground with 2000 SiC paper and polished with diamond paste through 3 and 1 µm.

The final polishing was made with a chemical polishing solution of 10 ml HNO3, 30 ml acetic acid, 40 ml H2O and 120 ml ethanol for 5s. Another etching was made with a solution of 5 g picric, 100 ml ethanol, 10 ml glacial acetic acid and 5ml distilled water for about 10s in order to reveal twins and grain boundaries.

The electron backscattered diffraction (EBSD) technique has been used to analyze twins and to determinate the twin orientation relationships. For this, it was performed a final polishing using a colloidal silica solution OP-S. The step size used in EBSD measurements was between 0.5 and 1 µm (depending on the grain size of the measured specimen).

Results and Discussion

Microstructure evolution

The initial microstructure of as-received material consists mainly of α-Mg with a recrystallized microstructure. The initial microstructure and inverse pole figure (IPF) map of as-received sheet are shown in Fig. 1. The initial microstructure consists of equiaxed grains, without mechanical twins. The mean grain size was about 13± 2 µm at the sheet surface. Granular eutectic precipitates rich in Al and Mn are heterogeneously distributed on the solid solution.

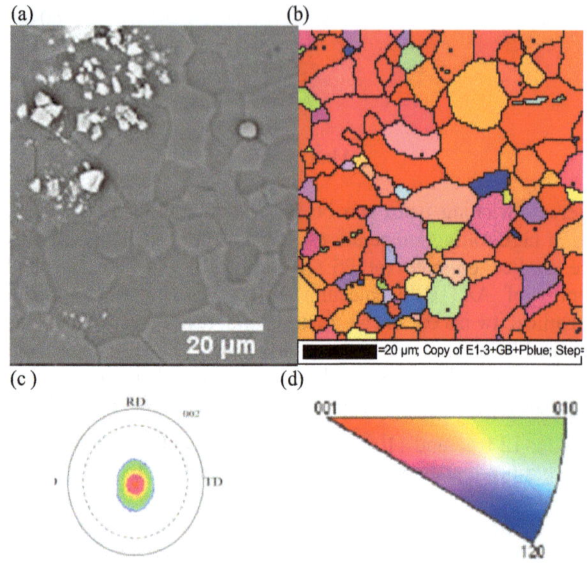

Figure 1 (a) Micrograph of initial microstructure, (b) EBSD image showing the inverse pole figure map (c) pole figure and (d) inverse pole figure obtained from as-received AZ31 plate. Grains in red have (0001) planes aligned with the rolling direction.

Figure 2 shows morphology of twins produced on the microstructure of warm and cold rolled samples. Looking at the twin evolution on micrographs, it is possible to see that both tensile and compressive twins occur more preferably on large grains.

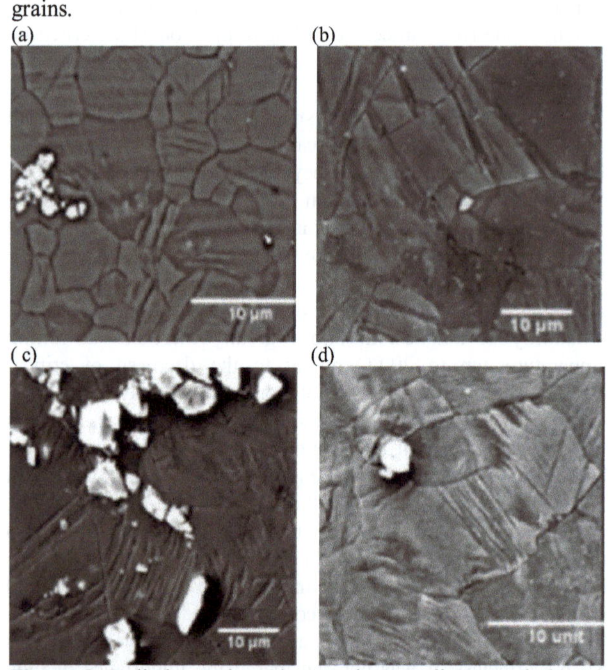

Figure 2. Rolled samples micrographs revealing morphology of contraction and extension twins formed when was applied a thickness reduction (a) 10%, (b) 15% on cold rolling and (c) 20%, (d) 30% of thickness reduction on warm rolling (at 250 ºC).

Comparing twin morphologies on micrographs of cold rolled samples (Fig. 2a and 2b), it can be noted that many thin twins were formed inner side of deformed grains on sample with low strain rate. Since the initial microstructure had a strong basal texture, these twins correspond to contraction twins. Their almost immobile twin boundaries likely contributed to their thin forms (less than 1 μm). However, some shear bands formed by big tension twins can be seen clearly in large grains. This kind of twins is preferably formed in strong textured wrought Mg alloys. Since it is well known that rolled Mg sheets have a strong basal-type texture (the c-axes are parallel to the normal direction, ND). These tension twinning was enhanced at high strain rates and low temperatures. Sample with 15% thickness reduction showed many shear bands formed by big tension twins. This kind of twins was predominantly formed because of contraction twins are activated under low critical resolved shear stresses (CRSS) between 20 and 35 MPa [12,16]. Moreover, increasing the strain rate over 10% of thickness reduction on cold rolling samples, twinning activity increased quickly, until the microstructure become saturated by tension and secondary twins. At that point, basal slips were severely hindered, limiting basal slips activities and generating failures in the microstructure due to abrupt lattice rotations.

On micrographs of warm rolling samples (Fig. 2c and 2d), it was evidenced the presence of tensile, compression and secondary twins on deformed grains, especially on sample with high thickness reduction. When the c-axis of Mg alloy grain was placed under compression, both contraction and tension twinning and non-basal slip were activated. Twinning under such conditions takes place mostly in $\{10\bar{1}1\}$ planes, causing a lattice reorientation of approximately 56°around the $\langle 11\bar{2}0\rangle$ axis. Occasionally, areas that have undergone contraction twinning were in a suitable orientation for subsequent extension twinning. Apparently, $\{10\bar{1}2\}$ tensile twins has a tendency to consume the entire volume of grains, probably due to the high mobility of their twin boundaries [5,14]. Other thing that can be evidenced is referred to the deformation degree, twinning concentration and twin boundary sizes increased with thickness reduction. It was probably as a result of multiple nucleation events followed by growth and coalescence of twin boundaries [8,11,13]. The consecutive occurrence of these two twinning mechanisms gives rise to secondary twinning [5]. The misorientation angle of the twin boundary surrounding a secondary-twinned area is approximately 38° around the $\langle 11\text{-}20\rangle$ axis. This mechanisms was develop because $\{1\bar{1}01\}$ compression twins are activated under high CRSS, between 30 and 100 MPa [13], comparable to that required to activate pyramidal slip (between 30 and 80 MPa) [14,15]. In this sense, compression twins are in competition with pyramidal and prismatic slips at high rolling temperatures. Their morphology typically appears to be relatively narrow with wavy boundaries (Fig 2d), these wavy boundaries indicate that twin boundaries can also change their orientations when they coalescence and growth at high rolling temperature. Meanwhile, it is interesting to noted that some $\{10\bar{1}2\}$ tension twins still persisted at high rolling temperature (250°C), acting as contributors to the strain accommodation, apparently, low concentration of tension twins does not inhibit the formability on warm- rolled samples.

The shear bands formed at high strain and on high rolling temperature evolved from multiples kind of twins. Note also that shear bands contain groups of parallel compression twins and tension twins with curved boundaries and secondary twins.

It is well know that compression twins produce a re-orientation of ~ 56° in a *c*-axis. This new orientation may be favorable for subsequent contraction twinning.

Secondary twins were not easy to identify on EBSD maps. However, it is well know that they reorient the c-axis by an additional re-orientation, which results in a final misorientation between 38° and 22°. The double twins can be identified on SEM micrographs as a set of twins with common interface bisecting two another twins. The double–twin interactions may mitigate the growth rates of the extension twin boundaries and the dislocation movement could be also limited.

Texture evolution

The {0001} basal pole figure of as-receive sheet exhibits an egg-shaped distribution, which is typical for materials with intense basal textures. The basal pole distribution indicates that the c-axis of grains is nearly normal to the sheet plane and a large volume fraction of basal planes are oriented parallel to the rolling plane.

The EBSD map of the as-received material showed a recrystallized microstructure with an intense basal texture where most part of basal planes was aligned with the rolling direction (Fig. 1). The initial texture played an important role in the twin formation during rolling process because intense basal texture facilitated the mechanical twin as a main deformation mechanism, leading basal plane reorientations that can modify the initial texture.

The figure 3 displays the EBSD maps illustrating the different type of twins formed during rolling processes, and also shows misorientations between grain and twin boundaries. For this, it was defined that high-angle grain boundaries have misorientations larger than 15°, and the low-angle grain boundaries have misorientations between 2 to 15°. Extension twin boundaries have about 86 ±5°, and compression twin boundaries have about 56 ±5°. Misorientations less than 2° were neglected. In order to facilitate distinction from untwined areas on EBSD maps, twin boundaries were highlighted using different colors. In blue were identified $\{10\bar{1}1\}$ contraction twins, in green were identified $\{10\bar{1}2\}$ extension twins and in red the double twins, which commonly were located next or inside the contraction twins.

The texture and microstructure quickly changed upon deformation. Twinning activity was evident during cold rolling processes. Approximately, a thickness reduction of 10% caused most of the blue and green grains to contain tensile twinning. At a strain of 15%, twinning grew very fast and the entire region underwent twinning. As the deformation degree increased, the EBSD inverse pole figure contained more of a non-indexed region, as shown in black.

On EBSD maps, it can be seen that twin boundaries decreased when the thickness reduction increased, though SEM micrographs showed the contrary. This was because the index quality decreased on samples rolled with high strains (over 10% of thickness reduction). However, in order to overcome this problem, twin volume fractions were also calculated on micrographs by a point counting technique (ASTM E562-02). In this sense, it can be seen that strain rate had a strong influence on the volume fraction of tension and contraction twins, higher strain rates give rise to higher volume fractions. The volume fraction produced with 15% thickness reduction was 20% higher than sample rolled with 10% thickness reduction. This was because the twin boundaries were widening, leading the appearance of new twins.

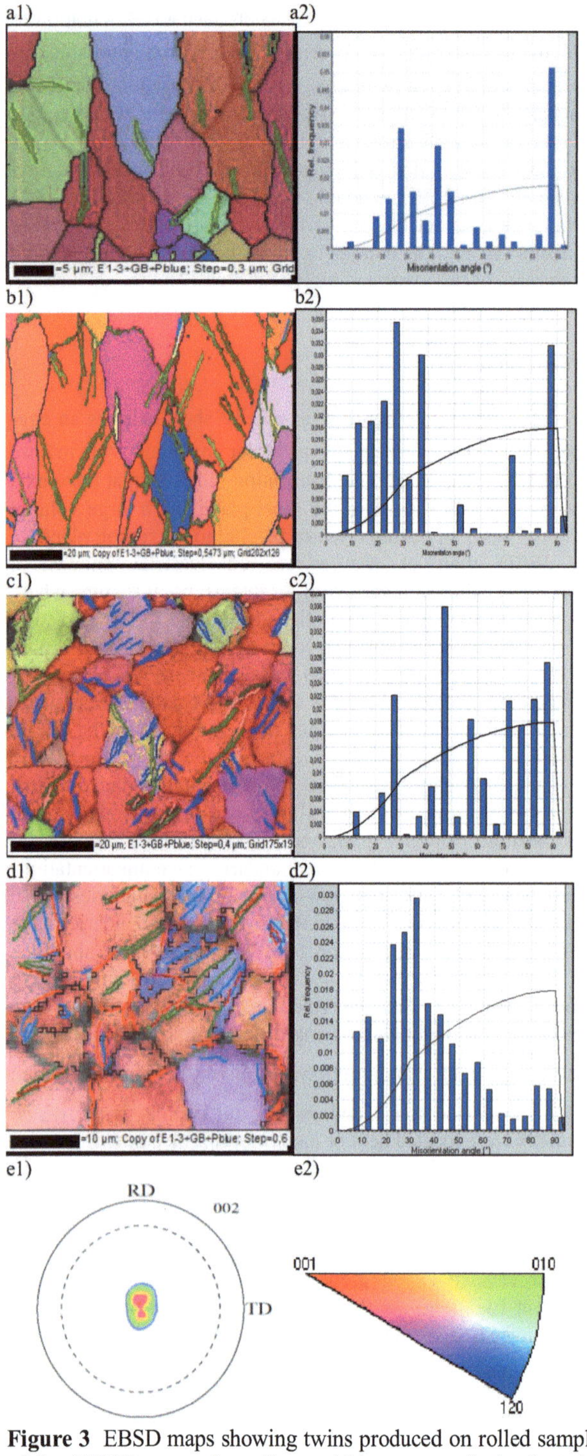

Figure 3 EBSD maps showing twins produced on rolled samples of the AZ31B magnesium alloy and histograms of misorientation distribution obtained from samples with thickness reduction of (a) 10%, 25°C; (b) 15%, 25°C; (c) 20%, 250°C and (d) 30%, 250°C (e) common form of (0002) basal pole figure.

The volume fraction of tension twins approached the saturation on high strains (15%), leading cracks began to be propagated on laterals of cold rolled sample. Thus, tension $\{10\bar{1}2\}$ twinning can contribute to the fatigue failure, especially when is generated $\{10\bar{1}1\}\{10\bar{1}2\}$ double twinning, they cause the formation of large surface steps, cracks and eventual failure.

Rolling temperature also affects the twin volume fractions significantly. Tension twins decreased with increasing temperature. At 250 °C, all types of twins were difficult to be determined on SEM micrographs. It was only considered the total number of twin points. The compression twin fraction was almost 35% in both thickness reductions of warm rolling samples.

Since warm rolled samples generated a mixture of different kind of twins, it was not possible to identify all misorientation angles on EBSD maps because of map qualities became poorer; the unindexed area became larger when increased the deformation degree. The figure 3 revealed that most misorientation angles are next to the theoretical values (table 1). Thus, The maps confirm that most of the twins observed on the SEM microscope, at high strain were $\{10\bar{1}2\}$ $<10\bar{1}1>$ extension twins in Fig.3 (a). The little difference was probably due to double twinning (38° and 22°).

On histograms of cold rolled samples, it can be noted that the frequency of extension twin boundaries increased almost twice more, applying high reduction (15%). The growth and combination of twin boundaries increased the twin boundaries fraction and the fraction of low-angle grain boundaries with misorientation angles below 15° also increased.

On warm rolling samples with 20% of thickness reduction, the fractions of contraction twin boundaries increased over 46%. Meanwhile, notice that the volume fraction of tension twin boundaries decreased more, the reasons can be attributed to twin boundaries, which were re-combined together during warm rolling and the new recrystallized grains restored the deformed microstructure.

The results reveal that the paired twins frequently occur at low misorientation grain boundaries. A twin impacts a grain boundary to stimulate the nucleation of a twin in the neighbouring grain. It is closely related with the misorientation angle between the neighbouring grains.

The volume fraction tension twins is lower on warm rolling samples and the low-angle grain boundaries take the majority during high deformation degree.

The texture of all rolled samples still possessed a strong basal component. However, the (0002) basal pole figures showed double peak with a tilts of ± 15 degrees away from the normal direction towards rolling direction. The transition from the single-peak to the double-peak type with rolling deformations was believed to be caused by the variation of shear strain and by the greater incidence of twinning during rolling processes.

CONCLUSIONS

The microstructural developments showed that twinning is the dominant deformation mechanism in both cold and warm rolling processes. However, the volume fraction of different twins is influenced by the grain size, rolling temperature, strain rate as well as initial texture, and the larger amount of twins is produced preferably on bigger grains.

On cold rolled samples. It has been found that strain rate enhances dramatically the activation of extension $\{10\bar{1}2\}$ twinning, while contraction and secondary twinning are not significantly influenced.

The $\{10\bar{1}1\}$ contraction twins appear under higher deformation degrees (over 20% of thickness reduction) and they are thinner than $\{10\bar{1}2\}$ tension twins.
Warm rolling samples still produce extension twins and formation of dense dislocation pile-ups within initial grains.
Contraction twinning may enhance ductility by reorienting the crystal lattice, which helps activate additional non-basal slip systems
Twinning behavior depends on the grain size as well as texture, with finer grained specimens exhibiting lower volume fraction of twins; the formation of twins was enhanced by grain coarsening.
In the undeformed microstructure, most of the basal poles were aligned near parallel to rolling direction. Increasing deformation, {0002} pole figure appear to be distributed towards RD

Acknowledgements

The authors are grateful to the Central Analítica-UFC/CT-INFRA/MCTI-SISNANO/Pró-Equipamentos CAPES. (Project code - 2014PP32)

REFERENCES

[1] B L Mordike, T. Ebert Magnesium properties applications potential. Mater Sci Eng A, 2001, A302: 37-45.
[2] M.R. Barnett: Materials Science and Engineering A, Vol. 464 (2007) No. 1-2, p.1.
[3] L. Jiang, J.J. Jonas, A.A. Luo, A.K. Sachdev and S. Godet: Scripta Materialia, Vol. 54 (2006) No. 5, p.771
[4] Yoo MH. Metall Trans A 1981;12A:10.
[5] P.E. Krajewski, Soc. Auto. Eng. 301 (2001) 21.
[6] F.K. Abu-Farha, M.R. Khraisheh, J. Mater. Eng. Perform. 16 (2007) 192.
[7] S.R. Agnew, O. Duygulu, Int. J. Plast. 21 (2005) 1161.
[8] A. Jain, S.R. Agnew, Mater. Sci. Eng.: A 462 (2007) 29.
[9] S.R. Agnew, C.N. Tome, D.W. Brown, T.M. Holden, S.C.Vogel, Scr. Mater. 48 (2003) 1003.
[10] N. Stanford and M.R. Barnett: Journal of Alloys and Compounds, Vol. 466 (2008) No. 1-2, p.182.
[11] X. Li, P. Yang, L.N. Wang, L. Meng and F. Cui: Materials Science and Engineering A, Vol. 517 (2009) No. 1-2, p.160.
[12] Kelley EW, Hosford WF. Trans Metall Soc AIME. 1968;242(1–2):5–13.
[13] Molnar P, Jager A, Lejˇcek P. Scr Mater. 2012;67(5):467, 470.
[14] Kitahara T, Ando S, Tsushida M, Kitahara H, Tonda H. Eng Mater. 2007;345–346(1–2):129132
[15] Knezevic M, Levinson A, Harris R, Mishra RK, Doherty RD, Kalidindi SR. Acta Mater. 2010;58(19):6230–6242.
[16] Ando S, Harada N, Tsushida M, Kitahara H, Tonda H. Eng Mater. 2007;345–346(1–2):101–104.
[17] D. Brown, S.R. Agnew, M. Bourke, T. Holden, S.C. Vogel, and C. Tome. Materials Science and Engineering: A, 399(1-2):1–12, 2005.

Magnesium Technology 2015
Edited by: Michele V. Manuel, Alok Singh, Martyn Alderman, and Neale R. Neelameggham
TMS (The Minerals, Metals & Materials Society), 2015

DISLOCATION-TWIN INTERACTIONS IN MAGNESIUM ALLOY AZ31

Fulin Wang and Sean Agnew
Materials Science and Engineering, University of Virginia, Charlottesville, Virginia, 22904-4745, USA

Keywords: TEM, extension twin, dislocation, stacking fault

Abstract

Transmission electron microscopy (TEM) was performed to study the defect structures in a cold deformed AZ31 magnesium alloy. The materials were compressed at room temperature along the rolling direction to promote the formation of {10-12} extension twins. Dislocation character and configuration in both the twin and adjacent matrix were investigated. It was found that basal <a> dislocations are dominant in the matrix, though some dislocations immediately adjacent to the twin boundary have <c> component. An abundance of dislocations with a <c> component were observed inside the twin, even at relatively low strain level of 2%. Defects on the twin boundary and in the proximity were further investigated, including stacking faults on the basal plane which appear to have resulted from partial dislocations emanating from the twin boundary.

Introduction

The $\{10\bar{1}2\}$ extension twinning is an important mechanism of plastic deformation in magnesium alloys. It is able to produce extension along the c-axis during e.g. in-plane compression of the textured material. When twinning occurs, the material exhibits a characteristic plateau and subsequent rapid strain hardening as observed in the S-shaped uniaxial compression flow curve presented in Figure 1. Because of the basal texture in the rolled plate, compression along the RD promotes compression nearly perpendicular to the c-axis in a large volume fraction of the grains. Such a straining direction promotes the activation of extension twinning. The deformation mechanisms prior to and at the plateau regime have been studied extensively. Basal slip and extension twinning have been confirmed as the dominant deformation modes at room temperature in polycrystalline AZ31, and non-basal slip modes was reported to occur at boundaries where the stress state is perturbed [1].

Many in-situ neutron diffraction studies have shown that highly textured Mg alloys begin $\{10\bar{1}2\}$ extension twinning by 0.5% strain (e.g. [2]), as indicated by a strong texture evolution corresponding to the near 90° reorientation of c-axes. Through the use of in-situ neutron diffraction experiments and elasto-plastic [3] and elasto-viscoplastic [4] crystal plasticity modeling, it has also become clear that slip of <a> type dislocations on basal planes precedes the onset of twinning. In fact, recent molecular dynamics simulations of a pile-up of basal dislocations on a low angle boundary was shown to result in the nucleation of a {10.2} extension twin [5]. The end of the plateau corresponds to the completion of Lüder's band propagation through the gage of the sample according to the picture proposed by Barnett et al. [6]. In short, a specified volume fraction of twinning has been produced within the entire gage region at this point. If further deformation (by twinning or otherwise) is to occur, higher stress levels are required. According to the twinning-detwinning (TDT) model of Wang et al. [7], this effect can be incorporated into the crystal plasticity models as the point at which further twinning becomes difficult.

Apparently, the distinct orientation of the twin relative to the loading direction, and the twin boundary have an influence on the deformation process. However, the interactions between slip and twinning which affect the strain hardening behavior are less understood. In addition to obstructing the dislocation slip, the coherent nature of the twin boundary allows the transmutation of dislocations [8] [9], which has been invoked as an explanation for the high hardening rate within the twin interior [10], and hence, the surrounding polycrystal. According to this argument, in addition to the fact that extension twins do not generate any orientation favorable for subsequent slip of the soft, basal mechanism inside the twin, twins are additionally hardened by high density of dislocations [11] [12]. The goal of the present study is to determine the dislocation character and arrangements in the twin and the adjacent matrix.

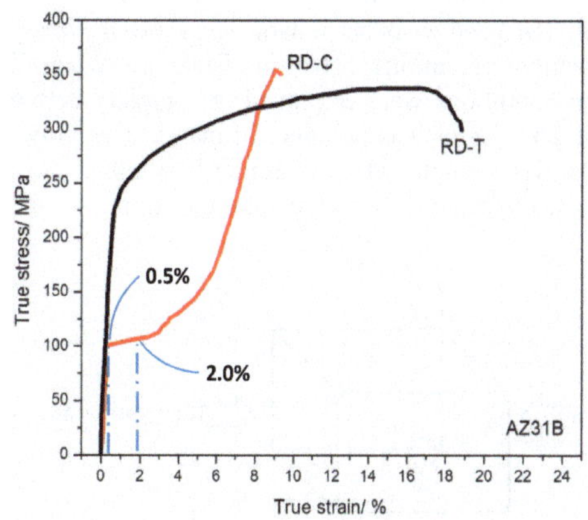

Figure 1. Flow curves obtained for loading along the rolling direction under tension (RD-T) and compression (RD-C). Figure reproduced from [13].

Experimental methods

The material used in this work is from a commercially rolled plate of magnesium alloy AZ31B with H24 temper (strain hardened and partially annealed). The nominal composition of the alloy is 3 wt% Al, 1 wt% Zn, and balance Mg). The rolled plate was subsequently compressed along the rolling direction (RD) at room temperature (RT). Compression samples with 0.5% and 2% strains were used to represent the start and end of the twinning-dominated strain hardening plateau.

For TEM analysis, thin sections parallel to the transverse plane were cut from the cuboidal compression sample using diamond saw (Figure 2). This orientation was selected to permit the twins to be observed primarily along the $\{11\bar{2}0\}$ direction. The sections

Table 1. Diffraction conditions for TEM analysis. The three zone axes (ZA) are 30° apart along the (0002) Kikuchi band. The diffraction vectors are grouped by their respective ZAs, and (0002) is available at all ZAs.

	ZA			$[\bar{1}2\bar{1}0]$		$[\bar{1}100]$		$[\bar{2}110]$	
Mode	b/ $\vec{g}$	(0002)	$(10\bar{1}0)$	$(\bar{1}011)$	$(10\bar{1}1)$	$(11\bar{2}0)$	$(01\bar{1}0)$	$(0\bar{1}11)$	$(01\bar{1}1)$
<a>	$1/3[11\bar{2}0]$	0	1	1	-1	2	-1	-1	1
	$1/3[1\bar{2}10]$	0	0	0	0	-1	1	1	-1
	$1/3[\bar{2}110]$	0	-1	-1	1	-1	0	0	0
<c+a>	$1/3[11\bar{2}3]$	2	1	0	2	2	1	0	2
	$1/3[1\bar{2}13]$	2	0	1	1	-1	-1	2	0
	$1/3[\bar{2}113]$	2	-1	2	0	-1	0	1	1
	$1/3[11\bar{2}\bar{3}]$	-2	1	-2	0	2	1	-2	0
	$1/3[1\bar{2}1\bar{3}]$	-2	0	-1	-1	-1	-1	0	-2
	$1/3[\bar{2}11\bar{3}]$	-2	-1	0	-2	-1	0	-1	-1
[c]	[0001]	2	0	1	1	0	0	1	0

were mechanically ground to the thickness of ~200 μm and then punched into 3 mm disks. The disks were further thinned by 1200 grit SiC paper to ~80 μm, prior to electropolish in a solution of 500 ml ethanol, 100 ml butyl cellosolve, 5.6 g magnesium perchlorate and 2.3 g lithium chloride, at 95 V and -50°C maintained by a solution of liquid nitrogen and methanol within a Fischione double-jet polisher. After perforation, the disks were rinsed by ethanol.

TEM observations were performed on a JEOL 2000FX TEM with a LaB_6 cathode operating at 200 kV. Images of the dislocation and fault structure in and around extension twins were obtained using bright field (BF) and weak beam dark field (WBDF) imaging with the diffraction conditions listed in Table 1. When all such diffraction conditions were employed, a complete determination of the Burgers vector was possible. In many cases only selected reflections were employed to distinguish whether the Burgers vector of the dislocation is <a> type or c-containing.

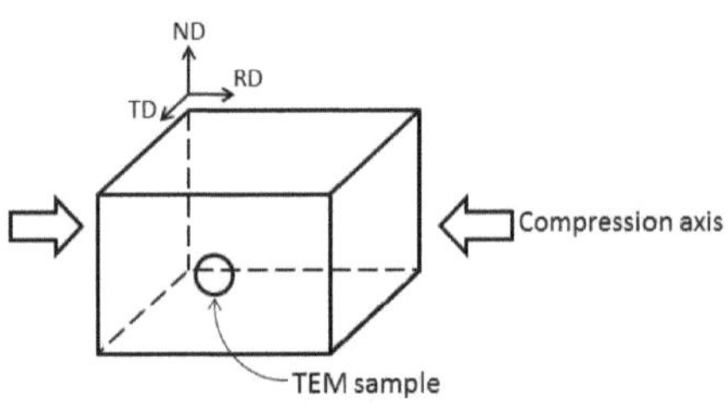

Figure 2. Orientation of the TEM disk in the compressed sample, the normal of the disk is parallel to the transvers direction (TD) of the rolled plate.

Results and discussion

TEM observations of dislocations

Regarding dislocations in the matrix at the onset of twinning (considered as 0.5% strain in this work), it is found that they are exclusively of <a> type, as indicated by their visibility in the $[1\bar{1}01]$ and $[1\bar{1}00]$ diffraction conditions and absence in the [0002] diffraction condition, Figure 3(d). Furthermore, stereological considerations indicated that the majority of these dislocations are lying within the basal plane (i.e., they are not prismatic or pyramidal dislocations). More specifically, it is noted that they are long and relatively straight when looking down the (0001) basal planes (near edge-on perspective in Figure 3). This is not surprising, given the low stress level that is required to produce a strain of 0.5% in compression along the rolling direction, and the high stress level that is required to activate prismatic slip (e.g., see high flow stress in RD-tension Figure 1). However, it is yet an important observation since the immediate vicinity of the twin is likely under stress concentration associated with the demands of twin accommodation. Some authors (e.g. Jonas et al. [14]) have emphasized the role of prismatic <a> dislocations in accommodating the twin strain within surrounding matrix grains. The present results suggest that this may not be necessary in the immediate matrix grain at low strain levels.

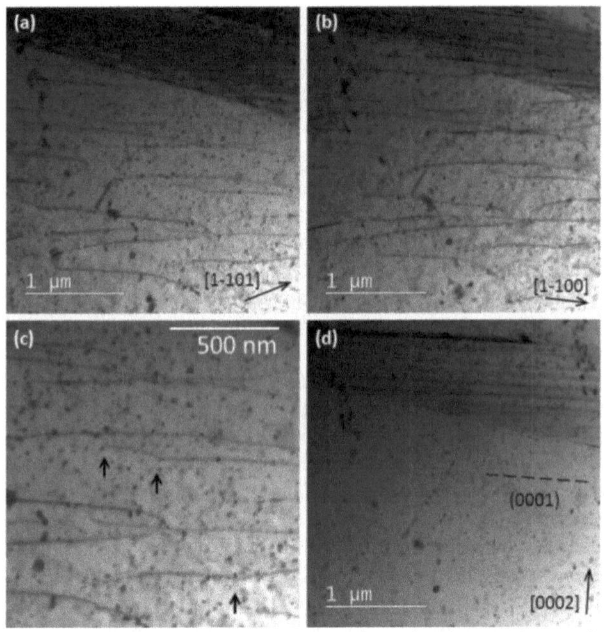

Figure 3. Straight <a> type dislocations in the 0.5% sample. Dislocation bow-out around precipitates is indicated by short arrows in (c). The imaged area is close to a twin in the upper-right corner in (a, b and d).

Also note-worthy is the occurrence of dislocation bow-out at precipitates in Figure 3(c). As reported in [15], the manganese rich particles in AZ31 are suggested to contribute to precipitation hardening and hence an increase in tensile yield strength of this alloy. The alloy is not considered age hardenable, since there is no practical heat treatment schedule that leads to appreciable hardening. Presumably, the precipitates are either too easily bypassed or too sparse in the matrix.

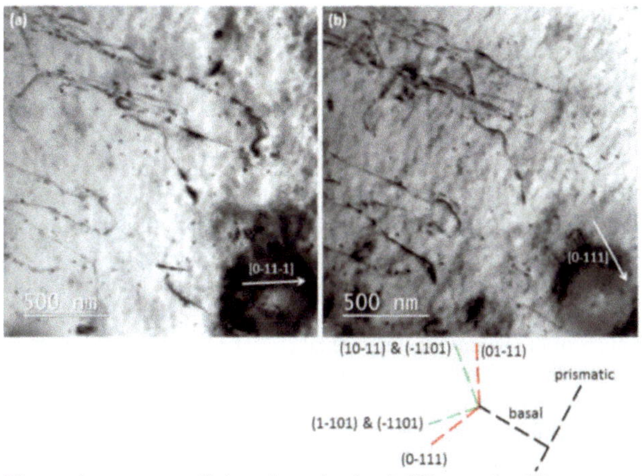

Figure 4. <a> type dislocations in the 0.5% sample. Plane traces are shown in (b).

The <a> type dislocations were occasionally observed to be operating on non-basal planes in the 0.5% sample. In Figure 4, dislocations are mainly on the basal plane while short segments are aligned with non-basal planes. It is observed, at least in this region, that the <a> dislocations appear to be more readily operating on the first order pyramidal plane $\{10\bar{1}1\}$ than on the prismatic plane $\{10\bar{1}0\}$. This has to be confirmed by examining the possibility of cross-slip of the <a> type dislocations. The presence of non-basal dislocations in both $[0\bar{1}1\bar{1}]$ and $[0\bar{1}11]$ reflections indicates that the <a> dislocations have the Burgers vector of either $[a_2]$ or $[a_3]$. In order for a screw dislocation to cross slip from the basal plane onto the pyramidal planes, the intercept of the basal and pyramidal planes has to be parallel to the Burgers vector. So $(10\bar{1}1)$, $(\bar{1}101)$, $(1\bar{1}01)$ and $(\bar{1}101)$ are the possible gliding planes for the $[a_2]$ or $[a_3]$ dislocations in the figure. In fact, close inspection shows that the traces of these four planes align closely with most out of basal dislocation segments.

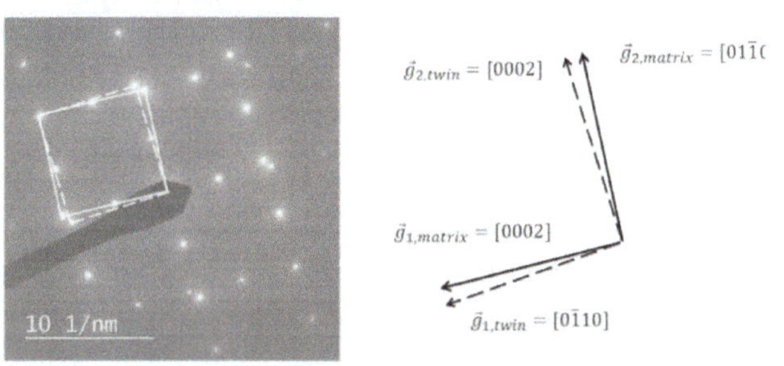

Figure 5. Selected area diffraction pattern showing the matrix-twin orientation relationship (86°). The extension twinning plane is determined as $(01\bar{1}2)$. The close alignment of some diffraction vectors from twin and matrix is illustrated on the right.

Extension twins assume a larger fraction of the microstructure at 2.0% strain. Dislocation analysis was focused in the vicinity of twins. The nearly 90° orientation relationship of directions from the matrix and the twin was utilized to compare the dislocation structure on both sides of a twin boundary, since it is possible to obtain a near two-beam diffraction condition on both the matrix and twinned regions. When the [0002] vector in the matrix is used as the two-beam condition for imaging, the condition in the twin occurs naturally as close to$[01\bar{1}0]$, and vice versa. This relationship is schematically illustrated in Figure 5. The [0002] reflection will only give contrast to the [c] dislocations, and the $[01\bar{1}0]$ reflection will give contrast to the <a> containing dislocations. If a dislocation in either matrix or twin is visible in both diffraction conditions, it has both <a> and [c] components.

At 2.0% strain, the most dislocations in the matrix are <a> type, with the exception of a few in the vicinity of the twin boundary. Figure 6(c) shows segments of dislocation with [c] component in the matrix connected to the twin boundary. While they are present also in the condition of g=$[01\bar{1}0]$ (Figure 6(b)), it is suggested that these are <c+a_2> or <c+a_3> dislocations, whose presence might be ascribed to the perturbed local stress state near the twin boundary or hint the emission from the twin boundary. Similar dislocations are also visible in Figure 7(b) on the right side and highlighted by the dashed line box. As is discussed in the previous paragraph, the $[01\bar{1}0]$ vector in the twin naturally leads to the [0002] reflection in the adjacent matrix, which gives contrast to the dislocation having the [c] component.

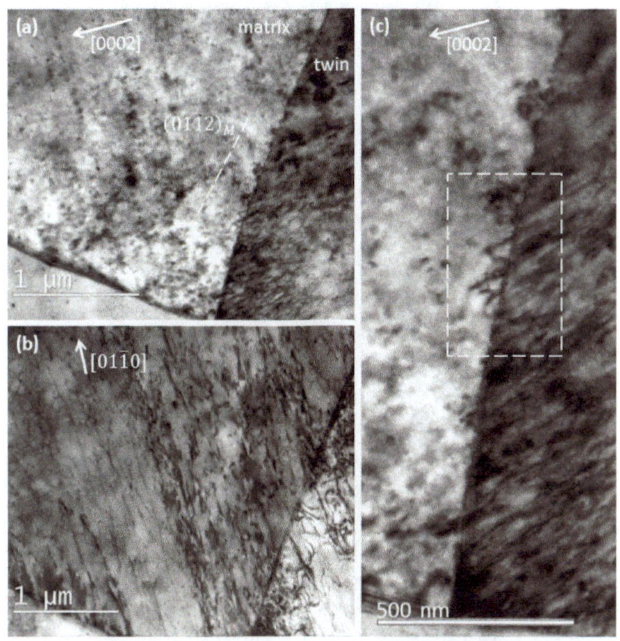

Figure 6. Dislocation structure in the twin and the matrix in the 2.0% sample. The diffraction vectors are those from the matrix. The trace of the twinning plane is indicated in (a) by the dashed line. The dashed line box in (c) highlights the dislocations with [c] component in the matrix connected with the twin boundary.

The primary <a> type dislocations in the matrix are accompanied by abundant [c] containing dislocations in the twin, as indicated by the dislocation contrast in both Figure 6(b) and (c). Furthermore, the coexistence of reflections $[01\bar{1}0]$ in the matrix and [0002] in the twin reveals in the same figure the $[a_2]$, $[a_3]$ dislocations on the basal plane in the matrix and [c] containing dislocations in the twin. The dislocation transmutation has been suggested involving two <a> type dislocations penetrating the $\{10\bar{1}2\}$ twin boundary and producing a <c+a> dislocation leaving a residual dislocation on the boundary [16] [17]. According to the correspondence matrix developed by M. Niewczas [9], a possible reaction is $[a_3]$ dislocations on the basal plane in the matrix produces a $[c + a_2]$ dislocation on the prismatic plane in the extension twin.

The dislocations inside the extension twin are displayed in Figure 7. The high density of dislocations and their large Burgers vectors ([c] containing) yield strong and overlapped strain field and therefore weak image contrast. Therefore, weak beam dark field (WBDF) technique is helpful, to reveal the contrast of individual dislocations. In terms of the morphology, most dislocations have shorter segment on the basal plane and longer segment on the non-basal planes, which is in contrast to the dislocations in the matrix of the 0.5% sample having longer segments on the basal plane (Figure 4).

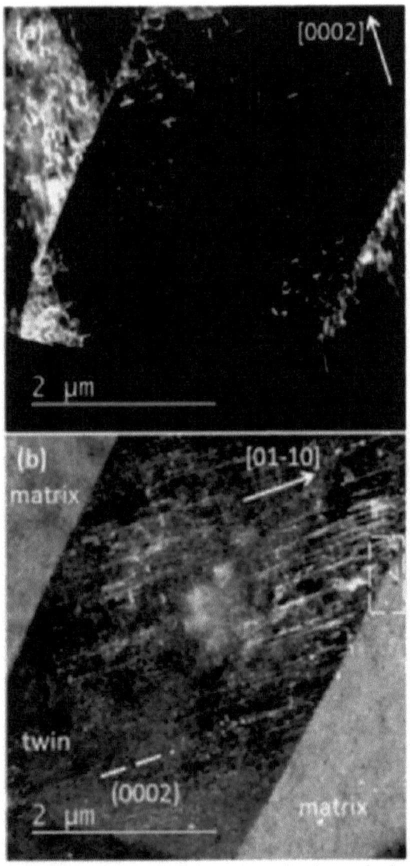

Figure 7. WBDF images of dislocations in the twin by using two diffraction vectors. The dashed line box in (b) highlights the [c] containing dislocation in the matrix.

According to the invisibility criterion for reflections [0002] and $[01\bar{1}0]$, in Figure 8, dislocations in (a) are either pure [c] or <c+a>, and those in (b) are either a_2, a_3 or $[c+a_2]$, $[c+a_3]$. Therefore, those dislocations present in both reflections as indicated by the short arrows have the Burgers vector of $[c+a_2]$ or $[c+a_3]$, and they are parallel to the basal plane; the non-rectilinear dislocations visible in Figure 8(a) but not in (b) have the Burgers vector of [c] or $[c+a_1]$ and are operating on the non-basal planes. It should be noted that the <c+a> dislocation could assume a dissociated configuration on the basal plane according to [18], therefore different contrast pattern follows than considering only the perfect dislocations. Analysis was focused here on the out of basal plane dislocations.

The existence of the possible Burgers vectors [c] or $[c+a_1]$ for non-basal dislocations was further examined by Schmid factor analysis. The external loading direction is closely aligned with the basal plane of the matrix due to the pronounced texture, and is approximately aligned with the c-axis in the extension twin by the characteristic 86° rotation. The Schmid factors of the possible slip systems in the twin are hence calculated assuming the loading direction along the [0001] direction and are listed in Table 2. For <c+a> dislocation, both the first and second order pyramidal planes are possible glide planes, and the Schmid factors for glide on the two planes are similar. The point of this section of the paper is to highlight the low probability that the contrast is due to c-type dislocations, rather than <c+a>, so only the Schmid factors for <c+a> on $\{2\bar{1}\bar{1}2\}$ are listed for comparison. The absolute values of the Schmid factors are taken since negative value merely means dislocation gliding on the opposite direction. It is readily revealed that the activation of pure [c] dislocation in the analyzed twin is unfavorable, whereas the <c+a> dislocations are necessary in terms of accommodating the imposed deformation judging from the near maximum Schmid factor. By this analysis, the Burgers vectors of the out of basal, non-rectilinear dislocations are $[c+a_1]$ and $[c-a_1]$.

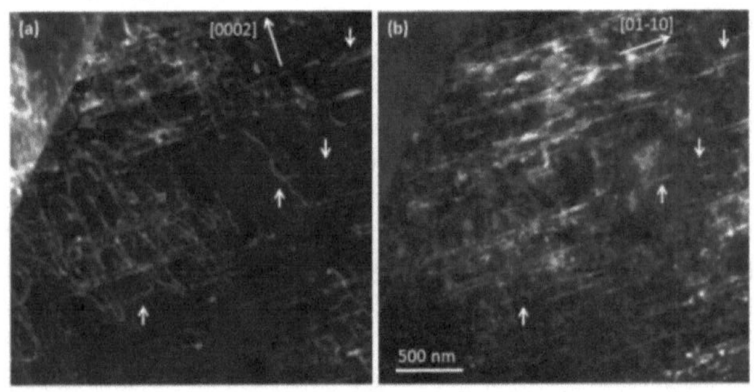

Figure 8. One area from Figure 7 to compare the visibility of dislocations in two diffraction conditions. The short arrows indicate the contrast from the same dislocation segment in both images.

Table 2. Schmid factors (*m*) for possible dislocation types. The loading direction is taken as [0001]. And the absolute values of *m* are taken.

Type	b	Slip plane	*m*
$[c-a_1]$	$1/3[\bar{2}113]$	$(2\bar{1}\bar{1}2)$	0.45
$[c+a_1]$	$1/3[\bar{2}113]$	$(\bar{2}11\bar{2})$	0.45
[c]	[0001]		0

TEM observation of stacking faults associated with twins

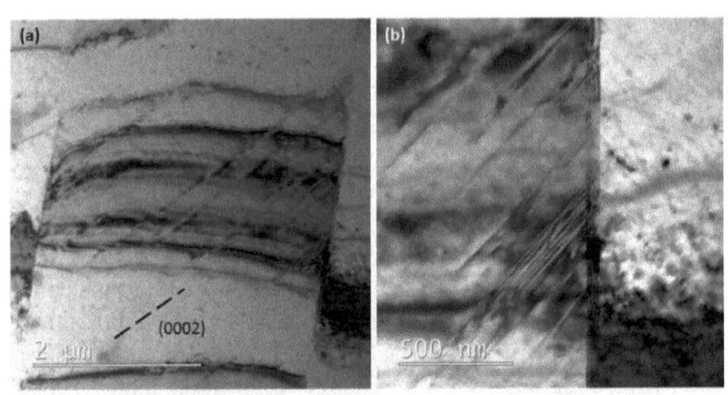

Figure 9. Stacking faults emanating from twin boundary in an extension twin in the 0.5% sample.

At 0.5% strain, there are very few dislocations of any type within the twins (Figure 9). On the other hand, there are wide stacking faults lying in the basal plane, being either I_1 or I_2 type. Although in the figure SFs seem to terminate in the interior of the twin, this is ascribed to the loss of contrast outside the bend contour, which is the dark horizontal band in the middle of Figure 9(a). At the bottom of this figure where a dark band exists, the SF is again visible. Considering the extended width of a few SFs at the center of Figure 9 (a), it is valid to conclude the SFs may span the entire twin. SFs traversing the twin were also reported by other authors, e.g. [19]. This is consistent with the twin growth (boundary migration) mechanism that has been proposed by J. Tu et al. [20]. Via molecular dynamic modeling, they suggest that the twin growth (migration of extension twin boundary) is accomplished by the migration of either the coherent twin boundary (CTB) or the prismatic/ basal interface (PB/ BP), both of which result in the emission of a leading Shockley partial dislocation on the basal plane from the twin boundary, leaving a SF in the swept region, which is I_2 type.

Figure 10 displays the SFs emitted from a twin boundary. Noteworthy is the lines with faint contrast parallel to the basal plane in the matrix, and these lines are connected with the SFs in the twin at the twin boundary. In the meantime, the contrast on the twin boundary superimposing on the boundary fringes is indicative of interface or twin dislocations. Further examination of these associated defects is required to reveal or prove the mechanism of SFs emission from twin boundary.

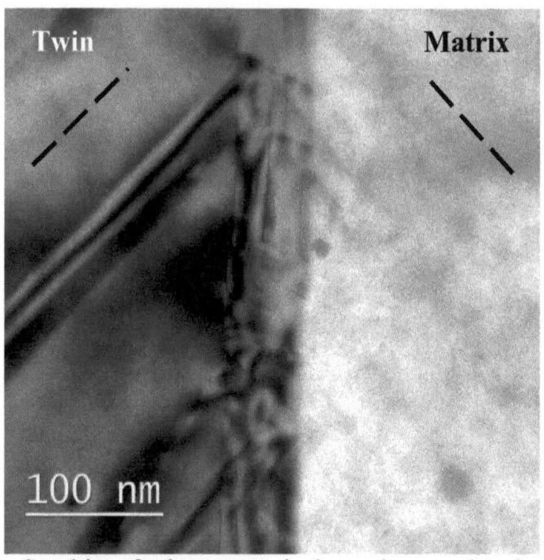

Figure 10. Stacking faults at a twin boundary and twin boundary dislocations in the 0.5% sample. The dashed lines indicate the trace of basal plane .

Finally, it is noted that no stacking faults are apparent within the matrix at the magnification employed in this study up to the strain of 2.0%. This is not surprising, even though the I_2 stacking fault energy is relatively low, because the equilibrium spacing of Shockley partials is only about 3 to 5 Burgers vectors or 6 to 16 Angstroms, as predicted by *ab initio* calculations [21].

Conclusions

By TEM analysis of the defect structure in the AZ31 RD-compression samples, specifically in the extension twin and the proximate matrix, the following conclusions can be drawn:

1. It is confirmed that in the matrix basal slip occurs firstly, as reported by the in-situ neutron diffraction and crystal plasticity modeling [3] [4].
2. At 0.5% strain where extension twinning begins to occur, the matrix is dominated by <a> dislocation on the basal plane, and dislocation density is low in the twin interior. However, SFs on the basal plane are present. And the morphology is consistent with the SFs emission mechanism from twin boundaries proposed by MD modeling [20].
3. At 2.0% strain, there is a higher density of <a> dislocations in the matrix. Dislocations with [c] component are observed emanating from the twin boundary into the matrix.
4. An abundant <c+a> dislocations were identified in the extension twin at 2.0% strain. These <c+a> dislocations have shorter length aligned with the basal plane and longer non-rectilinear segments on the non-basal planes. Their presence can be related to the higher hardening rate of the twin.

In addition to the above conclusions, several issues remain to be investigated in the future:

1. The <a> dislocation transmutation across an extension twin boundary might result in <c+a> dislocation in the twin. This provides a mechanism and explanation of higher dislocation density in the twin and the consequently high hardening rate.
2. Alternatively, the twin boundary may serve as a source mechanism for <c+a> dislocations apart from transmutation. Perhaps it is aided by the stress concentration provided by <a> dislocation pile-ups.
3. The twin boundary defect structure, specifically the interface dislocations can provide insights on the emission mechanism of stacking faults into the twin and the matrix.

Acknowledgements

This work was sponsored by the National Science Foundation grant number CMMI 1235259 (Mary Toney).

References

[1] J. Koike, "Enhanced deformation mechanisms by anisotropic plasticity in polycrystalline Mg alloys at room temperature," *Metall. Mater. Trans. A,* vol. 36, p. 1689–1696, 2005.

[2] D.W. Brown et al., "Internal strain and texture evolution during deformation twinning in magnesium," *Mat. Sci. Eng. A,* vol. 399, no. 1-2, pp. 1-12, 2005.

[3] S.R. Agnew, D.W. Brown, and C.N. Tome, "Validating a polycrystal model for the elastoplastic response of magnesium alloy AZ31 using in situ neutron diffraction," *Acta Mater.,* vol. 54, no. 18, pp. 4841-4852, 2006.

[4] H. Wang et al., "Study of lattice strains in magnesium alloy AZ31 based on a large strain elastic-viscoplastic self-consistent polycrystal model," *Int. J. Solids Struct.,* vol. 49, pp. 2155-2167, 2012.

[5] J. Wang, J. Beyerlein, and C.N. Tome, "An atomic and probabilistic perspective on twin nucleation in Mg," *Scripta Mater.,* vol. 63, no. 7, pp. 741-746, 2010.

[6] M.R. Barnett, M.D. Nave, and A. Ghaderi, "Yield point elongation due to twinning in a magnesium alloy," *Acta Mater.,* vol. 60, no. 4, pp. 1433-1443, 2012.

[7] H. Wang et al., "A constitutive model of twinning and

detwinning for HCP polycrystals," *Mater. Sci. Eng. A,* vol. 555, pp. 93-98, 2012.

[8] L. Remy, "Twin-slip interaction in f.c.c. crystals," *Acta Metallurgica,* vol. 25, no. 6, pp. 711-714, 1977.

[9] M. Niewczas, "Lattice correspondence during twinning in hexagonal close-packed crystals," *Acta Mater.,* vol. 58, no. 17, p. 5848–5857, 2010.

[10] Z.S. Basinski et al., *Revue de Metallurgie,* vol. 94, p. 1037, 1997.

[11] M.R. Barnett et al., "Influence of grain size on the compressive deformation of wrought Mg–3Al–1Zn," *Acta Mater.,* vol. 52, no. 17, p. 5093–5103, 2004.

[12] S.R. Agnew et al., "Study of slip mechanisms in a magnesium alloy by neutron diffraction and modeling," *Scripta Mater.,* vol. 48, no. 8, pp. 1003-1008, 2003.

[13] L. Wu et al., "The effects of texture and extension twinning on the low-cycle fatigue behavior," *Mater. Sci. Eng. A,* vol. 527, pp. 7057-7067, 2010.

[14] J.J. Jonas et al., "The role of strain accommodation during the variant selection of primary twins in magnesium," *Acta Mater.,* vol. 59, no. 5, pp. 2046-2056, 2011.

[15] N. Stanford, D. Atwell, "The Effect of Mn-rich Precipitates on the Strength of AZ31 Extrudates," *Metall. Mater. Trans. A,* vol. 44, pp. 4830-4843, 2013.

[16] S. R. Agnew, O. Duygulu, "Plastic anisotropy and the role of non-basal slip in magnesium alloy AZ31B," *International Journal of Plasticity,* vol. 21, p. 1161–1193, 2005.

[17] S. Morozumi, M. Kikuchi, and H. Yoshinaga, "Electron Microscope Observation in and around {1-102} Twins in Magnesium," *Transactions of the Japan Institute of Metals,* vol. 17, no. 3, pp. 158-164 , 1976.

[18] S.R. Agnew, L. Capolungo, and C.A. Calhoun;, "Connections between the basal I1 "growth" fault and <c+a> dislocations," *Acta Mater.,* 2014. Accepted.

[19] B. Li et al., "Transmission electron microscopy study of stacking faults and their interaction with pyramidal dislocations in deformed Mg," *Acta Materialia,* vol. 58, pp. 173-179, 2010.

[20] J. Tu et al., "Structural characterization of {10-12} twin boundaries in cobalt," *Applied Physics Letters,* vol. 103, p. 051903 , 2013.

[21] J.A. Yasi et al., "Basal and prism dislocation cores in magnesium: comparison of first-principles and embedded-atom-potential methods predictions," *Modelling Simul. Mater. Sci. Eng.,* vol. 17, p. 055012, 2009.

Magnesium Technology 2015
Edited by: Michele V. Manuel, Alok Singh, Martyn Alderman, and Neale R. Neelameggham
TMS (The Minerals, Metals & Materials Society), 2015

Role of Tensile Twinning on Fracture Behavior of Magnesium AZ31 Alloy

Subrahmanya Prasad Narla[1], Narasimhan R[1], Satyam Suwas[2]

[1]Department of Mechanical Engineering, Indian Institute of Science, Bangalore 560012, India. Tel: +91-80-22932959
[2]Department of Materials Engineering, Indian Institute of Science, Bangalore 560012, India. Tel: +91-80-22933245

Keywords: Magnesium alloy, Mg AZ31, Fracture behavior, Tensile twinning, EBSD.

Abstract

In this work, the mode I fracture response of a rolled Mg AZ31 alloy is investigated by conducting experiments using notched compact tension specimens. It is found that the notched fracture toughness $J_c \sim$ 42 N/mm which is moderately high. Detailed examination using EBSD and optical metallography indicates that profuse tensile twinning occurs almost over the entire uncracked ligament ahead of the notch root. This leads to considerable hardening as evidenced by slope of the load versus displacement curve and toughness enhancement by plastic dissipation. Strong texture changes are also observed near the extended crack tip. The fracture morphology involves dimpling at the specimen mid-plane and shear lip formation close to the free surfaces. The fracture resistance of the Mg alloy is discussed in the context of the above observations.

Introduction

Transportation industry always prefers to use light weight materials so as to increase the fuel efficiency. In this context, Mg alloys are attractive because of their low density and high specific strength. Further, among wrought Mg alloys, rolled Mg AZ31 has the greatest potential for structural application and has received considerable attention in recent years [1]. To use wrought magnesium alloys as structural materials, it is important that they have a good combination of mechanical properties - viz., ductility, tensile strength, strain hardening and fracture toughness.

Plastic deformation in Mg AZ31 is due to both slip and twinning. Deformation along the basal plane is accommodated by basal as well as non-basal slip systems (prismatic and pyramidal $\langle a \rangle$). Out of the four different types of slip systems, only pyramidal $\langle c+a \rangle$ can accommodate plastic deformation along c-axis. However, since the critical resolved shear stress (CRSS) of this slip system is very high, it is difficult to activate at room temperature. The only other mechanism that can account for deformation along c-axis at room temperature is deformation twinning. Extension along the c-axis of the crystal is caused by $\{10\bar{1}2\}$ or tensile twins (TTs). TTs form when tensile stress is applied along c-axis or compressive stress is applied perpendicular to c-axis. The interaction between these twin and slip systems, has a strong influence on the mechanical response of Mg alloys [2, 3].

There have been extensive studies on compression and tension behaviour of Mg AZ31 alloy. In contrast, there are relatively few reports on fracture response. The effect of texture on fracture toughness in a rolled Mg AZ31 alloy plate was investigated by Somekawa and Mukai [4]. Following ASTM-E399 procedure, fracture toughness tests were conducted using three point bend (TPB) specimens cut along RD, TD and 45° to RD from a rolled plate with thickness of 6 mm having a grain size of 65 μm. Dimples were observed on the fracture surface and the fracture toughness was found to be 17.6-20.7 MPa $m^{1/2}$.

Somekawa et al. [5] reported a fracture toughness value of 21.5 MPa $m^{1/2}$ for an extruded AZ31 bar with an average grain size of 51.6 μm. Tensile twins formed near the crack tip before crack initiation. At peak load, the crack propagated along the boundary between the TTs and the matrix without blunting resulting in brittle fracture. By decreasing the grain size, the tendency for twinning decreases as reported by Meyers et al. [6]. Somekawa and Mukai [7] found that fracture toughness increases from 12.7 MPa $m^{1/2}$ to 17.8 MPa $m^{1/2}$ by reducing the grain size from 55 μm to 1 μm in extruded pure magnesium. The fracture mechanism of fine-grained extruded Mg-Zn binary alloy with a grain size of 1-3 μm was studied by Somekawa et al. [8]. Subgrained structures were observed near the crack tip instead of deformation twins which lead to higher fracture toughness. On the fracture surface, dimple pattern was observed which indicates the occurrence of ductile fracture as opposed to brittle failure in the case of coarse-grained Mg AZ31 alloy. In a very recent work, Kondori and Benzerga [9] investigated the effect of notch root radius on tensile ductility of Mg AZ31 alloy and found that the fracture mechanism changes from micro-void coalescence to twin-sized micro-cracking as triaxiality increases. More recently, Kaushik et al. [10] have performed mode-I fracture tests using three point bend specimens of magnesium single crystals. In one of the orientations considered in their experiments, the notch front was aligned along c-axis which is similar to the basal-textured rolled polycrystalline Mg AZ31. From the EBSD maps on the free surface, it was observed that TTs nucleate near the loading edge and as load increases, extend towards notch root leaving a small region around the notch root free of twins. It was concluded that plastic dissipation due to profuse tensile twinning can enhance the fracture toughness although the crack eventually propagates along the prominent twin-matrix boundary and gets deflected at twin-twin intersections.

In this paper, mode-I fracture experiments were conducted at room temperature using compact tension (CT) specimens cut from a rolled Mg AZ31 plate having nearly basal texture. In one set of specimens, load was applied parallel to rolling direction (RD specimen) and the other was loaded along transverse direction (TD specimen). Since there was not much difference in behaviour exhibited by the two specimen sets, only results corresponding to RD specimen are presented in this paper. It was found that for the present alloy, notched fracture toughness $J_c \sim 42$ N/mm which is much higher compared to the fracture toughness values determined from fatigue precracked specimens [4]. The fracture surface shows crack tunneling in the specimen interior and shear lips on the free surface. SEM fractographs reveal dimple fracture. Extensive plastic deformation is noticed prior to crack initiation as manifested by nonlinearity in the load-displacement curves and is mediated by profuse tensile twinning. TEM investigation shows also strong dislocation activity in addition to tensile twinning. The moderately high notched fracture toughness displayed by the Mg alloy is rationalized by a combination of work of separation in the fracture process zone near the notch root associated with micro-void coalescence and dissipation in the surrounding plastic zone due to tensile twinning. Another important observation is occurrence of significant texture change near the extended crack tip following reorientation in the twinned regions. The implication of hardening associated with tensile twinning on the operative fracture mechanism is discussed.

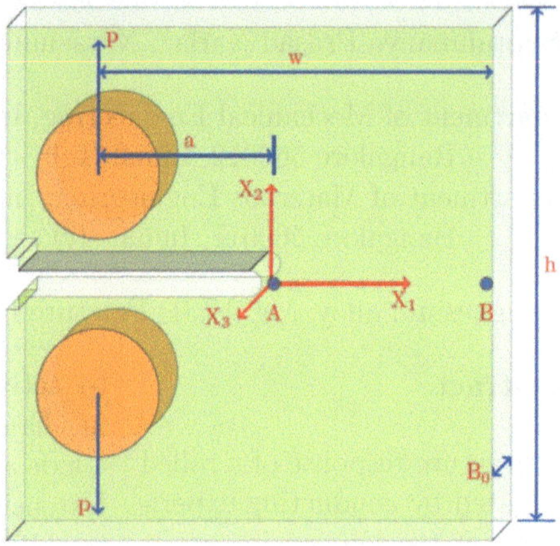

Figure 1: Schematic of compact tension (CT) specimen with various dimensions.

Experimental Procedure

The material used for the present experiments is a hot rolled magnesium alloy AZ31 with an average grain size of 14 μm. The CT specimens employed in the experiments are pre-notched using Electric Discharge Machining (EDM) to produce notches of width 90 μm parallel to TD and loaded along RD (denoted as X_1 and X_2 in Fig. 1). Referring to Fig. 1, the other dimensions of the specimen are a =22 mm, w=54 mm, h=60 mm and $B_0 = 9.6$ mm.

Specimens were loaded in displacement-control mode at a constant displacement rate of 0.3 mm/min in an Instron 8502 universal testing machine. In order to understand the evolution of microstructure with load level, some of the specimens were unloaded before crack initiation. Samples were extracted from the deformed specimens by EDM near the locations A and B in Fig. 1 for optical metallography, EBSD analysis and fractography studies.

For optical metallography, samples were mechanically polished followed by an etching process and observed using Zeiss Axio Vert.A1 microscope. In-order to get high quality Kikuchi patterns for EBSD analysis, samples were mechanically polished followed by electro-polishing. EBSD scans were performed with a step size of 0.8 μm in ESEM Quanta 200, FEI machine and the patterns were indexed using TSL software.

Results

The load (P) versus load line displacement (Δ) for the RD specimen is shown in Fig. 2. It can be seen

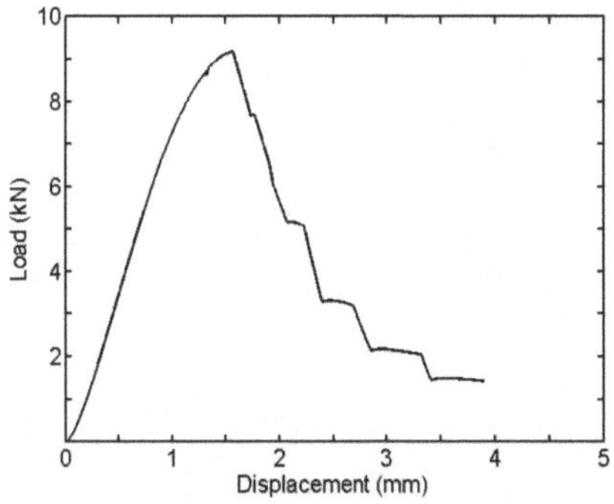

Figure 2: Load versus displacement curve for RD specimen.

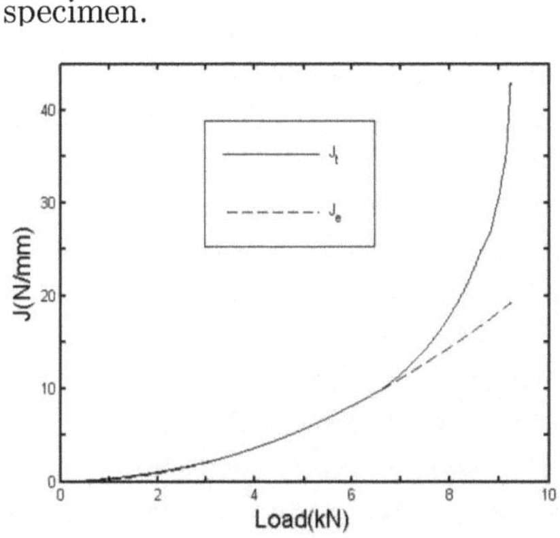

Figure 3: Variation of energy release rate J with load deduced from the load-displacement curve along with J_e versus load predicted by linear elastic analysis.

from this figure that there is a deviation from linearity at a load of 5 kN. The material displays high strain hardening which is evident from the continued strong increase in load with displacement beyond 5 kN. This will be discussed in the following as an outcome of profuse tensile twinning. Crack initiation was observed on the free surface at the peak load of 9.1 kN which was followed by stable crack growth.

The energy release rate J was computed from the P-Δ curve presented in Fig. 2 following the procedure described in ASTM E813 – 88 standards [11]. The total value of ${J_t}^i$ corresponding to a point (Δ^i, P^i) of the load versus displacement curve is obtained as:

$$ {J_t}^i = {J_e}^i + {J_p}^i \tag{1} $$

where ${J_e}^i$ and ${J_p}^i$ are elastic and plastic components of J^i. Assuming an isotropic response with Young's modulus E=45 GPa and Poisson's ratio $\nu = 0.29$, ${J_e}^i$ is calculated using an empirical expression for

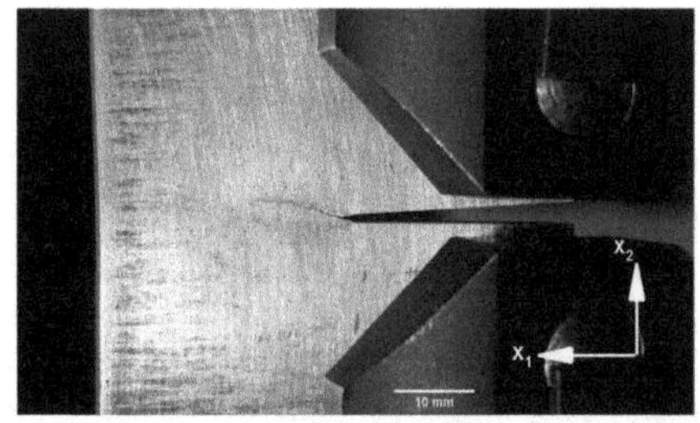

Figure 4: Photograph showing extended crack from the pre-notch. Crack trajectory appears inclined due to formation of shear lips adjacent to the specimen free surfaces.

the stress intensity factor as described in ASTM E813 – 88 standards [11]. Plastic component of ${J_p}^i$ is determined as

$$ {J_p}^i = \frac{\eta\, {A_p}^i}{B_0\, b}, \tag{2} $$

where $\eta = 2$ is a shape factor, $b = W - a$ is the uncracked ligament length, B_0 is the specimen thickness and ${A_p}^i$ is the area under load versus plastic displacement curve defined as in ASTM E813 – 88 standards [11]. The total energy release rate J determined by the above procedure is plotted against load P in Fig. 3 up to crack initiation stage along with the elastic estimate J_e. Both the curves J_e and J_t coincide up to $P \sim 5$ kN. Thereafter the J_t curve increases above J_e which indicates considerable plastic dissipation before crack initiation. At peak load, the value of J_t is 42.9 N/mm and the equivalent value of K_I is 45.9 MPa $m^{1/2}$. By contrast, Somekawa and Mukai [4] have reported values of K_{IC} in the range of 17.6 - 20.7 MPa $m^{1/2}$ for a specimen with a fatigue pre-crack, cut from rolled Mg AZ31 alloy sheet having basal texture. However, it must be mentioned that in their experiments, they could not meet the specimen size requirements stipulated by ASTM K_{IC} test procedure. Therefore, they employed the measured stretch zone height to estimate K_{IC}.

During the experiments, the region around the notch was carefully imaged using a digital camera mounted in front of the testing machine. It can be seen from Fig. 4, that the crack appears to grow along a curved path from the notch root due to the formation of shear lips. Crack tunneling occurs at the mid-thickness of the specimen while shear lips form adjacent to the free surfaces. SEM images of the fracture surface in the flat tunneled region and on the shear lips are shown in Fig. 5(a) and (b), respectively. From Fig. 5 (a), it can be seen that there is a combination of large and small voids which could have nucleated from second phase particles like

Mn, Al-Mn, MgO [9]. Hydrostatic stress is higher at mid-thickness which responsible for void initiation, growth and coalescence. In contrast to Fig. 5(a), the voids are shallow and sheared along an inclined direction in Fig. 5 (b).

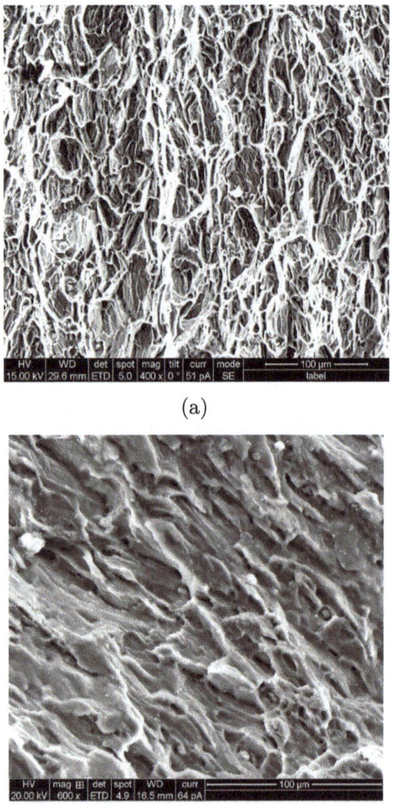

(a)

(b)

Figure 5: SEM fractographs obtained (a) at mid-thickness and (b) from shear lip region.

It can be seen from the macroscopic load-displacement and J versus load curves, there is considerable non-linear hardening behaviour for load higher than 5 kN. Microscopic analysis was conducted to understand the mechanisms responsible for the high macroscopic hardening. Specimes were unloaded before crack initiation at load levels of 8.1 and 8.7 kN. Optical metallographs and EBSD scans were taken from samples extracted at both locations A and B shown in Fig. 1 for the two load levels.

Inverse pole figure (IPF) taken near location B corresponding to a load of 8.7 kN is shown in Fig. 6. It can be seen that some grains such as those marked as A and B have multiple parallel twins. It is clear from the almost 90 degree lattice rotation within the twinned region (with respect to the matrix) that

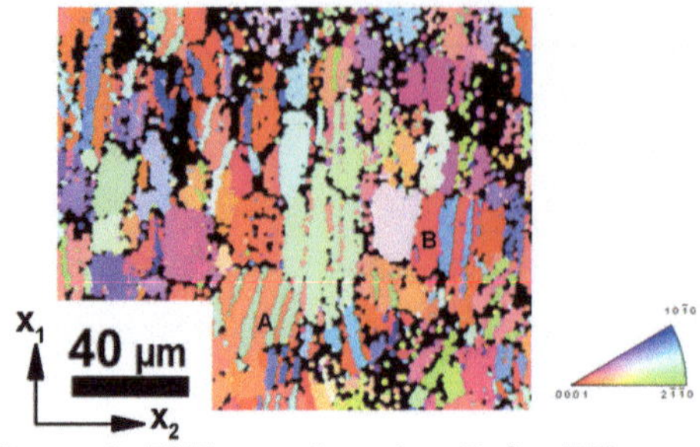

Figure 6: IPF near location B for RD specimen.

these are twins of $\{10\bar{1}2\}$ type. An optical metallograph at a distance of 1 mm in front of location B corresponding to a load of 8.7 kN is shown in Fig. 7. Most of the twins of $\{10\bar{1}2\}$ type are aligned along the central line of the notch. Twin growth can occur by either twin-twin coalescence of same twin variants or by lengthening and widening of individual twins [10]. Some of the grains such as marked as C, D and E in Fig. 7 have multiple thick parallel twins which are likely to undergo coalescence as load increases. Twin-twin intersections can be seen in grains marked as F and G. Thus, the lengthening of one twin is obstructed by another twin variant which was also observed by Kaushik et al. [10] in their fracture experiments on Mg single crystals. If there is no obstruction, the twins extend from one end to the other end of the grain. Twins extend across the grains marked as H and I. Twinning near the specimen edge as observed above can be attributed to the formation of plastic hinge in the uncracked ligament ahead of the notch tip [10]. The normal stress on the plane ahead of the notch tip changes from tensile to compressive beyond the hinge point which induces out-of-plane bulging (i.e., in [0001] direction) and triggers tensile twinning to accommodate it.

Optical metallograph obtained at location A corresponding to a load of 8.7 kN is shown in Fig. 8. Profuse tensile twinning can be seen near the notch tip from this figure. Kaushik et al. [10] in their fracture experiments on Mg single crystal, did not observe tensile twinning near notch root when the c-axis is along the notch front. The occurrence of TTs near the notch root in the present polycrystalline specimens is intriguing because it would lead to bulging along c-axis which in most grains is along the specimen thickness direction. Intrestingly, TTs were also observed on the surface of tension specimens loaded along RD by Koike et al. [2]. They named these twins as anomalous tensile twins. It was found that basal slip gets easily activated in grains having slight deviation of c-axis from the ND of the rolled sheet. The pile up of these basal dislocations at grain boundaries, creates a stress concentration and also gives rise to strain incompatibility with adjacent grains.

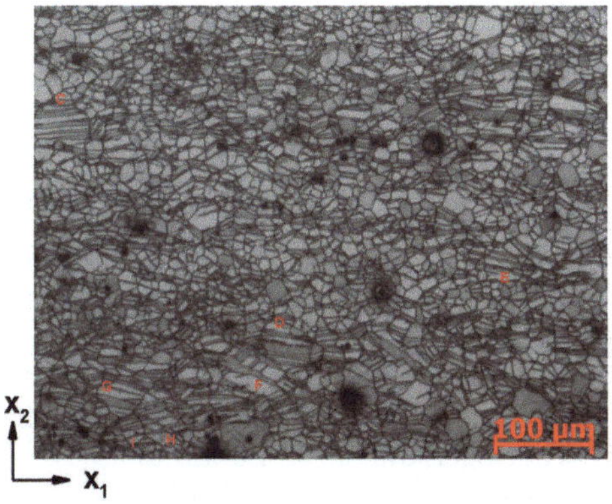

Figure 7: Opical metallograph at a distance of 1 mm from location B along the center-line of the notch for RD specimen.

Koike et al. [2] argued that in such circumstances, anamolous TTs form to accommodate the strain incompatibility and are triggered by the anisotropic stress concentration field created by the basal dislocation pile-up. Thus, as opposed to the global bending field mentioned earlier, it is the local stress and strain field at boundaries between soft and relatively harder grains that trigger TTs near the notch root.

Discussion

From the above observations, moderately high fracture toughness obtained in the present experiments could be due two reasons- viz., notch root effect and energy dissipation by profuse twinning in the background plastic zone. Fatigue pre-cracked specimens are expected to have less fracture toughness as compared to notched specimens. For Al alloys, Kamat and Eswara Prasad [12] have studied the effect of notch root radius on fracture toughness using compact tension specimens and found a threshold notch root radius below which toughness is constant and equal to that determined from fatigue pre-cracked specimens. No such study has been reported for Mg alloys in the literature. The results of the present work and those of Kondori and Benzerga [9] from notched tensile specimens indicate that when stress triaxiality is moderate, the fracture mechanism in AZ31 alloy is ductile void growth and coalescence. On the other hand, with reduction in notch root radius or in fatigue pre-cracked specimens, the fracture mechanism may transition to brittle due to occurrence of twin-size micro-cracks [5, 9]. This could cause reduction in toughness. Further, it is important to recognize that TT formation near the notch root as seen in Fig. 8 can contribute to strain hardening by Hall-Petch effect, Basinski mechanism and texture hardening. In the first, the effective slip length decreases due to the formation of TTs in the matrix region which can cause dislocation pile up at twin boundaries and lead to hardening. According to the second, glissile dislocations change to sessile within the twined regions thus making them much harder than the matrix [13]. Further, as evidenced by Figs. 6-8, since TTs can engulf the grain quickly, the whole grain can re-orient to a harder orientation which leads to texture hardening [3]. In this connection, it must be mentioned that significant texture changes were noticed around the extended crack tip in the present experiments. The combined influence of these mechanisms is reflected in the high slope of the load-displacment curve shown in Fig. 2 well after it deviates from linear elastic behavior. The hardening thus imparted by tensile twinning near the notch root has important consequences on the operative micro-void growth and coalescence fracture mechanism. Indeed, it has been shown by Thomason [14] through extension of Rice and Tracey analysis that growth of a spherical void in an isotropic plastic solid is retarded when it exhibits hardening. Thus, fracture initiation is expected to be delayed due to the influence of TTs occuring in the region around the fracture process zone where micro-void growth and coalescence takes place. Further, Kaushik et al. [10] showed that there is a correlation between evolution histories of J and average TT volume fraction from their experiments on Mg single crystals, thereby establishing role of TTs in energy dissipation and toughening. The same is expected to apply in the polycrystalline Mg alloy as well in view of the profuse tensile twinning seen in the uncracked ligament (Figs. 6-8). Thus, the moderately high fracture toughness is due to energy dissipation by both dislocation slip and twinning in the background plastic zone as well as the high work of separation in the fracture process zone (due to micro-void growth and coalescence and shearing in the shear-lip region).

Summary and Conclusions

In this work, mode I fracture experiments have been conducted at room temperature using CT specimens cut from a rolled Mg AZ31 alloy and loaded parallel to RD. The main observations / conclusions from the present work are summarized below.

Before crack initiation at peak load, there is considerable non-linear hardening behaviour observed from load-displacement curve. The notched fracture toughness value is $J_c \sim 42.9$ N/mm which is moderately high as compared to data reported in the

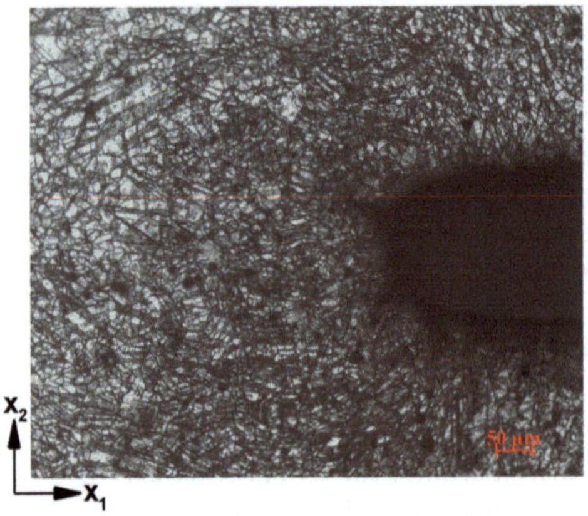

Figure 8: Opical metallograph obtained at location A.

literature for fatigue pre-cracked specimens. Crack propagation occurred by tunneling in the specimen interior and shear lips on the free surface. Fractographs show ductile void growth and coalescence at the mid-thickness and shear type failure in the shear lip region. The higher fracture resistance seen in the present notched specimens is attributed to two important effects caused by tensile twinning - substantial dissipation in the background plastic zone and retardation of void growth in the fracture process zone due to hardening.

Acknowledgments

R. Narasimhan would like to thank Department of Science and Technology (Government of India) for financial support under the JC Bose Fellowship scheme.

References

[1] C. Blawert, N. Hort, and K. U. Kainer, "Automotive applications of magnesium and its alloys," *Transactions of the Indian Institute of Metals*, vol. 57, pp. 397–408, 2004.

[2] J. Koike, Y. Sato, and D. Ando, "Origin of the anomalous $10\bar{1}2$ twinning during tensile deformation of Mg alloy sheet," *Materials Transactions*, vol. 49, pp. 2792–2800, 2008.

[3] M. Knezevic, A. Levinson, R. Harris, R. K. Mishra, R. D. Doherty, and S. R. Kalidindi, "Deformation twinning in AZ31: Influence on strain hardening and texture evolution," *Acta Materialia*, vol. 58, pp. 6230–6242, 2010.

[4] H. Somekawa and . T. Mukai, "Fracture toughness in a rolled AZ31 magnesium alloy," *Journal of Alloys and Compounds*, vol. 417, pp. 209–213, 2006.

[5] H. Somekawa, A. Singh, and T. Mukai, "Fracture mechanism of a coarse-grained magnesium alloy during fracture toughness testing," *Philosophical Magazine Letters*, vol. 89, pp. 2–10, 2009.

[6] M. A. Meyers, O. Vohringer, and V. A. Lubarda, "The onset of twinning in metals: A constitutive description," *Acta Materialia*, vol. 49, pp. 4025–4039, 2001.

[7] H. Somekawa and T. Mukai, "Effect of grain refinement on fracture toughness in extruded pure magnesium," *Journal of Alloys and Compounds*, vol. 53, pp. 1059–1064, 2005.

[8] H. Somekawa, K. Nakajima, A. Singh, and T. Mukai, "Ductile fracture mechanism in fine-grained magnesium alloy," *Philosophical Magazine Letters*, vol. 90, pp. 831–839, 2005.

[9] B. Kondori and A. A. Benzerga, "Effect of stress triaxiality on the flow and fracture of Mg alloy AZ31," *Metallurgical and Matrials Transactions A*, vol. 45A, pp. 3292–3307, 2014.

[10] V. Kaushik, R. Narasimhan, and R. K. Mishra, "Experimental study of fracture behaviour of magnesium single crystals," *Material Science & Engineering A*, vol. 590, pp. 174–185, 2014.

[11] *ASTM E813-88, 1988. Standard test method for* J_{IC}*, a measure of fracture toughness. American society for testing and materials.* (American society for testing and materials, 1988).

[12] S. V. Kamat and N. E. Prasad, "Effect of notch root radius on fracture toughness of an 8090 aluminium-lithium alloy," *Scripta Metallurgica et Materialia*, vol. 25, pp. 1519–1523, 1991.

[13] A. A. Salem, S. R. Kalidindi, R. D. Doherty, and S. L. Semiatin, "Strain hardening due to deformation twinning in α-Titanium: Mechanisms," *Metallurgical and Matrials Transactions A*, vol. 37A, pp. 259–268, 2006.

[14] P. F. Thomason, *Ductile fracture of metals.* (PERGAMON Press, Oxford, England, 1990), 1989.

Magnesium Technology 2015

Deformation II

Magnesium Technology 2015
Edited by: Michele V. Manuel, Alok Singh, Martyn Alderman, and Neale R. Neelameggham
TMS (The Minerals, Metals & Materials Society), 2015

New Model Predicting the Unusual Buckling Behavior of AZ31 Mg

Nitin Chandola[1], Oana Cazacu[1]*, Benoit Revil-Baudard[1]
[1]University of Florida, Dept. of Mechanical and Aerospace Engineering,
REEF,1350 N. Poquito Road Shalimar, FL 32579, USA

Keywords: Plastic buckling; strength differential effects; anisotropy; magnesium

Abstract

In this paper, for the first time it is demonstrated that only by accounting for the combined effects of the tension-compression asymmetry and strong anisotropy of AZ31 Mg it is possible to predict the very unusual buckling of this material, and most importantly capture the energy absorbed prior to buckling. It is shown that by using an elastic/plastic model based on Cazacu et al. [1] criterion that accounts for both anisotropy and strength differential effects, a strong influence of the direction of crushing is predicted. However, if Hill's [2] criterion is used i.e. strength differential effects are not taken into account, the predictions indicate that the orientation of the tube has no influence on the final geometry of the crushed specimen. Furthermore, the location and cross section of the buckle are not correctly captured by the Hill [2] yield criterion.

1. Introduction

Recently, the effect of plastic anisotropy on the buckling characteristics of cylindrical specimens subjected to axial compressive loads have been investigated for fcc materials (e.g. [3]). In comparison, the plastic buckling of hcp materials has received little attention. In this paper, for the first time fundamental issues related to plastic buckling of highly anisotropic AZ31 Mg cylindrical tubes subjected to axial compression are addressed. The outline of the paper is as follows. First, the constitutive models used to describe the plastic flow of AZ31Mg are discussed briefly. Next, FE simulations based on the correctly identified Cazacu et al. [1] yield criterion and the traditionally used Hill [2] yield criterion are carried out for two different loading orientations, rolling direction (RD) and normal direction (ND). The importance of the consideration of the specificities of the plastic flow of AZ31 Mg, namely its tension-compression asymmetry (TCA) on buckling predictions is demonstrated.

2. Macroscopic Elastic/Plastic Modeling

An elastic-plastic modeling approach in the framework of the mathematical theory of plasticity is considered to model the deformation response of AZ31 Mg. The total rate of deformation **D** is the sum of the plastic part $\mathbf{D}^P$ and the elastic part ($\mathbf{D}-\mathbf{D}^P$). The elastic stress-strain relationship is given by,

$$\overset{\Delta}{\sigma} = C^e : \left(D - D^P\right), \tag{1}$$

where, $\overset{\Delta}{\sigma}$ is the Green-Naghdi rate of the stress tensor σ, $\mathbf{C}^e$ is the isotropic fourth-order stiffness tensor. Associated plastic flow is assumed, i.e. the plastic part of the rate of deformation tensor is given by:

$$D^P = \dot{\lambda}\frac{\partial F}{\partial \sigma}, \tag{2}$$

where, F is the yield function and $\dot{\lambda} \geq 0$ is the plastic multiplier. Strain hardening is considered isotropic and governed by the accumulated plastic strain. Thus, the plastic potential F in Equation (2) is of the general form:

$$F\left(\sigma, \bar{\varepsilon}^P\right) = \bar{\sigma}\left(\sigma, \bar{\varepsilon}^P\right) - Y\left(\bar{\varepsilon}^P\right), \tag{3}$$

where, $\bar{\sigma}$ is the effective stress associated with the yield criterion, $\bar{\varepsilon}^P$ the corresponding work-equivalent accumulated plastic strain and $Y = Y\left(\bar{\varepsilon}^P\right)$ is a Voce-type hardening law:

$$Y\left(\bar{\varepsilon}^P\right) = A_0 - A_1 \exp\left(-A_2 \bar{\varepsilon}^P\right), \tag{4}$$

For AZ31 Mg, these parameters, identified from the uniaxial stress-strain curve in the rolling direction (RD) are: A_0 = 315.4 MPa, A_1 = 140.6 MPa, and A_2 =16.3.

<u>2.1 Cazacu et al. (2006) yield criterion</u>

According to the orthotropic form of Cazacu et al. [1] criterion the effective stress $\bar{\sigma}$ is:

$$\bar{\sigma} = B\left[\left(|\Sigma_1| - k\Sigma_1\right)^a + \left(|\Sigma_2| - k\Sigma_2\right)^a + \left(|\Sigma_3| - k\Sigma_3\right)^a\right]^{\frac{1}{a}}, \tag{5}$$

where, k is an internal variable, *a* is a homogeneity constant (usually a = 2) and Σ_1, Σ_2, Σ_3 are the principal values of the transformed stress deviator given as,

$$\Sigma = C : \sigma', \tag{6}$$

In Equation (6), σ' is the deviator of σ, and **C** is a fourth-order symmetric tensor characterizing the orthotropy of the material, while B is a constant defined such that $\bar{\sigma}$ reduces to the tensile flow stress in RD. Thus, the constant B is expressed in terms of the anisotropy coefficients C_{ij} ,with i, j = 1...3 and the material parameter *k* as follows:

$$m = 1/\sqrt{\left(|\Phi_1| - k\Phi_1\right)^2 + \left(|\Phi_2| - k\Phi_2\right)^2 + \left(|\Phi_3| - k\Phi_3\right)^2}$$

where $\Phi_1 = \left(2C_{11}\text{-}C_{12}\text{-}C_{13}\right)/3$, $\Phi_2 = \left(2C_{12}\text{-}C_{22}\text{-}C_{23}\right)/3$;

$\Phi_3 = \left(2C_{13}\text{-}C_{23}\text{-}C_{33}\right)/3$.

The yield criterion involves eight anisotropy coefficients, and a parameter k associated with strength-differential effects in plastic flow. For untextured (i.e. isotropic materials) **C= $\mathbf{I_4}$**, where $\mathbf{I_4}$ is the 4th order identity tensor, so the yield criterion involves only one material parameter, k. It is to be noted that all the parameters involved in the criterion can be determined from a few uniaxial tension and compression tests.

2.2 Hill (1948) yield criterion

For comparison purposes, Hill's [2] orthotropic yield criterion was also employed. In the axes of orthotropy, (*x*, *y*, *z*), (x-along RD, y along TD and z along ND) Hill's equivalent stress is:

$$\bar{\sigma}_{Hill} = \sqrt{\begin{array}{l} F(\sigma_{yy}-\sigma_{zz})^2 + G(\sigma_{zz}-\sigma_{xx})^2 + H(\sigma_{xx}-\sigma_{yy})^2 + \\ 2L(\sigma_{yz})^2 + 2M(\sigma_{zx})^2 + 2N(\sigma_{xy})^2 \end{array}}, \quad (7)$$

where F, G, H, L, M and N being material constants.

3. Application of Macroscopic Modeling Approach to Mg AZ31

The material parameters used in the above yield criteria are determined from the mechanical test results in uniaxial tension and compression reported by Khan and collaborators [4]. To capture the evolving anisotropy, the material parameters were considered to evolve with $\bar{\varepsilon}^p$. Figure 1 shows the AZ31 Mg yield surfaces according to the yield criteria for individual levels of $\bar{\varepsilon}^p$ (up to 0.1) while the value of their parameters are listed in Tables 1, 2. The values corresponding to any given level of plastic strain $\bar{\varepsilon}_j^p \le \bar{\varepsilon} \le \bar{\varepsilon}_{j+1}^p$ is obtained by linear interpolation.

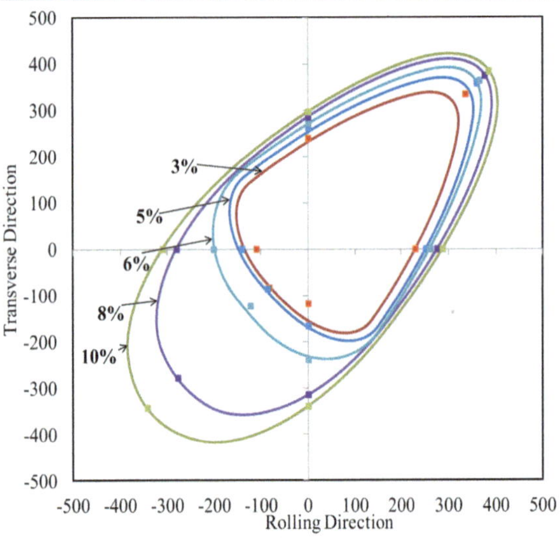

(a)

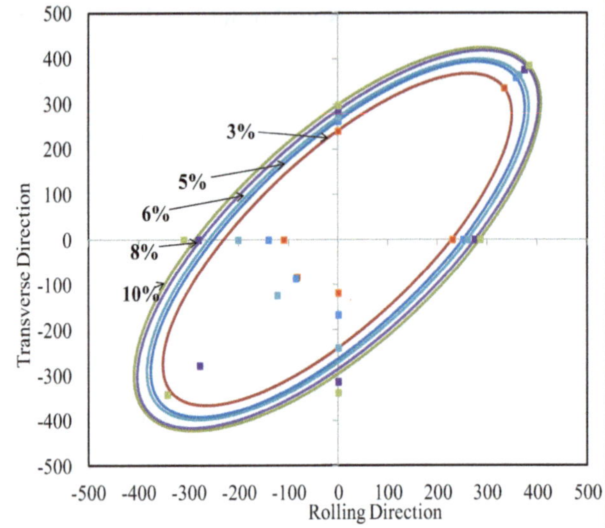

(b)

Figure 1. Evolution of the yield surfaces according to the (a) Cazacu et al. [1] yield criterion (b) Hill [2] yield criterion.

Table 1. Material parameters involved in Cazacu et al. [1] orthotropic yield criterion for AZ31 magnesium alloy corresponding to different values of plastic strain, $\bar{\varepsilon}^p$; C_{11} is set to unity.

$\bar{\varepsilon}^p$	C_{22}	C_{33}	C_{12}	C_{13}	C_{23}	C_{66}	*k*
0.03	1.09	3.34	-0.17	0.10	0.24	7.74	-0.62
0.05	1.09	3.34	-0.17	0.10	0.24	7.74	-0.62
0.06	1.07	2.90	-0.59	-0.28	-0.10	11.02	-0.52
0.08	1.10	1.44	-0.82	-0.52	-0.35	11.95	-0.21
0.10	1.08	0.88	-0.76	-0.66	-0.51	11.21	-0.17

Table 2. Orthotropy coefficients involved in Hill [2] orthotropic yield criterion (Equation (7)) anisotropy coefficients for AZ31 magnesium alloy corresponding to different values of plastic strain.

$\bar{\varepsilon}^p$	F	G	H	N
0.03	0.19	0.28	0.72	2.09
0.05	0.21	0.29	0.71	2.17
0.06	0.22	0.30	0.70	2.18
0.08	0.23	0.30	0.70	2.17
0.10	0.25	0.31	0.69	2.17

4. FE simulation of axial crushing of Mg AZ31 tube

All the axial crushing simulations are carried out using the commercial FE code ABAQUS and a user material routine that was developed for the anisotropic elastic-plastic model given by Equations (1)-(7). The geometry of the specimen and FE mesh (2736 quadratic hexahedral elements with full integration for 1/4th specimen) is shown in Figure 2. The lower nodes (z = 0) were pinned (no displacement allowed), while the upper nodes (z = L_0) were tied to a rigid tool. Axial displacement was imposed by the translation of this tool along the tube axis.

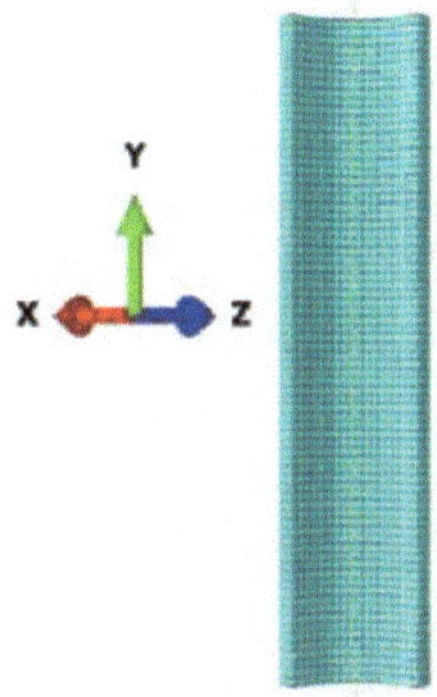

Figure 2. Typical mesh used for the finite element simulation.

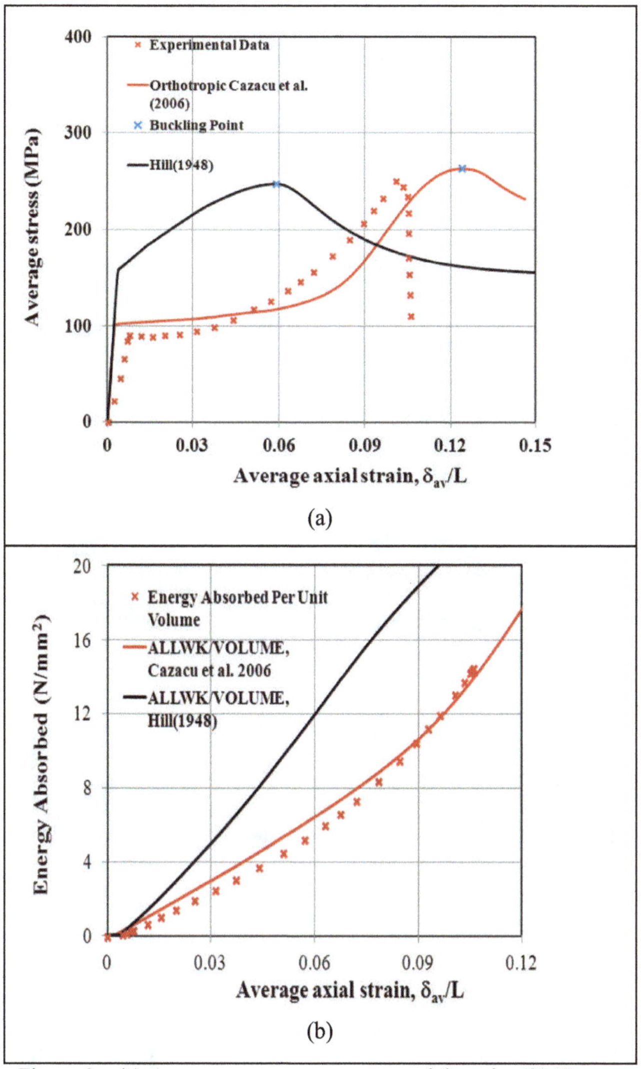

Figure 3. (a) Average stress vs. average axial strain; (b) Energy absorbed per unit volume vs. average axial strain obtained from tests (x), FE simulation using Hill [2] criterion (black line) and FE simulation using Cazacu et al. [1] criterion (red line) for AZ31 Mg tube with axis along RD.

The initial minimal section was meshed with 70 layers of elements and two elements were used along the wall thickness, h=2 mm. From Figure 3 we note that only the Cazacu et al. [1] yield criterion based model reproduces the characteristic S-shape of the average stress vs. average axial strain response. Additionally, the prediction of the Cazacu et al. [1] yield criterion in terms of energy absorbed by the material is in very good agreement with the experimental data, while using Hill [2] criterion the energy absorbed prior to buckling is overpredicted. The final profiles for the RD compression test simulated using both models are in given in Figures 4-5. The axial crushing of an Mg AZ31 tube oriented along ND is also studied, although no experimental data were available for verification. Figure 6 shows the predicted average stress vs. average strain response for both Cazacu et al. [1] and Hill [2] yield criterion. Figure 7 shows the predicted energy absorbed by the material per unit volume according to both criteria. From Figure 8 it is worth noting that the prediction of both models is very close for the axial crushing of a tube along ND in terms of the final profile. Both predict that the buckle will be located close to the boundary, and the final cross-section remains almost circular (Figure 8).

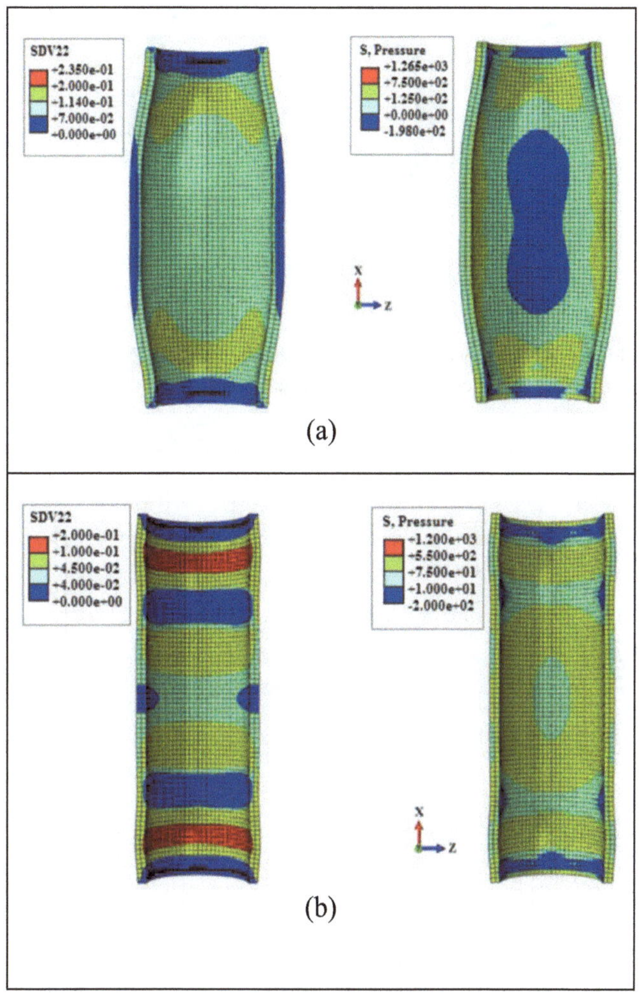

Figure 4. Deformed profile with isocontours of equivalent plastic strain and pressure in RD compression in the X-Z plane, predicted by (a) Cazacu et al. [1] criterion (b) Hill [2] criterion.

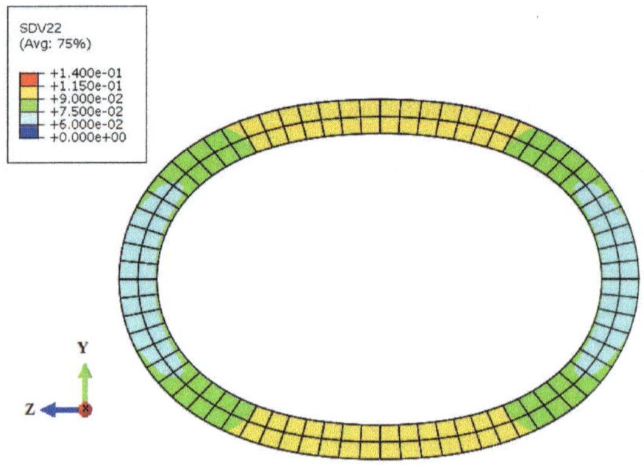

(a)

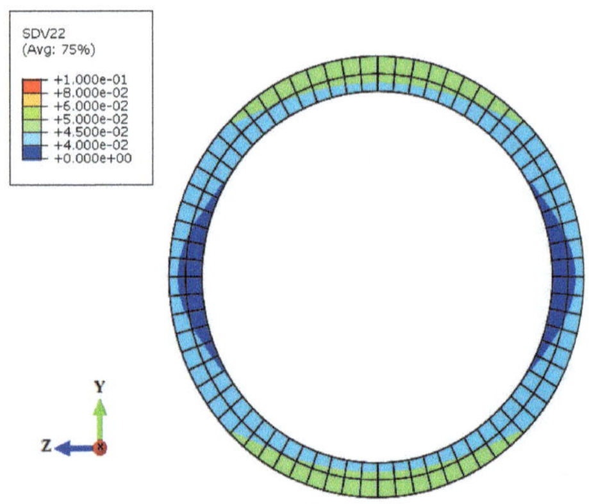

(b)

Figure 5. Deformed section in RD compression (Y-Z plane): (a) Cazacu et al. [1] criterion (b) Hill [2] criterion.

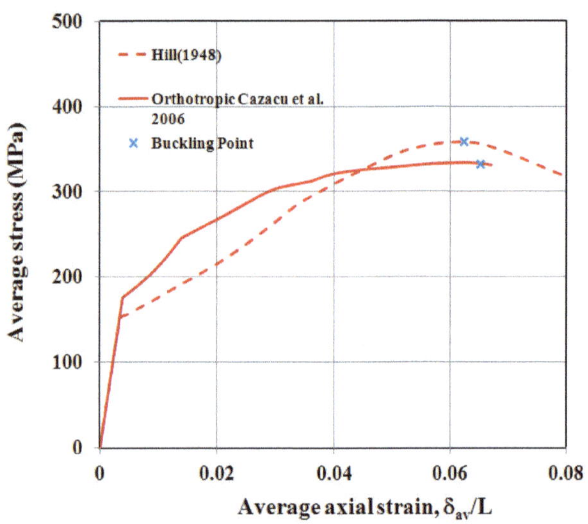

Figure 6: Average stress vs. average axial strain for crushing along ND according to the two criteria.

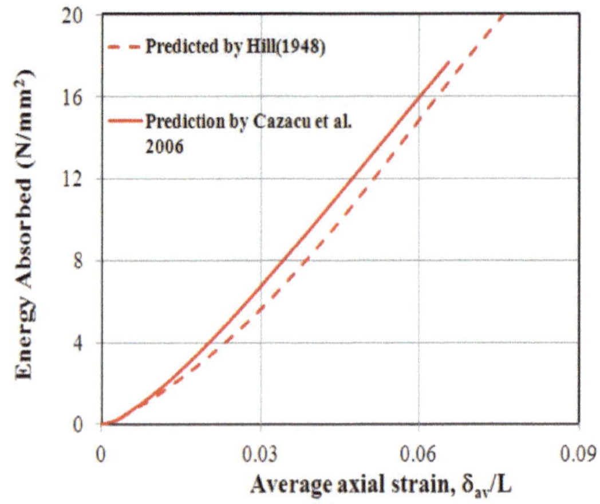

(b)

Figure 7. Energy absorbed per unit volume vs. average axial strain predicted by Cazacu et al. [1] and Hill [2] along ND

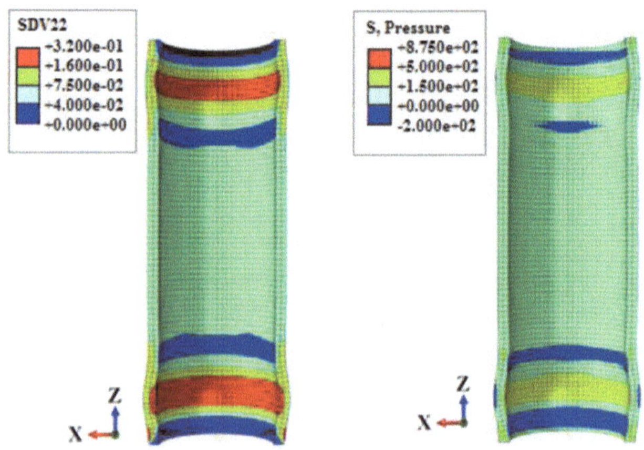

(a)

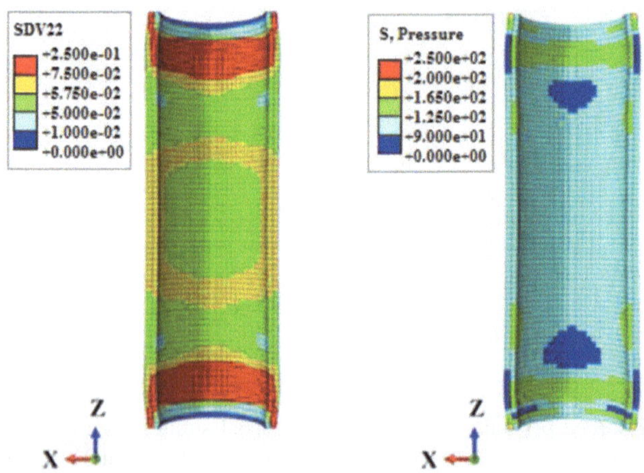

(b)

Figure 8. Deformed profile (inside surface) with isocontours of equivalent plastic strain and pressure in ND compression in the X-Z plane, predicted by (a) Cazacu et al. [1] yield criterion (b) Hill [2] yield criterion.

5. Conclusions

It was shown that Hill's [2] yield criterion, which does not account for tension-compression asymmtery is unable to predict the load vs. deformation behavior and the critical stress at buckling, irrespective of the direction of loading. While with the Hill [1] yield criterion, the orientation of the tube has no influence on the final geometry of the specimen (compare Figure 4(b) with Figure 7(b)), with the Cazacu et al. [1] yield criterion, a strong influence of the direction of crushing was predicted, the location and the final cross-section of the buckle being completely different (compare Figure 4(a) with Figure 4(b)). Most importantly, the very unusual behavior of Mg specimen when is subject to axial crushing along the RD direction was explained. At the buckling state, an outward buckles is observed in the plane (RD, ND), but an inward buckle occurs in the (RD, TD) plane. Furthermore, the unusual buckling behavior was correlated with the unusual uniaxial compression behavior in RD direction (unusual S-shape curve instead of a typical concave-down curve) (see also Figure 3). For the specimen with long axis along ND direction, no data were available. A completely different buckling behavior than for the RD specimen is predicted using Cazacu et al. [1] model. For the ND specimen, the buckles are located close to the lower and upper boundaries of the specimen, and the final cross-section remain almost circular.

In conclusion, only by accounting for the tension-compression asymmetry in the material both the load vs. deformation behavior and the critical stress at buckling of AZ31 Mg can be captured. Most importantly, the energy absorbed by the material prior to buckling can be correctly described only if the strength differential effects and anisotropy of the material are taken into consideration.

References

1. Cazacu, O., Plunkett, B., Barlat, F. Orthotropic yield criterion for hexagonal closed packed materials. Int. J. Plast. 22(2006), 1171-1194.

2. Hill, R., 1950. The mathematical theory of plasticity. Oxford University Press.

3. Kyriakides, S., Bardi, F. C., Paquette, J. A.. Wrinkling of circular tubes under axial compression: Effect of anisotropy. Transactions of the ASME-E-J.Appl. Mech., 72(2005), 301-305.

4. Khan, A., Pandey, A., Gnaupel-Herold, T., Mishra, R.K. Mechanical response and texture evolution of AZ31 alloy at large strains for different strain rates and temperatures. Int. J. Plast. 27(2011), 688-706.

Magnesium Technology 2015
Edited by: Michele V. Manuel, Alok Singh, Martyn Alderman, and Neale R. Neelameggham
TMS (The Minerals, Metals & Materials Society), 2015

WHY DO MAGNESIUM ALLOYS DEVELOP SHARP TEXTURES UPON DYNAMIC RECRYSTALLIZATION?

Haitham El Kadiri[1,2], Christopher D. Barrett[1,2], Aidin Imandoust[1,2], Sean R. Agnew[3], Mohammed Cherkaoui[4], Kaan Inal[5]
[1]Department of Mechanical Engineering, Mississippi State University, Starkville, MS 39762, United States
[2]Center for Advanced Vehicular System, Mississippi State University, Starkville, MS 39762, United States
[3]Materials Science and Engineering Department, University of Virginia, Charlottesville, VA 22904, United States
[4]School of Mechanical Engineering, Georgia Institute of Technology, Atlanta, Ga 30332, United States
[5]Mechanical and Mechatronics Engineering, University of Waterloo, Waterloo, ON N2L 3G1, Canada

Keywords: Grain boundary, Magnesium, Dynamic Recrystallization

Abstract

Using molecular dynamics (MD) simulations, we show that the maintenance of strong deformation texture during dynamic recrystallization of magnesium deformed at high temperature is controlled by the existence of energy landscape cusps and mobility spikes associated with specific grain boundaries generated by active deformation modes during the polygonization process. As the subgrains rotate to a final recrystallized state, they settle at misorientations with particular tilt axes that give out a best combination between low energy and high mobility of the associated grain boundary. Having identified the grain boundary types induced by extrusion of a commercial Mg alloy, AM30, we reconstructed them by MD simulations (notably of pure Mg) to identify their intrinsic defect structure via interfacial defect theory. These analyses enabled formal understanding of their excess potential energy variation with misorientation angles and their associated mobility. It is hypothesized that the change in the energy landscape and intrinsic defect structure induced by RE element additions, which highly segregate to the GBs, permits the subgrains to settle at more diverse misorientation angles, thus causing a weakening in the final texture.

Introduction

Texture sharpening during continuous dynamic recrystallization (CDRX) of traditional Mg alloys has been long identified as the main source of the low ductility and poor impact resistance associated with the mechanical behavior of these alloys. In fact, a sharp texture behaves more or less in the same way as a single crystal that, by virtue of the high propensity of twinning and large disparities in strength of active deformation modes in Mg, exhibits strong anisotropy and asymmetry and decreased total tensile elongation as temperature decreases and strain rate increases. Substantial research has been undertaken throughout the world, hoping to alter the inherent mechanisms of CDRX responsible for spawning sharp textures. The approach that is most widely supported consists of micro-alloying with rare-earth (e.g. Ce, La, Y, Gd) and/or alkaline element (e.g. Ca, Sr) additions. Additions of these elements enabled a certain degree of texture weakening commensurate with a marked reduction of both asymmetry and anisotropy and, consequently, some degree of increase in total elongation.

These prior studies did not, however, provide a clear understanding of why texture weakening occurs with addition of rare-earth elements. Although several hypotheses have been advanced to explain the texture randomization, including particle stimulated nucleation, Zener drag, grain growth, GB segregation, twinning, and lately, promotion of prismatic ⟨a⟩ cross-slip [1]-[2], a clear and consistent explanation for the 'rare earth' effect on texture has not yet been obtained, though GB segregation and solute drag have been indicated [1]–[7].

Grain boundaries (GBs) arguably represent the most critical microstructural feature in materials design principles for a wide range of properties [8], [9]. As a simple example, GB energy and mobility are the most fundamental mechanisms, which regulate the grain growth behavior, and it is now well established that both quantities are strongly dependent upon GB character (e.g. misorientation and interface angle) [10]–[14]. The effect of GBs however, are not phenomenologically accounted for in computational models above the atomistic scale, and most empirical attempts at GB modeling fall far short of sufficiently predicting GB behavior.

From a computational modeling perspective, GB mobility is probably the most controversial of all GB design relevant properties [12], [14]. This controversy emanates from the diversity of MD simulation methods, which do not agree with each other nor do they agree with experimental data, irrespective of the GB character [12]. Experimental measurements are themselves accompanied by significant uncertainties due to an unclear effect of impurities even in trace concentrations [15]–[20]. To several authors [12], [15], [16], the dependence of the mobility on the driving force (whether that be imposed deformation or simply capillarity effects of boundary energy and curvature) underscores a clear effect of the intrinsic defect structure on GB motion. They state that this problem "remains an open question" and requires a clear understanding of the atomic structure of GBs and how interfacial defects drive GB migration under different driving forces [12]. Khater et al. [21] investigated stressed-induced couple shear and migration of some symmetrical GBs in Zr with and without (pristine) grain boundary dislocations (GBDs). They show that the stress for motion is much higher in the pristine case than in the case when disconnections are present. Therefore, identification of the intrinsic defect structures of GBs is sine-qua-non for rigorous understanding and quantifying their mobilities. Hirouchi and Shibutani [15] recently demonstrated a discrepancy between results of electron-back scattered diffraction (EBSD) characterization and multi-phase field (M-PF) modeling of grain growth, which captured both GB energy and GB mobility. The authors attributed this discrepancy to a neglect of internal stress fields of interfacial defects by the current M-PF models.

Based on these observations, it is natural to hypothesize that in order to properly characterize the maintenance of strong texture during recrystallization, one must invoke the evolution of the intrinsic defect structure of the grain boundaries. Likewise, the direct effect of rare-earth elements on this defect structure would be expected to clarify the weaker textures observed when these elements are added to Mg. Following this hypothesis, we have performed experimental and modeling studies jointly to provide a new theory to explain the formation of Mg sharp textures.

Methodology

For our experimental investigations, we used a Mg alloy with the following composition in weight percentages: 2.54% Al, 0.4% Mn, 0.018% Zn, 0.03% Fe, 0.008% Si, 0.011% Cu, 0.025% Ce, 0.005% Ni, with the balance being Mg. The specimens were prepared by a sequence of extrusions from an initial diameter of 46cm. to an intermediate diameter of 33cm, and a final diameter of 23cm at 570°C and a speed of 0.3m/s. Following this, samples were machined from sections normal and parallel to the extrusion direction. Traditional grinding and polishing processes were performed, followed by electropolishing.

EBSD analyses were performed with a ZEISS SUPRA 40 FE-SEM machine. Scans were made in various 400 μm by 300 μm large regions with a step size of 200 μm. Identification of fully recrystallized grains and parent grains was performed by calculating the grain orientation spread (GOS) for each grain, and identifying low GOS with fully recrystallized grains (1° to 2°) and parent grains with high GOS (>10°). A few intermediate cases of GOS were observed with a range of 6°.

For our Molecular Dynamics studies, grain boundaries were relaxed using a combination of low-temperature equilibration and conjugate gradient minimization in LAMMPS [22] using the Liu EAM potential [23]. Periodic boundary conditions were applied to dimensions spanned by the grain boundaries with free surfaces normal to it at a distance of approximately 20nm. The energy and structure of the boundaries was determined after an equilibration and minimization routine. To study mobility of the boundary, the periodic condition normal to the tilt axis along the grain boundary was removed and a simple shear was applied in that direction on the grain boundary plane. The structure was strained until a constant stress of 40MPa was reached. This stress was maintained throughout the rest of the simulation.

Results and Discussion

EBSD measurement of the GOS showed two distinct peaks at 0.5° and 10.5°, as shown in Figure 1.

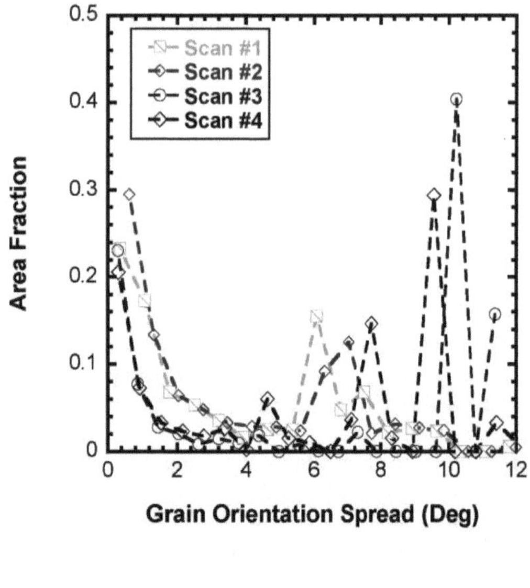

Figure 1: Grain orientation spread from four wide area EBSD scans. The data indicates that there are two major peaks near 0.5° and 10.5°, with a smaller peak around 7°. These peaks are correlated to fully recrystallized grains, parent grains, and partially recrystallized grains.

Thus we were able to unambiguously separate parent and recrystallized grains. The parent grains elongated in the extrusion direction (ED), and adopted a nearly exact ED$||\langle 10\bar{1}0\rangle$ fiber, referred to hereafter as a $\langle d\rangle$-axis fiber. The recrystallized grains adopted two different fiber orientations. One was a ED$||\langle 11\bar{2}0\rangle$ fiber, which is hereafter referred to $\langle a\rangle$-axis fiber, and the other was a $\langle d\rangle$-axis fiber.

Further examination of the recrystallized grains indicated that the $\langle a\rangle$-axis fiber was composed of grains with a consistent and nearly exactly 30° misorientation about the $[0001]$ pole with respect to their parent grain. Moreover, it was indicated that these boundaries have a tilt character. Analysis of the geometrically necessary dislocation content near these boundaries indicated a high density of prismatic-$\langle a\rangle$ type dislocations with a $[0001]$ line direction. This density indicates the likelihood of formation of the recrystallized grains by the CDRX mechanism of gradual rotation of the lattice by prismatic-$\langle a\rangle$ type dislocation accumulation along a subgrain boundary eventually leading to full-fledged grain boundary formation.

Similar examination of the $\langle d\rangle$-axis fiber indicated that it contains grains with boundaries residing within two major domains and a minor third domain. These domains coincide with misorientations of 23°-30° for the first domain, 53°-60° for the second domain, and 70° for the third domain. The first domain appears to relate to grains which are not fully recrystallized, as their GOS is 5° to 7°, in comparison to the unrecrystallized parent grains which have a GOS of about 12°, and the second and third domains, which have a GOS of less than 1°, indicating that they are fully recrystallized. CDRX formation of these grains arises from activity of basal-$\langle a\rangle$ and/or 2nd order pyramidal-$\langle c+a\rangle$ GNDs by gradual rotation around their common $\langle 10\bar{1}0\rangle$ characteristic Taylor axis.

In comparison of the $\langle a\rangle$-axis and $\langle d\rangle$-axis fibers, it was notable that the $\langle a\rangle$-axis fibers grew much larger on average than the $\langle d\rangle$-axis ones. This demonstrates that the 30° misorientation boundary with a $[0001]$ pole has a high mobility, particularly in comparison to the $\langle d\rangle$-axis grain boundaries.

In broad terms, these EBSD results indicate that the dramatic sharp texture adopted by Mg during DRX are a result of continuous grain boundary rotation being interrupted when certain misorientations are reached, leading to high populations of certain grain boundaries. In order to examine what might cause these boundaries to stop further grain rotation, we performed molecular dynamics (MD) simulations investigating the relationship between misorientation and the core structure and energy of the symmetric tilt boundaries on these two axes.

Some previous work has been done examining structure and energy of the $[0001]$ axis tilt boundaries and $[10\bar{1}0]$ axis tilt boundaries using MD [21],[24]. Since the $[10\bar{1}0]$ axis boundaries relationship between misorientation and grain boundary energy has been thoroughly explored by Wang and Beyerlein, we have not examined all of these boundaries. However, a similar plot for the [0001] axis boundaries was produced in this research, as shown in Figure 2.

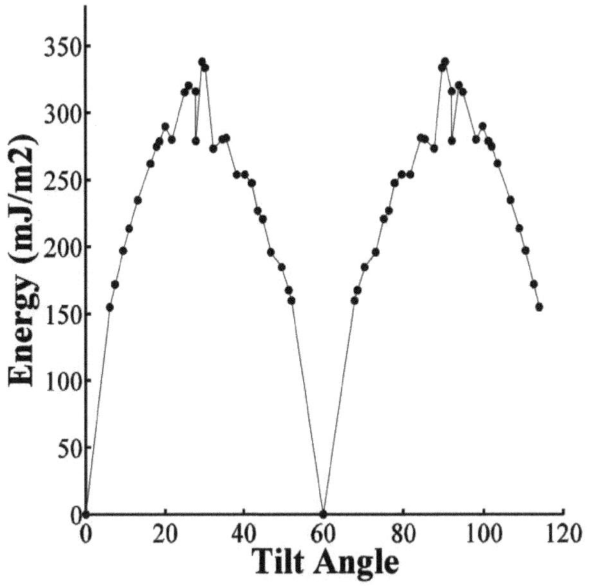

Figure 2: Tilt angle versus grain boundary energy for [0001] axis boundaries. The [0001] axis is symmetric about the 120° misorientation, and antisymmetric about the 60° misorientation. Unlike the $[10\bar{1}0]$ axis boundaries which have four major cusps, the [0001] axis boundaries have only one corresponding to the $\{4\bar{3}\bar{1}0\}$ twin at 28°.

Examination of the low-angle [0001] axis boundaries showed a structure which can be validly described as a tilt dislocation wall formed from prismatic-$\langle a \rangle$ dislocations. The energy increased proportionally with the dislocation density until the dislocation cores became too close together for a dislocation-based description to be effective. Figure 1 shows that there are three small cusps in the energy at misorientations of approximately 22°, 28°, and 34°.

The cusp at 28° is more dramatic than the other two. This boundary forms from adjoining planes of a $\{4\bar{3}\bar{1}0\}$ character. The boundary is structurally equivalent to introducing prismatic-$\langle a \rangle$ dislocations into a single crystal with a density of one every other prismatic plane to make a symmetric dislocation wall. The low energy, 275 mJ/m^2, distinguishes this boundary from such a description. It is better described as a $\{4\bar{3}\bar{1}0\}$ twin boundary because it satisfies the requirements of symmetry that the two grains are related by a reflection on the $\{4\bar{3}\bar{1}0\}$ plane, which is the K_1 plane, and because it lies at the bottom of an energy cusp, which means that the symmetry is not merely a geometric artifact, but is correlated to a low-energy core structure particular to this boundary. In confirmation of this description, we have observed that under applied simple shear, the boundary acts as a deformation twin and nucleates disconnections with a very wide core (~20a), and a Burgers vector of $a/\sqrt{13}$, where a is the lattice parameter. The width of the core and the inverse of the shear have been considered the best quantitative indicators at the atomic scale of the disconnection mobility [25]-[26]. This disconnection accommodates a moderate shear of 0.58, as seen in Figure 3. We correlate this boundary to the EBSD results indicating a 30° misorientation boundary with the same tilt axis, as it has a prominently low energy, the wide core predicts an unusually high mobility.

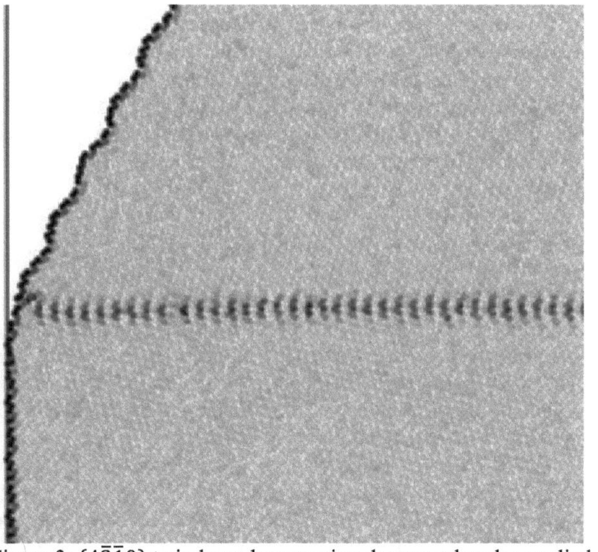

Figure 3: $\{4\bar{3}\bar{1}0\}$ twin boundary moving downward under applied shear stress. Coloring is by potential energy. The twin activates disconnections and migrates at an applied stress less than 40MPa.

Unlike the [0001] axis, the $[10\bar{1}0]$ axis is already known to host several prominent deformation twin boundaries. By reproducing some of the work of Wang and Beyerlein, we have confirmed that the $\{2\bar{1}\bar{1}1\}$ and $\{2\bar{1}\bar{1}2\}$ twin boundaries match the description given by the EBSD results of unique boundaries capable of stopping the continuous rotation of recrystallizing grains. These boundaries have low energies of 71 mJ/m^2 and 124 mJ/m^2 respectively, and while they are not highly mobile in Mg compared to other twin boundaries, simulation results indicate that they are mobile [27].

Thus the MD results for both the [0001] axis and $[10\bar{1}0]$ axis boundaries indicate that the boundary misorientation increase during CDRX is interrupted when a significant grain boundary cusp is reached. They also indicate this cusp corresponds closely to the point at which the cores of the dislocations begin to overlap. While these results are intuitive, they lead to far deeper insight into why Mg retains such sharp textures during DRX, which might otherwise be weakened. FCC and BCC grain boundaries tend to largely follow Frank's approximation for grain boundary energy. HCP boundaries, on the other hand, have significantly lower symmetry, which leads to unusual symmetric states forming at certain tilt boundaries. These points of greater symmetry usually correspond to twin boundaries, with both a lower energy, and a higher mobility than the surrounding grain boundary landscape.

Formation of twin boundaries as a result of applied stress has been deeply studied, though many questions still remain regarding their nucleation. Formation of twin boundaries by gradual rotation of the lattice into the twin relationship, however, is not discussed. This is a natural prediction, though, of what would happen if polygonization is carried out to its logical end with rotation about the appropriate axis and a sufficiently high dislocation density. Naturally, dislocations collect into subgrains because the total elastic energy of the domain is minimized by such an arrangement [28]. However, when the dislocation density increases enough to collapse into a twin boundary, this energy minimizing routine no longer holds because the core configuration has reached a high symmetry state which drastically reduces the energy of the grain boundary.

This stable boundary formation process is highly dependent on the unique symmetries of the resulting twin boundaries. Thus, it is hypothesized that the result of adding rare-earth elements is to upset this symmetry, while stabilizing higher energy boundaries, with the overall result of flattening the energy landscape in Mg. This effectively randomizes the texture during CDRX as the grain boundaries no longer have pitfalls to interrupt their gradual rotation.

Conclusions

While further investigation is clearly called for, in particular transmission electron microscopy of the boundaries common in sharp textures, and examination of the effects of rare-earth elements on grain boundary core structure and energy, we are confident that the theory we propose here correctly explains the retention of sharp textures during DRX of Mg.

DRX occurs as lattice dislocations collect into walls forming subgrains which slowly increase in density. The increase in density causes the recrystallizing grain to rotate and the dislocation wall takes on the character of a low-angle tilt boundary. This process continues to rotate the grain until the tilt boundary suddenly reaches a low energy cusp corresponding to a highly symmetric twin boundary. At this point, the rotation is interrupted and the absorption of dislocations by the boundary is much less favorable. In the case of the first domain in the $\langle d \rangle$-axis fiber, sufficiently high dislocation densities existed to overcome the barrier of that energy cusp and continue the grain rotation to the next energy cusp. A similar process probably occurred with the $\langle a \rangle$-axis fiber, but due to the higher symmetry on that axis, the grain would continue to rotate until it became identical in orientation to the parent grain, and thus unobservable.

Once grain rotation stops, there is an uptick in mobility associated with the high symmetry and lower defect density on the twin boundaries, fully corroborating the behavior observed in our EBSD data.

Acknowledgements

This work was sponsored by the National Science Foundation grant numbers CMMI 1235259 and 1235009 (Mary Toney as program manager).

References

[1] J. P. Hadorn, K. Hantzsche, S. Yi, J. Bohlen, D. Letzig, and S. R. Agnew, "Effects of Solute and Second-Phase Particles on the Texture of Nd-Containing Mg Alloys," *Metall. Mater. Trans. A*, vol. 43, no. 4, pp. 1363–1375, 2012.

[2] J. P. Hadorn, K. Hantzsche, S. Yi, J. Bohlen, D. Letzig, J. A. Wollmershauser, and S. R. Agnew, "Role of solute in the texture modification during hot deformation of Mg-rare earth alloys," *Metall. Mater. Trans. A*, vol. 43, no. 4, pp. 1347–1362, 2012.

[3] J. P. Hadorn and S. R. Agnew, "A new metastable phase in dilute, hot-rolled Mg–Nd alloys," *Mater. Sci. Eng. A*, vol. 533, pp. 9–16, 2012.

[4] J. P. Hadorn, R. P. Mulay, K. Hantzsche, S. Yi, J. Bohlen, D. Letzig, and S. R. Agnew, "Texture weakening effects in Ce-containing Mg alloys," *Metall. Mater. Trans. A*, vol. 44, no. 3, pp. 1566–1576, 2013.

[5] K. Hantzsche, J. Bohlen, J. Wendt, K. U. Kainer, S. B. Yi, and D. Letzig, "Effect of rare earth additions on microstructure and texture development of magnesium alloy sheets," *Scr. Mater.*, vol. 63, no. 7, pp. 725–730, 2010.

[6] J. D. Robson, "Effect of Rare-Earth Additions on the Texture of Wrought Magnesium Alloys: The Role of Grain Boundary Segregation," *Metall. Mater. Trans. A*, pp. 1–8, 2013.

[7] J. D. Robson, A. M. Twier, G. W. Lorimer, and P. Rogers, "Effect of extrusion conditions on microstructure, texture, and yield asymmetry in Mg–6Y–7Gd–0.5 wt% Zr alloy," *Mater. Sci. Eng. A*, vol. 528, no. 24, pp. 7247–7256, 2011.

[8] A. P. Sutton and R. W. Balluffi, "Interfaces in crystalline materials," 1995.

[9] A. P. Sutton and R. W. Balluffi, *Grain boundaries in crystalline materials*. Oxford Sci., Oxford, 1996.

[10] T. Frolov, S. V. Divinski, M. Asta, and Y. Mishin, "Effect of interface phase transformations on diffusion and segregation in high-angle grain boundaries," *Phys. Rev. Lett.*, vol. 110, no. 25, p. 255502, 2013.

[11] T. Frolov, D. L. Olmsted, M. Asta, and Y. Mishin, "Structural phase transformations in metallic grain boundaries," *Nat. Commun.*, vol. 4, p. 1899, 2013.

[12] Y. Mishin, M. Asta, and J. Li, "Atomistic modeling of interfaces and their impact on microstructure and properties," *Acta Mater.*, vol. 58, no. 4, pp. 1117–1151, 2010.

[13] A. E. Lobkovsky, A. Karma, M. I. Mendelev, M. Haataja, and D. J. Srolovitz, "Grain shape, grain boundary mobility and the Herring relation," *Acta Mater.*, vol. 52, no. 2, pp. 285–292, 2004.

[14] M. I. Mendelev, C. Deng, C. A. Schuh, and D. J. Srolovitz, "Comparison of molecular dynamics simulation methods for the study of grain boundary migration," *Model. Simul. Mater. Sci. Eng.*, vol. 21, no. 4, p. 045017, 2013.

[15] T. Hirouchi and Y. Shibutani, "Explicit Distinctions between 2D MPF Grain Growth Simulations and EBSD Analyses to Determine Driving Mechanism of Grain Growth," *Mater. Trans.*, vol. 54, no. 10, pp. 1884–1893, 2013.

[16] T. Hirouchi, T. Tsuru, and Y. Shibutani, "Grain growth prediction with inclination dependence of< 110> tilt grain boundary using multi-phase-field model with penalty for multiple junctions," *Comput. Mater. Sci.*, vol. 53, no. 1, pp. 474–482, 2012.

[17] M. Winning, G. Gottstein, and L. S. Shvindlerman, "Stress induced grain boundary motion," *Acta Mater.*, vol. 49, no. 2, pp. 211–219, 2001.

[18] M. Winning, G. Gottstein, and L. S. Shvindlerman, "On the mechanisms of grain boundary migration," *Acta Mater.*, vol. 50, no. 2, pp. 353–363, 2002.

[19] G. Gottstein and L. S. Shvindlerman, *Grain boundary migration in metals: thermodynamics, kinetics, applications*. CRC press, 1999.

[20] U. Czubayko, V. G. Sursaeva, G. Gottstein, and L. S. Shvindlerman, "Influence of triple junctions on grain boundary motion," *Acta Mater.*, vol. 46, no. 16, pp. 5863–5871, 1998.

[21] H. A. Khater, A. Serra, R. C. Pond, and J. P. Hirth, "The disconnection mechanism of coupled migration and shear at grain boundaries," *Acta Mater.*, vol. 60, no. 5, pp. 2007–2020, 2012.

[22] S. Plimpton, "Fast parallel algorithms for short-range molecular dynamics," *J Comput Phys*, vol. 117, no. 1, pp. 1–19, 1995.
[23] X. Y. Liu, J. B. Adams, F. Ercolessi, and J. A. Moriarty, "EAM potential for magnesium from quantum mechanical forces," *Modell Simul Mater Sci Eng*, vol. 4, p. 293, 1996.
[24] J. Wang and I. J. Beyerlein, "Atomic structures of symmetric tilt grain boundaries in hexagonal close packed (hcp) crystals," *Model. Simul. Mater. Sci. Eng.*, vol. 20, no. 2, p. 024002, 2012.
[25] A. Serra and D. J. Bacon, "Computer Simulation of Twinning Dislocations in Zirconium and Other HCP Metals," *Mater. Sci. Forum*, vol. 126, pp. 69–72, 1993.
[26] A. Serra, R. C. Pond, and D. J. Bacon, "Computer simulation of the structure and mobility of twinning disclocations in HCP Metals," *Acta Met. Mater*, vol. 39, no. 7, pp. 1469–1480, 1991.
[27] C. D. Barrett, H. El Kadiri, and M. A. Tschopp, "Breakdown of the Schmid law in homogeneous and heterogeneous nucleation events of slip and twinning in Magnesium," *J Mech Phys Solids*, vol. 60, no. 12, pp. 2084–2099, 2012.
[28] J. P. Hirth and J. Lothe, "Theory of dislocations," *John Wiley Sons Inc 1982*, p. 857, 1982.

Magnesium Technology 2015
Edited by: Michele V. Manuel, Alok Singh, Martyn Alderman, and Neale R. Neelameggham
TMS (The Minerals, Metals & Materials Society), 2015

CRYSTAL PLASTICITY MODELING OF THE DYNAMIC BEHAVIOR OF MAGNESIUM ALLOY, WE43-T5, PLATE

Jishnu Bhattacharyya[1], Sean Agnew[1], Peidong Wu[2], Wilburn Wittington[3], Haitham El Kadiri[3]
[1]Materials Science and Engineering, University of Virginia, Charlottesville, VA 22904, U.S.A.
[2]Mechanical Engineering, McMaster University, Hamilton, Ontario L8S 4L7, Canada
[3]Mechanical Engineering, Mississippi State University, Starkville, MS 39762, USA.

Keywords: VPSC, plastic anisotropy, r-value

Abstract

Understanding the high strain rate behavior of Mg alloys is of interest for applications ranging from armor to automobile crash worthiness. Toward this end, the viscoplastic self-consistent (VPSC) polycrystal plasticity code, including the recently developed twinning-detwinning (TDT) model, is used to describe the homogeneous plastic flow of the rare earth element containing Mg alloy, WE43-T5, plate. The model accounts for the presence of an initial, moderate texture and its evolution as during deformation. It reveals that the moderate texture is responsible for the difference between the plate through-thickness and in-plane behaviors. The model also helps to reconcile why the in-plane response is nearly isotropic, despite the presence of orthotropic (not radially symmetric) texture. Note that a single set of parameters was used to fit the entire set of results, i.e. it is a model, which can describe all of the observed strength, strain, and strain hardening anisotropies and asymmetries.

Introduction

Magnesium alloys are among the lightest structural materials. Hence, a considerable amount of research has been carried out in the recent years to examine the high strain rate behavior in quest of suitable armor materials [1–7]. One candidate material, is the rare earth containing precipitation hardenable Mg alloy designated WE43 (containing 3.7–4.3 wt.% yttrium, 2.4–4.4 wt.% neodymium and at least 0.4 wt.% zirconium) has demonstrated potential as an armor material in its direct-chill cast and hot-rolled plate form, given its good ballistic performance and higher energy absorbance capacity relative to competing lightweight metallic materials, such as Al-Mg alloy 5083 [8]. In a recent study [5], which explored the constitutive behavior of this commercially available plate alloy during high-strain-rate deformation, it was found that the yield strength at quasistatic as well as at dynamic strain rate was nearly isotropic even though a moderate crystallographic texture was present in the plate. However, the alloy still exhibits some tension–compression strength asymmetry and anisotropy in strain-hardening behavior. The goal of the current study was to determine the role of individual slip modes as well as the effect of crystallographic texture on the observed mechanical behavior at dynamic and quasistatic strain rate regimes. One of the approaches to study the grain level deformation mechanism is to use crystal plasticity modeling which allows an assessment of the strength and strain hardening behavior of individual slip systems. In this study, a visco-plastic self-consistent (VPSC) model [9] has been employed to examine the mechanical response of this alloy. Recent studies on precipitation hardenable Mg alloys, based on theoretical calculations of Orowan bowing mechanism [10–13] as well as elasto-plastic self-consistent (EPSC) modeling [14] have shown that the precipitates have different influences on the strength of different deformation modes. The crystal plasticity modeling employed in this study will also reveal the profound effect of the precipitates in this alloy.

Procedures

1. Materials and microstructure

The material used in this study is a 38 mm thick plate of WE43B alloy in T5 (artificially peak aged at 210 °C for 48 h) condition. The details of the processing are described elsewhere [5]. Optical metallography and crystallographic texture, as measured using X ray diffraction has also already been reported [5]. Notably, the c axes of the Mg alloy crystals tend to align with a direction near the plate normal direction (ND) but with a tilt of 25 ° towards the rolling direction (RD).

2. Mechanical testing

Split Hopkinson pressure bar testing was used assess the uniaxial tensile and compression behavior along RD, ND and transverse direction (TD) at a nominal strain rate of 1000/s. The experimental set up, sample geometry and data analysis are discussed in detail in an earlier publication [5]. The quasistatic uniaxial compression and tensile tests were carried out using a universal testing machine. The strain was recorded using a full scale contact extensometer and the r values discussed later in the paper are defined as [15]:

$$r_{RD} = \varepsilon_{TD}/\varepsilon_{ND} \tag{1}$$

$$r_{TD} = \varepsilon_{RD}/\varepsilon_{ND} \tag{2}$$

$$r_{ND} = \varepsilon_{RD}/\varepsilon_{TD} \tag{3}$$

For quasistatic tests, the samples were unloaded at a specific strain level, and the strains along the transverse and thickness directions were measured using a digital caliper. The constancy of volume condition ($\varepsilon_{ii} = 0$) was checked in all cases to ensure accurate measurements. For the high strain rate tests, the r values were obtained by measuring the lateral strains from a region away from the fracture surface on the tested samples.

3. Polycrystal plasticity modeling

VPSC with the twinning detwinning (TDT) model was used to model the deformation behavior of the WE43-T5 plate at dynamic (1000/s) and quasistatic (0.001/s) strain rates. The details of the model have been described elsewhere [16,17]. In short, the initial microstructure is simulated as a collection of crystal orientations with associated volume fractions representing the bulk texture. Each grain (defined by its crystallographic orientation) is considered to be a visco-plastic anisotropic ellipsoidal inclusion embedded in a visco-plastic anisotropic Homogeneous Effective Medium (HEM). The strain hardening behavior is described by the phenomenological Voce hardening law:

$$\tau_c^s = \tau_0^s + (\tau_1^s + \theta_1^s\Gamma)\left\{1 - exp\left(-\frac{\theta_0^s\Gamma}{\tau_1^s}\right)\right\} \qquad (4)$$

where, Γ is the accumulated shear strain in a grain, τ_c^s is the critical resolved shear stress (CRSS) of the sth slip system, the τ_0 and τ_1 are the initial and back-extrapolated increase in the threshold stress, while θ_0 and θ_1 are the initial and final slope of the hardening curve, respectively. These four parameters per deformation mode constitute the fitting parameters of the overall model.

The TDT model describes the deformation twinning by four processes: a) twin nucleation b) twin growth c) twin shrinkage and d) re-twinning [16]. Two empirical parameters *A1* and *A2* (< 1) control the volume fraction evolution of the twins.

$$\left(V^{th} = min\,(1.0, A1 + A2\frac{V^{eff}}{V^{acc}}\right) \qquad (5)$$

Where, V^{th} is the threshold volume fraction, V^{acc} is the accumulated twin fraction and V^{eff} is the effective twinned fraction. *A1* determines the level of strain which a grain can undergo before the twinning begins to be exhausted. *A2* controls the rate at which this exhaustion takes place once it has started. The combination of *A1* and *A2* control the length of the plateau and the steepness of the subsequent strain hardening. The main impact of these parameters is to prevent a grain from twinning 100% and to control the onset and rate of rapid hardening which follows prolific twinning.

Results

1. Initial texture

The initial texture of the WE43-T5 plate was obtained from averaging the experimentally (XRD) measured texture data obtained at 1/6th plane, 1/3rd plane and midplane and was modeled as 2000 discrete orientations with volume fractions representing the experimental data as described in reference [5].The resulting (0001), $(10\bar{1}0)$ and $(10\bar{1}1)$ pole figures are plotted using POLE8 as shown in Figure 1. As mentioned earlier, the plate exhibits a near basal texture, though the basal poles are not exactly aligned with ND but are tilted by ~25° towards RD, and with peak intensity of ~ 4 multiples of a random distribution (mrd).

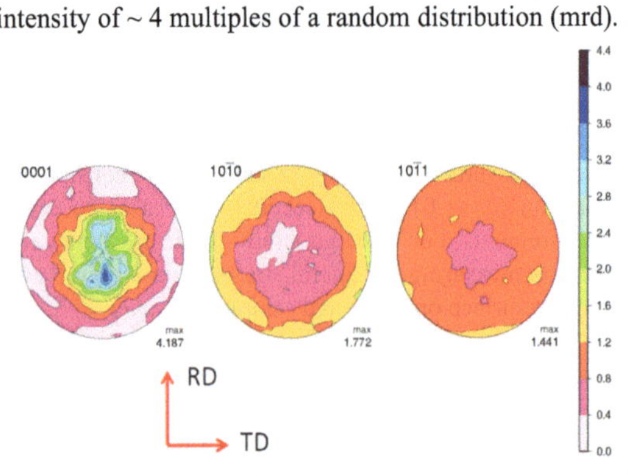

Figure 1: The initial texture of WE43T5 plate incorporated in the simulation. The (0001), $(10\bar{1}0)$ and $(10\bar{1}1)$ recalculated pole figures obtained from 2000 discrete orientations from the experimental data.

2. Crystal plasticity modeling

Uniaxial tensile and compression curves along RD and compression along ND were used to obtain initial estimates of the Voce hardening parameters of the VPSC model. Subsequently, the evolution of plastic anisotropy (r value) data was used to refine and constrain the model parameters to best fit the experimental data. For the present study, 3 slip modes and 1 twinning mode were considered: basal, prismatic, 2nd order pyramidal and tensile twinning. Thus, for the four deformation modes employed in the present study, there are a total of 16 primary fitting parameters. It is admitted that there are also the latent hardening parameters, which describe the potency of the interactions between the various slip and twinning modes. However, these were all assumed to be equal to one and were not adjusted. For twinning, the parameters *A1* and *A2* were set to 0.95 and 0.99 respectively, based on the rather linear nature of the flow curves where twinning is expected to be dominant.

Concerning strain rate sensitivity, the reference slip/twinning rate, $\dot{\gamma}_0$, of the viscoplastic constitutive law:

$$\left(\dot{\gamma^\alpha} = \dot{\gamma}_0 \left|\frac{\tau^\alpha}{\tau_c^\alpha}\right|^{1/m} sgn(\tau^\alpha)\right) \qquad (6)$$

was set equal to the actual rate of the mechanical tests: 1000/s for the high strain rate and 0.001/s for quasistatic tests. The rate sensitivity *m* was set to 0.05 for all slip systems in both cases. This is a bit higher than is observe experimentally. However, the stiffness of the matrix correlating stress and strain rate (based on equation 6), demands that higher m values be used in order to obtain a numerical solution. The affine linearization of equation (5) was employed throughout the study because it was found to give the best overall performance [18].

During RD and TD tension, by virtue of the texture, the most active slip systems are basal and prismatic slip of <a> type dislocations. Hence, the tensile data are used to determine the appropriate parameters corresponding to these two modes. Next, the ND compression data is used to determine the parameters which describe pyramidal <c + a> slip. Again, given the texture of the material, the c-axes of many grains are near parallel to the compression axis, which inhibits prismatic slip and tensile twinning. Thus, majority of the strain is accommodated by basal during the initial straining stage followed by <c + a> slip at higher strains. It is to be noted that since the stress strain response during the initial straining stages of the high strain rate tests are inaccurate [19] the 1% offset strains have been chosen to obtain the CRSS parameters for the slip systems. Once the initial round of fitting is carried out, the r value of RD tension is used to constrain the ratio between the CRSS values τ_{prism} : τ_{basal} and $\tau_{<c+a>}$: τ_{basal}. An increase in the ratio of prismatic to basal CRSS leads to a decrease in the r value. On the other hand, increasing the ratio of the <c+a> to basal CRSS values would increase the r value [20]. After adjusting the CRSS values, the hardening behaviors of these two modes are adjusted to fit the polycrystal hardening response, while making sure that the relative activities are still appropriate to model the r values.

Finally, compression along RD data were used to obtain the parameters for $\{10\bar{1}2\}$ tensile twinning. Based on the experimental data obtained, twinning is modeled with a linear hardening. The parameters obtained from the above procedure are listed in Table 1 for 1000/s and in Table 2 for 0.001/s.

The experimental and simulated flow curves for tension and compression along RD, TD and ND during dynamic deformation are shown in Figure 2. Two sets of experimental flow curves are presented to confirm the reproducibility of the tests. It can be seen that along RD, there is some tension-compression yield strength asymmetry and the model is able to capture this phenomenon. The model also captures the near linear twin dominated portion of the compression curve. For TD, there is very little yield asymmetry as seen from the experimental data. However, the model predicts a similar level asymmetry, as was predicted for the RD case. In the case of ND deformation, the experimentally observed yield asymmetry is the highest. If one recalls that a rigid-viscoplastic model will not well-represent the elasto-plastic transition, it can be concluded the ND compression curve is described well (at higher strain levels). The strength of the VPSC prediction of the tension behavior is slightly overestimated. Figure 3 shows the relative activities of the 3 slip modes and 1 twinning mode considered. As expected, during in-plane tension, the 2 primary modes are basal and prismatic slip, with prismatic dominating at larger strain levels. For compression, the basal slip mode is still the majority strain carrier, but the tensile twinning is also a major contributor a low strains. After twinning is exhausted, the activity of <c+a> slip increases.

The experimental and predicted flow curves at quasistatic strain rate of 0.001/s and the relative activities of different slip modes are shown in Figure 4 and 5 respectively. These results show that for a given strain rate, a single set of parameters can describe all of the flow behavior, strain hardening anisotropies and asymmetries. The yield strength flow curve shapes are similar to those observed at dynamic rates, while the flow strength asymmetries are generally lower. The deformation mode trends observed for the quasistatic strain rate predictions are also similar to those observed in the predictions of dynamic loading. The lower asymmetry does correlate with slightly lower twin activity.

Table 1: Voce hardening parameters for simulating flow curves at 1000/s

	τ_0	τ_1	θ_0	θ_1
Prismatic	165	18	600	0
Basal	60	45	1000	0
<c+a> 2nd order	200	70	1500	0
Tensile twin	85	0	200	200

Table 2: Voce hardening parameters for simulating flow curves at 0.001/s

	τ_0	τ_1	θ_0	θ_1
Prismatic	130	18	600	80
Basal	60	45	1000	30
<c+a> 2nd order	200	40	1000	0
Tensile twin	95	0	200	200

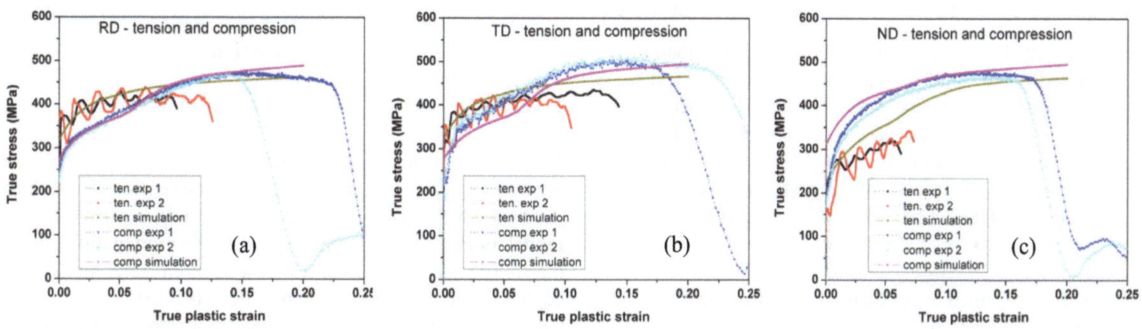

Figure 2: Experimental and simulated flow curves for tension and compression at 1000/s along (a) RD (b) TD and (c) ND

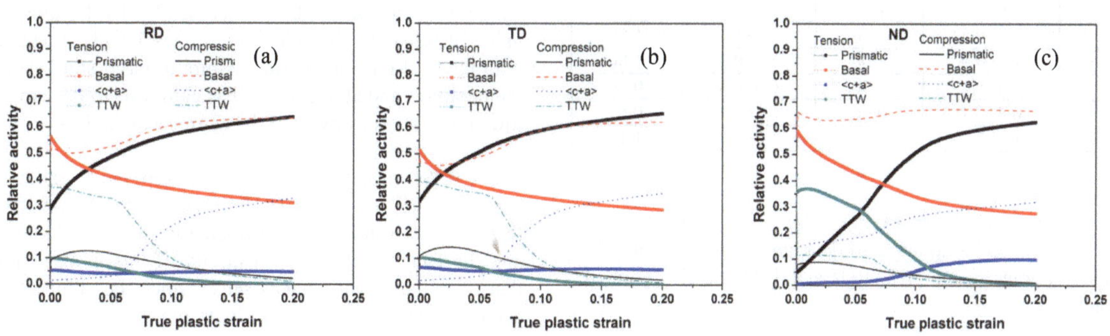

Figure 3: The relative activities of individual slip modes during tension and compression at 1000/s along (a) RD (b) TD and (c) ND

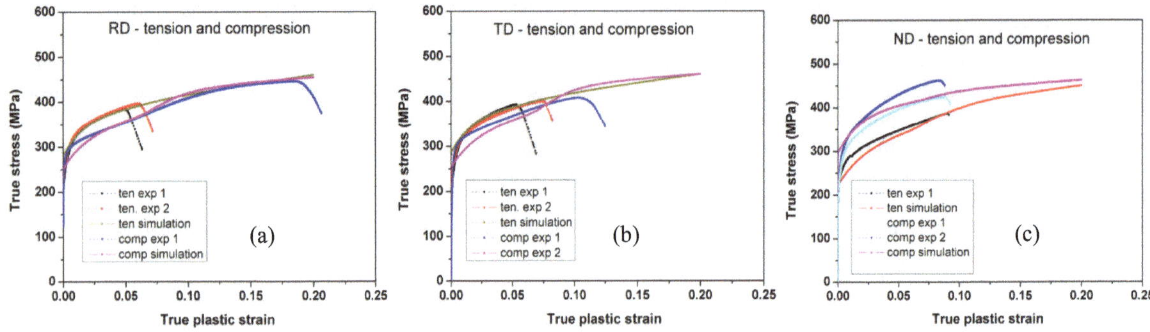

Figure 4: Experimental and simulated flow curves for tension and compression at 0.001/s along (a) RD (b) TD and (c) ND

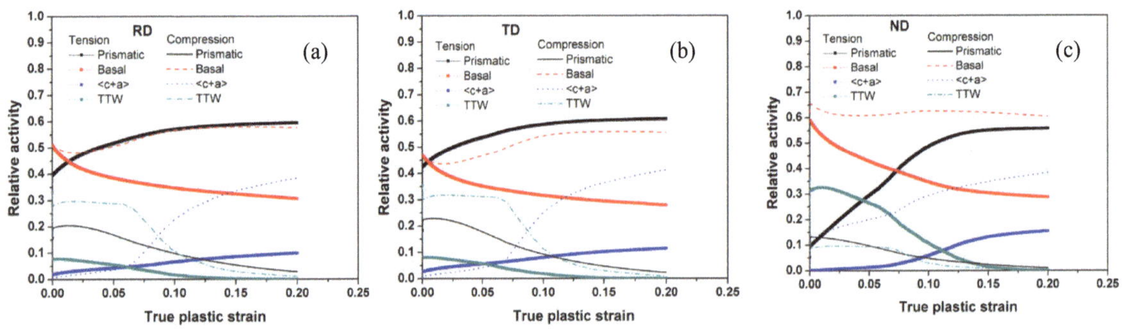

Figure 5: The relative activities of individual slip modes during tension and compression at 1000/s along (a) RD (b) TD and (c) ND

The experimental as well as simulated r values for all the loading conditions examined in this study are presented in Table 3. It is seen that the simulated r values correlate with the experimentally measured values, especially if one takes into account the rather large experimental uncertainties associated with these measurements. This level of agreement suggests that the model adequately captures the relative activities of the individual slip modes.

Table 3: The r values obtained experimentally and from the VPSC model at the same level of strain.

Loading condition	at 1000/s		at 0.001/s	
	Experimental	Simulated	Experimental	Simulated
RDt	1.6±0.5	1.40	1.3±0.5	1.58
TDt	1.8±0.5	1.56	1.8±0.3	1.56
NDt	1.5±0.9	1.00	1.2±0.3	1.06
RDc	0.4±0.2	0.50	0.5±0.3	0.56
TDc	0.5±0.2	0.56	0.5±0.4	0.55
NDc	1.7±0.7	1.15	1.2±0.7	1.15

Table 4: Strain rate sensitivities along different loading directions. Calculated based upon all data, up to 5000 1/s, or only the cylinder samples tested at 0.001/s and 1000/s and presented within brackets (#.###).

Strain level → Loading condition	0.01	0.05	0.10
RDt	0.008	0.006	-
TDt	0.009	0.004	-
NDt	-0.009	-0.011	-
RDc	0.009 (0.000)	0.008 (0.008)	0.016 (0.010)
TDc	0.006	0.009	0.014
NDc	(0.000)	(0.003)	(0.005)

Discussion

1. Rate sensitivities of individual slip/twin modes

The experimentally observed strain rate sensitivities, *m*, are presented in Table 4, including the compression data which was published previously [5]. Even though a rate sensitive constitutive model is employed for slip system selection within the crystal plasticity code (Eq. 6), the use of a different reference rate in that equation allows isolation of the rate sensitivity effects within the CRSS values themselves. This permits isolation of the strain rate sensitivities of the individual deformation modes. For example, it is found that all the experimental data, at both high and quasistatic strain rate, can be modeled successfully by keeping the CRSS value of basal slip constant (see Tables 1 and 2). This correlates with the well-known athermal nature of basal slip [21] at ambient temperatures and above, as well as conforms to the experimentally low strain rate sensitivity of the aggregate (Table 4)..

The CRSS values of the prismatic slip mode show some rate sensitivity $m = 0.017$ (0.015 if $(\tau_0 + \tau_1)$ is considered). The thermally activated nature of prismatic slip is well known [22,23] and the fact that the rate sensitivity of this mode is higher than that of the aggregate makes sense, since the strain is also being accommodated by modes which are less rate sensitive. For instance, basal slip appears to be rate insensitive and carries a significant fraction of the strain during RD and TD tension.

On the other hand, it is found that the CRSS values of the <c+a> slip mode are essentially rate insensitive. If, instead, the back extrapolated flow stress $(\tau_0 + \tau_1)$ is considered, the rate sensitivity of <c+a> slip is $m = 0.0085$. These values correlate well with the experimentally observed rate sensitivity during ND compression [5], which is the direction along which <c+a> slip is most active.

In fact the high strain rate sensitivity of the <c+a> mode appears to be slightly higher than the aggregate, which again makes sense since basal slip is still the majority strain carrier.

In case of twinning, the modeling actually suggests a negative rate sensitivity (i.e. higher CRSS values at lower strain rates). Note that the slight in-plane tension – compression yield asymmetry (Figs. 2 and 4) observed experimentally as well as in the simulated flow curves is due to tensile twinning. The relative activity plots (Fig. Figs. 3 and 5) show that tensile twinning is active during the in plane compression which leads to lower strength as compared to in plane tension where macroscopic yield is dominated by prismatic slip. The level of in plane tension-compression asymmetry is lower under quasistatic loading conditions than it is under dynamic conditions. This further provides evidence that tensile twinning is less active at lower strain rates. Nevertheless, this point merits further investigation and experiments are presently being performed in order to obtain the volume fraction of twins at the two strain rates examined to see how the volume fraction of twins changes with strain rate.

2. Plastic anisotropy and asymmetry

The r values obtained for ND compression and tension are greater than 1 indicating larger straining along RD than TD. This is due to the initial tilt of the basal poles towards RD, which enables basal slip to accommodate a larger strain along RD than TD. In case of uniaxial compression within the plane of the plate (i.e. along RD and TD) the r values are < 1 which correlates with the occurrence of tensile twinning. The r values for uniaxial tension within the plane of the plate are greater than 1, but there is essentially no difference between the RD and TD as far as plastic anisotropy in concerned, unlike strongly textured AZ31 sheets/plates where a significantly higher r value is observed in case of TD tension as compared to RD [20].

Among the 3 directions viz. RD TD and ND, the ND exhibits the maximum amount of tension compression asymmetry at both dynamic and quasistatic strain rates. In this case, the sample deformed in tension, exhibits lower flow stress due to tensile twinning. The simulation of ND tension may be viewed as a sort of validation study since it will test whether or not the twinning parameters (which control yielding) and the prismatic slip parameters (which control higher strain plasticity in this case) are correct. The results presented in Fig. 2 show that, at high strain rate, the model overestimates the ND tension flow stress and thus under predicts the tension-compression asymmetry. However, in the quasistatic regime, it can be seen that the ND tension flow stress is reasonably well predicted by the model (Fig. 4). Further study is being carried out to quantify the volume fraction of twins which will help to confirm these results.

It is to be noted that the normal direction asymmetry is a critical result of the study. Both the low strength and the low ductility evidenced along this direction result in a low energy absorption potential for the plate loaded in tension along the through thickness direction. Initial ballistic tests performed at ARL on Mg alloy WE43-T5 were encouraging [8], but they evidenced large spallation failures, which correlate with a weakness with respect to tensile loading parallel to the plate normal direction.

3. Role of precipitation and solute hardening – stronger basal slip

Previous studies have found that precipitation hardening is one of the most potent strengthening mechanism in Mg alloys. In Mg alloys containing rare earth elements, the main strengthening precipitates form as prismatic plates [24,25] and is found to be very effective in blocking basal slip [10,11]. For the current study, the CRSS of basal slip is found to be 60 MPa, which is considerably higher than the values obtained for AZ31 alloys.[26,27] Simple calculations based on Orowan bowing of dislocations against non–shearable precipitates have shown that these obstacles can lead to significant strengthening of the basal slip. The increase in the CRSS $\Delta\tau$ is given by [10]:

$$\Delta\tau = \frac{Gb\ln\left(\frac{0.886\sqrt{d_t t_t}}{b}\right)}{2\pi\sqrt{1-\nu}\left(0.825\sqrt{\frac{d_t t_t}{f}}-0.393d_t-0.886t_t\right)} \quad (7)$$

where, G is the shear modulus of the Mg matrix phase, b is the magnitude of the Burgers vector of the slip dislocations (in this case basal <a> type), ν is the Poisson's ratio, and f is the volume fraction, d_t is the diameter and t_t is the thickness of the precipitates. By considering a precipitate volume fraction of 5% for this T5 plate, as an initial estimate that is slightly lower than the value of 7% observed previously measured T6 casting [14]and an estimated aspect ratio (d_t/t_t) of 5 [10], an increase in the CRSS of basal slip by ~ 46 MPa can be achieved. Currently experiments are being carried out to quantify the exact volume fraction and geometry of the precipitates present in this alloy in T5 condition.

Recent first principles calculations [29] have revealed the potency of several elements as solid solution strengthener of the basal slip. From this study it is found that Y has an optimum solubility and strength potency to harden the basal slip considerably. The WE43 alloy contains 3.7-4.3 wt% Y along with 2.5 wt% Nd. Considering that almost all the Nd is used up in forming the precipitates in T5 condition, and the solution hardening comes only from Y, calculations show that the present amount in this alloy would lead to an increase of ~ 12 MPa of the basal CRSS.

Thus, a high CRSS of 60 MPa for basal slip as determined from the modeling seems to stem from the combined effects of precipitation hardening by the prismatic plates and solute strengthening by the rare earth elements viz. Y present in this alloy. Similar analyses are underway to describe the strengthening effects of the solute and precipitates on the other deformation mechanisms.

Conclusions

The crystal plasticity modeling results establishes that the initial texture of the plate is responsible for the plate through-thickness anisotropy and the near in-plane isotropy. The model also provides the mechanisms (slip modes) responsible for the observed flow response and evolution of the plastic anisotropy (r value) with straining. Finally, the rate sensitivities of individual slip systems can be obtained from the model. Is shown that prismatic slip is the most rate sensitive, followed by <c+a>. On the other hand, basal slip and tensile twinning are rate insensitive. Further investigation is being carried out to understand the role of thermally activated dislocation motion which is responsible for the observed rate sensitivities of each mode. Also, experiments are being carried out to quantify the texture evolution during these mechanical tests which would provide another means of constraining the crystal plasticity model, e.g., it will enable quantification of the volume fraction of twins present in the material.

Acknowledgements

The research at U.V.A. and M.S.U. was sponsored by the United States Army Research Office under contract number W911NF-12-1-0455 monitored by Drs. Suveen Mathaudhu and David Steppe. The views and conclusions contained in this document are those of the authors and should not be interpreted as representing the official policies, either expressed or implied, of the Army Research Laboratory or the U.S. Government. The U.S. government is authorized to reproduce and distribute reprints for government purposes notwithstanding any copyright notation hereon.

References

1. T. Mukai, M. Yamanoi, and K. Higashi, Mater. Sci. Forum **350-351**, 97 (2000).
2. K. Ishikawa, H. Watanabe, and T. Mukai, Mater. Lett. **59**, 1511 (2005).
3. K. Ishikawa, H. Watanabe, and T. Mukai, J. Mater. Sci. **40**, 1577 (2005).
4. I. Ulacia, N. V. Dudamell, F. Gálvez, S. Yi, M. T. Pérez-Prado, and I. Hurtado, Acta Mater. **58**, 2988 (2010).
5. S. Agnew, W. Whittington, A. Oppedal, H. El Kadiri, M. Shaeffer, K. T. Ramesh, J. Bhattacharyya, R. Delorme, and B. Davis, Jom **66**, 277 (2014).
6. T. Mukai, T. Mohri, M. Mabuchi, M. Nakamura, K. Ishikawa, and K. Higashi, Scr. Mater. **39**, 1249 (1998).
7. K. Eswar Prasad, B. Li, N. Dixit, M. Shaffer, S. N. Mathaudhu, and K. T. Ramesh, Jom **66**, 291 (2014).
8. J. Hamilton, S. T. Brennan, Y. Sohn, B. Davis, R. Delorme, and K. Cho, *Microstructural Characteristics of High Rate Plastic Deformation in Elektron ^{TM} WE43 Magnesium Alloy* (2012), pp. 433–438.
9. R. A. Lebensohn and C. N. Tomé, Acta Metall. Mater. **41**, 2611 (1993).
10. J. F. Nie, Scr. Mater. **48**, 1009 (2003).
11. J. D. Robson, N. Stanford, and M. R. Barnett, Acta Mater. **59**, 1945 (2011).
12. J. D. Robson, N. Stanford, and M. R. Barnett, Metall. Mater. Trans. A **44**, 2984 (2012).
13. J. Robson, Metall. Mater. Trans. A **45**, 5226 (2014).
14. S. R. Agnew, R. P. Mulay, F. J. Polesak, C. a. Calhoun, J. J. Bhattacharyya, and B. Clausen, Acta Mater. **61**, 3769 (2013).
15. A. Jain and S. R. Agnew, Mater. Sci. Eng. A **462**, 29 (2007).
16. H. Wang, P. D. Wu, C. N. Tomé, and J. Wang, Mater. Sci. Eng. A **555**, 93 (2012).
17. H. Wang, P. D. Wu, J. Wang, and C. N. Tomé, Int. J. Plast. **49**, 36 (2013).
18. H. Wang, B. Raeisinia, P. D. Wu, S. R. Agnew, and C. N. Tomé, Int. J. Solids Struct. **47**, 2905 (2010).
19. B. A. Gama, S. L. Lopatnikov, and J. W. Gillespie, Appl. Mech. Rev. **57**, 223 (2004).
20. S. R. Agnew and Ö. Duygulu, Int. J. Plast. **21**, 1161 (2005).
21. H. Conrad, L. Hays, G. Schoeck, and H. Wiedersich, Acta Metall. **9**, 367 (1961).
22. A. Akhtar and E. Teghtsoonian, Acta Metall. **17**, 1351 (1969).
23. B. Raeisinia, S. R. Agnew, and A. Akhtar, Metall. Mater. Trans. A **42**, 1418 (2010).
24. J. F. Nie and B. C. Muddle, Acta Mater. **48**, 1691 (2000).
25. J. . Nie, X. . Xiao, C. . Luo, and B. . Muddle, Micron **32**, 857 (2001).
26. A. Jain, O. Duygulu, D. W. Brown, C. N. Tomé, and S. R. Agnew, Mater. Sci. Eng. A **486**, 545 (2008).
27. S. R. Agnew, in *Magnes. Technol. 2002*, edited by H. I. Kaplan (TMS (The Minerals, Metals & Materials Society), 2002, 2002), pp. 169–174.
29. J. a. Yasi, L. G. Hector, and D. R. Trinkle, Acta Mater. **58**, 5704 (2010).

Magnesium Technology 2015
Edited by: Michele V. Manuel, Alok Singh, Martyn Alderman, and Neale R. Neelameggham
TMS (The Minerals, Metals & Materials Society), 2015

Micromechanical Modeling of Evolving Anisotropy in AZ31 Mg for Various Strain Paths

Nitin Chandola[1], Oana Cazacu[1]*, Raja. K. Mishra[2]
[1]University of Florida, Dept. of Mechanical and Aerospace Engineering,
REEF,1350 N. Poquito Road Shalimar, FL 32579, USA
[2] General Motors Research and Development Center, 30500 Mound Road, Warren, MI 40890-9055, USA

Keywords: Simple shear; Twinning; Texture evolution; Visco-plastic Self Consistent Model

Abstract

Accurate description of the mechanical response of AZ31 Mg requires consideration of its strong anisotropy both at the single crystal and polycrystal level and their evolution with plastic deformation. In this paper, we apply the self-consistent mean field model developed by Lebensohn and Tomé [1] and the VPSC code to the description of the behavior of a rolled AZ31 plate. The predictive capabilities of the model are demonstrated. It is shown that with the same set of parameters, which were identified based on a few mechanical tests, it is possible to predict with great accuracy both the mechanical stress-strain response and texture evolution for loading paths that were not used for calibration, such as off-axis loadings, and simple shear loadings.

1. Introduction

There has been a large body of literature on AZ31 Mg. Experimental data reported include the macroscopic stress-strain response and the microstructure evolution under various loadings such as uniaxial tension and compression along different orientations [1-5], simple shear [4-5], etc. In the framework of crystal plasticity, the VPSC model [1] has been used to provide understanding of the active deformation mechanisms. For example., Agnew and Duygulu [2] calibrated the VPSC model using as input the initial texture and uniaxial tension and compression stress-strain curves and further used it to demonstrate the importance of <c + a> slip on texture evolution along the same loading paths for a rolled plate. Using VPSC, Yi et al. [3] demonstrated that basal <a> slip and tensile twinning have a significant effect on anisotropy in uniaxial tension while <c+a> slip activity must be considered along with tensile twinning activity to describe the key features of the response in uniaxial compression of an extruded bar. Walde and Reidel [6] used the VPSC model to understand how the starting texture affects earing formation during cup drawing. In this paper, we demonstrate that using the same set of parameters for the VPSC model, calibrated using few tests, it is possible to describe with accuracy both the shear response of an AZ31 rolled plate. Comparison between the experimental and predicted textures as well as twin volume fractions are also presented. As mentioned, there are many studies devoted to the investigation of Mg AZ31 using VPSC, however the capabilities of this model to describe the in-plane anisotropy of the mechanical response of the material, in particular the stress-strain for in-plane orientations other than the axes of orthotropy were not assessed. Here, for the first time the material's response in uniaxial loading along the 45 in-plane direction (DD) is simulated and compared with experimental data, both in terms of stress-strain response and texture evolution.

2. VPSC Model

The VPSC model is only briefly presented in what follows, a detailed description can be found in the review article by Tomé and Lebensohn [8]. A self-consistent approach is used to describe the interaction of a grain with its surroundings. Every grain is treated as an anisotropic ellipsoidal inclusion embedded in a homogeneous effective medium. Plastic deformation in each grain occurs by activation of both slip and twinning modes. The grain level constitutive behavior relating the Cauchy stress deviator and the viscoplastic strain rate **d** is:

$$\mathbf{d}(\mathbf{x})=\sum_{s}\mathrm{m}^{s}(\mathbf{x})\dot{\gamma}^{s}(\mathbf{x}),\quad \dot{\gamma}^{s}=\dot{\gamma}_{0}\left|\frac{\mathbf{m}^{s}:\boldsymbol{\sigma}'}{\tau_{C}^{s}}\right|^{1/n}\operatorname{sgn}\left(\mathbf{m}^{s}:\boldsymbol{\sigma}'\right), \tag{1}$$

where the sum is over all slip and twin systems, $\mathbf{m}^s$ is the Schmid tensor, $\dot{\gamma}^s$ is the shear rate on the system (s) given by a power law with $\dot{\gamma}_0$, n, and τ_C being the reference slip rate, rate-sensitivity exponent, and resistance of the slip/twin system (s), respectively. A Voce-type hardening law is used to describe the evolution of the critical resolved shear stress as a function of accumulated shear strain:

$$\tau_{c}^{s}=\tau_{0}^{s}+\left(\tau_{1}^{s}+\theta_{1}^{s}\Gamma\right)\left(1-\exp\left(-\frac{\theta_{0}^{s}\Gamma}{\tau_{1}^{s}}\right)\right), \tag{2}$$

where τ_0^s, τ_1^s, θ_0^s and θ_1^s are constants, and Γ is the accumulated shear in all active deformation systems. In addition, self and latent hardening are captured using the coupling coefficients $h^{ss'}$, which empirically account for the obstacles that new dislocations (or twins) associated with system *s'* create for the propagation of dislocations (or twins) on system *s*. The increase in the threshold stress over the time interval Δt is calculated as:

$$\Delta\tau_{o}^{s(r)}=\frac{d\tau^{*s(r)}}{d\Gamma^{(r)}}\sum_{k'}h^{ss'}\dot{\gamma}^{s'(r)}\Delta t, \tag{3}$$

Twinning is modeled using the Predominant Twin Reorientation Scheme (PTR) ([7]). In the PTR scheme, twinning differs from slip in its directionality, which means activation is allowed only if the resolved shear stress is positive.

3. Modeling of the Mechanical Response of AZ31 Mg Using the VPSC Model

3.1 Calibration of VPSC Parameters for AZ31 Mg

In the following the room temperature response under quasi-static conditions (strain rate of 10^{-4}/s) All the data used for calibration were either obtained at GM and/or reported in [5].

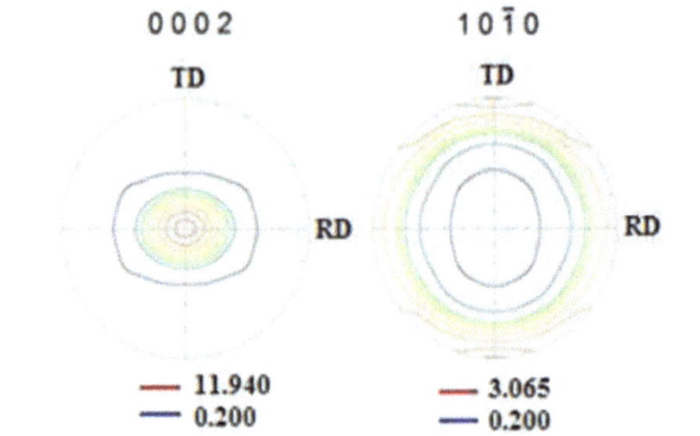

Figure 1. Pole figures showing the initial texture of the AZ31 Mg rolled plate. RD and TD designate the rolling and transverse in-plane directions, respectively.

The initial texture of the material was determined by the authors from the entire EBSD scan (784x808 μm) to generate 2762 weighted orientations (see Figure 1). Note that the initial texture of the material is strongly basal with an almost equal fraction of grains having the c-axis slightly tilted away from the sheet normal towards +RD and –RD, respectively (around ±20^0).

For calibration of the VPSC model, we propose a step-by-step procedure in which we isolate as much as possible the potential systems for accommodating the imposed plastic strain. We start by determining the parameters that characterize the basal and prismatic <a> slip systems using the stress-strain curve in uniaxial tension along RD. Since the c-axis of most grains is aligned with the normal direction or deviates by a very small angle from it, in the ND compression, activation of pyramidal <c + a> slip is expected. Thus, from the stress-strain response in uniaxial compression along ND, we identify the parameters associated with this system. A characteristic ~86° reorientation of the basal poles during uniaxial compression along RD has been observed in experimental tests and is attributed to the activation of the {10-12} extension twinning system [1-4]. Hence, from the uniaxial compression test along RD, we identify the parameters for this twinning system and further estimate the validity of the identification of the <c + a> slip system obtained using the ND compression test data. Simple shear test data can be used to identify the pyramidal <a> slip system parameters. Table 1 gives the values of the parameters of the model.

Table 1: VPSC parameters determined for simple shear tests.

Mode	τ_0	τ_1	θ_0	θ_1
Basal, $\{0001\}\langle 1\bar{2}10\rangle$	17.5	5	3000	35
Prismatic <a>, $\{1\bar{1}00\}\langle 11\bar{2}0\rangle$	85	33	550	70
Pyramidal <a>, $\{10\bar{1}1\}\langle \bar{1}2\bar{1}0\rangle$	100	30	30	10
Pyramidal II <c + a>, $\{10\bar{1}1\}\langle \bar{1}\bar{1}23\rangle$	148	50	8500	0
Tensile Twinning, $\{10\bar{1}2\}\langle \bar{1}101\rangle$	52	0	0	0

In Figure 2 is shown the comparison between the only reported deformation textures (see [5]) measured in uniaxial RD tension and RD compression, respectively along with the textures predicted by VPSC. We also verify that the predicted microstructure evolution contributing to plastic deformation corroborate with observed deformation trends (see Figures 3 and 4). In all the stress-strain plots the experimental data is represented by symbols while the VPSC prediction is represented by solid line.

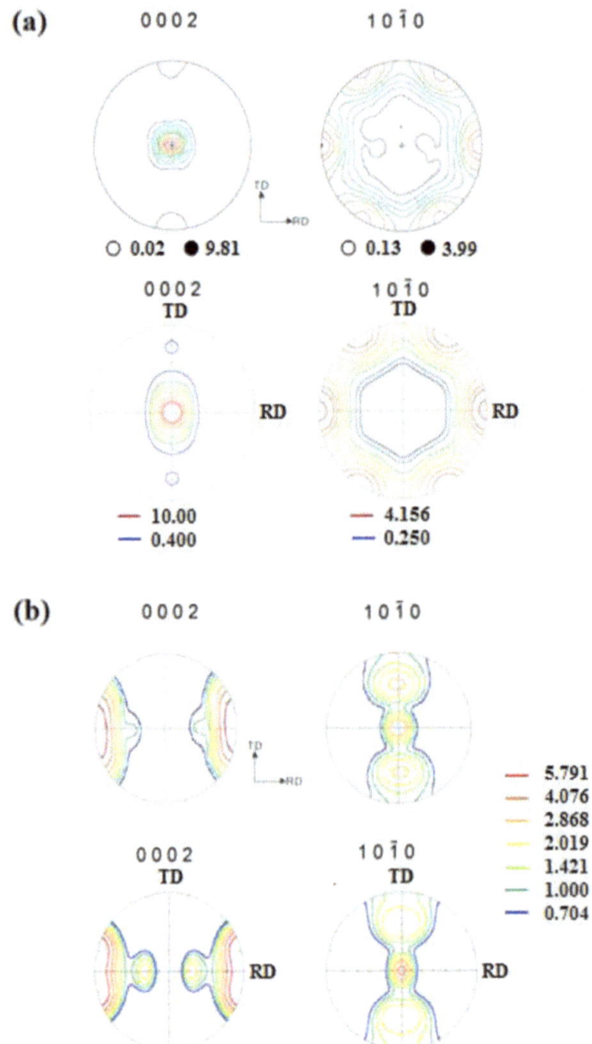

Figure 2. Pole figures showing measured texture (top) and predicted texture (bottom) in: (a) uniaxial tension along RD at 13% strain (failure) (b) uniaxial compression along RD at 8% strain.

While Figures 3 (a)-(b) show simulations for strain paths that were used to calibrate the parameters of the VPSC model, Figure 4(a) shows the model predictions for ND tension. It should be noted that this peculiar "S-shape" of the ND tension curve predicted by VPSC was also observed experimentally by Guo et al.[9] for an AZ31 Mg plate of similar texture. In the next subsection, the predictions of the VPSC model along other in-plane directions for both uniaxial tension and compression are presented.

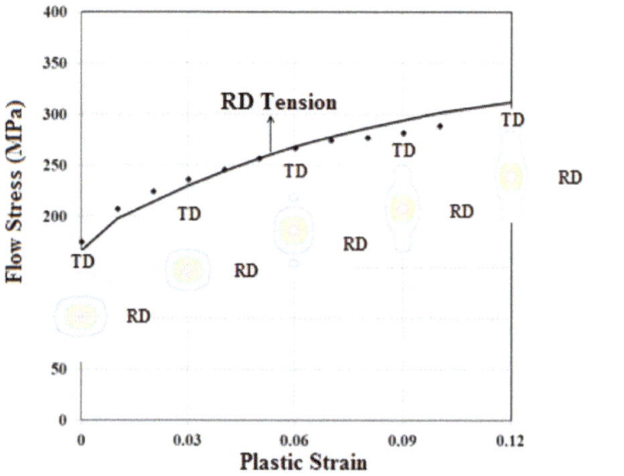

(a)

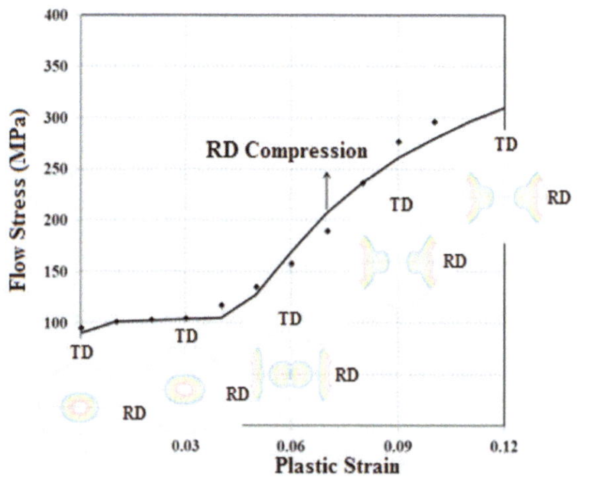

(b)

Figure 3. Stress-strain response and evolution of microstructure ((0 0 0 2) pole figure) predicted by the VPSC model in (a) RD tension, (b) RD compression.

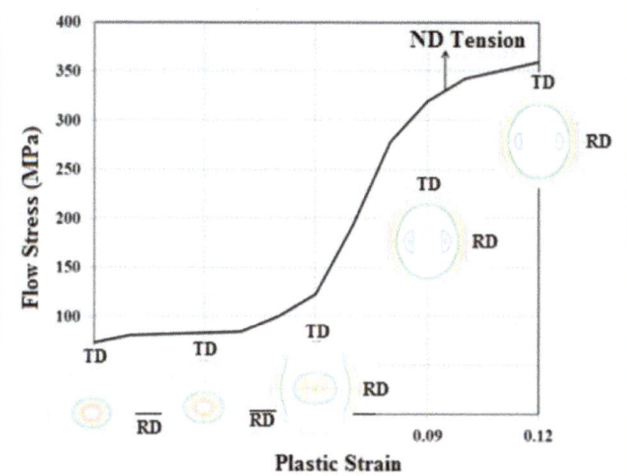

(a)

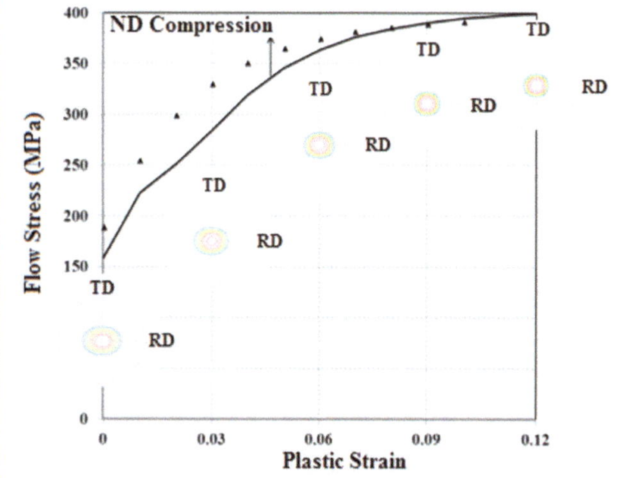

(b)

Figure 4. Stress-strain response and evolution of microstructure ((0 0 0 2) pole figure) predicted by the VPSC model in (a) ND tension, (b) ND compression.

3.2. VPSC Predictions of the In-Plane anisotropy: Simulations for Off-axis Uniaxial Loadings

The VPSC model is now used to predict the material behavior in uniaxial tension and compression for loadings along 45^0 to the rolling direction in the plane of the sheet (DD orientation) as well as in the in-plane 90 or TD direction.

Note the excellent agreement between VPSC and data for the TD loadings. Note that for TD compression the model predicts that at higher strains (i.e. above 4%), the c-axis of the majority of grains is rotating towards the transverse direction and becomes aligned with TD at ~ 9%. To simulate the response for the off-axis DD orientation, we rotated the starting texture by 45^0 about the through-thickness direction (i.e. ND) and imposed strain along the corresponding direction, while enforcing zero average stress along the two lateral directions of the sample. Note the excellent agreement between model and the stress-strain data. It is important to note that the initial yield and subsequent hardening rate predicted by the model for DD tension and compression tests are higher than that predicted for the RD direction and lower than those predicted for the TD. This trend matches well the experimentally observed in-plane anisotropy of the stress-strain response (see [5]). No texture measurements were reported for the DD orientation. The simulated texture evolution is strictly predictive. Note that the RD-TD axes shown in the pole figures (Fig.5) are in fact the original RD-TD axes rotated by 45^0 about ND. Note also the very strong texture evolution in DD compression, with drastic changes occurring between 4% and 9% strains. While here we show only the 45^0 direction simulations, for each in-plane compression test the model predicts a unique S-shape stress-strain curve (see Figures 5(b), 5(d)), which indicates that tensile twinning activity should be observed in uniaxial compression tests irrespective of the in-plane orientation. This finding needs to be further verified/validated experimentally.

3.3. VPSC Predictions for Shear Loading

The simple shear tests were carried out using an MTS tension/torsion system and specially designed fixtures with gage length of 38.2 mm, width of 6 mm and a thickness of 2 mm ([5]).

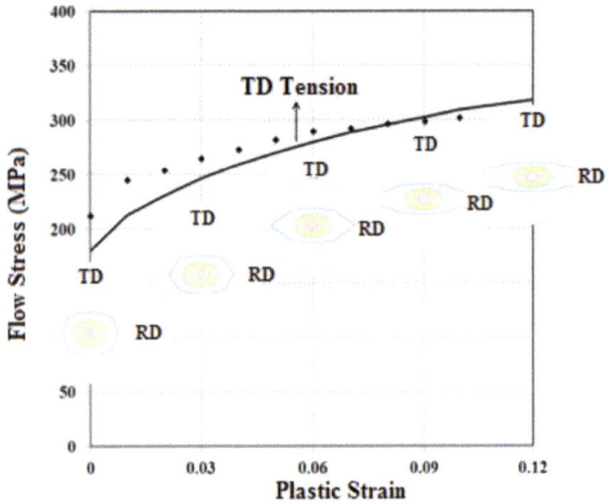

(a)

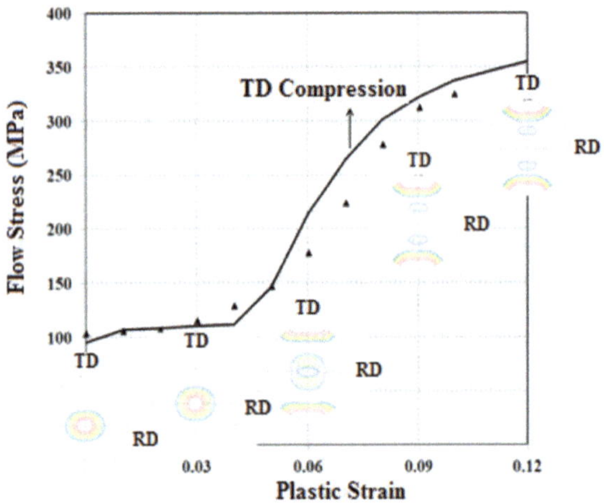

(b)

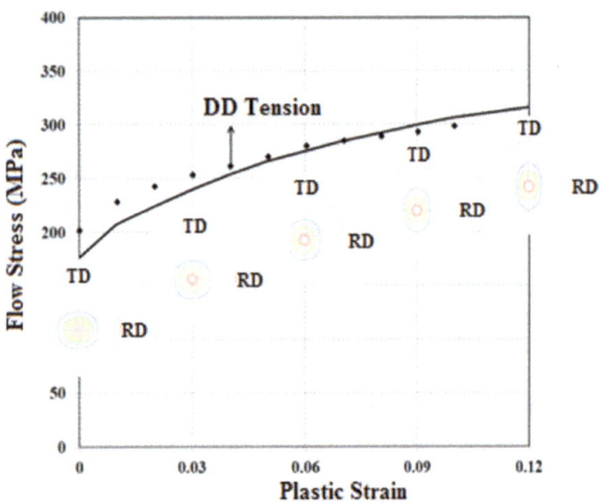

(c)

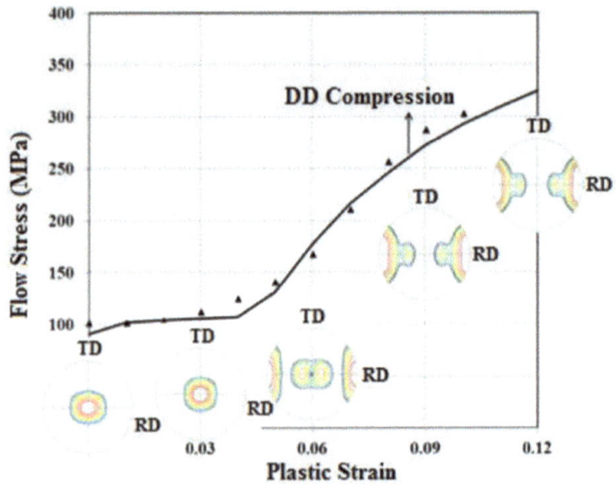

(d)

Figure 5. Comparison of experimental (points) and predicted stress-strain response (lines) in (a) TD tension, (b) TD compression, (c) DD tension and (d) DD compression. Note predicted texture evolution as shown by (0 0 0 2) pole figures.

Neutron diffraction analysis was carried out on cube specimens cut from as-received sheet specimens and deformed specimens using the BT8 residual stress diffractometer [5]. Simulation results for simple shear along RD (see Figure 6-7) show that the model predicts the stress-strain response with great accuracy. The deformation mode activity plot (Fig. 6(b)) shows that tensile twinning is initially activated and then saturates at higher strains. It is highest at approximately 3.5% von Mises equivalent strain where a slight knee in the stress-strain response can be observed indicating a change in slope. It saturates at around 10% equivalent strain and correspondingly the slope of the stress-strain curve remains constant thereafter. It is important to note that Lou et al. [4] also observed a knee in the stress strain curve in simple shear at approximately 2.5% equivalent strain, which was attributed to tensile twinning. The final texture observed in the RD shear is very close to that predicted by VPSC, as shown in Figure 7(b).

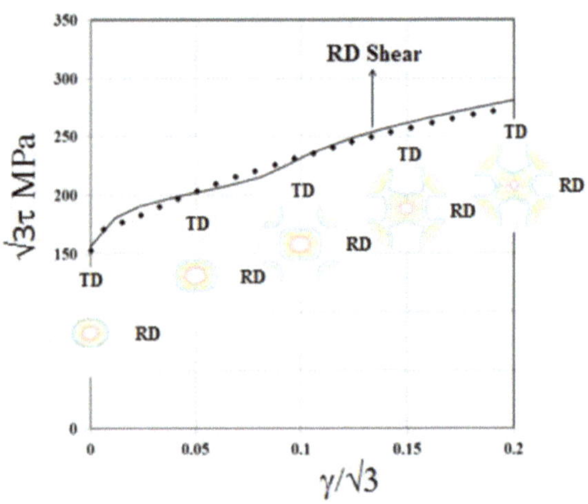

(a)

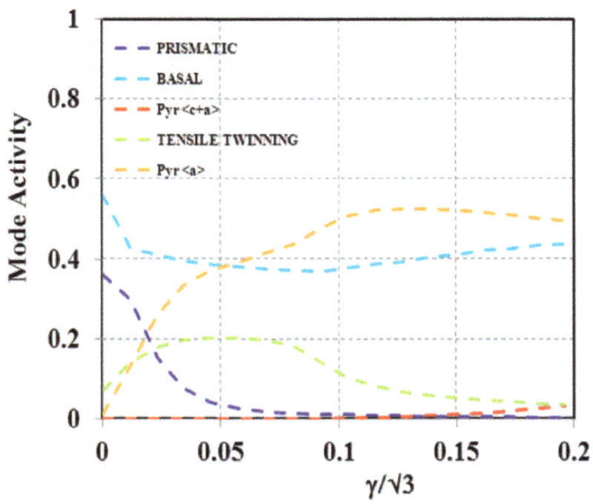

(b)

Figure 6. VPSC predictions in RD shear: (a) stress-strain response compared with data (symbols) along with predicted texture evolution with accumulated strain ((0 0 0 2) pole figures), (b) relative deformation mode activity.

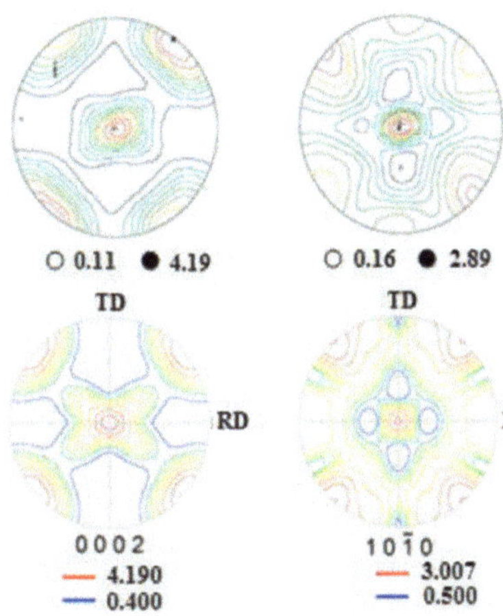

(a)

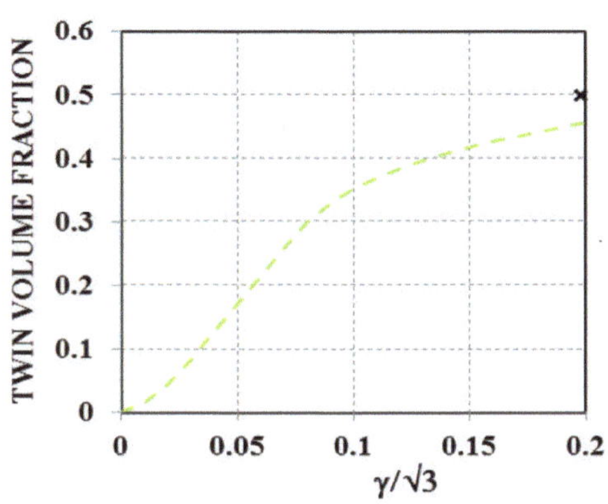

(b)

Figure 7. VPSC predictions for RD shear: (a) observed texture (top) and predicted (bottom) texture at equivalent strain 20%, (b) tensile twinning activity in RD shear; the symbol shows the only available measured twin volume fraction reported in [5].

The deformation mode activity plot (Fig. 6(b)) shows that tensile twinning is initially activated and then saturates at higher strains. It is highest at approximately 3.5% von Mises equivalent strain

4. Conclusions

Using a step-by-step approach in conjunction with mechanical data in uniaxial tension along RD, uniaxial compression in the RD and ND, and shear in RD reported by Khan et al. [5] the VPSC model was identified. It was shown that using the VPSC model thus calibrated it is possible to explain and accurately predict the stress-strain behavior and texture evolution of an AZ31 Mg rolled plate for all uniaxial loading directions. Moreover, the response for off-axis loadings (45 for RD) was for the first time simulated and compared with experimental data. The peculiar S-shape of the stress-strain curve in ND tension observed by Guo et al. [9] for an AZ31 Mg plate of similar initial texture was also captured. Moreover, it was shown, that by correctly accounting for slip system and twinning activity, it is possible to concomitantly describe the stress-strain response, the microstructure evolution and the twin volume fraction at 20% (Mises) equivalent strain for simple shear along the rolling direction.

Acknowledgement

O. Cazacu and N. Chandola wish to thank Dr. Ricardo Lebensohn of Los Alamos National Laboratories for providing the VPSC code and insightful discussions.

References

1. Lebensohn, R. A., Tome, C. N. A self-consistent anisotropic approach for the simulation of plastic deformation and texture development of polycrystals: Application to zirconium alloys. Acta Metall. et Mater. 41(1993), 2611-2624.
2. Agnew, S.R., Duygulu, O. Plastic Anisotropy and the role of non-basal slip in magnesium alloy AZ31B. Int. J. Plast. 21 (2005), 1161-1193.
3. Yi, C., Davies, H.J., Brokmeier, H.-G., Bolmaro, R.E., Kainer, K.U., Homeyer, J. Deformation and texture evolution in AZ31 magnesium alloy during uniaxial loading, Acta Mater., 54 (2006), 549-562.
4. Lou, X.Y., Li, M., Boger, R.K., Agnew, S.R., Wagoner, R.H. Hardening evolution of AZ31B Mg sheet. Int. J. Plast. 23(2007),44-86.
5. Khan, A., Pandey, A., Gnaupel-Herold, T., Mishra, R.K. Mechanical response and texture evolution of AZ31 alloy at large strains for different strain rates and temperatures. Int. J. Plast. 27(2011), 688-706.
6. Walde, T., Riedel, H. Simulation of earring during deep drawing of magnesium alloy AZ31. Acta Mater., 55 (2007), 867-874.
7. Van Houtte, P. Simulation of the rolling and shear texture of brass by the Taylor theory adapted for mechanical twinning. Acta Metallurgica, 26 (1978), 591-604.
8. Tomé, C.N., Lebensohn, R., 2004. "Self-consistent homogenization methods for texture and anisotropy" in: *Continuum Scale Simulation of Engineering Materials: Fundamentals, Microstructures*, Process Applications. D. Raabe, F. Roters, F. Barlat and L.-Q. Chen (Eds.), Willey, 352-378.
9. Guo X.Q.,Wu W., Wu P.D., Qiao H., An K., Liaw P.K. On the Swift effect and twinning in a rolled magnesium alloy under free-end torsion. Scripta Mater. 69 (2013), 319–322.

Magnesium Technology 2015
Edited by: Michele V. Manuel, Alok Singh, Martyn Alderman, and Neale R. Neelameggham
TMS (The Minerals, Metals & Materials Society), 2015

RECRYSTALLIZATION BEHAVIOR OF THE MAGNESIUM ALLOY ZE20

Xianfeng Ma[1], Zhihua Huang[1], Mei Li[2], John E. Allison[1]
[1]Department of Materials Science and Engineering, University of Michigan, Ann Arbor, MI 48109, USA
[2]Research and Advanced Engineering Laboratory, Ford Motor Company, Dearborn, MI 48124, USA

Keywords: Recrystallization, Magnesium, Grain orientation spread

Abstract

Understanding and predicting the evolution of recrystallized microstructures is an important part of the Integrated Computational Materials Engineering (ICME) toolset required for magnesium alloys. In this study, the recrystallization behavior of a new magnesium-rare earth alloy, ZE20, was investigated at typical temperatures (i.e. 375 °C, and 425 °C). The characteristics of recrystallization in ZE20 were analyzed via microscopic observations. The recrystallization microstructure was characterized and quantified using a Grain Orientation Spread (GOS) technique based on electron back scatter diffraction (EBSD) scans. The evolution of texture and grain size during recrystallization at 425 °C was examined by EBSD analysis. Static recrystallization kinetics at both temperatures was obtained experimentally and modeled using the JMAK relationship.

Introduction

Magnesium alloys are attractive materials for developing light-weight vehicles because of their high strength-weight ratio [1-3]. However, the low formability limits their wide application due to limited slip systems and strong deformation texture formed during processing. Recrystallization is one means to refine the microstructure [4] and improve the formability of magnesium [4, 5]. Hence, understanding and predicting the evolution of recrystallized microstructures forms an important part of the ICME toolset for magnesium alloys.

Recently, Luo et al. [2] reported that a new magnesium alloy ZE20 (Mg-2%Zn-0.2%Ce) exhibits outstanding ductility and strength. This was rationalized based on the solid solution strengthening effect of Zn, combined with the influence of Ce to reduce the formation of strong texture during processing. Limited research has been conducted on the dynamic recrystallization of magnesium alloys such as AZ31 [1, 6], AZ31B [7] and AM30 [8]. However, very little attention has been paid to the static post-deformation recrystallization behavior during annealing of magnesium alloys [4, 5].

In this study, the recrystallization behavior of the magnesium alloy ZE20 was investigated at two typical temperatures (i.e. 375 °C, and 425 °C). The characteristics of recrystallization were analyzed via microscopic observations and electron back-scatter diffraction (EBSD) measurements. The recrystallization kinetics was obtained by quantification of recrystallization fraction using the Grain Orientation Spread (GOS) technique.

Materials and Experimental

The ZE20 material used in this study had a nominal composition (wt. %): 2.4 Zn, 0.23 Ce, 0.10 Si, 0.02 Mn, 0.01 Al, <0.005 Fe, <0.005 Cu, < 0.003 Ni with Mg comprising the balance. The material was received as a cast billet with 180 mm diameter and 50mm height. Cylindrical samples (Φ10mm x H12mm) were machined from the edge region of as-cast billet, to insure a consistent starting microstructure in all samples. Cold deformation was produced via compression tests performed on an Instron testing machine. Samples were compressed to 25% height reduction (ε = 0.29) at 0.3mm/min speed. To minimize friction effects between the sample and platen, teflon tape was used with polished platens. The recrystallization heat treatments were performed in a Carbolite box furnace at 375 °C and 425 °C with times ranging from 5 minutes to 14 hours. The samples were water quenched immediately after removing from the furnace. Vickers microhardness was measured using a CM-400 AT Clark Microhardness Tester at a force of 200 gram. Five points were measured and the average value was used in analysis.

Samples for optical metallography were prepared using a standard metallographic preparation procedure, i.e. grinding to 1200 grit and polishing to 1 μm, then etching in an etchant of 4 g picric acid, 10 mL acetic acid, 10 mL DI water, and 100 mL ethanol for 5 sec. The samples for EBSD study were polished to 0.05 μm, and finished by etching in an acetic-nitric solution (60mL ethanol, 20mL water, 15 mL acetic acid and 5mL nitric acid) for 3~5 sec. Accelerating voltage of 15~20 keV was used for all scans and a step size between 0.5~2 μm was selected according to the scan region. Optical images were taken using a Nikon AZ100 Optical Microscope. A Tescan MIRA scanning electron microscope (SEM) with an EDAX Octane Plus EBSD detector was used.

Results and Discussion

Figure 1 shows an SEM back-scattered electronic image of the starting microstructure of an as-cast ZE20 sample. Mg-Zn-Ce particles (identified by energy dispersive spectrum analysis) were distributed primarily along the grain boundaries. The EBSD scan in Fig. 2 shows the grain morphology of as-cast ZE20, with the measured average grain size to be 390 μm. EBSD study also indicated that the initial texture of as-cast billet was close to random as shown in Fig. 2.

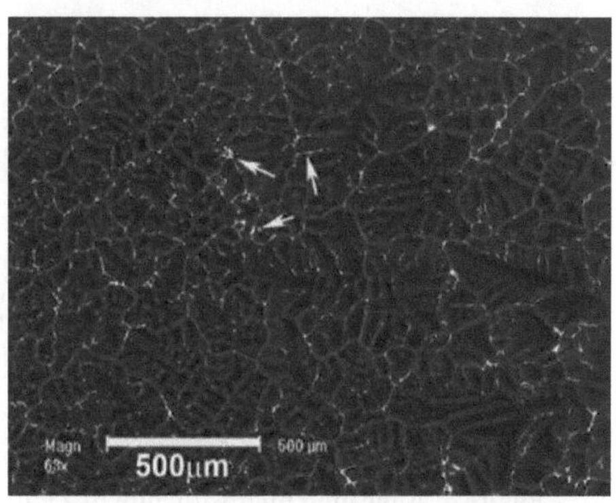

Figure 1. The as-cast microstructure of ZE20 billet (arrows indicate the Mg-Zn-Ce particles)

Figure 2. The EBSD micrograph of as-cast microstructure of ZE20 billet (at edge)

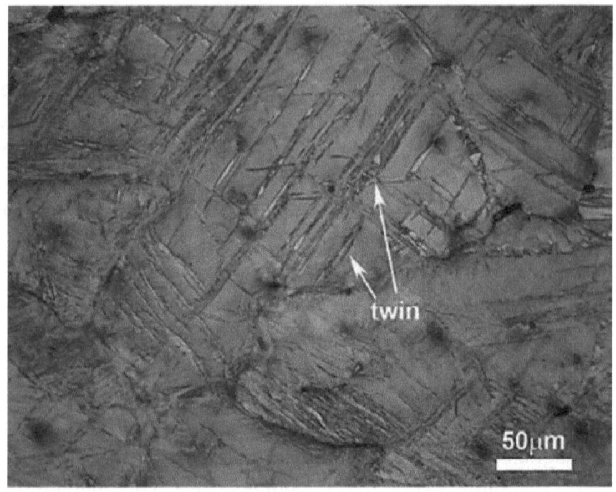

Figure 3. Optical microstructure of ZE20 after cold compressed to 25% height reduction (on longitudinal section)

Figure 3 shows an example of the deformed microstructure of ZE20 after 25% cold compression. The cold deformed microstructure shows a large number of deformation twins, which is common for Mg alloys at ambient temperature [4, 5], due to the limited number of slip systems [9]. Deformation bands were occasionally observed in some samples.

Figure 4. Annealing microstructure of ZE20 alloy on longitudinal section: (a) 425 °C, 10min, (b) 375 °C, 10 min, (c) 425 °C, 14 h (d) 375 °C, 14 h

The representative optical microstructures after annealing at 425 °C and 375 °C were shown in Fig. 4. A high amount of recrystallization was observed after 10 minutes annealing at 425 °C (Fig. 4a). Recrystallized grains appear smaller than the starting microstructures. In Fig. 4b, small recrystallized grains were also observed after annealing at 375 °C for 10min. Deformation twins remained due to slower recrystallization process at 375 °C, examples of which are observed in Fig. 4b. After 14 hours annealing at 425 °C, a microstructure composed of equiaxed grains was formed (Fig. 4c). Fig. 4d shows the microstructure at 375 °C for 14 hours, in which recrystallization was not fully completed and large deformed grains containing twins were evident.

It is difficult to quantify recrystallization from optical images, in part due to the presence of twins. Hence, both the fractional softening method and the EBSD Grain Orientation Spread techniques were evaluated. The Vickers microhardness was measured and used to analyze the softening fraction [10] using the following relationship:

$$X_{rex} \approx X_H = (H_\varepsilon - H_t)/(H_\varepsilon - H_0) \qquad (1)$$

where H_ε, H_t, and H_0 are the hardness values for as-deformed structures evolved at various strains (ε), for intermediate annealing at time (*t*) and for a fully annealed condition at longer times, respectively. This equation is the measurement of softening factor X_H, which is approximately close to recrystallization fraction assuming that recrystallization is the only source of softening.

Figure 5 shows the Vickers hardness measurements on samples annealed at 425 °C and 375 °C. For samples at both temperatures, it is seen that the hardness dropped remarkably in the first 10 min and then remained relatively constant at times up to 840 min. Generally, the samples show slightly lower hardness values after annealing at 425 °C than those at 375 °C.

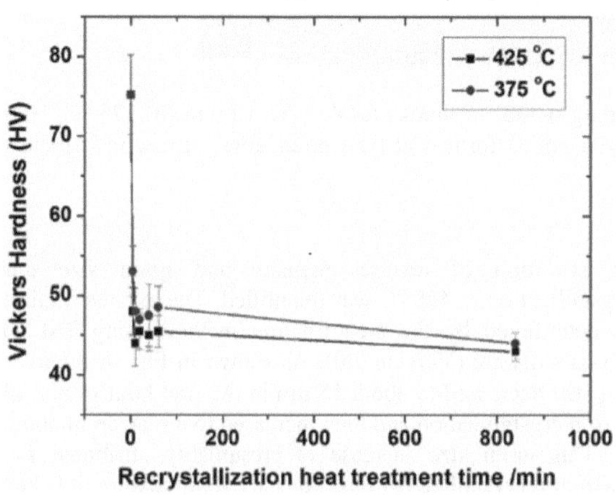

Figure 5. Vickers hardness results versus heat treatment time for ZE20 alloy

The recrystallization fraction was estimated using Eq. (1), show in Fig. 6. In this calculation, we take H_0 as the hardness of the sample annealed at 425 °C for 14 hour, assuming that it is close to fully recrystallization state.

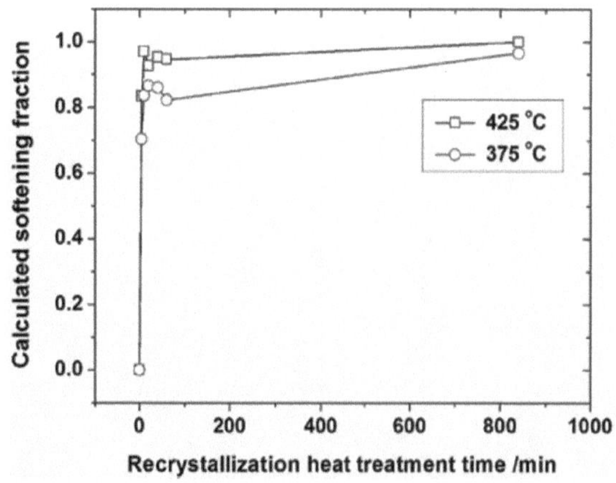

Figure 6. Softening fraction calculated from Vickers hardness change for samples at different recrystallization heat treatment time

EBSD measurements were also used to quantify the fraction of recrystallized grains. EBSD scans (1 mm by 1 mm) were obtained at the very center of the longitudinal section of each sample, where based on finite element analysis of the cylindrical sample, the strain is uniform to within 3% dispersion. Grain orientation spread analysis (GOS) was used to identify the recrystallized grains (1 degree threshold of misorientation) (see Fig. 7). The recrystallized grains are highlighted in white and the non-recrystallized grains are in different colors according to the intragranular misorientation. Fig. 7a (425 °C, 10 min) shows well-developed recrystallized grains and some deformed parent grains (the fraction of recrystallized grains was 65~70%). Recrystallized grains generated along deformation twin are noted with arrows in Fig. 7a. It is seen more clearly in Fig. 7b (375 °C, 10 min) that twin boundaries can be preferred nucleation sites for recrystallization. This is consistent with a previous study in AZ31 magnesium alloy [5]. In Fig. 7b, the fraction of recrystallized grains was determined to be 35%. The sample after 14 hour annealing at 425 °C (Fig. 7c) exhibited a fine equiaxed recrystallization microstructure with occasional parent grain, indicating it is close to fully recrystallized. At 375 °C after 14 hours annealing, recrystallization was not completed (~69% recrystallized). In Fig. 7d, the parent grains can be observed (marked by an arrow). The results obtained by EBSD-GOS analysis are consistent with the optical observations in Fig. 4.

Quantification of the recrystallization fraction can be obtained by image analysis of the GOS results in Fig. 7. The non-indexed points in the grain map (Fig. 7), e.g. particles and grain boundary neighborhood, were not taken into account when calculating the area fraction recrystallized. The fraction at each annealing time measured by EBSD-GOS was summarized in Fig. 8 (solid line), in which the estimated recrystallization fraction by hardness value (dotted line) was also plotted for comparison. The fraction recrystallized is higher at 425 °C than that at 375 °C. This is expected due to the fact that higher annealing temperature promotes the thermally activated process controlling recrystallization. The fraction softening method appears to overestimate the recrystallization fraction, presumably due to the recovery which also occurs during the annealing heat treatment.

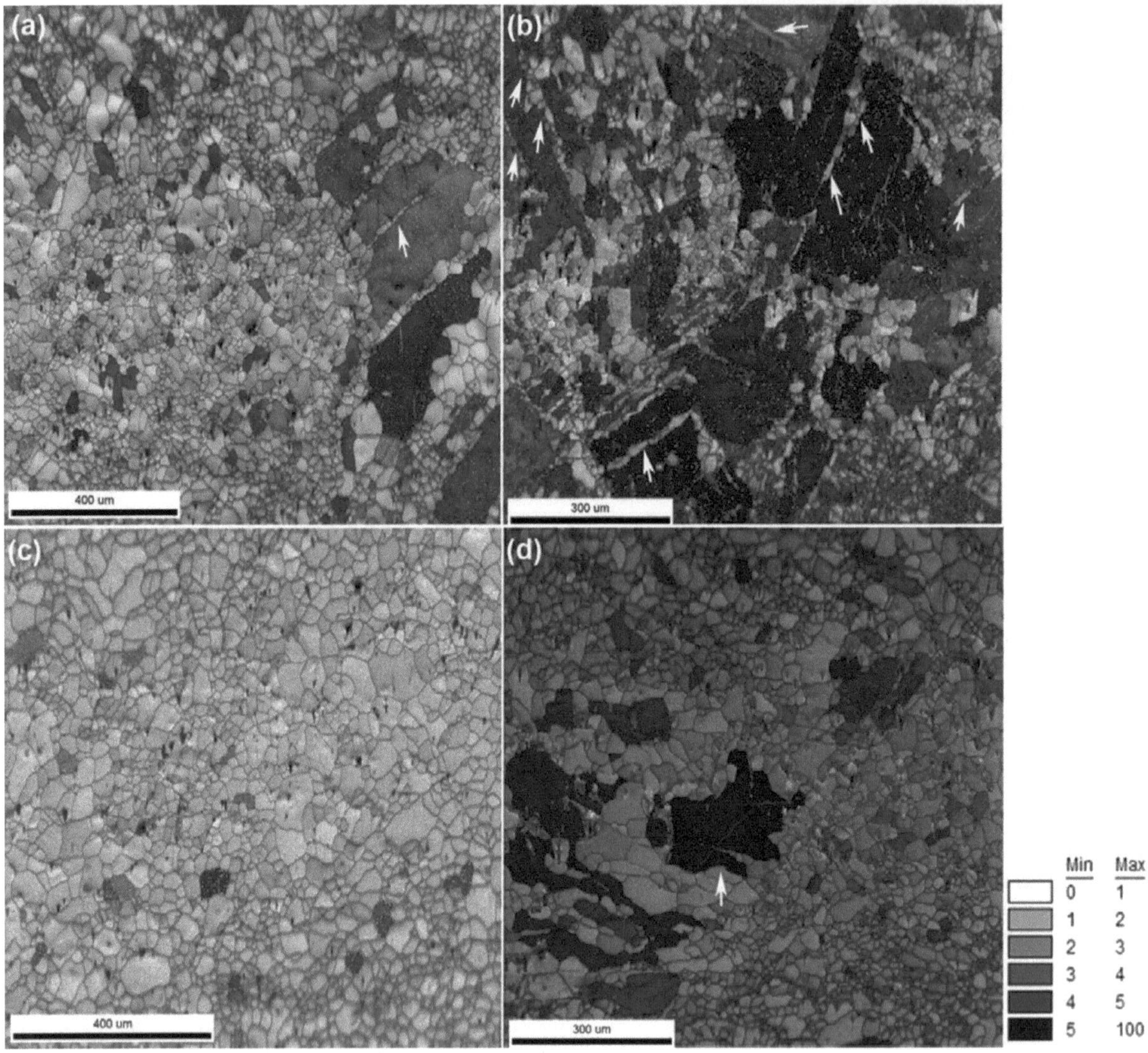

Figure 7. Analysis of recrystallization fraction using grain orientation spreading (GOS) technique:(a) 425 °C, 10 min (b) 375 °C, 10 min; (c) 425 °C, 14 h (d) 375 °C, 14 h (Arrows in (a) and (b) show recrystallized grains formed at twin boundaries. Arrow in (d) shows a deformed grain)

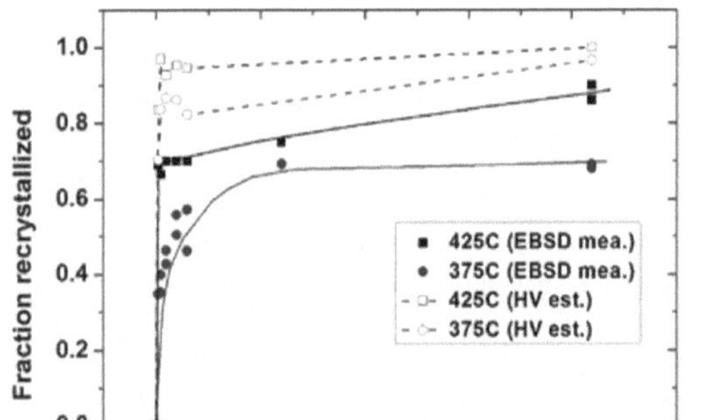

Figure 8. Static recrystallization kinetics of ZE20 alloy: area fraction of recrystallized grains versus heat treatment time

The evolution of texture intensity and grain size during recrystallization at 425 °C was quantified. The average grain size was determined by the area fraction method using TSL OIM analysis software (Version 7.0). As shown in Fig. 9, the average grain size decreased to about 15 μm in the first hour of annealing due to recrystallization and then increased to a plateau of about 25 μm. This grain size increase is presumably attributed to the growth of recrystallized grains. For the overall texture, it is known that during compression, the basal poles tend to align along the compression axis and gradually this basal texture weakens during recrystallization [11]. Fig. 10 shows an example of the texture of a recrystallized sample (425 °C, 14 hours). The texture intensity decreases substantially during the first hour (red points in Fig. 9), due to the replacement of textured deformed grains by newly formed recrystallization grains. The result indicated that after 425 °C annealing for 14 hours, a uniform microstructure with fine recrystallization grain as well as low texture intensity (~4 times random) can be obtained. This refined microstructure is expected to benefit the formability and strength of ZE20 alloy, as has been reported in other Mg alloys [12, 13].

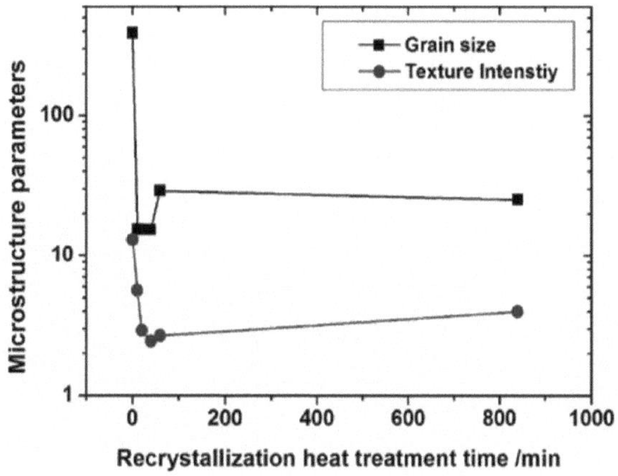

Figure 9. Grain size and texture intensity versus recrystallization heat treatment time at 425 °C

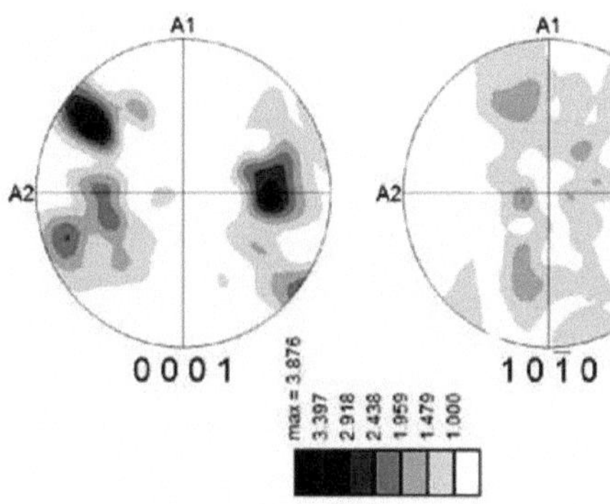

Figure 10. Pole figure of ZE20 sample annealed at 425 °C for 14 hour ('A2' is the compression direction, and 'A1' is the radial direction of cylinder sample)

A commonly used approach for capturing this recrystallization behavior is the JMAK relationship [11, 14]:

$$X_{rex} = 1 - \exp\left[-k\left(\frac{t}{t_{0.5}}\right)^n\right] \tag{2}$$

where X_{rex} is the fraction recrystallized, t is recrystallization time, $t_{0.5}$ is the time for 50% recrystallization, k is a material constant, and n is the Avrami exponent.

In the current study, the experimental EBSD GOS-determined fraction recrystallized versus annealing time in Fig. 8 was used to calibrate the JMAK model (Eq. (2)). n is obtained by plotting $ln(ln(1/(1\text{-}X_{rex})))$ versus $ln(t)$, which follows a linear relationship with a slope of n. k is taken as 0.693 to fit the model at $X_{rex} = 50\%$ at $t_{0.5}$ [15]. $t_{0.5}$ is estimated by interpolating or extrapolating the experimental data in Fig. 8, which yields a value of 1 min and 40 min for 425 °C and 375 °C, respectively. Thereby, the static recrystallization kinetics of ZE20 was modeled using Eq. (2), as shown in Fig. 11. It is seen that this model captures the fraction of recrystallized grains over the range of annealing times investigated.

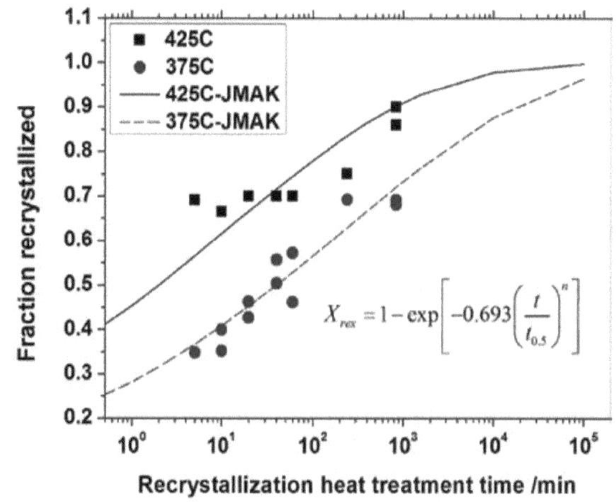

Figure 11. Recrystallization kinetics of ZE20 and associated JMAK relationship

Conclusion

The static recrystallization behavior of ZE20 alloy has been studied after annealing at 425 °C and 375 °C. Significant microstructure changes occurred by recrystallization within the first 10 minutes at both temperatures. At 425 °C, ZE20 showed an almost fully recrystallized microstructure after 14 hour annealing. Cold working and recrystallization annealing appears to be an effective way of refining the microstructure of ZE20 alloy reducing the average grain size from 390 µm to ~25 µm. At 375 °C, even after 14 hour annealing, recrystallization was not fully completed. Twin boundaries were observed to be one of the preferred nucleation sites for recrystallization. The recrystallization kinetics of ZE20 at both 425 °C and 375 °C were obtained using the EBSD-GOS method, whereas the fractional softening method overestimated the recrystallization fraction. The recrystallization kinetics can be captured using a JMAK relationship.

Acknowledgements

The authors gratefully acknowledge the support of US Automotive Materials Partnership and Department of Energy (DOE). This material is based upon work supported by the DOE National Energy Technology Laboratory under the Award Number DE-EE0005660. This paper was prepared as an account of work sponsored by an agency of the United States Government. Neither the United States Government nor any agency thereof, nor any of their employees, makes any warranty, express or implies, or assumes any legal liability or responsibility for the accuracy, completeness, or usefulness of any information, apparatus, product, or process disclosed, or represents that its use would not infringe privately owned rights. Reference herein to any specific commercial product, process, or service by trade name, trademark, manufacturer, or otherwise does not necessarily constitute or imply its endorsement, recommendation, or favoring by the United States Government or any agency thereof. The views and opinions of authors expressed herein do not necessarily state or reflect those of the United States Government or any agency thereof. The authors would also like to acknowledge E. Deda, A. Trump, A. Murphy, and T. Berman at the University of Michigan for their assistance.

References

[1] Huang S. Y., Li M., Allison J. E., Zhang S. R., Li D. Y., Peng Y. H. (2012) Effects of Direct Extrusion Process on Microstructure, Texture Evolution and Yield Strength of Magnesium Alloy AZ31. *Magnesium Technology*,349-354.

[2] Luo A. A., Mishra R. K., Sachdev A. K. (2011) High-ductility magnesium-zinc-cerium extrusion alloys. *Scripta Mater.* **64**,410-413.

[3] Kojima Y., Aizawa T., Kamado S. (2000) Magnesium Alloys Development towards the 21st Century. *Mater Sci Forum* **350-351**,19-30.

[4] Jain J., Poole W. J., Sinclair C. W. (2006) A study on the static recrystallization of cold rolled magnesium alloy AZ80. *Magnesium Technology 2006*,147-152.

[5] Levinson A., Mishra R. K., Doherty R. D., Kalidindi S. R. (2013) Influence of deformation twinning on static annealing of AZ31 Mg alloy. *Acta Mater.* **61**,5966-5978.

[6] Fatemi-Varzaneh S. M., Zarei-Hanzaki A., Beladi H. (2007) Dynamic recrystallization in AZ31 magnesium alloy. *Mater. Sci. Eng. A* **456**,52-57.

[7] Al-Samman T., Gottstein G. (2008) Dynamic recrystallization during high temperature deformation of magnesium. *Mater. Sci. Eng. A* **490**,411-420.

[8] Ma Q., Li B., Whittington W. R., Oppedal A. L., Wang P. T., Horstemeyer M. F. (2014) Texture evolution during dynamic recrystallization in a magnesium alloy at 450 degrees C. *Acta Mater.* **67**,102-115.

[9] Barnett M. R., Nave M. D., Bettles C. J. (2004) Deformation microstructures and textures of some cold rolled Mg alloys. *Materials Science and Engineering: A* **386**,205-211.

[10] Yang X. Y., Okabe Y., Miura H., Sakai T. (2012) Annealing of a magnesium alloy AZ31 after interrupted cold deformation. *Mater Design* **36**,626-632.

[11] Humphreys F. J., Hatherly M. (2004) *Recrystallization and Related Annealing Phenomena (Second Edition).* in, Elsevier, Oxford.

[12] Mukai T., Yamanoi M., Watanabe H., Higashi K. (2001) Ductility enhancement in AZ31 magnesium alloy by controlling its grain structure. *Scripta Mater.* **45**,89-94.

[13] Koike J. (2005) Enhanced deformation mechanisms by anisotropic plasticity in polycrystalline Mg alloys at room temperature. *Metall. Mat. Trans. A* **36**,1689-1696.

[14] Jensen D. J. (1995) Growth-Rates and Misorientation Relationships between Growing Nuclei Grains and the Surrounding Deformed Matrix during Recrystallization. *Acta Metall. Mater.* **43**,4117-4129.

[15] Liu J., Cui Z., Ruan L. (2011) A new kinetics model of dynamic recrystallization for magnesium alloy AZ31B. *Mater. Sci. Eng. A* **529**,300-310.

Magnesium Technology 2015
Edited by: Michele V. Manuel, Alok Singh, Martyn Alderman, and Neale R. Neelameggham
TMS (The Minerals, Metals & Materials Society), 2015

BALLISTIC CHARACTERIZATION OF THE SCALABILITY OF Mg ALLOY AMX602

Tyrone L. Jones[1], Katsuyoshi Kondoh[2], David Moore[3], Isamu Otsuka[4], Alan Annis[3], Hiroto Nakazawa[5], Yoshinori Ohori[5], Ryo Numasawa[4], Masamichi Takahashi[6]
[1] *US Army Research Laboratory; RDRL-WMP-E, APG, MD 21005 USA*
[2] *Joining and Welding Research Institute; Osaka University, 11-1 Mihogaoka, Ibaraki 567-0047 OSAKA JAPAN*
[3] *Taber Extrusions; 915 S. Elmira; Russellville, AR 72802 USA*
[4] *Epson Atmix; 4-44 Kaigan, Kawaragi, Hachinohe, Aomori 039-1161 JAPAN*
[5] *Pacific Sowa; 3-17,1-Chome Shinkawa, Chuo-ku Tokyo 104-0033 JAPAN*
[6] *National Material L.P. 1965 Pratt Boulevard, Elk Grove Village, IL 60007 USA*

Keywords: SWAP AMX602, Ballistic Analysis, Scalability

Abstract

The US Army Research Laboratory (ARL) and the Osaka University Joining and Welding Research Institute (JWRI) formed a collaborative partnership with Taber Extrusions, Epson, Pacific Sowa, Kurimoto, and National Material LP to domestically reproduce and scale-up military grade magnesium alloy AMX602 at the Taber Extrusion manufacturing facility in Gulfport, MS. AMX602 material was provided in the form of 38.1-mm (1.5-inch) wide bars, 101.6-mm (4-inch) wide plate, and 152.4-mm (6-in) wide plate. The ARL and the JWRI conducted mechanical analysis and dynamic impact examination to evaluate the lateral dimension scale-up of AMX602. The results were parametrically analyzed and compared to conventionally processed AZ31B-H24 and AA5083-H131. Details of the scalability of the AMX602 alloy are provided.

Introduction

The US Army Research Laboratory's (ARL) ballistic analysis of magnesium (Mg) alloys over the last 8 years has led to an increased understanding of the material's failure mechanisms and relationship between Mg alloy strength and ductility requirements for lightweight armor applications [*1*]. While Mg alloys have been used for military structural applications since WWII, very little research has been done to improve its mediocre ballistic performance [*2*]. The highest strength commercial Mg alloy available in plate form, AZ31B (1.78 g/cm^3), has proven to be a very good substitute armor material for AA5083 (2.66 g/cm^3) against armor-piercing projectiles on an equal weight basis [*3*]. It is also an adequate substitute armor material against fragment simulating projectiles (FSPs) within an areal density range that is FSP dependent [*4*]. However, shearing and scabbing are dominant features during ballistic impact. The ballistic data generated by ARL was used to develop the first set of Mg alloy acceptance standards, MIL-DTL-323333 (MR), titled, "Armor Plate, Magnesium Alloy, AZ31B, Applique" [*5*]. Ultimate tensile strength (UTS), tensile yield strength (YTS), ductility, and grain size are all important parameters in determining the ballistic performance of metals. The bulk material properties of AZ31B are shown in Table 1. Figure 1 correlates impact energy absorption (J) vs. Mg armor alloy AZ31B grain size (μm) [*6*].

In 2009, ARL collaborated with the Joining and Welding Research Institute (JWRI) of Osaka University under contract through the International Technology Center-Pacific to develop and evaluate high-strength, high-ductility Mg alloy plate for structural applications that would exceed the ballistic performance of AA5083. Initial evaluation of conventionally rolled AZ31B plate versus powder formed AZ31B plate showed that grain refinement is one parameter that is needed to significantly improve the viability of Mg armor alloy plate [*7*].

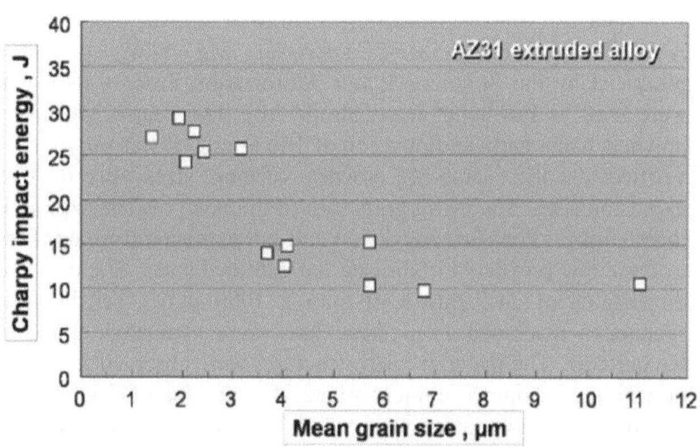

Figure 1. AZ31B Grain Size versus Impact Energy Absorption.

New fundamental Mg alloying is needed to increase the impact energy and thus the performance of Mg alloy plates. The result showed a research opportunity to make Mg alloys viable armor materials that could compete with current aluminum armor alloy solutions [*4*]. Based on the preliminary material and ballistic analysis, the ARL/JWRI program set goals to develop Mg alloys with the mechanical properties shown on the lower row of Table 1.

Table 1. Mechanical Properties/Goals of Mg Alloys

	Ultimate Tensile Strength (MPA)	Tensile Yield Strength (MPA)	Elongation to Failure (%)
AZ31B [*7*]	245	150	7
New Mg Alloy	400	350	20

Clearly, there were two potential paths forward towards achieving these set goals:

1. Discover new chemical compositions to create high strength, high ductility Mg alloys, while not compromising the desirably low density of 1.78 g/cm^3.
2. Improve grain refinement through novel processing techniques to produce high-strength, high-ductility Mg alloys.

As a result, ARL and JWRI collaboratively created two new experimental Mg alloys, AMX602 and ZAXE1711, in extruded 40-mm bars starting with an advanced metallurgical powder process. The successful ballistic and corrosion evaluation of each material [8], expanded our partnership into a coalition to reproduce and scale up the lateral dimensions of AMX602 bars into 305-mm x 305-mm (12-inch x 12-inch) plates for commercial production in the United States. Cost was the driver in our selection of AMX602 over ZAXE1711. However, once scaling up of AMX602 is achieved, the same processing methodology can be applied to reproduce ZAXE1711. Details of the scale up process and ballistic evaluation of AMX602 are discussed in the next sections.

Material Exploration

AMX602 (Mg-6Al-0.5Mn-2Ca/mass%) Mg alloy powders produced by the Spinning Water Atomization Process (SWAP) were used as raw input materials [9, 10]. The coarse Mg alloy powders had a particle dimension of 1 to 5-mm. It was previously verified that the coarse Mg powders of these sizes were of low explosion risk. The α-Mg grain size of the raw powders was less than 0.5 μm. Powder compaction and hot extrusion were applied to these raw powders to fabricate the extruded bars. The bar had dimensions of 24.5 mm x 40 mm x 1000 mm. Tensile test specimens machined from these bars were evaluated at room temperature. The material microstructures were observed using an optical microscope. The microstructural evaluation of ZAXE1711 is withheld from discussion until the patent application is processed.

Experimental Evaluation of Raw Materials

In SWAP powder preparation, schematically illustrated in Figure 2 (a), non-combustive AMX602 magnesium alloy ingots were melted at 1053 K in a ceramic crucible covered by a protective inert gas. The molten metals were directly streamed inside the spinning water chamber from a crucible nozzle. Table 2 shows the chemical composition of AMX602 alloy powders prepared by SWAP. The calcium is necessary because it reduces the combustive properties of the magnesium alloys. The impurity content of Fe and Cu is controlled to less than 0.005% because they are known corrosive elements in magnesium alloys. As shown in Figure 2 (b), the length of the coarse AMX602 powders prepared by SWAP is ~1 - 4 mm, and they are of irregular shape. A cast ingot with the same composition was also prepared as a reference input material.

Powder Metallurgy

The powder was consolidated at room temperature using a 2000kN hydraulic press machine to fabricate the green compact. The green compact had a relative density of 85% and 42 mm diameter. The columnar compact and cast ingot were heated at ~573 - 673K for 180 seconds in an argon gas atmosphere, then immediately consolidated into full density material by hot extrusion. An extrusion ratio of 37 and an extrusion speed 1 m/s were used in this study.

Table 2. Chemical Compositions of Non-Combustive Magnesium Alloy Powders.

Al	Zn	Mn	Fe	Si	Cu	Ca	Mg
6.01	0.007	0.26	0.002	0.038	0.004	2.09	Bal.

(a)

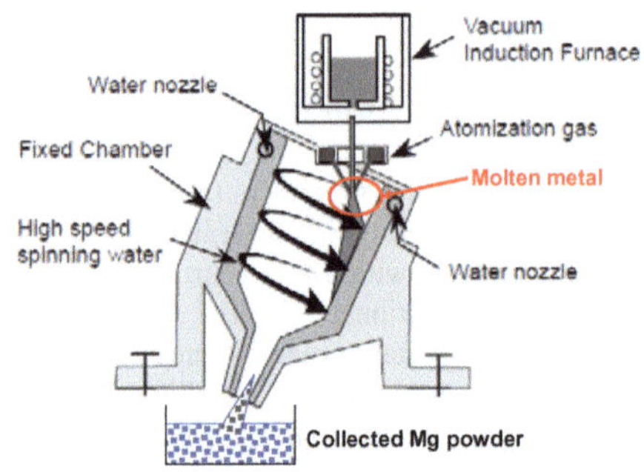

(b)

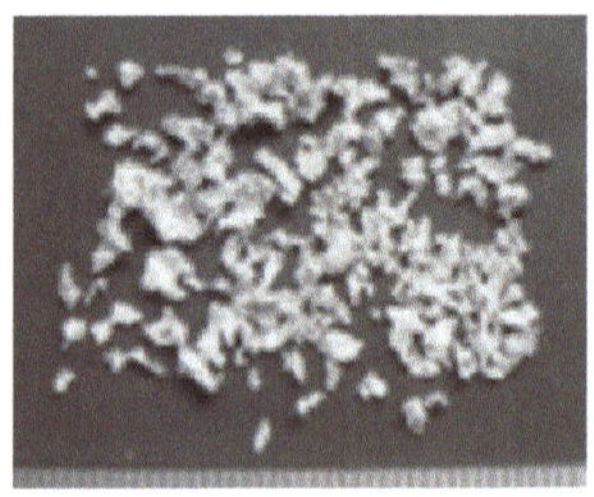

Figure 2. (a) Schematic Illustration of SWAP Equipment to Produce Rapidly Solidified Mg Alloy Powders; (b) Morphology of Coarse Magnesium Alloy Powder Prepared by SWAP.

Fabrication Procedure

Plate A and Plate B extrusions were performed with a 2722 tonne (3,000 ton) press, 279-mm (11-inch) container with the 25.4-mm x 101.6-mm (1-inch x 4-inch) die. The conditions are detailed in Table 4.

The extrusion of Plate A had surface flaws. The cause was thought to be that the extrusion speed was too fast and the compaction too short. Hence, Plate B underwent a longer compaction time and a slower extrusion speed. The material surface and cross-section of Plate B was greatly improved based on visual inspection of the extrusion.

Three experiments were conducted using 2 Kurimoto billets on the 1633 tonne (1800 ton) press 178mm (7-inch) container to produce 25.4-mm x 101.6-mm (1-inch x 4-inch) plates, and then a 25.4-mm x 152.4-mm (1-inch x 6-inch) plate on the 2722 tonne (3,000 ton) press 279-mm (11-inch) container. Results are shown in Table 5.

The material surface and cross-section of Plate C, Plate D, and Plate E looked optically homogeneous. The material surface of Plate E had the best surface appearance.

Table 4. Processing Specifications of the Taber Billets

Plate	A	B
Powder (kg)	25	27
Container Temp (F)	800	800
Die Temp (F)	799	800
Billet Length (inch)	8.5	9.0
Nose Billet Temp (F)	596	565
Butt Billet Temp (F)	740	600
Extrusion Speed (m/min)	2.44	1.52

Table 5. Processing Specifications of the Kurimoto Billets

Plate	C	D	E
Die Temp (F)	700	550	700
Billet Length (inch)	12.0	12.0	17.0
Nose Billet Temp (F)	600	500	550
Butt Billet Temp (F)	Not Measured	550	693
Extrusion Speed (m/min)	1.83	1.22	1.83

The extruded samples were sent to Osaka University for mechanical property measurement. The results will be discussed in the next section.

Mechanical Property Analysis

As mentioned above, all extruded plate-specimens were fabricated by using the mass-production equipment under various pre-heating and extrusion conditions. Figure 3 shows no cracking and fracture of the specimen surfaces. In addition, Fig. 4 indicates that the mean UTS values of the specimens were ~355-365 MPa, and the scatter was small. This mean value was almost same as that of the small-scale specimens fabricated in JWRI. YS and elongation are also similar to those of the small-scale ones. This means that extrusion process was very stable and the mass-production process could be employed in consolidation of SWAP-Mg alloy powders.

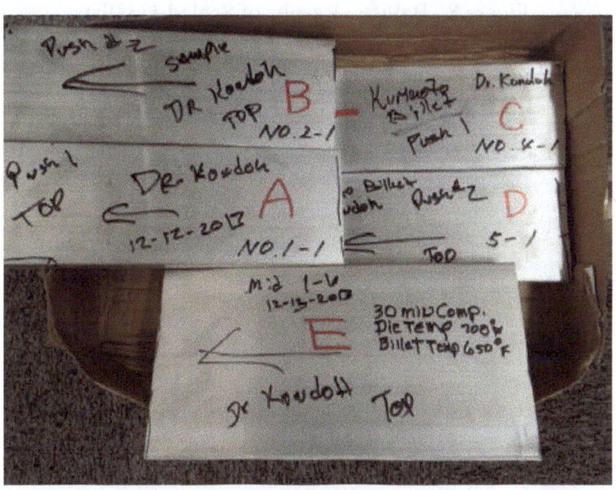

Figure 3. Extruded Mg AMX602 Plate

L-Direction	(MPa)	(MPa)	(%)
	YS	UTS	Elongation
A-L direction	287.3	365.2	14.0
B-L direction	293.3	362.9	12.5
C-L direction	281.1	355.7	14.8
D-L direction	278.9	357.2	16.1
E-L direction	285.6	362.7	14.1

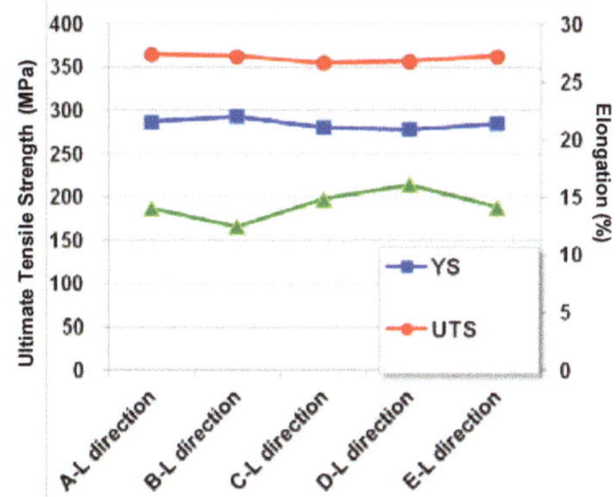

Figure 4. Mechanical Properties of Extruded AMX602 Plate

Ballistic Experimental Procedures

The extruded SWAP magnesium alloy AMX602 was ballistic characterized for 38-mm (1.5-inch) wide, 101.6mm (4-inch) wide, and (152.4mm) 6-inch wide plate.

The ballistic threat was selected in a manner that would allow for direct comparison to other metal alloy armor standards, particularly to the aluminum alloy 5083 (AA5083) armor plate standard. The weldable, AA5083 is currently specified for use in many vehicle armor systems [11]. The test projectile selected was the 0.30-caliber FSP because it provides for quantifiable macro-mechanism features to be observed between magnesium and aluminum alloys at the 45 kg/m2 (9.2 pound per square foot) areal density range.

Ballistic testing of all magnesium alloy plate samples was performed by ARL at Aberdeen Proving Grounds, Maryland in accordance with MIL-STD-662F, issued 18 December 1997 [12]. Ballistic results were characterized using the standard V_{50} test methodology, also documented in MIL-STD-662F.

The specific ballistic threat used to test the magnesium alloy plate samples was the 0.30-cal FSP designed in accordance with MIL-DTL-46593B (MR), issued 6 July 2006, as depicted below in Figure 6 [13]. The weight and hardness specifications are shown in Table 3.

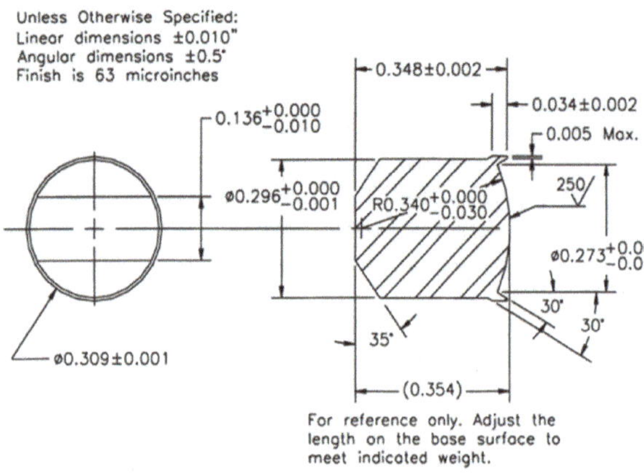

All Units Are in Inches.

Figure 6. The 0.30-cal. FSP Schematic Diagram.

Table 3. Projectile Weight and Hardness Specifications

FSP Type	Weight (g)	Rockwell Hardness C
0.30-cal.	2.9	30±2

Ballistic Experimental Results

Table 4 compares the ballistic limits of the 38.1-mm wide AMX602 bars manufactured at Taber Extrusions to 40-mm AMX602 bars manufactured by JWRI, ZAXE1711 bars manufactured by JWRI, AZ31B (reference material) manufactured by Magnesium Elektron North America, and AA5083 in terms of equivalent areal densities (i.e., mass per unit surface area) and in terms of the actual plate thicknesses. The AA5083 plate was evaluated in 2014. The V_{50} ballistic limits of samples B1 and B2 were estimated due to the limited material available. The hardness of the Mg plates was measured on a Brinell 500-kg scale, while the aluminum plate was measured on the 3000-kg scale.

Table 4. V_{50} Ballistic Limits versus the 0.30-cal FSP

Metal Alloys	Manufacturer	Plate Thickess	Plate Width	Hardness	Ballistic Limit
		mm	mm	HBN	m/s
AZ31B-H24	Mg-Elektron	25.400	304.8	61	843
AA5083 (2014)	--	17.04	304.8	99	1026
AMX602-1 (JWRI)	JWRI/KURIMOTO	25.190/25.235	40	80	1061
AMX602-2 (JWRI)	JWRI/KURIMOTO	25.171/25.210/25.197	40	80/80/83	1092
AMX602-3 (JWRI)	JWRI/KURIMOTO	25.171/25.178	40	80	1105
AMX602-B1 (Taber)	Taber Extrusions	25.489	38.1	99	1122
AMX602-B2 (Taber)	Taber Extrusions	25.483	38.1	99	1121
ZAXE1711-B (JWRI)	JWRI/KURIMOTO	25.248	40	80	1111
ZAXE1711-C (JWRI)	JWRI/KURIMOTO	25.229/25.241	40	77/83	1117
ZAXE1711-D (JWRI)	JWRI/KURIMOTO	25.210/25.279	40	80	1140

Figure 7 compares the reproduced 38.1-mm wide AMX602 bars by Taber to the original 40-mm bars produced by JWRI. The performance of AMX602-B1 and AMX602-B2 are similar.

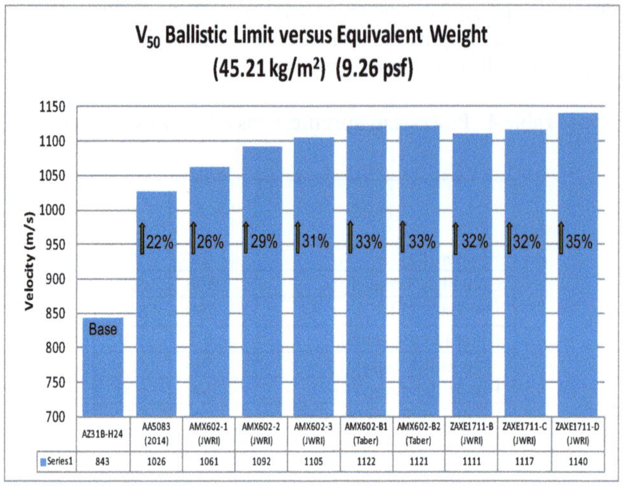

Figure 7. Ballistic Results of 38-mm wide Plate.

The 38.1-mm wide AMX602-B2 bar was evaluated again with the scaled up AMX602 plates. The performance of 101.6-inch wide AMX602 plates yielded the highest V_{50} ballistic limits as shown in Figure 8. Figure 9 illustrates the damage of the 25.4-mm x 101.6-mm (1-inch x 4-inch) plate and the 25.4mm x 152.4mm (1-inch x 6-inch) plate.

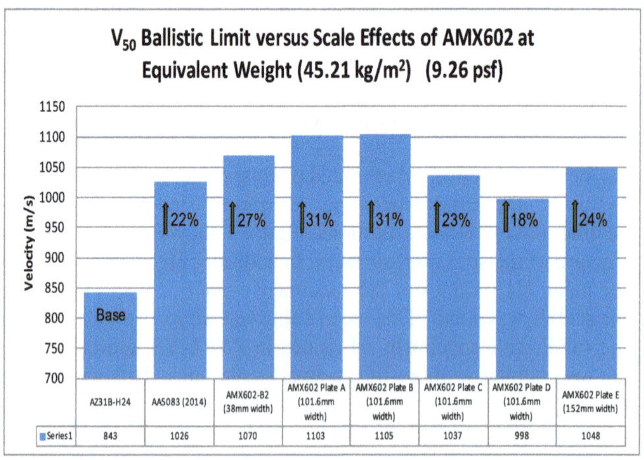

Figure 8. Ballistic Results of Scaled-Up Plate.

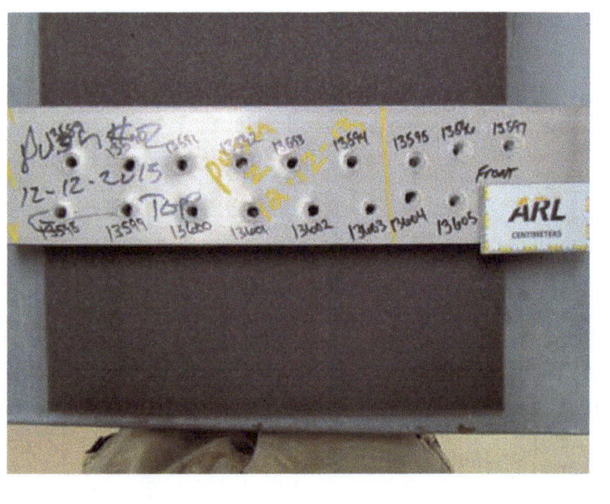

Entry of Impact - 25.4-mm x 101.6-mm (1-inch x 4-inch)

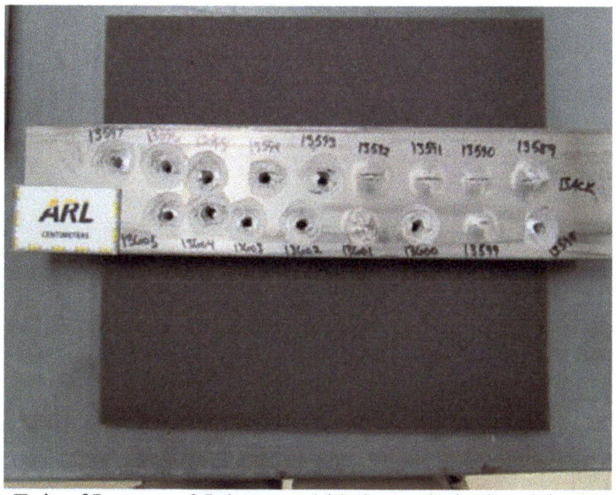

Exit of Impact - 25.4-mm x 101.6-mm (1-inch x 4-inch)

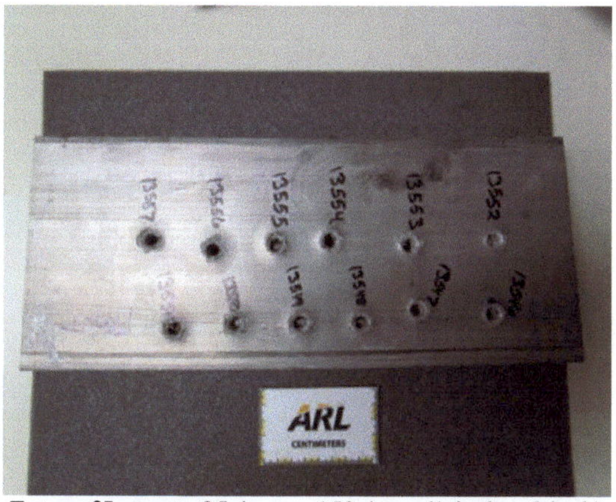

Entry of Impact - 25.4mm x 152.4mm (1-inch x 6-inch)

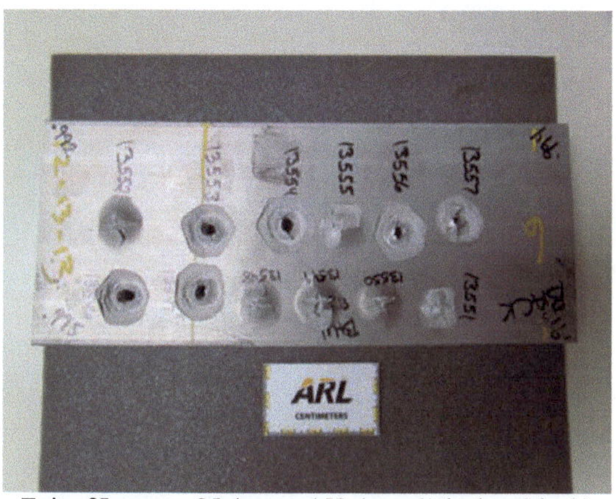

Exit of Impact - 25.4mm x 152.4mm (1-inch x 6-inch)

Figure 9. Post-Ballistic Images of the AMX602 Plates

Conclusions

Successful reproduction of the ballistic performance of AMX602 was achieved on widths from 38.1-mm wide bars thru 101.6-mm wide plates produced by Taber. There was a slight decrease in ballistic performance in the 152.4-inch width plates of AMX602 compared to the narrower plates. One possible reason for this decrease is the reduced yield strength. Comparable penetration resistance across all the plates was observed. There was comparable localized damage for the 38.1-mm width bars thru the 152.4-mm width plates of AMX602. The plates exhibited good scab containment at the back of the material. Plate A revealed good mechanical properties, especially good ductility. Future research should focus on understanding how to improve yield strength (transverse and longitudinal) of the plate. Minimizing the loss of yield strength may prove vital to the penetration resistance of any magnesium alloy.

References

1. Jones, T.; DeLorme, R.; Burkins, M.; Gooch, W. *Ballistic Evaluation of Magnesium Alloy AZ31B*; ARL-TR-4077; U.S. Army Research Laboratory: Aberdeen Proving Ground, MD, April 2007.
2. Mathaudhu, S.; Nyberg, E.; *Magnesium Alloys in Army Applications: Past, Current and Future Solutions*; Proceedings of the 2010 Minerals, Metals and Materials (TMS) Annual Symposium, Seattle, WA, February 2010.
3. Jones, T.; DeLorme, R.; *Development of a Ballistic Specification for Magnesium Alloy AZ31B*; ARL-TR-4664; U.S. Army Research Laboratory: Aberdeen Proving Ground, MD; December 2008.
4. Jones, T.; DeLorme, R.; *A Comparison of the Ballistic Performance Between Rolled Plate in AZ31B-H24 Magnesium and 5083-H131 Aluminum*; 24th International Symposium on Ballistics, New Orleans, LA; September 2008.
5. MIL-DTL-32333 (MR); Armor Plate, Magnesium Alloy, AZ31B, Applique; 2009.
6. Liao, J.; Hotta, M.; Kaneko, K.; Kondoh, K.; *Enhanced Impact Toughness of Magnesium Alloy by Grain Refinement*; Scripta Materialia 2009, 61; 208–211.
7. Jones, T.; Kondoh, K.; *Initial Evaluation of Advanced Powder Metallurgy Magnesium Alloys for Dynamic Applications*; ARL-TR-4828; U.S. Army Research Laboratory, Aberdeen Proving Ground, MD; May 2009.
8. Jones, T.; DeLorme, R.; *Ballistic and Corrosion Analysis of New Military-Grade Magnesium Alloys AMX602 and ZAXE1711 for Armor Applications*; ARL-TR-5931; U.S. Army Research Laboratory: Aberdeen Proving Ground, MD; February 2012.
9. Sakamoto, M.; Akiyama, S.; Hagio, T.; Ogi, K.; *Control of Oxidation Surface Film and Suppression of Ignition of Molten Mg-Ca Alloy by Ca Addition*; Journal of Japan Foundry Engineering Society; 69 (1997) 227-233.
10. Nishida, S.; Motomura, I.; *Estimation of Heat Transfer Coefficient and Temperature Transition on Melt Drag Process of AZ31 Magnesium Alloy by Heat Transfer and Solidification Analysis*; Journal of Japan Institute of Light Metals; 58 (2008) 439-442.
11. MIL-DTL-46027K (MR); Armor Plate, Aluminum Alloy, AA5083, Weldable; 2013.
12. MIL-STD-662F; V50 Ballistic Test for Armor; 1997.
13. MIL-DTL-46593B (MR); Projectile, Calibers .22, .30, .50, and 20MM Fragment-Simulating; 2006.

Magnesium Technology 2015
Edited by: Michele V. Manuel, Alok Singh, Martyn Alderman, and Neale R. Neelameggham
TMS (The Minerals, Metals & Materials Society), 2015

Deformation Response of Mg-Y Alloys under Dynamic Loading

Toshiji Mukai[1], Masaki Nagao[1], Tomofumi Terada[1], Hidetoshi Somekawa[2], Alok Singh[2]

[1]Mechanical Engineering Department (Kobe University); 1-1 Rokkodai-cho, Nada; Kobe, Hyogo 657-8501, Japan
[2] National Institute for Materials Science, Sengen, Tsukuba, Ibaraki 305-0047, Japan

Keywords: magnesium, yttrium, high strain rate, compression, deformation twin, dynamic recovery

Extended Abstract

Weight reduction of automobiles and aircrafts improves fuel economy and reduces greenhouse gas emissions. Use of Mg alloys may allow weight reduction because of their low densities, but adoption is hindered because they exhibit limited ductility at ambient temperatures [1]. In a previous study of fracture toughness in a Mg alloy, crack readily propagated near twin boundaries and resulted in poor durability [2]. It has been shown that pile-up of dislocations at the interface between the matrix and deformation twins caused stress concentration to form cracks [2]. Another study suggested that the ductility of Mg alloys is further limited under dynamic loading due to lowered activity of dislocations [3]. It has also been reported that a Mg-Al-Mn alloy had pronounced mechanical anisotropy at high strain rates of around 1.0×10^3 s^{-1} [4]. Therefore, the mechanical properties of Mg alloys should be evaluated accurately for applications involving possible dynamic loading.

It has been reported that addition of rare earth elements such as yttrium may improve mechanical properties over conventional Mg-Al-Zn alloys by minimizing intensity of the basal texture [5–9]. However, the deformation behavior of Mg alloys containing yttrium is not sufficiently understood at strain rates exceeding 1×10^3 s^{-1}. This study examines a binary Mg-Y alloy at high strain rates in compression to clarify the effect of yttrium addition on the mechanical properties. Details have been reported elsewhere [10].

This study investigated a binary Mg-0.6at.%Y alloy. A billet was fabricated by gravity casting into a steel mold. The cast binary alloy was homogenized at 773 K, followed by quenching in water. It was then extruded at 673 K at a ratio of 25:1. The extruded rod having a diameter of 8 mm was annealed at 673 K for 16.5 h to form an equiaxed grain structure with an average grain size of 42 μm. As a reference material, extruded magnesium with 99.95% purity and an average grain size of 50 μm was prepared. Cylindrical specimens having a height of 8 mm and a diameter of 4 mm were machined parallel to the extrusion direction. To examine the mechanical anisotropy, cylindrical specimens of height 6.5 mm and diameter 4 mm were also prepared along two directions, parallel and perpendicular to the extrusion direction. High strain rate compression tests were performed by using a split Hopkinson pressure bar (SHPB) at ambient temperature (298 K). The average strain rate was measured to be 1.4×10^3 s^{-1} for a specimen of 8.0 mm height and to be 1.7×10^3 s^{-1} for a specimen of 6.5 mm height. The shape of the specimen during deformation was detected as a series of images by an ultra-high speed camera (Shimadzu HPV-1) with a sampling time of 4 μs.

By investigation of the Mg-0.6 at.% Y alloy and comparing it with a commercial purity magnesium sample, we have found that yttrium solute contributed to enhanced compressive ductility, reduced strain hardening rate, and minimized deformation asymmetry in magnesium, even under dynamic loading. Although $\{10\bar{1}2\}$ twins were the predominant deformation mechanism for the Mg-0.6Y alloy in the early-stage of deformation, cracks did not initiate near these twin boundaries, unlike in the conventional Mg alloy AZ31. This suggests occurrence of stress relaxation during deformation. Subgrain formation and c-axis rotation possibly release stress concentrations during high strain rate deformation of the Mg-Y alloy.

This work was supported in part by Toyota Motor Corp. and by a Grant-in-Aid for Scientific Research (No. 25246012) from the Ministry of Education, Culture, Sports, Science and Technology of Japan.

Reference

1. T. Mukai, H. Watanabe, K. Higashi, Mater. Sci. Technl., 2000, vol. 16, 11-12, pp. 1314-1319.
2. H. Somekawa, A. Singh and T. Mukai: Philo. Mag. Lett., 2009 vol. 89, 1, pp.2-10.
3. T. Mukai, M. Yamanoi, H. Watanabe and K. Higashi: Mater. Trans., 2001, vol. 42, 7, pp.1177-1181.
4. P. Mao, Z. Liu and C. Wang: Mater. Sci. Eng.A 2012, vol. 539, pp.13-21.
5. E.A. Ball and P.B. Prangnell, Scripta Metall. Mater., 1994, vol. 31, pp. 111-116.
6. S.R. Agnew, M.H. Yoo, C.N. Tome, Acta Mater., 2001, vol. 49, pp. 4277-4289.
7. S. Miura, S. Imagawa, T. Toyoda, K. Ohkubo and T. Mohri, Mater. Trans., 2008 vol. 49, 5, pp.952-956.
8. J. Bohlen, M. R. Nurnberg, J. W. Senn, D. Letzing, S. R. Agnew, Acta Mater., 2007, vol. 55, 6, pp. 2101-2112.
9. J. Geng, Y. B. Chun, N. Stanford, C. H. J. Davies, J. F. Nie, M. R. Barneet, Mater. Sci. Eng.A, 2011, vol. 528, 10-11, pp.3659-3665.
10. M. Nagao, T. Tereda, H. Somekawa, A. Singh, T. Mukai, JOM, 2014 vol. 66, pp.305-311.

LARGE STRAIN BEHAVIOUR OF ZEK100 MAGNESIUM ALLOY AT VARIOUS STRAIN RATES

Julie Lévesque[1], Srihari Kurukuri[2], Raja Mishra[3], Michael Worswick[2], Kaan Inal[2]
[1]Department of Mechanical Engineering, Université Laval; 1065 Avenue de la Médecine; Québec, G1V 0A6, Canada
[2]Department of Mechanical and Mechatronics Engineering, University of Waterloo; 200 University Avenue West; Waterloo, Ontario, N2L 3G1, Canada
[3]General Motors Research and Development Center; 10500 Mound Road; Warren, Michigan, 48090, USA

Keywords: Magnesium alloys, crystal plasticity, twinning

Abstract

A constitutive framework based on a rate-dependent crystal plasticity theory is employed to simulate large strain deformation in hexagonal closed-packed metals that deform by slip and twinning. The model allows the twinned zones and the parent matrix to rotate independently. ZEK100 magnesium alloy sheets which significant texture weakening compared to AZ31 sheets are investigated using the model. There is considerable in-plane anisotropy and tension compression asymmetry in the flow behavior of ZEK100. Simulations of uniaxial tension in different directions at various strain rates and the accompanying texture evolution are performed and they are in very good agreement with experimental measurements. The effect of strain rate on the activation of the various slip systems and twinning show that differences in the strain rate dependence of yield stress and R-values in ZEK100 have their origin in the activation of different deformation mechanisms.

Introduction

Due to their large potential for weight saving and fuel economy, magnesium alloys can play a significant role in the reduction of vehicle weight which will directly contribute to increase vehicle efficiency (in terms of fuel consumption). The low formability of magnesium alloys is one of the limitations for their increased usage in auto manufacturing. This is due to their hexagonal closed-packed structure, which offers a limited number of slip systems that are active at room temperature. Rolled sheets of common alloy AZ31 possess a very strong basal texture [1], resulting in a low formability.

One way to improve the formability of magnesium alloy at room temperature is the addition of rare earth. ZEK100 alloy sheets have a more random texture and possesses a better formability than AZ31 [2]. However, there is still considerable anisotropy in this material. In this work, the behavior of ZEK100 sheet material during uniaxial tension at different strain rate, as published by Kurukuri et al. [3] is studied using a Taylor-type polycristal plasticity model. This model is based on the formulation introduced by Kalidindi [4] and is an extension of the model used in previous work by Lévesque et al. [5].

Constitutive model

The plasticity model for polycrystals formulated by Asaro and Needleman [6] and modified by Kalidindi (1998) is employed in this work. Accordingly, the total deformation of a crystallite is taken to be the result of three distinct physical mechanisms: crystallographic slip, twinning and elastic lattice distortion. The systems that are considered in this work are presented are basal prismatic, pyramidal <a> and <c+a> slip and extension twinning. Extension twinning is active only when the c-axis is parallel to a positive stress.

Plastic deformation is envisaged to occur as a set of simple shears along the various slip and twinning systems, leaving the lattice and the slip system vectors ($s_{(\alpha)}$, $m_{(\alpha)}$) not only essentially undistorted, but also un-rotated. Next, the material and lattice are considered to deform elastically and rotate rigidly from the plastically deformed state to the current configuration. The deformation gradient F is written as

$$F = F^* F^P \tag{1}$$

where F^P consists of crystallographic slip and twinning, while the elastic deformation and any rigid body rotation are embodied in F^*. Since this is a Taylor-type polycrystal plasticity models, the same deformation gradient is used for every grain within the polycrystal aggregate. In order to be more realistic, however, different plastic and elastic deformation gradients are used within the grains, for the twinned and untwined regions. From (1), the spatial gradient of velocity can be written as

$$L = \dot{F}F^{-1} = L^* + L^P \tag{2}$$

where

$$L^* = \dot{F}^* F^{*-1},\ L^P = F^*\left(\dot{F}^P F^{P^{-1}}\right)F^{*-1} \tag{3}$$

Using Eq. (2), the elastic and the rigid body rotation of velocity gradient is written as follows

$$L^* = L - L^P \tag{4}$$

It is proposed in this formulation that, L^* in the matrix and in twinned regions can be considered independently and computed in each time step. Thus, they are written as

$$L^*_{mt} = L - L^P_{mt} \tag{5}$$

$$L^*_{tw\beta} = L - L^P_{tw\beta} \tag{6}$$

Using velocity gradients as formulated in Eq. (5) and Eq. (6) leads to independent rotation for the twinned regions and the parent matrix.

Taking symmetric and antisymmetric parts of the above relations leads to the elastic and plastic strain-rates D^* and D^P, the so-called plastic spin Ω^P, and the spin Ω^* associated with the rigid lattice rotation (Eq. 7).

$$D = D^* + D^P\ ,\ \Omega = \Omega^* + \Omega^P \tag{7}$$

The vectors $s^{(\alpha)}$ and $m^{(\alpha)}$ are regarded as lattice vectors so that they stretch and rotate by

$$s^*_{(\alpha)} = F^* s_{(\alpha)}\ ;\ m^*_{(\alpha)} = m_{(\alpha)} F^{*-1} \tag{8}$$

By introducing the following symmetric and skew-symmetric tensors for each slip system α,

$$P_{(\alpha)} = \frac{1}{2}\left[s^*_{(\alpha)} \otimes m^*_{(\alpha)} + m^*_{(\alpha)} \otimes s^*_{(\alpha)}\right] \quad (9)$$

$$W_{(\alpha)} = \frac{1}{2}\left[s^*_{(\alpha)} \otimes m^*_{(\alpha)} - m^*_{(\alpha)} \otimes s^*_{(\alpha)}\right] \quad (10)$$

the plastic strain-rate and spin for the crystal can be, respectively, written as

$$D^P = \left(1 - \sum_{\beta=1}^{N^T} f_{(\beta)}\right)\left[\sum_{\alpha=1}^{N^S} P_{(\alpha)}\dot{\gamma}_{(\alpha)} + \sum_{\beta=1}^{N^T} P_{(\beta)}\dot{f}_{(\beta)}\gamma^{tw}\right] + \sum_{\beta=1}^{N^T}\left[f_{(\beta)}\sum_{\alpha=1}^{N^S}\left(P^{tw\beta}_{(\alpha)}\dot{\gamma}^{tw\beta}_{(\alpha)}\right)\right] \quad (11)$$

and

$$\Omega^P = \left(1 - \sum_{\beta=1}^{N^T} f_{(\beta)}\right)\left[\sum_{\alpha=1}^{N^S} W_{(\alpha)}\dot{\gamma}_{(\alpha)} + \sum_{\beta=1}^{N^T} W_{(\beta)}\dot{f}_{(\beta)}\gamma^{tw}\right] + \sum_{\beta=1}^{N^T}\left[f_{(\beta)}\sum_{\alpha=1}^{N^S}\left(W^{tw\beta}_{(\alpha)}\dot{\gamma}^{tw\beta}_{(\alpha)}\right)\right] \quad (12)$$

where $\dot{\gamma}_{(\alpha)}$ is the shear rate on the slip system α, $f_{(b)}$ is the volume fraction of the grains that have been twinned to the system β, γ^{tw} is the amount of shear associated with twinning and $\dot{f}_{(\beta)}$ is the rate of twinning of the system β. The superscript *twβ* indicates that the parameter considered is calculated for a region twinned according to the system β. N^S and N^T represent the number of slip systems and number of twinning systems, respectively. In Eqs. (8) and (9), the contribution of twinning to the deformation is multiplied by the untwinned volume fraction, since new twinning is considered to happen in the untwinned region of the grain. This approach is different than that presented by Kalidindi (1998), but leads to similar results.
The slip and twinning rates to be substituted into Eq. (11) and (12) are assumed to be governed by the following power-law expressions:

$$\dot{\gamma}_{(\alpha)} = \dot{\gamma}_{(0)}\, sgn\,\tau_{(\alpha)}\left|\frac{\tau_{(\alpha)}}{g_{(\alpha)}}\right|^{1/m} \quad (13)$$

$$\dot{f}_{(\alpha)} = \frac{\dot{f}_{(0)}}{\gamma^{tw}}\, sgn\,\tau_{(\alpha)}\left|\frac{\tau_{(\alpha)}}{g_{(\alpha)}}\right|^{1/m} \quad (14)$$

where $\dot{\gamma}_{(\alpha)}$ and $\dot{f}_{(\alpha)}$ are reference shear and twinning rates taken to be the same for all the slip and twinning systems, $\tau_{(\alpha)} = P_{(\alpha)} : \sigma$ is the resolved shear stress on the slip system α, $g_{(\alpha)}$ is its hardness and m is the strain-rate sensitivity index. The term $g_{(\alpha)}$ characterizes the current strain-hardened state of all slip systems. The rate of increase of the function $\dot{g}_{(\alpha)}$ is defined by the hardening law:

$$\dot{g}_{(\alpha)} = \sum_{\beta} h_{(\alpha\beta)}\left|\dot{\gamma}_{(\beta)}\right| \quad (15)$$

where $g_{(\alpha)}(0)$ is the initial hardness, taken to be a constant τ_0 for each slip system, and $h_{(\alpha\beta)}$ are the hardening moduli. The form of these moduli is

$$h_{(\alpha\beta)} = q_{(\alpha\beta)} h_{(\beta)} \quad (\text{no sum on } \beta) \quad (16)$$

where $h_{(\beta)}$ is a single slip hardening rate and $q_{(\alpha\beta)}$ is the matrix describing the latent hardening behaviour of the crystallite. In this work, latent hardening parameters are all set to one.

A Voce-type hardening law is used to represent the hardness of the various slip and twinning systems during deformation. An expression accounting for the hardening caused by the twin boundaries is also incorporated in the formulation.

$$h_{(\alpha)} = h_1 \exp\left(h_2 \gamma_a\right) + h_{TB} \quad (17)$$

Where h_1 and h_2 are constants and γ_a is the sum of the accumulated slip on all the slip systems; i.e.

$$\gamma_a = \int_0^t \sum_{\alpha=1}^{N^S}\left|\dot{\gamma}_{(\alpha)}\right| dt \quad (18)$$

The term h_{TB} represents the hardening caused by the twin boundaries which restrict the dislocation motion. The density of twin boundaries increases as the volumetric fraction of the twins increases, but as twinning continues, the twins consolidate and the twin boundaries disappear, leading to a decrease in their density. Accordingly, the value of h_{TB} increases up to a certain amount of twinning, where the density of the twin boundaries within a grain is the highest, and then decreases.

The Cauchy stress, σ, over a grain can be obtained by averaging the stresses in the twinned and un-twinned parts of the grain;

$$\sigma = \left(1 - \sum_{\beta=1}^{N^T} f_{(\beta)}\right)\sigma^{mt} + \sum_{\beta=1}^{N^T} f_{(\beta)}\sigma^{tw\beta} \quad (19)$$

where σ^{mt} is the Cauchy stress in the matrix (un-twinned regions of the grain) and $\sigma^{tw\beta}$ is the Cauchy stress in the β-twinned region.

Results and discussion

The material investigated in this study is rolled sheets of magnesium ZEK100 alloy. Stress-strain curves in tension in different directions and various strain rates, as well as texture data, have been provided by Kurukuri et al. [3]. As can be observed in Figure 1, basal planes in the material are mostly aligned parallel to the rolling direction of the sheet, while prismatic planes are perpendicular to the rolling direction.

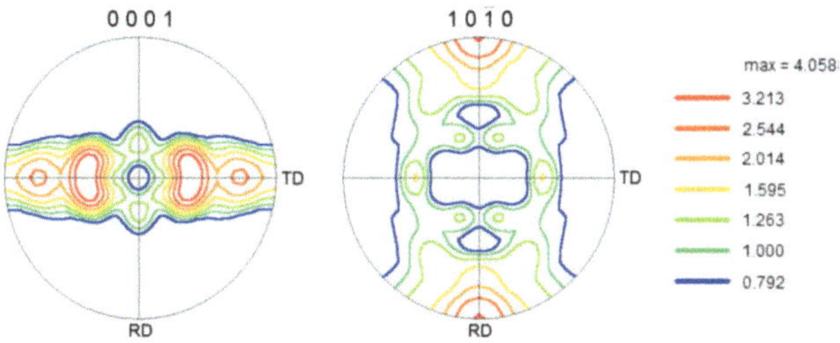

Figure 1: Initial texture of ZEK100 sheets

The model has been calibrated for strain rates of $0.001 s^{-1}$ and $0.1 s^{-1}$. As can be seen in figure 2, a very good correlation is obtained in all directions. In all cases, parameters for a given strain rate are kept the same for simulations in the 3 directions. The only difference in input is that the initial texture is being rotated.

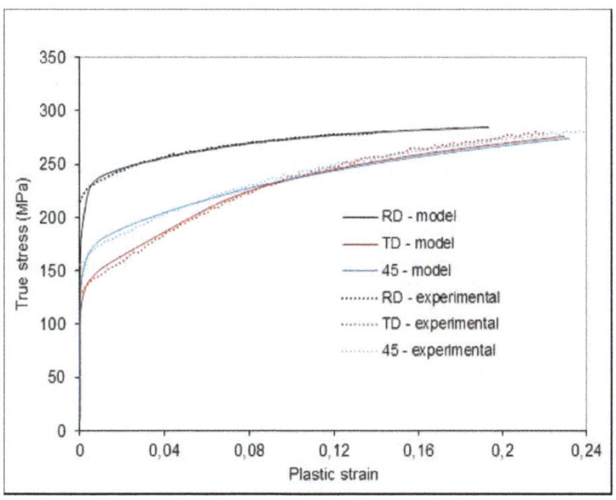

Figure 2: Stress-strain curves obtained from the model (solid lines) and from experiments (dotted lines) in the rolling (RD) and transverse (TD) direction and at a 45 degrees angle from the rolling direction, in the sheet plane (45), at a strain rate of 0.001/s.

Figure 3 shows the evolution of twinning in the different directions. Twinning is a lot more active in the transverse direction than in the rolling directions. This is due to the initial texture of the sheet. Lattices are in a favorable orientation for twinning only when their c-axes are parallel to the tensile stress direction.

The relative accumulated slip for the various slip system families are presented in Figures 4, 5 and 6. The difference in the activation of the deformation mechanisms when loading along different direction can clearly be assessed. In the rolling direction, prismatic slip is the most active, and twinning is very low. Due to that low amount of twinning, pyramidal <c+a> slip is also very active, since it is the most important system that can accommodate deformation along the c-axis. In directions where twinning is more active, like in the transverse direction and 45 degrees orientation, basal and prismatic slip are the most active. Less <c+a> slip is present, since twinning itself can accommodate deformation along the c-axis.

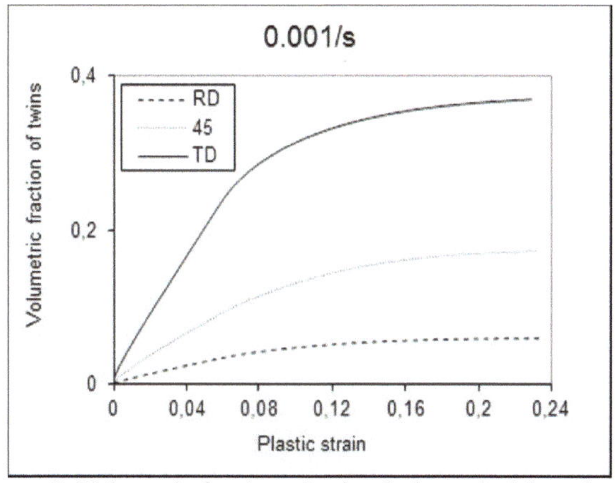

Figure 3: Volumetric fraction of twins during uniaxial tension at 0.001/s in the rolling (RD) and transverse (TD) direction and at a 45 degrees angle from the rolling direction, in the sheet plane (45).

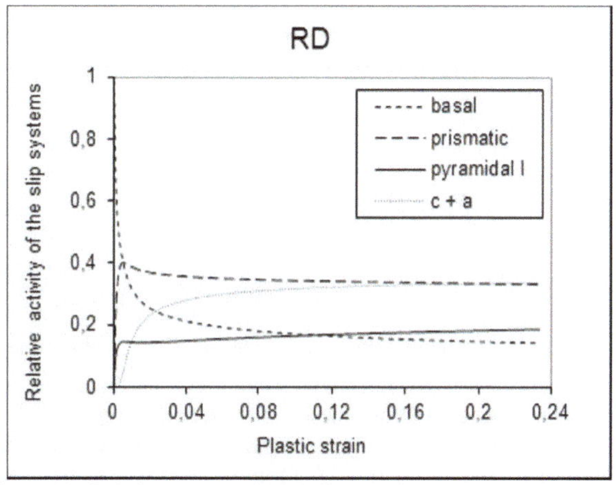

Figure 4: Relative accumulated slip on the various slip systems families during uniaxial tension at 0.001/s in the rolling direction.

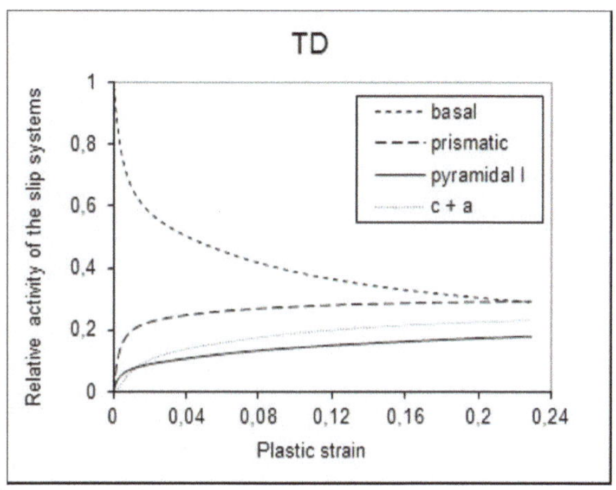

Figure 5: Relative accumulated slip on the various slip systems families during uniaxial tension at 0.001/s in the transverse direction.

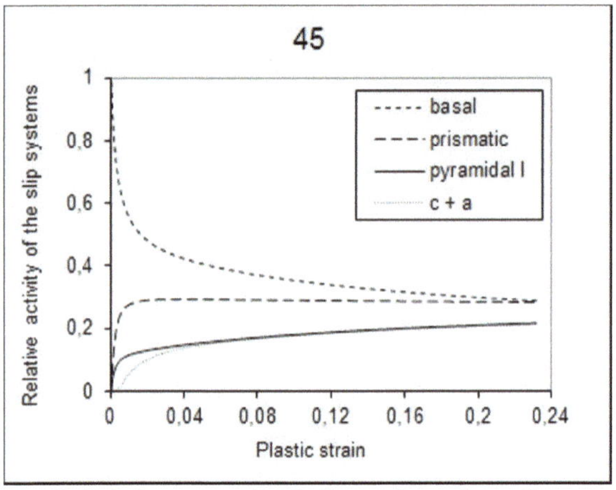

Figure 6: Relative accumulated slip on the various slip systems families during uniaxial tension at 0.001/s at a 45 degrees angle from the rolling direction, in the sheet plane.

The model has also been calibrated for uniaxial tension at $0.1s^{-1}$. Coefficients C have then been calculated for logarithmic functions of the type:

$$P_{\dot{\varepsilon}} = P_{\dot{\varepsilon}(R)}\left(1+C\ln\left(\frac{\dot{\varepsilon}}{\dot{\varepsilon}_{(R)}}\right)\right) \qquad (20)$$

Where P is the parameter to be calibrated, C is a constant, $\dot{\varepsilon}$ is the strain rate, $\dot{\varepsilon}_{(R)}$ is the reference strain rate and $P_{\dot{\varepsilon}(R)}$ is the value of the parameter at the reference strain rate. In this case, the parameters that need to be adjusted are the critical resolved shear stress (CRSS) for the slip systems and twinning systems and the values of h_l and h_{tw} in Eq. 17. More detailed explanations and functions can be found in Lévesque et al. [5]. Figures 7 shows the curves fitted for a strain rate of $0.1s^{-1}$.

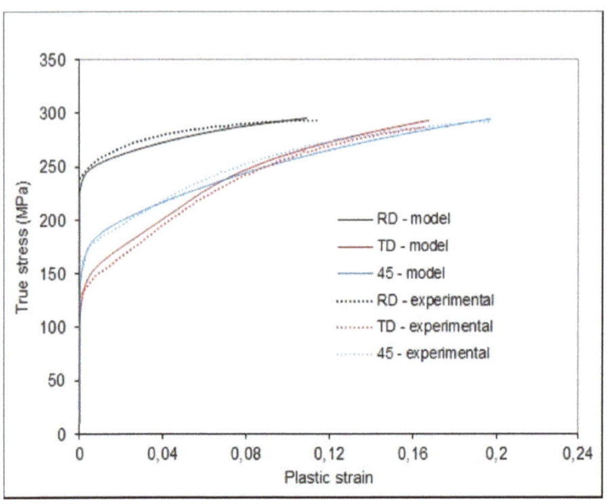

Figure 7: Stress-strain curves obtained from the model (solid lines) and from experiments (dotted lines) in the rolling (RD) and transverse (TD) direction and at a 45 degrees angle from the rolling direction, in the sheet plane (45), at a strain rate of 0.1/s.

The values for the constants C that have been found after calibration at a strain rate of $0.1s^{-1}$ for the CRSS of slip systems, the CRSS for twinning systems, the h_l value and the h_{tb} value are 0.01, -0.217, 0.0418 and 0.11, respectively. Using these values, parametsrs have been calculated and used to simulate uniaxial tension at a strain rate of 0.01s-1. Obtained stress-strain curves are shown in Figure 8. The predicted curves are in fairly good agreement with the experimental curves.

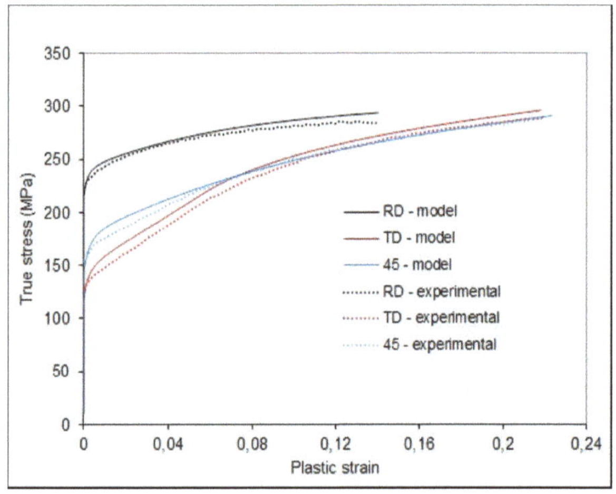

Figure 8: Stress-strain curves obtained from the model (solid lines) and from experiments (dotted lines) in the rolling (RD) and transverse (TD) direction and at a 45 degrees angle from the rolling direction, in the sheet plane (45), at a strain rate of 0.01/s.

Figure 9 presents the texture after 11.4% deformation in the rolling direction. There is not a big texture evolution during this test, since twinning is not very active. However, in transverse direction, twinning is more active and texture evolution is more important, as can be seen in figure 10.

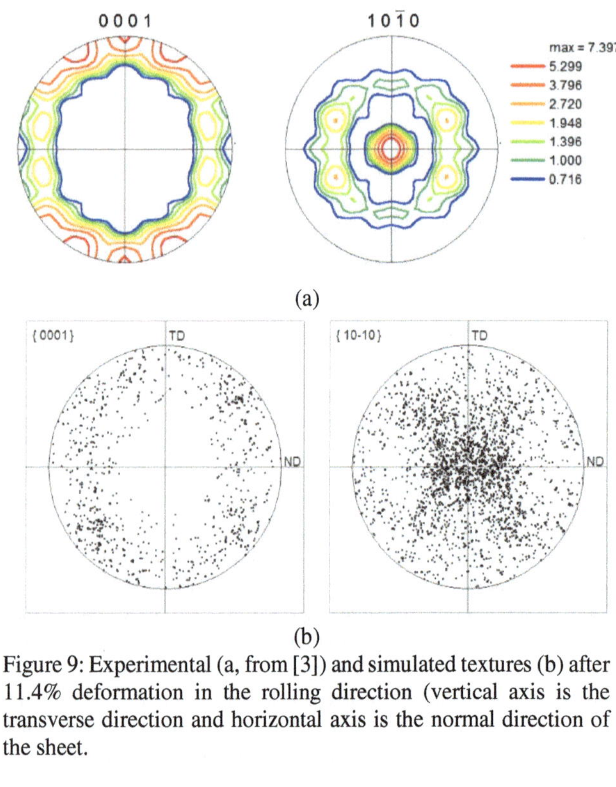

(a)

(b)

Figure 9: Experimental (a, from [3]) and simulated textures (b) after 11.4% deformation in the rolling direction (vertical axis is the transverse direction and horizontal axis is the normal direction of the sheet.

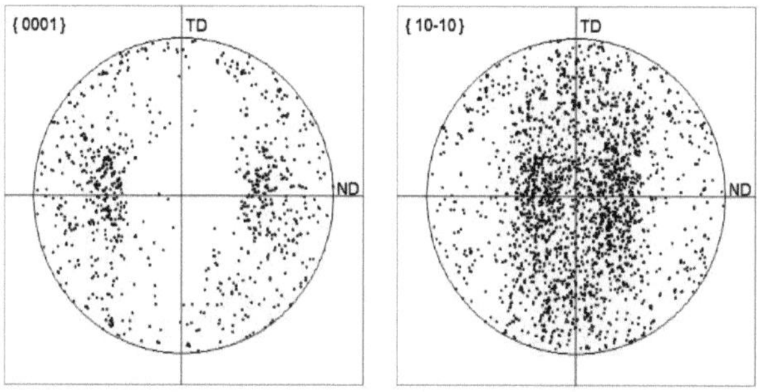

Figure 10: Simulated texture after 11.4% deformation in the transverse direction

Conclusion

Results show that even if ZEK100 sheets have a weaker textures than AZ31, loading direction has an important impact on its deformation behavior. Using an extended Taylor-type model, it is possible to simulate uniaxial tension along rolling direction, transverse direction, and at a 45 degrees angle. The use of simple logarithmic functions allows to find model parameters for different strain rates. The model can be useful in studying the deformation mechanisms that are the most active under different loading path and strain rates.

Acknowledgements

This work was supported by the Natural Sciences and Engineering Research Council of Canada (NSERC), the research program of the Magnesium Network (MagNET) and General Motors of Canada.

References

1. Khan, A., Pandey, A., Gnäupel-Herold, T. and Mishra, R.K., 2011, Mechanical response and texture evolution of AZ31 alloy at large strains for different strain rates and temperatures. International Journal of Plasticity 27: 688-706.
2. Boba, M., Worswick, M.J., Mishra, R., Carter, J., 2012, Formability of AZ31 and ZEK100 magnesium alloy sheets at elevated temperatures, Gm2012: 9th Internatonal Conference on Magnesium Alloys and their Applications.
3. Kurukuri, S., Worswick, M.J., Bardelcik, A., Mishra, R., Carter, J., 2014, Constitutive behavior of commercial grande ZEK100 magnesium alloy sheet over a wide range of strain rate, Metallurgical and materials transactions A, 45: 3321-3337.
4. Kalidindi, S.R., 1998. Incorporation of Deformation Twinning in Crystal Plasticity Models. Journal of the Mechanics and Physics of Solids 46: 267-290.
5. Lévesque, J., Inal, K., Neale, K.W., Mishra, R.K. (2010). Numerical modeling of formability of extruded magnesium tubes. International journal of plasticity. 26: 65-83.
6. Asaro, R.J., Needleman, A., 1985. Texture Development and Strain Hardening in Rate Dependent Polycrystals. Acta Metallurgica 33, 923-953.

Magnesium Technology 2015
Edited by: Michele V. Manuel, Alok Singh, Martyn Alderman, and Neale R. Neelameggham
TMS (The Minerals, Metals & Materials Society), 2015

Effect of Solute Segregation on Fracture Behavior of Mg Alloy

Tomoaki Kawa[1], Masatake Yamaguchi[2], Naoko Ikeo[1], Toshiji Mukai[1]
[1]Mechanical Engineering Department (Kobe University); 1-1 Rokkodai-cho, Nada; Kobe, Hyogo 657-8501, Japan
[2]Center for Computational Science and e-Systems (Japan Atomic Energy Agency); 2-4 Sirakata Sirane, Tokai–mura; Naka-gun, Ibaraki, 319-1195, Japan

Keywords: magnesium alloy, fracture behavior, impact three-point bending

Abstract

Improving mechanical properties of magnesium and understanding fracture behavior under impact loading are necessary to apply magnesium alloys to structural components of automobiles. We have investigated the fracture behavior of binary magnesium alloys by three-point bending experiment and conducted a first principle calculation to estimate the effect of solute segregation on fracture energy. In this paper, we have focused on experimental result of impact three-point bending test for Mg-0.3at.%Y alloy and the results of the test were compared with that of AZ31 commercially available alloy [1]. As a result, the crack propagation speed of Mg-0.3at.%Y was found to be slower than that of AZ31 alloys. Moreover, the absorbed energy of Mg-0.3at.%Y was more than twice as high as that of AZ31 alloys. These results suggested that yttrium solute in magnesium improved the fracture toughness of magnesium under impact loading. Then, fracture surface was observed by SEM to consider the effect of microstructure on crack propagation speed.

Introduction

The application of magnesium in structural components is desirable because of its low density. In addition to the lightness, magnesium has many advantages, for example, good machinability, high damping capacity, good recyclability and so on. However, magnesium is not generally used for structural components in automobiles because their ductility and/or toughness are low at room temperature. Then various magnesium alloys has been investigated in order to improve static mechanical properties [2-4]. For example, alloying with yttrium enhances ductility of magnesium due to weakened intensity of basal texture and activation of non-basal slip [5]. Moreover, twin boundaries in Mg-Y alloy is not likely to be crack initiation point [6].

For adoption of magnesium alloys as structural materials in the automotive industry, understanding impact fracture behavior is also important to ensure safety. Charpy impact test is often used to evaluate impact fracture toughness [7]. However, impact velocity in this method is limited (~5 m/s) and crack propagation behavior is difficult to observe. Thus, Yokoyama *et al.* proposed a novel impact bending test procedure using split Hopkinson pressure bar technique as an evaluation method of dynamic fracture toughness [8]. This method enables to deform specimen under higher impact velocity than Charpy impact test and observe crack propagation behavior.

In this research, crack propagation speed and energy absorption capability of Mg-0.3at.%Y alloy under impact loading were evaluated by impact three-point bending test and compared with commercially available magnesium alloys. Further, fracture surface was observed by SEM in order to consider the effect of yttrium solute and microstructure on impact fracture behavior of magnesium.

Experimental Procedure

Impact three-point bending test was carried out in this study. The schematic illustration of experimental set up is shown in Figure 1 (a). The incident bar had a diameter of 16 mm and a length of 1300 mm. The contact end of incident bar with specimen had a point angle of 30 degrees, which is similar to the striker for Charpy impact test. The diameters and lengths of two transmission bars were 16mm and 1100 mm, respectively. The contact end of transmission bar with specimen was designed as an anvil for Charpy impact test. These three elastic bars were made of SKD11 tool steel. Two strain gauges are installed on each of incident bar and the transmission bars at fixed positions. A brass bar having a diameter of 16 mm and a length of 420 mm was used as a striker bar.

The Charpy V-notch specimen was used in this study. The configuration of the specimen is shown in Figure 1 (b). The notch had a depth of 2 mm and a point angle of 45 degrees. As-extruded Mg-0.3at.%Y alloy was used in this research. The extrusion ratio and temperature of this alloy were 12.6 and 453 K, respectively. Specimens were collected parallel to the extruding direction (ED) and notched perpendicular to ED. For comparison, as-cast and as-extruded AZ31 commercially available alloys were also used. The extrusion ratio and temperature of as-extruded AZ31 alloy were 13.6 and 623 K, respectively. The notch of as-extruded AZ31 alloy was also perpendicular to ED.

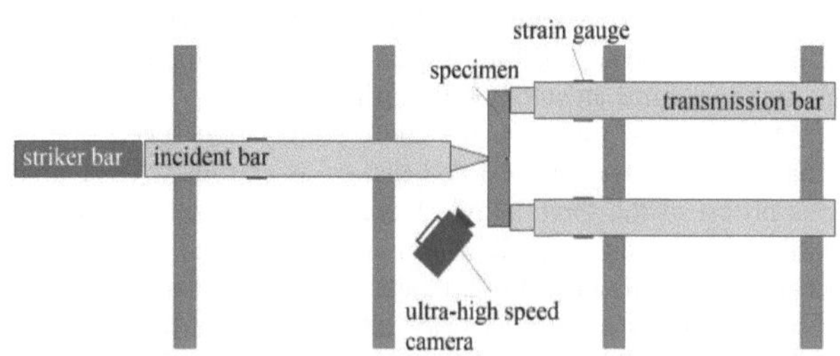

(a)

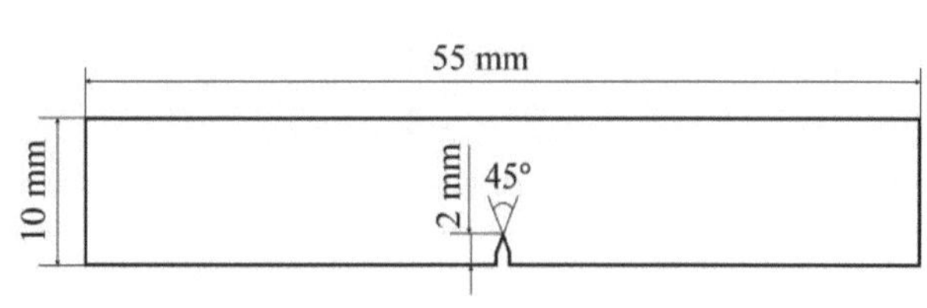

(b)

Figure 1. (a) Schematic illustration of impact three-point bending machine and (b) configuration of V-notch specimen.

In the impact three-point bending test, the striker bar was accelerated by the compressed air and impacted of the incident bar. Then stress wave were generated at the back end of incident bar. The specimen loaded when the stress wave reached contact end of the incident bar. After that a part of the stress wave which was needed to deform specimen was transmitted to the transmission bars. Incident and transmitted stress waves were detected by strain gauges. The load was calculated by following equation.

$$\varepsilon = (2 \times V_g)/(G \times V \times K)$$

$$P = \varepsilon \times E \times A$$

Here, V_g is the voltage output from an oscilloscope, G is the gain of the oscilloscope. V is the input voltage, K is the gauge factor, E is the Young's modulus of the transmission bars, respectively. The load applied to the specimen is the sum of the loads applied to the two transmission bars.

Also, deformation and fracture behavior of specimen was recorded by an ultra-high speed camera. The frame rate was 250,000 fps. From a series of detected images, the impact velocity of the incident bar and crack length variation with time were measured. Moreover, average crack propagation speeds from 150 to 200 μs after impact were calculated from the slope of crack length-time curves by least-square method for each alloy.

Result and Discussion

Optical microscope images of three alloys were shown in Figure 2. The grains was columnar in the as-extruded Mg-0.3at.%Y alloy. On the other hand, as-cast AZ31 alloy had coarse grains (>500 μm) and as-extruded AZ31 alloy had bimodal grain structure.

Figure 3 is an example of image detected by ultra-high speed camera during impact three-point bending test of as-cast AZ31 specimen. It can be seen from the image, the crack was initiated at the notch tip. In other specimens the crack was also initiated at the notch tip. Impact velocities in all test were measured to be about 12 m/s. It can be said that the impact velocities in these tests were enough to assume an automobile crash.

As shown in Figure 4, the crack propagation speed of Mg-0.3at.%Y alloy was slowest and that of as-extruded AZ31 alloy was fastest of the three alloys. Moreover, the absorbed energies until crack length reached 2 mm were calculated by sectional measurement of load-displacement curves. As a result, the absorbed energy of Mg-0.3at.%Y alloy was more than twice as high as AZ31 alloys as shown in Figure 5. Figure 5 also shows that the absorbed energies of as-cast and as-extruded AZ31 alloys were almost same.

Fracture surface was observed by SEM to inspect the fracture behavior. From the SEM image shown in Figure 6(a), it was found that Mg-0.3at.%Y specimen exhibited ductile fracture with many dimples. On the other hand, Figures 6(b) and 6(c) suggest that the crack propagated near twin boundaries in as-cast and as-extruded AZ31 alloys because there were a lot of parallel linear edges spreading over the region almost same size with the matrix grains in those fracture surface. Hence, it can be said that the slower crack propagation in Mg-0.3at.%Y specimen was caused by higher ductility and reduced crack initiation at twin boundary. Further, when compared with as-extruded AZ31 alloy, the crack propagation speed was slower in as-cast AZ31 alloy. These results were affected by the difference in grain size. The grains of as-cast AZ31 alloy was very coarse (>500 μm). Therefore, the formation of deformation twin in as-cast AZ31 alloy may require longer time than that in as-extruded AZ31 alloy. Consequently, this difference contributed to the slower crack propagation.

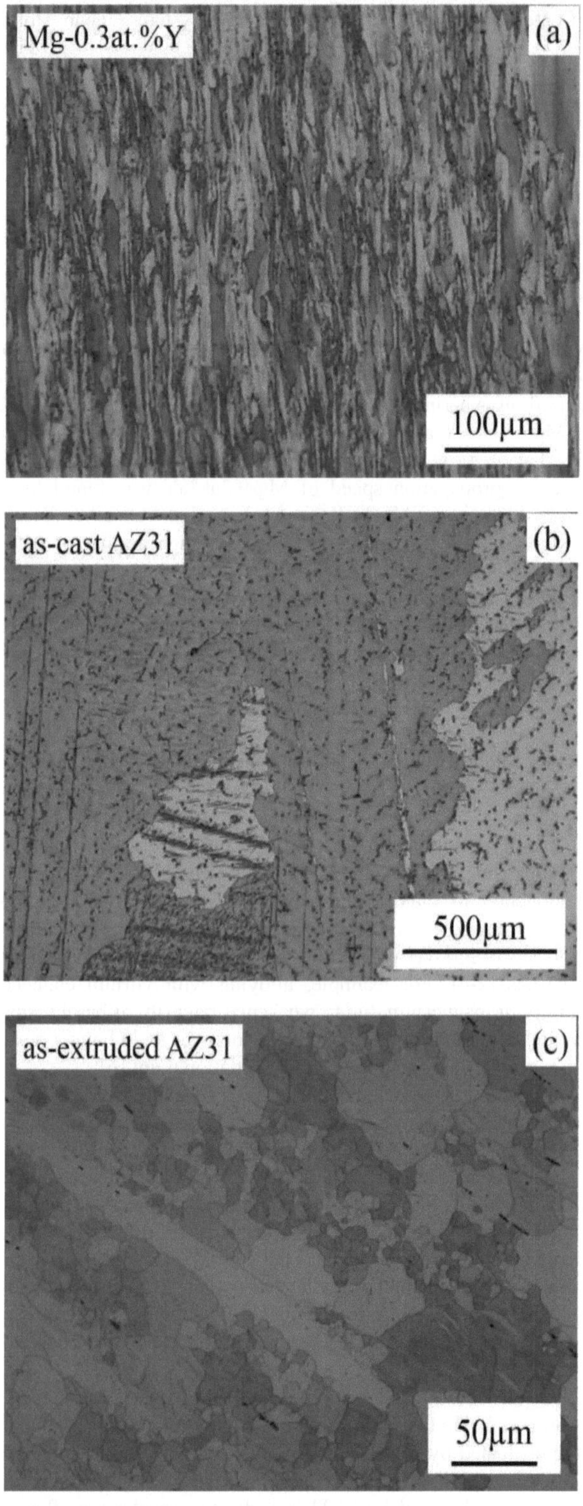

Figure 2. Optical microscope image of (a) Mg-0.3at.%Y, (b) as-cast AZ31 and (c) as-extruded AZ31.

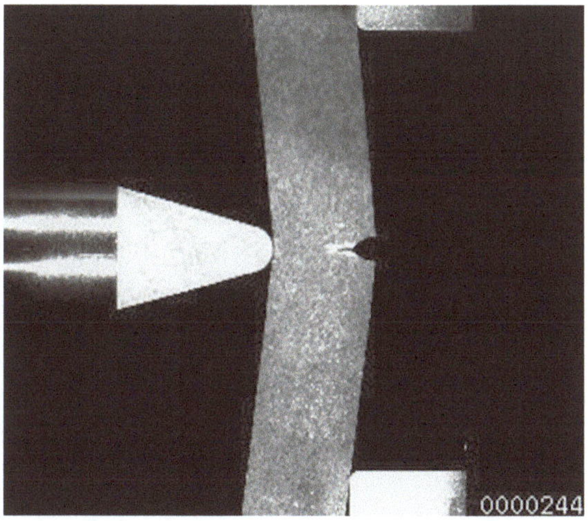

Figure 3. An example of detected image by ultra-high speed camera during impact three-point bending test.

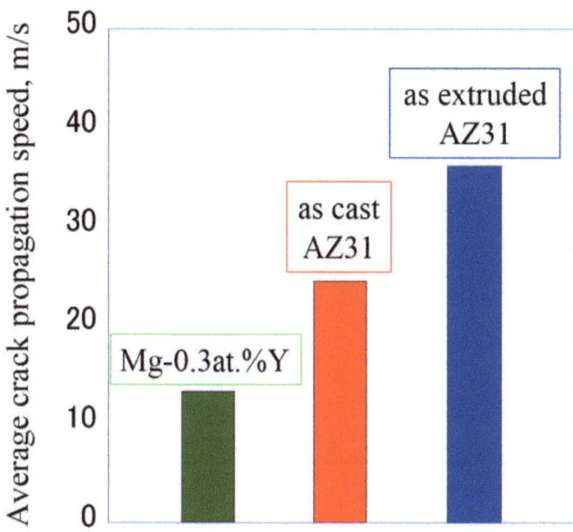

Figure 4. Comparison of average crack propagation speed.

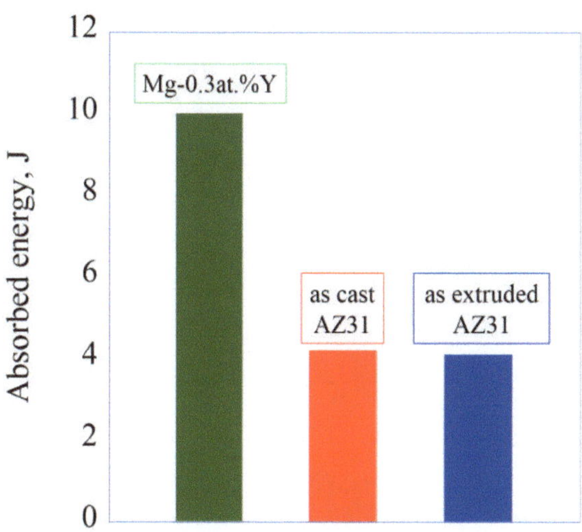

Figure 5. Comparison of absorbed energy calculated from load-displacement curves.

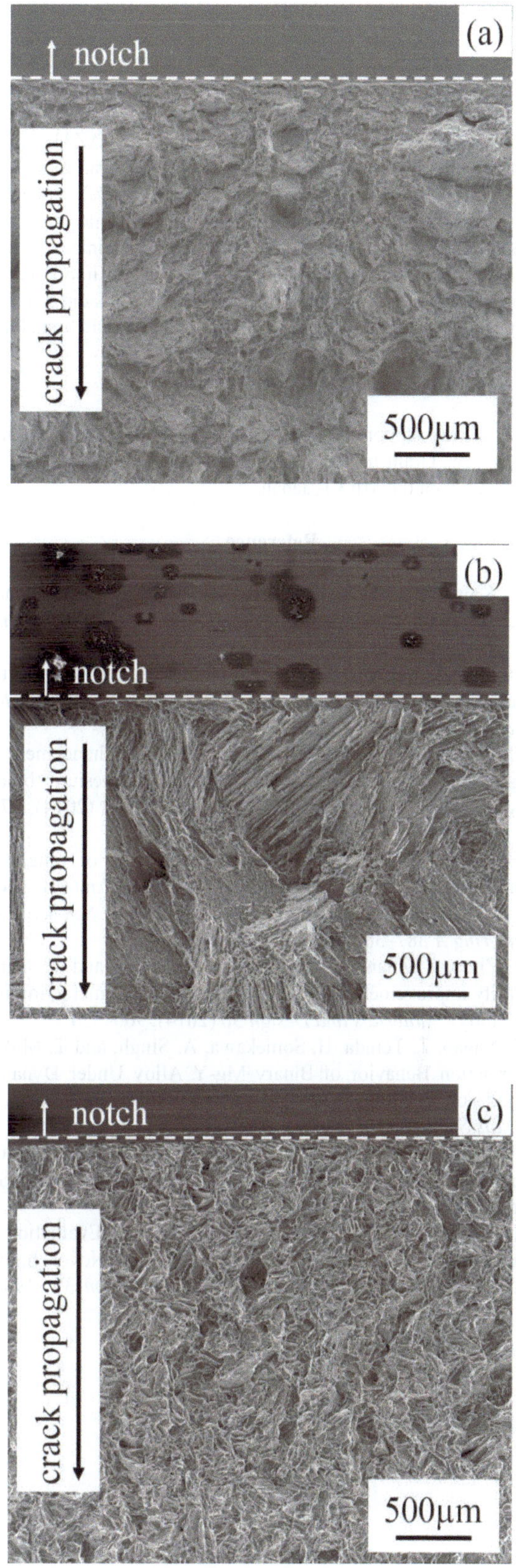

Figure 6. Fracture surface of (a) Mg-0.3at.%Y, (b) as-cast AZ31 and (c) as-extruded AZ31.

Summary

Impact three-point bending test was carried out for Mg-0.3at.%Y alloy and the results of the test were compared with that of commercial AZ31 alloys. The crack propagation speed of Mg-0.3at.%Y specimen was slower than those of AZ31 alloys. Moreover, the absorbed energy of Mg-0.3at.%Y was more than twice as high as that of AZ31 alloys. Mg-0.3at.%Y specimen exhibited ductile fracture with many dimples while the crack propagated near twin boundaries in AZ31 specimens. These results suggested that yttrium solute in magnesium improved the fracture toughness of magnesium under impact loading due to enhanced ductility and reducing crack initiation at twin boundary.

Acknowledgements

This work was supported in part by Toyota Motor Corp. Funding, in part by Grant-in-Aid for Scientific Research (A) (no. 25246012) from the MEXT, Japan.

Reference

1. T. Kawa, M. Nagao and T. Mukai, "Evaluation of impact fracture behavior of AZ31 magnesium alloy," *Applied Mechanics and Materials* 566 (2014), 316–321
2. H. Somekawa, and T. Mukai, "High strength and fracture toughness balance on the extruded Mg–Ca–Zn alloy," *Materials Science and Engineering A* 459 (2007), 366–370
3. H. Somekawa, A. Singh, and T. Mukai, "Enhancement of toughness by grain boundary control in magnesium binary alloys," *Materials Science and Engineering A* 612 (2014), 172–178
4. E. Gariboldi and A. Lo Conte, "Evaluation of toughness characteristics of a die-cast Mg–Al–Mn alloy Am60B under different microstructural conditions," *Materials Science and Engineering A* 387–389 (2004), 34-40
5. N. Zhou, Z. Zhang, L. Jin, J. Dong, B. Chen, and W. Ding, "Ductility improvement by twinning and twin–slip interaction in a Mg-Y alloy," *Materials and Design* 56 (2014), 966–974
6. M. Nagao, T. Terada, H. Somekawa, A. Singh, and T. Mukai, "Deformation Behavior of Binary Mg-Y Alloy Under Dynamic Compression Loading," *The Journal of The Minerals, Metals & Materials Society* 66 (2) (2014), 305-311
7. H. Kuhn and D. Medlin, ASM Handbook Volume 08: Mechanical Testing and Evaluation (Materials Park, ASM International, 2000)
8. T. Yokoyama, K. Kishida, and M. Nakano, "Evaluation of Dynamic Fracture-Initiation Toughness using a Novel Impact Bending Test Procedure," *J. Soc. Mater. Sci. Japan* 39 (1990), 1637-1643. (in Japanese)

Magnesium Technology 2015

Wrought

Magnesium Technology 2015
Edited by: Michele V. Manuel, Alok Singh, Martyn Alderman, and Neale R. Neelameggham
TMS (The Minerals, Metals & Materials Society), 2015

DEVELOPMENT OF THIN-WALLED MAGNESIUM ALLOY EXTRUSIONS FOR IMPROVED CRASH PERFORMANCE BASED UPON TEXTURE CONTROL

Bruce W. Williams[1], Sean R. Agnew[2], Robert W. Klein[2], Jonathan McKinley[1]
[1] CanmetMATERIALS, Natural Resources Canada, 183 Longwood Road South, Hamilton, Ontario L8P 0A5, Canada
[2] Materials Science & Engineering, University of Virginia, 395 McCormick Road, Charlottesville, Virginia 22904-4745, USA

Keywords: extrusion, texture, magnesium, crystal plasticity

Abstract

Recent investigations suggest that it is possible to achieve dramatic modifications to both strength and ductility of magnesium alloys through a combination of alloying, grain refinement, and texture control. The current work explores the possibility of altering the texture in extruded thin-walled magnesium alloy tubes for improved ductility during axial crush in which energy is absorbed through progressive buckling. The texture evolution was predicted using the viscoplastic self-consistent (VPSC) crystal plasticity model, with strain path input from continuum-based finite element simulations of extrusion. A limited diversity of textures can be induced by altering the strain path through the extrusion die design. In some cases, such as for simple bar extrusion, the textures predicted can be connected with simple shape change. In other cases, a subtle influence of strain path involving shear-reverse-shear is predicted. The most promising textures predicted for a variety of strain paths are selected for subsequent experimental study.

Introduction

One limitation to the increased utilization of wrought magnesium alloys for body and structural components in the automotive industry is poor room temperature bend ductility [1,2]. Magnesium alloys currently available for production would not necessarily meet the crash energy absorption needed for critical areas in the automotive structure. Development of a magnesium alloy and processing protocol which would enable adequate energy absorption during crash conditions would be a significant achievement.

Severe plastic deformation (SPD) techniques have been used to improve both strength and ductility of magnesium alloys [3-6]. Previous experimental work had demonstrated the wide variety of textures possible if the strain path is varied through the equal channel angular extrusion (ECAE) multi-pass process [6]. More recent studies [7] have also confirmed those earlier studies of texture and influence of texture on mechanical properties of magnesium alloys. For example, Chino et al. [8] examined the texture of another novel extrusion process and explicitly showed that texture is responsible for enhanced ductility by annealing the material afterward. Orlov et al. [9] has shown that SPD can be used to produce a ZK60 alloy bar with about 2.5 times the elongation along with improved strength, as compared to a conventionally processed alloy. This was attributed to a fine-grained microstructure, which promotes strength, and a texture which enhances ductility by promoting basal slip in the hexagonal-closed packed (hcp) alloy. Other investigations have shown that addition of alloying elements, such as rare-earth elements, can also improve mechanical properties, especially ductility. Though the key to the development of a thin-walled magnesium alloy extrusion that can absorb energy during crash at room temperature will likely involve a combination of alloying, grain refinement, and texture control, the current work focuses on the influence of the texture.

Due to the various extrusion methods that could be utilized to produce a thin-walled magnesium alloy component, it was deemed necessary to first utilize computational tools to predict texture under a variety of possible strain paths and identify an ideal flow path for the magnesium alloy during extrusion of thin-walled components. The computational method used is similar to the method detailed by Beyerlein et al. [10] in which the velocity gradient predicted from finite element models is used as input to crystal plasticity predictions. Agnew et al. [6] has shown that VPSC can reasonably reproduce textures observed from experiment for hcp magnesium alloys. Though the focus of the current work is computational, brief details of concurrent experimental work are provided below to provide some insight into the geometries selected for the computational study.

Extrusion Experiments

A vertical extrusion press at CanmetMATERIALS is being utilized to extrude thin-walled magnesium alloy components. The extrusion tooling for this system is depicted in Figure 1, in which an extrusion billet 88.9 mm in diameter can be extruded into a thin-walled component. A 50.8 mm outer diameter tube with 2.5 mm thickness can be extruded using a bridge-die, which results in extrusion seams [11]. There is some concern that the tubular extrusion would fail at the extrusion seams during subsequent deformation. Additionally, a 40 mm wide x 20 mm high x 2 mm thick C-channel extrusion was produced from an 88.9 mm diameter billet. The advantage of the C-channel die is that there is no extrusion seam, though two C-channels will need to be welded together to produce a square tube for subsequent crash tests. The extrusions were produced from an extruded AZ31B rod at a temperature of 400 °C and a ram speed of 254 mm/min.

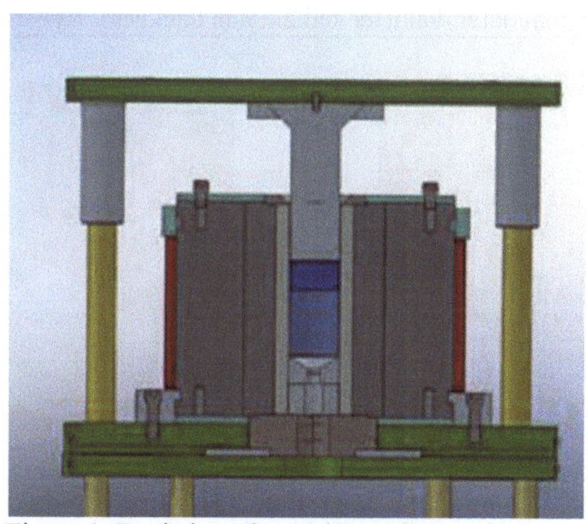

Figure 1. Depiction of experimental extrusion setup

Computational Method

Continuum-Based Finite Element Simulations

Continuum-based finite element simulations were performed using an Arbitrary Lagrangian-Eulerian (ALE) adaptive meshing technique in ABAQUS. The flow stress of the material was specified to be 100 MPa and perfectly-plastic, which is approximately the flow stress of AZ31B at 300 °C at a strain-rate of 0.1 s^{-1} [12]. The coefficient of friction between the flowing material and the stationary surfaces was specified to be zero in all of the simulations, which prevented any dead zones from being predicted during extrusion. For now, the goal is to simply explore the possibility to create distinctions in the deformation texture evolution from very distinct die designs. Future work on promising die designs should employ a simulation method that is capable of accounting for the influence of high levels of friction and dead zones during the extrusion simulation. Eight extrusion profiles were considered, which are shown in Figure 2, and referred to as Profiles #1-8 and a description of the eight profiles is provided in Table 1.

Table 1: Description of extrusion simulation profiles

Profile	Description	Extrusion Velocity (mm/min)
#1	Axisymmetric bar extrusion – diameter 88.9 mm to 38.1 mm	508
#2	Plane strain strip extrusion – 88.9 mm thick to 38.1 mm thick	508
#3	Diameter 38.1 mm bar extruded to tube with 57.2 mm outer diameter and 2.5 mm thickness – gap thickness = 3 mm	254
#4	Diameter 38.1 mm bar extruded to tube with 57.2 mm outer diameter and 2.5 mm thickness – gap thickness = 10 mm	254
#5	Diameter 38.1 mm bar extruded to tube with 57.2 mm outer diameter and 2.5 mm thickness – transition zone ~ 20 mm	254
#6	Tube with 57.2 mm outer diameter and 15 mm thickness extruded to tube with 57.2 mm outer diameter and 2.5 mm thickness	508
#7	Tube with 88.9 mm outer diameter and 19 mm thickness extruded to tube with 88.9 mm outer diameter and 2.5 mm thickness	254
#8	Tube with 88.9 mm outer diameter and 19 mm thickness extruded to tube with 55.8 mm outer diameter and 2.5 mm thickness	254

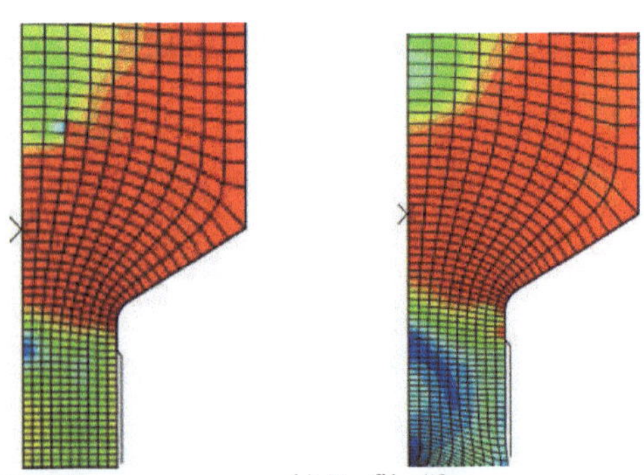

a) Profile #1 b) Profile #2

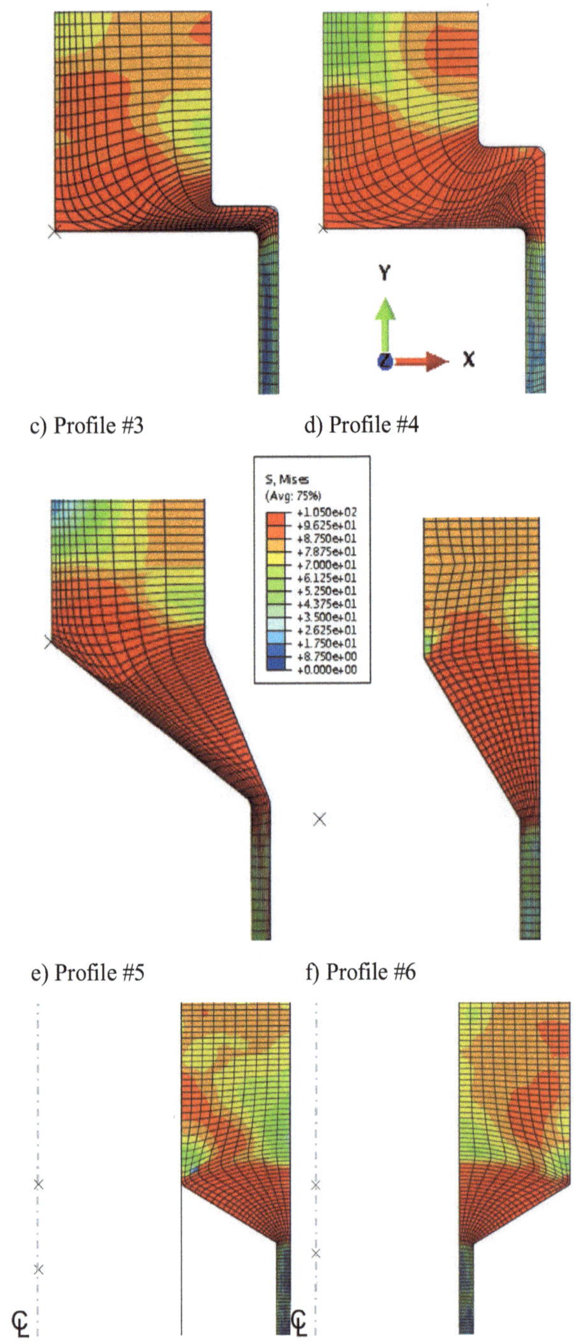

c) Profile #3 d) Profile #4

e) Profile #5 f) Profile #6

g) Profile #7 h) Profile #8

Figure 2: Simulated extrusion geometries (contours of von Mises effective stress)

All of the simulations were performed with 2D axisymmetric elements, except Profile #2, which was performed with 2D plane strain elements. Profiles #1 and #2 were performed with simple strain paths known to produce a specific texture, i.e. axisymmetric bar extrusions with $\langle 10\bar{1}0 \rangle$ fiber parallel to the extrusion direction (ED) and plane strain strip extrusion with $\langle 0001 \rangle$ parallel to plate normal direction, respectively, to help validate the simulation technique.

The strain path of a typical element during the extrusion simulation was output in the form of the deformation gradient, F_{ij}, was converted to velocity gradient according to,

$$L_{ij} = dF_{ij}/dt\, F_{ij}^{-1} \quad (1)$$

where there are five components output from the primarily 2D axisymmetric simulations, L_{11}, L_{12}, L_{33}, L_{12}, and L_{21} with the 1, 2, and 3 components corresponding to the x, y, and z directions of Figure 2, respectively. The velocity gradient was then input to the VPSC crystal plasticity simulations. The velocity gradients were calculated using the extrusion velocity given in Table 1, which is representative of the displacement of the extrusion ram during experiment, such that reasonably accurate strain-rates should be captured in the subsequent crystal plasticity simulations.

VPSC Crystal Plasticity Simulations

Crystal plasticity simulations were performed using VPSC 7c using a similar method to that detailed by Beyerlein et al. [10]. Agnew et al. [6] has previously demonstrated that VPSC can predict reasonably accurate textures in wrought magnesium alloys. In addition to the velocity gradient, the critical resolved shear stress (CRSS) ratio of basal:prismatic:<c+a> slip was an important input to the model. Initial simulations were performed with an initial ratio of 1:8:6, which was the set of slip system strength ratios that seemed to best reproduce the texture evolution of a typical magnesium alloy, AZ31, during ECAE [4]. If there were difficulties (lack of convergence during the simulations) in producing predictions then the ratio would be reduced to 1:6:6. Larger values of prismatic and <c+a> slip would be expected to lead to stronger texture. Lowering the CRSS ratio could weaken the texture and is similar to an alloying or temperature effect by permitting more non-basal slip, which can slow the plastic rotation below the level when basal slip dominates the strain accommodation. An initially random texture was used to represent the initial state of the material prior to extrusion. The texture data output from the VPSC simulations are provided as (0001) and $(10\bar{1}0)$ pole figures. As is typical for these types of simulations, the strength of the texture can be over-predicted but qualitative features of the pole figures are well predicted. One source of additional spread in experimental textures is the variation in orientation that develops within grains, whereas the present simulations assume the grains remain as single orientations.

Results of Finite Element Analysis

Due to large distortions of the mesh that occur during extrusion simulations, an Arbitrary Lagrangian-Eulerian (ALE) method was utilized which combines features of both a Lagrangian (mesh and material move together) and Eulerian analyses (material moves through mesh). In this method, the mesh and material initially move together, but to prevent the elements from becoming too distorted, remeshing was employed at regular intervals. The strain in a specific element can fluctuate as the location of the element is constantly changing due to both distortion and remeshing. Consequently, material point tracking was utilized to accurately describe the state of strain for a specified point. This was accomplished using the post-processing engine, EnSight, in which the velocity and displacement data of the ABAQUS output were used to calculate the position of a material point through the simulation [13]. The strain components, taken as the deformation gradient, were then queried along the material point path and plotted at different distances travelled for the material point. Figure 3 shows the path of the tracer particle during simulations of Profiles #3 and #8. The corresponding deformation and velocity gradients obtained for the tracer particle are provided in Figures 4 and 5 for Profiles #3 and #8, respectively.

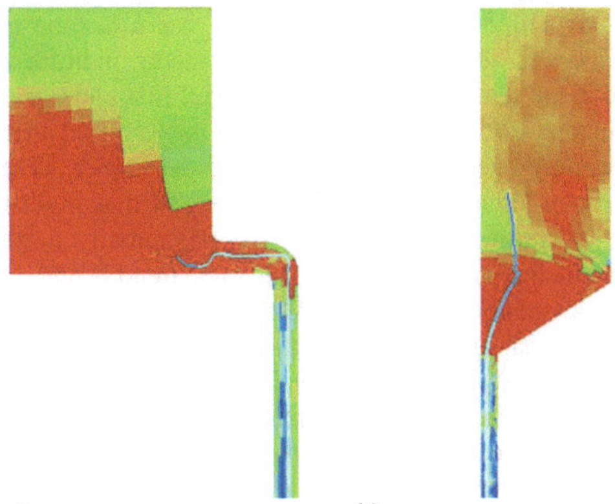

a) b)

Figure 3. Path line utilized to track deformation gradient during extrusion simulations from a) Profile #3 and b) Profile #8 (contours of von Mises effective stress, max.=105 MPa)

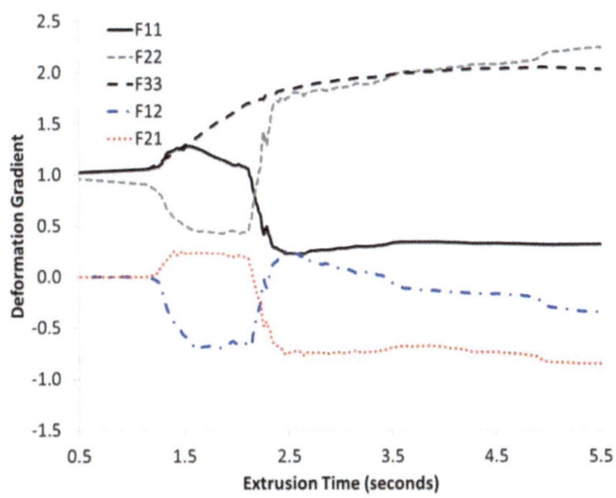

a)

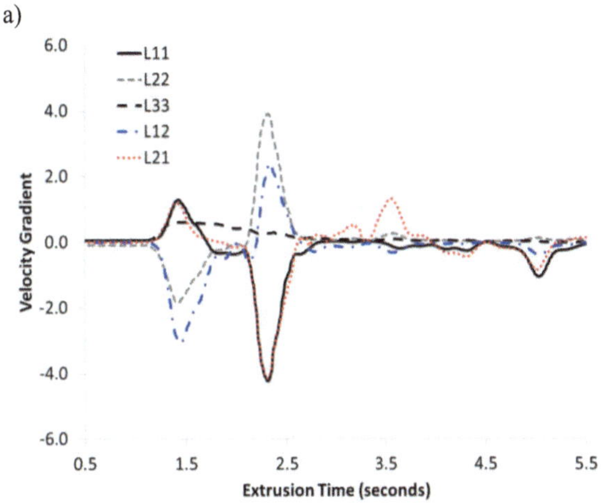

b)

Figure 4. Time evolution of the components of a) the deformation and b) the velocity gradient predicted from the simulation of Profile #3

The evolution of the components of the deformation gradient for Profile #3 show that the material initially undergoes a compression in the y-direction (F_{22} in Figure 4a) as the material flows into the 3 mm gap (Figure 2c), followed by tension as the material flows around the sharp corner. The opposite trend is predicted for F_{11}. F_{33} indicates that the material undergoes tension in the z-direction as the material flows from a bar to a tube of larger circumference. The F_{12} and F_{21} components indicate a strong reversal in the sense of shear as the material first undergoes strong shear in one direction as it flows into the gap and then shears in the opposite sense as it flows around the corner. The velocity gradient in Figure 4b shows that the largest magnitudes, both negative and positive, occur as the material initially flows into the gap and then around the corner. This same trend is present in all of the tube extrusion simulations (#3-#8), though the level of shear and reverse shear are different for the different profiles. For example, the trends are slightly different in Profile #8. Figure 5a indicates that the material undergoes a large tensile strain in the y-direction (F_{22}) as the material is compressed in the x-direction (F_{11}). F_{33} indicates compression as the material flows from the pierced billet of larger outer circumference to a tube of reduced circumference. Figure 5b indicates that the magnitudes of the velocity gradients are largest for the shear component L_{21} as the material flows through the transition zone (Figure 3b).

a)

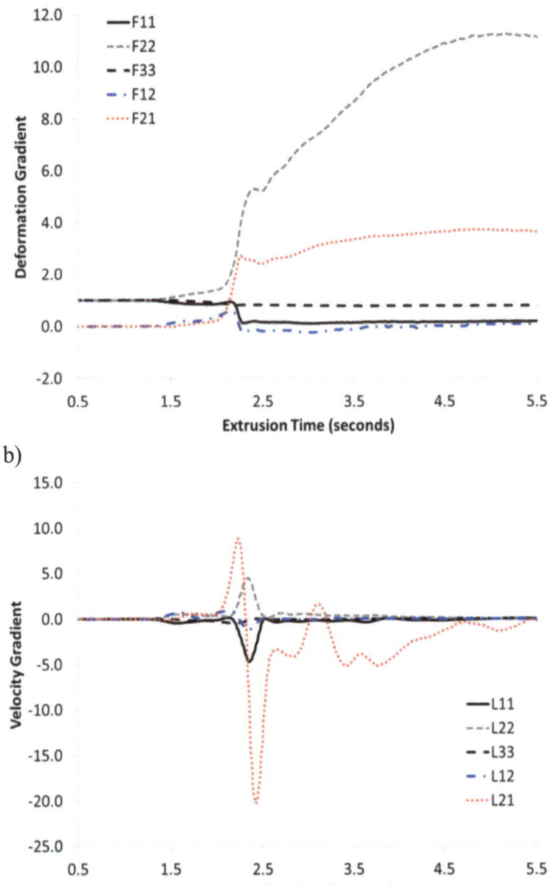

Figure 5. Time evolution of the components of a) the deformation and b) the velocity gradient predicted from the simulation of Profile #8

Results of Deformation Texture Predictions

Axisymmetric Bar and Plane Strain Strip Extrusion

Axisymmetric bar and plane stain strip extrusion simulations were performed to produce predictions of known textures in magnesium alloys. The predicted texture for axisymmetric bar extrusion, Profile #1, is shown in Figure 6. As expected, a so-called "rod texture" is predicted with basal plane normal vectors all directed radially outward. The slight clockwise rotation of the simulated texture about the azimuthal (3) direction is in part due to the shear imposed by the extrusion die. The predicted texture from Profile #2 (similar to plane-strain compression) is shown in Figure 7. The texture is similar to that observed in cold-rolled magnesium alloys. More interestingly, a similarity exists between the predicted texture in Figure 7 and those observed in actual extruded tubes of alloys AM30 and AZ31 [14], which suggests the primary deformation in those processes is plane strain compression.

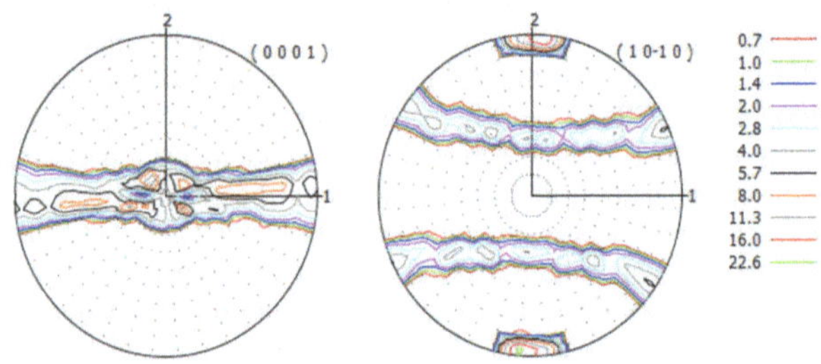

Figure 6. Basal and prismatic pole figure resulting from extrusion Profile #1 – axisymmetric bar– extrusion direction vertical and radial directions along the equatorial plane of the pole figures.

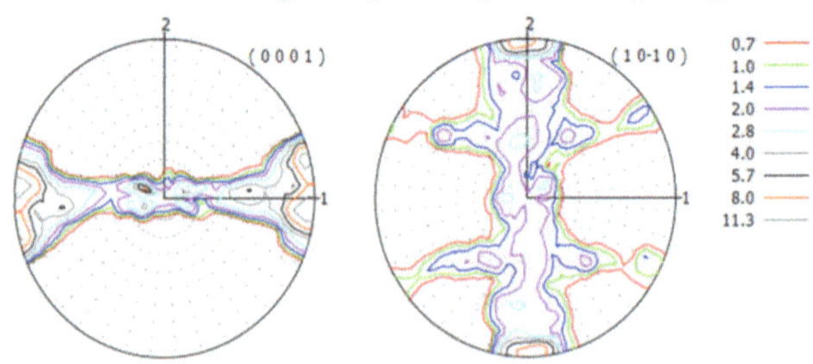

Figure 7. Basal and prismatic pole figure resulting from extrusion Profile #2 – similar to plate rolling – extrusion direction vertical and plate "normal" direction to the left.

Bar to Tube Extrusions

Figures 8-10 show the results of VPSC simulations from Profiles #3-5. Note that the simulation of Profile #3 was performed with a CRSS ratio of 1:6:6 (Figure 8) and indicates a strong compression texture placing <0001> nearly parallel to direction 1. When the results from Profile #3 are compared with the subsequent profiles, it will be shown that this is one profile in which there is a clear tilt in the pole figures about the azimuthal direction seen in Figure 8, which is a direct result of the shear inducted during extrusion. The predicted texture is similar to that detailed in [9] which also involved shear-reverse-shear geometries. The weaker component in the center of the (0001) pole figure are due to grains in the simulated texture which were essentially perfectly oriented to accommodate the imposed deformation with prismatic slip.

Figures 2c and 4a show that the material undergoes compression in the vertical (y-) direction as the material flows into the 3 mm gap. As a slight variation of this profile geometry, a die with a gap height of 10 mm was simulated (Figure 2d). The predicted texture for this geometry (Profile #4) is shown in Figure 9, which was also performed with a CRSS ratio of 1:6:6 for ready comparison. The result is similar to that obtained with Profile #3 (with the smaller gap), with a rigid body rotation that renders the texture components more in line with the sample coordinates. In fact, it is suggested that because the magnitude of the shear-reverse-shear is more balanced in this case, this leads to less net rigid body rotation of the material.

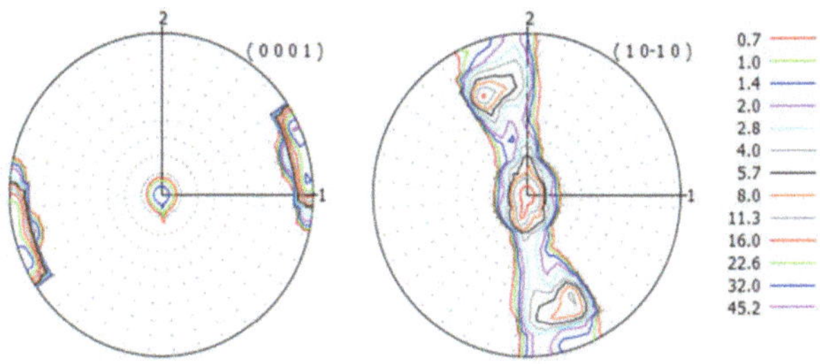

Figure 8. Basal and prismatic pole figure resulting from extrusion Profile #3 with CRSS ratio of 1:6:6 (extrusion direction (ED) vertical, radial direction (RaD) right, and azimuthal direction (AzD) into the page)

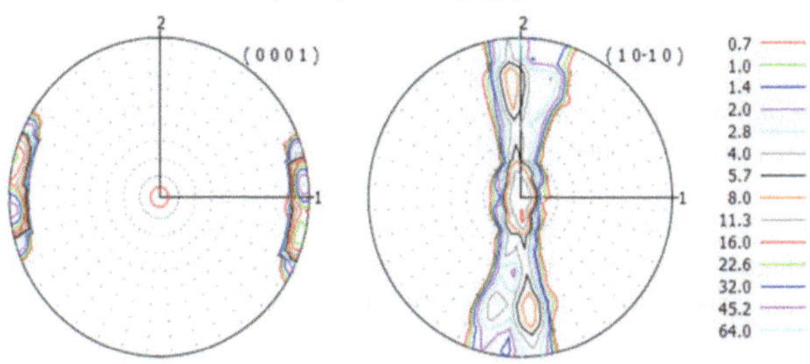

Figure 9. Basal and prismatic pole figure resulting from extrusion Profile #4 with CRSS ratio of 1:6:6 (same orientation as Fig. 8)

The next geometry studied was based on a gradual transition of the material into the shear zone (Figure 2e). By introducing a gradual transition, the material would undergo straining in both the x- and y- directions, with some strain in the z-direction as well. The predicted texture for Profile #5 is shown in Figure 10. The results show a similar texture to that predicted from Profile #4 (Figure 9), in that the ⟨0001⟩ fiber is parallel to the radial direction. However, the $\langle 10\bar{1}0\rangle$ fiber is parallel to ED for Profile #5 (Figure 10) whereas it is parallel to the AzD in Profile #4 (Figure 9). Such a difference in the orientation of the $\langle 10\bar{1}0\rangle$ is not expected to have a significant influence on tube properties.

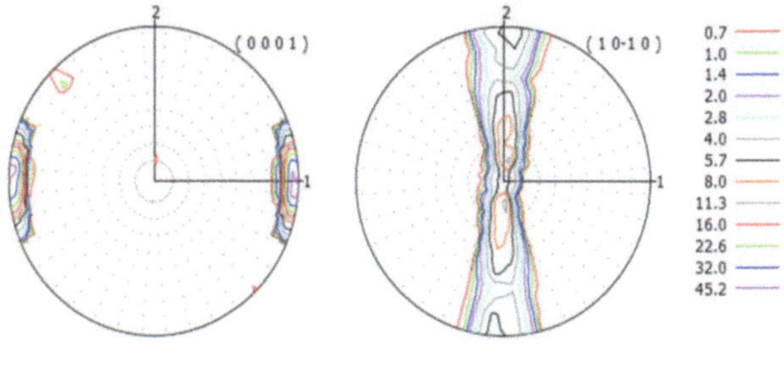

Figure 10. Pole figure resulting from extrusion Profile #5 with CRSS ratio of 1:6:6 (same orientation as Fig. 8)

Pierced Billet Extrusions

The next set of simulations (Profiles #6-8 in Figure 2) represent an extrusion operation where an initially thick tube is extruded to a thin-walled tube, beginning with a pierced billet. In some ways, it resembles a plate rolling operation in the sense that one dimension (radial or thickness) is strongly reduced while one dimension is strongly increased (extrusion or rolling). One distinction is that the dimension along azimuthal direction is not fixed, as is case in the transverse direction during rolling. Depending on the geometry, the azimuthal direction may be strained in tension or compression. The texture predicted for Profiles #6-8 which were shown in Figures 2f-2h are given in Figures 11 to 13. CRSS ratios of 1:6:6 were employed in each simulation.

The results shown in Figures 11 and 12, for Profiles #6 and #7, respectively, are similar to each other albeit with a slight distinction due to the amount of shear imposed. This makes sense, because the overall type of deformation is similar. Only the sharpness of the transition is distinct. Further, these two cases result in textures which are similar to most of the preceding texture (Figures 7 - 10). For all of the axisymmetric cases, Profiles #3 – 7, the inner diameter is being expanded. Practically speaking, these cases would be difficult to achieve as they would impose high loads on a mandrel and/or bridge-die.

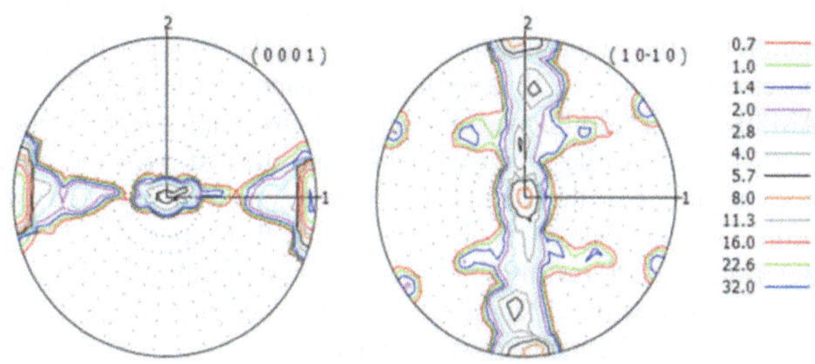

Figure 11. Basal and prismatic pole figure resulting from extrusion Profile #6 with CRSS ratio of 1:6:6

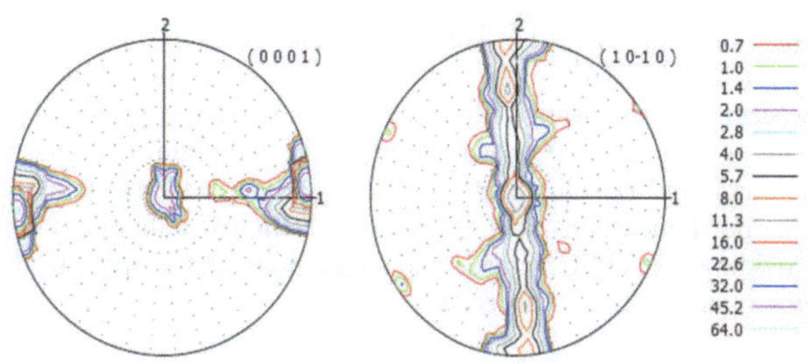

Figure 12. Basal and prismatic pole figure resulting from extrusion Profile #7 with CRSS ratio of 1:6:6

The final profile, Profile #8, is of greater interest practically as it can be produced in a straightforward way, using a mandrel extrusion system. It results in a texture type which is intermediate to that of axisymmetric bar extrusion (Figure 6) with a $\langle 10\bar{1}0\rangle$ fiber oriented parallel to the extrusion and all of the tube expansion cases (Figures 7-12), which have a <0001> fiber nearly parallel to the radial direction (RaD). As shown in Figure 13, Profile #8 has both the $\langle 10\bar{1}0\rangle$ || ED and the <0001> nearly parallel to RaD.

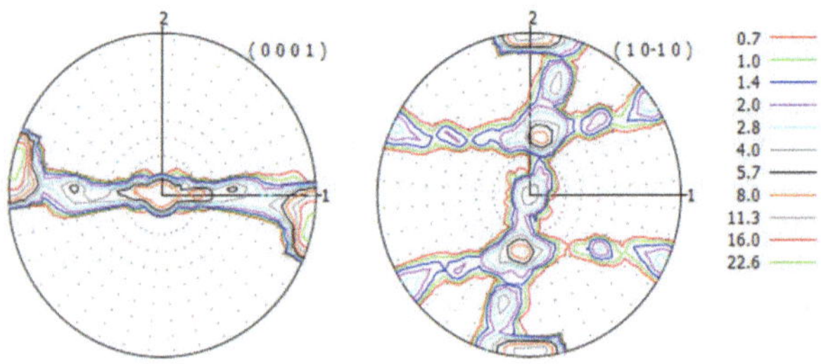

Figure 13. Basal and prismatic pole figure resulting from extrusion Profile #8 with CRSS ratio of 1:6:6

Conclusions

Texture evolution in thin-walled magnesium alloy extrusions was predicted using the viscoplastic self-consistent (VPSC) crystal plasticity model, with strain path input from continuum-based finite element simulations of the extrusion process. In the case of simple strain paths, the predictions were in reasonable agreement with the experimentally observed textures. Interesting textures were predicted for cases of bar-to-tube and thick-to-thin walled tube extrusions of a hypothetical magnesium alloy where the CRSS ratios for basal:prismatic:<c+a> slip were set to 1:6:6. For the most readily achieved case (Profile #8), the texture is a rational combination of that observed in axisymmetric bar extrusion, which results in $\langle 10\bar{1}0\rangle$ parallel to the extrusion axis, and plane strain compression, which places the <0001> nearly parallel to the radial direction. This is similar to what has been observed experimentally after tube extrusion of alloys AZ31 and AM30 [14]. The influence of strain path involving shear-reverse-shear could also be predicted, which is reflected by some rigid body rotations of the dominant texture components between one profile and another. Overall, it was found that range of possible deformation textures in thin-walled magnesium alloys is limited. Experiments with different types of die designs are presently underway to validate the model predictions.

Acknowledgements

The authors would like to thank Lucian Blaga of CanmetMATERIALS for assistance in carrying out the experimental extrusions. Funding for this work was provided by Natural Resources Canada through the Program of Energy Research and Development. Thanks to Drs. Irene Beyerlein and Ricardo Lebensohn of Los Alamos National Laboratory for providing the VPSC 7c code and suggestions with respect to the present application of the code.

References

1. D.A. Wagner et al., "Test results and FEA predictions from magnesium AM30 extruded beams in bending and axial compression," Proceedings of Magnesium Technology 2009, 541-547.

2. J. Rossiter et al. "Numerical modeling of the failure of magnesium tubes under compressive loading," Journal of Materials Engineering and Technology, 134 (2012).

3. T. Mukai, H. Watanabe and K. Higashi, "Application of superplasticity in commercial magnesium alloy for fabrication of structural components," Materials Science and Technology, 16 (2000), 1314-1319.

4. S.R. Agnew et al., "Enhanced Ductility in Strongly Textured Magnesium Produced by Equal Channel Angular (ECA) Processing," Scripta Materialia, 50 (2004), 377-381.

5. S.R. Agnew et al., "Texture Evolution of Five Wrought Magnesium Alloys During Route A Equal Channel Angular Extrusion: Experiments and Simulations" Acta Materialia, 53 (2005), 3135-3146.

6. S.R. Agnew et al., "Crystallographic Texture Evolution of Three Wrought Magnesium Alloys During Equal Channel Angular Extrusion," Materials Science and Engineering A, 408 (2005), 72-78.

7. S. Seippa et al., "Microstructure, crystallographic texture and mechanical properties of the magnesium alloy AZ31B after different routes of thermo-mechanical processing," International Journal of Plasticity, 35 (2012), 155–166.

8. Y. Chino, K. Sassa, and M. Mabuchi, "Enhancement of tensile ductility of magnesium alloy produced by torsion extrusion," Scripta Materialia, 59 (2008), 399-402.

9. D. Orlov et al., "Improvement of mechanical properties of magnesium alloy ZK60 by integrated extrusion and equal channel angular processing," Acta Materialia, 59 (2011), 375-385.

10. I.J. Beyerlein et al., "Non-uniform microstructure and texture evolution during equal channel angular extrusion," Philosophical Magazine, 85 (2005), 1359–1394.

11. N.H. Alharthi and W.Z. Misiolek, "Extrusion Welding in a Magnesium Alloy (AM30) Hollow Section Extrudate," Advanced Materials & Processes, February (2014), 16-17.

12. J. Liu, Z. Cui, and C. Li, "Modelling of flow stress characterizing dynamic recrystallization for magnesium alloy AZ31B," Computational Materials Science, 41 (2008), 375-382.

13. EnSight – CSM and CFD Post processing, https://www.ceisoftware.com, [Accessed August 2014].

14. L. Jiang et al., "Twinning and texture development in two Mg alloys subjected to loading along three different strain paths," Acta Materialia, 55 (2007), 3899-3910.

Magnesium Technology 2015
Edited by: Michele V. Manuel, Alok Singh, Martyn Alderman, and Neale R. Neelameggham
TMS (The Minerals, Metals & Materials Society), 2015

ROLE OF Zr IN THE MICROSTRUCTURE EVOLUTION IN Mg-Zn-Zr BASED WROUGHT ALLOYS

T. Bhattacharjee[1,2], T.T. Sasaki[1], B. C. Suh[3], T. Nakata[4], S. Kamado[4], N.J. Kim[3], K. Hono[1,2]

[1] National Institute for Materials Science (NIMS), 1-2-1 Sengen, Tsukuba, Ibaraki, 3050047, Japan
[2] University of Tsukuba, 1-1-1 Tennodai, Tsukuba, Ibaraki, 3050006, Japan
[3] Pohang Institute of Science and Technology (POSTECH), Hyoja-dong, Pohang, 790784, Republic of Korea
[4] Nagaoka University of Technology, 1603-1 Kami-Tomioka Machi, Nagaoka, Niigata, 9402137,Japan

Keywords: Magnesium alloy, ZK60, Zirconium,

Abstract

We investigated the role of Zr in the microstructure evolution in Mg-Zn-Zr based wrought alloys fabricated by twin-roll casting and hot rolling (TRC-HR) process and extrusion process. In the as-cast condition, the Mg(Zn,Zr) precipitates are observed in the as-cast Mg-6.2Zn-0.5Zr (ZK60) alloys indicating that the Mg(Zn,Zr) precipitates are formed during melting, solidification or cooling process. The ZK60 alloy exhibited finer grain structure compared to the Mg-6.2Zn (Z6) alloy in both as-cast and solution treated conditions since the Mg(Zn,Zr) inhibited the grain growth. The Mg(Zn,Zr) particle also contributed to the fine-grain structure formation during the extrusion and TRC-HR processes. During the wrought processing, the recrystallization starts to occur from the grain boundaries of the homogenized sample, and spread to the grain interior. Since the Mg(Zn,Zr) particle inhibited the grain growth, the fine recrystallized grain structure is formed in the ZK60 wrought alloy.

Introduction

There are strong demands for the development of high strength wrought alloys applicable to transportation vehicles to reduce their weight. Mg-6.2Zn-0.5Zr (ZK60) alloy is one of the most commonly used wrought alloys because of the relatively high yield strength over 260 MPa, satisfactory large ductility and excellent damping capacity [1].

The trace amount of Zr has been added as a grain refiner of the cast alloys. Zr-enriched cores formed during melting have been believed to provide nucleation sites for the magnesium grains leading to the grain refinement [2-5]. Recent studies have reported some overlooked roles of Zr in the Mg-Zn based alloys [6-10]. Bhattacharjee et al. reported that Mg(Zn,Zr) precipitates are formed in a ZK60 cast alloy subjected to a heat treatment at 350°C [8]. The structure of the Mg(Zn,Zr) precipitates were identical to $MgZn_2$ type phases of β_1' and β_2' phases, which are typically observed in the Mg-6Zn (Z6) alloy subjected to aging. Since the β_1' and β_2' phases in the aged Mg-6Zn (Z6) alloy are dissolved by heat treatment at 350 $^{\circ}$C, they concluded that Zr stabilized the β_1' and β_2' precipitates at elevated temperature. These thermally stable Mg(Zn,Zr) precipitates critically affected the grain structure of the wrought alloys. In both extruded bars [7, 10] and twin-roll cast and hot-rolled (TRC-HR) sheets [9], much finer grain structure was formed in the ZK60 alloy compared to the Z6 alloy.

Although the presence of the Mg(Zn,Zr) phase was confirmed in the ZK60 alloy subjected to the heat treatment, it still remains unclear how the Mg(Zn,Zr) phase is formed since there has been no detailed characterization work on the as-cast ZK60 alloy. On the other hand, Oh-ishi et al reported that the neck-lace type recrystallization process caused the formation of bimodal grain structure consisting of fine recrystallized grains and coarse unrecrystallized grains during the extrusion process of a ZK60 alloy microalloyed with Ag and Ca [7]. In this work, we have compared the microstructure of the as-cast and heat treated ZK60 alloys to investigate the formation of the Mg(Zn,Zr) precipitates, and discussed the role of the Mg(Zn,Zr) particle on the microstructure evolution during the extrusion and TRC-HR processes in Mg-6.2Zn based alloy.

Experimental Procedure

Alloy ingots with nominal compositions (in wt%) of Mg–6.2Zn (Z6), Mg–6.2Zn–0.6Zr (ZK60) were prepared by induction melting of high purity Mg, Zn and Mg-34.6wt.%Zr master alloy in an argon atmosphere using a steel crucible. The alloy compositions and their nomenclatures are shown in Table 1 in both atomic and weight per cent. The as-cast samples were then heat treated at 350 $^{\circ}$C for 48 h (homogenization treatment) in a muffle furnace.

The ingots were then extruded at 350 $^{\circ}$C with an extrusion ratio of 20:1 at a ram speed of 0.1 mm^{-1}. To fabricate the twin-roll cast and hot-rolled (TRC-HR) sample, the ingots were remelted under a mixture of CO_2 and SF_6 and transferred into a preheated tundish held at 680–700 $^{\circ}$C followed by twin roll casting. The roll gap was set at 1.8 mm and roll speed was 4 m/min. The cast sheets were then homogenized at 330 $^{\circ}$C for 2 h and then hot rolled at 300 $^{\circ}$C for three passes with a total reduction of 50%.

TEM observations were conducted using FEI Tecnai T20 and Titan G2 80-200 microscopes operating at 200 kV. TEM specimens were prepared using twin jet electro-polishing using a solution of 300 ml 2-butoxy ethanol, 15.9 g lithium chloride, 33.5 g of magnesium perchlorate in 1500 ml methanol at a temperature of -45°C as well as ion-polishing using Gatan Precision Ion Polishing System (PIPS).

Results and Discussions

Table 1: The chemical compositions and alloy nomenclatures of the samples used in this study

Nomenclature	at.%	wt.%
Z6	Mg-2.4Zn	Mg-6.2Zn
ZK60	Mg-2.4Zn-0.16Zr	Mg-6.2Zn-0.6Zr

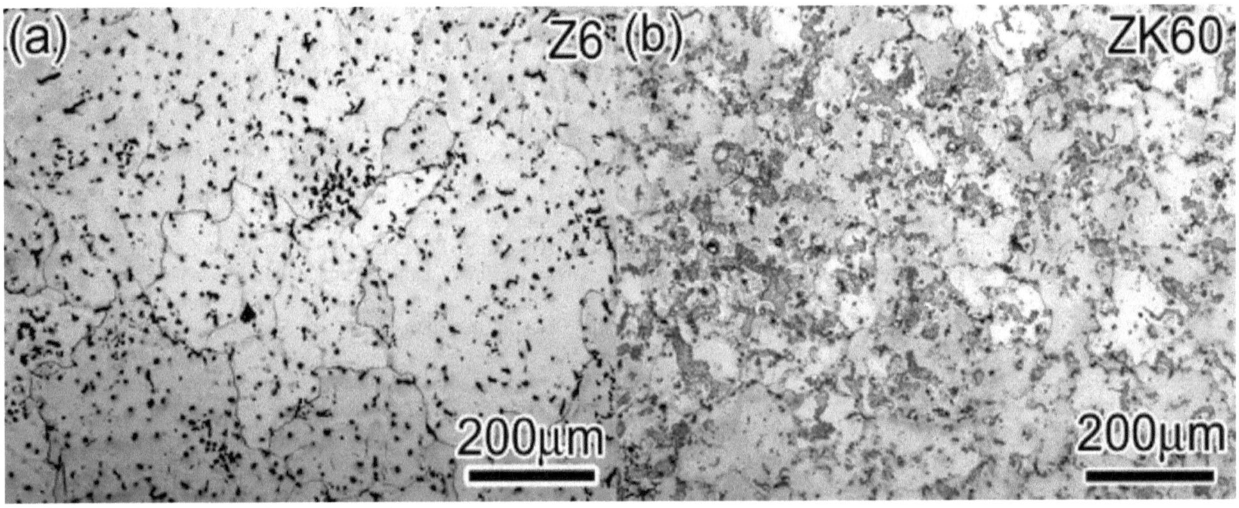

Figure 1: Optical micrographs taken from as-cast (a) Z6 and (b) ZK60 alloys

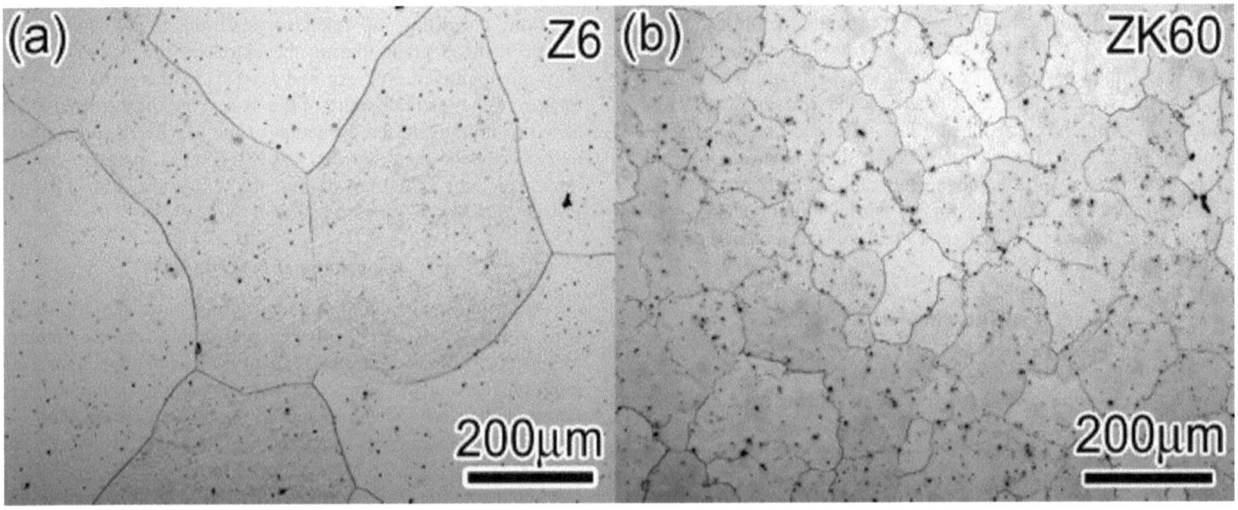

Figure 2: Optical micrographs taken from heat treated (a) Z6 and (b) ZK60 alloys

Figure 1 shows the optical micrographs of the as-cast Z6 and ZK60 alloys. The ZK60 alloy has finer grain structure compared to the Z6 alloy. The average grain size of the as-cast Z6 and ZK60 alloys is 340 $\pm$ 100and 100 $\pm$ 20 mm, respectively. There are higher volume fraction of secondary phase along the grain boundaries in the ZK60 alloy than the Z6 alloy.

The grain growth during the heat treatment at 350 ^{0}C is significantly restricted in the ZK60 alloy compared to the Z6 alloy. Figure 2 shows the optical micrographs of the heat treated Z6 and ZK60 alloys. In the ZK60 alloy, the grains grew from 100 $\pm$ 20 to 140 $\pm$ 40 μm by the heat treatment as shown in Fig. 2 (b) while those in the Z6 alloy significantly coarsened to 600 $\pm$ 200 μm during the heat treatment (Fig. 2 (a)). Most of the coarse secondary phase particles, which can be observed in the as-cast optical micrographs are dissolved into the magnesium matrix by the heat treatment. Although previous work has reported that the fine nanoscale β_1' and β_2' phase still remains within the grains of ZK60 alloy [8].

Figure 3 (a) shows the bright field TEM image of the as-cast ZK60 alloy taken from the zone axes of $[11\bar{2}0]$. Two different shapes of secondary phases are observed within the grain; one is the rod-like phase growing along [0001] direction, and the other is the plate like precipitate forming on the (0001) plane. The rods growing along the [0001] direction are the β_1' phase, and the β_1' phase have the size of 620 ± 150 nm in length and 30 ± 10 nm in diameter. The plates are the β_2' phase [6], and they have the size of 30 ± 10 nm in thickness and 10 ± 5 nm in diameter. Figure 3 (b) through (d) are the high angle annular dark field scanning TEM (HAADF-STEM) image and EDS elemental maps of Zn and Zr taken from the zone axis of [0001]. The EDS elemental maps show that the Zr is enriched within the precipitates as well as Zn.

The β_1' and β_2' phases observed in the as-cast sample are much coarser than those observed in the heat treated sample [8]. However, EDS elemental maps clearly confirms the presence of the Mg(Zn,Zr) phase. In the as-cast condition, the fine Mg(Zn,Zr) are formed, and the β_1' and β_2' phases consisting of Mg and Zn are formed in contact with the Mg(Zn,Zr) phase similar to the β_1' and β_2' phases formed in the low temperature aged ZK60 alloy [8]. In this work, it was hard to find Zr-rich cores, which have been believed to cause the refinement of the magnesium grains. However, this work shows that the Mg(Zn,Zr) precipitates are formed during melting, solidification or cooling processes. In addition, the fine

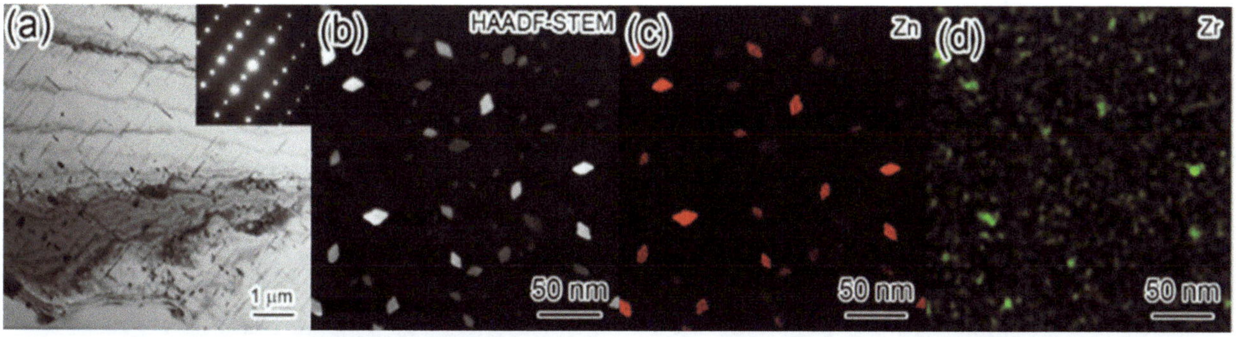

Figure 3: (a) is Bright field TEM image obtained from the as-cast ZK60 alloy. Note that the image was recorded from the zone axis of $[11\bar{2}0]$. (b) through (d) are HAADF-STEM image and EDS elemental maps of Zn and Zr taken from the zone axis of [0001].

precipitates are also dispersed along the grain boundary in the as-cast alloy indicating that the grain boundary is pinned by these particles as shown in Fig. 3 (a). Therefore the Mg(Zn,Zr) particles can also be considered to work as a grain refiner in the ZK60 cast alloy together with the Zr-rich particles which works as an inoculant.

Figure 4 and 5 show optical micrographs of the microstructure evolution during the extrusion process in the Z6 and ZK60 alloys, respectively. In both Z6 and ZK60 alloys, the grain structure at 3 cm above the die entrance shows the similar structure to the homogenized samples as shown in Fig. 4 (a) and Fig. 5 (a). At 1 cm from the die entrance, twin formation is obvious in the Z6 alloys while it is not the case in the ZK60 alloy, Fig. 4 (b) and Fig. 5 (b). The recrystallized grains are observed along the grain boundaries (Fig. 4 (b) and Fig. 5 (b)). Near the die entrance, the recrystallized area spreads into the grain interior in both Z6 and ZK60 alloys, Fig. 4 (c) and Fig. 5 (c). However, after the extrusion, the ZK60 alloy exhibits bimodal grain structure consisting of fine equi-axed recrystallized grains with the size of 10±5 µm and coarse unrecrystallized grains elongated along the extrusion direction while the grain structure of the Z6 alloy is uniform with the size of 30±15 µm as shown in Fig. 4 (d) and Fig. 5 (d). Since the Mg(Zn,Zr) phase remain after the heat treatment at 350 ^{O}C [8], the Mg(Zn,Zr) particles caused the difference in grain structure of the extruded sample between the Z6 and ZK60 alloys. As shown in Fig. 4 (b) and Fig. 5 (b), the recrystallized grains mainly nucleate along the grain boundaries of the homogenized sample at the beginning of extrusion process. Then, the recrystallized area spreads from the surface to interior of the grain of the homogenized sample. In the Z6 alloy, the secondary phases are dissolved in the magnesium matrix by the homogenization prior to the extrusion at 350 ^{O}C. This means that single phase is kept in the Z6 alloy as there is no obstacles to impede the recrystallization during the extrusion of the Z6 alloy, a uniform but coarse grain structure is formed. However, since the Mg(Zn,Zr) particles are dispersed throughout the grains in the homogenized ZK60 alloy, these particles worked as obstacles to spread the recrystallized area into the grain interior. This caused the formation of the unrecrystallized grains. In addition, the recrystallized grains in the ZK60 alloy has much finer size compared to those in the Z6 alloy. The homogenized ZK60 alloy consisted of finer grain structure of 140±40 µm compared to the Z6 alloy, 600±200 µm as shown in Fig. 4 (a) and Fig. 5 (a). There are more nucleation sites for the recrystallized grains in the ZK60 alloy. Since the growth of these recrystallized grains are pinned by the Mg(Zn,Zr) particles, the finer recrystallized grain structure is formed in the ZK60 alloy.

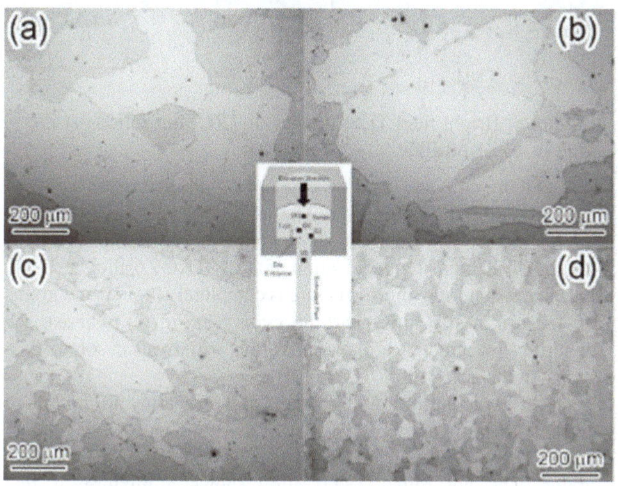

Figure 4: Optical micrographs showing the microstructure evolution during the extrusion process in Z6 alloy. (a): Top of the homogenized sample, (b) 1 cm above the die entrance, (c): near the die entrance and (d): extruded part. Note that inset figure shows the position of each figures.

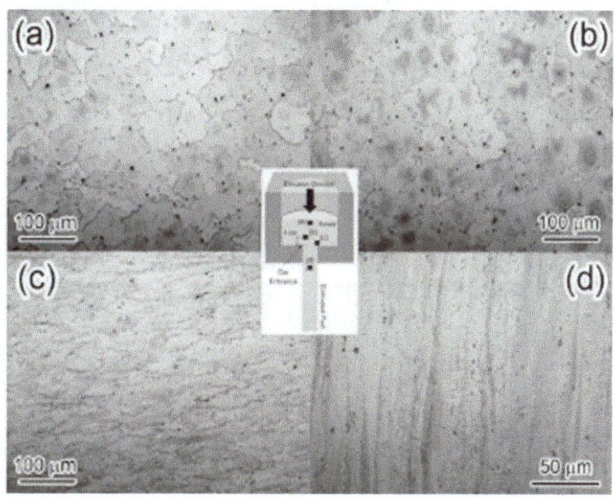

Figure 5: Optical micrographs showing the microstructure evolution during the extrusion process in ZK60 alloy. (a): Top of the homogenized sample, (b) 1 cm above the die entrance, (c): near the die entrance and (d): extruded part. Note that inset figure shows the position of each figures.

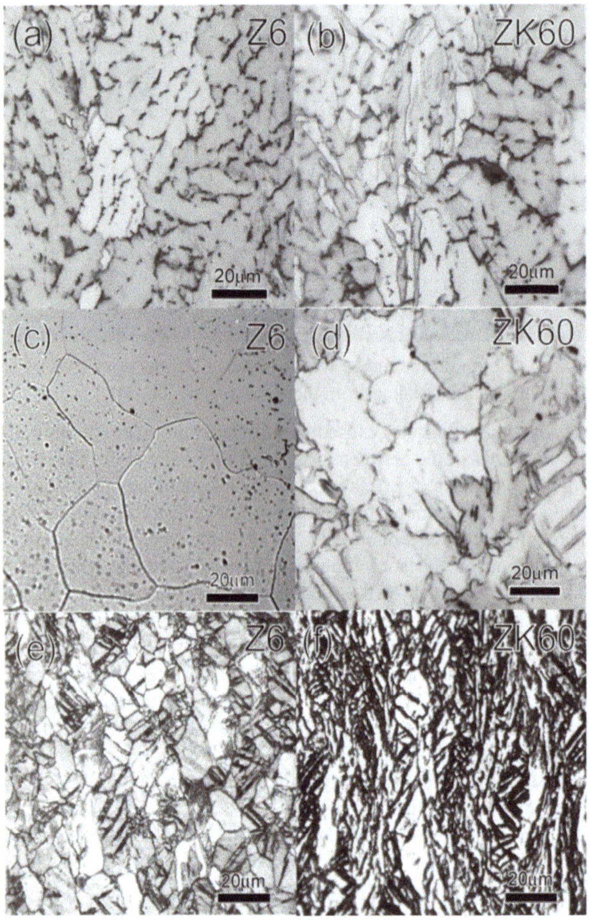

Figure 6: Optical micrographs of (a) As-TRC Z6, (b) As-TRC ZK60, (c) homogenized Z6, (d) homogenized ZK60, (e) 3rd-pass- hot rolled Z6, (f) 3rd-pass- hot rolled ZK60

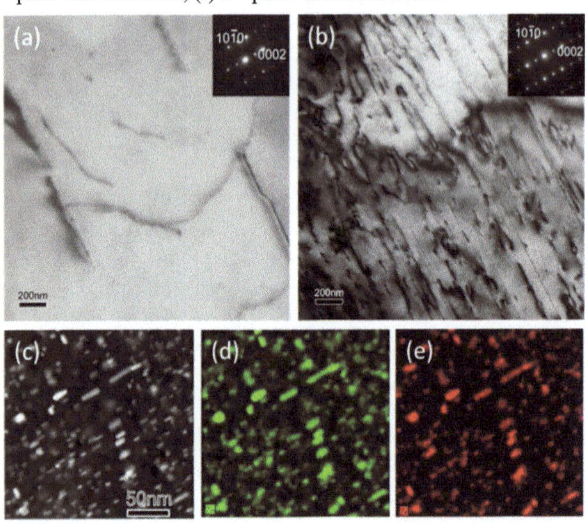

Figure 7: Bright field TEM images of homogenized (a) Z6 and (b) ZK60 alloy. (c) HAADF-STEM image of homogenized ZK60 alloy and corresponding EDS map of (d) Zn and (e) Zr. (Electron beam is approximately parallel to the <11$\bar{2}$0> direction)

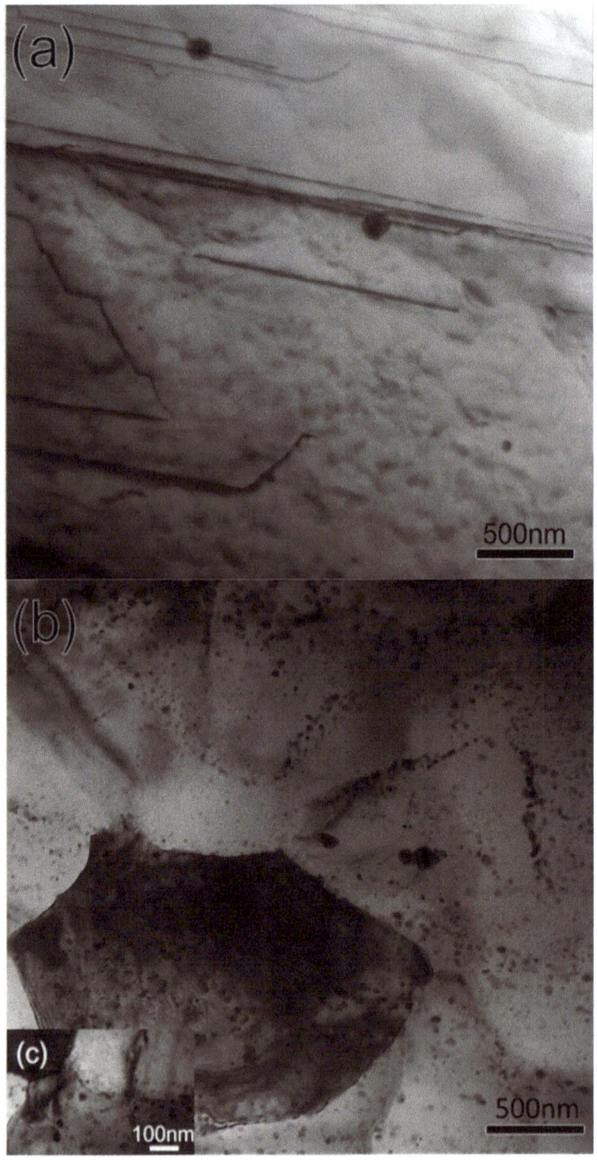

Figure 8: Bright field TEM images of TRC- hot rolled (a) Z6 and (b, c) ZK60 alloy.

Figure 6 (a) and (b) shows the optical microstructures of twin roll cast Z6 and ZK60 alloys. The as-TRC sheets have dendritic microstructure with twins in some grains in both samples (Fig. 6 (a) and (b)). Upon homogenization of the as-TRC Z6 alloy at 330ºC for 2h, the dendritic microstructure is completely broken and grains of ~50 μm in size forms with no twins (Fig. 6 (c)), whereas the dendritic-like structure still remains along with twins in some of the grains in the ZK60 alloy (Fig. 6 (d)).

After the hot rolling, the grains in the Z6 alloy are refined to around ~20 μm (Fig. 6 (e)). Most of the grains are equiaxed and dynamically recrystallized. Twin can be also observed in some grains. The hot rolled ZK60 alloy shows a mixture of fine recrystallized grains of around 1-4 μm with unrecrystallized regions elongated along the rolling direction with size greater than 20 μm containing twins (Fig. 6 (f)).

The bright field TEM images of Z6 and ZK60 alloy homogenized at 330ºC for 2 h after being twin roll cast are shown in Figure 7 (a) and (b) respectively. The Z6 alloy has very few dislocations compared to ZK60 alloy which has a very high dislocation density. Figure 7 (c) is the HAADF-STEM image showing the dispersion of the fine rod like precipitates with the size of 25 ± 20 nm in length and 8 ± 2 nm in width within the matrix of the ZK60 alloy. The EDS elemental maps of Zn and Zr shows that these precipitates are enriched with Zn and Zr (Fig. 7 (d) and (e)) and confirms the presence of Mg(Zn,Zr) precipitates similar to those observed in the heat treated ZK60 alloy [8].

The bright field TEM image of hot rolled Z6 alloy shows the presence of few spherical Mg-Zn particles with a size of above 100 nm (Fig. 8 (a)), whereas the ZK60 alloy shows a very high density of heterogeneously dispersed fine spherical Mg(Zn,Zr) particles with the size of 30 ± 25 nm (Fig. 8 (b)). The shapes of the fine Mg(Zn,Zr) particles becomes spherical upon hot rolling as compared to their fine rod and plate like shape in the homogenized condition (Fig. 7(c)). The presence of these fine precipitates in the grains tends to retard the recrystallization and also prevent grain growth of the fine recrystallized grains by pinning them (Fig. 8(c)) giving rise to the bimodal microstructure in the ZK60 alloy.

Summary

In this work, we compared the microstructure of the as-cast and heat treated Mg-6.2Zn (Z6) and Mg-6.2Zn-0.6Zr (ZK60) alloys. This work shows that the Mg(Zn,Zr) particles are already formed in the as-cast sample. The ZK60 alloy shows finer grain structure compared to the Mg-6.2Zn (Z6) alloy in both as-cast and solution treated conditions since the Mg(Zn,Zr) inhibits grain growth. The Mg(Zn,Zr) phase also critically effect the recrystallization behavior during the wrought processing of ZK60 alloy leading to its characteristic bimodal microstructure whereas in Z6 alloy the microstructure consists of equiaxed and coarse recrystallized grains.

Acknowledgement

This work was supported by JSPS, Graint-in-Aid for Young Scientists (B), 24760605, Grant-in-Aid for Scientific Research on Innovative Area, "Bulk Nano Metal", and JST, Advanced Low Carbon Technology Research and Development Program, (ALCA).

References

[1] M.M. Avedesian, H. Baker, eds. Magnesium and Magnesium Alloys ASM Specialty Handbook. ASM International, Materials Park, OH, 1999.

[2] E.F. Emley, Principles of Magnesium Technology, Pergamon Press, Oxford, 1966.

[3] Y.C. Lee, A.K. Dahle and D.H. StJohn, "The Role of Solute in Grain Refinement of Magnesium", Metallurgical and Materials Transactions, 31A (11) (2000) 2895-2905.

[4] Y. Tamura, N. Kono, T. Motegi, E. Sato, "Grain refining of pure magnesium by adding Mg-Zr master alloy", J. Japan Inst. Light Metals 47 (1997) 679-684.

[5] M. Qian, D. H. StJohn and M. T. Frost, "Zirconium alloying and grain refinement of magnesium alloys", Magnesium Technology 2003, ed. by H. I. Kaplan, TMS, Warrendale PA, USA (2003) 209-214.

[6] H. Watanabe, K. Moriwaki, T. Mukai, T. Ohsuna, K. Hiraga, K. Higashi, "Materials Processing for Structural Stability in a ZK60 Magnesium Alloy", Mater. Trans. 44 (4) (2003) 775-781.

[7] K. Oh-ishi, C.L. Mendis, T. Homma, S. Kamado, T. Ohkubo, K. Hono, Acta "Bimodally grained microstructure development during hot extrusion of Mg–2.4 Zn–0.1 Ag–0.1 Ca–0.16 Zr (at.%) alloys", Acta Materialia 57 (2009) 5593-5604.

[8] T. Bhattacharjee, C.L. Mendis, T.T. Sasaki, T. Ohkubo and K. Hono, "Effect of Zr addition on the precipitation in Mg–Zn-based alloy", Scripta Materialia, 67 (12) (2012) 967-970

[9] T. Bhattacharjee, B.C Suh, T.T. Sasaki, T. Ohkubo, N.J. Kim, K. Hono, "High strength and formable Mg–6.2Zn–0.5Zr–0.2Ca alloy sheet processed by twin roll casting", Materials Science and Engineering A, 609 (2014) 154-160.

[10] T. Bhattacharjee, T. Nakata, T.T. Sasaki, S. Kamado, and K. Hono, " Effect of microalloyed Zr on the extruded microstructure of Mg–6.2Zn-based alloys", Scripta Materialia, 90-91 (2014) 37-40

Magnesium Technology 2015
Edited by: Michele V. Manuel, Alok Singh, Martyn Alderman, and Neale R. Neelameggham
TMS (The Minerals, Metals & Materials Society), 2015

Effect of Alloy Composition on Microstructure and Strength of Fine Grained Extruded Mg-Zn-Y Alloys Containing Quasicrystal Phase

Alok Singh[1], Yoshiaki Osawa[1], Hidetoshi Somekawa[1], Toshiji Mukai[2], Catherine J. Parrish[3], Donald S. Shih[4]

[1]Structural Materials Unit, National Institute for Materials Science, Tsukuba 305-0047, Japan
[2]Department of Mechanical Engineering, Kobe University, Kobe 657-8501, Japan
[3]Boeing Research and Technology, The Boeing Company, Huntington Beach, CA, USA
[4]Boeing Research and Technology, The Boeing Company, St. Louis, MO, USA

Keywords: Magnesium alloys, Mg-Zn-Y, quasicrystal, mechanical properties, yield anisotropy

Abstract

Earlier, we reported very high yield strengths of 400 MPa accompanied by total elongations of over 12% in Mg-6xZn-xY alloys, where x=1.0, 0.5, 0.35 and 0.2 in at%, dispersed with quasicrystalline phase and processed by extrusion. Here we report on the effect of low amount of alloying elements zinc and yttrium on the texture and anisotropy of the Mg-Zn-Y alloys. With decreasing amount of the alloying elements, the yield strength of the extruded alloys remains close to 400 MPa with total elongation to failure of about 12%, while the anisotropy keeps increasing. Thus very high tensile strengths with moderate ductility are achieved even in very dilute alloys. With lesser amount of alloying elements the degree of dynamic recrystallization is reduced and the intensity of the basal texture is increased, resulting in higher yield anisotropy.

Introduction

As the usage of automobiles globally expands rapidly, it has become an urgent priority to find ways to conserve fuel and limit environmental impact. One of the basic methods is to reduce the weight of the automobiles, especially since other technologies such as hybrid add more weight. More use of the lightest structural metals, i.e., magnesium alloys, thus sound inevitable. It is thus important to develop magnesium alloys with better mechanical properties. Common techniques to improve mechanical properties are refinement of grain size, precipitation and dispersion of intermetallic phases. Grain refinement is particularly effective in magnesium alloys because of high slope of Hall-Petch plots [1]. Grain refinement occurs simultaneously with wrought processes such as rolling and extrusion. Wrought processing also introduces a basal texture which results in yield anisotropy in which tensile yield strength (YS) is larger than the compressive YS. The texture can be controlled to change the balance between strength and ductility. Finer grains obtained by wrought processing are associated with weaker basal texture and show better ductility and less yield anisotropy.

One of the strengthening methods of great interest is the dispersion of quasicrystalline icosahedral (i-) phase. A thermodynamically stable ternary i-phase of composition Mg_3Zn_6RE occurs in Mg-Zn-RE alloys where RE are rare earth elements Y, Gd, Tb, Dy, Ho or Er. This ternary phase occurs in direct equilibrium with α-Mg, forming a two phase field [2, 3]. Development of magnesium alloys by a dispersion of this phase by wrought processing has been studied extensively [4, 5, 6, 7, 8, 9, 10, 11, 12, 13, 14, 15, 16, 17, 18]. The i-phase stimulates dynamic recrystallization during wrought processing, resulting in finer grain size and weaker basal texture [12, 15]. Thus these alloys exhibit high strength in combination with good ductility, resulting in high fracture toughness [19].

In a significant result, Singh et al. [17] showed that micron-size grains can be obtained in Mg-Zn-Y alloys by direct extrusion of chill cast alloys, exhibiting YS of $\approx$400 MPa with total elongation to failure of over 12%. The alloys contained nano-sized precipitates of the order of about 15 nm, precipitated dynamically from a supersaturated matrix during extrusion. The microstructure (grain size, texture and precipitation) and mechanical properties (tensile and compression strength and ductility) of the alloys have been analyzed in detail by varying the alloy composition and microstructure [20]. The alloys showed a texture similar to those reported for Mg-RE alloys. The textures were weaker for more concentrated alloys and for lower extrusion temperatures. Dispersion of i-phase produces finer grain size by stimulating dynamic recrystallization during wrought pro-

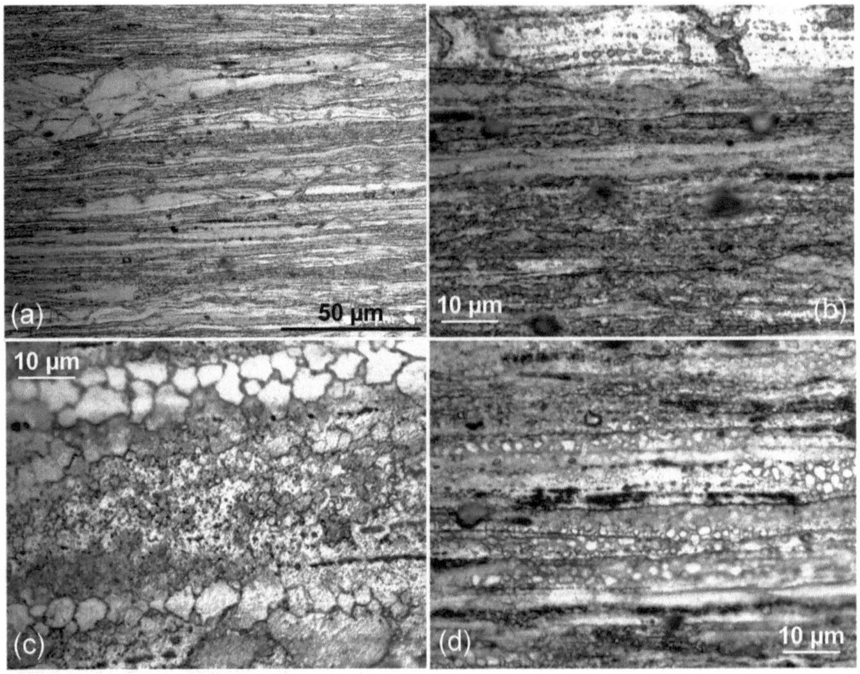

Figure 1: Optical micrographs of as extruded (a,b) ZW31, (c) ZW61 and (d) ZW82 alloys.

cessing, resulting in high strengths. Simultaneously, the texture is modified, resulting in higher dutility.

We report on development of very dilute Mg-Zn-Y alloys, containing a minimum amount of 0.2at% yttrium, which were chill cast and then extruded at temperatures close to 250°C. Preliminary results were reported in a recent publication [21]. Here we present more detailed information and analysis of the final microstructure by presenting new information on recrystallized microstructure and texture, to explain the effect of alloy composition on the deformation behavior of these alloys in tension and compression.

Experimental Procedure

Three alloys of composition Mg-6x Zn-x Y, where x = 0.2, 0.35 and 0.5 in at%, designated as ZW31, ZW61 and ZW82 by their wt% composition, were prepared by melting of commercial purity metals in an electric furnace under a cover gas. The alloys were cast in to a steel chill cast mold of diameter 46 mm. Billets of 40 mm diameter were machined from the casting and extruded at temperatures in the range 235 and 260°C (ZW31 at 250°C, ZW61 at 260°C and ZW82 at 235°C) into 8 mm diamter rods (extrusion ratio 25:1). Microstructural observations were made on sections along the extrusion direction, by optical microscopy, scanning electron microscopy (SEM) and transmission electron microscopy (TEM). SEM observations were performed on JEOL JSM7001F microscope. TEM investigations were performed on JEOL 2000FX-II microscope, after thinning the samples by precision ion milling. Grain sizes were determined by multiplying the mean intercept by a factor of 1.74 to compensate for stereological effects [22]. Grain structure and textures were also determined by electron backscattered diffraction (EBSD) equipped on SEM with orientation imaging microscopy (OIM) software by TexSEM Laboratories of EDAX. Mechanical properties were determined by tension and compression tests on an Instron testing machine at an initial strain rate of 10^{-3}/s on samples machined along the length of the extruded rods. Tensile specimens had a gauge length of 15 mm with a diameter of 3 mm. Compression specimens were 8 mm long with a diameter of 4 mm.

Results

The cast structure of the as cast alloys was dendritic, with i-phase solidifying in the interdendritic spaces [21]. The dendrite size was about 2 mm in case of ZW31 alloy, and about 600 μm in case of

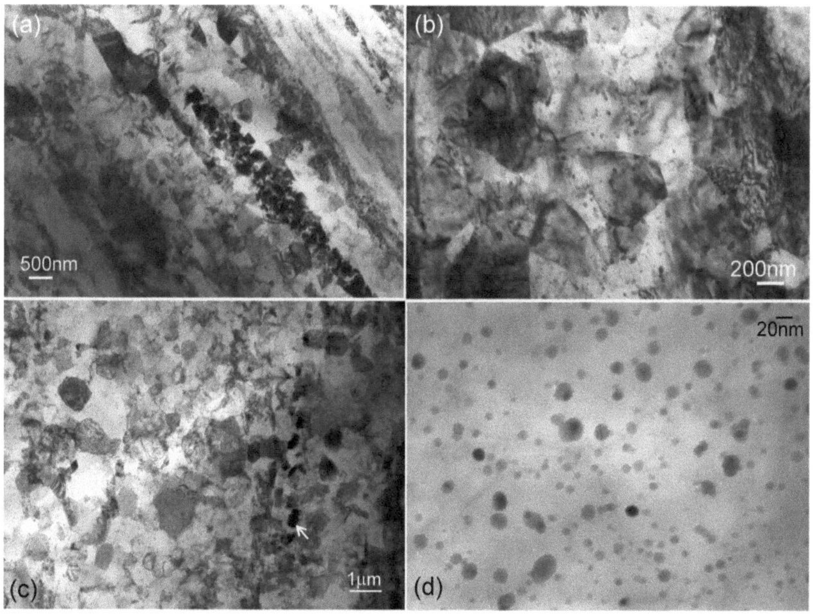

Figure 2: Bright Field TEM micrographs from as extruded (a,b) ZW31 and (c,d) ZW61 alloys. An i-phase particle is pointed out by an arrow in (c).

ZW61 and ZW82 alloys. The thickness of the dendrite arms was estimated to be about 30 μm in all the alloys. The regions near the eutectic i-phase were supersaturated in zinc and yttrium, as determined by EDS.

Upon extrusion, the eutectic phase was distributed along the extrusion direction. Volume percent of the dispersed phase was estimated to be 3.8, 8.1 and 9.1 in the alloys ZW31, ZW61 and ZW82, respectively [21]. Although the average grain size was reported to be very fine, of about a micron in size [21], the grain size is not in fact uniform in all the alloys. There are various degrees of recrystallization depending upon the alloy composition. Fig. 1 shows optical micrographs of the extruded alloys. The extrusion direction is horizontal in all these micrographs. Fig. 1(a) shows a relatively lower magnification micrograph of the extruded ZW31 alloy. Among the bands of fine grain size there are large islands of unrecrystallized regions in which apparent shear bands criss-cross each other. Fig. 1(b) is a larger magnification micrograph from the same sample. Fine grains in the flow along the extrusion direction are not fully resolved. However, there appear to be recrystallized grains of about a micron in size, in the unrecrystallized region near the top. A rough estimate of grain size in the range 1.7 and 1.9 μm was made. In case of ZW61 alloy in Fig. 1(c), banded rows of large grains of average size 7.4±1.4 μm are observed. Outside these, smaller grain size of average 2.4±1.0 μm occur. Grains were much more uniform in case of extruded ZW82 alloy, as shown in Fig. 1(d). An average was measured to be 2.0±0.32 μm.

TEM micrographs are shown in Fig. 2. Fig. 2(a) is a micrograph from extruded ZW31 alloy. Diagonally near center is a dispersion of the ternary eutectic (i) phase. Towards its right appears to be an unrecrystallized region. It is however banded with elongated subgrains. In the center are very fine recrystallized grains. Most do not have clearly defined grain boundaries. A more detailed bright field micrograph in Fig. 2(b) shows a region with cold-work structure in which some grains with sharp grain boundaries are present. Presence of nano-sized precipitates can also be noticed in this micrograph. Fig. 2(c) shows grain structure in extruded ZW61 alloy. Fine grains of about a micron in size with clear grain boundaries are observed. Precipitation in this alloy is shown in the high magnification micrograph of Fig. 2(d). As reported, these precipitates are less than about 25 nm in size, most of them of about 10 nm [21].

The samples were examined by EBSD. Fig. 3 shows inverse pole figure maps and pole figures for the three extruded alloys. In all the samples there are grains elongated along the extrusion direction, as well as discrete grains of about one micron size. The remaining

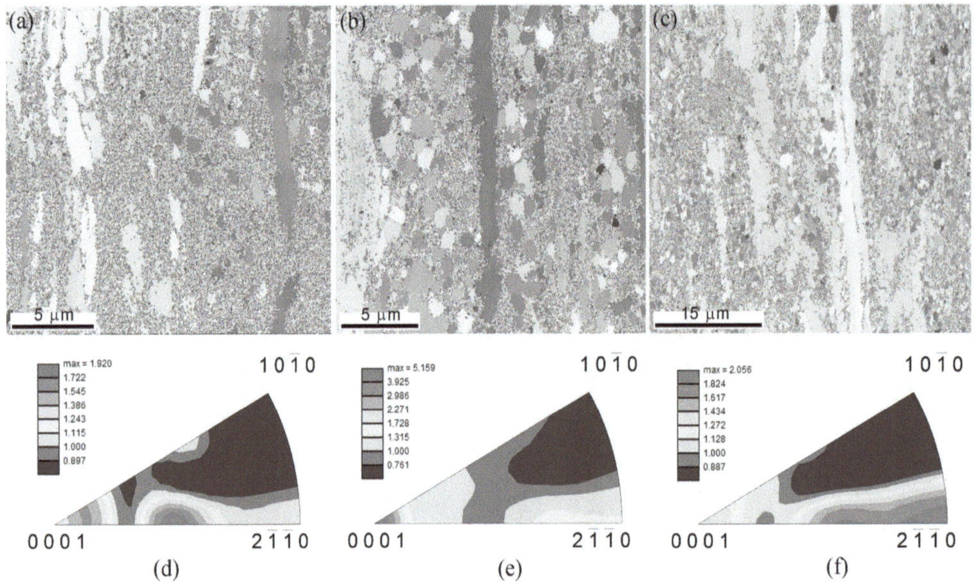

Figure 3: EBSD inverse pole figure maps and pole figures of extruded (a,d) ZW31, (b,e) ZW61 and (c,f) ZW82 alloy samples.

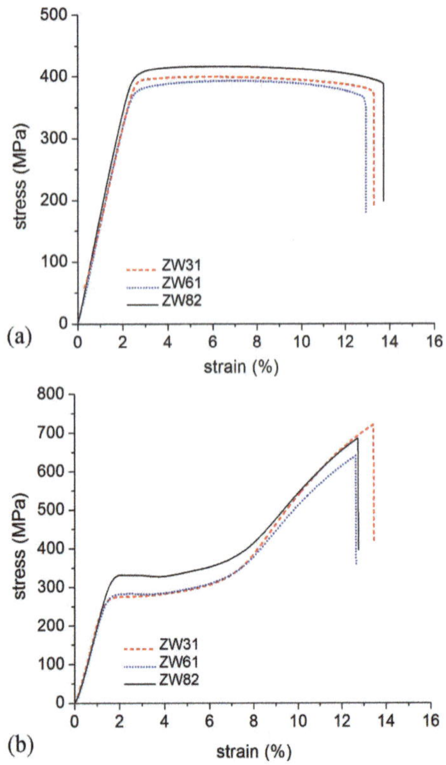

Figure 4: Stress-strain plots of (a) tensile and (b) compression tests of extruded ZW31, ZW61 and ZW82 alloys [21].

regions are believed to have high strain. In the pole figures the ZW61 alloy shows the highest intensity of basal texture. However, since the observations were made in small regions of the sample, this may not be representative of the whole samples.

Fig. 4 shows the tensile and compression test plots. The tensile YS of the three alloys are in a close range of 377 to 405 MPa [21]. The compressive YS shows a larger range of 266 to 327 MPa. The total elongation to failure are 12 to 13%.

Discussion

The micrographs in Fig. 1 show that there are large unrecrystallized areas in extruded ZW31 alloy. These areas are elongated along the extrusion direction, which indicates that these are related to single grains in the cast structure. Grains which are oriented with basal planes parallel to the extrusion direction undergo least deformation and therefore are likely to remain unrecrystallized. Evidence of such orientation of unrecrystallized grains in these alloys was presented earlier [21]. Since this alloy has the least amount of second phase particles in the form of i-phase, it undergoes the least amount of recrystallization. The ZW61 alloy does not have such prominently unrecrystallized regions, but show rows of large sized grains (Fig. 1(c)). These large grains appear to have been produced by recrystallization.

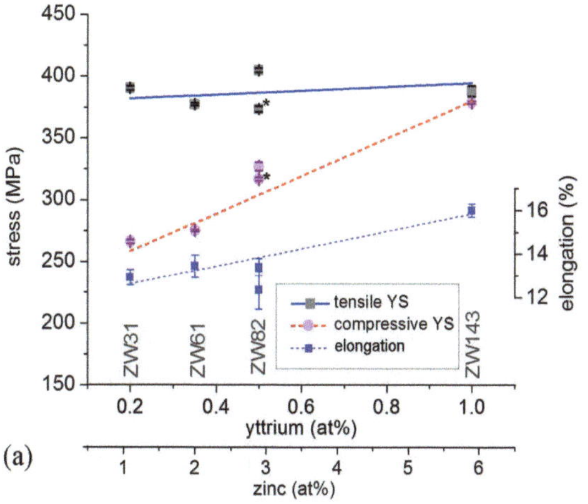

(a)

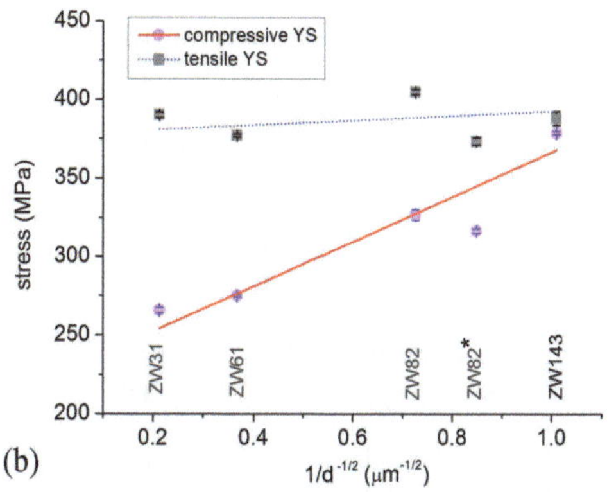

(b)

Figure 5: (a) Tensile and compression YS and elongations plotted against zinc and yttrium concentration of the extruded alloys. (b) Hall-Petch plots of the extruded alloys in tension and compression. Data of the alloy ZW82* is from reference [20] and ZW143 from [17].

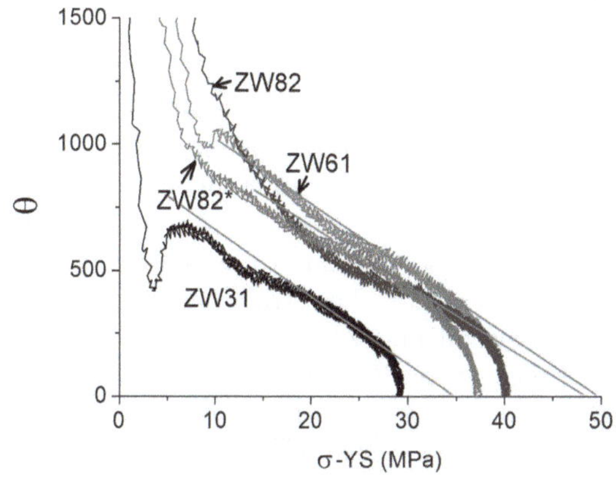

Figure 6: Plots of strain hardening rate θ against σ-YS. Slopes of stage III hardening are shown by straight lines. Plot of the alloy ZW82* is from reference [20].

Therefore they are not likely to have strained structure, unlike the unrecrystallized grains which have low angle boundaries and subgrain structure. This may be a cause that the tensile and compression yield strengths of ZW61 alloy are lower than those of the ZW31 alloy (Fig. 4). Despite this, the yield assymmetry ratio (YAS ratio, ratio of compression YS to tensile YS) of the alloys shows a progressive reduction with decreasing alloying content, being 0.68, 0.73 and 0.81 for alloys ZW31, ZW61 and ZW82, respectively [21]. YAS ratio of near 1.0 can be achieved in ZW143 alloy [17, 20]. Accordingly, YS in tension and compression and elongations are plotted against alloying concentration in Fig. 5(a). Fairly linear plots are obtained. By extrapolation, for diminished alloying contents, the tensile YS of 379 MPa and compression YS of 234 MPa are obtained. Elongations are between 12 and 16%.

Grain size of the extruded alloys were reported to be about one micron, corresponding to the recrystallized grain sizes. Accordingly, the tensile YS are in a close range within 28 MPa. However, the compression YS varies with the composition. The subgrains and low angle boundaries within the unrecrystallized grains are effective against dislocation slip, but not against twinning. Therefore unrecrystallized regions do not affect the tensile strength as much as the compressive strength. A Hall-Petch plot is plotted in Fig. 5(b) in which the largest grain size were taken from the optical micrographs in Fig. 1. A reasonably linear plot for compression strength is obtained. The slope k of this plot is 142.9±1.7 MPaμm$^{1/2}$ and intercept σ_0 223.7 MPa. In case of tensile strength, since all values are in a narrow range, they are expected to show linearity. The effective grain size in this case will include subgrains.

To understand the overall effect of microstructural elements on the deformation behavior, the work hardening rate θ, $d\sigma/d\epsilon$, where σ and ϵ are true stress and true strain, respectively, were calculated. Fig. 6(a) shows plots of θ against σ-YS, the net flow stress (the dislocation contribution), also known as Kocks-Mecking plots [23]. An elasto-plastic region, which is a sharp drop in the curve, leads directly to stage III hardening, which is characterized by decreasing work hardening rate and multiple slip. In case of ZW82 alloy the transition from the elasto-plastic region to stage III is smooth and extended. In case of

ZW61 and ZW31 this transition is sharp, with a rapid rise in the hardening rate at the beginning. This is indicative of a strong basal texture [24]. From the intensity of this sharp transition, it can be said that the texture in ZW31 is much stronger than in ZW61. The ZW61 alloy has the most linear or stable plot in stage III.

The deformation behavior appears to be sensitive to the extrusion temperature. The strain hardening rate plots in Fig. 6 also include ZW82 alloy from reference [20], extruded at 260°C, referred to here as ZW82*. Its strain hardening rate curve shows a continuous transition to stage III hardening, similar to ZW82, but the transition is not as extended. The plot of ZW81* is similar to that of ZW61, which is also extruded at 260°C, with a major difference that the elasto-plastic transition to stage III hardening of ZW61 is sharp, which suggests a very different basal texture compared to ZW82*.

Conclusions

Mg-Zn-Y alloys chill cast and extruded in the temperature range 235°C to 260°C showed very high tensile YS in the range 377 to 405 MPa with elongations of over 12%. As the concetration of alloying elements is reduced, the recrystallization effect is reduced and the basal texture intensity increases, resulting in lower yield asymmetry ratio. With alloy composition of very low amount of alloying elements, compression YS of over 234 MPa is obtained.

References

[1] R. Armstrong, I. Codd, M. Douthwaite and N.J. Petch, Philos. Mag. 1962, 7, 45.

[2] A.P. Tsai, Y. Murakami and A. Niikura, Phil. Mag. A 80 (2000) 1043.

[3] N. Lebrun, A Stamou, C. Baetzner and J. Robinson: "Magnesium-Yttrium-Zinc" in Ternary Phase Diagrams (eds. G. Petzow and E. Effenberg), VCH, 1988, pp. 702-710.

[4] D.H. Bae, S.H. Kim, D.H. Kim, W.T. Kim, Acta Mater. 50 (2002) 2343.

[5] D.H. Bae, M.H. Lee, K.T. Kim, W.T. Kim, D.H. Kim, J. Alloy Compounds 342 (2002) 445.

[6] A. Singh, M. Nakamura, M. Watanabe, A. Kato, A.P. Tsai, Scripta Mater. 49 (2003) 417.

[7] H. Taniuchi, H. Watanabe, H. Okumura, S. Kamado, Y. Kojima, Y. Kawamura, Mater. Sci. Forum 419-422 (2003) 255.

[8] A. Singh, M. Watanabe, A. Kato, A.P. Tsai, Mater. Sci. Engg. A 385 (2004) 382.

[9] Y. Zhang, X. Zeng, L. Liu, C. Lu, H. Zhou, Q. Li and Y. Zhu, Mater. Sci. Engg. A 373 (2004) 320.

[10] M.Y. Zheng, X.G. Qiao, S.W. Xu, K. Wu, S. Kamado and Y. Kojima, J. Mater. Sci. 40 (2005) 2587.

[11] D.K. Xu, L. Liu, Y.B. Xu and E.H. Han, Mater. Sci. Engg. A 443 (2007) 248.

[12] A. Singh, H. Somekawa and T. Mukai, Scripta Mater. 56 (2007) 935.

[13] A. Muller, G. Garces, P. Perez and P. Adeva, J. Alloys Comp. 443 (2007) L1.

[14] E. Mora, G. Garces, E. Onorbe, P. Perez and P. Adeva, Scripta Mater. 60 (2009) 776.

[15] J.Y. Lee, H.K. Lim, D.H. Kim, W.T. Kim and D.H. Kim, Mater. Sci. Engg. A 491 (2008) 349.

[16] H. Somekawa, A. Singh, Y. Osawa and T. Mukai, Mater. Trans. 49 (2008) 1947.

[17] A. Singh, Y. Osawa, H. Somekawa and T. Mukai, Scripta Mater. 64 (2011) 661.

[18] A. Singh, Y. Osawa, H. Somekawa and T. Mukai, in "Magnesium Technology 2011" (eds. W.H. Sillekens, S.R. Agnew, N.R. Neelameggham and S.N. Mathaudhu), TMS, Warrendale (PA), 2011, p. 239.

[19] H. Somekawa, A. Singh and T. Mukai, Scripta Mater. 56 (2007) 1091.

[20] A. Singh, Y. Osawa, H. Somekawa, T. Mukai, Mater. Sci. Eng. A 611 (2014) 242.

[21] A. Singh, Y. Osawa, H. Somekawa, T. Mukai, C.J. Parrish and D.S. Shih, Metall. Mater. Trans. A 45A (2014) 3232.

[22] A. W. Thompson, Metallography 5 (1972) 366.

[23] U.F. Kocks and H. Mecking, Prog. Mater. Sci. 48 (2003) 171.

[24] J.A. del Valle, F. Carreno, O.A. Ruano, Acta Mater. 54 (2006) 4247.

Magnesium Technology 2015
Edited by: Michele V. Manuel, Alok Singh, Martyn Alderman, and Neale R. Neelameggham
TMS (The Minerals, Metals & Materials Society), 2015

An Extruded and Peak Aged Mg-5Gd-3Y-1Zn-Zr Alloy with High Strength

D. Wu [a], J.L. Li [a], M. Hong [b], W.H. Wang [c], R.S. Chen [a,*], E.H. Han [a], W. Ke [a]
[a] The Group of Magnesium Alloys and Their Applications, Institute of Metal Research, Chinese Academy of Sciences, 62 Wencui Road, Shenyang 110016, China
[b] Shenyang Aerospace University, Shenyang 110036, China
[c] Shenyang Ligong University, Shenyang 110159, China.※

Keywords: Mg-Gd-Y-Zn alloy, Extrusion, Aging, Mechanical properties

Abstract

This article presents an extruded Mg-5Gd-3Y-1Zn-Zr alloy produced by conventional ingot metallurgy, exhibiting high-strength and excellent ductility at room and elevated temperatures, with a low density of 1.84g/cm^3. The 0.2% yield strength (YS), ultimate tensile strength (UTS) and elongation-to-failure (EL) at room temperature, of samples in T5 condition, can reach 324MPa, 382MPa and 9.3%, respectively. At 250℃, the UTS can be still higher than 317MPa, with EL of 21%. The excellent mechanical properties were attributed to the high density of the static aging prismatic precipitates in the matrix, dynamic β phase precipitated on the dynamic recrystallization grain boundaries, and bimodal grain size distribution with fine recrystallized grains and large deformed grains with intense basal texture and LPSO phase inside the grain.

Introduction

Now for saving energy as well as for reducing weight, urgent development of new lightweight structural materials for aerospace, aircraft and automobile industries are strongly desired [1]. As the lightest metal Magnesium alloys, especially Mg-RE(rare earth) based alloys which have high strength at both room temperature and elevated temperature, have been extensively studied by researchers over the world [2, 3]. It was reported that Mg-Gd-Y system alloys, such as Mg-15Gd-2Y-0.5Zr/Mg-10Gd-3Y-0.4Zr alloys, showed excellent mechanical properties at room temperature and good creep resistance at elevated temperature compared with conventional Mg alloys [4-8], including WE54, which had the best high temperature properties among the existing commercial magnesium alloys. And the high strength of Mg-Gd-Y-Zr based alloys at high temperature is mainly attributed to the additions of gadolinium and yttrium elements. It has been known that Mg-Gd-Y-Zr alloys exhibits a three-stage precipitation sequence during ageing at 200 ℃ , i.e. α'-Mg(S.S.S.S.)→β'' ($D0_{19}$)→β'(cbco)→β(fcc) [8,9].

Subsequently, T Honma [10]et al. found the novel long-period stacking ordered (LPSO) structures formed in both hot extruded or ingot metallurgy Mg-Gd-Y-Zn alloys. By studying the structural transformation of Mg-10Gd-3Y-xZn-0.4Zr alloys in the process of heat treatment, Li [11] and Liu [12] et al. revealed that the addition of zinc can significantly enhance alloy's precipitation strengthening. Moreover, Shao [13]et al. found that LPSO phase was prone to kinking during deformation, and the fined LPSO caused by deformation significantly increased the high temperature strength of alloy. For instance [14,15], the cast ingots of Mg-8.2Gd-3.8Y-1.0Zn-0.3Zr (wt.%) alloy with hot extrusion and aging process had excellent properties at room temperature, whose tensile strength is up to 470MPa, and elongation is about 9%. However, high gadolinium content not only corresponds to high cost but also means high density, which are generally not expected.

Recently, Homma researched the mechanical properties and microstructure of hot extruded and aged Mg-1.8Gd-1.8Y-0.7Zn-0.2Zr (at.%), and found the tensile strength, yield strength and elongation of it was 542MPa, 473MPa and 8%, respectively [11]. The good strength and high toughness properties of this ultrahigh strength magnesium alloy bars are mainly due to fine dynamic precipitation and aging precipitation. In order to obtain an alloy with low density, high strength and high toughness at both room temperature and elevated temperature, this paper studied an extruded Mg-5Gd-3Y-1Zn-Zr alloy produced by conventional ingot metallurgy, which had a low density of 1.84g/cm^3.The relationship between the microstructures and corresponding mechanical properties was investigated.

Experimental procedures

The material Mg-5Gd-3Y-1Zn-0.5Zr (wt.%) alloy (GWZ531) with a diameter of 50mm prepared for hot extrusion was produced by conventional ingot metallurgy. The cylinder-shaped specimens were homogenized (as-initial) and then hot extruded at 400℃ with an extrusion ratio of 11:1 at the speed of 5 mm/min. Subsequently the as-extruded specimens were aged at 200℃ for various periods of time, from 0.01h to 120h. The microstructures of the specimens were observed using optical microscope (OM), and transmission electron microscope (TEM). For optical observation the specimens were etched with acetic picral (2g of picric acid, 5ml of acetic acid, 5ml of water and 25ml of ethanol) for 15-30s. Foils for TEM observation were polished to a thickness about 0.1mm, and then thinned to perforation by ion milling. Electron back-scatter diffraction (EBSD) observation was employed for the metallographic analyses by scanning electron microscopy on a JEOL JSM-5600 machine operating at 25KV. Orientation imaging microscopy was measured at a step size of 0.6μm. The macro texture was measured with a Bruckre D8 DISCOVVER diffract meter along extruding direction (ED) direction. Vickers hardness testing was performed using 500g load and a holding time of 10s. No fewer

※**Corresponding author**: Rongshi Chen; Tel: +86-24-23926646; Fax: +86-24-23894149;
E-mail: rschen@imr.ac.cn

than 8 measurements were taken in each specimen. Tensile tests were carried out at a strain rate of $1\times10^{-3}s^{-1}$ at room temperature and 250℃ with gauge length of 10mm along ED(extruded direction).

Results and discussion

Age hardening: Figure 1 shows the hardness curve of the extruded GWZ531 alloy during isothermal ageing at 200℃. As the aging time increases from 0.01h to 50h, there appears to be little hardness increase, while at 18h and 50h, the hardness drops to some extent. This is because of the incubation period of aging precipitates, and the recovery effect may result in the softening of matrix. Further aging leads to the hardness beginning to increase rapidly, and the peak hardness (HV99) was obtained at about 60h. The alloy then shows a hardness plateau from 60h to 96h with the Vickers hardness still higher than HV98, followed by a slight decline up to 120h.

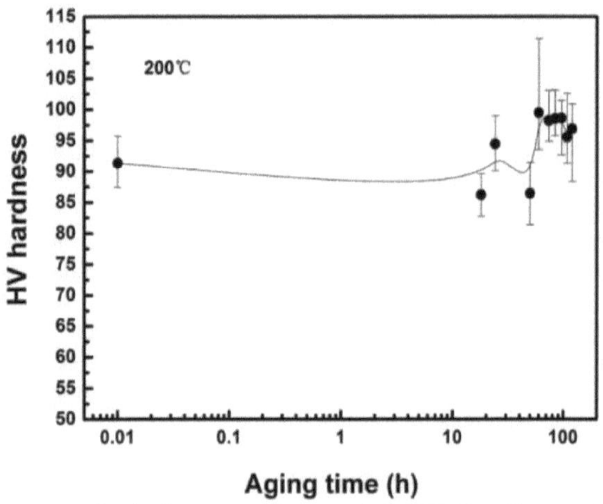

Fig.1 Age-hardening curves for extruded GWZ531 alloy aged at 200℃

Microstructures The microstructures of GWZ531 alloy in solutionized and extruded condition are shown in Fig.2. A lot of lamellar LPSO phases can be observed in the initial equiaxed grain with the grain size of about 88μm (Fig.2a). Fig.2b shows the optical microstructure of the as-extruded alloy observed on planes parallel to ED, where the ED is parallel to the scale bar. As it can be seen, it exhibits bimodal microstructure of large deformed grains and fine dynamic recrystallized (DRXed) grains. In unDRXed area the grains are elongated visibly along the extrusion, with 20~50μm in thickness and several hundreds of micrometers in length. Near the deformed grain boundaries, there are a large amount of fine dynamic recrystallized grains, of which the average size is about 4μm calculated with a line-intercept method. Moreover, the LPSO phases in the large deformed grain are obviously bent or event folded, as indicated by red arrows in Fig.2b, which will release the strain concentration near these phases and coordinate deformation [16]. Fine DRXed grains distributed among these LPSO phase and the vicinity of the original grain boundaries, where the strain concentration provided the energy for the activation of the DRX.

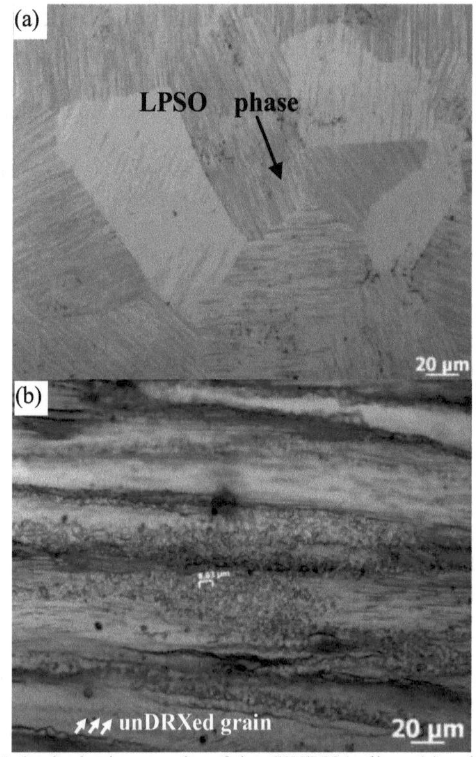

Fig .2 Optical micrographs of the GWZ531 alloy: (a)as-initial (b)as-extruded.

The TEM images of the as-aged alloy are shown in Fig.3. Fig3a shows the TEM bright-field(BF) image of large deformed grains area with LPSO phase and corresponding selected area electron diffraction (SAED) pattern, Fig. 3b shows the BF image of a DRXed region and corresponding selected area electron diffraction (SAED) pattern along $[-111]_{\beta}$ direction. It is worth noting that there are considerably irregular phases precipitated in the DRXed grain boundaries and also inside the grains, of which the size is about half of the re-crystallized grains. Wherein, the prismatic precipitated β' phase ($Mg_{15}(Gd+Y)_3$) in the matrix with a micro-morphology of disc or convex lens shape is still thermally stable until 250℃[17]. Moreover, the phases on the boundaries are β phase caused by dynamic recrystallization are shown in Fig3c. Fig3d shows there are also have lamellar LPSO phases in fine grain, which may be one reason benefit for DRXed. Fig.4 shows the EBSD results of the samples of as-extruded GWZ531 alloy observed on the planes parallel to ED measured. Large grains with (0002) basal planes approximately parallel to the extruding surface and fine DRXed grains with random orientation can be found. Furthermore, conspicuous color variation caused by the sub-grain and high density of dislocations can be observed in the deformed grains shown. Because the LPSO phase precipitated on (0002) basal planes of the α-Mg lattice [18], the kink deformation in the LPSO phases occur due to the stress concentration also led to the variation of the grain orientation.

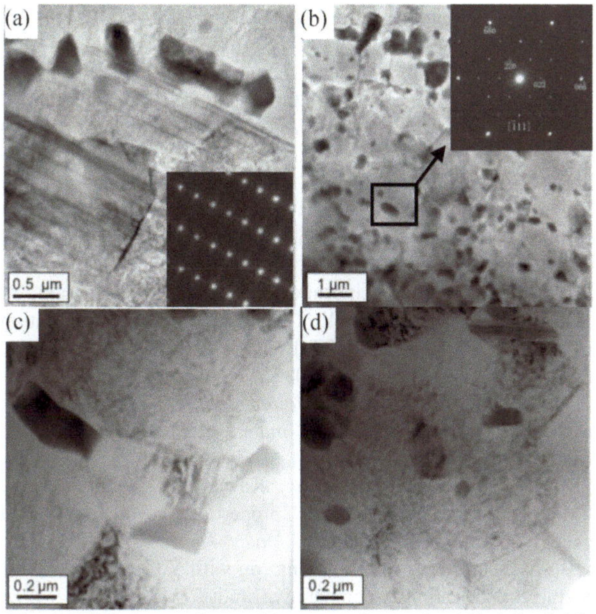

Fig. 3 TEM observations of the as-aged samples:(a)bright-field (BF)image of a large deformed grains area and corresponding SAED pattern with incident electron beam// $[11\text{-}20]_{\alpha\text{-}Mg}$(b)bright-field image of a DRXed area and corresponding SAED pattern with incident electron beam//$[-111]_{\beta}$ (c)BF image of precipitated β' phase (d)BF image of precipitated β phase

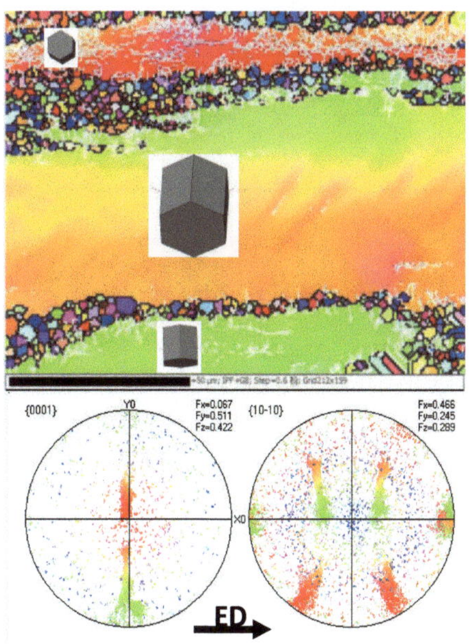

Fig. 4 EBSD orientation maps of as-extruded alloy

Texture Fig. 4 focuses on the micro-texture of the samples of the as-extruded GWZ531 alloy. It can be seen that the as-extruded samples exhibit a basal texture with the (0002) basal planes parallel to the ED. In addition, the macro-texture of the extruded alloy is also tested by XRD, as shown in Fig.5, and which is similar to the results obtained by EBSD that the extruded samples have a basal texture along the ED. Furthermore, from the XRD result, it can be seen that the basal pole maximum intensity broadened perpendicular to the ED, which means the texture is weakened in some degree. It was supposed that the RE elements addition to the Mg alloys weakens the deformation basal texture [19].

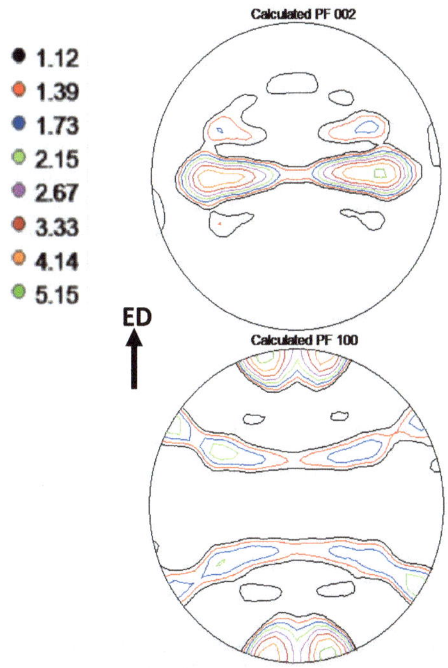

Fig.5 Macro-texture of the as-extruded alloy

Mechanical Properties Fig. 6 shows the engineering stress-strain curves of the as-extruded and peak aged alloy at room temperature (RT) and at 250℃ along the ED. The extruded GWZ531 alloy exhibits high-strength and excellent ductility at room and elevated temperatures. The 0.2% yield strength (YS), ultimate tensile strength (UTS) and elongation-to-failure (EL) at room temperature, of the peak aged samples can reach 324MPa, 382MPa and 9.3%, respectively, little higher than the as-extruded one. Moreover, the UTS of the peak aged samples can be still higher than 310MPa, with EL of 21% at 250℃. As shown in Fig. 3, the precipitated prismatic β' phase and dynamic β phase, which are thermally stably, play an important rule for strengthening at room temperate and high temperature. In addition, the good performance of this alloy is also attributed to the bimodal grain size distribution with fine recrystallized grains and large deformed grains with intense basal texture and LPSO phase inside the grain.

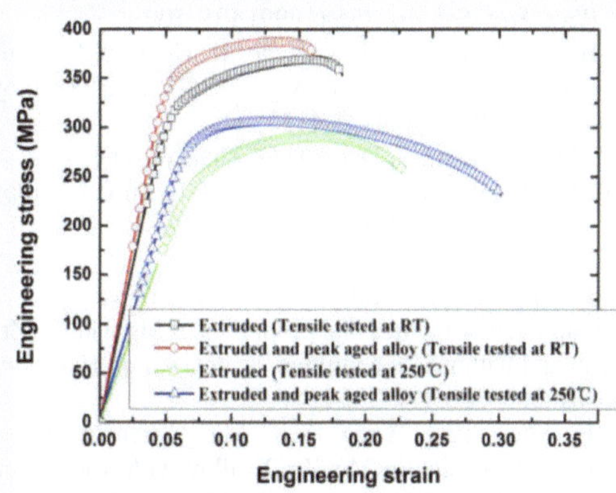

Fig.6 Engineering stress-strain curves of the as-extruded and peak aged alloy at room temperature and at 250°Calong the ED

Conclusion

The extruded GWZ531 alloy exhibits a bimodal microstructure with fine dynamic recrystallized grains and large deformed grains. After peak ageing, it shows a UTS of 382MPa and 317MPa at room temperature and 250℃, respectively, with a satisfactory ductility. The high strength can be attributed to the high density of the static aging prismatic precipitates in the matrix, dynamic β phase precipitated on the dynamic recrystallization grain boundaries, fine recrystallized grains in sub-micrometer and large deformed grains with intense basal texture and LPSO phase inside the grain.

Acknowledgements

This work was funded by the National Basic Research Program of China (973 Program)through project No. 2013CB632202, and National Natural Science Foundation of China (NSFC) through projects No. 51105350 and No. 51301173, respectively.

References

[1]. K. Yamada, et al., "Alloy development of high toughness Mg-Gd-Y-Zn-Zr alloys," *Materials transactions,* 2006, 47(4): 1066.

[2]. T. Honma, T. Ohkubo, K. Hono, S. Kamado, "Chemistry of nanoscale precipitates in Mg-2.1Gd-0.6Y-0.2Zr (at.%) alloy investigated by the atom probe technique ," *Mater. Sci. Eng. A,* 2005,395(1-2): 301-306.

[3]. S.M. He, etal., "Microstructure and strengthening mechanism of high strength Mg-10Gd-2Y-0.5Zr alloy," *J Alloys Compd,* 2007, 427(1-2): 316-323.

[4]. B.L. Mordike, "Creep-resistant magnesium alloys," *Mater. Sci. Eng.A,* 324 (2002) 103–112.

[5]. I.A. Anyanwu, S. Kamado, Y. Kojima, "Aging characteristics and high temperature tensile properties of Mg-Gd-Y-Zr alloy," *Mater. Trans,* 42 (2001) 1206–1211.

[6]. I.A. Anyanwu, S. Kamado, Y. Kojima, "Creep properties of Mg-Gd-Y-Zr alloy," *Mater. Trans,* 42 (2001) 1212–1218.

[7]. P. Vostry, B. Smola, I. Stulikova, F.V. Buch, B.L. Mordike, " Microstructure evolution in isochronally heat treated Mg- Gd alloys, " Phys. Stat. Sol.(a), 175 (1999) 491–500.

[8].B. Smola, I. Stulikova, F.V. Buch, B.L. Mordike, " Structural aspects of high performance Mg alloys design," Mater. Sci. Eng. A, 324(2002) 113–117.

[9] P. Vostry, B. Smola, I. Stulikova, F.V. Buch, B.L. Mordike, " Microstructure evolution in isochronally heat treated Mg- Gd alloys," *Phys. Stat. Sol. (a),* 175 (1999) 491–500.

[10]. T. Honma, T. Ohkub, S. Kamado, K. Hono, "Effect of Zn additions on the age hardening of Mg-2.0Gd-1.2Y-0.2Zr alloys," *Acta Materialia,* 2007(55):4137-4150

[11]. D.J. Li, X.Q. Zeng, J. Dong, C.Q. Zhai, W.J. Ding, " Microstructure evolution of Mg-10Gd-3Y-1.2Zn-0.4Zr alloy during heat-treatment at 773K," *Journal of Alloys and Compounds,* 2009, 468: 164-169

[12]. X.B. Liu , R.S. Chen, E.H. Han, " Effects of ageing treatment on microstructures and properties of Mg–Gd–Y–Zr alloys with and without Zn additions," *Journal of Alloys and Compounds,* 2008, 465(1): 232-238.

[13]. X.H. Shao, Z.Q. Yang, X.L. Ma, "Strengthening and toughening mechanisms in Mg–Zn–Y alloy with a long period stacking ordered structure," *Acta Materialia,* 2010, 58(14): 4760-4771.

[14]. Y.Q. Chi, M.Y. Zheng, C. Xu, et al., " Effect of ageing treatment on the microstructure, texture and mechanical properties of extruded Mg-8.2 Gd-3.8Y-1Zn-0.4 Zr (wt.%) alloy," *Materials Science and Engineering: A,* 2012.

[15]. C. Xu, M.Y. Zheng, S.W. Xu, et al, "Ultra high-strength Mg–Gd–Y–Zn–Zr alloy sheets processed by large-strain hot rolling and ageing," *Materials Scienceand Engineering: A,* 2012, 547: 93-98.

[16]. K. Hagihara, A. Kinoshita, Y. Sugino, M. Yamasaki, Y. Kawamura, H.Y. Yasuda. Y. Umakoshi, "Plastic deformation behavior of Mg97Zn1Y2 extruded alloys," *Intermetallics* ,18(2010)1079-1085.

[17]. J.F. Nie, *"Effects of precipitate shape and orientation on dispersion strengthening in magnesium alloys,"* Scripta Materialia, 2003, 48, 1009-1015.

[18]. T. Itoi, T. Suzuki, Y. Kawamura, M. Hirohashi, "Microstructure and mechanical properties of Mg-Zn-Y rolled sheet with a Mg12ZnY phase," *Mater. Trans,* 51(2010).

[19]. N. Stanford, "Micro-alloying Mg with Y,Ce,Gd and La for texture modification-A comparative study," *Mater. Sci. Eng. A* , 527 (2010) 2669–2677.

Magnesium Technology 2015
Edited by: Michele V. Manuel, Alok Singh, Martyn Alderman, and Neale R. Neelameggham
TMS (The Minerals, Metals & Materials Society), 2015

STRENGTHENING IN THERMOMECHANICALLY PROCESSED MAGNESIUM ALLOYS

B. Mansoor[1], R.F. Decker[2], S.E. LeBeau[2]

[1]Mechanical Engineering Program. Texas A&M University at Qatar, Doha, Qatar
[2]Thixomat Inc., nanoMag LLC., 13753 Otterson Court, Livonia, Michigan 48150, USA

Keywords: Thixomolding®, Mg, Thermomechanical Processing, AZ61, AZ70

Abstract

Commercial Mg alloys, compared to other engineering materials such as steels or aluminum materials have inferior strengths (Y.S. = ~120 MPa), limited ductility and poor formability. Furthermore, due to high costs their use in structural applications for transportation industry is still rather limited. Therefore, there is significant interest in developing microstructure modification routes to produce novel Mg base alloys with an attractive combination of strength and ductility at room temperature as-well as warm temperature formability. In order to promote use of such microstructurally engineered Mg materials, better understanding of the relationship between microstructure, texture etc. with mechanical properties must be developed for a range of different alloys. In this work, microstructure evolution and mechanical response of two thermomechanically processed Mg alloys AZ61L and AZ70-TH were investigated. Initial findings of this work are presented here. The processed materials exhibited a good combination of strength and tensile ductility at room temperature that was further enhanced (Y.S. > 250 MPa, El. % > 10%) by low temperature (180°C) annealing treatment for 1 hr. The ductility and in-plane anisotropy in mechanical property was found to be related to basal texture formation in the sheet plane. In addition to the Hall-Petch strengthening due to near ultrafine grain size, β-particles from as-molded microstructure, complement strengthening by sub-dividing and possibly solutionizing/re-precipitating into nano-sized, well-dispersed, obstacles to dislocation motion and grain growth.

Introduction

The compelling need for energy-efficient, sustainable transportation systems has motivated significant R&D efforts to promote wider spread use of ultra-light weight magnesium and Mg-based alloys [1]. Magnesium (Mg) and Mg-based alloys have drawn significant interest as attractive lightweight structural materials with high specific mechanical properties. However, Mg wrought alloys due to their moderate strength, limited ductility and high costs find limited commercial application [2]. Grain refinement down to nano/ultrafine grain size, dispersion of secondary phases and control of precipitates by thermo-mechanical processing can significantly improve mechanical properties of metallic alloys at ambient and elevated temperatures [3-6]. Severe plastic deformation is an effective method for grain refinement of bulk metallic materials. Some noteworthy severe deformation processes include: Equal Channel Angular Extrusion ECAE [4], High Pressure Torsion HPT [3], Accumulated Roll Bonding [5] and Friction Stir Processing [6-9]. Ultra-fine grained microstructure with < 1 μm grain size can be produced by such methods. Thixomolding®, is a semisolid, injection molding process that in addition to near-net shape parts produces low porosity Mg plates [10,11]. The details of the thixomolding process have been reported extensively in the past [13-14]. In addition to near-net shape parts, thixomolding routinely produces low porosity Mg plates. Microstructure of thixomolded alloys typically consists of 3-15 vol.% α_p-Mg globules (~ 30-100 μm) in a matrix of homogenous and fine, secondary α_s-Mg grains (4-5 μm). Furthermore, the α_s-Mg grains are surrounded by a eutectic aggregate (1-2 μm) that consists of α_e-Mg and an intermetallic compound β ($Mg_{17}Al_{12}$) [15]. Commercially available, wrought Mg alloys in hot-rolled or extruded condition are characterized by coarse and highly bimodal microstructures [15]. In comparison, thixomolded alloys with fine and relatively homogeneous microstructure are therefore, suitable candidates for secondary thermomechanical processing to enhance mechanical properties by further grain refinement and homogenization. Thixomat® Inc., has developed a novel method to Thixomold plates then Thermomechanically Process (TTMP) them to produce Mg sheets with near ultrafine grain size, high specific stiffness and strength [9]. This study is aimed to reveal microstructure and texture evolution of TTMP-processed magnesium-zinc alloys, AZ61L and AZ70-TH. In addition, tensile properties of processed material before and after annealing are studied. The eventual goal is to develop high strength Mg sheets capable of being formed into aero/automotive structural parts for light-weighting to improve fuel efficiency and reduce CO_2 emissions.

Experimental

Materials

The materials used in this research were semi-solid molded AZ61L (nominal composition, wt. %: 6.5 Al, 0.46 Zn, 0.14 Mn, 0.01 Si, 0.03 Fe and Mg in bal.) and AZ70-TH (nominal composition, wt. %: 7.4 Al, 0.47 Zn, 0.2 Mn, 0.003 Si, 0.003 Fe and Mg in bal.). They were provided by Thixomat® Inc., in the form of 3.25 mm thick as-molded plates and 1.5 mm thick thermo-mechanically processed sheets obtained from the as-molded material. Details of the proprietary thermomechanical processing procedure used by Thixomat Inc., to obtain the 1.5 mm sheets from the as-molded plates has been reported elsewhere [10,12] and will not be discussed in this paper.

Microstructure

Microstructure examination was carried out on as-molded, FSP processed and aged Mg alloys. The as-molded samples were sectioned in a plane containing the normal direction (ND) and molding/rolling direction (RD). FSP samples were sectioned in a plane containing plate normal direction (ND) and transverse direction (TD). The metallographic samples were mechanically polished by the standard methods and etched by acetic-picral solution (4.2 g picric acid, 70 ml ethanol, 10 ml acetic acid and 10 ml distilled water). A Quanta FEG 250 SEM operating at 20-30 KeV was employed to observe the very fine microstructures. Grain sizes were measured from the micrographs by an area-based approximation. In order to determine the optimum conditions for aging, low temperature annealing treatments were performed at 150°C, 180°C, 220°C, 250°C, and 300°C for 0.5, 1 and 1.5 hrs.

Mechanical Properties and Texture

Room-temperature tensile tests were performed on samples with 12.5 mm long × 3 mm wide gauge sections in a computer-controlled, 5982 Instron machine at a constant crosshead speed of 0.5 mm/min. Vickers hardness was measured at an indentation load of 300g for 30 seconds loading time. Crystallographic texture measurements were made for {0002} basal poles on the plate/sheet rolling plane in a Rigaku D/MAX-B rotating anode X-ray diffractometer using CuKα radiation at 40 kV and 100 mA.

Results and Discussions

Microstructure of Thixomolded Mg

In Figure 1 and 2, microstructure of as-Molded AZ61L and AZ70-TH Mg show un-melted, equiaxed, primary α_p-Mg (15-18 % vol., for both alloys) grains dispersed in a continuous matrix of solidified, equiaxed, fine, secondary, pro-eutectic α_s-Mg grains, delineated by a network of eutectic component i.e. an intermetallic compound β ($Mg_{17}Al_{12}$).

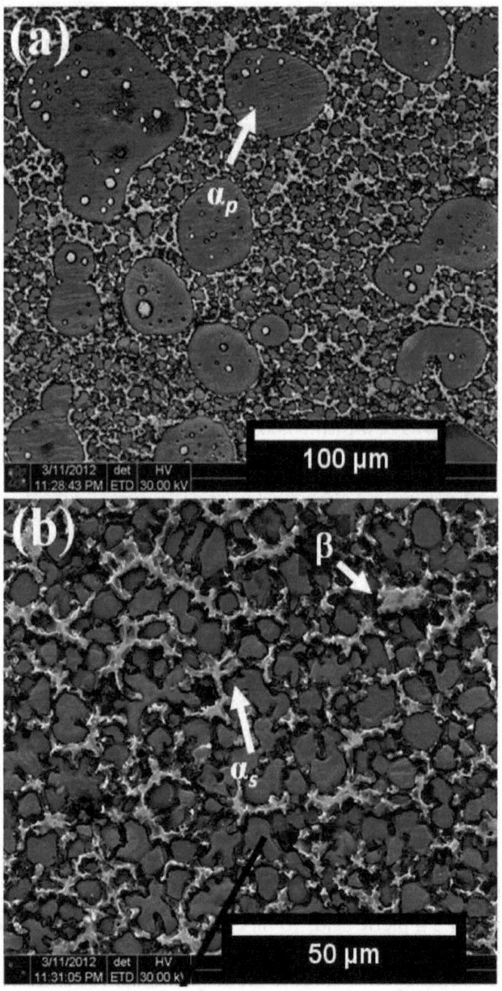

Figure. 1 The microstructure of as-molded AZ61L Mg alloy.

The avg. grain size of α_s-Mg grains is ~ 5 µm while the eutectic intermetallic film has a thickness of ~1 µm. No evidence of coupled eutectic growth was found in this work [10, 11].

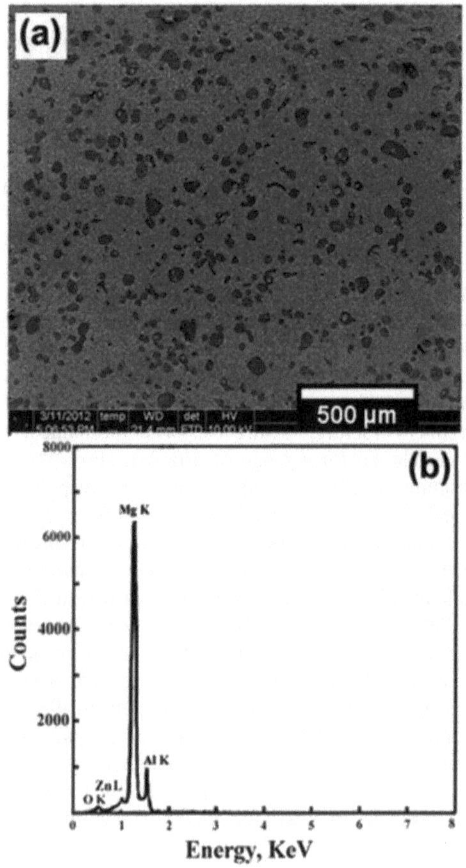

Figure. 2 (a) The microstructure of as-molded AZ70TH Mg alloy. (b) EDS spot analysis of α_s-Mg grain.

Microstructural of Thermomechanically Processed alloys

In Figure 3, the as-TTMP microstructure for (a-c) AZ61 and (d-e) AZ70 Mg shows significant grain refinement down to near ultrafine grain size ~ 1-2 µm. The eutectic component is dispersed in the form of a string of broken down particles while no shear-banding or profuse twinning is observed. The high magnification image in (c) shows broken down, dispersed eutectic β-particles (~50-100 nm, indicated by arrows) of different morphologies. This was confirmed by EDS point analysis as shown in (f). In Figure 3 (d-e), some relatively coarse and elongated grains, reminiscent of α_p-Mg grains from as-molded condition, delineated by fine grains (indicated by arrows) are observed while no eutectic film or broken down β-particles are seen in this region. The presence of fine grains decorating the grain boundaries indicates dynamic recrystallization as mechanism of grain sub-division process [15, 16]. In annealing experiments, (micrographs not shown here) no appreciable change in microstructure was noticed after 180°C exposure for 1 hr. However, some recrystallization was seen at >220 °C while at ≥ 250 °C, the material initially retained fine grain microstructure but grain coarsening was visible after 30 mins.

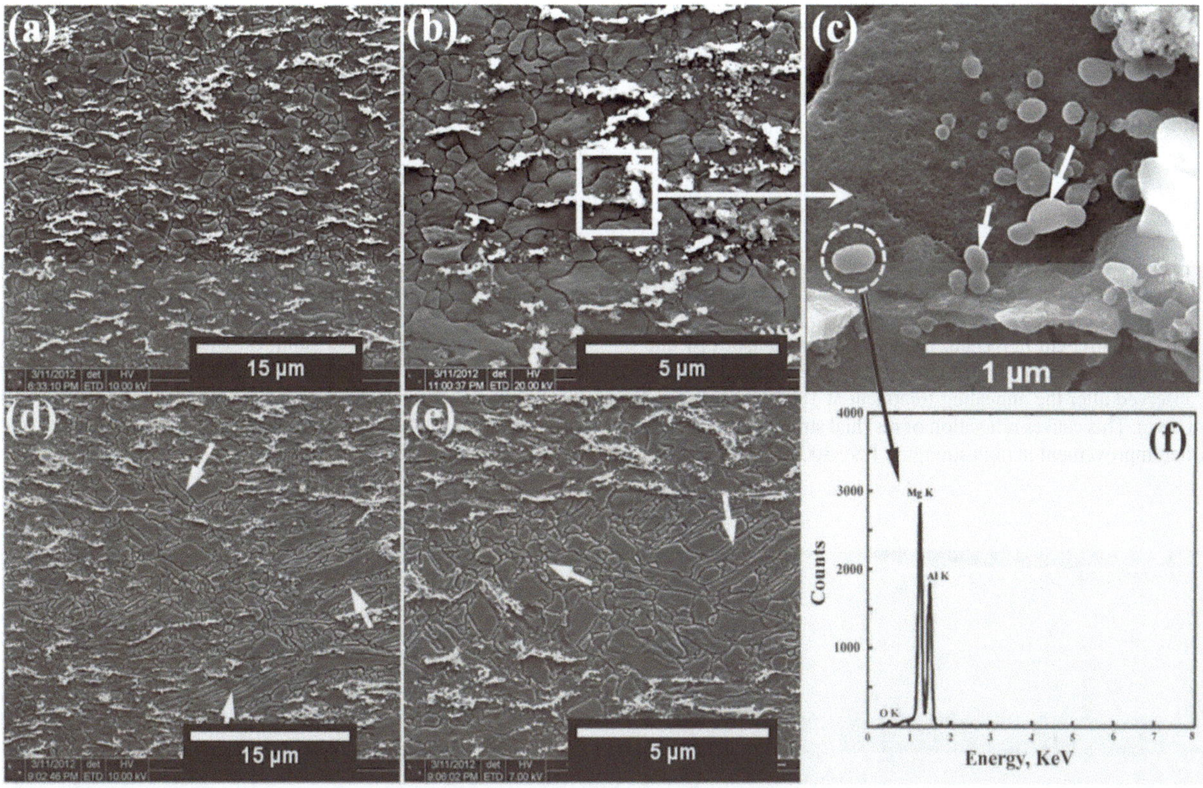

Figure. 3 The microstructure evolution in as-processed (a-c) AZ61L and (d-e) AZ70-TH Mg. (c) Different morphologies (arrows) of sub-divided (~50-100 nm) intermetallic β-particles ($Mg_{17}Al_{12}$), that were confirmed by EDS spot analysis in (f).

Mechanical Behavior

The tensile response and texture evolution of as-processed material is correlated in this section. In as-molded condition, the strain incompatibility between the (bcc) eutectic β-intermetallics and the (hcp) αs-Mg matrix leads to brittle interfaces and therefore low ductility (Y.S for as-molded condition ~ 125-150 MPa, EL. 6-13 %) [14]. In as-TTMP condition, both materials exhibit higher strength, accompanied with improvement in tensile ductility as shown in Figure 4. The Y.S of the as-processed material is ~1.5 times that of the as-molded condition and the UTS of the annealed material is ~ 1.7 times that of the as-molded condition. On the other hand, the tensile elongation of the as-molded and as-processed are drastically different with the processed material exhibiting significant improvement in ductility. Both processed alloys exhibit in-plane anisotropy and the Y.S and EL%., is found to be inferior in TD. Pole figures for {0002} plane are shown in Figure 5. Random texture is observed in as-molded condition while texture evolution after thermomechanical processing causes the basal planes to become parallel with the sheet face, which is typical secondary processed Mg sheets involving a finish rolling step [18]. The intensity of basal poles increases after thermomechanical processing and AZ61 has highest intensity of ~4.2 multiples of a random distribution (MRD), while for AZ70 highest basal pole intensity is 5.5 MRD. In both cases a spread of basal poles in the RD exists that can explain the in-plane anisotropy of mechanical response. The density of basal poles is distributed around ND with the maximum located at ND. AZ70 is found to be more ductile and it corresponds well with relatively higher intensity of basal poles for AZ70. The stress strain curves show a very different strain hardening behavior for the three conditions. The as-processed and annealed condition exhibits the highest strain hardening and the greatest UTS while the as-processed (TD) condition exhibits the lowest strain hardening and the lowest UTS.

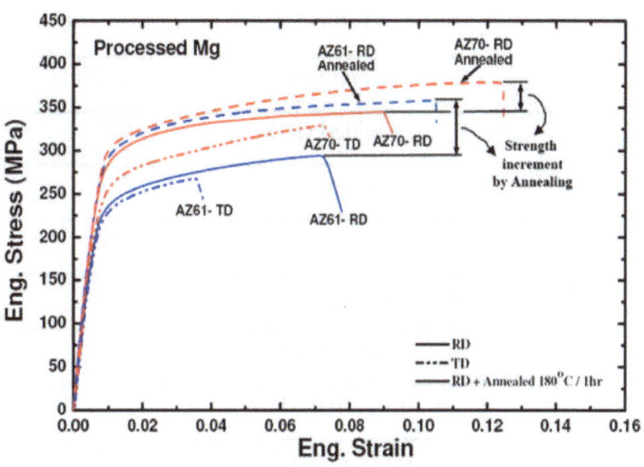

Figure. 4 The engineering stress strain curves for the material under investigation at room temperature. The TTMP processed material after annealing shows balanced mechanical properties.

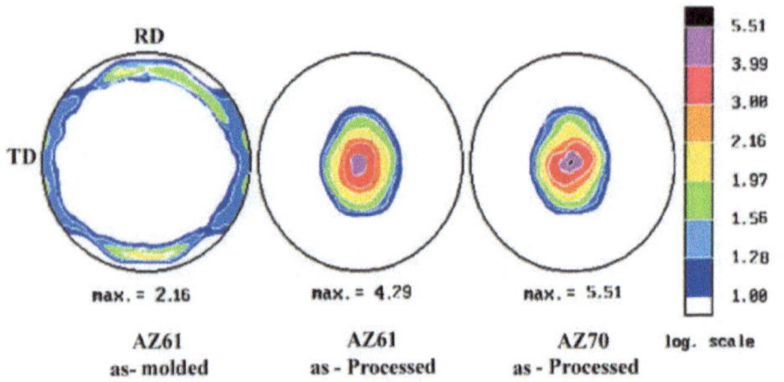

Figure. 5 {0002} pole figures for as-molded and as-processed conditions.

This improves strength in RD but leads to in-plane anisotropy due to hcp crystal structure of Mg. Since no significant texture changes were observed after the annealing treatment at 180°C for 1 hr (not shown here). This causes relaxation of residual stresses which leads to further improvement in both strength (T.S >300 MPa) and tensile ductility (>8 %). This could be explained by the fact that the short time annealing near recrystallization temperature of Mg (200°C), results in a lower dislocation density compared to the as-processed condition. Also, the high dislocation content in the processed alloy may lead to re-precipitation of some β-precipitates solutionized during thermomechanical treatment. In addition to the Hall-Petch strengthening due to near ultrafine grain size [19, 20], β particles from as-molded microstructure complement strengthening by sub-dividing and possibly solutionizing/re-precipitating into nano-sized, well-dispersed obstacles to the dislocation motion and grain growth. Furthermore, thermal treatments help optimize tensile response. Figure 6 shows the fracture surfaces of tensile tested samples for both TTMP materials after annealing treatment. The SEM images in (b) and (c) show regions that corresponds to the middle of the tensile tested samples. Both samples largely show ductile dimples in annealed condition but the fracture surface for the AZ61 Figure 6 (a) shows an area with sharp, brittle characteristics. Some precipitates are also visible in the back scatter images of the fracture surfaces but cannot be fully resolved.

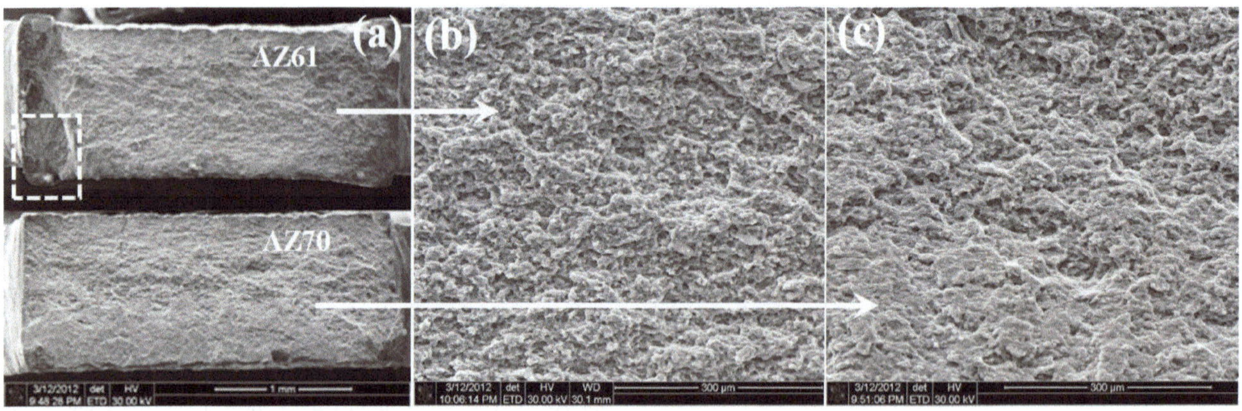

Figure 6 (a) SEM images show the fracture surface after tensile test for TTMP (a) AZ61L, (b) AZ70-TH after annealing treatment. The highlighted area for AZ61 shows some sharp, brittle fracture region. The higher magnification, representative SEM images in (b) and (c) show largely ductile fracture for both alloys.

Table-1 Mechanical properties of the magnesium alloys under investigation.

Alloy	Condition	YS (MPa)	Std. Dev.	UTS(MPa)	Std. Dev.	El. %	Std. Dev.
AZ61	molded (RD)	135	5	200	4	3.5 %	0.2
	processed (RD)	222	12	276	11	7.0 %	0.5
	processed (TD)	210	9	255	8	3.2 %	0.4
	processed + annealed (RD)	250	5	332	3	10.5 %	0.2
AZ70	molded (RD)	150	2	215	1	5.5 %	0.1
	processed (RD)	265	15	325	9	9.2 %	0.6
	processed (TD)	230	8	305	6	7.1 %	0.5
	processed +annealed (RD)	285	5	365	4	12.5 %	0.1

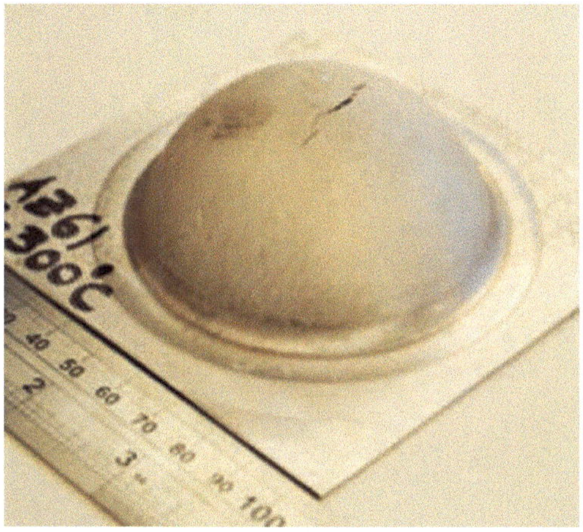

Figure. 7 A photograph of a dome obtained by bulge forming test. This depicts the potential of processed material for elevated temperature forming applications. This test was carried out at a forming pressure of 1MPa at 300°C in ~10 mins

Formability improvement:

Figure 7 shows a typical hemispherical dome in AZ61 obtained by gas pressure at 300°C under biaxial forming condition. The fracture region shown is oriented in the rolling direction. When we compared the dome heights achieved for two materials under different forming temperatures. The TTMP alloy with homogeneously distributed ultra-fine grain size forms for much longer time to larger dome heights compared to the as-molded condition that has a larger grain size with bimodal distribution and cast structure. Another important observation is that for TTMP alloys, increase in forming temperature results in larger dome heights which may be explained on the basis of improved strain rate sensitivity at elevated temperatures. It should be noted that dome superior dome heights were observed in TTMP AZ70 as compared to AZ61 material. Most notably, a dome height of 30.5 mm was achieved in AZ70 at 300°C after forming for approximately ~ 7 mins and a limit strain of 73 pct. is measured near dome's pole. In this particular case, the average value of biaxial stress was calculated to be 44 MPa by using the membrane equation $\sigma = pr/2t$, where p is gas pressure, r is dome radius, and t is sheet thickness. The preliminary evidence from bulge forming tests presented in this work suggests that formability can be enhanced in these two materials by secondary deformation processing.

Conclusions

The following are the conclusions of this work:

1. The as-processed AZ61L and AZ70-TH alloys both showed near ultrafine grain size (~1-2 μm) with dispersed eutectic nano-scale (~50-100 nm) intermetallic β-particles. No shear-banding or profuse twinning was seen. The mechanism of grain refinement was found to be recovery and DRX.

2. In processed condition, AZ70 alloy had superior mechanical properties as compared to AZ61. Both AZ61L and AZ70-TH alloys showed Y.S > 225 MP with a good combination of ~ 7-10 % elongation. Additionally, low temperature annealing treatment for 1hr further enhanced the high strength (UTS > 300 MPa), as well as improved the ductility (>10%).

3. In-plane anisotropy in mechanical property was found and it was related to the dominant basal texture in as-processed condition.

4. The improved mechanical properties including forming response are believed to be due to Hall-Petch strengthening resulting as well as the resistance offered to dislocation motion by presence of nano-sized intermetallic β-particles. Fracture surfaces for both alloys have ductile characteristics but AZ61 alloy shows some brittle features as-well and that is consistent with low ductility observed in this alloy as compared to AZ70.

5. Present results demonstrate the mechanical property improvements due to microstructural engineering down to nano grain structure by secondary processing in ultra-lightweight Mg materials. Further studies on warm formability and creep are required to promote industrial use of Mg based materials.

Acknowledgements

This research was performed with support from the Qatar Foundation under the National Priorities Research Program grant# NPRP 4 - 1063 - 2 - 397 and NPRP 4 - 1411 - 2 - 555. The authors acknowledge this financial support with gratitude. The authors also acknowledge the support from NanoMag LLC and Thixomat, Inc. for providing the magnesium materials.

References

1. Pollock TM. Weight Loss with Magnesium Alloys. Science. 2010;328:986-7.
2. Mathaudhu SN, Nyberg EA. Magnesium alloys in US military applications: past, current and future solutions In: Agnew SR, R. NN, A. NE, H. SW, editors. Magnesium Technology 20102010. p. 27-32.
3. Valiev RZ, Korznikov AV, Mulyukov RR. Structure and properties of ultrafine-grained materials produced by severe plastic deformation. Materials Science and Engineering: A. 1993;168:141-8.
4. Segal VM. Materials processing by simple shear. Materials Science and Engineering: A. 1995;197:157-64.
5. Saito Y, Tsuji N, Utsunomiya H, Sakai T, Hong RG. Ultra-fine grained bulk aluminum produced by accumulative roll-bonding (ARB) process. Scripta Materialia. 1998;39:1221-7.
6. Ma ZY, Mishra RS, Mahoney MW, Grimes R. High strain rate superplasticity in friction stir processed Al–Mg–Zr alloy. Materials Science and Engineering: A. 2003;351:148-53.
7. A. Albakri, B. Mansoor, H. Nassar, M. Khraisheh, Thermo-mechanical and metallurgical aspects in friction stir processing of AZ31 Mg alloy—A numerical and experimental investigation, Journal of Materials Processing Technology, 213 (2013) 279-290.
8. B. Mansoor, A. Ghosh, Microstructure and tensile behavior of a friction stir processed magnesium alloy, Acta Materialia, 60 (2012) 5079-5088.

9. B. Mansoor, S. Mukherjee, A. Ghosh, Microstructure and porosity in thixomolded Mg alloys and minimizing adverse effects on formability, Materials Science and Engineering: A, 512 (2009) 10-18.

10. Campbell J. Rheocasting and Thixocasting - a review of progress to-date. Foundry Trade Journal. 1975;138: 291-5.

11. Decker RF, Carnahan R, Babij E, Mihelich J, Spalding G, Thompson L. Magnesium semi-solid metal forming. Materials Park, OH: ASM International; 1996.

12. Chen Z, Huang J, Decker R, Lebeau S, Walker L, Cavin O, et al. The Effect of Thermomechanical Processing on the Tensile, Fatigue, and Creep Behavior of Magnesium Alloy AM60. Metallurgical and Materials Transactions A. 2011;42:1386-99.

13. Park SHC, Sato YS, Kokawa H. Microstructural evolution and its effect on Hall-Petch relationship in friction stir welding of molded Mg alloy AZ91D. Journal of Materials Science. 2003;38:4379-83.

14. Berman TD, Donlon W, Decker R, Huang J, Pollock TM, Jones JW. Microstructure Evolution in AZ61L During TTMP and Subsequent Annealing Treatments. Magnesium Technology 2011: John Wiley & Sons, Inc.; 2011. p. 599-603.

15. Ponge D, Gottstein G. Necklace formation during dynamic recrystallization: mechanisms and impact on flow behavior. Acta Materialia. 1998;46:69-80.

16. Sellars CM. Annealing Process-Recovery, Recrystallization and Grain Growth. Proceedings of 7th Ris Int Symp on Metallurgy and Materials Science Ris National Laboratory, Roskilde, Denmark1986. p. 69-80.

17. Lü YZ, Wang QD, Ding WJ, Zeng XQ, Zhu YP. Fracture behavior of AZ91 magnesium alloy. Materials Letters. 2000;44:265-8.

18. Foley DC, Al-Maharbi M, Hartwig KT, Karaman I, Kecskes LJ, Mathaudhu SN. Grain refinement vs. crystallographic texture: Mechanical anisotropy in a magnesium alloy. Scripta Materialia. 2011;64:193-6.

19. Hall EO. The Deformation and Ageing of Mild Steel: III Discussion of Results. Proceedings of the Physical Society Section B. 1951;64:747.

20. Petch NJ. Cleavage strength of polycrystals. Iron and Steel Institute- Journal. 1953;174 25-8.

Magnesium Technology 2015
Edited by: Michele V. Manuel, Alok Singh, Martyn Alderman, and Neale R. Neelameggham
TMS (The Minerals, Metals & Materials Society), 2015

MICROSTRUCTURES AND MECHANICAL PROPERTIES OF Mg-1at%X ALLOYS PROCESSED WITH HIGH-PRESSURE TORSION

Hiroyuki Kawabata[1], Shigeru Kuramoto[1], Keiichiro Oh-ishi[1]
[1]Toyota Central R&D Laboratory; 41-1 Yokomichi, Nagakute, Aichi, 480-1192, Japan

Keywords: Magnesium alloy, High-pressure torsion, Solute concentration, Grain refinement

Abstract

A number of researchers have reported the mechanical properties of Mg alloys processed with high-pressure torsion (HPT), which is a typical method of severe plastic deformation. However, the effect of alloying elements on the mechanical properties of HPT-processed Mg alloys was unclear.

In this study, to clarify the influence of the additional elements on the microstructures and mechanical properties of magnesium alloys, Mg-1at%x alloys (x=Al, Zn, Sn, Bi, Y) were produced by HPT processing. The microstructures and mechanical properties were investigated, and the effects of the various alloying elements were compared. The specimens were heat treated to maximize the amount of solute content in the matrix, and then subjected to HPT processing. Significant grain refinement was achieved by HPT prpcessing, with solute segregation at the grain boundaries. The hardness of the alloys was increased by the HPT processing, and this effect cannot be explained by grain refinement alone. Finally, the different strengthening effects of the various alloying elements are discussed.

Introduction

To achieve significant weight reduction in automobile manufacturing, magnesium alloys are very effective, as these are the lightest practical metals for use in automotive components. Mg-Al alloy castings and highly heat resistant magnesium alloy (for example, Mg-Al-Ca, Mg-Al-R.E., Mg-Al-Si, Mg-Al-Sr, etc) castings are already used in automobiles. High-strength and high-ductility magnesium alloys are being actively investigated for uses such as interior door panels, and the number of parts for which magnesium alloys are applicable is expected to expand greatly in the near future.

High-pressure torsion (HPT) processing can achieve high-strength and high-ductility magnesium alloys. A number of researchers have reported the microstructures and mechanical properties of Al alloys subjected to severe plastic deformation with HPT processing. For example, it has been reported that nanoclusters are formed in Al-Zn-Mg alloys after huge distortions induced by HPT processing, and that this contributes to the formation of high-strength alloys [1]. On the other hand, it has also been reported that most of the precipitation occurs not at dislocations or within grains but at the grain boundaries of Al-Mg-Si alloys that are aged after HPT processing [2]. This effect is not limited to Al alloys; it has been reported that HPT processing increases the strength of pure magnesium [3] and AZ31 alloy [4]. In other words, fine grains should be expected in magnesium alloys after HPT processing. Furthermore, HPT processing can provide a solution and precipitation state of the solute element that is unavailable via general processing and heat-treatment methods.

In order to clarify the influence of the additional elements on the microstructures and mechanical properties of magnesium alloys, Mg-1at%x alloys (X=Al, Zn, Sn, Bi, Y) were produced in this study by HPT processing, and their microstructures and mechanical properties were investigated.

Experimental

Samples

Mg-1at%x alloys were investigated, where X is an alloying element to magnesium. Al, Zn, Sn, Bi, and Y were selected as elements that have a high solid-solution limit according to the respective Mg-x binary phase diagrams. A pure Mg ingot was melted in a steel crucible, and then pure Al, Zn, Sn, Bi, or Y was added to the molten Mg at 973 K to prepare the Mg-1at%x alloy. The molten alloys were poured into a mold to prepare castings. The castings were heat treated at 673 K for 48 h to maximize the solute content of the matrix. After heat treatment, $\phi 10\times 1$ mm test pieces were scraped out of the castings and subjected to HPT processing.

HPT processing

Figure 1 shows a schematic illustration of HPT, and Table I shows the HPT processing conditions employed in this study. The rotational anvil was turned for 10 times at a speed of 1 rpm under a compressive stress of 2 GPa, so that the specimens were given huge distortions by the HPT processing. Torque curves were measured during the HPT processing.

Evaluation

The chemical composition of each alloy was measured by inductively coupled plasma optical emission spectrometry (ICP-OES) analysis. The alloy microstructures before HPT processing were observed by scanning electron microscopy (SEM). The microstructures of the alloys after HPT processing were observed by bright-field scanning transmission electron microscopy (BF-STEM) and high-angle annular dark field scanning transmission electron microscopy (HAADF-STEM). The equivalent circle diameter, which were calculated from the areas of the each grain measured by image analysis of BF-STEM images, was defined as the grain size. The solute concentration at the grain boundaries was analyzed by energy dispersive X-ray spectrometry (EDX). Finally, the Vickers hardness of the alloys before and after HPT processing was measured under a 100 g load applied for a period of 15 seconds.

Results and Discussion

Table II shows the chemical compositions of the Mg-1at%x alloys. The chemical compositions of Zn, Sn, and Bi were very close to the target concentrations. The amount of Al of the alloy was slightly more than the target value. The yield of Y in Mg-1at%Y was about 80%. The impurity concentrations were on the order of ppm, and were below the detection limit in each alloy.

Figure 2 shows torque curves obtained during the HPT processing of Mg-1at%x alloys. The torque at 9 rotations decreased in the order: Mg-1at%Bi, Mg-1at%Al, Mg-1at%Sn, Mg-1at%Y, Mg-1at%Zn. As the number of revolutions increased, the torque tended to decrease initially and increase after 4 rotations in Mg-1at%Al. For Mg-1at%Y, the torque tended to decrease initially and increase after 6 rotations in the same way. On the other hands, Mg-1at%Sn, Mg-1at%Zn, and Mg-1at%Bi, the torque had a peak torque, and the number of revolutions at which this peak torque was observed differed for each alloy. The peak occurred at 8 rotations for Mg-1at%Sn and 7 rotations for Mg-1at%Zn. The Mg-1at%Bi torque peaked at 4 rotations, which was the fewest, and then decreased as additional rotations were applied.

Figure 3 shows the microstructures of the Mg-1at%Al alloys. Figure 3(a) shows the alloy microstructures before HPT processing, Figures 3(b) and (c) show BF-STEM images, and Figure 3(d) shows HAADF-STEM images of the alloys after HPT processing. Figure 4, 5, 6, and 7 show the microstructures of the Mg-1at%Zn, Mg-1at%Sn, Mg-1at%Bi, and Mg-1at%Y each. The microstructures observed before HPT processing indicated that Al, Zn, and Y form a solid solution with the α-matrix in each alloy. Sn and Bi formed not only a solid solution but also the intermetallic compounds Mg_2Sn and Mg_3Bi_2. These compounds are not expected to exist according to the binary phase diagrams. It is considered that the heat-treatment condition was insufficient. Table III shows the grain sizes of the Mg-1at%x alloys. The grain sizes of all alloys were reduced by HPT processing, with or without the formation of intermetallic compounds. The effects of grain refinement were especially prominent for Mg-1at%Al, Mg-1at%Zn, Mg-1at%Sn, and Mg-1at%Y, and the grain sizes of these alloys were reduced to less than 200 nm by the HPT processing. On the other hand, the grain size of Mg-1at%Bi after HPT processing was approximately 470 nm, which was slightly larger than that of the other alloys. Therefore, the refinement effect in the HPT processing of Mg-1at%Bi is weak.

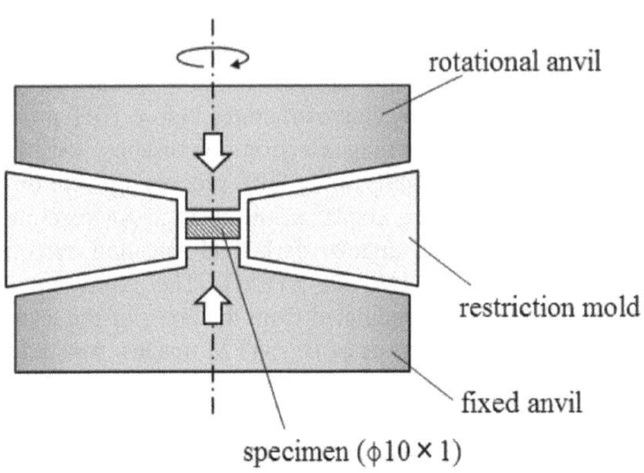

Figure 1. Schematic illustration of high-pressure torsion (HPT).

Table I. HPT processing conditions.

Temperature	R.T.
Rotational speed	60rpm
Rotation	10 rotations
Pressure	2GPa

Table II. Chemical compositions of the magnesium alloys.

alloy	element x		Cu	Ni	Fe
	at%	wt%	ppm	ppm	ppm
Mg-1at%Al	1.18	1.31	< 20	32	51
Mg-1at%Zn	1.02	2.70	< 20	32	< 50
Mg-1at%Sn	0.98	4.61	< 20	28	< 50
Mg-1at%Bi	0.99	7.88	< 20	37	< 50
Mg-1at%Y	0.83	2.96	< 20	37	< 50

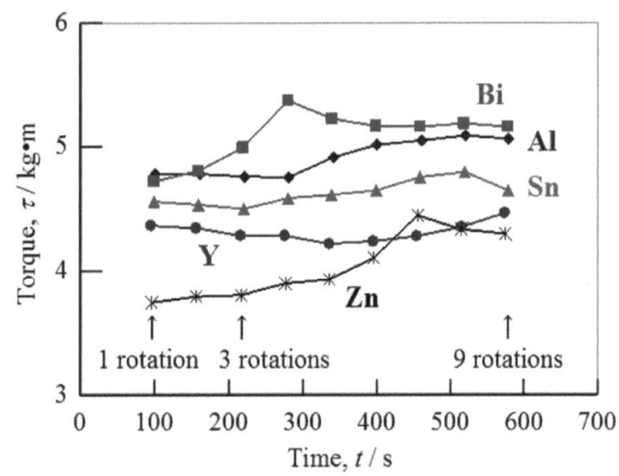

Figure 2. Torque curves obtained during the HPT processing of Mg-1at%x alloys (x=Al, Zn, Sn, Bi, Y).

As shown in Figure 3(d), solute segregation at the grain boundaries was observed in all alloys. Especially, solute segregation in Mg-1at%Sn alloy was observed clearly. Therefore, solute segregation of Mg-1at%Sn alloy was investigated in more detail. Figure 8 shows the results of TEM analysis of the alloy. Sn was detected at a grain boundary (#2), but almost no Sn was detected in the grain boundary neighborhood (#1, #3). Figure 8(a) shows that there was no intermetallic compound (Mg_2Sn) at the grain boundary. Therefore, Sn was concentrated at the grain boundary.

Figure 9 shows the Vickers hardness values of Mg-1at%x alloys before and after HPT processing. The Vickers hardness of Mg-1at%Al, Mg-1at%Zn, Mg-1at%Sn, and Mg-1at%Bi samples before HPT processing (samples heat treated after casting to maximize the amount of solute content in the matrix) was approximately 40 HV, and that of Mg-1at%Y was 48 HV. The samples after HPT processing were classified as alloys with a Vickers hardness of less than 60 HV (Mg-1at%Sn, Mg-1at%Bi), alloys with a Vickers hardness above 80 HV (Mg-1at%Al, Mg-

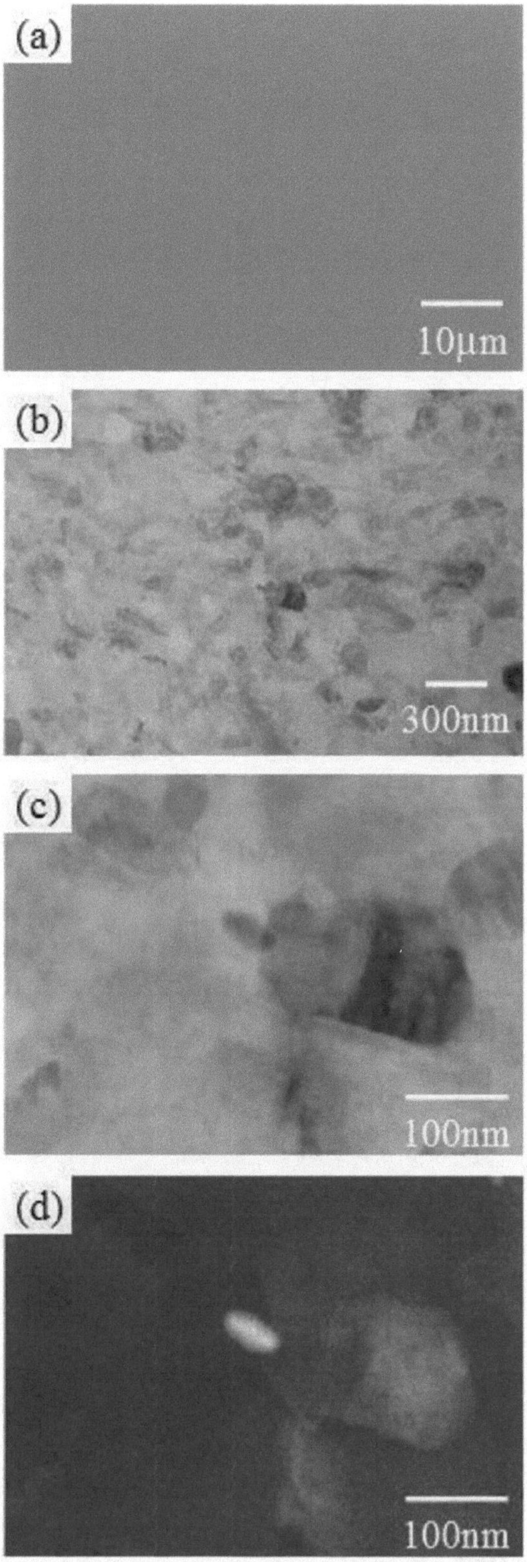

Figure 3. Microstructures of Mg-1at%Al alloy. (a) SEM-BSE images before HPT processing, (b) and (c) BF-STEM images after HPT processing, (d) HAADF-STEM images after HPT processing.

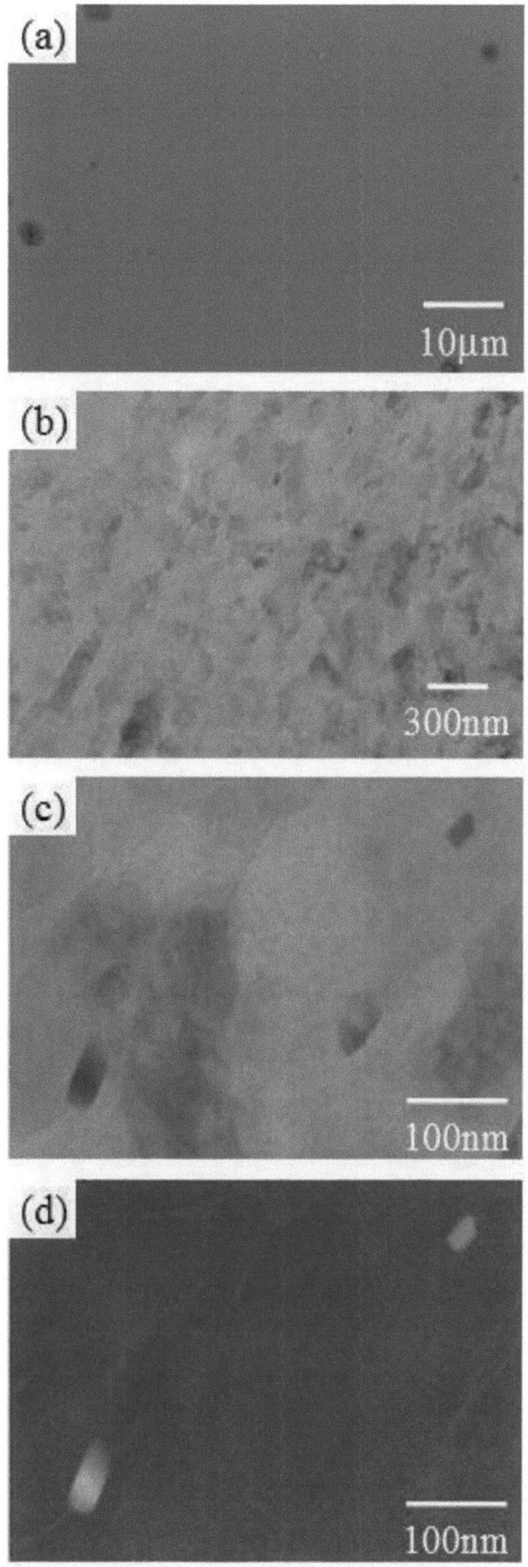

Figure 4. Microstructures of Mg-1at%Zn alloy. (a) SEM-BSE images before HPT processing, (b) and (c) BF-STEM images after HPT processing, (d) HAADF-STEM images after HPT processing.

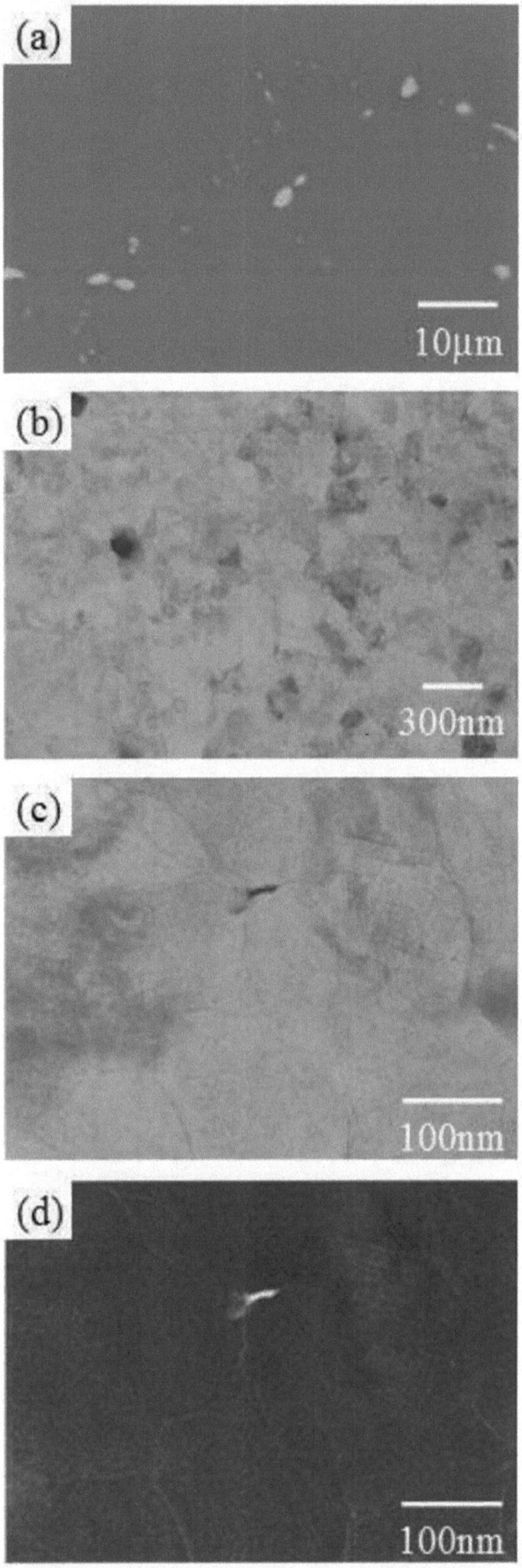

Figure 5. Microstructures of Mg-1at%Sn alloy. (a) SEM-BSE images before HPT processing, (b) and (c) BF-STEM images after HPT processing, (d) HAADF-STEM images after HPT processing.

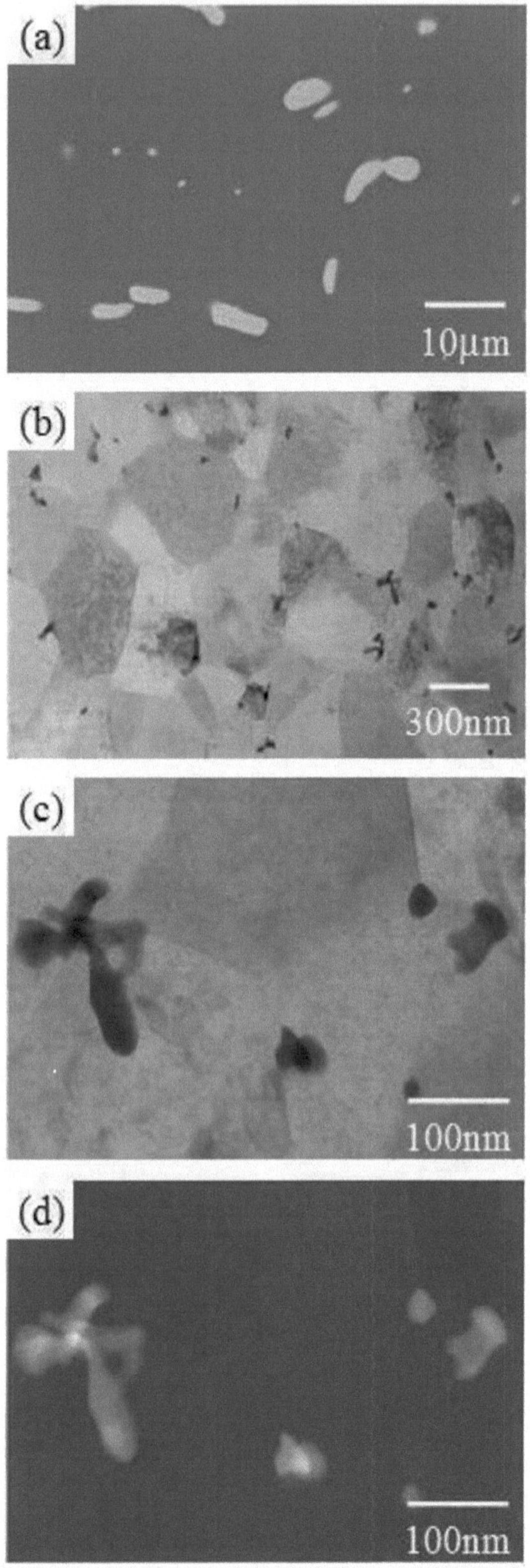

Figure 6. Microstructures of Mg-1at%Bi alloy. (a) SEM-BSE images before HPT processing, (b) and (c) BF-STEM images after HPT processing, (d) HAADF-STEM images after HPT processing.

lat%Y), and alloys with a Vickers hardness of approximately 110 HV (Mg-1at%Zn).

As mentioned above, HPT processing of a supersaturated solid solution of a Mg-x alloy results in fine grains and a concentration of solute x at the grain boundary. In general, by applying a heat treatment to a supersaturated solid solution of metals, precipitates are often formed in grains and at grain boundaries, but solute concentration in grain boundaries are rarely formed. Therefore, HPT processing can result in a unique microstructure. On the other hand, HPT processing enhanced the hardness of the alloys, and the amount of increase varied by alloy. The grain sizes of Mg-1at%x alloys (x=Al, Zn, Sn, Y) after HPT processing were almost the same. Therefore, solute concentration at the grain boundaries was supposed as one of the causes of the differences in alloy hardness. It is considered that the further inspection of this point is necessary.

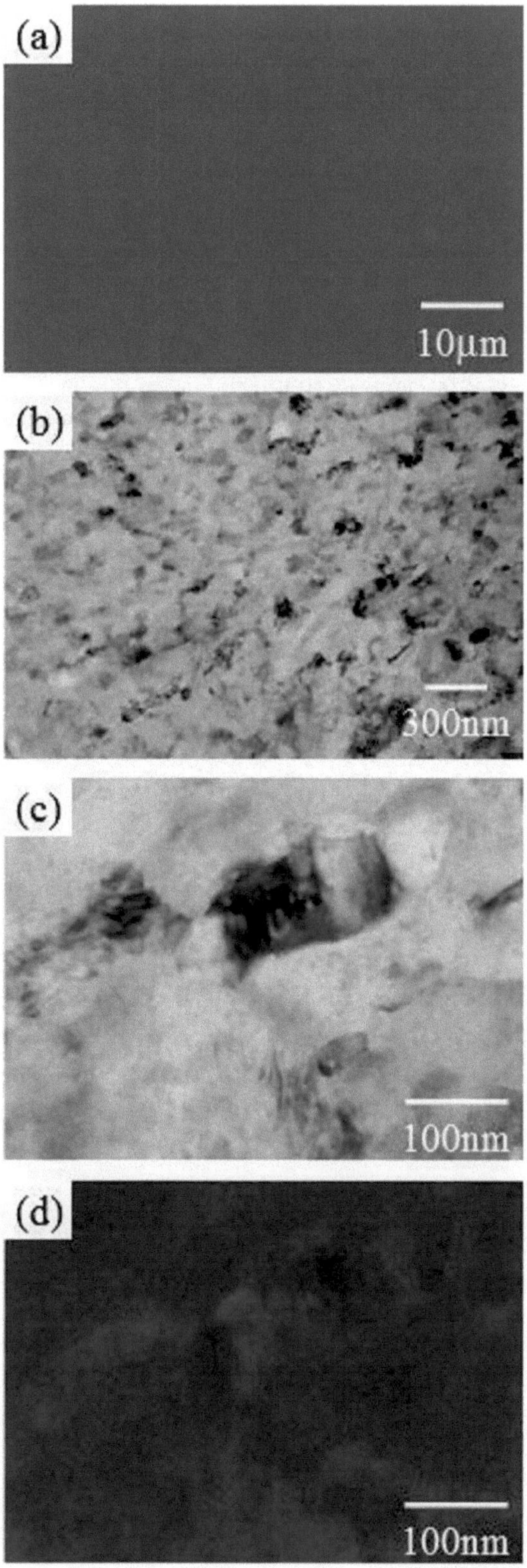

Figure 7. Microstructures of Mg-1at%Y alloy. (a) SEM-BSE images before HPT processing, (b) and (c) BF-STEM images after HPT processing, (d) HAADF-STEM images after HPT processing.

Table III. Grain sizes of the magnesium alloys.

alloy	grain size (nm)
Mg-1at%Al	160
Mg-1at%Zn	140
Mg-1at%Sn	180
Mg-1at%Bi	470
Mg-1at%Y	140

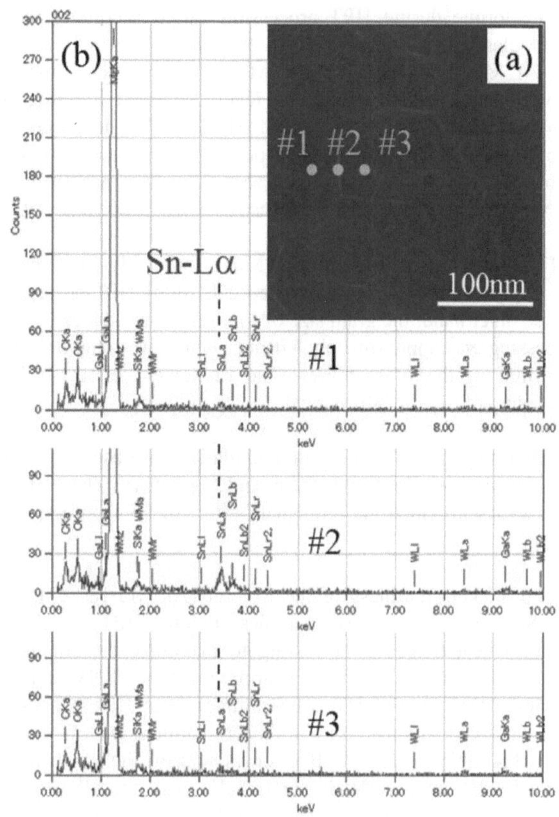

Figure 8. TEM analysis results. (a) DF-STEM image, (b) point analysis of the Mg-1at%Sn alloy.

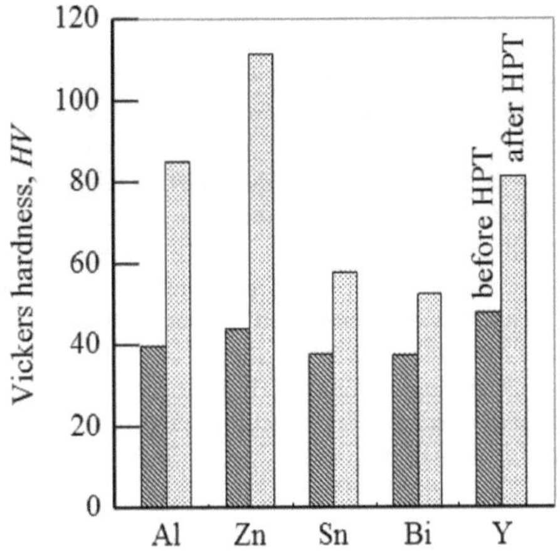

Figure 9. Vickers hardness of Mg-1at%x alloys before and after HPT processing (x=Al, Zn, Sn, Bi, Y).

Summary and Conclusion

Mg-1at%x alloys (x=Al, Zn, Sn, Bi, Y) were produced by HPT processing, and their torque behaviors, microstructures, and mechanical properties were investigated. The results obtained are summarized as follows:

1. The torque during HPT processing varied according to the added element, and decreased in the order: Mg-1at%Bi, Mg-1at%Al, Mg-1at%Sn, Mg-1at%Y, Mg-1at%Zn. As the number of revolutions increased, the torque level decreased initially, and then increased in Mg-1at%Al and Mg-1at%Y. The number of revolutions required to reach the peak torque was different for each alloy: 8 rotations for Mg-1at%Sn, 7 for Mg-1at%Zn, and 4 for Mg-1at%Bi, which was the fewest.

2. The grain sizes of Mg-1at%Al, Mg-1at%Zn, Mg-1at%Sn, and Mg-1at%Y were reduced to less than 200 nm by HPT processing. On the other hand, the grain size of Mg-1at%Bi produced by HPT processing was approximately 470 nm, which was slightly larger than that of the other alloys. The refinement effect in HPT processing of Mg-1at%Bi is weak.

3. Solute segregation at the grain boundaries was observed in Mg-1at%Al, Mg-1at%Zn, Mg-1at%Sn, and Mg-1at%Y. These are the same alloys in which the refinement effect of HPT processing is strong.

4. The Vickers hardness of the Mg-1at%Al, Mg-1at%Zn, Mg-1at%Sn, and Mg-1at%Bi samples before HPT processing (samples heat treated after casting to maximize the amount of solute content in the matrix) was approximately 40 HV, and that of Mg-1at%Y was 48 HV. The samples after HPT processing were classified as alloys with a Vickers hardness of less than 60 HV (Mg-1at%Sn, Mg-1at%Bi), alloys with a Vickers hardness above 80 HV (Mg-1at%Al, Mg-1at%Y), and one alloy with a Vickers hardness of approximately 110 HV (Mg-1at%Zn).

Acknowledgement

This work was supported by Grant-in-Aid for Scientific Research No.25630313 from Japan Society for the Promotion of Science (JSPS).

Reference

1. P.V. Liddicoat et al., "Nanostructural Hierarchy Increases the Strength of Aluminium Alloys," Nat. Commun., 1:63, doi: 10.1038/ncomms1062, (2010).
2. Katsumi Watanabe et al., "Aging behavior and microstructure of aged excess Mg type Al–Mg–Si alloys after HPT processing," *Journal of Japan Institute of Light Metals*, 63(2013),406-412.
3. K. Edalati et al., "High-pressure torsion of pure magnesium: Evolution of mechanical properties, microstructures and hydrogen storage capacity with equivalent strain," *Scripta Materialia*, 64(2011),880-883.
4. J. Vratna et al., "Investigation of microhardness and microstructure of AZ31 alloy after high pressure torsion," *TMS*, Magnesium Technology 2011(2011),589-594.

Magnesium Technology 2015
Edited by: Michele V. Manuel, Alok Singh, Martyn Alderman, and Neale R. Neelameggham
TMS (The Minerals, Metals & Materials Society), 2015

DEPENDENCE OF COMPRESSION-TENSION LOADING ON TWINNING IN WROUGHT Mg ALLOY

Daria Drozdenko[1], Jan Bohlen[2], Sangbong Yi[2], Dietmar Letzig[2], František Chmelík[1], Patrik Dobroň[1]

[1]Charles University in Prague, Faculty of Mathematics and Physics, Department of Physics of Materials, Ke Karlovu 5, Prague 2, 12116, Czech Republic
[2]Helmholtz-Zentrum Geesthacht, Zentrum für Material- und Küstenforschung GmbH, Max-Planck-Straße 1, Geesthacht, D21502, Germany

Keywords: Magnesium, Twinning, Detwinning, EBSD

Abstract

Wrought magnesium alloy AZ31 was pre-compressed and subsequently subjected to tensile loading. Due to a strong fibre texture of the alloy the level of pre-compression stress significantly influences the following tensile test. The results of acoustic emission (AE) measurement are correlated with the stress-time curves and differences in the AE count rate were used to reveal changes in underlying deformation mechanisms.
First, the compression-tension test was recorded by using the AE technique and then the test was repeated to specific points on the deformation curve, where the AE count rate exhibits strong changes in its activity. A pronounced S-shape of the tensile curve and AE peaks were observed. It highlights the role of detwinning and the nucleation of new twins during the tensile test. The subsequent analysis of microstructure using electron back scattered diffraction (EBSD) method opens a detailed insight into microstructural changes.

Introduction

The complexity of the deformation of magnesium (Mg) is related to its hexagonal close packed (*hcp*) structure. The relatively low crystal symmetry gives a reduced number of active slip systems at room temperature [1-4]. In this case, twinning, as an additional deformation mechanism, becomes active in Mg alloys [5, 6]. It produces either tensile or compressive strains along the crystallographic *c*-axis in *hcp* lattice. The most common twinning in Mg alloys is the {10.2} <10.1> tensile twinning [7], which results in a tensile strain parallel to the *c*-axis (hence the pseudonym: "Tensile twins") or a compression strain perpendicular to the *c*-axis [8-10].
This type of twin provides strain along the *c*-axis, while basal and prismatic <*a*> glide cannot contribute to straining in this direction. Furthermore, for an activation of <c+a> pyramidal slip system a higher stress is typically required [11].
Moreover, various "compression" twinning modes (either of the {10.1} or {10.3} planes) become active when applying compression strain parallel to the *c*-axis [4, 12].
Therefore, the twin formation depends on the preferred grain orientation (texture) and the loading direction. The texture of wrought Mg alloys is characterized by the preferential orientation of the *c*-axis perpendicular to the extrusion direction (ED), so tensile twins are preferentially activated in compression along the extrusion axis rather than in tension [3].
The twinning mechanisms are usually investigated during reversal loading tests with respect to the *c*-axis of an *hcp* lattice [13-15]. The influence of twins induced by pre-straining on the reloading behavior of magnesium alloys can be found in [6, 16-18], where the phenomenon of detwinning during a reversal loading was reported. This process is characterized by a thickness reduction or disappearance of existing twin lamellae, i.e. twin lamellae are rotating back to the matrix (parent) orientation [19].
Bohlen et al. [13] studied the dependence of the grain size on the activation of twins in magnesium extrusions. It was found that twins are preferentially formed in larger grains during compression loading and only with increasing stress appear also in smaller grains. Moreover, a strong fibre texture and the level of pre-compression stress significantly influence the following tensile test. Detailed grain by grain analysis of activation of different variants of twins in Mg alloys can be also found in [19, 20].
The acoustic emission (AE) technique is an *in situ* method for the detection of transient elastic waves, which are generated within the material due to sudden localized structural changes. Based on differences of the amplitude of the respective signal, the AE method could be used to reveal different types of deformation processes [21].
Electron backscattering diffraction (EBSD) was applied to obtain local orientation images in order to reveal twinning–untwinning processes [13, 20, 22].
Numerous applications require knowledge of the deformation behavior of materials under cyclic loading. It is essential to study the twinning in wrought Mg alloys, since this deformation mechanism is important for straining along the *c*-axis, hence tensile-compression asymmetry and planar anisotropy.
The main aim of the present work is to analyze the detwinning processes and nucleation of various types of twins in partly twinned material. The concurrent AE measurement is used to distinguish between twin nucleation, twin growth and detwinning processes.

Experimental procedures

AZ31 magnesium alloy (Mg + 3 wt%Al + 1 wt%Zn + 0.3 wt%Mn) was indirectly extruded at 300^{0}C with an extrusion rate (profile exit speed) of 5 m/min. The extrusion ratio was 1:30. Specimens (gauge length of 15 mm, diameter of 8 mm) with screws on both ends were machined from the round bar parallel to the ED. Deformation test (compressive loading followed by tension) was carried out at room temperature and a constant strain rate of 10^{-3} s^{-1} using a universal testing machine Zwick Z050.
A computer-controlled PCI-2 device (Physical Acoustic Corporation) was used to monitor the AE activity during the mechanical testing. The threshold level of detection was set as 26 dB to avoid the background noise. A preamplifier with gain of ~40 dB was used. The sensor was attached to the specimen surface using silicon grease and a spring.
Electron Back Scattering Diffraction technique in a field emission gun scanning electron microscope (Zeiss Ultra 55, EDAX/TSL EBSD system and Hikari detector) was used to perform texture

analysis. Mapping was done on longitudinal sections of the samples with the step size of 0.4 μm. The specimens for EBSD investigations were polished by standard method down to OPS colloidal silica suspension and subsequently electropolished by means of the Struers AC-2-II electrolyte. The EBSD data were processed by the data acquisition software TSL-OIM Analysis 7.0.

Results and discussion

The deformation curve and the concurrent AE count rate as the functions of measuring time for cyclic test are shown in Figure 1a.

a)

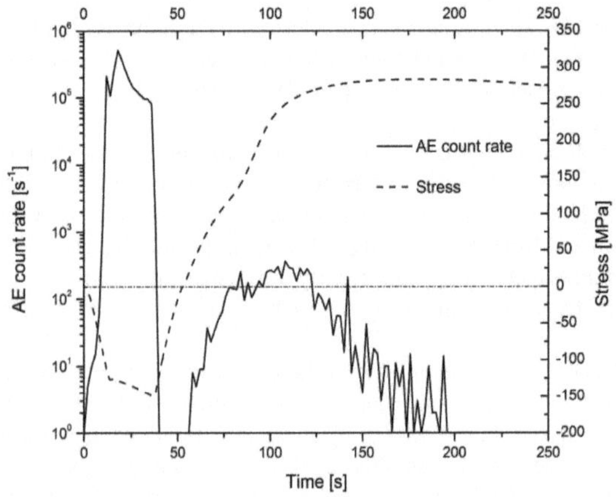

b)

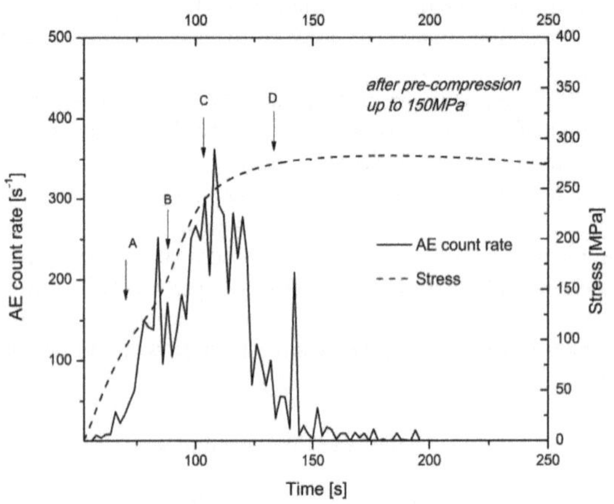

Figure 1. Stress and AE count rate *vs.* time curves for a) cyclic test, b) tensile part of cyclic test after pre-compression up to 150 MPa.

In order to achieve a microstructure with {10-12}<10-11> type of twins, the first part of the cyclic test was compression up to 150 MPa (the compression yield point is 125 MPa). Subsequent tensile loading was performed until fracture. The AE signal was recorded during the whole compression-tension loading. The test was repeated and stopped at distinct points on the deformation curve, where the AE count rate exhibits strong changes in its activity in order to investigate changes in microstructure (Figure 1b). For microstructural analysis, samples after pre-compression and tension up to 95 MPa, 150 MPa, 230 MPa, 275 MPa (denoted as sample A, B, C, D) were chosen.

The stress-time curve for the tensile part of the cyclic test (Figure 1b) exhibits an S-shape with few inflection points, which could be explained by activation the different types of deformation processes. A similar deformation behavior was found in [13], where with increasing pre-compression level a more pronounced S-shape of deformation curve was observed.

The AE count rate is significantly lower for the tensile than for the compression part (see Figure 1a) of the test. The AE count rate during tensile loading (Figure 1) exhibits few pronounced peaks, which strongly correspond to inflection points on the deformation curve and are followed by a decrease up to the end of the test.

The initial texture of the extruded alloy exhibits a typical basal character, where the *c*-axis of the *hcp* structure is perpendicular to the ED. In our previous work [13], it was shown that microstructure of extruded AZ31 after pre-compression up to 150 MPa contains tensile twins.

After the pre-compression loading, the *c*-axes of the crystals are aligned with the loading direction due to the 86.6° reorientation, which is caused by tensile twinning. Therefore, the basal pole figure shows the intensity peak in the middle.

During subsequentl tension, this intensity peak weakens, which is linked with the reorientation to the <10-10> parent orientation. A similar texture development was observed in Mg extrusions [4, 13, 23, 24].

Figure 2 shows the pole figures in the ED of samples A, B, C and D.

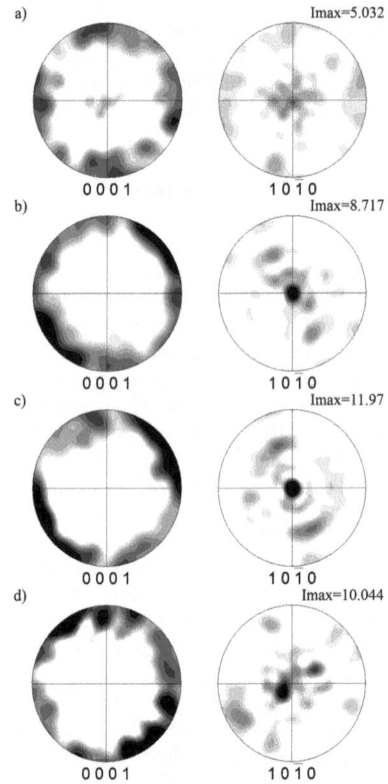

Figure 2. Pole figures in the extrusion direction of samples after pre-compression and subsequent tension up to a) 95 MPa (A), b) 150 MPa (B), c) 230 MPa (C), d) 275 MPa (D). The measured pole figures were obtained by EBSD.

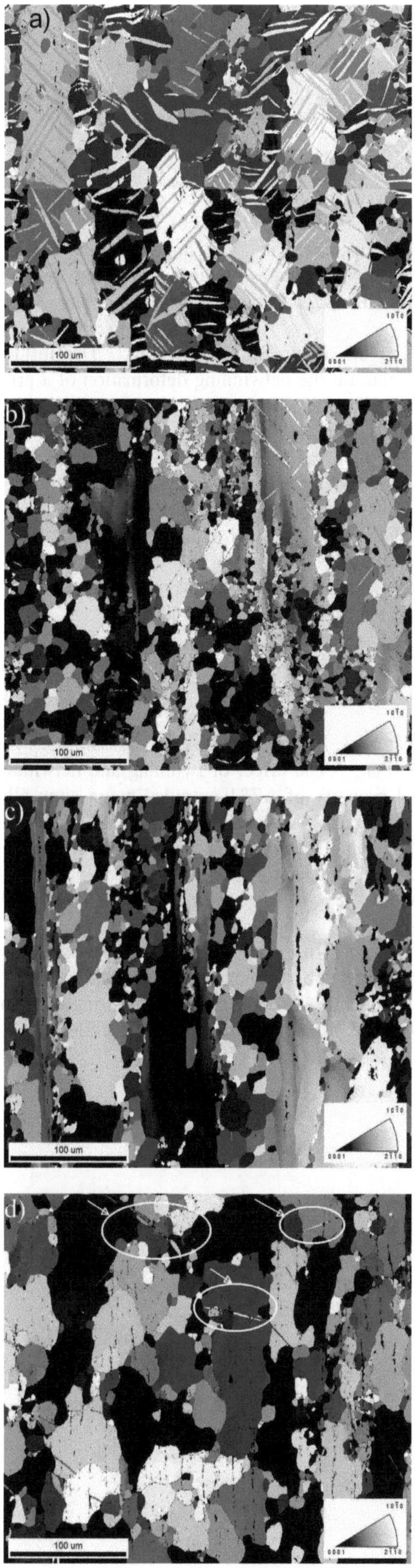

Figure 3. Orientation maps of the longitudinal section of extruded round bar after pre-compression and subsequent tension up to a) 95 MPa (A), b) 150 MPa (B), c) 230 MPa (C), d) 275 MPa (D).

The basal pole of the sample A still contains the peak in the middle. Due to the relatively small analyzed area of samples, few grains contribute by peaks in the radial part of basal pole. Intensity of the prismatic pole becomes stronger with increasing tensile stress up to 230 MPa.

Figure 3a shows the orientation map of longitudinal section of the sample A (pre-compression up to 150 MPa and tensile loading up to 95 MPa). The reoriented fractions which are spread all over the microstructure are identified as {10-12} tensile twins. Smaller grains are free from twins and twin boundaries are identified in large grains. The same grain size effect was observed by Bohlen et al [13]. The sample A was subsequently loaded in tension and therefore it could be supposed that existing twins had already shrunk. A continuous progress of the detwinning process could be seen in Figure 3b, where the tensile twins are still observed in only a few grains. The analysis of the misorientation distribution demonstrates the absence of a particular misorientation of 86 ± 5^0, which corresponds to tensile twinning. Based on this we could assume that tensile loading up to 230 MPa gives completely twin free microstructure, see Figure 3c.

It was experimentally observed that twinned grains in Mg alloy can be easily detwinned during reversed loading [14-17]. This mechanism requires less stress due to the fact that the twin boundaries already exist, and no nucleation is necessary [6].

It is generally accepted that twin growth and shrinkage do not give AE response [21]. The reorientation of the matrix during detwinning increases the free path of moving dislocations, which leads to higher AE activity. Therefore, the increase in the AE count rate during loading from 150 MPa to 230 MPa could be associated with the activity of dislocation slip system.

A pronounced AE count rate peak arises with a subsequent increase of tensile load. The EBSD orientation map of the sample D (tension up to 275 MPa) shows the existence of compression twins (misorientation angle of 57 ± 5^0), which are marked by circles with arrows in Figure 3d. A rapid increase in the AE count rate can be explained by nucleation of twins. As was reported in [21, 27], twin formation is an excellent source of AE in contrary to twin thickening (no AE signal). The AE signals, observed after the microscopic yield point, are produced by strong dynamic processes whose AE activity is not yet reduced by strain hardening.

The activation of compression twins at later stage of deformation supports the idea that the stress, which is required for nucleation of new twins, is higher than the stress required to move a twin boundaries, i.e. to propagate the twin boundaries to enlarge or shrink the twins [9, 26]. Moreover, the critical resolved shear stress (CRSS) for compression twins is much higher than for other deformation mode (prismatic and pyramidal slip) and therefore compression twins appear at later stage of plastic deformation.

Conclusions

During a reversed loading along the extrusion direction of wrought magnesium alloy AZ31, detwinning is a significant deformation mechanism and is preferred over the nucleation and growth of new twins. After a pre-compression up to 150 MPa and subsequent tension up to 230 MPa twin free microstructure was observed. After finishing the detwinning process at a later stage of plastic deformation, compression twins appear. The twinning-detwinning process and nucleation of additional type of twins was directly confirmed by electron backscatter diffraction. The strong AE activity indicates collective dislocation motion in detwinned grains and formation of new twins in already twinned grains.

Acknowledgements

This work was supported by the Czech Science Foundation under the grant 13 - 19812S, by the Grant Agency of Charles University under the grant No. 1882314 and also by the grant SVV-2014-260091. The work of P.D. is a part of activities of the Charles University Research Center "Physics of Condensed Matter and Functional Materials".

References

1. M.R. Barnett, Z. Keshavarz, A.G. Beer, D. Atwell, "Influence of grain size on the compressive deformation of wrought Mg–3Al–1Zn," *Acta Materialia*, 52 (17) (2004), 5093.
2. C.H.J. Davies, S. Yi, J. Bohlen, K.U. Kainer, H.G. Brokmeier, "Synchrotron radiation investigation of twinning in extruded magnesium alloy AZ31," *Materials Science Forum* 495-497 (2005), 1633.
3. J. Bohlen, S.B. Yi, J. Swiostek, D. Letzig, H.-G. Brokmeier, K. U. Kainer, "Microstructure and texture development during hydrostatic extrusion of magnesium alloy AZ31," *Scripta Materialia* 53 (2005), 259-264.
4. D. Sarker, D.L. Chen, "Detwinning and strain hardening of an extruded magnesium alloy during compression," *Scripta Materialia* 67(2) (2012) 165–168.
5. O. Muransky, D.G. Carr, P. Sittner, E.C. Oliver "In situ neutron diffraction investigation of deformation twinning and pseudoelastic-like behaviour of extruded AZ31 magnesium alloy," *International Journal of Plasticity*, 25 (2009), 1107–1127.
6. G. Proust, C.N. Tomé, A. Jain, S.R. Agnew, "Modeling the effect of twinning and detwinning during strain-path changes of magnesium alloy AZ31," *International Journal of Plasticity*, 25 (2009), 861–880.
7. D.W. Brown, S.R. Agnew, M.A.M. Bourke, T.M. Holden, S.C. Vogel, C.N. Tome, "Internal strain and texture evolution during deformation twinning in magnesium," *Materials Science and Engineering: A*, 399 (2005), 1–12.
8. M.R. Barnett, "Twinning and the ductility of magnesium alloys: Part I: "Tension" twins," *Materials Science and Engineering: A* 464 (1–2) (2007), 1–7.
9. P.G. Partridge, "The crystallography and deformation modes of hexahonal close-packed metals," *Metall. Review* 118, (1967), 169.
10. E.W. Kelley, W.F. Hosford, "Plane-strain compression of Magnesium and Magnesium Alloy Crystals," *Trans. Metall. Soc. AIME*, 242 (1968), 5–13.
11. W. Wu, H. Qiao, K. An, X. Guo, P. Wu, P.K. Liaw, "Investigation of deformation dynamics in a wrought magnesium alloy", *International Journal of Plasticity* 62 (2014), 105–120.
12. M.R. Barnett, "Twinning and the ductility of magnesium alloys: Part II. "Contraction" twins," *Materials Science and Engineering: A*, 464 (1–2) (2007), 8-16.
13. J. Bohlen, P. Dobron, L. Nascimento, K. Parfenenko, F. Chmelik, D. Letzig, "The Effect of Reversed Loading Conditions on the Mechanical Behaviour of Extruded Magnesium Alloy AZ31," *Acta Physica Polonica, A*. 122 (3) (2012), 444.
14. Y.N. Wang, J.C. Huang, "The Role of Twinning and Untwinning in Yielding Behavior in Hot-Extruded Mg-Al-Zn Alloy," Acta Materialia. 55 (2007), 897-905.
15. X.Y. Lou, M. Li, R.K. Boger, S.R. Agnew, R.H. Wagoner, "Hardening evolution of AZ31B Mg sheet," *International Journal of Plasticity* 23 (1) (2007), 44–86.
16. C.H. Caceres, T. Sumitomo, M. Veidt, "Pseudoelastic behaviour of cast magnesium AZ91 alloy under cyclic loading–unloading," *Acta Materialia*, 51 (2003), 6211–6218.
17. S Kleiner, P.J Uggowitzer, "Mechanical anisotropy of extruded Mg–6% Al–1% Zn alloy," *Materials Science and Engineering: A*, 379 (1–2) (2004), 258-263.
18. A. Jain, S.R. Agnew "Effect of twinning on the mechanical behavior of a magnesium alloy sheet during strain path changes," *Magnesium Technology*, 2006 (2006), 219.
19. A. Chapuis, Y. Xin, X. Zhou, Q. Liu, "{10-12} Twin variants selection mechanisms during twinning, re-twinning and detwinning," *Materials Science and Engineering: A* 612 (2014), 431–439.
20. Y. Xin, X. Zhou, L. Lv, Q. Liu, "The influence of a secondary twin on the detwinning deformation of a primary twin in Mg–3Al–1Zn alloy," *Materials Science and Engineering: A* 606 (2014), 81–91.
21. C.R. Heiple, S.H. Carpenter, "Acoustic emission produced by deformation of metals and alloys – A review: Part II.," J. Acoustic Emission 6 (1987) 215-237.
22. S.H. Park, J.H. Lee, Y.H. Huh, S.-G. Hong, "Enhancing the effect of texture control using {1 0−1 2} twins by retarding detwinning activity in rolled Mg–3Al–1Zn alloy," *Scripta Materialia* 69(11-12) (2013), 797–800.
23. Y. Li, Y. Cui, H. Bian, S. Sun, N. Tang, Y. Chen, B. Liu, Y. Koizumi, A. Chiba, "Detwining in Mg alloy with a high density of twin boundaries," *Sci. Technol. Adv. Mater.* 15 (2014), 1-8.
24. L. Wang, G. Huang, Q. Quan, P. Bassani, E. Mostaed, M. Vedani, F. Pan, " The effect of twinning and detwinning on the mechanical property of AZ31 extruded magnesium alloy during strain-path changes," *Materials & Design* 63 (2014), 177–184.
25. T. Richeton, P. Dobron, F. Chmelik, J. Weiss, F. Louchet, "On the critical character of plasticity in metallic single crystals," *Materials Science and Engineering: A*, 424 (1–2) (2006), 190-195.
26. R.E. Reed Hill, R. Abbaschian, *Physical Metallurgy Principles*, (PWS Publ. Company, Boston 1994).
27. J.P. Toronchuk, "Acoustic emission during twinning of zinc single crystals," *Material Evaluation,* (1977), 51-53.

Magnesium Technology 2015
Edited by: Michele V. Manuel, Alok Singh, Martyn Alderman, and Neale R. Neelameggham
TMS (The Minerals, Metals & Materials Society), 2015

EFFECT OF Mn CONTENT ON MICROSTRUCTURES AND MECHANICAL PROPERTIES OF Mg-Al-Ca-Mn ALLOYS FABRICATED BY HIGH-SPEED EXTRUSION

Taiki Nakata[1], Kazunori Shimizu[2], Yasunobu Matsumoto[2], Satoru Hanaki[2], Shigeharu Kamado[1]
[1]Nagaoka University of Technology, 1603-1, Kamitomioka, Nagaoka, Niigata 940-2188, Japan
[2]Sankyo Tateyama, Inc. Sankyo Material-Company, 8-3, Nagonoe, Imizu, Toyama 934-8515, Japan

Keywords: Mg-Al-Ca-Mn alloy, high-speed extrusion, age hardening, texture, tensile property

Abstract

Mg-0.3Al-0.2Ca-xMn (x=0.1, 0.2, 0.4 at.%) alloys were extruded at an extraordinarily high speed of 60 m/min, and the effect of Mn content on the extrudability, age-hardening response, and tensile properties was investigated. An increase in Mn content improves the surface quality of the high-speed extruded alloys. Moreover, the as-extruded Mg-0.3Al-0.2Ca-0.4Mn alloy shows much higher strengths than the as-extruded Mg-0.3Al-0.2Ca-0.1Mn alloy and also exhibits enough elongation of about 25%; the proof stress was improved from 135 MPa to 190 MPa with an increase of Mn content from 0.1% to 0.4%. Whereas the Mg-0.3Al-0.2Ca-0.4Mn alloy shows a minor age-hardening by T5-treatment compared to the Mg-0.3Al-0.2Ca-0.1Mn alloy, the proof stresses of both alloys are increased about 20 MPa. The high proof stress despite of its minor age-hardening is attributed to the strong basal texture observed in the high Mn containing alloy.

Introduction

Global warming is becoming one of the serious problems in the world due to a recent increase in greenhouse gas emissions which is partially caused by transportation vehicles. From this point of view, wrought Mg alloys have a great potential as structural materials for weight reduction of transportation vehicles to improve their fuel efficiencies. However, commercial Mg alloys such as AZ alloy series show poor extrudability and lower strength compared to 6xxx series Al alloys [1, 2]. Thus industrial companies are eager for the development of Mg alloys with good extrudability and high strengths. The extrudability of Mg alloys can be enhanced by lowering the alloying elements such as Al and Zn [2-4]. However, Mg alloys containing a small amount of Al and Zn shows poor strengths due to low solid solution strengthening [2], and little precipitation hardening as well as the AZ31 alloy. On the other hand, Mg-0.3Al-0.3Ca (at.%) and Mg-0.3Zn-0.3Ca (at.%) alloys show a remarkable age-hardening because of the precipitation of Guinier-Preston zones lying on the basal plane of Mg matrix regardless of their low content of alloying elements [5, 6]. Considering these results, we believe that the dilute Mg-Al-Ca and Mg-Zn-Ca alloy systems are the promising materials which can provide better extrudability and higher strengths for the practical use. Besides, a trace amount of Mn addition to Mg-Al alloys produces nano-scale Al-Mn particles in Mg matrix; improvement in strengths by precipitation of Al-Mn particles is achieved [7]. Furthermore, Mn addition leads to a formation of strong basal texture and fine grains in wrought Mg alloys [8, 9]; even only a small amount of Mn improves the strength of wrought Mg alloys. In this work, we have extruded the Mg-0.3Al-0.2Ca dilute alloys which contains 0.1, 0.2, and 0.4 at.% Mn at high speed, and investigated the effect of Mn content on the microstructures and tensile properties of the high-speed extruded alloys.

Experimental Procedure

Alloy ingots with chemical compositions of Mg-0.3Al-0.2Ca-0.1Mn (0.1Mn alloy), Mg-0.3Al-0.2Ca-0.2Mn (0.2Mn alloy), and Mg-0.3Al-0.2Ca-0.4Mn (0.4Mn alloy) in at.% were prepared by direct chill (DC) casting. Table 1 summarizes the nomenclature and the chemical compositions in both at.%, and wt.% in parentheses.

Table 1: Nomenclatures and chemical compositions of the investigated alloys

at.% (wt.%)

Nomenclature	Al	Ca	Mn	Mg
0.1Mn	0.30(0.33)	0.21(0.34)	0.11(0.24)	Bal.
0.2Mn	0.31(0.34)	0.19(0.32)	0.19(0.43)	Bal.
0.4Mn	0.30(0.27)	0.19(0.31)	0.37(0.84)	Bal.

Before extrusion, homogenization heat-treatment was carried out for all the DC billets at 450 °C for 1 h in an Ar-atmosphere followed by water quenching. The homogenized alloys were extruded at 400 °C at a die-exit speed of 60 m/min with an extrusion ratio of 20. The extruded samples were subjected to an aging treatment at 200 °C; the age-hardening responses were measured using Vickers hardness tester (Mitsutoyo, HM-221) with a load of 500 g. Tensile properties of as-extruded and peak-aged (T5) alloys were evaluated by a tensile test at an initial strain rate of 10^{-3} s^{-1}, using the samples with a gauge length of 20 mm and a diameter of 4 mm. Microstructures of as-extruded alloys were characterized by an optical microscope (OLYMPUS, BX51M) and a scanning electron microscope (JEOL, FE-SEM7001F) equipped with an electron backscattered diffraction apparatus (TSL, Orientation Imaging Microscopy software).

Results and discussion

Figure 1 shows the bar-surfaces of all the alloys extruded at 400 °C at a die-exit speed of 60 m/min. Surface cracks are observed in the 0.1Mn and 0.2Mn alloys. In contrast, the 0.4Mn alloy exhibits a crack-free surface in spite of extruding at an extraordinarily high-speed of 60 m/min.

Figure 2 (a) shows the tensile stress-strain curves of the as-extruded alloys. The tensile strength is improved from 203 MPa to 229 MPa when the Mn content is increased from 0.1 at.% to 0.4 at.%. The 0.2 % proof stress is also enhanced with increasing the Mn content; the 0.2 % proof stress of the 0.1Mn, 0.2Mn and 0.4Mn alloys are 134 MPa, 167 MPa and 190 MPa, respectively. Figure 2 (b) shows the age-hardening curves of all the as-extruded alloys. The Vickers hardness of the 0.1Mn alloy is increased about 4 HV after the aging treatment for 8 h and reached to the peak hardness of 48 HV. The 0.2Mn and 0.4Mn alloys show the relatively high Vickers hardness of about 49 HV and 50 HV, respectively even just after the extrusion. However, these two alloys show a minor increase in Vickers hardness compared to the 0.1Mn alloy. Figure 2 (c) shows the tensile stress-stain curves of the T5-treated alloys. Both tensile strength and 0.2% proof stress of the T5-treated alloys are also improved with an increase in Mn content. The 0.2 % proof stress of the 0.1Mn alloy is improved about 14 MPa by the peak aging. The 0.1Mn alloy shows a higher increase in Vickers hardness by the peak aging compared to the 0.2Mn and 0.4Mn alloys as shown in Figure 2(a). However, the 0.2Mn and 0.4Mn alloys show relatively high enhancement in the 0.2 % proof stress of 19 MPa and 20 MPa, respectively, compared to the 0.1Mn alloy.

Figure 3 shows the inverse pole figure (IPF) maps, average recrystallized grain sizes, (0001) pole figures, and average (0001) $<11\bar{2}0>$ Schmid factors of the as-extruded alloys. These images were taken from the plane parallel to the extrusion direction. The 0.1Mn alloy shows fully-recrystallized structure. In contrast, the 0.2Mn and 0.4Mn alloys contain a small fraction of non-recrystallized grains. The average recrystallized grain sizes of the 0.1Mn, 0.2Mn and 0.4Mn alloys are 23 μm, 11 μm, and 13 μm, respectively. The alloy containing 0.2 at.% and 0.4 at.% Mn shows a significant reduction in the grain size. Whereas the 0.1Mn and 0.2Mn alloys show homogeneous grains at the recrystallized regions, both coarse grains of about 20 μm, which are elongated to the extrusion direction, and fine grains of about 2 -3 μm coexist in the 0.4Mn alloy. The average (0001) $<11\bar{2}0>$ Schmid factor of the 0.1Mn, 0.2Mn and 0.4Mn alloys are 0.31, 0.27, and 0.25, respectively; an increase in Mn content leads to a strong basal texture.

Figure 4 shows the optical microscopic images of the as-extruded alloys. These images were taken from the plane parallel to the extrusion direction. At the grain boundaries, black particles can be observed in all the alloys. The particles are considered to be an Al-Mn compound, and the numbers of this compound increase with increasing the Mn content of the alloys.

In this work, we have investigated the effect of Mn content on the microstructures and tensile properties of the high-speed extruded Mg-0.3Al-0.2Ca alloys. A high amount of Mn addition improves the quality of the extruded bar surfaces; the surface cracking vanished in the 0.4Mn alloy extrudate (Figure 1). Al-Mn compounds are formed in an alloy containing Al and Mn [7, 10], and the numbers of Al-Mn compounds increase with increasing the Mn content in the investigated alloys (Figure 4). The formation of Al-Mn compounds decreases the solid solubility of Al and Ca, and the extrudability of Mg alloys can be enhanced with decreasing the solid solution alloying elements [2-4]. Thus the Mg-Al-Ca-Mn alloy with high Mn content shows the good surface quality (Figure 1). Also, an increase in Mn content leads to significant refinements in the recrystallized grains (Figure 3). When the extrudates are exposed to a high temperature, secondary grain growth is expected to occur after extrusion. However, fine particles can suppress the grain growth due to Zener drag effect [11], and the numbers of Al-Mn compounds increase with an increase in the Mn content (Figure 4). As a result, the 0.2Mn and 0.4Mn alloys show finer grain sizes than the 0.1Mn alloy by a suppression of grain growth after extrusion. Huppman [12] reported that grains coarsened by static recrystallization shows weak basal texture. The 0.1Mn alloy has relatively large grains, which are expected to be coarsened after extrusion, compared to the 0.2Mn and 0.4Mn alloys (Figure 3). In contrast, the 0.4Mn alloy possesses high fraction of fine grains obtained by a suppression of grain growth due to Al-Mn compounds (Figure 3 and 4). Thus the 0.4Mn alloy exhibits strong basal texture while the texture of the 0.1Mn alloy is weakened. In the as-extruded condition, the 0.4Mn alloy shows the highest tensile strength and 0.2% proof stress among all the investigated alloys. The strengthening mechanisms for conventional extruded Mg alloys are mainly divided into solute-strengthening, fine-grain strengthening, and texture strengthening [7, 13]. Because of their low contents of solute elements, we can ignore the solute-strengthening effect. In addition, considering the Hall-Petch coefficient is 2.61 MPa $mm^{1/2}$ in an Mg alloy [14], the difference of the 0.2% proof stresses between the 0.1Mn and 0.4Mn alloys is estimated to be 6 MPa which is significantly lower than the experimental value. Yi et al. reported that the activity of basal slip plays an important role in initiating strain in an extruded AZ31 alloy sheet [15]. When tensile test carried out parallel to the extrusion direction, the plastic deformation begins with activation of basal slip even with its low Schmid factor. The relative activity of basal slip reaches to more than 0.9 up to the equivalent strain of 0.01. Thus, we conclude that the highest 0.2% proof stress of the 0.4Mn alloy among all the as-extruded alloys is attributed to the strong basal texture (Figure 3). Both tensile strength and 0.2 % proof stress are increased by T5-treatment. Jayaraj [5] reported that monolayer ordered Guinier-Preston zones are formed in an aged Mg-Al-Ca alloy system. The investigated alloys used in this study have similar compositions as mentioned by Jayaraj. Thus the enhancement in both tensile strength and 0.2 % proof stress is attributed to the formation of monolayer ordered Guinier-Preston zones in the Mg-matrix. Although the 0.4Mn alloy shows the lowest increment in Vickers hardness, this alloy shows the highest improvement in the 0.2% proof stress, and the 0.2% proof stress attain to 210 MPa. The highest increase in the 0.2% proof stress is also attributed to the strong basal texture of the 0.4Mn alloy. The Orowan strengthening in tensile stress, $\Delta\sigma$, is estimated from [7]:

$$\Delta\sigma = \Delta\tau / M$$

where $\Delta\tau$ is the increment in CRSS by precipitates and M is the Schmid factor of metals. The 0.4Mn alloy shows the lowest (0001) $<11\bar{2}0>$ Schmid factor among all the investigated alloys. Thus the enhancement in the 0.2 % proof stress by T5-treatment is highest despite of its minor age-hardening.

Summary

In summary, high-speed extrusion is performed using Mg-Al-Ca-Mn alloys containing different amount of Mn. Addition of large amount of Mn to Mg-0.3Al-0.2Ca (at.%) alloy improves the surface quality of the extruded alloys, refines the recrystallized grain size due to the pinning effect by fine Al-Mn compounds, and sharpens the basal texture. Consequently, the tensile strength and 0.2 % proof stress are enhanced due to the formation of strong basal texture in the high Mn containing alloy, and enough elongation of more than 20% is also obtained because of the fine grains. Particularly, the T5-treated 0.4Mn alloy shows the highest 0.2 % proof stress of 210 MPa.

Acknowledgement

This work was supported by JSPS Grant-in-Aid for Scientific Research (A) Grant Number 25249101, and JST, Advanced Low Carbon Technology Research and Development Program (ALCA).

References

1. D. L. Atwell and M. R. Barnett, "Extrusion Limits of Magnesium Alloy," *Metall. Mater. Trans. A*, 38 (2007), 3032-3041.

2. C. Davis and M. Barnett, "Expanding the Extrusion Limits of Wrought Magnesium Alloys," *Journal of Minerals, Metals, and Materials Society*, 56 (November, 2004), 22-24

3. T. Murai et al., "Effects of zinc and manganese contents on extrudability of Mg-Al-Zn alloys", *Journal of Japan Institute of Light Metals*, 53 (2003), 27-31

4. T. Murai, "Extrusion of magnesium alloys", *Journal of Japan Institute of Light Metals*, 54 (2004), 472-477

5. J. Jayaraj et al., "Enhanced precipitation hardening of Mg-Ca alloy by Al addition", *Scripta Materialia*, 63 (2010), 831-834

6. K. Oh-ishi et al., "Age-hardening response of Mg-0.3Ca at.% alloys with different Zn contents", *Material Science and Engineering A*, 526 (2009), 177-184

7. N. Stanford and D. Atwell, "The Effect of Mn-rich Precipitates on the Strength of AZ31 Extrudates", *Metallurgical and Materials Transactions A*, 44 (2013), 4830-4843

8. N. Stanford, "The effect of calcium on the texture, microstructure and mechanical properties of extruded Mg-Mn-Ca alloys", *Materials Science and Engineering A*, 528 (2010), 314-322

9. L. B. Tong et al., "Effect of Mn addition on microstructure, texture and mechanical properties of Mg-Zn-Ca-alloy", *Materials Science and Engineering A*, 528 (2011), 3741-3747

10. A. Jansson, "A Thermodynamic Evaluation of the Al-Mn System", *Metallurgical Transactions A*, 23 (1992), 2953-2962

11. F. J. Humphreys and M. Hatherly, RECRYSTALLIZAION and Related Annealing Phenomena Second Edition (United Kingdam, UK: Elsevier Society, (2004), 109

12. M. Huppmann et al., "Changes of the texture and mechanical properties of the extruded Mg alloy", *Materials Science and Engineering A*, 528 (2010), 342-354

13. L. Zhang et al., "Microstructure and strengthening mechanism of a thermomechanically treated Mg-10Gd-3Y-1Sn-0.5Zr alloy", *Material Science and Engineering A*, 565 (2013), 262-268

14. Y. Wang and H. Chool, "Influence of texture on Hall-Petch relationships in an Mg alloy", *Acta Materialia*, 81 (2014), 83-97

15. S.-B.Yi et al., "Deformation and texture evolution in AZ31 magnesium alloy during uniaxial loading", *Acta Materialia*, 54 (2006), 579-562.

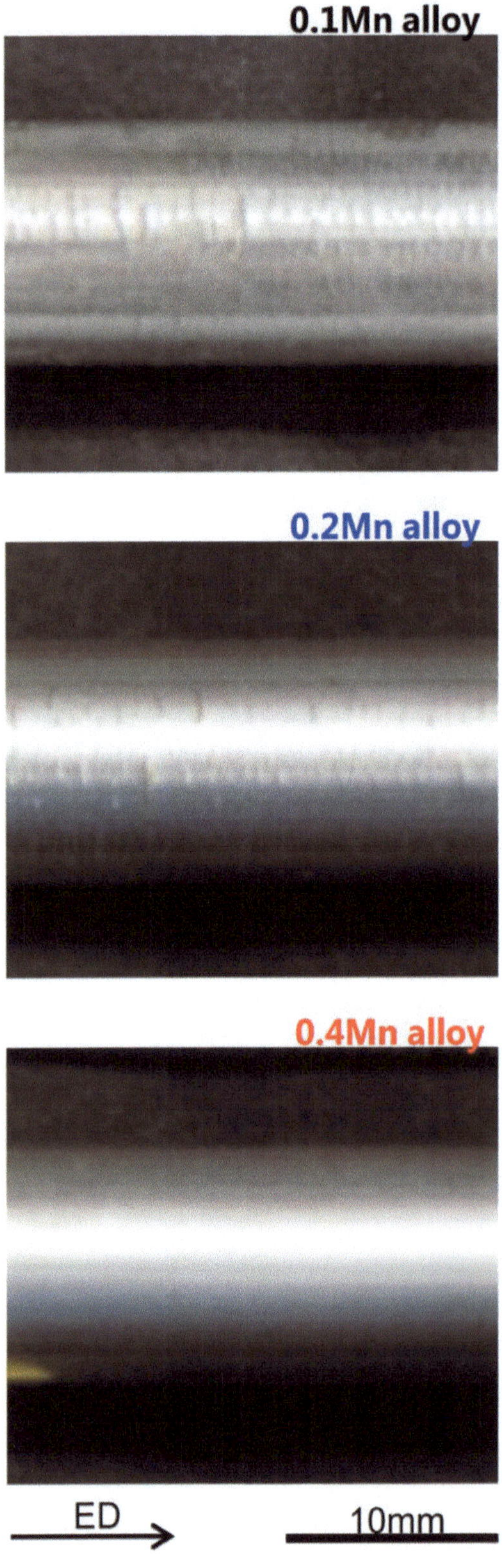

Figure1: Bar surfaces of the alloys extruded at 400 °C at a high die-exit speed of 60 m/min.

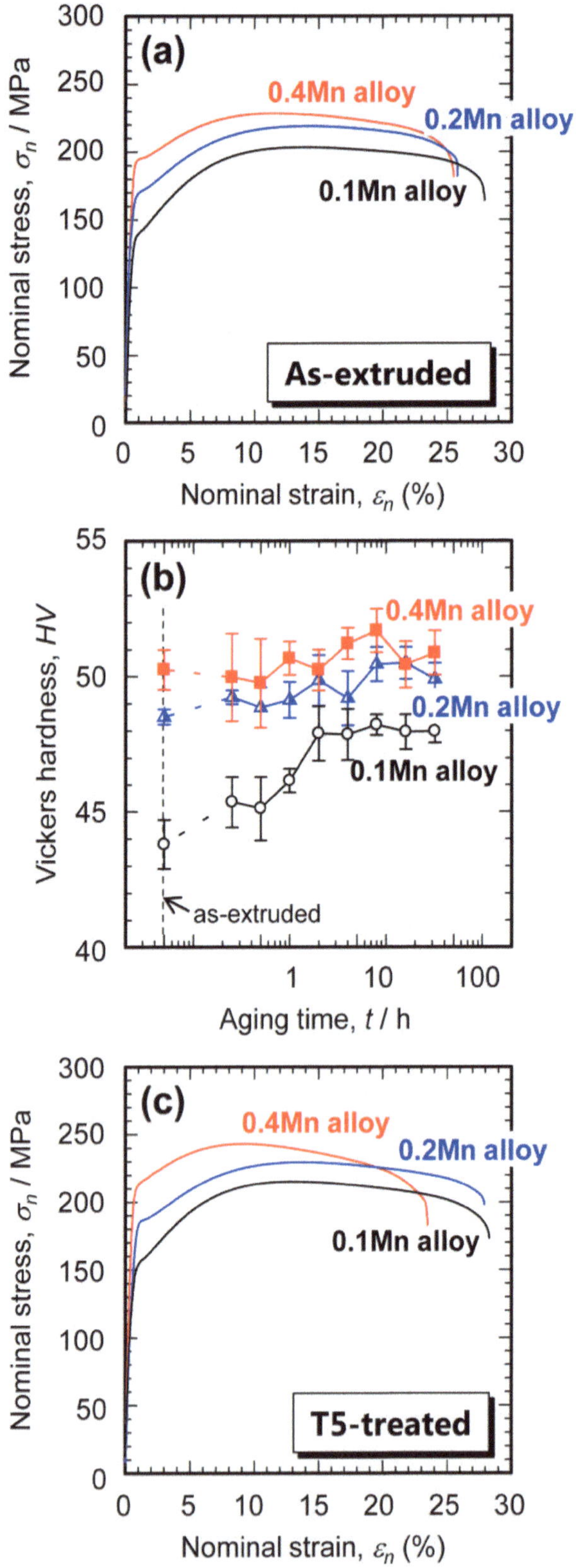

Figure 2: (a) Tensile stress-strain curves of the as-extruded alloys. (b) Age-hardening curves of the alloys. (c) Tensile stress-strain curves of the T5-treated alloys.

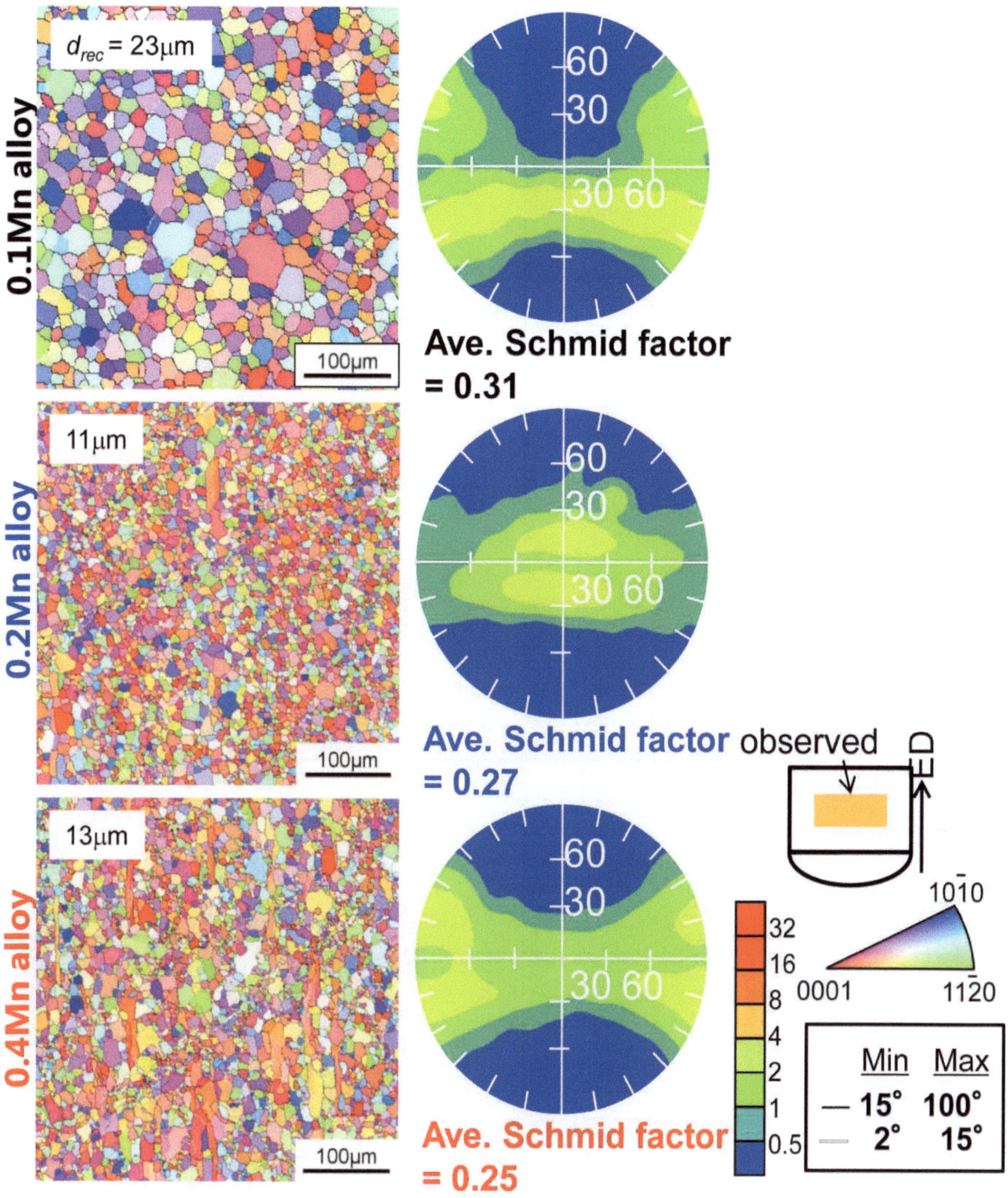

Figure 3: Inverse pole figure (IPF) maps, average recrystallized grain sizes, (0001) pole figures, and (0001) <11$\bar{2}$0> Schmid factors of the as-extruded alloys. Observed surfaces are parallel to the extrusion direction.

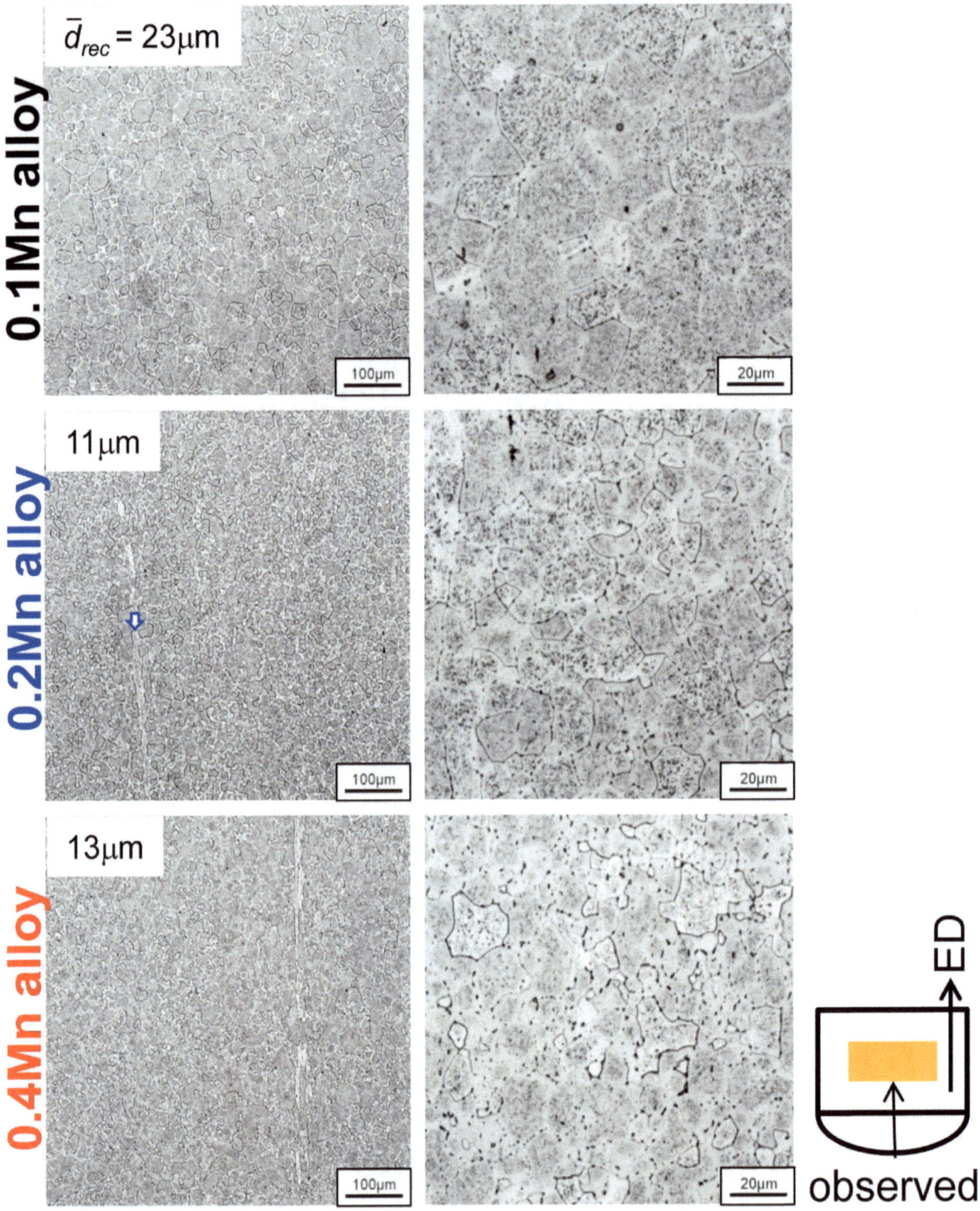

Figure 4: Optical microscopic images and the recrystallized grain sizes of the as-extruded alloys. Observed surfaces are parallel to the extrusion direction.

Magnesium Technology 2015
Edited by: Michele V. Manuel, Alok Singh, Martyn Alderman, and Neale R. Neelameggham
TMS (The Minerals, Metals & Materials Society), 2015

EFFECTS OF EXTRUSION PROCESSING AND HEAT TREATMENT ON MECHANICAL PROPERTY AND HEAT DISSIPATION PERFORMANCE OF Mg-2.5Nd-1.0Zn-0.5Zr ALLOY

Jixu Wang, Jieyu Zhang, Guangxin Wu, Wangjun Peng
Shanghai Key Laboratory of Modern Metallurgy & Materials Processing, Shanghai University

Keywords: Extrusion, Heat treatment, Mechanical property, Dissipation performance

Abstract

As heat dissipation material, magnesium alloys have smaller specific gravity and bigger specific heat capacity than copper alloys and aluminum alloys. The object of this study focused on the Mg-2.5Nd-1.0Zn-0.5Zr (wt.%) alloy. A research on Mg-xNd-yZn-0.4Zr (x=1.25-3.25,y=0-2.0,wt.%) alloys had been carried out systematically, but there was few reports about the research on heat dissipation performance of Mg-Nd based alloys. In this work, the effects of hot extrusion and heat treatment on mechanical properties and thermal performance of Mg-2.5Nd-1.0Zn-0.5Zr (wt.%) alloy were studied. The results show that the heat treatment and extrusion processing can improve the ultimate tensile strength (UTS) from 188.2Mpa to 231.2MPa. While the breaking elongation is also improved from 1.1% to 2.3%. Then, a patented way[1] to evaluate the dissipation performance is used in this work. And results reveal that with proper solution treatment, aging treatment and extrusion processing, the heat dissipation performance of this alloy can also be improved obviously.

1. Introduction

Magnesium alloys are one of the most promising candidates for practical application in heat dissipation field. As its smaller specific gravity and bigger specific heat capacity, it is quite suitable to be used as heat dissipation material, especially for electronic products such as mobile phones and computers. Studies[2-3] have shown that working temperature of LED increases from 63°C to 74°C, the service life of the LED decreases by 3/4. So the research and development of heat dissipation materials and finding the methods to improve the thermal conductivity is very important[4].

Magnesium alloys with different composition have quite different properties. Therefor the composition of Magnesium alloys is very important for practical application. Nd is a rare earth element, with maximum solubility in solid Mg of 3.6 wt% at eutectic temperature which is showed in the phase diagram. Li Mingzhao[5] studied the effects of Nd on AZ31 alloy with Scanning Electron Microscope (SEM) , Energy Dispersive Spectrometer (EDS) , X-Ray Diffraction (XRD) etc. The results show Nd can improve the microstructure and mechanical properties. A small amount of Zn can also contribute to strength due to solid solution strengthening. Zr is an effective grain refining agent in magnesium alloys without Al. The contribution of Zr to the strengthening due to the formation of fine grains by grain boundary strengthening. Heat treatment and extrusion are typical processes that can refine grains and improve mechanical properties. In the present work, the mechanical property and heat dissipation of Mg-2.5Nd-1.0Zn-0.5Zr(wt.%) alloy with or without extrusion and heat treatment were studied.

2. Experimental

Mg-Nd-Zn-Zr as-cast alloy is used as raw material, and the chemical components are showed in table I.

Table I. Mg-Nd-Zn-Zr alloy composition (wt. %)

	Nd	Zn	Zr	Mg
composition	2.8	0.85	0.64	95.71

Four specimens with different treatments which are showed in table II were analyzed. The four samples are named A, B, C and D. (A = Not Extruded and Not Heat Treated, B = Not Extruded and Heat Treated, C =Extruded and Not Heat Treated, and D = Extruded and Heat Treated). The extrusion ratio of sample C and sample D is 16, and the heat treatment of sample B and sample D is 4 hours solution treatment at 525 ℃ and 50 hours aging treatment at 200 ℃.

Table II. Processing condition of the specimens

Specimens	Extrusion ratio16	Solution treatment (525℃)	Aging treatment (200℃)
A	×	0h	0h
B	×	4h	50h
C	√	0h	0h
D	√	4h	50h

The microstructure and the appearance of fracture of the 4 specimens A, B, C and D were observed by SEM. The phase composition was identified by XRD.

The specimens in different conditions possess different heat dissipation performance, in this work, a patented way[1] was used to evaluate it.

The tensile tests were carried out at room temperature with a rate of 1.0mm/min. The schedule of tensile test specimen is showed in figure 1. The extrusion processing is carried out, and the ratio is 16 at 400℃, and the rate is 10mm/s.

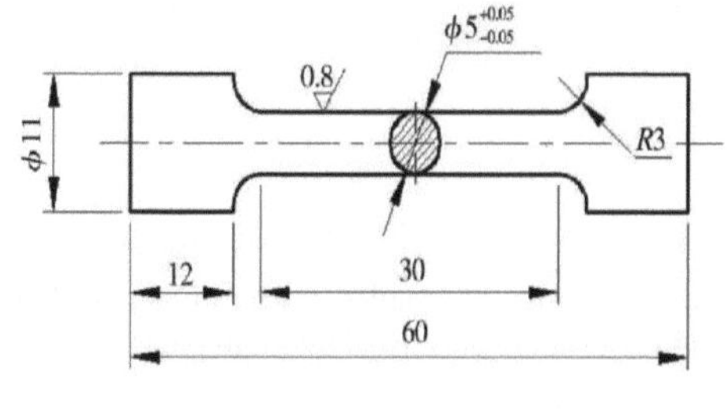

Fig.1 Schedule of the tensile test specimen.

3. Results and discussion

3.1 Microstructure

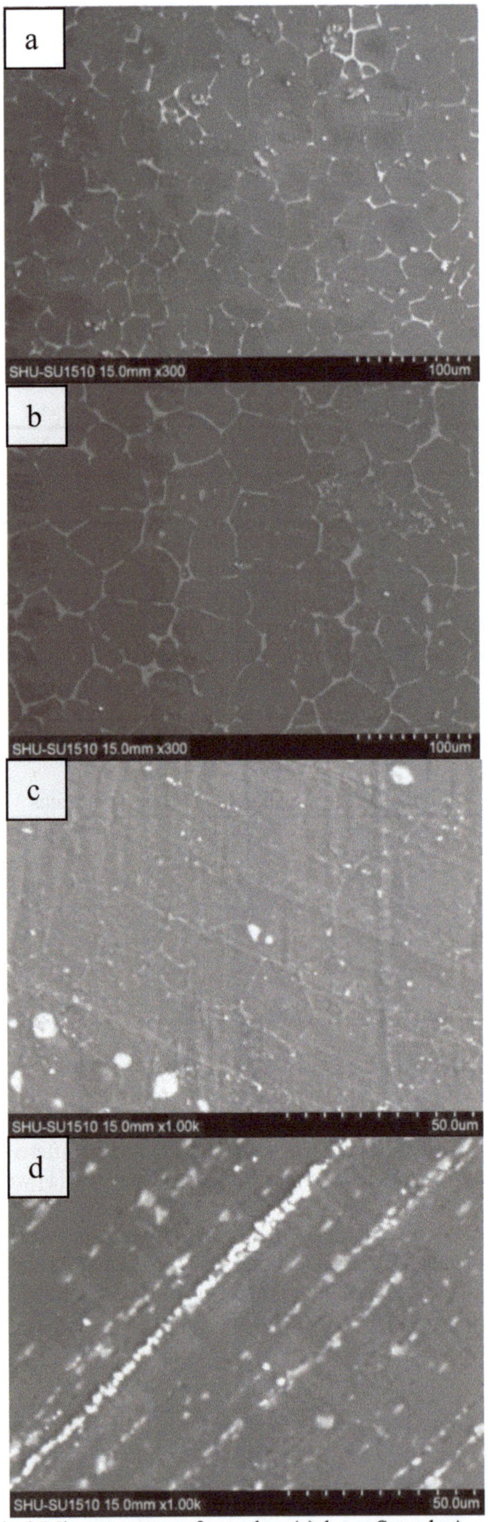

Fig.2 Microstructure of samples. (a)shows Sample A, not extruded and not heat treated, (b)shows Sample B, not extruded and heat treated, (c)shows Sample C, extruded and not heat treated (d)shows Sample D, extruded and heat treated .

The microstructure of the 4 specimens was shown in Fig. 2. It can be found that the alloy is mainly composed by α-Mg and the second phase as shown in Fig.2a. Compare Fig.2 a. and b, it shows that the grains grow up after the heat treatment and the grain boundary is more clearly.

The line scanning result of EDS as shown in Fig.3 indicates that the Nd and Zn mainly exist in the second phase. Zr distributed evenly.

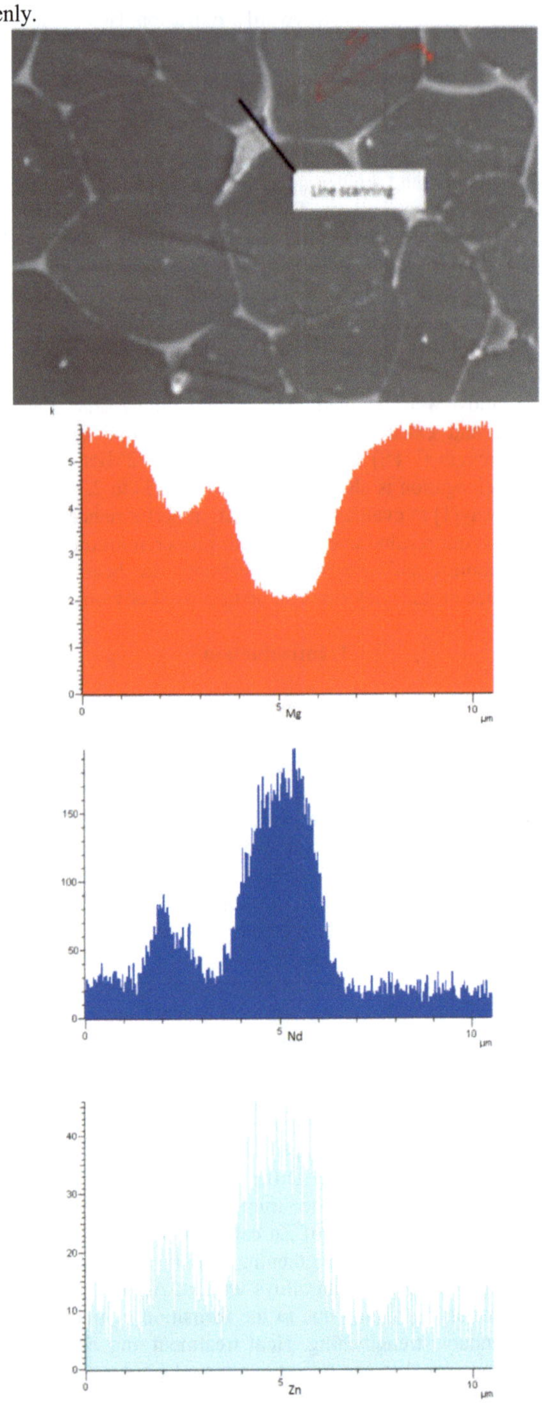

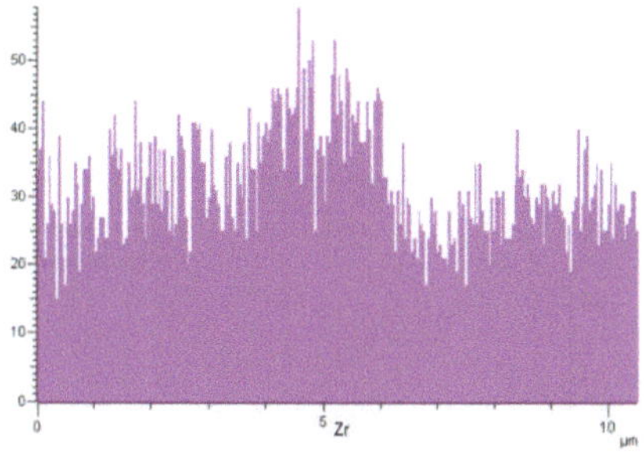

Fig.3 Line scanning result of EDS of the sample A.

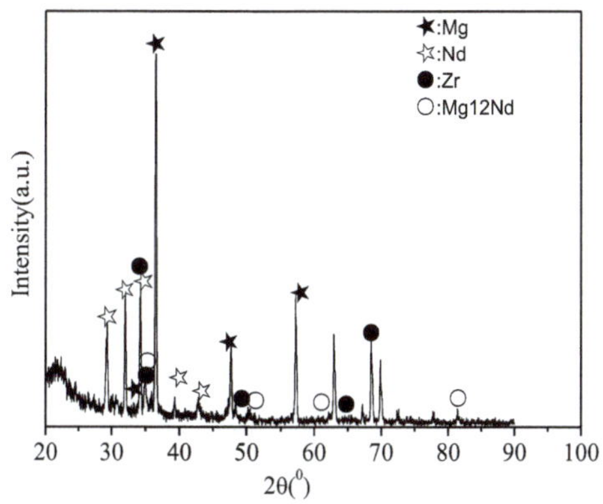

Fig.4 XRD pattern of sample A.

The XRD result indicates the α-Mg and $Mg_{12}Nd$ phases are in the alloy matrix as shown in Fig.4. The metallographic structure of extruded alloy is showed in Fig.2(c), the grain size was significantly decreased. The microstructure of extruded alloy with heat treatment is showed in Fig. 2(d), the grain size is smaller than the extruded alloy without heat treatment.

3.2 Mechanical properties

The tensile curve is showed in Fig.5, and the mechanical properties is showed in Table III. The results indicate that solution treatment and aging treatment can obviously improve the mechanical performance of the alloy without extrusion processing. During the aging treatment many solution element and precipitated phase precipitated out, so the precipitation of the solution element and precipitated phase contributes to the mechanical property improvement.

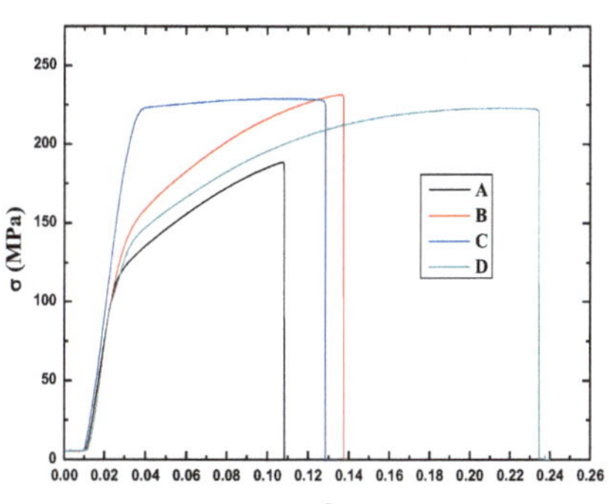

Fig.5 Tensile curve of the specimens.

Table III Mechanical properties

Samples	UTS(MPa)	YS(MPa)	Elongation (%)
A	188.2	111.8	1.1
B	231.2	133.3	1.4
C	229.9	227.3	1.3
D	223.5	131.2	2.3

It can be found that the extrusion processing can also improve the mechanical property. The extruded alloy without heat treatment has a stable stress yielding process shows that it's brittle rupture. And the extruded alloy with heat treatment has better plasticity, the elongation is significantly improved. But at the same time the yield strength decreased a little. Compare the as-cast alloy, the as-cast alloy with heat treatment and the extruded alloy without heat treatment, the extruded alloy with heat treatment has a better mechanical property and plastic performance.

The appearance of fracture is showed in Fig.6. Magnesium alloys have close-packed HCP structure with low symmetry and its c/a is 1.623, close to the ideal close-packed c/a of 1.633 and less slip systems at room temperature. Its fracture behavior is generally cleavage fracture or quasi-cleavage fracture. A typical fracture appearance of the as-cast alloy stretch is shown in Fig. 6(a). Cast alloy fracture can obviously see a large tear edge, cleavage facets and grain boundaries. Cast alloy is constituted by α-Mg matrix and $Mg_{12}Nd$ phase which is the second eutectic phase and $Mg_{12}Nd$ is composed brittle and complex shapes, therefor during the stretching process, the stress accumulated at the grain boundaries, resulting in local strength concentration of $Mg_{12}Nd$ and where is easy to product microcracks. Since there is no strengthening phase, after the microcracks come out through the second phase, the crack will cross the grain and it will be cleavage fracture. It can also extend through the second phase, so the fracture is structured by the cracked second phase and cleavage plane. Overall it is a quasi-cleavage fracture.

The sample appearance of sample B, C and D are also quasi-cleavage fracture as shown in Fig.6. Compared with the sample A, B and C, sample D has more torn edges, it explained why sample D has better plastic performance.

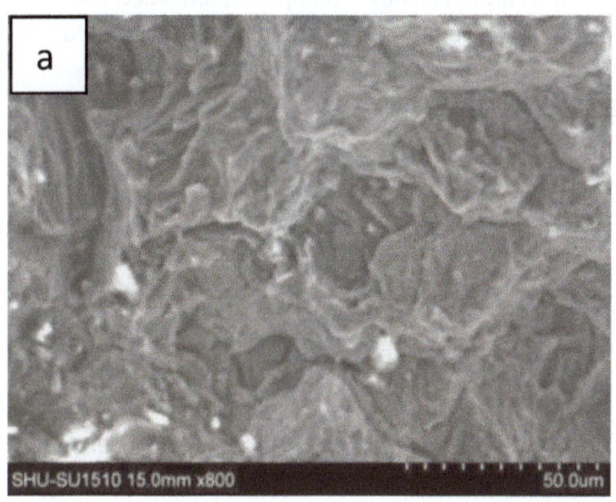

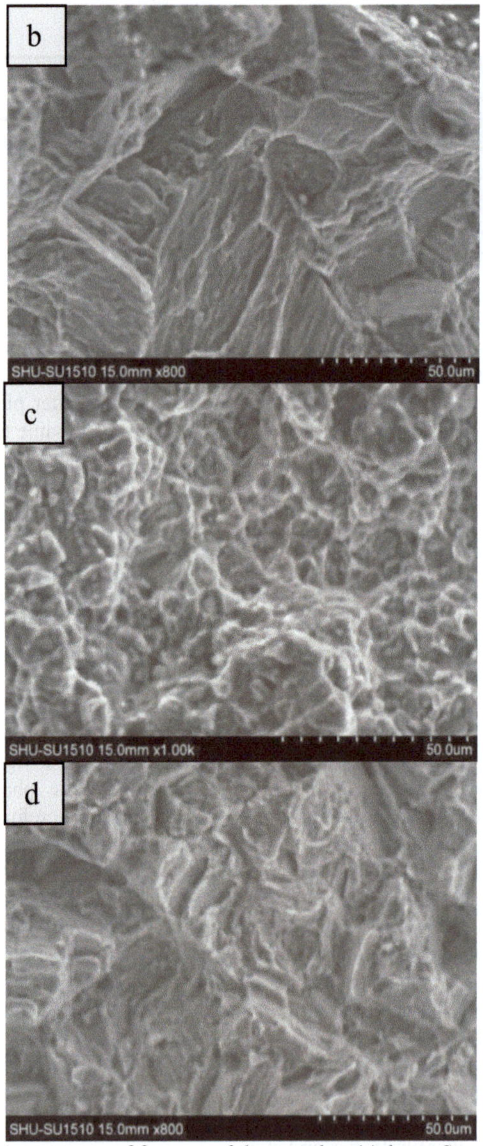

Fig.6 Appearance of fracture of the samples. (a)shows Sample A, not extruded and not heat treated, (b)shows Sample B, not extruded and heat treated, (c)shows Sample C, extruded and not heat treated (d)shows Sample D, extruded and heat treated .

3.3 Thermophysical property analysis

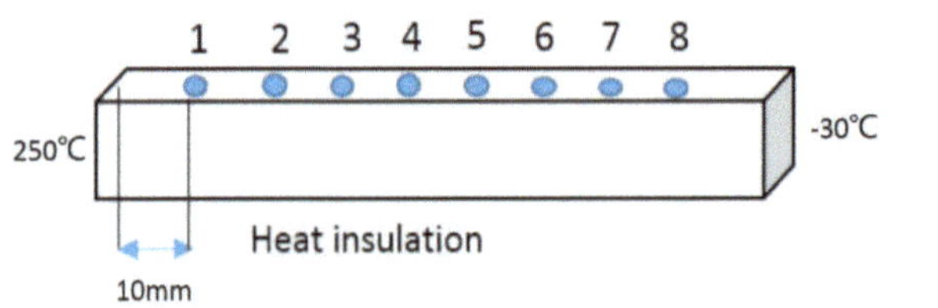

Fig.7 Sketch of one-dimensional heat transfer boundary condition.

The heat transfer sketch is shown in Fig.7, the hot side is kept at 250℃, and the cold side is kept at -30℃. Other surfaces are heat insulation. Since the extruded alloy has better plastic processing performance, the thermophysical property of sample C and sample D is compared in this work. The temperature distribution is shown in Fig.8. In the same boundary condition, the sample with heat treatment can make the cold side temperature lower, it proved that the heat treatment can improve the heat dissipation performance of the alloy.

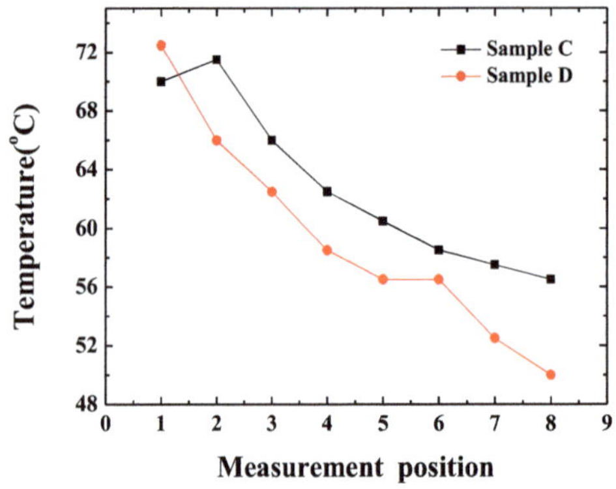

Fig.8 The temperature distribution curve of the sample C (extruded and not heat treated), and sample D(extruded and heat treated).

Conclusion

The themophysical property and mechanical property of Mg-2.5Nd-1.0Zn-0.5Zr (wt.%) alloy with or without extrusion and heat treatment were investigated. Main results obtained in this study are following:

(1)The microstructure of the extruded alloy is refined obviously.

(2)The grain refinement and precipitation strengthening contributes to the mechanical properties improvement of the extruded alloy.

(3)Heat treatment on the extruded alloy improve the elongation significantly but reduces the strength slightly.

(4)The heat treatment can also improve the heat dissipation performance of the extruded alloy.

References

[1]Jieyu Zhang, Jixu Wang. Device and method to evaluate heat dissipation performance. China Patent No. 201410286343.9

[2]H.J. Yu, and H. Y. You. “Development orientation of LED thermal conductive board,” China Illuminating Engineering Journal, 1-2 (2010), 76-80.

[3]M. Arik, C. Becker, S. Weaver et al. “Thermal management of LEDs: package to system” .Proceedings of the SPIE-The International Society for Optical Engineering, 5187(2004), 64-75.

[4]B. Zhang. “Cooling with Al-Mg-Si alloy microstructure and properties researc,” HuNan: Central South University, 2003.

[5] E Z.L. Ning , H. Wang , H.H. Liu ， et al ”Effects of Ndon microstructures and properties at the elevated temperature of a Mg–0.3Zn–0.32Zr alloy,” Materials and Design, 31 (2010), 4438–4444.

Magnesium Technology 2015

Forming and Alloy Design

Magnesium Technology 2015
Edited by: Michele V. Manuel, Alok Singh, Martyn Alderman, and Neale R. Neelameggham
TMS (The Minerals, Metals & Materials Society), 2015

EFFECT OF INITIAL MICROSTRUCTURES ON COLD-ROLLABILITY OF RE-FREE MAGNESIUM ALLOYS

Young Min Kim[1,2], Su Mi Jo[1], Bong Sun You[1,2]
[1]Advanced Materials Engineering, Korea University of Science and Technology, 217 Gajungro, Yuseong-gu, Daejeon 305-350, Korea
[2]Light Metal Division, Korea Institute of Materials Science, 797 Changwondaero, Seongsan-gu, Changwon 642-831, Korea

Keywords: magnesium alloys, cold rolling, recrystallization, work hardening exponent

Abstract

The effects of alloying elements and processing parameters on cold-rollability of RE-free magnesium alloys were investigated. Several representative alloys with different microstructural features such as solution or precipitation hardening were selected and extruded to control the initial microstructures by changing extrusion speed and temperature before cold-rolling. The results show that the uniform elongation and the work hardening exponent reflecting inclusive microstructural features of grain size, precipitates and dislocation density is linearly correlated with cold-rollability. It can be concluded that an alloy with weak basal texture can possess improved cold-formability depending on microstructural features effectively affecting work hardening exponent besides basal texture intensity. In particular the fraction of precipitates and deformed grains sensitively affected work hardening exponent and thus the cold-rollability of magnesium alloys. It was confirmed that poor cold-rollability of the precipitation-hardened alloys such as TAZ531 and TAZ711 could be improved by removing precipitates and deformed grains via a homogenization treatment.

Introduction

Magnesium alloys with the lightest density among structural metals and high specific strength have been already used for portable electronic devices via die-casting. Recently potential application of sheet-type of magnesium alloys to transportation vehicles such as automotive, high-speed train and aerospace is attracting more attention to obtain considerable weight saving and thus reduce CO_2 production. Many efforts have been made extensively on the development of new wrought magnesium alloys with high strength and excellent formability. However, due to the poor room temperature formability of magnesium alloys resulted from their hexagonal crystal structure and the associated insufficient independent slip systems, they are necessary to be processed at elevated temperatures leading to a considerable increase in the cost of sheet production against competing metals such as aluminum and steels.

It has been known that a considerable decrease in basal texture intensity through the activation of non-basal slip and twining is essential for enhancing cold-formability of magnesium alloys [1-7]. The addition of high amount of rare earth elements and the application of high temperature processing and heat treatment were also reported to improve formability and rollability at room temperature [2-6]. Hantzsche et al. [5] and Wu et al. [6] investigated the influence of rare earth elements on texture weakening. X. Huang et al. [4] increased the rolling temperature from 350°C to 525°C and found a significant decrease in basal texture intensity of AZ31 and resultantly enhanced cold deep drawability. With respect to the relation of cold-formability with microstructural features, Kang et.al. [8] found a linear relation of their results of limit dome height (LDH) test with work hardening exponent in magnesium alloys and suggested that magnesium alloys with larger grains had higher LDH values. On the contrary, Huang et al. [3] found the opposite effect of grain size from their results that a grain-coarsening with maintaining basal texture intensity led to a deterioration of both stretch formability and deep drawability at a room temperature [3]. Besides the grain size effect on cold-formability which is still questionable, the effects of other microstructural factors such as precipitates and dislocation density have not been understood so far.

The cold-rolling behavior in magnesium alloys is basically different from the deformation mode of deep drawing; in other words, the cold-rolling is governed by the plane strain tension, but stretching and drawing are governed by the bi-axial tension mode. Similar to cold-formability, nevertheless the cold-rollability of magnesium alloys can be also affected by various microstructural features. However, there has been little information on the relation of microstructural features and cold-rollability in various alloy systems, especially RE-free alloy systems, with different microstructural features, while the deformation behavior and the associated texture change of AZ31 alloy during cold rolling have been well understood [7]. In the present study, we have systematically investigated the influences of initial microstructures on cold rollability of RE-free magnesium alloys. It is necessary to understand how to control initial microstructures for the improvement in cold rollability of magnesium alloys without expensive rare earth elements and special rolling processes, which cause considerable cost increase.

Experimental Procedure

Table 1. The extrusion conditions of the initial materials in the present study.

Alloys	Extrusion conditions		
	Extrusion temperature (T_E)	Ram speed (V_B)	Extrusion ratio (ER)
AZ61 ZK60 TAZ531 TAZ711	250°C, 300°C	0.5, 3.0 mm/s	30:1

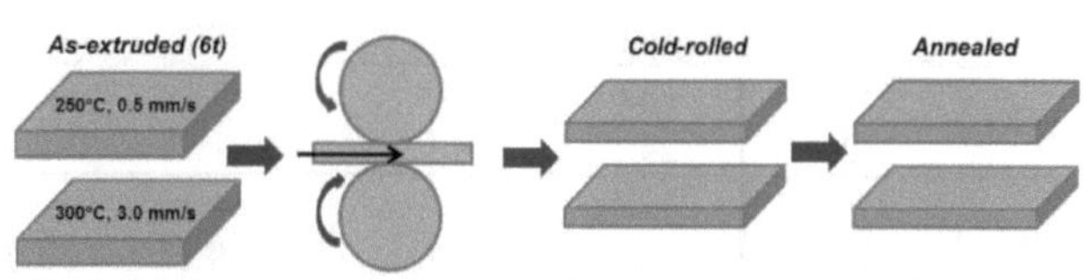

Fig. 1. Schematic illustration of cold-rolling and subsequent annealing

Mg-6Al-0.7Zn-0.2Mn (AZ61), Mg-6Zn-0.5Zr (ZK60), Mg-5Sn-1Al-1Zn (TAZ531) and Mg-7Sn-1Al-1Zn (TAZ711) alloys were selected to produce different types of microstructural features such as grain size and precipitates showing thus different strain hardening behavior. For a precise control of initial microstructure, the extruded sheets with the dimensions of 28 mm width × 6 mm thickness produced by different conditions as shown in Table 1 were prepared for cold rolling experiments. Fig. 1 shows overall experimental procedure of rolling and annealing. The rolling was carried out at room temperature with a rolling mill in which twin rolls with the diameter of 280 mm were rotated at a constant speed of 5 mpm (meter per minute) or 0.595 rad/s. The thickness reduction of each pass was approximately 3% and the rolling was repeated until the samples were broken. We defined the rollability at room temperature (i.e., cold-rollability) as an accumulated rolling reduction just before such a fracture occurrence. The rolled sheets were annealed at 250°C for 20 minutes in order to obtain recrystallized microstructures. Specimens for microstructure observation were prepared by mechanical polishing with a polycrystalline diamond suspension glycol based solution. The optical micrographs were taken after etching with a solution of picric acid (5 g), acetic acid (5 ml), distilled water (10 ml), and ethanol (100 ml). All microstructural analyses of the rolled sheets were performed on the plane normal to the transverse direction. Average grain size, area fraction of recrystallized grains and maximum intensity of (0002) pole figure were measured using EBSD. Sheet-type samples for tensile test with a gage width of 6 mm and a gage length of 25 mm were prepared in the longitudinal direction of the rolled sheets, and were tested at room temperature at a crosshead speed of 1.5 mm/min.

Results and Discussion

The initial microstructures of as-extruded sheets before cold rolling are shown in Fig. 2, and the detailed microstructural features are summarized in Table 2. The alloy sheets extruded under the condition of relatively fast ram speed (3.0 mm/s) and higher temperature (300°C) have microstructures characterized as homogeneous structures with coarse grains and larger fraction of recrystallized grains, compared with those extruded slowly (0.5 mm/s) at low temperature (250°C). In the cases of TAZ531 and TAZ711 which has large amount of Mg_2Sn precipitates [9], the extruded sheets under 250°C-0.5 mm/s tend to have large non-recrystallized grains elongated along the extrusion direction and resultantly higher maximum intensity of (0002) pole figure. Fig. 3 shows strain-stress curves of the extruded sheets with different microstructural features as shown in Table 2 where the tensile data measured from the curves are also listed. The alloys used in the present study are divided by two groups in terms of yield ratio (YR, yield strength/tensile strength); that is, AZ61 and TAZ531 alloys have lower yield ratio indicating that they are capable of accommodate external plastic deformation. On the contrary, ZK60

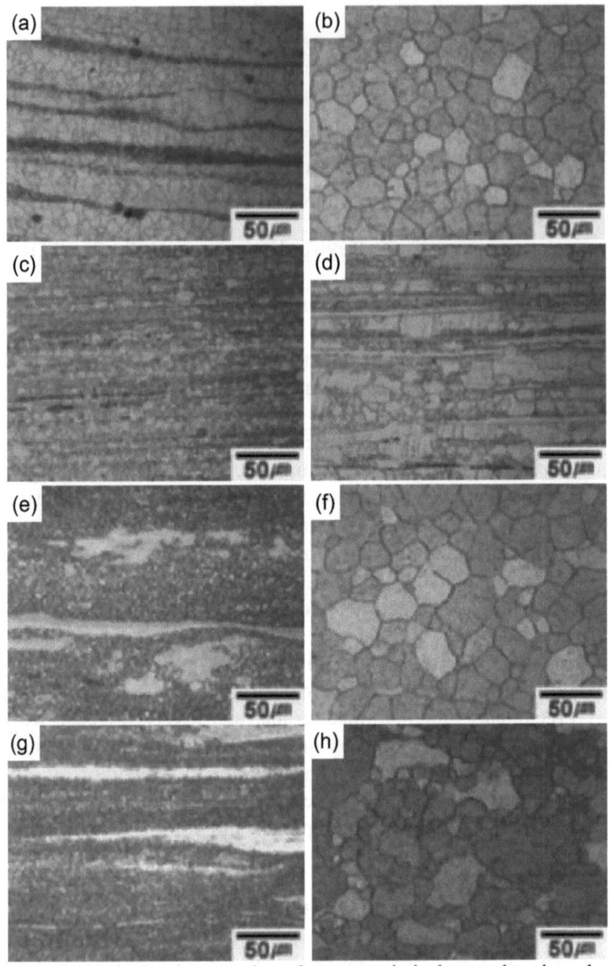

Fig. 2. Optical micrographs of as-extruded sheets showing the initial microstructures before cold rolling: (a and b) AZ61, (c and d) ZK60, (e and f) TAZ531, (g and h) TAZ711 alloys. The left figures (a, c, e and g) were taken from the samples extruded under the condition of 250°C-0.5 mm/s and the right ones (b, d, f and h) were taken under the extrusion condition of 300°C-3.0 mm/s.

Table 2. Microstructural features, tensile properties and work hardening exponent of the extruded samples

Alloy	T_E - V_B	AGS (μm)	$I_{max,(0002)}$	vol. frac. of RXed grain	vol. frac. of ppt	YS (MPa)	UTS (MPa)	El. (%)	U.El. (%)	YR (%)	N
AZ61	250-0.5	8.4	7.3	87	3.1	167	304	27	22.2	55	0.249
	300-3	16.0	9.2	97	0.4	155	291	24	21.7	53	0.267
ZK60	250-0.5	13.6	10.4	65	1.6	210	290	24	13.3	72	0.127
	300-3	24.1	8.1	71	1.4	209	303	27	16.5	69	0.183
TAZ531	250-0.5	12.4	16.1	84	9.3	184	312	12	12.1	59	0.186
	300-3	28.4	7.7	98	5.2	159	279	14	16.1	57	0.208
TAZ711	250-0.5	8.1	11.4	81	15.8	222	309	11	10.4	72	0.148
	300-3	24.5	8.4	90	18.3	189	287	13	10.9	66	0.159

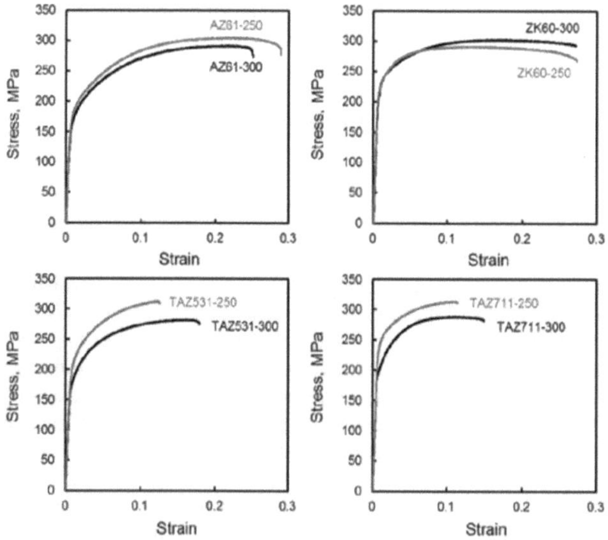

Fig. 3. Strain-stress curves of the extruded sheets with different microstructures.

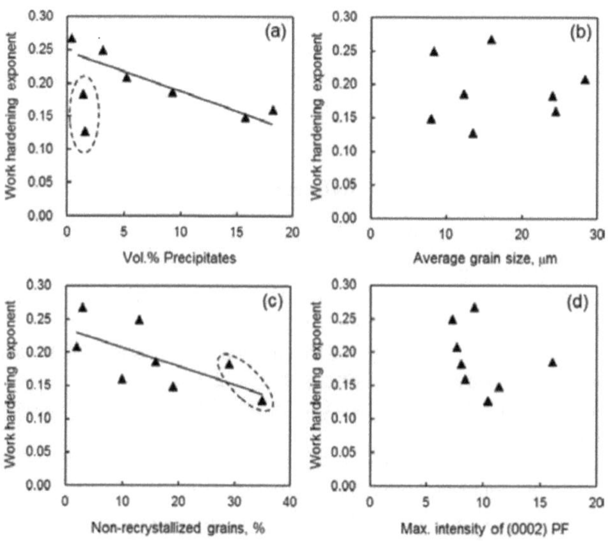

Fig. 4. The relationship of work hardening exponent (N) and microstructural features: (a) volume fraction of precipitates, (b) average grain size, (c) area fraction of non-recrystallized grains, and (d) maximum intensity of (0002) pole figure.

and TAZ711 alloys have higher yield strength and yield ratio due to solid solution and precipitation hardening. It should be also noted that yield ratio is well corresponded to the work hardening exponent (N) shown in Table 2. It is well known that microstructural features such as grain size, solute concentration, precipitates and dislocation density are inclusively reflected to the work hardening exponent. Therefore, it is difficult to evaluate quantitatively an individual influence of each microstructural feature on work hardening exponent. Nevertheless, from the relationship between work hardening exponent and microstructural features as shown in Fig. 4, a relative contribution of each microstructural factor could be evaluated. In the present study, it is obvious that the fraction of precipitates and non-recrystallized grains contributed more influentially to the change

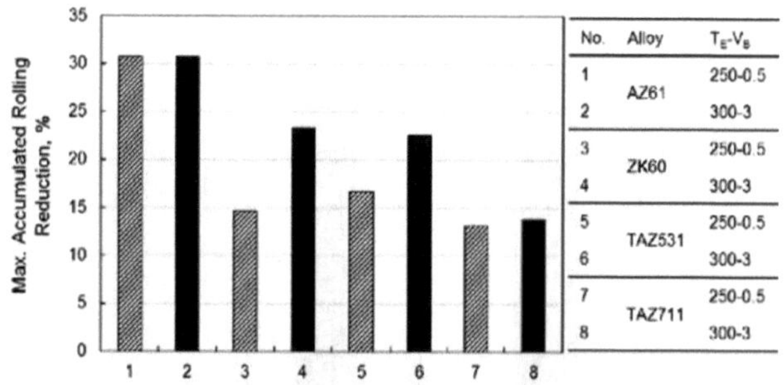

Fig. 5. Maximum accumulated cold-rolling reduction of the four alloys subjected to different extrusion conditions.

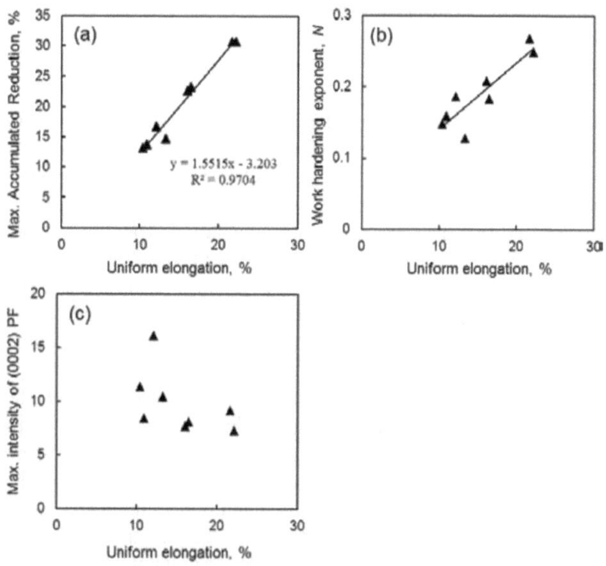

Fig. 6. The relation of (a) maximum accumulated rolling reduction – uniform elongation, (b) uniform elongation – work hardening exponent, and (c) uniform elongation – maximum basal texture intensity

in work hardening exponent than grain size and basal texture intensity. Two points out of the regression line inside the dashed circle in Fig. 4(a) are corresponded to the samples of ZK60-250-0.3 and ZK60-300-3.0 extruded sheets which have much larger fraction of non-recrystallization grains, i.e. deformed grains, than the others. As deformed grains have in general large density of dislocations, they cause a decrease in work hardening capability. With regard to maximum basal texture intensity no clear relation with work hardening exponent can be found in Fig. 4(d).

The maximum accumulated rolling reduction at room temperature using the extruded sheets with the initial microstructures listed in Table 2 is shown in Fig. 5. In addition, Fig. 6(a) shows that the cold-rollability has a linear correlation with uniform elongation of the extruded sheets before cold-rolling. Interestingly, the result of a regression analysis indicates that the maximum accumulated rolling reduction can be simply estimated at 1.5 times the uniform elongation of initial materials. It is also seen in Figs. 6(b) and (c) that the uniform elongation can be related to work hardening exponent not maximum intensity of basal texture. Similar to the situation as shown in Fig 4(d) there is no correlation between maximum intensity of basal texture and uniform elongation. This

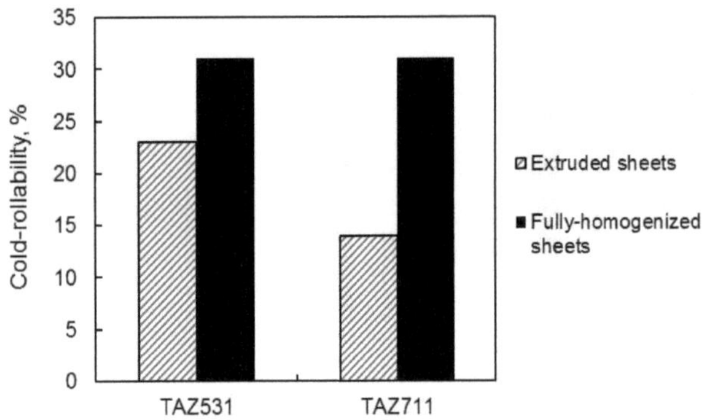

Fig. 7. The improvement in cold-rollability by homogenization heat treatment in the TAZ531 and TAZ711 alloy sheets

is because the cold-rolling behavior of the extruded sheets with relatively low basal texture intensity in the range of 7 to 9 is governed by microstructural factors such as amount of precipitates and fraction of deformed grains, while the ones with strong basal texture lead to a decrease in uniform elongation. Therefore, it can be concluded that an alloy with strong basal texture definitely has low cold-rollability, on the contrary an alloy with weak basal texture can possess excellent cold-formability depending on microstructural features effectively affecting work hardening exponent.

It is obviously seen that the alloy sheets extruded under the conditions of 300°C and 3.0 mm/s show overall better cold-rollability. This is because larger grain size and low basal texture intensity are attributed to the improved cold-rollability of the sheets. Regardless initial extrusion conditions, the AZ61 alloy sheets with the initial microstructures of larger fraction of recrystallized grains and small amount of precipitates have the best cold-rollability. In the cases of ZK60 alloy sheets, they have relatively large fraction of non-recrystallized grains, i.e. deformed grains, after extrusion process, and resultantly they have lower cold-rollability than AZ61 alloy sheets. In the cases of the precipitate-hardened TAZ531 and TAZ711 sheets with the initial condition of 300°C-3.0mm/s, they have limited cold-rollability despite large grain size of > 20 μm and low texture intensity < 9. Since an increase in the amount of precipitates decreases in work hardening exponent as shown in Table 2, TAZ711 has lower work hardening exponent than TAZ531 and thus the former could be cold-rolled up to less accumulated reduction than the latter. In the present study, it is obvious that among various microstructural factors the volume fraction of non-crystallized grains and the amount of precipitates more sensitively affect work hardening exponent than grain size. Therefore, the best way of microstructure control to get improved cold-rollability is to increase grain size and decrease volume fraction of precipitates and deformed grains corresponding to the microstructural feature of the AZ61 sheet extruded under the initial condition of 300°C-3.0 mm/s, indicating the microstructure fully-homogenized in a Mg single phase region can satisfy the above requirements.

In order to confirm the effect of homogenization treatment in TAZ531 and TAZ711 sheets which have large amount of Mg_2Sn precipitates and thus low cold-rollability, the extruded sheets were homogenized before cold-rolling at the temperatures that only Mg single phase exist and they were rolled at room temperature until the samples were broken. Fig. 7 shows the improved cold-rollability of the homogenized sheets. It can be confirmed that the cold-rollability of TAZ531 and TAZ711 sheets increased 8% and 17% more.

Conclusions

The effects of initial microstructures on cold-rollability of RE-free magnesium alloys were investigated in the present study. It was found that the uniform elongation and the work hardening exponent reflecting inclusive microstructural features of grain size, precipitates and dislocation density have a linear correlation with cold-rollability. Although a weak texture is well-known to lead to the improved formability at room temperature, the cold-rollability of magnesium alloys can change depending on microstructural features corresponding to work hardening exponent besides basal texture intensity. In particular the fraction of precipitates and deformed grains sensitively affected work hardening exponent and thus the cold-rollability of magnesium alloys. It was confirmed that poor cold-rollability of the ppt-hardened alloys such as TAZ531 and TAZ711 could be improved by removing precipitates and deformed grains via a homogenization treatment. In order to achieve current goals of low cost production and further improvement in mechanical properties of sheet-type magnesium alloys, the results of the present study can be helpful for understanding how to control initial microstructures for the improvement in cold rollability without expensive rare earth elements and special rolling processes.

References

1. M.R. Barnett, M.D. Nave, and C.J. Bettles, Materials Science and Engineering, A 386 (2004) 205-211.
2. M. Kohzu, K. Kii, Y. Nagata, H. Nishio, K. Higashi, and H. Inoue, Materials Transactions, 51 (2010) 749-755.
3. X. Huang, Y. Chino, M. Mabuchi, and M. Matsuda, Materials Science and Engineering, A 611(2014) 152-161.
4. X. Huang, K. Suzuki, Y. Chino, and M. Mabuchi, Materials Science and Engineering, A 565 (2013) 359-372.
5. K. Hantzsche, J. Bohlen, J. Wendt, K.U. Kainer, S.B. Yi, and D. Letzig, Scripta Materialia, 63 (2010) 725-730.
6. D. Wu, R.S. Chen, and E.H. Han, Journal of Alloys and Compounds, 509 (2011) 2856-2863.
7. A. Styczynski, C. Hartig, J. Bohlen, and D. Letzig, Scripta Materialia, 50 (2004) 943-947.
8. D.H. Kang, D.W. Kim, S. Kim, G.T. Bae, K.H. Kim, and N.J. Kim, Scripta Materialia, 61 (2009) 768-771.
9. S.H. Park, Y.M. Kim, C.D. Yim, H.S. Kim, and B.S. You, Magnesium Technology (2012) 439-442.

Magnesium Technology 2015
Edited by: Michele V. Manuel, Alok Singh, Martyn Alderman, and Neale R. Neelameggham
TMS (The Minerals, Metals & Materials Society), 2015

Studies on the Magnesium Alloys Cladding in the Plastic Forming Processes (Die Forging and Extrusion) Using as the Clad Layer Corrosion Resistant Aluminum Alloys

Piotr Korczak MSc, IMN OML Skawina, POLAND; Bartłomiej Płonka Ph.D. , IMN OML Skawina, POLAND; Dariusz Leśniak Ph.D., AGH Kraków, POLAND; Marek Nowak MSc, IMN OML Skawina, POLAND; Krzysztof Remsak, MSc, IMN OML Skawina, POLAND; Sonia Boczkal Ph.D. IMN OML Skawina, POLAND

Keywords: metal cladding, magnesium alloys, extrusion

Abstract

The studies was conducted in order to improve the corrosion properties of magnesium alloys by forming on the surface a thin layer of aluminum alloy. As the feedstock material MgAlZn alloy (AZ series) have been used. Cladding process was realized in direct and indirect extrusion process. As feedstock to extrusion process billets with sleeve pipe (cladding material) pressed on them were used. The resulting claded rods have been used for forging process trials, which allowed to determine their susceptibility to further plastic forming. After forging micro-and macrostructure of samples were tested, paying particular attention to the welding zone of both materials (claded alloy and clad layer). In the final stage of work forgings used in trials in salt spray chamber which allowed to verify the corrosion properties of technology being developed.

Introduction

Owing to their inherent advantages, magnesium alloys are becoming more and more popular in the world. The undeniable fact is their growing use in numerour sectors of the industry. Some of the users are completely satisfied with the parameters that the cast materials can offer, others, more demanding, tend to use the elements subjected to plastic working [1,2]. And yet, no matter which material is used, one of the most serious drawbacks of these alloys often comes to light, and it is their low resistance to corrosion. Currently, in the world, this problem is solved using the following methods:

- protecting the surface with layers produced by electrochemical methods (e.g. chromate layers)
- painting the surface
- applying a layer of other metals such as aluminium (e.g. by sputtering methods)
- laser Al+Al_2O_3 claddign [3,4]

All of the above mentioned methods are applied to the finished items. This article will discuss the possibility of applying protective layers of aluminium alloys onto magnesium alloys still at the stage of the manufacturing process (i.e. extrusion or forging).

Test Procedure

Tests were carried out on an MgAlZnAZ81 alloy; the plating material was the EN AW-6060 aluminium alloy (the choice was dictated by the fact that this alloy is readily available and offers good corrosion resistance combined with high workability). Tables 1 and 2 show chemical compositions of both alloys. The research started with proper preparation of the feedstock for extrusion trials. It was decided to use feedstock in the form of 94mm diameter AZ80A alloy billets with 98mm diameter sleeves made of the EN AW-6060 alloy press fitted onto the billet. Because the surface of both materials is easily oxidised, wire brushing was applied first and then pressing so tight as to ensure the closest fit possible. The aim of this operation was to reduce the risk of the penetration of atmospheric gas and water vapour between the two "working" surfaces, which might result in their re-oxidation. Figures 1 and 2 show illustration and image of the feedstock preparation.

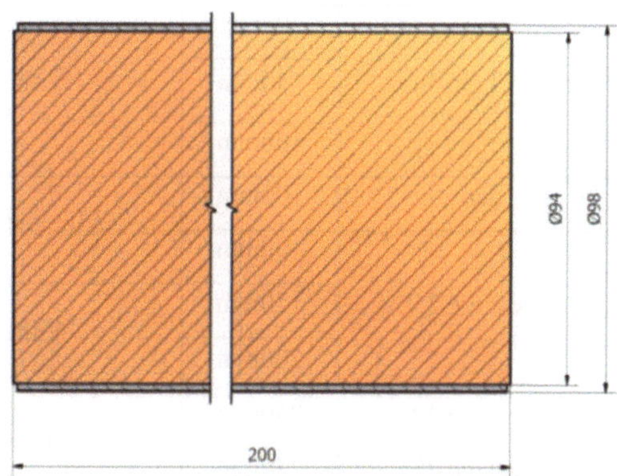

Fig 1. Dimensions of the billet-sleeve feedstock.

Table 1. Chemical composition of EN AW – 6060 alloy.

Element	Al	Mg	Si	Fe	Mn	Zn	Cr	Ti	Cu
Content (%)	Remainder	0,35-0,6	0,30-0,6	0,10-0,30	max 0,10	max 0,15	max 0,05	max 0,10	max 0,10

Table 2. Chemical composition of AZ80A alloy

Element	Mg	Al	Zn	Mn
Conten (%)	Remainder	8,5	0,5	0,15

Fig. 2 The AZ80A billet and EN AW-6060 sleeve, photo taken before fitting process.

Extrusion tests

The extrusion tests were performed on a horizontal direct-indirect 500T press. The feedstock, the press container and tools were preheated to 350-380°C. Conical and flat dies with Ø26mm diameter (the elongation $\lambda = 14$) providing dead zones of different shapes were used in the tests. Detailed test procedures are given in Table 3.

Three special billets were used in the tests. Their ends were sharpened so as to make them fit the die angles of 20°, 25°, 30°, respectively. Then, after press fitting of the sleeve onto the billet, its end was spun to make it adhere very tightly to the angled face of the billet (Fig. 3). The aim of this operation was to eliminate the formation of dead zones in which the build-up of magnesium alloy would interfere with the uniform flow of the sleeve material.

To check the flow of material depending on the tool geometry, feedstock geometry and the extrusion system, each extrusion butt was sectioned, etched, and then photographed. The results are shown in Table 4.

Table. 3 Extrusion test procedure.

No.	Die type and angle	Billet length [mm]	Type of billet and shell fitting process	Extrusion system	Velocity [mm/s]	Temperature material /tools
1	conical die 30°	200mm	Without thermal process	Direct	1mm/s	380°C /360°C
2		150mm				
3		100mm				
11	flat	100mm	With thermal process 380°/10h	Indirect	1,5mm/s	
12	conical die 25°	100mm				
13		150mm	Without thermal process			
14		200mm				
21	conical die 30°	250mm	Without thermal process + angle faced billet	Direct	0,75mm/s	
22	conical die 25°	250mm				
23	conical die 20°	250mm				

Table 4. Extrusion process results.

No.	Die type /angle	Billet length [mm]	Extrus-ion System	Cross-section
1	30°	200	Dierct	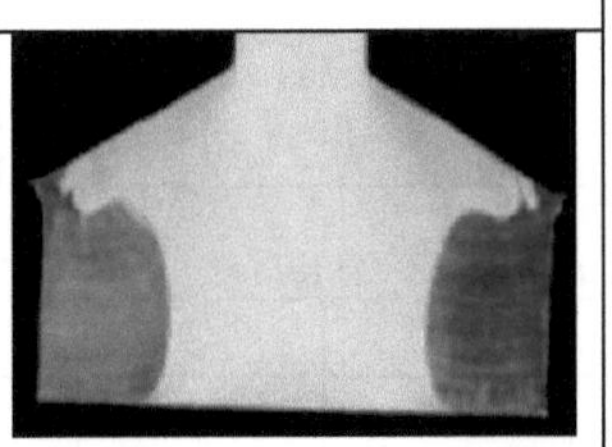
2	30°	150		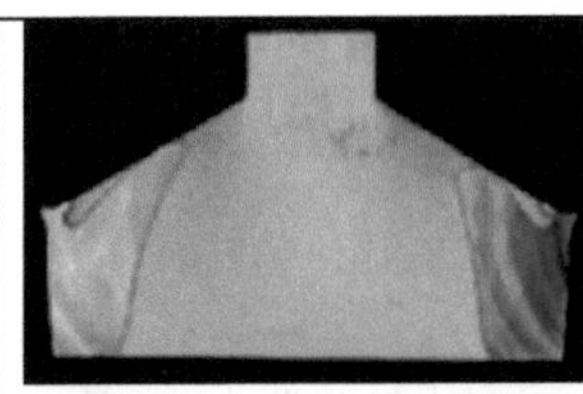
3	30°	100		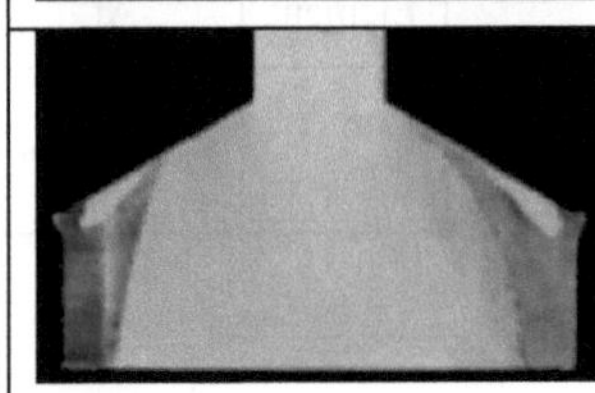

11	flat	100	Indirect	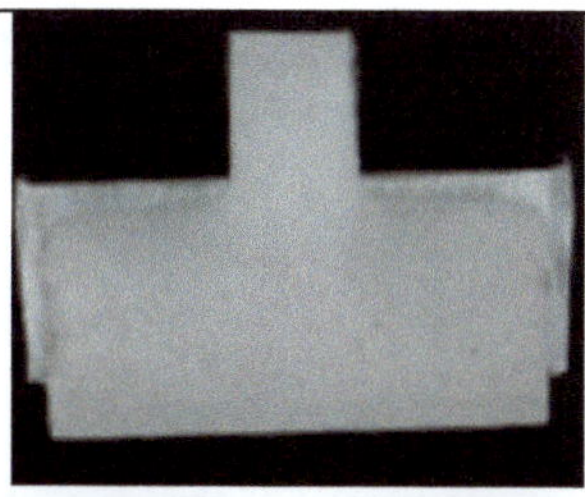
12	25°	100		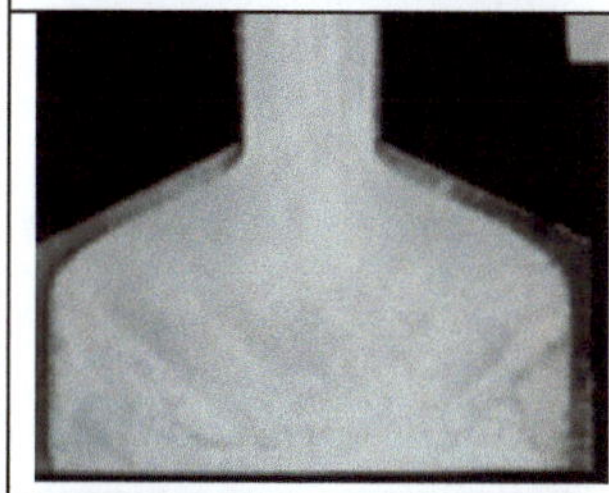
13	25°	150		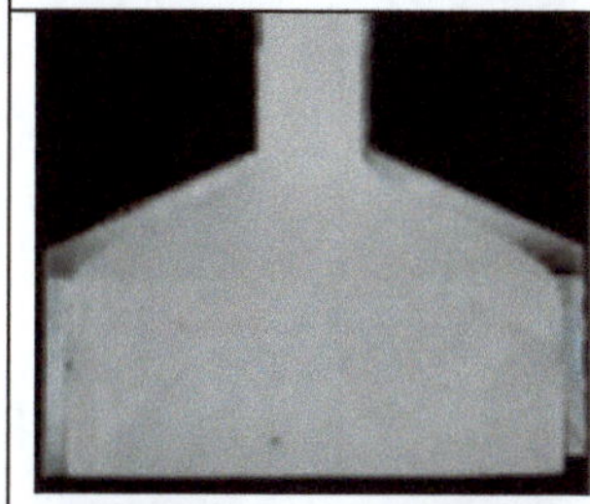
14	25°	200		
21	30°	250	Direct + angle faced billet	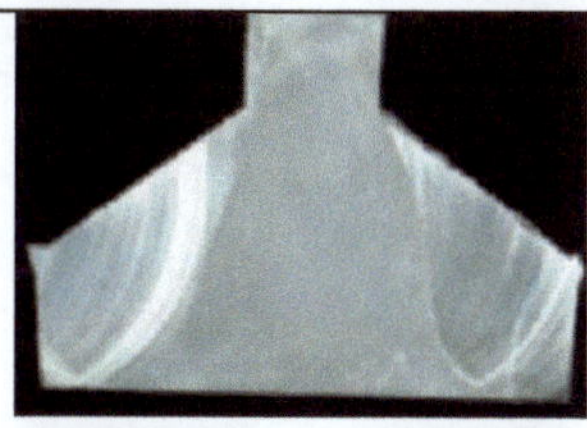
22	25°	250		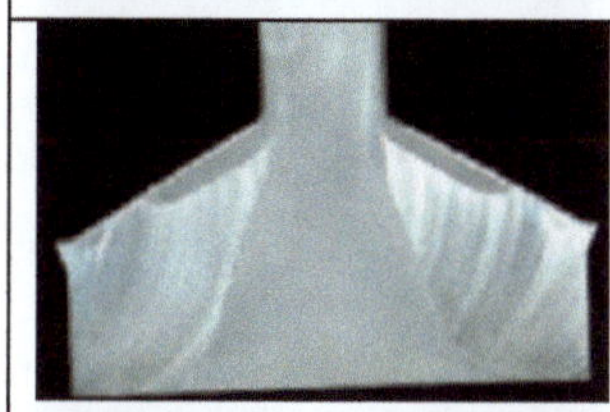
23	20°	250		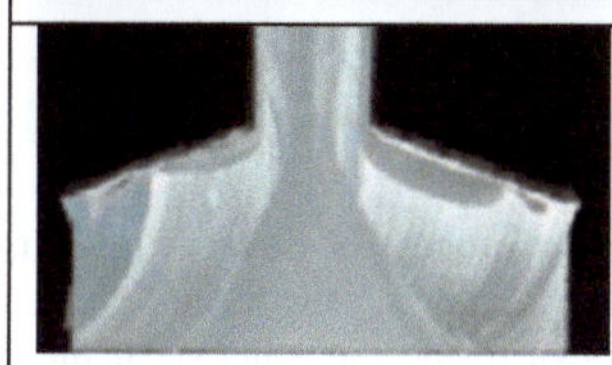

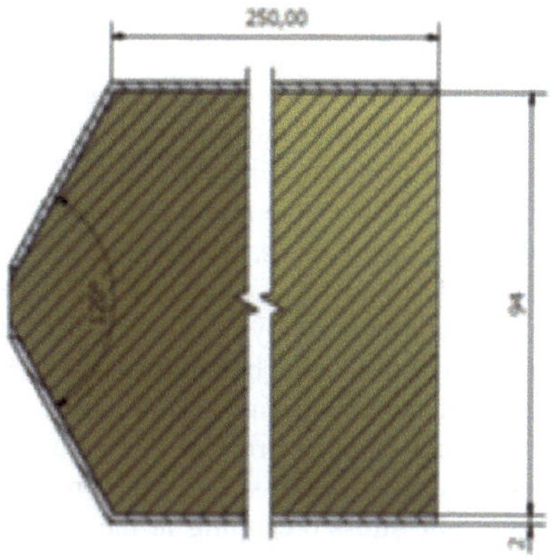

Fig. 3. Angle-face billet.

Examining the cross sections of the extrusion butts, particular attention should be paid to differences in the metal flow, depending on the extrusion system. Definitely preferred is here the indirect extrusion mode, in which the sleeve material is flowing evenly and without evidence of being "sucked" inside. The direct extrusion system, due to the formation of dead zones in which at the initial stage of extrusion the magnesium alloy builds up, while at the end of the process the "sucking in" effect occurs, does not favour the formation of a uniform plated layer (Fig. 4) [2]. The situation slightly changed with the use of sleeves spun on the conical front face of magnesium billet but only for a very narrow range of the die angles (the angle must be very sharp to eliminate as much as possible the formation of dead zones). It was also observed that, despite the rigorous procedure of press fitting the sleeve onto the billet, cases occurred when air was trapped in between the walls. The result was the formation of gas blowholes under the layer of plating (Fig. 4c).

To determine the quality of bond formed between the Mg-Al alloy and Al alloy at the billet–plated layer interface, and to exactly measure the thickness of the plated layer, microstructures of the resulting extruded rods were examined. The obtained microstructures are shown in Table 5.

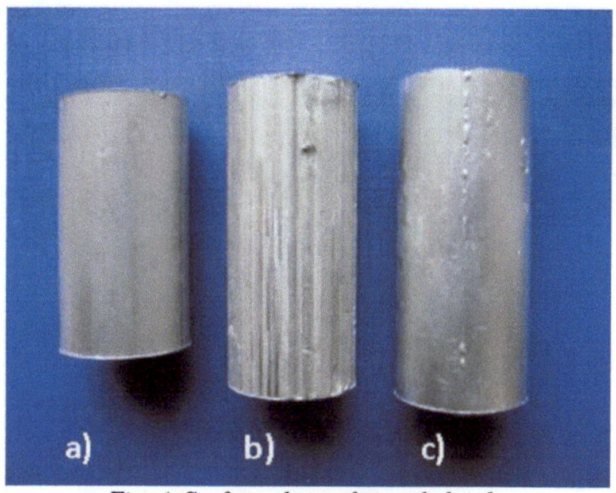

Fig. 4. Surface photo of extruded rods.
a) proper layer of plate
b) dead zones effect
c) gas blowholes under the layer of plating material

Careful analysis of these microstructures shows clear effect of the multi-layered arrangement of alloys obtained in the case of direct extrusion system. This is due to the fact that the billet front face was not coated with a layer of the plating material. At the start of the extrusion process, the maximum flow rate had the layers of material concentrated in the axis of the billet. On approaching the billet surface, the increasing volumes of material created the, so-called, dead zones, where the forces of extrusion did not operate and the material remained undeformed. At the very beginning of the process, only magnesium alloy was flowing along the conventional dead zone boundary (therefore the starting portion of the rod remained uncoated with a layer of the plating material). Aluminium alloy (forming a plated layer on the product) started flowing at a later stage of the process. Due to the favourable hydrostatic stress prevailing in the deformation zone, aluminium alloy was welded to the magnesium core, this also including the magnesium alloy in the dead zone. The result was magnesium alloy penetrating through the plated layer (on the circumference of the extruded rod). This is an undesirable effect and results from the specific nature of the direct extrusion process and from the fact that it is not possible to completely eliminate the formation of dead zones. Even when conical dies were used, some residual dead zones remained and moved to the inside of the billet (the container of the press).

Table 5. Microstructures of extruded bars.

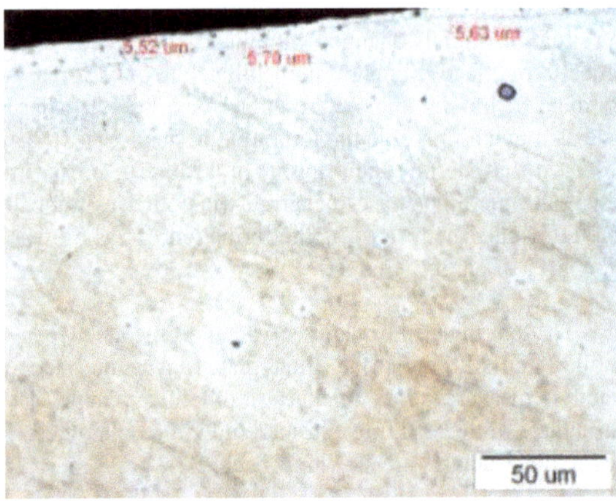

No. 1 Die angle 30° L_{billet}=200mm direct extrusion

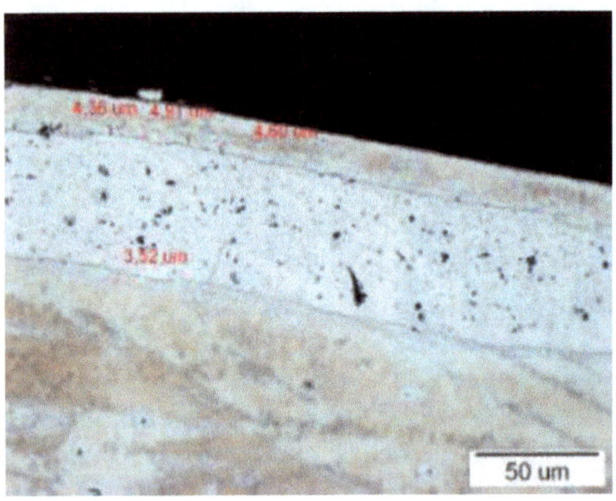

No. 2 Die angle 30° L_{billet}=150mm direct extrusion

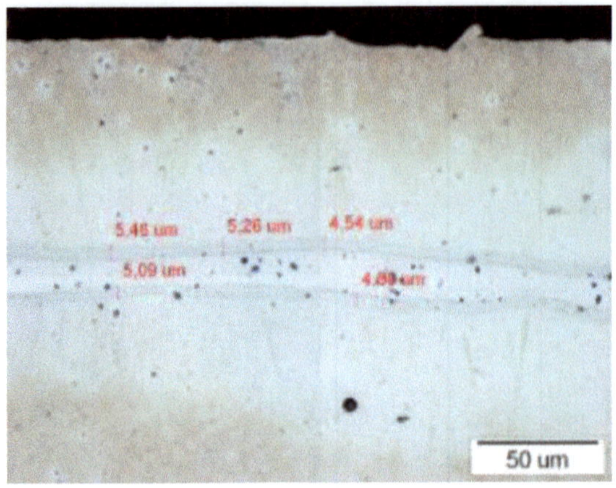

No. 3 Die angle 30° L_{billet}=100mm direct extrusion

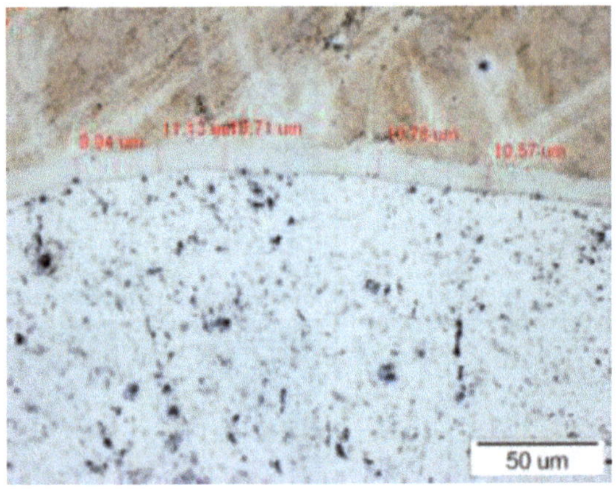

No. 11 Flat die L_{billet}=100mm indirect extrusion

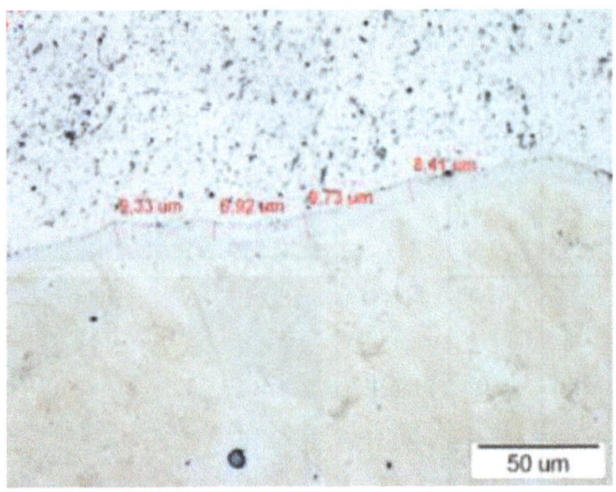

No. 12 Die angle 25° L_{billet}=100mm indirect extrusion

In the process of direct extrusion, there is the effect of back-pull of material, which means that the outer layers of the billet are drawn inside the billet (near the billet axis) where, as mentioned above, the material flow velocity is the highest. As a consequence, the plating material appears in the interior of the extruded rod which is also considered a type of defect. This problem can be eliminated only by leaving larger butt end, but this will have a negative impact on the overall yield of the extrusion process.

The problems described above did not occur in the indirect extrusion. The plating material was applied evenly. In the aluminium alloy-magnesium alloy phase boundary, an $Al_{12}Mg_{17}$ phase of 9 to 15 μm thickness was formed. In some areas of the bonding phase, cracks appeared. Most probably they were due to various stresses formed in the extrusion process or to a difference in the flow velocity of magnesium alloy layers and aluminium alloy forming a layer of plating.

Forging tests

Rods produced in the extrusion trials were next used in die forging tests conducted on a vertical 250T press. The geometry of connecting rod was selected for the tests as a most representative one. To improve the flow of material at the edges of forging, the feedstock was prepared in two versions - as a standard cylinder and as a cylinder with drilled bases (Fig. 2). In the first stage of forging, the depth of drilling was 15 mm, but after analysis of the results it was found that this did not guarantee a sufficient allowance of material at the edges of forgings. The next trials were made on cylinders drilled to a depth of 20 and 25mm.

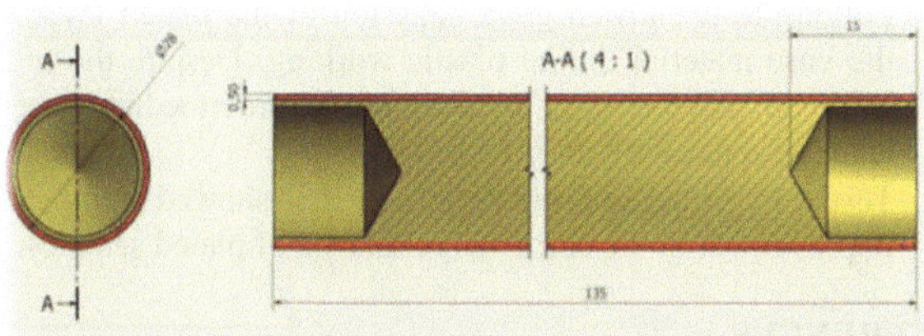

Fig. 5. Feedstock geometry.

The forging process parameters are shown in Table 6. Figure 7 shows a photograph of produced forgings.

Table 6. Forging parameters.

Alloy / parameter	*AZ80A*	*EN AW-2519*
Feedstock temperature	*350-380 ℃*	*400-450 ℃*
Die temperature	*300 ℃*	*280 ℃*
Velocity	*2 mm/s*	*2 mm/s*

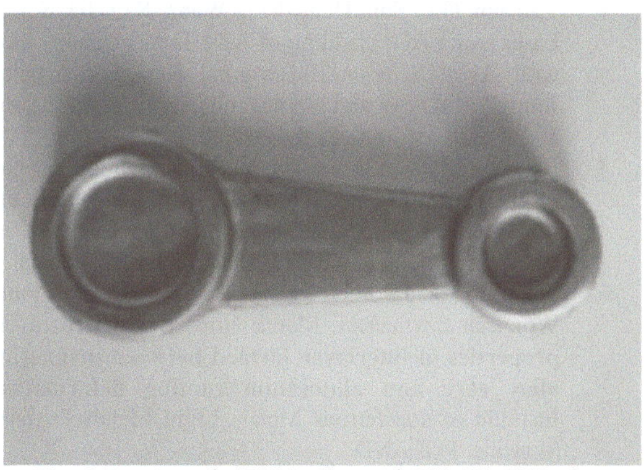

Fig. 6 Forging of AZ80A alloy with EN AW-6060 plate.

As a next step, the resulting forgings were cut along the axis and subjected to etching to expose the structure and uniformity of the plating. As seen in the photos below (Figs. 7 - 10), in places of the highest deformation, some discontinuities have appeared in the layer of plating.

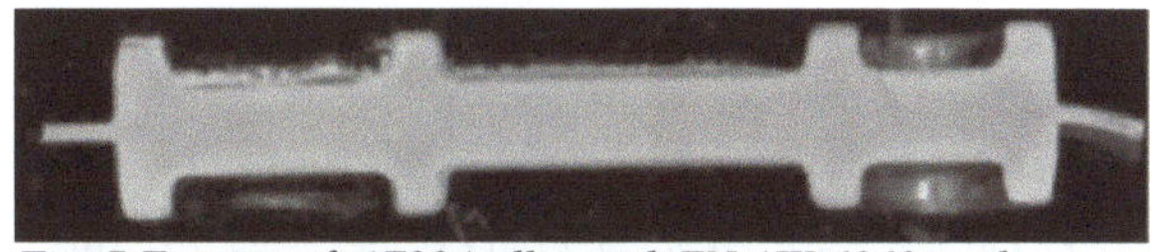

Fig. 7 Forging of AZ80A alloy with EN AW-6060 made form full cylinder feedstock

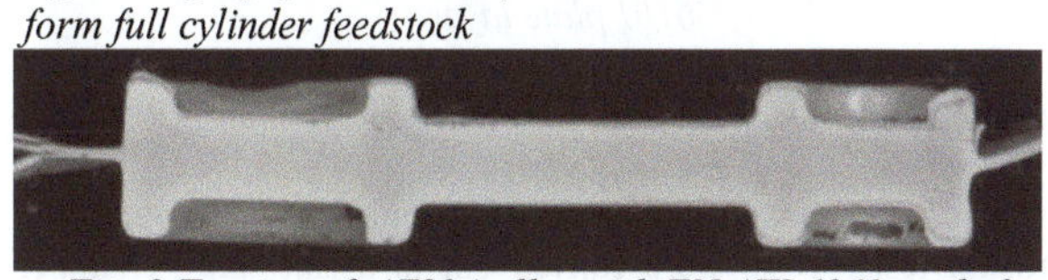

Fig. 8 Forging of AZ80A alloy with EN AW-6060 made form with drilled bases (15mm. deep)

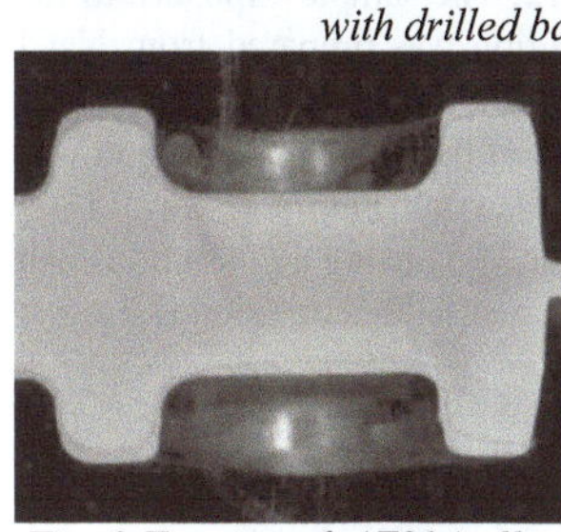

Fig. 9 Forging of AZ80A alloy with EN AW-6060 made form full cylinder feedstock

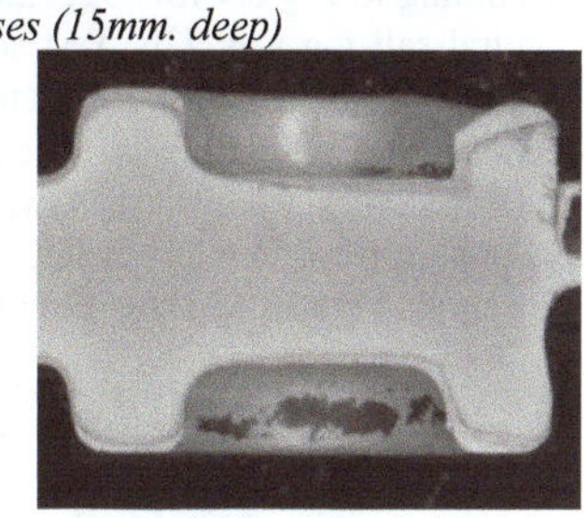

Fig. 10 Forging of AZ80A alloy with EN AW-6060 made form with drilled bases (15mm. deep)

The analysis of the results showed that plating on the samples made from the drilled feedstock had better flow properties, although even in these cases, some discontinuities occurred in the places of the most severe plastic deformation. Therefore, in the next series, tests were performed under the same conditions as the conditions used in the previous test, but using the feedstock drilled to a depth of 20 and 25 mm.

Examining these photographs it is easy to note that the plating has much better continuity. Unfortunately, when the feedstock was drilled to a depth too large, the volume of the available material became insufficient and unfilled sections appeared. After analysis of all the photos, it was found that the optimal feedstock is a cylinder drilled to a depth of 20 mm. Tests made on cylinders drilled deeper showed the occurrence of very adverse effects such as unfilled sections, and the situation could only improve the use of slug guaranteeing more optimal geometry of the feedstock with proper allowances of material both plated and plating.

To check the quality of bond between the material plated and plating, the after-forging microstructure of the material was examined. Photo in Figures 11 show a very satisfactory bond produced between both materials, the place of the weld is clear, and the weld itself, in spite of high hardness (and hence brittleness), is continuous and uniform which should result in a fast bond between the plating material and the base stock.

The photo on figure 11 show a characteristic black scratch appearing in the weld which might suggest cracking and delamination. However, this is the effect of the high hardness of the weld, which differs in the abrasion properties or even breaks off during preparation [3,4].

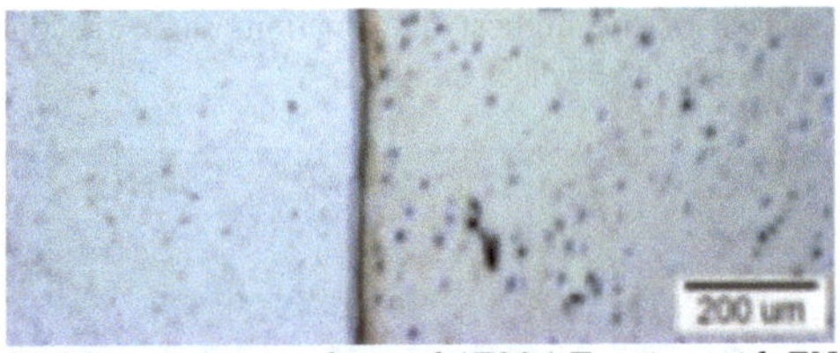

Fig. 11 Microstructure photo of AZ80A Forging with EN AW-6101 plate layer.

Corrosion test

In a next step, the Mg-Al samples were subjected to corrosion tests. Tests were carried out in a salt spray chamber according to PN-EN ISO 9227:2012. The sample exposure to the neutral salt fog was 72h. Test solution was prepared from NaCl (50g/l) of analytical purity and demineralised water with a conductivity of 1.52 μS/cm. The pH of thus prepared solution was 7.19. Samples were washed with acetone before testing. Tests were performed in a test chamber at 35°C using a continuous salt fog spray. Figures 12 show samples before the corrosion testing.

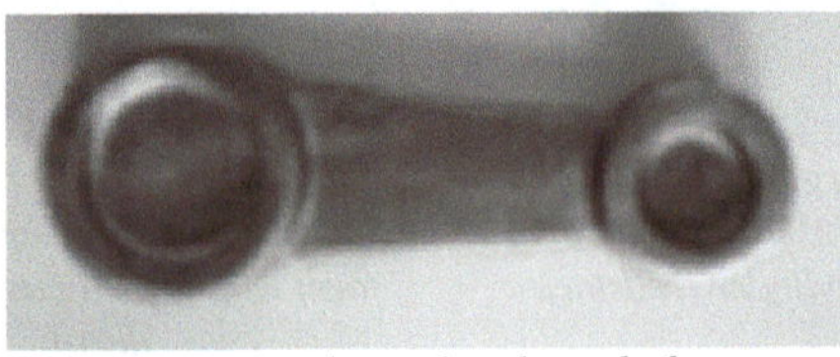

Fig. 12 AZ80A Forging without plate layer before corrosion tests.

After the test, the samples were subjected to visual inspection to see the progress of corrosion. Samples made of aluminium showed signs of spot corrosion. The sample of magnesium alloy without a sleeve was characterised by deep pitting corrosion. Magnesium in aluminium sleeve had traces of pitting corrosion but only in the side edges of the sample. The sample of magnesium was covered with corrosion products as early as after 24 h of the corrosion test duration. The corrosion products formed on the sample surface are shown in Figure 13.

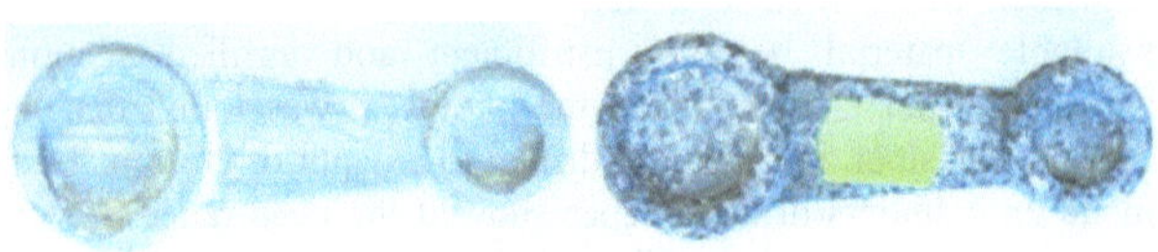

Fig.13 *AZ80A Forging with (left) and without (right) plate layer after corrosion tests.*

The results of the sample weight loss are given in Table 6. Analysing the results of gravimetric tests it was found that the AZ80A alloy without a sleeve exhibited poor corrosion resistance in environments containing chloride ions. The sample weight loss was 6,867 g and thus exceeded 11 times the weight loss observed in the sample with an aluminium sleeve. Photo of sample after the corrosion test are shown in Figure 14.

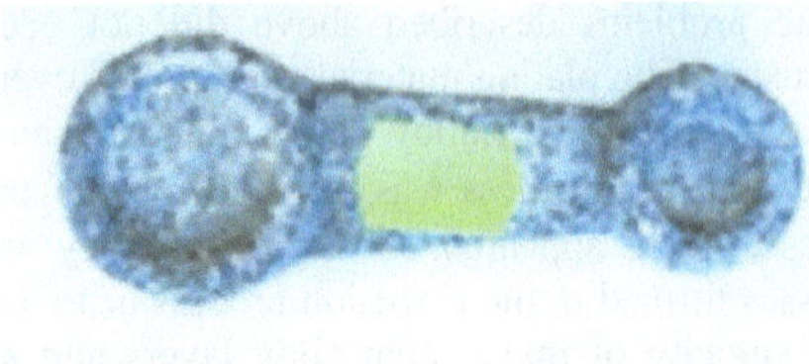

Fig. 14. *AZ80A Forging without plate layer after corrosion tests (with corrosion products removed).*

Summary

In subsequent stages of the research, numerous tests of extrusion, forging and heat treatment were carried out. At all stages, the quality of bond in a base material – plated layer system was examined along with the continuity of protective layer. At the end of the work, tests were conducted in a salt spray chamber to verify the degree of protection that can be achieved when applying the tested technology. In extrusion tests, much better results were obtained when an indirect system was used. The reason is much smaller proportional share of dead zones, which have been shown to cause a significant non-uniformity in the plated layer.

- Forging tests showed good adhesion of the plated layer to the base material during plastic working. Despite the high hardness and hence brittleness of the intermediate layer, cracks did not form during forging.
- Tests carried out in a salt spray chamber showed significant improvement in the corrosion resistance of plated samples.

REFERENCES

1. ASM Speciality Handbook, Magnesium and Magnesium Alloys, ASM International Materials Park, OH, 2004.
2. Metals Handbook Ninth Edition, Properties and Selction Nonferroue Alloys and Pure Metals, Vol. 2, Metals Park, Ohio, 1979, p.707-832.
3. M. Hazra, A.K. Mondal, S. Kumar, C. Blawert, Narendra B. Dahotre Laser surface cladding of MRI 153M magnesium alloy with (Al+Al2O3)Original Research Article / Surface and Coatings Technology, Volume 203, Issue 16, 30 May 2009, Pages 2292-2299
4. Yao Jun, G.P. Sun, Hong-Ying Wang, S.Q. Jia, S.S. Jia Laser (Nd:YAG) cladding of AZ91D magnesium alloys with Al + Si + Al2O3Original Research Article / Journal of Alloys and Compounds, Volume 407, Issues 1–2, 5 January 2006, Pages 201-207
5. Hiroaki Matsumo, Sadao Watabane, Shuji Hanada **"Fabrication of pure Al/Mg-Li Alloy clad plate and its mechanical properties",** Institute for materials research, Tohoku university, Sendai 980-8577, Japan
6. Sonia Boczkal, Piotr Korczak, Bartłomiej Płonka, Wojciech Szymański, Marek Nowak, **"Structure and properties of interlayer formed between magnesium alloy core and aluminium coating deformation"** Institute of non-ferrous Metals Light Metals Division, Skawina, POLAND

The research was carried out under Project No. POIG.01.03.01-00-015/09 entitled: "Advanced materials and technologies for their production" co-financed from the structural fund; the project implementation period is 2010-2013.

INNOWACYJNA GOSPODARKA
NARODOWA STRATEGIA SPÓJNOŚCI

UNIA EUROPEJSKA
EUROPEJSKI FUNDUSZ ROZWOJU REGIONALNEGO

Magnesium Technology 2015
Edited by: Michele V. Manuel, Alok Singh, Martyn Alderman, and Neale R. Neelameggham
TMS (The Minerals, Metals & Materials Society), 2015

THE EFFECTS OF PLASTIC ANISOTROPY IN WARM AND HOT FORMING OF MAGNESIUM SHEET MATERIALS

Eric M. Taleff[1,2], Aravindha R. Antoniswamy[2,a], Alexander J. Carpenter[1,b] and Emre Yavuz[1,c]

1. The University of Texas at Austin, Mechanical Engineering, 204 E. Dean Keeton St., C2200, Austin, TX 78757
2. The University of Texas at Austin, Materials Science & Engineering, 204 E. Dean Keeton St., C2201, Austin, TX 78757

a. Currently with Intel Corp., Assembly & Test Technology Development, 5000 W. Chandler Blvd., Chandler, AZ 85226
b. Currently with Southwest Research Institute, Engineering Dynamics, 6220 Culebra Rd., San Antonio, TX 78238
c. Currently with Turkish Petroleum Corporation, Söğütözü Mahallesi, 2180. Cadde, No: 86 06100 Çankaya, Ankara, Turkey

Keywords: Magnesium, plastic anisotropy, warm forming, hot forming, sheet forming

Abstract

Mg alloy sheet materials often exhibit plastic anisotropy at room temperature as a result of the limited slip systems available in the HCP lattice combined with a commonly strong basal texture. Less well studied is plastic anisotropy developed at the elevated temperatures associated with warm and hot forming. At these elevated temperatures, particularly above 200 °C, the activation of additional slip systems significantly increases ductility. However, plastic anisotropy is also induced at elevated temperatures by a strong crystallographic texture, and it can require an accounting in material constitutive models to achieve accurate forming simulations. The type and degree of anisotropy under these conditions depend on both texture and deformation mechanism. The current understanding of plastic anisotropy in Mg AZ31B and ZEK100 sheet materials at elevated temperatures is reviewed in this article. The recent construction of material models that can support accurate simulations of simple hot forming cases is also reviewed with strategies to account for plastic anisotropy in forming simulations.

Introduction

Mg-based alloys are recently of particular interest to manufacturers of vehicles because of their potential to reduce vehicle mass, a goal very important to improving vehicle performance. While the use of cast Mg alloys is relatively well established in current vehicle manufacturing processes, particularly for the die casting of components, the use of wrought Mg alloys is in its infancy. One of the more significant barriers to increased use of wrought Mg alloys is a generally poor ductility at room temperature, which makes deformation processing of wrought material into complex components a difficult challenge. The source of low ductility is the limited number of slip systems available at room temperature in the hexagonal-close-packed (HCP) crystal structure of Mg, which evidences only two unique basal systems and one unique non-basal system that requires high stress to activate. Although twinning can assist this limited number of slip systems, it is still insufficient to meet the Taylor criterion for ductile polycrystals [1]. However, increasing temperature enables the activation of additional slip systems in Mg and its alloys [2]. Because of these additional slip systems, ductility increases significantly with increasing temperature. This is demonstrated in Figure 1, which shows a fit to data from AZ31B sheet for tensile elongation as a function of temperature at a strain rate of 10^{-2} s^{-1}; see Ref. [3] for original data that produced this fit. The drop in elongation at the highest temperatures, which follows a steady increase in elongation with temperature, is because of early rupture arising from cavitation [3]. At quite high temperatures, superplastic forming is even possible in some Mg alloys. For these reasons, forming at elevated temperatures is an approach that can make possible the production of complex component geometries from wrought Mg alloys. One example is the recent success in hot gas-pressure forming of a production component from AZ31B sheet [4].

It is usual for wrought Mg sheet materials to possess a strong crystallographic texture, most commonly of the basal type. A ramification of this strong texture and the anisotropy of slip in the HCP crystal is significant plastic anisotropy. Plastic anisotropy in Mg sheet materials is generally greatest at room temperature, but it also persists to elevated temperatures. While plastic anisotropy decreases with increasing temperature and the commensurate increase in available slip systems, it remains sufficiently pronounced to necessitate an accounting of it in the design and implementation of elevated-temperature forming processes. A material constitutive model was recently introduced that accounts for the effects of plastic anisotropy in Mg alloy AZ31B sheet during deformation at 450 °C [5], the temperature used to form the production component described in Ref. [4]. The methodologies used to construct that model [5] and to extend it to lower temperatures [3] are reviewed in this article in the context of AZ31B and ZEK100 wrought Mg alloy sheet.

Plastic Anisotropy in Sheet Materials

Plastic anisotropy is quite common for materials used in sheet forming operations. The methods used to account for anisotropy in simulating sheet forming are well established

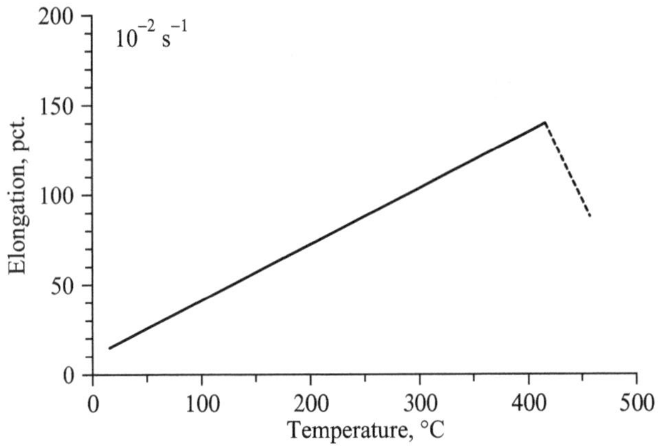

Figure 1: The tensile elongations of AZ31B are shown as a function of temperature for tests conducted at 10^{-2} s^{-1} [3]. Data used in fitting are available in Ref. [3].

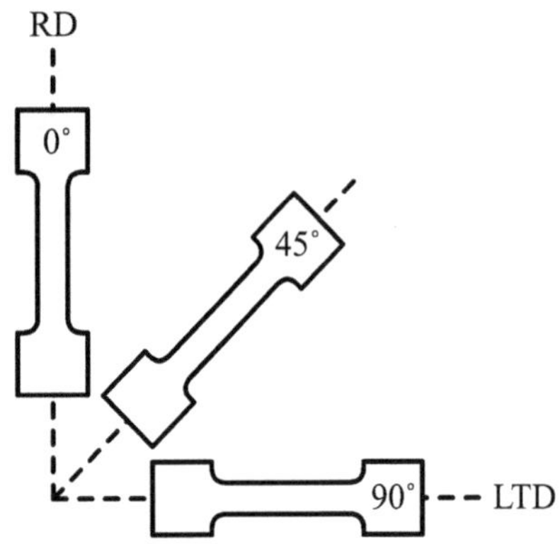

Figure 2: Tensile coupons are shown in the three orientations typically used to evaluate planar anisotropy in sheet materials. RD is the rolling direction and LTD is the long-transverse direction.

and typically rely upon measuring the Lankford coefficient, also called the R-value. The R-value can be readily measured from a tensile coupon as,

$$R = \frac{\varepsilon_w}{\varepsilon_t} \quad , \tag{1}$$

where ε_w is the width strain and ε_t is the thickness strain. For the general case of planar anisotropy, tensile coupons should be tested in three orientations relative to the sheet rolling direction, RD, as indicated in Figure 2. These tests provide R-values along each of the three directions, typically 0, 45 and 90° to the RD. If these three R-values are identical, then anisotropy is reduced to the simpler case of normal anisotropy. If all three are identical and equal to unity, $R = 1$, this is the case of isotropic behavior. The case of normal anisotropy may be dealt with using a simple flow rule that assumes a flow stress of σ_f within the sheet plane as,

$$\sigma_f^2 = \sigma_1^2 + \sigma_2^2 - \frac{2R}{1+R}\sigma_1\sigma_2 \quad , \tag{2}$$

where σ_1 and σ_2 are the principal stresses in the sheet plane. The sheet normal stress is assumed negligible, $\sigma_3 \approx 0$. When the R-values are only slightly different among the three orientations shown in Figure 2, then normal anisotropy may be sufficient to approximate plasticity in the sheet, and an average R-value can be calculated as,

$$\overline{R} = \frac{R_0 + 2R_{45} + R_{90}}{4} \quad . \tag{3}$$

When significant differences occur between these directional values, then the more complicated case of planar anisotropy must be addressed. The trends presented in this review were all fit to data for R-values either measured or corrected to a reference true strain of $\varepsilon_{ref} = 0.1$.

Creep Flow and Plastic Anisotropy

Creep Mechanisms

The increase in ductility with increasing temperature shown in Figure 1 for AZ31B occurs initially because of the introduction of slip on non-basal planes. However, as the recrystallization temperature is exceeded, creep flow begins to play an important role in increasing ductility. Upon initially exceeding the recrystallization temperature, the primary mechanism contributing to creep deformation is dislocation-climb (DC) creep. The fundamental DC creep mechanism is controlled by lattice self-diffusion and is often referred to as five-power creep. For the case of steady-state creep flow, the situation best described by current creep theory, five-power creep can be effectively described through the following application of the phenomenological equation for creep [6],

$$\dot{\varepsilon}_{DCL} = A_{DCL}\left(\frac{D_L}{b^2}\right)\left(\frac{\sigma}{E}\right)^5 \quad , \tag{4}$$

where $\dot{\varepsilon}_{DCL}$ is the true-strain rate attributed to this mechanism, A_{DCL} is a material constant of approximately 10^{11} for materials of high stacking-fault energy [7,8], D_L is the diffusion coefficient for lattice self-diffusion, b is the magnitude of the Burgers vector, σ is the flow stress and E is the temperature-dependent unrelaxed Young's modulus. The stress exponent of this mechanism is typically near five, as shown in Eqn. 4, and is the reason that this dislocation creep mechanism is often referred to as five-power creep. Because creep deformation under this mechanism results from dislocation slip, albeit rate controlled by diffusion during the dislocation climb step of recovery, it is subject to the same slip-controlled sources of anisotropy evident at lower temperatures. Thus, a strong recrystallization texture can produce plastic anisotropy under the DC creep mechanism.

Figure 3 demonstrates the change in R-value for AZ31B sheet as temperature increases for a tensile strain rate of

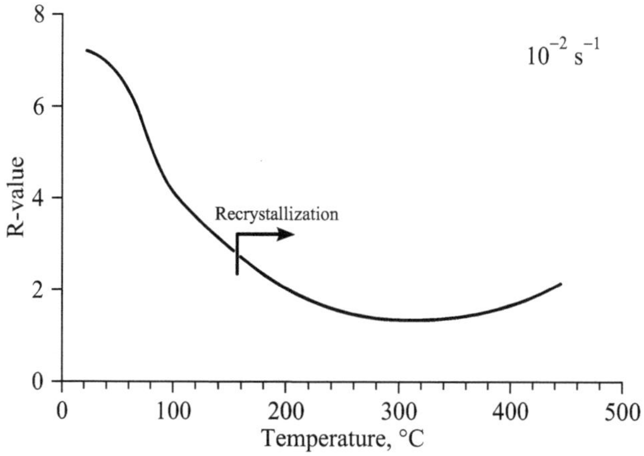

Figure 3: A fit to the R-values of AZ31B is shown as a function of temperature for tests conducted at 10^{-2} s^{-1}. Data that produced this curve were normalized to an approximate true strain of 0.1 [3]. Data used in fitting are available in Ref. [3].

10^{-2} s^{-1} [3]. The plastic anisotropy of this material is well approximated as normal anisotropy with a single R-value for a given test/material condition. The high R-value of approximately seven near room temperature rapidly drops with increasing temperature. After exceeding the recrystallization temperature, indicated by an arrow in Figure 3, DC creep begins to contribute to plastic flow. At the lowest temperatures of creep, dislocation climb is typically first accommodated by dislocation pipe diffusion, instead of lattice diffusion, because of its low activation energy. This form of dislocation-climb creep is reasonably described in the steady-state condition by the following relation [7,9],

$$\dot{\varepsilon}_{\mathrm{DCP}} = A_{\mathrm{DCP}} \left(\frac{D_P}{b^2}\right)\left(\frac{\sigma}{E}\right)^7 \quad , \qquad (5)$$

where $\dot{\varepsilon}_{\mathrm{DCP}}$ is the true-strain rate attributed to this mechanism, A_{DCP} is a material constant of approximately 5×10^{12} for materials of high stacking-fault energy [8] and D_P is the diffusion coefficient for pipe diffusion. The increase in the stress exponent to seven in Eqn. 5 is because of the dependence of dislocation density on stress. This form of DC creep is likewise subject to plastic anisotropy arising from the slip process. Note from Figure 3 that the R-value remains greater than unity until temperature increases to 300 °C.

As temperature further increases, the fine-grained AZ31B materials become susceptible to grain-boundary-sliding (GBS) creep, the mechanism fundamentally associated with superplastic deformation [10,11]. GBS creep accommodated by lattice self-diffusion obeys the following relationship under steady-state deformation,

$$\dot{\varepsilon}_{\mathrm{GBS}} = A_{\mathrm{GBS}} \left(\frac{D_L}{d^2}\right)\left(\frac{\sigma}{E}\right)^2 \quad , \qquad (6)$$

where $\dot{\varepsilon}_{\mathrm{GBS}}$ is the true-strain rate attributed to this mechanism, A_{GBS} is a material constant of approximately 6.4×10^9 for materials of high stacking-fault energy [8] and d is the grain size. By its nature, GBS creep is largely insensitive to crystallographic texture and tends to produce an isotropic plastic response. Another unique aspect of GBS creep deformation is its tendency to randomize crystallographic texture [12–14], which also promotes isotropic deformation. However, large strains are required to note this randomization effect. It is GBS creep that is responsible for the minimum R-value shown in Figure 3.

The two DC creep mechanisms and single GBS creep mechanism described above are all competitive, which means that they can act together toward producing strain and strain rate [6]. The total strain rate observed is then the sum of the contributions from all active creep mechanisms,

$$\dot{\varepsilon} = \sum_{i=1}^{N} \dot{\varepsilon}_i = \dot{\varepsilon}_{\mathrm{DCP}} + \dot{\varepsilon}_{\mathrm{DCL}} + \dot{\varepsilon}_{\mathrm{GBS}} \quad , \qquad (7)$$

where the sum of strain rates from N different competitive mechanisms involves the contribution from each mechanism i. The plastic anisotropy displayed is then also the result of contributions from each of these active creep mechanisms. Thus, as deformation changes from domination by DC creep (anisotropic) to GBS creep (isotropic) with increasing temperature, the R-value reduces to the minimum shown in Figure 3. The modest rise in the R-value with temperature beyond that minimum suggests a reduction in GBS creep because of increasing temperature. This is true, but it is a secondary effect from grain growth at the highest temperatures. Because GBS creep depends strongly on grain size and grain growth becomes significant during deformation at the highest temperatures, the rise in the R-value is most directly a result of increasing grain size decreasing the contribution to strain from GBS creep. This change in the contribution from GBS creep with a change in grain size may be calculated using Eqn. 6. Thus, the rise in R-value at the highest temperatures will depend strongly on the $\varepsilon_{\mathrm{ref}}$ at which R is measured because larger strains allow more grain growth.

Effective Plastic Anisotropy

When the relative contributions from all active creep mechanisms are known, then the effective plastic anisotropy may be calculated. For the case of DC creep mechanisms, both of which produce similar levels of normal anisotropy in AZ31B, competing with the isotropic deformation of GBS creep, the following calculations are appropriate [5,15]. First, the fraction of deformation associated with the DC creep mechanisms is calculated as,

$$f_{\mathrm{DC}} = \frac{\varepsilon_{\mathrm{DC}}}{\varepsilon_{\mathrm{DC}} + \varepsilon_{\mathrm{GBS}}} = \frac{\dot{\varepsilon}_{\mathrm{DC}}}{\dot{\varepsilon}_{\mathrm{DC}} + \dot{\varepsilon}_{\mathrm{GBS}}} \quad , \qquad (8)$$

where the strains and strain rates for DC creep are summed from the contributions of both DC creep mechanisms. The

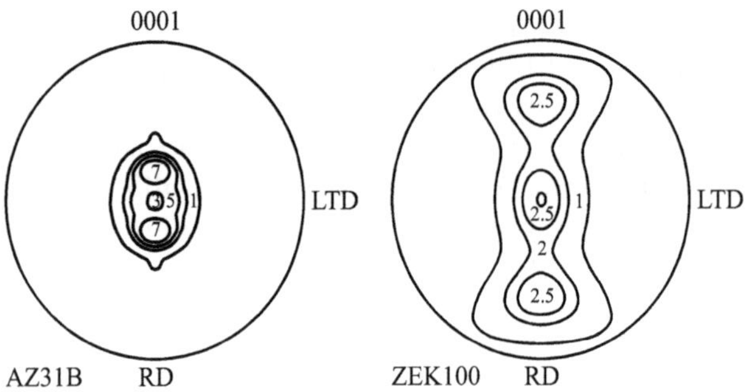

Figure 4: Pole figures are shown for (left) AZ31B and (right) ZEK100 Mg alloy sheet materials [3]. The numbers in the figures indicate approximate multiples of random.

effective R-value observed in the material will then be,

$$R_{\text{eff}} = \frac{R_{\text{GBS}}(1+R_{\text{DC}})(1-f_{\text{DC}})+R_{\text{DC}}(1+R_{\text{GBS}})f_{\text{DC}}}{(1+R_{\text{DC}})(1-f_{\text{DC}})+(1+R_{\text{GBS}})f_{\text{DC}}} \quad (9)$$

These equations provide a means to combine Eqns. 4 through 6 into an effective material constitutive model suitable for implementation in finite-element-method (FEM) software to predict plastic deformation of AZ31B at elevated temperature. The efficacy of this approach was demonstrated in Refs. [3,5].

The Case of Planar Anisotropy

Eqn. 9 is particular to the case of normal anisotropy, which is appropriate for wrought AZ31B sheet [5]. However, wrought sheet of Mg alloy ZEK100, which contains modest additions or rare-earth elements, is a case where planar anisotropy must be considered. The reason for this is a change from the typical basal texture of AZ31B to a split-basal texture in ZEK100. Examples of pole figures from both these materials are shown in Figure 4 [3]. For deformation at 450°C and 10^{-2} s^{-1}, ZEK100 sheet produces R values of $R_0 = 0.9$, $R_{45} = 0.86$ and $R_{90} = 0.81$. These values differ sufficiently to require accounting for planar anisotropy to achieve accurate FEM simulations of forming [3]. However, the coarser grain size of ZEK100 can eliminate the GBS creep mechanism, leaving only the two DC creep mechanisms to dominate deformation.

Conclusions

The methodologies used to construct accurate material constitutive models for predicting deformation of wrought AZ31B and ZEK100 sheet materials at elevated temperatures were reviewed. The theory of steady-state creep provided a description of the creep mechanisms that contribute to deformation at elevated temperatures. The DC creep mechanisms contribute to plastic anisotropy, and GBS creep promotes isotropic deformation. Quantifying the relative contributions of these mechanisms makes the calculation of effective plastic anisotropy possible. The basal texture of wrought AZ31B sheet produces plastic anisotropy that is essentially normal in behavior and can be described using a single effective R-value. The split-basal texture in wrought ZEK100 sheet, however, produces anisotropy that varies with direction within the sheet, a case of planar anisotropy. Thus, FEM predictions of hot deformation in ZEK100 require a material constitutive model that accounts for planar anisotropy.

References

[1] G. I. Taylor, *J. Inst. Metals*, 62 (1938), pp. 307–324.

[2] P. W. Flynn, J. Mote, and J. E. Dorn, *Trans. AIME*, 221 (1961), 1148–1154.

[3] A. R. Antoniswamy, *The Construction and Use of Physics-Based Plasticity Models and Forming-Limit Diagrams to Predict Elevated Temperature Forming of Three Magnesium Alloy Sheet Materials.* (PhD thesis, The University of Texas at Austin, 2013).

[4] GM News, 23 October 2012. `http://media.gm.com/media/us/en/gm/news.detail.html/content/Pages/news/us/en/2012/Oct/1023_GM_Magnesium.html`

[5] A. J. Carpenter, A. R. Antoniswamy, J. T. Carter, J. Hector, Louis G., and E. M. Taleff, *Acta Mater.*, 68 (2014), 254–266.

[6] O. D. Sherby and P. M. Burke, *Prog. Mater. Sci.*, 13 (1968), 325–390.

[7] O. A. Ruano, A. K. Miller, and O. D. Sherby, *Mater. Sci. Engrg.*, 51 (1981), 9–16.

[8] O. A. Ruano and O. D. Sherby, *Revue Phys. Appl.*, 23 (1988), 625–637.

[9] H. Luthy, A. K. Miller, and O. D. Sherby, *Acta Metallurgica*, 28 (1980), 169–178.

[10] O. D. Sherby and J. Wadsworth, *Prog. Mater. Sci.*, 33 (1989), 169–221.

[11] T. G. Nieh, J. Wadsworth, and O. D. Sherby, *Superplasticity in Metals and Ceramics*, Ch. 5: Fine-Structure Superplastic Metals, (Cambridge, U.K.: Cambridge University Press, 1997), 40–53.

[12] M. Perez-Prado, T. McNelley, D. Swisher, G. Gonzalez-Doncel, and O. Ruano, *Mater. Sci. Engrg. A*, 342 (2003), 216–230.

[13] T. R. McNelley, K. Oh-oishi, A. P. Zhilyaev, S. Swaminathan, P. E. Krajewski, and E. M. Taleff, *Metall. Mater. Engrg. A*, 39 (2008), 50–64.

[14] T. R. McNelley, K. Oh-Ishi, S. Swaminathan, J. R. Bradley, P. E. Krajewski, and E. M. Taleff, *Mater. Sci. Forum*, 735 (2013), 43–48.

[15] M. R. Barnett, A. Ghaderi, I. Sabirov, and B. Hutchinson, *Scripta Mater.*, 61 (2009), 277–280.

Magnesium Technology 2015
Edited by: Michele V. Manuel, Alok Singh, Martyn Alderman, and Neale R. Neelameggham
TMS (The Minerals, Metals & Materials Society), 2015

Effect of Heat Treatment on Microstructure and Damping Capacity of Twin Roll Cast ZK60 Strip

Hongmei Chen[1], Qianhao Zang[1], Jing Zhang[2], Jaehyung Cho[3], Yunxue Jin[1], and Yuke Shi[1]
[1]Provincial Key Lab of Advanced Welding Technology, Jiangsu University of Science and Technology, Zhenjiang, 212003, China
[2]School of Metallurgical and Materials Engineering, Jiangsu University of Science and Technology, Zhangjiagang, 215600, China
[3]Korea Institute of Materials Science, Changwon, 641010, South Korea

Keywords: magnesium alloy; twin roll casting; microstructure; damping capacity; damping peak

Abstract

ZK60 strip with 4mm thickness was fabricated by Twin Roll Casting technology. The effect of annealing, solid solution and artificial aging heat treatment on microstructures, mechanical properties and damping capacities were discussed in this paper, and the optimized heat treatment parameters was also determined. Fine equiaxed structure was obtained after annealing treatment. The room temperature damping value (Q^{-1}) of annealed ZK60 strip was 0.0085. There was a damping peak at 279.7°C after 350°C annealing treatment; this was caused by the grain boundaries sliding at a higher temperature. The room temperature damping value (Q^{-1}) of ZK60 strip was 0.0071 after solid solution heat treatment. A very broad damping peak appeared from 280°C to 330°C. The T6 treatment parameters were solution at 375°C for 3hrs, and then artificial aging at 175°C with different holding time. The best damping capacity and mechanical properties were obtained after artificial aging at 175°C for 1hr.

Introduction

Magnesium alloys have a number of unique advantages such as low density, high specific strength, high specific elastic modulus and high damping capacity[1-5]. With the development of modern industry and transportation, noise pollution has become a serious environmental problem. In order to reduce the vibration and noise in industry and transportation fields, the applications of high strength and high damping magnesium alloys were considered[6]. Pure magnesium has the best damping capacity among various metallic materials[7]. However, poor mechanical properties of pure magnesium limit its more widespread applications.

The twin roll cast (designated as TRC in short) technique has been applied to manufacture magnesium alloy sheets with mechanical properties equivalent to or better than those after hot rolling of CC ingots. Compared to normal casting[8-11], twin roll casting produces strips with higher solid supersaturation, finer microstructure, dispersed compounds and lighter element segregation.

Thus, in this paper, the effect of high temperature annealing, solution and aging on the microstructure, mechanical properties and damping capacities of TRC ZK60 strip are investigated in detail. The optimized heat treatment parameters have been determined by analyzing the microstructure, mechanical properties and damping capacities.

Experimental Procedure

The material used in this study was the TRC ZK60 strip. The chemical composition of the ZK60 alloy was analyzed as 6.72Zn-0.343Zr-Mg balance in weight percent by Optical Emission Spectrometer (SPECTRO MAXx) apparatus.

For optical examination, the samples were sectioned, cold mounted, polished by silicon dioxide paste with grain size of 1μm and finally etched in nitric acid and alcohol mixed solution, and then examined by using a ZEISS optical microscope. The phase compositions were determined by an X-ray diffraction instrument (XRD, Shimadzu XRD-6000). The mode for XRD examination is continuous scan, with a scan speed of 4°/min within a range of 20°-80°.The value reported for HV represents the average of ten separate measurements taken at randomly selected points using a load of 100gf for 10s. The tensile tests were performed on a SANS universal testing machine with a test speed of 1.25 mm/min. Damping samples were machined to dimensions of 35mm×12mm×1mm. Damping capacity was measured by dynamic mechanical analyzer (NETZSCH DMA-242C) in single cantilever deformation mode. The measurement frequency is 0.5 Hz, 1Hz, 5 Hz and 10 Hz, respectively. The temperature range was from 25°C to 400°C and the heating rate was 5°C/min.

Results and analysis

<u>Microstructure of twin roll cast ZK60 strip</u>

Microstructural change of ZK60 alloy strip at different conditions are shown in Fig.1. The microstructure of TRC strip consisted of dendrite structure, with eutectics and intermetallic compounds located in the interdendritic region (Fig.1a). The microstructure of ZK60 strip was homogenized but dendrite structure was retained after annealing heat treatment (Fig.1b). After solution treatment (Fig.1c), the intermetallic compounds were dissolved, which could reduce the volume fraction of the second phase. The solution treatment also induces an obvious grain growth. After solution treatment followed by aging at 175°C for 1hr (Fig.1d), the grain size and distribution were not changed significantly.

<u>X-ray diffraction analysis</u>

At the annealed condition, the eutectics and intermetallic compounds in the interdendritic region mainly contained Mg and Zn elements. It mainly consists of $MgZn_2$ phase. At T4 heat-treatment state, the solution treatment results in decreasing volume fraction of the second phase. The $MgZn_2$ phase were dissolved at T4 heat-treatment at 375°C for 3hrs. Most of the $MgZn_2$ phase precipitated again during the aging treatment at 175°C for 3hrs.

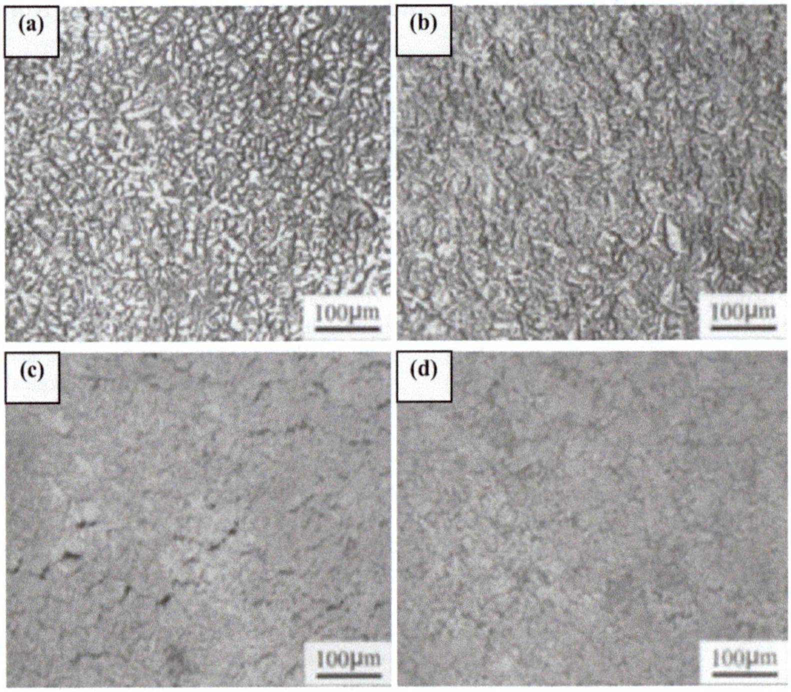

Fig.1. Microstructures of TRC ZK60 alloys
(a) as twin roll cast strip; (b) annealed 350°C×1hr; (c) T4 375°C×3hrs; (d) T6 375°C×3hrs+175°C×1hr

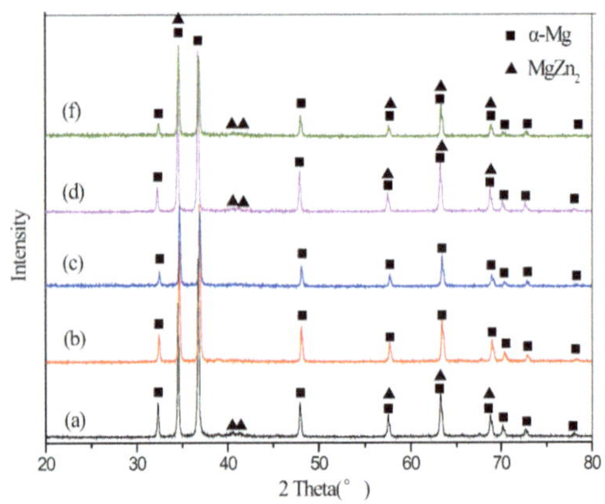

Fig.2 XRD patterns of TRC ZK60 alloys in various heat treatment conditions: (a) annealed 350°C×1hr; (b) T4 375°C×3hrs; (c) T6 375°C×3hrs+175°C×1hr; (d) T6 375°C×3hrs+175°C×3hrs; (f) T6 375°C×3hrs+175°C×18hrs.

Mechanical properties of TRC ZK60 strips

The tensile properties of TRC ZK60 strip tested at room temperature are shown in Fig.3a.The as-cast strips were given solution treatment at 375 °C for 3hrs, and then aged at 175°C for 1hr, 3hrs, 6hrs, 10hrs and 18hrs. The tensile properties of the aging treated strip were much higher than that of as-cast strip. This is attributed to the formation of precipitation during aging process. At aging for 1hr, the tensile properties were increased slightly in comparison with those aging for other durations. Fig. 3b shows the aging curve of TRC ZK60 strips at the aging temperature of 175°C with different aging durations ranging from 1hr to18hrs. After aging for 3hrs, the hardness increased to the peak value of 79.17 HV. With increasing aging time, the hardness decreased again. According to the XRD patterns, many $MgZn_2$ phase precipitated after aging for 3hrs this induced increasing of hardness.

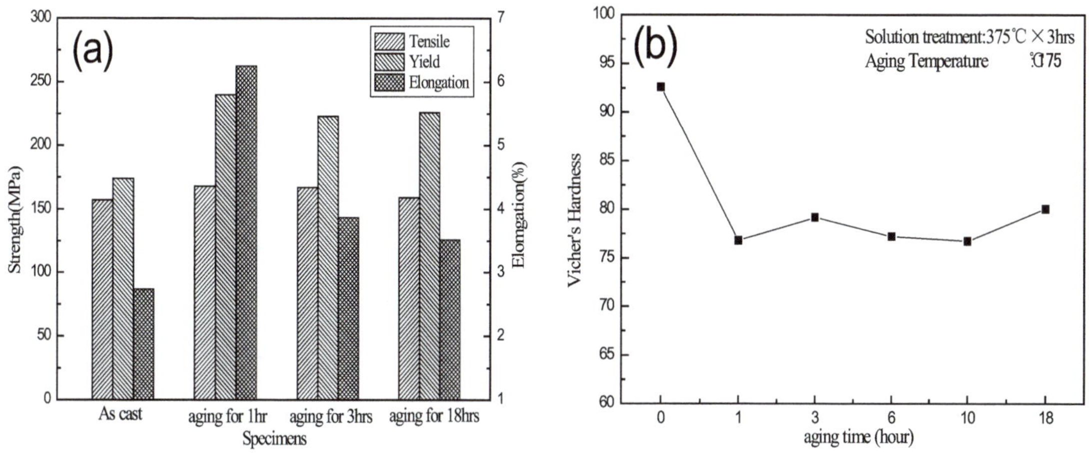

Fig.3 Mechanical properties of TRC ZK60 strips: (a) effect of aging time on tensile properties; (b) aging curve

Damping capacity of TRC ZK60 strip

Damping capacity of TRC ZK60 strip at room temperature are shown in Fig.4. At the annealed condition, the room temperature damping values (Q^{-1}) of ZK60 strip were 0.0079, 0.0085 and 0.0096, respectively. The damping values were mainly higher than that of the T4 treated specimens as shown in Fig.4. The best damping value of TRC ZK60 strip is 0.0098 after T6 treatment (solution treated at 375°C for 3hrs and then aging treated at 175°C for 1hr). There are many factors that can affect the room temperature damping values such as grain size, second phase and impurity atoms. After aging for 3hrs and even longer time, the second phase precipitates again; the distance of the distribution of dislocations is reduced. The damping value of aging for 1hr is higher than that of aging for 3hrs and long time.

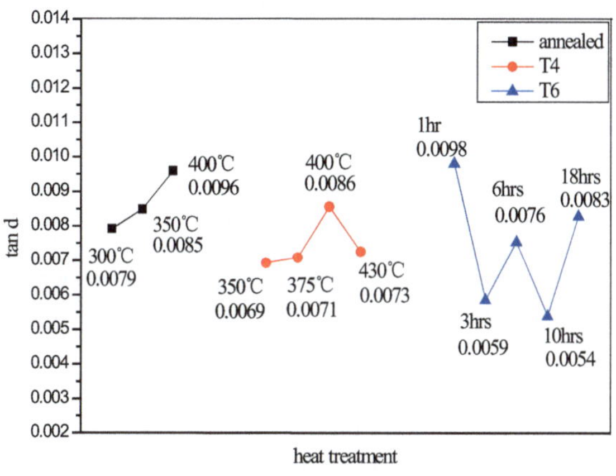

Fig.4 Damping capacity of TRC ZK60 strips at room temperature

Temperature-dependent damping capacity of heat treated TRC ZK60 strips are shown in Fig.5. At annealed condition (Fig.5a), there were damping peaks at 279.3°C, 351.5°C and 302.5°C after 300°C, 350°C and 400°C annealing treatment, respectively. The damping peaks are caused by the grain boundaries sliding at medium temperature ranging from 215°C to 247°C (the medium temperature ranging from 0.35Tm to 0.4 Tm)[8]. The Tm of TRC ZK60 strip in this paper was 616.27°C. The temperatures of the three damping peaks present were beyond medium temperature range. This was mainly because it was very sensitive to the distribution of impurity atoms in the grains and dislocations[9]. At T4 treatment condition (Fig.5b), a very broad damping peak appeared from 280°C to 330°C. This wide damping peak was attributed to many overlapping small damping peaks. At T6 treatment condition (Fig.5c), there were damping peaks at 283°C and 286°C after aging for 6hrs and 18hrs, respectively. After aging for 1hr, 3hrs and 10hrs, these peaks disappeared. These broad damping peaks might be due to the distribution of activation energy or relaxation time [10].

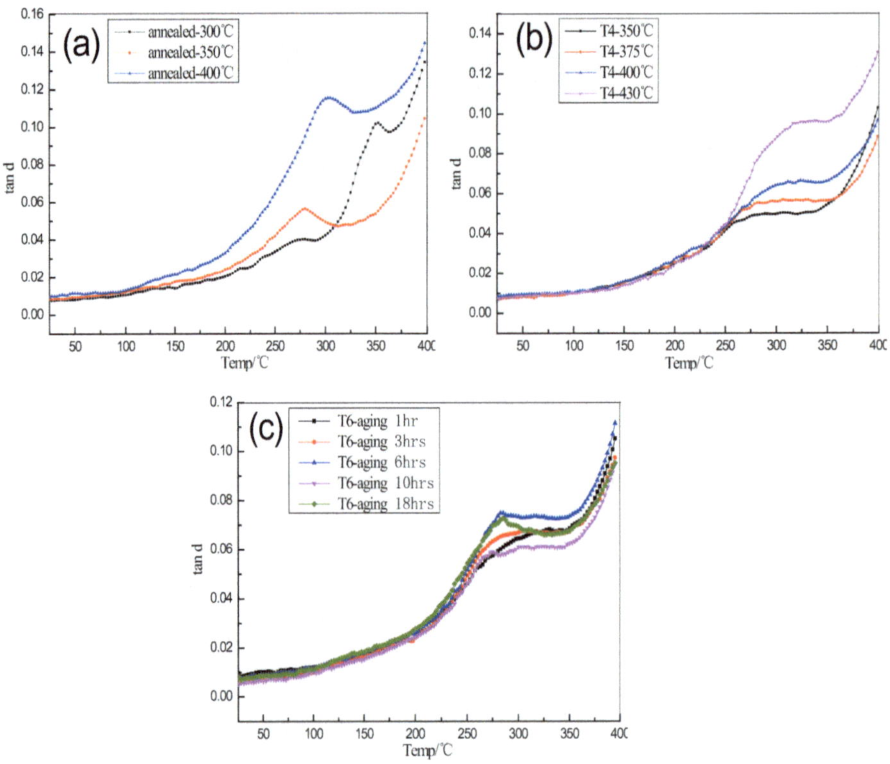

Fig.5 Temperature-dependent damping capacity of TRC ZK60 strips: (a) annealed treatment; (b) T4 treatment; (c) T6 treatment

Conclusions

(1) The microstructure of twin roll cast ZK60 strip consisted of dendrite structure. Several $MgZn_2$ phase precipitated again during the aging treatment at 175°C for 3hrs; it resulted in increasing of the hardness.

(2) Fine equiaxed structure was obtained after annealing treatment. The room temperature damping value (Q^{-1}) of annealed ZK60 strip was 0.0085. There was a damping peak at 279.7°C after 350°C annealing treatment.

(3) The room temperature damping values (Q-1) of T6 heat treated ZK60 strip were 0.0098, 0.0059, 0.0076, 0.0054 and 0.0083, respectively. The best damping capacity and mechanical properties were obtained after artificial aging at 175°C for 1hr.

Acknowledgement

This work was supported by National Natural Science Foundation (Project No:51301077) and Natural Science Foundation of Jiangsu Province (Project No:BK20130470).

References

[1] B.L. Mordike, T. Ebert, Magnesium properties-applications-potential. Materials Science and Engineering A 2001, 302:37-45.

[2] Y. Xiong, Q. Yu, Y.Y. Jiang. An experimental study of cyclic plastic deformation of extruded ZK60 magnesium alloy under uniaxial loading at room temperature. International Journal of Plasticity, 2014,53:107-124.

[3] T. Bhattacharjee, C.L. Mendis, T.T. Sasaki, T. Ohkubo,et al. Effect of Zr addition on the precipitation in Mg–Zn-based alloy. Scripta Materialia, 2012, 67(12): 967-970

[4] M. Shahzad, L. Wagner. Microstructure development during extrusion in a wrought Mg–Zn–Zr alloy. Scripta Materialia, 2009, 60(7): 536-538

[5] A. S. Khan, A. Pandey, T. Gnäupel-Herold, R. K. Mishra. Mechanical response and texture evolution of AZ31 alloy at large strains for different strain rates and temperatures. International Journal of Plasticity, 2011, 27(5): 688-706

[5] N. Srikanth, X.L. Zhong, M. Gupta. Enhancing damping of pure magnesium using nano-size alumina particulates. Materials Letters 2005, 59: 3851-3855.

[7] R. Schaller, Metal matrix composites: a smart choice for high damping materials. Journal of Alloys and Compounds 2003, 355: 131–135.

[8] H. Watari, K. Davey, M.T. Rasgado, T. Haga, et al. Semi-solid manufacturing process of magnesium alloys by twin-roll casting. J Mater Process Technol 2004, 96:1662-1667.

[9] S.B. Kang, J.H. Cho, H.W. Kim, Y.M. Jin. Effect of heat treatment on microstructure and mechanical properties in ZK60 alloy sheet. Materials Science Forum 2008, 567-568: 361-364.

[10] H.M. Chen, S.B. Kang, H.S. Yu, et al. Effect of heat treatment on microstructure and mechanical properties of twin roll cast and sequential warm rolled ZK60 alloy sheets. Journal of Alloys and Compounds, 2009, 476: 324-328.

[11] M. SHAHZAD, L. WAGNER. Microstructure development during extrusion in a wrought Mg-Zn-Z0r alloy. Scripta Materialia, 2009, 60: 536-538.

[12] C. Belamri, S. Belhas, A. Rivière. Damping of a High-purity Aluminum Single Crystal at High Temperatures. Materials Science and Engineering A 2006, 442: 142~146.
[13] C.L. Mendis, J.H. Bae, N.J. Kim. Microstructures and tensile properties of a twin roll cast and heat-treated Mg–2.4Zn–0.1Ag–0.1Ca–0.1Zr alloy. Scripta Materialia 2011, 64: 335–338.
[14] D.Q. Wan, J.C. Wang, G.F. Wang, et al. High damping properties of Mg-Si binary hypoeutectic alloys. Materials Letters 2009, 63: 391–393.

Magnesium Technology 2015
Edited by: Michele V. Manuel, Alok Singh, Martyn Alderman, and Neale R. Neelameggham
TMS (The Minerals, Metals & Materials Society), 2015

DEFORMATION BEHAVIOR OF ROLLED MAGNESIUM SLABS AND TWIN ROLL CAST STRIPS STUDIED BY THE ACOUSTIC EMISSION TECHNIQUE

Patrik Dobroň[1*], Daria Drozdenko[1], Jan Bohlen[2], Dietmar Letzig[2], František Chmelík[1]

[1]Charles University in Prague, Faculty of Mathematics and Physics, Department of Physics of Materials, Ke Karlovu 5, 12116, Prague 2, Czech Republic
[2]Helmholtz-Zentrum-Geesthacht, Zentrum für Material- und Küstenforschung GmbH, Max-Planck-Straße 1, D21502 Geesthacht, Germany

Keywords: Magnesium alloy sheet, Mechanical anisotropy, Acoustic emission, Texture

Abstract

The influence of texture on deformation behavior has been investigated for conventionally rolled magnesium slab and rolled twin roll cast (TRC) magnesium strip in form of sheets. The Mg-Zn-Rare earth alloy sheets were deformed at room temperature with the tensile axis oriented in the rolling (RD), 45° and the transversal direction (TD). The basal pole intensity aligned with the sheet normal direction is lower for the conventionally rolled sheet by comparison with the rolled TRC strip. Higher yield strength (YS) in the RD than in the TD for conventionally rolled sheet and higher YS for TD than RD in rolled TRC strip was observed. The acoustic emission (AE) analysis correlates the microstructure and the stress-time curves to the active deformation mechanisms. This aspect relates to the activity of the basal slip and tensile twinning, particularly during TD tension, which corresponds with the AE activity.

Introduction

Mechanical properties (especially strength, ductility and formability) of rolled magnesium alloys having hexagonal close packed (hcp) lattice are significantly influenced by strong textures [1]. The texture development and the deformation behavior of magnesium alloys depend on the activation of both dislocation slip and twinning.

Rolled sheets of the Mg-based alloys exhibit a typical basal texture with orientation of basal planes almost parallel to the sheet plane [2-4]. Such a texture favors deformation twinning if a compressive stress is applied along the rolling direction (RD) [5]. In the case of the tensile in-plane test, deformation can be realized practically without (near-zero) tensile twinning [6, 7]. Bohlen et al. [4] have recently reported the texture development in six Mg-Zn alloys with dilute additions of Y, Nd, Ce and Mn. It should be mentioned that the angular distribution of the basal poles in ZE10 is significantly broader to the transversal direction (TD); it is not affected in the RD. On the other hand, the prismatic plane poles are mostly oriented parallel to the RD. Therefore the strong basal texture is responsible for tension – compression asymmetry and anisotropy.

Twin roll casting (TRC) of magnesium alloys is a cost-effective process and has beneficial effects on resulting microstructure (reduction of segregation, homogeneous microstructure etc.). To improve the strength and the formability of TRC magnesium sheets, it is necessary to obtain fine grained and homogeneous microstructure with well- distributed in-plane texture [8, 9]. Thus, solidification and recrystalization during TRC process have strong impact on resulting mechanical properties. Simulation of such processes is a way how to better understand and subsequently influence the final microstructure [10, 11]. Most research deals with TRC of Mg-Al based alloys [12-14], where a relative good knowledge of the production of conventionally rolled AZ-magnesium alloys can be applied. Recent studies are also focused on other Mg alloys such as Mg-Zn-Rare earth alloys, which are promising materials for industrial applications.

For the study of dynamic processes involved in plastic deformation of Mg alloy sheets, the acoustic emission (AE) technique can be efficiently used. It is based on sensitive detection of the transient elastic waves generated within the material during deformation due to sudden localized structure changes [15]. Detected AE signals can be parameterized according to their amplitude, number of counts, duration, energy or frequency. A direct correlation of the AE parameters with the stress-strain curve yields information on the dynamic changes indicating e.g. activation of various deformation mechanisms during plastic deformation of Mg alloys [16, 17].

The aim of the present work is to examine deformation mechanisms occurring during tensile testing of conventionally rolled magnesium slab and rolled TRC strip by the AE method. The time dependence of the stress and *in situ* AE were used in order to identify deformation mechanisms during loading. The AE activity, represented by the AE count rate and the amplitudes of the individual AE events, reveals dynamic changes in collective dislocation processes and possible twin nucleation during in-plane tensile tests under given conditions.

Experimental

Magnesium alloy ZE10 (Mg + 1.3 wt.% Zn+ 0.2 wt.% Ce + 0.1 wt.% La) was received in the form of slabs and twin roll cast strips. They were subsequently subjected to conventional rolling into sheets of the thickness of 1.5 mm. Tensile samples (width 6 mm, gauge length 35 mm) were machined with loading axis in RD, 45° and TD. Deformation tests were performed in a universal testing machine Zwick Z50 at room temperature and at a constant strain rate of 10^{-3} s^{-1}.

Global texture information about the sheets in the as-received condition was obtained using the X-ray diffraction. A Panalytical X-ray diffractometer setup using CuKα radiation was employed to measure pole figures on polished samples in reflection geometry to a sample tilt of 70°. The selected samples were ground to their midsection, polished mechanically and etched in a solution of HNO_3 (65%) : ethylene glycol : water = 11 : 66 : 23 for 10 min.

AE during deformation tests was monitored by computer-controlled PCI-2 device (Physical Acoustic Corporation) using a piezoelectric transducer with flat response between 100 and 600 kHz and a preamplifier giving a gain of 40 dB. The threshold level of the detection was set above the peak noise level to 26 dB.

Results

Both, conventionally rolled slab and rolled TRC strip exhibit a typical microstructure consisting of elongated grains into the RD with an average grain size of 20±2 μm and 17±2 μm, respectively (Fig. 1). The initial textures represented by the distribution of basal (0001) and prismatic poles are shown in Fig. 2. The (0001) pole figure of both rolled slab and TRC strip display texture with the orientation of basal planes tilted away from the plane of the sheet (normal direction (ND)). For the rolled slab, the intensity of basal planes in the (0001) pole figure significantly varies with respect to the RD and the TD. The characteristic angular spread of basal planes with a double peak towards the TD is observed. Such a texture is different from a texture observed in rolled TRC strip (Fig. 2b).

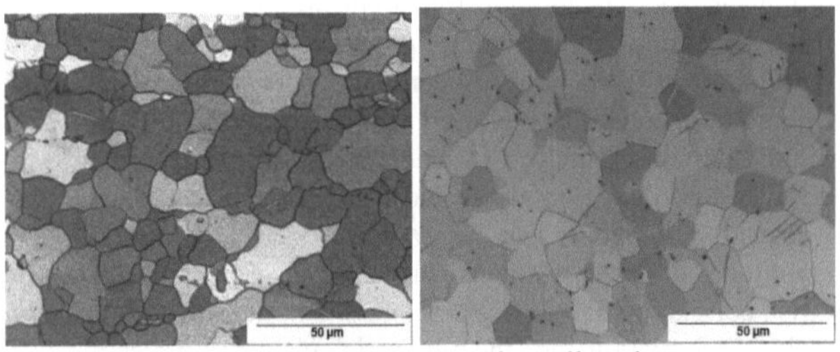

Fig. 1. Microstructure of ZE10 magnesium alloy sheet a) conventionally rolled slab, b) rolled TRC strip.

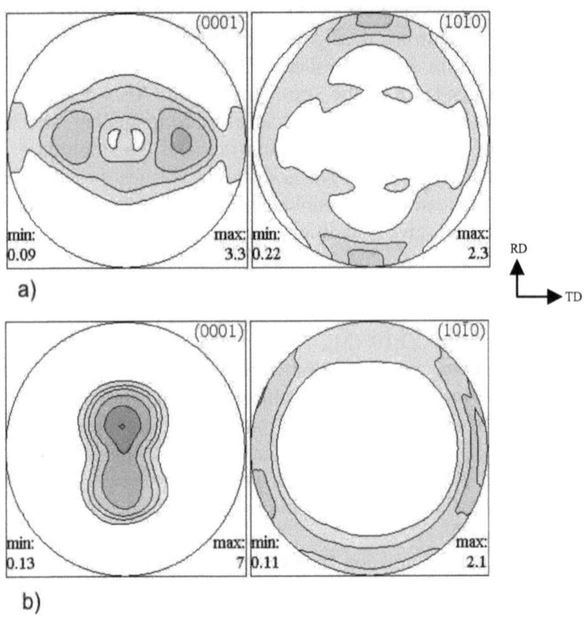

Fig. 2. Initial texture of a) conventionally rolled slab, b) rolled TRC strip.

In rolled TRC strip, the basal planes are split towards the RD rather than to the TD, is the latter being a typical texture for conventional magnesium alloy sheets, such as AZ31.

Figs. 3a,b,c show the stress–time curves with corresponding AE response for tension along the RD, 45° and the TD of the conventionally rolled slab, respectively. The mechanical behavior is significantly influenced by texture. In-plane anisotropy of the yield strength (YS) for all three orientations of sample is clearly seen; the highest value of the YS is found in RD sample and the lowest YS is in the TD. Likewise, differences in the AE response are observed. Depicted AE parameters (AE count rate and amplitudes of individual AE events) exhibit a similar tendency for each sample orientation. In the RD sample, the maximum of the AE activity is connected with the macroscopic yield point and after achieving the YS the AE parameters significantly decrease. The 45° and TD sample display a second AE maximum after the YS, which is lower than the first one observed at the microscopic yield point.

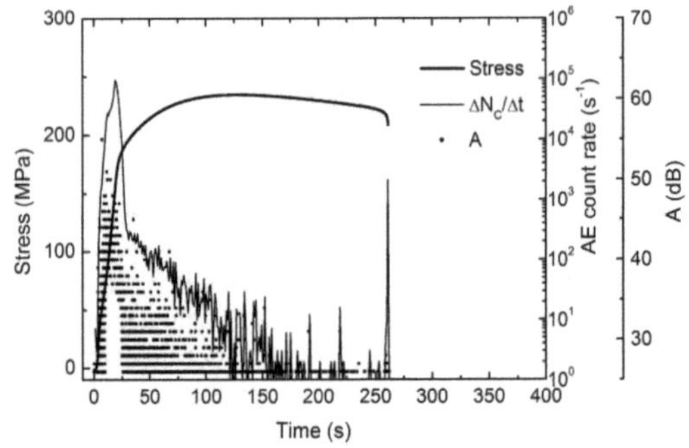

a)

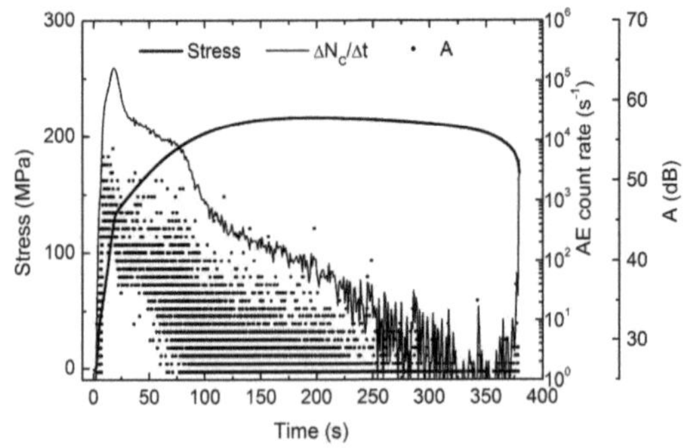

b)

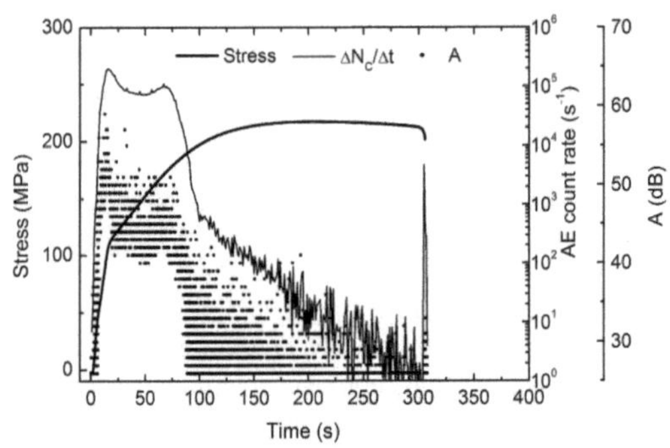

c)

Fig. 3. Stress, the AE count rate and the amplitude for individual AE events vs. time curves for a) RD b) 45° and c) TD of conventionally rolled slab.

A considerable difference in initial texture of the rolled slab and TRC strip is also manifested in the deformation behavior of both sheets. Deformation curves with the AE response for tension along the RD, 45° and the TD for rolled TRC strip are presented in Figs. 4a,b,c, respectively. It can be seen that the highest value of the YS is in the TD sample and decreases with the orientation to the RD, which is an opposite tendency by comparison with the deformation behavior of the rolled slab (Fig. 3).

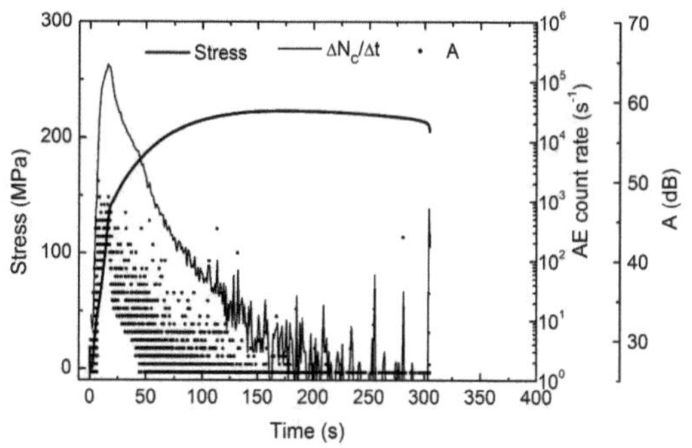

a)

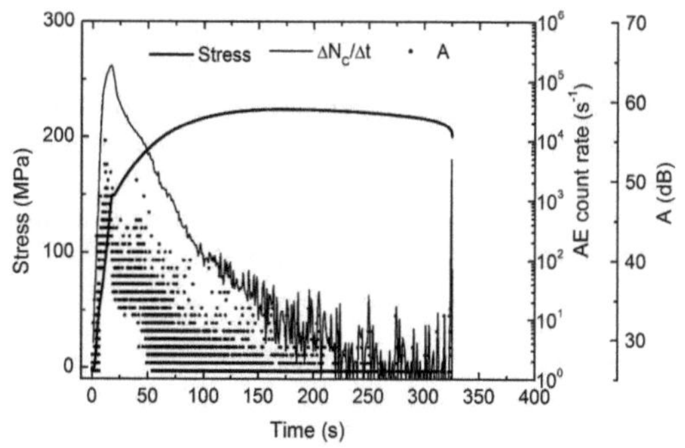

b)

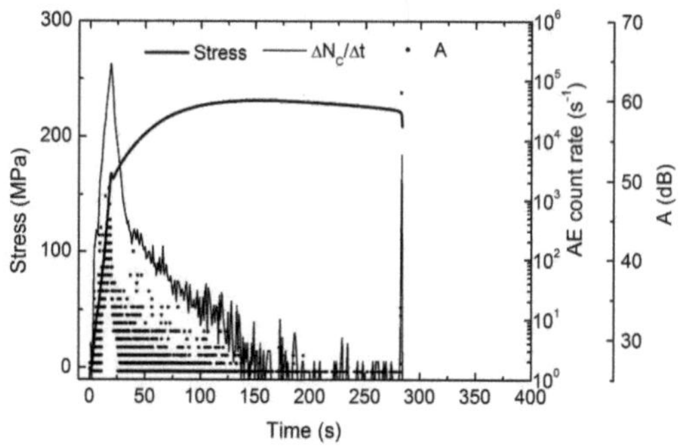

c)

Fig. 4. Stress, the AE count rate and the amplitude for individual AE events vs. time curves for a) RD, b) 45° and c) TD of rolled TRC strip.

Moreover, a variation in the YS for all three orientations of rolled TRC strip is also considerable lower than for rolled slab. The AE parameters reach their maxima at the YS and the following decrease is more pronounced in the TD sample than in the other two samples. Comparing to rolled slab, no second AE maxima during deformation hardening were observed.

Discussion

The conventional rolling process led to a fine grain structure, in comparison with the rolled TRC strip with "untypical" texture for Mg alloy sheets, i.e. with basal planes split towards the TD. A similar texture was observed in rolled and annealed Mg-Zn-Ce alloy sheet [18, 19]. Rolling of the TRC strip should bring a benefit in the weak texture and homogeneous microstructure. Nevertheless, the texture is relative strong with basal planes tilted into the RD.

The results of deformation tests indicate different micromechanisms responsible for the deformation behavior of the rolled slab and rolled TRC strip. In-plane anisotropy of the YS due to texture is clearly seen for both rolled materials. The AE measurement during straining is a useful tool indicating the microstructural evolution (dislocation evolution, twinning) and it helps to estimate deformation micromechanisms. The AE results show that collective dislocation processes start at a low stress, even below the microscopic yield point, in both materials. The characteristic AE peak at the YS is usually explained by a massive dislocation multiplication when basal and non-basal slip systems are activated. In the case of rolled slab, the narrow AE peak observed for samples deformed in the RD can be assumed as a consequence of the activation of the primary (basal) and secondary slip systems (higher work hardening than in the TD). The subsequent strong decrease in the AE activity can be explained by increasing number of sessile dislocations, which reduce the free path of moving dislocations. The second AE peak, observed in 45° and TD sample, can be attributed to twinning activity. Twin nucleation is an excellent source producing high amplitude AE signals (burst type) [20]. It has been reported [20, 21] that the twin growth does not produce AE contrary to the twin nucleation. Therefore, the second peak in the count rate, as well as the persistence of high amplitude signals, following the peak in both directions is consistent with twin nucleation. Likewise, spread of basal poles into TD and S - shape of the deformation curve, as a typical sign of twinning activity, support this explanation.

Moreover, Dobroň et al. [19] have proven with the help of EBSD measurements that an increase in the AE count rate despite of a strong strain hardening in ZE10 magnesium sheet is caused by the twinning activity.

In the case of the rolled TRC strip, the in-plane anisotropy of the YS is smaller than for the rolled slab what is in agreement with observed initial texture. The spread of basal planes into the RD is responsible for the lowest value of the YS in this direction (Fig. 4). The plastic deformation starts in basal planes and continues in non-basal slip systems. The increasing dislocation density and transition from primary (basal) to secondary (non-basal) slip mechanisms significantly reduce the AE activity. The absence of the second AE peak in rolled TRC strip can be ascribed to unfavorable orientation of grains for a massive twin nucleation. Broader AE count rate peak in the RD and 45° samples, in comparison to the TD, seems to be connected with collective dislocation processes.

This assumption is also supported by an occurrence of lower amplitudes for individual AE events in rolled slab than in rolled TRC strip during strengthening after the YS. As was already mentioned above, the nucleation of twins is linked with a release of high amplitude AE signals.

Conclusions

Tensile tests at room temperature revealed that the different initial texture between conventionally rolled slab and rolled TRC strip has a great effect on dominant deformation processes (slip and twinning) during the early stage of straining. In rolled slab, the basal planes are split towards the TD and with a help of the AE parameters it was shown that the second AE maximum during tensile loading of 45° and TD samples indicates the nucleation of the tensile twins. In rolled TRC strip, the basal texture is more pronounced in the RD and the absence of high AE signals after the microscopic yield point seems to correspond to the transition from the initial basal to a secondary non-basal slip and therefore, lower in-plane anisotropy in rolled TRC strip was observed.

Acknowledgement

This work received support from the Czech Science Foundation under grant 13 - 19812S. The work of P.D. is a part of activities of the Charles University Research Center "Physics of Condensed Matter and Functional Materials". D.D. is grateful for financial support from the Grant Agency of Charles University under grant No. 1882314 and also for the grant SVV-2014-260091.

References

[1] C.J. Bettles, M.A. Gibson, "Current wrought magnesium alloys: strengths and weakness," *JOM - J. Min. Met. Mat. S.*, 57 (2005), 46-49.

[2] K. Hantzsche, J. Bohlen, J. Wendt, K.U. Kainer, S.B. Yi, D. Letzig, "Effect of rare earth additions on microstructure and texture development of magnesium alloy sheets," *Scripta Mater.*, 63 (2010), 725-730.

[3] A. Jain, S.R. Agnew, "Modeling the temperature dependent effect of twinning on the behavior of magnesium alloy AZ31B sheet," *Mater. Sci. Eng. A*, 462 (2007), 29-36.

[4] J. Bohlen, M.R. Nürnberg, J.W. Senn, D. Letzig, S.R. Agnew, "The texture and anisotropy of magnesium–zinc–rare earth alloy sheets," *Acta Mater.*, 55 (2007), 2101-2112.

[5] R.E. Reed-Hill, "Role of deformation twinning in determining the mechanical properties of metals: the inhomogeneity of plastic deformation," *Am. Soc. Metals*, 285 (1973), 311.

[6] S.R. Agnew, O. Duygulu, "Plastic anisotropy and the role of non-basal slip in magnesium alloy AZ31B," *Int. J. Plasticity*, 21 (2005), 1161-1193.

[7] J. Bohlen, F. Chmelík, P. Dobroň, F. Kaiser, D. Letzig, P. Lukáč, K.U. Kainer, "Orientation effects on acoustic emission during tensile deformation of hot rolled magnesium alloy AZ31," *J. Alloy Compd.*, 378 (2004), 207-213.

[8] Y. Wang, S.B. Kang, J. Cho,"Microstructure and mechanical properties of Mg–Al–Mn–Ca alloy sheet produced by twin roll casting and sequential warm rolling," *J. Alloy Compd.*, 509 (2011), 704-711.

[9] E. Essadiqi, I.H. Jung, M.A. Wells, "7 - Twin roll casting of magnesium,"*Advances in Wrought Magnesium Alloys* (2012), 272-303.

[10] J. Zeng, R. Koitzsch, H. Pfeifer, B. Friedrich, "Numerical simulation of the twin-roll casting process of magnesium alloy strip," *J. Mater. Process. Tech.*, 209 (2009), 2321-2328.

[11] S. Chen, J. Chen, J. Yu, L. Lv, "Simulation of microstructures in solidification of aluminum twin-roll continuous casting," *Procedia Engineering*, 31(2012), 509-514.

[12] B.S. You, Ch.D. Yim, S.H. Kim, "Solidification of AZ31 magnesium alloy plate in a horizontal continuous casting process," *Mater. Sci. Eng. A*, 413–414 (2005), 139-143.

[13] T. Mino, M. Asakawa, D. Lee, T. Fujiwara, K. Matsuzaki, M. Kobayashi, "Twin-roll strip casting of AZ61 magnesium alloy and improvement of formability by structure-control rolling," *J. Mater. Process. Tech.*, 177 (2006), 534-538.

[14] N. Stanford, D. Phelan, "The formation of randomly textured magnesium alloy sheet through rapid solidification," *Acta Mater.*, 58 (2010), 3642-3654.

[15] C.R. Heiple, S.H. Carpenter, "Acoustic emission produced by deformation of metals and alloys – A review: Part I.," *J. Acoustic Emission*, 6 (1987), 177-204.

[16] P. Dobroň, J. Bohlen, F. Chmelík, P. Lukáč, D. Letzig, K.U. Kainer, "Mechanical anisotropy of AZ31 magnesium alloy sheet investigated by the acoustic emission technique," *Kovove Mater.*, 45 (2007), 129-133.

[17] A. Vinogradov, D. Orlov, A. Danyuk, Y. Estrin, "Effect of grain size on the mechanisms of plastic deformation in wrought Mg–Zn–Zr alloy revealed by acoustic emission measurements," *Acta Mater.*, 61 (2013), 2044-2056.

[18] L.W.F. Mackenzie, M.O. Pekguleryuz, "The recrystallization and texture of magnesium–zinc–cerium alloys," *Scripta Mater.*, 59 (2008), 665-668.

[19] P. Dobroň, J. Balík, F. Chmelík, K. Illková, J. Bohlen, D. Letzig, P. Lukáč, "A study of mechanical anisotropy of Mg–Zn–Rare earth alloy sheet, "*J. Alloy Compd.*, 588 (2014), 628-632.

[20] C.R. Heiple, S.H. Carpenter, "Acoustic emission produced by deformation of metals and alloys – A review: Part II.," J. Acoustic Emission 6 (1987) 215-237.

[21] J.P. Toronchuk, "Acoustic emission during twinning of zinc single crystals," *Mater. Eval.*, (1977), 51-53.

Magnesium Technology 2015
Edited by: Michele V. Manuel, Alok Singh, Martyn Alderman, and Neale R. Neelameggham
TMS (The Minerals, Metals & Materials Society), 2015

EFFECTS OF TEXTURE AND ALLOYING ELEMENTS ON STRETCH FORMABILITY OF Mg ALLOY SHEETS

Byeong-Chan Suh[1], Jae H. Kim[1], Jun Ho Bae[2], Nack J. Kim[1]
[1]Pohang University of Science and Technology (POSTECH), Pohang 790-784, Korea
[2]Korea Institute of Materials Science (KIMS), Changwon, 641-831, Korea

Keywords: Mg alloys, Formability, Texture, Twinning.

Abstract

A study has been made on the effects of texture and alloying elements on stretch formability of Mg alloys. AZ31, AT31, and ZX11 alloys having different textures have been subjected to uni-axial tensile loading and bi-axial stretch forming tests, and their microstructures and textures before and after deformation have been analyzed. It shows that the deformation of AT31 alloy is mainly controlled by the activation of prismatic <a> slip besides the basal slip. On the other hand, there is a significantly high activity of tension twinning in ZX11 alloy regardless of tensile loading directions, resulting in the smallest Lankford values in ZX11 alloy among three alloys, followed by AT31 and AZ31 alloys. Nevertheless, the results of Erichsen cup tests in these alloys show the different behaviors such that AT31 alloy has the best stretch formability, followed by ZX11 and AZ31 alloys. It has been shown that the excellent stretch formability of AT31 alloy comes from an increased activity of prismatic <a> slip and its characteristic texture counterbalanced by broadened distribution of basal poles along the transverse direction and split basal poles along the rolling direction. Although ZX11 alloy shows an increased activity of tension twinning during stretch forming, its orthotropic texture would lead to significant anisotropic deformation at a late stage of stretch forming, resulting in a premature failure.

Introduction

The development of magnesium sheet alloys is essential to support the growing needs of the automotive industry for lightweight components. For the feasible application of Mg sheet products, however, it is necessary that Mg sheets can be readily formed into complicated shapes at room temperature. It is well known that Mg alloy sheets typically exhibit strong basal texture after thermomechanical treatment and such strong basal texture has an adverse effect on the formability at room temperature. Therefore, there have been numerous studies on modifying the texture of Mg alloy sheets to a weaker and more random distribution of basal poles by the addition of alloying elements such as rare earth (RE) and Ca [1-6]. Although its maximum basal pole intensity is weak, the textures of these RE or Ca containing alloys are typically characterized by the orthotropic or asymmetrical texture, i.e., broadened distribution as well as splitting of basal poles along the same direction. Considering that biaxial tensile stress is operative during stretch forming, it is expected that such asymmetrical texture of RE or Ca containing alloys would result in anisotropic in-plane deformation [7] and poor formability. Nevertheless, these alloys all show good formability [1, 8, 9], which has been ascribed mainly to their weak texture intensity without detailed understanding of their detailed deformation behavior. The deformation behavior of Mg alloys is difficult to understand since not just the texture but also the alloying elements play important roles in controlling various modes of deformation. It is needed to separate the effect of alloying elements from the dominant effect of texture on deformation behavior to understand their intrinsic effects. In the present study, three alloys with different texture and alloy composition have been subjected to uni-axial tension and bi-axial stretch forming tests and their microstructure and texture evolution during forming have been investigated.

Experimental Procedure

Chemical compositions of these Mg alloys investigated in the present study are Mg-3Al-1Zn, Mg-3Al-1Sn, and Mg-1Zn-1Ca (wt.%), which are referred to as 'AZ31', 'AT31', and 'ZX11', respectively. AT31 and ZX11 alloys were fabricated by a laboratory twin-roll caster at POSTECH. Details of the TRC process are described elsewhere [5, 10, 11]. After twin-roll casting (TRC), they were homogenized at 673 K for 3 h for AT31 alloy and at 713 K for 1 h for ZX11 alloy, followed by water quenching. The homogenized AT31 and ZX11 alloys were hot rolled at 473 K and 573 K, respectively, for 5 passes with a total reduction of 50 % (13% reduction per pass). After the final rolling, the sheets were given final annealing at 473 K for 3 h for AT31 alloy and at 713 K for 30 min for ZX11 alloy, followed by water quenching. AZ31 alloy sheets were obtained from the Research Institute of Industrial Science and Technology (RIST), Pohang, Korea. As-received AZ31 alloy was annealed at 473K for 3 h. Microstructure of the alloys was observed by optical microscopy and electron backscattering diffraction (EBSD). The texture was analyzed at the midsection of the specimens with Co K_α radiation and the pole figures were obtained using the Schulz reflection method. Tensile tests were conducted using the specimens with a gauge length of 12.5 mm, a gauge width of 5 mm, and a gauge thickness of 1 mm at the strain rate of 6.4×10^{-4} s^{-1}. Stretch formability was measured by the Erichsen cup test. The punch diameter and speed used were 27.5 mm and 0.5 mm s^{-1}, respectively. Silicon oil was used as a lubricant.

Results and Discussion

Microstructure and Texture

The microstructures in the RD-ND plane and (0001) pole figure of as-annealed AZ31, AT31, and ZX11 alloys are shown in Fig. 1. The average grain sizes are 5.9 μm, 5.2 μm, and 10.8 μm for AZ31, AT31, and ZX11 alloys, respectively. In case of AT31 alloy, small equiaxed grains coexist with relatively large elongated grains, which probably were not dynamically recrystallized during rolling process. It shows that the characteristics of texture vary quite a lot with alloy compositions. The intensities of basal poles are 9.9, 6.8, and 2.0 for AZ31, AT31, and ZX11 alloys, respectively. While AZ31 alloy shows a typical strong basal texture, AT31 alloy shows a texture with broadened angular distribution of basal poles from the normal direction (ND) towards the transverse direction (TD) but the basal poles tilted 20 – 30° from the ND towards the rolling direction (RD). It is expected that anisotropy would not be so significant in such case due to counterbalancing effect of broadened basal poles along the TD and split basal poles along the RD. On the other hand, the texture of ZX11 alloy is weak and can be characterized by a broadened angular distribution as well as a splitting (30 – 60°) of basal poles from the ND towards the TD.

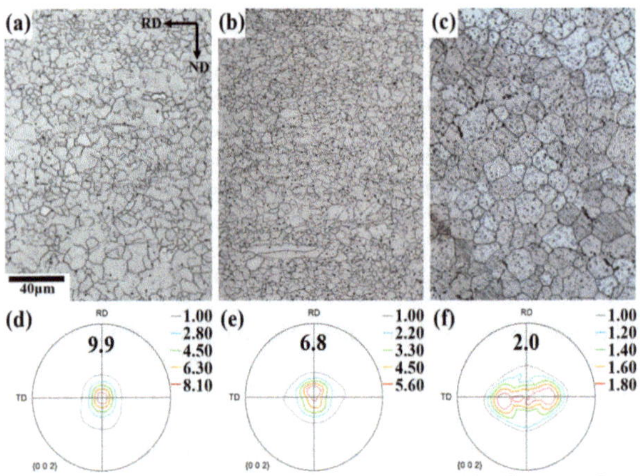

Figure 1. Optical micrographs and (0001) pole figures of as-annealed (a, d) AZ31, (b, e) AT31, and (c, f) ZX11 alloys.

As shown in Fig. 1, all the alloys show somewhat equally broadened angular distribution of basal poles, over the range of 0-50°, from the ND towards the RD. Therefore, when tensile loaded along the RD, it gives similar Schmid factors (SFs) for basal slip which is a dominant deformation mode in Mg alloys at room temperature. The calculated average SFs for basal slip under tensile loading along the RD are 0.221, 0.237, and 0.241 for AZ31, AT31, and ZX11 alloys, respectively. The average SFs for basal slip have also been calculated for tensile loading along the TD, and are 0.199, 0.227, and 0.298 for AZ31, AT31, and ZX11 alloys, respectively. It shows that the SFs show the same trend along the RD and TD, i.e., the smallest in AZ31 alloy, followed by AT31 and ZX11 alloys. It also shows that the SFs of ZX11 alloy is much larger along the TD than the RD, which can be easily expected from its (0001) pole figure showing the broadened and split basal poles from the ND towards the TD.

Mechanical Properties

Tensile properties of the alloys are summarized in Table I and their representative engineering stress-strain curves are shown in Fig. 2a. It shows that yield strength (YS) is highest in AZ31 alloy, followed by AT31 and ZX11 alloys. This is in agreement with the results of texture analyses, which show that the maximum basal pole intensity is highest and average SF is lowest in AZ31 alloy, followed by AT31 and ZX11 alloys. It also shows that although the difference is not large, both AZ31 and AT31 alloys show lower YS and ultimate tensile strength (UTS), and larger elongation in RD than in TD, in agreement with the calculated average SFs. On the other hand, ZX11 alloy shows a rather large difference in tensile properties depending on tensile loading directions. Unlike the cases of AZ31 and AT31 alloys, ZX11 alloy shows much higher YS and UTS in RD than in TD. Such behavior undoubtedly comes from its unique texture (Fig. 1f), which gives higher SFs for basal slip along the TD than the RD. As can be seen in Fig. 2a, the stress-strain curve of ZX11 alloy shows the sigmoidal shape, particularly when tested along the TD, suggesting the onset of tension twinning at an early stage of deformation, which is also responsible for lower YS in TD than in RD.

Table I. Tensile properties and IE values of the alloys.

Alloy	Testing direction	YS (MPa)	UTS (MPa)	ε_u (%)	ε_t (%)	IE (mm)
AZ31	RD	186.8	272.9	14.1	24.7	4.8
	TD	205.6	276.3	12.8	23.3	
AT31	RD	179.2	258.8	16.7	27.4	10.2
	TD	185.7	261.0	17.2	25.6	
ZX11	RD	154.9	230.6	11.6	12.3	8.8
	TD	114.0	214.4	13.2	13.6	

Erichsen cup tests of the alloys give quite interesting results. It shows that AT31 alloy shows the highest index Erichsen (IE) value, followed by ZX11 and AZ31 alloys. Although the smallest IE value of AZ31 alloy is not surprising due to its strong basal texture, the higher IE value of AT31 alloy than that of ZX alloy is against the prediction based on the characteristics of microstructure and texture, i.e., stretch formability increases with increasing grain size and weakening texture [1, 12-14]

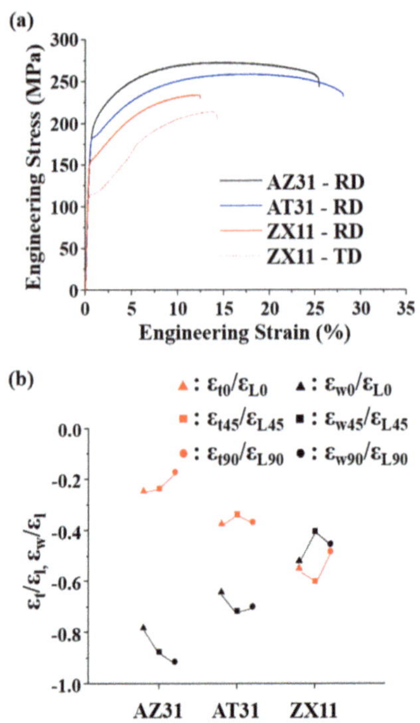

Figure 2. (a) Engineering stress-strain curves of the alloys along the RD (the curve along the TD for ZX11 alloy is also shown) and (b) The ratio of the thickness strain to the longitudinal strain ($\varepsilon_t/\varepsilon_l$) and the ratio of the width strain to the longitudinal strain ($\varepsilon_w/\varepsilon_l$) at the strain of 0.1.

The Lankford values (r-values) of three alloys have been measured at the strain of 0.1. It shows that ZX11 alloy has the smallest average r-value ($\bar{r}$) of 1.28, followed by AT31 (1.97) and AZ31 (4.01) alloys. Although the difference is small, on the other hand, AT31 alloy shows the smallest planar anisotropy (Δr) of 0.33, followed by ZX11 (0.42) and AZ31 (0.56) alloys. Considering that stretch formability of Mg alloys is mainly determined by their abilities to deform through thickness direction, the above-mentioned differences in r-values among the alloys suggest that ZX11 alloy would show better stretch formability (i.e., higher IE value) than AT31 alloy, which does not agree with the experimental results. The components of r-value, i.e., thickness strain and width strain, were compared with the longitudinal strain and the results are plotted in Fig. 2b. It shows that as expected from r-values, ZX11 alloy shows the largest ratio of thickness strain to longitudinal strain ($\varepsilon_t/\varepsilon_l$) and the smallest ratio of width strain to longitudinal strain ($\varepsilon_w/\varepsilon_l$) and AZ31 alloys shows the opposite behavior. Although the stretch formability of Mg alloys is mainly determined by their ability to deform through thickness direction as mentioned previously, too much thickness strain at an early stage of deformation would eventually lead to early failure during stretch forming, i.e., small IE value, which might be case for ZX11 alloy showing the largest $\varepsilon_t/\varepsilon_l$ at the given longitudinal strain.

Deformation Behavior

Uni-axial Deformation

Fig. 3 shows the distribution of misorientation angles of the alloys after deformation to 10% strain. It shows that there is a development of new peak at the misorientation angle of about 86° after deformation in AZ31 and ZX11 alloys, indicating the formation of tension twins [15, 16]. Analyses of misorientation profile clearly show that the activity of tension twins in ZX11 alloy is much higher than in AZ31 alloy. On the other hands, AT31 alloy does not show significant difference in the distribution of misorientation angles before and after deformation, indicating that the activity of tension twinning in AT31 alloy is much less than in other alloys. It is interesting to note that there is a rather large amount of tension twins formed in AZ31 alloy, although its texture is not favorable for the formation of tension twins since the c-axes of matrix grains are mostly perpendicular to the tensile loading direction. As pointed out by Suh et al. [17], it is mainly due to the operation of compress stress along the direction perpendicular to the tensile loading direction due to constraint imposed by the deformation of neighboring grains. It can also explain why the ZX11 alloy having the orthotropic texture shows the similar fraction of tension twins after tensile deformation along the TD and the RD.

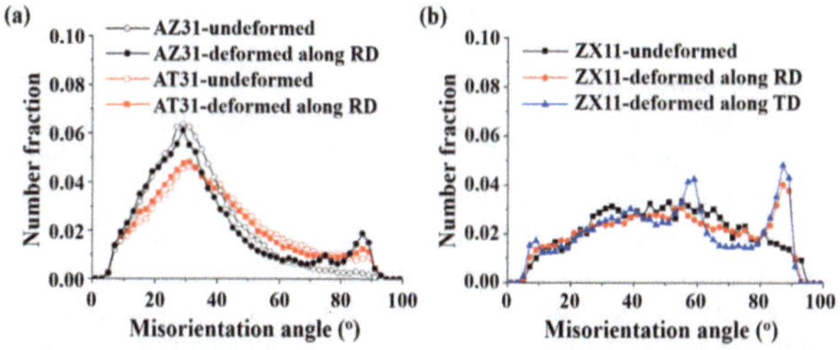

Figure 3. Misorientation profiles of (a) AZ31 and AT31 alloys before and after deformation along the RD and (b) ZX11 alloy before and after deformation along the RD and TD.

To identify the deformation modes of the alloys, polycrystal plasticity simulation based on visco-plasticity self-consistent (VPSC) model has been conducted [18]. It shows that besides the basal slip, tension twinning and the prismatic <a> slip are the dominant deformation modes for ZX11 and AT31 alloys, respectively. The operation of prismatic <a> slip in AT31 alloy has also been confirmed by TEM analysis [18].

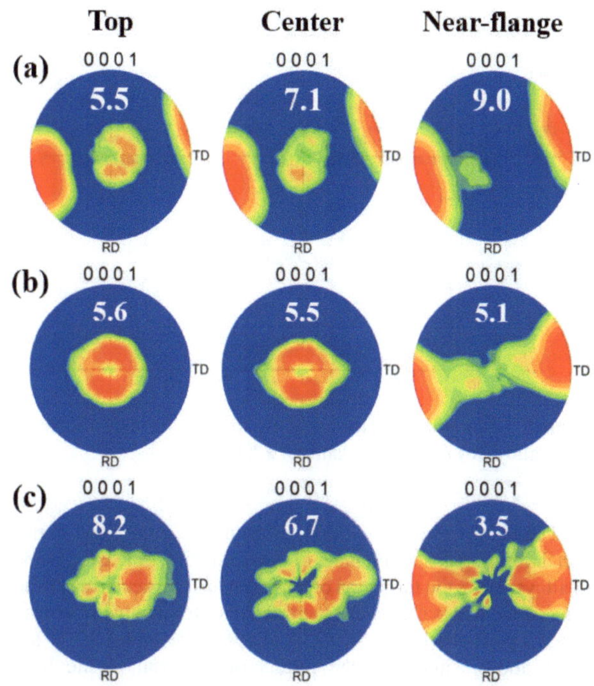

Figure 4. (0001) pole figures of the stretch formed alloys taken from the mid-thickness TD-RD plane of the sheet; (a) AZ31 after fracture, (b) AT31 after fracture, and (c) ZX11 after fracture.

Bi-axial Stretch Forming

To observe the microstructural and texture evolution during stretch forming, three different parts of the sheets were analyzed; top, center, and near-flange areas of stretch formed sheet. Fig. 4a shows the (0001) pole figures of AZ31 alloy taken from the mid-thickness TD-RD plane of the sheet after fracture (IE of 4.8 mm). There is a formation of numerous tension twins in AZ31 alloy, whose c-axes are parallel to the TD, as clearly evidenced by the newly developed basal poles near the TD in Fig. 4a. It can also be noticed that the fraction of tension twins is almost same regardless of the area (Fig. 5a).

In the case of AT31 alloy having the large IE value of 10.2 mm (Fig. 4b), top and center areas do not show the presence of tension twins in contrast to the case of AZ31 alloy (Fig. 5b). However, tension twins having their c-axes parallel to the TD are frequently observed in near-flange areas. Comparison of Fig. 5a and Fig. 5b shows that the formation of tension twins is more active during stretch forming of AZ31 alloy than that of AT31, similar to the results of uni-axial tensile tests. In case of ZX11 alloy, it can be seen that the distributions of basal poles in the top and center areas after stretch forming are quite similar to the initial texture having broadened and split basal poles from the ND towards the TD. The formation of tension twins is seldom observed at the top and center areas of ZX11 alloy. On the other hand, a fairly large amount of tension twins are found in near-fringe area (Fig. 5c). It should be noted that while AZ31 alloy shows the similar fraction of tension twins regardless of the area, both AT31 and ZX11 alloys show large fractions of tension twins in the near-flange area, but virtually no tension twins in the top and center areas.

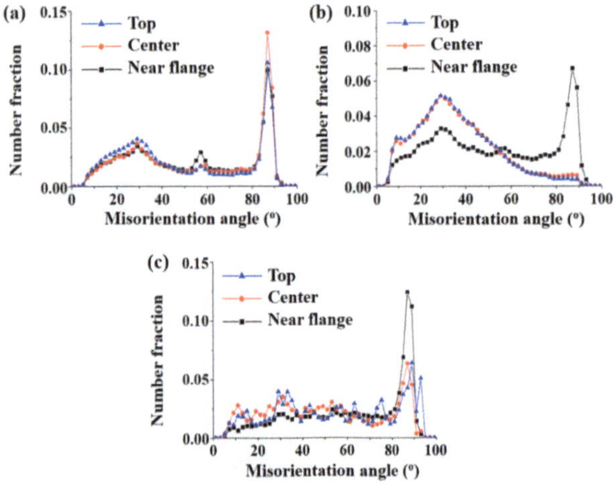

Figure 5. Misorientation profiles of the stretched formed alloys; (a) AZ31 alloy, (b) AT31 alloy, and (c) ZX11 alloy.

The above results show that although the stress state during stretch forming is not favorable for the formation of tension twins, all alloys shows the formation of tension twins, particularly in near-flange area. It also shows that the fraction of tension twins in near-flange area is the largest in ZX11 alloys followed by AZ31 and AT31 alloys, similar to the results of uni-axial tensile tests. However, the fraction of tension twins at the near-top area of the dome is the largest in AZ31 alloy, followed by ZX11 and AT31 alloys. In the case of uni-axial tensile test, tension twins can form in the grains unfavorably oriented for the formation of tension twins by the operation of compressive stress perpendicular to their c-axes due to the constraint by neighboring grains. However, such compressive stress is generally smaller than the imposed tensile stress during bi-axial stretch forming, making the formation of tension twins difficult in the basal textured grains during stretch forming.

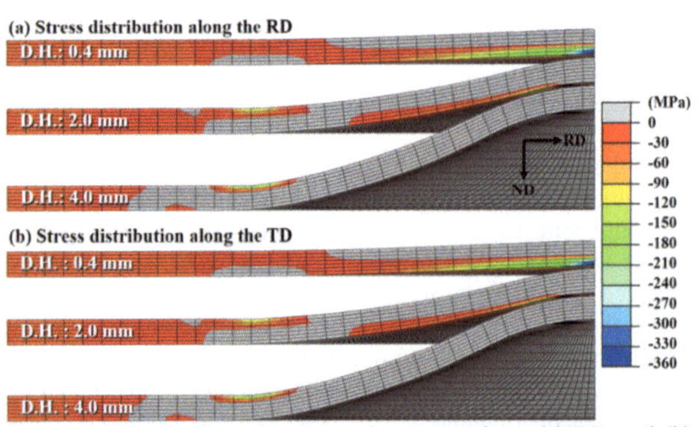

Figure 6. Distribution of compressive stress along (a) RD and (b) TD at the dome height of 0.4, 2.0, and 4.0 mm in AT31 alloy.

Nevertheless, the formation of tension twins is observed in AZ31 alloy throughout the sheet, from the near-flange to the top area of stretch formed sheet. To understand the reason behind such formation of tension twins, finite element method (FEM) based on ABACUS simulation has been conducted. Fig. 6 shows the variations in compressive stress distributions along the RD and TD with the dome height of AT31 alloy (tensile stress is represented as a gray color). It shows that the compressive stress operates at the lower surface of the sheet at an early stage of stretch forming. With an increase in dome height, most of sheet will be subjected to tensile stress but the near-flange area will still be subjected to compressive stress until the imposed tensile stress becomes larger than the compressive stress. In particular, the lower part of near-flange areas will be subjected to compressive stress due to bending of the sheet [10]. Even under such circumstances, however, AT31 alloy show the smallest fraction of tension twins among the alloys, similar to the results of uni-axial tensile tests. Although tension twinning is promoted at an early stage of stretch forming regardless of alloy systems, however, the deformation behavior of alloys would be different at a late stage of stretch forming when there is no in-plane compressive stress. Comparing the behavior of AZ31 and ZX11 alloys, one can notice that there is still a significant fraction of grains with their c-axes inclined towards the TD in AZ31 alloy (Fig. 4a), while c-axes of most grains are oriented towards the ND in ZX11 alloy (Fig. 4c). Analyses of misorientation angles also show that the fraction of tension twins does not change with the location in AZ31 alloy (Fig. 5a), but there is virtually no tension twins in the top and center areas of ZX11 alloy (Fig. 5c). It indicates that while the deformation of AZ31 alloy occurs mainly by basal slip, there is an active detwinning in ZX11 alloy due to reversed stress state at a late stage of stretch forming. Such enhanced activity of tension twinning (as well as detwinning) results in better stretch formability for ZX11 alloy than that for AZ31 alloy.

Although the above results show that tension twinning can be beneficial for the improvement of stretch formability, comparison of ZX11 and AT31 alloys shows that the initial texture of alloy is also important in determining the stretch formability. As shown in Fig. 1f, ZX11 alloy shows an orthotropic texture whose basal poles are oriented along the TD. In such a case, deformation during stretch forming is not homogeneous across the plane since stretching would be easier along the TD than other directions, resulting in a severe anisotropic deformation although the alloy's active tension twinning would help relieving the applied stress. On the other hand, AT31 alloy show a texture counterbalanced by broadened distribution of basal poles along the TD and split basal poles along the RD. Then, the alloy would show excellent stretch formability if it can deform by additional deformation mode besides basal slip. This is the case of AT31 alloy, which shows an increased activity of prismatic slip by the addition of Sn.

Summary

In the present study, three alloys with different texture and alloy compositions, AZ31, AT31, and ZX11 alloys, have been subjected to uni-axial tensile tests and bi-axial stretch forming tests. It shows that the predictions made based on the results of uni-axial tensile tests do not agree with the experimental results of stretch forming tests. While ZX11 alloy shows the smallest r-values, the best stretch formability has been obtained in AT31 alloy. Detailed analyses of deformed alloys show that ZX11 alloy has the smallest average r-value due to its weak texture and an enhanced activity of tension twinning induced by Ca addition. However, its orthotropic texture, i.e., broadening and splitting of basal poles along the TD, would result in a severe anisotropic deformation at a late stage of stretch forming, which might be responsible for its worse stretch formability than that of AT31 alloy. On the other hand, AT31 alloy has the more ideal texture counterbalanced by broadened distribution of basal poles along the TD and split basal poles along the RD. Such texture as well as an enhanced prismatic slip induced by Sn addition improve the stretch formability of AT31 alloy over those of other alloys

Acknowledgement

This work was supported by the WPM Program funded by the Ministry of Industry, Trade and Energy, Korea.

References

1. D. Wu, R.S. Chen, E.H. Han, "Excellent room-temperature ductility and formability of rolled Mg–Gd–Zn alloy sheets", *Journal of Alloys and Compounds*, 509 (2011) 2856-2863.
2. S.R. Agnew, M.H. Yoo, C.N. Tome, "Application of texture simulation to understanding mechanical behavior of Mg and solid solution alloys containing Li or Y", *Acta Materialia*, 49 (2001) 4277-4289.
3. J. Bohlen et al., "The texture and anisotropy of magnesium-zinc-rare earth alloy sheets", *Acta Materialia*, 55 (2007) 2101-2112.
4. T. Al-Samman, X. Li, "Sheet texture modification in magnesium-based alloys by selective rare earth alloying", *Materials Science and Engineering: A*, 528 (2011) 3809-3822.
5. D.W. Kim et al., "Texture Evolution in Mg-Zn-Ca Alloy Sheets", *Metallurgical and Materials Transactions A*, 44 (2013) 2950-2961.
6. J.-Y. Lee et al., "Comparison of static recrystallization behavior in hot rolled Mg–3Al–1Zn and Mg–3Zn–0.5Ca sheets", *Journal of Alloys and Compounds*, 589 (2014) 240-246.
7. B.-C. Suh et al., "Current issues in magnesium sheet alloys: Where do we go from here?", *Scripta Materialia*, 84–85 (2014) 1-6.
8. K.H. Kim et al., "Microstructure and texture evolution of Mg alloys during twin-roll casting and subsequent hot rolling", *Scripta Materialia*, 63 (2010) 716-720.
9. Y. Chino et al., "Influence of Zn concentration on stretch formability at room temperature of Mg–Zn–Ce alloy", *Materials Science and Engineering: A*, 528 (2010) 566-572.
10. S.S. Park et al., "Microstructure and tensile properties of twin-roll cast Mg–Zn–Mn–Al alloys", *Scripta Materialia*, 57 (2007) 793-796.
11. S.S. Park, Y.S. Park, N.J. Kim, "Microstructure and Properties of Strip Cas AZ91 Mg Alloy", *METALS AND MATERIALS International*, 8 (2002) 551-554.
12. D.H. Kang et al., "Relationship between stretch formability and work-hardening capacity of twin-roll cast Mg alloys at room temperature", *Scripta Materialia*, 61 (2009) 768-771.
13. C.E. Dreyer et al., "Formability of a more randomly textured magnesium alloy sheet: Application of an improved warm sheet formability test", *Journal of Materials Processing Technology*, 210 (2010) 37-47.
14. Y. Chino, M. Kado, M. Mabuchi, "Enhancement of tensile ductility and stretch formability of magnesium by addition of 0.2wt%(0.035at%)Ce", *Materials Science and Engineering: A*, 494 (2008) 343-349.
15. J.W. Christian, S. Mahajan, "Deformation Twinning", *Progress in Materials Science*, 39 (1995) 1-157.
16. M.D. Nave, M.R. Barnett, "Microstructures and textures of pure magnesium deformed in plane-strain compression", *Scripta Materialia*, 51 (2004) 881-885.
17. B.-C. Suh et al., "Twinning behavior of Mg–4Zn–1Gd alloy sheet during longitudinal tensile deformation", *Scripta Materialia*, 69 (2013) 465-468.
18. B.-C. Suh, N.J. Kim, "Effect of initial texture and alloying elements on deformation behavior of Mg alloys", in preparation (2014).

Magnesium Technology 2015
Edited by: Michele V. Manuel, Alok Singh, Martyn Alderman, and Neale R. Neelameggham
TMS (The Minerals, Metals & Materials Society), 2015

Progress in Thermodynamic Database Development for ICME of Mg Alloys

Rainer Schmid-Fetzer[1]

[1]Institute of Metallurgy, Clausthal University of Technology, Robert-Koch-Str. 42, D-38678 Clausthal-Zellerfeld, Germany

Keywords: Computational thermodynamics, Calphad, Multicomponent Mg alloys.

Abstract

Integrated thermodynamic software packages are a powerful tool to analyze phase formation in multicomponent Mg alloys, providing key data for focused alloy and process development. The thermodynamic database of the alloy system is the indispensable basis for this direct application, but also needed for any further step towards Integrated Computational Materials Engineering. The database quality, which may vary widely, is crucial for the reliability of such applications.

Progress in development and extension of the large thermodynamic database for magnesium alloys is reported, with emphasis on consistency, coherency and quality assurance. One aspect is the coherent description of large solid solution ranges of intermetallic phases in multicomponent Mg alloys with rare earths. Further addition of Zn may produce complex icosahedral, long-period stacking order (LPSO) or other hexagonal phases. All these may form highly efficient secondary phase precipitates in the (Mg) matrix. Examples are given for Mg-Ce-La, Mg-Ce-Y, and Mg-Gd-Nd-Y-Zn-Zr alloys.

Introduction

Integrated Computational Materials Engineering (ICME) is still growing as a strong and wide field in Materials Science and Engineering. One corner stone of ICME is Computational Thermodynamics (CT). The author's view of CT is depicted in Fig. 1. The key point, the application of "Calphad-generated" databases in Materials Science & Engineering, is unique in the sense that it allows to proceed to truly **multi**phase and **multi**component phase equilibria. This goes beyond calculation of phase diagrams. It also enables simulation of phase formation in alloys, ceramics and other complex materials. Moreover, chemical reactions and processes may be simulated, but that is not in the focus of this work.

Central to this approach is the Calphad method [1, 2]. The quintessence is depicted in Fig. 2. Thermodynamic data, such as enthalpy, heat capacity, vapor pressure, etc, are usually available from experimental studies and partly also from *ab initio* studies. In a different class of data we may have phase equilibrium/transformation data at our disposal. These are currently only available experimentally with sufficient accuracy. The key point of the Calphad method is the consistent combination of these two classes of data. That is done by generation of a unique set of Gibbs energy functions, one for each phase φ, with the variables temperature, pressure and phase composition. The multiphase and multicomponent phase equilibria are then calculated by minimizing the Gibbs energy of the system at constant temperature, pressure and alloy composition. That is done by smart distribution of atoms on the available stable or metastable phases in any of the current thermodynamic software packages, Pandat [3], Thermocalc [4], or Factsage [5]. Thus, the formerly separated classes of data are combined into a consistent picture which also exploits the strong thermodynamic rules governing the interdependence of all data.

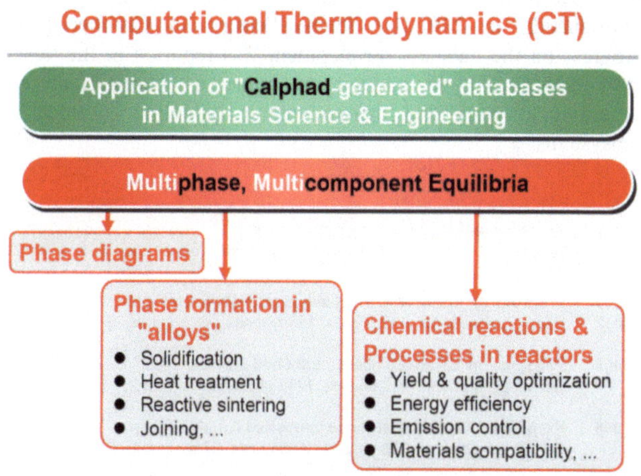

Figure 1: Overview of Computational Thermodynamics and its main applications.

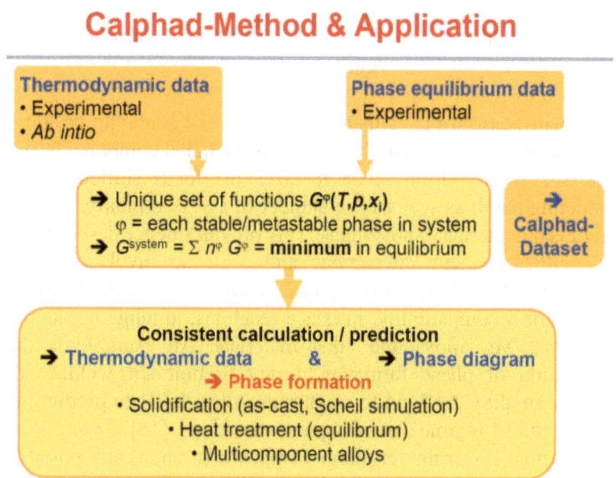

Figure 2: Overview of the Calphad-method and specific applications in the simulation of phase formation.

The generated "Calphad dataset" (thermodynamic description) of the system comprises, for each phase, both the Gibbs energy models and the pertinent thermodynamic parameters. In the whole, for large systems, this constitutes a thermodynamic database, such as the Mg alloy database in the focus of this work.

Also depicted in Fig. 2 is the typical direct application. It encompasses the internally consistent calculation of any thermodynamic property and phase diagrams with the ability to extrapolate into unknown regions in temperature/composition space and into multicomponent systems. Moreover, crucial data concerning phase formation in the alloy system may be calculated or predicted. Specifically for magnesium alloys this approach has been demonstrated as a powerful tool in focused alloy design and process optimization [6-8]. The present work will focus on specific aspects of a thermodynamic Mg alloy database

Thermodynamic Mg alloy database development

Focused work on the development of the dedicated thermodynamic Mg alloy database started in 1995 in an ongoing effort in the authors' group by key experiments combined with Calphad modeling as sketched in Fig. 2. Milestones in that development are highlighted in Fig. 3. Results were first summarized in 1998 by Pisch and Schmid-Fetzer [9] and subsequently in 2001 by Schmid-Fetzer and Gröbner [10]. Progress on the database development was reported in the years 2005-2013[11-15], in addition to numerous publications on experimental work and thermodynamic modeling of specific subsystems.

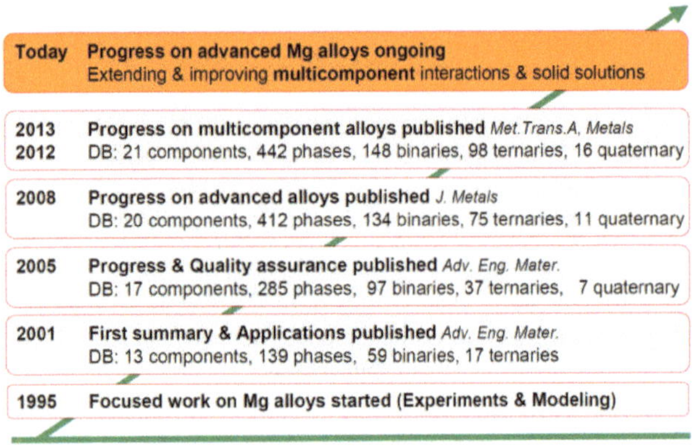

Figure 3: Milestones in development of the thermodynamic Mg alloy database

It is evident from Fig. 3 that during the initial stage the progress was mainly on increasing the number of included components and assessed phases. From 2008 to 2013 the focus shifted to the modeling depth as seen by the number of assessed binary, ternary and even quaternary subsystems. This provides better descriptions of multicomponent interactions and also the involvement of complex solid solution phases. That enables exciting applications to materials compatibility, interface reactions, joining, and so on, involving Mg alloys and other materials. An example is the simulation of phase formation during friction stir welding of aluminum alloy AA5454 to magnesium alloy AZ91 to predict the temperature regime of incipient melting [16]. For such applications covering very large composition ranges any possible phase that might occur within the multicomponent system must have been properly modeled. That applies also to the widely used Scheil simulation of the as-cast structure because the composition of the residual liquid often changes dramatically by segregation and moves far away from the initial Mg-rich alloy. This consistent incorporation of multicomponent interactions is also the current focus of the ongoing progress in the Mg alloy database development.

As the basis the interactions need to be assessed in the complete composition range of binary subsystems. The current status, as frozen in the version PanMg2013 (www.computherm.com) of the database, is depicted in Fig. 4, where the color coding indicates the reliability of the description. That reliability is often limited by the unavailability of more reliable experimental input data on thermodynamics and phase equilibria, as required for the Calphad modeling depicted in Fig. 2. In order to proceed realistically into multicomponent systems the systematic assessment of binary and especially also of crucial ternary subsystems is essential, 98 of which are included in the current status.

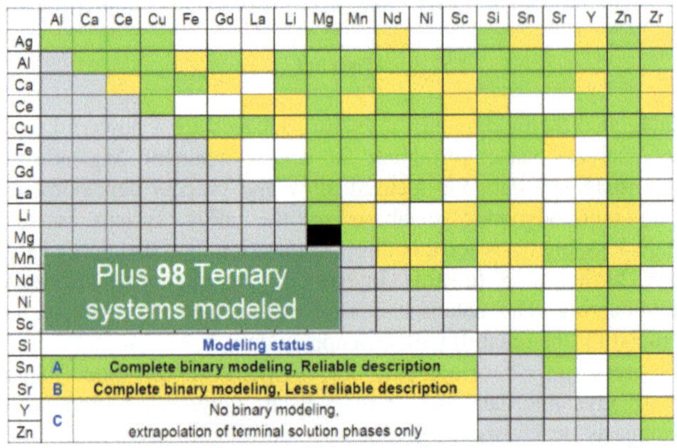

Figure 4: Status of assessed binary and ternary systems in the version PanMg2013 of the Mg database.

With growing size the key issues arising are consistency, coherency and quality assurance. These issues also concern extension, maintenance and updating of the database [15]. It has been shown [17] that the key issue *quality assurance* involves addressing four quite distinct questions concerning the thermodynamic descriptions. It should be verified if these are correct, reasonable, accurate, and safe.

The issues *consistency* and *coherency* can be exemplified by considering the two ternary systems Mg-Ce-La and Mg-Ce-Y as follows. Consistency demands that, generally, if any newly assessed ternary system is incorporated in the database it may need adjustment to the phase models of the binary systems in the database. Conversely, if any binary system is updated the check of any higher order system and probably an adjustment is required, especially for the ternary systems.

Coherency needs very careful consideration because binary compounds may be coherent in ternary or higher order systems if crystallizing in the same structure. For example, the four binary Mg-Ce intermetallic compounds shown at the Mg-Ce edge of the phase diagram in Fig. 5 are $CeMg_{12}$, Ce_5Mg_{41}, $CeMg_3$, and CeMg. Three of them, $CeMg_{12}$, $CeMg_3$, and CeMg, have a stable counterpart at the Mg-La edge, $LaMg_{12}$, $LaMg_3$, and LaMg, respectively, with the same crystal structure. The protrusion of the binary compounds by solid solution into the ternary system is highlighted by red arrows in Fig. 5, which is calculated from the data assessed in [18].

It is crucial that such phases with same crystal structure are unified in the thermodynamic modeling and that unique phase names are assigned. They must be consistent and should be concise and intuitive, such as RMg_3 for the solid solution phase range $CeMg_3$-$LaMg_3$. That phase is modeled by intermixing of Ce and La on the first sublattice of the single phase $(Ce, La)_1Mg_3$. In the entire Mg database the phase RMg_3 with crystal structure $cF16$-BiF_3 is modeled as $(\mathbf{Ce},\mathbf{Gd},\mathbf{La},\mathbf{Nd},Y)_1(Li,\mathbf{Mg},Zn)_3$. The major constituents are highlighted by bold font whereas Y, Li, and Zn are minor constituents indicating that these elements exhibit limited solubility only in the phase RMg_3. Therefore, RMg_3 exists as stable compound in four of the binary edge systems, as $CeMg_3$, $GdMg_3$, $LaMg_3$, or $NdMg_3$, respectively.

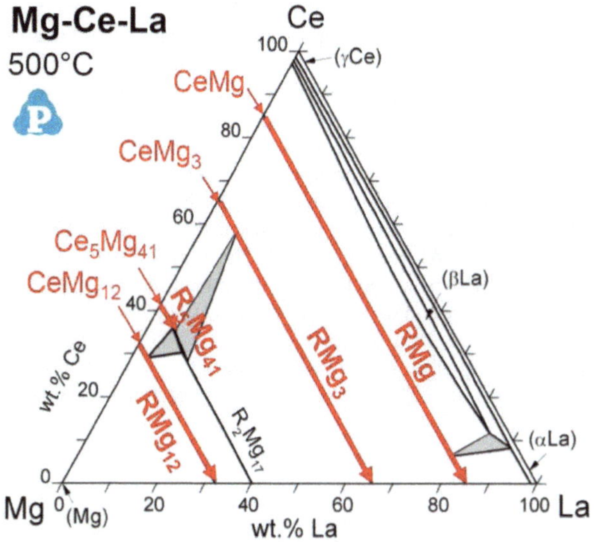

Figure 5: Calculated Mg-Ce-La isothermal phase diagram section at 500°C. The protrusion of four binary Mg-Ce intermetallic compounds in the ternary system is highlighted by red arrows.

The phase protrusion may, of course also be viewed vice versa from the Mg-La to the Mg-Ce edge in Fig. 5. In the ternary Mg-Ce-La example - instead of modeling the six compounds $CeMg_{12}$, $CeMg_3$, CeMg, $LaMg_{12}$, $LaMg_3$, and LaMg separately - only three (but more complex) phases, RMg_{12}, RMg_3, and RMg, are assessed in the database. Therefore, the number of assessed phases may decrease with increasing modeling depth.

The phase diagram Mg-Ce-Y at 500°C in Fig. 6 is calculated from the thermodynamic description assessed in [19]. Again, we see the protrusion of four binary Mg-Ce intermetallic compounds into the ternary system. However, with dissolution of Y only the CeMg extends as RMg phase all the way to the Mg-Y edge whereas the three compounds $CeMg_{12}$, Ce_5Mg_{41}, and $CeMg_3$, show limited solubility only; the corresponding Mg-Y compounds are not stable.

A special case is the RMg_2 phase which might be mistaken as a ternary phase if considering just the 500°C equilibria. However, it originates as the $CeMg_2$ phase from the Mg-Ce edge which is stable at higher temperature only in the binary system but stabilized by dissolution of Y in the $(Ce, Y)Mg_2$ phase. Another peculiarity is that a phase with the same general composition $(Ce,Y)Mg_2$ protrudes from the binary Mg-Y edge. However, the binary compound Mg_2Y possesses a different crystal structure, C14, compared to the binary $CeMg_2$, C15. Therefore, even though the sublattice description $(Ce,Y)Mg_2$ is formally the same, these are modeled as separate phases and named separately as RMg_2(C15) and Mg_2Y(C14). In the entire Mg database these two phases are modeled as follows with major constituents highlighted by bold font:

RMg_2(C15) as $(\mathbf{Ce},\mathbf{Gd},\mathbf{La},\mathbf{Nd},\mathbf{Y})_1(\mathbf{Al},\mathrm{Cu},\mathbf{Mg},\mathrm{Zn})_2$, which is stable in eight binary systems as $CeMg_2$, $LaMg_2$, $NdMg_2$, $CeAl_2$ $GdAl_2$, $LaAl_2$, $NdAl_2$, YAl_2;

Mg_2Y(C14) as $(\mathbf{Mg})_2(\mathrm{Ce},\mathrm{Gd},\mathrm{Nd},\mathbf{Y})_1$, which is stable only in the binary Mg-Y system as Mg_2Y with limited solubility of Ce, Gd, and Nd.

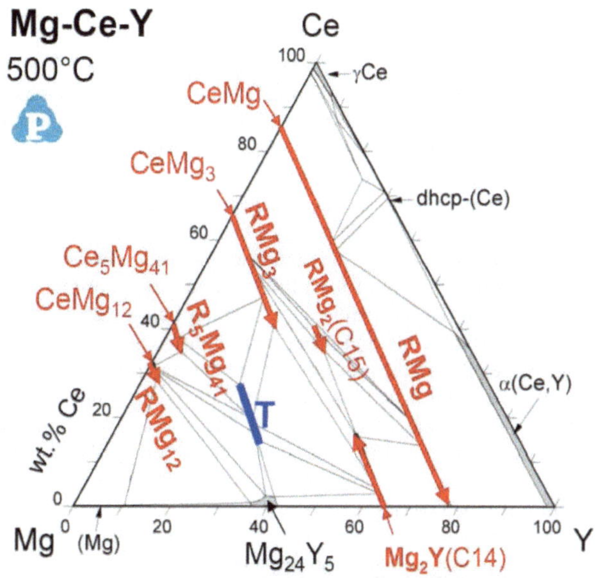

Figure 6: Calculated Mg-Ce-Y isothermal phase diagram section at 500°C. The protrusion of the **five** binary Mg-Ce intermetallic compounds in the ternary system is highlighted by red arrows.

This leaves the phase T, marked blue in Fig. 6, as the only truly ternary phase in the Mg-Ce-Y system. It does not connect to any binary edge, not even at temperatures different from 500°C. This example highlights that phase identification in the microstructure of such Mg alloys based only on local composition analysis, such as SEM/EDS or microprobe, may be hard to interpret without viewing the wider picture of the entire phase diagram.

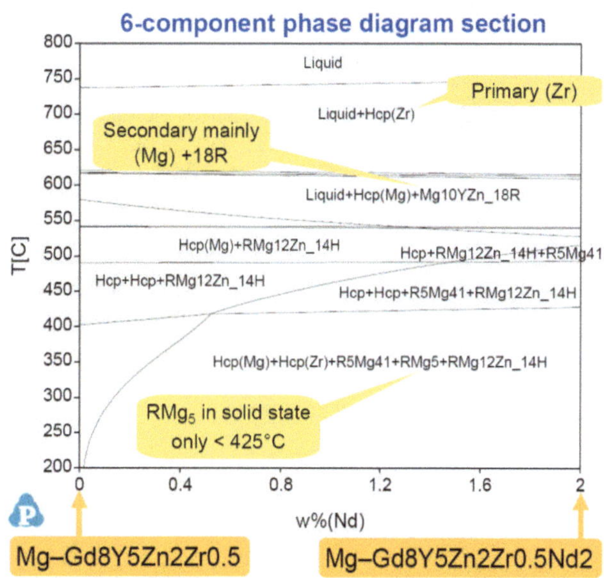

Figure 7: Calculated phase diagram section from alloy Mg–Gd8Y5Zn2Zr0.5 to alloy Mg–Gd8Y5Zn2Zr0.5Nd2.

Application to Mg-Gd-Y-Zn-Zr alloys with Nd addition

The effect of Nd addition on the microstructure and mechanical properties of Mg–Gd8–Y5–Zn2–Zr0.5 (wt.%) alloy was studied by Zhang et al. [20]. Their experimental data are chosen as an example to highlight the potential of applying CT in this work to predict phase formation in real six-component Mg alloys.

An overview of the phases expected by adding 0-2 wt.% Nd to the alloy Mg–Gd8Y5Zn2Zr0.5 is given by the phase diagram section from the base alloy into the six-component system towards alloy Mg–Gd8Y5Zn2Zr0.5Nd2 calculated by Pandat and PanMg2013 in Fig. 7.

Three main messages are given in this phase diagram for the cooling path of molten alloys: Close to 750°C the primary crystallizing phase is (Zr), the secondary phases are mainly (Mg) and 18R, which is one of the LPSO phases in that system. The RMg_5 phase forms in a solid state reaction below 425°C under these equilibrium conditions; that will be different in the as-cast state. As detailed in the previous section the concise phase names cannot reflect the actual composition if they are solid solution phases. That is the case for six of the seven phases in the equilibrium region of Fig. 7 as shown in Table I. Only the 18R phase is modeled as stoichiometric as opposed to the other complex solution phases.

Table I. Phase names and constituents pertinent to Fig. 7

Phase name	Phase constituents
(Zr), Hcp	(Mg,Gd,Nd,Y,Zn,**Zr**)
(Mg), Hcp	(**Mg**,Gd,Nd,Y,Zn, Zr)
18R	$Mg_{10}Y_1Zn_1$
14H	$(Gd,Y)_1Mg_{12}Zn_1$
RMg5	$(Gd,Nd,Y,Zn)_1Mg_5$
RMg3	$(Gd,Nd,Y,Zn)_1Mg_3$
R5Mg41	$(Nd,Y,Zn)_1Mg_5$

The as-cast condition of such alloys is much better reflected by simulations based on Scheil conditions as shown for just the two alloys with the minimum and maximum Nd-content, Mg–Gd8Y5Zn2Zr0.5 and Mg–Gd8Y5Zn2Zr0.5Nd2 in Figs. 8 and 9, respectively.

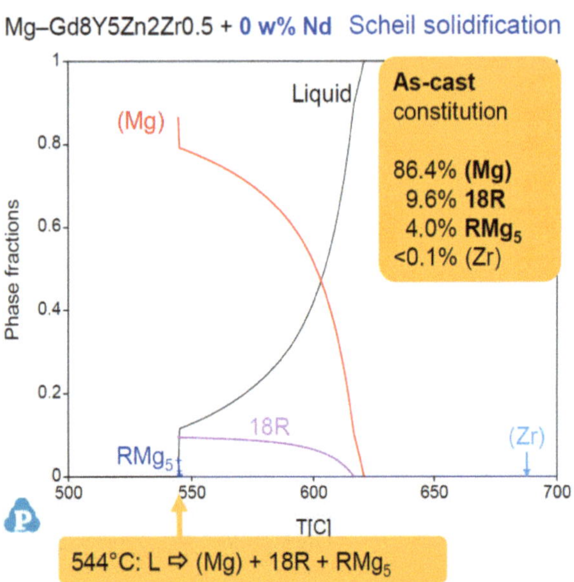

Figure 8: Phase fractions developing during Scheil solidification simulation in alloy Mg–Gd8Y5Zn2Zr0.5.

The classic Scheil equation [21] cannot be applied to real-world scenarios of multicomponent alloys with different types of solid phases precipitating. Significant progress has been made by combining Calphad-type generated thermodynamic databases with the basic Scheil approximations. This solidification simulation using the "Scheil approximation" in context of multicomponent multiphase solidification is explained elsewhere [22]. It often provides a good estimate on the as-cast constitution or microstructure if solid-state diffusion is negligible.

The amount of primary phase (Zr) is very small, the corresponding curve almost coincides with the abscissa in Figs. 8 and 9. However, this amount is sufficient to act as a well known grain refiner in Al-free Mg alloys. The joint crystallization of (Mg) and 18R starts at about 620°C and extends down to 544°C in the Nd-free alloy and to 520°C in the alloy with 2 wt.% Nd. This precipitate phase, 18R, is clearly evidenced by TEM in the as-cast alloy by Zhang et al. [20]. It is also interesting that RMg_5 phase forms from the melt, close to the reaction terminating solidification.

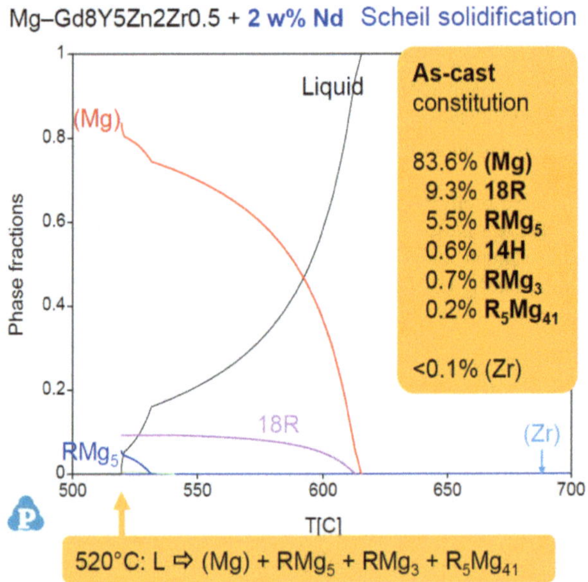

Figure 9: Phase fractions developing during Scheil solidification simulation in alloy Mg–Gd8Y5Zn2Zr0.5Nd2.

The expected phase formation during heat treatment may be read from the phase diagram, shown in the magnification above 500°C in Fig. 10. At 520°C only the phases (Mg) + 14H are stable. The fractions of these phases cannot be read from this multicomponent vertical section because the tie lines are not in that plane.

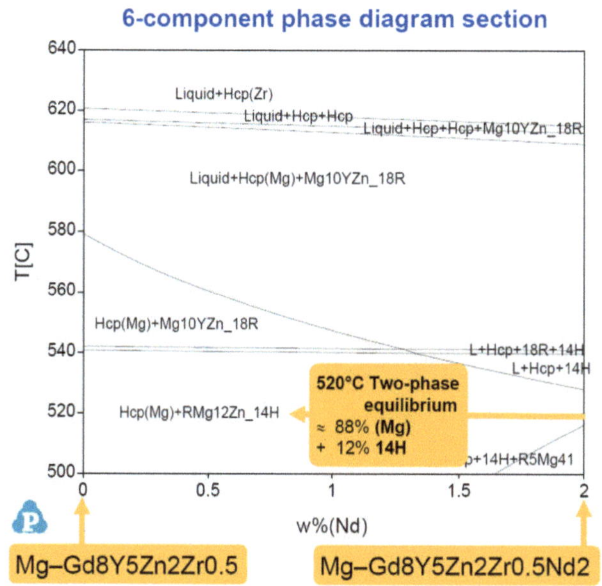

Figure 10: Magnified region of the phase diagram section in Fig. 7.

However, equilibrium calculation along the composition line at 520°C from 0-2 wt.% Nd reveals that the molar phase fractions are almost independent on the alloy composition in that range and values of about 88% (Mg) + 12% 14H are obtained. In fact, the transformation of as-cast phase 18R to the phase 14H in the heat treated sample is clearly evidenced by TEM [20]. A direct (congruent) transformation of 18R → 14H is not possible because their compositions differ in the 6-component phase diagram. This transformation, which is not an equilibrium reaction, involves at least the (Mg) solution phase, the fraction of which also changes as seen from the calculated data. However, the fractions of all other non-equilibrium phases in the as-cast constitution summarized in Figs. 8 and 9 also changes in pursuit of the equilibrium constitution. Coining that into a reaction equation may be misleading. It is more precise to simply compare the calculated phase fractions at start (as-cast) and end (equilibrium) of the "solution heat treatment" to see which phase amount is expected to shrink and which to grow.

References

1. L. Kaufman, H. Bernstein, "Computer Calculations of Phase Diagrams", Academic Press: New York, NY, 1970.

2. H. L. Lukas, S. G. Fries, B. Sundman, "Computational thermodynamics: The Calphad Method", Cambridge University Press, 2007.

3. W. Cao, S. Chen, F. Zhang, K. Wu, Y. Yang, Y.A. Chang, R. Schmid-Fetzer, W.A. Oates, "PANDAT software with PanEngine, PanOptimizer and PanPrecipitation for multi-component phase diagram calculation and materials property simulation", *Calphad* 33 (2009) 328–342.

4. J.O. Andersson, T. Helander, L. Höglund, P. Shi, B. Sundman, "Thermo-Calc & DICTRA, computational tools for materials science", *Calphad*, 26 (2002) 273–312.

5. C.W. Bale, E. Bélisle, P. Chartrand, S.A. Decterov, G. Eriksson, K. Hack, I.-H. Jung, Y.-B. Kang, J. Melançon, A.D. Pelton, C. Robelin, S. Petersen, "FactSage thermochemical software and databases — recent developments", *Calphad*, 33 (2009) 295-311

6. A.A. Luo, P. Fu, L. Peng, X. Kang, X., Z. Li, T. Zhu, "Solidification microstructure and mechanical properties of cast magnesium-aluminum-tin alloys", *Met. Mater. Trans.*, 43A (2012) 360–368.

7. A.A. Luo, B.R. Powell, A.K. Sachdev, "Computational phase equilibria and experimental investigation of magnesium-aluminum-calcium alloys", *Intermetallics*, 24 (2012) 22–29.

8. A.C. Hänzi, F.H. Dalla Torre, L. Zhao, P. Gunde, R. Schmid-Fetzer, M. Kuehlein, J.F. Löffler, P.J. Uggowitzer, "Design strategy for microalloyed ultra-ductile magnesium alloys", *Phil. Mag. Lett.*, 89 (2009) 377–390.

9. A. Pisch, R. Schmid-Fetzer, "Development of a thermodynamic Mg-alloy database by experiments and modeling", in "Magnesium Alloys and their Applications", B.L. Mordike, K.U. Kainer (eds.), Int. Conf., Wolfsburg, April 28-30, 1998, Werkstoff-Informationsgesellschaft mbH, Frankfurt, (1998) 139-144.

10. R. Schmid-Fetzer, J. Gröbner, "Focused Development of Magnesium Alloys using the Calphad Approach", *Advanced Engineering Materials,* 3 (2001) 947-961.

11. R. Schmid-Fetzer, A. Janz, J. Gröbner, M. Ohno, "Aspects of quality assurance in a thermodynamic Mg alloy database", *Advanced Engineering Materials*, 7 (2005) 1142-1149.

12. R. Schmid-Fetzer, J. Groebner, D. Mirkovic, A. Janz, A. Kozlov, "Constitution of Magnesium Alloys", in "Magnesium Technology 2007" R. Beals, M. Pekguleryuz, N. Neelameggham (eds.), TMS (The Minerals, Metals & Materials Society), (2007) 339-344.

13. J. Gröbner, A. Janz, A. Kozlov, D. Mirkovic, R. Schmid-Fetzer, "Phase diagrams of advanced magnesium alloys containing Al, Ca, Sn, Sr and Mn", *J. Metals*, No.12 (December), (2008) 32-38.

14. R. Schmid-Fetzer, J. Gröbner, "Thermodynamic database for Mg alloys — progress in multicomponent modeling", *Metals*, 2, (2012) 377-398 (Special Issue "Magnesium Technology"), Open Access, www.mdpi.com/2075-4701/2/3/377, DOI: 10.3390/met2030377

15. J. Gröbner, R. Schmid-Fetzer, "Key issues in a thermodynamic Mg alloy database" *Metall. Mater. Trans. A*, A44, (2013) 2918-2934.

16. O. Klag, J. Gröbner, G. Wagner, R. Schmid-Fetzer, D. Eifler, "Microstructural and Thermodynamic Investigations on Friction Stir Welded Mg/Al-Joints", *Int. J. Materials Research (Z. Metallkunde)*, 105 (2014) 145-155.

17. R. Schmid-Fetzer, D. Andersson, P. Y. Chevalier, L. Eleno, O. Fabrichnaya, U.R. Kattner, B. Sundman, C. Wang, A. Watson, L. Zabdyr, M. Zinkevich, "Assessment techniques, database design and software facilities for thermodynamics and diffusion". *Calphad*, 31 (2007) 38-52.

18. J. Gröbner, M. Hampl, R. Schmid-Fetzer, M.A. Easton, S.M. Zhu, M.A. Gibson, J.F.Nie, "Phase Analysis of Mg-La-Nd and Mg-La-Ce Alloys", *Intermetallics*, 28 (2012) 92–101.

19. J. Gröbner, R. Schmid-Fetzer, "Thermodynamic modeling of the Mg-Ce-Gd-Y system", *Scripta Materialia*, 63 (2010) 674-679.

20. Xingguo Zhang, Linggang Meng, Canfeng Fang, Peng Peng, Fei Ja, Hai Hao, "Effect of Nd on the microstructure and mechanical properties of Mg–8Gd–5Y–2Zn–0.5Zr alloy". *Materials Science and Engineering: A*, 586 (2013) 19–24.

21. E. Scheil, "Bemerkungen zur Schichtkristallbildung", *Z. Metallkd.*, , 34 (1942) 70-72.

22. R. Schmid-Fetzer, "Phase diagrams - the Beginning of Wisdom", *J. Phase Equil. Diffusion*, (2014) online-first, Open Access DOI: 10.1007/s11669-014-0343-5.

SIMULATION OF CONCURRENT PRECIPITATION OF TWO STRENGTHENING PHASES IN MAGNESIUM ALLOYS

Weihua Sun[1], Chuan Zhang[2], Andrew D. Klarner[1], Weisheng Cao[2], Alan A. Luo[1, 3]
[1]Department of Materials Science and Engineering, The Ohio State University, Columbus, OH 43210, USA
[2]CompuTherm LLC, Madison, WI 53719, USA
[3]Department of Integrated Systems Engineering, The Ohio State University, Columbus, OH 43210, USA

Keywords: Magnesium alloys, Precipitation simulation, CALPHAD

Abstract

The precipitation kinetics and microstructure in Mg-Sn binary and Mg-Al-Sn ternary alloys are simulated using PanPrecipitation coupled with Mg thermodynamic database and a newly established mobility database of the Mg-Al-Sn ternary system. Both Mg_2Sn and $Mg_{17}Al_{12}$ precipitates are considered in this work. The obtained kinetic parameters for these two precipitates can be used in the simulation of both individual and concurrent precipitations of $Mg_{17}Al_{12}$ and Mg_2Sn in Mg-Al-Sn alloys. The simulated microstructure evolution, such as the particle size and number density, are in agreement with experimental data.

Introduction

Magnesium alloys are considered as promising lightweight materials in the transportation industry due to their low density. Currently, the most widely used Mg alloys in industry are Mg-Al based alloys, but their usage is limited to room temperature applications because of their poor creep resistance at elevated temperatures. The poor performance of current Mg-Al based alloys at temperature above 120°C is caused by the softening of grain boundary $Mg_{17}Al_{12}$, which has a low melting point of 450°C [1]. Luo et al. [2] systematically investigated the microstructure and mechanical properties of as-cast Mg-Al-Sn alloys and suggested that Mg-7Al-2Sn and Mg-7Al-5Sn (wt.%) offer good balance of mechanical properties. The Mg-Al-Sn system consists of two aging hardening systems, i.e., Mg-Al and Mg-Sn systems. By precipitation of fine $Mg_{17}Al_{12}$ and/or Mg_2Sn particles in (Mg) matrix during aging, the yield strength of Mg-Al-Sn alloys could be improved, as reported from the investigation of Mg-7Al-2Sn (wt.%) alloy by Shi et al. [3]. Moreover, the precipitation of Mg_2Sn with a melting point of 774°C could improve the creep resistance of the alloys at elevated temperatures [4].

In order to develop precipitation hardening in Mg-Al-Sn alloys, both experimental and simulation work should be carried out to understand the microstructure evolution during aging process and its influence on the mechanical properties of the alloys. Previously, Zhang et al. [5] has simulated the precipitation of AZ91 alloy during aging process with PanPrecipitation of Pandat software [6]. In the present work, previous work on AZ91 alloy [5] will be extended to simulate the precipitation of Mg_2Sn in Mg-Sn binary alloys and concurrent precipitation of $Mg_{17}Al_{12}$ and Mg_2Sn in Mg-Al-Sn ternary alloys during aging. The microstructure evolution information, such as number density, average particle size etc., will be predicted.

Methodology

The general flowchart of precipitation simulation is shown in Fig. 1. In order to perform a quantitative simulation, reliable thermodynamic and mobility databases for Mg-Al-Sn system should be available for extracting accurate thermodynamic and mobility data, such as driving force and diffusivity. In the present simulation, the PanMagnesium thermodynamic database [7] is directly utilized. The atomic mobility database of Mg-Al-Sn system is developed in this study based on assessment of diffusion data in literature using the CALPHAD approach (CALculation of PHAse Diagrams) [8] as a first step. The parameters including nucleation site parameter and interfacial energy of Mg_2Sn/(Mg) are then calibrated from the simulation of the two Mg-Sn alloys in this study. The simulated microstructure evolution information, such as number density and size of precipitate will be compared with experimental data in literature. Finally, by combing the parameters for Mg_2Sn/Mg in the current work and the modified parameters for $Mg_{17}Al_{12}$/Mg based on previous work [5], the concurrent precipitation of two precipitates in Mg-Al-Sn alloys is predicted. The basic theories for atomic mobility assessment and precipitation simulation are briefly introduced as follows.

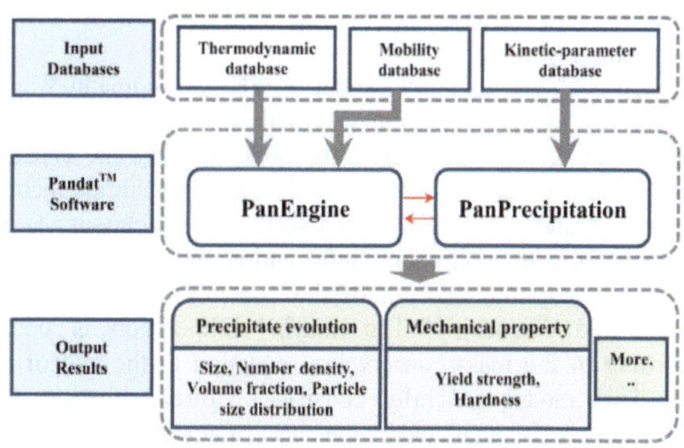

Figure 1. The architecture of PanPrecipitation [5]

<u>Atomic mobility assessment</u>
According to the absolute rate theory arguments, the atomic mobility of the element B, M_B can be divided into a frequency factor M_B^0 and an activation enthalpy Q_B [9, 10]. M_B can be expressed as [9]:

$$M_B = M_B^0 \exp\left(\frac{-Q_B}{RT}\right)\frac{1}{RT} = \exp\left(\frac{RT\ln M_B^0}{RT}\right)\exp\left(\frac{-Q_B}{RT}\right)\frac{1}{RT} \qquad (1)$$

Where R is the gas constant and T is the temperature. The parameters $RT\ln M_B^0$ and Q_B are dependent on temperature, composition and pressure. In CALPHAD approach, the composition dependency of these two parameters can be expressed by a linear combination of the values at each endpoint of the composition space and a Redlich-Kister polynomial [11]:

$$\Phi_B = \sum_i x_i \Phi_B^i + \sum_i \sum_{j>i} x_i x_j \left[\sum_{r=0}^{m} {}^r\Phi_B^{i,j}(x_i\text{-}x_j)^r\right] + \sum_i \sum_{j>i} \sum_{k>j} x_i x_j x_k \left[\sum_s v_{ijk}^s \; {}^s\Phi_B^{i,j,k}\right] \; (s = i, j, k) \quad (2)$$

Where x_i is the molar fraction of element i, Φ_B^i is for pure i and represents the value of one endpoint in the composition space. ${}^r\Phi_B^{i,j}$ and ${}^s\Phi_B^{i,j,k}$ are the binary and ternary interaction parameters. The parameter v_{ijk}^s is given by

$$v_{ijk}^s = x_s + (1\text{-}x_i\text{-}x_j\text{-}x_k)/3 \; (s = i, j \text{ or } k) \quad (3)$$

Assuming the monovacancy atomic exchange as the main diffusion mechanism, the tracer diffusivity D_i^* is related to the atomic mobility M_i by the Einstein's relation:

$$D_i^* = RTM_i \quad (4)$$

For a substitutional solution phase, the interdiffusion coefficient in terms of the volume fixed reference frame can be given by the following expression:

$$\tilde{D}_{kj}^n = \sum_i (\delta_{ik} - x_k) \cdot x_i \cdot M_i \cdot \left(\frac{\partial\mu_i}{\partial x_j} - \frac{\partial\mu_i}{\partial x_n}\right) \quad (5)$$

Where δ_{ik} is the Kronecker delta (δ_{ik}=1 if i=k, otherwise δ_{ik}=0), μ_i is the chemical potential. Based on the relations above, the parameter Φ_B can be numerically assessed by fitting to the experimental diffusion coefficients including self diffusivity, tracer diffusivity and chemical diffusivity.

Precipitation simulation

The classical precipitation model KWN (Kampmann-Wagner-numerical) model [12] implemented in PanPrecipitation is selected for the current work. This model features the capability of modeling nucleation, growth and coarsening simultaneously. The precipitates are divided into a series of size classes. At each time step, the microstructure information in each size class is updated taking into account of the concurrence of nucleation, growth and coarsening. The matrix composition is updated according to the mass conservation equation at the end of each step. This model generally consists of three equations. The nucleation rate is calculated by the classic nucleation theory:

$$J = N_0 Z\beta \exp(-\frac{\Delta G^*}{kT})\exp(-\frac{\tau}{t}) \quad (6)$$

N_0 is the nucleation site density, *Z* is the Zeldovich factor and β is the rate at which the atoms from the matrix are attached to the critical nucleus. ΔG* is the Gibbs energy of formation of a critical nucleus. k is the Boltzmann constant, T is the temperature. τ is the incubation time for nucleation, t is the time.

The growth of precipitates is expressed by a model proposed by Morral and Purdy [13]. This model for multicomponent alloys assumes a diffusion-controlled process and takes into account the Gibbs-Thomson effect.

$$\frac{dR}{dt} = \frac{K}{R}\left(\frac{1}{R^*} - \frac{1}{R}\right) \quad (7)$$

$$K = \frac{2\alpha V_m}{(C^{\alpha\beta}][M]^{-1}[C^{\alpha\beta})} \quad (8)$$

R is the radius of the interface and R* is the radius of the critical nucleus. V_m is the molar volume of the precipitate. $(C^{\alpha\beta}]$ and $[C^{\alpha\beta})$ are the row and column vector of the solute concentration difference between matrix phase and precipitate. [M] is the chemical mobility matrix. The flux of precipitates in and out the each size is expressed by the following continuity equation:

$$\frac{\partial N(r,t)}{\partial t} = -\frac{\partial(N(r,t)v(r,t))}{\partial r} + \delta(r - r_{cr}(t))j(t) \quad (9)$$

N(r,t) is the number density of precipitate and v(r,t) is the growth velocity. δ is the Kronecker delta. r_{cr} is the critical radius.

Results and discussion

Atomic mobility assessment

There are just a few experimental diffusion coefficients related to the hcp(Mg) phase of Mg-Al-Sn system, as listed in Table I. It can be seen from Fig. 2 that there is discrepancy among the experimental interdiffusion coefficients from Ref. [19-21]. The data reported by Kulkarni and Luo [21] are less concentration dependent and more self-consistent, while the data from Kammerer et al. [19] and Brennan et al. [20] are more concentration dependent. More diffusion experiment is needed to clarify the discrepancy.

Table I. Summary of experimental data for hcp (Mg) of Mg-Al-Sn alloys.

System	Date Type		Temp. range(K)	Ref.
Mg	Self diffusion	D_{Mg}^{Mg}	741-908	[14]
			775-906	[15]
Mg-Al	Impurity diffusion	D_{Al}^{Mg}	573-673	[16]
			635-693	[17]
			623-723	[18]
	Inter-diffusion	D_{AlAl}^{Mg}	623-723	[19]
			573-673	[20]
			653-693	[21]
Mg-Sn	Impurity diffusion	D_{Sn}^{Mg}	748-903	[22]

The atomic mobility parameters are listed in Table II and only the parameters of nine end members are used without introduction of any binary or ternary interaction parameter. The Q_{Mg}^{Mg} is directly obtained from experimental self diffusion coefficient of Mg [15]. Q_{Sn}^{Mg} is from experimental impurity diffusion coefficient of Sn in hcp(Mg) [22] because no other data are reported. Since the impurity diffusion coefficients of Al in hcp(Mg) from Brennan et al. [16] are in agreement with the data from Das et al. [17], their data [16] are used for Q_{Al}^{Mg}. It should be mentioned that no interaction parameter is introduced for Mg-Al system since in the optimization process, the atomic mobility of Mg-Al system consisted of only four end member parameters could describe the interdiffusion coefficient of Ref. [21] , as shown in Fig. 2. The introduction of interaction parameters has no significant improvement of the fitting to the experimental data. The hypothetical end member parameter Q_{Al}^{Al} is from the estimation of Cui et al. [23]. Q_{Sn}^{Sn} is estimated using the empirical equation [24].

$$Q_B = RT_m(K + 1.5V) \quad (10)$$

$$D_0 = 2.506 \times 10^9 Q_B a^2 \quad (11)$$

Q_B is the activation energy and D_0 is the frequency factor. K denotes the crystal structure factor, V is the valence, and a is the lattice constant. For metastable Sn with hcp structure, the melting point is calculated to be 499K [7], K=15.5 for hcp phase [24], V=2.5 [24], a=3.282Å [25], so Q_{Sn}=79862 J/mol and $D_0=2.15576\times10^{-5}m^2/s$. The other hypothetical end members Q_{Mg}^{Sn} and Q_{Al}^{Sn} are assumed to be equal to the mobility parameter Q_{Sn}^{Sn} while Q_{Mg}^{Al} and Q_{Sn}^{Al} are equal to Q_{Al}^{Al}.

Table II Summary of the atomic mobility for Mg, Al and Sn in hcp(Mg) of the Mg-Al-Sn system assessed in the present work.

Mobility	Parameter
Mg	$Q_{Mg}^{Mg} = -138164 + R*T*LN(1.75E-4)$
	$Q_{Mg}^{Al} = -79790 + R*T*LN(2.38E-5)$
	$Q_{Mg}^{Sn} = -79862 + R*T*LN(2.15576E-5)$
Al	$Q_{Al}^{Mg} = -155000 + R*T*LN(3.9E-3)$
	$Q_{Al}^{Al} = -79790 + R*T*LN(2.38E-5)^a$
	$Q_{Al}^{Sn} = -79862 + R*T*LN(2.15576E-5)$
Sn	$Q_{Sn}^{Mg} = -149887 + R*T*LN(4.8251E-4)$
	$Q_{Sn}^{Al} = -79790 + R*T*LN(2.38E-5)$
	$Q_{Sn}^{Sn} = -79862 + R*T*LN(2.15576E-5)$

a: From assessment by Cui et al. [23]

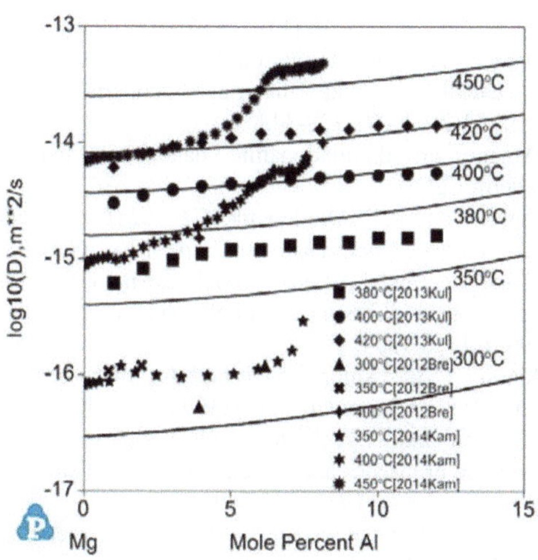

Figure 2. Calculated interdiffusion coefficients in the Mg rich region of Mg-Al system along with the experimental data from Ref. [19-21].

Precipitation simulation

The shape of Mg_2Sn precipitate is not spherical according to the experimental observations [26]. By assuming shape-preserved growth of the precipitate, the shape of the precipitate could be treated as sphere [5]. The conversion method described by Zhang et al. [5] is used to calculate the equivalent particle size. By coupling to the thermodynamic database and the atomic mobility parameters described above, the precipitation of Mg_2Sn at 200°C in two supersaturated Mg-Sn alloys is simulated in the first step. Figure 3 shows the calculated number density for the two Mg-Sn alloys are in agreement with the experimental data [26]. Figure 4 shows the predicted radius for Mg-1.9 at.% Sn alloy at 240 hours is 128nm, which is consistent with the experimental value of 112nm [26]. However, for the Mg-1.3 at.% Sn alloy, the simulated radius at 1000 hours is 160nm, which is lower than the experimental value of 197 nm [26], but still within the experimental error range. Table III shows the parameters used for the simulation. The nucleation site parameter and interfacial energy are two adjustable parameters in the simulation.

Table III Parameters used in the simulation.

Parameter	Value
Nucleation site parameter	1×10^{-14}
Interfacial energy of Mg/Mg_2Sn interface	$0.25\ J/m^2$
Molar volume of hcp(Mg) matrix	$1.4\times10^{-5}m^3/mol$
Molar volume of Mg_2Sn precipitate	$1.55\times10^{-5}m^3/mol$

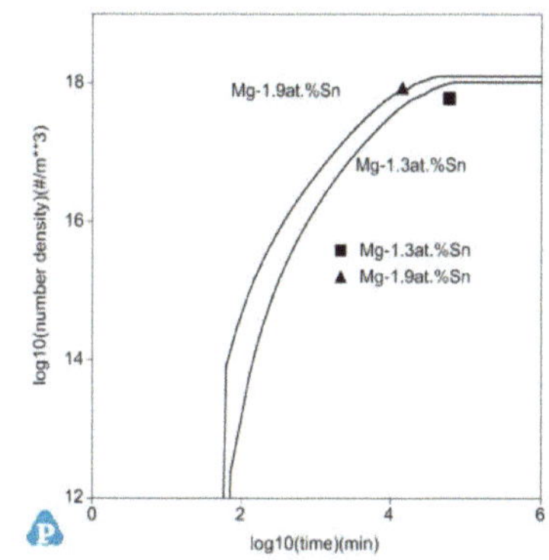

Figure 3. Predicted number density of Mg_2Sn precipitate in Mg-1.3 at.% Sn and Mg-1.9 at.% alloys along with the experimental data [26].

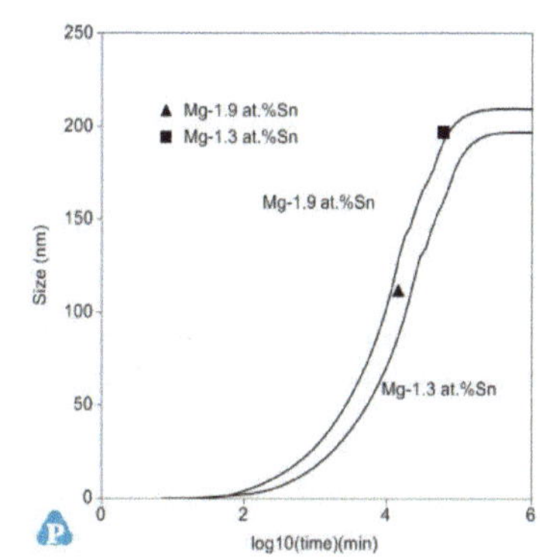

Figure 4. Predicted radius of Mg_2Sn precipitate in Mg-1.3 at.% Sn and Mg-1.9 at.% alloys along with the experimental data [26].

By combining the parameters for Mg_2Sn and $Mg_{17}Al_{12}$, the concurrent precipitation of $Mg_{17}Al_{12}$ and Mg_2Sn at 200°C in Mg-7Al-2Sn (wt.%) alloy is simulated. Figure 5 shows the volume fractions of of Mg_2Sn and $Mg_{17}Al_{12}$. After about 100 hours, the fraction of $Mg_{17}Al_{12}$ reaches equilibrium value while the fraction of Mg_2Sn is considerably low. There are two reasons, the nucleation of $Mg_{17}Al_{12}$ is considerably stronger than that of Mg_2Sn as shown in Fig. 6 and the growth of $Mg_{17}Al_{12}$ is faster as can be seen in Fig.7. According to the simulation, the precipitation of $Mg_{17}Al_{12}$ is much stronger than that of Mg_2Sn. However, the present simulation utilizes the parameters from separate simulations of AZ91 alloy [5] and Mg-Sn alloys. The results will be clarified by the ongoing TEM work to investigate how Al could promote the precipitation of Sn in the ternary alloys. In addition, by using the microstructure evolution information, such as particle size distribution, number density and solute concentration in (Mg) as input, the precipitation hardening and solid solution hardening could be predicted from empirical equations. Thus, the yield strength is obtained and the hardness could also be deduced from the relationship between yield strength and hardness.

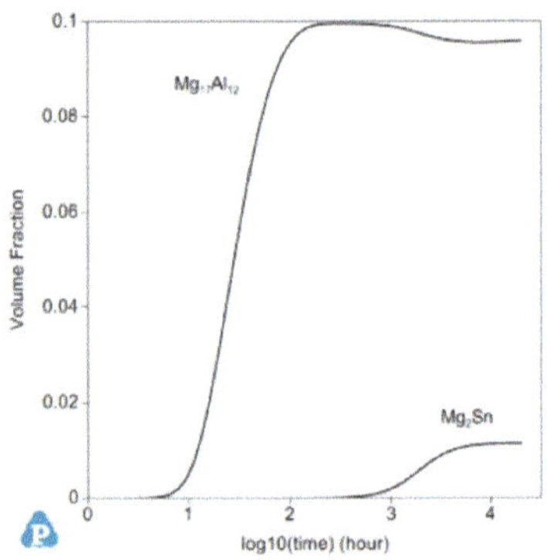

Figure 5. Predicted volume fraction of $Mg_{17}Al_{12}$ and Mg_2Sn precipitates in Mg-7Al-2Sn (wt.%) alloy at 200°C.

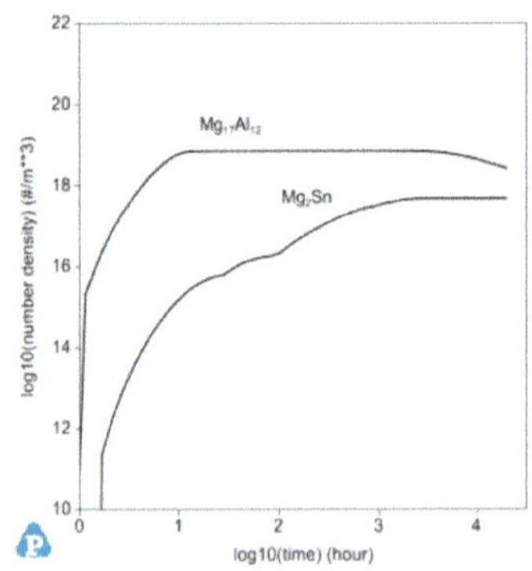

Figure 6. Predicted number density of $Mg_{17}Al_{12}$ and Mg_2Sn precipitates in Mg-7Al-2Sn (wt.%) alloy at 200°C.

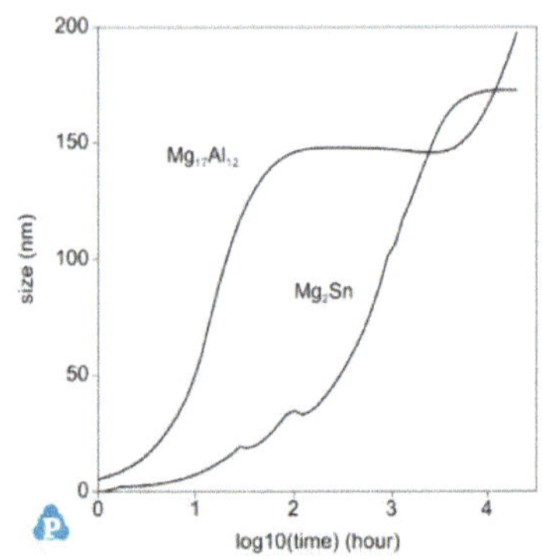

Figure 7. Predicted average size of $Mg_{17}Al_{12}$ and Mg_2Sn precipitates in Mg-7Al-2Sn (wt.%) alloy at 200°C.

Conclusions

Using CALPHAD approach, the atomic mobility parameters of Mg-Al-Sn system are established based on assessment of literature experimental data. By coupling to thermodynamic and atomic mobility parameters, the microstructure evolution in the aging process of two Mg-Sn alloys and concurrent precipitation of $Mg_{17}Al_{12}$ and Mg_2Sn in Mg-7Al-Sn (wt.%) alloy are simulated. It is predicted that the precipitation of $Mg_{17}Al_{12}$ is much strong than that of Mg_2Sn.

Acknowledgements

This work is sponsored by the U.S. Department of Energy under project DE-EE0006450.

References

1. A.A Luo, "Recent magnesium alloy development for elevated temperature applications", *Int. Mater. Rev.*, 49(1) (2004) 13-30.
2. A.A Luo et al., "Solidification microstructure and mechanical properties of cast magnesium-aluminum-tin alloys", *Metall. Mater. Trans A.*, 43(1) (2012), 360-368.
3. X.Y. Shi et al., "Microstructure and mechanical properties of Mg-7Al-2Sn alloy processed by super vacuum die-casting", *Metall. Mater. Trans A.*, 44(10) (2013), 4788-4799.
4. W. Ke et al., "Investigations in the magnesium-tin system", *Mater. Sci. Forum*, 488-489 (2005), 135-138.
5. C. Zhang et al., "Precipitation simulation of AZ91 alloy", *JOM*, 66(3) (2014), 389-396.
6. W. Cao et al., "PANDAT software with PanEngine, PanOptimizer and PanPrecipitation for multi-component phase diagram calculation and materials property simulation", *CALPHAD*, 33(2) (2009), 328-342.
7. PanMagnesium thermodynamic database (Madison, WI: CompuTherm LLC)
8. N. Saunders and A.P. Miodownik, *CALPHAD (Calculation of Phase Diagrams): A Comprehensive Guide* (Elsevier, 1998)
9. J.O. Andersson and J.Ågren, "Models for numerical treatment of multicomponent diffusion in simple phases", *J. Appl. Phys.*, 72(4) (1992), 1350-1354.
10. B. Jönsson, "Assessment of the mobility of carbon in fcc C-Cr- Fe-Ni alloys", *Z.Metallkd.*, 85(7) (1994), 502-509.
11. O. Redlich and A.T. Kister, "Algebraic representation of thermodynamic properties and the classification of solution", *Ind. Eng. Chem.* Res. 40(2) (1948) 345-349.
12. R. Kampmann and R. Wagner, "Kinetics of precipitation in metastable binary alloys-theory and application to Cu-1.9 at % Ti and Ni-14 at % Al", *Decomposition of Alloys: the early stages, Proceedings of the 2nd Acta-Scripta Metallurgica Conference*, ed. P. Haasen et al. (Pergamon Press, 1983), 91-103.
13. J.E. Morral and G.R. Purdy, "Particle coarsening in binary and multicomponent alloys", *Scripta Metall. Mater.*, 30(7) (1994), 905-908.
14. P.G. Shewmon, "Self-diffusion in magnesium single crystals", *Trans. Metall. Soc, AIME*, 206 (1956), 918-922.
15. J. Combronde and G. Brebec, "Anisotropy for self diffusion in magnesium", *Acta Metall.*, 19(12) (1971), 1393-1399.
16. S. Brennan et al., "Aluminum impurity diffusion in magnesium", *J. Phase Equilib. Diff.*, 33(2) (2012), 121-125.
17. S.K. Das et al, "Anisotropic diffusion behavior of Al in Mg: diffusion couple study using Mg single crystal", *Metall. Mater. Trans A.*, 44(6) (2013), 2539-2547.
18. C. Kammerer et al., "Al and Zn impurity diffusion in binary and ternary magnesium solid-solutions," *Magnesium Technology 2014*, ed. M. Alderman et al., TMS, 2014, 407-411.

19. C. Kammerer et al., "Impurity diffusion coefficients of Al and Zn in Mg determined from solid-to-solid diffusion couples", *Magnesium Technology 2014*, ed. M. Alderman et al., TMS, 2014, 505-509.
20. S. Brennan et al., "Interdiffusion in the Mg-Al system and intrinsic diffusion in β-Mg_2Al_3", *Metall. Mater. Trans. A*, 43(11) (2012), 4043-4052.
21. K.N. Kulkarni and A.A. Luo, "Interdiffusion and phase growth kinetics in magnesium-aluminum binary system", *J. Phase Equilib. Diff.*, 34(2) (2013), 104-115.
22. J. Combronde and G. Brebec, "Diffusion of Ag, Cd, In, Sn and Sb in magnesium", *Acta Metall.*, 20 (1972), 37-44.
23. Y.W. Cui et al., "Study of diffusion mobility of Al–Zn solid solution", *J. Phase Equilib. Diff.*, 27(4) (2006), 333-342.
24. John Askill, *Tracer diffusion data for metals, alloys, and simple oxides* (New York, NY, Plenum Publishing Corporation, 1970), 19-26.
25. C. Yu et al., "Ab initio calculation of the properties and pressure induced transition of Sn," *Solid State Commun.*, 140(11-12) (2006), 538-543.
26. C.L. Mendis et al., "Refinement of precipitate distributions in an age-hardenable Mg-Sn alloy through microalloying", *Phil. Mag. Let.*, 86(7) (2006), 443-456.

Magnesium Technology 2015
Edited by: Michele V. Manuel, Alok Singh, Martyn Alderman, and Neale R. Neelameggham
TMS (The Minerals, Metals & Materials Society), 2015

STRENGTHENING MECHANISMS IN $Mg_{97}Zn_1Y_2$ ALLOYS

Zhiqing Yang, Hengqiang Ye
Shenyang National Laboratory for Materials Science, Institute of Metal Research, Chinese Academy of Science; 72 Wenhua Rd; Shenyang, 110016, China

Keywords: Mg alloy, Strength, Ductility, Long-periodic stacking ordered phase, Dislocation, Twinning, Stacking-fault

Abstract

We elucidate the strengthening mechanisms of Mg-Zn-Y alloys containing long-periodic stacking ordered (LPSO) structures, based on comprehensive electron microscopy investigations. Kinking of the LPSO structures is not only an important way to accommodate plastic deformation effectively, but also simultaneously strengthens the alloy as a result of microstructural refinement. In addition, kink boundaries in the LPSO structures can effectively restrict propagation of microcracks, benefiting both the alloy's strength and ductility. Using atomic-resolution imaging, we found that the stacking-faults with Zn and Y segregation and the dynamic dislocation-solute interactions in Mg matrix also play important roles in strengthening the alloys, besides the LPSO structures. The stacking-faults can hinder the generation and propagation of $\{10\bar{1}2\}$ deformation twins, reducing the potential nucleation sites for microcracks. Interactions between solute atoms with dislocations promote the dissociation of both "**a**" and "**a** + **c**" dislocations, leading to nanometer-sized structures similar to GP zones that can act as obstacles for dislocation motion in Mg matrix. Both the wide stacking-faults with Zn and Y segregation formed during solidification, and nanometer-sized stacking-faults produced by dislocation dissociation have significant contributions to the mechanical properties at elevated temperatures.

Introduction

Mg alloys are attractive light-weight engineering materials owing to their low density, high specific strength, excellent damping and ease of recycling [1]. However, the relatively low strength and poor ductility of Mg alloys limit their widespread engineering applications. Improving the strength and ductility is a central goal for widespread application of Mg alloys. In addition, development of Mg alloys with good creep resistance at elevated temperatures is of general interest and importance [2]. Great efforts have been made in alloy design and processing in order to improve their mechanical properties. Mg alloys with long-periodic stacking ordered (LPSO) structures are found to show good strength at both ambient and elevated temperatures. A tensile yield strength of ~ 600 MPa and elongation of ~ 5.0% were achieved by rapidly solidified powder metallurgy processing of a $Mg_{97}Zn_1Y_2$ (at.%) alloy [3]. Hot-extruded $Mg_{97}Y_2Cu_1$ (at.%) alloy exhibited tensile strength and elongation of 377 MPa and 8.1% at ambient temperature, and 344 MPa and 16.3% at 200 ℃ [4]. The high tensile strength at 200 ℃ in Mg alloys strengthened by LPSO structures was not observed in Mg alloys consisting of other types of secondary strengthening phases. For example, the tensile strength of extruded AZ31B alloy decreased significantly from about 300 MPa to about 100 MPa when raising the temperature from ~25 ℃ to 200 ℃ [4].

The aim of our study is to reveal the microscopic strengthening mechanisms of Mg alloys with LPSO structures, by using a $Mg_{97}Zn_1Y_2$ (at.%) alloy as the model material. Previous studies were mainly focused on the strengthening effect of the LPSO structures. However, the superior mechanical properties of this kind of Mg alloys should not be solely attributed to the LPSO structures distributed along grain boundaries. The Mg matrix should also play a role in the high strength, especially at elevated temperatures, since a highly softened Mg matrix is expected to result in remarkable decrease in the material's strength.

Experiments

A $Mg_{97}Zn_1Y_2$ (at.%) alloy was prepared using a high frequency induction furnace under argon atmosphere protection. Mechanical tests were performed at 300 ℃ at a strain rate of 1.0×10^{-3} s^{-1}, and then quenched into water to minimize thermo-driven microstructural recovery after deformation. A Tecnai G2 transmission electron microscope (TEM) and a Nion UltraSTEM100 dedicated scanning TEM (STEM) were used to characterize the lattice defect microstructures. Atomic resolution Z contrast images were recorded using a collection half-angle range of 86-200 mrad to reduce the effect of strain contrast on images around lattice defects [5,6].

Results and discussion

Figure 1 is scanning electron microscopy (SEM) and TEM micrographs of the as-cast and deformed samples. The alloy has a duplex microstructure. The phase with bright contrast in the SEM images is the LPSO phase that is enriched in Zn and Y, and the black phase is Mg matrix. The LPSO phase form a three-dimensional network at Mg grain boundaries. The darker parallel lines observed in the LPSO region are thin slices of Mg, Figure 1a. The LPSO plates are straight and several hundred micrometers in width in the as-cast state. Obvious kinking of most LPSO plates were observed in the deformed samples, as indicated by white arrows in Figure 1b. Figure 1c is a TEM image of a region with several kinked LPSO plates (the dark phase) in a sample deformed to a strain of about 20%. The width of kink bands in LPSO phase decreases with the deformation strain, and can be less than one micron in samples deformed to a strain of 60%, showing continuous refinement of the LPSO phase with increasing the deformation strain. Figure 1d is a high-resolution TEM image recorded at a kink boundary. The kink boundary is composed of basal edge dislocations with the same sign and a Burgers vector of $b = 1/3\langle 11\bar{2}0\rangle$. Fourier transformation analysis shows that the mean distance between neighboring dislocations is about 0.6 nm, corresponding to a tilt angle of about 20°, which is consistent with the TEM observation shown in Figure 1c. Systematic high-resolution TEM analyses found that neighboring low-angle kink boundaries were composed of dislocations with opposite signs [7]. Therefore, kinking of the LPSO phase should be attributed to the generation, slip and rearrangement of dislocation pairs of opposite sign on basal planes during plastic deformation.

Kinking is a common deformation behavior of materials with layered structures [8]. On the other hand, kinking can lead to local hardening as a result of microstructural refinement by producing high density of kink boundaries (Fig. 1), since the kink boundaries can act as barriers for dislocation motion, following the Hall-Petch relationship. We found that the Vickers microhardness of the LPSO region was increased from 78 to 112 by compressing the as-cast alloy to a strain of 60% [7]. In addition, the newly formed kink boundaries can prevent the propagation of microcracks produced along Mg nano-slices sandwiched between LPSO plates, which is beneficial to both strength and damage tolerance of the alloy.

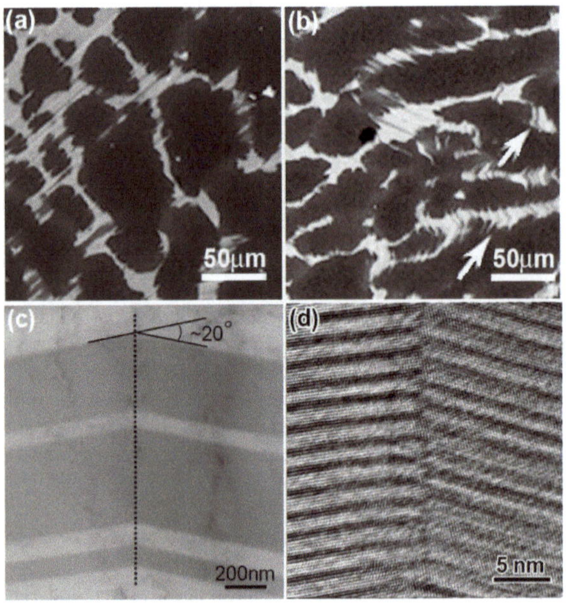

Figure 1. (a, b) Back-scattered electron SEM images for as-cast and deformed samples. (c) A TEM image showing a kink boundary in LPSO phase in deformed samples. (d) A high-resolution TEM image showing fine structure of the kink boundary shown in (c). The dotted line in (c) marks the kink boundary, and the kink angle is indicated.

Twinning is an important deformation mode of Mg and its alloys, due to the limited number of slip systems in hexagonal structures. However, deformation twin could only be scarcely observed in Mg matrix in $Mg_{97}Zn_1Y_2$ alloys, as shown in Figure 2a. This is a {10$\bar{1}$2} deformation twin, according to electron diffraction analysis and HRETM observation [9]. It is interesting to note that the width of the twin band keeps decreasing from the middle part to its two ends, and one end of the twin band stops at a stacking-fault (SF) at the upper left corner of Figure 2a. This implies that the SF in Mg matrix might play an important role in preventing the propagation and widening of the deformation twins. Figure 2b shows a typical high-resolution TEM image for the {10$\bar{1}$2} twin boundaries. The misorientation between basal planes in the matrix and twin is measure to be about 84°. Additionally, the twin boundary is curved and extended, instead of an ideal lattice plane of {10$\bar{1}$2} as in pure Mg. This indicates that the twin boundary

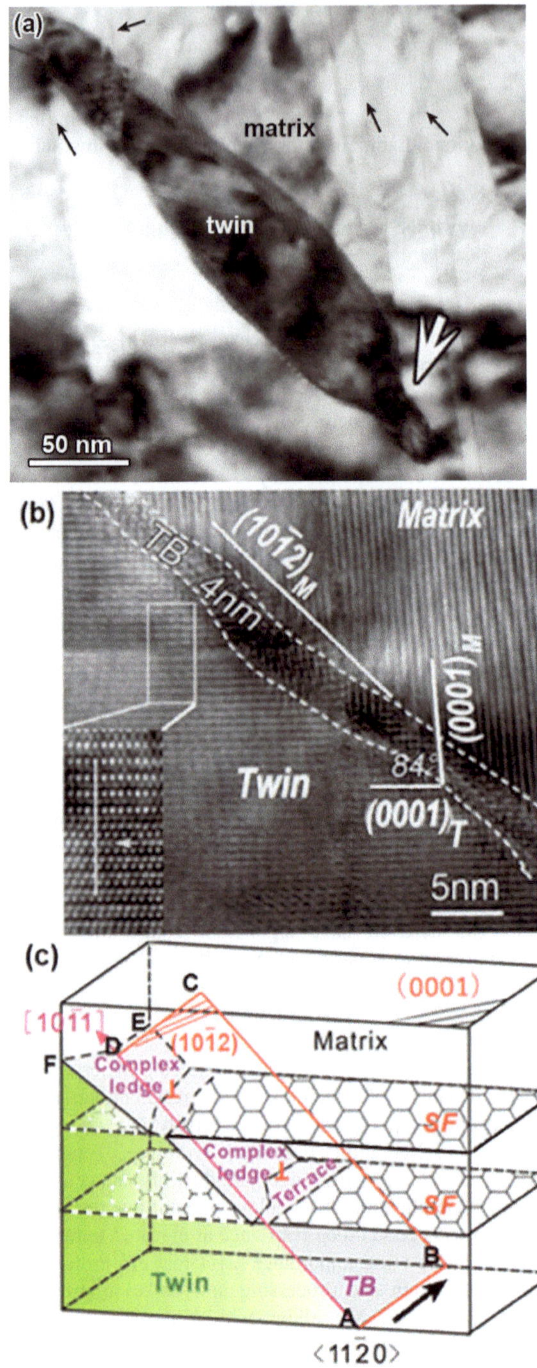

Figure 2. (a) TEM image showing shrinkage of a deformation twin in Mg matrix with basal stacking-faults. Black arrows indicate SFs in Mg matrix. An white arrow points to local shrinkage of the twin upon interacting with a SF. (b) High-resolution TEM image showing fine structure of a twin boundary. The inset in (b) is an enlarged image for a SF emitted from the boundary. (c) Schematic drawing describing the mechanism for the shrinkage of deformation twins by interacting with stacking-faults. The twin boundary is curved and deviated from the [10$\bar{1}$1] direction and {10$\bar{1}$2} plane, resulting in shrinking of the twin. Hexagons in the twin were drawn using dashed lines, since the original basal planes were not there anymore after twinning.

deviates slightly from the $\{10\bar{1}2\}$ along both the $[10\bar{1}1]$ and $\langle 11\bar{2}0\rangle$ directions upon propagation of the twin boundary. The SFs in the matrix lattice were still present, but could not be found on the twin side. This shows that the SF was recovered within the twin band during the deformation twinning process. Recovery of a section of the SFshould leave a partial dislocation at the twin boundary, which should tilt the boundary from its ideal habit plane of $\{10\bar{1}2\}$.

A proposed mechanism accounting for the deviation and extension of twin boundaries is schematically shown in Figure 2c, based on the above analysis. All the Burgers vectors of basal partial dislocations are not parallel to the $\langle 11\bar{2}0\rangle$ direction, in which one is perpendicular to, and the other two are inclined to $\langle 11\bar{2}0\rangle$ by 30°. Twin boundary partial dislocations with a Burgers vector perpendicular to $\langle 11\bar{2}0\rangle$ result in simple rotation of the twining direction around the $\langle 11\bar{2}0\rangle$ axis, while the 30° partials lead to deflection of the twinning plane around both $\langle 11\bar{2}0\rangle$ and $[10\bar{1}1]$ axes. Since there is a high density of SFs in the Mg matrix, an array of partial dislocations appear at the twin boundaries associated the formation and propagation of deformation twins, resulting in the deviation of the twin boundary from a $\{10\bar{1}2\}$ plane indicated by "ABCD" to an irregular curved plane "ABEF", as shown in Figure 2c. The presence of extra partial dislocation arrays is responsible for the deviation of twin boundaries from their habit $\{10\bar{1}2\}$ planes, leading to the shrinkage of the twin finally. New emission of deformation SFs was occasionally observed in deformation twins, as shown in Figure 2b.

In addition, we found that nanometer-thick LPSO plates and thin slice of Mg consisting of SFs with inter-spacing of several nanometers could result in sharp necking of the deformation twin. The critical resolved shear stress to recover a basal stacking fault in Mg grains of the $Mg_{97}Zn_1Y_2$ alloy was estimated to be above 26.5 MPa [10], corresponding to a shear stress above 36.2 MPa along the $[10\bar{1}1]$ direction in the habit $\{10\bar{1}2\}$ twinning plane. The simultaneous recovery of three consecutive basal SFs would require local shear stress in the $\{10\bar{1}2\}$ habit plane above 108.6 MPa which is higher than the yield stress of the $Mg_{97}Zn_1Y_2$ alloy at 300 ℃. Therefore, the scarce formation of deformation twins in the $Mg_{97}Zn_1Y_2$ alloy is plausibly related to the strong interaction and reaction between SFs and deformation twin. The critical shear stress for twinning in Mg is only several MPa [11], twinning-mediated deformation will result in yielding of Mg alloys at lower applied stress. In addition, it has been found that twin boundaries are potential nucleation sites for microcracks which may lead to premature failure of Mg alloys [12]. The formation and propagation of microcracks can result in significant decrease of flow stress during plastic deformation of materials. Therefore, decreasing the possibility of deformation twinning is expected to benefit the strength of Mg alloys.

Figure 3 shows Z-contras STEM images of SFs in Mg grains in as-cast and deformed specimens. The Zn/Y enriched growth SFs have widths of the same magnitude of the Mg grain diameters of several hundred micrometers, as shown in Figure 3a. No dislocation was observed in the as-cast Mg grains, except those partial dislocations at the end of SFs terminated in the Mg grains (Fig. 3a). Many nanometer-sized white features were observed in the deformed Mg grains, as indicated by arrows in Figure 3b. High-resolution Z-contrast imaging demonstrated that all those nanometer-sized white features were SFs resulted from dissociation of "a" or "a + c" dislocations, and Zn/Y segregation on the newly produced SFs, as shown in Figures 4 and 5.

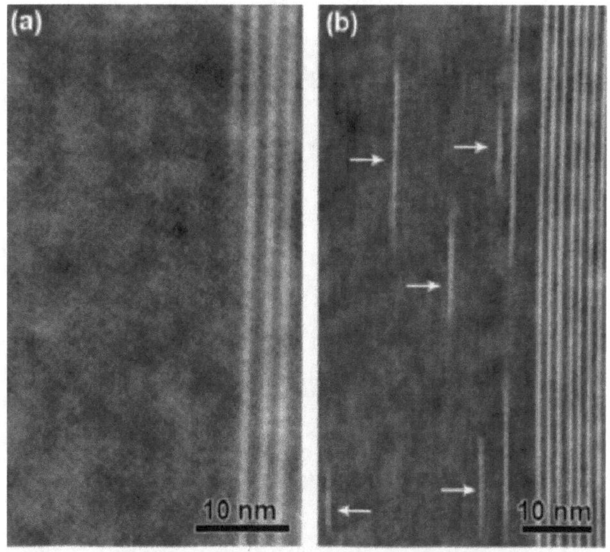

Figure 3. Z-contrast STEM images for quite long growth SFs and Mg matrix in as-cast sample (a), and newly formed nanometer-sized SFs in Mg of the hot deformed samples.

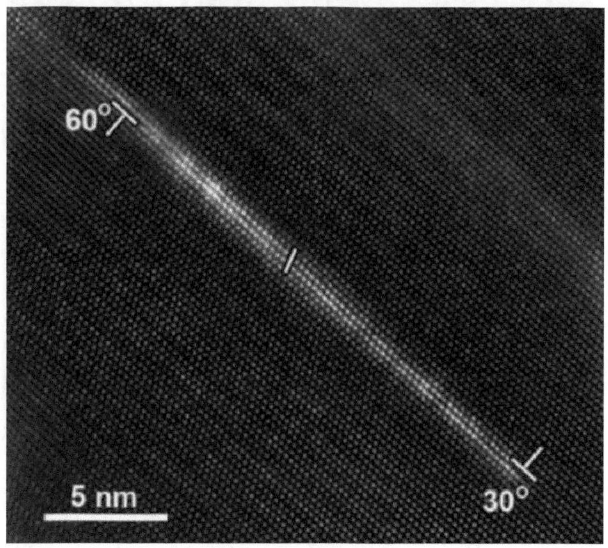

Figure 4. High-resolution Z-contrast images showing the dissociation of "**a**" dislocations and segregation of Zn/Y at SF. A 60° "**a**" dislocation dissociated into a 30° partial (left) and a 60° partial (right), producing a four-layer deformation SF.

Figure 4 is a high-resolution Z-contrast image for a SF of about 30 nm wide as a result of the dissociation of 60 "**a**" dislocations. The higher intensities of atomic columns along the SFs are due to Zn and Y segregation. The dissociation of "**a**" dislocations produced "ABCA" four-layer faulted structures, due to the glide of the

Shockley partials. Four-layer SFs formed by dissociation of screw dislocations were also observed in the hot deformed Mg alloys. Since those growth and deformation planar defects are bounded with partial dislocations, it is proper to refer them as stacking faults, rather than a kind of precipitates [13], according to the dislocation theory [10,14]. The enrichment of Zn and Y solutes at the stacking faults is known as Suzuki segregation associated with chemical interactions between stacking faults and solutes [15].

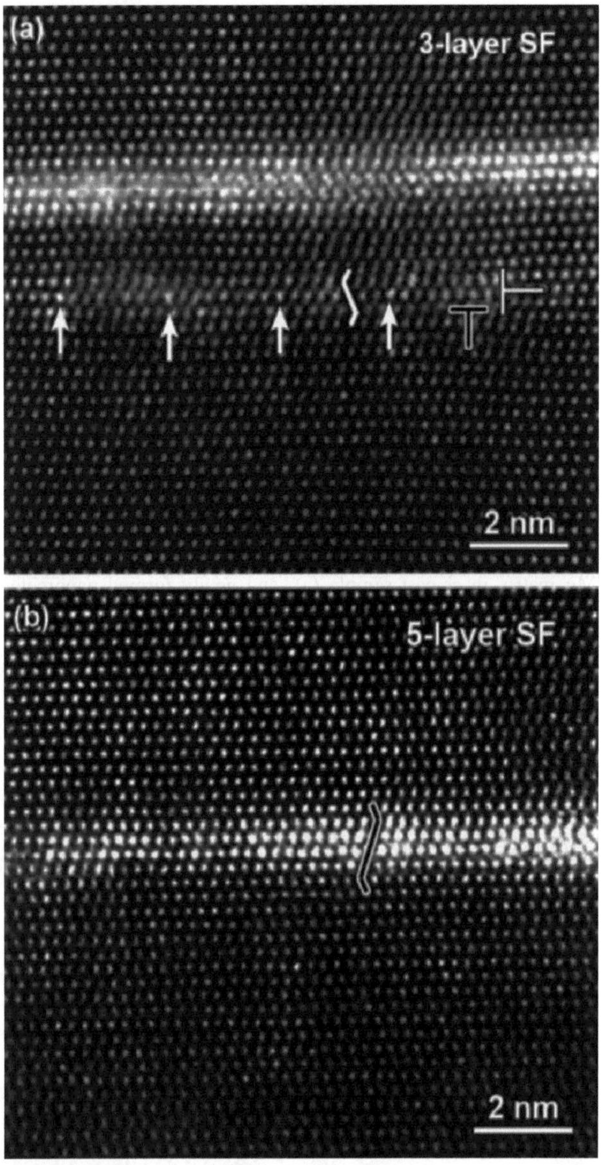

Figure 5. High-resolution Z-contrast images showing SFs produced by dissociation of "**a** + **c**" dislocations and segregation of Zn/Y at SFs. (a) Dissociation of an "**a** + **c**" dislocation with a 60° "**a**" component. The atomic columns show higher brightness due to Zn/Y segregation, as indicated by white arrows as examples. The 3-layer SF is indicated white bars. The black and white "⊥" indicate the approximate positions of the Shockley partial and the Frank partial ½[0001], respectively. (b) Dissociation of an "**a** + **c**" dislocation with a screw "**a**" component, producing a 5-layer SF, as indicated by black bars.

Besides those four-layer faulted structures produced by the dissociation of basal "**a**" dislocations, three-layer and five-layer faulted structures were also observed in the deformed Mg grains, as shown in Figures 5a and 5b respectively. Burgers circuits surrounding some newly formed short SFs show closure failure of "**a** + **c**" for three-layer faults or of "**c**" for five-layer faults on high-resolution Z-contrast images. Close analyses of the images in combination with theory of dislocation dissociation showed that both the three-layer and five-layer faulted structures were originated from the dissociation of "**a** + **c**" dislocations [10,14]. The three-layer SF shown in Figure 5a was produced by the dissociation of a "**a** + **c**" dislocation with a 60 "**a**" component, while the dislocation producing the five-layer fault had a screw "**a**" component. An "**a** + **c**" dislocation dissociated into two Shockley partials and two Frank partials . The Shockley partials glided on the basal planes, while the Frank partials with a Burgers vector $1/2[0001]$ needed to climb in opposite directions, during the widening of the SFs [10].

There is obvious difference in the concentration of Zn/Y segregation among the newly formed SFs, as shown in Figures 4 and 5. Also, there are changes in Zn/Y concentration from one end to the other end within the 5-layer faulted structure (Figure 5b). The variation of Zn/Y concentration should be a result of fluctuation in solute concentration within the Mg matrix, since the surrounding Mg matrix is the supplier of solute atoms. Suzuki segregation is believed to have a strong influence on the mechanical properties of alloys. For example, the strength of cold-rolled Co-Ni-based superalloys is believed to be further increased by ageing at intermediate temperatures due to the Suzuki segregation at SFs [16]. Suzuki segregation is also considered responsible for other phenomena, such as serrated flow, positive temperature dependence of strength, dynamic strain ageing (abnormal strain rate dependences of flow stress), and deformation twinning in alloys. Our high-resolution Z-contrast imaging confirmed the occurrence of Suzuki segregation on the atomic level for the first time. The Suzuki segregation at the newly formed nanometer-wide SFs hinders their further motion, and can also serve as nanometer-sized blocking centers, acting like precipitates or GP zones [17,18], for other dislocations. This is believed to be responsible, at least partly, for the superior strength at elevated temperatures of the Mg-Zn-Y alloys. Both the strong interaction between Zn and Y atoms in Mg and the reduction of elastic strain produced by Zn/Y substitutional atoms in the lattice are believed responsible for the occurrence of co-segregation of Zn and Y at SFs.

Conclusion

In summary, we studied the strengthening mechanisms of $Mg_{97}Zn_1Y_2$ alloys from the micrometer, nanometer down to the atomic scale through comprehensive electron microscopy investigations. The Mg-Zn-Y alloy containing LPSO shows superior mechanical strength at elevated temperatures. TEM and STEM studies show that not only the large LPSO strengthening phase, but also the wide growth SFs with Zn/Y segregation across the Mg grains and those nanometer-sized SFs formed dynamically as a result of dislocation dissociation and Suzuki segregation, play important roles in strengthening the alloy. Main conclusions are as follows:

On the one hand, kinking of LPSO phases is an important way to afford considerable plastic strain, benefiting the ductility of the alloy; on the other hand, dynamic microstructural refinement resulted from LPSO kinking can harden the alloy according to the Hall-Petch relationship. Furthermore, kink boundaries can block the microcracks propagating along basal planes, which also benefits the material's strength. Producing layered structure is thus believed effective in improving Mg alloys' mechanical performance.

Growth SFs in Mg matrix can hinder the formation and propagation of deformation twins, reducing potential nucleation sites for microcracks. This plays a role in preventing the alloy from premature failure, consequently improving its strength. The high shear stress required to recover the SFs to normal hexagonal stacking plays an important role in hindering deformation twinning in the present Mg-Zn-Y alloy.

Newly formed nanometer-sized SFs with Suzuki segregation of Zn and Y atoms in Mg matrix during deformation can act like GP zones in pinning dislocation motion, which strengthens the Mg matrix, playing an important role in strengthening the alloy at elevated temperatures.

Acknowledgement

We thank Dr. X. H. Shao, J. F. Liu and W. W. Hu for assistance in sample preparation, mechanical tests and electron microscopy experiments. We thank the STEM group in Oak Ridge National Laboratory for using the UltraSTEM100 microscope. Z.Q.Y. thanks Dr. M. F. Chisholm at ORNL and Prof. S. J. Pennycook at University of Tennessee for valuable discussions. The research was financially supported by the NSFC (51171189, 51371178 and 51390473).

Reference

[1] M. M. Avedesian, H. Baker. Magnesium and magnesium alloys. Maerials Park (OH): ASM international, 1999.
[2] M. Bamberger, G. Dehm. Annu. Rev. Mater. Res. 38, 2008, 505.
[3] A. Inoue et al. J. Mater. Res. **16**, 2001, 1894.
[4] Y. Kawamura et al. Scripta Mater. **55**, 2006, 453.
[5] S. J. Pennycook. Ultramicroscopy **30**, 1989, 58.
[6] S. J. Pennycook. Annu. Rev. Mater. Sci. **22**, 1992, 171.
[7] X. H. Shao et al. Acta Mater. **58**, 2010, 4760.
[8] E. Orawan. Nature **149**, 1942, 463.
[9] X. H. Shao et al. Philos. Mag. Lett. **94**, 2014, 150.
[10] Z. Q. Yang et al. Acta Mater. **61**, 2013, 350.
[11] X. Y. Yang et al. Acta Mater. Sinica **45**, 2009, 1303.
[12] P. Cizek et al. Scripta Mater. **59**, 2008, 959.
[13] Y. M. Zhu et al. Script Mater.**60**, 2009, 980.
[14] J. P. Hirth, J. Lothe. Theory of dislocations. New York: Wiley, 1982.
[15] H. Suzuki. Sci. Rep. Res. Inst. Tohoku Univ. **A4**, 1949, 455.
[16] Y. Koizumi et al. Acta Mater. **60**, 2012, 2901.
[17] Y. L. Zhao et al. Acta Mater. **61**, 2013, 1624.
[18] M. Karlik et al. Acta Mater. **45**, 1997, 3251.

Magnesium Technology 2015

Corrosion, Coatings, Fatigue, and Fracture

Magnesium Technology 2015
Edited by: Michele V. Manuel, Alok Singh, Martyn Alderman, and Neale R. Neelameggham
TMS (The Minerals, Metals & Materials Society), 2015

FATIGUE AND CORROSION FATIGUE OF COLD DRAWN WE43 WIRES

Adam J. Griebel[1], Jeremy E. Schaffer[1]

[1]Research & Development, Fort Wayne Metals Research Products Corp., Fort Wayne, IN, 46809, USA

Keywords: wire drawing, magnesium wire, cold drawn magnesium, WE43, fatigue, rotary beam fatigue

Abstract

Interest in magnesium alloys as temporary implant materials continues to grow. However, significant processing and performance gaps exist between material needs for medical devices and traditional magnesium processing technology. To help close this gap, a process to enable cold-drawn magnesium alloy wires has been developed. Cold drawing offers processing and property enhancements including strengthening through work-hardening and grain refinement. Tensile strengths exceeding 600 MPa have been achieved in WE43 wires after significant cold reduction. Understanding the mechanical fatigue performance of these wires will be critical when designing an absorbable implant. In this work, WE43 wires were drawn to a diameter of 127 microns with 90% cold work, and then either aged or annealed to effect various grain sizes. Rotary beam fatigue testing was performed on cold-drawn and thermally treated wires in both ambient air and body-temperature saline with a goal to inform the relationship between processing, environment, and fatigue performance.

Introduction

Magnesium and its alloys have generated considerable interest as a temporary biomedical implant material in recent years[1–3]. Patients requiring devices such as cardiovascular stents, orthopedic pins and screws, or soft tissue fixation materials like staples and sutures may benefit from a material that is absorbed by the body after it has served its purpose. Mg alloys are a promising material solution, yet many processing and performance gaps exist between current medical device raw materials and traditional Mg processes. Most metallic, non-degrading versions of the aforementioned devices are generally made from raw material that has been manufactured using some amount of cold-deformation. For example, raw material forms such as tubing used in the manufacture of stents, rods for screws, or wire for staples are heavily dependent on various cold-deformation manufacturing processes.

Cold wire drawing is an established, robust, and versatile manufacturing method. In addition to being a very efficient and scalable process, cold drawing allows many property enhancements. Beneficial property improvements include the gain in strength through work-hardening, and substantial grain refinement through repeated drawing and annealing steps. Excellent dimensional control and surface finish are also possible and enabled by drawing through single crystal natural diamond dies with proper lubrication.

Further, cold-drawn magnesium alloys may be well-suited for absorbable devices. Problematically, to date, cold drawn magnesium alloys have not been produced on a commercial scale. While recent reports have demonstrated feasibility drawing some HCP magnesium alloys with moderate levels of cold work [4–7], achieving quality results remains a challenge due to limited room-temperature deformation mechanisms.

WE43 is a relatively high-strength and corrosion-resistant magnesium alloy containing nominally 4 wt% yttrium, 3 wt% Rare Earths (primarily neodymium), and 0.5 wt% zirconium. It has been proposed as an absorbable implant material, and has generally shown an acceptable biological response both *in vitro* [8,9] and *in vivo* [10–12].

Mechanical fatigue is a phenomena whereby materials fail due to repetitive loading, often at levels significantly less than the yield or ultimate tensile strength of the material. Fatigue and corrosion-fatigue are recognized as important modes of device failure today [13–16]. Understanding the impact of fatigue, and especially corrosion-fatigue, of absorbable implant materials will be critical to successful device design. Several studies have investigated fatigue of magnesium alloy implant materials [8,17], and also cold-drawn ferrous materials [18], but none have investigated cold-drawn magnesium alloy wire.

With this in mind, the purpose of this study was to investigate the tensile properties and fatigue performance of fine, cold-drawn WE43 wires with varying thermal treatments in ambient air and body-temperature saline environments.

Experimental

Wire Production

Raw material used in the study was 3.2 mm diameter WE43B in the as-extruded condition. This material was repeatedly drawn and annealed as needed to reduce the diameter to 0.4 mm in preparation for final drawing steps. Interpass and final annealing steps were conducted at a temperature of 400-500°C at an adequate dwell time to ensure full recrystallization. Wire were cold drawn by conventional industrial processes, using natural diamond drawing dies and oil-based lubricants. After the anneal at 0.4 mm diameter, wires were drawn without annealing to a diameter of 0.127 mm, corresponding to a cold work level of 90% (Eq. 1).

$$Cold\ Work = \left(1 - \frac{d_f^2}{d_i^2}\right) * 100\% \quad (1)$$

At this final 0.127 mm diameter, wires were subjected to various thermal treatments. Specifically, thermal treatments consisted of treating the as-drawn wires at 250, 375, and 450°C, holding dwell time constant. These temperatures elicited an aging response (250°C), a stress relief (375°C), and a recrystallized microstructure (450°C).

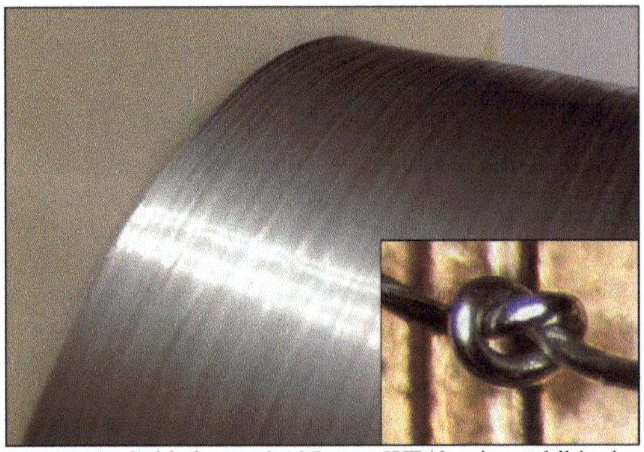

Figure 1. Cold drawn 0.127 mm WE43 wire exhibited a bright finish and good formability (inset).

Tensile Analysis

Wires in each of the four conditions (As Drawn, 250°C, 375°C, and 450°C) were tensile tested (N = 4) to assess mechanical strength and available elastic strain for fatigue testing. Tensile testing was conducted using an Instron 5966 with a 10 N load cell. Gauge length was 127 mm. Strain rate was $0.00167s^{-1}$, and sampling rate was 20 Hz.

Fatigue Testing

Rotary beam fatigue testing (Figure 2) was employed to assess fatigue behavior. Rotary beam fatigue testing allows for fully-reversed loading by axially rotating wires through a defined radius, effecting a known strain level. Testing equipment (Positool, Inc., Brunswick, OH) rotated wires at 60 Hz in ambient (22°C) air at strains ranging from 0.25 to 1.0 percent. Tests were completed at N = 3 per strain level. Additionally, samples of the 375°C and 450°C heat treatments were tested in Dulbecco's phosphate buffered saline (PBS) held at 37 ± 1°C.

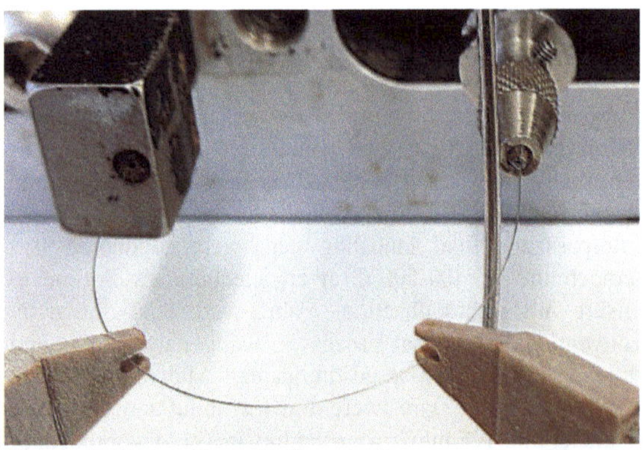

Figure 2. Rotary Beam Fatigue testing apparatus. The wire is held at a predetermined strain by a rotating chuck, right, and a bushing, left.

Results

Tensile Analysis

Tensile testing of the 4 wire conditions revealed a wide range of available properties (Table 1, Figure 3). The As Drawn condition, with 90% cold work, exhibited a tensile strength exceeding 500 MPa and high elasticity, with a 0.2% offset yield strain of 1.44%. Overall elongation to failure was relatively low, at 2.4%, but this did not translate into a brittle material, as the As Drawn wire could still be formed into a knot (Figure 1, inset).

Table 1. Mechanical properties of WE43 wire samples (mean ± standard deviation).

Condition	*0.2% YS (MPa)*	*UTS (MPa)*	*Elongation (%)*	*Modulus (GPa)*	*0.2% Yield Strain (%)*
As Drawn	**493** ± 2	**521.6** ± 0.4	**2.4** ± 0.2	**34.3** ± 0.2	**1.44** ± 0.01
250°C	**579** ± 4	**608.3** ± 0.5	**2.8** ± 0.2	**35.4** ± 0.3	**1.64** ± 0.02
375°C	**479** ± 1	**495.1** ± 0.6	**5.9** ± 0.7	**35.2** ± 0.2	**1.36** ± 0.01
450°C	**298** ± 3	**325** ± 2	**11** ± 2	**32.4** ± 0.3	**0.92** ± 0.01

Thermally treating the cold worked wire at 250°C resulted in a substantial increase in both strength, surpassing 600 MPa, and elasticity. The 375°C treatment slightly reduced strength, but dramatically increased toughness, with elongation to failure more than doubling. The wire treated at 450°C had a substantially reduced strength, but high elongation, as is typical of annealed metals.

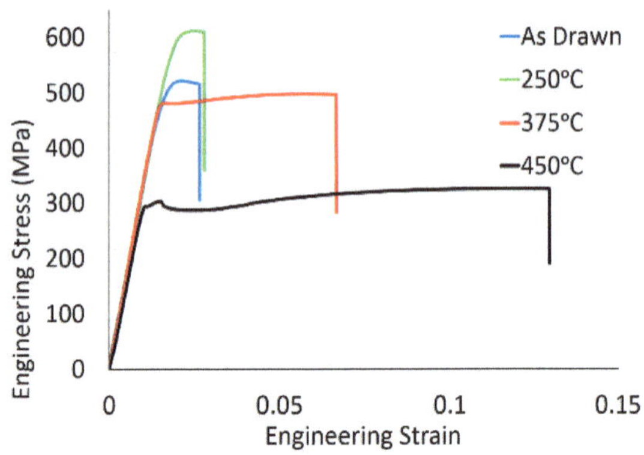

Figure 3. Representative tensile curves for WE43 wires used in this study.

Fatigue Testing

Fatigue results for each test can be seen in Figure 4. In the As Drawn condition, the minimum number of cycles to fracture of the three tests increased from 1,008 at 1.0% strain to 251,388 at 0.25% strain.

The 250°C samples performed slightly better than the As Drawn wires, although testing was only conducted from 1.0

to 0.4% strain, with minimum cycles to failure of 2,376 and 58,932, respectively.

The 375°C treatment enabled the best performance in high-cycle fatigue, with all three 0.4% strain samples achieving the run-out limit of 10 million cycles. The high strain tests were improved versus the As Drawn condition, but less than the 250°C condition.

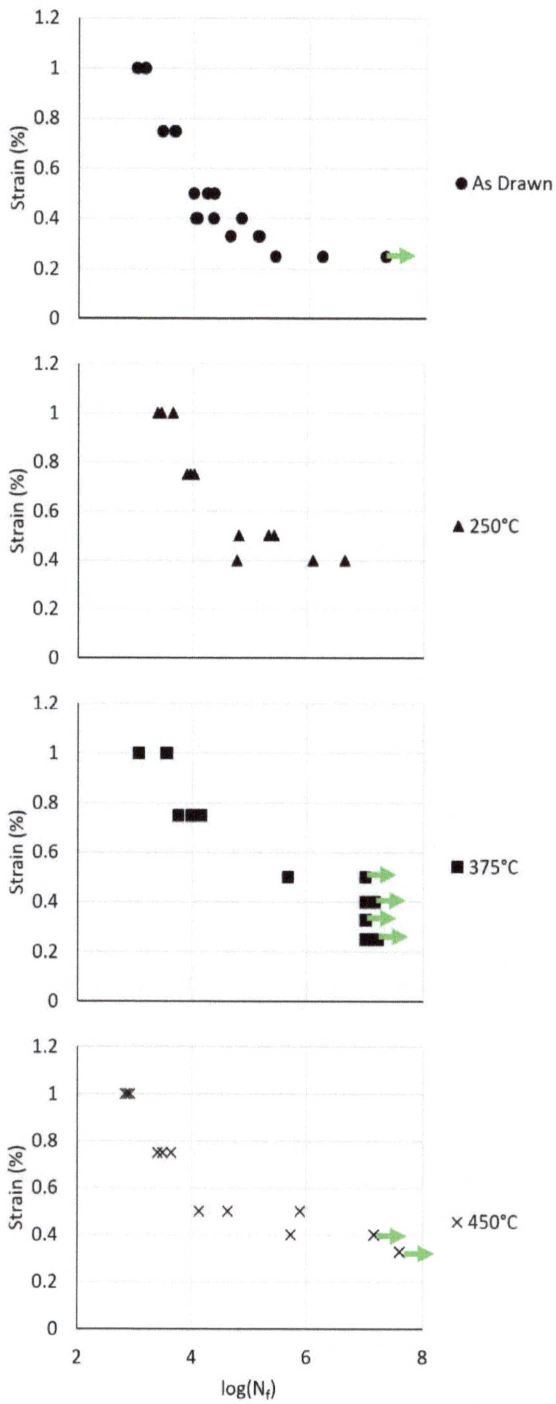

Figure 4. Fatigue results in air for each condition. Green arrows indicate fracture has not occurred.

The wires treated at 450°C had the shortest cycles to failure at high strain levels (684 minimum at 1%), but performed well in the high-cycle regime. Two of the three 0.4% samples achieved 10^7 run out, as well have all 0.33% strain samples.

The addition of a corrosive, body-temperature PBS environment reduced fatigue life (Figure 5). The influence of corrosion becomes more prominent in the high-cycle regime. For example, the 375°C wire tested in air achieved complete 10^7 runout at 0.4% strain and below. In PBS, the maximum cycles achieved at 0.4% strain was 320,688. At 0.25% strain, the minimum cycles to failure was 330,948 and the mean was 2,991,044 cycles. The 450°C wire had a substantially reduced fatigue life in PBS, and did not achieve greater than 486,972 cycles at any strain level.

Discussion

The purpose of the present study was to offer a first look into the tensile and fatigue performance of novel WE43 wires produced through cold drawing and subsequent thermal treatments. While more testing is required and ongoing, the results presented herein are encouraging.

The tensile strengths achieved in the As Drawn, 250°C, and 375°C wires are very high for magnesium-based materials, and the annealed 450°C wire exhibits reasonably good strength for WE43. These wires also exhibited a high degree of elasticity, in some cases greater than 1.5%. Wires such as these may have use in self-expanding medical device applications such as braided stents. Plastic strain to failure was fairly low in the As Drawn and 250°C conditions. However, this phenomenon is common in many drawn materials, such as stainless steel, and the wires are still easily formed without fracture. They may, however, be more susceptible to fatigue crack propagation, and this notion is supported in the fatigue data.

Rotary beam fatigue testing is a strain-governed method, and there is a relation between the yield strain of the wires in tension and the low-cycle, high-strain fatigue regime results.

For example, the 250°C wires had the highest amount of elastic strain and the highest minimum life at 1% and 0.75% strain. Conversely, the 450°C wires were experiencing plastic deformation at 1% strain, and thus performed relatively poorly in fatigue at this level.

It may not be possible to achieve a 1:1 *in vitro* – *in vivo* correlation by accelerated means. This testing, conducted at 60 Hz, may not accurately represent physiological loading conditions. For example, a stent may experience flexure at 72 beats per minute (1.2 Hz). Accelerated corrosion fatigue testing may not allow for sufficient corrosion to occur, and may also prevent crack passivation. However, the testing conducted in body-temperature PBS offers a first look into the interactive effects of fatigue and corrosion on WE43 wires. It serves as an indication and reminder that fatigue in the corrosive body environment may be significantly reduced when compared to results in air.

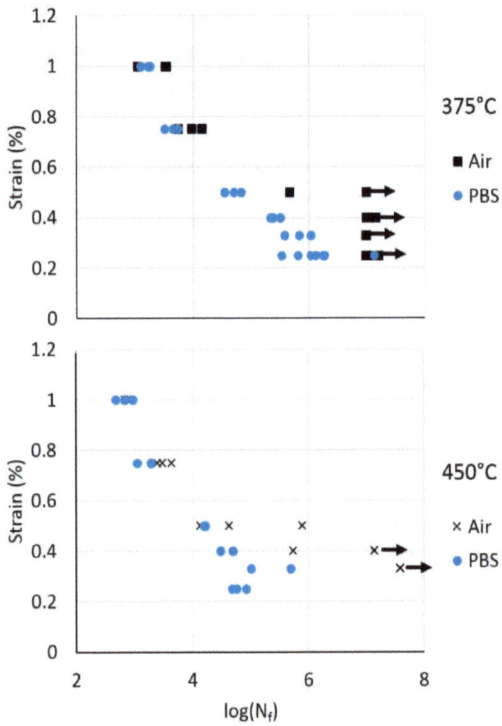

Figure 5. Fatigue life was reduced in air. 375 and 450°C wire samples were fatigue tested in a 37°C phosphate-buffered saline solution.

The wire that was cold drawn and then subjected to the 375°C treatment may offer an optimal balance of strength, toughness, and fatigue resistance for vascular implant applications. It has been suggested that an absorbable braided wire stent used to treat peripheral arterial disease must withstand 10-20 million cycles (3-6 months at 72 beats per minute) at up to 0.25% strain [19]. The 375°C (and 450°C) wire meets this criteria in air, but falls short in this initial test.

Current work is ongoing to address some of the limitations of the present study and broaden the scope. While fatigue data presented herein are only N=3, testing is continuing, and ultimately each strain level will be tested at N=7. A microstructural analysis in underway to determine grain size, precipitates, and crystal texture in each of the conditions as appropriate. Scanning electron microscopy (SEM) has been used to analyze fatigue fracture faces to determine the cause of crack initiation [20], and a similar analysis with these materials will shed light on the limiting factor of fatigue performance, be it nonmetallic inclusions originating during the melt, surface quality, or other factors.

Conclusions

The results of the study indicate that heavily-cold-worked WE43 wire achieved high tensile strength, and may be thermally treated to effect a wide range of mechanical properties. These properties can have a large influence on fatigue performance, and with the correct process, a fatigue strength exceeding 0.4% strain at 10 million cycles may be achieved in air.

Acknowledgements

The assistance of David Snider and Drew Finan in wire production and fatigue testing, respectively, is gratefully acknowledged. The authors also thank Fort Wayne Metals leadership for continued support.

References

[1] M. P. Staiger, A. M. Pietak, J. Huadmai, and G. Dias, "Magnesium and its alloys as orthopedic biomaterials: A review," *Biomaterials*, vol. 27, no. 9, pp. 1728–1734, Mar. 2006.

[2] Y. F. Zheng, X. N. Gu, and F. Witte, "Biodegradable metals," *Mater. Sci. Eng. R Rep.*, vol. 77, pp. 1–34, Mar. 2014.

[3] F. Witte, J. Fischer, J. Nellesen, H.-A. Crostack, V. Kaese, A. Pisch, F. Beckmann, and H. Windhagen, "In vitro and in vivo corrosion measurements of magnesium alloys," *Biomaterials*, vol. 27, no. 7, pp. 1013–1018, Mar. 2006.

[4] H. Chao, Y. Yu, and E. Wang, "ACHIEVING ULTRAFINE GRAIN SIZE IN Mg-Al-Zn ALLOY BY COLD DRAWING," *Int. J. Mod. Phys. B*, vol. 23, no. 06n07, pp. 927–933, Mar. 2009.

[5] H. CHAO, H. SUN, and E. WANG, "Working hardening behaviors of severely cold deformed and fine-grained AZ31 Mg alloys at room temperature," *Trans. Nonferrous Met. Soc. China*, vol. 21, Supplement 2, pp. s235–s241, Aug. 2011.

[6] T. Bajor, Z. Muskalski, and M. Suliga, "Research on the drawing process with a large total deformation wires of AZ31 alloy," *J. Phys. Conf. Ser.*, vol. 240, no. 1, p. 012107, Jul. 2010.

[7] A. Milenin and P. Kustra, "Numerical and Experimental Analysis of Wire Drawing for Hardly Deformable Biocompatible Magnesium Alloys / Numeryczna I Doswiadczalna Analiza Ciagnienia Drutu Z Trudno Odkształcalnych Biozgodnych Stopów Magnezu," *Arch. Metall. Mater.*, vol. 58, no. 1, Jan. 2013.

[8] X. N. Gu, W. R. Zhou, Y. F. Zheng, Y. Cheng, S. C. Wei, S. P. Zhong, T. F. Xi, and L. J. Chen, "Corrosion fatigue behaviors of two biomedical Mg alloys – AZ91D and WE43 – In simulated body fluid," *Acta Biomater.*, vol. 6, no. 12, pp. 4605–4613, Dec. 2010.

[9] D. Liu, Y. Ding, T. Guo, X. Qin, C. Guo, S. Yu, and S. Lin, "Influence of fine-grain and solid-solution strengthening on mechanical properties and in vitro degradation of WE43 alloy," *Biomed. Mater.*, vol. 9, no. 1, p. 015014, Feb. 2014.

[10] F. Witte, V. Kaese, H. Haferkamp, E. Switzer, A. Meyer-Lindenberg, C. J. Wirth, and H. Windhagen,

"In vivo corrosion of four magnesium alloys and the associated bone response," *Biomaterials*, vol. 26, no. 17, pp. 3557–3563, Jun. 2005.

[11] C. Di Mario, H. Griffiths, O. Goktekin, N. Peeters, J. Verbist, M. Bosiers, K. Deloose, B. Heublein, R. Rohde, V. Kasese, C. Ilsley, and R. Erbel, "Drug-Eluting Bioabsorbable Magnesium Stent," *J. Intervent. Cardiol.*, vol. 17, no. 6, pp. 391–395, Dec. 2004.

[12] H. Windhagen, K. Radtke, A. Weizbauer, J. Diekmann, Y. Noll, U. Kreimeyer, R. Schavan, C. Stukenborg-Colsman, and H. Waizy, "Biodegradable magnesium-based screw clinically equivalent to titanium screw in hallux valgus surgery: short term results of the first prospective, randomized, controlled clinical pilot study," *Biomed Eng Online*, vol. 12, p. 62, 2013.

[13] S. H. Teoh, "Fatigue of biomaterials: a review," *Int. J. Fatigue*, vol. 22, no. 10, pp. 825–837, Nov. 2000.

[14] M. Niinomi, "Fatigue characteristics of metallic biomaterials," *Int. J. Fatigue*, vol. 29, no. 6, pp. 992–1000, Jun. 2007.

[15] R. A. Antunes and M. C. L. de Oliveira, "Corrosion fatigue of biomedical metallic alloys: Mechanisms and mitigation," *Acta Biomater.*, vol. 8, no. 3, pp. 937–962, Mar. 2012.

[16] R. K. Singh Raman, S. Jafari, and S. E. Harandi, "Corrosion fatigue fracture of magnesium alloys in bioimplant applications: A review," *Eng. Fract. Mech.*

[17] S. Jafari and R. K. Singh Raman, "Corrosion Fatigue Behaviour of a Common AZ91D Magnesium Alloy in Modified Simulated Body Fluid," *Adv. Mater. Res.*, vol. 891–892, pp. 267–272, Mar. 2014.

[18] J. E. Schaffer, E. A. Nauman, and L. A. Stanciu, "Cold-Drawn Bioabsorbable Ferrous and Ferrous Composite Wires: An Evaluation of Mechanical Strength and Fatigue Durability," *Metall. Mater. Trans. B*, vol. 43, no. 4, pp. 984–994, Aug. 2012.

[19] J. E. Schaffer, "Development and characterization of vascular prosthetics for controlled bioabsorption," *Theses Diss. Available ProQuest*, pp. 1–274, Jan. 2012.

[20] J. E. Schaffer and D. L. Plumley, "Fatigue Performance of Nitinol Round Wire with Varying Cold Work Reductions," *J. Mater. Eng. Perform.*, vol. 18, no. 5–6, pp. 563–568, Aug. 2009.

Magnesium Technology 2015
Edited by: Michele V. Manuel, Alok Singh, Martyn Alderman, and Neale R. Neelameggham
TMS (The Minerals, Metals & Materials Society), 2015

DEFORMATION AND FAILURE MODES OF RAPIDLY SOLIDIFIED, ULTRA-FINE GRAIN, AMX602 AND ZAXE1711 MAGNESIUM ALLOYS

John F. Chinella
U.S. Army Research Laboratory, Aberdeen Proving Ground, MD 21005

Keywords: Magnesium Alloys, Wrought Magnesium, Mechanical Properties, Plastic Flow, Fracture

Abstract

This paper characterizes microstructure, mechanical properties, and the plastic flow and fracture behaviors of the high strength and high ductility, ultra-fine grain (UFG), magnesium alloys AMX602 (Mg-5.4Al-0.39Mn-1.77Ca) and ZAXE1711 (Mg-7.28Al-1.20La-1.18Ca-0.93Zn). Rapidly solidified processed (RSP) powders produced by Spinning Water Atomization (SWAP) were formed into compacts that were subsequently extruded into flat bars. Specific characteristics of the mechanical properties, and deformation and failure modes and features are revealed in response to loading conditions in tension, compression, shear, and impact. The material behaviors are shown with numerical and stress-strain test data, and power law hardening constitutive plastic flow parameters. Fracture modes and features are shown with optical and SEM microscopy. The ZAXE1711 alloy reveals lower yield strength, a high rate of strain hardening, greater ultimate strength and impact toughness, and more homogeneous less localized plastic flow and fracture.

Introduction

Plastic flow and fracture behaviors, through mechanical properties in tension, compression, and Charpy impact tests, are described of magnesium-aluminum-manganese-calcium materials AMX602 (Mg-5.4Al-0.39Mn-1.77Ca) and magnesium-zinc-aluminum-calcium-rare earth lanthanum ZAXE1711 (Mg-0.93Zn-7.28Al-1.180Ca-1.20La). These wrought, high strength and creep resistant materials were prepared from the Spinning Water Atomization Process (SWAP) powders, compacted to green density, then hot extruded [1-3]. The SWAP powder process provides an extraordinary high rate of solidification, about 10^6 K/s, an ultra-fine grain size, and a refined, homogeneous structure of end product unattainable by casting or wrought processing of ingots. The SWAP refined microstructures provides a means to improve hardness, strengthening, and toughness, and to avoid deleterious micro-mechanical effects, e.g. microcrack nucleation from particles that promote fracture during either quasi-static or high strain rate deformation. Identification of the deformation and fracture modes of these SWAP alloys may help reveal which specific mechanical and material properties that may better contribute to safety and protection for vehicles.

Magnesium alloys are now of interest as a means for weight savings in vehicles to meet objectives of improved fuel efficiency, thereby lowering energy costs, and minimizing air pollution as in densely populated urban environments. SWAP powder material may produce materials with multiple roles of improved resistance to creep, environmental durability, strength, and toughness. For either a lightweight structure or automotive component, Mg alloys may help provide crash-protection for vehicles and occupants, either by resisting deformation, distributing forces, or absorption of energy.

The Army has interest in Mg alloys as a protection and component material in vehicles for weight savings to improve fuel efficiency and mobility. Wrought medium-high strength Mg alloy extrudate and plate may provide efficient designs. It is important to be aware that with increased strength and hardness of materials designed for protection, there is greater tendency for the failure modes to transition from those dominated by plastic flow to more localized modes of deformation and fracture. It follows the fracture is a more brittle behavior with less energy absorption, with result of less protection. With increasing strength the protection material failures modes are: 1) strain-hardening, ductile plastic flow; 2) shear localization; and 3) dynamic toughness [4].

Deformation by ductile plastic flow in modern metals allows lightweight and damage-tolerant structural designs. Based on classical flow stress curve concept, strength of a ductile metal is decided by the relation of plastic strain, ε, and flow stress, σ_{max}, and the relation of ε, and the fracture resistance stress, σ_{crit}. When rate of increase of σ_{max} is higher than σ_{crit} with ε, fracture occurs when $\sigma_{max} = \sigma_{crit}$. The σ_{max} - $\bar{\varepsilon}_{max}$ curves may be described as: 1) σ_{max} directly reaching cleavage stress (σ_{crit}) level; 2) σ_{max} reaching σ_{crit} following deformation; 3) σ_{crit} reaches a critical strain $\bar{\varepsilon}_{crit}$, causing ductile fracture. During failure, a ductile fracture (crack) may change into a brittle type because of stress concentration and strain hardening [5].

Microstructure, thickness, and constraint influence fracture behavior of protection materials. Because ballistic protection occurs at high rates, it is often assumed that thermal softening plays a dominant role, yet mechanical softening effects such as void nucleation, growth, and coalescence, or texture initiate the onset of localization of plastic flow and shear [4]. Void nucleation and growth in metals [4] and Mg alloys [6-7] occurs in states of dominate tension or shear. In compression with shear, stress concentrations and elastic-incompatibility generate localized bursts of dislocation generation or microcracking around defect [8] alloy constituent inclusions, and precipitates. These localized deformations may coalesce together and propagate to a critical macro size scale during fracture.

Mg at room temperature is difficult to form, with limiting effect of crack failure. Dislocation slip is limited by few slip planes; mainly, slip occurs on the basal plane, twinning on pyramidal planes. With stress parallel to basal planes, twinning is only possible in compression; with stress perpendicular to basal planes twinning is only possible in tension [9]. Typically, there is a strong crystallographic texture in sheet in which grains are oriented in a direction difficult to form, resulting in high flow stresses, modest work hardening, and low values of uniform elongation [10]. Hot-work processing of rare earth (RE) alloyed Mg alloys may enhance formability. During hot work processing RE dispersoids stimulate static recrystallization and RE solute enhances non-basal slip and restricts recrystallization. The product microstructure, in conventional alloys with minimal dispersoids, has more random weakened texture and a grain refined microstructure that enhances plasticity with more uniform and greater elongation, ductility, and work hardening. However, in specific high alloy Mg-RE materials the particles may deleteriously localize fracture by void nucleation or microcracking during deformation at room temperature [2, 6-7]. Mg protection materials which control effects of texture, void nucleation, or microcracking may better resist a brittle manner of fracture and the localization of plastic flow and fracture, to provide improved, more consistent levels of protection.

Experimental Materials and Test Procedures

The experimental alloys have been described previously [1-3]. Table I reveals the chemical composition of the alloys. The process temperatures conditions for extrusion ranged from 200-350 °C, (see Table II) with an extrusion speed of 1 m/s and extrusion ratio of 37. The extrusion of SWAP powder compacts at temperatures less than 300 °C provide a supersaturated wrought processed material without precipitation of solutes, and minor amounts precipitation of Al_2Ca, and $Al_{11}La_3$ compounds at extrusion temperatures less than 350 °C [1-2].

Table I. Chemical composition of the experimental materials

AMX602, weight percent (w%)							
Al	Ca	Mn	C	Si	H	Zn	La
5.42	1.77	0.39	0.031	0.028	0.022	0.011	0.007
ZAXE1711, w%							
Al	Ca	La	Zn	H	Si	C	Mn
7.28	1.20	1.180	0.930	0.038	0.029	0.028	0.012

Results from Luvak Inc. analytical report 0-60103

Table II. Billet Temper designations and extrusion temperatures

Alloy	Temper Designation (TD)	Temperature (°C)
AMX602	1	350
	2	300
	3	250
ZAXE1711	B	350
	C	250
	D	200

Characterizations of extruded bar materials ZAXE1711-D, AMX602-3 were conducted by optical and scanning electron microscopy (SEM), chemical analysis, tension and compression, and Charpy V-notch (CV) and un-notched (UN) sub-size impact mechanical test specimens. The tension specimen gage length is 25.4-mm with a diameter of 6.401-mm; compression specimen length is 16.51-mm with 8.255-mm diameter. The UN Charpy impact specimens have a sub-sized width of 8.89-mm, the CV specimens have section area geometry of 10.00 x 10.0-mm.

Experimental Results and Discussion

Results of microhardness tests (see Table III) and extrusion temperature (see Table II) of the bar materials reveal: 1) the ZAXE1711 is less sensitive to process temperature than the AMX602 alloy; 2) hardness values of the low temperature process temper AMX602 are greater than the low temperature processed ZAXE1711.

Test results in engineering compression and tension (see Tables IV, V, and Figures 1-4) reveal these materials have high levels of strength and elongation (or reduction). Proof stresses are near identical in compression and tension. Compression proof stress in Mg tends to low values when twinning and texture dominate [9]. Fracture occurred near 45° to the compression axis, localized in a thin zone and the fracture propagated in a rapid brittle manner (see Figure 1). After plastic deformation to high levels of strain with high levels of flow stress, the final fracture of the tension specimens occurred in a somewhat brittle manner during final fracture, see Figure 4); there was very little drop in load by the plastic deformation after maximum load. Tension test results are similar to previous reports for cold compacted, extruded SWAP powders [3]; i.e., here the ZAXE1711 has 6-MPa greater Y.S., 3-MPa less U.T.S., and 1.2% less El.%; the AMX602 has 16-MPa greater Y.S., 11-MPa greater U.T.S., and 1.7% less El.

Mechanical properties are superior to intermetallic particle containing, extruded cast billet AMX602 and ZAXE1711 [1, 2]. ZAXE1711 has 80-MPa greater Y.S., 90-MPa greater U.T.S., but 1.2% less ductility. AMX602 has 77-MPa greater Y.S., 64-MPa greater U.T.S., and equal ductility. Recently, SWAP AMX602 and ZAXE1711 achieved greater levels [1-2] in strength and ductility. The ZAXE1711 and AMX602 exceed tensile requirements for ASTM B107-13 WE54-T5 and WE43C–T5 extruded bar, or yield strengths of sand cast WE54 and WE43 [9]. In comparison to advanced Mg-RE alloys, strengths exceed differential speed rolled (DSR), 723 K annealed, WE43 sheet [7]; there is lower strength but greater ductility and strain hardening than Mg-5.72Al-2.96Zn-6.05Nd RSP powder extrudate [6].

In strength performance comparison to industrial Al 6061-T651 [9] in tension, (see Table V) the ZAXE1711 has Y.S. 20-MPa less, U.T.S. 48-MPa greater, and ductility 2.2% less. The AMX602 Y.S. and U.T.S. exceed the strength of 6061-T651, respectively by 27-, and 32-MPa, with 3.6% less ductility.

The AMX602 plastic flow stress appears to have little strain hardening either in compression up to high levels of strain over 0.05, or in tension below strains of 0.02 (see Figures 2-3). The ZAXE1711 demonstrates strong strain hardening in compression over 0.03 strain and in tension strain hardening begins from the onset of plastic flow. The absence of strain hardening in AMX602 may be considered deleterious for toughness and damage tolerance, as any local microscale defect based fracture process may be more easily initiated and propagated by localized plastic flow or microcracking, early in deformation or loading.

Despite having lower yield strength, the cumulative strain hardening of ZAZE1711 results in flow stress meeting and then exceeding AMX602 flow stress beyond 0.06 strain for both compression and tension (see Figures 2-3). Note, the SWAP ZAXE1711 material has demonstrated greater levels of ballistic protection than AMX602 [3].

Table III. Vickers microhardness test results by temper, kg/mm^2

	602-1	602-2	602-3
Avg.	84	89	91
Std. Dev.	0.53	0.35	0.67
	ZAX-B	ZAX-C	ZAX-D
Avg.	88	87	88
Std. Dev.	0.89	0.65	0.66

1-kg load, average of 10-readings.

Table IV. Engineering properties in compression.

Alloy	Statistic	E (GPa)	Y.S (MPa)	U.C.S. (MPa)	Re. (%)
ZAXE1711-D	Ave.	47.5	284	520	10.6
	S.D.	2.8	1.6	22.6	0.44
AMX602-3	Ave.	46.7	333	479	9.2
	S.D.	2.2	2.8	25.7	0.64

3-specimen averages, L-direction; E = elastic modulus; Y.S. = 0.2% offset yield strength; U.C.S. = ultimate strength in compression; Re = reduction %; Ave. = average; S.D. = standard deviation.

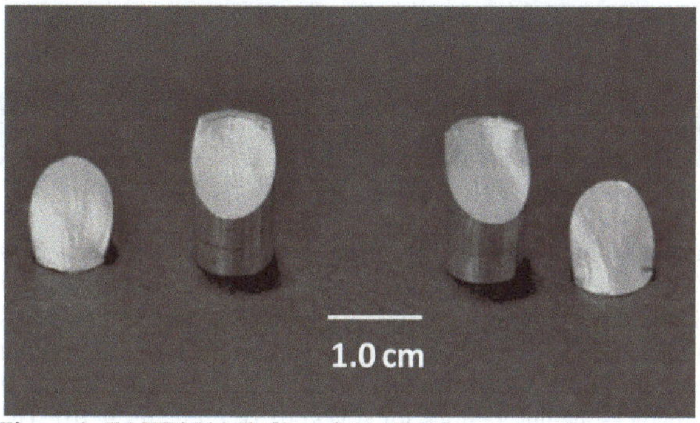

Figure 1. ZAXE1711 (left) and AMX602 compression specimens.

Table V. Engineering properties in tension

Alloy	Statistic	E (GPa)	Y.S. (MPa)	U.T.S. (MPa)	El . (%)	RA (%
ZAXE1711	Ave.	43.1	280	385	17.5	18.2
	S.D.	4.1	1.4	1.4	1.9	2.8
AMX602	Ave.	45.3	327	369	16.1	15.7
	S.D.	0.82	3.5	3.1	2.4	4.7

3-specimen averages, L-direction; Y.S = 0.02% offset yield strength; U.T.S = ultimate strength in tension; El. = elongation; R.A. reduction of area.

Table VI presents properties of true stress and true strain in tension. At maximum load ZAXE 1711 strain hardens to higher levels of true uniform stress, σ_u, than AMX602, 445-MPa, versus 419-MPa; however true strain values, ε_u, at maximum load for both alloys are near identical, ZAXE1711 with 0.14 ε_u versus 0.13 ε_u for AMX602. Similarly the values of necking strain, ε_n, from maximum load to final fracture are near-identical for both alloys, 0.06 for ZAXE1711 and 0.05 for AMX602.

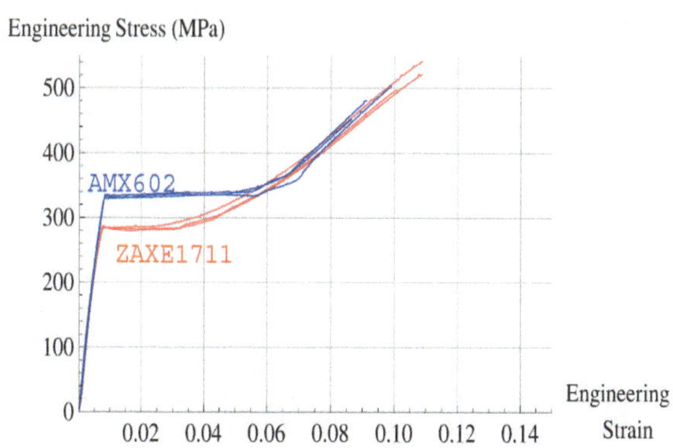

Figure 2. Plots of engineering stress versus strain in compression.

Table VII reveals strain hardening behavior with the *average rate of strain hardening*, dσ/dε, determined between true yield stress and the σ_u, and by the *strain hardening exponent*, n, of power law hardening, $\sigma = K\varepsilon^n$, for true stress-strain plastic flow in the uniform plastic deformation region. The n is the slope of a log true stress versus log true strain plot for the region of uniform plastic flow. AMX602 with 0.13-n has moderate strain hardening; e.g. greater than 6061-T651. ZAXE1711 has 0.18-n, high strain hardening similar to 2024-T351 Al. The n-values greatly exceed 0.052-n, for RSP Mg-Al-Zn-Nd alloy [6]. DSR WE43 sheet, annealed at 723 K [7], has 0.23-0.24-n, and approximate equal ε_f and ε_u to ZAXE1711.

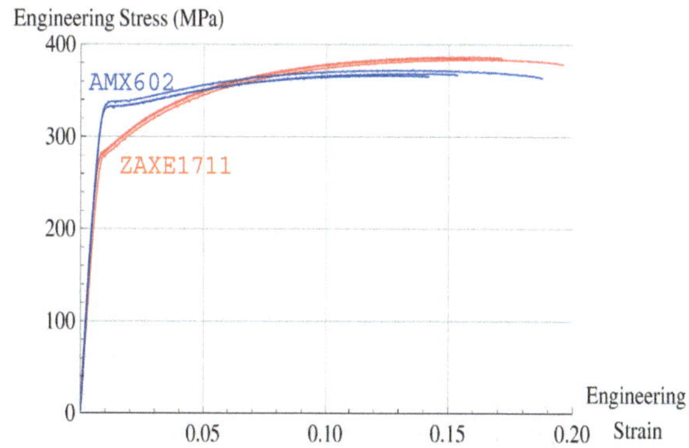

Figure 3.Plots of engineering stress versus strain in tension.

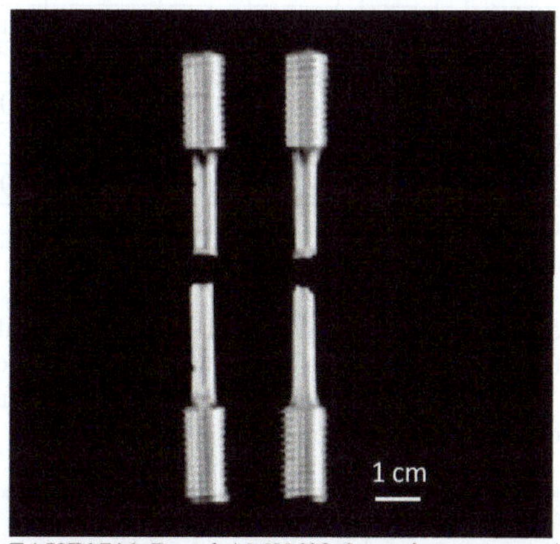

Figure 4. ZAXE1711-D and AMX602-3 tension test specimens.

Table VI. True stress and strain in tension

Alloy & Statistic	σ_u (MPa)	ε_u	ε_n	σ_f (MPa)	ε_f
ZAXE1711					
Ave.	445	0.14	0.06	467	0.20
S.D.	2.76	0.01	0.03	11.8	0.03
AMX602					
Ave.	419	0.13	0.05	434	0.18
S.D.	4.75	0.004	0.05	22.4	0.05

σ_u = true stress at maximum load $\sigma_u = P_{max}/A_u$; ε_u = true uniform strain = ln A_0/A_u; ε_n = true local necking strain = ln A_u/A_f; σ_f = (uncorrected) true strength at fracture = P_f/A_f; ε_f = true strain at fracture = ln A_0/A_f.

Table VII. Strain hardening and plastic flow; dσ/dε rate, and $\sigma = K\varepsilon^n$ power law hardening parameters K, n

Alloy &	dσ/dε	K	n	ε range fit
Statistic	(MPa)	(MPa)		
ZAXE1711				
Ave.	1204	636.0	0.18	≥ 0.02
S.D.	31.56	0.683	0.003	
AMX602				
Ave.	760.2	546.6	0.13	≥ 0.05
S.D.	11.38	5.47	0.003	

Notes: dσ/dε = average rate of hardening to maximum load σ_u; Power law hardening to maximum load: K = strength coefficient, n = strain hardening exponent.

The materials of high strain hardening theoretically may better resist the condition of plastic instability in tension, dσ/dε = σ, of material that contains microscale flaws or zones of weakness; the material surrounding defects may strain harden to maintain load.

The values of Charpy impact energies (see Table VIII) of these alloys are low; less than austempered ductile iron (ADI) values of 7.3-, and 133 J respective VN and UN impact energy. However, by mass-specific impact energy, J/ρ, the energy absorption per unit mass meets or exceeds ADI energy. Similarly, the mass-specific impact toughness values appear similar to those for low alloy high strength martensitic carbon steels.

Table VIII. Charpy impact toughness, L-S orientation

Alloy	Statistic	UN	VN	UN	VN
		(J)	(J)	(ft-lbs)	(ft-lbs)
ZAXE1711	Ave.	37.1	5.3	27.4	3.9
	S.D.	4.61	0.18	3.4	0.13
AMX602	Ave.	31.9	4.8	23.5	3.5
	S.D.	2.37	0.12	1.8	0.08

VN = V-notch specimen; UN = Un-notched specimen.
UN: impact energy/cm^2

Figures 5-6 reveal failure modes in tension with a ductile failure mode of fracture and the dynamically recrystallized grain structure of the alloys. The failure includes some features of microcracking and decohesive failure.

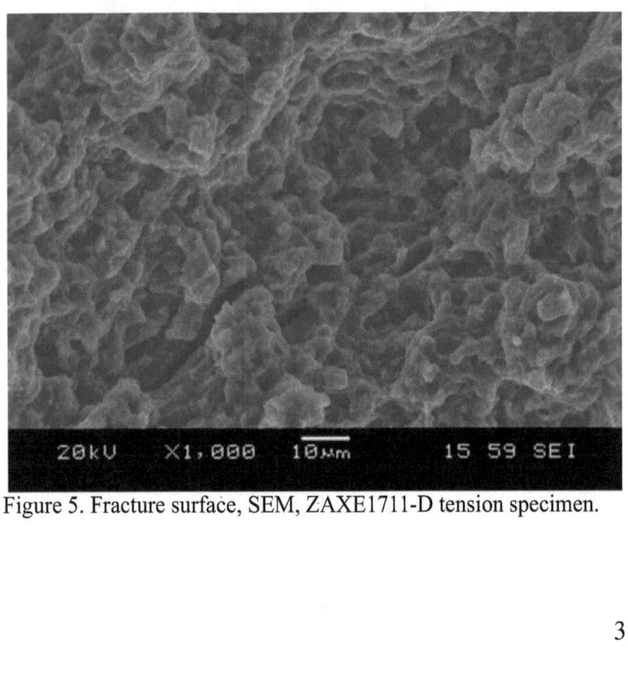

Figure 5. Fracture surface, SEM, ZAXE1711-D tension specimen.

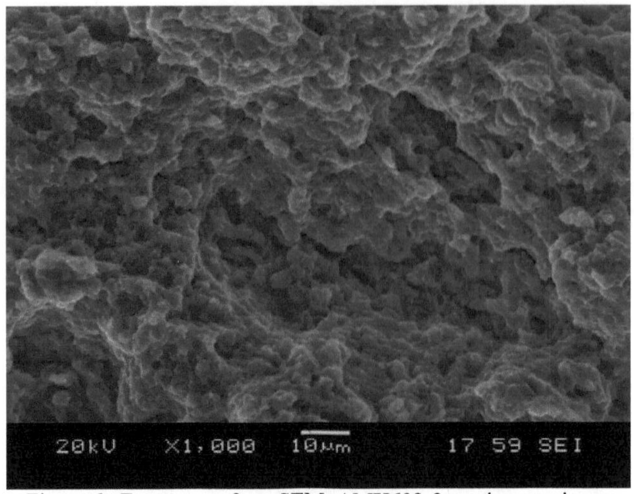

Figure 6. Fracture surface, SEM, AMX602-3 tension specimen.

Fracture features of the CV impact specimens (see Figures 7-8) reveal minor amounts of shear fracture along edges and absence of lateral expansion of material of the hinge-area. The specimens fail under Mode I in tension as observed by fracture features that propagate from the notch for ZAXE1711 and by localized microcrack-opening regions for AMX602 (see Figures 9-10). The size of the cracks developed in the AMX602 UN Charpy fracture zone range from 50μ to 800 μ.

Figures 11-12, optical, 13-16 SEM, reveal deformation and fracture feature of the UN (bend) Charpy specimens, which include a compression region with lateral expansion and region of tension with lateral contraction. The compression region of expansion includes localized crack opening and shear displacements, with the AMX602 showing greatest localization of fracture. The AMX602 material clearly reveals more closely-spaced and amounts of inhomogeneous deformation. For ZAXE1711 strains, average lateral compressive expansion = 0.09, and average lateral tensile contraction = 0.04; AMX602 strains averaged 0.07 lateral expansion and 0.03 contraction. There was between 0.01 to 0.02 strain in thickness contraction for all specimens, near 0.02 contraction for ZAXE1711; near 0.01 contraction for AMX602.

Figure 7. ZAXE1711-D CV impact specimen.

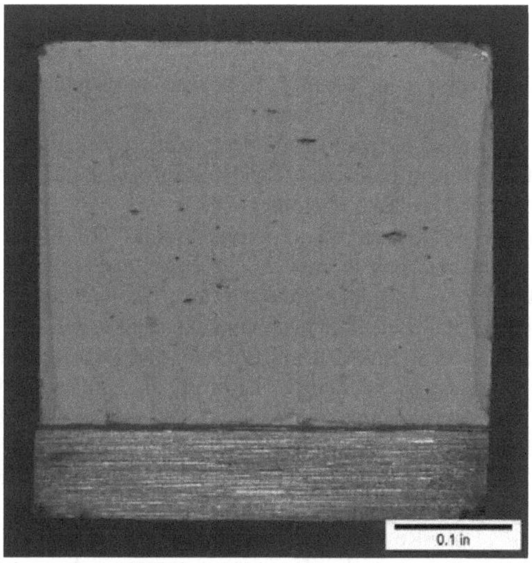

Figure 8. AMX602-3 CV impact specimen.

Figure 9. Fracture surface, SEM, CV specimen ZAXE1711-D.

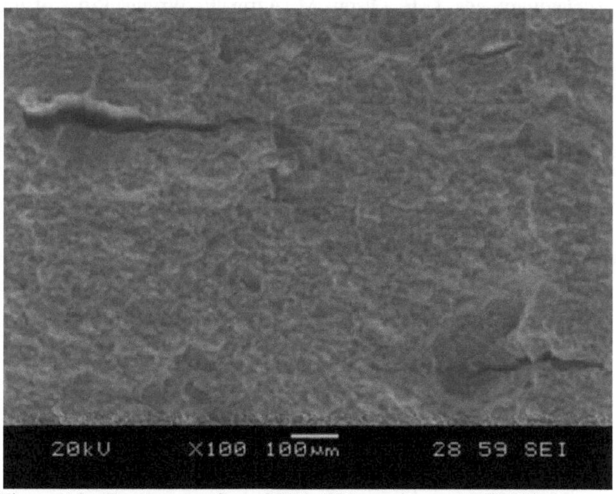

Figure 10 . Fracture surface, SEM, CV specimen AMX602-3.

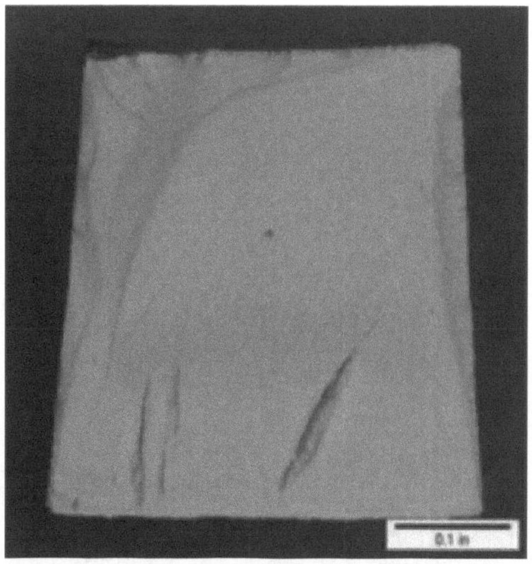
Figure 11. ZAXE1711-D UN impact specimen with deformations of bend-tension, bend-compression, and shear, and crack opening modes. Impact and compression load (bottom).

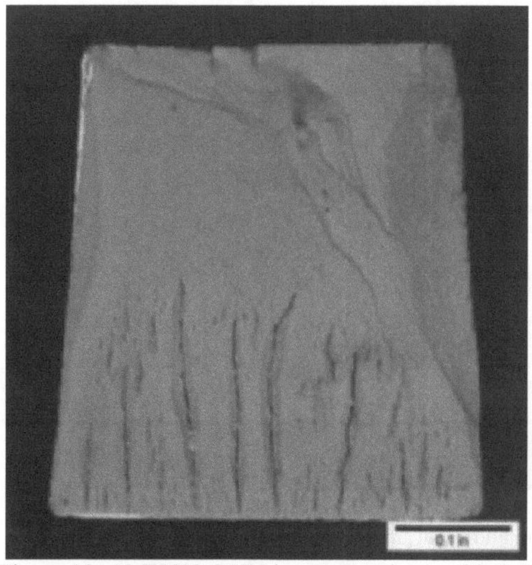
Figure 12. AMX602-3 UN impact specimen with deformations of bend-tension, bend-compression, and shear, and crack opening modes. Impact and compression load (bottom).

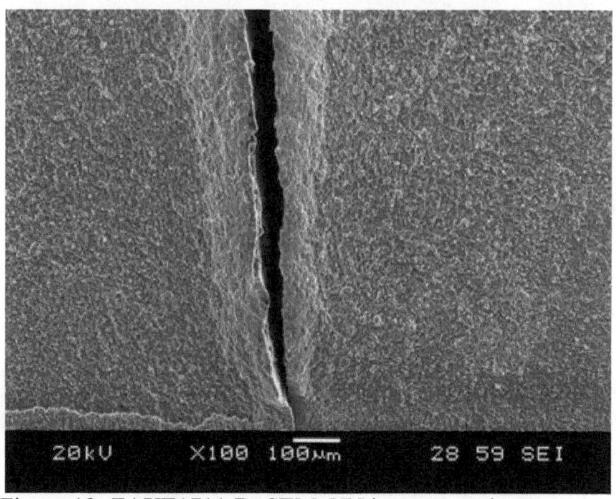

Figure 13. ZAXE1711-D, SEM, UN impact specimen.

Figure 14. AMX602-3 UN Charpy fracture topography, impact side bottom.

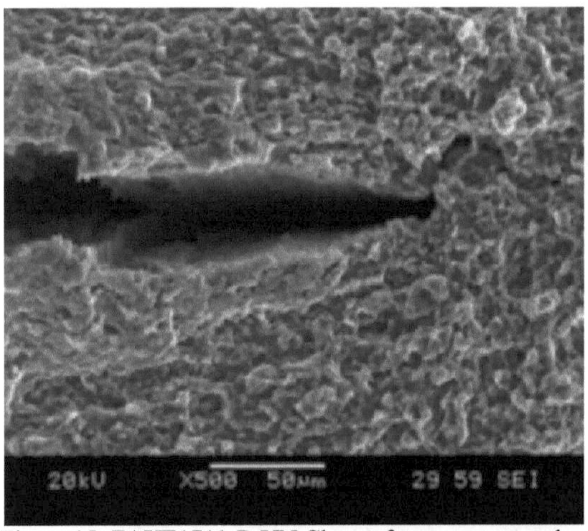

Figure 15. ZAXE1711-D UN Charpy fracture topography with crack opening and shear, impact side left.

Figure 16. AMX602-3 UN Charpy fracture topography with crack opening and shear, impact side bottom.

Conclusions

1. The SWAP powder, green compact, hot extrusion manufacture method provide wrought Mg material with high strength and ductility. Mechanical behaviors and microstructures suggest the absence of constituent particles lessens deleterious effects that lower ductility and promote fracture.
2. In comparison to DSR WE43 alloy annealed at 723 K, the ZAXE1711 material has greater engineering yield strength, greater ultimate strength, and equal ductility. These materials have near equal values of strain hardening exponent n, and ε_u.
3. The low levels of strain hardening of the AMX602 appear deleterious to damage tolerance. The material demonstrated proclivity for localized crack and shear mode fractures. The spacing of cracks appears to suggest low levels of fracture toughness and critical flaw size. The amount of shear on the surface area is dependent on the cracks formed during fracture.
4. The ZAXE1711 Mg-Re material demonstrates strong strain hardening behavior, greater levels of impact energy, and greater resistance to formation of localized crack and shear modes in the fracture zone during impact loading.
5. The ZAXE1711 has less yield strength than AMX602, but has demonstrated ballistic protection levels greater than AMX602. The characteristics of high strain hardening of ZAXE1711 appear beneficial for applications of protection.

References

[1] A. Elsayed, K. Kondo, H. Imai, J. Umeda, "Microstructure and mechanical properties of hot extruded Mg-Al-Mn-Ca alloy produced by rapid solidification powder metallurgy." *Materials and Design* 31 (2010) 2444-2453.

[2] A. Elsayed, J. Umeda, K. Kondo, "Application of rapid solidification powder to the fabrication of high-strength, high ductility Mg-Al-Zn-Ca-La alloy through hot extrusion." *Acta Materialia* 59 (2011) 273-282.

[3] T.L. Jones, K. Kondoh, "Ballistic Analysis of New Military Grade Magnesium Alloy for Armor Applications;" *Magnesium Technology 2011*, eds. W.H. Sillekins, S.R. Agnew, N.R. Neellameggham, S.N. Matthandhu; TMS (The Mining, Metals, & Materials Society), 2011.

[4] J.F. Mescal, H. Rodgers, "The Role of Shear Instability in Ballistic Penetration." *Innovations in Ultrahigh-Strength Steel Technology*, eds. G.B. Olson, M. Azrin, E. Wright, Proc. of the 34th Sagamore Conf. Aug. 30-Sept. 3, 1987, Lake George, NY.

[5] T. Kobayashi, *Strength and Toughness of Materials*. Springer, Tokyo, 2004.

[6] T.S. Srivatsan , S. Vasudevan, M. Petrorali, "An investigation of the quasi-static fracture behavior of a rapidly solidified magnesium alloy." *J. of Alloys and Compounds 460 (2008) 386–391.*

[7] X. Huang, K. Suzuki, Y. Chino, "Static recrystallization and mechanical properties of Mg–4Y–3RE magnesium alloy sheet processed by differential speed rolling at 823 K" *Materials Science and Engineering A 538 (2012) 281–287.*

[8]. M.F. Ashby, S.H. Gelles, L.E. Tanner, "The Stress at Which Dislocations are Generated at a Particle-Matrix Interface." *Philos. Mag.* 1969, 19, 757.

[9] I.J. Polmear, *Light Alloys From Traditional Alloys to Nanocrystals*, fourth edition, Elsevier; London, UK, 2006.

[10] J. Bohlen, M.R. Nürnberg, J.W. Senn, D. Letzig, S.R. Agnew, "The texture and anisotropy of magnesium–zinc–rare earth alloy sheets." *Acta Materialia 55 (2007) 2101–2112.*

Magnesium Technology 2015
Edited by: Michele V. Manuel, Alok Singh, Martyn Alderman, and Neale R. Neelameggham
TMS (The Minerals, Metals & Materials Society), 2015

EFFECT OF Sn:Zn RATIO ON CORROSION BEHAVIOR OF Mg-aSn-bZn EXTRUSIONS

Chang Dong Yim[1,2], Sang Kyu Woo[2], Bong Sun You[1,2]
[1]Korea Institute of Materials Science, 797 Changwondae-ro, Seongsan-gu, Changwon, 642-831, Republic of Korea
[2]University of Science and Technology, 217 Gajungro, Yuseong-gu, Daejeon, 305-350, Republic of Korea

Keywords: Magnesium, Sn:Zn Ratio, Corrosion Rate, Hydrogen Evolution

Abstract

The effect of Sn:Zn ratio on corrosion behavior of magnesium alloys containing Sn and Zn was evaluated systematically by potentiodynamic polarization and immersion tests. The hydrogen evolution rate during cathodic polarization and the average corrosion rate measured by immersion test increased with increasing Sn content and Sn:Zn ratio. The changes of fraction of second phase with more positive corrosion potential and solute contents in the matrix phase were responsible for the change of corrosion behavior according to Sn content and Sn:Zn ratio. Mg_2Sn particle accelerated the corrosion by formation of micro-galvanic cell, which resulted in faster corrosion rates of Mg-5Sn-xZn alloys than those of Mg-2Sn-xZn alloys. The increase of Zn in the matrix was harmful to the corrosion resistance because Zn accelerated the hydrogen evolution although ZnO was beneficial to the passivity of surface film.

Introduction

The precipitation hardening is the most effective method to increase the strength of wrought magnesium alloys. The Mg-Sn system is known as a precipitation system, which has a relatively high solubility limit at 834K and a low solubility at ambient temperature. Even though the Mg-Sn system has great potential as a high strength wrought alloy via precipitation hardening, thus far its slow age hardening response has limited the practical application of Mg-Sn system alloys. Adding alloying elements as a means to enhance the age hardening response of Mg-Sn system alloys has been studied extensively. Zn has been an effective alloying element for improving the age hardening response of the Mg-Sn system alloys and for a more uniform distribution of fine Mg_2Sn particles. According to the previous studies, the Mg-Sn-Zn system has great potential to produce a high strength and highly corrosion resistant wrought alloy. Although many studies on the mechanical behaviours of this alloy system have been conducted, few studies focused on the corrosion behaviour of the Mg-Sn-Zn system have been carried out. In this study, the corrosion behaviour of a magnesium alloy containing Sn and Zn was evaluated via potentiodynamic polarization and immersion tests. The change of corrosion behavior according to the composition of the alloy was analyzed with respect to its microstructure.

Experimental Procedure

2 kg of pure magnesium ingot were put into a carbon crucible and heated to 700°C by induction heating under a protective atmosphere of CO_2 and SF_6 gases. Pure tin granule and pure zinc ingot were added into the melt according to the given composition. The melt was poured into a cylindrical mold preheated to 200 °C. The compositions of the alloys were analyzed by induction coupled plasma (ICP) as shown in table 1.
A homogenization heat treatment was carried out for 24 h at 450 °C to homogenize the microstructure of the as-cast billets. After homogenization, the billets were machined into billets with dimensions of 78 mm in diameter and 145 mm in length for extrusion. The machined billets were preheated to 350 °C for 3 h and then extruded directly through a rectangular die having dimensions of 4 mm in thickness and 50 mm in width. The die and container were also preheated to 350 °C. The extrusion ratio and speed were 25:1 and 1.5 m/min, respectively.

Table I. Composition of Mg-aSn-bZn Alloys

Nominal Composition	Designation	ICP Results (wt%)		
		Mg	Sn	Zn
Mg-2Sn-0.4Zn	TZ204	97.68	1.90	0.42
Mg-2Sn-0.8Sn	TZ208	97.28	1.88	0.84
Mg-2Sn-1.2Zn	TZ212	97.08	1.74	1.18
Mg-5Sn-1.0Zn	TZ510	94.45	4.57	0.98
Mg-5Sn-2.0Zn	TZ520	93.72	4.43	1.85
Mg-5Sn-3.0Zn	TZ530	92.50	4.60	2.90

The specimens prepared for microstructure observation were cut from the as-extruded plates and mounted using epoxy resin. Each mounted specimen was mechanically ground using #240 ~ #2400 grit SiC papers and then polished using paste containing diamond particles of 3 μm, 1 μm and 0.25 μm. Each polished specimen was etched using a solution of 3 g picric acid, 20 ml acetic acid, 20 ml distilled water, and 50 ml ethanol to reveal the microstructure more clearly. The microstructures of as-extruded Mg-aSn-bZn alloys were observed using an optical microscope (Nikon Ephiphot 200). The constituent phases of the as-extruded Mg-aSn-bZn alloys were analyzed by measuring the X-ray diffraction pattern using an X-ray diffractometer (D/Max-2500VL/PC). The scanning range of 2θ was from 20° to 80° and the scanning speed was 2 °/min.
The electrochemical behavior of the as-extruded Mg-aSn-bZn alloys was evaluated via potentiodynamic test. Specimens of 16 mm in diameter and 2 mm in thickness were machined from the as-extruded plates and then ground mechanically using #240 ~ #2400 grit SiC papers. Each specimen was inserted into a holder and then put into a flask containing a 3.5 wt% NaCl solution at 25 °C. A three-electrodes array was applied to the potentio-dynamic test. The working electrode, reference electrode, and counter electrode were the specimen, a saturated calomel electrode, and a carbon rod, respectively. The open circuit potential (OCP) was measured over 3600 s, and then the external voltage was changed from -0.25 V to 0.5 V with respect to an OCP at rate of 0.5 mV/s. The potentiodynamic test was stopped when the current density reached 0.01 mA/cm^2, even though the external voltage did not reach 0.5 V with respect to the OCP.
The average corrosion rates were measured using immersion test. Specimens of 28 mm in width, and 36 mm in length, and 3mm in thickness were machined from the as-extruded plates and then ground mechanically using #240 ~ #2400 grit SiC papers. The specimens for the immersion test were put into a 3.5 wt% NaCl solution, and the temperature of the solution was maintained isothermally at 25°C using an isothermal control chamber. After

immersion for 72 h, the corrosion products on the surface were removed using a solution of 200 g/L CrO_3 and 10 g/L $AgNO_3$. Weight loss before and after immersion test was measured and converted to the average corrosion rate according to ASTM G1-03. The corrosion rate was determined as the average value of the three test results.

Results and Discussion

Figure 1 shows the microstructures of as-extruded TZ208 and TZ520 alloys. The grain structure showed a typical equiaxed grain structure by dynamic recrystallization during hot extrusion. The grain size decreased remarkably with increasing of Sn content, which seemed to be resulted from the difference of initial grain size before extrusion and existence of second phase particle. Table 2 shows the average grain size before and after extrusion.

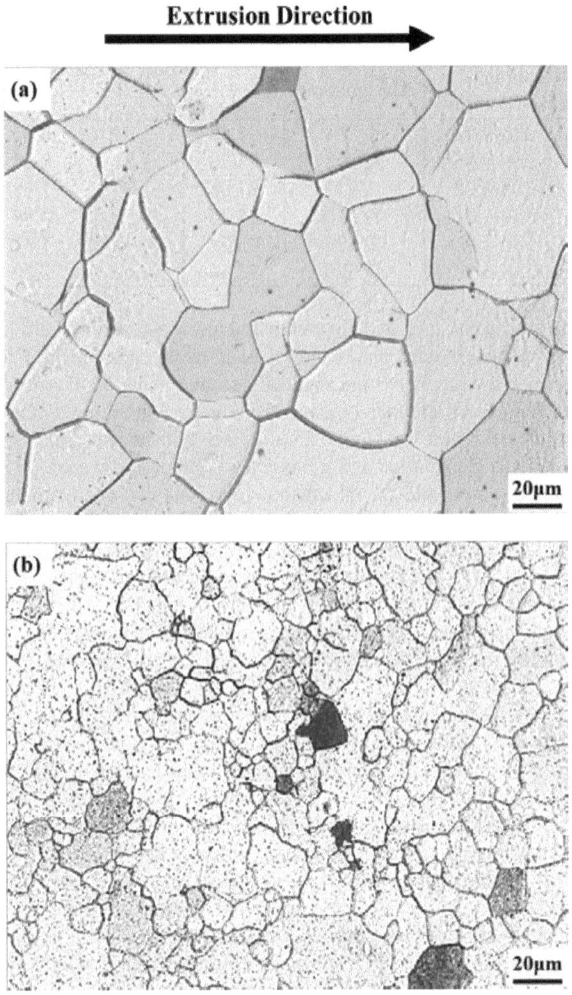

Figure 1. Microstructures of as-extruded Mg-aSn-bZn alloys; (a) TZ208 (b) TZ520.

Takigawa [1] reported that the size of grains formed by dynamic recrystallization during hot deformation of magnesium alloys decreased with decreasing the initial grain size. It was reported [2-3] that the grain size of as-cast Mg-xSn binary alloys decreased with increasing Sn content due to higher GRF (growth restriction factor). In this study, the grain size of as-cast Mg-aSn-bZn alloys decreased with increasing Sn content, which was similar to the previous reports [2-3]. The grain size of Mg-aSn-bZn alloys after homogenization also decreased with increasing Sn content. Therefore, it is reasonable to consider that the difference of initial grain size is one of the reasons for change of grain size formed by dynamic recrystallization according to Sn content.

Cheng [4] reported that fine Mg_2Sn particles could form from supersaturated matrix of Mg-xSn binary alloys during hot extrusion by dynamic precipitation and they could suppress the growth of grain formed by dynamic recrystallization. According to thermodynamic calculation, it was predicted that the formation of Mg_2Sn particles would not occur in Mg-2Sn-xZn alloys but it would occur in Mg-5Sn-xZn alloys during preheating and hot extrusion at 623K, which is well matched to the XRD patterns of as-extruded Mg-aSn-bZn alloys as shown in figure 2. Therefore, it seemed that the formation of fine Mg_2Sn particles played a role in suppression of grain growth.

Table II. Average Grain Size of Mg-aSn-bZn Alloys

Alloy Designation	Grain Size (μm)	
	Before Extrusion	After Extrusion
TZ204	452	51
TZ208	467	56
TZ212	479	61
TZ510	348	32
TZ520	325	25
TZ530	311	21

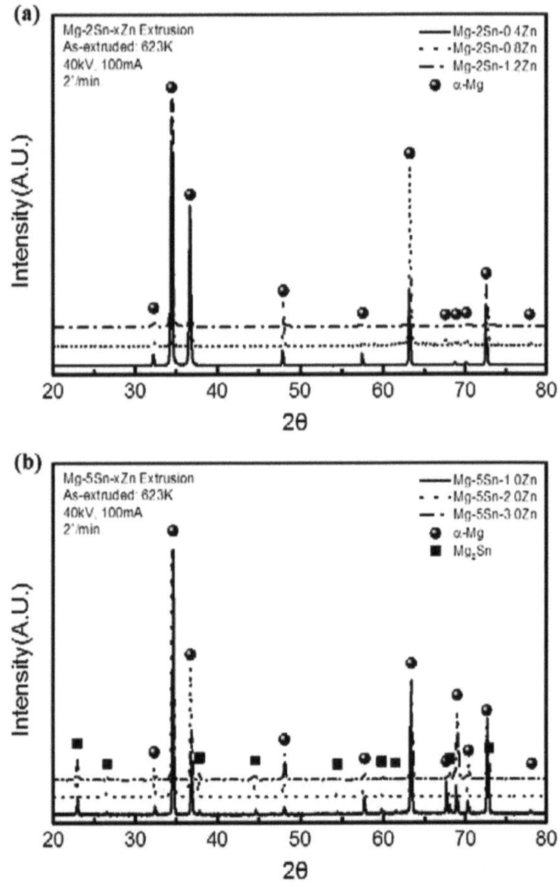

Figure 2. XRD patterns of as-extruded Mg-aSn-bZn alloys; (a) Mg-2Sn-xZn (b) Mg-5Sn-xZn.

Figure 3 shows the change of OCPs and potentiodynamic polarization curves of as-extruded Mg-aSn-bZn alloys. As shown in figure 3(a), OCP increased with increasing Sn content and Sn:Zn ratio, which resulted from the increase of passivity of surface film by addition of more Sn and Zn [5-6]. The corrosion potential, E_{corr}, was not well matched to OCPs due to modification of surface during cathodic polarization. The current density at cathodic polarization region increased with increasing Sn content and Sn:Zn ratio. It means that the hydrogen evolution rate increases with Sn content and Sn:Zn ratio.

Figure 4 shows the average corrosion rates of as-extruded Mg-aSn-bZn alloys which were measured by immersion test. The average corrosion rate increased with increasing Sn content and Sn:Zn ratio, which was well matched to the hydrogen evolution rate as shown in figure 3(b).

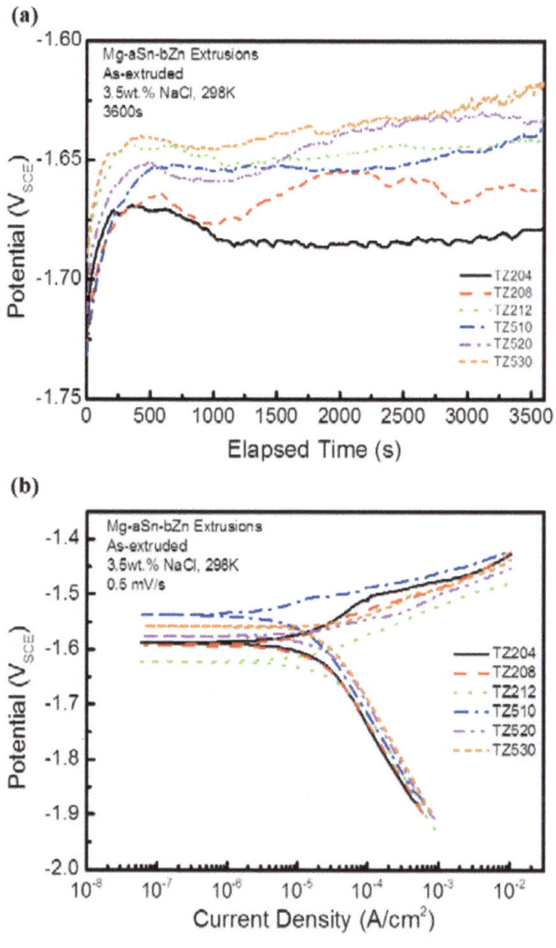

Figure 3. Electrochemical behaviors of as-extruded Mg-aSn-bZn alloys; (a) OCP (b) Potentiodynamic polarization.

The electrochemical and corrosion behaviors of metallic materials are strongly dependent on the microstructural factors including sort and amount of second phase, solute content in matrix phase, grain size and texture, etc. In case of magnesium alloys, the effect of second phase on corrosion behavior may be most important because most of second phases in magnesium alloys have more positive potential than matrix phase and therefore a micro-galvanic cell between matrix and second phases accelerates the corrosion of magnesium alloys. As shown in figure 2, there was no Mg_2Sn particle in Mg-2Sn-xZn alloys, which resulted in slower corrosion rates of Mg-2Sn-xZn alloys than those of Mg-5Sn-xZn alloys.

It was also reported [6] that Mg_2Sn particle accelerated the hydrogen evolution, which was well matched to the cathodic polarization curves as shown in figure 3(b). In case of Mg-2Sn-xZn alloys, Zn content in the matrix phase increased because all of Zn added into the alloy was dissolved in the matrix phase. The increase of Zn content in the matrix phase resulted in the increase of passivity of surface film by formation of more ZnO. But, ZnO has higher conductivity than MgO and $Mg(OH)_2$, which results in larger hydrogen evolution.

The average grain size also affected the cathodic polarization curve. It was reported [6] that the current density at cathodic region decreased with increasing grain size, which was well matched to the result shown in figure 3(b).

In case of magnesium alloys, it seems that hydrogen evolution has more dominant effect on overall corrosion resistance than the passivity of surface film. Therefore, the corrosion rate increased with increasing Sn:Zn ratio in Mg-2Sn-xZn alloys due to higher hydrogen evolution.

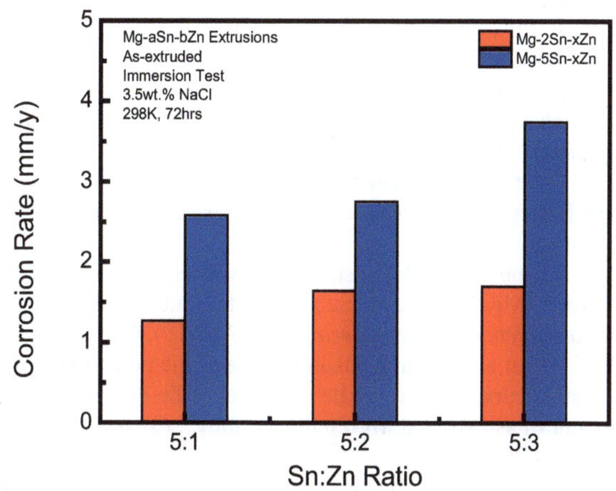

Figure 4. Average corrosion rates of as-extruded Mg-aSn-bZn alloys.

In case of as-extruded Mg-5Sn-xZn alloys, the fraction of Mg_2Sn particle increased with increasing Sn:Zn ratio according to thermodynamic calculation as shown in figure 5. The Zn content in the matrix phase also increased with Sn:Zn ratio. Therefore, the corrosion rate of Mg-5Sn-xZn alloys increased with increasing Sn:Zn ratio due to higher fraction of Mg_2Sn particle and higher Zn content in the matrix phase.

Conclusions

As-extruded Mg-aSn-bZn alloys showed a typical equiaxed grain structure by dynamic recrystallization during hot extrusion. The formation of Mg_2Sn particle was observed in Mg-5Sn-xZn alloys although there was no Mg_2Sn particle in Mg-2Sn-xZn alloys. The formation of Mg_2Sn particle resulted in faster corrosion rate by formation of micro-galvanic cell. The increase of Sn:Zn ratio also accelerated the corrosion of as-extruded Mg-aSn-bZn alloys due to higher Zn content in the matrix phase and/or higher fraction of Mg_2Sn particle.

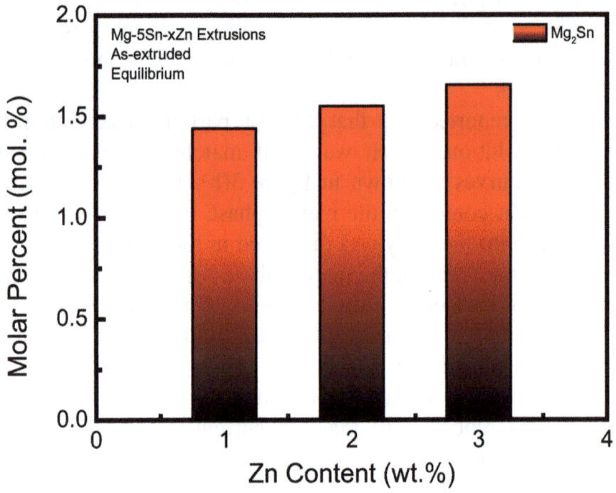

Figure 5. Change of fraction of Mg_2Sn particle in as-extruded Mg-5Sn-xZn alloys calculated by Pandat 7.0.

References

1. Y. Takigawa et al., "Effect of Initial Grain Size on Dynamically Recrystallized Grain Size in AZ31 Magnesium Alloy," *Materials Transactions*, 49 (2008), 1979-1982.
2. H. Liu et al., "The Microstructure, Tensile Properties, and Creep Behavior of As-cast Mg-(1-10)%Sn Alloys," *Journal of Alloys and Compounds*, 440 (2007), 122-126.
3. J.W. Fu et al., "Formation of the Solidified Microstructure in Mg-Sn Binary Alloy," *Journal of Crystal Growth*, 322 (2011), 84-90.
4. W.L. Cheng et al., "Microstructure and Mechanical Properties of Binary Mg-Sn Alloys Subjected to Indirect Extrusion," *Materials Science and Engineering A*, 527 (2010), 4650-4653.
5. H.-Y. Ha et al., "Limitations in the use of the potentiodynamic polarisation curves to investigate the effect of Zn on the corrosion behaviour of as-extruded Mg-Zn binary alloy," *Corrosion Science*, 75 (2013), 426-433.
6. H.-Y. Ha et al., "Influences of metallurgical factors on the corrosion behaviour of extruded binary Mg-Sn alloys," *Corrosion Science*, 75 (2014), 369-379.

Magnesium Technology 2015
Edited by: Michele V. Manuel, Alok Singh, Martyn Alderman, and Neale R. Neelameggham
TMS (The Minerals, Metals & Materials Society), 2015

STRESS CORROSION CRACKING OF ZEK100 MAGNESIUM ALLOY FOR AUTOMOTIVE APPLICATIONS

Xin Pang[1], Chao Shi[1], Renata Zavadil[1]
[1]CanmetMATERIALS, Natural Resources Canada; 183 Longwood Road South; Hamilton, Ontario, L8P 0A5, Canada

Keywords: Wrought Magnesium Alloy, Stress Corrosion Cracking, Slow Strain Rate Testing.

Abstract

ZEK100 magnesium alloy has attracted considerable interest for automotive body structure applications in light-weight vehicles due to its excellent formability at room temperature. However, the intrinsic tendency of magnesium alloys to corrode under wet conditions has been a concern. Reports on the corrosion resistance and, in particular, the stress corrosion cracking (SCC) susceptibility of ZEK100 under automotive service conditions have been scarce. In this work, the SCC of ZEK100 magnesium alloy was characterized by slow strain rate testing method. The effects of microstructure and corrosion environment on the SCC resistance of the alloy have been investigated. The results represent outcomes from a US-Canada-China collaborative research and development project - Magnesium Front End Research and Development (MFERD) currently underway.

Introduction

Magnesium and its alloys are light-weight structural materials that possess a high strength-to-weight ratio and have attracted considerable attention for lightweight applications [1]. Use of magnesium alloys in automobiles offers great potential in vehicle lightweighting and significant environmental benefits due to the increased fuel economy and reduced greenhouse gas emission. However, their tendency to corrosion under road spray conditions has been a concern and a limiting factor for their more widespread applications. In addition to general corrosion and galvanic corrosion, magnesium alloys are susceptible to environmentally assisted corrosion, including stress corrosion cracking (SCC) and corrosion fatigue, since many components of automobiles are often subjected to stress in a corrosive environment.

SCC has long been recognized as a major cause of failure of engineering components and structures. Despite the fact that SCC has been widely studied and well-documented [2-5], there is still a lack of understanding regarding the operating mechanism. Moreover, in the case of magnesium, new alloys, processing technologies, as well as surface treatment and coating techniques are under active development. The influences of these new processes and treatments on the SCC of magnesium alloys are unknown. Therefore it is of fundamental importance and practical interest to study the SCC of magnesium alloys.

Since 2007, CanmetMATERIALS has engaged in a US-Canada-China collaborative project known as Magnesium Front End Research and Development (MFERD). The overall objective of the project is the development of enabling technologies and engineering knowledge base to permit increased deployment of magnesium alloys in automotive body structures, with specific attention to the front end substructure [6-8]. Demonstration front end subassemblies have been designed and fabricated, consisting of AM60 shock tower (die cast), AZ31 upper rail (warm formed sheet) and AZ31 or AM30 lower rail (extrusion). These structures have been studied in terms of joining methods, strength, durability and corrosion. In the later phases of the project, ZEK100 has been added to the scope of study and its formability and corrosion resistance are being characterized for applications in automotive body structures [9].

The addition of rare-earth elements such as Nd, Ce, Y and Gd has been shown to improve the room-temperature formability of magnesium alloys, by modifying and weakening their basal texture and thus reducing their room temperature anisotropy [10-12]. Yet very little reports can be found in literature which studied the corrosion properties, especially the SCC susceptibility of this type of magnesium alloys. In this work, the SCC of rare-earth alloyed ZEK100 was characterized by slow strain rate testing (SSRT) method. The microstructure and fracture surface of the alloy were analyzed using optical and scanning electron microscopy (SEM). SCC mechanism of ZEK100 in different corrosion environments was discussed.

Experimental Procedures

A coil of ZEK100 rolled sheet was obtained from Magnesium Elektron. The nominal chemical composition of the alloy is 1.0 wt% Zn, 0.2 wt% Zr, 0.12 wt% Nd, and the balance Mg. The sheet was used as fabricated and the nominal thickness is 1.5 mm. For microstructure analysis using optical microscopy, the specimen cut from the ZEK100 was cold-mounted in epoxy with the surface of the rolled sheet exposed for examination. The polished specimen was etched in a solution comprising 4.2 g picric acid, 10 ml acetic acid, and 100 ml ethanol.

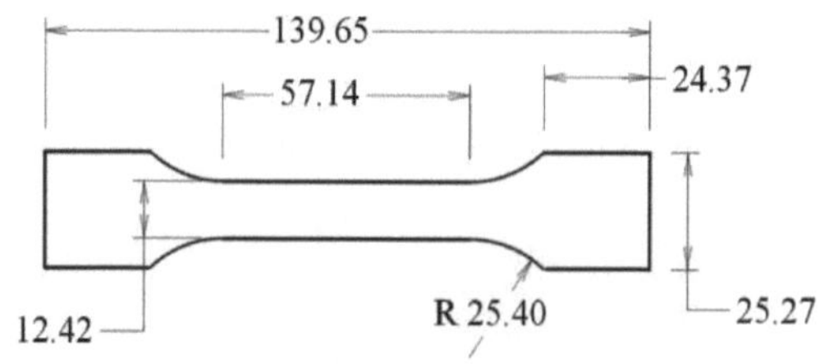

Figure 1 Specimen dimension for SSRT (mm).

The SSRT specimens were cut out of the sheet with their longitudinal axis parallel to the rolling direction. The specimen dimension is shown in Figure 1. SSRT was conducted per ASTM G129 standard using a Cortest SSRT/CERT frame at a nominal strain rate of 1×10^{-6} s^{-1} in air, distilled water, or a salt solution. Two replicates were tested for each condition, and a good reproducibility was observed. The typical results were presented in this paper. The salt solution was the electrolyte as specified in SAE J2334 standard [13], containing 0.5% NaCl, 0.1% $CaCl_2$, and 0.075% $NaHCO_3$. After test, the time to fracture (T_f), elongation to fracture (E_f), reduction of area (RA), and ultimate tensile stress (UTS) of each specimen were recorded.

It must be noted that the specimen elongation recorded on this equipment was measured by two linear variable displacement transducers (LVDTs) attached at two sides of the specimen, which doesn't represent the precise strain level of the specimen, and hence is referred to as "apparent strain". The fracture surfaces of the SSRT specimens were analyzed using a Nova NanoSEM scanning electron microscope.

Results

The optical micrograph of the rolled ZEK100 alloy presented in Figure 2 reveals very fine grains with plenty of twins.

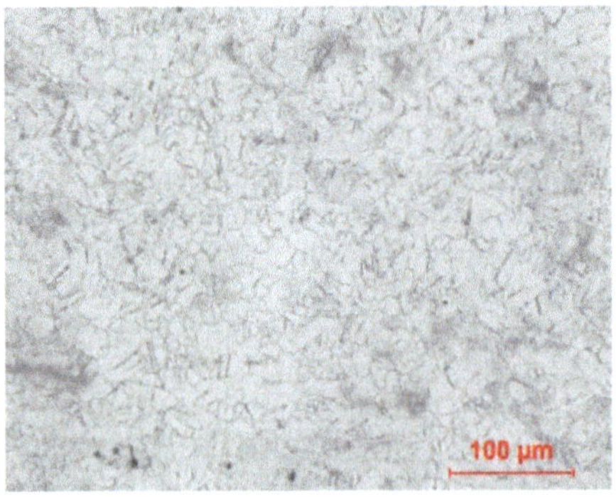

Figure 2 Optical micrograph of ZEK100.

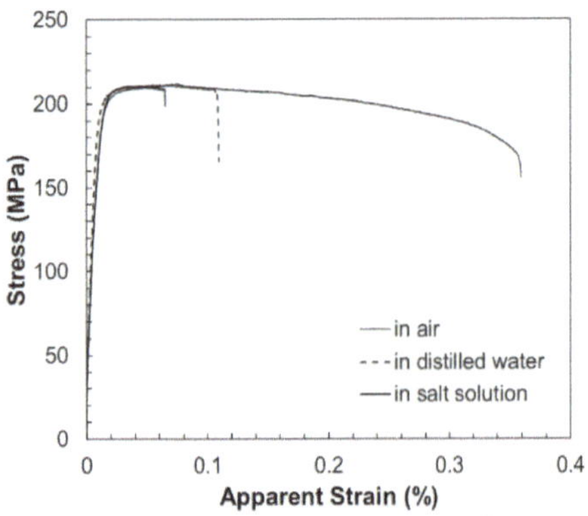

Figure 3 Stress vs. apparent strain curves for ZEK100.

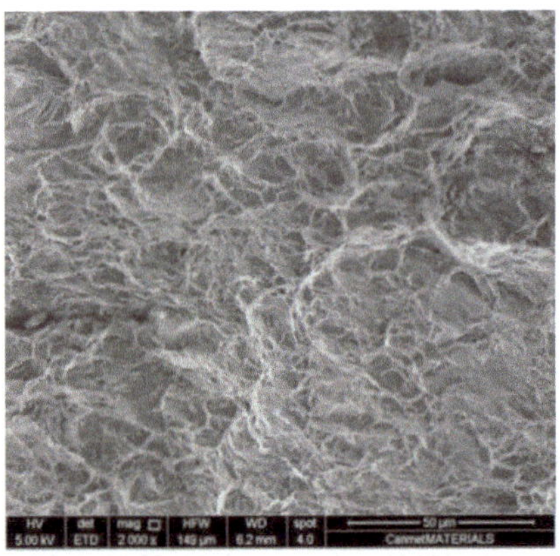

Figure 4 Fracture surfaces of ZEK100 specimens in air.

Figure 3 shows the stress-strain curves of ZEK100 SSRT specimens tested in air, distilled water, and salt solution. SAE J2334 is a cyclic corrosion test standard commonly adopted by researchers and automotive original equipment manufacturers (OEMs) to characterize the corrosion resistance of materials in a sequence of dry and wet environments. In this work, a salt solution as per SAE J2334 was used for SSRT to represent the corrosive environment of real vehicle service conditions.

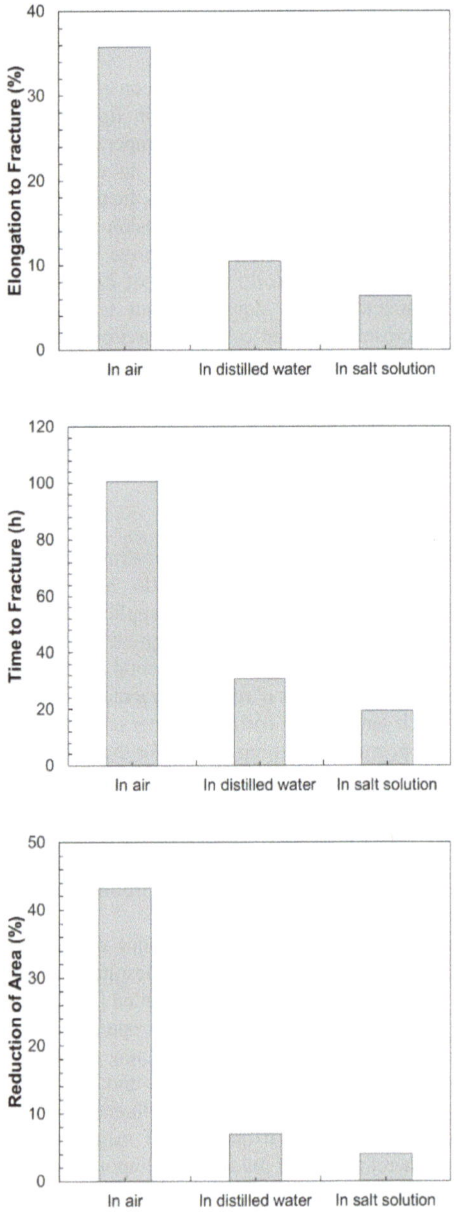

Figure 5 Effects of the test environment on (a) E_f, (b) T_f, and (c) RA of the ZEK100 SSRT specimens.

Specimen tested in air showed signs of ductility, as indicated by the fracture surface (Figure 4) and the large E_f, T_f and RA values (Figure 5). The fracture surface of the specimen tested in air shows a typical dimpled structure of a transgranular ductile rupture (Figure 4).

In contrast, the results indicated that the specimens tested in distilled water and salt solution had suffered corrosion, pitting, and SCC. The E_f, T_f and RA of the specimen tested in distilled water decreased significantly compared to those for the specimen tested in air, and further decrease was obtained for the specimen tested in the salt solution (Figure 5). No pronounced decrease in UTS was observed for both specimens tested in distilled water and salt solution compared to that tested in air (Figure 3).

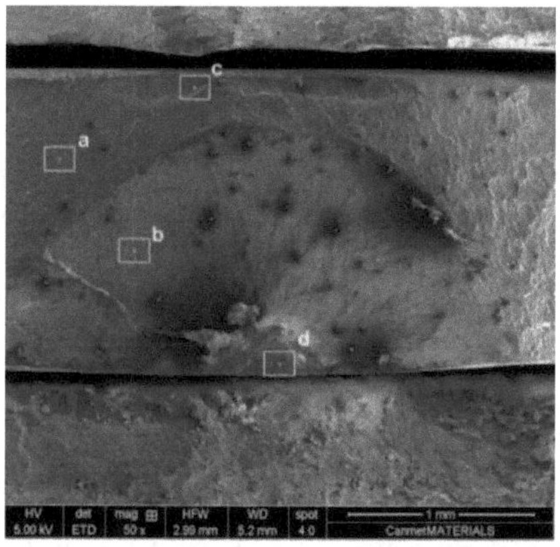

Figure 6 Fracture surface of ZEK100 in distilled water.

It can be concluded from the SSRT results that the ZEK100 alloy is more susceptible to SCC in the salt solution and less susceptible to SCC in distilled water, although the difference is not large. This is probably caused by the aggressive ions such as Cl^- in the salt solution which promote both anodic dissolution of magnesium and hydrogen production by cathodic reaction [14,15].

The fracture surface of ZEK100 specimen tested in distilled water, shown in Figures 6 and 7, can be interpreted as mixed interganular and quasi-cleavage transgranular fracture [15]. The specimen exhibited a large area of flat, cleavage-like facets showing characteristic of brittle rupture (Figure 7a,b). Examination under a higher magnification also revealed macroscopically flat fracture surface regions at the gauge surface (Figure 7c) associated with SCC. It can be seen that the fracture was initiated from a defect site at the specimen surface (Figures 6, section d). The defect was caused by the corrosion and embrittlement of the magnesium alloy in water, as indicated by the corrosion products and intergranular cracks observed by SEM at a higher magnification (Figure 7d).

For the specimen tested in the salt solution, corrosion of magnesium is more severe and multiple large pits formed at the specimen surface, from which brittle rupture was initiated, as shown in the fracture surface in Figure 8.

The corrosion of magnesium in aqueous solutions can be described by the following reactions [16]:

$Mg \rightarrow Mg^{2+} + 2e^-$ anodic reaction (1)

$2H_2O + 2e^- \rightarrow 2OH^- + 2[H]$ cathodic reaction (2)

$2\,[H] \rightarrow H_2 \uparrow$ hydrogen combination reaction (3)

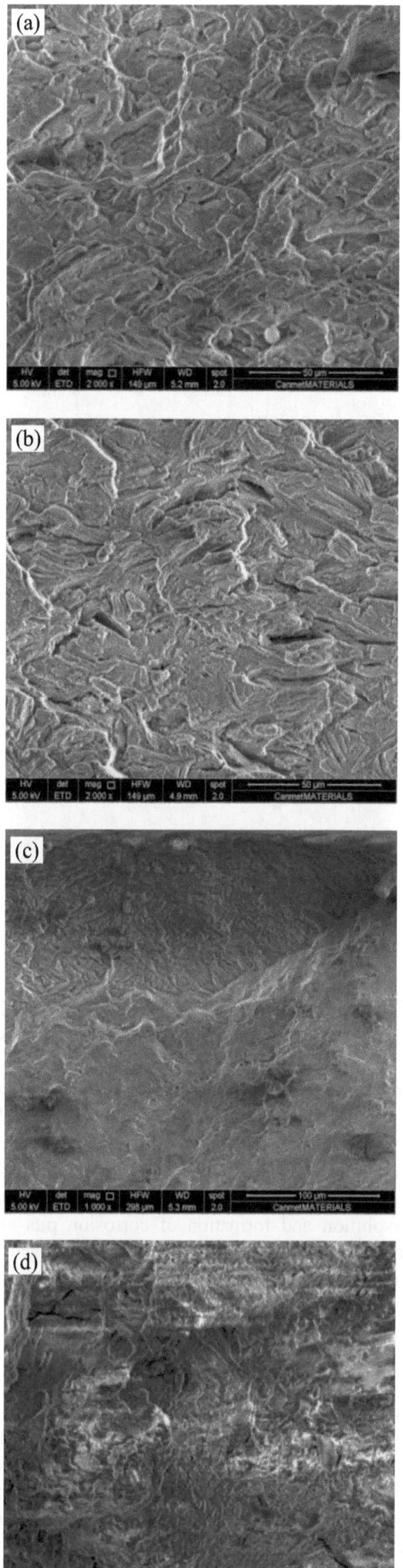

Figure 7 Micrographic images of the (a) section a, (b) section b, (c) section c, and (d) section d in Figure 6.

$$Mg^{2+} + 2OH^- \rightarrow Mg(OH)_2 \quad \text{corrosion product formation} \quad (4)$$

The overall corrosion reaction is:

$$Mg + 2H_2O \rightarrow Mg(OH)_2 + H_2 \uparrow \quad (5)$$

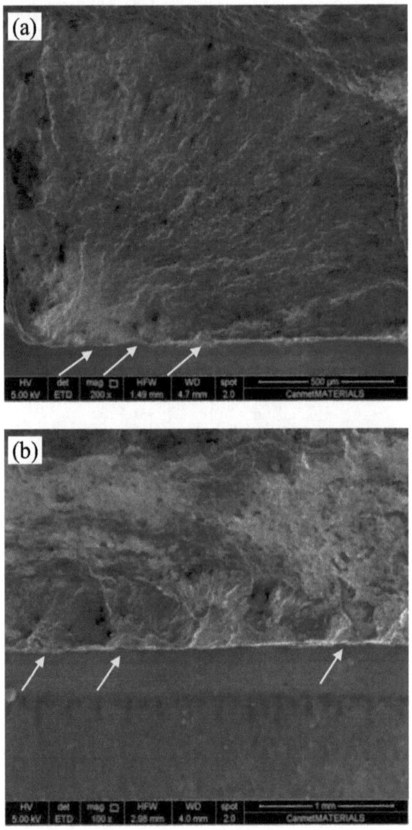

Figure 8 Micrographic images of fracture surface of ZEK100 tested in the salt solution. Arrows denote pits at the surface.

Part of the hydrogen atoms generated by cathodic reaction would penetrate into the magnesium alloy and cause embrittlement mechanism. It is suggested that hydrogen embrittlement is the principal operating mechanism of SCC in the ZEK100 alloy in distilled water and the salt solution. It must be pointed out that the anodic dissolution and formation of corrosion pits at the metal surface provide defect sites with increased hydrogen production and hydrogen entry into the material, and hence promote crack initiation and SCC of the magnesium alloy.

Conclusion

ZEK100 sheet magnesium alloy is susceptible to SCC in distilled and the salt solution, and the SCC behavior of ZEK100 tested in these two environments is similar. The hydrogen generated by cathodic reaction can penetrate into the metal and reduce the cohesive strength of the magnesium alloy, hence resulting in SCC caused by hydrogen embrittlement. The aggressive ions in the salt solution promote anodic dissolution of magnesium and formation of surface defects such as corrosion pits, which in turn promote the SCC of ZEK100 alloy.

References

1. B. L. Mordike, T. Ebert, Mater. Sci. and Eng. A, 2001, 302, 37-45.
2. D. Najjar, T. Magnin, T. J. Warner, Mater. Sci. and Eng. A, 1997, 238, 293-302.
3. N. Winzer, A. Atrens, G. Song, E. Ghali, W. Dietzel, K. U. Kainer, N. Hort, C. Blawert, Adv. Eng. Mater., 2005, 7, 659-693.
4. T. D. Burleigh, Corrosion, 1991, 47, 89-98.
5. K. Sieradzki, R. C. Newman, J. Phys. Chem. Solids, 1987, 48, 1101-1113.
6. A. A. Luo, J. F. Quinn, R. Verma, Y. -M. Wang, T. M. Lee, D. A. Wagner, J. H. Forsmark, X. Su, J. Zindel, M. Li, S. D. Logan, S. Bilkhu, R. C. McCune, Light Metal Age, 2012, 70 (2), 54-58.
7. R. C. McCune, J. Forsmark, B. Schneider, A. A. Luo, H. Gu, W. Schumacher, X. Chen, F. Vartolas, SAE Int. J. Mater. Manf., 2013, 6, 242-247.
8. J. H. Forsmark, M. Li, X. Su, D. A. Wagner, J. Zindel, A. A. Luo, J. F. Quinn, R. Verma, Y. -M. Wang, S. D. Logan, S. Bilkhu, R. C. McCune, Magnesium Technology 2014, 517-524.
9. S. Kurukuri, M. J. Worswick, A. Bardelcik, R. K. Mishra, J. T. Carter, Metall. Mater. Trans. A, 2014, 45A, 3321-3337.
10. J. Bohlen, M. R. Nürnberg, J.W. Senn, D. Letzig, S. R. Agnew, Acta Mater., 2007, 55 (6), 2101-2012.
11. J. Bohlen, S. Yi, D. Letzig, K. U. Kainer, Mater. Sci. Eng. A, 2010, 527 (26), 7092–7098.
12. L. Jiang, J. J. Jonas, R. K. Mishra, Mater. Sci. Eng. A, 2011, 528 (21), 6596–6605.
13. SAE International Surface Vehicle Standard, SAE Standard J2334, Rev. Dec. 2003.
14. G. L. Song, A. Atrens, Adv. Eng. Mater., 2003, 5 (12), 837-858.
15. R. G. Song, C. Blawert, W. Dietzel, A. Atrens, Mater. Sci. Eng. A, 2005, 399, 308-317.
16. G. L. Song, A. Atrens, Adv. Eng. Mater., 1999, 1, 11-33.

Magnesium Technology 2015
Edited by: Michele V. Manuel, Alok Singh, Martyn Alderman, and Neale R. Neelameggham
TMS (The Minerals, Metals & Materials Society), 2015

EFFECT OF Ca AND Y ON CORROSION BEHAVIOR OF EXTRUDED AZ SERIES Mg ALLOYS

Sang Kyu Woo[1], Chang Dong Yim[1,2], Young Min Kim[1,2], Bong Sun You[1,2]
[1] Advanced Materials Engineering, University of Science and Technology, 217 Gajungro Yuseong-gu, Daejeon, 305-350, Korea
[2] Light Metal Division, Korea Institute of Materials Science, 797 Changwondaero, Seongsan-gu, Changwon, Gyeongnam, 642-831, Korea

Keywords: Magnesium alloys, Corrosion, Calcium, Yttrium, Microstructure

Abstract

It is well known that ignition temperature increases dramatically through combined addition of Ca and Y to AZ series Mg alloys. Therefore, the industrial application of Mg alloys is expected to be expanded gradually. However, studies on corrosion behavior of these alloys have been carried out rarely even though the evaluation and understanding of corrosion behavior are necessary for wide application. In this study, the changes of microstructure and corrosion behavior of AZ series alloys according to the content of Ca and Y were systemically evaluated by immersion and salt spray tests. The corrosion resistance of AZ series alloys was improved dramatically by combined addition of Ca and Y without degradation of mechanical properties. It implies that the AZ series alloys containing optimized amounts of Ca and Y have the excellent combination of ignition resistance, corrosion resistance and mechanical properties.

Introduction

Magnesium alloys have many advantages such as high specific strength and castability but they also have some disadvantages such as high reactivity with oxygen and a poor corrosion resistance. These problems must be overcome for widespread application of magnesium alloys. Many researchers have been studied to improve ignition resistance. One of methods which can improve ignition resistance is the addition of alloying element. Among various elements, Ca is the most effective one which can increase ignition temperature dramatically [1-4]. Unfortunately, the addition of Ca causes a decrease of mechanical properties. Therefore, many studies have been carried out to solve this problem. According to previous researches, the combined addition of Ca and Y to AZ series Mg alloys can increase ignition temperature dramatically, without degradation of mechanical properties [5-6]. However, the evaluation and understanding of corrosion behavior of these alloys are necessary for a widespread application. In this study, the changes of microstructure and corrosion behavior of AZ series alloys with the content of Ca and Y were systemically evaluated by immersion and salt spray tests in NaCl solution.

Experimental Procedure

Mg-3Al-1Zn and Mg-9Al-1Zn ternary alloys with the addition of Ca or the combined addition of Ca and Y were investigated in this study. 2.5 kg of commercial AZ series alloy ingot was put into a graphite crucible and heated to 973 K under a protective atmosphere of CO_2 and SF_6 gases using induction heating furnace. Granule type of pure calcium and pure yttrium were added into the melt according to the given compositions. The melt was held for 10 minutes at 973 K for complete melting the granules, and then it was poured into a cylindrical steel mold which was preheated to 473 K. Before extrusion, a homogenization heat treatment was carried out for 24 hours at 673K. After heat treatment, the billets were machined into the billets of 78 mm in diameter and 145 mm in length for extrusion. The machined billets were preheated to 573 K and then indirectly extruded through a rectangular die. The die and container were also preheated to 573 K. The extrusion ratio and ram speed were 30:1 and 1 mm/sec., respectively. Table 1 shows the composition of AZx1-aCa-bY alloys and extrusion conditions of these alloys.

Table 1. Composition and extrusion conditions of AZx1-aCa-bY alloys

Nominal Composition (wt.%)	Extrusion conditions		
	Pre-heating temp.	Ram speed	Extrusion ratio
AZ31	573K	1.0mm/sec	30:1
AZ31-0.75Ca			
AZ31-0.5Ca-0.25Y			
AZ91			
AZ91-0.25Ca			
AZ91-0.15Ca-0.1Y			

The specimen for microstructural observation was cut from the extruded plate and mounted by epoxy resin. The mounted specimen was mechanically ground using SiC papers up to #2400 grit and then polished using pastes containing diamond particles of 3 μm, 1 μm and 0.25 μm. The polished specimen was etched using a solution of 3 g picric acid, 10 ml acetic acid, 10 ml distilled water, and 100 ml ethanol. The microstructures of as-extruded AZx1-aCa-bY alloys were observed using an optical microscope (Nikon, Epiphot 200). The constituent phases of the as-extruded AZx1-aCa-bY alloys were analyzed by measurement of X-ray diffraction (Rigaku, D/Max-2500VL) pattern. A scanning range of 2θ was from 20° to 80° and a scanning speed was 2°/min. A scanning electron microscope (SEM) (Jeol, JSM-6610LV equipped with an energy dispersive X-ray spectroscopy) was also used to observe microstructure more detail and identification of each particle was carried out using energy dispersive X-ray spectroscopy (EDS) analysis.
The corrosion behaviors of as-extruded AZx1-aCa-bY alloys were evaluated by immersion and salt spray tests. The specimen of 28 mm in width, 36 mm in length, and 6mm in thickness was machined from the as-extruded plate and then ground mechanically using SiC papers up to #2400 grit. In case of immersion test, specimens were immersed in 3.5wt.% NaCl for 72 hours and temperature was maintained at 298 K by temperature control chamber. In case of salt spray test, specimens were put in cyclic corrosion test chamber (ASCOTT, CC450ip), and a salt fog was sprayed for 72 hours at 308 K according to ASTM B117.

After immersion and salt spray tests during given periods, the specimens were cleaned through immersion in a mixed solution (170 g/L CrO_3 + 10 g/L $AgNO_3$ + 20 g/L $Ba(No_3)_2$) to remove the corrosion products, and then the weight loss of the specimen was measured.

Results and Discussion

Figure 1 shows the microstructures of as-extruded AZx1-aCa-bY alloys taken by an optical microscope. As shown in figure 1, the microstructures of as-extruded AZx1-aCa-bY alloys showed a typical equi-axed grain structure by dynamic recrystallization during hot extrusion. It seems that there was little difference in grain size of as-extruded AZx1-aCa-bY alloys by addition of Ca or combined addition of Ca and Y. But sort and volume fraction of constituent phases were changed.

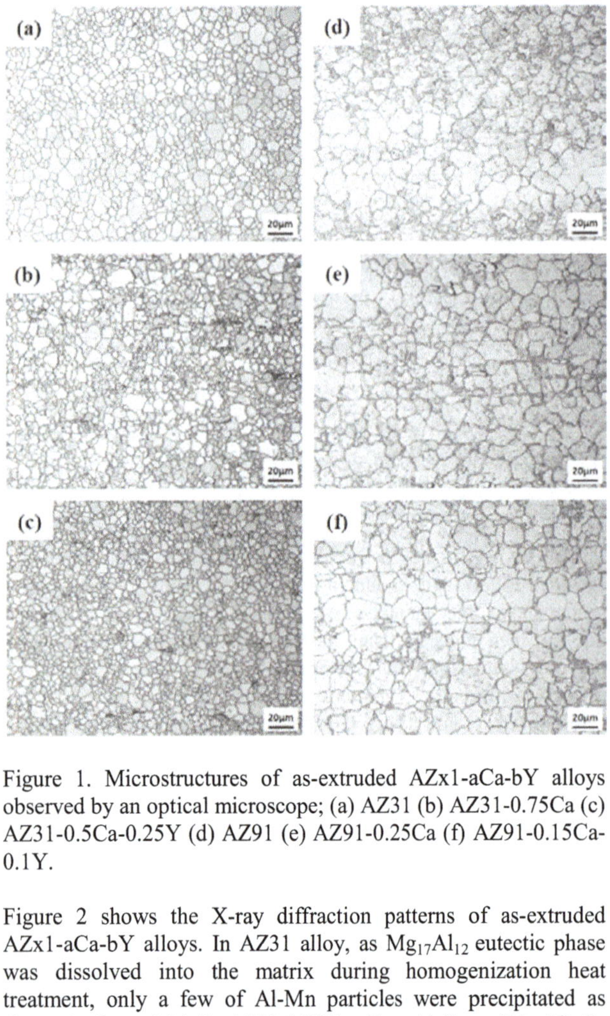

Figure 1. Microstructures of as-extruded AZx1-aCa-bY alloys observed by an optical microscope; (a) AZ31 (b) AZ31-0.75Ca (c) AZ31-0.5Ca-0.25Y (d) AZ91 (e) AZ91-0.25Ca (f) AZ91-0.15Ca-0.1Y.

Figure 2 shows the X-ray diffraction patterns of as-extruded AZx1-aCa-bY alloys. In AZ31 alloy, as $Mg_{17}Al_{12}$ eutectic phase was dissolved into the matrix during homogenization heat treatment, only a few of Al-Mn particles were precipitated as shown in figure 3(a). In AZ31-0.75Ca alloy, Al_2Ca or $(Mg,Al)_2Ca$ phase was precipitated with Al_8Mn_5 phase. Similarly, in AZ31-0.5Ca-0.25Y alloy, most of second phases were Al_2Ca and $(Mg,Al)_2Ca$ but Al_8Mn_4Y phase formed by addition of Y instead of Al_8Mn_5 phase. The changes of second phases in AZ91 series alloys by addition of Ca or combined addition of Ca and Y were similar to those of AZ31 series alloys in which a large amount of β-$Mg_{17}Al_{12}$ phases were precipitated.
The morphology and composition of phases in as-extruded AZx1-aCa-bY alloys observed by SEM were shown in figures 3 and 4.

As shown in figure 3(a), Al_8Mn_5 particles with various sizes from sub-micron to 5 μm were uniformly distributed both at the grain boundaries and inside grains in the AZ31 alloy. When Ca was added, a Mg-Al-Ca phase with a long rod shape which is expected as C36-$(Mg,Al)_2Ca$ phase [7] were aligned along the extrusion direction and these particles are relatively larger than Al_8Mn_5 particles as shown in figure 3(b). As shown in figure 3(c), the addition of yttrium led to the formation of polygonal Al_8Mn_4Y particles with various sizes instead of Al_8Mn_5 phase. In case of AZ91 series alloys, similar changes of second phases were also observed according to alloy composition. In the figure 4(a), it is shown that AZ91 alloy has mainly fine β-$Mg_{17}Al_{12}$ particles precipitated along grain boundaries with small amount of Al_8Mn_5 particles. As shown in figure 4(b), when Ca was added, Mg-Al-Ca particles formed as a type of long band aligned in the extrusion direction, while volume fraction of β-$Mg_{17}Al_{12}$ particles decreased. It is clearly shown that Mg-Al-Ca particles are much larger than β-$Mg_{17}Al_{12}$ particles, because the former remains undissolved in a matrix during homogenization heat treatment at 673 K due to its thermal stability. When Ca and Y were added to AZ91 alloy, Mg-Al-Ca and Al_8Mn_4Y particles can be observed in the manner similar to the AZ31 alloy containing Ca and Y. It could be also noted that volume fraction of β-$Mg_{17}Al_{12}$ particles in AZ91-0.15Ca-0.1Y alloy was smaller than that of AZ91 alloy.

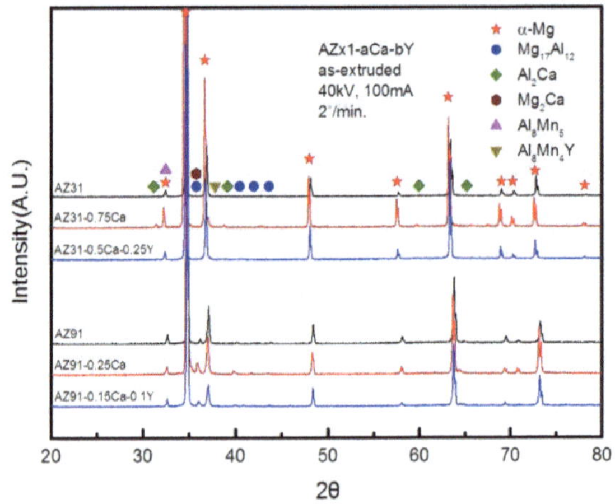

Figure 2. X-ray diffraction patterns of as-extruded AZx1-aCa-bY alloys.

Figure 5 shows the average corrosion rates of as-extruded AZx1-aCa-bY alloys which were calculated from weight loss after immersion and salt spray tests. Overall trends of the corrosion rate according to alloy composition were similar in both tests, but the difference of absolute value was larger in immersion than that in salt spray test. It seemed to be resulted from different corrosion environment. In case of immersion test, electrolyte always contact a substrate while time for electrolyte to contact substrate is limited during salt spray test. Therefore, the mechanism and controlling factors of corrosion are different according to the environmental situation, which results in difference of absolute value of corrosion rate.
When Ca was added, the average corrosion rate increased in both tests. It seems that the increase of corrosion rate results from a galvanic corrosion effect by precipitated Al_2Ca or $(Mg,Al)_2Ca$ phases. It was reported that the corrosion rates of as-cast AZ series alloys decreased with increasing Ca contents up to 1.0wt.% [8] because the Ca-containing second phases formed semi-continuously along grain boundaries acted as a barrier against

corrosion propagation. It is known that coarse Ca-containing second phases undissolved during homogenization heat treatment were fragmented to fine particles during extrusion. Therefore, they could not act as a barrier against corrosion propagation and thus accelerated the corrosion by means of the formation of micro-galvanic cell with the matrix.

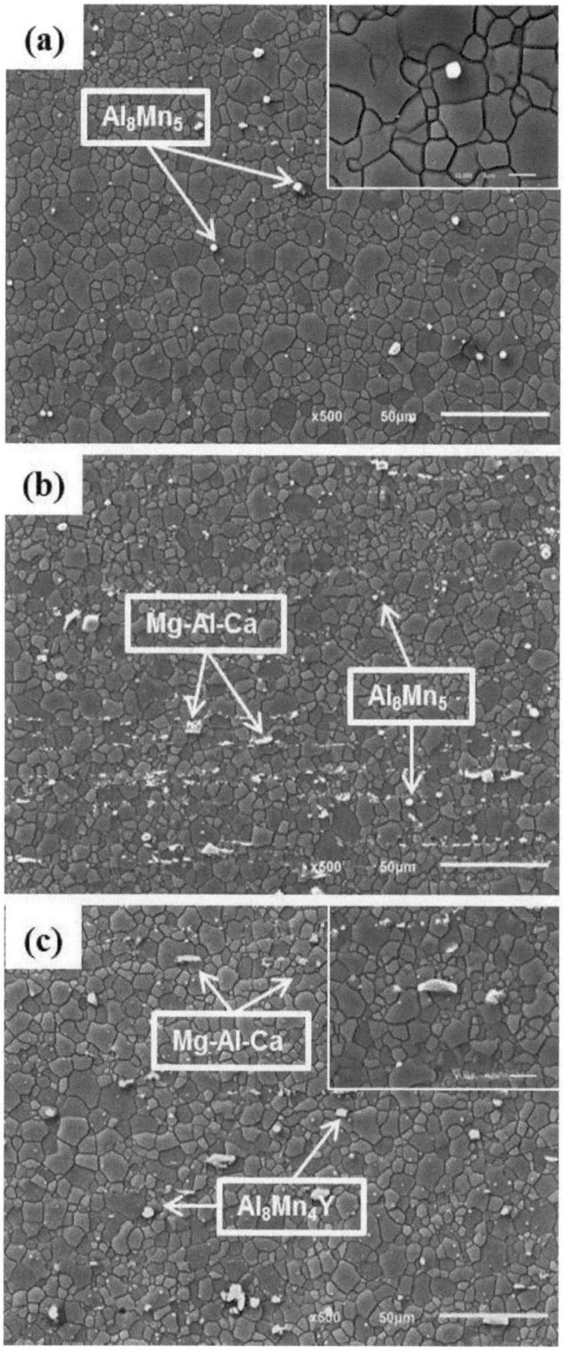

Figure 3. Microstructures of as-extruded AZx1-aCa-bY alloys observed by scanning electron microscope; (a) AZ31 (b) AZ31-0.5Ca (c) AZ31-0.5Ca-0.25Y.

On the other hands, the corrosion rates of the alloys containing Ca and Y were lower than or similar to those of AZx1 alloys. It seems that various factors affect the corrosion behaviors of AZx1-aCa-bY alloys. One factor is the volume fraction of Mg-Al-Ca phase which accelerates the corrosion. The volume fraction of Mg-Al-Ca particles decreases by the substitution of Ca by Y while remaining total amount of Ca and Y. That is beneficial to improve the corrosion resistance of AZx1-aCa-bY alloys. Another factor is the formation of Al_8Mn_4Y phases. The formation of Al_8Mn_4Y phases instead of Al_8Mn_5 phases can contribute to improve corrosion resistance through attenuation of micro-galvanic effect.

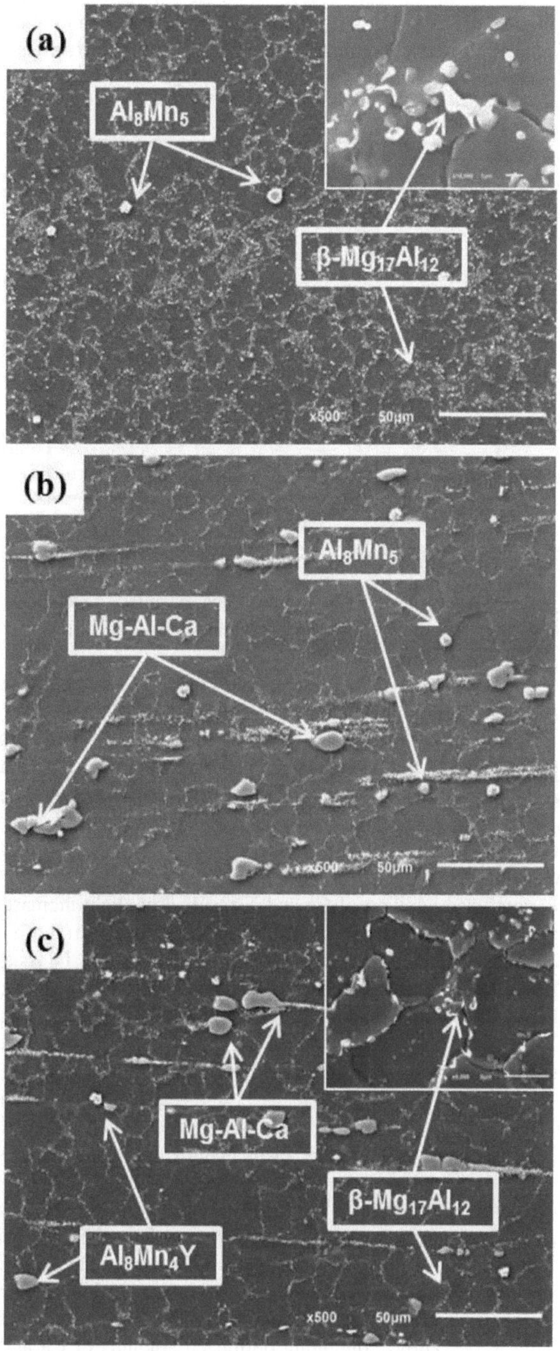

Figure 4. Microstructures of as-extruded AZx1-aCa-bY alloys observed by scanning electron microscope; (a) AZ91 (b) AZ91-0.25Ca (c) AZ91-0.15Ca-0.1Y.

Also, according to Kim et al. [5], protective double layered structures which consist of MgO-CaO + CaO-Y_2O_3 was formed on surface of melt of AZx1 alloys with combined addition of Ca and Y during casting. This layer contributes to improve the ignition resistance of AZx1-aCa-bY alloy. In this way, changes of surface layer can contribute to improve the protectiveness of surface film of as-extruded AZx1-aCa-bY alloy. However, the corrosion rate of AZx1-aCa-bY alloy was similar to these of AZx1 alloy in salt spray test which is strongly affected by properties of surface film. It seems that the microstructural change of phases is more dominant factor than the protectiveness of surface film on corrosion behaviors.

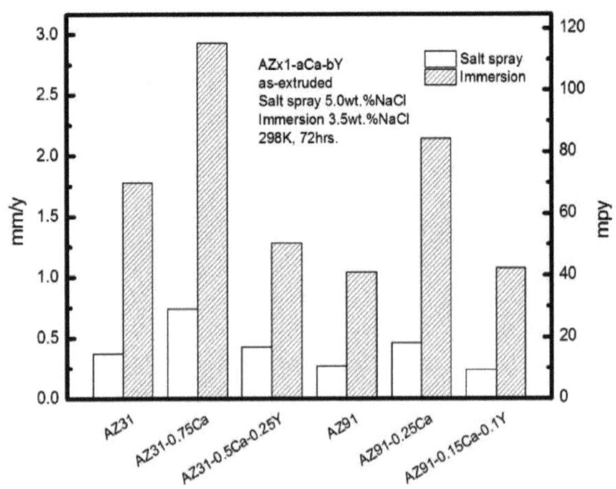

Figure 5. Average corrosion rates of as-extruded AZx1-aCa-bY alloys after salt spray and immersion tests in NaCl solution.

The corrosion rates of AZ91 series alloys are slower than those of AZ31 series alloys because β-$Mg_{17}Al_{12}$ phase acts as barrier to propagation of corrosion [9]. In immersion test, some improvement of the corrosion rate observed between AZ31 and AZ31-0.5Ca-0.25Y alloy, whereas comparable between AZ91 and AZ91-0.15Ca-0.1Y alloy. From this result, it is assumed that the precipitation of β-$Mg_{17}Al_{12}$ phase weaken the beneficial effect of change of phases by addition of Ca and Y.
The combined addition of Ca and Y resulted in slight increase of yield strength. The effect of combined addition of Ca and Y on elongation was different according to the base alloy. In case of AZ31 alloy, the elongation increased but it decreased in AZ91 alloy by the combined addition of Ca and Y.

Conclusions

In this study, the corrosion behaviors of as-extruded AZx1-aCa-bY alloys were systemically evaluated by immersion and salt spray tests. The sort and volume fraction of second phases were changed by addition of Ca or combined addition of Ca and Y and then these microstructural changes affected the corrosion behaviors. The results of this study show that the corrosion resistance of AZ series alloys can be improved by combined addition of Ca and Y instead of addition of Ca only. It means that the AZ series alloys containing optimized amounts of Ca and Y have the excellent combination of ignition resistance, corrosion resistance and mechanical properties.

References

1. M. Sakamoto et al., "Suppression of Ignition and Burning of Molten Mg Alloys by Ca Bearing Stable Oxide Film," *Journal of Materials Science Letters*, 16 (12) (1997), 1048-1050.
2. B.S. You et al., "Effect of Calcium Additions on the Oxidation Behavior in Magnesium Alloys," *Scripta Materialia*, 42 (11) (2000), 1089-1094.
3. B.H. Choi et al., "Effect of Ca Addition on the Oxidation Resistance of AZ91 Magnesium Alloys at Elevated Temperatures," *Metals and Materials International*, 9 (4) (2003), 395-398.
4. F. Czerwinski, "Controlling the Ignition and Flammability of Magnesium for Aerospace Applications," *Corrosion Science*, 86 (2014) 1-16.
5. Y.M. Kim et al., "Mechanical Properties and High-Temperature Oxidation Behavior of Mg-Al-Zn-Ca-Y Magnesium Alloys" (Paper presented at the Magnesium Technology 2012 - TMS 2012 Annual Meeting and Exhibition, Orlando, FL, U.S.A., 11-15 March 2012), 217-219.
6. B.S. You et al., "Oxidation and Corrosion Behavior of Non-Flammable Magnesium Alloys Containing Ca and Y" (Paper presented at the Magnesium Technology 2014 - TMS 2014 Annual Meeting and Exhibition, San Diego, CA, U.S.A., 16-20 February 2014), 325-329
7. Y.M. Kim et al., "Effect of the Ca Content on the Microstructural Evolution of Ca Containing AZ31 Cast Alloys," Metals and Materials International, 17 (2011), 583-586.
8. C.D. Yim et al., "Effect of Ca Addition on the Corrosion Resistance of Gravity Cast AZ31 Magnesium Alloy," *Materials Transactions*, 48 (5) (2007), 1023-1028.
9. G.L. Song, A. Atrens, "Understanding Magnesium Corrosion," *Advanced Engineering Materials*, 5 (12) (2003), 837-858.

CHARACTERIZATION OF COATINGS ON STEEL SELF-PIERCING RIVETS FOR USE WITH MAGNESIUM ALLOYS

Robert C. McCune[1], Joy H. Forsmark[2], Vinod Upadhyay[3] and Dante Battocchi[3]
[1]Robert C. McCune & Associates, LLC, West Bloomfield, MI, 48322, USA
[2] Ford Motor Company, Dearborn, MI, 48124, USA
[3]North Dakota State University, Fargo, ND, 58105, USA

Keywords: magnesium, galvanic corrosion

Abstract

Incorporation of magnesium alloys in self-pierce rivet (SPR) joints poses several unique challenges among which are the creation of spurious galvanic cells and aggravated corrosion of adjacent magnesium when coated steel rivets are employed. This work firstly reviews efforts on development of coatings to steel fasteners for the diminution of galvanic corrosion when used with magnesium alloys. Secondly, approaches, based on several electrochemical methods, for the measurement of the galvanic-limiting effect of a number of commercially-available coatings to hardened 10B37 steel self-piercing rivets inserted into alloy couples incorporating several grades of magnesium are reported. Electrochemical impedance spectroscopy (EIS), zero-resistance ammeter (ZRA), corrosion potential and potential-mapping visualization methods (e.g. scanning vibrating electrode technique – SVET) are illustrated for the several rivet coatings considered.

Introduction

<u>The Magnesium Front End Research and Development Project</u>

This work contributes to the U.S. Automotive Materials Partnership LLC (USAMP) Magnesium Front End Research and Development Project ("MFERD") which has previously been reviewed [1-3]. The project, begun in 2006, has the goal of developing both the knowledge base and enabling technologies that could permit greater deployment of magnesium alloys in automotive body substructures – of which the "front end" is exemplary. The project, with support of the U.S. Department of Energy, also includes an international collaboration involving organizations in Canada and the People's Republic of China.

<u>The Use of Steel Fasteners in Lightweight Auto Body Construction Incorporating Magnesium Alloys</u>

Of the challenges in developing "magnesium-intensive" substructures, the confluence of joining, corrosion and durability is notable, particularly with mixed materials of construction including steel and aluminum. Most automotive body construction employs fasteners in addition to weldments or adhesive joints. These may be various threaded fasteners, rivets or other point-attaching devices, typically made of hardened steel, for reasons of economy and strength. The self-piercing rivet (SPR) is typical.

Figure 1 shows a cross-sectional picture of a coated steel self-piercing rivet used to join die-cast AM60B magnesium to a 6082-T4 aluminum extrusion for this project. The technology of self-piercing rivets has recently been reviewed [4] and recent advances include the use of external heating to improve ductility of the magnesium [5] and development of finite-element methods to understand the local metal deformation process accompanying riveting [6]. For the particular case shown in Figure 1, no supplemental heating of the joint was necessary and material stacks can thus be judiciously designed with magnesium alloys to avoid this complication.

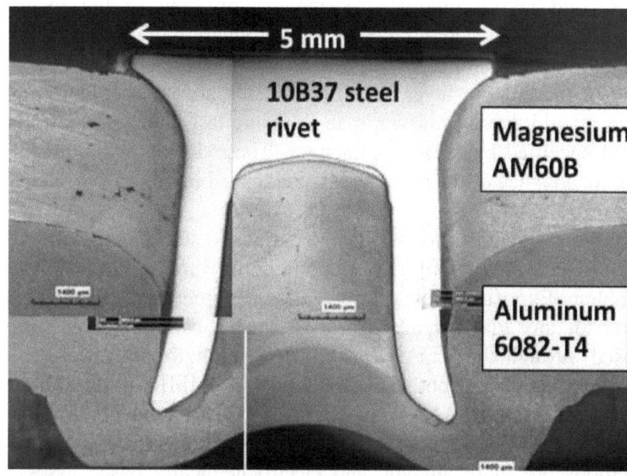

Figure 1. Cross-sectional image of a 5 mm self-piercing steel rivet joining an AM60B die casting to aluminum 6082-T4 extrusion.

Hardened and tempered steel (typically 10B37) remains the material of choice for SPRs; however, consideration has been given to the use of high-strength aluminum alloys such as 7278-T6 [7] for joining aluminum sheets, thereby reducing the degree of galvanic coupling expected with a steel-based rivet. One strategic option with steel fasteners is the use of coatings which then thwart the natural tendency for iron to act as a vigorous cathode when electrically connected to magnesium in the presence of an electrolyte. The means for assessment of the performance of candidate coatings to steel rivets in galvanic couples with magnesium is the focus of this work.

<u>Coating Technologies for Steel Fasteners</u>

Coatings to steel fasteners are usually intended to protect the steel of the fastener itself from corrosion. This is normally achieved through coatings that are either sacrificial relative to the steel of the fastener or provide a corrosion barrier. Such coatings don't necessarily provide the isolation and galvanic minimization desired for effective joints with magnesium, owing to the extreme sensitivity of magnesium to even very small amounts of iron capable of sustaining hydrogen reduction as the cathodic reaction. In various studies exploring conventional coatings to steel fasteners to minimize galvanic corrosion of adjacent magnesium alloys [8-10], insulating layers are usually more effective than metallic coatings in reduction of galvanic coupling. USAMP, in a prior project [11], explored the production and use of highly-insulating coatings achieved by physical or plasma-enhanced chemical vapor deposition coatings to steel substrates in order to achieve this end for use with magnesium.

Measurement, Analysis and Experiment Background

Galvanic corrosion in the vicinity of SPRs or other coated steel fasteners in magnesium can take several forms. In the most severe case (*cf.* Fig. 2), erosion of the adjacent magnesium can occur with attendant reduction in joint strength due to loss of section thickness and clamping support. In a study of strength loss for steel SPR rivets joining aluminum sheets, Howard and Sunday [12] actually observed little, if any, loss in lap-shear strength accompanying exposure of the test plates to 3.5% NaCl aqueous solution in the ASTM G44-75 procedure over a definitive period.

A second metric involves "cosmetic" corrosion of the forms shown in Fig. 3. Although basic joint strength may still be maintained, the appearance of white corrosion product and undercutting of the coating layer by filiform attack are unacceptable defects should they occur within an exposure test time and environment specified by each manufacturer. These defects would occur well in advance of the material loss evidenced in Fig.2.

Comparative studies for assessing coatings to steel fasteners used with magnesium [8-10], remain a straightforward approach to the problem. Exposure cycles can be tailored to meet acceptance criteria for any given manufacturer. A quantification scheme proposed through the VDA (Verband der Automobilindustrie) [13] has been employed both for studies within "MFERD" as well as the companion USAMP project [11] for visual and gravimetric measure of magnesium loss. Beyond such basic comparative methods are more analytic and electrochemically based assessments.

A monograph from ASTM [14] includes reviews of numerical methods emerging in the late 1980's including finite difference, finite element and boundary element methods to determine galvanic current flows for particular systems of materials and geometries.

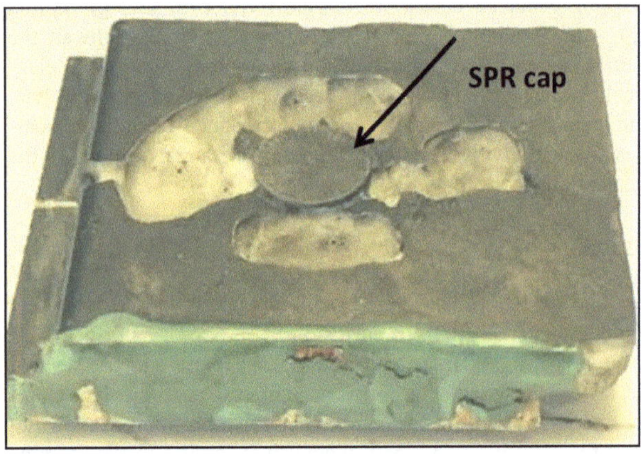

Figure 2. Loss of magnesium by galvanic attack near a self-piercing rivet head in a coated Mg-Al assembly.

More recent treatments of numerical methods aimed specifically at magnesium couples are by Verbrugge [15], Deshpande [16] and Hoeche and Isokovic [17]. Hoeche [18] recently showed a model specifically for an aluminum self-piercing rivet embedded in magnesium. Numerical methods ultimately rely on high-quality electrochemical data for quantification of the boundary conditions at the electrode surfaces.

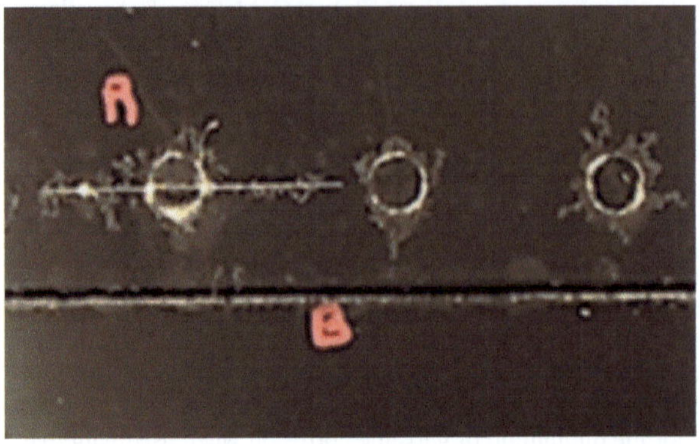

Figure 3. Cosmetic and filiform undercutting corrosion adjacent to self-piercing rivet caps in a polymer-coated overlap joint with AM60B magnesium surrounding the rivet head. Rivet marked with 'A' included a scribe through the paint layers above and near the rivet head.

Experimental approaches have been developed to probe the local environments where galvanic attack is occurring *in situ*. Microscopic probes of electrochemical parameters in the vicinity of steel fasteners in aluminum were demonstrated by Mizukami [19]. A segmented electrode technique was developed by Jia, Song and Atrens [20] to both determine corrosion currents at bimetal junctions as well as assess agreement with computational prediction. The scanning vibrating electrode technique (SVET) is among the more dramatic means for visualizing microsocopic current flows and electric fields above corroding metals, including galvanic couples *in situ* [21]. Deshpande [22] has demonstrated the capability of this approach for magnesium-steel couples as an adjunct to the numerical methods.

In this work, conventional electrochemical techniques (e.g. DC polarization, zero-resistance ammeter and impedance spectroscopy) as well as SVET were employed to assess both the nature of the coated rivet as an electrode and the degree of galvanic coupling between the rivets and surrounding magnesium. The remainder of this work discusses the rivet materials themselves, selected coatings, experimental procedures and results.

Experimental Procedures

Materials and Surface Treatments.

The rivets were provided by Henrob Corp. (New Hudson, MI), and were of a 5mm head diameter, constructed of 10B37 boron steel, heat treated to hardness of 47 R_C (*cf* Fig. 1). The coatings studied are listed in Table I along with test designations. (All testing at North Dakota State University was conducted 'blind.')

The 'baseline' coating supplied by Henrob for many applications is a Zn-Sn alloy applied by barrel (impact) plating. For this study, two versions of aluminizing were explored with the intent of producing largely inert coatings to the steel, either as pure aluminum or with supplemental surface treatments as described to improve its passivity. One version was prepared using ion-vapor deposition (IVD) aluminum supplied by Titanium Finishing Co. (East Greenville, PA) [23] to a nominal ('Level I') coating thickness of approximately 12 microns. IVD coatings of aluminum on fasteners are typically "tumbled" to densify the surface layers, and "thicker" ('Level II') coating layers are also possible at greater treatment times. The thicker coatings in this study were not found to offer improvements in galvanic corrosion over the basic coating. The second aluminizing process was provided by AlumiPlate Inc. (Minneapolis, MN). The AlumiPlate® process is a non-aqueous electroplating of aluminum to nominal thicknesses of approximately 25 microns. The coating may also be provided with a chromate conversion coating to increase passivity of the surface [24].

Both IVD and AlumiPlate®-coated rivets had versions which were further treated through an inorganic electrophoretic coating process "Alodine® EC² ™" (Electro-Ceramic Coating) provided by Henkel Corp. (Madison Hts., MI) [25]. This process, which is electrical in nature, promotes the development of a ceramic-like and highly insulating layer on the surface of aluminum, for which it was designed. This then necessitates the pre-coating of the steel rivet with aluminum to act as a 'platform' for deposition of the EC² ™ layer. A final layer of a polymeric infiltrant may then be incorporated to assure closure of any porosity in the ceramic-like layer, thereby limiting exposure of any base metal. A schematization of the layer sequences for this type of composite surface treatment is shown in Figure 4. This arrangement would be typified by sample codes 'C' and 'H' in Table I. Operationally, this layering is intended to produce a highly insulating surface coating as was envisioned in the prior USAMP study [11].

Table I. Summary of rivet coating technologies and sample coding designations.

Code	Description
A	AlumiPlate® with chromate conversion coating
B	AlumiPlate® coating only
C	AlumiPlate® coating with Henkel EC² layer and sealant
D	Uncoated (bare) rivet, as heat treated
E	Henrob Zn-Sn alloy baseline barrel coating
F	Ion-vapor Deposited (IVD) Aluminum coating only
G	IVD Aluminum coating with Henkel EC² layer
H	IVD Aluminum coating with Henkel EC² layer and sealant

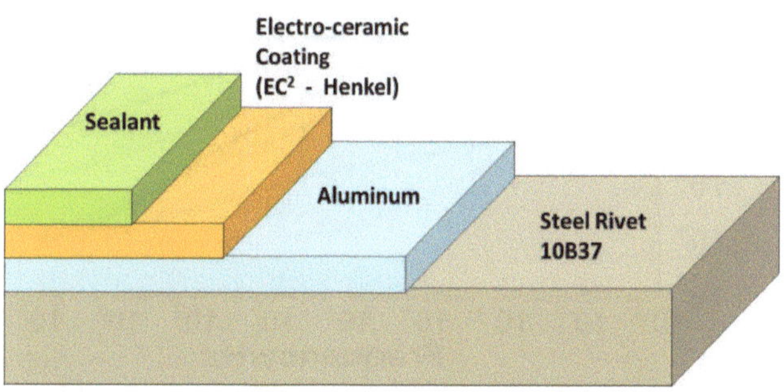

Figure 4. Schematization of composite insulating surface layer on aluminized steel rivets.

Electrochemical Methods

All electrochemical testing was conducted at the Center for Surface Protection of North Dakota State University. Four basic protocols aimed primarily at the steel rivet coating behaviors are described below.

Potentiodynamic Scanning and Electrochemical Impedance Spectroscopy EIS). Potentiodynamic scanning and EIS were performed using a conventional three-electrode cell where the rivet acted as the working electrode, a platinum mesh as the counter electrode, and a saturated calomel electrode (SCE) was the reference electrode. The electrolyte used was aqueous 3.5 wt. % NaCl. The aim of these measurements was to assess the degree of metallic character and corrosion potential of the rivet itself relative to known references and influence of any coatings.

A Gamry IFC 1000 Potentiostat/Galvanostat/ZRA was used for all the electrochemical measurements. Data were acquired using Gamry software, framework version 6.21. A Perspex™ glass cylinder mounted on a plastic board and clamped with an O-ring insert was used to hold the electrolyte. For EIS [26], impedance response from 10^{-2} Hz to 10^{5} Hz was measured at a potential perturbation of 10mV vs open circuit potential (OCP). Data were acquired at the rate of 10 points per decade. Potentiodynamic scans (PDS) were obtained by scanning the potential of the working electrode from -0.5 to +1.5 Volts with respect to the OCP at a scan rate of 1mV/sec. PDS data was used primarily as an indicator of the electrode characteristics for the rivets, which were always cathodic relative to the surrounding magnesium.

Zero-resistance Ammeter (ZRA). The galvanic corrosion current measurement between rivet and die-cast magnesium used the two-compartment arrangement shown in Figure 5. One compartment consisted of a beaker filled with the electrolyte (also 3.5% NaCl) into which the SPR of interest was placed as a working electrode (WE-1) along with a saturated calomel electrode (SCE) as reference. Care was taken to permit only the exterior cap and shank portion of the rivet to be exposed to electrolyte. A second compartment consisted of a die-cast AM60B Mg plate as working electrode (WE-2) mounted with a Perspex™ cylinder and clamped with an O-ring insert to contain the electrolyte. An agar salt bridge was used to join the two compartments for purposes of directing current flow through the ammeter, while permitting communication of potential and ionic exchange between the half cells. The potentiostat was operated using the zero resistance ammeter (ZRA) mode [27], ensuring that the two working electrodes behave as if they were connected by a zero resistance shunt, thereby facilitating the monitoring of galvanic current and combined potential of the couple.

Scanning Vibrating Electrode Technique (SVET). SVET experiments were performed to measure the current density (CD) distribution over the surface of the metal/electrolyte interface at the free corrosion potential. This technique uses a vibrating electrode to scan a sample surface and then detects the ohmic potential drop produced by local currents generated from actively corroding surfaces immersed in an electrolyte [21]. From the variation in potentials, regions of varying current densities can be located and measured. For these experiments, the rivet of interest was potted into a pre-machined flat piece of AM30 wrought magnesium, to produce a coplanar surface with rivet head and surrounding magnesium. Wrought AM30 was chosen in this case for greater homogeneity in comparison to the die-cast AM60B.

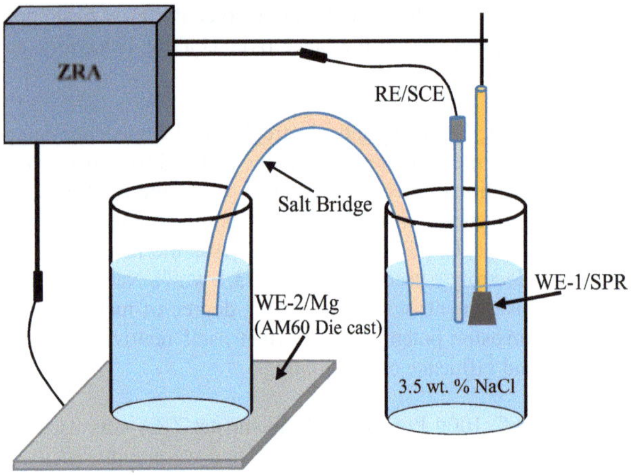

Figure 5. Schematization of the Zero-resistance Ammeter (ZRA) method employed for measurement of corrosion current between coated self-piercing rivets and AM60B die-cast magnesium.

SVET instrumentation was from Applicable Electronics, (Sandwich, MA, USA). The scanning probe was a Pt-Ir microelectrode with a 10μm diameter tip that was platinized to a 20μm sphere. Schematization of the approach is shown in Figure 6. The probe was placed 150 μm above the sample surface and vibrated with amplitude of 20 μm along the X and Y direction. A pair of platinized Pt wires was used as a reference and bath ground electrode respectively. Scans were initiated immediately after immersion in an equal volume mixture of 0.05 wt. % NaCl and 0.05 wt. % $MgSO_4$.The probe made a 20x20 measurements in each scan that lasted 6 minutes approximately. An area of approximately 3 x 3 mm was exposed for scanning, that consisted of part of rivet, Mg bulk, and the rivet-Mg interface. The unscanned region was masked with nail polish and transparent tape on top to define the area trapped by the thin layer of electrolyte.

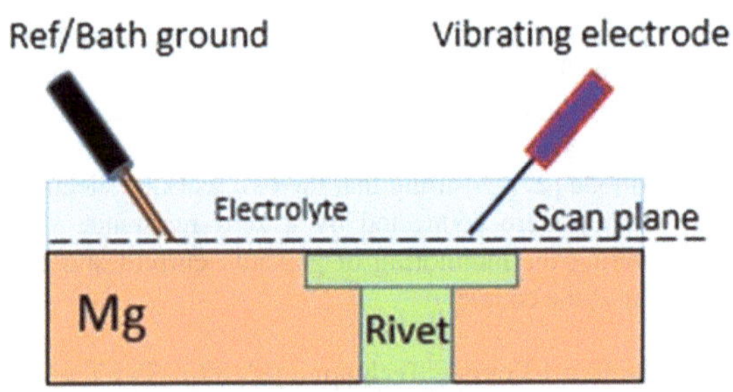

Figure 6. Schematization of the SVET technique.

Results and Discussion

Open Circuit and DC Polarization

Open-circuit (corrosion) potentials for the metallic surfaces of interest (including reference materials), using the single cell arrangement with reference electrode are compared in Table II for both rivets permitted to equilibrate at the open circuit (corrosion potential), and the Ecorr deduced from DC polarization scans using fitting routines. Data from aluminum (99.9%) and magnesium (AM60B die cast) reference materials are also included.

Table II. Open circuit and Ecorr values for selected rivets and coatings.

Sample Designation	Description	OCP (V vs SCE) (steady state)	E_{corr} (V vs SCE) (polarization)
B	steel + AlumiPlate	-1.16	-1.00 ± 0.07
C	steel+ AlumiPlate+ EC^2 + sealant	-1.09	-1.05 ± 0.09
D	uncoated steel	-0.64	-0.67 ± 0.17
E	Zn-Sn-coated steel	-1.04	-1.22 ± 0.09
F	steel + IVD Al coat	-0.73	-0.71 ± 0.04
aluminum	reference material	- 0.78	N/A
magnesium	reference (AM60B)	- 1.59	N/A

Electrochemical Impedance Spectra

The impedance Bode plots (i.e. impedance modulus vs frequency) for the various rivet coatings from Table I are shown in Figure 7. The AlumiPlate® -treated rivets with EC^2 layer and sealant showed the most dielectric-like behavior, evidenced from the low-frequency impedance in excess of 10^8 Ω. Although similarly processed, the IVD aluminum sequence showed convergence with this only at the highest frequency, suggesting a "leakier" capacitance at low frequencies. As suspected, the uncoated rivet and baseline Zn-Sn coating were most metallic-like in character, showing least frequency dependence. Comparison of the 'Level I' IVD coated rivet (F) with the same rivet receiving only the EC^2 layer suggested a diminution of low-frequency impedance characteristic of some breakdown of the protective capability of the aluminization. This was also confirmed by observation of corrosive attack to the underlying steel during EC^2 processing by Henkel. Such breakdown of the aluminizing layer was not observed when EC^2 was applied to the AlumiPlate® coating, suggestive of the thicker and less porous layer.

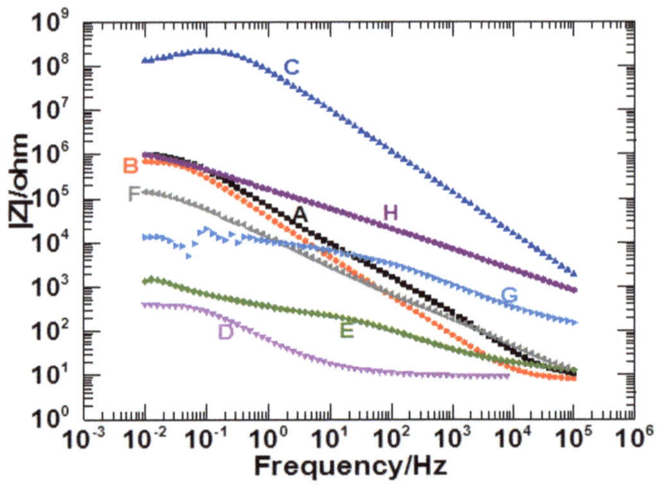

Figure 7. Summary of EIS spectra (Bode Plot) for rivet coatings summarized in Table I.

Galvanic Coupling Current

The galvanic currents at steady state between the various coated rivets and AM60B magnesium test coupon are summarized in Table III. Additionally, the corrosion appearances of example magnesium plates under the cathodic influence of particular rivets, at the end of exposure are shown in Figure 8, wherein least corrosion occurred for the composite coating 'C' and greatest for the case of the uncoated rivet 'D.' The galvanic coupling current for coating 'C' was found to be almost seven decades lower than for the bare steel rivet. Coating 'H' which included the same process steps as 'C' but with the more porous aluminizing did not exhibit the same degree of insulating behavior. For coating 'C', the cell potential shows a value near the open circuit potential for AM60B suggesting the cell voltage to be dominated by the anodic reaction of the magnesium corrosion, the cathode in this case being mostly inert.

Table III Summary of galvanic coupling currents between rivet (coating) and AM60B magnesium plate.

Sample Designation	Steady-state Galvanic Current (A)	Steady-state cell potential V vs SCE
A	**1.23 E-04**	**-1.11**
B	**1.46 E-04**	**-1.13**
C	**3.12 E-10**	**-1.56**
D	**4.25 E-04**	**-1.13**
E	**7.19 E-05**	**-1.38**
F	**1.53 E-04**	**-1.09**
G	**3.25 E-04**	**-1.19**
H	**1.07 E-04**	**-1.26**

Figure 8 illustrates the appearance of the anodic component of the galvanic couple (i.e. AM60B magnesium) following the test run couples to the rivet of interest. Sample 'C' shows a minimally corroded surface consistent with the highly insulating coating to the rivet, while 'D' shows the worst case galvanic coupling to an uncoated steel rivet.

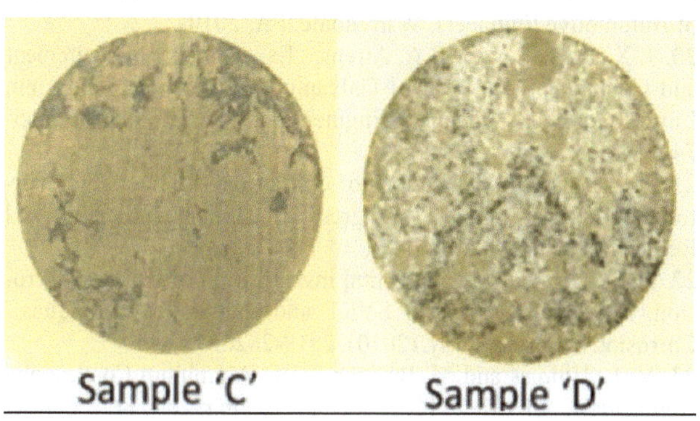

Figure 8. Appearance of the AM60B anodic surfaces (magnesium) following galvanic coupling to the particular rivet configurations shown.

SVET Observations

Figure 9 illustrates the approach to the SVET measurements of current densities in the electrolyte for the coated rivets as potted into wrought AM30 magnesium plate. The inset illustrates the data collecting matrix and image of the vibrating probe tip for the baseline Zn-Sn coated rivet. As corrosion of the magnesium progresses, hydrogen bubbling is also observed primarily on the cathodic (rivet) surface, but also at the junction with magnesium. SVET indicated some anodic activity on the rivet itself, as well as the adjacent magnesium. Comparison of the Zn-Sn (code 'E') coating with the bare rivet (code 'D') indicated almost a 50% reduction in anodic current at the magnesium surface when the coating is employed.

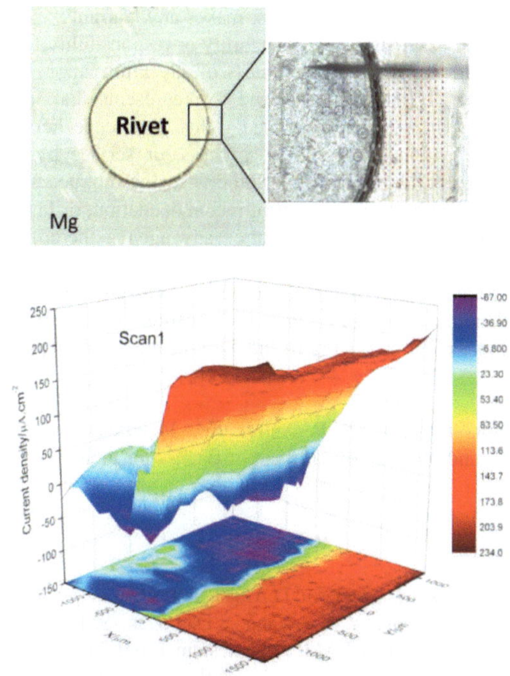

Figure 9. Images of selected area for SVET probe, test matrix points (inset) and map of current density in the electrolyte above the metal surface (negative values are cathodic, positive are anodic).

Summary and Conclusions

Of the several approaches taken here for characterization of coatings to steel rivets for the purpose of defusing galvanic corrosion when in proximity to magnesium, the following points are underscored:

a.) The "open circuit" or free corrosion potential of the coated rivet itself can give an estimation of the metallic cathode potential for the electrolyte selected – e.g. whether it is more like iron, aluminum or other alloy (e.g. Zn).

b.) EIS of the rivet coating is most useful in assessing the degree to which such coating creates a dielectric or insulating layer relative to more metallic conductors. This feature of the rivet coating is seen to be advantageous for the application. Additional kinetic studies of the breakdown of the dielectric layer are possible.

c.) The galvanic coupling current (density) as measured by the ZRA configuration provides an estimate of the severity of worst-case coupled corrosion for the separated electrodes. The more insulating or galvanic-limiting coatings show measurable reduction in this mode.

d.) The SVET mode gives a visualization of the degree of galvanic attack for the particular geometry of the embedded rivet and magnitude of galvanic current through the thin layer of electrolyte.

Acknowledgments

This material is based upon work supported by the Department of Energy, National Energy Technology Laboratory under Award Number No. DE-EE0005660. This report was prepared as an account of work sponsored by an agency of the United States Government. Neither the United States Government nor any agency thereof, nor any of their employees, makes any warranty, express or implied, or assumes any legal liability or responsibility for the accuracy, completeness, or usefulness of any information, apparatus, product, or process disclosed, or represents that its use would not infringe privately owned rights. Reference herein to any specific commercial product, process, or service by trade name, trademark, manufacturer, or otherwise does not necessarily constitute or imply its endorsement, recommendation, or favoring by the United States Government or any agency thereof. The views and opinions of authors expressed herein do not necessarily state or reflect those of the United States Government or any agency thereof. Such support does not constitute an endorsement by the Department of Energy of the work or the views expressed herein.

The authors thank the U.S. Automotive Materials Partnership (USAMP) and the U.S. Department of Energy for their support of this work.

References

1. J.H. Forsmark, M. Li, X. Su, D.A. Wagner, J. Zindel, A.A. Luo, J.F. Quinn, R. Verma, Y.-M. Wang, S.D. Logan, S. Bilkhu, and R.C. McCune, "The USAMP Magnesium Front End Research and Development Project – Results of the Magnesium "Demonstration Structure", Magnesium Technology 2014, eds. Martyn Alderman, Michele V. Manuel, Norbert Hort, and Neale R. Neelameggham, TMS (The Minerals, Metals and Materials Society), Warrendale, PA, 2014) 517-524.

2. A. A. Luo, J. F. Quinn, R. Verma, Y.-M. Wang, T.M. Lee, D. A. Wagner, J.H. Forsmark, X. Su, J. Zindel, M. Li, S.D. Logan. S. Bilkhu and R. C. McCune, "The USAMP Magnesium Research and Development Project," Light Metal Age, 2012, (4), 54-58.

3. A.A. Luo, E.A. Nyberg, K. Sadayappan and W. Shi, "Magnesium Front End Research and Development: A Canada-China-USA Collaboration", Magnesium Technology 2008, eds. M.O. Pekguleryuz, et al, TMS, Warrendale, PA, 2008), 3-10.

4. A. Chrysanthou and X. Sun, eds., Self-piercing riveting - Properties, processing and applications, (Oxford, Woodhead Publishing Ltd. 2014).

5. Y. Durandet, R. Deam, A. Beer, W. Song, S. Blacket, "Laser assisted self-pierce riveting of AZ31 magnesium alloy strips", Materials and Design, 31, SUPPL. 1, 2010, S13-S16.

6. Y. Abe, T. Kato and K. Muri, "Joinability of aluminum alloy and mild steel sheets by self piercing rivet," J. Materials Processing Technology,177 (2006), 417-421.

7. N.-H. Hoang, R. Porcaro , M. Langseth , A.-G. Hanssen, "Self-piercing riveting connections using aluminium rivets," International Journal of Solids and Structures, 47, (2010), 427-439.

8. David Tawil, Magnesium Elektron Corp. Presentation: "The Principles of Magnesium Corrosion Protection," (web site) http://www.magnesium-elek-ek-tron.com/data/downloads/Corrosion%20Protection%20Principles.pdf

9. David Hawke and Thomas Ruden, "Galvanic Compatibility of Coated Steel Fasteners with Magnesium," SAE Transactions: Journal of Materials & Manufacturing, 104, (1995), 358-362. (SAE Paper #950429).

10. Jan Ivar Skar, Malte Isacsson and Zheng Tan, "Galvanic Corrosion of Die Cast Magnesium Exposed on Vehicles and in Accelerated Laboratory Tests," SAE Paper No. 2006-01-0255, Society of Automotive Engineers, Warrendale, PA, USA (2006).

11. http://energy.gov/sites/prod/files/2014/04/f14/4_usamp_cooperative_research.pdf

12. R. M. Howard and S.P. Sunday, "The Corrosion Performance of Steel Self-Piercing Rivets When Used with Aluminum Components," SAE Paper 831816, (Society of Automotive Engineers, Warrendale, PA), 1983.

13. Verband der Automobilindustrie (VDA) Test sheet: "Checking of the contact corrosion behavior of screwed joints in magnesium alloys," Proof copy, June 1999.

14. Harvey P. Hack, ed., "Galvanic Corrosion" ASTM Special Technical Publication 978, (American Society for Testing and Materials, Philadelphia, PA, 1988).

15. M. Verbrugge, "Galvanic Corrosion over a Semi-infinite Planar Surface," Corrosion Science, 48, (2006), 3489–3512.

16. K. Desphande, "Validated numerical modelling of galvanic corrosion for couples: Magnesium alloy (AE44)–mild steel and AE44–aluminium alloy (AA6063) in brine solution, Corrosion Science, 52, (2010), 3514-3522.

17. D. Hoeche and J. Isakovic, "Level-set modeling of galvanic corrosion of magnesium," in Mg2012 Conference, Vancouver, BC, Canada, July 8-13, 2012 Proceedings: Symposium Simulation and Modeling, 2012.

18. D. Hoeche, "Towards the simulation based design of Mg/Al hybrid component joints in terms of corrosion prevention," Proceedings EuroCORR Conference, ISBN: 978-989-98850-0-4, European Federation of Corrosion Event 343, September 1 - 5, Estoril, Portugal 2013.

19. T. Mizukami, "Simulation of Galvanic Corrosion of Aluminum Materials for Vehicles," SAE Paper 2010-01-0724, (Society of Automotive Engineers, Warrendale, PA, 2010).

20. J. X. Jia, G.Song and A. Atrens, "Experimental Measurement and Computer Simulation of Galvanic Corrosion of Magnesium Coupled to Steel," Advanced Engineering Materials,9, (2007), 65-74.

21. H. S. Isaacs, "Applications of current measurement over corroding metallic surfaces," Progress in Clinical and Biological Research, 210, (1986), 37-44.

22. K. Deshpande, "Experimental investigation of galvanic corrosion: Comparison between SVET and immersion techniques," Corrosion Science, 52 [9], (2010), 2819-2826.

23. V. L. Holmes and M. Bridger, "IVD Aluminum Coating and Application of the Process at Boeing – St. Louis," http://www.ipsenusa.com/Documents/Ipsen-USA/Technical-Papers/IVD-Paper.pdf

24.http://preview.alumiplate.com/wp-content/uploads/AlumiPlate-Brochure-.pdf

25.http://www.henkelna.com/us/content_data/59949_lt4958alodineec2electroceramiccoatings.pdf

26. F. Mansfeld, "Use of electrochemical impedance spectroscopy for the study of corrosion protection by polymer coatings," J. Appl. Electrochem., 25 [3], (1995),187–202.

27. J. Wolstenholme, "Inexpensive Zero-Resistance Ammeter for Galvanic Studies", British Corrosion Journal, 9 [2],(1974) 116 – 117.

Magnesium Technology 2015
Edited by: Michele V. Manuel, Alok Singh, Martyn Alderman, and Neale R. Neelameggham
TMS (The Minerals, Metals & Materials Society), 2015

Investigation of Coating and Corrosion Mitigation Strategies in Magnesium/Mixed Metal Assemblies

Joy H. Forsmark[1], Robert C. McCune[2], Terry Giles[3], Michelle Audette[3], Jasmine Snowden[3], Jeff Stalker[4], Matthew Morey[5], Matt O'Keefe[6], and Carlos Castano[6]
[1] Ford Motor Company, Dearborn, MI, 48121, USA
[2] Robert McCune and Associates, West Bloomfield, MI 48322, USA
[3] Henkel North America, Madison Heights, MI 48071, USA
[4] PPG Industries, Inc., Allison Park, PA 15101, USA
[5] Atotech Worldwide, Rock Hill, SC 29730, USA
[6] Missouri S&T, Rolla, MO 65409, USA

Keywords: Magnesium alloys, Mg corrosion, automotive lightweighting

Abstract

The US Automotive Materials Partnership through the Magnesium-Intensive Front End Development Project (MFERD) is currently investigating a number of joining, coating and corrosion mitigation strategies to incorporate magnesium components into the automotive body-in-white with the ultimate goal of decreasing vehicle curb weight, thus improving fuel economy. Because Mg is anodic to all other structural metals, this is a key hurdle to Mg component implementation in vehicles. This paper will discuss the results of a study to examine the effectiveness of different corrosion mitigation strategies in joined plate assemblies and provide some insight into the systems challenges of incorporation of Mg parts into a vehicle. Details of a statistically-designed experiment developed to explore the interaction of several materials of construction (magnesium, steel and aluminum), pretreatment and topcoatings, joining methods and standardized test protocols including SAE J-2334 and ASTM B-117 are discussed. A number of avenues have emerged from this study as potential strategies for corrosion mitigation.

Introduction

The US Automotive Materials Partnership (USAMP), the materials arm of the US Council for Automotive Research (USACAR), is a consortium of Chrysler Group, Ford Motor Company, and General Motors Company with the purpose of promoting collaborative activities on "pre-competitive" materials research. A major focus of the research in recent years has been in the area of light-weight materials: specifically aluminum and magnesium alloys. In 2007, USAMP initiated a project to examine the challenges and provide strategies to utilize Mg alloy components in automotive body applications, the Mg Front End Research and Development Program (MFERD). The first two phases of the project examined issues associated with an all-Mg front end structure by examining design strategies that accounted for the different mechanical and physical properties of Mg from traditional automotive structural materials[1].

One outcome of Phases 1 and 2 of MFERD was the acknowledgement that incorporation of Mg components into the body-in-white, at least in the near term, would be a mixed materials exercise. Therefore, a major focus of the MFERD Phase 3 project has been to investigate the issues associated with joining Mg to dissimilar materials. Because the standard reduction potential of Mg is far below that of Al and Fe, the tendency for galvanic corrosion at joint areas is very strong. Much of the focus in this program has been to examine different combinations of joining and coating methods. Isolation strategies included coating the base metal and the fastener; the role of adhesives in corrosion protection was also examined. This paper describes the development and test results of a coupon matrix used for initial investigation of joining, coating, and corrosion mitigation strategies for multi-material structures incorporating Mg.

Background

Current Practice of Corrosion Mitigation in Automotive Mg Components

There are several Mg alloy components currently in service in applications where corrosion is a concern because they are external parts that are exposed to the environment. In the case of the Mg suspension cross member for the 2006 Corvette Z06, the components were installed uncoated on the vehicle and the primary galvanic corrosion mitigation strategy was the use of external isolation at the joining interfaces. Aluminum alloy washers and mylar sheet were used as isolation techniques in this application [2]. This part was joined to the body-in-white after much of the vehicle had been built and painted.

Other Mg components such as the 2004 F150 Grill Opening Reinforcement bracket [3], the Mg spare tire carrier on the 2007 Jeep Wrangler or the 2011 Lincoln MKT rear liftgate door inner use a protective coating as part of the corrosion mitigation strategy. The coating process consists of a surface cleaning and pretreatment followed by the application of a powder epoxy layer. The parts are then bolted using Al isolators and additionally, in the case of the door inner, adhesively bonded to the rest of the vehicle. The coating systems for Mg parts perform three important functions: 1) provides additional corrosion protection in service for the part, 2) provides a "Class A" cosmetic surface appearance in components that are externally visible and 3) isolates the Mg component from the rest of the paint shop for components that must be installed on the body-in-white prior to painting. The third point is critical because a conventional automotive paint shop is designed primarily for steel and utilizes cleaning steps that are incompatible with Mg, potentially damaging the component and contaminating the pretreatment bathes.

Design Approach in the MFERD Program

Thus far, all of the external Mg component applications that are in production are single component castings that are bolted on to the body-in-white using fasteners with isolators. However, if Mg components are to make further inroads into body applications, there will need to be more use of alternative, high-volume, joining methods that cannot allow for isolating washers and the part itself will be expected to go through a traditional paint shop which may or may not be compatible with the Mg alloy.

Recently, self piercing riveting (SPR) and friction stir welding (FSW) techniques have had commercial applications within the body-in-white, chassis and closure applications in aluminum alloys [4-6]. These techniques, along with adhesive, also show some promise in applications using Mg alloys and were utilized successfully in joining dissimilar Mg components in MFERD phase 2 [1]. Both of these joining techniques are being investigated in the current program.

Automotive bodies are painted, in part, to mitigate corrosion of the base metal and joined surfaces. An idealized corrosion protection system for Mg is shown in Figure 1. In general, a base metal will go through a cleaning step, pretreatment, polymeric primer (e.g. electrocoat), and then finally application of topcoats. The polymer topcoat and paint steps require a short, low temperature cure or "bake". If adhesive is to be applied to any surface, the pretreatment must be applied to the bare metal surface in order for sufficient adhesion to take place. The entire painting process must be compatible with all materials in an assembly. Thus the goals of the coupon matrix experiments were twofold: 1) examine commercially-available pretreatment systems and protocols for their compatibility with multi-material assemblies that include Mg alloys and 2) examine various joining/coating strategies for corrosion mitigation capability those assemblies.

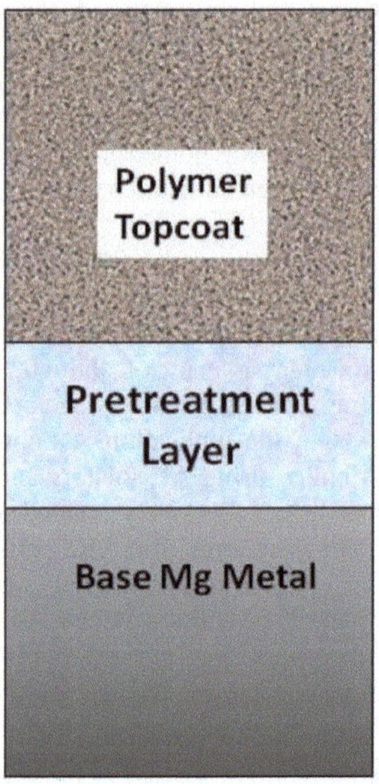

Figure 1. Idealized corrosion protection system showing the Mg base material, a pretreatment layer to improve adhesion for an adhesive or topcoat, and a polymer topcoat layer for environmental protection or surface quality

Experimental Procedure

To examine different coating and corrosion mitigation strategies, a matrix of plate assemblies was created (Table 1). Two and three plate assemblies were created from the following base plates (dimensions 101 mm x 152 mm): a) AM60 high pressure die cast plates produced by ACT Test Panels with a 3.0 mm nominal thickness; b) Al alloy 6082 extruded plates produced by Kaiser Aluminum with a nominal thickness of 3.0 mm and used in the T4 condition; c) Al alloy 6022 sheet with a nominal thickness of 1.5 mm in the T4E40 condition; and d) HSLA 350 electrogalvanized steel sheet with a nominal thickness of 1.0 mm.

Three joining methods were examined: self piercing rivets (SPR), friction stir linear welding (FSLW) and conventional breakstem rivets with predrilled holes (both Al and Steel rivets examined). In addition, Henkel Terokal® 5089 one-part adhesive was applied to some samples as part of the SPR or FSLW joining process; all break stem rivet joints utilized adhesive. The plate assemblies consisted of AM60 joined to 6082 Al using SPRs (produced at Henrob Corp. in Wixom, MI), AM60 joined to HSLA galvanized steel using adhesive and break stem open rivets (produced at Stanley Engineered Fastening, Novi, MI) and AM60 joined to 6022 Al using FSLW (produced at Hitachi NA, Farmington, MI). The full matrix is shown in Table 1. Figures 2a, b, and c indicate the plate assembly types and stack-ups. The overlap of the joint area for the plate assemblies was 30 mm. As indicated in Table 1, Groups A and B were essentially similar with the exception of the SPR coating. Group D assemblies utilized an aluminum standard rivet at the top of the joint and a steel standard rivet at the lower joint area. These samples enabled an evaluation of rivet material and coating on the corrosion mitigation of Mg assemblies.

Four different chromium-free, zirconium containing pretreatments provided by Atotech, Henkel, and PPG were investigated as initial coatings for the assemblies. In this work, they are designated, in no particular order, as J, K, L, and M. The assemblies then had a final top coat of PPG 590-534 electrocoat (ecoat) or ProTech ES542N49 Epoxy Powder coat. Table 2 shows the matrix of treatments and plate assembly designation. The plate assemblies were all treated after final joining. Pretreatment J is only designated for use with Mg and Al alloys so it was only applied to the 2-plate assemblies. The other pretreatments are also compatible with electrogalvanized steel so they were used on all assembly combinations including Group D (3-plate assemblies). All of the pretreatment operations were performed at the supplying companies and the plate assemblies were then transferred to PPG Industries, Allison Park, PA (ecoat) or Almond Products, Spring Lake, MI (powder coat) for final top coating. Plate assemblies that included adhesive required an initial pretreatment prior to joining to ensure that the adhesive would adhere to the surface. Care was taken to ensure that samples were reprocessed using the same pretreatment after assembly.

Table 1. Plate assembly designations and joints

Designation	Plate Order	Joint #1	Joint #2
A	Mg/Al	SPR (Zn/Sn coating)	NA
B	Mg/Al	SPR (Al coating)	NA
C	Mg/Al	SPR (Al coating) + adhesive	NA
D	Al/Mg/galvanized Steel	SPR (Al coating)	Breakstem Rivets + adhesive
E	Al/Mg	FSW	NA
F	Al/Mg	FSW + adhesive	NA

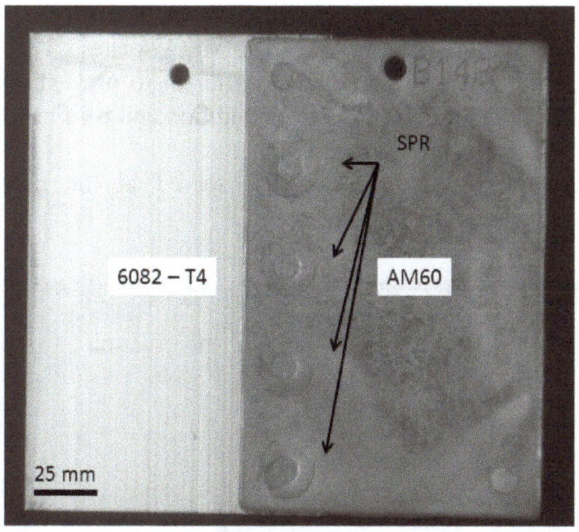

Figure 2a. Uncoated 2 plate assembly using SPR (note: AM60 is the top plate)

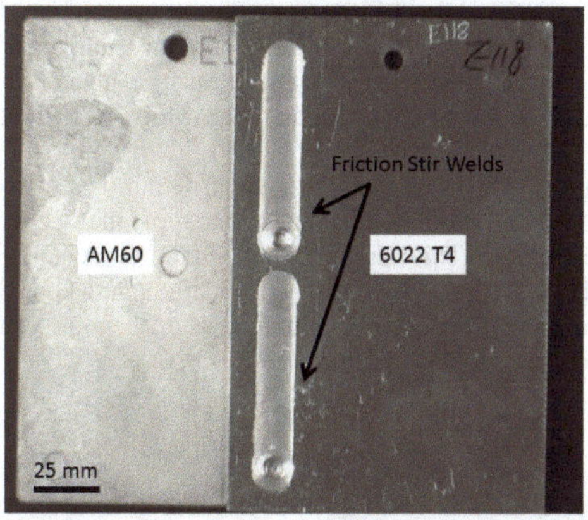

Figure 2b. 2 plate assembly using FSW (note: Al 6022 is the top plate)

Figure 2c. 3 plate assembly configuration consisting of 6082-T4 extruded Al plate, AM60 die cast Mg plate, and 350G electrogalvanized HSLA steel sheet

Samples that utilized adhesive also went through a curing heat treatment of 180 °C for 30 minutes. The assemblies were placed vertically on a rack approximately 10 to 15 mm apart and heat treated in batches.

Once the manufacture of the plate assemblies was complete, three replicate samples of each condition were sent to Henkel and PPG Industries for testing under SAE J2334 [7] or ASTM B-117 [8] laboratory corrosion protocols, respectively. SAE J2334 is a cyclic corrosion test, designed to expose a sample under accelerated automotive conditions seen in the field. Under this protocol, all samples were placed on a rack at the same orientation with the rivet head or FSLW keyhole facing on the angled side up. The racks were placed in a humidity controlled chamber at 50°C and once every 24 hours the samples were exposed to a humidity cycle for 6 hours, removed and immersed in a salt solution for 15 minutes and then placed back in the chamber where they were exposed to a dry cycle for 17 hours and 45 minutes. Each cycle was conducted once a day for five days, with a dry soak over the weekend. A total of 60 cycles were run. Each plate was also scribed according to ASTM standard D1654[9] with a horizontal scribe mark (one rivet was also scribed). One replicate from each condition was subjected to a scribe evaluation to examine the quality of the coating process.

ASTM B-117 is a severe salt fog exposure test that has been used to examine corrosion in joint areas. In this test, the samples were continuously exposed to a fog of 5% NaCl solution inside a chamber. This test was extremely severe and all plate assemblies were pulled from the test chamber after 500 hours of exposure.

All joint areas were examined and documented for corrosion at the fasteners and joints and also at the overlap of the plates and the interface areas. Samples from the SAE J2334 tests were sent to Missouri S&T for further evaluation and sectioning.

Table 2. Plate assembly coatings. EC= electrocoat, PC = powder coat

Designation	Pretreat before Assembly?	Pretreatment	Topcoat
A	No	J, K, L, M	EC, PC
B	No	J, K, L, M	EC, PC
C	Yes	J, K, L, M	EC, PC
D	Yes	K, L, M	EC, PC
E	No	J, K, L, M	EC, PC
F	Yes	J	EC, PC

Results

Effectiveness of Corrosion Mitigation Strategies on Base Metal Corrosion

Several interesting results were observed from this analysis. First, the scribe measurements gave an indication of the feasibility of multi-material coating strategies on the metal substrates. Figure 3 shows the scribe creepback measurements made on a subset of plate assemblies of the AM60 Mg plates after 60 cycles were run on SAE J2334 cyclic corrosion testing. The results indicate that corrosion did undercut the coating layers in most cases. Group B showed the highest levels of scribe creepback and ecoat versus power coat did not appear to make a strong difference in performance. One Zr-based pretreatment was much worse than the other pretreatments in the 2 plate assemblies (Group B and Group C). All of the Zr-based pretreatments appeared to perform the same in the 3 plate assemblies (Group D) that included the steel panel. In all cases, the Al panels indicated no creepback from the scribe. Additional measurements on the steel panels in Group D indicated a consistent scribe creepback measurement of 5-7 mm for all cases.

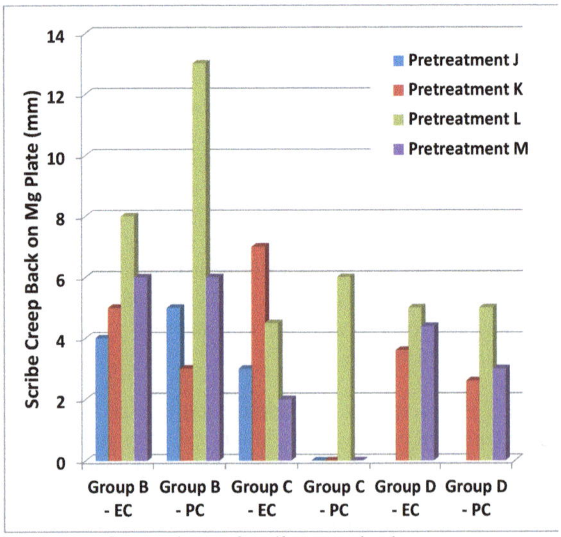

Figure 3. Comparison of scribe creepback measurements on Mg plates measured at the conclusion of 60 cycles of SAE J2334 corrosion testing.

Effectiveness of Corrosion Mitigation Strategies in Joint Regions

The joint areas were also examined for all the plate assemblies. In the cases of the SPR joints, some evidence of corrosion was apparent around the surface of the rivet head as can be seen in Figure 4. This was especially true for the ecoated samples but was also evident in some of the powder coated samples. Groups A, B, C, and D all showed white corrosion product around the SPR head. This occurred in both the conventional Zn/Sn coated rivets and for the Al coated rivets and also appeared regardless of pretreatment. The powder coating did appear to protect the SPR head from corrosion a little better than the ecoat. However, if the powder coat was scratched, white corrosion product formed. Figure 4 also shows that there is some corrosion evident at the plate overlap edges for the ecoated samples.

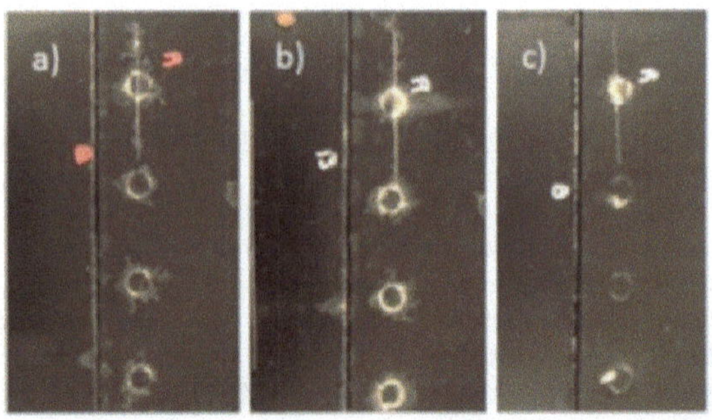

Figure 4. Examples of typical corrosion product observed around SPRs and at plate edge overlap areas in tested plate assemblies at 60 cycle SAE J2334. a) Group A with pretreat L, ecoat; b) Group B with pretreat L, ecoat; c) Group B with pretreat L, powder coat.

Figure 5 shows cross-sections of the rivets. The cross-sections confirm that most of the corrosion activity occurred at the rivet head/Mg plate interface. There was little to no corrosion occurring at the interface between the Mg/Al plates. The coating process, therefore, appeared to be sufficient to prevent corrosion internal to the lap joint for this test condition and configuration.

Some corrosion product is also observed at the plate edge interface for all samples.

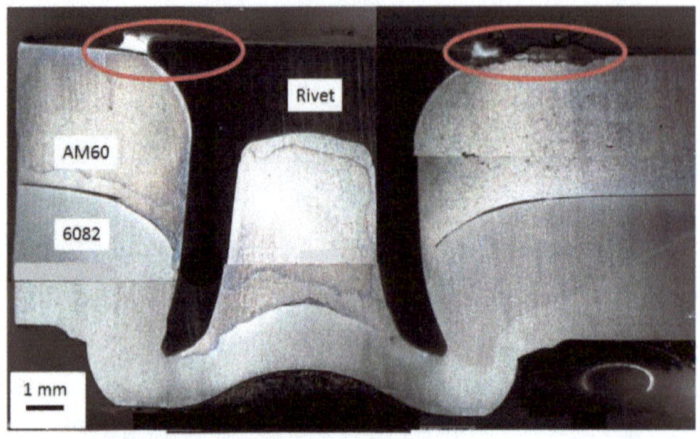

Figure 5. Cross-section of typical rivet in this study indicating corrosion product and pitting in AM60 surface at and near rivet head (red circles) but no corrosion at the AM60/6082 Al interface.

Observations of the FSLW joint areas in Group E also indicated white corrosion product on the surface of the weld in the "keyhole" region. This can be seen in Figure 6. Again, this was observed under all coating combinations although the powder coat did appear to perform better on other areas of the joint away from the keyhole and at the overlap area.

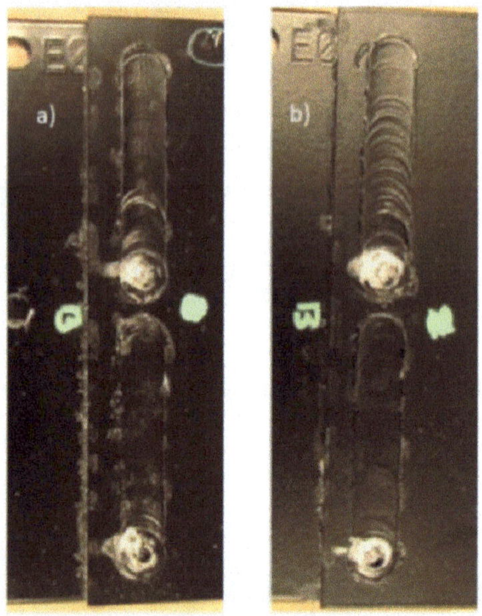

Figure 6. Examples of typical corrosion product observed in tested FSLW plate assemblies at 60 cycle SAE J2334. a) Group E with pretreat J, and ecoat and b) Group E with pretreat J and powder coat.

An examination of the Group D plate assemblies showed similar behavior to Group B in the AM60/6082 joint area. In the part of the assembly where the Mg and the steel were joined, corrosion product was observed surrounding the steel rivet (Figure 7). However, the aluminum rivet appeared to be free of corrosion product and the surface was clean in that region. The adhesive that was applied during the joining process did appear to protect the edges from corrosion where it squeezed out of the interface region. Edges where the adhesive did not squeeze out did appear to have some corrosion evident. This was observed on both the front and the back side of the coupon assembly. Later results in B117 indicated very pronounced corrosion on the edges where there was not adhesive indicating that the coating process is poor in this region for all combinations.

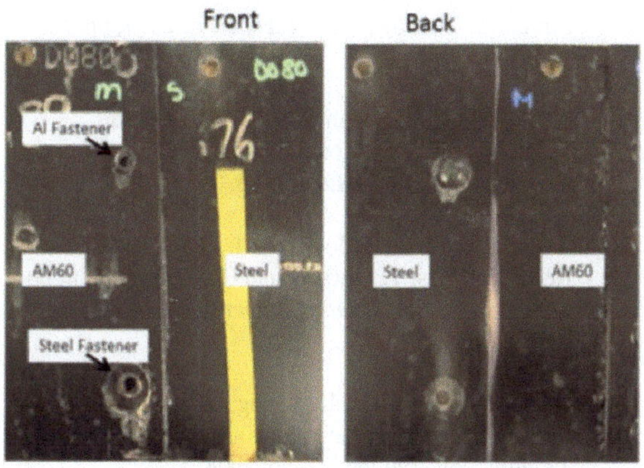

Figure 7. Examples of typical corrosion product observed in the Mg/Steel joint area of tested Group D plate assemblies at 60 cycle SAE J2334. Pretreatment M plus Ecoat

Discussion

The primary focus of this research was to examine the feasibility of joining multi-materials including Mg prior to painting an assembly. This strategy involves a complete change in the traditional paint shop for cleaning and pretreating assemblies and vehicles. Four Zr-based pretreatment systems were examined for feasibility in coating multi-material assemblies. In the SAE J2334 testing, the results indicated that while there appeared to be some differences between the pretreatments in terms of scribe corrosion results, none of the coating strategies appeared to mitigate corrosion entirely. The three manufacturers of the pretreatments were invited to pretreat these assemblies using their standard process. Since the cleaning requirements for Mg, Al, and galvanized steel are all different, a compromise in the cleaning steps was needed to allow for treatment of all three materials at the same time. This is actually an enormous challenge because optimized cleaning and coating preparation methods can be different when these materials are treated in isolation. The scribe measurements indicated that while all the pretreatment/topcoating systems did well with the Al plates, the Mg plates and the galvanized steel plates were not protected as well. Both of those materials showed scribe creep back readings that were higher than the 2 mm value generally allowable in SAE J2334. Further investigation will be required to understand whether this is a result of the cleaning process or the pretreatment chemistry itself. Both the powder coating and the ecoat appeared to protect the surface of the base metal from corrosion as long as the coating stayed intact.

The joint areas showed macroscopically visible cosmetic corrosion in all cases. In the case of the SPR joints, corrosion appeared to start at the edge interface of the rivet head and the AM60 plate and then undercut the pretreatment and topcoat layer on the surrounding Mg plate, creating pitting in the material surrounding the rivet head. This occurred regardless of pretreatment of the plate assembly and SPR rivet coating. Rings of white corrosion product were almost always evident around the SPRs on the ecoated samples. Powder coating did appear to offer some protection. However, if there were defects or scribes on the rivet that could disturb the powder coat, corrosion product was formed. The mechanism of corrosion is shown schematically in Figure 8. The powder coat probably offers some protection because it is about 5 times thicker than the ecoat (100 μm range versus 20 μm). The edges around the rivet head may be sealed a little better under powder coating. Once corrosion attack begins, however, the pretreatment and topcoat layer is undercut, resulting in very severe pitting corrosion in the area of Mg in the vicinity of the rivet head.

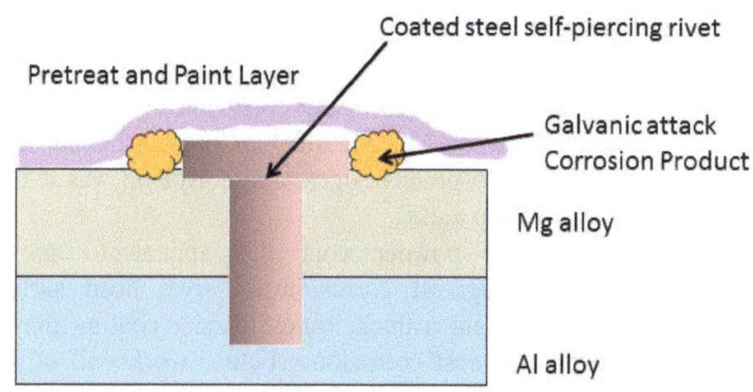

Figure 8. Schematic diagram of SPR head corrosion

One of the objectives of this study was to investigate if an alternative coating material for the rivet itself would protect against corrosion. The results indicate that SPRs treated with an aluminum coating fared no better than SPRs treated with a standard Zn/Sn coating. This may be due to the thin layer of coating (approx. 15 μm). In the case of Group D, an all Al fastener was used to join the Mg plate to the galvanized steel plate and that fastener showed very good corrosion resistance. Further investigation is required in this area.

FSLW samples also showed corrosion in the keyhole area and, in some of the ecoated samples, on the surface of the weld itself. This occurred despite the fact that this side of the plate assembly was Al and base material scribe measurements indicated that this material performed well under cyclic corrosion testing. This is possibly caused by issues of coating these areas. There was some flash on the surface of the weld and uneven surface in the interior of the keyhole. This may have been difficult to clean and coat and may have resulted in poorly protected material.

Finally, the study also wanted to investigate the role of structural adhesive in mitigating corrosion. Examination of the overlap areas in Groups C and D that contained adhesive showed that adhesive that come out between the plates did appear to mitigate corrosion by protecting the edges of the plates from contacting each other but in the absence of adhesive, corrosion did occur and was very severe in Group D.

This study did show some promising coating strategies for corrosion mitigation but also some weaknesses. Galvanic corrosion will result from contact between the bare metals so they must be protected. Future work will need to include the application of a sealant product to the rivet heads, FSLW keyholes, and overlap edges of the dissimilar materials involved in the assembly.

Conclusions

- Multi-material systems including Mg can be pretreated and coated in the same line although further refinement to the cleaning and pretreatment process will be required to ensure coating adhesion and resistance to undercutting, particularly for Mg and steel parts processed together.
- Joint areas are still a concern: mechanical fasteners should still be investigated for suitable coatings and/or materials to be compatible with Mg
- FSLW joints did indicate corrosion in the keyhole area of weld. This is possibly due to inadequate coating of the base Al material in this area because of the very rough surface quality.
- An adhesive can mitigate corrosion by providing a barrier between Mg and steel. Coatings appear to be inadequate to protect edges, and corrosion was evident at the overlap areas.
- Undisrupted powder coat did appear to provide protection against corrosion on rivet head surfaces suggesting that a thick, robust barrier coating may be effective against corrosion. Future work will utilized sealants on the rivet heads, FSLW keyhole regions, and component edges.

Acknowledgements

This material is based upon work supported by the Department of Energy National Energy Technology Laboratory under Award Number DE-EE0005660, with the United States Automotive Partnership LLC (USAMP). This report was prepared as an account of work sponsored by an agency of the United States Government. Neither the United States Government nor any agency thereof, nor any of their employees, makes any warranty, express or implied, or assumes any legal liability or responsibility for the accuracy, completeness, or usefulness of any information, apparatus, product, or process disclosed, or represents that its use would not infringe privately owned rights. Reference herein to any specific commercial product, process, or service by trade name, trademark, manufacturer, or otherwise does not necessarily constitute or imply its endorsement, recommendation, or favoring by the United States Government or any agency thereof. The views and opinions of authors expressed herein do not necessarily state or reflect those of the United States Government or any agency thereof.

References

[1] J. H. Forsmark, M. Li, X. Su, D. A. Wagner, J. W. Zindel, A. A. Luo, J. F. Quinn, R. Verma, Y.-M. Wang, S. D. Logan, and R. C. McCune, "The USAMP Magnesium Front End Research and Development Project - Results of the Magnesium 'Demonstration' Structure," in *Magnesium Technology 2014*, pp. 517–524.

[2] G. S. Cole, "Magnesium Corrosion Protection Techniques in the Automotive Industry," in *Corrosion Prevention of Magnesium Alloys*, 2013, pp. 489–506.

[3] J. S. Balzer, P. K. Dellock, M. H. Maj, G. S. Cole, D. Reed, T. Davis, T. Lawson, and G. Simonds, "Support Assembly," *SAE Technical Paper Series.*,#2003-01-0186, 2003.

[4] X. He, I. Pearson, and K. Young, "Self-Pierce Riveting for sheet materials: State of the art," *J. Mater. Process. Technol.*, vol. 199, pp. 27–36, 2008.

[5] S. Ohhama, T. Hata, T. Yahaba, T. Kobayashi, T. Miyahara, and M. Sayama, "Application of an FSW Continuous Welding Technology for Steel and Aluminum to an Automotive Subframe," *SAE Technical Paper Series.*,#2013-01-0372, 2013.

[6] T. Pan, "Friction Stir Spot Welding (FSSW) – A Literature Review," *SAE Technical Paper Series.*,#2007-01-1702, 2007.

[7] "Laboratory cyclic corrosion test," *SAE Standard J2334*, SAE International Surface Vehicle Standard , 2003.

[8] "Standard Practice for Operating Salt Spray (Fog) Apparatus 1," ASTM Standard B117-11.

[9] "Standard Test Method for Evaluation of Painted or Coated Specimens subjected to corrosive environments," ASTM Standard D1654-08, 2005.

Magnesium Technology 2015
Edited by: Michele V. Manuel, Alok Singh, Martyn Alderman, and Neale R. Neelameggham
TMS (The Minerals, Metals & Materials Society), 2015

INFLUENCE OF PULSE TIME ON THE STRUCTURAL AND TRIBOLOGICAL PROPERTIES OF MICRO ARC OXIDIZED AZ91D MAGNESIUM ALLOY

Deniz Kilic, Faiz Muhaffel, Yakup Yurekturk, Murat Baydogan
Istanbul Technical University, Department of Metallurgical and Materials Engineering, ITU Ayazaga Campus, Sarıyer, Istanbul-Turkey

Keywords: Magnesium alloy, Micro arc oxidation, Oxide coating, Wear.

Abstract

Micro arc oxidation process is recently developed as a promising surface modification technique applying high voltage and current density onto the light metals. The magnesium alloys exhibit nearly the lowest density among metallic structural materials but its poor characteristics in corrosion resistance, wear resistance, hardness and so on, limit its wide-range of applications. Through micro arc oxidation, thick and wear-resistant ceramic coatings are directly formed on the surface of magnesium alloys. In this study, oxide coatings were formed on AZ91D magnesium alloy in a silicate-based electrolyte by micro arc oxidation (MAO) process. MAO process was applied in an alkaline electrolyte with different pulse time and the contribution of applied pulse time in microstructural and wear resistance was systematically investigated. Structure, composition and tribological characteristics of the coatings were studied by scanning electron microscope (SEM), X-ray diffraction (XRD) and dry sliding reciprocating wear tests.

Introduction

Magnesium and its alloys have gained significant attention for applications requiring high strength to weight ratio, low density, high dimensional stability, good electromagnetic shielding and damping characteristics, high stiffness and good machining [1–3]. Therefore, magnesium alloys are very attractive engineering materials for the applications such as automotive, construction, aerospace, communication fields, sporting goods and household equipment. However, the use of magnesium-based materials is limited in the industry due to insufficient wear resistance against steel, low corrosion resistance and low hardness [4]. Hence, surface modification techniques including sol-gel, conversion coating, electroless plating, anodizing and micro arc oxidation (MAO) processes are attempted to overcome these drawbacks [5–7].

The MAO process, which is the most favorable one among these methods, is an environmental friendly surface modification treatment that generates a plasma state by applying a high voltage (>200 V) in an alkaline electrolyte. MAO is a rapid and high voltage anodization process which eliminates columnar structure of anodization step and forms crystalline oxide layer on the magnesium-based substrates. With respect to the mentioned techniques above, micro arc oxidized magnesium alloys obtain better adhesion, improved wear resistance, hardness and corrosion resistance depend on the parameters of the treatment. It is well documented in the literature that the properties of MAO coatings are mainly affected by applied electrical parameters such as voltage, current density, frequency, pulse time, duty cycle and processing time [7–11]. Among them, pulse time of the process acts as an important parameter on the formation of the oxide layer on magnesium alloy [12].

The current work is an attempt to understand the effect of processing pulse time in modifying the phase structure, surface and cross-sectional morphology, elemental composition and wear behavior of the oxide coating formed in silicate-based electrolyte.

Experimental procedure

Commercially available AZ91D magnesium alloy discs (25 mm in diameter and 4 mm in thickness) were used as the substrate. Samples were ground with SiC sandpaper up to 1200#. They were then cleaned with acetone and distilled water, successively, and dried at room temperature in prior to MAO process.

The MAO device consists of an AC power supply (constant voltage and constant current density modes are provided), a stainless steel electrolyte bath, an air stirring system and a cooling chamber system. The samples were used as anode while the stainless steel container was used as cathode in the electrolytic bath. The temperature of the electrolyte was kept at 20 ± 2 °C by the cooling system during MAO process in order to prevent evaporation of the solution and regulate the chemical reaction rate in the electrolyte. In this study, the constant voltage mode was selected. Samples were oxidized in an aqueous electrolyte containing 15 g/L of sodium silicate (Alfa Aesar) and 2 g/L of potassium hydroxide (WVR BDH Prolabo) with applied voltage range of 400 V in the positive half cycle and 80 V in the negative half cycle. Pulse frequency and duty cycle were changed from 417 to 1000 Hz and from 20 to 66%, respectively. Each sample was oxidized for 5 min and then cleaned ultrasonically in acetone and distilled water. The samples are coded according to the pulse time as shown in Table 1.

Table 1. Applied pulse time of the MAO samples.

Sample	Positive pulse time (μs)	Negative pulse time (μs)	Stop time (μs)
MAO-100	100	100	400
MAO-200	200	200	400
MAO-400	400	400	400
MAO-800	800	800	400

Morphology and mean elemental composition of the oxide layers were examined by an energy dispersive spectrometer (EDS, Oxford Instruments) equipped scanning electron microscope (SEM, Hitachi, TM-1000). Phase composition of the coatings was identified by X-ray diffractometer (XRD, GBC, MMA 027) using Cu Kα radiation (λ = 0.154 nm) at 35 kV and 28.5 mA between 20-80°. Mean surface roughness (Ra) of the surface of the samples was examined by surface profilometer (Veeco Dektak 6M) under 5 mg load, with a scan distance of 2000 μm. Ten examinations were done for each sample. Coating thickness of the samples was evaluated from the cross-sectional SEM micrographs. Ten measurements were done for each sample throughout the coating.

The friction and wear properties of the uncoated and MAO applied substrates investigated at dry sliding condition using a computer controlled reciprocal wear tester. 6 mm alumina ball was used as abrasive tool and sliding speed was 15 mm/s. The sliding distance, length of the wear track and load were 50 m, 5 mm and 2 N, respectively.

Results and Discussion

Fig. 1 shows XRD patterns of MAO treated samples in silicate-based electrolyte. The coating mainly composed of cubic MgO and $MgSiO_3$ phase. Due to high penetration of X-rays also magnesium peaks were observed on the pattern of the oxidized sample. It is suggested that the relatively less crystalline MgO phase has changed into cubic MgO with the increase of the processing pulse time due to high temperature and pressure during the sparking stage of the MAO process [7]. The presence of $MgSiO_3$ formation revealed that the silicate ions in the electrolyte favored the formation of $MgSiO_3$ in the phase composition of the coatings for all the applied pulse regimes [12]. It is apparent that intensity of $MgSiO_3$ peaks increases as the processing pulse time increases with a corresponding decrement of MgO phase.

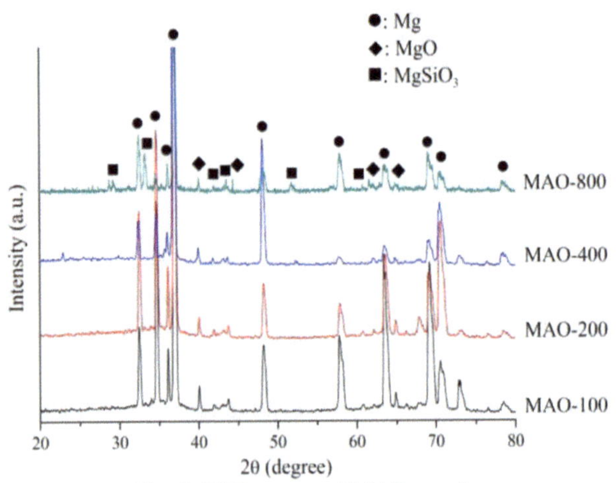

Fig. 1. XRD spectra of MAO samples.

Surface SEM micrographs of the MAO samples with different applied pulse time regimes are shown in Fig. 2. The oxidized surfaces presented homogenous distribution of pores throughout the surface. Some cracks were visible on the surface of MAO-800 sample, which was oxidized for the highest applied pulse time in the present study. The size of the pores was mainly dependent on the applied pulse time and it is clear that an increase in the processing pulse time let to widening of the pores up to 20 μm in diameter. These pores were generated by intensive sparking during MAO process [11]. Longer pulse time caused high temperature and pressure on the surface and enlarged the micro-discharge channels during MAO process which seen on the surface as pores.

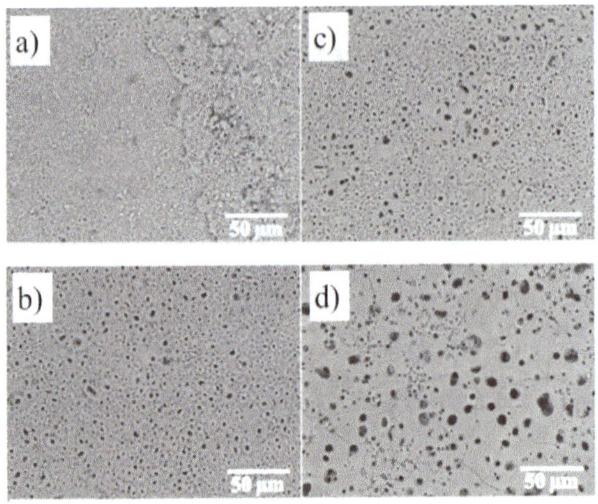

Fig. 2. Surface SEM micrograph of (a) MAO-100, (b) MAO-200, (c) MAO-400 and (d) MAO-800 samples.

Cross-sectional SEM morphologies of MAO coatings obtained with different applied pulse time are shown in Fig. 3. MAO-100 sample which has lowest pulse time has a very thin coating due to insufficient discharging during MAO process. In the case of MAO-200 and MAO-400 samples, they exhibited a thicker coating morphology and major part of the coatings was relatively compact. On the other hand, the MAO-800 sample showed a wavy and defective microstructure.

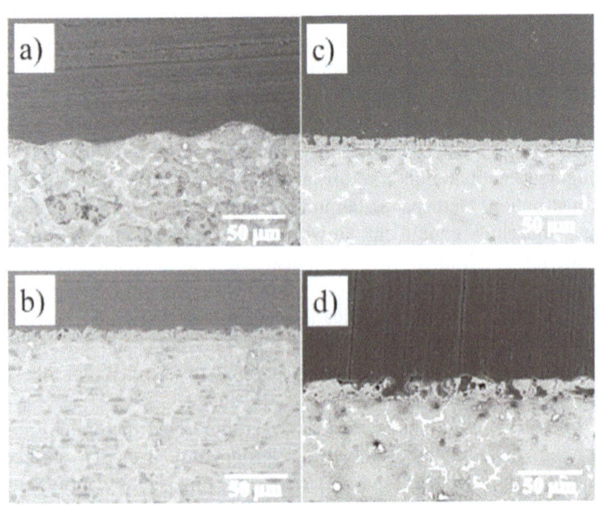

Fig. 3. Cross-sectional SEM micrograph of (a) MAO-100, (b) MAO-200, (c) MAO-400 and (d) MAO-800 samples.

EDS analyses were performed on the surface of the coating formed with different applied pulse time and the results are presented in Table 2. Elemental composition analyses revealed that the coatings mainly consisted of Mg and Si with low amount of Al (from substrate), K and Na (from electrolyte). Even though the coating comprised of oxide phases, oxygen was not detected due to the inadequacy of EDS equipment in determination of small elements from the samples. It was found that the wt.% content of Mg from the substrate generally followed a decreasing trend with the increasing applied pulse time. This was attributed that less Mg coming from the substrate could be detected in the application of longer pulses producing thicker coatings. In contrast to Mg, Si content coming from the coating itself exhibited almost a constant variation within the applied pulse time intervals. The contents of Na and K species (present in electrolyte), on the other hand, generally showed a slight increment which is the consequence of thicker coatings having elements coming from the electrolyte when the applied pulse time was longer.

Table 2. Elemental composition of MAO samples.

Sample	Elemental Composition (wt.%)				
	Na	Mg	Al	Si	K
MAO-100	1.9	40.7	4.1	51.1	2.1
MAO-200	3.4	34.9	3.5	54.8	3.4
MAO-400	5.1	32.1	2.5	55.8	4.4
MAO-800	4.7	32.5	3.9	52.6	6.3

The variations of average surface roughness and coating thickness of MAO coatings formed with different applied pulse time are shown in Fig. 4. It is obvious that the increment of pulse time favored thicker and rougher coating morphology. While the sparking period was extended, the oxidation process induced roughening and thickening of the oxide layer. Moreover, formation of oxide coating with large pores is also responsible for the increase of the surface roughness. Therefore, the surface roughness results are consistent with the SEM morphologies of these coatings which are shown in Fig. 2. However, applying 800 µs pulse time let to deterioration of the oxide coating as shown in Fig 3d.

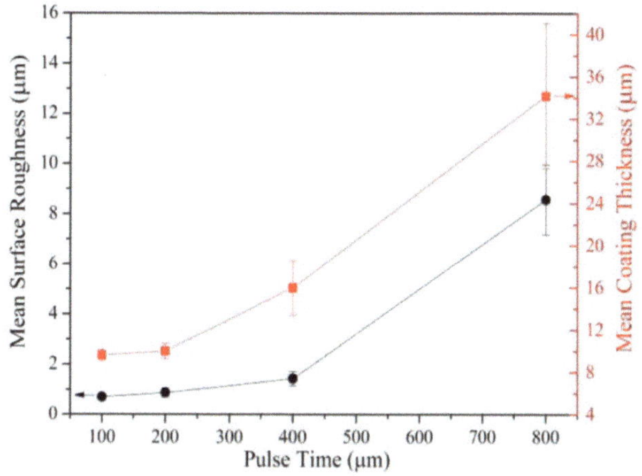

Fig. 4. Variation of surface roughness and coating thickness of MAO coatings.

Variation of friction coefficient and sliding distance for AZ91D magnesium substrate and coating of MAO-200 and MAO-400 samples is given in Fig. 5. It is seen that the friction coefficient of AZ91D substrate showed a lower value but due to severe delamination on the surface, the friction coefficient was in a wavy nature. However, MAO coatings exhibited a higher friction coefficient along with a smooth increment in friction coefficient value.

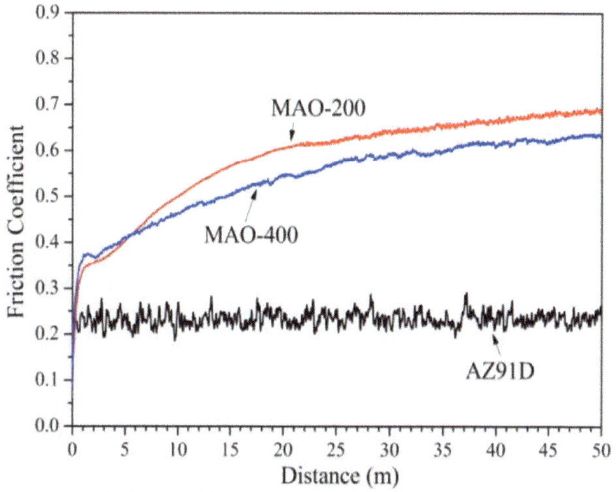

Fig. 5. Friction coefficient of AZ91D substrate, MAO-200 and MAO-400 samples.

Fig. 6 shows the relationship between the relative wear resistance of AZ91D magnesium substrate and coatings of MAO-200 and MAO-400 samples. It can be seen that the relative wear resistance increases rapidly when the substrate was coated by micro arc oxidation. MAO-400 exhibited better wear resistance as compared to MAO-200 sample. This was mainly due to thicker oxide coatings on the surface of MAO-400 sample. Increasing crystallinity of the oxide phases with increasing applied pulse time may be another factor contributing higher wear resistance and lower friction coefficient of MAO-400 sample.

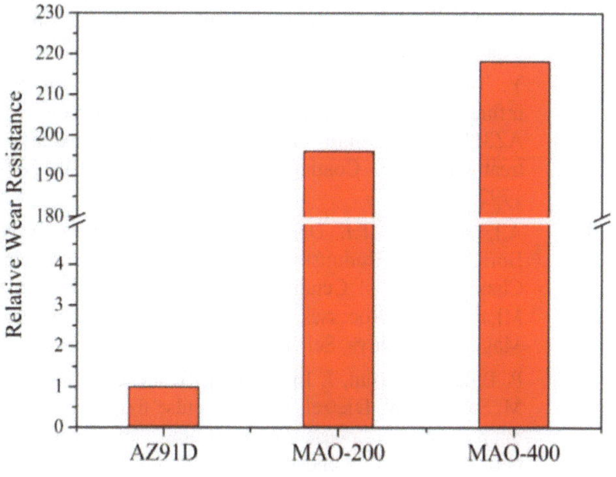

Fig. 6. Relative wear resistance of AZ91D substrate, MAO-200 and MAO-400 samples.

Conclusion

An oxide coating consisted of magnesium oxide and magnesium silicate was generated on AZ91D magnesium alloy. Applying low pulse on time (MAO-100) during MAO process let to a very thin and less crystalline microstructure while applied high pulse time (MAO-800) induced an undesirable microstructure due to high roughness and porosity in the oxide coating. Under dry sliding conditions, MAOed samples exhibited much higher wear resistance than untreated AZ91D alloy. Results also showed that application time of a pulse during micro arc oxidation can also affect the wear resistance of AZ91D alloy.

Amongst the micro arc oxidized samples investigated, MAO-400 sample exhibited the best tribological performance as compared to MAO-200 sample, in terms of higher wear resistance and lower friction coefficient mainly due to thicker oxide coatings on the surface.

References

[1] M. Laleh, A.S. Rouhaghdam, T. Shahrabi, A. Shanghi, Effect of alumina sol addition to micro-arc oxidation electrolyte on the properties of MAO coatings formed on magnesium alloy AZ91D, J. Alloys Compd. 496 (2010) 548–552.

[2] M. Laleh, F. Kargar, a. S. Rouhaghdam, Investigation of rare earth sealing of porous micro-arc oxidation coating formed on AZ91D magnesium alloy, J. Rare Earths. 30 (2012) 1293–1297.

[3] J. Liang, P.B. Srinivasan, C. Blawert, M. Störmer, W. Dietzel, Electrochemical corrosion behaviour of plasma electrolytic oxidation coatings on AM50 magnesium alloy formed in silicate and phosphate based electrolytes, Electrochim. Acta. 54 (2009) 3842–3850.

[4] M. Weiyi, H. Yong, Study on Micro-Arc Oxidized Coatings on Magnesium in Three Different Electrolytes, Rare Met. Mater. Eng. 39 (2010) 1129–1134.

[5] P. Shi, W.F. Ng, M.H. Wong, F.T. Cheng, Improvement of corrosion resistance of pure magnesium in Hanks' solution by microarc oxidation with sol–gel TiO2 sealing, J. Alloys Compd. 469 (2009) 286–292.

[6] W. Zhang, C. He, Electroless Ni-WP/Ni-B duplex coatings on AZ91D magnesium alloy, in: Mech. Autom. Control Eng., 2010: pp. 3895 – 3899.

[7] Y. Tang, X. Zhao, K. Jiang, J. Chen, Y. Zuo, The influences of duty cycle on the bonding strength of AZ31B magnesium alloy by microarc oxidation treatment, Surf. Coatings Technol. 205 (2010) 1789–1792.

[8] A.I. Mamaev, T.I. Dorofeeva, V.A. Mamaeva, V.N. Borikov, Y.Y. Budnitskya, A.A. Makarov, Voltammetry Characteristics of Ceramic Coatings Obtained by Pulse Microplasma Processes on Aluminium , Titanium and Magnesium Alloys, Sci. York. (2000) 340–342.

[9] P. Bala Srinivasan, J. Liang, R.G. Balajeee, C. Blawert, M. Störmer, W. Dietzel, Effect of pulse frequency on the microstructure, phase composition and corrosion performance of a phosphate-based plasma electrolytic oxidation coated AM50 magnesium alloy, Appl. Surf. Sci. 256 (2010) 3928–3935.

[10] L. Chang, F. Cao, J. Cai, W. Liu, Z. Zhang, J. Zhang, Influence of electric parameters on MAO of AZ91D magnesium alloy using alternative square-wave power source, Trans. Nonferrous Met. Soc. China. 21 (2011) 307–316.

[11] P. Bala Srinivasan, J. Liang, C. Blawert, M. Störmer, W. Dietzel, Effect of current density on the microstructure and corrosion behaviour of plasma electrolytic oxidation treated AM50 magnesium alloy, Appl. Surf. Sci. 255 (2009) 4212–4218.

[12] R.O. Hussein, D.O. Northwood, X. Nie, The influence of pulse timing and current mode on the microstructure and corrosion behaviour of a plasma electrolytic oxidation (PEO) coated AM60B magnesium alloy, J. Alloys Compd. 541 (2012) 41–48.

Magnesium Technology 2015
Edited by: Michele V. Manuel, Alok Singh, Martyn Alderman, and Neale R. Neelameggham
TMS (The Minerals, Metals & Materials Society), 2015

ELECTROLESS Ni-P/NANO-SiO_2 COMPOSITE PLATING ON DUAL PHASE MAGNESIUM-LITHIUM ALLOY

Y. Zou, Z. W. Zhang*, M. L. Zhang
Key Laboratory of Superlight Materials and Surface Technology, Ministry of Education, College of Materials Science and Chemical Engineering, Harbin Engineering University, Harbin 150001, P R China

Keywords: Magnesium-Lithium alloy, electroless plating, Ni-P/nano-SiO_2 composite coating, corrosion resistance

Abstract

The application of Mg-Li alloys is restricted in practice due to mainly poor corrosion resistance and wear resistance. Electroless nickel plating is one of the common and effective ways to protect alloys from corrosion. In this study, nano-SiO_2 particles with Ni-P matrix have been successfully co-deposited onto dual phase Mg-8Li base alloy through electroless plating, generating homogeneously Ni-P/nano-SiO_2 composite coating. The morphology, elemental composition and structures of coatings were investigated. Coating performances were evaluated using hardness tests and electrochemical analysis. The results indicate that the Ni-P/nano-SiO_2 composite coating can significantly improve the wear and corrosion resistance.

Introduction

Magnesium-lithium (Mg-Li) alloys possess the lowest density among metallic structural materials [1]. Mg-Li alloys have many excellent properties, such as high strength-to-weight ratio, good formability, damping capacity and shock resistance ability et al. [1, 2]. These properties make Mg-Li alloys the promising candidates in a wide range of fields including aeronautics, automotive industries, military and consumer electronic industries [3-5]. However, the application of Mg-Li alloys is restricted in practice due to some undesirable properties, mainly poor corrosion and wear resistance [6-9]. Electroless Ni-P coatings are widely used in various industries. Co-depositing another metallic or non-metallic elements or abrasive/lubricate particles or combination of them in binary Ni-P matrix can further enhance corrosion and wear resistance [10-14]. It can be rationally anticipated that the inclusion of nano-sized particles in electroless Ni-P alloy coating would broaden the scopes of the coating in engineering, because various types of nano-particles have special natures. However, the so-called composite coatings are prone to some problems, such as non-uniform particulates distribution and numerous defects, because of the high surface energy and activity of nano-particles in the plating bath [14, 15]. Therefore, excellent comprehensive properties of the electroless Ni-P based composite coatings are highly dependent on the preparation techniques, specifically stable dispersion of the nano-particles in plating bath.

In this study, electroless Ni-P/nano-SiO_2 composite plating on dual phase Mg-8Li alloy was developed. The effects of nano-SiO_2 on microstructure, structure, wear resistance and corrosion resistance of the composite coatings were investigated.

Experimental

Materials

Dual phase Mg-Li base alloy with the composition (wt. %) of Mg-8Li-2Al-0.7La was studied. The duplex structure is composed of α phase (a solid solution of Li in Mg) and β phase (a solid solution of Mg in Li). The specimens with dimensions of $20\times20\times3$ mm^3 were abraded with 600# SiC paper.

Plating process

The plating process consisted of pre-grinding, washing, activation, pre-treatment and electroless Ni-P/nano-SiO_2 plating. The formulations and process conditions for different steps are summarized in Table I.

Table I. Formulations and process conditions for various processing operations.

Stage	Processing	formulation and condition
1	Cleaning	Acetone with ultrasonic assistance, RT, 10min
2	Activation	10mL/L HNO_3 and 60mL/L H_3PO_4, RT, 20s
3	Pre-treatment	0.8 g/L $Ce(NO_3)_3$ and 25 g/L $KMnO_4$, RT, 15min
4	Plating	16 g/L $NiSO_4\cdot6H_2O$, 15 g/L $NaH_2PO_2\cdot H_2O$, 13 g/L $NaC_2H_3O_2$, 8 g/L NH_4HF_2, 12 ml/L HF, 5 g/L nano-SiO_2, pH= 6.5, 68°C, with ultrasonic assistance.

The electroless Ni-P/nano-SiO_2 plating process was divided to two steps. Electroless Ni-P plating was first conducted for 20min. Then SiO_2 nano-particles were added to the nickel electroless plating bath and ultrasonically dispersed for 15min, followed by co-depositing for 2 h.

Characterization

The morphology and chemical composition of nanocomposite deposits were characterized by Scanning Electron Microscopy (SEM) (JSM-6480A, Japan Electronics) and Energy Dispersive X-ray Spectroscopy (EDS) (INC250, Japan Electronics). Phase identification of the deposits was performed by X-ray diffraction on Rigaku D/max-TTR-III diffractometer using Cu Kα. The hardness of the coatings was measured using Vickers hardness tester HXS-1000Z with a load of 300 g for 20 s. The corrosion performances of the deposits were evaluated by the electrochemical impedance spectroscopy (EIS) and potentiodynamic polarization in 3.5 wt% NaCl solution. All electrochemical experiments were carried out using a electrochemical workstation (CHI660D, China) based on the conventional three-electrode system. A saturated calomel electrode (SCE, +0.242 V versus SHE) as the reference electrode and a platinum sheet as the counter electrode. The sample was embedded in epoxy resin leaving an area of 10 mm×10 mm as the working electrode. Prior to the EIS tests, the sample was immersed in the corrosion solution for 45 min to obtain a steady state at open circuit potential (OCP). The EIS data were collected in the frequency range from 100 kHz to 10 mHz with a 5 mV amplitude perturbation. The analysis of the EIS data was performed by fitting the experimental results to equivalent circuits using ZSimpWin 3.10 software. After EIS measurement, each sample was settled for 5min to attain its stable OCP again, and then the potentiodynamic polarization curves were recorded at a scanning rate 1mV/s. All electrochemical corrosion tests were performed in static conditions at 25±0.5℃ under environment atmosphere.

Results and discussion

Morphology, compositions and structures of Ni-P/nano-SiO_2 composite coatings

Figure 1 shows the optical/SEM morphologies before and after electroless Ni-P/nano-SiO_2 plating of Mg-Li base alloy. Figure 1(a) depicts a typical dual-phase microstructure, consisting of α phase (white) and β phase (gray). Figure 1(b) and (c) show that a compact and uniform Ni-P/nano-SiO_2 composite coating was successfully deposited onto the Mg-Li base alloy. It can be seen from Figure 1(c) that many nodular protrusions uniformly distributed over the coating. No obvious flaws or apertures can be observed.

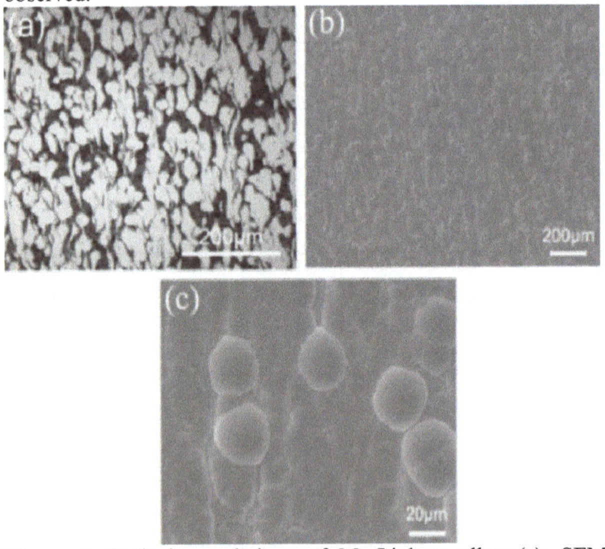

Figure 1. Optical morphology of Mg-Li base alloy (a), SEM images of electroless Ni-P/nano-SiO_2 coating with two magnifications (b, c).

To determine the composition of electroless coating, specifically the concentration of P, the composition of coating matrix was focused on and the corresponding EDS spectrum is shown in Figure 2. The EDS results indicate that the coatings are composed of nickel and phosphorous (Figure 2), which manifest that the surface of Mg-Li base alloy was thoroughly covered by the Ni-P coatings. The average phosphorus concentration of the coating is ~ 7.2 wt%. It is generally accepted [16] that the Ni-P coatings containing 1~5 wt% phosphorus (low phosphorus) have a crystalline structure and 6~9 wt% phosphorus (moderate phosphorus) have a two-phase structure with a mixture of crystalline and amorphous phases. While 10~13 wt% phosphorus (high phosphorus) leads to an amorphous structure. Thus, the Ni-P/nano-SiO_2 composite coating can be regarded as a mixed structure with crystalline and amorphous.

Figure 3 shows the XRD pattern of Ni-P/nano-SiO_2 composite coating. It can be seen that the diffraction pattern of composite coating demonstrates a broad peak at ~ 45° with preferred orientation of Ni(111), confirming a mixture of polycrystalline and amorphous phases. This is in consistence with the EDS results. However, besides the broad peak around 45°, there are some other sharp peaks which can be indexed as Ni_3P and SiO_2, indicating the formation of Ni-P/nano-SiO_2 composite in the coating.

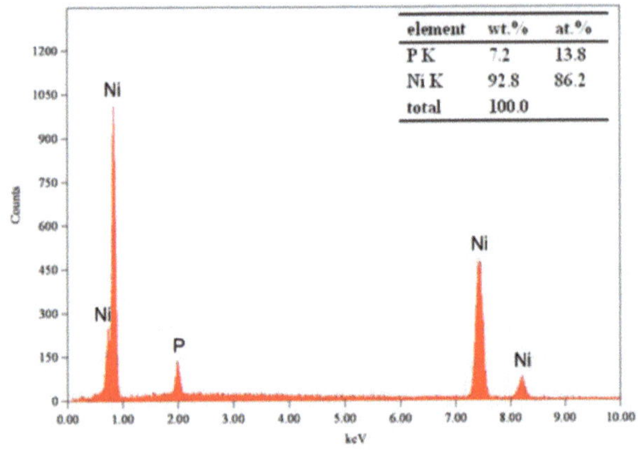

Figure 2. EDS spectrum of electroless Ni-P/nano-SiO_2 composite coating.

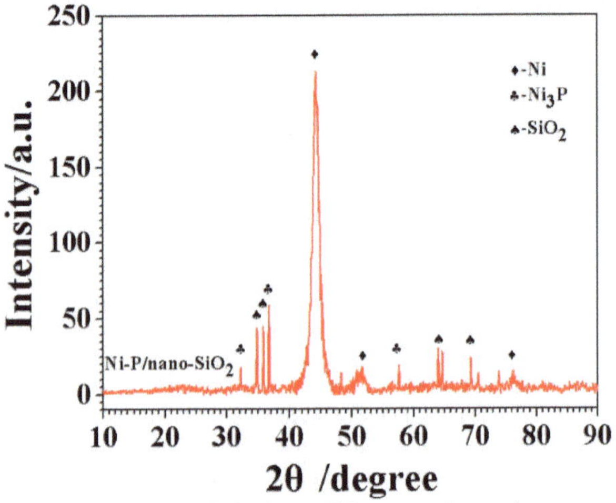

Figure 3. XRD pattern of Ni-P/nano-SiO_2 composite coating.

Hardness of the Ni-P/nano-SiO_2 composite coating

Hardness of coating is the resistance to deformation caused by the indentation, which is an important indicator of coating performance, and to a certain extent, reflects the wear resistance of coatings. The Vickers hardness values of the coatings are shown in Figure 4. The hardness of the as-deposited Ni-P/nano-SiO_2 composite coating is about 678HV, which is ~855% times higher than that of the Mg-Li base alloy (about 71HV), implying a significant improvement of wear resistance through electroless Ni-P/nano-SiO_2 coating.

Corrosion performance of the Ni-P/nano-SiO_2 composite coating

The corrosion performance of the Ni-P/nano-SiO_2 composite coated Mg-Li alloy was studied by electrochemical impedance spectroscopy (EIS) and potentiodynamic polarization in 3.5 wt.% NaCl solution. The plots of Nyquist, Module and phase Bode for bare alloy and Ni-P/nano-SiO_2 composite coating are shown in Figure 5. The Nyquist plot (Figure 5a) shows that the Mg-Li substrate exhibits a high-frequency capacitive loop and one low-frequency inductance loop. The capacitive loop in the high frequency region represents the double layer capacitance and charge transfer resistance at the electrode/electrolyte interface. The inductance loop in the low frequency domain is mainly ascribed to the initial stage of the localized corrosion [17]. While there are no inductance loops for the Ni-P/nano-SiO_2 coated samples, indicating that the as-obtained coating shows more resistant to pit corrosion. Besides, the Ni-P/nano-SiO_2 coating has a much larger capacitive reactance arc than the bare alloy, implying that the Ni-P/nano-SiO_2 coated samples have better anti-corrosion performance.

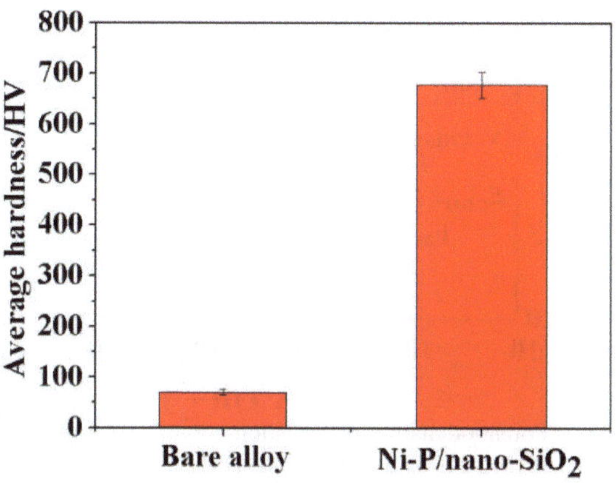

Figure 4. Hardness of bare alloy and Ni-P/nano-SiO_2 composite coating.

To account for corrosion behavior of Mg-Li base substrate and Ni-P/nano-SiO_2 composite coating, equivalent circuits that were used for EIS data analysis are shown in Figure 6. Because there exist inductance loops for the Mg-Li base substrate but no inductance loops for the Ni-P deposited samples, two equivalent circuits were used, one for Mg-Li base substrate (Figure 6a) and another for Ni-P deposited samples (Figure 6b). In Figure 6a, R_S represents the solution resistance, R_a is charge-transfer resistance, the resistance R_b and inductance L can be used to specify the inductive behavior. In Figure 6b, R_S also represents the solution resistance, R_1 is the porosity resistance on the coating surface, R_2 is the charge-transfer resistance. The constant phase element (CPE, designated as Q) is used instead of ideal capacitance (Cdl) to account for non-ideal behavior of double layer due to surface inhomogeneity, roughness and adsorption effects [18].The impedance of CPE can be expressed in:

$$Z_{CPE}=1/Q(jw)^{-n}$$

where Q is the CPE constant, and j is the imaginary number, ω is the angular frequency (in rad s^{-1}) and n ($0 < n < 1$) is a CPE exponent which can be used as a gauge of the heterogeneity or roughness of the surface. The polarization resistance, R_p, can be calculated as $R_p = R_1 + R_2$. The EIS data were fitted by ZSimpWin 3.10 software using the given equivalent circuits. EIS parameters are summarized in Table II. It is found that the polarization resistance, R_p of the Ni-P/nano-SiO_2 coating (2914 Ω cm^2) is ~ 2900% times higher than that of bare alloy (96.12 Ω cm^2), indicating that the corrosion protective ability have been significantly improved by Ni-P/nano-SiO_2 coating.

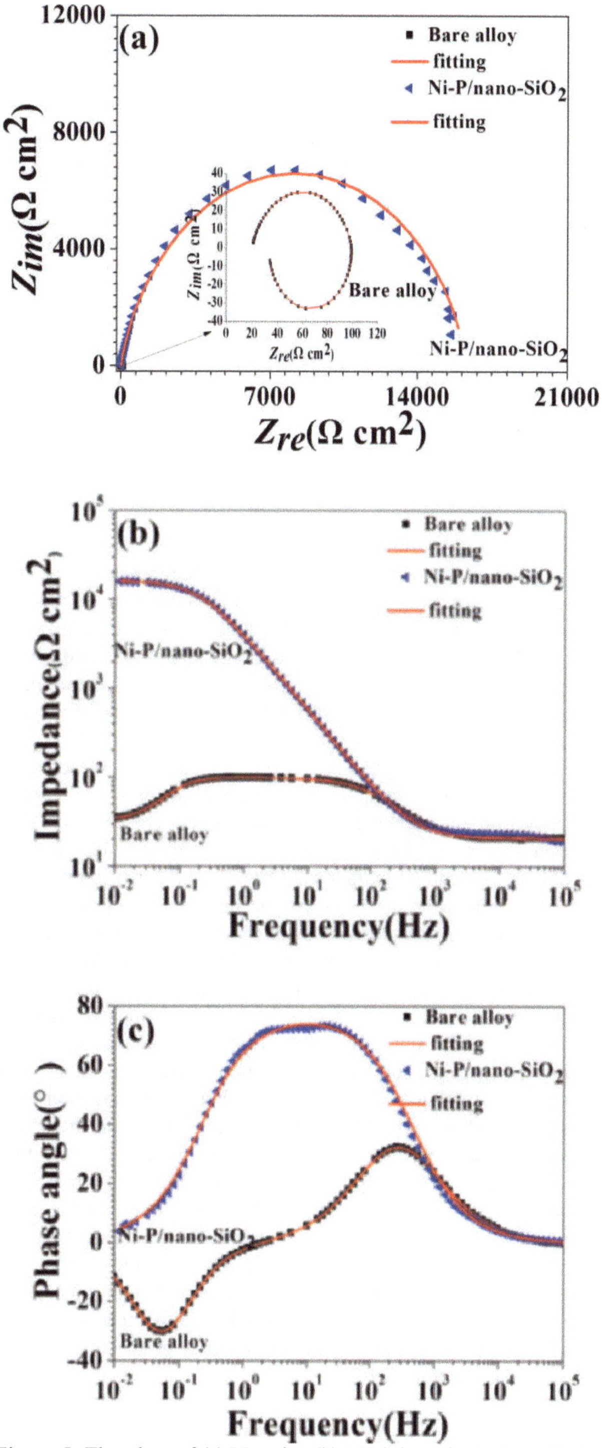

Figure 5. The plots of (a) Nyquist, (b) Module and (c) phase Bode for bare alloy and Ni-P/nano-SiO_2 composite coating determined from electrochemical impedance spectroscopy (EIS) in 3.5 wt.% NaCl solution.

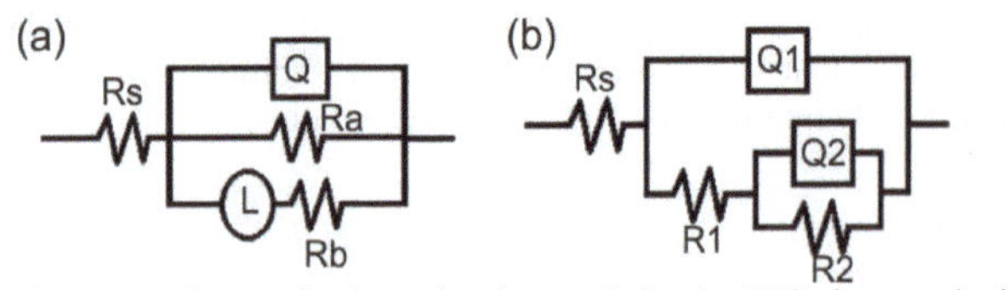

Figure 6. The equivalent circuits used for the EIS data analysis. (a) for Mg-Li bare and (b) for Ni-P/nano-SiO_2 coated samples.

Table II. EIS parameters for the bare alloy and Ni-P/nano-SiO_2 coated samples in 3.5 wt.% NaCl solution.

Sample	Bare alloy		Ni-P/nano-SiO_2	
	fitting data	%error	fitting data	%error
$R_s(\Omega cm^2)$	19.72	0.04132	20.64	0.4541
$CPE_1(S\ cm^{-2}s^n)$	6.193×10^{-5}	0.2093	4.820×10^{-5}	2.574
n_1	0.8194	0.03602	0.8653	0.763
$R_a/R_1(\Omega cm^2)$	80.08	0.02815	1631	9.451
L	163.1	0.05925		
$CPE_2(S\ cm^{-2}s^n)$			3.984×10^{-6}	7.033
n_2			0.981	9.13
$R_b/R_2(\Omega cm^2)$	16.04	0.133	1283	1.76
$R_p(\Omega cm^2)$	96.12		2914	

Figure 7 shows the potentiodynamic polarization in 3.5 wt.% NaCl solution. And the calculated electrochemical corrosion data determined from the potentiodynamic polarization curves are summarized in Table III. We note that the anodic polarization curves do not clearly indicate a Tafel region, and the anodic Tafel slopes are not necessarily robust for the extrapolated determination of corrosion current density (i_{corr}) alone, due to the presence of a passive window for Ni-P/nano-SiO_2 coated specimens and a very low slope for uncoated samples. Hence the corrosion potential and corrosion current density are derived directly from the cathodic polarization curves. The cathodic Tafel slope (β_c) is derived from the fitted linear region. After the slope is obtained, it can be extrapolated back to the corrosion potential (E_{corr}) to obtain the corrosion current density (i_{corr}) [19]. For the Ni-P/nano-SiO_2 composite coatings, the corrosion potentials (E_{corr}) were shifted significantly toward the positive direction and the corrosion current density (i_{corr}) was dramatically decreased from 1.59×10^{-4} A cm^{-2} to 1.08×10^{-5} A cm^{-2}, as compared with that of the Mg-Li alloy substrate. This indicates a significant improvement in corrosion protection after Ni-P/nano-SiO_2 composite plating. Moreover, an obvious passivation was observed with a noticeable passive range ~550 mV in the anodic branch of the Ni-P/nano-SiO_2 coatings (Figure 7) due to a passivation process [20]. The passivation film can protect the sample from corrosion during anodic polarization in the NaCl solution. In fact, the high corrosion resistance can be attributed to the mixed polycrystalline-amorphous structures and the passivation of the coating. It was reported that amorphous structure offer better corrosion protective ability than equivalent polycrystalline because of their free grain or grain boundaries as well as glassy nature [21-23].

Conclusions

Nano-SiO_2 particles with Ni-P matrix have been successfully co-deposited onto dual phase Mg-8Li base alloy through nickel electroless plating, generating homogeneously Ni-P/nano-SiO_2 composite coatings. The Ni-P/nano-SiO_2 composite coatings have a compact and uniform microstructure with a mixture of polycrystalline-amorphous phases. The hardness of Ni-P/nano-SiO_2 composite coating is ~ 678HV, 855% times higher than that of the Mg-Li base alloy (~ 71HV), implying a significant improvement in wear resistance. The results obtained from EIS and polarization measurements illustrate that the surface corrosion protection ability can be significantly improved by the formation of Ni-P/nano-SiO_2 composite coatings. The polarization measurements indicated that the corrosion resistance of Ni-P/nano-SiO_2 composite coating is improved by the formation of passive film. The combination of good wear and corrosion resistance should have a potential use in specific applications.

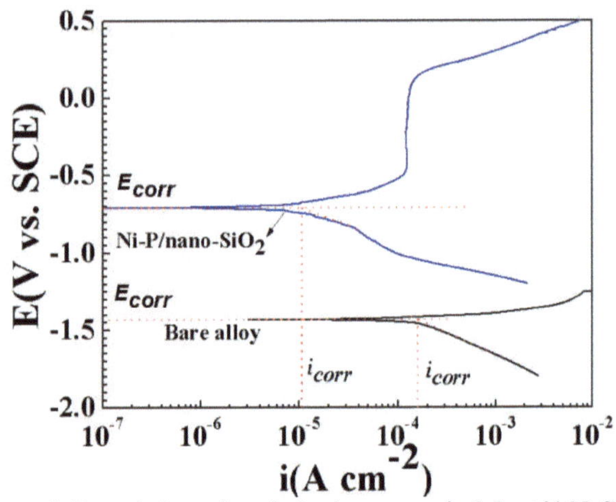

Figure 7. Potentiodynamic polarization curves in 3.5 wt.% NaCl solution of Mg-Li bare alloy and Ni-P/nano-SiO_2 composite coatings.

Table III. The electrochemical parameters of Mg-Li bare alloy and Ni-P/nano-SiO_2 composite coatings obtained from the potentiodynamic polarization curves

Sample	E_{corr}(V)	β_c(V decade^{-1})	i_{corr} (A cm^{-2})
bare alloy	1.432	-0.256	1.59×10^{-4}
Ni-P/nano-SiO_2 coatings	0.708	-0.185	1.08×10^{-5}

Acknowledgements

This work was supported by the Fundamental Research Funds for the Central Universities (HEUCFZ1308, HEUCFT1306), the NSFC Funding (51171081, 51371062 and U1460102), NSFHLJ (ZD201411) and the Project 2013DB04 at NPL, CAEP. YZ is benefited from the International Exchange Program of Harbin Engineering University for Innovation-oriented Talents Cultivation.

References

[1] A. Sharma, R. Uma Rani, and S. Mayanna, "Thermal studies on electrodeposited black oxide coating on magnesium alloys," *Thermochimica acta*, 376 (2001), 67-75.

[2] J. M. Song, T. X. Wen, and J. Y. Wang, "Vibration fracture properties of a lightweight Mg-Li-Zn alloy," *Scripta materialia,* 56 (2007), 529-532.
[3] A. Sanschagrin et al, "Mechanical properties and microstructure of new magnesium-lithium base alloys," *Materials Science and Engineering: A*, 220 (1996), 69-77.
[4] Z. Trojanová et al., "Deformation behaviour of Mg-Li alloys at elevated temperatures," *Materials Science and Engineering: A*, 410 (2005), 148-151.
[5] S. Wang, et al., "Microstructures and mechanical properties of 5 wt.% Al_2Y p/Mg-Li composite," *Materials Letters*, 60 (2006), 1863-1865.
[6] M. C. Zhao et al., "An exploratory study of the corrosion of Mg alloys during interrupted salt spray testing," *Corrosion Science*, 51 (2009), 1277-1292.
[7] L. Liu, M. Schlesinger, "Corrosion of magnesium and its alloys," *Corrosion Science*, 51 (2009), 1733-1737.
[8] J. Gray, B. Luan, "Protective coatings on magnesium and its alloys-a critical review," *Journal of alloys and compounds*, 336 (2002), 88-113.
[9] R. Ambat, W. Zhou, "Electroless nickel-plating on AZ91D magnesium alloy: effect of substrate microstructure and plating parameters," *Surface and Coatings Technology*, 179 (2004), 124-134.
[10] A. Grosjean, M. Rezrazi, and J. Takadoum, "Hardness, friction and wear characteristics of nickel-SiC electroless composite deposits," *Surf Coat Technol*, 137 (2001), 92-96.
[11] J. N. Balaraju, T. S. N. Sankara-Narayanan, and S. K. Seshadri, "Electroless Ni-P composite coatings," *Appl Electrochem,* 33 (2003), 807-816.
[12] R. Wang et al., "Preparation and characterization of nanodiamond cores coated with a thin Ni-Zn-P alloy film," *Mater Character* 59 (2008), 108-111.
[13] T. Rabizadeh, and S. R. Allahkaram, "Corrosion resistance enhancement of Ni-P electroless coatings by incorporation of nano-SiO_2 particles," *Materials and Design*, 32 (2011), 133-138.
[14] D. Dong et al., "Preparation and properties of electroless Ni-P-SiO_2 composite coatings"*Applied Surface Science*, 255 (2009), 7051-7055.
[15] S. Alirezaei et al., "Wear behavior of Ni-P and Ni-P-Al_2O_3 electroless coatings," *Wear,* 262 (2007), 978-985.
[16] K. Keong, W. Sha, "Crystallisation and phase transformation behaviour of electroless nickel-phosphorus deposits and their engineering properties," *Surface Engineering,* 18 (2002), 329-343.
[17] Y. Song et al., "Corrosion characterization of Mg–8Li alloy in NaCl solution," *Corrosion Science*, 51 (2009), 1087-1094.
[18] H.P. Liu et al, "Effect of organic additives on the corrosion resistance properties of electroless nickel deposits," *Thin solid films*, 516 (2008), 1883-1889.
[19] H. Yang et al., "A homogenisation pre-treatment for adherent and corrosion-resistant Ni electroplated coatings on Mg-alloy AZ91D," *Corrosion Science*, 79 (2014), 41-49.
[20] J. Lian et al., "Electroless Ni-P deposition plus zinc phosphate coating on AZ91D magnesium alloy," *Surface and Coatings Technology*, 200 (2006), 5956-5962.
[21] H. Ashassi-Sorkhabi, B. Shaabani, and D. Seifzadeh, "Corrosion inhibition of mild steel by some Schiff base compounds in hydrochloric acid," *Applied Surface Science*, 239 (2005), 154-164.
[22] J. Balaraju, T. S. Narayanan, and S. Seshadri, "Evaluation of the corrosion resistance of electroless Ni-P and Ni-P composite coatings by electrochemical impedance spectroscopy," *Journal of solid state electrochemistry*, 5 (2001), 334-338.
[23] R. Abdel Hameed, and A. Fekry, "Electrochemical impedance studies of modified Ni-P and Ni-Cu-P deposits in alkaline medium," *Electrochimica Acta*, 55 (2010), 5922-5929.

Magnesium Technology 2015

Casting and Metal Matrix Composites

Magnesium Technology 2015
Edited by: Michele V. Manuel, Alok Singh, Martyn Alderman, and Neale R. Neelameggham
TMS (The Minerals, Metals & Materials Society), 2015

Process-Structure-Property Correlations for HPDC AM60B

Pouya Sharifi[1], Ying Fan[1], Kumar Sadayappan[2], Gabriel Birsan[2], Jeff Wood[1]

[1]Department of Mechanical and Materials Engineering, University of Western Ontario, London,

[2]Canmet MATERIALS, Hamilton, ON, Canada

Keywords: High pressure die casting, Modelling and simulation, AM60, Tensile properties

Abstract

In this paper, we characterize the local microstructure and mechanical properties of an AM60 magnesium alloy high-pressure die-casting in order to further develop predictive process-structure-property models. The casting process is simulated using the commercial code ProCAST™ in order to extract information regarding mold-filling and local solidification conditions including solidification rate and expected levels of shrinkage porosity. This data is used to predict the presence of knit lines, local porosity and the through-thickness variation of grain size (i.e. determination of skin and core thickness). Microstructural data is then used to predict upper and lower bounds on local mechanical properties including yield strength, tensile strength and ductility. A high degree of correlation between predicted and measured mechanical properties is found.

Introduction

Due to the excellent strength-to-weight ratio of magnesium alloys **[1]**, these alloys have been identified as potential materials enabling efficient weight reduction in the automobile industry. High-pressure die-casting (HPDC) is the preferred manufacturing for producing automobile components. The increasing use of magnesium alloy die-castings requires a better understanding of the relationship between process, microstructure and mechanical properties. The variations in mechanical properties have been found to depend upon features of the casting process, such as porosity, inclusions, oxide skins, bimodal distribution of grain size through the casting thickness **[1-3]**. This paper is aimed at developing predictive capabilities of mechanical performance of die-cast magnesium alloy from the microstructural features.

Background

Determination of skin thickness

The HPDC process produces different grain sizes across the thickness of the casting. This results in what are referred to as the skin and core regions. The skin region is comprised of fine grains due to a high solidification rate near the mold wall as compared to the interior. The core region contains a coarser grained microstructure, a higher percentage of defects and large externally solidified grains (ESGs) which are formed in the shot sleeve **[2,4]**. Some investigations **[4-6]** show the yield strength of HPDC Mg-Al alloy is controlled by the skin region. In order to predict mechanical properties, it is critical to determine the skin thickness or area fraction of that region. Recently, a model **[6]** has been developed to predict skin thickness using heat transfer equations and cooling rate–grain size relationship. An exact analytical solution to the Stefan problem was used to compute the one dimensional heat transfer for a section of the casting and to calculate its temperature distribution along the cross section. The cooling rates are then calculated with respect to its distance from the mold wall of the casting. The transition from the skin region to the core region can be identified by a grain size threshold value of

$$R_{10}=490\ (°C/s)$$

$$R_{11}=375\ (°C/s)$$

10 or 11 μm. The value of critical cooling rate was calculated from empirical relationship between cooling rate and grain size. Where, R_{10} and R_{11} are cooling rate values corresponding to the average grain size at 10 and 11μm, respectively. The point at which this threshold cooling rate straight line intersects with the cooling rate versus distance from surface, will give the thickness of the skin layer. In this study, the cooling rate value at central point was extracted from simulation results.

Modified Hall-Petch equation

According to the distribution of grains in microstructure of a die cast magnesium alloy, previous investigation **[7]** proposed the modified Hall-Petch equation, as:

$$\sigma_y = \sum f_i(\sigma_0 + Kd^{-\frac{1}{2}}) \qquad (1)$$

Where k and σ_0 are Hall-Petch parameters, f_i is the fraction of the sample thickness of each of the skin and core regions, and d is the grain size. This model can consider the effect of local grain size on the overall yield strength as compared to original Hall-Petch equation.

Failure model

Numerous studies have been conducted on the influence of porosity on the tensile elongation of die cast magnesium alloys. Ghosh **[8]** proposed the failure model which is able to predict the effect of internal porosity on the elongation. This model is well suited to microstructures of HPDC that contain pre-existing pores. Under the assumption that the material containing internal porosity f, exists in axial local equilibrium under tensile loading, the conventional equation for stress distribution can be expressed in terms of load carrying area as the following equation:

$$(\varepsilon_i)^n(1-f)\exp(-\varepsilon_i) = (\varepsilon_h)^n \exp(-\varepsilon_h) \quad (2)$$

Where n is the strain-hardening coefficient of the material and ε_i is the critical strain in the region surrounding the spherical pore and ε_h is fracture strain. Recently this model was modified by Weiler et al **[9]** who suggested a modified form of failure model which contains the strain concentration factor which considers the location parameter for the distribution of porosity. This modified failure model for HPDC alloys is given as:

$$\varepsilon_f = 0.016 f^{-0.3}\left[0.18Ln\left(\frac{d_t - r}{t}\right) + 1.17\right] \quad (3)$$

Where, d_t is the distance from the center of the equivalent pore to the edge of the sample; r is the radius of the equivalent pore and t is the thickness of the specimen. This model is able to clarify the role of porosity location on the fracture strain. In this study, the modified failure model was used to estimate the range of tensile elongation for corresponding area fraction of porosity.

Experimental and simulation methods

Experimental method

The alloy examined in this work is AM60 whose nominal chemical compositions are given in **Table 1**. Top hat automotive component of magnesium alloy is produced by high pressure die casting (HPDC) at Canmet MATERIALS in Hamilton, Ontario, Canada. During filling, a high vacuum in the die is used to reduce the level of gas porosity. The process parameters used for in this investigation, are shown in **Table 2**. To examine the effect of flow pattern on the microstructure and mechanical properties of die cast magnesium alloy, twelve tensile coupons are selected as shown in **Figure 1**.

Figure 1: Pictures showing locations of 12 tensile specimen cut from a die-cast Top hat

Sample preparation for microstructure examination was carried out by grinding with different grits of sand papers from 180 to 1200. The samples were further polished with a polishing cloth in slurries of 1μm, and finally 0.5μm diameter alumina. The grain structure was revealed by selectively etching in 2% Nital. Optical microscope and digital image analysis (DIA) software were then used to measure the average grain size and area fraction of porosity. In this study, three different terms were used to evaluate amount of porosity: bulk porosity, fractographic porosity and shrinkage porosity. The bulk porosity was obtained by measuring the area fraction of porosity into the surface which is close to the fracture surface. The fractographic porosity was expressed as the area ratio of porosity to the entire area of a fracture surface. The area of porosity on fracture surface was measured by scanning electron microscope (SEM) according to the quantitative fractographic technique. The shrinkage porosity was predicted by the commercial code ProCAST™.

Tensile testing was performed on an Instron servo hydraulic load frame with 250kN load cell at a constant strain rate of $0.05s^{-1}$. The load was monotonically increased until failure. The tensile specimens were machined to the dimensions specified in ASTM B557M.

Table 1: Chemical composition of AM60 alloy

Alloy	**Mg**	**Al**	**Zn**	**Mn**	**Si**	**Cu**	**Fe**	**Ni**	**Be**
AM60	93.6	6.1	-------	0.32	0.005	0.001	0.001	0.001	0.001

Table 2: The process parameters (boundary condition) used in the ProCAST™ simulation.

Process parameters	First stage shot speed (m/s)	Second stage shot speed (m/s)	Intensification pressure (bar)	Die temperature (^{0}C)	Melt temperature (^{0}C)	Vacuum (mbar)
Top hat	0.8-1	4.5-5	550-650	160	700	<30

Table 3: The values of thermophysical variables were used in the ProCAST™, simulation.

Parameters	T_l(°C)	T_s(°C)	α(m²/s)	k(W/m °C)	ρ(kg/m³)	c(J/kg °C)	L_f(kg/J)
	615	565	0.044	101.8	1800	1333.8	370000

Simulation method

In this study, the commercial software ProCAST™, was used to optimize the process parameters before conducting HPDC process. This commercial software is a finite-element program that solves thermal and fluid equations in casting systems. The element sizes used for the mesh ranged from 0.9–1.3mm for the casting, and from 7.0–14.0mm for the die sections. The thermo-physical properties used for this simulation are presented in **Table 3**. The process parameters used in this simulation are also shown.

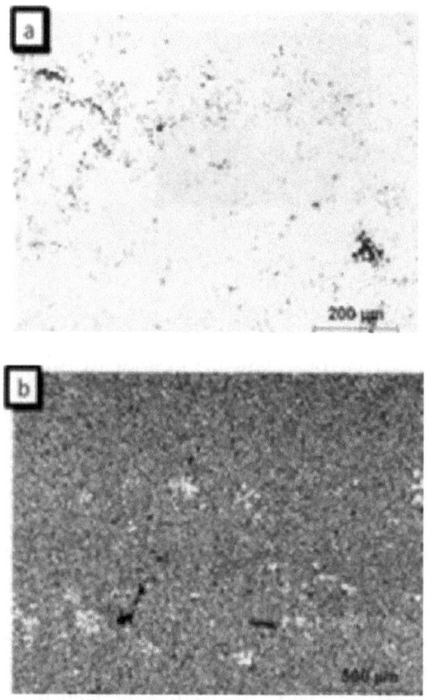

Figure 2: Optical pictures of AM60 HPDC: a) As-polished b) Etched. The presence of porosity and shrinkage can be observed in (a).The variation of grain size between skin and core region are shown in (b).

Result and discussion

Figure 2 shows the typical as-polished microstructures for a sample of the Top hat casting. The area fraction of porosity was measured by optical metallography; fractographic porosity was measured by the SEM and the predicted shrinkage porosity was predicted via ProCAST™, for each sample. Typical SEM images of the fracture surface of die cast AM60 alloy are presented in Figure 3. The filling and solidifying process at different times are shown in **Figure 4**. After filling into cavities the casting changes from liquid state to solid state with the decrease of temperature. In this model, fluid flow simulation starts in the sleeve and comprises the moving boundary of the plungers as indicated in Table 2. Note that we assumed moving liquid metal into cavity towards the gate, preferably without entrapping the air in the cold chamber. This means the simulation is able to predict only shrinkage porosity. However, the simulation results give us an opportunity to identify the locations which have specific filling or solidification conditions such as: last-to-solidify region, last-to-fill region and knit line. These terms are discussed in the Ref. **[4]**.

The fracture surfaces contain shrinkage porosity as well as gas porosity. In the case of Top hat, there is no notable variation in the solidification time as shown in **Figure 4**. As a result there is neither last-to-solidify region nor last–to-fill region close to the overflow. As can be seen in **Figure 4**, simulation results indicate several knit lines occur close to the gate. These regions have a greater chance of porosity and oxide formation due to the fact that two molten metal fronts meet.

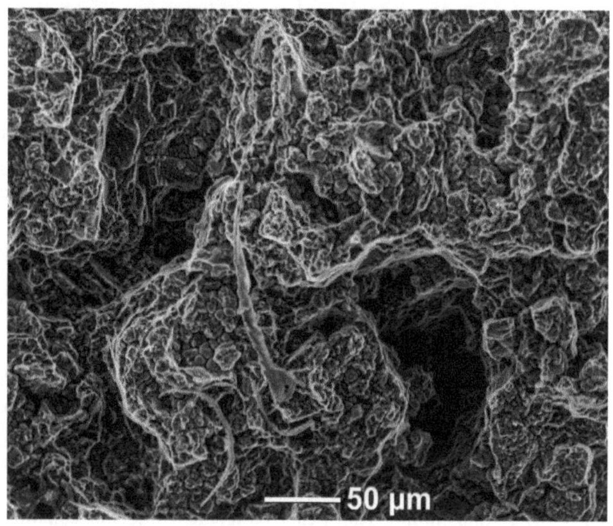

Figure 3: SEM image of a fracture surface of the HPDC AM60 alloy.

Figure 2 shows the typical microstructure of sample which is extracted from Top hat. It is observed that the ESGs occupy significantly larger fractions of the volume at the core region whilst farther away from the center the ESGs are more dilute. It is also noted that the fraction of ESGs decreases with distance from the gate. This morphology of grains formed in the shot sleeve prior to inserting the die cavity and the migration of them to the center as a result of lift forces acting on the solid. The ESGs increase the average grain size and consequently should have a

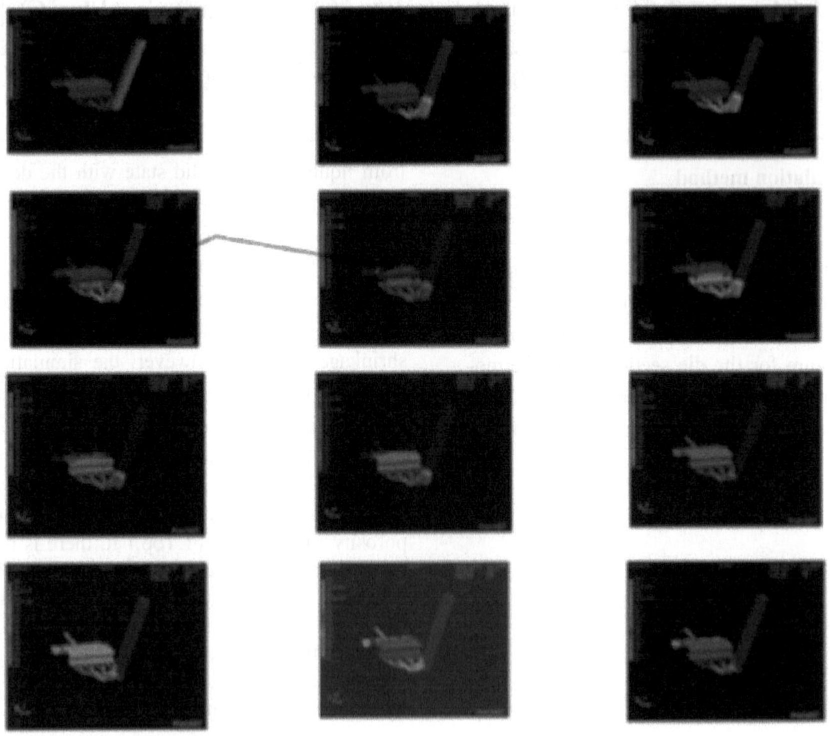

Figure 4: Filling sequence and solidification conditions predicted in Pro-cast. Arrows show the location of knit lines

deleterious impact on the yield strength. The literature **[4]** shows that fine grains of α-magnesium are 2 times harder than ESGs. Thus, either the lack of ESCs in the skin region or the decreased presence of ESGs in the core region implies higher hardness and yield strength.

To quantitatively determine the effects of grain size on the yield strength, the modified Hall-Petch equation was applied. In a previous paper by our research group, we used microindentation testing to determine the Hall-Petch equation for average grain sizes from 4.8-75 μm and found $\boldsymbol{k}$ = 152.9 $\boldsymbol{MPa\text{-}\mu m^{-1/2}}$, and $\boldsymbol{\sigma_o}$ = 62.1 ***MPa***. Substituting Hall-Petch parameters into the equation (1), gives.

$$\sigma_y = \sum f_t (62.1 + 152.9 d^{-\frac{1}{2}})$$

The data measured in this study, average grain sizes range from 5 to 9 μm in the skin region, and from 12 to 17 μm in the core region. It should be noted that the average grain size for core region without ESGs ranges from 9 to 11 μm and the sizes of ESGs are from 38 to 40 μm. Thus, the area fractions of three different regions were measured to be used in the modified Hall-Petch equation. **Figure 5** compares the yield strength obtained by experiments and prediction using modified and conventional Hall-Petch equations. The errors for the predictions are 1.9% and 7% for the modified and conventional Hall-Petch equation, respectively. This finding confirms that the yield strength predicted by the modified Hall-Petch is in very good agreement with the experimental results, better than the predictions of the conventional Hall-Petch equation.

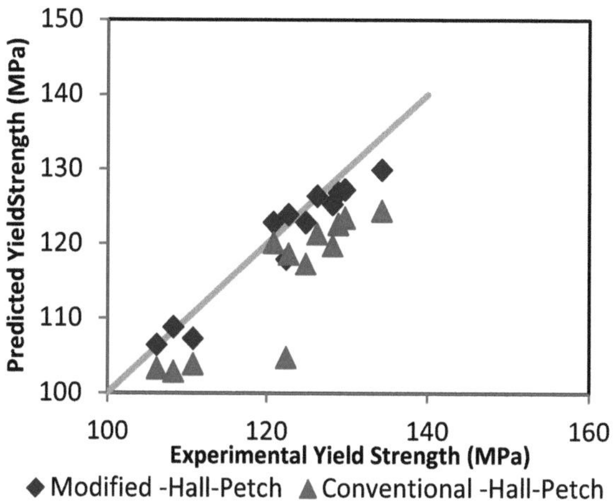

Figure 5: Accuracy of predicted yield strengths of specimens cut from a Top-Hat die casting

The modified failure model discussed above was investigated to estimate the experimental tensile elongation of the HPDC magnesium alloys using the measured porosity data. This model requires the inputs of the area fraction and the location of the porosity. Due to the skin region of HPDC magnesium alloys is typically free from porosity, we assume a minimum tensile elongation when the pore is located adjacent to the skin, while a maximum tensile elongation occurs for pores located at the central point of the specimen. The thickness of the skin region then can be used as a minimum distance from a possible pore location to the specimen surface, substituting for the parameter expression: (d_t- r) in Eq. (4). The skin thickness has been calculated using the approach described above. The simulation results show that the solidification time is similar for all locations. Thus, the calculated skin thickness is the same for all locations. Finally, we calculated the skin thickness of 400 μm for the die-cast Top hat with a thickness of 2.6 ***mm*** (±0.1 ***mm***).

The predicted values of tensile elongation (lower bound or upper bound) using the modified models are presented in **Figure 6**, in comparison with the experimental results. The area fraction of porosity was measured with three methods for each sample: bulk porosity, fractographic porosity and predicted shrinkage porosity. In the case of fractographic porosity, the tensile elongations generally fit well with the prediction while the shrinkage porosity shows considerable variation with the predicted range. This is attributed to the simulation which neglects the gas porosity in the measurement of area fraction of porosity. In the case of bulk porosity, the average error percentage is 7.1% as compared to the predicted range. **Figure 6** also exhibits there is slightly difference between the predicted elongation versus fractography and the tensile elongation. It should note that the area fraction of porosity used in the Ghosh model or the modified model is based on the porosity measurement of pre-facture. Thus, the mentioned variation might be attributed to the growth of internal defects during tensile testing, especially in higher elongation values. However, the tensile elongation versus fractography predicted by the modified failure model is in very good agreement with experimental results. This model gives us a possibility to estimate the local elongation across a die-cast component with the knowledge of local area fraction of porosity and skin thickness.

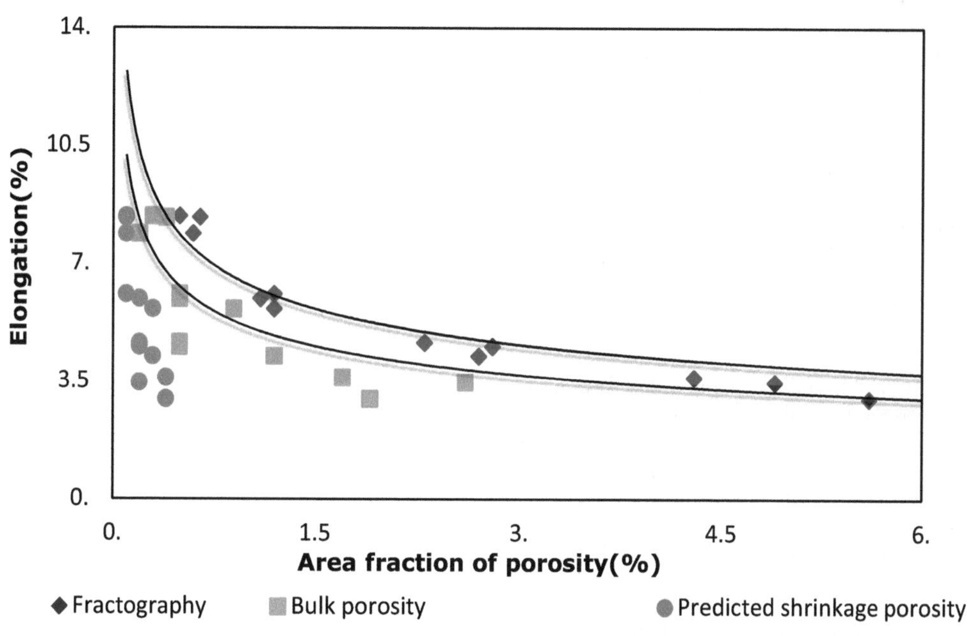

Figure 6 Predicted tensile elongations as a function of area fraction of porosity using the modified analytical model compared with experimental tensile elongations

Conclusion

This paper described the HPDC process simulation for AM60 Top hat and the correlation between microstructural features and tensile properties. Simulation results give us the opportunity to characterize the location of knit lines, local porosity and the through-thickness variation of grain size. It has been demonstrated that the tensile yield strength is dependent on the local grain size distribution and the skin thickness. The macroscopic ductility is dependent upon the local area fraction of porosity with a lower-bound dictated by the combination of pore fraction and skin thickness.

ACKNOWLEDGMENTS

The work is jointly supported by NSERC - Automotive Partnership Canada program and NRCan's Office of Energy R&D through the Program on Energy R&D. Technical support from the Magnesium Front End R&D (MFERD) research team at USAMP is appreciated. The authors of this paper would also like to thank the casting group at Canmet MATERIALS for their assistance in performing casting trials.

References

[1] A.K. Dahle et al, "Solidification of cast magnesium alloy" J. Light Metals 1 99–103(2001).

[2] J.P. Weiler et al., "Relationship between internal porosity and fracture strength of die-cast magnesium AM60B alloy", Materials Science and Engineering, A395 (2005), 315-322.

[3] A.M. Gokhale and G.R. Patel, "Origins of variability in the mechanical properties of AM60 magnesium alloy castings", Magnesium Technology 2001, ed. J. Hryn (Warrendale, PA:TMS, 2001), 195-199.

[4] J.P. Weiler et al., "Variability of skin thickness in an AM60B magnesium alloy die-casting," Materials Science and Engineering A, A419 (2006), 297-305.

[5] P.Sharifi et al., " Predicting the flow stress of high pressure die cast magnesium alloys," Journal of Alloys and Compounds 605 (2014) 237–243.

[6] P.Sharifi et al., " A novel non-experimental prediction of skin thickness in magnesium alloys using heat-transfer model" submitted to Alloys and compounds (2014).

[7] J.P. Weiler et al.," Structure -property realtionship For die-cast magnesium alloys ", Magnesium Technology 2010, ed. J. Hryn (Warrendale, PA:TMS, 2010), 413-417.

[8] A.K. Ghosh, "Tensile instability and necking in materials with strain hardening and strain-rate hardening," Acta Materialia, 25 (1977), 1413-1424.

[9] J.P. Weiler et al., "Modeling fracture properties in a die-cast AM60B Magnesium alloy II – the effects of the size and location of porosity determined using finite element simulations,"Materials Science and Engineering A, A527 (2009), 32-37.

[10] S. Biwaz et al. " Relationship Between the 3D Porosity and b-Phase Distributions and the Mechanical Properties of a High Pressure Die Cast AZ91 Mg Alloy" Metallurgy And Materials Transactions A, Volume 44A(2013).

Magnesium Technology 2015
Edited by: Michele V. Manuel, Alok Singh, Martyn Alderman, and Neale R. Neelameggham
TMS (The Minerals, Metals & Materials Society), 2015

INFLUENCE OF ALLOY COMPOSITION ON CAST CRACKING AND HEAT RESISTANCE OF Mg-Al-Ca CAST ALLOY

Hajime Kato[1], Hiroyuki Kawabata[1], Shuji Inoue[2], Masaaki Kondo[3]
[1]Toyota Central R&D Labs., Inc.; 41-1 Yokomichi, Nagakute, Aichi 480-1192, Japan
[2]Aisin Seiki Co., Ltd.; 2-1 Asahi-Machi, Kariya, Aichi 448-8650, Japan
[3]Toyota Motor Corp.; 1 Toyota-Cho, Toyota, Aichi 471-8572, Japan

Keywords: Magnesium, Casting, Castability, Cast Cracking, Heat Resistance

Abstract

Magnesium alloys are important materials since they can be used to significantly reduce the weight of automobiles, but their application has been limited thus far. One possible way to extend the usage of magnesium alloys is applying them to peripheral parts of the engine that are large, have complicated shapes, and are exposed to high temperatures under the loaded condition. However, heat-resistant magnesium alloys always have poor castability, since such alloys have a lot of intermetallic compounds with poor ductility.

In this study, hot cracking tests of Mg-Al-Ca cast alloys were performed in a high-pressure die-casting machine using a die with an I-shaped cavity to investigate the castability. At the same time, the heat resistance of the cast alloys was evaluated using a bolt load retention test. It was found that adding tin to heat-resistant Mg-Al-Ca improved its hot cracking behavior without reducing its excellent heat resistance.

Introduction

Magnesium is the lightest of all the practical metals. To achieve significant automobile lightening, it is expected that the usage of magnesium alloys will be extended to applicable parts. Currently, magnesium alloys are used for indoor parts under relatively low thermal load conditions. In order to extend the usage of magnesium alloys, they must be applied to parts under high-temperature and high-load conditions, which require large and complex shapes. To realize this, magnesium alloys require higher heat-resistance and better castability than general-purpose magnesium alloys.

Adding calcium to Mg-Al based alloys enhances their heat resistance, and several reports have been published on the heat resistance of Mg-Al-Ca alloys [1-3]. However, a method of quantitatively evaluating the castability of these alloys has not been established.

Cast cracking is classified into "hot tearing" and "hot cracking". "Hot tearing" means the rupture prior to final solidification, and "hot cracking" means the rupture of the fully solid casting during cooling. Both types of cast cracking in aluminum alloys have been investigated. Hot tearing has been investigated by high-pressure die casting with a T-shaped cavity [4, 5]. Hot cracking has been investigated by high-pressure die casting with an I-shaped cavity [4]. On the other hand, the evaluation of cast cracking in heat-resistant Mg-Al-Ca alloys cast by gravity casting [6] and the solidification transferability of the alloys [7] have been investigated. However, the evaluation of cast cracking for alloys cast by high-pressure die casting has hardly been reported on.

Al_2Ca intermetallic compound, which solidifies during the solidification process of Mg-Al-Ca alloys [8, 9], decreases the ductility of the alloys [10]. Therefore, cast cracking occurs due to the restraint provided by the casting molds after solidification. Namely, the principal factor of cast cracking in Mg-Al-Ca alloys is hot cracking.

In this study, to achieve excellent castability and high heat resistance, the effect of the aluminum and calcium content on the cast cracking of Mg-Al-Ca cast alloy was investigated by the high-pressure die-casting of a die with an I-shaped cavity. Moreover, the effect of the addition of tin on cast cracking was also investigated.

Experiment

Evaluation of castability

Mg-(5–9%)Al-(0–3%)Ca-0.3%Mn-0.5%Si alloys were cast with ingots of pure Mg(99.99%), pure Al(99.99%), and Mg-3.3%Mn and grains of pure Ca(99.5%) and pure Si(99.5%). These alloys, which were 1.2 kg per batch, were melted in a steel crucible. The casting temperature was 50 K above the liquidus temperature, the shot pressure was 64 MPa, and the shot velocity was 0.4 m/s. Figure 1 shows a schematic illustration of the die with an I-shaped cavity used in this research. The parallel length was varied in the range of 20–90 mm. Thermal insulating material was adhered to the center of the parallel part. The molten metal was injected from the bottom to the top of the cavity. Then, solidification progressed from the upper and lower parts toward the center of the I-shaped cavity, and final solidification took place at the center of the parallel part. The alloys were cast by high-pressure die casting, and the parallel part length was changed. The parallel lengths (x) without cast cracking were investigated, and the maximum length was defined as the maximum parallel length (L).

Evaluation of heat resistance

Alloys melted using the method mentioned above were cast by high-pressure die casting into a mold with a rectangular cavity of dimensions 20×22×50 mm^3. Cylindrical specimens were prepared for a bolt load retention test by scraping the casts. The specimens were fastened with bolts, which applied a compressive load of 15 kN, and were then kept at 175°C for 300 hours in a furnace. The fraction of the retained bolt load was calculated according to the ratio between the bolt load after the test and the initial bolt load.

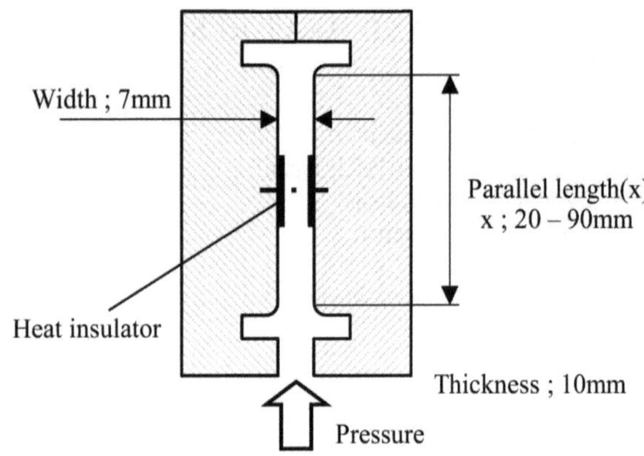

Figure 1. Schematic illustration of die with I-shaped cavity.

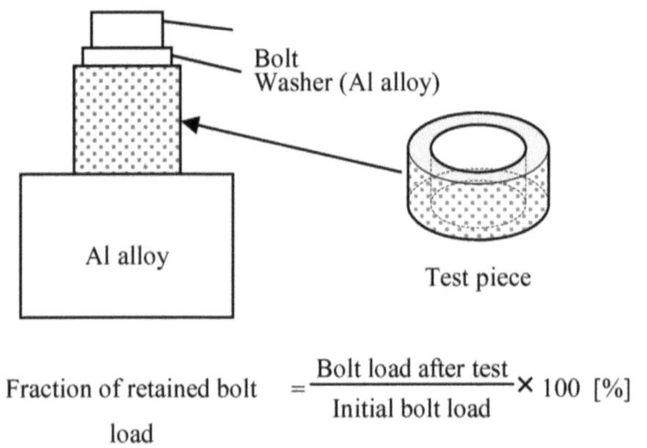

Figure 2. Schematic illustration of bolt load retention test method.

Results and Discussion

Mg-Al-Ca alloy castability

Figure 3 shows the relationship between the maximum parallel length and the amount of calcium and aluminum in Mg-(5–9)%Al-(0–3)%Ca-0.5%Si-0.3%Mn alloys. In the alloys containing less than 0.5% calcium, the maximum parallel length was 80 mm regardless of the aluminum content. In the alloys containing more than 0.5% calcium, the maximum parallel length shortened, and cast cracking easily occurred with increasing calcium content regardless of the aluminum content. Specifically, in the alloy containing 9% aluminum, increasing the calcium content from 2% to 3% sharply made the maximum parallel length short and the castability poor. In addition, as the Al content decreased, the maximum parallel length of the alloys with the same Ca content shortened.

Figure 4 shows the microstructures of the center of the parallel part for Mg-9%Al-3%Ca-0.5%Si-0.3%Mn alloy (whose maximum parallel length was shortest) and AZ91D alloy (whose maximum parallel length was longest). Al_2Ca intermetallic compounds were observed in Mg-9%Al-3%Ca-0.5%Si-0.3%Mn alloy, which included a high percentage of calcium. The cracking propagated along the eutectic structures of the alloy. The quantity of Al_2Ca compounds increased with an increase in the quantity of Ca in Mg-Al alloys, ductility became poor, and cast cracking occurred [8]. On the other hand, β phase ($Mg_{17}Al_{12}$ intermetallic compound) and small cracks are observed in the microstructure of the AZ91D alloy. It is considered that the small cracks are shrinkage cavities formed by tensile stress on the residual liquid phase just before the completion of solidification. Because these shrinkage cavities were formed, the tensile stress from the solidification shrinkage relaxed, and the maximum parallel length increased. In Figure 3, because the area of the eutectic phase decreased with a decrease in Al content and the effect of relaxing the tensile stress from the solidification shrinkage decreases, the maximum parallel length of the alloys with the same content of Ca decreases with a decrease in Al content.

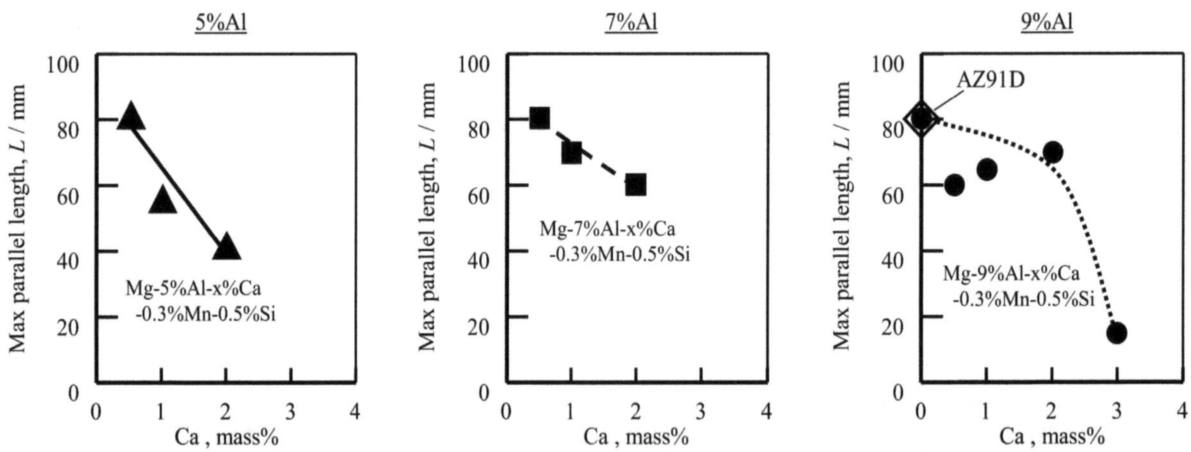

Figure 3. Relationship between the maximum parallel length and the amount of calcium and aluminum in Mg-(5–9)%Al-(0–3)%Ca-0.5%Si-0.3%Mn alloys.

In order to increase the maximum parallel length (that is, to improve cast cracking), it is very effective to relax the solidification shrinkage stress by forming shrinkage cavities. Therefore, in Mg-Al-Ca alloys, tin was chosen as the additional element, which increases the area of the eutectic phase without deteriorating the heat resistance. Figure 5 shows the effect of the tin addition on the maximum parallel length. The maximum parallel length increased by adding tin. Figure 6 shows the appearance of the casting and the microstructure of the center of the parallel part for Mg-5%Al-2%Ca-0.5%Si-0.3%Mn-0.2%Sn alloy. Cast cracking did not occur in the specimen, but small cracks were observed in the microstructure of the specimen. It was considered that cast cracking was improved due to the generation of small shrinkage cavities.

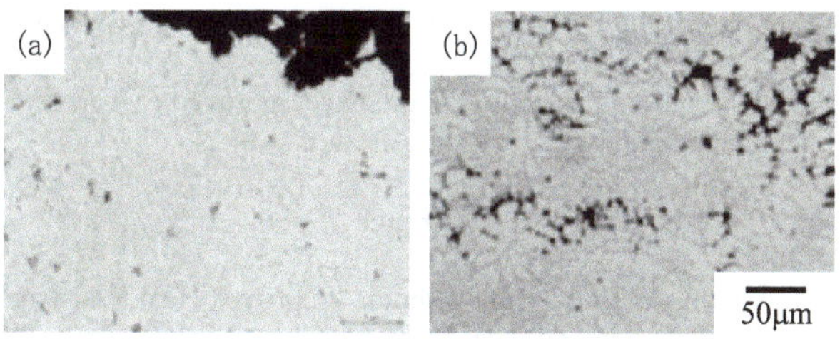

Figure 4. Microstructure of test pieces: (a) Mg-9%Al-3%Ca-0.5%Si-0.3%Mn alloy, (b)AZ91D.

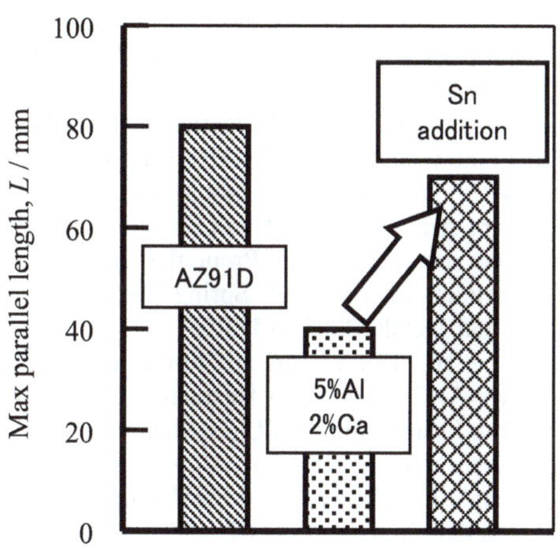

Figure 5. Effect of Sn addition on the maximum parallel length.

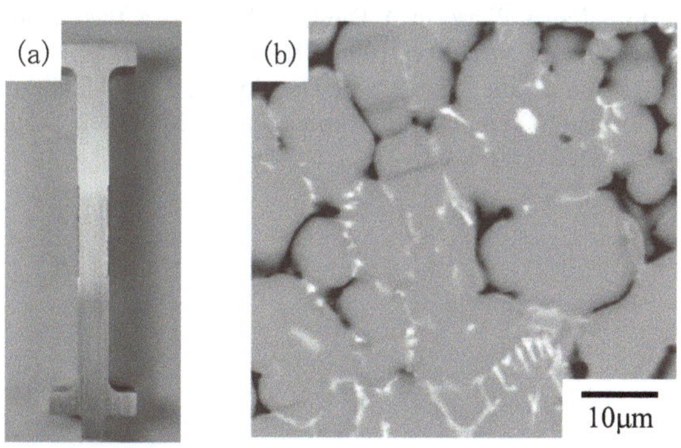

Figure 6. (a) Appearance and (b) microstructure of Mg-5%Al-2%Ca-0.5%Si-0.3%Mn-0.2%Sn alloy test pieces.

Heat resistance of Mg-Al-Ca alloy

Figure 7 shows the microstructure and electron probe microanalysis (EPMA) maps of Mg-5%Al-2%Ca-0.5%Si-0.3%Mn-0.2%Sn alloy. For the compounds containing Ca, the Ca-Al compound (Al_2Ca) and the Ca-Sn compound were used. Namely, the tin added to Mg-Al-Ca alloy is spent on the formation of the Ca-Sn compound, while Al_2Ca structures (which can exhibit excellent heat resistance for the alloys [1]) are maintained. Simultaneously, because the Ca-Sn compounds formed eutectic structures, a residual liquid phase was formed just before the completion of solidification and relaxed the solidification shrinkage stress. In other words, it can be assumed that the tin addition can improve the cast cracking tendency by maintaining the heat resistance of Mg-Al-Ca alloy.

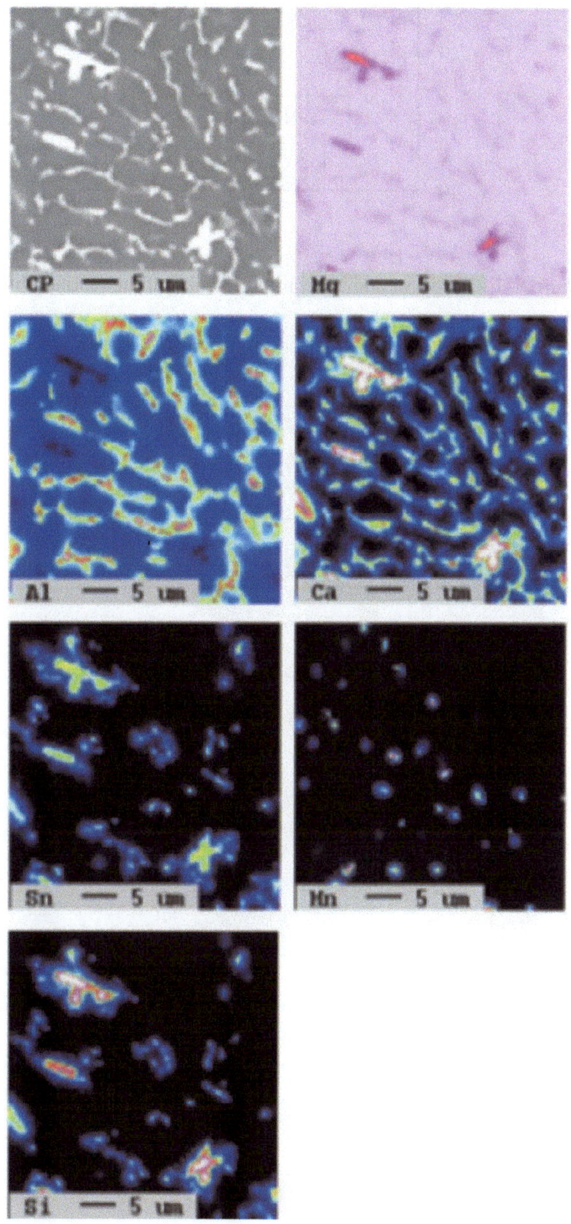

Figure 7. Microstructure and electron microprobe analysis maps of Mg-5%Al-2%Ca-0.5%Si-0.3%Mn-0.2%Sn alloy.

Figure 8 shows the result of the bolt load retention test. The fraction of the retained bolt load of Mg-6%Al-2.2%Ca-0.5%Si-0.3%Mn-0.1%Sn alloy was bigger than that of AZ91D alloy. As mentioned above, the tin addition can improve the cast cracking tendency while maintaining the heat resistance of Mg-Al-Ca alloy.

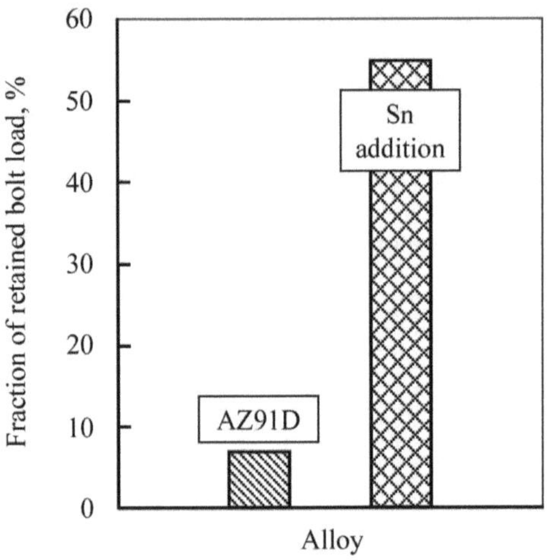

Figure 8. Effect of Sn addition on the fraction of the retained bolt load.

Summary and Conclusion

Cast cracking of Mg-Al-Ca alloys was investigated by high-pressure die casting with a die having an I-shaped cavity. Moreover, the fraction of the retained bolt load for the alloys was investigated. The results obtained are summarized as follows:

1. Increasing the calcium content and decreasing the aluminum content shortened the maximum parallel length and reduced cast cracking.
2. Adding tin to Mg-Al-Ca alloy improved the cast cracking of the alloys.
3. Cast cracking was improved by the relaxation of the solidification shrinkage stress due to the formation of shrinkage cavities.
4. Adding tin to Mg-Al-Ca alloys maintained the formation of the Al_2Ca intermetallic compound and formed a Ca-Sn based intermetallic compound.
5. For these reasons, the alloys could have improved cast cracking while maintaining excellent heat resistance.

Reference

1. R. Ninomiya, T.Ojiro and Kubota, "Improved Heat Resistance of Mg-Al Alloys by The Ca Addition," Acta Metall. Mater., 43 (1995), 669-674.
2. S. M. Zhu, B. L. Mordike and J. F. Nie, "Creep Properties of A Mg-Al-Ca Alloy Produced by Different Casting Technologies," Mater. Sci. Eng., A483-484, (2008), 585-586.
3. J. Bai et al., "Effect of Al Contents on Microstructures, Tensile and Creep Properties of Mg-Al-Sr-Ca Alloy," J. Alloys Comp., 437 (2007), 247-253.
4. T. Horie et al., "Hot Crack Evaluation in High Pressure Die-Casting of Light Weight Alloys," (Abstracts of the 2005 Spring Meeting of The Japan Institute of Light Metals, 2005), 155-156.
5. H. Kawahara et al., "Effect of Iron Content on The Hot Tearing Characteristics and Mechanical Properties in Al-Mg-Mn Cast Alloys," (Abstracts of The 2005 Spring Meeting of The Japan Institute of Light Metals, 2005), 157-158.
6. Y. Nagano et al., "Hot Tearing and Solidification Structures in Sr Added Mg-Al-Ca Alloys," (Abstracts of The 2014 Spring Meeting of The Japan Institute of Light Metals, 2014), 125-126.
7. H. Kawabata et al., "Effect of Alloy Composition on Heat Resistance and Castability of Mg-Al_2Ca(Mg) Quasi-Binary Alloys," J. JILM, 60 (2010), 596-601.
8. S.M. Liang et al., "Thermal Analysis and Solidification Pathways of Mg-Al-Ca System Alloys," Mater. Sci. Eng., A480 (2008), 365-372.
9. A. Suzuki et al., "Structure and Transition of Eutectic $(Mg,Al)_2$Ca Laves Phase in A Die-Cast Mg-Al-Ca Base Alloy," Scripta Materialia, 51 (2004), 1005-1010.
10. E. Aghion et al., "The Art of Developing New Magnesium Alloys for High Temperature Applications," Materials Science Forum, 419-422 (2003), 407-418.

Magnesium Technology 2015
Edited by: Michele V. Manuel, Alok Singh, Martyn Alderman, and Neale R. Neelameggham
TMS (The Minerals, Metals & Materials Society), 2015

MICROSTRUCTURAL SCALE EFFECTS ON THERMAL EXPANSION BEHAVIOUR OF CAST AZ91D

Hoda Dini, Nils-Eric Andersson, Ehsan Ghassemali, Anders E. W. Jarfors
Jönköping University, School of Engineering, Department of Materials & Manufacturing, P.O. Box 1026, SE-551 11, Jönköping, Sweden

Keywords: Microstructure, AZ91D, $Mg_{17}Al_{12}$, Magnesium, Thermal expansion

Abstract

The effect of microstructure on thermal expansion of AZ91D cast alloy was studied. Samples with equiaxed grains and a controlled secondary dendrite arm spacing (SDAS) were fabricated using gradient solidification. SDAS was chosen to represent the range of microstructural scale found in sand castings down to that of high pressure die casting. Optical microscopy and electron backscatter diffraction (EBSD) were used for microstructural characterization. The relation between thermal expansion and microstructural scale of existing phases precipitated, in particular grain size, SDAS and fraction of $Mg_{17}Al_{12}$ was analyzed.

Introduction

Cast AZ91D (Mg–9wt.%Al–1.0wt.%Zn) has applications in automotive industry, and in hand-held tools and equipment. Examples of these are for instance crank case housings, due to the high strength-to-weight ratio, castability and ductility of AZ91D. The crank case is commonly exposed to thermal cycling in the temperature range from 293 K to 473 K. The main application limiting factor of AZ91 is its rapid loss of strength with increasing temperature as well as limited creep resistance [1-3].

Mechanical properties especially creep resistance is extensively studied, mostly in die cast conditions [4-7]. On the other hand, some studies show that AZ91 produced by different methods can exhibit significantly different properties. These differences are attributed to different microstructural features such as α-grain size and fraction of $Mg_{17}Al_{12}$ [8-10]. Hence, in order to study the high temperature properties of the alloy and assessing stress and strain, it is important to know the correlation between microstructure and thermal expansion coefficient (CTE) as this, in many cases, is the cause of stress that generate creep.

The correlation of microstructure with thermal properties of AZ91 is a subject which has not been studied. This paper focuses on the influence of the microstructure on the thermal expansion properties of AZ91D alloy.

Experimental procedure

Preparation of material

Commercial purity AZ91D, Table 1, was melted in a mild steel crucible under a dry air–0.5% SF_6 gas mixture. The melt was sucked into 1000 mm long and 10 mm in diameter mild steel tubes, preheated to 523 K. The cast rods were cut into lengths of 170 mm and cleaned. Re-melting was made in Bridgman furnace under a dry air–0.5% SF_6 gas mixture. At a temperature of 923 K the solidification with a controlled rate was achieved by moving the furnace upwards away from the stationary sample which resulted in a well-controlled solidification rate and well-defined secondary dendrite arm spacing (SDAS). The pulling rate of the furnace was varied from 0.3mm/s up to 6mm/s.

Table 1. The chemical composition of the AZ91D alloy (wt.%)

Mg	Al	Zn	Mn	Fe	Ni	Cu	Si
Bulk	8.5	0.67	0.25	0.0023	0.0017	0.021	0.017

Microstructural characterization

Longitudinal cross-sections were prepared to investigate the different microstructures under optical microscopy. Sample preparation consisted of mechanical grinding and polishing by alcohol based diamonds suspension of 9 µm, 3 µm and 1 µm. For optical microscopy samples were etched with 10 ml HF (48%) in 90 ml water to darken $Mg_{17}Al_{12}$ [11]. When determining the particle size (diameter), two transversal lengths having perpendicular directions were drawn for each particle in the micrograph. Normally, at least 100 measurements were made in each sample. The area fraction of phases were defined by using an Olympus Stream motion 1.8 image analysis system, using contrast based recognition and a particle size discrimination. Five representative images from each sample were studied for this evaluation. The shape factor of particles were obtained from $4\pi A/p^2$ relation where A and p are area and perimeter of particles respectively. The $Mg_{17}Al_{12}$ phase precipitates along triple junction grain boundaries and cross-section based measures capture the degree of connectivity poorly. The secondary dendrite arm spacing (SDAS) was determined using light-optical microscope with measuring more than 50 dendrites for each specimen for sufficient accuracy. To assess actual grain size, electron backscattered diffraction (EBSD) was applied.

Coefficient of thermal expansion

The coefficients of thermal expansion (CTE) of AZ91D was determined using a NETZSCH 402PC dilatometer at 5 K/min heating rate and with 1.2 min^{-1} Ar gas flow. Displacement of the 12 mm long AZ91 samples as a function of temperature (318 K to 430 K) was measured using an alumina probe. CTE was then calculated from the displacement data.

Differential scanning calorimetric

Disk shaped samples having 42.1 mg weight were prepared for differential scanning calorimetric (DSC) measurements with five times temperature cycle from 300 K to 430 K. Heating rate was 5 K/min. and cycles included static or dynamic temperature steps in order to achieve desired temperature. The measurements were carried out with NETZSCH 404 C calorimeter.

Results and discussion

Microstructure

Figure 1 shows AZ91D optical microstructures for fast and slow solidification speed. The observed equiaxed microstructures contain α -Mg (white phase) surrounded by $Mg_{17}Al_{12}$ particles (dark gray phase). Intermetallic precipitated in different shapes at grain boundaries as well as inter-dendritic regions. The sizes of the $Mg_{17}Al_{12}$ particles are measured to be mostly about several micrometers (see Table 2). Although the shape factor of this phase for different solidification speed showed the same type of wide distribution ranging from very low to very high value, area fraction value was quite different. Increasing the solidification speed leads to lower area fraction of $Mg_{17}Al_{12}$ intermetallic. Table 2 is shown the microstructural scale for cast AZ91D with different solidification speed. The EBSD orientation mapping of the samples are illustrated in Figure 2. The maps show random orientation of the grains suggesting that the sample manufacturing did not result in any texture and grains have no preferred directionality. The assumption of isotropic behavior on sample level could thus be made. Besides, it is observed that increasing the furnace drawing rate leads to increasing the average grain size. This is due to the fact that the nucleation was different in the slow and fast drawing rate.

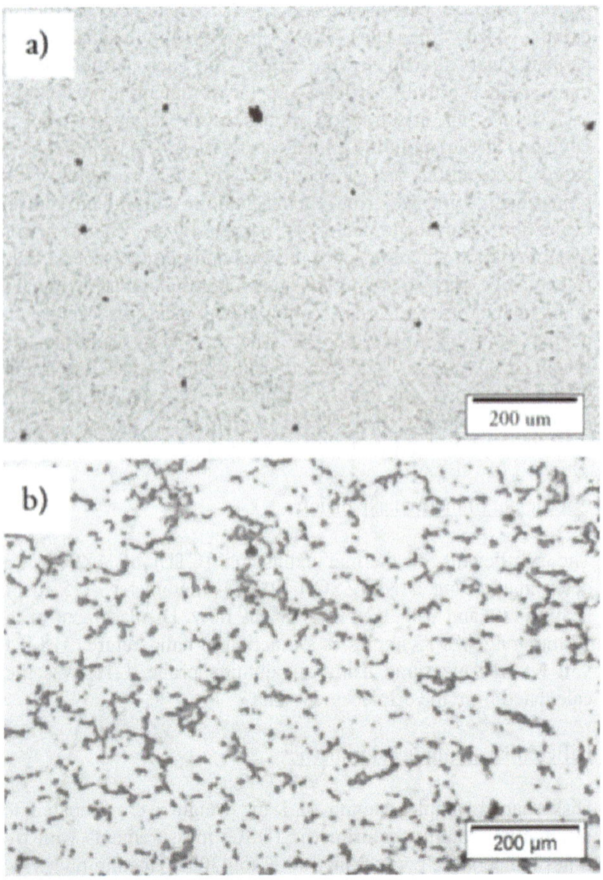

Figure 1. Optical micrograph illustrating SDAS (a) (5.4±2.1 µm) for samples drawn at 6 mm/s and (b) (26.9±2.4 µm) for samples drawn at 0.3 mm/s.

Table 2. Experimental results for microstructural scale obtained by using 6 and 0.3 mm/s drawing speed

Solidification speed (mm/s)	6	0.3
SDAS (µm)	5.4±2.1	26.9±2.4
Minimum diameter of $Mg_{17}Al_{12}$ (µm)	0.8	1.3
Maximum diameter of $Mg_{17}Al_{12}$ (µm)	16.5	24.2
Area fraction of $Mg_{17}Al_{12}$ (%)	7.1±1.1	11.3±1.4
$Mg_{17}Al_{12}$ shape factor $4\pi A/p^2$	0.64±0.4	0.67±0.5
Grain size (µm)	206±6	90±7

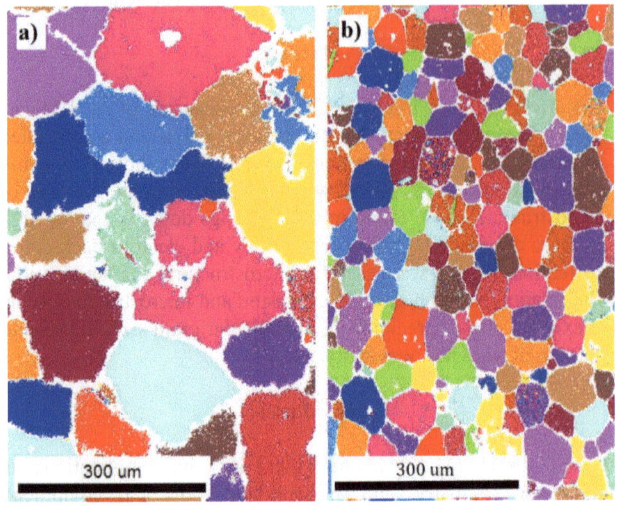

Figure 2. Illustration of EBSD orientation map for (a) (206±5.6 µm) for samples drawn at 6 mm/s and (b) 90.3±7.3 µm) for samples drawn at 0.3 mm/s. Non-indexed regions (white regions) around the grain boundaries represent the gamma intermetallic phases. Non-indexed reigns inside the grains indicate defects and/or deformed area.

Coefficient of thermal expansion

Figure 3 shows the relative elongation of alloy from 323 to 423 K for five cycles. It can be seen that the cast alloy with low fraction of intermetallic expanded more than alloy containing higher intermetallic phase fraction.
The coefficient of thermal expansion (CTE) at a particular temperature is calculated as follows:

$$\alpha = \frac{\partial}{\partial T}\left(\frac{\Delta l}{l}\right) \quad \text{Eq 1}$$

where l is the actual length of specimen at T.

The mean linear thermal expansion coefficient can be derived as follows:

$$\bar{\alpha} = \frac{1}{l_0}\left(\frac{\Delta l}{\Delta T}\right) \quad \text{Eq 2}$$

where l_0 is the initial specimen length, Δl is the change in length over a temperature interval ΔT.

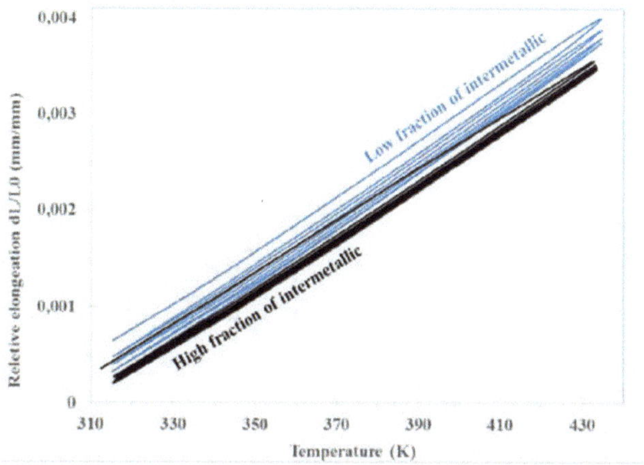

Figure 3. Relative elongation of AZ91D cast alloy containing different level of intermetallic fraction.

The differential linear thermal expansion coefficient of AZ91D for different microstructure in the temperature range of 323 to 420 K is represented in Figure 4. It is indicated that the value of CTE increases with increasing temperature. In the beginning of temperature range of 323 to 348 K the rate of thermal expansion increased faster, which can be attributed to the residual stress. Residual stress due to the phases in and around grains is released by increasing temperature [12].

It is revealed that CTE values for sample with lower fraction of $Mg_{17}Al_{12}$ are higher than the value for samples with higher fraction of intermetallic (see Figure 4). As it was showed earlier by optical microstructure investigation the fast solidification speed (6 mm/s drawing rate) has lower amount of intermetallic in microstructure (7.1±1.1%). Hence, it is reasonable to assume that there is limited interconnectivity of intermetallic in this microstructure. This assumption is in a good agreement with the fact indicated by Jarfors *et al.*[13] that intermediate connectivity of particles starts at 8% and by 11% particle fraction there is a full connectivity between particles. Hence, one can assume that the effective CTE upper bound (when there is a limited particle connectivity) is suitably equals to CTE value for microstructure with low level of intermetallic fraction. The upper bound of linear thermal expansion coefficient, $(\alpha_{eff})_{up}$, of an isotropic two-phase composite can be obtained by Eq 3, and lower bound, $(\alpha_{eff})_{low}$, Eq 4, can be obtained if indices 1 and 2 are interchanged [14].

$$(\alpha_{eff})_{up} = \alpha_2 - f_1(\alpha_2 - \alpha_1)\frac{K_1(3K_2 + 4G_2)}{K_2(3K_1 + 4G_2) + 4f_1G_2(K_1 - K_2)}$$

Eq 3

$$(\alpha_{eff})_{low} = \alpha_1 - f_2(\alpha_1 - \alpha_2)\frac{K_2(3K_1 + 4G_1)}{K_1(3K_2 + 4G_1) + 4f_2G_1(K_2 - K_1)}$$

Eq 4

where, α, f, K and G are linear thermal expansion coefficient, phase fraction, bulk modulus and shear modulus of phase 1 and phase 2 respectively. In Eq 3 indices 1 is representing intermetallic properties and 2 is representing matrix properties. While, in Eq 4 the indices 1 and 2 are representing matrix and intermetallic properties respectively.

The effective CTE upper bound fit is shown in Figure 4. If in Eq 3 the $(\alpha_{eff})_{up}$ equals to actual experimental value of liner thermal coefficient obtained for 6 mm/s drawing rate (microstructure with lower value of intermetallic), and having value of α_1, K_1, G_1 and f_1 corresponded to CTE, bulk, shear and area fraction of intermetallic the α_2, K_2, G_2 corresponding to matrix values are obtainable. The temperature dependency of intermetallic elastic modulus (bulk and shear modulus) were applied after work of Zhang *et al.* (2010) [15] where the elastic stiffness are obtained from First-principles calculations based on the stress–strain method. For more detail the reads are referred to the original reference [15]. Accordingly, the actual matrix's CTE value of the upper bound fit as a function of temperature can be derived as a model with second order equation. The graphs showing the changing of CTE for matrix and intermetallic are presented in Figure 5. Furthermore, the temperature dependency of bulk modulus and shear modulus for matrix was derived in the same way and compared with the available results for intermetallic modulus [15]. The elastic modulus of matrix and intermetallic are represented in figure 6.

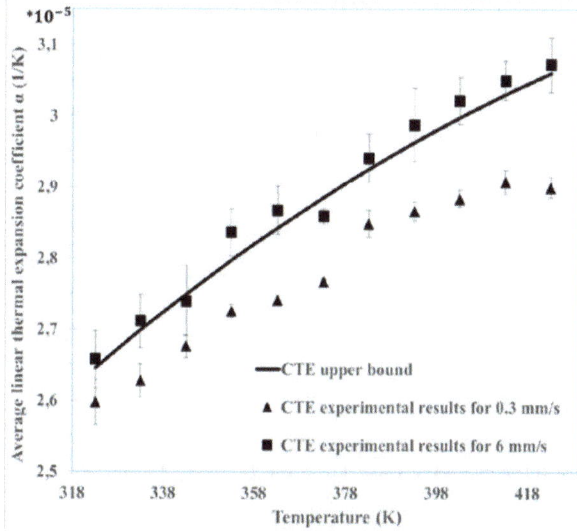

Figure 4. Illustration of CTE for AZ91D cast alloy applying 6 mm/s and 0.3 mm/s drawing speed a function of temperature. The error bars are 95% confidence intervals. The line shows the upper bound fit for calibration of AZ91D thermal expansion coefficient considering no connectivity between intermetallic. The upper bound fit obtained from actual CTE value by 6 mm/s drawing rate leading to low fraction of intermetallic.

The calculated values of CTE, bulk and shear modulus for matrix and intermetallic were applied in order to obtain corresponding CTE lower bound of AZ91D with high fraction of intermetallic. The lower bound of linear thermal expansion coefficient $(\alpha_{eff})_{low}$ of an isotropic two-phase composite can be obtained by Eq 4. In Eq 4 α_1, K_1, G_1 and f_1 are matrix value for CTE, bulk modulus, shear modulus and phase fraction respectively and α_2, K_2, G_2 are corresponding to intermetallic CTE and stiffness modulus, accordingly the value of CTE lower bound was obtained. Experimental results for this alloy including correlated upper and lower bound of CTE are presented in Figure 7.

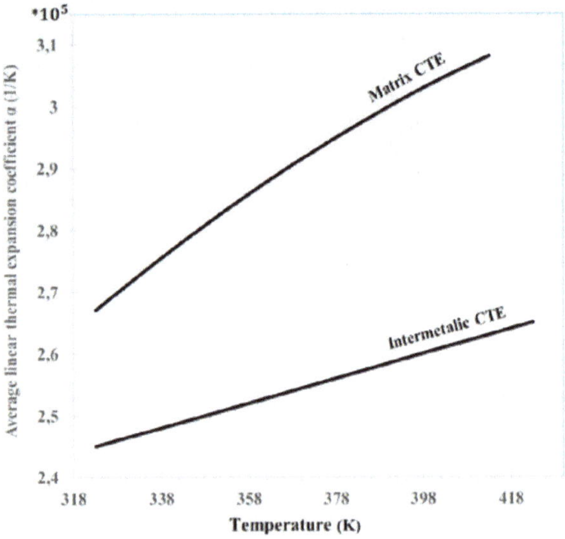

Figure 5. Illustration of temperature dependency of thermal expansion coefficient of matrix phase and intermetallic phase. Matrix CTE value are obtained using the CTE upper bound fit. Intermetallic values are obtained after reference [15].

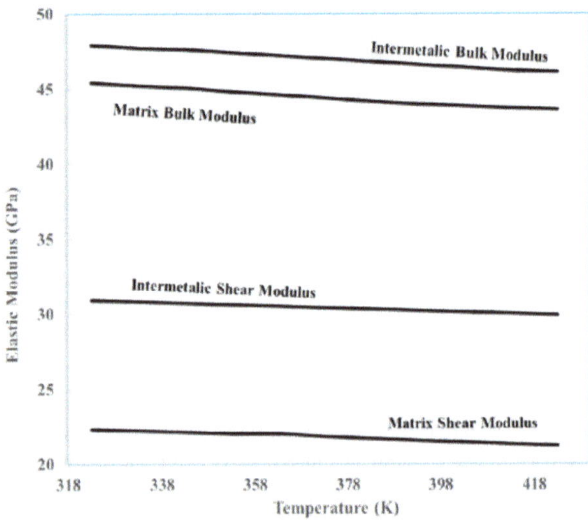

Figure 6. Comparing the temperature dependency of bulk and shear modulus for matrix phase of AZ91D and intermetallic ($Mg_{17}Al_{12}$) as a function of temperature.

The results strongly suggest the idea that there is a correlation between the connectivity of intermetallic in the microstructure and the thermal expansion of the alloy. It is revealed that higher fraction of intermetallic phase results in decreasing CTE of the alloy. By increasing the fraction of intermetallic, $Mg_{17}Al_{12}$ phase particles start to impinge each other, forming a network and the network starts to be continuous especially between dendrites. The CTE of the alloy results from interaction of CTE of matrix and intermetallic phase. Thermal expansion coefficient of intermetallic is lower than matrix leading to less expansion by increasing temperature. Consequently, the material with higher fraction expands less than alloy having a lower fraction of $Mg_{17}Al_{12}$ phase.

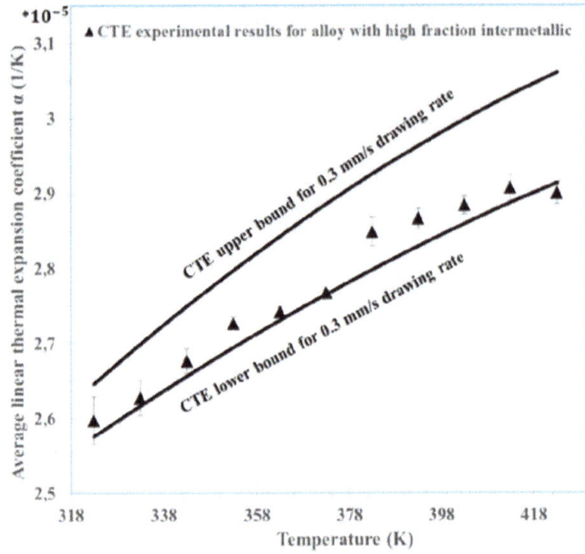

Figure 7. The upper and lower bound of thermal expansion for AZ91D cast applying 0.3 mm/s furnace drawing rate. The experimental results are shown with 95% confidence interval.

It is worth mentioning that the idea of a continuous network of intermetallic in the slow speed solidification samples was previously realized through mechanical testing by the authors [16]. It was confirmed, that the yield strength of AZ91D alloy was primarily determined by the network of Mg17Al12. This is in good agreement with the findings of Cao *et al.* [17], that concluded that an increase in fraction of $Mg_{17}Al_{12}$, increased yield strength. $Mg_{17}Al_{12}$ (E=71.9 GPa [18]) is more rigid than the matrix (E=45 GPa [18]). Hence, $Mg_{17}Al_{12}$ will take up higher load than the matrix. Lü *et al.* [19], concluded that $Mg_{17}Al_{12}$ is a brittle phase, but under the current experimental conditions, the difference in CTE between matrix and intermetallic would only result in strain levels in the elastic regime only, Figure 5.

Differential scanning calorimetry

In order to rule out non-equilibrium effects that may occur during heating DSC was used. Figure 8 shows the DSC curves obtained for the as-cast samples during heating from room temperature up to 430 K at a heating rate of 5 K/min. In this temperature range (300-420 K) the shape of DSC curves are in good agreement with literature [20, 21].There was no significant difference the shape of the DSC signal between two materials. This suggested that the difference in thermal expansion was not due to non-equilibrium effects originating from sample manufacturing. The conclusion is that the reason for the lower CTE values for the sample, with higher fraction of intermetallics, is the connectivity of the $Mg_{17}Al_{12}$ phase.

Conclusion

In present work, samples having controlled microstructures and showing isotropic properties, were prepared concluding that the manufacturing technique is suitable. The correlation between microstructure and thermal expansions of AZ91D was studied. The results demonstrated that the connectivity of intermetallic phase was contributing strongly to the thermal expansion of the alloy. A continuous network of intermetallic was formed by increasing and impinging of the $Mg_{17}Al_{12}$ particles.

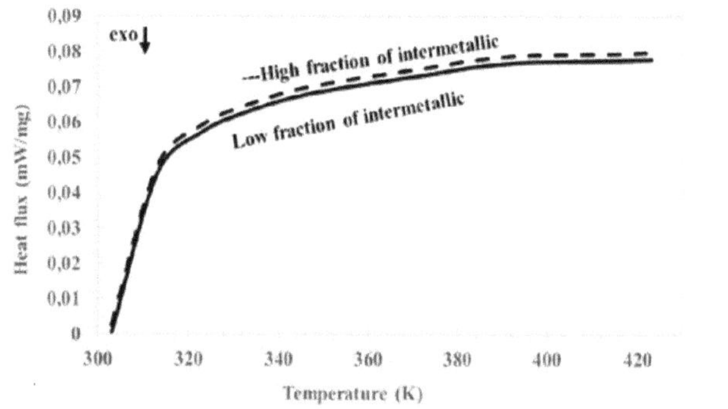

Figure 8. Curves of DSC of AZ91D alloy with different fraction level of intermetallic during heating process.

This conclusion is in a good agreement with available mechanical test results which showed that the higher intermetallic phase fraction contributes to the higher yield strength of AZ91D. The lower CTE bound suggests the possibility of intermetallic ($Mg_{17}Al_{12}$) network formation for the high fraction of intermetallic (11.3±1.2 %). The CTE curve on the upper bound corresponded to low fraction of intermetallic (7.1±0.9%) and possibility of network formation was considered to be almost zero. The lower CTE of intermetallic causes the reduced sensitivity to increasing temperature.

References

[1] Luo A, Renaud J, Nakatsugawa I, Plourde J. Magnesium castings for automotive applications. Jom. 1995;47:28-31.

[2] Powell BR, Luo AA, Rezhets V, Bommarito JJ, Tiwari BL. Development of creep-resistant magnesium alloys for powertrain applications: Part 1 of 2. SAE Technical Paper; 2001.

[3] Eliezer D, Aghion E, Froes FS. Magnesium science, technology and applications. Advanced Performance Materials. 1998;5:201-12.

[4] Regev M, Aghion E, Rosen A. Creep studies of AZ91D pressure die casting. Materials Science and Engineering: A. 1997;234:123-6.

[5] Gutman E, Unigovski Y, Levkovitch M, Koren Z. Influence of porosity and casting conditions on creep of die-cast Mg alloy. Journal of materials science letters. 1998;17:1787-9.

[6] Blum W, Li Y, Zeng X, Ing PZD-, Von Grossmann B, Ing CHD-. Creep deformation mechanisms in high-pressure die-cast magnesium-aluminum-base alloys. Metallurgical and Materials Transactions A. 2005;36:1721-8.

[7] Blum W, Watzinger B, Zhang P. Creep of Die-Cast Light-Weight Mg-Al-base Alloy AZ91hp. Advanced Engineering Materials. 2000;2:349-55.

[8] Kleiner S, Beffort O, Wahlen A, Uggowitzer P. Microstructure and mechanical properties of squeeze cast and semi-solid cast Mg–Al alloys. Journal of Light Metals. 2002;2:277-80.

[9] Spigarelli S, Cabibbo M, Evangelista E, Talianker M, Ezersky V. Analysis of the creep behaviour of a thixoformed AZ91 magnesium alloy. Materials Science and Engineering: A. 2000;289:172-81.

[10] Koren Z, Rosenson H, Gutman E, Unigovski YB, Eliezer A. Development of semisolid casting for AZ91 and AM50 magnesium alloys. Journal of Light Metals. 2002;2:81-7.

[11] Handbook A. Metallography and microstructures. Edited by GF Vander Voort, ASM Intenational. 2004;9.

[12] Chen C, Wang Q, Yin D. Thermal properties of Mg–11Y–5Gd–2Zn–0.5 Zr (wt.%) alloy. Journal of Alloys and Compounds. 2009;487:560-3.

[13] Jarfors AE, Torgny Carlsson, Anders Eliasson, Hans Keife, Cornel-Mihai Nicolescu, Bengt Rundqvist, Björn Sandberg. Tillverkningsteknologi. Lund: Studentlitteratur; 2006.

[14] Rosen BW, Hashin Z. Effective thermal expansion coefficients and specific heats of composite materials. International Journal of Engineering Science. 1970;8:157-73.

[15] Zhang H, Shang S, Wang Y, Saengdeejing A, Chen L, Liu Z. First-principles calculations of the elastic, phonon and thermodynamic properties of Al< sub> 12</sub> Mg< sub> 17</sub>. Acta Materialia. 2010;58:4012-8.

[16] Dini H, Andersson NE, Jarfors A. Effects of Microstructure on Deformation Behaviour of AZ91D Cast Alloy. TMS 2014 Supplemental Proceedings. 2014:565-72.

[17] Cao H, Wessén M. Effect of microstructure on mechanical properties of as-cast Mg-Al alloys. Metallurgical and materials transactions A. 2004;35:309-19.

[18] Huang Z-w, Zhao Y-h, Hou H, Zhao Y-h, Niu X-f, Han P-d. Structural, thermodynamics and elastic properties of Mg 17 Al 12, Al 2 Y and Al 4 Ba phases by first-principles calculations. Journal of Central South University. 2012;19:1475-81.

[19] Lü Y, Wang Q, Ding W, Zeng X, Zhu Y. Fracture behavior of AZ91 magnesium alloy. Materials Letters. 2000;44:265-8.

[20] Lindemann A, Schmidt J, Todte M, Zeuner T. Thermal analytical investigations of the magnesium alloys AM 60 and AZ 91 including the melting range. Thermochimica acta. 2002;382:269-75.

[21] Braszczyńska-Malik K. Thermal analytical investigations of the magnesium alloy AZ91. Archives of Foundry Engineering. 2007;7.

Magnesium Technology 2015
Edited by: Michele V. Manuel, Alok Singh, Martyn Alderman, and Neale R. Neelameggham
TMS (The Minerals, Metals & Materials Society), 2015

PRECIPITATION SEQUENCE IN A Mg-Sm-Zn-Zr ALLOY

Xiangyu Xia[1*], Amirreza Sanaty-Zadeh[1,2], Ran Chen[3], Xiaoqin Zeng[3], Alan A. Luo[2,4], Donald S. Stone[1,2]

[1] Materials Science Program; University of Wisconsin – Madison, 1509 University Ave., Madison, WI 53706, USA
[2] Department of Materials Science and Engineering, University of Wisconsin-Madison, 1509 University Ave., Madison, WI 53706, USA
[3] Department of Materials Science and Engineering, Shanghai Jiaotong University, 800 Dongchuan Road, Shanghai, China
[4] Department of Materials Science and Engineering, The Ohio State University, 116 W. 19th Ave, Columbus, OH 43210, USA

Keywords: Mg-Sm-Zn-Zr, precipitation, heat treatment, mechanical property, transmission electron microscopy

Abstract

In this research, aging behavior and precipitation sequence in an age-hardenable Mg-4Sm-1.3Zn-0.4Zr (wt%) alloy were investigated, using Differential Scanning Calorimeter (DSC) and Transmission Electron Microscopy (TEM) techniques. The precipitation sequence was determined to be Super Saturated Solid Solution (S.S.S.S) → solute atom clusters → γ'' → γ' → γ, which changes drastically compared to those reported for Zn-free Mg-Sm-Zr alloys. Among them, γ'', with a GP zone structure, lies on $\{0001\}_\alpha$ basal planes, and is the dominant precipitate responsible for age hardening effect. γ' has a hexagonal structure, with a=0.556nm, c=0.521nm.

Introduction

Magnesium and its alloys have remarkable potential in reducing energy consumption in transportation industries due to their low density, high specific stiffness, excellent castability, and other favorable attributes [1-4]. However, the relatively poor mechanical properties of magnesium alloys have limited their applications in critical structural components. Additions of rare-earth (RE) elements to Mg alloys have been reported to significantly enhance strength properties and creep resistance by solid solution strengthening, precipitation hardening, and/or texture randomization [5, 6]. Studies on correlation between mechanical properties and microstructure evolution in several Mg-RE alloys, including yttrium (Y), cerium (Ce), neodymium (Nd), and gadolinium (Gd), have been reported [7-12].

Samarium (Sm), with relatively high solid solubility in magnesium matrix near eutectic point and ability to form similar intermetallic phases to other Mg-RE systems [13, 14], is considered as promising alloying elements for Mg alloy optimization. Nishijima et al [15] reported age-hardening response during isothermal aging at 200°C in Mg-0.99at%Sm. The precipitation sequence was proposed to be Super Saturated Solid Solution (S.S.S.S) → β'' ($D0_{19}$, a=0.642nm, c=0.521nm) → β'_1 (orthorhombic, a=0.64 nm, b=3.334nm, c=0.52nm) → β_1 (FCC, a=0.73nm), and the main strengthening precipitates was β'_1 [15]. In this work, all notations are named after ref [16] for convenience unless otherwise noted. Recently, Zhang et al [17], and Su [18], acquired a precipitation sequence of S.S.S.S → β''→ β_1. The addition of Zr is for grain refinement but does not otherwise affect precipitation sequence in Mg alloys [19]. The main strengthening precipitate, in contrary to the report in [15], is β''. Although there are some disagreements in precipitation sequence, all of these studies show that addition of Sm introduces β series of precipitates, which, lying on prismatic planes of α(Mg) matrix, acts as obstacles for dislocation motion and thus increases hardness.

Ternary addition of Zn to binary Mg-RE alloys changes β series of precipitates to γ series [16]. In an early study of precipitates in Mg-2.8Nd-1.3Zn alloys [20], the precipitation sequence was determined as S.S.S.S → low-temperature reaction → γ'' (HCP, a=0.55nm, c=1.56nm) → γ (FCC, a=0.72nm). γ'' plates lie on basal plane of α(Mg) matrix, with an orientation relationship of $(0001)_{\gamma''}//(0001)_\alpha$, $[2\text{-}1\text{-}10]_{\gamma''}//[10\text{-}10]_\alpha$. γ rods lie on prismatic planes, $(011)_\gamma//(0001)_\alpha$, $[\text{-}1\text{-}11]_\gamma//[2\text{-}1\text{-}10]_\alpha$. Ping [21], and Wilson [22] investigated Mg-2.4RE-0.4Zn-0.6Zr, and modified γ'' to be γ', having an ordered G.P. zone structure, with a=0.556nm. Yuan and Zheng [23, 24] investigated the effect of Zn on the aging behavior of Mg-xZn-3Sm-0.5Gd alloys (x=0, 0.3, 0.6), and found that with increase of Zn, the dominant strengthening precipitate changes from β'' to β' (orthorhombic, a=0.64nm, b=2.223nm, c=0.52nm), and then to γ' (HCP, a=0.556nm).

Although addition of Zn changes the precipitation sequence and further affects the mechanical properties drastically, a clear understanding on precipitation sequence and nature of γ precipitates is not available, which is crucial to illustrate the role of Zn in change of microstructure evolution in Mg-RE-Zn-Zr alloys. This paper presents a systematical study on precipitation sequence in high-Zn Mg-Sm-Zn-Zr alloys, aiming to provide a comprehensive understanding of the microstructure evolution, properties of metastable precipitates, and kinetics of precipitation in these alloys, for further alloy design and processing optimization of Mg alloys.

Experimental procedure

The as-cast alloy sample with a target composition of Mg-4Sm-1Zn-0.4Zr were prepared using commercially pure Mg, Zn (>99.9%), Mg-20Sm and Mg-30Zr master alloy ingots. The alloying ingredients were melted in a steel crucible and cast into a steel mold preheated to 230°C under protection of SF_6/CO_2, and then degassed by bubbling pure Argon for about 20min at 740°C. At last, the slag was removed prior to casting. The chemical composition of the cast alloy, measured using Inductively Coupled Plasma Mass Spectrometry (ICP-MS) was Mg-4.36Sm-1.27Zn-0.39Zr. For protection against oxidation during heat treatment, all samples were wrapped with tantalum foil and sealed in quartz tubes back-filled with high-purity He atmosphere (>99.999%) before heating.
Specimens were solid solution treated at 520°C for eight hours, followed by quenching in room temperature (RT) water. The solid

solutionized (S.S) samples were then aged at 200°C for different lengths of time (0.5h, 1h, 2h, 4h, 8h, 10h, 16h, 32h, 64h, 200h). Aged alloy samples were ground and polished with distilled water to achieve clean and even surfaces for microhardness measurements, which were conducted on a Wilson Tukon 1102/1202 microhardness tester, with 200kg load. For each sample, at least 8 indents of around 50*50μm diamond shape were analyzed and average values were used to improve precision. Phase identifications were performed by X-ray diffraction (XRD), using a General Area Diffraction Detector System (GADDS), with Cu Kα radiation (40kV, 50mA). A FEG LEO-1530 Scanning electron microscope (SEM) was used to study the microstructures of as-cast alloy samples.

DSC tests were performed on S.S alloy samples, on a DQ100 Differential Scanning Calorimeter from TA instruments. Samples were cut to around 20mg, then polished and cleaned with ethanol. Standard Aluminum pans were used as references as well as to encapsulate the sample alloys. Samples underwent cyclic thermal treatment from 40°C to 450°C, for three cycles, with 4 different heating rates of 5, 10, 15 and 20°C/min. The third cycle (over aged condition) was used to help establish a background, which was then subtracted from the first cycle (S.S condition). To identify metastable precipitate phases formed in this process, S.S samples were heated at 10°C/min to the peak temperatures measured in DSC curves, held for 3 minutes, and then quench in RT water. These samples were then analyzed by Transmission Electron Microscopy (TEM) (Tecnai TF30) to identify the precipitate types, morphology and diffraction patterns. For sample preparation, 1000μm thick slices were cut from these samples and ground down to ~50μm, using Allied Multiprep machine with diamond lapping films of different grid size from 15μm to 0.1μm. The polished samples were then ion milled with Fischione 1050, (3.0kV ion gun energy and 5° milling angle). A final 0.6kV ion gun energy was used to clean up surface.

Results and discussion

<u>As-cast and solid solution treated samples</u>

Fig. 1a shows back-scattered electron (BSE) Scanning Electron Microscopy (SEM) images of as-cast samples, in which two bright intermetallic phases separating dendrites of a dark matrix. Based on an Energy Dispersive X-ray Spectroscopy (EDS) analysis, the dark matrix is α(Mg) matrix, the gray intermetallic is $Mg_{41}Sm_5$, and the bright intermetallic is Mg_3Sm, corresponding well to the phase diagrams in [13, 14]. XRD spectra of as-cast and S.S. samples are shown in Fig. 1b. In the as-cast alloy, all peaks can be attributed to α(Mg), Mg_3Sm, or $Mg_{41}Sm_5$, being consistent with the SEM analysis and the phase diagrams. After solid solution treatment all the peaks of intermetallic phases vanish, indicating dissolution of all intermetallic phases into the α(Mg) matrix. The change in relative intensity of α(Mg) peaks might come from the mechanical polish prior to XRD analysis and does not affect conclusion on dissolution of intermetallic phases. Since the focus of this work is on the precipitation sequence in this alloy, no further investigation on as-cast or S.S samples will be reported.

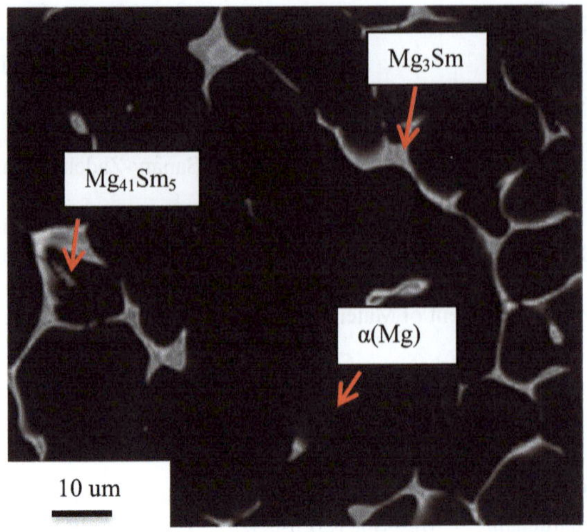

(a)

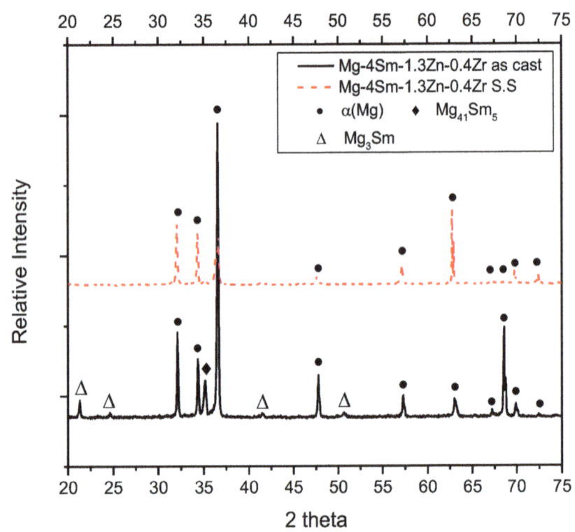

(b)

Fig.1. (a): SEM BSE images of as-cast sample alloys Mg-4Sm-1.3Zn-0.4Zr (b): XRD spectra of as cast and S.S sample alloys

<u>Precipitation Hardening</u>

The S.S. samples were held in refrigerator at about 5°C for 12h, and then isothermally aged at 200°C for various times. The pre-aging treatment is known to promote clustering in the solid solutionized samples, resulting in higher densities of nuclei and precipitates [25]. The variation of hardness with aging time is shown in Fig. 2 with error bars indicating standard deviations in the measurements. It is seen that this alloy exhibits typical age hardening response. At initial stage, the hardness increases with increasing aging time, representing an incubation period. The peak hardness is then reached at around 16h, after which the hardness gradually decreases, as a result of over-aging coming from precipitates coarsening or formation of a new precipitate phase.

To identify the precipitates phase responsible for hardness increase, TEM analysis was performed on samples aged at 16h

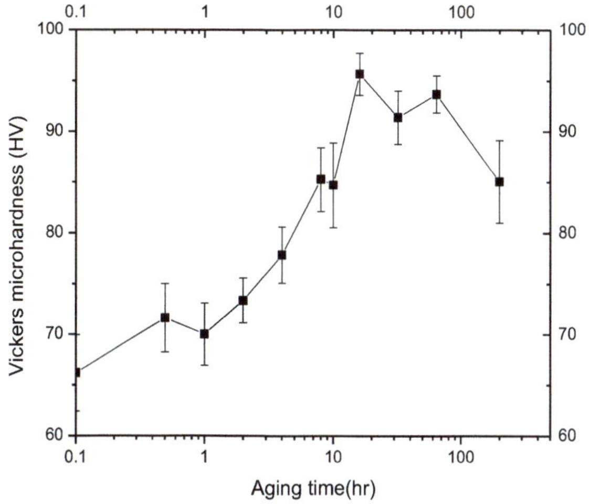

Fig.2. Isothermal age hardening response of Mg-4Sm-1.3Zn-0.4Zr alloys at 200°C

(peak-aged). Figs. 3a and b show microstructure and diffraction patterns of peak-aged sample, taken near zone axis $[01\text{-}10]_\alpha$ and $[2\text{-}1\text{-}10]_\alpha$, respectively. Both of them reveal 10-30nm long and one atom layer thick needle-like precipitates, lying on $(0001)_\alpha$ basal planes. When zone axis is on $[01\text{-}10]_\alpha$, streaks are parallel to $\{0001\}_\alpha$ at $1/3\{\text{-}2110\}_\alpha$ and $2/3\{\text{-}2110\}_\alpha$. And similar streaks are found along $\{0001\}_\alpha$ when zone axis is parallel to $[2\text{-}1\text{-}10]_\alpha$. Thus these precipitates are determined to have a G.P. zone structure, being the same as γ' in [21]. In this work, however, the precipitates are named γ''. Based on the diffraction patterns, the orientation relationship between γ'' and α(Mg) matrix is such that $[10\text{-}10]_{\gamma''}//[2\text{-}1\text{-}10]_\alpha$. Lying on basal planes, these γ'' precipitates hinder dislocation movements and contribute to the hardening of Mg-4Sm-1.3Zn-0.4Zr alloy, act as main strengthening precipitates during aging [16], regardless of its high aspect ratio (>15). Besides γ'', some dim linear features lie on prismatic planes of α(Mg) matrix. Very weak streaks can be found in diffraction pattern perpendicular to $(0001)_\alpha$, implying these features might belong to β group of precipitates [16], though the specific type remains unknown.

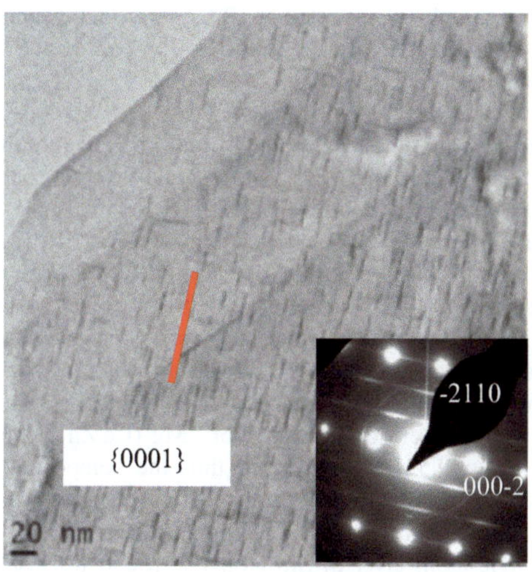

(a)

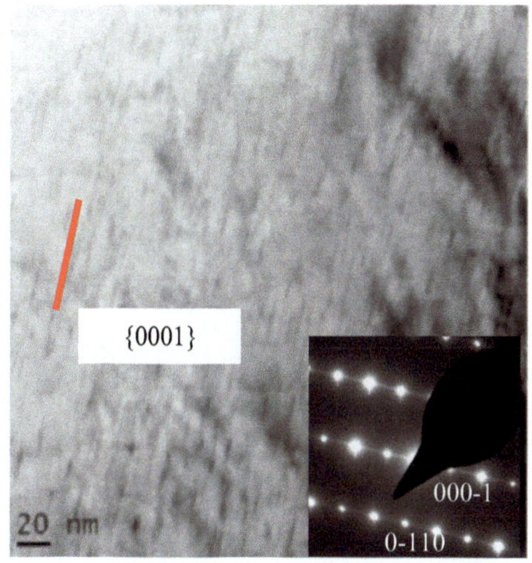

(b)

Fig.3. TEM Bright Field image and corresponding Diffraction Patterns of sample alloy aged 200°C for 16h, (a): $B//[01\text{-}10]_\alpha$; (b): $B//[2\text{-}1\text{-}10]_\alpha$

Precipitation Sequence

Despite the findings for peak-aged precipitates, details in early precipitation stage and precipitation evolution remain unknown. For this purpose DSC experiments were conducted. Fig. 4 shows the DSC curve acquired from S.S sample at a heating rate of 10°C/min. Exothermal peaks are found at 150.8, 291.4, 353.0°C, representing three different precipitation reactions. To identify these transition reactions, S.S samples were heated at a rate of 10°C/min, kept at specific peak temperature (151, 290, 353°C, respectively) for 3 min, and then quenched in RT water. Figs. 5a and b show HRTEM and corresponding FFT images of sample alloy at peak temperature 151°C, on zone axis $[0001]_\alpha$ and $[2\text{-}1\text{-}10]_\alpha$, respectively. HRTEM image near $[0001]_\alpha$ show dark areas randomly distributed in the grain without specific diffraction features, tentatively identified as clusters of solute atoms [11]. When zone axis is near $[2\text{-}1\text{-}10]_\alpha$, Fig. 5b exhibits that solute atoms separately form multiple atom layers, resulting in discontinuous streaks in FFT image. Compared to the G.P. zone structure of γ'', these are likely the precursors of γ''.

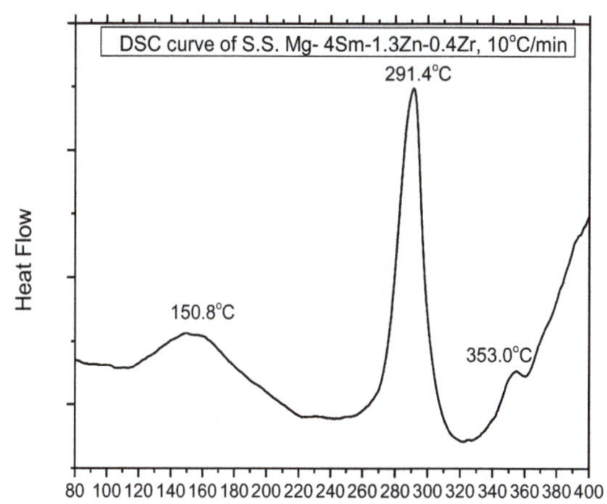

Fig.4. DSC curve of S.S sample alloy at 10°C/min

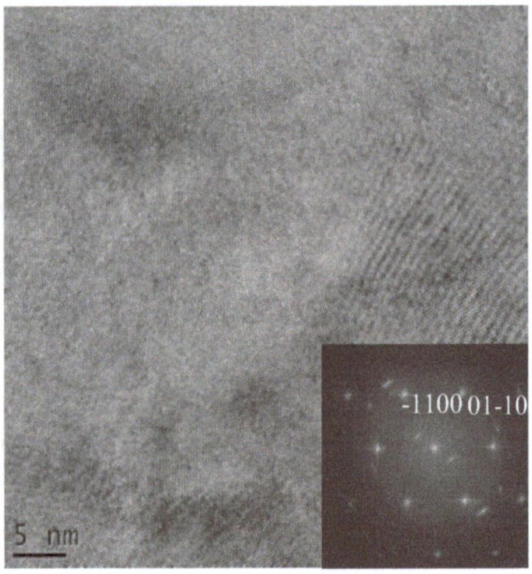

(a)

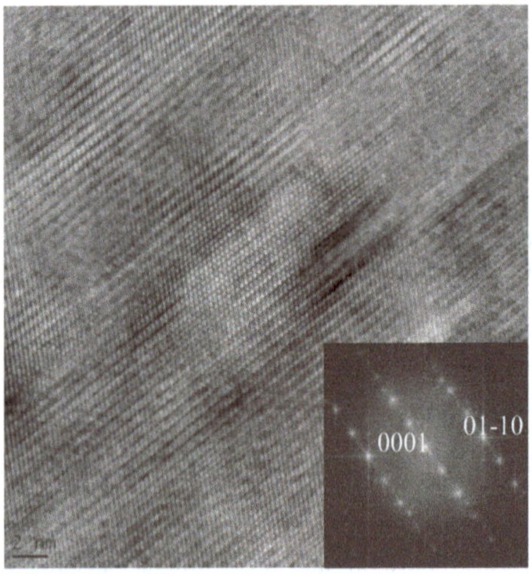

(b)

Fig.5. TEM Bright Field image and corresponding Diffraction Patterns of sample alloy heated at 10°C/min to 151°C and held for 3 minutes before cooling (a): B//$[0001]_{\alpha}$; (b): B//$[2\text{-}1\text{-}10]_{\alpha}$

Microstructures and corresponding diffraction patterns of the sample alloy annealed at 290°C are shown in Figs. 6a-b, taken near zone axis $[01\text{-}10]_{\alpha}$ and $[2\text{-}1\text{-}10]_{\alpha}$, respectively. From Fig. 6b, microstructures and diffraction patterns of precipitates are clearly identical to those in Fig.3 a, and thus the 290°C phase transition peak is designated to the formation of γ'' precipitates. An HRTEM image of γ'' lying in α(Mg) matrix is shown in Fig. 5d, near zone axis $[2\text{-}1\text{-}10]_{\alpha}$. A high-contrast Sm single atom layer lie on the basal plane of α(Mg) matrix, and the strains in the nearby lattice are released by low-contrast Zn atom layers, as demonstration of G.P. zone structure of γ''.

Figs. 6a-b show bright field images and corresponding diffraction patterns of sample alloy at 353°C, with beam parallel to $[01\text{-}10]_{\alpha}$ and $[2\text{-}1\text{-}10]_{\alpha}$, respectively. From diffraction pattern when B//$[01\text{-}10]_{\alpha}$, streaks at $1/3\{\text{-}2110\}_{\alpha}$ and $2/3\{\text{-}2110\}_{\alpha}$ suggest that precipitate in this sample also has a hexagonal structure with a=0.556nm. Without any features visible in bright field image,

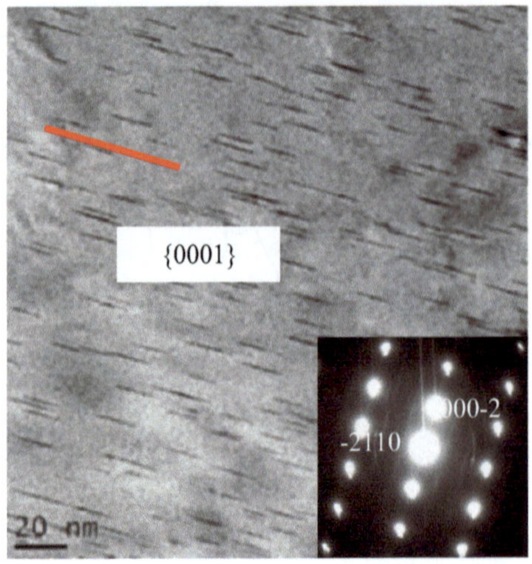

(a)

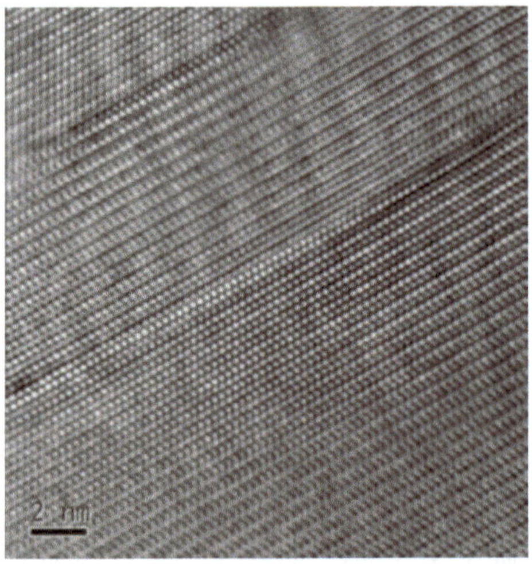

(b)

Fig.6. TEM Bright Field image and corresponding Diffraction Patterns of sample alloy heated at 10°C/min to 290°C and held for 3 minutes before cooling (a): B//$[01\text{-}10]_{\alpha}$; (b): HRTEM of the same sample, B//$[2\text{-}1\text{-}10]_{\alpha}$

it is concluded that the habit plane is $\{0001\}_{\alpha}$. The morphology of these precipitates changes from single-layered needles to multiple-layered plates, with 30-80nm in length and 5-15nm in width. In addition, the strong reflections at $1/3\{\text{-}2110\}_{\alpha}$ and $2/3\{\text{-}2110\}_{\alpha}$ positions in streaks parallel to $\{0001\}_{\alpha}$ implies that this precipitate has an independent crystal structure instead of merely adding layers to the G.P. zone structure. Since no outstanding precipitate reflections can be found at $1/n\{0001\}_{\alpha}$, c is equal to that of α(Mg), ie 0.521nm. This structure (hexagonal, a=0.556nm, c=0.521nm) is close to that of $Mg_5(Ce,Zn)$ (hexagonal, a=0.571nm, c=0.521nm, [16]) and thus the composition of this precipitate, named as γ', is proposed to be $Mg_5(Sm,Zn)$. An HRTEM of γ' is shown in Fig.8d near zone axis $[2\text{-}1\text{-}10]_{\alpha}$. Instead of the single atom layer of Sm atoms surrounded by smaller Zn atoms, as that of γ'', Sm layers segregate together and Zn atoms migrate into these layers, forming a hexagonal structure coherent with the α(Mg) matrix.

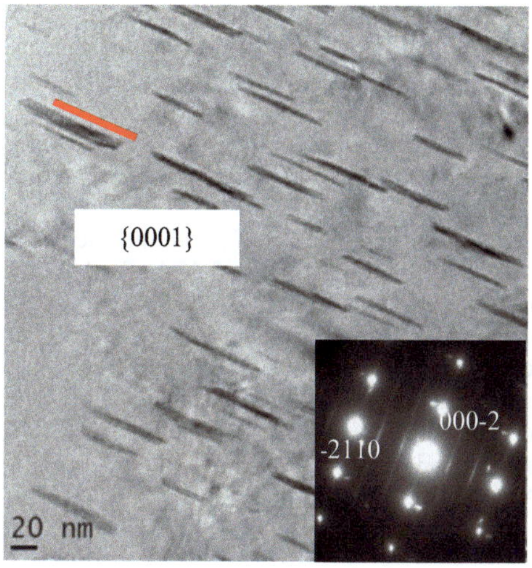

(a)

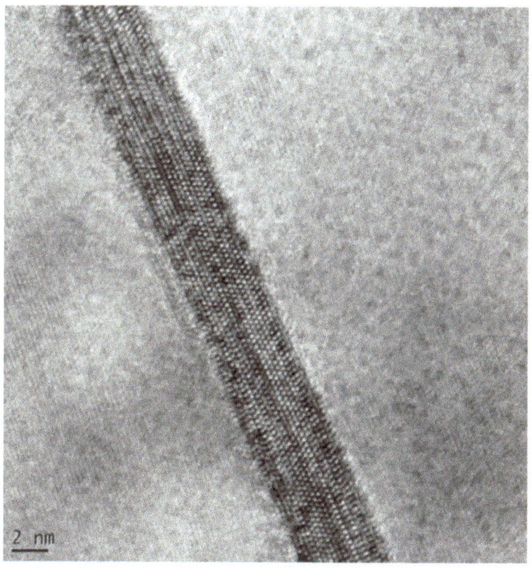

(b)

Fig.7. TEM Bright Field image and corresponding Diffraction Patterns of sample alloy heated to 353°C at 10°C/min (a): B//[01-10]$_{\alpha}$; (b): HRTEM of the same sample, B//[2-1-10]$_{\alpha}$

Based upon DSC and TEM analyses, the precipitation sequence in Mg-4Sm-1Zn-0.4Zr can be determined as solute atoms clusters → γ'' → γ'. Crystal structures and morphologies of these precipitates are listed in Table 1. γ was reported to be the final stage in the precipitation sequence. The DSC curves of S.S samples at four heating rates are presented in Fig. 8. As shown, peak temperatures increase with increasing heating rates, indicating these precipitation reactions are thermally activated and kinetically controlled. Further kinetic analysis is under study.

Table.1. structures and kinetic parameters of metastable phases in Mg-4Sm-1.3Zn-0.4Zr

Phase	Crystal structure	Morphology
Solute atom clusters	-	-
γ''	G.P. zone a=0.556nm	Needle-like 10-30nm long One atom layer thick
γ'	Hexagonal a=0.556nm c=0.521nm	Plate 30-80nm long 5-15nm thick

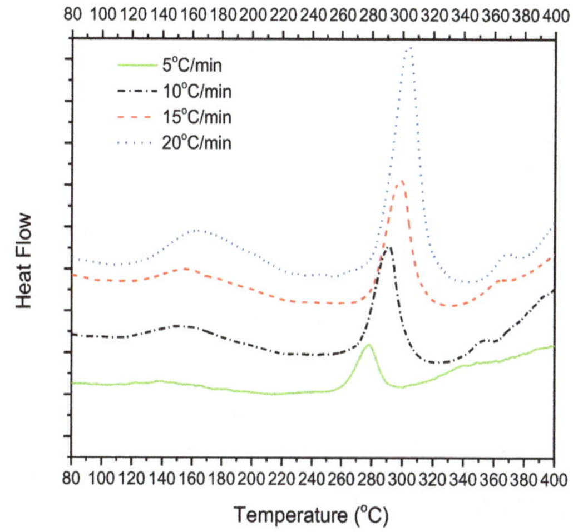

Fig.8. DSC curves of S.S sample alloys at 5, 10, 15, 20^{0}C/min

Conclusion

Aging behavior and precipitation sequence in a Mg-4Sm-1.3Zn-0.4Zr alloy were researched. Results exhibit that it shows classical age hardening response, as a result of precipitation hardening. The precipitation sequence in high-Zn content Mg-Sm-Zn-Zr alloys are determined to be S.S.S.S → solute atom clusters → γ'' → γ' → γ. Lying on {0001}$_{\alpha}$ basal planes, γ'' has a G.P zone structure of single solute atom layers, and is dominant in precipitation hardening. γ' has a hexagonal structure, a=0.556nm, c=0.521nm. Due to the structure similarity the chemical composition of γ' is proposed to be Mg_5(Sm,Zn), and further study is undergoing.

Acknowledgement

The authors acknowledge the financial support from National Science Foundation (Award No. 1005762). We would also like to thank Mr. Xin Su, Mr. Humberto Melgarejo, and Ms. Nayomi Plaza for their helpful discussions.

References

[1]. A.A Luo, "Recent magnesium alloy development for elevated temperature applications", International Materials Review. 49 (2004)13-30

[2]. M.O Pekguleryuz, A.A Kaya, In: K.U.Kainer, editor. Magnesium alloys and their Applications. Germany, Wiley-VCH Verlag GmbH, 2003, pp.74

[3]. M. Bamberger, G. Dehm, "Trends in the Development of New Mg Alloys", Annu. Rev. Mater.Res.38 (2008)505-33

[4]. A.A Luo, R.K Mishra, B.R Powell, A.K Sachdev, "Magnesium Alloy Development for Automotive Applications", Materials Science Forum. 706-709 (2012)69-82

[5]. L.L Rokhlin, Magnesium Alloys Containing Rare Earth Metals: Structure and Properties. CRC Press, 2003

[6]. I.P Moreno, T.K Nandy, J.W Jones, J.E Allison, T.M Pollock, "Microstructure and Creep Behavior of a Die Cast Magnesium-Rare Earth alloy", Magnesium Technology 2002, pp.111-116

[7]. K. Hantzsche, J.Bohlen, J.Wendt, K.Kainer, S.Yi, D.Letzig, "Effect of rare earth additions on microstructure and texture development of magneiusm alloy sheets", Scripta Mater. 63 (2010)725-730

[8]. R.K. Mishra, A.K.Gupta, P.R.Rao, A.K.Sachdev, A.Kumar, A.Luo, "Influence of cerium on the texture and ductility of magnesium extrusions", Scripta Mater. 59 (2008) 562-565

[9]. D.Wu, R.S.Chen, E.H.Han, "Excellent room-temperature ductility and formability of rolled Mg-Gd-Zn alloys sheets", J. Alloys Comp. 509 (2011) 2856-2863

[10]. T.J. Pike, B. Noble, "The formation and structure of precipitates in a dilute mageniusm-neodymium alloy", J. the Less-Common Met. 30 (1973) 63-74

[11]. A.Sanaty-Zadeh, X. Xia, A. A. Luo, D. S. Stone. "Precipitation evolution and kinetics in a magnesium-neodymium-zinc alloy", J. Alloys Comp. 583(2014) 434-440

[12]. L. Ma, R. K. Mishra, M. P. Balogh, L. Peng, A. A. Luo, A.K. Sachdev, W. Ding, "Effect of Zn on the microstructure evolution of extruded Mg-3Nd(-Zn_-Zr(wt%) alloys", Mater. Sci. Eng. A 543 (2012) 12-21

[13]. X. Xia, A Sanaty-Zadeh, C Zhang, X Q. Zeng, A A. Luo, Y A. Chang, D S. Stone, "Thermodynamic modeling and experimental investigation of the magnesium-zinc-samarium alloys", J Alloys Comp, 593 (2014) 71-78

[14]. Drit M.E, Rokhlin L.L, Abrikina N.P, Kinzhibalo V.V, Tyvanchuk A.T, "PHASE EQUILIBRIA IN THE MG-SM-ZN SYSTEM", Izvestiya Akademii Nauk SSSR. Metally. 6 (1985)194-200

[15]. M Nishijima, K Hiraga, M Yamasaki, Y Kawamura, "Characterization of Precipitates in Mg-Sm Alloy Aged at 200 C, Studied by High-Resolution Transmission Electron Microscopy and High-Angle Annular Detector Dark-Field Scanning Transmission Electron Microscopy", Materials transactions, 50(07) (2009) 1747-1752.

[16]. J Nie, "Precipitation and Hardening in Magnesium Alloys", Metallurgical and Materials Transtractions A, 43A (2012) 3891-3939

[17]. Zhenyan Zhang, phd thesis, "Study on the microstructure, mechanical properties and precipitated phase transformations of Mg(-Gd)-Sm-Zr alloys", Shanghai Jiaotong University, 2009

[18]. Xin Su, master thesis, "First-principle calculation, microstructures and properties of Mg-RE alloys", Shanghai Jiaotong University, 2013

[19]. E. F. Emley, Principle of Magnesium Technology, Pergamon Press Ltd, 1966

[20]. P.A.Nuttall, T.J.Pike, B.Noble, "Metallography of dilute Mg-Nd-Zn alloys", Metallography, 13 (1980) 3-20

[21]. D.-H.Ping, K.Hono, J.F.Nie, "Atom Probe Characterization of plate-like precipitates in a Mg-RE-Zn-Zr casting alloy", Scripta Materialia, 48 (2003) 1017-1022

[22]. R.Wilson, C.J.Bettles, B.C.Muddle, J.F.Nie, "Precipitation hardening in Mg-3wt%Nd(-Zn) casting alloys", Materials Science Forum, 419-422 (2003) 267-272

[23]. M Yuan, Z Zheng, "Effects of Zn on the microstructure and mechanical properties of Mg-3Sm-0.5Gd-xZn-0.5Zr (x=0, 0.3, 0.6) alloy", J Alloys Comp, 590 (2014) 355-361

[24]. M Yuan, Z Zheng, "Microstructure and mechanical properties of a Mg-Sm-Gd-Zn-Zr Alloy processed by heat treatment", Metallorgr. Microstruct. Anal. 3 (2014) 131-137

[25]. A. Luo, D. J. Lloyd, A. Gupta, W. V. Yourdelis, "Precipitation and dissolution kinetics in Al-Li-Cu-Mg alloy 8090", Acta Mater. 41 (1993) 769-776

Magnesium Technology 2015
Edited by: Michele V. Manuel, Alok Singh, Martyn Alderman, and Neale R. Neelameggham
TMS (The Minerals, Metals & Materials Society), 2015

THE EXOMET PROJECT: EU/ESA RESEARCH ON HIGH-PERFORMANCE LIGHT-METAL ALLOYS AND NANOCOMPOSITES

W.H. Sillekens
European Space Agency; PO Box 299; 2200 AG Noordwijk, Netherlands

Keywords: magnesium alloys, chemical grain refinement, metal-matrix nanocomposites, external fields

Extended Abstract

The performance of structural materials is commonly associated with such design parameters as strength and stiffness relative to their density; a recognized means to further enhance the weight-saving potential of low-density materials is thus to improve on their mechanical attributes. The European Community research project ExoMet that started in mid-2012 targets such high-performance aluminum- and magnesium-based materials by exploring novel grain-refining and nanoparticle additions in conjunction with melt treatment by means of external fields (electromagnetic, ultrasonic, mechanical). These external fields are to provide for an effective and efficient dispersion of the additions in the melt and their uniform distribution in the as-cast material. The consortium of 27 companies, universities and research organizations from eleven countries integrates various scientific and technological disciplines as well as application areas – including automotive and (aero)-space.

This paper gives an overview of the project, including its scope for development and its organization. Focusing on magnesium, exemplary results are presented; the application perspective is considered as well. Being based on a recent overview [1], the paper will also include latest notable results. The project's website is an additional source of information [2].

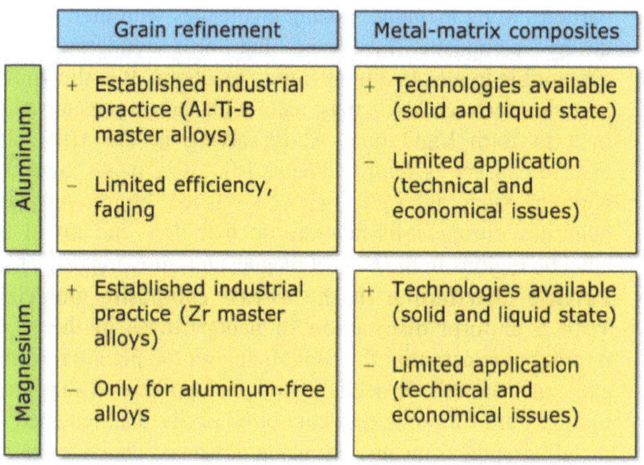

Figure 1. State of the art for the topics considered.

To set the scene for the research topics at hand, the current situation is simply represented in **Figure 1**. Chemical grain refinement – or grain refinement by inoculation – consists of adding (typically micrometer-sized) particles to the metal melt that act as crystal nuclei upon solidification, leading eventually to refined microstructures with improved mechanical response. Metal-matrix composites consist of metals in which particles are included as reinforcements; these particles can be of various types, shapes, and sizes (usually micrometer-sized and above). The emergence of nano-sized particle powders has given a new impulse to the research on the latter, although the issues with the proper incorporation of such particles into these so-called nanocomposites are even more of a challenge.

Departing from this situation, the ExoMet project focuses on some distinct research and development areas with their associated anticipated technological benefits.

With respect to the materials, the addition of ceramic (and other) particles to conventional alloys is considered in relatively minor concentrations. This targets the following.

- **Chemical grain refinement** (GR), the controlling mechanisms of which entail both crystal nucleation and crystal growth during solidification. Grain refinement in metallic materials improves both strength (according to the Hall-Petch relationship) and ductility of finished components such as shape castings. Given that microstructural coarsening (grain growth) will set in upon exceeding some threshold in-service temperature, mechanical property enhancement by means of grain refinement is of prime concern to room-temperature applications. Further, inoculation is especially relevant for casting processes that feature low solidification rates such as sand casting and permanent-mold casting, as well as direct-chill billet casting. The last-mentioned process is an initial step in wrought products' manufacturing, in which grain refinement reduces hot-tearing susceptibility. In addition, it improves subsequent forming (e.g., extrudability) and promotes a uniform fine-grained recrystallized microstructure – and hence better and more consistent mechanical properties – in the finished product.
- **Metal-matrix nanocomposites** (MMNCs), in which the embedded particles act as local reinforcements (yet may induce some grain refinement during solidification as well). This is based on so-called Orowan strengthening, that arises when dislocation movement is impeded by such shear-resistant obstructions in the glide plane. The mechanism is most effective when these particles are in the nanometer range, and densely and uniformly dispersed; under these conditions, the particles are reported not to negatively affect ductility as well. Assuming the use of thermally stable additions, the effect extends to elevated-temperature strength and creep resistance, and is therefore of particular relevance for high-temperature applications such as gearboxes. Further, such nanoparticle additions are assumed to not substantially offset fluidity/castability and to be equally beneficial for different casting processes. Typical compounds that are being suggested for inclusion, such as SiC, Al_2O_3, MgO, ZrO_2 and AlN, are generally strong/hard and of high melting point versus the metal matrix. Production capacities and associated costs of the reinforcements as nanopowders have meanwhile made it to a point that additions can be economically viable if of limited concentrations in the composite materials (e.g., to a few percent).

With respect to the proper inclusion of these (nano)-particles in the material, it is recognized that this requires appropriate melt-treatment and consolidation processes that use **external fields** rather than to rely on processes that imply internal modes of mixing (such as natural convection) only. Issues to be overcome are the de-agglomeration and wetting of the particles as well as their uniform distribution in the crucible and finally in the mold. The principles explored for that are electromagnetic, ultrasonic, and mechanical stirring. Such external fields have been implemented before for certain melt-treatment operations. Electromagnetic stirring – which involves the application of inductors (coils) around the crucible that impose stationary and alternating magnetic fields in the melt leading to forced convection and/or vibrations – has amongst others been explored for the production of thixotropic feedstock by means of continuous casting. Ultrasonic stirring – which involves the dipping and operation of an ultrasonic vibration device (called also "probe" or "sonotrode") in the melt – has also been studied as a means of modifying solidification and hence the microstructure of the as-cast material. And finally mechanical stirring – which involves the use of impellors, screws and other moving rotors along with non-moving stators – has been studied in a variety of embodiments, as of late in an implementation with a pair of co-rotating intermeshing screws.

It is generally thought that a very intensive stirring of the melt with the particles is required for the effective synthesis of MMNCs. This synthesis can in principle be applied on the level of master alloys (with high concentrations of particles, to be diluted with base metal upon shape casting of the actual components in the foundry) or on the level of feedstock that is ready for use (with low concentrations of particles, being either pre-manufactured as ingots or being prepared as a melt in the foundry). Similarly, it is considered that such rigorous stirring will be instrumental in enhancing the incorporation of grain refiners and thus will yield a better grain-refining efficiency.

The ExoMet research project is being conducted as a so-called large-scale integrating project within the 7th Framework Program of the European Commission. The project aims for a substantial enhancement of material properties, new microstructures, and modest implementation costs into production technology lines. Relating to the project's full title ("Physical processing of molten light alloys under the influence of external fields"), the scientific and technological objectives are as follows.

- Understanding of underlying mechanisms and resolution of current technical constraints (with agglomeration, wetting, chemical reactions and so on) of particle dispersion in aluminum and magnesium melts.
- Substantial enhancement of (mechanical) attributes as strength, ductility and creep resistance over conventional aluminum and magnesium alloys, but without compromising other essential features such as castability and corrosion resistance.
- Technical solutions for field-assisted particle addition in industrial practice (master alloys and/or shape casting).
- Demonstration of technical viability of the approach on the basis of selected prototype components.

The project has a duration of 50 months (from June 2012 to July 2016). Its resources include a 19.2 M€ budget (of which 9.5 M€ EC contribution), about 116 person-years of manpower, and some 90 researchers and engineers being structurally engaged in its practical implementation. The consortium consists of 27 companies, research organizations and universities from eleven countries on three continents (Australia, France, Germany, Hungary, Italy, the Netherlands, Norway, Russia, Spain, Sweden, and the United Kingdom).

With the project being well on its way as of writing of this paper, results from the research and development efforts are gradually becoming available and are published in as far as it concerns non-confidential outcomes. Arranged from upstream to downstream manufacturing, exemplary findings from distinct project activities are as follows.

- **Nanoparticle production and characterization.** Nanoparticles can be manufactured with a variety of methods, of which the plasma-chemical method (based on thermochemical decomposition of liquid-sprayed agents in a high-temperature heat-transfer medium) and electro-explosion of wire (based on electrically atomizing a metal in a controlled atmosphere) are explored amongst others [3]. Important in the characterization of these particles – that typically appear as powders with a low bulk density – are the chemical composition of the main constituent and any side constituents, but also their size distribution and shape as well as their specific surface area which derives from that.
- **Mixing patterns in metal melts.** As noted before, metal melts can be treated with several types of external fields, each with their own characteristics as well as practical implications. With specific designs, it has been shown that acoustic waves (implying vigorous agitation of the melt) can be generated by electromagnetic induction [4]. In an equipment using a so-called traveling magnetic field, efforts are also reported to arrive at grain refinement and nanoparticle dispersion [5].
- **Interface reactions between metal and particles.** One of the distinguishing features of MMNCs compared to conventional composites is the extensive surface area of the particles to be included, posing challenges with their wetting during composite synthesis, but also rendering them susceptible to elevated temperature reactions. It is shown that particles such as Al_2O_3 may react with the magnesium-alloy melt to form MgO (and Al dissolving in the melt), that instead can act as reinforcements [6]. Also, it is shown that particles such as Y_2O_3 may be pushed by the growing dendrites during solidification, so that they end up in the eutectic regions in the as-cast magnesium-alloy matrix [7].
- **Particle distribution in the as-cast composite materials.** With a uniform distribution of nanoparticles in the metal matrix being essential for capitalizing on the potential of this class of materials, it needs also to be appreciated that these particles are of a size that cannot be easily distinguished by metallographical means. By using advanced characterization means such as X-ray tomography and small-angle neutron scattering, however, spatial distributions of particles such as Y_2O_3 have been visualized and quantified [8].
- **Mechanical properties of the as-cast composite materials.** With high-strength light-metal alloys being commonly based on precipitation strengthening, the stability of the materials' mechanical attributes is secured only up to a certain (long-term) exposure temperature. MMNCs here would have the principal advantage of reinforcements that would not soften and/or dissolve upon high-temperature exposure. In line with this, it has been demonstrated that an addition of 1 wt.% AlN particles can improve creep resistance by a factor of roughly three up to quite harsh testing conditions (of 240 °C of

temperature and 140 MPa of compressive stress), as reported in the earlier mentioned overview paper [1].

Modeling is an integral part of the project activities and relates to various technological issues. While simulations on the macro-scale are proving useful to describe mixing patterns in metal melts as outlined above, mesa- and micro-scale models are being developed to enhance scientific understanding as well. This includes such aspects as the behavior of particles in a flow and during solidification, the onset of cavitation, the calculation of interfacial energies and the assessment of favorable conditions for de-agglomeration of particles in a melt. For instance, phase-field modeling has been used to study the parameters affecting the pushing or engulfment of particles by an advancing solidification front [9]. Yet another example is in further unraveling the mechanisms of grain nucleation and grain growth in support of the chemical grain-refinement developments [10].

As for the application perspective, it was already stated that there is a general drive toward aluminum- and magnesium-based materials with increased mechanical performance to compete with other materials for structural lightweight applications, especially when such enhancements are not compromising manufacturability, costs, and/or other (technical) attributes. As such, possible applications of these materials may tap on the current markets for aluminum and magnesium alloys. More in particular, the prospective application areas as considered by and included in the project consortium are outlined in **Table I**. This inventory is limited to transport, being the predominant sector if it comes to structural lightweight applications. Further, it is restricted to shape castings (wrought applications are omitted for simplicity's sake, but yet not to be overlooked overall).

Table I. Global end-user application features.

SECTOR		PRODUCTION VOLUME*	WEIGHT-SAVING DRIVE*	MANUFACTURING PROCESS**	DEVELOPMENT ROUTE
Automotive	Passenger cars	very high	medium	HPDC, PMC	(GR), MMNCs
	Utility vehicles	high	medium	HPDC, PMC	(GR), MMNCs
(Aero)-space	Aircrafts	medium	high	SC	GR, MMNCs
	Space vehicles	low / one-off	very high	SC, M	GR, MMNCs

* Relative to other application areas

** HPDC = high-pressure die casting, PMC = permanent-mold casting, SC = sand casting, M = machining

Acknowledgement

The author wishes to acknowledge financial support from the ExoMet project, which is co-funded by the European Commission in the 7th Framework Program (contract FP7-NMP3-LA-2012-280421), by the European Space Agency, and by the individual partner organizations. Acknowledged is also the support of all ExoMet project participants who contributed with their results to this paper.

References

1. W.H. Sillekens, D.J. Jarvis, A. Vorozhtsov, V. Bojarevics, C.F. Badini, M. Pavese, S. Terzi, L. Salvo, L. Katsarou, H. Dieringa, "The ExoMet project: EU/ESA research on high-performance light-metal alloys and nanocomposites", *Metallurgical and Materials Transactions A*, 45 (2014), 3349–3361

2. N.N., ExoMet project website (http://www.exomet-project.eu)

3. S. Vorozhtsov, I. Zhukov, A. Vorozhtsov, A. Zhukov, D. Eskin, A. Kvetinskaya, "Synthesis of micro- and nanoparticles of metal oxides and their application for reinforcement of Al-based alloys", *Advances in Materials Science and Engineering* (2014), in press

4. G. Djambazov, V. Bojarevics, B. Lebon, K. Pericleous, "Contactless acoustic wave generation in a melt by electromagnetic induction", *Light Metals 2014*, TMS (2014), 1379–1382

5. M. Garrido, Y. Fautrelle, L. Davoust, V. Bojarevics, K. Pericleous, M. Megahed, O. Koeser, "Grain refinement and nanoparticle dispersion using traveling magnetic field", *Light Metals 2014*, TMS (2014), 1405–1410

6. M. Mounib, M. Pavese, C. Badini, W. Lefebvre, H. Dieringa, "Reactivity and microstructure of Al_2O_3-reinforced magnesium matrix nanocomposites", *Advances in Materials Science and Engineering* (2014), 476079

7. M. Mounib, W. Lefebvre, "Advanced characterization of metal matrix nano-composites", *Magnesium Technology 2014*, TMS (2014), 461–464

8. S. Terzi, R. Daudin, J. Villanova, P. Srirangam, P. Lhuissier, L. Salvo, E. Boller, R. Schweins, P. Lindner,J.-J. Blandin, P. Lee, H. Lemmel, "X-ray tomography and small-angle neutron scattering characterization of nano-composites: Static and in situ experiments", *Light Metals 2014*, TMS (2014), 1389–1393

9. T. Pusztai, L. Rátkai, A. Szállás, L. Gránásy, "Phase-field modeling of solidification in light-metal matrix nanocomposites", *Magnesium Technology 2014*, TMS (2014), 455–459

10. A. Prasad, L. Yuan L., P.D. Lee, D.H. StJohn, "The interdependence model of grain nucleation: A numerical analysis of the nucleation-free zone", *Acta Materialia*, 61/16 (2013), 5914–5927

PHASE EVALUATION OF Sr AND CaO ADDED Mg-Al-Si ALLOYS

Young-Gil Jung[1,2], Young-Ok Yoon[1], Shae K. Kim[1], Hyun Kyu Lim[1], and Do Hyang Kim[2]
[1]Rare Metal R&BD group, Korea Institute of Industrial Technology, Incheon, Korea
[2]Dept. of Metallurgical Engineering, Yonsei University, Seoul, Korea

Keywords: Mg-Al-Si alloys, CaO, CaMgSi, Mechanical properties

Abstract

In the previous study, to modify the morphology of Mg_2Si phase, we added CaO in Mg-Al-Si alloy. However, CaMgSi phase was formed by the addition of CaO in Mg-Al-Si alloy and the morphology of CaMgSi phase is needle type. Therefore, CaMgSi phase would deteriorate the mechanical properties. On the other hand, when adding Sr in CaO added Mg-Al-Si alloy, the modified CaMgSi phase was confirmed.

This study is aimed at the evaluation of formed phase by Sr and CaO addition in Mg-Al-Si alloys and the investigation of mechanical properties and modification mechanism on the modified CaMgSi phase by the addition of Sr in CaO added Mg-Al-Si alloy.

Introduction

Recently, to decrease fuel consumption and emission in the automobile industries, Mg alloys are widely used for the reduction of weight. Especially, the application of Mg alloys on the powertrain parts has the largest potential for weight reduction [1]. However, because powertrain parts in automobiles are exposed at the high temperature atmosphere for a long time, in order to use Mg alloys on these parts, the superior creep resistance at high temperature should be required. For this reason, the various Mg alloys have been developed for high temperature applications. Currently, the most widely used high temperature Mg alloys are AJ (Mg-Al-Sr), MRI (Mg-Al-Sr-Ca-xx), and AX (Mg-Al-Ca) series, which offer a good combination of castability and high creep resistance. However, these alloys are expensive. Moreover, AX series is prone to limited castability, and AJ series is required for higher casting temperature [2-3]. Among these alloys, the advantage of AS (Mg-Al-Si) alloys is inexpensive and the improved creep performance exhibited rather than AZ (Mg-Al-Zn), AM (Mg-Al-Mn) series. However, molten AS alloys easily ignite and are oxidized without melt protective gases during melting and casting processes due to their high reactivity. Therefore, the most disadvantage of AS alloys are the limited castability. Furthermore, the coarse Chinese script type Mg_2Si phase will deteriorate the mechanical properties of Mg alloys. Therefore, in order to modify the Chinese script type Mg_2Si phases in the AS alloys, many methods have been studied such as the addition of alloying elements [4]. Recently, Ha et al. [5] reported that addition of CaO to Mg alloys gives the similar effect of the addition of Ca because CaO gets reduced by reacting with the liquid Mg. This finding is important from the viewpoint of industry because CaO is more cost effective and less hazardous to handle than Ca. Thus, in the previous studies, to solve the disadvantage of AS alloys, CaO added in AS alloys, and then ignition and oxidation resistance of molten AS alloys improved and the morphology of Mg_2Si was changed [6]. On the other hand, with the addition of CaO in AS alloys, needle type CaMgSi phase was formed. These needle type phases are similar to the morphology of Chinese script type Mg_2Si. Therefore, these phases have a bad effect on the mechanical properties. For this reason, to modify needle type CaMgSi, Sr element was selected. Because it was reported that positive effect on the morphology of Si containing phase in Mg alloys provided [7]. Through the preliminary test, when the addition of Sr in CaO added AS alloy, the modified CaMgSi phases were observed. Therefore, this study has been carried out to investigate modification mechanism on the modified CaMgSi phase and mechanical properties by the addition of Sr in CaO added AS alloys. We used "O" as a code letter to be distinguished from alloys with metallic Calcium whose code letter is "X".

Experimental Procedure

The experimental alloys were listed in Table 1. The as-cast alloys were prepared by melting with commercial-pure Mg, Al, Mg-10Si, Mg-5CaO, and Al-10Sr master alloys in a mild steel crucible under the protection of a mixed gas atmosphere of SF_6 and CO_2. The melt was held at 690 °C for several minutes and then poured into permanent mold. The microstructural analysis was carried out using an optical microscopy (OM) and scanning electron microscopy (SEM) equipped with energy dispersive X-ray spectrometer (EDS) after etching with a solution of 2 vol.% nitric acid + ethyl alcohol. Also, the specimens for the tensile test were milled into a dog-bone design for mechanical testing based on the ASTM B557-84 (gauge length: 30 mm, diameter: 6.0 mm). The tensile tests were performed at room and high temperatures (150 °C). The condition of creep test was carried out at a stress level well below the tensile yield strength. The conditions are as follows: Given temperature is 125 °C, 150 °C, 175 °C, and 200 °C.

Results and Discussion

Microstructure

Fig. 1 shows the microstructures of the as-cast AS51, ASO5105, and ASJO51005 alloys. The tested alloys were consisted of α-Mg matrix and the secondary phases. Fig. 2 shows SEM images of the tested alloys. According to the EDS analysis results, the microstructure of as-cast AS51 alloy was consisted of Chinese script type Mg_2Si phases with β-$Mg_{17}Al_{12}$ phases. When adding CaO in AS51 alloy, needle type CaMgSi phase was observed in ASO5105 alloy.

Table 1. Nominal chemical compositions of experimental alloys (wt.%)

Alloy ID	Mg	Al	Si	Ca*	Sr
AS51	Bal.	5.0	0.7	-	-
ASO5105	Bal.	5.0	0.7	0.5	-
ASJO51005	Bal.	5.0	0.7	0.5	0.3

(*: Ca contents reduced from CaO)

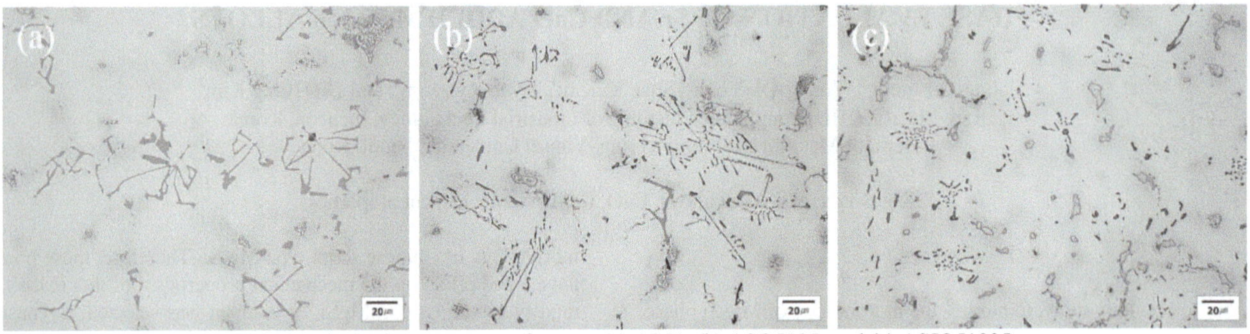

Fig. 1. Optical microstructures of as-cast (a) AS51, (b) ASO5105, and (c) ASJO51005

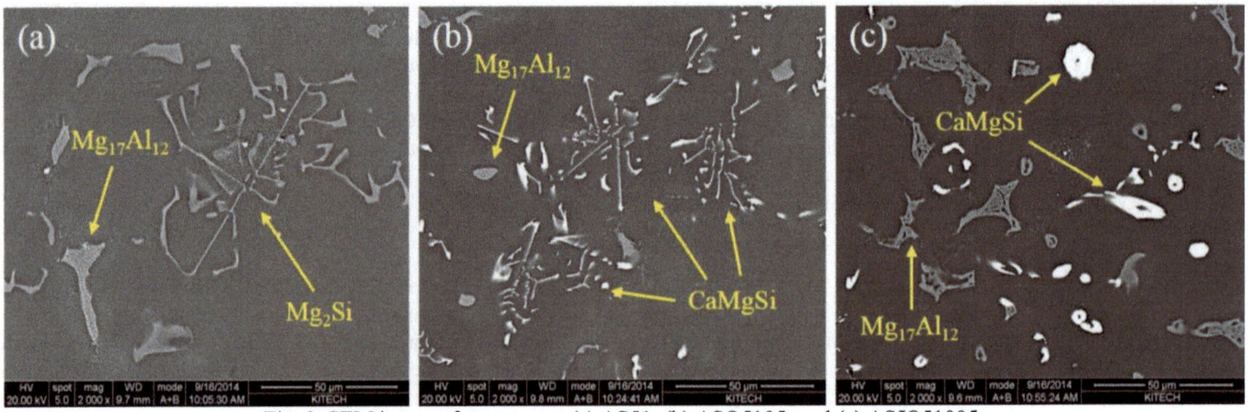

Fig. 2. SEM image of as-cast cast (a) AS51, (b) ASO5105, and (c) ASJO51005

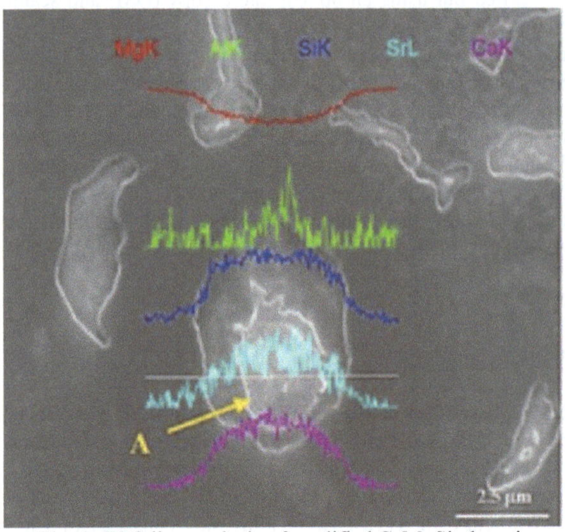

Fig. 3. EDS line analysis of modified CaMgSi phase in ASJO51005 alloy

Table. 2. EDS point analysis result of area A

Area	Mg	Al	Si	Ca	Sr
A	53.58	8.52	25.71	9.09	3.09

Comparing with the microstructure of as-cast AS51 and ASO5105 alloys, the morphology needle type CaMgSi phase in ASO5105 alloy is similar to that of Chines script type Mg_2Si in AS51 alloys. Especially, by the addition of Sr in ASO5105, the morphology of CaMgSi phase of ASJO51005 alloy was distinctly different from that of ASO5105 alloys. Fig. 3 shows the EDS line analysis result of modified CaMgSi phase in ASJO51005 alloy. From the EDS linescan, Al and Sr atoms in modified CaMgSi phase in ASJO51005 alloy was detected. The quantitative EDS analysis results listed in Table 2. From the SEM analysis results, Al-Sr compound in CaMgSi phase was observed. In previous studies, it has been reported that the formed Al-Sr compound by the addition of Sr element in Mg-Al based alloys was Al_4Sr phase [7, 8].

In the process of nucleus formation, the boundary energy of the heterogeneous nucleus with the crystallization phase has an effect on nucleus formation and it depends on the structure of the two contacting crystalline faces. Bramfitt suggested the mathematical model of the two-dimensional planar disregistry for the case of crystallographically dissimilar atomic arrangement between the substrate and the nucleated solid [9-10]. Bramfitt's disregistry is equation and one criterion of heterogeneous nucleation is that the disregistry of nucleant planes is less than 15 %. Fig. 4 shows the relationship of some possible crystal faces of (1 0 0) Al_4Sr to (1 1 0) CaMgSi, (1 0 1) CaMgSi, and (0 1 1) CaMgSi, respectively. Table 3 shows the results of the calculation of planar disregistry for some possible crystallographic orientations for CaMgSi nucleation on the Al_4Sr particles. It can be seen that when the orientations relationship between CaMgSi phase and Al_4Sr phase is (1 0 0) Al_4Sr // (1 0 1) CaMgSi, the disregistry is the lowest 1.0 %, which is less than 15 %. Therefore, Al_4Sr phase can act as the heterogeneous nucleation for CaMgSi phase by this orientation relationship.

Mechanical properties

The tensile properties at room temperature of AS51, ASO5105, and ASJO51005 alloys are listed in Table 4. The tensile yield strength at room temperature of the tested alloys exhibit the similar value.

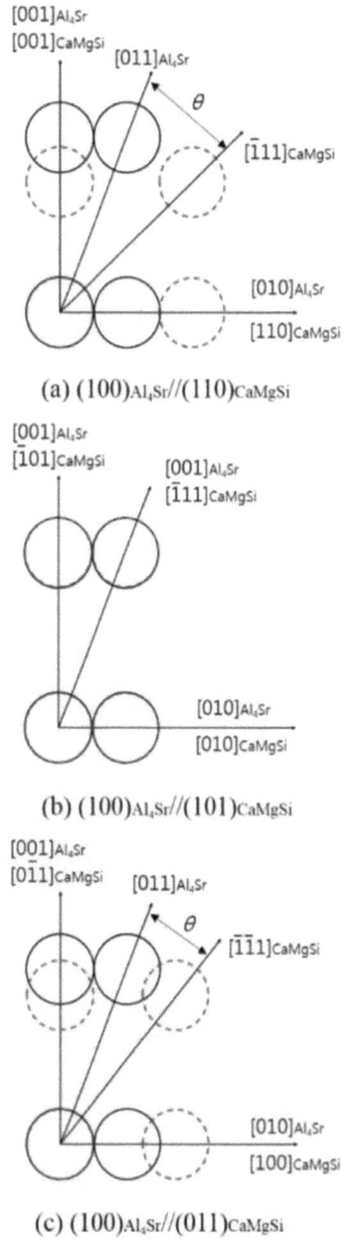

Fig. 4. Schematics of crystallographical relationships between Al_4Sr and CaMgSi phases

Among the tested alloys, the elongation at room temperature of ASJO51005 alloy is higher than that of AS51 and ASO5105 alloys, because the morphology of needle type CaMgSi phase was modified. Also, among the tested alloys, the tensile yield strength at high temperature (150 °C) of ASJO51005 alloy is higher than those of AS51 and ASO5105 alloys. Because modified CaMgSi phases is thermally stable at high temperature, this intermetallic phase at high temperature does not diffuse easily in the α-Mg [11].

According to results of tensile tests, ASJO51005 can exhibit the good creep resistance among tested alloys.

Fig. 5 shows the creep curve of the tested alloys at 60 MPa. Given the temperature condition is under the 125 °C, 150 °C, 175 °C, and 200 °C. Although all tested alloys occurred creep fracture at the 200 °C with the rapid increase of creep strain, the creep strain and secondary creep rate at 125 °C, 150 °C, and 175 °C of ASO5105 and

Table 3. Disregistry and crystallographical relationships between Al_4Sr and CaMgSi phases

Case	[h k l] Al_4Sr	[h k l] CaMgSi	θ	$\delta^{[hkl]Al4Sr}_{[hkl]CaMgSi}$
(1 0 0) Al_4Sr // (1 1 0) CaMgSi	[0 0 1]	[0 0 1]	0	30.6 %
	[0 1 1]	[$\bar{1}$ 1 1]	24.372	
	[0 1 0]	[1 1 0]	0	
(1 0 0) Al_4Sr // (1 0 1) CaMgSi	[0 0 1]	[$\bar{1}$ 0 1]	0	1.0 %
	[0 1 1]	[$\bar{1}$ 1 1]	0	
	[0 1 0]	[0 1 0]	0	
(1 0 0) Al_4Sr // (0 1 1) CaMgSi	[0 0 1]	[0 $\bar{1}$ 1]	0	21.0 %
	[0 1 1]	[$\bar{1}$ $\bar{1}$ 1]	16.597	
	[0 1 0]	[1 0 0]	0	

Table 4. The tensile test results at the room and high temperature (150 °C)

Alloy ID	Room temperature			at 150 °C		
	TYS (MPa)	UTS (MPa)	E (%)	TYS (MPa)	UTS (MPa)	E (%)
AS51	88.9	178.1	4.7	82.1	152.9	11.6
ASO5105	90.5	189.0	5.6	83.0	159.5	11.3
ASJO51005	89.8	185.0	6.2	88.2	164.5	10

ASJO51005 alloys is lower than those of AS51 alloy. From the creep test results, ASJO51005 alloy exhibited best creep properties at the all condition.This means that modified CaMgSi phase can play a significant role in increasing creep resistance, because of its high thermal stability. Also, from the gradient of the secondary creep in creep curve, the secondary creep rate can be calculated and the results are listed in Table 5. Fig. 6 shows the temperature dependence of secondary creep rate at 60 MPa. From the calculated secondary creep rate, activation energy (Q) value of AS51, ASO5105, and ASJO51005 alloys are 104.18 kJ/mol, 94.05 kJ/mol, and 98.42 kJ/mol, respectively. The range of 92-135 kJ/mol reported for the activation energy for dislocation glide in basal planes of pure magnesium [12]. Therefore, the dominant creep mechanism of tested alloy is dislocation glide at all conditions.

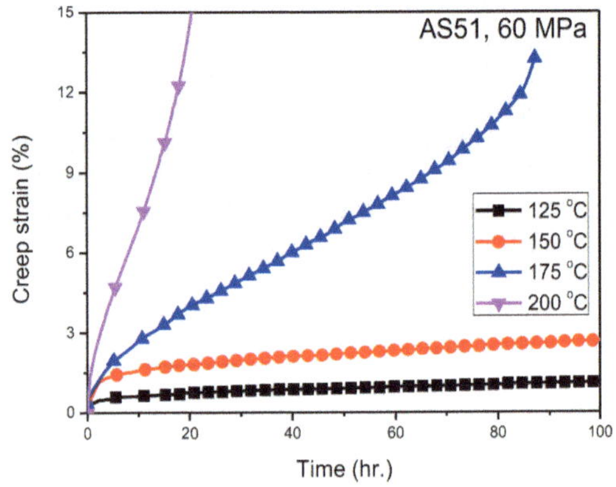

Fig. 5 Creep curves of tested alloys at 60 MPa: under the 125 °C, 150 °C, 175 °C, and 200 °C

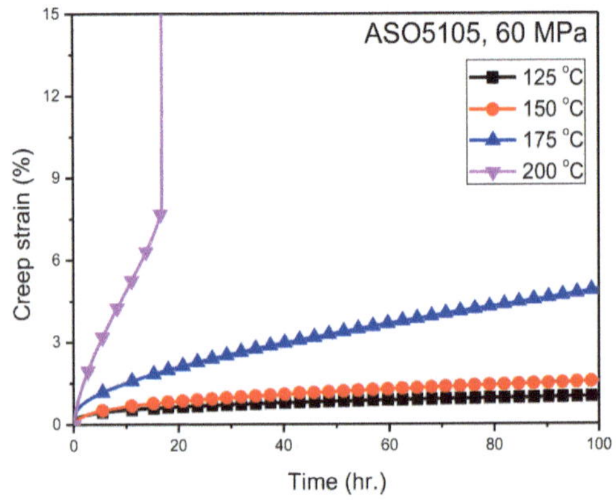

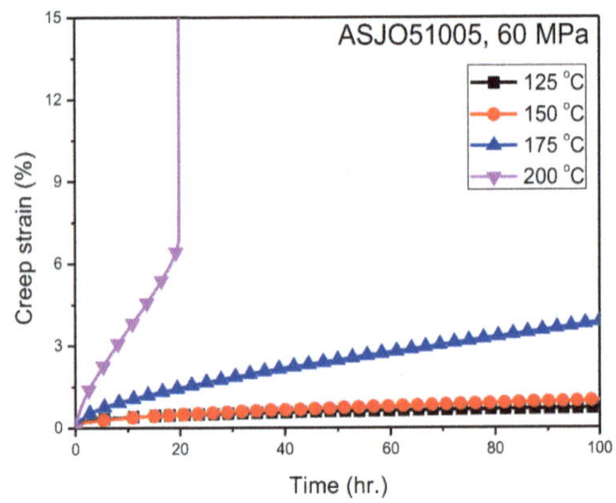

Fig. 5 (Continued) Creep curves of tested alloys at 60 MPa: under the 125 °C, 150 °C, 175 °C, and 200 °C

Table 5. Secondary creep rate and creep strain at 60 MPa: under the 125 °C, 150 °C, 175 °C, and 200 °C (*: Fractured)

Alloys	Creep strain at 100 hr (%)			
	125 °C	150 °C	175 °C	200 °C
AS51	1.11	2.69	13.64*	18.87*
ASO5105	1.01	1.57	4.49	7.98*
ASJO51005	0.71	1.00	3.86	7.94*
Alloys	**Secondary creep rate (/s)**			
	125 °C	150 °C	175 °C	200 °C
AS51	1.23E-08	2.76E-08	2.95E-07	1.49E-06
ASO5105	1.01E-08	2.20E-08	8.82E-08	1.02E-06
ASJO51005	6.48E-09	1.58E-08	7.74E-08	7.65E-07

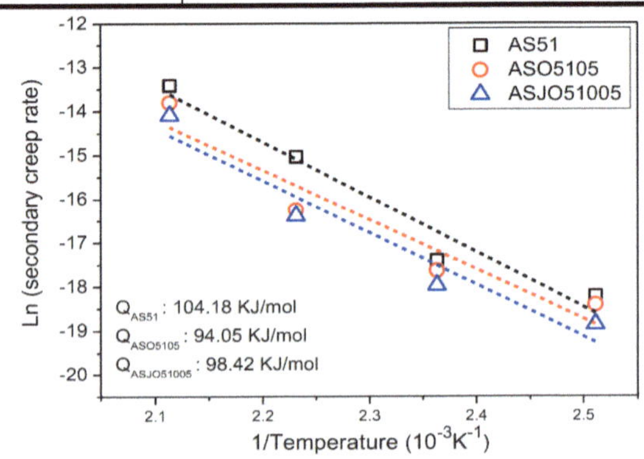

Fig. 6. Temperature dependence of secondary creep rate at 60 MPa

Conclusion

In the present study, mechanical properties and modification mechanism were investigated.

The conclusions are as follows:

(1) The morphology of CaMgSi phase is modified from needle type to round type, because the formed Al_4Sr phase by adding Sr in ASO5105 alloy acts as the heterogeneous nucleation site for the CaMgSi phase.

(2) The elongation at room temperature of ASJO51005 alloy is improved by modified CaMgSi phase. Among the tested alloys, ASJO51005 alloy exhibits good mechanical properties at room and elevated temperatures.

(3) From the creep test results, ASJO51005 alloy exhibited best creep properties at the all condition.

References

[1] Chris D., "MAGTECH1: Mg alloy and processing technologies for lightweight transport application", *DTI global watch mission report, September-October* (2004)

[2] M. M. Avedesian, H. Baker, Eds., Magnesium and Magnesium Alloys, ASM International, Metals Park (1999)

[3] M. pekguleryuz, P. Labelle, D. Argo, E. Baril, "*Magnesium diecasting alloy AJ62X with superior creep resistance, ductility and diecatability*", Magnesium Technology 2003 (2003) 201-206

[4] J. J. Kim, D. H. Kim, K. S. Shin, and N. J. Kim. "*Modification of Mg_2Si morphology in squeeze cast Mg-Al-Zn-Si alloys by Ca or P addition*" Scripta Mater, 41 (1999) 333-340

[5] S. H. Ha, J. K. Lee, and Shae K. Kim, "Effect of CaO on oxidation resistance and microstructure of pure Mg, Material Transactions, Vol. 49, No.5 (2008) 1081-1083.

[6] Y. G. Jung, H. K. Lim, Y. O. Yoon, Shae k., and D. H. Kim, "*Effects of alloying elements and cooling rate on morphology of phases in CaO Added Mg-Al-Si Alloys*", Magnesium Technology 2013 (2013) 333-339

[7] Nam K Y, Song D H, Lee C W, Lee S W, Park Y H, Cho K M, and Park I M., "*Modification of Mg_2Si morphology in as-cast Mg-Al-Si alloys with strontium and antimony*", Mater. Sci. Forum (2006) 510-511

[8] Tang S. Q., Zhou J. X, Tian C. W., and Yang Y. S., "*Morphology modification of Mg_2Si by Sr addition in Mg-4%Si alloy*", Tran. Nonferrous Met. Soc. China 21 (2011) 1932-1936

[9] B. L. Bramffit, "*The effect of Carbide and Nitride additions on the heterogeneous nucleation behavior of liquid iron*", Metall. Trans, vol. 1 (1969)

[10] Yuan G., Liu M., Ding W., and Akihisa I. "*Microstructure and mechanical properties of Mg-Zn-Si-based alloys*", Mater. Sci. Eng. A357 (2003) 314-320

[11] Y. Carbonneau, A. Couture, A. Van Neste, and R. Tremblay, "*On the observation of a new ternary MgSiCa phase in Mg-Si alloys*", Metall. Trans. A., 29A (1998) 1759

[12] B. Jing, S. Yangshan, X. Shan, X. Feng, and Z. Tianbai, "*Microstructure and tensile creep behavior of Mg-4Al based Magnesium alloys with alkaline-earth elements Sr and Ca addition*", Mater. Sci. Eng. A419, (2006) 181-188

Magnesium Technology 2015
Edited by: Michele V. Manuel, Alok Singh, Martyn Alderman, and Neale R. Neelameggham
TMS (The Minerals, Metals & Materials Society), 2015

HIGH PERFORMANCE Mg6Zn NANOCOMPOSITES FABRICATED THROUGH FRICTION STIR PROCESSING

Jiaquan Xu[1], Chezheng Cao[1], Shamiparna Das[2], Lianyi Chen[3], Chao Ma[3], Rajiv S. Mishra[2], Xiaochun Li[3]
[1]Department of Materials Science and Engineering, University of California, Los Angeles, CA 90095, USA
[2]Center for Friction Stir Processing, Department of Materials Science and Engineering, University of North Texas, TX 76203, USA
[3]Department of Mechanical and Aerospace Engineering, University of California, Los Angeles, CA 90095, USA

Keyword: Mg6Zn nanocomposites, semi-solid mechanical mixing, friction stir processing

Abstract

Mg-based metal matrix nanocomposites (MMNCs) are expected to provide significant enhancement of properties by introducing thermally stable ceramic nanoparticles into the metal matrix. However, it is extremely difficult to achieve a uniform distribution of nanoparticles in the magnesium matrix for the predicted significant property enhancement. In this study an unprecedented uniform distribution and dispersion of 6 vol.% SiC nanoparticles in Mg6Zn matrix was obtained through a novel scalable processing method that combines semi-solid mechanical mixing and solid-state friction stir processing. The resulting Mg6Zn nanocomposites exhibit tremendous hardness and strength enhancement. The results provide a viable pathway for the development and production of high performance Mg-based nanocomposites for numerous applications.

Introduction

Mg-based metal matrix nanocomposites (MMNCs) are expected to provide significant enhancement of properties by introducing thermally stable ceramic nanoparticles into the metal matrix. However, it is extremely difficult to efficiently incorporate a large amount of nanoparticles into metals and to achieve a uniform distribution of nanoparticles in the magnesium matrix for the predicted significant property enhancement [1,2,3,4,5]. The poor wettability and high specific surface areas of the nanoparticles often hinder the incorporation and dispersion of nanoparticles into Mg melt. Moreover, the van der Waals force between nanoparticles induces agglomeration, clustering and further floating, settling in molten metals.

After tremendous effort, two major routes for MMNCs processing have been established – solid-state processing (e.g. powder metallurgy) and liquid state solidification processing. In the powder metallurgy route, high shear stress generated by high-energy ball milling is applied to enforce nanoparticles into matrix metal and shear nanoparticle clusters. However, the low nanoparticles loading (less than 2 vol.%) for effective dispersion limited this method for high performance Mg nanocomposites [6,7]. And the strict conditions to avoid oxidations during high-energy ball milling also make this method not economic for scalable processing of Mg nanocomposites [8,9]. In the liquid-state solidification processing, firstly, non-oxide nanoparticles are easy to burn during their feeding at high temperature (above the liquidus temperature of the metal); secondly, nanoparticles form clusters in molten metal due to the attractive van der Waals force and the lack of a repulsive force; thirdly, nanoparticles or clusters tend to float or settle without effective incorporation into the molten metal due to a poor wettability with the molten metal; and fourthly, nanoparticles are normally pushed to the grain boundary by solidification fronts during solidification [10].

Here we utilized the semi-solid mixing for effective SiC nanoparticle incorporation into Mg6Zn matrix. The low temperature at semi-solid state could reduce the nanoparticles burning and oxidation effectively, while a high viscosity of the metal at semi-solid state would trap nanoparticles inside the matrix metal. Moreover, the micro clusters of incorporated SiC nanoparticles were friction stir processed (FSP) to achieve a uniform dispersion and distribution inside the solid Mg6Zn matrix. The microhardness of Mg6Zn + 6 vol.% SiC was increased by 178% after FSP. Tensile tests were also conducted to test the mechanical properties of Mg6Zn + 6 vol.% SiC nanocomposites.

Materials and method

Mg6Zn with 6 vol.% SiC were fabricated by semi-solid processing. The schematic of the experimental setup is in Figure 1. Pure Mg (99.93%) of 365.7 g and pure Zn of 21.94 g were melt together at 700 °C in a furnace under SF_6/CO_2 flow gas protection. The metal melt were then set at 600 °C at semi-solid state. Mechanical stir with a four blades (25.4 mm in diameter) at 1400 rpm were applied at the 1/3 height of the liquid melt. The vortex were generated by the rotating blade in a SiC crucible with a diameter of 70 mm. 41.3 g SiC nanoparticles were loaded and stirred into and maintained inside the melt with the help of the high viscosity of the semi-solid melt. The melt were then cooled to room temperature in the air. The solidified nanocomposites were re-melted inside a plate mold inside the furnace at 700 °C under the same gas flow protection. The nanocomposites were solidified to plate shape and cooled to room temperature inside the mold in the furnace.

The nanocomposites plates were friction stir processed with a H13 steel pin. The pin length is 6 mm and the pin diameter is 7.8 mm with a shoulder diameter 25.5 mm. The rotational speed is 1500 rpm. The traverse speed is 25.4 mm/min. The tilt angle is 3°. The materials were processed for four passes back and forth. Pure Mg6Zn was processed under same conditions for comparison.
The friction stir processed nanocomposites were wire cut into ASTM type V tensile bar. The tensile tests were conducted on the INSTRON 5966 tensile and compression testers. Vickers hardness tests were conducted using a 200 gf load for 10 s. The sample was grinded and ion milled to expose the microstructure and embedded nanoparticles. The microstructure of the samples was studied with powder X-ray diffraction (XRD) and Field Emission Scanning Electron Microscopy (SEM) (Navo Nano 230) with energy dispersive spectrometry (EDS). XRD was conducted using a Bruker D8 Advance Powder diffractometer with Cu Kα radiation.

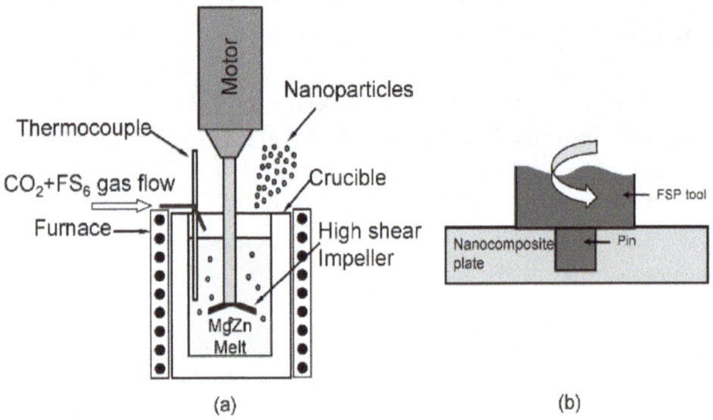

Figure 1. The schematic experimental setup: (a) semi-solid processing (b) friction stir processing.

Results and discussion

XRD of the nanocomposites

The XRD of the nanocomposites after FSP were shown in Figure 2. Pure Mg, SiC and MgZn phases are identified in this sample. No reaction products from Mg6Zn and SiC or oxidation of nanoparticles were identified. The XRD pattern indicates the stability of both matrix materials and nanoparticles through the whole procedure.

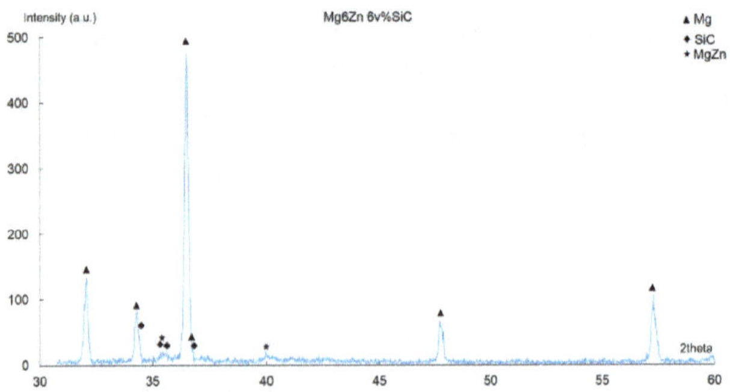

Figure 2. XRD pattern of the nanocomposites: Mg, SiC and MgZn were identified in this sample.

Microstructures

The microstructure of the nanocomposites after semi-solid mixing, but before FSP, were checked by optical microscope, as shown in Figure 3. Micro clusters of SiC nanoparticles are distributed in Mg6Zn matrix. The micro clusters indicate that high loading of SiC nanoparticles were successfully loaded into Mg6Zn at semi-solid state with the help of mechanical stirring.

The microstructure of both pure Mg6Zn and nanocomposites after FSP revealed by SEM were shown in Figure 4. For pure Mg6Zn in Figure 4(a) and (b), the surface is clean with ripples induced by ion milling. For nanocomposites in Figure 4(c) and (d), the surface is quite rough and SiC nanoparticles (dark spots as represented by arrow points) were well dispersed and distributed inside Mg6Zn matrix. No clusters were identified. The rough surface of Mg6Zn were believed to be introduced by the different milling rate of well dispersed SiC nanoparticles and Mg6Zn matrix during ion milling. And the dispersion SiC nanoparticles were from the disaggregation of micro clusters in Figure 3 by the shear force during friction stir.

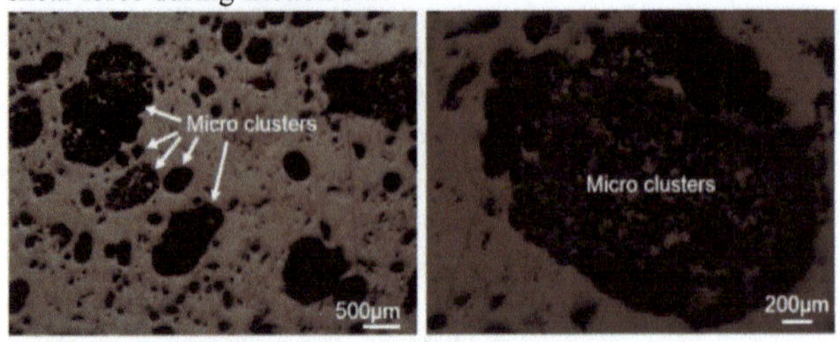

Figure 3. Micro clusters of SiC in Mg6Zn after semi-solid nanoparticle incorporation before FSP.

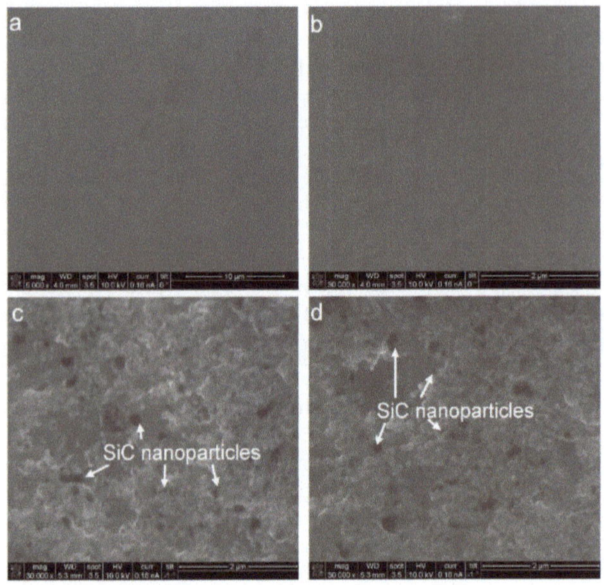

Figure 4. Microstructures after FSP: (a)(b)SEM images of pure Mg6Zn after FSP; (c)(d) SEM images of dispersed SiC nanoparticles in Mg6Zn after FSP (dark spots as represented by arrows points are SiC nanoparticles).

Mechanical properties

The microhardness of the nanocomposites before and after FSP is shown in Figure 5. The hardness (Vickers Hardness) of Mg6Zn + 6 vol.% SiC nanoparticles before FSP is 38. After FSP, the microhardness is 107, increased by 187%. Also compared with the hardness of pure Mg6Zn after FSP, the microhardness of nanocomposites after FSP has an 84% increase. The microhardness indicates FSP dispersed SiC nanoparticles in the matrix, leading to the tremendous increase of the microhardness.

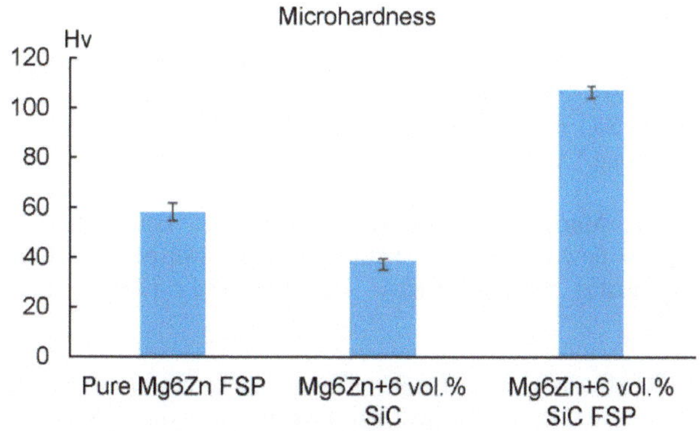

Figure 5. Microhardness of different samples.

The tensile test results of Mg6Zn + 6 vol.% SiC (data collected from five tensile) were shown in Table 1. The average yield strength and the ultimate tensile strengthen are 265.6 MPa and 303.7 MPa, respectively. The average elongation is about 2.11%.

Table 1. Tensile test results.

Yield strength (MPa)	Ultimate tensile strength (MPa)	Elongation
265.6	303.7	2.11%

Discussion

Master nanocomposites with a high percentage loading of nanoparticles are achieved through the efficient incorporation at semi-solid state. At semi-solid state, the viscosity of the semi-solid metal depends on the volume fraction of solidified metal, which depends on the temperature. A suitable viscosity can be found to prevent the nanoparticles from floating or settling by tuning the fraction of solidified metal through temperature control. At a suitable temperature, the nanoparticles were stirred into and maintained inside the slurry. Also the low temperature at semi-solid state reduce the reaction and oxidation of nanoparticles, enhancing the quality of incorporated nanoparticle. XRD pattern indicates there is no detectable reaction products and oxidation. The efficient incorporation at semi-solid state provide an economic method for high loading of nanoparticle incorporation. When comparing with liquid state solidification processing, the incorporation at semi-solid state improves the quality of incorporated nanoparticles.

Friction stir processing was used to disperse nano-reinforcements for metal matrix composites in several studies, especially in cases of Al- and Mg based MMCs and MMNCs [11,12,13,14,15,16,17]. The dispersion of the particles and the interfacial bonding between the particles and the Al- and Mg matrix after friction stir processing were found to be excellent [1, 18 , 19 , 20]. The trajectory of nanoparticles moving during friction stir processing were revealed in previous studies[18]. The significant particle movement induced by friction stir processing provides a powerful tool for dispersion of nanoparticles. Compared with the liquid-state solidification processing, high viscosity of matrix metal during friction stir processing effectively hindered clusters formation, floating or settling of nanoparticles in molten metal; Moreover, the solid state friction stir processing avoids the pushing of nanoparticles to the grain boundary by solidification fronts during solidification.

The strengthening effect of nanoparticles in the nanocomposites is characterized by the Orowan equation, since the second phases are ceramics, thus are all non-shearable by dislocations. The strengthening equation is $\Delta\sigma_{Orowan} = \frac{0.13G_m b}{d_p[(\frac{1}{2V_p})^{1/3}-1]}\ln\frac{d_p}{2b}$, where G_m is the Young's modulus 45 GPa and $b=(1/3)[2\bar{1}\bar{1}0]$ is the Burger's vector at room temperature. d_p is the diameter of nanoparticles and V_p is the volume fraction of nanoparticles. Theoretical Orowan strength contributed by 6 vol.% SiC in Mg6Zn matrix is 68.9 MPa. It corresponds to 27.7 increase in microhardness [21,22].

Compared with pure Mg6Zn after FSP, Mg6Zn nanocomposites are believed to benefit from the refined microstructure of matrix materials by Zener pinning effect[21] in addition to the Orowan strengthening contribution. The dispersed nanoparticles will apply pining pressures on the moving grain boundaries to restrict the grain growth during friction stir to gain more refined grains.

Conclusion

Here we combined the semi-solid mixing for effective SiC nanoparticle incorporation and friction stir processing to achieve high performance Mg6Zn+6vol.%SiC nanocomposites. The low temperature and high viscosity at semi-solid state could effectively improve the nanoparticles loading and reduce the oxidation/reaction. Moreover, friction stir processing effectively breaks micro clusters of incorporated SiC nanoparticles to achieve a uniform dispersion and distribution inside Mg6Zn matrix. The microhardness and tensile tests were conducted and the mechanical properties of Mg6Zn + 6 vol.% SiC nanocomposites are significantly enhanced. This method provides an economic pathway for processing of high performance Mg nanocomposites.

Acknowledgement

This work was partially supported by National Institute of Standards and Technology (NIST) through its Technology Innovation Program.

References

1. C. J. Lee, J. C. Huang, and P. J. Hsieh. "Mg based nano-composites fabricated by friction stir processing." *Scripta Materialia* 54.7 (2006): 1415-1420.

2. L.-Y. Chen, et al. "Novel nanoprocessing route for bulk graphene nanoplatelets reinforced metal matrix nanocomposites." Scripta Materialia 67.1 (2012): 29-32.

3. L.-Y. Chen, et al. "A novel manufacturing route for production of high-performance metal matrix nanocomposites." *Manufacturing Letters* 1.2 (2013): 62-65.

4. L.-Y. Chen, et al. "Achieving Uniform Distribution and Dispersion of a High Percentage Nanoparticles in Mg18Sn Matrix by Solidification Processing."*Magnesium Technology 2014* (2014): 465-470

5. L.-Y. Chen, et al. "Achieving uniform distribution and dispersion of a high percentage of nanoparticles in metal matrix nanocomposites by solidification processing." *Scripta Materialia* 69.8 (2013): 634-637.

6. C. Capdevila and H. KDH. Bhadeshia, "Manufacturing and Microstructural Evolution of Mechanuically Alloyed Oxide Dispersion Strengthened Superalloys*." *Advanced Engineering Materials* 3.9 (2001): 647.

7. H. Ferkel, and B. L. Mordike. "Magnesium strengthened by SiC nanoparticles."*Materials Science and Engineering: A* 298.1 (2001): 193-199.

8. K. S. Tun, and M. Gupta. "Improving mechanical properties of magnesium using nano-yttria reinforcement and microwave assisted powder metallurgy method." *Composites Science and Technology* 67.13 (2007): 2657-2664.

9. C. S. Goh, et al. "Development of novel carbon nanotube reinforced magnesium nanocomposites using the powder metallurgy technique."*Nanotechnology* 17.1 (2006): 7.

10. J. Q. Xu, et al. "Theoretical study and pathways for nanoparticle capture during solidification of metal melt." *Journal of Physics: Condensed Matter* 24.25 (2012): 255304.

11. R. S. Mishra, Z. Y. Ma, and Indrajit Charit. "Friction stir processing: a novel technique for fabrication of surface composite." *Materials Science and Engineering: A* 341.1 (2003): 307-310.

12. R. S. Mishra, and Murray W. Mahoney, eds. *Friction stir welding and processing*. ASM International, 2007.

13. J. Q. Xu, et al. " Urchin-like AlOOH nanostructures on Al microspheres grown via in-situ oxide template " *Materials Science and Engineering: B* 188 (2014): 89.

14. J. Q. Xu, et al. " Assembly of metals and nanoparticles into novel nanocomposite superstructures." *Scientific Reports* 3(2013): 1730.

15. M. Mohammad, K. Besharati, G. Barmouz, and J. Seyfi. "On the role of processing parameters in producing Cu/SiC metal matrix composites via friction stir processing: Investigating microstructure, microhardness, wear and tensile behavior." *Materials characterization* 62.1 (2011): 108-117.

16. Q. Liu, et al. "Microstructure and mechanical property of multi-walled carbon nanotubes reinforced aluminum matrix composites fabricated by friction stir processing." *Materials & Design* 45 (2013): 343-348.

17. A. Dolatkhah, et al. "Investigating effects of process parameters on microstructural and mechanical properties of Al5052/SiC metal matrix composite fabricated via friction stir processing." *Materials & Design* 37 (2012): 458-464.

18. Y. Morisada, et al. "Three-dimensional visualization of material flow during friction stir welding by two pairs of X-ray transmission systems." *Scripta Materialia* 65.12 (2011): 1085-1088.

19. Y. Morisada, et al. "Effect of friction stir processing with SiC particles on microstructure and hardness of AZ31." Materials Science and Engineering: A433 (2006) 50-54.

20. C. I. Chang, et al. "On the Hardening of Friction Stir Processed Mg-AZ31 Based Composites with 5-20% Nano-ZrO~ 2 and Nano-SiO~ 2 Particles."*Materials transactions* 47.12 (2006): 2942.

21. Z. Zhang, and D. L. Chen. "Consideration of Orowan strengthening effect in particulate-reinforced metal matrix nanocomposites: A model for predicting their yield strength." *Scripta Materialia* 54.7 (2006): 1321-1326.

22 Z. Zhang, and D. L. Chen. "Contribution of Orowan strengthening effect in particulate-reinforced metal matrix nanocomposites." *Materials Science and Engineering: A* 483 (2008): 148-152.

Magnesium Technology 2015
Edited by: Michele V. Manuel, Alok Singh, Martyn Alderman, and Neale R. Neelameggham
TMS (The Minerals, Metals & Materials Society), 2015

SYNTHESIS AND CHARACTERIZATION OF NOVEL MAGNESIUM MATERIALS CONTAINING COPPER-TITANIUM BASED ($Cu_{50}Ti_{50}$) AMORPHOUS ALLOY PARTICLES

S. Sankaranarayanan [1], N. Agrawal [2], S. Jayalakshmi [1, 3], Q.B. Nguyen [1], M. Gupta [1]
[1] Department of Mechanical Engineering, National University of Singapore, Singapore 117576
[2] Department of Ceramic Engineering, National Institute of Technology, Rourkela, India 769008
[3] Department of Mechanical Engineering, Bannari Amman Institute of Technology, Satyamangalam, India 638401

Keywords: Magnesium-Metal Matrix Composites, Amorphous reinforcement, Microwave Sintering, Microstructure, Mechanical Properties

Abstract

This study elucidates about the development and characterization of novel magnesium (Mg) materials containing amorphous Copper-Titanium based ($Cu_{50}Ti_{50}$) alloy particles. It involves the incorporation of $Cu_{50}Ti_{50}$ amorphous alloy powder (prepared using high-energy ball milling process) in pure Mg matrix using microwave sintering assisted powder metallurgy technique followed by hot extrusion. The microstructure and mechanical properties of the developed Mg/$Cu_{50}Ti_{50}$ composite were evaluated. Microstructural studies revealed the retention of amorphous structure of reinforcement, its uniform distribution in Mg-matrix without interfacial reaction products. Mechanical properties under indentation loads indicated hardness improvement due to amorphous $Cu_{50}Ti_{50}$ particle addition. Under compression loads, the developed composites exhibited enhanced strength with negligible change in ductility. The observed mechanical properties are discussed using the process-structure-property relationship.

Introduction

Magnesium is an excellent candidate for weight critical applications, primarily in the automotive and aerospace sectors. It exhibit desirable properties such as low density, high strength to weight ratio and excellent damping capacity which makes it highly suitable for the above mentioned engineering applications [1-3]. However, it is rarely used in its pure form because of its limitations such as poor ductility, low elastic modulus and detrimental properties at elevated service temperatures [3-5]. For mechanical property improvement, it is conventionally alloyed with other metallic elements such as Al, Mn, Zn, Zr, Ag and Rare Earth metals [1-7]. While those conventional Mg-alloys exhibited better mechanical properties when compared to pure Mg, they often suffer from poor thermal stability. On the other hand, the Mg-based metal matrix composites prepared through the incorporation of hard ceramic reinforcement generally exhibit better strength properties alongside thermal stability. However, they suffer from extreme brittleness [7].

In this regard, there exists a great demand to explore other, alternate reinforcements which could possibly enhance the properties of Mg for its extensive commercial utilization. One plausible type of novel reinforcement technology is to utilize bulk metallic glass (BMG) particles which are a class of unique materials having a dis-ordered crystal structure. They exhibit superior mechanical properties such as high strength, fracture toughness and high elastic limit at ambient temperature [8]. Since, the BMG particle reinforcements are metallic in nature, better interfacial properties are anticipated (due to strong mechanical bonding) which would impart superior strength to Mg-matrix, when compared to those conventional ceramic reinforcements [7]. The study of available literature reveals that only limited studies have been carried out so far to investigate the effects of metallic glass addition on the properties of Mg [9, 10]. Hence, an attempt is made in this research work to investigate the influence of amorphous $Cu_{50}Ti_{50}$ particulates addition on the microstructural and mechanical properties of Mg. Pure Mg was selected as the matrix material in order to isolate other microstructural factors imposed by alloying elements. Mg materials required for the current study were synthesized through the microwave sintering assisted powder metallurgy technique followed by hot extrusion [11]. The materials thus synthesized were examined for microstructural, physical and mechanical properties. Particular emphasis was placed on correlating the effect of $Cu_{50}Ti_{50}$ as reinforcement with the microstructural, thermal and tensile properties of pure magnesium.

Experimental procedure

Preparation of Amorphous Alloy Reinforcements

Amorphous alloy powder with composition $Cu_{50}Ti_{50}$ (in atomic percent, at.%) was prepared by mechanically alloying powder mixtures of elemental Cu and Ti metals. The powder mixture was milled at room temperature in air for 20 h, using a Retsch PM400 planetary ball mill with a ball-to-powder ratio of 3:1 and milling speed of 200 rpm [10]. The structural analysis of the amorphous powder was conducted using X-ray diffraction (XRD, Shimadzu LAB-XRD-6000, Cu Kα; λ = 1.54056 Å) and a scanning electron microscope (Hitachi FESEMS4300).

Preparation of Mg-composites

To produce Mg-10wt.%$Cu_{50}Ti_{50}$ composite, elemental Mg-powder (98.5% purity, particle size: <300 μm) was blended with the required amount of $Cu_{50}Ti_{50}$ amorphous powder for a duration of 1 h, and consolidated at room temperature at 450 MPa for 1 min. The compacted cylindrical billets of 36 mm diameter were microwave sintered at 100% power level (900 W) for 13 min, to achieve a temperature of 510 °C (based on prior calibration). The sintered billets were soaked at 400 °C for 1 h, and hot extruded into rods of 10 mm diameter at 350 °C. The extruded rods were used for characterization and testing.

Characterization

X-ray diffraction studies: For X-ray analyses, polished samples from the extruded rods were subjected to x-ray radiation using an automated Shimadzu LAB-XRD-6000 diffractometer (Cu, Kα= 1.54056 Å), with a scanning speed of ~2°/min. Phase identification was carried out by matching the Bragg angle and peak intensities with the standard peaks of Mg, Cu, Ti and other related phases.

Microstructural characterization: Microstructural characterization studies were conducted on the extruded pure Mg and Mg-amorphous particle reinforced composite to examine the particle distribution and for the formation of intermetallic phases. Hitachi FESEMS4300 field emission scanning electron microscope was used for the study. Scion image analysis software was used to analyze the captured micrographs to identify the average amorphous particle size and aspect ratio.

Mechanical properties: Mechanical properties of the developed Mg-materials were measured under indentation and compression loads. Indentation measurements were carried out on as-polished samples of the extruded Mg-materials using a Shimadzu HMV automatic digital microhardness tester (rather macro-hardness, as the average matrix grain size of the developed Mg-materials is expected to be few microns diameter) with a Vickers indenter, under a test load of 25 gf and a dwell time of 15 s. The tests were conducted on three samples for each composition for 10–15 repeatable readings. Uniaxial compression tests were performed at room temperature using a Materials Test System (MTS 810) in accordance with ASTM test method ASTM: E9–09. The tests were conducted at a constant strain rate of 8.3 x 10-4 s_1on cylindrical samples with aspect ratio, l/d ~ 1.5. A minimum of 5 tests were conducted so as to obtain repeatable values.

Results and Discussion

Structure and thermal properties of the amorphous powder

Structural changes during the ball-milling process leading to the amorphization of elemental Cu and Ti powder (composition ratio -50:50 at.%) can be understood by carefully observing the x-ray diffraction pattern after different ball-milling time as shown in Fig. 1a. The distinct crystalline Cu and Ti peaks (of high intensity) observed in the raw materials were seen to gradually transform into amorphous halo upon ball-milling. The observed diffused halo characteristics observed in the XRD pattern is typical of the amorphous structure which confirms the amorphous state of the produced powder [10]. The results of image analysis conducted on FESEM micrograph (Fig. 1b) showed that the particles are less than ~40 μm in size with a mean diameter of ~25 μm.

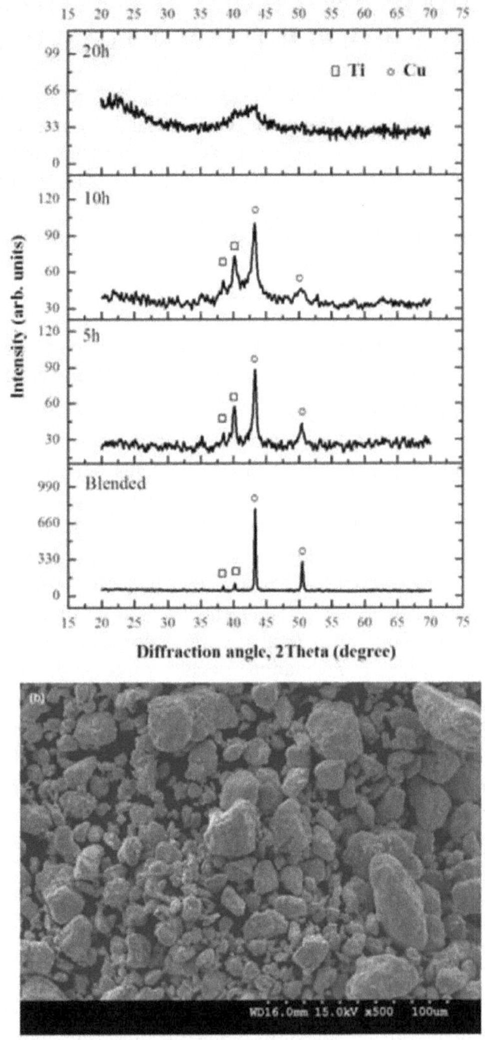

Figure 1 - (a) X-ray diffraction pattern of Cu and Ti after ball milling, with a ball-to-powder (BPR) ratio of 3:1 at 200 rpm. The amorphous structure was obtained after ~20 hours and (b) FESEM image of amorphous particles.

Synthesis of pure Mg and Mg-$Cu_{50}Ti_{50}$ amorphous particle reinforced composite

Synthesis of pure Mg and Mg-10wt.%$Cu_{50}Ti_{50}$ amorphous particle reinforced composite were successfully carried out by bi-directional microwave sintering followed by hot extrusion. The visual observation of sintered material did not reveal any macro-defects such as radial cracks. The surface of extruded rods was smooth and free from pores. The high sintering temperatures and short holding time during microwave sintering ensured efficient sintering with dense structure.

X-ray diffraction studies: The results of X-ray diffraction studies conducted on the extruded rods of the pure Mg and that of $Cu_{50}Ti_{50}$ amorphous particle reinforced Mg composite is shown in Fig. 2a. It shows prominent peaks corresponding to the crystalline Mg phase. In addition, the amorphous halo is also observed [10]. Further, peaks corresponding to other related phases are not seen which indicates the absence of any reaction between Mg-matrix and the amorphous reinforcement particles and the retention of reinforcements' amorphous structure even after sintering and extrusion processes. Available literature suggests that the addition of Cu to Mg would generally result in the formation of Mg_2Cu and Cu_2Mg phases [12, 13]. However, the relatively short duration microwave sintering technique employed in this study allowed the retention of amorphous structure of the reinforcement powder and thereby, prevents the formation of any intermetallic phase [13].

Microstructural characterization

Distribution of amorphous particles in Mg-matrix: The micro-structural characteristics of the developed Mg-composite are shown in Fig. 2(b-c). It can be seen from Fig. 2b that the amorphous reinforcement particles are distributed fairly uniformly in the Mg-matrix, without clustering or agglomeration. Further, the high magnification image (Fig. 2c) shows good bonding (without any interfacial cracks, micro-pores and de-bonding etc.) at the matrix/particle interface which is free of any interfacial reaction products, thus confirming the observations made in x-ray analysis.

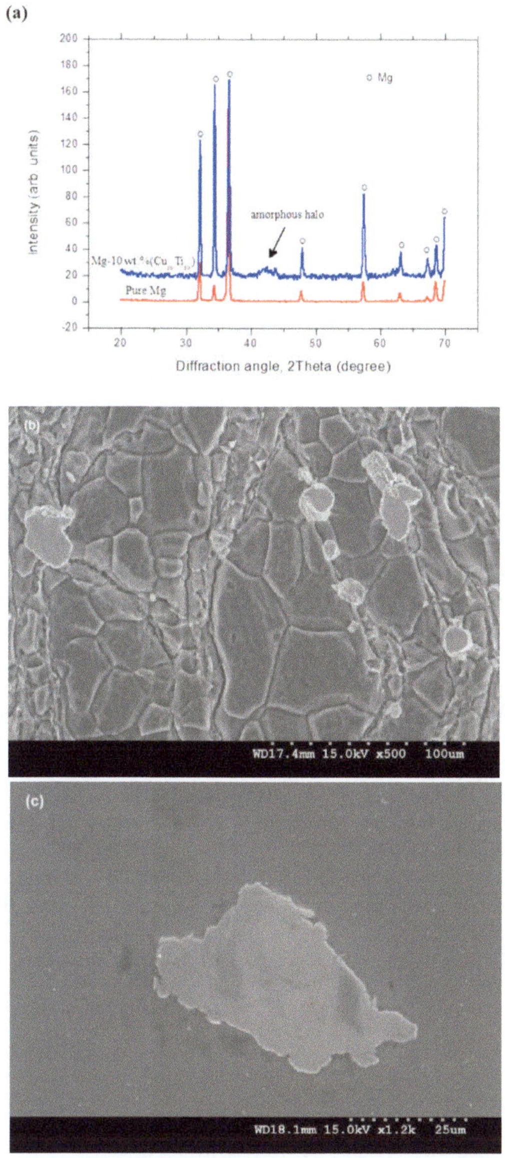

Figure 2 - X-ray diffractograms of pure Mg and Mg-10wt.%$Cu_{50}Ti_{50}$ amorphous alloy particle reinforced composite. SEM micrographs showing (b) distribution of $Cu_{50}Ti_{50}$ amorphous particles and (c) clear interface without interfacial reaction products in Mg-10wt.%$Cu_{50}Ti_{50}$ 10wt.%$Cu_{50}Ti_{50}$ amorphous alloy particle reinforced composite.

Mechanical properties

Micro-hardness: The results of micro-hardness measurements conducted on pure Mg and its composite containing amorphous $Cu_{50}Ti_{50}$ particles are shown in Table 1. It indicates an increase in the mean hardness value of Mg-composite (by ~25 %) due to amorphous $Cu_{50}Ti_{50}$ particle addition. The observed increment in micro-hardness value (Table 1) can be attributed primarily to the higher hardness of the amorphous reinforcement (920±35 Hv, measured). The high hardness value of the amorphous particle reinforcements impart higher constraint to localized matrix deformation during indentation and increases the matrix work-hardening, thereby resulting in higher hardness of the composite [14, 15].

Compressive properties: The results of compressive properties measurements conducted on the extruded samples of pure Mg and its Mg-10wt.%$Cu_{50}Ti_{50}$ composite are listed in Table 1. The results indicate that the incorporation of amorphous $Cu_{50}Ti_{50}$ particles have resulted in a significant improvement in the compressive strength properties (0.2% offset yield strength by ~45% and ultimate strength by ~40%) of Mg-10wt.%$Cu_{50}Ti_{50}$ composite when compared to its monolithic counterpart. The observed increase in the compressive strength properties due to amorphous $Cu_{50}Ti_{50}$ particle addition is mainly due to the high load bearing capacity of the reinforcement particles, (i.e. inherent high strength of the reinforcement which effectively bear the load transferred from the matrix to the particle) [15]. In the present case, the amorphous reinforcement has inherently high strength (compressive strength of amorphous alloys/BMG > 2GPa), that eventually results in higher composite strength properties. In addition to the load bearing capacity, the amorphous reinforcement would also contribute towards the strengthening of Mg-matrix by dislocation strengthening due to the generation and interaction of the geometrically necessary dislocations to accommodate the mismatch in (a) elastic modulus and (b) thermal expansion coefficient [5,14,15]. The difference in the coefficients of thermal expansion and elastic modulus between the matrix and reinforcing particles can cause residual plastic strain in the matrix around the particles, resulting in the above-mentioned enhanced dislocation density strengthening effect. Further, the very fine particles (sub-micron size) would contribute towards the matrix strengthening by dislocation bowing. It is interesting to note that the compressive ductility of the composite has been retained. This positive effect of ductility retention can be attributed to the unique properties of high strength and high elastic strain limit of the amorphous particles, which is unlike that seen in conventional MMCs, wherein the addition of ceramic particles usually resulted in poor ductility [16].

Table 1 - Results of Mechanical Property Measurements

S. No.	Materials	Micro-hardness [H_v]	Compressive Properties		
			0.2% Yield Strength [MPa]	Ultimate Compressive Stress [MPa]	Ductility (%)
1	Pure Mg	51 ± 2	61 ± 1	283 ± 5	18.7 ± 1.1
2	Mg-10wt.%$Cu_{50}Ti_{50}$	63 ± 4	93 ± 3	403 ± 5	17.5 ± 0.5

Conclusions

Novel Mg-composite containing 10wt.% $Cu_{50}Ti_{50}$ amorphous alloy particles was synthesized using microwave assisted rapid sintering technique, followed by hot extrusion. The developed Mg-composite was investigated for its structural and mechanical properties with respect to pure Mg. Based on the structure-property correlation, the following conclusions are drawn.

1. Microwave sintering followed by hot extrusion can successfully synthesize magnesium composites containing $Cu_{50}Ti_{50}$ amorphous reinforcement without altering its amorphous structure.
2. The addition of $Cu_{50}Ti_{50}$ amorphous reinforcement has significantly increased the mean hardness of pure Mg by ~25%.
3. Under compressive loads, the incorporation of $Cu_{50}Ti_{50}$ amorphous particles has significantly enhanced the strength of pure Mg (by ~40%) without largely affective the ductility.
4. The enhancement in mechanical properties due to amorphous reinforcement addition can be attributed to: (i) the inherent superior properties of high strength and high elastic strain limit of the $Cu_{50}Ti_{50}$ amorphous reinforcement and (ii) excellent matrix/particle interface, which together contribute to efficient load transfer and various matrix strengthening mechanisms.

Acknowledgement

The financial support of Ministry of Education Academic Research Funding (WBS #R-265-000-493-112) is gratefully acknowledged.

References

1. M. M. Avedesian and H. Baker, *ASM specialty handbook: magnesium and magnesium alloys* (ASM international 1999), 274.
2. B. Mordike and T. Ebert, "*Magnesium: Properties—applications—potential*" *Materials Science and Engineering: A* 302 (1) (2001), 37-45.
3. K. U. Kainer and F. Kaiser, *Magnesium alloys and technology* (Wiley 2003).
4. I. J. Polmear, "*Magnesium alloys and applications*" *Materials science and technology* 10 (1) (1994), 1-16.
5. G. E. Dieter, *Mechanical metallurgy (*New York: McGraw-Hill 1976), Vol. 3.
6. Z. Yang et al., "*Review on research and development of magnesium alloys*" *Acta Metallurgica Sinica (English Letters)* 21 (5) (2008), 313-328.
7. M. Gupta and S. M. L. Nai, *Magnesium, magnesium alloys, and magnesium composites* (Wiley 2011).
8. A. Inoue, *Bulk amorphous alloys* (Springer: 2001).
9. D. Dudina et al., "*A magnesium alloy matrix composite reinforced with metallic glass*" *Composites Science and Technology* 69 (15), (2009), 2734-2736.
10. S. Jayalakshmi et al., "*Development of novel Mg–$Ni_{60}Nb_{40}$ amorphous particle reinforced composites with enhanced hardness and compressive response*" *Materials & Design* 53, (2014), 849-855.
11. M. Gupta, M. Lai, D. Saravanaranganathan, "*Synthesis, microstructure and properties characterization of disintegrated melt deposited Mg/SiC composite*". *Journal of materials science* 35 (9), (2000), 2155-2165.
12. W. Wong and M. Gupta, "*Development of Mg/Cu nanocomposites using microwave assisted rapid sintering*" *Composites Science and Technology* 67 (7), (2007), 1541-1552.
13. W. Wong, "*PhD Thesis: Development of Advanced Materials Using Microwaves*" (National University of Singapore 2007).
14. W. D. Callister, D. G. Rethwisch, *Materials science and engineering: an introduction*. (New York: John Wiley & Sons Inc. 2007).
15. D. J. Lloyd, *Particle reinforced aluminium and magnesium matrix composites. International Materials Reviews* 39 (1), (1994), 1-23.
16. K. K. Chawla, N. Chawla, *Metal-Matrix Composites. (New York: Wiley 2004).*

Magnesium Technology 2015

Biomedical Applications

Mechanical and Corrosive Properties of Two Magnesium Wires: Mg4Gd and Mg6Ag

Petra Maier[1], Gabor Szakács[2], Marcin Wala[1], Norbert Hort[2]

[1]University of Applied Sciences Stralsund, Zur Schwedenschanze 15, D-18435 Stralsund, Germany
[2]Helmholtz-Zentrum Geesthacht, Magnesium Innovation Center, Max-Planck-Straße 1, D-21502 Geesthacht, Germany

Keywords: magnesium wire, bending, pitting corrosion

Abstract

The mechanical and corrosive properties of two Magnesium wires are studied in the field of microhardness, tension-, compression- and 3-point-bending tests, corrosion and its influence on the mechanical properties. Due to recrystallization during their complex forming processes (casting, extrusion, wire drawing), both wires show a fine grained microstructure resulting in high strength and ductility. However, the ductility is mostly evaluated by 3-point bending and compression; due to the notch effect in the clamp area, the maximum tensile strength and elongation under tension cannot be measured. Both alloys show a tensile-compressive yield asymmetry. Even RE-elements are known to reduce this asymmetry, the Mg-Gd alloy shows 100 MPa higher strength in tension than compression. The asymmetry of the Mg-Ag wire is similar. Overall the wires show very high strength and hardness, Mg6Ag slightly higher compared to Mg4Gd. Strong pitting corrosion is found and reduces strongly the tensile and bending strength.

Introduction

Mg wires have recently gained interest in the area of biomedical application: either for pins, rods and clips for temporary fixation of bone fracture in an mm range or for forming stents; here much smaller diameters will be needed. Due to recrystallization during or after their complex forming processes, metal wires show fine grained microstructure resulting in high strength and ductility. A study by Hort et al. [1] shows that addition of Gd, as one example of RE elements, improves the mechanical properties and corrosion rate of Mg. The work of Feyerabend et al. [2] describes that the acute toxicity of Gd is only moderate. Tests regarding the cytotoxicity in osteoblast-like cells show that Gd could be a suitable alloying element. Tie et al. [3] indicate that Mg-Ag alloys have satisfactory corrosion properties under cell culture condition and much better mechanical properties than pure Mg. Binary Mg-Ag alloys are designed as antibacterial material to treat infections in an implant site. Recent publications by Griebel et al. [4] (concerning cold-drawn ZM21 and WE43 wires) and Chatterton et al. [5] (studying the texture evolution during RE-Mg wire drawing) underline the growing interest in Mg wires. Since these wires are exposed to bending, before operation and during implantation or by additional shaping into stents the strength influenced by corrosion as well as the corrosion influenced of work hardening is of interest.

Experimental

Production

Out of two different binary alloys (Mg4Gd with 3.9 wt.% Gd and Mg6Ag with 6.4 wt.% Ag) wires of a diameter of 1.6 mm were produced from cylindrical gravity castings followed by solution heat treatment (Mg4Gd: 500 °C / 6 h and Mg6Ag: 440 °C / 6 h) via direct extrusion (Mg4Gd: 430 °C and Mg6Ag: 350 °C) and drawing. Wire drawing itself was done at 350 °C using up to 9 drawing steps with an average deformation degree of 0.3. A short annealing (Mg4Gd: 470 °C / 15 min and Mg6Gd: 430 °C / 15 min) followed to release forming stresses.

Testing

For microscopy examinations, the samples were ground with SiC papers to a grit size of 2500, then polished with a mixture of water free OP-S and 1 µm diamond suspensions and finally etched with an etchant solution of 4g picric acid, 7ml acetic acid, 40 ml distillate water and 140 ml ethanol. The microstructures were analysed with a DMI 5000M light microscope. The grain sizes were determined with line interception method.

Mechanical testing was conducted on a TIRA28100 universal testing machine. Samples for tensile tests had a length of 50 mm, for bending 30 mm (span of 25 mm, deformation speed 1 mm/min). Compression samples had a ratio of height to diameter of 2:1 and were carefully prepared to guarantee parallel top and bottom faces. Ringer-Acetate solution was used in the corrosion tests due to its similarity to human blood. Three wires were corroded at once in 500 ml electrolyte at 37 °C over 1, 2 and 3 days. Hardness measurements (HV1) were conducted on a Zwick Indentation Hardness Tester ZHU/Z2.5.

Materials

Microstructure

As-cast material consists of rather coarse grained microstructure, up to 920 ± 160 µm in case of Mg6Ag and 878 ± 235 µm for Mg4Gd, respectively. As a result of recrystallization during extrusion both wires show much smaller grain sizes after extrusion, 13 ± 4.5 µm in case of Mg4Gd and 45 ± 18.5 µm for Mg6Ag. After drawing the average grain size increases a little, up to 30 ± 10.3 µm of the Mg4Gd wire (see Figure 1a) and up to 80 ± 17.1 µm of the Mg6Ag (see Figure 1b). The grains of the Mg6Ag wire are extensively twinned and crossed by second phase alignments in extrusion and drawing direction (here seen as vertical lines and more often in the middle of the wire).

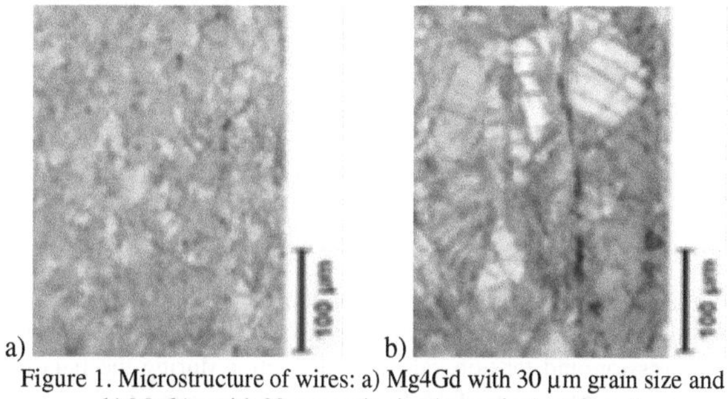

Figure 1. Microstructure of wires: a) Mg4Gd with 30 µm grain size and b) Mg6Ag with 80 µm grain size (extensively twinned)

Mechanical Properties: hardness and 3-Point-Bending

To evaluate the mechanical property changes during processing, the hardness HV1 of the Mg4Gd and Mg6Ag wires of different processing steps has been measured, see Table I. The changes are a result of processing and grain size modifications. The hardness of the as-cast material is slightly higher for Mg4Gd compared to Mg6Ag. T4 solution heat treatment does not change the hardness and grain size significantly for both alloys. However, extrusion increases hardness stronger in Mg6Ag. From the increase in hardness during drawing it can be seen again, that Mg6Ag is acting stronger to the (warm) forming processes: hardness increases by x1.6 (compared to x1.3 for Mg4Gd), even Mg6Ag has the larger grain size (but extensive twinning).

Table I. Hardness HV1 of Mg4Gd and Mg6Ag wires of different processing steps

	Mg4Gd	**Mg6Ag**
as-cast	**47.1** ± 4.9	**40.3** ± 2.6
T4 solution heat treatment	**46.8** ± 3.7	**38.0** ± 4.9
extruded -longitudinal-	**50.1** ± 1.5	**50.6** ± 3.9
extruded -cross section-	**51.8** ± 0.5	**49.7** ± 2.2
drawn -longitudinal-	**62.9** ± 1.3	**81.0** ± 4.4
drawn -cross section-	**68.7** ± 3.2	**79.1** ± 2.8

Figure 2 shows the bending stress-outer strain curves of the Mg4Gd (black) and the Mg6Ag (gray) wires at two different deformation speeds: 1 mm/mm and 10 mm/min. The higher deformation speed causes slightly higher strain hardening, here more significant at the Mg4Gd wire. However, concerning the overall strain hardening ability: the curves of Mg6Ag show higher bending yield stress (~480 MPa) and a slightly higher stress-strain slope within the plastic deformation region before necking. In Mg6Ag the twin boundaries act additionally to the grain boundaries as blockages for dislocation moving, causing higher strain hardening.

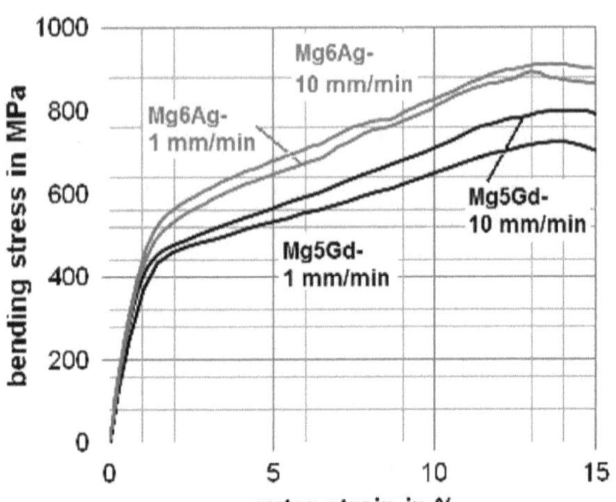

Figure 2. Bending stress - outer strain curves of Mg4Gd (black) and Mg6Ag (gray) wires at two different deformation speeds: 1 mm/mm and 10 mm/min

Wires have been pre-bent to the region of necking and used for hardness measurement to evaluate possible differences between compression and tension side. The deflection during the bending test at an outer strain of 15 % was 10 mm, causing a bend wire with an inner angle less than 90°. Since no crack initiation has been seen, both wires show very ductile behaviour.

Figure 3 shows the hardness increase as a result of bending at a deformation speed of 1 mm/min. The solid line shows the hardness increase at the tensile side and the dotted line at the compression side for Mg4Gd (black curves) and Mg6Ag (gray curves). Mg4Gd reached at the outer strain of 15 % a bending stress of 700 MPa and Mg6Ag of 870 MPa, respectively. The indents have a distance of 500 μm; indent 11 represents the location of the maximum bending moment.

A general hardness increase can be seen, towards the left and right the average hardness of each wire is starting to show. Followed from Table 1, Mg6Ag shows higher (as-)drawn hardness and higher hardness increase during bending, which agrees with the higher bending stress in Figure 2. Furthermore it can be seen, that the hardness at the compression side in both wires is higher than at tension side. This is known for small thicknesses and only turns around at higher sample thickness as the result of shifting the neutral axis towards the compression side.

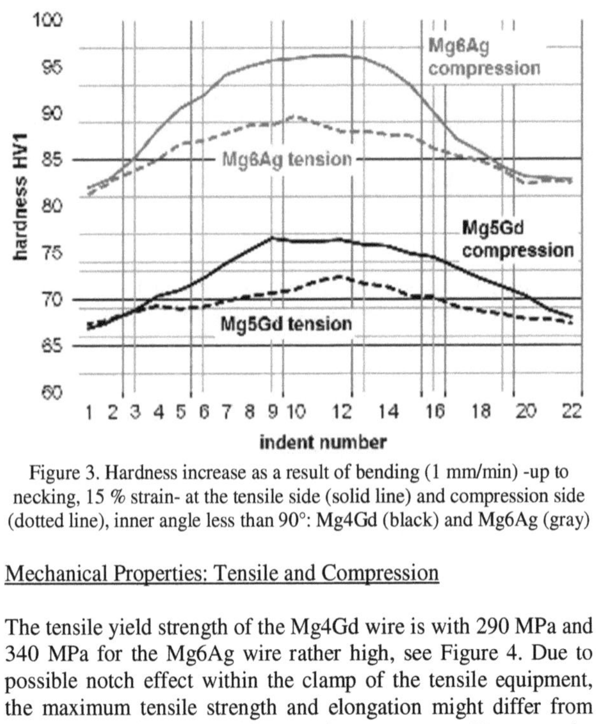

Figure 3. Hardness increase as a result of bending (1 mm/min) -up to necking, 15 % strain- at the tensile side (solid line) and compression side (dotted line), inner angle less than 90°: Mg4Gd (black) and Mg6Ag (gray)

Mechanical Properties: Tensile and Compression

The tensile yield strength of the Mg4Gd wire is with 290 MPa and 340 MPa for the Mg6Ag wire rather high, see Figure 4. Due to possible notch effect within the clamp of the tensile equipment, the maximum tensile strength and elongation might differ from tests done with longer wires or other test setup. However, the elongation up to 4 % of the Mg4Gd wire results in a reduction of initial area by 40 % (see Figure 5a), and by 30 % area reduction of the Mg6Ag wire with 1.5 % elongation.

From additional compression tests a tensile-compressive yield asymmetry (Δ 100 MPa) can be seen for both wires (Figure 4). RE elements are known to reduce the tensile-compressive yield asymmetry by reducing the texture [6], but 3.9 wt.% of Gd seems not enough effective. Binary Mg-10Gd in extruded condition only hardly shows any texture and yield asymmetry [7]. From the compression tests the ductility of the wires can be seen again. The samples failed under 45 ° shear fracture.

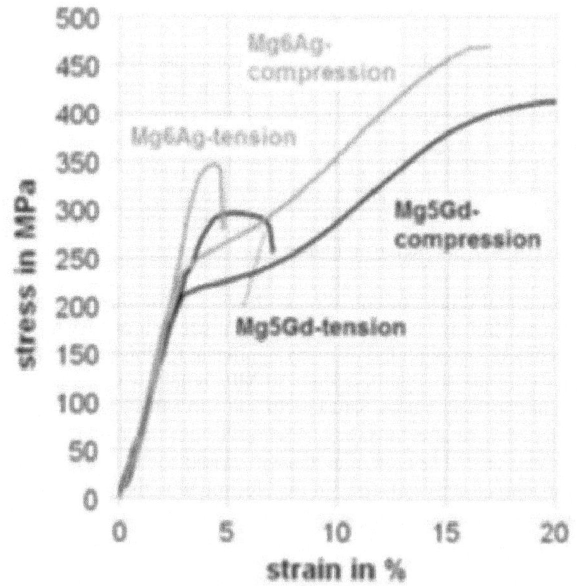

Figure 4. Tensile and compression stress - strain curves of Mg4Gd (black) and Mg6Ag (gray) wires

Figure 5b shows the ductile fracture behaviour of Mg4Gd wire. The morphology of voids and dimples (some make second phases within visible) looks similar to the Mg6Ag fractured surface after tensile test; just the reduction in initial area is slightly smaller due to its lower elongation at fracture.

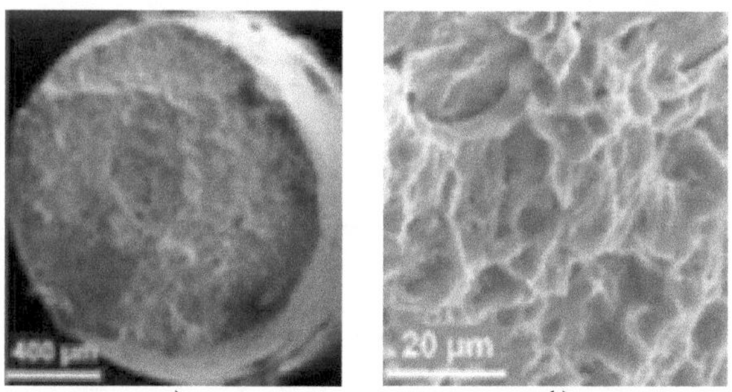

a) b)

Figure 5. SEM Fracture surface after tensile test of Mg4Gd, a) overview showing initial area reduction and b) detail showing ductile voids

Results and Discussion

Corrosion

To correlate the corrosion damage to the loss of mechanical properties the corrosion behaviour itself has been investigated. The wires were exposed to 500 ml Ringer solution (37 °C) for 1, 2 and 3 days (each time 3 wires). The rather high amount of electrolyte kept the pH-value down in neutral level.

Figure 6 (Mg4Gd) and Figure 7 (Mg6Ag) show the corroded wires after: a) 1 day, b) 2 days and c) 3 days. Both wires show very local corrosion. Mg4Gd forms mainly the white corrosion product $Mg(OH)_2$, which partly remains and partly falls off. After 3 days the wire still has an increased weight of plus 10%. Mg6Ag corrodes under forming $Mg(OH)_2$, $CaCO_3$ as well as AgCl and shows weight loss and strong pitting corrosion. By looking at Figure 7c no mechanical properties are expected.

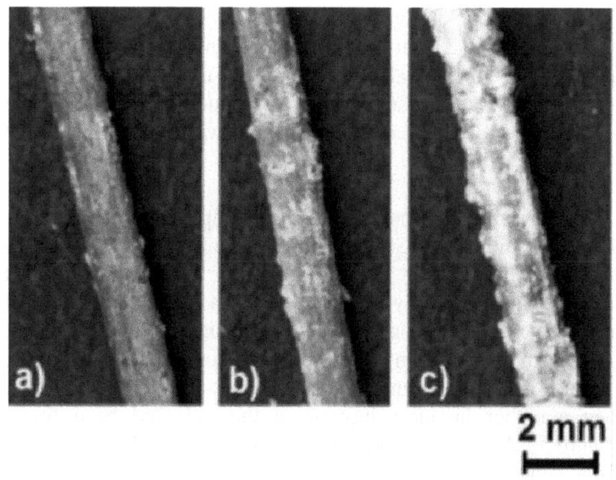

Figure 6. Photograph of Mg4Gd wires after corrosion: a) 1 day, b) 2 days and c) 3 days

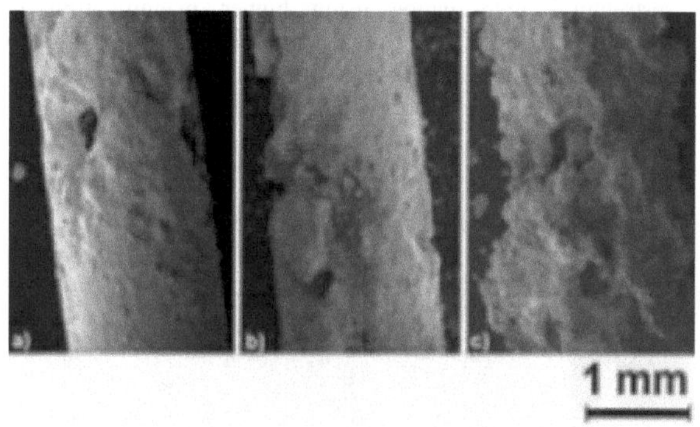

Figure 7. SEM of Mg6Ag wires after corrosion: a) 1 day, b) 2 days and c) 3 days

Because of a partly remaining passive layer, the pitting corrosion of the Mg4Gd wire can be pointed out better by micrographs (longitudinal), see Figure 8a and b, which show a corroded wire after 1 day exposure. Pitting corrosion starts off with an elliptical shape (see Figure 8a left pit) and transforms in subsurface or undercutting (see Figure 8a right pit) or even horizontal pit shape (see Figure 8b). Pitting corrosion does not only reduce the wire diameter, it also causes increased stress intensity (notch effect).

a) 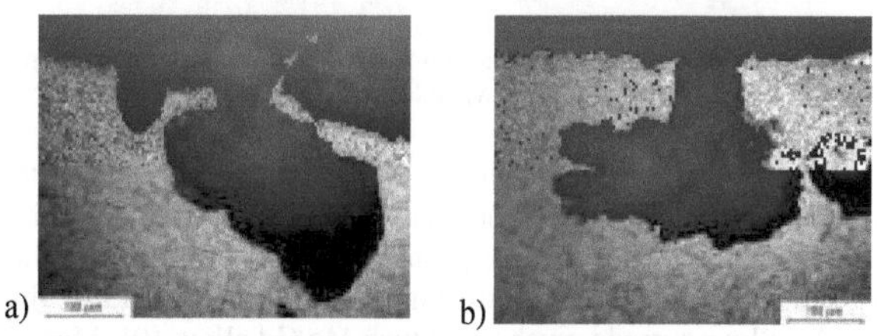 b)

Figure 8. Micrograph of corrosion pits in Mg4Gd wires after 1 day corrosion, shaped as a) elliptical and subsurface, b) horizontal

Influence of pre-bent wires on corrosion

Experiments of pre-bending the wires to the region of necking (see hardness measurements on them in Figure 3) have been repeated to find out the influence of plastic stress/strain on the corrosion. The outside of the bent wire undergoes tensile stresses during 3-point-bending and the inside compressive stresses, respectively. Straight after deformation the wires were exposed to corrosion. Since the wires have not been hold in their "elastic-plastic distortion" during corrosion (that would need a clamping device), elastic and even plastic back-deformation by pit initiation takes place. The residual stress situation at the out- and inside of the wire will have changed to lower values. However, tensile strength is still assumed at the outside wire and compressive strength at the inside.

Figure 9a shows the longitudinal macrograph of a Mg4Gd wire after 2 days corrosion. Strong pitting can be seen at the outside of the wire, where tensile stresses occur. The corrosion process is therefore accelerated by open pores within the forming passive layer, which is inhomogenous even without overlaying stresses. Tensile stress is known to support crack initiation and growth. There is hardly seen any corrosion attack at compression side, even a higher hardness increase at this side has been found (Figure 3). An increase in hardness caused by a higher density of dislocation leads to high anodic dissolution [8]. However, that seems to act less major compared to the beneficial compressive residual stresses not allowing corrosion pits to develop. Microstructural investigation using higher magnification to discuss the corrosion microstructure (influence of twins, grain boundaries and second phases) is in progress. However, the general shape of the pits agrees with the micrographs in Figure 8.

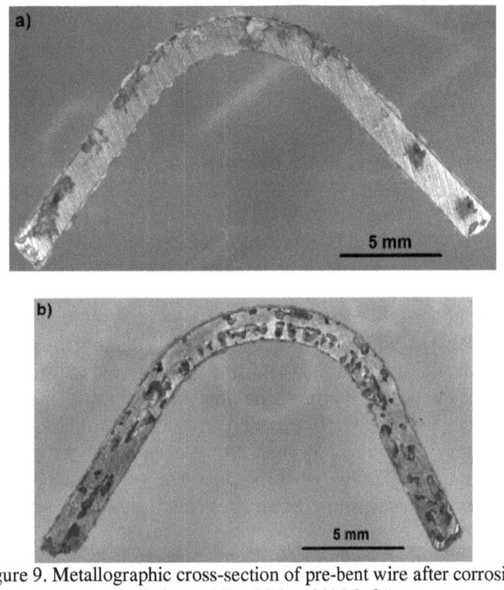

Figure 9. Metallographic cross-section of pre-bent wire after corrosion over 2 days: a) Mg4Gd and b) Mg6Ag

Looking at Figure 9b, a different "picture" is shown. In the Mg6Ag wire pitting corrosion can be seen all around the wire. Degradation tests also in phosphate buffered saline and in other physiological conditions on extruded materials have shown that Mg-Ag alloys corrode much faster than Mg-Gd alloys [9]. Mg6Ag has larger and heavily twinned grains. Especially twinned grains (increased density of lattice defects) accelerate corrosion [10]. Therefore the higher corrosion rate of Mg6Ag alloy seems to overlay the different stress situations; even compressive residual stresses are not beneficial. The holes formed along the middle of the wire are a result of ongoing corrosion at upper or lower parts of the wire. The second phase alignment in extrusion and drawing direction seems to lead the corrosion expanding trough the middle of the wire: here microgalvanic corrosion between matrix (α-Mg) and novel second phases (see micrograph in Figure 1b).

<u>Influence of corrosion on tensile properties</u>

Being aware of the strong local corrosion, the influence of the pitting corrosion on the tensile strength has been evaluated. Corrosion exposure only took place at the gauge length (the wire parts, which were clamped in the machine were protected against corrosion). Figure 10a and b show the stress-strain curves of Mg4Gd (Figure 10a) and Mg6Ag (Figure 10b) as a function of corrosion time - 1, 2 and 3 days - compared to the reference wire, not corroded, see therefore Figure 4.

After 1 and 2 days of corrosion both wires show similar retained strength. But by taking the higher initial strength of Mg6Ag into account, the strength loss is more significant for this alloy. After 3 days a tensile strength less than 50 MPa is left, whereas Mg4Gd reaches strength values between 30 and 140 MPa. Looking at the corroded wire after 3 days in Figure 7c, low structural stability was expected. The corrosion damage caused some wires to break apart during corrosion test (in the electrolyte without mechanical implementation). Furthermore it is noticeable that with increasing corrosion time data scattering also increases.

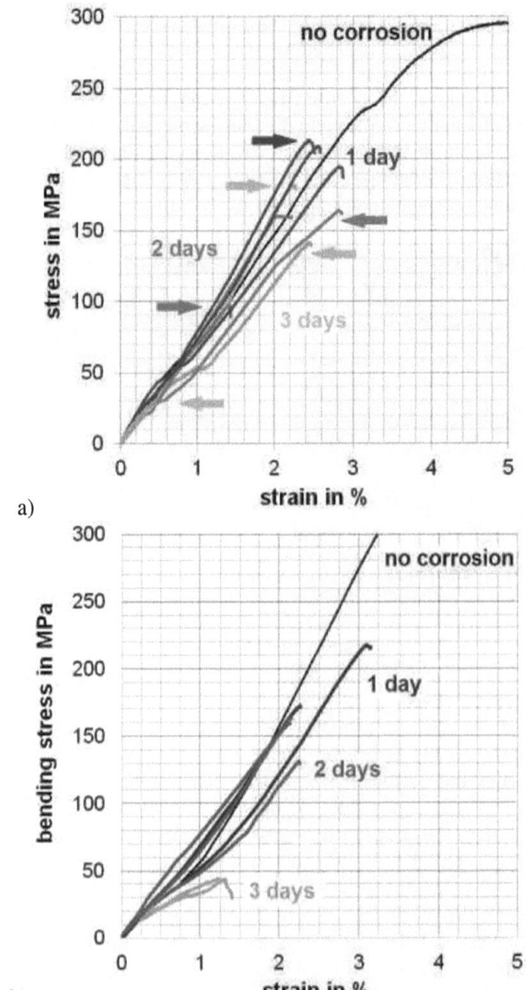

Figure 10. Tensile stress - strain curves as a function of corrosion time: without corrosion and 1, 2, 3 days of a) Mg4Gd and b) Mg6Ag wires

Indicating the strength loss as a function of time is one possible way to evaluate the influence of corrosion. However, the pits forming during corrosion varying in size and shape even after identical time frame, see here again Figure 8. To correlate the strength loss more specific to ongoing corrosion the remaining load bearing area should be taken into account. Without micro-CT imaging this area can only determined after testing by either fractography or micrographs of fractured samples (this will be explained in the next section "Influence of corrosion on bending properties").

Figure 11 shows the fracture surface of a Mg6Ag wire after 3 days corrosion, showing extensive pitting corrosion. Mostly the upper right fractured area appears as the non corroded initial area before tensile test. Overall not more than 25 % of area remained after 3 days, causing a significantly loss of strength down to 40 MPa. Taking a maximum tensile strength of 350 MPa as the reference value (see Figure 4), only 10 % of initial strength remains.

Figure 11. Fracture surface of Mg6Ag wire after 3 days of corrosion showing extensive pitting corrosion

Influence of corrosion on bending properties

The bending stress - outer strain curves in Figure 12a (Mg4Gd) and 12b (Mg6Ag) reveal in comparison with the non-corroded wire a significant reduction in bending yield and maximum bending strength as well as ductility after corrosion. The loss in mechanical properties is higher for Mg6Ag, see data for 1 day.

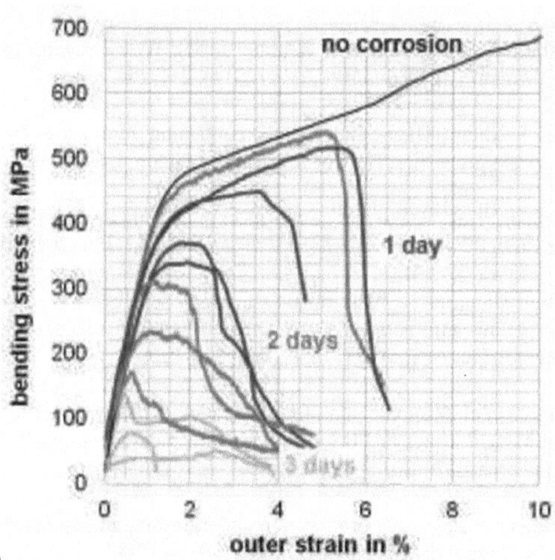

a)

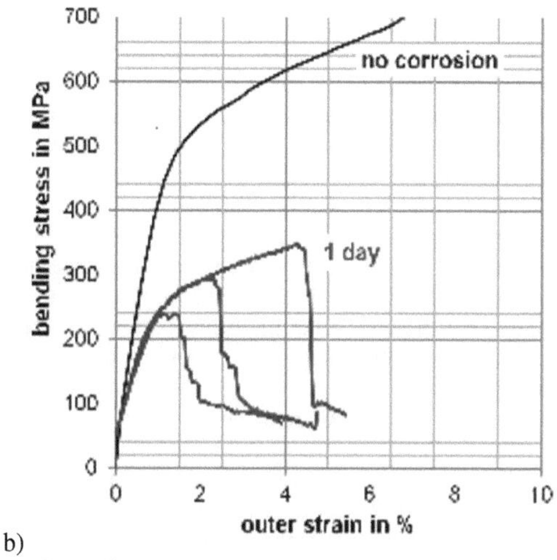

b)

Figure 12. Bending stress - outer strain curves as a function of corrosion time: without corrosion and 1, 2, 3 days of a) Mg4Gd and b) Mg6Ag

Since the device to measure the deflection during 3-point-bending applies a small force to the wire, the damage in the Mg6Ag alloy after 2 days of corrosion proceeded in such a way that these wires could not be tested due to its fragility. However, the Mg4Gd wire shows again a high data scattering at the same duration of corrosion: after 1 day the maximum bending strength varies from 520 MPa to 340 MPa and from 540 MPa to 170 MPa for 2 days, respectively. Form these data scattering it is concluded that once local corrosion takes place it causes a notch effect and the material fails due to stress peaks.

The distribution in remaining strength depends on the size and shape of the corrosion pits, which have been found very different, see again Figure 8. With increasing corrosion time failure occurs more often "off-centre of the supporting pins" (meaning not opposite the upper loading pin). That means that some pits cause due to their larger size a high enough stress intensity as a result of the notch effect by pitting to overcome the highest bending moment at mid span and act as crack initiation points.

Figure 13 shows a micrograph of a cracked Mg4Gd wire under bending after corrosion over 2 days. The marked left side (black line) indicates the initial corrosion pit before fracture. The opening toward the right is the final crack. The pit had a depth of ~ 100 µm and was ~ 80 µm wide. It was shaped elliptical, with the depth exceeding the width.

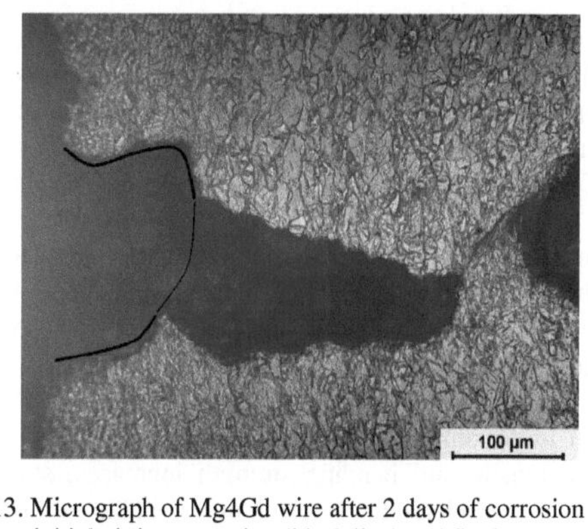

Figure 13. Micrograph of Mg4Gd wire after 2 days of corrosion showing initial pitting corrosion (black line) and final crack

Knowing from notched samples the V notch factor is depending on its geometry: its angle, radius and length. Applying these geometrical parameters to the shape of the corrosion pits narrow, deep pits will increase the stress intensity more than wide, shallow pits. However, the corrosion pit in Figure 12 has caused a loss in bending yield strength of 100 MPa.

The fracture toughness will depend on the microstructure. Since the non-corroded wire did not show any cracking under bending, this value could not been evaluated. The results of the corroded wires show that the area under the bending stress – outer strain curves decreases with increasing corrosion time, indication decreasing toughness.

Conclusion

This study of two Mg wires for consideration of biomedical application (Mg4Gd for its moderate corrosion rate and good mechanical properties and Mg6Ag for having an antibacterial effect) gives the following key findings:

(1) after the final process step "wire drawing" a rather fine grained microstructure can be found: Mg4Gd with 30 μm and Mg6Ag with 80 μm (but heavily twinned)

(2) hardness measurements reveal higher strain hardening ability in Mg6Ag (counts for extrusion, wire drawing and deformation by bending), Mg6Ag shows larger grains, but higher hardness (caused by the twins)

(3) especially 3 point bending shows the high ductility of these wires, they do not crack under bending

(4) bending yield strength of Mg6Ag with 480 MPa slightly higher than of Mg4Gd with 400 MPa

(5) a tensile-compression yield asymmetry can be seen from the tensile and compression tests, compression tests result in sheared fractures under 45 °

(6) tensile tests lack of determination of elongation at fracture due to notch effect by gripping the sample, but reduction of initial area and fractography present a ductile fracture behavior (voids and dimples, some with second phases enclosed)

(7) corrosion: twinned and larger grained microstructure of Mg6Ag corrodes faster than Mg4Gd, both alloys show strong local corrosion: pitting corrosion

(8) second phases (Mg_5Gd in Mg4Gd and $Mg_3Ag/Mg_{54}Ag_{17}$ in Mg6Ag), mostly aligned in extrusion and drawing direction, will cause microgalvanic corrosion, an additional solution heat treatment will reduce second phases and residual stresses, the effect of T4 on mechanical properties remains to find out

(9) corrosion pits act as crack initiation point when also exposed to mechanical loading: both in tension and bending, strong decrease in strength and ductility has been found after corrosion in Ringer solution at 37 °C

(10) with increasing corrosion time, data scattering of yield and maximum tensile and bending strength increases, strength loss should not be correlated to corrosion duration, but to the corrosion pit size and shape

(11) corrosion pits have been found to develop from small elliptical pits to subsurface and even horizontal pits, however, even an elliptical pit of a size of 100 μm deep and 80 μm wide causes a strength loss of 100 MPa, which is 75% of initial bending yield strength.

To summarize: when applying mechanical load, more important than the value of corrosion rate (in weight or geometry loss) is the understanding of corrosion damage/form, especially when pitting corrosion occurs. After all, all these findings are a result of corrosion in Ringer solution. Other media will act different. But so far some pitting corrosion always occurs in magnesium corrosion, so further work bringing the pit size and shape together with the loss of mechanical properties makes the ongoing corrosion independent of media and time. However, the properties of the material itself, like microstructure, strength and ductility will influence the remaining properties, like ongoing crack propagation. Since the mechanical properties itself are very good, but the degradation is not homogenous, a surface coating might improve long term stability for possible biomedical application.

Acknowledgement

The authors appreciate the support of J. Briesemeister (UAS Stralsund) in carrying out tree-point-bending tests. The assistance of H. Habeck (UAS Stralsund) during corrosion tests is greatly appreciated. The group of C. Vogt (Hannover University, Germany) is thanked for analyzing the corrosion products of Mg6Ag wires, including SEM imaging.

References

1. N. Hort et al. "Magnesium alloys as implant materials - Principles of property design for Mg-RE alloys," *Acta Biomaterialia* 6 (2010), 1714-1725.

2. Feyerabend et al. "Evaluation of short-term effects of rare earth and other elements used in magnesium alloys on primary cells and cell lines" *Acta Biomaterialica* 6 (2010), 1834-1842.

3. D. Tie et al. "In vitro mechanical and corrosion properties of biodegradable Mg–Ag alloys" *Materials and Corrosion* 64 (2013), 1-7.

4. A.J. Griebel et al. "Cold-drawn ZM21 and WE43 wires exhibit exceptional strength and ductility" *European Cells and Materials* (2014), in press

5. Chatterton et al. "Texture Evolution during Wire Drawing of Mg-RE Alloy" *Magnesium Technology 2014* (2014), 251-256.

6. L.L. Rokhlin, Magnesium alloys containing rare earth elements – structure and properties, (Taylor & Francis, London, UK 2003).

7. P. Maier et al. "Influence of Nd in extruded Mg10Gd base alloys on fatigue strength" *Materials Science Forum*, Vol. 783-786 (2014) 419-424.

8. G. Ben Hamu et al. "The relation between severe plastic deformation microstructure and corrosion behaviour of AZ31 magnesium alloy," *Journal of Alloys Compound* 468 (2009), 222–229.

9. I. Marco et al. "Comparison of the degradation behaviour of binary Mg-Ag and Mg-Gd alloys in PBS and in physiological conditions" *European Cells and Materials* (2014), in press

10. P. Maier et al. "Crack propagation under bending in cast Mg10GdxNd-T4 alloys" *Magnesium Technology 2014* (2014), 77-82.

Magnesium Technology 2015
Edited by: Michele V. Manuel, Alok Singh, Martyn Alderman, and Neale R. Neelameggham
TMS (The Minerals, Metals & Materials Society), 2015

DEGRADATION BEHAVIOR OF Mg-Ca NAIL AFTER PENETRATION INTO ARTIFICIAL BONE

Naoko Ikeo[1], Junichi Shimizu[1], Chihiro Ishigaki[1], Yuya Sano[2], Yoshinaka Shimizu[3], Toshiji Mukai[1]
[1]Department of Mechanical Engineering, Kobe University, 1-1 Rokkodai-cho, Nada-ku, Kobe, 657-8501, Japan
[2] Division of Advanced Prosthetic Dentistry, Tohoku University Graduate School of Dentistry, 4-1 Seiryo-cho, Aoba-ku, Sendai, 980-8575, Japan
[3] Division of Oral and Craniofacial Anatomy, Tohoku University Graduate School of Dentistry, 4-1 Seiryo-cho, Aoba-ku, Sendai, 980-8575, Japan

Keywords: Biodegradable materials, biocorrosion, residual strain, magnesium nail, magnesium-calcium alloy

Abstract

For applying a magnesium alloy to nails in the guided bone regeneration (GBR) method, sufficient strength and appropriate degradation speed are required to fix a membrane. In the case, the nails are exposed to body fluid after penetrating the alveolar bone. Therefore, in this research, degradation behavior after penetration of magnesium-calcium alloy which is expected to possess high biocompatibility was investigated. As a result, the Mg-Ca alloy nails were degraded inhomogeneously after immersion for 4 weeks in simulated body fluid. The degraded portions corresponded to the distribution of residual strain estimated by finite element analysis. Mg-Ca nails without precipitates possessed comparatively gradual degradation rate. It can be also confirmed that the region with residual strain degraded preferentially when compared the CT images and the residual strain distribution after penetration.

Introduction

When permanent teeth are lost by disease or accidents, mastication function can be recuperated by reconstruction of teeth using artificial dental root. However, these titanium implants cannot be inserted into the mouth if there is a large defect by diseases or accidents in alveolar bone which supports the dental roots. In these cases, reconstruction of alveolar bone is conducted by using guided bone regeneration (GBR) method before implantation of artificial dental implant [1].
In GBR method, at first, bone substitute which promotes bone growth fills up the defect area. To preserve this medication in the defects and avoid invasion of other soft tissue, the affected area is covered by the membranes. Finally, these membranes are fixed by driving nails into surrounding mandible bone. In current systems, polytetrafluoroethylene and titanium is utilized for membranes and nails respectively. Since they are insoluble in human body, removal surgery is carried out after reconstruction of alveolar bone.
Because there is physical and mental suffer due to the removal surgery, it is desirable to substitute biodegradable nails and membranes for conventional such devices. Therefore, various researches of application of biodegradable materials such as magnesium for these methods are conducted in dental field [2, 3].
For development of biodegradable metals of tools for penetration in GBR method, sufficient strength is required. However, strength of commercially pure magnesium for penetration is not sufficient as biomaterials for surgery of hard tissue, it is essential to strengthening of magnesium. For practical application, these nails also must maintain sufficient fixation capability of membrane in spite of degradation until the alveolar bone recovers. In former research, magnesium alloy strengthened by both alloying with calcium, one of essential elements for human body, and grain refinement [4]. As a result, nails made of magnesium with high yield strength could be penetrated into artificial bones and pig mandible bone[5]. Though nails of Mg-Ca alloy were successfully penetrated in former research, it can be expected that residual strain remains because of deformation during penetration into mandible bone. Strain field due to penetration may affect the surface energy of nails. This means that distribution of residual strain will be dependent on degradation behaviors [6-8]. Therefore, in this research, degradation behaviors of magnesium-calcium nails for GBR method were clarified by immersion in simulated body fluid after penetration to reveal the effect of residual strain.

Experimental and Analytical Procedures

Mg-0.2at.%Ca alloy was prepared by gravity casting. Matrix grains in the alloy were refined via double extrusion. Extrusion temperatures and extrusion ratio of first extrusion was 350 ℃, and 20:1, respectively. Extrusion temperatures and extrusion ratio of secondary extrusion was 250 ℃ and 25:1, respectively. The microstructures in the cross sections of extruded bar was observed by TEM. Grain size was calculated by the average linear intercept length.
The nails were machined from the extruded bars. The dimensions of head diameter, shaft diameter, total length, head height and acral angle was 2 mm, 1 mm, 5 mm, 1 mm and 30°, respectively as shown in Fig. 1. Solid Rigid Polyurethane 50 pcf (Sawbones) fabricated according to ASTM F 1839 was used as substitution of human mandible bone because they replicate mechanical properties of human bone. They were machined to cylinders with a diameter of 8 mm and height of 3 mm. The penetration test of these nails to artificial bone was performed at a displacement rate of 0.002 mm/s.

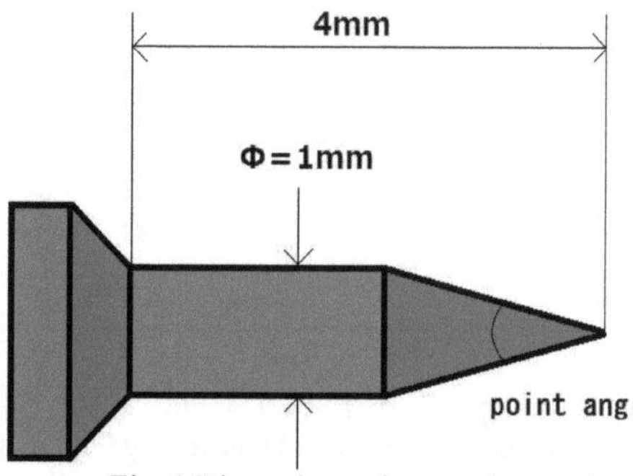

Fig. 1 Dimensions of magnesium nails.

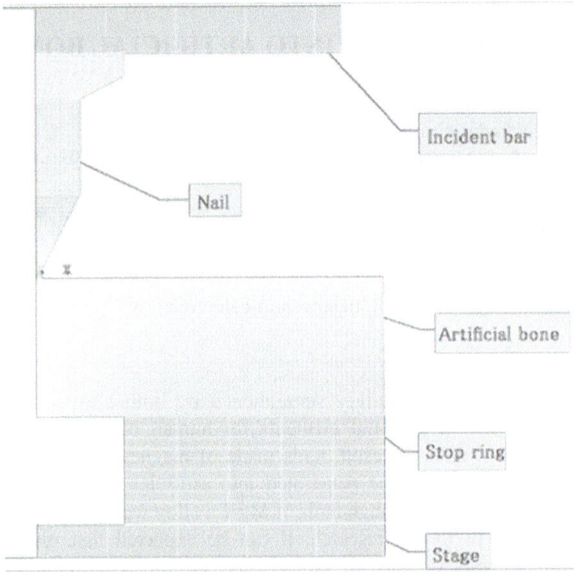

Fig. 2 FE model for penetration of magnesium nails into an artificial bone.

Since the strain field may affect the degradation rate at the surface of magnesium, a finite element (FE) analysis was conducted with ANSYS/LS-DYNA to clarify the distribution of residual strain after penetration. The axial symmetry model of meshes using square eight nodes for an incident bar, a stop ring, a nail and artificial bone was used as illustrated in Fig. 2. The dimensions of models were the same as those of the actual nails and bones. Young's moduli and Poisson ratios of magnesium nail and artificial bone were 45 GPa, 1.5 GPa, 0.3 and 0.5, respectively. The coefficient of static and dynamic friction between nails and the artificial bone was assumed to be 0.3 and 0.2, respectively. The responses of plastic stress to plastic strain of magnesium nails and artificial bones were derived from the nominal stress strain curves obtained by quasi-static compression test shown in Fig. 3.

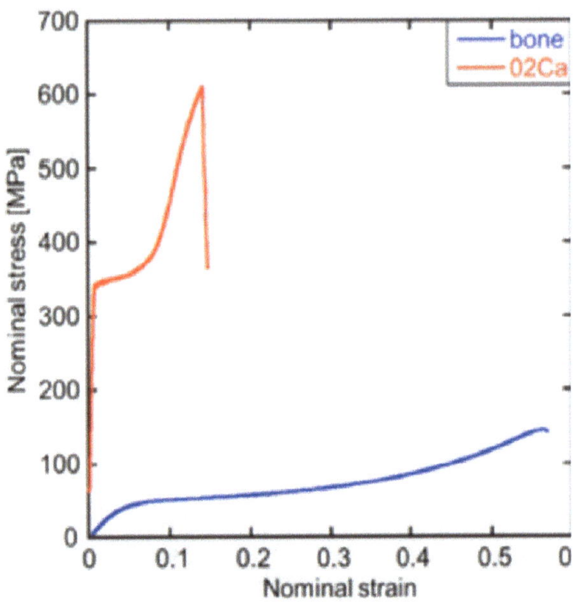

Fig. 3 Stress-strain curves obtained by quasi-static compression of magnesium-calcium alloy and artificial bone.

The diameter and height of magnesium specimen were 4 mm and 4 mm, respectively. Though these compression tests were carried out at ambient temperature and at a strain rate of $1x10^{-3}$ s^{-1}, the penetration speed of 12 mm/s was used for the FE analysis due to for the limitation of analysis time. Since the load displacement curves of the FE analysis analyses were consistent with the penetration tests in the former research [5], the compressive mechanical properties were found to be suitable for the penetration analysis. During the analysis, the mesh which satisfied the fracture criteria defined by the fracture strain shown in Fig. 3 was eliminated.

After the penetration of a nail to an artificial bone, the nails together with the bone was immersed into simulated human body fluid which consisted of Eagle's minimum essential medium (E-MEM; Nissui Pharmaceutical. Co. Ltd.) supplemented with 10 vol % fetal bovine serum (JRH Biosciences Inc.). To prevent contamination, 100U/ml penicillin and 100 μg/ml streptomycin were also added to body fluid. In addition to monitor the pH during immersion tests, phenol red was added to these fluids. These fluids were stirred at a speed of 200 rpm in a sterile glass container in humidified air containing 5 vol % CO_2 at 37 °C. The change in volume during degradation was quantified by micro-focus X-ray computerized tomography (micro-CT) images.

Results and Discussion

Fig. 4 shows the microstructure of Mg-0.2at.%Ca whose cross section was parallel to the extrusion direction. There are little precipitates in this figure. This corresponds to the solute limit of calcium into magnesium. Since the grain size of this alloy was about 500 nm, the high strength can be expected by ultra grain refinement because its compressive yield stress is comparable to that of Mg-0.3at.% alloy used in the former research [4].

Fig. 5 shows a load-displacement curve obtained by a penetration test of a nail made of Mg-0.2at.%Ca alloy which possessed microstructure shown in Fig. 4. The load increased as the penetration depth.. During penetration, no drop of load was found during the penetration. The maximum value of load showed the fully penetration of the nail. Therefore, this curve indicated that no fracture or buckling of the nail occurred during penetration. The maximum load was about 170 N. Therefore, the compressive stress in the parallel portion of the nail can be estimated to be 216 MPa. The compressive yield stress obtained by quasi-static compression test at 1×10^{-3} s^{-1} was higher than 300 MPa. Therefore, it can be expected that a magnesium alloy having compressive yield stress over 216 MPa contributes to penetration without buckling.

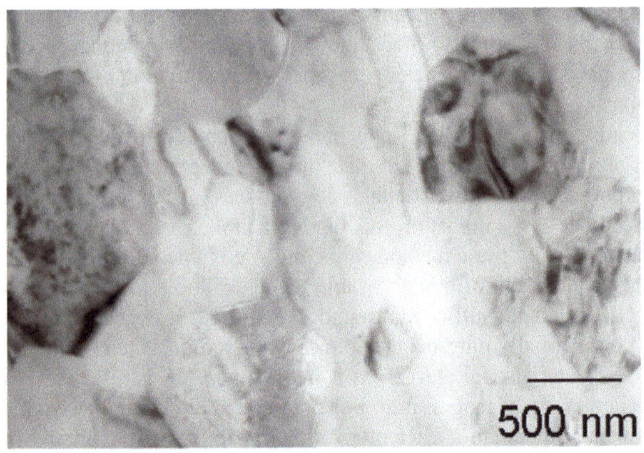

Fig. 4 Microstructures of Mg-0.2at.%Ca alloy.

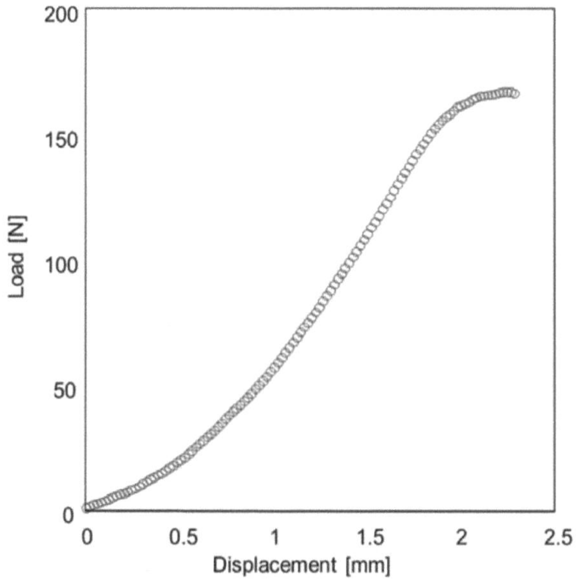

Fig. 5 Load-displacement curve obtained by actual penetration test with a magnesium nail into artificial bone.

Fig. 6 shows the cross section images of a nail parallel to the penetration axis inspected by micro-CT analysis. The cross sections before penetration and after the immersion test for 4 weeks are shown in Fig. 6 (a) and (b), respectively. This image indicates that there is no change in shape because of buckling and fracture of magnesium nail. This result corresponds to the mechanical response as shown in Fig. 5. Moreover, the density of the artificial bone around the penetrated portion was found to be higher than that of the undeformed state.

Fig. 6 (b) shows the degradation occurred inhomogeneously. The degraded area was the reentrant area which is boundary of head area and the parallel area. In addition, defects due to the degradation are also observed at the edge of nails and the upper contact boundary of the nail and the artificial bone. Gray-color area exists at the reentrant surface of the nail which may be a deposit layer generated by reaction between the Mg-Ca nail and simulated body fluid. Former research suggests that the deposit layer is expected to be $Mg(OH)_2$, Mg_2CO_3、$Mg_3(PO_4)_2$ [4]. The degradation rate was estimated to be 4% of volume degraded during immersion for 4 weeks by direct analysis with VG Studio MAX for the CT data. This result corresponds to former research that the gradual degradation occurred without precipitates. However, this degradation rate is too slow for the application to GBR method because of the duration for bone healing [1]. Therefore, it is essential to confirm the degradation rate by conduction of in vivo tests with this alloy and another Mg-Ca alloy containing precipitates.

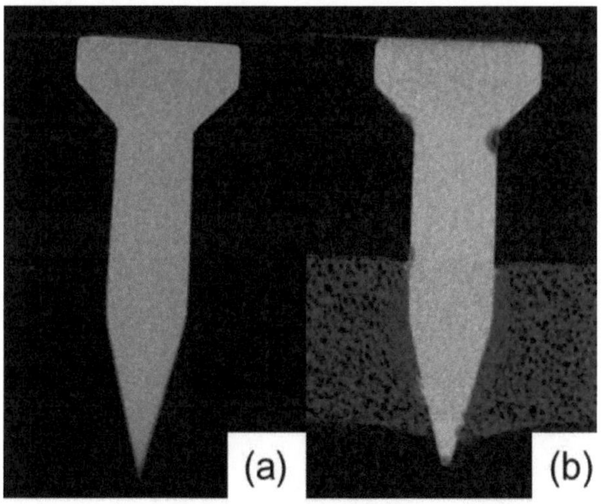

Fig. 6 Cross section images of a Mg-Ca nail inspected by micro-CT; (a) before penetration, (b) after immersion for 4 weeks in simulated body fluids

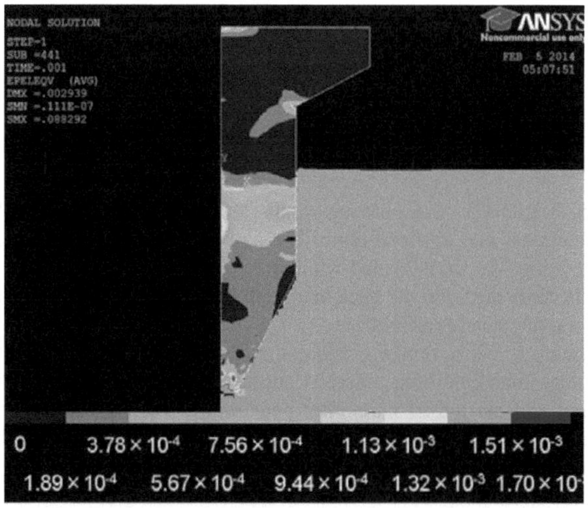

Fig. 7 The total Strain distribution of a Mg-Ca nail after penetration.

To clarify the effect of residual strain by penetration on degradation behavior, total strain distribution is displayed in Fig. 7. In this research, residual strain distribution introduced during machining was assumed to be ignored. The scale of strain from 0 to $1.7x10^{-3}$ is shown by a contour map. The darker area corresponds to the area with high residual stress. This area is confirmed at the edge of nails, and the boundary of head and parallel area. Therefore, when compared the contour map with the CT image in Fig. 6 (b), the degraded area corresponds to the area with higher residual strain. Therefore it can be summarized that the area with higher residual stress degraded preferentially after penetration to bone.

There are many researches which focused on the effect of residual strain rate or residual stress on corrosion resistance [6-8]. It has been reported that the degradation rate of Mg-Ca alloy without laser shock peening degraded was faster than that of the specimens with the process [7]. It is said that this result is due to the planer compressive stress. In contrast, other research reports that tensile residual stress affected the corrosion resistance [8]. Therefore, it is expected that the residual strain at the concave area was introduced by residual tensile stress or residual compressive stress without planner component. As a result, the penetration to bones may contribute the quick degradation of magnesium nails due to higher surface energy than that of other area.

In contrast, there is another strain field in the middle of nails. However, there is little degradation in the nails as shown in Fig. 6 (b). When focusing on the middle area of nail, it is confirmed that the artificial bone around the nails was densified by collapse of pores by penetration. Therefore, it is expected that the existence of densified bone layer functioned as protective layer of magnesium nails.

In addition, the edge of neck is concave area. This change in geometry also may affect the degradation behavior in the circulated liquid.

Conclusion

Mg-0.2at.%Ca nail having high strength was successfully penetrated into artificial bone. The nail degraded gradually since the alloy was strengthened without any precipitates. Moreover, it was confirmed that the neck area with a relatively higher residual strain degraded preferentially.
However, when thinking about the nails after penetration into human mandible bone, there is another metabolic system at the head and the tail of nails after penetration. This is because the head is located at the cortical bone and the tail is located at the cancellous bone. Therefore, it is desired to investigate the degradation behavior in vivo.

Acknowledgement

This work was supported by the Funding Program for Grant-in-Aid (Grant Number 23390451 & 25246012) for Scientific Research from the Japan Society for the Promotion of Science (JSPS).

References

[1] M. Retzepi, and N. Donos: "Guided Bone Regeneration: biological principle and therapeutic applications", *Clinical Oral Implants Research*, Vol 21, (2010), pp567-576.
[2] R. Rojaee, M. Fathi, K. Raeissi; "Controlling the degradation rate of AZ91 magnesium alloy via sol–gel derived nanostructured hydroxyapatite coating", Materials Science and Engineering: C, 33 (2013), pp 3817-3825
[3] Y. Shimizu, A. Yamamoto, T. Mukai, et al; 3rd International Symposium for Interface Oral Health Science, 2010, 318-320
[4] N. Ikeo et al.; "Degradation behavior of Mg-Ca binary alloys with ultrafine grain structures in simulated body fluid", Materials Science and Engineering: C, under review
[5] J. Shimizu et al.; "Mechanical Response of Mg-Ca Alloy Nails Penetrating the Bone Tissue and Their in vitro Degradation Behavior", paper presented at ISETS'2013, Nagoya.
[6] B. Denkena, A. Lucas; "Biocompatible magnesium alloys as absorbable implant materials – Adjusted surface and subsurface properties by machining processes", CIRP Annals - Manufacturing Technology, 56 (2007), pp. 113-116.
[7] Y. Guo, M.P. Sealy, C. Guo; "Significant improvement of corrosion resistance of biodegradable metallic implants processed by laser shock peening, CIRP Annals - Manufacturing Technology", CIRP Annals - Manufacturing Technology, 61 (2012), pp. 583-586.
[8] Z. Pu, G.L Song, S. Yang, J.C. Outeiro, O.W. Dillon Jr, D.A. Puleo, I.S. Jawahir; "Grain refined and basal textured surface produced by burnishing for improved corrosion performance of AZ31B Mg alloy", Corrosion Science, 57 (2012), pp. 192-201.

Magnesium Technology 2015
Edited by: Michele V. Manuel, Alok Singh, Martyn Alderman, and Neale R. Neelameggham
TMS (The Minerals, Metals & Materials Society), 2015

IMPROVING THE CORROSION RESISTANCE OF BIODEGRADABLE MAGNESIUM ALLOYS BY DIFFUSION COATING PROCESS

Galit Katarivas Levy and Eli Aghion
Department of Materials Engineering, Ben-Gurion University of the Negev, P.O. Box 653, Beer-Sheva 84105, Israel

Keywords: Magnesium, Diffusion coating, Neodymium, Biodegradable Implants

Abstract

Magnesium alloys suffer from accelerated corrosion in physiological environment and hence their use as a structural material for biodegradable implants is limited. The present study focuses on a diffusion coating treatment that amplifies the beneficial effect of Neodymium on the corrosion resistance of magnesium alloys. The diffusion coating layer was obtained by applying 1 μm Nd coating on EW10X04 magnesium alloy using Electron-gun evaporator and PVD process. The coated alloy was heat treated at 350°C for 3 hours in a protective atmosphere of N_2+0.2%SF_6. The microstructure characteristics were evaluated by SEM, XRD, and XPS; the corrosion resistance was examined by potentiodynamic polarization and EIS analysis. The corrosion resistance of the diffusion coated alloy was significantly improved compared to the uncoated material. This was related to: (i) formation of Nd_2O_3 in the outer scale, (ii) integration of Nd in the MgO oxide layer, and (iii) formation of secondary phase $Mg_{41}Nd_5$ along the grain boundaries of α-Mg.

Introduction

For more than a hundred years, researchers have been investigating the possibility of using magnesium and its alloys as structural material for biomedical applications [1-3]. Lately, those applications relate to orthopedic implants [4-7], cardiovascular stents [8,9], and tissue engineering scaffolds [10,11]. The potential of magnesium and its alloys as implant material exceeds that of other materials due to their unique advantages in terms of biocompatibility and bio-degradability [12-14]. In the field of orthopedics and in particular bone fixation, magnesium is considered a preferable structural material due to its mechanical strength, which is relatively similar to the typical strength of bones and the fact that it stimulates bone growth [2,15]. The main problem with magnesium alloys relates to their reduced life span in terms of corrosion resistance, which can cause premature loss of structural integrity of the implant and the formation of hydrogen gas bubbles that can cause separation of tissues, and in rare cases gas embolism [16-18]. Those problems actually limit the use of Mg implants in practice.

Surface coating seems to be an attractive method to delay the accelerated corrosion of magnesium implants at the initial exposure phase. In addition, those coatings are considered to be easily controlled and cost effective [13,19]. Among various coating technologies, diffusion coating is considered one of the most promising methods. Diffusion coating has high adhesion strength due to the strong metallurgical bond between the coating and the substrate and good mechanical properties. Moreover, the diffusion coating improves not only the corrosion resistance but also the wear resistance due to intermetallic compounds that are formed during the heat treatment [20-22]. The intermetallic compounds contain Nd, which is considered a non-carcinogenic element with lack of cytotoxicity in the living body [23-25].

The present study aims to evaluate the properties and metallurgical characteristics of diffusion coating treatment applied to newly developed biodegradable magnesium alloy EW10X04 using Nd as an external coating prior to the heat treatment. Neodymium was selected as a coating substrate because small amounts of this element are considered to have lower toxicity [15] and they are barely absorbed by the surrounding tissue and body fluid [26]. In addition, Nd has a beneficial effect on the corrosion resistance and mechanical properties of magnesium alloys [11,23-28]. Hence, the overall motivation for this study was to explore the potential of diffusion coating treatment as a method to improve the corrosion resistance of biodegradable magnesium implants.

Experimental

The chemical composition of the selected alloy EW10X04 was: Mg-1.2%Nd-0.5%Y-0.5%Zr-0.4%Ca. The detailed metallurgical procedure for preparing this alloy as well as its environmental behavior in *in vitro* and *in vivo* conditions were introduced in previous studies [16,17]. Cylindrical disks of 15 mm diameter and 5 mm thickness were obtained by machining from as-cast ingots and used as test specimen substrates for the diffusion coating treatment. The cleaning of the specimen before PVD coating included ultrasonic cleaning in ethanol and subsequent cleaning in acetone.

PVD coating of 1 μm Nd was carried out using an electron gun evaporator with target plate made from 99.99%Nd under high vacuum of 2.7×10^{-7} mbar. The evaporation rate selected for this PVD process was about 4 Å/sec. Additional information on the PVD procedure used in this study is described in a previous paper [18]. The PVD coating was followed by heat treatment at 350°C for 3 hours in a protective atmosphere of N_2+0.2%SF_6 using a vacuum furnace (GSL1300X, MTI Corporation) to obtain a diffusion coating system.

The metallurgical characteristics of the diffusion coating layer were evaluated using scanning electron microscopy (SEM), X-ray diffraction (XRD), and X-ray photoelectron spectroscopy (XPS) analysis. The parameters of the XRD were CuKα wavelength and scanning rate of 2°/min. The sputtering rate used with XPS analysis was 0.07 nm/min.

The corrosion resistance of the biodegradable magnesium alloy with and without the diffusion coating layer was evaluated in terms of electrochemical corrosion behavior. This included potentiodynamic polarization and electrochemical impedance spectroscopy (EIS) analysis. Both tests used a saturated calomel electrode (SEC) as reference electrode and platinum as a counter electrode. The scanning rate for the potentiodynamic polarization analysis was 0.167 mV/s; the EIS measurements were recorded in open circuit potential with amplitude of 7 mV and frequency range between 100 kHz and 10 mHz. The immersion times used for the EIS measurements were up to 36 hours. All the electrochemical tests were carried out in 0.9% NaCl solution saturated with $Mg(OH)_2$ at room temperature. The use of

$Mg(OH)_2$ aims to eliminate the side effect of the corrosion products of Mg that can change the pH of the solution during the corrosion testing [29-30].

Results and Discussion

The microstructure at a cross-section of the tested alloy with Nd coating is shown in Figure 1. This revealed a bright layer of Nd coating that was obtained by the PVD process in the electron gun evaporation. The thickness of this layer was between 0.7 and 2.0 µm. Previous research [18] indicated that the Nd coating was composed of three sub-layers: the external layer (up to 0.1 µm from the surface) of Nd_2O_3 only, the internal layer (0.1-0.4 µm) composed of Nd and Nd_2O_3, and the deeper layer (above 0.4 µm) of pure Nd.

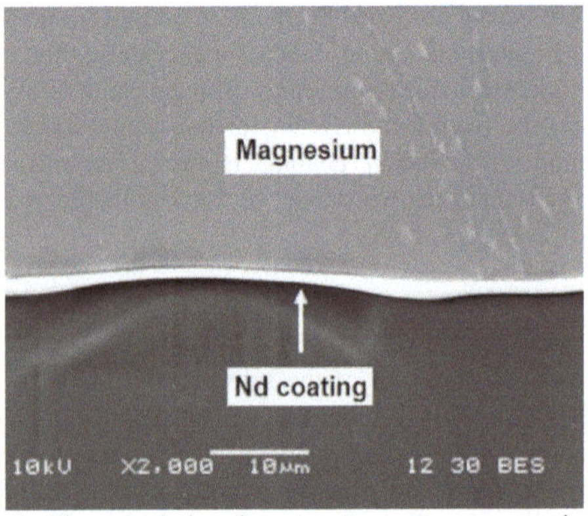

Figure 1: Characteristic microstructure at a cross-section after PVD Nd coating as obtained by SEM.

A close-up view of the microstructure at the interface layer close to the external surface after the diffusion coating treatment is shown in Figure 2. This clearly indicates that the grain boundaries were heavily decorated with the second phase $Mg_{41}Nd_5$ [18].

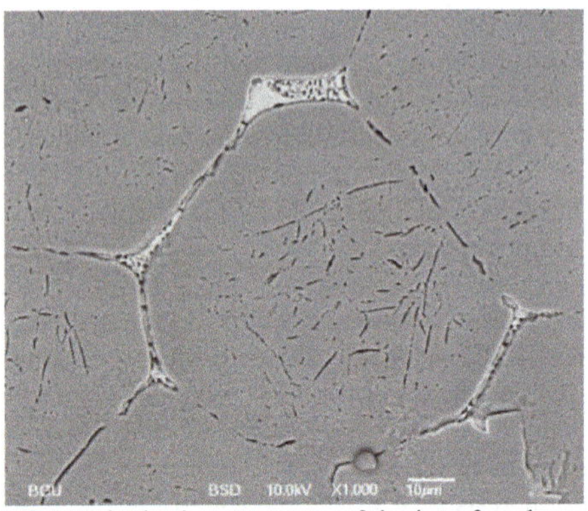

Figure 2: Typical microstructure of the interface layer after diffusion coating treatment as obtained by SEM.

Line profile measurements from the external surface of Nd, Mg, and Oxygen, obtained by X-ray photoelectron spectroscopy after the diffusion coating treatment are shown in Figure 3. This revealed that Mg was not present at the surface and its content was dramatically increased above a depth of 0.4 µm. In parallel, the content of Nd and oxygen beyond the depth of 0.4 µm decreased.

High-resolution XPS spectra for Nd-3d, O-1s, and Mg-1s after diffusion coating are shown in Figure 4(a), (b), and (c), respectively. This explored the layers composing the diffusion coating zone (apart from the inherent alloying elements of the alloy): external layer of Nd_2O_3, inner layer of Nd + Nd_2O_3 (up to a depth of about 0.4 µm), intermediate layer of Mg + MgO + Nd (up to a depth of about 0.7 µm), and matrix alloy subtract. A schematic illustration showing the different layers that composed the external surface after diffusion coating treatment is shown in Figure 5.

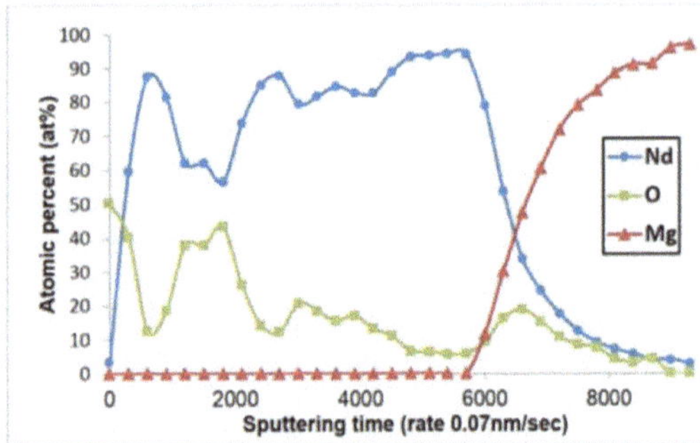

Figure 3: Depth profiles across the diffusion coating layer as obtained by XPS sputtering analysis.

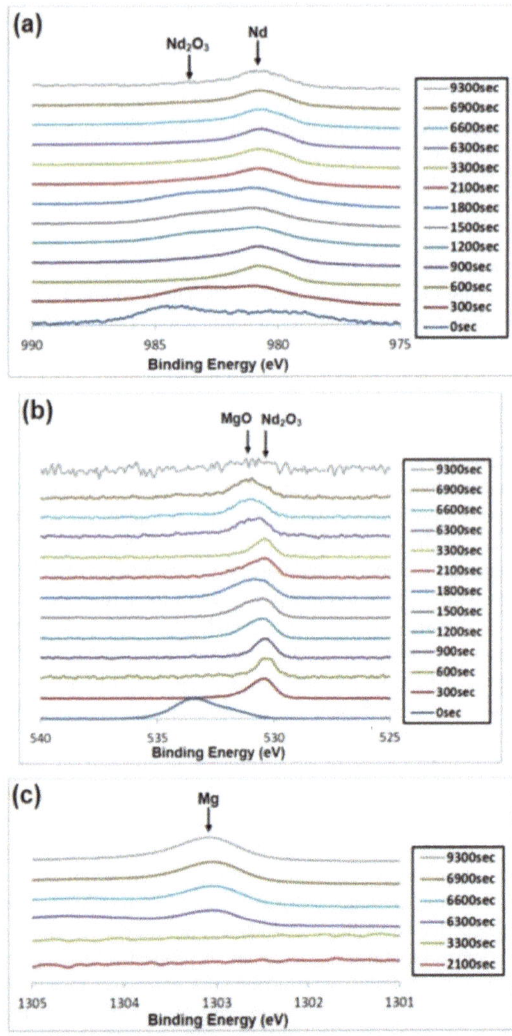

Figure 4: Diffusion coating layer's characteristics as obtained by high-resolution XPS spectra using different sputtering times (a) Nd spectra, (b) Oxygen spectra, (c) Mg spectra.

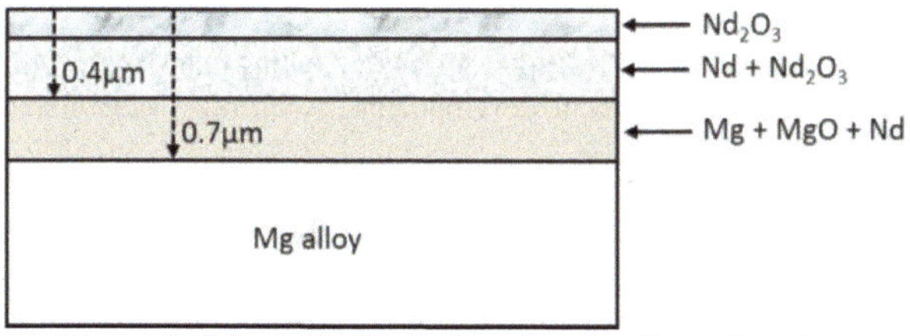

Figure 5: Schematic illustration showing the various layers obtained after the diffusion coating treatment.

The electrochemical behavior of the uncoated alloy and the diffusion-coated system obtained by potentiodynamic polarization and EIS analysis are shown in Figures 6 and 7, respectively. According to Tafel extrapolation obtained from the polarization curves, it is clearly evident that the corrosion resistance of the diffusion coating system was significantly improved compared to the uncoated alloy. In addition, the corrosion potential of the diffusion-coated system was -1.6 V compared to -1.74 V for the uncoated alloy. This also supports the evidence that the diffusion coated system was superior in terms of corrosion resistance. The EIS measurements showed that the polarization resistance of the diffusion-coated system was significantly higher compared to the uncoated alloy in the initial exposure phase as well as after 36-hour immersion. As the polarization resistance represents the corrosion resistance of the external film [18], this indicates that the corrosion resistance of the diffusion-coated system was higher than that of the uncoated alloy, which is in line with the results obtained by the potentiodynamic polarization analysis.

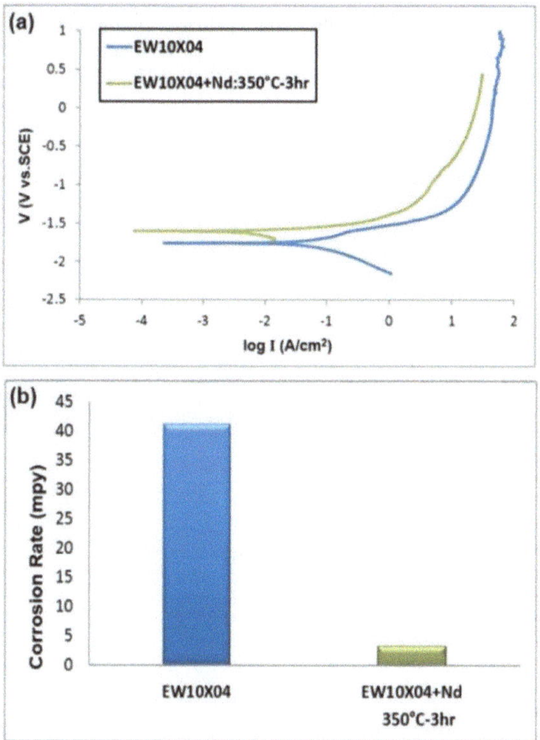

Figure 6: (a) Potentiodynamic polarization curves obtained in 0.9% NaCl solution saturated with $Mg(OH)_2$. (b) Corresponding corrosion rates calculated by Tafel extrapolation.

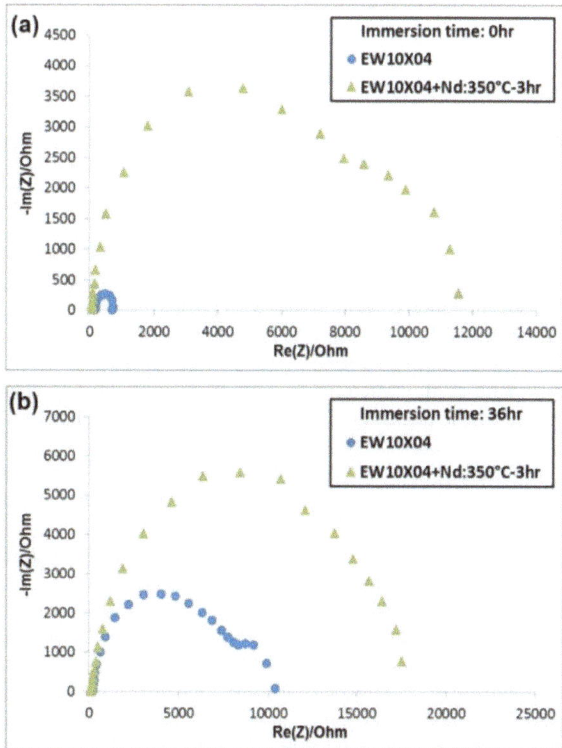

Figure 7: Nyquist plots in 0.9%NaCl solution saturated with $Mg(OH)_2$ obtained for the uncoated alloy and the diffusion-coated system after immersion time of (a) 0 hr (b) 36 hr.

The improved corrosion resistance of the diffusion-coated system as obtained by the electrochemical study (Figures 6 and 7) is related to the beneficial effect of Nd on the corrosion performance of the diffusion layer and external surface. This beneficial effect is mainly demonstrated by the formation of Nd_2O_3 film as shown in the XPS analysis (Figure 4) and by the increased amount of Nd at grain boundaries in the form of secondary phase $Mg_{41}Nd_5$ (Figure 2). The fact that the second phase has a significantly improved corrosion resistance compared to Mg alloy matrix and as it tends to form nearly continuously around the grains may suggest that this phase provides an effective corrosion resistance barrier to the Mg alloy matrix grains. Furthermore, as Nd is also integrated in the MgO layer located 0.4-0.7 µm from the external surface (Figure 5), it is believed that the enrichment of the MgO with Nd and probably some Nd-oxide phases also contribute to the corrosion resistance of the natural Mg oxide layer.

Conclusion

The obtained results clearly indicate that diffusion coating treatment with Nd can improve the corrosion resistance of EW10X04 alloy in simulated physiological electrolytes. This was explained by the beneficial effect of Nd on the oxide scale in terms of its corrosion resistance and by the nearly continuous precipitation of noble $Mg_{41}Nd_5$ phase of α-Mg boundaries.

Acknowledgment

The authors would like to thank Ms. Hila Pnini and Ms. Noa Taub from Ben-Gurion University for their assistance in the experimental activities.

References

1. F. Witte, "The history of biodegradable magnesium implants: A review," *Acta Biomater,* 6 (2010), 1680-1692.
2. L. Zhang et al., "The effect of selected alloying element additions on properties of Mg-based alloy as bioimplants: A literature review," *Frontiers of Materials Science,* 7 (2013), 227-236.
3. N. Kirkland et al., "Assessing the corrosion of biodegradable magnesium implants–A critical review of current methodologies and their limitations," *Acta Biomater,* 8 (2012), 925-936.
4. E. Aghion, A. Arnon, D. Atar, G. Segal, *US Patent,* PCT/IL 2007/000520 (2007).
5. F. Witte et al., "Microtomography of magnesium implants in bone and their degradation," *Proc. SPIE 6318, Developments in X-Ray Tomography V,* 2006, no. 6318.
6. M. Salahshoor and Y. Guo, "Biodegradable orthopedic magnesium-calcium (MgCa) alloys, processing, and corrosion performance," *Materials,* 5 (2012), 135-155.
7. R.A. Lindtner et al, "Comparative biomechanical and radiological characterization of osseointegration of a biodegradable magnesium alloy pin and a copolymeric control for osteosynthesis," *Journal of the Mechanical Behavior of Biomedical Materials,* 28 (2013), 232-243.
8. W. Zhou et al., "Mechanical property, biocorrosion and in vitro biocompatibility evaluations of Mg–Li–(Al)–(RE) alloys for future cardiovascular stent application," *Acta biomaterialia,* 9 (2013), 8488-8498.
9. W. Wu et al., "Experimental data confirm numerical modeling of the degradation process of magnesium alloys stents," *Acta Biomaterialia,* 9 (2013), 8730-8739.
10. K. Bobe et al., "In vitro and in vivo evaluation of biodegradable, open-porous scaffolds made of sintered magnesium W4 short fibres," *Acta Biomaterialia,* 9 (2013), 8611-8623.
11. X. Zhao et al., "A Comparison of Corrosion Behavior in Saline Environment: Rare Earth Metals (Y, Nd, Gd, Dy) for Alloying of Biodegradable Magnesium Alloys," *Journal of Materials Science & Technology,* 29 (2013), 781-787.
12. Y. Xin et al., "In vitro studies of biomedical magnesium alloys in a simulated physiological environment: A review," *Acta Biomaterialia,* 7 (2011), 1452-1459.
13. X. Lin et al., "Characterization of micro-arc oxidation coating post-treated by hydrofluoric acid on biodegradable ZK60 magnesium alloy," *Surface and Coatings Technology,* 232 (2013), 899-905.
14. C. Yigang et al., "In vitro and in vivo assessment of the biocompatibility of an Mg–6Zn alloy in the bile," *J Mater Sci: Mater Med,* 25 (2014), 471-480.
15. F. Witte et al., "In vivo corrosion of four magnesium alloys and the associated bone response," *Biomaterials,* 26 (2005), 3557-3563.
16. E. Aghion and G. Levy, "The effect of Ca on the in vitro corrosion performance of biodegradable Mg-Nd-Y-Zr alloy," *J Mater Sci,* 45 (2010), 3096-3101.
17. E. Aghion et al., "In vivo behavior of biodegradable Mg–Nd–Y–Zr–Ca alloy," *J Mater Sci Mater Med,* 23 (2012), 805-812.
18. G. Levy and E. Aghion, "Effect of diffusion coating of Nd on the corrosion resistance of biodegradable Mg implants in simulated physiological electrolyte," *Acta biomaterialia,* 9 (2013), 8624-8630.
19. X. Lin et al., "In vitro degradation and biocompatibility of a strontium-containing micro-arc oxidation coating on the biodegradable ZK60 magnesium alloy," *Appl Surf Sci,* 288 (2014), 718-726.
20. C. Zhong et al., "Protective diffusion coatings on magnesium alloys: A review of recent developments," *J Alloys Compounds,* 520 (2012), 11-21.
21. M. Zhang and P. Kelly, "Surface alloying of AZ91D alloy by diffusion coating," *J Mater Res,* 17 (2002), 2477-2479.
22. J.E. Gray and B. Luan, "Protective Coatings on Magnesium and Its Alloys - A Critical Review," *Journal of Alloys and Compounds,* 336 (2002), 88-113.
23. L. Xu et al., "In vivo corrosion behavior of Mg-Mn-Zn alloy for bone implant application," *Journal of Biomedical Materials Research Part A,* 83A (2007), 703-711.
24. I. Nakatsugawa et al., "Corrosion of magnesium alloys containing rare earth elements," *Corrosion Reviews (UK),* 16 (1998), 139-157.
25. F. Witte et al., "Degradable biomaterials based on magnesium corrosion," *Current Opinion in Solid State and Materials Science,* 12 (2008), 63-72.
26. Y. Ji et al., "The suppression effect of light rare earth elements on proliferation of two cancer cell lines." *Biomedical and environmental sciences: BES,* 13 (2000), 287.
27. F. Feyerabend et al., "Evaluation of short-term effects of rare earth and other elements used in magnesium alloys on primary cells and cell lines," *Acta Biomaterialia,* 6 (2010), 1834-1842.
28. A. Atrens et al., "Corrosion mechanism applicable to biodegradable magnesium implants," *Materials Science and Engineering: B,* 176 (2011), 1609-1636.
29. H.B. Yao et al., "An XPS investigation of the oxidation/corrosion of melt-spun Mg," *Applied Surface Science,* 158 (2000), 112-119.
30. E. Aghion and N. Lulu. "The corrosion performance of die-cast magnesium alloy MRI230D in 3.5%NaCl solution saturated with $Mg(OH)_2$," *Mater Charact,* 61 (2010), 1221-1226.

EFFECTS OF HEAT TREATMENT ON BIO-CORROSION PROPERTIES OF Mg-Zn-xMn (x= 0.5, 1.0, and 1.5 wt.%) ALLOYS AS BIODEGRADABLE MATERIALS

Wonseok Yang[1, 2], Young-Ok Yoon[1], Shae K. Kim[1], Hyun Kyu Lim[1] and Do Hyang Kim[2]
1 Rare Metal R&BD Group, Korea Institute of Industrial Technology, Incheon, Korea
2 Dept. of Metallurgical Engineering, Yonsei University, Seoul, Korea

Keywords: Biodegradable, Heat treatment, Magnesium alloys

Abstract

Mg alloys have unique characteristics such as high specific strength, low density, high corrosion rate, etc., as functional as well as structural materials. Mg-Zn alloys have good biocompatibility because Mg and Zn are abundant nutritional elements in the human's body. However, Mg alloys with multi-phase cause galvanic corrosion by corrosion potential differences among constituent phases. Therefore, the application of Mg alloys on bio-material parts are limited.

In this study, bio-corrosion properties of Mg-Zn-Mn alloys according to the solid solution and distribution of phases by various Mn contents and heat treatment condition were evaluated. The results of tensile and *in-vitro* corrosion tests indicated that tensile and bio-corrosion properties could be adjusted by controlling the Mn contents and heat treatment. The tensile yield strength (TYS), ultimate tensile strength (UTS) and Elongation of Mg-Zn-Mn alloys increased up to 0.5 wt.% and then slightly decreased with Mn contents. However, TYS, UTS, and Elongation of T4 treated Mg-Zn-Mn alloys increased with increasing Mn contents. When MgZn phase and Mn particle were dissolved in the matrix, bio-corrosion properties were improved in hank's solution. From the results of immersion test, the Mg-3Zn-0.5Mn alloy has good corrosion properties at 2 steps T4 treatment.

Introduction

Magnesium and magnesium alloys are light metals with characteristics including low density, high specific strength and high specific stiffness [1]. In particular, the elastic modulus and compressive yield strength of magnesium are close to natural bone, which makes them a promising substitute for biodegradable medical applications in the biomaterial field [2, 3]. Many clinical cases as well as *in-vitro* and *in-vivo* assessments have demonstrated that Mg alloys possess good biocompatibility [4-6]. However, currently, the developed Mg alloys are containing alloying elements such as aluminum (Al) and rare earth (RE) elements. The addition of Al or RE elements can improve the corrosion resistance and strength of Mg alloys [7-9]. However, Al is harmful to osteoblast and neuron [10, 11]. And RE elements are difficult to be absorbed and excreted by human bodies. In addition, it is found that the some RE elements can also lead to hepatotoxicity [12]. Apparently, Mg alloys containing alloying elements such as Al, RE are not suitable for biodegradable applications. Otherwise other Mg alloys containing harmless alloying element have attracted much attention. Mg-Ca, Mg-Zn, Mg-Zn-Ca, and Mg-Zn-Mn fall into this category, and the research shows that these Mg alloys are potential biodegradable materials with satisfactory mechanical properties. On the other hand, these alloys occurs to galvanic corrosion by formation of micro-galvanic cell, because of corrosion potential differences between α-Mg and secondary phases. Thus, it is difficult to apply for biomaterials. Recently, this phenomenon have been reported that corrosion resistance is improved by heat treatment at Mg-Zn alloys. However, this method decreases the mechanical properties.

In this study, the effect of the T4 treated in Mg-Zn-Mn alloys on the mechanical and bio-corrosion properties were studied to investigate interaction between microstructure and corrosion properties. The main approach of the present study was to carefully assess the effect of various Mn contents and solid solution on the corrosion properties of Mg-Zn-Mn alloys in Hank's solution. Also, the observation of the microstructures and examination of mechanical properties of the tested alloys were carried out.

Experimental procedure

Mg-Zn-Mn alloys were prepared with highly pure Mg, Zn, and commercial Mg-5Mn master alloy. Melting and alloying operations were carried out in a steel crucible under the protection of a mixed gas atmosphere of SF_6 and CO_2. Pure Zn and Mg-Mn master alloy were added into the pure Mg melts at 720 °C. The melts were kept for 15 min at 720 °C. And, all the required alloying elements were dissolved in the molten alloy, and then the melts were cooled down to 690 °C and poured into a steel mold of the cylindrical cavity with the dimension of 50 mm in diameter and 200 mm in height. Chemical compositions of as-cast alloys were analyzed by ICP-mass spectrometer and the results were listed in Table 1.

The as-cast billets were 2-step homogenized at the 400 °C for 12 hours + 520 °C for 12 hours. All billets were removed from the furnace and cooled to room temperature in the air. Indirect extrusion at 335 °C was performed with billets of 100 mm height and 50 mm length and the extrusion ratio was 19.6.

Table 1. ICP analysis results of as-cast alloys.

Alloy No.	Description		Chemical composition (wt.%)					
			Mg	Zn	Mn	Cu	Ni	Fe
1	Mg-3Zn-0.5Mn	Nominal	96.5	3	0.5	< 0.05	< 0.03	< 0.03
		Analyzed	Bal.	3.09	0.54	0.0003	0.0015	0.0015
2	Mg-3Zn-1.0Mn	Nominal	96	3	1.0	< 0.05	< 0.03	< 0.03
		Analyzed	Bal.	2.97	0.95	0.0003	0.0001	0.0016
3	Mg-3Zn-1.5Mn	Nominal	95.5	3	1.5	< 0.05	< 0.03	< 0.03
		Analyzed	Bal.	2.99	1.39	0.0004	0.0001	0.0022

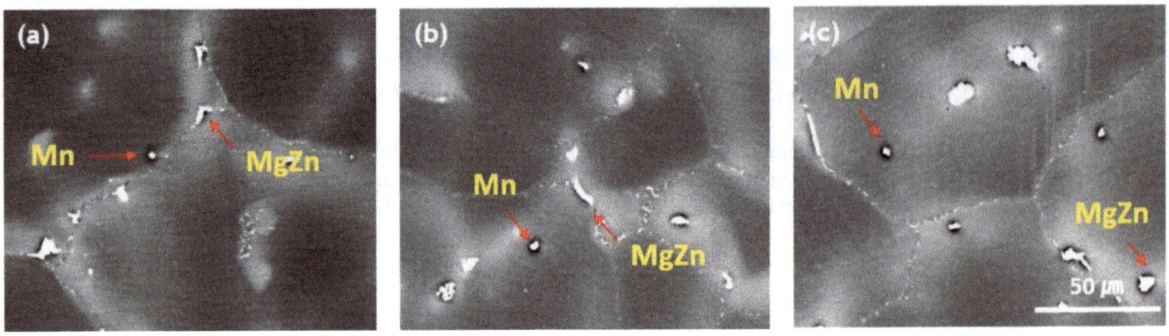

Fig. 1. SEM images of as-cast (a) Mg-3Zn-0.5Mn, (b) Mg-3Zn-1.0Mn, and (c) Mg-3Zn-1.5Mn alloys

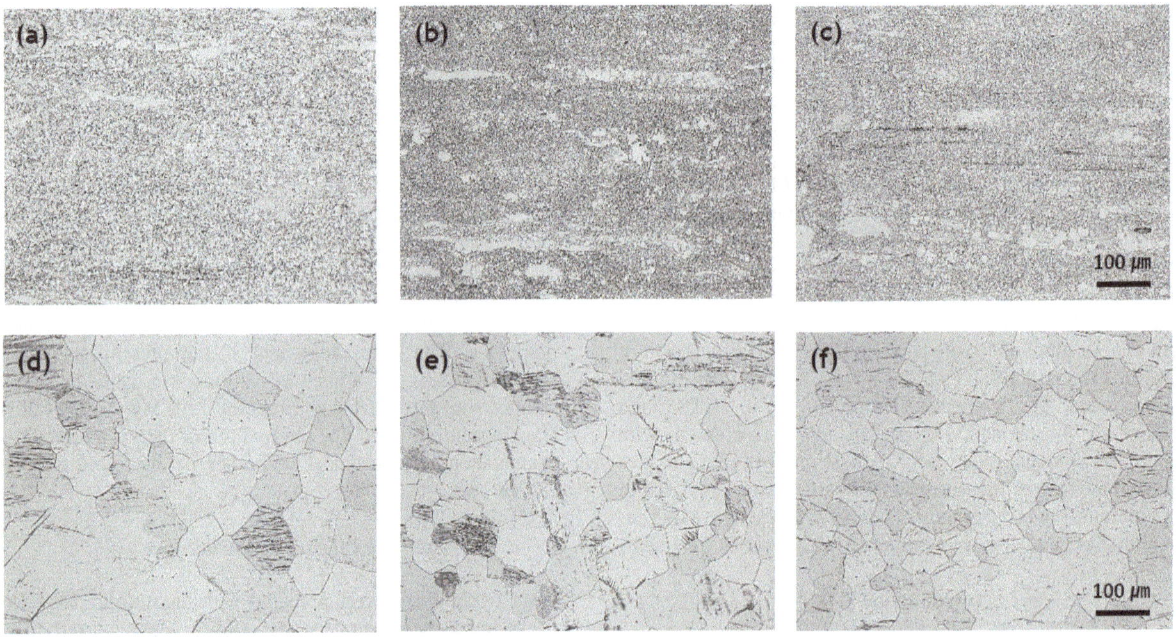

Fig. 2. Optical microstructures of as-extruded (a) Mg-3Zn-0.5Mn, (b) Mg-3Zn-1.0Mn, (c) Mg-3Zn-1.5Mn alloys, and 400 °C for 1 h + 520 °C for 0.5 h 2-step T4 treated (d) Mg-3Zn-0.5Mn, (e) Mg-3Zn-1.0Mn, and (f) Mg-3Zn-1.5Mn alloys.

The T4 treatments followed by water quenching was carried out at the 400 °C for 1 hour, 400 °C for 1 hour + 520 °C for 0.5 hour, and 400 °C for 1 hour + 520 °C for 1 hour. Mechanical testing samples were cut from the extruded bars and T4 treated extruded bars. A dog bone specimen with a gauge length of 25 mm, a width of 6 mm, and a thickness of 2 mm was selected for tensile test. Tensile test was carried out using and universal testing machine at a constant crosshead speed of 1.5 mm/min at room temperature. At least three samples were tested for each condition. Immersion test was performed in Hank's solution. The pH value of Hank's solution was adjusted with HCl and NaOH to 7.2-7.4, to avoid precipitation or formation of sediments in the solution. Test temperature was controlled around 37 ± 1 °C, which was equal to the human body normal temperature. Samples were removed after 7 days of immersion, rinsed with distilled water, and cleaned with chromic acid to remove the corrosion products. Changes in the surface morphologies and the microstructures of the samples before and after immersion were characterized by SEM. The hydrogen evolution rate was monitored during the soaking experiment. The immersion tests specimens with a size of 15 mm * 15 mm * 2 mm were cut and subsequently grounded with 360 ~ 1200 grit SiC papers.

Results and discussion

Microstructures

Generally, Mn forms intermetallic compound with Al and Fe of small amount in melts. And then, these particles sink to the bottom of the molten metal. However, the remaining Mn element presents only Mn particle, it exists at the grain boundaries or inter dendrite in the matrix after solidification on Mg-Zn alloys [13].

The microstructures of as-cast (a) Mg-3Zn-0.5Mn, (b) Mg-3Zn-1.0Mn, and (c) Mg-3Zn-1.5Mn alloys were shown in Fig. 1. MgZn phases and Mn particles of all alloys were observed around grain boundaries. Fig. 2 (a)-(c) showed optical microstructures of as-extruded Mg-Zn-Mn alloys along the longitudinal direction. The grain size was smaller than that of the as-cast specimen by the extrusion and dynamic recrystallization which happened during the extrusion process. However, with the increase of Mn contents, some coarse or un-recrystallized grains were observed. Fig. 2 (d)-(f) showed the microstructures of 400 °C for 1 h + 520 °C for 0.5 h 2-step T4 treated specimens. Compared with Fig. 2 (a) and (d), the grain size of T4 treated specimen was larger than that of the as-extruded specimen, because of grain growth during heat treatment.

Table 2. Tensile test results of Mg-Zn-Mn alloys

Alloys / heat treatment conditions		TYS (MPa)	UTS (MPa)	Elongation (%)
Ref.) Mg-3Zn	As-extruded	110.3	236.6	23.90
Mg-3Zn-0.5Mn	As-extruded	189.5	264.3	36.93
	T4 treated (@ 400 °C × 1 h)	137.2	235.4	25.72
	T4 treated (@ 400 °C × 1 h + 520 °C × 0.5 h)	123.5	207.6	11.92
	T4 treated (@ 400 °C × 1 h + 520 °C × 1 h)	118.2	204.7	9.72
Mg-3Zn-1.0Mn	As-extruded	169.7	258.2	36.76
	T4 treated (@ 400 °C × 1 h)	143.7	243.1	24.35
	T4 treated (@ 400 °C × 1 h + 520 °C × 0.5 h)	123.9	212.3	13.92
	T4 treated (@ 400 °C × 1 h + 520 °C × 1 h)	123.3	207.0	10.52
Mg-3Zn-1.5Mn	As-extruded	159.3	257.3	32.34
	T4 treated (@ 400 °C × 1 h)	148.3	248.4	25.58
	T4 treated (@ 400 °C × 1 h + 520 °C × 0.5 h)	130.2	223.5	20.08
	T4 treated (@ 400 °C × 1 h + 520 °C × 1 h)	128.6	222.1	18.32

However, in the case of microstructures of T4 treated alloys (Fig. 2 (d)-(f)), grain size decreased with the increase of Mn contents. This phenomenon might explain that grain growth was suppressed by supersaturated Mn or Mn particles.

Mechanical properties

Table 2 showed the tensile test results of as-extruded and T4 treated Mg-Zn-Mn alloys. The yield strength, ultimate tensile strength, and the elongation of as-extruded Mg-3Zn-0.5Mn alloy were 189.5 MPa, 264.3 MPa, and 36.9 %, respectively. And then, with the increase of Mn contents, all tensile properties were slightly decreased. This phenomenon could result from coarse or un-recrystallized grains, as shown in Fig. 2 (a)-(c). However, tensile properties of these alloys were higher than those of as-extruded Mg-3Zn alloy. Mn element increased tensile yield strength, ultimate tensile strength, and elongation in Mg-Zn alloys up to 0.5

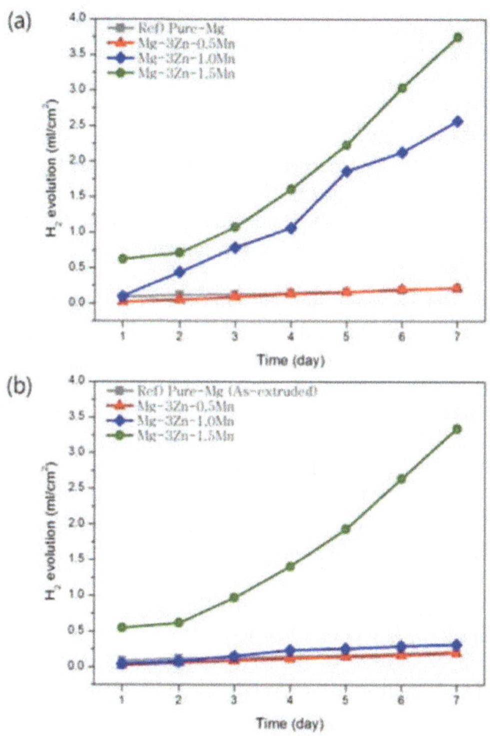

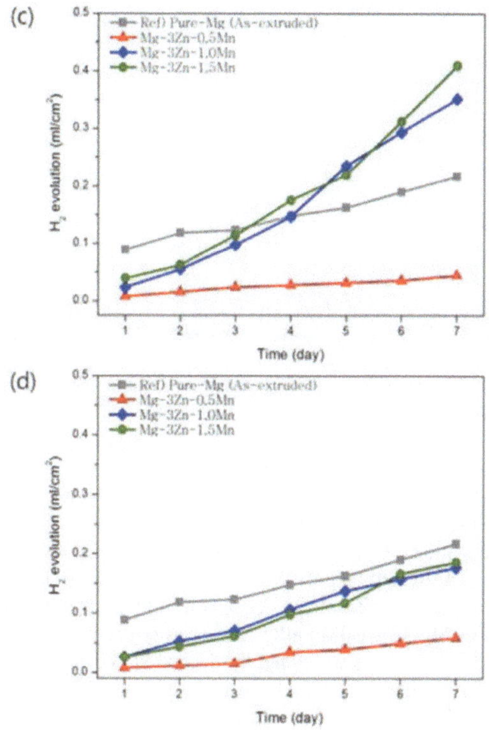

Fig. 4. H2 evolution of (a) as-extruded, and T4 treated (b) 400 °C for 1 h, (c) 400 °C for 1 h + 520 °C for 0.5 h, and (d) 400 °C for 1 h + 520 °C for 1 h

wt.%. However, over the 0.5 wt.%, all tensile properties were decreased. On the other hand, after T4 treatment, strength and elongation were lower than those of as-extruded alloys by grain growth during the heat treatment. However, with the increase of Mn contents, the strength reduction ratio was decreased due to the suppression of grain growth by supersaturated Mn or Mn particles, as shown in Fig. 2 (d)-(f).

Immersion test results

Fig. 4 showed the results of hydrogen evolution of as-extruded, 1 step T4 treated, and 2 steps T4 treated specimens. In the hydrogen evolution of as-extruded alloys, Mg-3Zn-0.5Mn alloy in Fig. 4 (a)

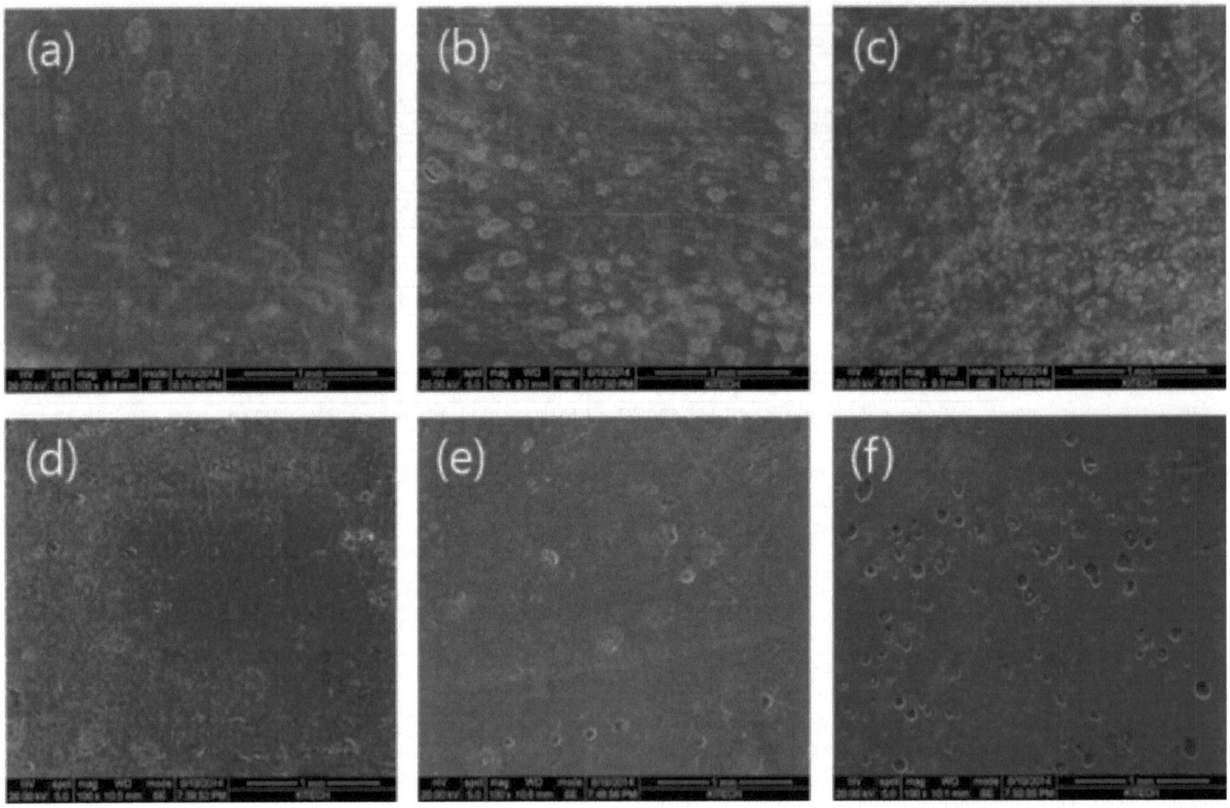

Fig. 5. Surface morphology after immersion test at 168 h of as-extruded (a) Mg-3Zn-0.5Mn, (b) Mg-3Zn-1.0Mn, (c) Mg-3Zn-1.5Mn, and 400 °C for 1 h + 520 °C for 0.5 h 2-step T4 treated (d) Mg-3Zn-0.5Mn, (e) Mg-3Zn-1.0Mn, and (f) Mg-3Zn-1.5Mn alloys.

exhibited the same hydrogen evolution level as the reference value of as-extruded pure Mg. The amount of hydrogen evolution of Mg-3Zn-1.0Mn and Mg-3Zn-1.5Mn alloys sharply increased more than Mg-3Zn-0.5Mn alloy. Fig. 4 (b) showed the test results of 1-step T4 treated Mg-Zn-Mn alloys to evaluate the solid solution effects of MgZn phase. As shown in Fig. 4 (b), hydrogen evolution of Mg-3Zn-1.0Mn alloy was sharply decreased. On the other hand, other alloys were slightly decreased. When compared with the result after 1-step T4 treatment, the hydrogen evolution of Mg-3Zn-1.5Mn alloy was sharply decreased by 2-step T4 treatment as shown in Fig. 4 (c). Fig. 4 (d) showed the results of hydrogen evolution when the 2nd heat treatment time expanded from 0.5 to 1 hour. The amounts of hydrogen of all specimens with longer 2nd heat treatment decreased compared with specimens which were heat-treated shorter 2nd heat treatment. In addition, the amount of hydrogen of these alloys generated less than that of reference specimen, which was as-extruded pure Mg due to reduce Mn particles by 2nd step solid solution treatment at 520 °C.

The SEM images of the surface of Mg-Zn-Mn alloy samples were immersed in hank's solution for 168 hours, as shown in Fig. 5 These images were similar to the results of immersion tests. Fig. 5 (a)-(c), the degree of corrosion increased from 0.5Mn to 1.5Mn in Mg-3Zn alloys due to increase Mn contents. However, after T4 treatment, the corroded surface area was decreased, as shown in Fig. 5 (e)-(f).

Conclusion

In the present study, mechanical properties and bio-corrosion properties of Mn added heat treated Mg-Zn-Mn alloys were investigated. The conclusions are as follows:

(1) The addition of 0.5 wt.% Mn in Mg-3Zn alloy exhibited the good mechanical properties, and then it decreased with the increase of Mn contents. However, after T4 treatment, strength reduce ratio was decreased because grain growth was suppressed with the increase of Mn contents.

(2) With the increase of Mn contents, H_2 evolution was increased. According to the distribution and solid solution of Zn and Mn in matrix by the T4 treatment, H2 evolution decreased. This phenomenon is clearly observed with the increase of Mn contents in Mg-3Zn alloys.

References

[1] M.M. Avedesian, and H. Baker., ASM Specialty Handbook, Magnesium and Magnesium alloys, ASM International Materials Park, USA, Ohio, 4

[2] M.P. Staiger, A.M. Pietak, J. Huadmai, and G. Dias., Biomaterials 27, 1728-1734. (2006)

[3] F. Witte, V. Kaese, H. Switzer, A. Meyer-Lindenberg, C.J. Wirth, and H. Windhag., Biomaterials 26, 3557-3563 (2005)

[4] G.L. Song, and S.Z. Song., Adv. Eng. Mater. 9(4), 298-302 (2007)

[5] Y.C. Xin, K.F. Huo, H. Tao, G.Y. Tang, and P.K. Chu., Acta Biomater. 4, 2008-2015 (2008)
[6] Y.C. Xin, T. Hu, and P.K. Chu., J. Electrochem. Soc. 157, 238-243 (2010)
[7] C. F. Li, Y.H. Liu, Q. Wang, L. Zhang, and D.W. Zhang., Mater. Charact. 61, 123-127 (2010)
[8] T. Rzychon, A. kielbus, J. Cwajna, and J. Mizera., Mater. Charact. 60, 1107-1113 (2009)
[9] X.M. Wang, X.Q. Zeng, S.S. Yao, G.S. Wu, and Y.J. Lai., Mater. Charact. 59, 618-623 (2008)
[10] J.A. Helson, and H.J. Breme (Eds.)., Metals as Biomaterials, John Wiley & Sons, England, 23 (1998)
[11] S.S. El-Rahman., Pharmacol. Res. 47, 189-194 (2003)
[12] N. Yumikom T. Yukari, T. Yasuhide, S. Tadashi, and I. Yoshio., Fundam. Appl. Tocicol. 37, 106-116 (1997)
[13] M.M. Avedesian, and H. Baker., ASM Specialty Handbook, Magnesium and Magnesium alloys, ASM International Materials Park, USA, Ohio, 15

Magnesium Technology 2015
Edited by: Michele V. Manuel, Alok Singh, Martyn Alderman, and Neale R. Neelameggham
TMS (The Minerals, Metals & Materials Society), 2015

EFFECTS OF Ti AND TiB_2 NANOPARTICULATES ON ROOM TEMPERATURE MECHANICAL PROPERTIES AND IN VITRO DEGRADATION OF PURE Mg

Ganesh Kumar Meenashisundaram, Mui Hoon Nai and Manoj Gupta *

Department of Mechanical Engineering,
National University of Singapore,
9 Engineering Drive 1,
Singapore - 117576, Singapore

mpegm@nus.edu.sg (Manoj Gupta)

Keywords: Mg 1 vol.% Ti, Mg 1 vol.% TiB_2, composites, in vitro degradation, tensile properties, compression properties.

ABSTRACT

Mg 1 vol.% Ti and Mg 1 vol.% TiB_2 composites containing Ti (30-50 nm) and TiB_2 (~ 60 nm) nanoparticulates were successfully synthesized using disintegrated melt deposition technique followed by hot extrusion. In vitro degradation of synthesized pure magnesium and composites were assessed by immersion testing in Dulbecco's Modified Eagle's Medium (DMEM) + 10% Fetal Bovine Serum (FBS) solution for a maximum duration of 28 days. Determination of corrosion rates by weight loss technique reveals that after 28 days of immersion testing, Mg 1 vol.% Ti exhibited the best corrosion resistance followed by pure magnesium and finally by Mg 1 vol.% TiB_2 composite. The room temperature mechanical properties of the synthesized composites were found to surpass those of pure magnesium. On tensile and compressive loading, substantial strengthening of pure magnesium was observed with 1 vol.% Ti addition whereas appreciable increase in tensile and compressive fracture strains of pure magnesium was observed with 1 vol.% TiB_2 addition.

1. INTRODUCTION

Higher expectations of human beings, their desire to live longer and ever growing population has increased the demand for biomaterials. Selection of biomaterials especially for load bearing applications such as orthopedic implants plays an important role. It was estimated that the total number of hip revision surgery is expected to increase by ~ 137% and knee revision surgery by ~ 607% between the years 2005 and 2030 which may further increase the demand for implants [1, 2]. Natural bone is a bio-nanocomposite made up of collagen fiber matrix and hydroxyapatite crystals (HAp) ($Ca_{10}(PO_4)_6(OH)_2$) [3]. Collagen fiber (or the organic phase) of the bone is a flexible material with high fracture toughness and HAp crystals (or the inorganic phase) are brittle materials with high hardness. Combination of two different materials with different properties makes the natural bone a bio-nanocomposite [3]. Stainless steel, titanium (Ti) and Ti alloys are the commonly used implant materials for orthopedic applications. Stainless steel has a high risk of Ni toxicity leading to allergic reactions within human body [4]. Both Ti and stainless steel implants requires second surgery to remove the implants after the wound healing process. Further, Ti does not bond with bone leading to implant loosening [3] and allergic responses due to Ti implants due to higher concentrations of Ti have been reported [5, 6].

Magnesium (Mg) is the second most abundant intracellular cation and total Mg in human body is distributed such that half is in soft tissues and the other half in the bone [7]. Human bone serves as Mg reservoir [8]. The recommended daily intake of Mg is ~ 240 to 420 mg which is ~ 17 to 50 times more than that of iron (8 to 18 mg) [9]. Mechanical properties of Mg based materials are closer to that of cortical bone with lower mouduli (41 to 45 GPa) than Ti based materials (110 to 117 GPa) thereby mitigates stress shielding and implant instability [10]. Further, it was observed that Mg safely degrades in vivo and its mechanical properties enable them to be used for load bearing applications [11]. For their effective utilization, further improvement in mechanical properties of pure Mg is very much essential and addition of nanoparticulate (NP) reinforcements to Mg materials is considered as a promising technique when compared to that of alloying and grain refinement [12]. Magnesium matrix nanocomposites (MMNCs) have been extensively studied in the recent years. Selection of suitable reinforcements and similarities between natural bone and MMNCs as "nanocomposites" make research on new MMNCs more exciting targeted towards biomedical applications. Ti does not form any intermetallic compound with Mg [13] and has a wide range of biomedical applications. High hardness of TiB_2 improves the wear resistance of Ti alloys used for biomedical applications [14]. Apart from its high hardness, titanium diboride (TiB_2) has a very good corrosion resistance and both Mg and TiB_2 have similarities in their crystal structures [15].

Mechanical suitability of biomaterials is generally investigated by studying their Young's modulus, tensile and compression properties [3]. After investigating on their mechanical properties, in vitro degradation properties of the nanocomposites are analyzed by immersing them within a physiological medium. Previously the authors have synthesized Mg (0.58, 0.97 and 1.98) vol.% Ti and Mg (0.58, 0.97 and 1.98) vol.% TiB_2 nanocomposites by using disintegrated melt deposition technique followed by hot extrusion [16, 17]. In the present study, authors compare the mechanical and in vitro degradation properties of pure Mg and low volume fraction (~ 1 vol.%) Ti and TiB_2 reinforced Mg nanocomposites. In vitro degradation properties of the nanocomposites are assessed by immersion in DMEM + 10% FBS solution up to 28 days and corrosion rates of the composites are determined by using weight loss technique as per ASTM: G31-72. Young's modulus of the synthesized Mg composites was measured by using nanoindentation technique and their mechanical properties are compared to that of natural bone.

2. EXPERIMENTAL PROCEDURE

In the present study, > 99.9% pure elemental Mg turnings supplied by ACROS organics, USA, were used as the base material. The required amount of Ti of size 30-50 nm supplied by U.S research and nanomaterials and TiB_2 of size ~ 60 nm supplied by NaBond Co. Ltd, China were used as the reinforcement phases. Pure Mg and Mg ~1 vol.% Ti, Mg ~1 vol.% TiB_2 nanocomposites were synthesized by using Disintegrated Melt Deposition technique followed by hot extrusion. The procedures involved in the synthesis of pure Mg and MMNCs are discussed briefly in previous works [16, 17]. The samples from hot extruded rods were used for characterization. Grain size of pure Mg and composite samples were measured by using an Olympus metallographic optical microscope and Scion image analysis software. X-ray diffraction studies investigated at the cross section of extruded samples were conducted by using an automated Shimadzu lab X-XRD-6000 diffractometer (Cu Kα, λ = 1.54056 Å). Young's modulus of the composites was determined by using a nanoindentation system with nanoscope IIIa controller (Digital Instruments, Santa Barbara, USA). Berkovich tip was used for indentation and the tip area function was calibrated using a standard quartz sample between contact depths of 50 and 300 nm. With ideal area function of C_0 = 24.5 and an indentation loading segment up to a maximum of 1500 μN consisting of a loading segment for 5 s, a holding segment for 2 s and an unloading segment for 5 s, each sample was indented for a minimum of 12 readings. Tensile tests were conducted as per ASTM: E8M-01on a dog-bone shaped round samples of 5 mm gauge diameter and 25 mm gauge length to determine the smooth bar tensile properties of the synthesized pure Mg and composites using a fully automated servo hydraulic mechanical testing machine MTS-810. The strain rate was set to 1.693 x 10^{-4} s^{-1} and an Instron 2630-100 series extensometer was used to measure the fracture strain. Compression tests were conducted as per ASTM: E9-89a using MTS-810 testing machine at a strain rate of 8.334 x 10^{-5} s^{-1}. The test specimens of 8 mm diameter and length to diameter ratio (l/d) ~ 1 were used. Microhardness tests were conducted as per ASTM: E384-08 on flat and mechanically polished samples using a Shimadzu HMV automatic digital microhardness tester with Vickers indenter. For immersion testing, pure Mg and composite samples were prepared with dimensions of 8 mm diameter and length to diameter ratio l/d ~1. Surface area and weight of the samples were carefully recorded before immersion. The specimens were thoroughly cleaned with acetone using a sonicated bath followed by UV sterilization for 30 min. After sterilization, the samples were immersed within a physiological medium containing Dulbecco's Modified Eagle's Medium (DMEM) + 10% Fetal Bovine Serum (FBS) + 1% Penicillin with an initial pH of 7.4 and placed in an incubator at a temperature of 37.4 °C (in a humidified atmosphere) with 5 % CO_2. Volume to surface area ratio of the physiological medium used in the immersion test was maintained at 20 ml cm^{-2} as per ASTM: G31-72. The specimens were removed after 1 week, 2 weeks, 3 weeks and 28 days of immersion, thoroughly rinsed with distilled water followed by ethanol and air dried. The difference in weight before and after immersion was recorded by using a weighing machine A&D GX600 with 0.001 gram resolution and degradation rates (in mm $year^{-1}$) of the composites were obtained according to ASTM: G31-72. Corrosion rate is given by:

$$\text{Corrosion rate (mm year}^{-1}) = \frac{(K * W)}{(A * T * D)} \qquad \text{eq. (1)}$$

Where, time conversion coefficient, K = 8.76 x 10^4, W is the weight difference before and after immersion (g), A is the sample area exposed to solution (cm^2), T is the exposure time (h) and D is the density of the material (g cm^{-3}). The pH value of the solution was also recorded after the immersion test.

3. RESULTS AND DISCUSSION

3.1 Synthesis of Pure Mg and Mg ~1 vol.% Ti, Mg ~1 vol.% TiB_2 Nanocomposites

Observations made on as-deposited pure Mg and Mg ~1 vol.% Ti, Mg ~1 vol.% TiB_2 nanocomposites indicated absence of macropores and blow holes. Further, there was no evidence of microstructural defects and circumferential cracks after hot extrusion. Density measurements of the nanocomposites measured using a gas pycnometer indicated that using the adopted fabrication methodology, near dense Mg materials can be synthesized. Porosities of Pure Mg and Mg ~1 vol.% Ti, Mg ~1 vol.% TiB_2 nanocomposites were found to be ~ 0.20%. The metallographically polished samples of nanocomposites exhibited uniform distribution of Ti and TiB_2 nano reinforcements in Mg matrix when analyzed using a field emission scanning electron microscope (FESEM).

3.2 X-Ray Diffraction Studies

X-ray diffractograms of pure Mg and Mg ~1 vol.% Ti, Mg ~1 vol.% TiB_2 nanocomposites are shown in Fig. 1. The diffractograms indicated that there is no phase formation between Mg and Ti. Further, a faint peak corresponding to Ti was visible in the diffractogram of Mg ~1 vol.% Ti nanocomposite but peaks corresponding to TiB_2 were not visible in Mg ~1 vol.% TiB_2 nanocomposite.

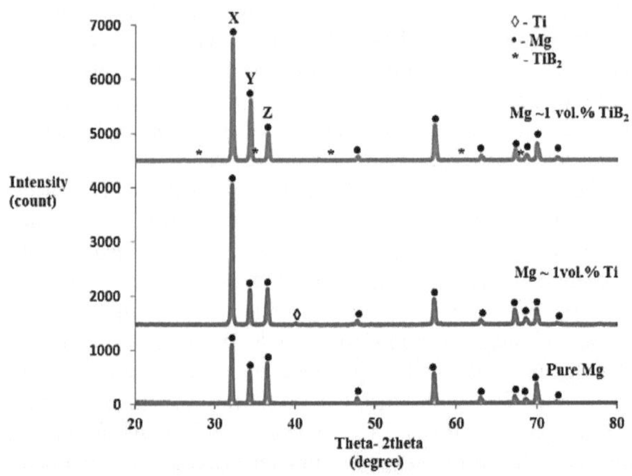

* indicates areas where TiB_2 peaks are expected

Fig.1: X-ray diffractograms of synthesized pure mg and Mg ~1 vol.% Ti, Mg ~1 vol.% TiB_2 nanocomposites

In the X-ray diffractograms of extruded pure Mg, the peaks observed at 2θ = 32°, 34° and 36° correspond to (10-10) prism, (0002) basal and (10-11) pyramidal planes of HCP Mg-crystal [17, 18], highlighted as X, Y and Z respectively in Fig.1. It is observed that the peak corresponding to prismatic plane (X = 32°) is the maximum for all the synthesized Mg materials. Further, the peak corresponding to basal plane (Y = 34°) of pure Mg is found to relatively increase with ~ 1 vol.% TiB_2 addition when compared to that of Mg ~1 vol.% Ti nanocomposite. The dominance of prismatic intensities (X = 32°) of synthesized Mg materials when analyzed along the cross section of the extruded rod indicate that any of the prismatic plane is perpendicular to the extrusion direction [17, 18].

3.3 Microstructure

Grain size analysis of pure Mg and Mg ~1 vol.% Ti, Mg ~1 vol.% TiB_2 nanocomposites are shown in Table 1.

Table 1: Results of Microstructure and Microhardness studies

Material	Reinforcement (in wt.%)	Grain size (x 10^{-6} m)	Microhardness (Hv)
Pure Mg	Nil	45 ± 2.4	52 ± 1.5
Mg ~1 vol.% Ti	2.5	2.5 ± 1.5	58 ± 1.5
Mg ~1 vol.% TiB_2	2.5	28 ± 3.6	69 ± 2

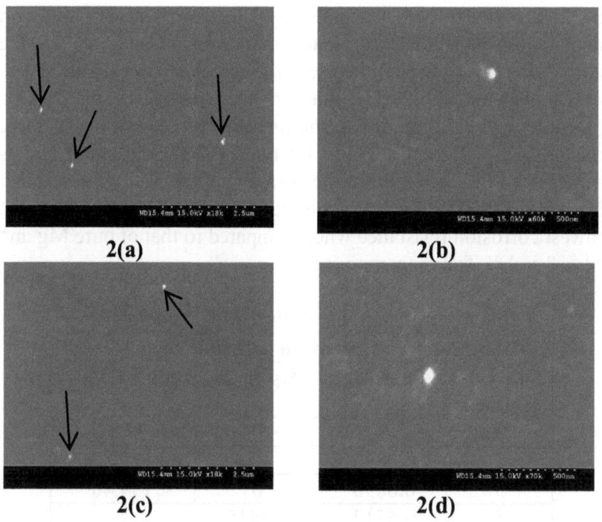

Fig. 2: Distribution of nano Ti and TiB_2 particulates in (a) Mg ~1 Ti and (b) Mg ~ 1 TiB_2 composites & Interfacial integrity of Mg-Ti and Mg-TiB_2 in (a) Mg ~1 Ti and (b) Mg ~ 1 TiB_2 composites

During recrystallization, small sized second phase particles were found to inhibit grain growth significantly when compared to that of large sized reinforcements [19]. With addition of ~ 1 vol.% Ti and TiB_2 nanoparticulates, the grain size of pure Mg decreased by ~ 94% and ~ 33% respectively. The reduction in grain size with addition of nanoparticulates may be due to: (a) presence of uniformly distributed Ti and TiB_2 nanoparticulates (as shown in Fig. 2) in pure Mg matrix and (b) pinning effect due to the presence of nanoparticulates [20, 21]. When compared with TiB_2, nano Ti particulates were found to significantly reduce the grain size of pure Mg which may be due to good wettability of Ti by molten Mg [19].

3.4 Mechanical Properties

3.4.1 Microhardness: The microhardness values of the composites are shown in Table 1. High hardness values of uniformly distributed Ti and TiB_2 nanoparticulates leads to constraint in the localized deformation of Mg matrix and thereby improves the hardness of pure Mg. A maximum hardness of ~ 69 Hv for Mg ~ 1 vol.% TiB_2 was observed which is ~ 33% and ~19% greater than that of pure Mg and Mg ~1vol.% Ti nanocomposite respectively.

3.4.2 Young's Modulus: The average reduced modulus of pure Mg and synthesized nanocomposites are obtained by using the nanoindentation technique. The actual Young's modulus of the composites is calculated by using Oliver and Pharr model [22]:

$$Es = (1-vs^2)\left(\frac{1}{Er} - \frac{1-vi^2}{Ei}\right)^{-1} \qquad \text{eq.(2)}$$

Where, *vs* and *vi* are the poisson's ratios of the specimen and indenter (0.07) respectively. Ei is the modulus of diamond indenter (1141 GPa) and Er is the reduced modulus of the synthesized Mg materials. In all the calculations, vs is considered to be poisson ratio of pure magnesium, which is 0.35. Previously, Polmear reported the Young's modulus of pure Mg in as-cast condition to be ~ 40 GPa [23]. In the present study, a marginal improvement in the modulus of pure Mg is observed with the addition of low volume fraction (~ 1 vol.%) Ti and TiB_2 nanoparticulates. Thus the modulus of pure Mg is not affected by the presence of low volume fraction Ti and TiB_2 nanoparticulates.

Table 2: Results of Young's modulus measurements using nanoindentation technique

Material	Average reduced Young's modulus [GPa]	Young's modulus [GPa]
Pure Mg	41.2	37.5
Mg ~ 1Ti	42.97	39.17
Mg ~ 1TiB_2	41.8	38.06

3.4.3 Tensile Properties: Room temperature tensile properties of the synthesized pure Mg and Mg ~1 vol.% Ti, Mg ~1 vol.% TiB_2 nanocomposites are shown in Table 3. From stress-strain curve (as shown in Fig. 3) , with the addition of ~ 1 vol.% of Ti and TiB_2, the 0.2% tensile yield strength (0.2 TYS) and ultimate tensile strength (UTS) of pure Mg were found to increase and a maximum improvements of ~ 47% and ~ 25% were observed with ~ 1 vol.% Ti addition. With the addition of ~ 1 vol.% TiB_2, the 0.2% TYS and UTS of pure Mg increased by ~ 20% and ~ 10% respectively and a maximum tensile fracture strain of 16% which is ~ 95% greater than that of pure Mg was observed. Thus, addition of ~ 1 vol.% Ti led to strengthening of pure Mg whereas presence of ~1 vol.% TiB_2 in pure Mg led to improvement of its ductility.

Significant increase in the strengths of Mg ~1 vol.% Ti nanocomposites may be due to: (a) Hall-Petch strengthening from significant grain refinement, (b) good wettability between Mg and Ti [24], (c) Taylor strengthening due to mismatch in modulus between Mg matrix (E_{Mg} = 40 GPa) and Ti nanoparticulates (E_{Ti} = 110.2 GPa) [25], (d) Forest strengthening due to mismatch in the coefficient of thermal expansion (CTE) values between Mg matrix

(~ 27 x 10^{-6}/K) and Ti nanoparticulates (9.1 x 10^{-6}/K) [25] and (e) Orowan strengthening by dislocation bowing due to ultrafine nanoparticulates. Presence of strong basal texture of extruded Mg materials, with basal planes (0002) aligned parallel to the extrusion direction, is the potential reason for their poor room temperature ductility.

Table 3: Results of room temperature tensile testing

Material	0.2% TYS [MPa]	UTS [MPa]	Fracture strain [%]	Energy absorbed [MJ/m^3]
Pure Mg	92±5	157±5	8.2±0.2	11±1
Mg ~ 1Ti	135±3	197±8	8.3±0.6	16±1
Mg ~ 1TiB_2	110±3	173±8	16±0.5	25±2

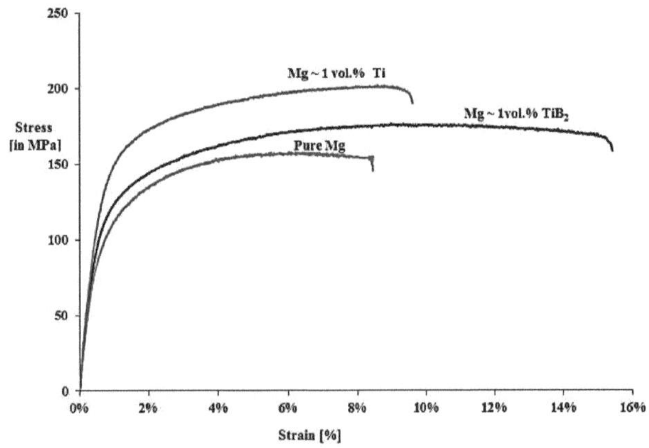

Fig. 3: Stress-strain curves of pure Mg and Mg ~1 vol.% Ti, Mg ~ 1 vol.% TiB_2 nanocomposite

High improvement in ductility of pure Mg with the addition of ~ 1 vol.% TiB_2 nanoparticulates may be due to the ability of TiB_2 nanoparticulates to modify the basal texture of hot extruded pure Mg. From the X-ray diffraction studies, the basal intensity of extruded pure Mg measured along the cross section was found to increase with the addition of ~ 1 vol.% TiB_2 which indicates that the basal planes may be tilted at an angle and no longer parallel to the extrusion direction. High yield strength Mg materials such as Mg ~1 vol.% Ti nanocomposite may lead to both difficulty in activating of basal slip and inability of twinning [26-28]. Restriction of twinning and delay in yielding contributes to high strength and low fracture strain of Mg ~1 vol.% Ti nanocomposite. Further, a maximum of ~ 125% increase in energy absorbed up to tensile fracture was observed with Mg ~1 vol.% TiB_2 nanocomposites.

3.4.4 Compression Properties: Room temperature compression properties of the synthesized pure Mg and Mg ~1 vol.% Ti, Mg ~1 vol.% TiB_2 nanocomposites are discussed in Table 4. A maximum of ~ 128% increase in the 0.2% compressive yield strength (0.2% CYS), ~ 25% increase in the ultimate compressive strength (UCS), and ~ 26% increase in energy absorbed up to compressive failure was observed with the addition of ~ 1 vol.% Ti nanoparticulates with inappreciable increase in the compressive fracture strain. Significant increase in the strengths of Mg ~1 vol.% Ti nanocomposites may be due to: (a) Hall-Petch strengthening from significant grain refinement, (b) good wettability between Mg and Ti [24] and thereby leading to sufficient transfer of load from Mg matrix to Ti reinforcements and (c) Orowan strengthening due to presence of uniformly distributed Ti particulates. With ~ 1 vol.% addition of TiB_2 nanoparticulates to pure Mg, maximum compressive fracture strain of ~ 21% was observed.

Table 4: Results of room temperature compressive testing

Material	0.2% CYS [MPa]	UCS [MPa]	Fracture strain [%]	Energy absorbed [MJ/m^3]
Pure Mg	57±3	332±10	18	34±2
Mg ~1Ti	130±8	413±15	18.5	43±2
Mg ~1TiB_2	75±7	333±9	21±1	42±2.5

The tensile-compression anisotropy "TCA", which is $\sigma_{y,t}$ / $\sigma_{y,c}$, where $\sigma_{y,t}$ and $\sigma_{y,c}$ are uniaxial tensile and compressive yield strengths respectively, for the synthesized pure Mg, Mg ~1 vol.% Ti and Mg ~1 vol.% TiB_2 nanocomposites is found to be 1.61, 1.038 and 1.466, respectively. Among the synthesized Mg materials, Mg ~1 vol.% Ti exhibited the lowest TCA value with maximum tensile and compressive yield strengths which may be attributed to higher load bearing capacity of Ti due to good wettability between molten Mg and Ti. Minimal tensile fracture strain of ~ 8.3 % which is ~ 48% lower than that of Mg ~1 vol.% TiB_2 was observed in Mg ~1 vol.% Ti nanocomposite.

3.5 Immersion Test

Corrosion rates of the synthesized pure Mg and Mg ~1 vol.% Ti, Mg ~1 vol.% TiB_2 nanocomposites immersed in DMEM + 10% FBS + 1% Penicillin are calculated at different soaking intervals (1, 2, 3, 5, 7, 10, 14, 21 and 28 days) for an exhaustive study on degradation behavior of the nanocomposites. Using eq. (1), the corrosion rates of the synthesized Mg materials are calculated and are shown in Table 5. Further, the change in pH of the physiological solution due to the presence of the synthesized Mg materials at different soaking intervals is also measured and is represented in Fig. 4. From the corrosion rates, it is observed that Mg ~1 vol.% TiB_2 degrades completely within 3 days and has the lowest corrosion resistance when compared to that of pure Mg and Mg ~1 vol.% Ti nanocomposite.

Table 5: Corrosion rate of Pure Mg and composites

Soaking time [Days]	Corrosion rate [mm year^{-1}]		
	Pure Mg	Mg ~1 Ti	Mg ~1 TiB_2
1	0	0	27.81
2	0.6890	0	129.87
3	0.6890	0	154.40
5	0.5512	0.3415	-
7	0.3937	0.0089	-
10	0.2756	0.1366	-
14	0.1968	0.0976	-
21	0.1312	0.0650	-
28	0.2214	0.0406	-

The corrosion rates of pure Mg and Mg ~1 vol.% Ti nanocomposite samples are compared in Fig.5. It is observed that up to 3 days of immersion, Mg ~ 1 vol.% Ti nanocomposite exhibited zero degradation where as in pure Mg, after 24 h of immersion, the corrosion rate increased to a maximum of ~ 0.69 mm year^{-1} and gradually decreased. Mg 1vol.% Ti nanocomposite samples exhibited the best corrosion resistance and retains its mechanical integrity. The average corrosion rate of Mg ~1 vol.% Ti nanocomposite was found to be ~ 0.1150 mm year^{-1} for 28 days which is ~ 71% lesser than that of pure Mg (0.3935 mm year^{-1}). But the pH of the physiological medium for the Mg ~1 vol.% Ti nanocomposite was found to increase steadily after 2 weeks of immersion whereas the pH of pure Mg increased after 3 weeks.

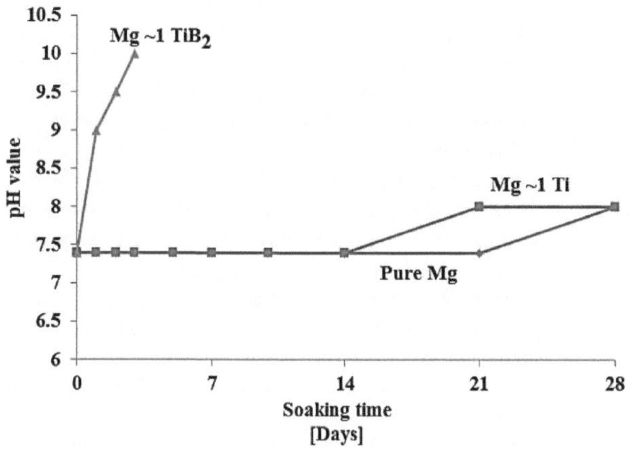

Fig 4: pH value of DMEM + 10% FBS solution incubating pure Mg and Mg ~1 vol.% Ti, Mg ~1 vol.% TiB_2 nanocomposites

Weakening of texture of pure Mg through particulate reinforcement addition leads to decrease in its corrosion properties and similar observations were visible with Mg HAP microcomposites [29]. Texture and porosity of the Mg-HAP microcomposites are the properties that were found to control their corrosion rates. In this study, all the tested Mg materials have almost equal porosity levels of ~ 0.20% which makes the comparison between the in vitro degradation properties of Mg materials more practical. Further, it shows that texture of Mg-TiB_2 nanocomposites may lead to its significant corrosion rate.

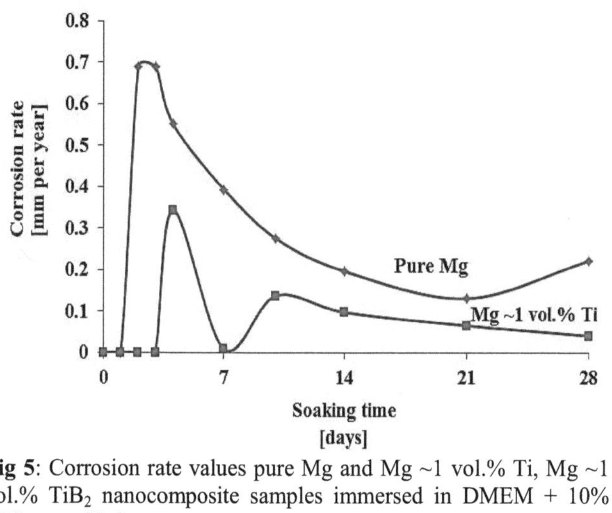

Fig 5: Corrosion rate values pure Mg and Mg ~1 vol.% Ti, Mg ~1 vol.% TiB_2 nanocomposite samples immersed in DMEM + 10% FBS up to 28 days

The mechanical properties of cortical and cancellous bones are compared to that of the synthesized Mg materials in Table 6. The 0.2% tensile yield strength (0.2% TYS) of pure Mg is found to be lesser than that of cortical bone. The mechanical properties of Mg ~ 1 vol.% Ti and Mg ~1 vol.% TiB_2 are greater than that of cortical and cancellous bones. Further, from the results of immersion testing in a physiological medium of DMEM+10% FBS, Mg ~ 1 vol.% Ti is found to surpass those of pure Mg and Mg ~1 vol.% TiB_2 nanocomposites. For further studies on suitability of Mg ~ 1 vol.% Ti nanocomposites as a biomaterial, a direct or indirect cytotoxicity test followed by more appropriate in vivo using animal models and later humans may be investigated.

Table 6: Mechanical properties of bone Vs synthesized Mg materials

Material	Modulus [GPa]	0.2% TYS [MPa]	UTS [MPa]	UCS [MPa]	Tensile fracture strain [%]
Cortical bone [30]	5-23	104.9-114.3	35-283	164-240	1.07-2.1
Cancellous bone [30]	10-1570 (MPa)	-	1.5-38	-	-
Pure Mg[a]	37.5	92±5	157±5	332±10	8.2±0.2
Mg ~1Ti[a]	39.17	135±3	197±8	413±15	8.3±0.6
Mg ~1TiB_2[a]	38.06	110±3	173±8	333±9	16±0.5

* [a] respresents Mg materials synthesized in the present study

4. CONCLUSIONS

The following are the primary conclusions of the present study:

a) Disintegrated melt deposition technique followed by hot extrusion can be utilized to synthesize Mg based nanocomposites targeting biomedical applications.
b) For the same volume fraction (~ 1) of Ti and TiB_2 nanoparticulates addition, significant improvements in the strengths of pure Mg both in tension and compression were observed with the addition of ~ 1 vol.% Ti metallic reinforcements.
c) Maximum tensile and compressive fracture strain of ~ 16% and ~ 21% was observed with the addition of ~ 1vol.% TiB_2 ceramic reinforcements.
d) A marginal improvement in the Young's modulus of the synthesized Mg materials was observed through nanoindentation studies.
e) Immersion test of the nanocomposites indicated significant decrease in the corrosion rate of pure magnesium by ~ 71% with the addition of ~ 1 vol.% Ti nanoparticulates when immersed within DMEM + 10% FBS up to 28 days. Eventhough TiB_2 has good corrosion resistance, addition of ~ 1 vol.% TiB_2 to pure Mg decreased its corrosion rate significantly.
f) The mechanical properties of the synthesized nanocomposites surpassed that of natural bone.

REFERENCES

1. Geetha, M., et al., *Ti based biomaterials, the ultimate choice for orthopaedic implants–a review.* Progress in Materials Science, 2009. **54**(3): p. 397-425.
2. Kurtz, S.M., et al., *Future clinical and economic impact of revision total hip and knee arthroplasty.* The Journal of Bone & Joint Surgery, 2007. **89**(suppl_3): p. 144-151.
3. Katti, K.S., *Biomaterials in total joint replacement.* Colloids and Surfaces B: Biointerfaces, 2004. **39**(3): p. 133-142.
4. Niinomi, M., *Recent metallic materials for biomedical applications.* Metallurgical and materials transactions A, 2002. **33**(3): p. 477-486.
5. Marchi, J., et al., *Physico-chemical characterization of zirconia–titania composites coated with an apatite layer for dental implants.* Dental Materials, 2013. **29**(9): p. 954-962.
6. Mohseni, E., E. Zalnezhad, and A. Bushroa, *Comparative investigation on the adhesion of hydroxyapatite coating on Ti–6Al–4V implant: A review paper.* International Journal of Adhesion and Adhesives, 2014. **48**: p. 238-257.
7. Elin, R.J., *Magnesium metabolism in health and disease.* Disease-a-month, 1988. **34**(4): p. 166-218.
8. Watson, R.R. and V.R. Preedy, *Bioactive food as dietary interventions for diabetes.* 2012: Academic Press.
9. Trumbo, P., et al., *Dietary reference intakes for energy, carbohydrate, fiber, fat, fatty acids, cholesterol, protein and amino acids.* Journal of the American Dietetic Association, 2002. **102**(11): p. 1621-1630.
10. Farraro, K.F., et al., *Revolutionizing orthopaedic biomaterials: The potential of biodegradable and bioresorbable magnesium-based materials for functional tissue engineering.* Journal of biomechanics, 2014. **47**(9): p. 1979-1986.
11. Staiger, M.P., et al., *Magnesium and its alloys as orthopedic biomaterials: a review.* Biomaterials, 2006. **27**(9): p. 1728-1734.
12. Ferguson, J., et al., *On the strength and strain to failure in particle-reinforced magnesium metal-matrix nanocomposites (Mg MMNCs).* Materials Science and Engineering: A, 2012. **558**: p. 193-204.
13. Murray, J.L., *Phase diagrams of binary titanium alloys.* 1987.
14. Abkowitz, S., et al., *CermeTi® discontinuously reinforced Ti-matrix composites: Manufacturing, properties, and applications.* Jom, 2004. **56**(5): p. 37-41.
15. Xiuqing, Z., et al., *The mechanical properties of magnesium matrix composites reinforced with (TiB< sub> 2</sub>+ TiC) ceramic particulates.* Materials Letters, 2005. **59**(17): p. 2105-2109.
16. Meenashisundaram, G.K. and M. Gupta, *Low volume fraction nano-titanium particulates for improving the mechanical response of pure magnesium.* Journal of Alloys and Compounds, 2014. **593**: p. 176-183.
17. Meenashisundaram, G.K., S. Sankaranarayanan, and M. Gupta, *Enhancing overall tensile and compressive response of pure Mg using nano-TiB2 particulates.* Materials Characterization, 2014. **94**: p. 178-188.
18. Seetharaman, S., et al., *Synthesis and characterization of nano boron nitride reinforced magnesium composites produced by the microwave sintering method.* Materials, 2013. **6**(5): p. 1940-1955.
19. Habibi, M.K., S.P. Joshi, and M. Gupta, *Hierarchical magnesium nano-composites for enhanced mechanical response.* Acta Materialia, 2010. **58**(18): p. 6104-6114.
20. Habibnejad-Korayem, M., R. Mahmudi, and W. Poole, *Enhanced properties of Mg-based nano-composites reinforced with Al< sub> 2</sub> O< sub> 3</sub> nano-particles.* Materials Science and Engineering: A, 2009. **519**(1): p. 198-203.
21. Sankaranarayanan, S., et al., *Effect of nanoscale boron carbide particle addition on the microstructural evolution and mechanical response of pure magnesium.* Materials & Design, 2014. **56**: p. 428-436.
22. Oliver, W.C. and G.M. Pharr, *An improved technique for determining hardness and elastic modulus using load and displacement sensing indentation experiments.* Journal of materials research, 1992. **7**(06): p. 1564-1583.
23. Avedesian, M.M. and H. Baker, *Magnesium and magnesium alloys.* 1999: ASM international.
24. Kondoh, K., et al., *Wettability of pure Ti by molten pure Mg droplets.* Acta Materialia, 2010. **58**(2): p. 606-614.
25. Smithells, C., *Smithells Metal Reference Book 7th end.* 1992, Boston: Butterworth-Heinemann.
26. Gehrmann, R., M.M. Frommert, and G. Gottstein, *Texture effects on plastic deformation of magnesium.* Materials Science and Engineering: A, 2005. **395**(1): p. 338-349.
27. Garcés, G., P. Pérez, and P. Adeva, *Effect of the extrusion texture on the mechanical behaviour of Mg–SiC p composites.* Scripta materialia, 2005. **52**(7): p. 615-619.
28. Garcés, G., et al., *Effect of volume fraction and particle size on the microstructure and plastic deformation of Mg–Y 2 O 3 composites.* Materials Science and Engineering: A, 2006. **419**(1): p. 357-364.
29. Campo, R.d., et al., *Mechanical properties and corrosion behavior of Mg-HAP composites.* Journal of the mechanical behavior of biomedical materials, 2014.
30. Witte, F., et al., *Degradable biomaterials based on magnesium corrosion.* Current opinion in solid state and materials science, 2008. **12**(5): p. 63-72.

MICROSTRUCTURE AND PROPERTIES OF MAGNESIUM ALLOY Mg-1Zn-1Ca (ZX11)

L. Katsarou[1], K. Suresh[2], K.P. Rao[2], N. Hort[1], C. Blawert[1], C.L. Mendis[1], H. Dieringa[1]

[1] MagIC – Magnesium Innovation Centre, Helmholtz-Zentrum Geesthacht, Max-Planck-Str. 1, 21502 Geesthacht, Germany
[2] Department of Mechanical and Biomedical Engineering, City University of Hong Kong, Tat Chee Avenue, Kowloon, Hong Kong

Keywords: magnesium alloy, ZX11, corrosion, creep, microstructure

Abstract

In recent years, some magnesium alloy systems have received attention to serve as potential materials for orthopedic implants due to their biocompatibility and biodegradability. Besides acceptable mechanical strength and corrosion rate, also non-toxicity is an important criterion in the development of these degradable magnesium alloys. Zinc and calcium are essential micro-nutrients in the body, therefore are not expected to be harmful, and positively influence strength by grain refinement and age hardening. To identify biomedical as well as other applications, the as-cast Mg-1Zn-1Ca (ZX11) material was tested for standard corrosion resistance as well as compression and creep strength, also at elevated temperatures. Microstructural investigations complete the determination of relevant characteristics for the use of ZX11. Grain size reduction is observed along the radius of the cylinder and SEM-EDX analysis reveals Mg_2Ca and $Mg_6Ca_2Zn_3$ phases have formed on the grain boundaries. Dislocation climbing seems to be the rate controlling deformation mechanism for creep. Compression strength increases with temperature gradually increased up to 100 °C, plateaus between 100 and 175 °C and decreases after that. Acceptable corrosion properties have been observed.

Introduction

Zinc and calcium are promising alloying elements for biodegradable and non-toxic magnesium alloys for medical applications. Zinc because of its high solid solubility (6.2 wt.-% at 635 °C), which strengthens magnesium alloys, and its positive effect on corrosion performance [1]. Zinc and calcium are essential micro-nutrients in the body and their bio-absorption is beneficial. In non-medical investigations, a number of studies have shown that Mg-Zn-Ca alloys appear to have an excellent combination of yield strength and formability [2]. It is also noted that the Mg-Ca system shows a lot of promise for the development of low cost creep resistant alloys as an alternative to rare earths, particularly due to the precipitation hardening effect [3,4,5]. This effect also improves the tensile strength and fracture properties of the alloys [6].

Experimental

Mg-1Zn-1Ca (ZX11) was cast in a modified permanent mold casting process, which results in a dense structure without blowholes and shrinkage porosity. Chemical analysis was performed by X-ray Fluorescence (XRF) analysis using Bruker Explorer S4. Table I gives the composition of alloy ZX11.

Table I. Composition of the ZX11

Element	Mg	Ca	Mn	Zn	Others
wt.-%	97.96	0.86	0.03	1.14	<0.01

Disks were cut from the casted cylinders and three samples were cut from the disk as shown in the following schematic representation (Figure 1). Each sample was embedded in resin, grinded, polished and etched for the optical microscope pictures and the samples were then used for hardness testing. Hardness tests were performed using the HV5 test standard with an EMCOTEST M1C 010 machine.

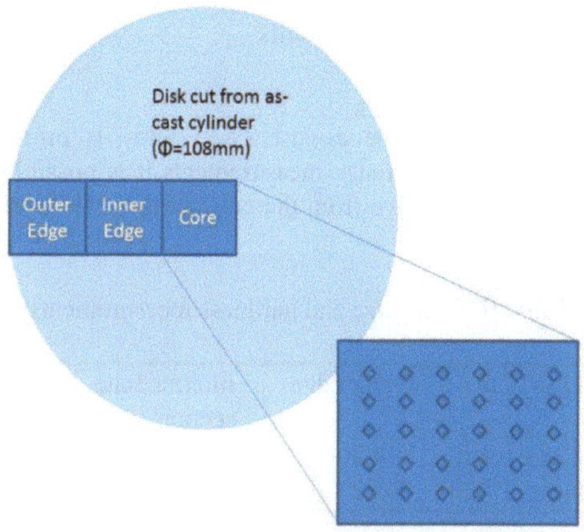

Figure 1: Schematic representation of how the samples were taken for microstructure and hardness testing

Pandat 8 thermodynamic software and the PanMg8 database was used to calculate the isothermal sections of Mg-Ca-Zn phase diagram at the Mg rich end and the Scheil solidification curve for Mg-1Zn-1Ca alloy.

Scanning electron microscopy of the polished but un-etched alloy was performed on a Carl Zeiss Ultra 55 scanning electron microscope (SEM) operating at 15 kV with an EDAX energy dispersive X-ray spectrometer (EDXS).

Compression creep tests were performed with cylinders having 6 mm diameter and 15 mm length. They were prepared with spark erosion machining. At a constant temperature of 200 °C, ATS lever creep tests systems apply constant stresses between 70 and 160 MPa. Stress is partially beyond the yield point, which allows conclusions about the deformation mechanisms during the transition from pure creep to the classical plastic deformation. Deformation is computed as a function of time, giving the creep curves. From first derivatives of those curves, giving the creep rate, minimum creep rates were determined.

For evaluating the yield and ultimate compressive strength of ZX11, cylindrical specimens of 10 mm in dia and 15 mm in height were machined from the slice of cast billet. Compression tests were conducted in the temperature range of 25-250 °C (steps of 25 °C) at a slow strain rate of 0.0001 s^{-1}. Graphite powder mixed with grease was used as the lubricant. The experimental load-stroke data were reformed into true stress-true strain curves by the standard equations. The ultimate compressive strength (UCS) was noted and the yield strength (YS) was calculated at 0.2% offset strain.

Information about the electrochemical behaviour of the Mg1Zn1Ca alloy was obtained by performing potentiodynamic polarization measurements in 0.5% NaCl solution (stirred, pH 6.5, RT, saturated with atmospheric oxygen). A typical three electrode cell (Ag/AgCl reference (with 3 mol/l KCl), Pt counter and the specimens as working electrode) was used. The exposed surface area of the specimens was 1.54 cm^2 and the volume of the electrolyte was 330 ml. After 30 minutes recording of the free corrosion potential, the polarization scan was started from –200 mV relative to the free corrosion potential with a scan rate of 0.2 mV/s. Additional information about the degradation behaviour was obtained from electrochemical impedance spectroscopy (EIS) measurements performed in the same electrolyte and the same cell as described above for the polarisation measurements. Latter tests were carried out at open circuit potential in the frequency range of 0.01 to 30,000 Hz using amplitudes of ±10 mV rms.

Results

Microstructure and hardness

The microstructure of the as-cast ZX11 alloy is presented in Figure 2. From the grain size measurements it is possible to see the difference as we move from the edge of the disk towards the center, see Table II.

Table II. Grain size and hardness measurements

Mg-1Zn-1Ca	Outer edge	Intermediate section	Central core
Grain size (μm)	124.20	140.70	164.60
Hardness (HV5)	44.82	43.13	42.98

The isothermal section at 350 °C of the Mg rich end of Mg-Ca-Zn phase diagram, Figure 3, shows that Mg-1Ca-1Zn alloy only contained α-Mg and Mg_2Ca phase, while the corresponding Scheil solidification calculations, Figure 4, show a small fraction of ternary $Mg_6Ca_2Zn_3$ phase in addition to the two phases expected from the equilibrium phase diagram.

The SEM micrograph of the as-cast Mg-1Zn-1Ca alloy show a semi-continuous distribution of intermetallic particles distributed along the dendritic boundaries, Figure 5, with a closer examination, Figure 6 showing two types of intermetallic particles along the boundaries. The chemical compositions of these particles are summarized in Table III. The first type of particles contained a higher concentration of Zn and some Ca with the Ca:Zn ratio being 0.55 and is likely to be $Mg_6Ca_2Zn_3$ phase. The second type of intermetallics particles contained a higher concentration of Ca compared to Zn with Ca:Zn ratio being ~8.1. This phase is likely to be Mg_2Ca phase. The SEM microstructures showed significant amount of $Mg_6Ca_2Zn_3$ intermetallic particles at the grain boundaries. This is different from the isothermal section shown in Figure 3 and those reported by Nie and Muddle [7] for an alloy with similar composition, where only Mg_2Ca was observed in the as-cast condition. However, it matches the Scheil calculations given in Figure 4.

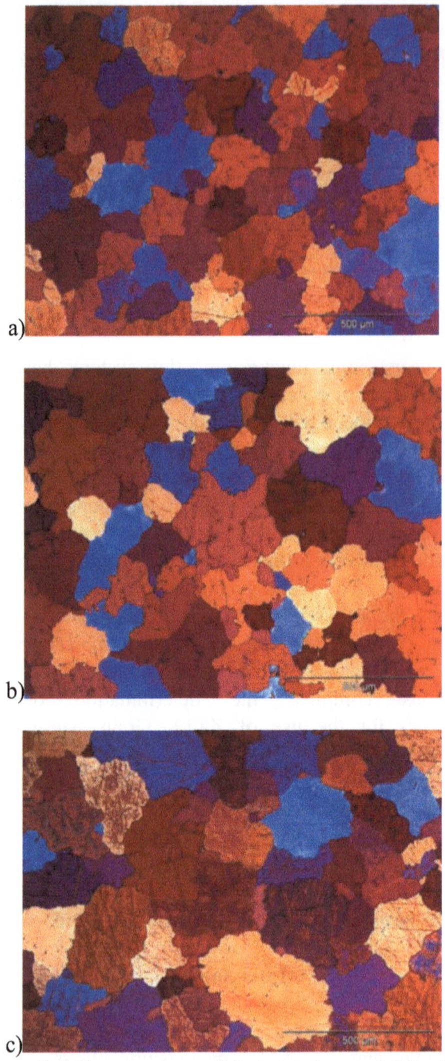

Figure 2: Microstructure of the casted material (optical microscopy). a) outer edge, b) intermediate section, c) central core as described in the previous schematic representation (Figure 1).

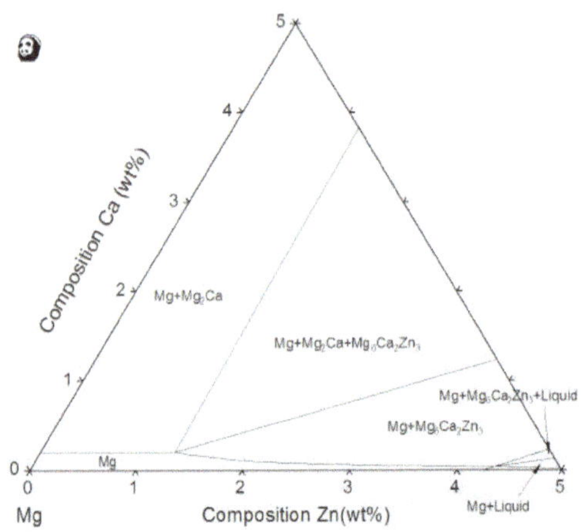

Figure 3: 350°C isothermal section of Mg rich end of the Mg-Zn-Ca phase diagram.

Table III. Composition of the intermetallic particles at the boundaries

Phase	Composition (at%)			Ratio
	Mg	Ca	Zn	Ca:Zn
Type 1	81.4±3.8	6.6±0.7	11.9±1.1	0.55
Type 2	75.9±4.5	21.5±1.2	2.6±0.7	8.14
Matrix	98.8±1.2	0.20±0.05	1.1±0.1	0.2

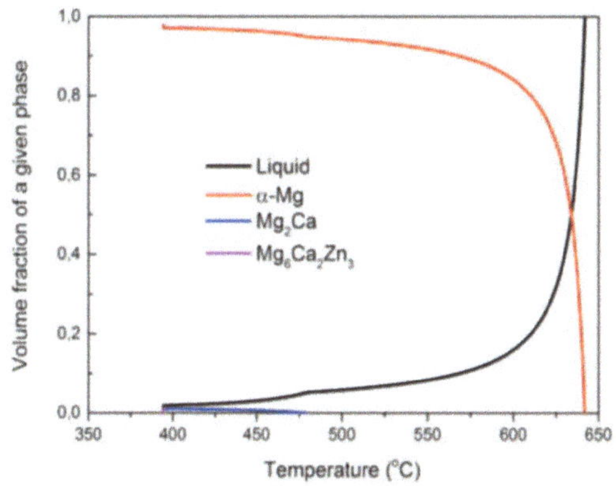

Figure 4: Scheil solidification curve for Mg-1Zn-1Ca alloy.

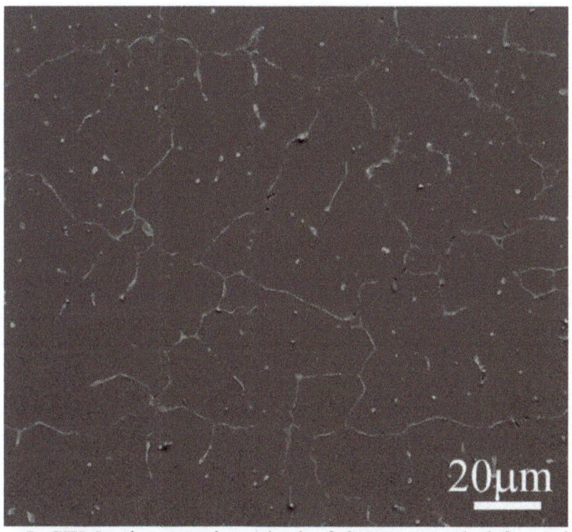

Figure 5: SEM micrograph typical of those recorded from the as-cast Mg-1Zn-1Ca alloy.

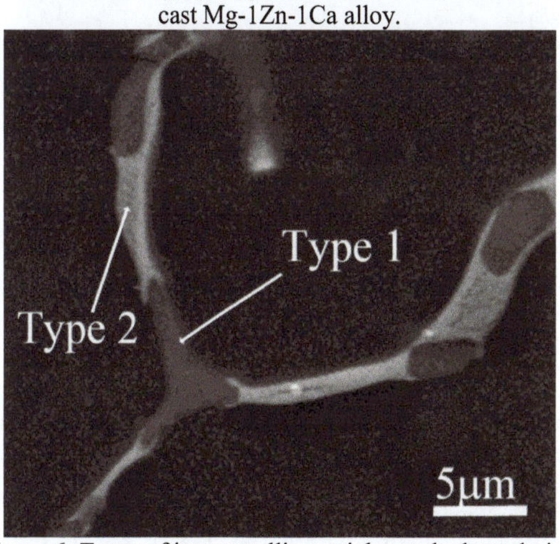

Figure 6: Types of intermetallic particles at the boundaries.

<u>Creep Behaviour</u>

Compression creep tests at 200 °C and stresses between 70 and 160 MPa were performed and creep curves of these tests are shown in Figure 7. Primary creep at the beginning of the curves yields in a regime of secondary creep having a constant deformation rate. This constant or at higher stresses minimum creep rate is plotted over the applied stress, see Figure 8.

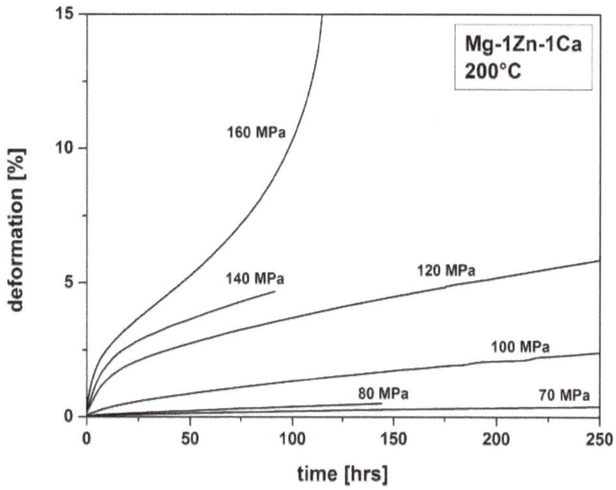

Figure 7: Creep curves from compression creep tests performed at 200°C and stresses between 70 and 160 MPa.

Temperature (T) and stress (σ) dependence of minimum creep rate $\dot{\varepsilon}_s$ (=$d\varepsilon_s/dt$) is usually given by the Norton-Equation (1):

$$\dot{\varepsilon}_s = \frac{ADGb}{kT}\left(\frac{\sigma}{G}\right)^n \qquad (1)$$

where A is a dimensionless constant that depends on the material, D the diffusion coefficient (=D_0 exp(-Q_c/RT). D_0 is the frequency factor, Q_c the activation energy for creep, R the gas constant, and T the absolute temperature), G is the shear modulus, b the Burgers vector, and k the Boltzmann constant.

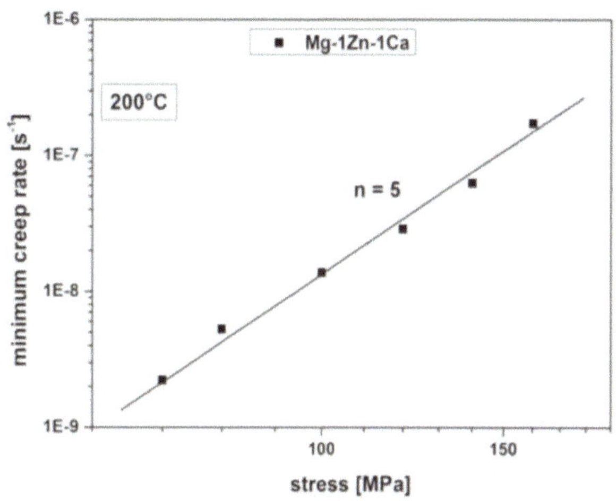

Figure 8: Minimum creep rates over applied stress from tests at 200 °C.

After performing several creep tests at one temperature and different applied stresses, it is possible to determine the stress exponent n from double-logarithmic plots of minimum creep rate from the tests as a function of applied stress. In accordance with equation (1), the slope of the fitted line is n. This plot can be seen in Figure 8. The stress exponent n is directly related to deformation mechanisms taking place during creep, n=5 is related to dislocation climbing at high temperatures [8, 9, 10].

Compression strength

The typical true stress–true strain curves obtained on cast ZX11 alloy at various temperatures are shown in Figure 9. The variation of UCS and YS of ZX11 with temperature is shown in Figure 10. The error estimated for the compression data is within 3%. The strength gradually decreases with increase in temperature from 25 to 100 °C, and reaches near plateau in the range of 100 to 175 °C. The alloy has shown significant softening upon further increase in temperature, with UCS approaching YS at 250 °C. The presence of intermetallic phases like Mg_2Ca and $Mg_6Ca_2Zn_3$ strengthens the grain boundaries and reduces their sliding due to pinning which may be responsible for the occurrence of plateau in the strength variation.

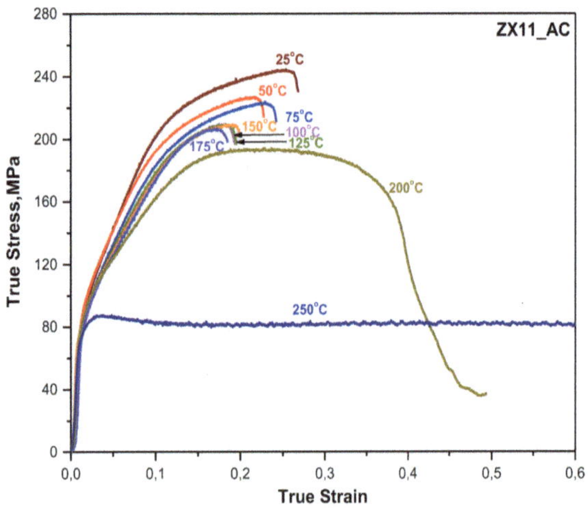

Figure 9: Flow curves for compression tests at different temperatures.

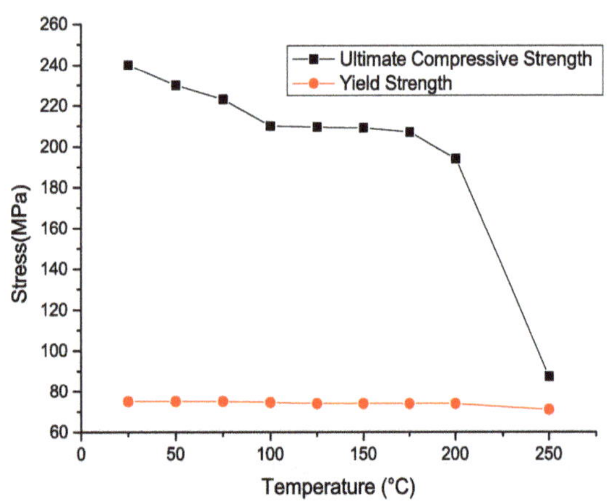

Figure 10: Yield strength and ultimate compressive strength results.

The calculated compressive YS at 0.2% offset of cast ZX11 alloy varied between 71-75 MPa, whereas the UCS varied between 88-240 MPa. It may be noted that the UCS of ZX11 in the temperature range of 25-200 °C is higher than other zinc and calcium containing cast magnesium alloys like AZ31, AZX311 [13], ABaX422 [14], TX32, TXZ321 and, TXAZ3211 [15].

At temperatures higher than 175°C, thermal energy overcomes the pinning effect and grain boundary sliding resumes. Guan et al. [11] reported that the $Mg_6Ca_2Zn_3$ controls the grain growth and it is helpful for the enhancement of mechanical properties in ZX alloys. Bamberger et al. [12] reported the existence of the above phases by thermodynamic calculation and SEM with X-ray mapping, and found that the precipitation of $Mg_6Ca_2Zn_3$ is responsible for the hardening of ZX alloys.

Corrosion

Polarization. The free corrosion potential of the alloy after 30 min of immersion to the 0.5% NaCl electrolyte is -1653 ± 8 mV and the corrosion rate is 1 ± 0.1mm/year. A slightly retarded anodic dissolution of the alloy can be observed with a breakdown potential of around -1500 mV (Figure 11).

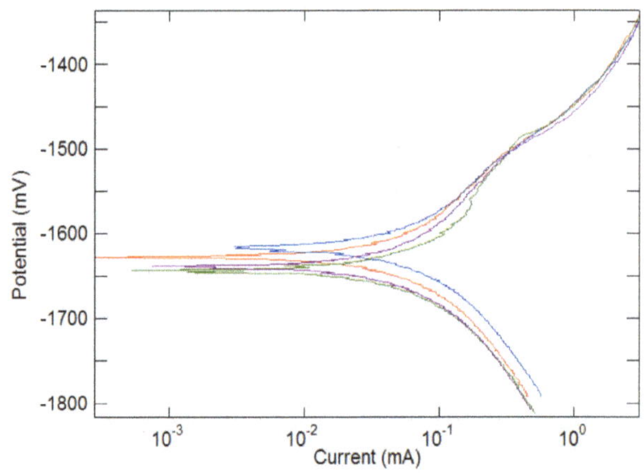

Figure 11: Potentiodynamic polarization curves for the Mg-1Zn-1Ca alloy (4 measurements).

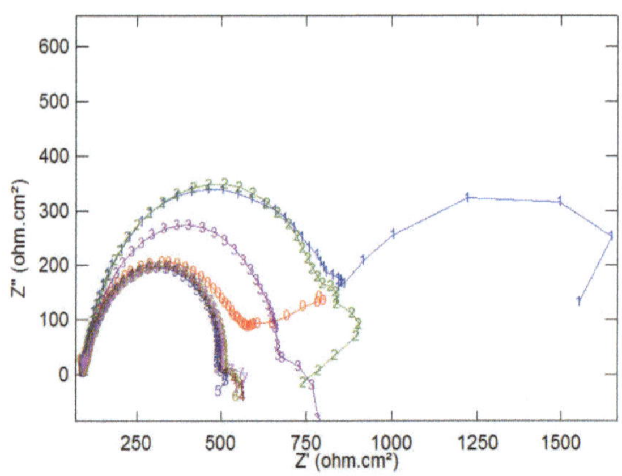

Figure 12: Nyquist plots for the Mg-1Zn-1Ca alloy after various immersion times in 0.5% NaCl solution. The numbers on the lines indicate the immersion time in the electrolyte until the measurement was performed, see here:

Number.	0	1	2	3	4	5	6	7
Time [hrs]	1/12	1	4	10	24	48	72	96

EIS. The results show that the alloy is not able to form a stable passive film in the electrolyte (Figure 12). Although having a passive film right after immersion (likely formed already in air) which is growing in thickness until 1 hour immersion the alloy fails to stabilize this film in the long term. Already the next measurement after 4 hours indicates the occurrence of localized corrosion attack visible by the inductive loop. After this breakdown the film is not recovering anymore, but the resistance remains finally at a stable value of around 400 $Ohmcm^2$.

Conclusions

Castings of ZX11 show smaller grains at the outer section and larger at the core. This is due to the higher cooling rate at the outer area, where the material was closer to the steel mold. Smaller grains cause higher hardness, as it is known from Hall-Petch relationship. SEM-EDX analysis shows two different phases at the grain boundaries. They are Mg_2Ca and $Mg_6Ca_2Zn_3$, respectively, matching Scheil-calculations, but are different from the isothermal section of ternary phase diagram that indicates only Mg_2Ca. Compression creep tests at 200 °C and stresses between 70 and 160 MPa were performed and minimum creep rates determined. A stress exponent of n=5 is found, which corresponds to dislocation climbing as rate controlling deformation mechanism during creep. Compression strength decreases with increase in temperature from 25 to 100 °C, and reaches near plateau in the range of 100 to 175 °C. The alloy shows significant softening upon further increase in temperature, with UCS approaching YS at 250 °C. Though Mg-1Zn-1Ca alloy in the present condition is not among the best performing Mg alloys in terms of corrosion resistance, it is at an acceptable level and appears to be relative stable in the long term.

Acknowledgements

Support from the Research Grants Council of Hong Kong through project no. G_HK006/12 is acknowledged, as well as support from the DAAD through project "Magnesiumlegierungen für biodegradable Anwendungen", project-ID 56203444.

References

[1] G. Song, A. Atrens, Corrosion mechanisms of magnesium alloys, Advanced Engineering Materials 1 (1999) 11-33.

[2] D.-W. Kim, B.-C. Suh, M.-S. Shim, J.H. Bae, D.H. Kim and Nack J. Kim, Texture evolution in Mg-Zn-Ca alloy sheets, Metallurgical and Materials Transactions A 44A (2013) 2950-2961.

[3] X. Gao, S.M. Zhu, B.C. Muddle, J.F. Nie, Precipitation-hardened Mg-Ca-Zn alloys with superior creep resistance, Scripta Matererialia 53 (2005) 1321-1326.

[4] J.F. Nie and B.C. Muddle, Precipitation hardening of Mg-Ca(-Zn) alloys, Scripta Materialia 37 (1997) 1475-1481.

[5] Y. Ortega, M.A. Monge, R. Pareja, The precipitation process in Mg-Ca-(Zn) alloys investigated by positron annihilation spectroscopy, Journal of Alloys and Compounds 463 (2008) 62-66.

[6] Y. Ortega, T. Leguey, R. Pareja, Tensile fracture behavior of aging hardened Mg-1Ca and Mg-1Ca-1Zn alloys, Materials Letters 62 (2008) 3893-3895.

[7] J.F. Nie and B.C. Muddle, Precipitation hardening of Mg-Ca(-Zn) alloys, Scripta Materialia 37 (1997) 1475-1481.

[8] O.D. Sherby, P.M. Burke, Mechanical behavior of crystalline solids at elevated temperature, Progress in Materials Science 13 (1968) 325-389.

[9] F.A. Mohamed., K.T. Park, E.J. Lavernia, Creep behavior of discontinuous SiC-Al composites, Materials Science and Engineering A150 (1992) 21-35.

[10] J. Weertman, Steady-State Creep of Crystals, Journal of Applied Physics 28, 4 (1957) 1185-1189.

[11] S. Guan, J. Hu, L. Wang, S. Zhu, H. Wang, H. Wang, W. Li, Z. Ren, S. Chen, E. Meng, J. Gao, S. Hou, B. Wang and B. Chen, Mg alloys development and surface modification for biomedical application in: Magnesium alloys – corrosion and surface treatments, ed. F. Czerwinski, InTech, Europe, (2011) 109-152.

[12] M. Bamberger, G. Levi, J.B.Vander Sande, Precipitation hardening in Mg-Ca-Zn alloys, Metallurgical and Materials Transactions A 37 (2006) 481-487.

[13] K. Suresh, K.P. Rao, Y.V.R.K. Prasad, N. Hort, and K.U. Kainer, Effect of calcium addition on the hot working behavior of as-cast AZ31 magnesium alloy, Materials Science and Engineering A 588 (2013) 272-279.

[14] K.P. Rao, H.Y. Ip, K. Suresh, Y.V.R.K. Prasad, C.M.L. Wu, N. Hort, and K.U. Kainer, Compressive strength and hot deformation mechanisms in as-cast Mg-4Al-2Ba-2Ca (ABaX422) alloy, Philosophical Magazine 93 (2013) 4364-4377.

[15] K. Suresh, K.P. Rao, Y.V.R.K. Prasad, N. Hort, and K.U. Kainer, Microstructure and mechanical properties of as-cast Mg-Sn-Ca alloys and effect of alloying elements, Transactions of the Nonferrous Metallurgical Society China 23 (2013) 3604-3610.

POWDER METALLURGICAL SYNTHESIS OF BIODEGRADABLE Mg-HYDROXYAPATITE COMPOSITES FOR BIOMEDICAL APPLICATIONS

Cesar Augusto Stüpp[a], Gábor Szakács[a], Chamini Lakshi Mendis[a], Felix Gensch[b], Sören Müller[b], Frank Feyerabend[a], Dachamir Hotza[c], Marcio Celso Fredel[c], and Norbert Hort[a]

[a.] Institute of Materials Research, Helmholtz-Zentrum Geesthacht, Max-Planck-Str. 1, D-21502 Geesthacht, Germany

[b] Forschungszentrum Strangpressen, TU Berlin, Gustav-Meyer-Allee 25, 13355 Berlin, Germany;
[c]Universidade Federal de Santa Catarina, Brazil

Keywords: hydroxyapatite, ball milling, Mg-MMC, bio-corrosion

Abstract

Magnesium alloys with acceptable or even controllable corrosion rates, where mechanical properties are not significantly modified or worsened, have been increasingly investigated in the last decade for use as biomaterials. This work shows an approach with a magnesium metal matrix composite (Mg-MMC), composed of ZK60 as base material and hydroxyapatite (HA) particles. The composite was produced by mechanical alloying followed by hot extrusion, as HA in contact with molten magnesium releases toxic gases such as phosphine (PH_3). This work will present the influence of different amounts of HA on corrosion behaviour and mechanical properties of the investigated composites. Compared to the ZK60 alloy, corrosion is expected to be delayed, without localized corrosion. The mechanical properties are not expected to be compromised with such composite during tissue's healing period.

Introduction

Non-toxicity, degradability, biocompatibility and mechanical properties close to those of bone are reasons why Mg is suitable for biological implant applications [1, 2]. Mg alloys undergo degradation during exposure to physiological conditions, making magnesium an ideal candidate for degradable implants [1]. Mg degradation products can be absorbed, consumed and excreted within a period of time in physiological systems [3]. Additionally, Mg is a common constituent in the human body, where at least half is stored in bone tissues [4]. The high reactivity and rapid corrosion rates observed in Mg can result in loss of mechanical integrity before the surrounded tissue has enough time to heal and bear loads [5].

One approach to solve this problem is the use of Mg metal matrix composite (Mg-MMC), where ceramic particles are distributed within the Mg matrix. Hydroxyapatite (HA) is one such ceramic that is already used in implant applications, as it is biocompatible and bioactive. HA is a constituent of bones and aids osseointegration and cell adhesion [6, 7, 8]. Previous studies have shown that addition of HA reduced the corrosion rates in Mg alloys. Witte et al. [5] reported that adding 20 wt.% HA to AZ91 decreased corrosion rate in artificial seawater. Feng and Han [9] reported a smaller weight loss for ZK60 in physiological saline when reinforced with calcium polyphosphate (CPP).

The biocompatibility of alloying additions is important when choosing alloys for biomedical applications. Both zinc and zirconium are biocompatible. Zinc deficiency causes many physiological functions to be perturbed, such as bone formation, causing metabolic disorders that can result in many diseases [10, 11, 12]. Zirconium is also found in the human body, but its functions are not yet clear [13]. ZK60 has been widely investigated for structural applications and its mechanical and corrosion properties were also investigated [9, 10, 14]. Therefore, ZK60 is a suitable material to be used for the development of biodegradable Mg composite containing HA.

In this contribution we report the findings from an investigation on the ball milled and extruded Mg-6%Zn-0.5%Zr (wt.%) (ZK60) alloy with varying amounts of HA (10 wt.% (6 vol%) and 20 wt.% (13 vol%)). Compression and immersion tests were used to evaluate mechanical properties and corrosion rates along with the microstructure. For easier understanding, the convention ZK60, ZK60-10 and ZK60-20 is adopted for the composite with 0 wt.%, 10 wt.% and 20 wt.% of HA, respectively.

Experimental procedure

ZK60 ingots were cast and machined into millimeter scale chips, and commercially available synthetic HA was used (particle size <42 μm, Alfa Aesar GmbH, Germany). The ZK60 alloy chips and HA were ball milled for 7 hours, at 250 rpm and with a ball to powder ratio (BPR) of 20:1. The powders were hot extruded to produce consolidated material at a temperature of 270 °C, with an extrusion ratio of 1:10 and speed of 9 mm/s.

Specimens from extruded composites were embedded in epoxy, ground with SiC papers with a grit size from 500 to 2500 and then polished with water-free 0.5 μm colloidal silica solution for optical microscopy and scanning electron microscopy (SEM). Samples for optical microscopy were etched in a solution of ethanol (70%), distilled water (20%) and acetic acid (5%) with addition of 8-9 g of picric acid. SEM investigations were conducted with a Tescan VEGA 3 SEM equipped with an energy dispersive X-ray spectrometer (EDXS).

The compression tests were performed on a Zwick 050 testing machine, with a strain gauge to measure elongations, at room temperature with an initial strain rate of 10^{-3} s^{-1}. The 5 compression specimens of each composite were machined from the extruded bars. Compression samples were 9.5 mm diameter and 17 mm length and prepared according to DIN 50106 [15].

Immersion tests specimens were 8.5 mm diameter and 3 mm thickness and were machined from the extruded rods. These samples were sterilized by autoclaving for 20 min at 121 °C. Immersion tests were conducted according to the standard ISO 10993-5 [16], for 72 h in 3 ml of Dulbecco's modified eagle medium (DMEM) with 10 vol.% fetal bovine serum (FBS). The tests were conducted at 37 °C, with a controlled test environment containing 5% CO_2, 20% O_2, and 95% relative humidity. Following immersion tests, the samples were cleaned in chromic acid (180 g/L) to remove corrosion products and were dried for 24 h at 50 °C. The weight of the specimens was determined before the immersion tests and after the removal of corrosion products as described above, to determine the corrosion rate for each of the alloys.

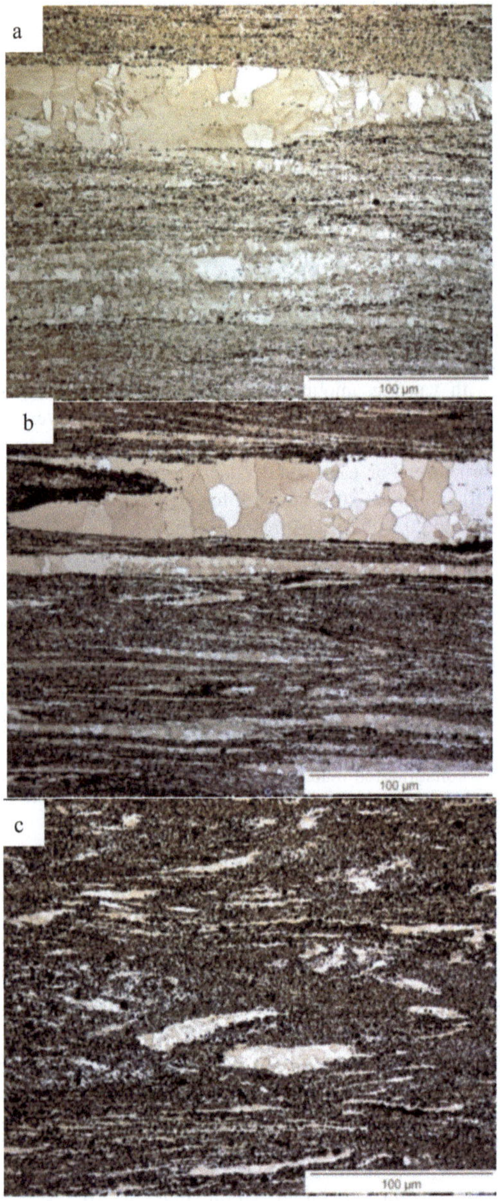

Figure 1: Optical micrographs typical of the ZK60 (a) without HA, (b) with 10 wt.% HA and (c) 20 wt.% HA. Images are parallel to the extrusion direction.

Results and discussion

The optical microstructures of ZK60 with 0, 10 and 20 wt% HA are shown in Figure 1. The microstructures contain a duplex distribution of grains with regions of large grains surrounded by fine grains. The large grained regions occur along the extrusion direction. With the addition of HA, regions of coarse grains decreased and only small regions of coarse grains are observed in ZK60-20. The coarse grain size also decreased with the increase in HA content. In ZK60 and ZK60-10, these grains range from 15 µm to 40 µm, while in ZK60-20, they are less than 20 µm. The fine grained region contained grains that are less than 1 µm. HA is homogeneously distributed throughout the samples, and this distribution causes the formation of fine grains. In ZK60 alloy the fine grained regions are due to the presence of oxide particles introduced during machining and ball milling.

The compression stress-strain curves are shown in Figure 2 and the pertinent features of the stress strain curves are summarized in Table 1. The compression stress-strain curves show that the ultimate strength and yield strength increase with increased HA content, while the elongation decreases. The 0.2% compressive yield strength is 277, 341 and 356 MPa for ZK60, ZK60-10 and ZK60-20 alloys respectively. The increase in yield strength due to HA additions was approximately 60 MPa with 20% HA additions. Elongation to failure changed from 9.6% for ZK60, 6.6% for ZK60-10 and 5.6% for ZK60-20. This is approximately 42% decrease in the elongation to failure from ZK60 with the addition of 20% HA.

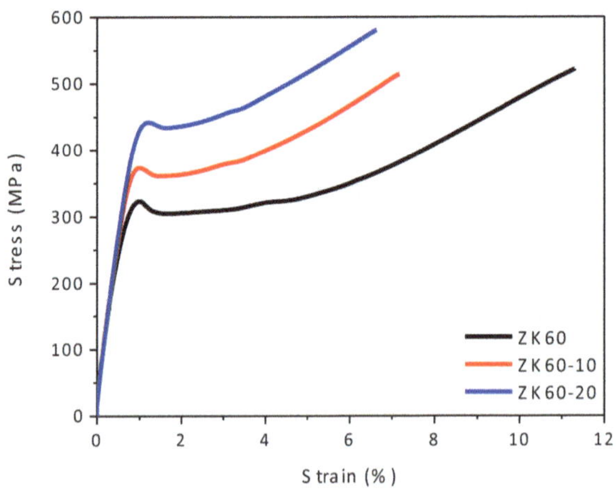

Figure 2: The compressive stress-strain curves of as extruded ZK60 magnesium alloy and ZK60-HA composites.

Table 1: Room temperature compressive properties of ZK60 alloy, ZK60-10 and ZK60-20 composites.

Material	0.2%CYS (MPa)	UCS(MPa)	Elongation (%)
ZK60	277 ± 22	485 ± 35	9.5 ± 1.0
ZK60-10	341 ± 20	529 ± 13	6.6 ± 0.4
ZK60-20	356 ± 54	554 ± 20	5.1 ± 0.4

Table 2: The corrosion rates of ZK60 alloy with 0, 10 and 20 wt% HA additions after immersion in DMEM+10%FBS solution.

Material	Corrosion rate (mm/year)
ZK60	0.50 ± 0.16
ZK60-10	0.47 ± 0.14
ZK60-20	0.41 ± 0.10

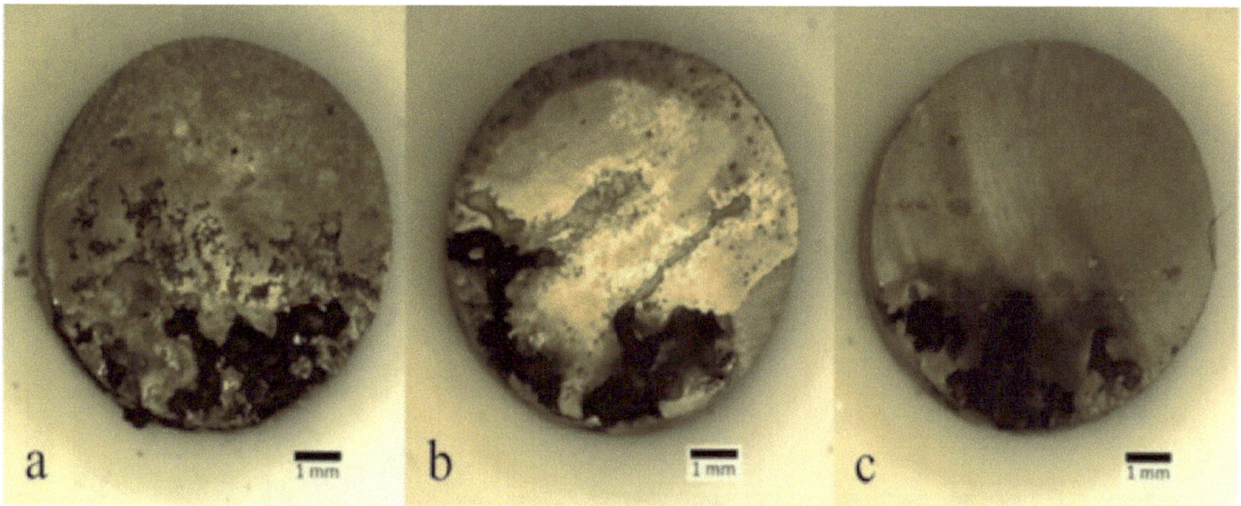

Figure 3: Mg-MMC samples after 72 h immersion in DMEM + 10 vol.% FBS: (a) ZK60, (b) ZK60-10 and (c) ZK60-20.

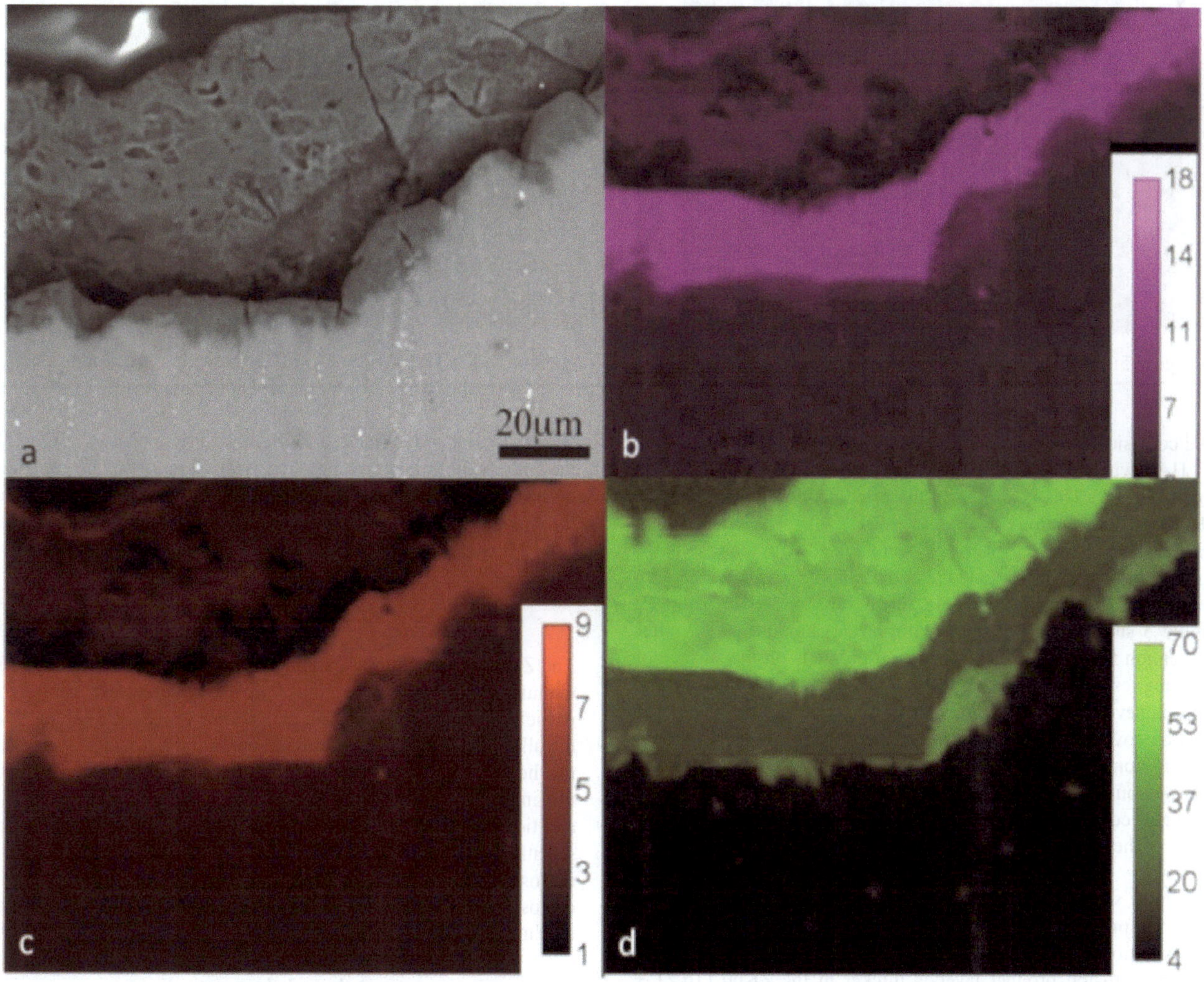

Figure 4: (a) SEM backscattered image of ZK60 alloy after 72h immersion in DMEM + 10 vol.% FBS, showing corrosion product formation and EDX elemental maps showing the (b) Ca, (c) P and (d) O distributions.

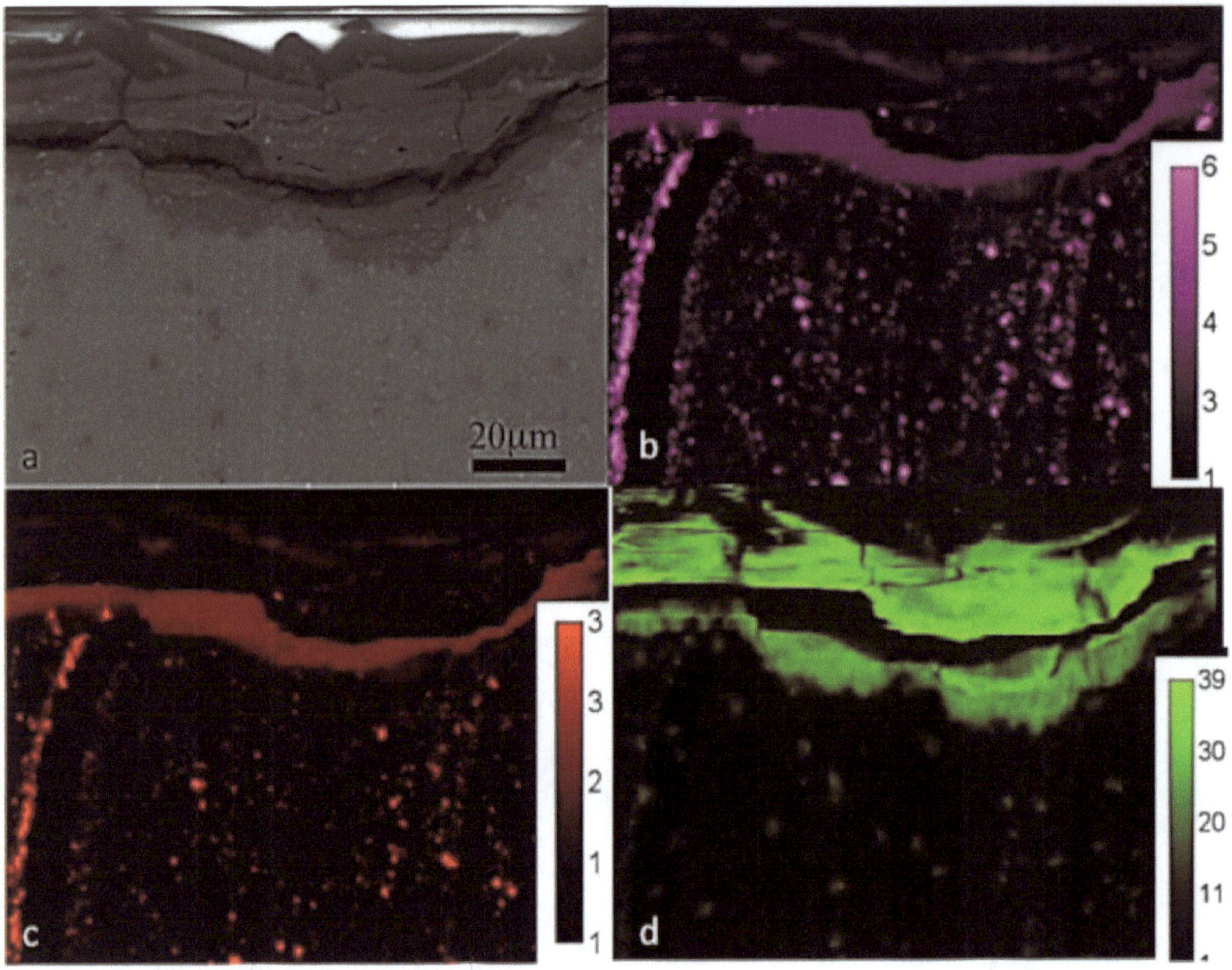

Figure 5: (a) SEM backscattered image of ZK60-10 alloy after 72h immersion in DMEM + 10 vol.% FBS, showing corrosion product formation and distribution of HA; and EDX elemental maps showing the (b) Ca, (c) P and (d) O distributions.

The corrosion rate of the ZK60 alloy decreased with the addition of HA, with 20% HA containing alloy showing the lowest corrosion rate. This is similar to the previous observations [9] on ZK60 alloy containing CPP and AZ91 with HA in artificial seawater [5]. However, this result is contrasted with the results for pure Mg where an increase in the corrosion rate was detected with the addition of HA [17]. These results are shown in Table 2. Figure 3 shows ZK60 samples with various amount of HA after immersion tests showing the corrosion products on the surface of the samples. The corrosion on these samples is not uniform and there is severe localized corrosion in all three samples. The area fraction covered by the localized corrosion seems to decrease with the addition of HA, with 20% HA containing sample showing the least amount of localized corrosion. The reasons behind the localized corrosion are not clear and thought to be due to break down of the very thin oxide layer that formed during autoclaving.

The cross sections corresponding EDX elemental maps showing corrosion product build up in samples immersed in DMEM + 10% FBS solution for ZK60 and ZK60-10 are shown in Figures 4 and 5. The corrosion product layer is thicker in the ZK60 (16.21 ± 2.00 µm approximately) compared with ZK60-10 (4.50 ± 1.35 µm approximately), Figures 4 (a) and 5 (a) respectively. The corrosion product observed for both ZK60 and ZK60-10 are similar with the layer closest to the Mg-MMC consists of O while the intermediate layer is enriched with Ca and P. The top most layers consisted of O. In ZK60-10 the distribution of HA particles through the composite as well as in the corrosion product layer can be observed, Figure 5.

Conclusions

The ZK60 magnesium alloy and its composites were successfully manufactured by ball milling followed by hot extrusion. HA particle size decreased substantially after processing, along with the alloy's grain size. Furthermore, particulates are closely bonded to the Mg matrix. The maximum compressive strength was obtained with ZK60-20 and the elongation decreased with the addition of HA. After immersion tests, it is clear that HA addition slightly decreases the corrosion rate of the composite. The corrosion is macro and localized, with the formation of a corrosion layer based on HA constituents (phosphate and calcium).

Acknowledgements

C.S. acknowledges the project BRAGECRIM for the provision of financial support in form of a stipendium provided by CAPES.

References

1. M.P. Staiger et al., "Magnesium and its alloys as orthopedic biomaterials: a review", *Biomaterials*, 27 (2006), 1728-1734.

2. W.F. Ng, M.H. Wong and F.T. Cheng, "Stearic acid coating on magnesium for enhancing corrosion resistance in hanks' solution", *Surface & Coatings Technology*, 204 (2010), 1823-1830.

3. F. Witte et al., "In vitro and in vivo corrosion measurements of magnesium alloys", *Biomaterials*, 27 (2006), 1013-1018.

4. G. Song, "Control of biodegradation of biocompatible magnesium alloys", *Corrosion Science*, 49 (2007), 1696–1701.

5. F. Witte et al., "Biodegradable magnesium-hydroxyapatite metal matrix composites", *Biomaterials*, 28 (2007), 2163-2174.

6. C.P.A.T. Klein et al., "Studies of the solubility of different calcium phosphate ceramic particles *in vitro", Biomaterials*, 11 (1990), 509-512.

7. M.J. Shen et al., "Effect of bimodal size SiC particulates on microstructure and mechanical properties of AZ31B magnesium matrix composites", *Materials and Design*, 52 (2013), 1011-1017.

8. Y. Yun et al., "Revolutionizing biodegradable metals", *Materials Today*, 12 (2009), 22-32.

9. A. Feng and Y. Han, "Mechanical and *in vitro* degradation behavior of ultrafine calcium polyphosphate reinforced magnesium-alloy composites", *Materials and Design*, 32 (2011), 2813-2820.

10. A. Feng and Y. Han, "The microstructure, mechanical and corrosion properties of calcium polyphosphate reinforced ZK60A magnesium alloy composites", *Journal of Alloys and Compounds*, 504 (2010), 585-593.

11. M. Yamaguchi, "Role of zinc in bone formation and bone resorption", *The Journal of Trace Elements in Experimental Medicine*, 11 (1998), 119-135.

12. A.S. Prasad and D. Oberleas, "Zinc: Human nutrition and metabolic effects", *Annals of Internal Medicine*, 73 (1970), 631-636.

13. X. Lin et al., "The in vitro degradation process and biocompatibility of a ZK60 magnesium alloy with a forsterite-containing micro-arc oxidation coating", *Acta Biomaterialia*, 9 (2013), 8631-8642.

14. X. Wang et al., "Microstructure refining and property improvement of ZK60 magnesium alloy by hot rolling", *Transactions of Nonferrous Metals Society of China*, 21 (2011), s242-s246.

15. DIN 50106 Testing of Metallic Materials; Compression Test (1978)

16. ISO10993-5, Biological evaluation of medical devices - Part 5: Tests for cytotoxicity: In vitro methods, International Standards Organization, Geneva (1993).

17. X. Gu et al., "Microstructure, mechanical property, bio-corrosion and cytotoxicity evaluations of Mg/HA composites", *Materials Science and Engineering C*, 30 (2010), 827-832.

Magnesium Technology 2015
Edited by: Michele V. Manuel, Alok Singh, Martyn Alderman, and Neale R. Neelameggham
TMS (The Minerals, Metals & Materials Society), 2015

CORRELATION BETWEEN MECHANICAL BEHAVIOUR AND MICROSTRUCTURE IN THE Mg-Ca-Si-Sr SYSTEM FOR DEGRADABLE BIOMATERIALS BASED ON THERMODYNAMIC CALCULATIONS

A Gil-Santos[1], G Szakacs[2], N Moelans[1], N Hort[2], O Van der Biest[1]

[1]Department of Materials Engineering, KU Leuven, Kasteelpark Arenberg, 44, 3001 Leuven, Belgium
[2]Magnesium Innovation Centre (MagIC), Helmholtz-Zentrum Geesthacht, Max-Planck-Strasse 1, 21502 Geesthacht, Germany

Keywords: Magnesium alloy, Mechanical properties, Microstructure, Phase diagrams.

Abstract

Magnesium and its alloys are being investigated worldwide for application in biodegradable implants, yet there are no commercially available alloys specifically developed for this application. Magnesium is an essential element in the human body and it would be logical then to explore the development of alloys that contain only elements which are known to be biocompatible because they are already part of the human body.

In this work, Ca, Si and Sr were selected to develop magnesium alloys for biomedical application due to their good biocompatibility. A number of ternary Mg-Ca-Si and Mg-Si-Sr alloy compositions were explored. The alloys were cast in steel molds and their phase content was measured and compared with thermodynamic predictions. The effect of the characterized phases on the hardness and compressive strength was evaluated.

Introduction

Magnesium rich compositions of ternary Mg-Ca-Si and Mg-Si-Sr were explored by casting a number of alloys and characterizing their phase content in terms of crystallographic structure, phase composition, volume fraction and distribution in the microstructure, in order to correlate these with mechanical properties. Due to the fact that the alloys are produced by casting their microstructure, and thus their mechanical properties, strongly depend on the solidification. Therefore, an understanding of the sequence of formation and evolution of the intermetallic phases during solidification in ternary and higher order systems is essential to achieve control over the microstructure of the alloys. Thermodynamic calculations offer a powerful tool to characterize the phase formation and evolution during the solidification path.

Thermodynamic calculations allow to predict the phase formation throughout a solidification process. Since the atomic diffusion process is very slow in the solid phase, equilibrium cooling is hardly achieved for most real cases. Often no diffusion in the solid phase and full mixing in the liquid is considered, according to the Scheil hypothesis. Depending on the alloy system and the experimental conditions, the phase composition obtained in real solidification experiments often demonstrates that a mixture of Scheil and equilibrium solidification took place. In fact, in order to account for the differences between the simulation and the experimental observation, one should consider equilibrium calculations and/or kinetic effects [1]. When considering the final phase fraction of solid phases from calculations, the Scheil solidification curves were in much better agreement with the experimental observations than the equilibrium calculations. Therefore only Scheil solidification is considered in here.

Element selection

Three elements, Ca, Si and Sr, were selected to develop Mg alloys for biomedical application. In addition to their good biocompatibility [2,3], those elements are naturally present in the human body reducing the risk of rejection and toxicity. Ca, Si and Sr are involved in essential functions and in bone regeneration processes. They also improve the mechanical properties of the alloys and/or facilitate the production process, as outlined in the following paragraphs.

Calcium is an essential element that performs essential functions in bone structures and calcium salts are considered nontoxic [4]. The human adult body contains about 1100g Ca, about 1.5% of the body weight, and the skeleton contains 99% of the total body Ca. The normal serum level is 0.919-0.993 mg/L [5]. Previous studies on Ca as an alloying element have shown that Mg-Ca alloys will be degraded within bone and have biocompatibility both in vivo and in vitro. The formation of the compound of $Ca_{10}(PO_4)_6(OH)_2$ (hydroxyapatite) and $Mg_3(PO_4)_2$ (magnesium phosphate) is very important and necessary when using the alloy in implant applications. Furthermore, ions released from Mg-Ca alloy systems are beneficial for bone healing [6]. Within Mg alloys, Ca, with a 0.68wt.% solubility in magnesium helps in grain size reduction and because of the low solubility of Ca in Mg the intermetallic Mg_2Ca is formed, improving the creep resistance. On the other hand, the intermetallic Mg_2Ca phase is brittle and highly reactive. Since it forms at the grain boundaries, it could possibly act as a crack source and, due to his high reactivity, it will increase the corrosion rate of the alloy.

Silicon plays an important role in the healing process and helps to build the immune system [7]. This is important for the growth and development of bone and connective tissue. Si plays an essential role in bone formation; indeed, silicon ions are known to be involved in the calcification process of young bones [8,9]. Thus, the presence of silicon has a significant effect in the osteogenesis process. In fact, it has been demonstrated that the incorporation of Si into apatites induces the formation of a higher amount of bone tissue as compared to undoped apatites. Moreover, silicon improves material bioactivity by inducing the formation of Si–OH groups on the material surface [10]. These groups trigger the nucleation and formation of apatite layers on the surface improving the material–bone bonding [11]. Silicon based devices, with Mg used in their components, have been tested as organic electronic implants [12]. They disintegrate and dissolve when previously immersed in deionized water. So, the idea of a complete disappearance via resorption by the human body seems to be within reach with Mg-Si alloys. Silicon has a 0.006wt.% solubility in Mg; the initial effect of very small additions of Si in the alloy is to give a pronounced increase in ductility, with the tensile strength remaining almost unaffected. With further Si additions, the hardness and tensile strength are improved by the presence of Mg_2Si precipitates [13]. Si helps the grain refinement. During casting, Si increases the fluidity of the molten alloy [14].

The human body contains about 320 mg of *strontium*, 99% contained in the skeleton and the rest distributed in soft tissues [4]. It was reported [15] that the presence of Sr in hydroxyapatite promotes osteoblast function and subsequent bone formation. Sr addition enhances the in vivo biocompatibility and the new bone formation inducing capacity. It has been demonstrated to increase bone mass through its stimulatory effect on osteoblasts (which increases bone formation), and its inhibitory effect on osteoclasts (which cause bone loss through bone resorption). At the recommended daily dosage of 2,000 mg, Sr has shown a significant reduction of vertebral fractures and hip fractures in clinical trials, with decreased back pain and decreased body height loss compared to a placebo group [16]. Adding Sr into the alloy, with a 0.35wt.% solubility in Mg, increases the compressive strength. Sr reduces the surface tension of Mg, improving the castability. In fact, Sr is added in gravity cast alloys to reduce micro-shrinkage porosity. However, the tenacious white oxide that forms on the melt may create problems during casting [14].

Experimental

The alloys were prepared at Magnesium Innovation Centre MagIC (Geesthacht, Germany) using permanent mold gravity casting by melting 99.95% pure Mg ingots, pure Sr, Si and Ca chips (98.9%, 99.9% 97.3%) and Mg10Ca master alloy. The raw materials were molten in an electrical resistance furnace under the protective atmosphere of Ar and SF_6. They were heated up to 700°C in about 40 minutes. When this mixture was molten, the Si and/or Ca chips preheated at 400°C, were added. Next, the melt was kept for about 45 minutes between 700°C and 750°C, and stirred with a steel paddle every 15 minutes for homogeneity. The molten alloy was then stabilized at 760°C, before pouring into a preheated boron nitride coated mild steel die at 400°C. After two minutes, the die was opened and the solidified alloys were cooled down to room temperature. 18 mm diameter and 20 cm length cylinders were produced.

Disk shaped specimens for microstructural characterization were cut with 2 mm height and 18mm diameter from the center of the 20 cm long cast specimens. These disk samples were then ground and polished with alcohol and alumina suspension. The composition of the alloys (tables I and II) was determined by optical emission spectrometry (ICP-OES). The surface morphology was determined by scanning electron microscopy (SEM) and energy dispersive X-ray spectroscopy (EDS). In order to analyze small secondary phases present in the alloys X-ray diffraction (XRD) as well as field emission electron probe micro microanalyzer (EPMA) were used for microstructural and crystallographic characterization.

Compression test samples were obtained by electric discharge machining (EDM). The specimens were first cut into 10 mm thick slabs and from each slab 3 cylinders were produced from the central part with a height of 10 mm and 6 mm diameter. Compression tests were performed on an Instron 4600 applying maximal load of 50Kg parallel with the axis of the cast cylinder. For each alloy from 3 to 6 cylinders were tested. Then HV 0.25 Vickers hardness tests were performed to compare the hardness depending on the concentration of alloying elements.

The presence of secondary phases and their fractions (vs T) were previously calculated with the ThermoCalc 4.0 and Pandat 8.1 software using commercially available Mg databases. The phase content was predicted assuming Scheil condition and afterwards experimentally measured with both X-ray diffraction and using an image analysis based method (See figure 2).

Results & Discussion

Thermodynamic prediction and experimental verification

The presence of the predicted phases (table I) at room temperature was confirmed for the Mg-Ca-Si ternary system (figures 1 and 2).

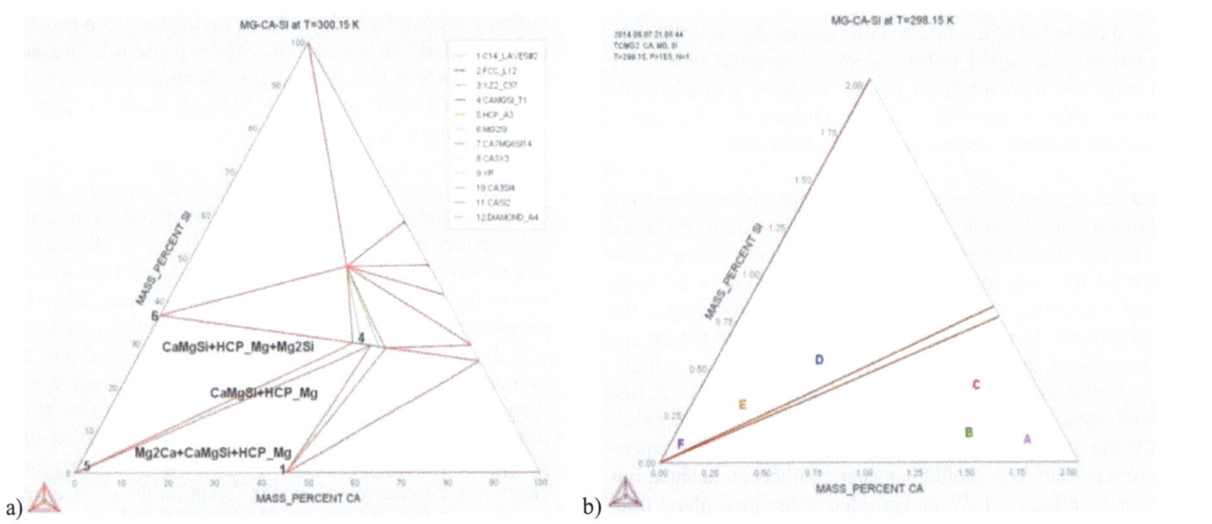

Figure 1. a) Isothermal section at room temperature for the Mg-Ca-Si system calculated with TCMG2 database in ThermoCalc 4. b) Magnification of the magnesium rich corner including the studied alloys.

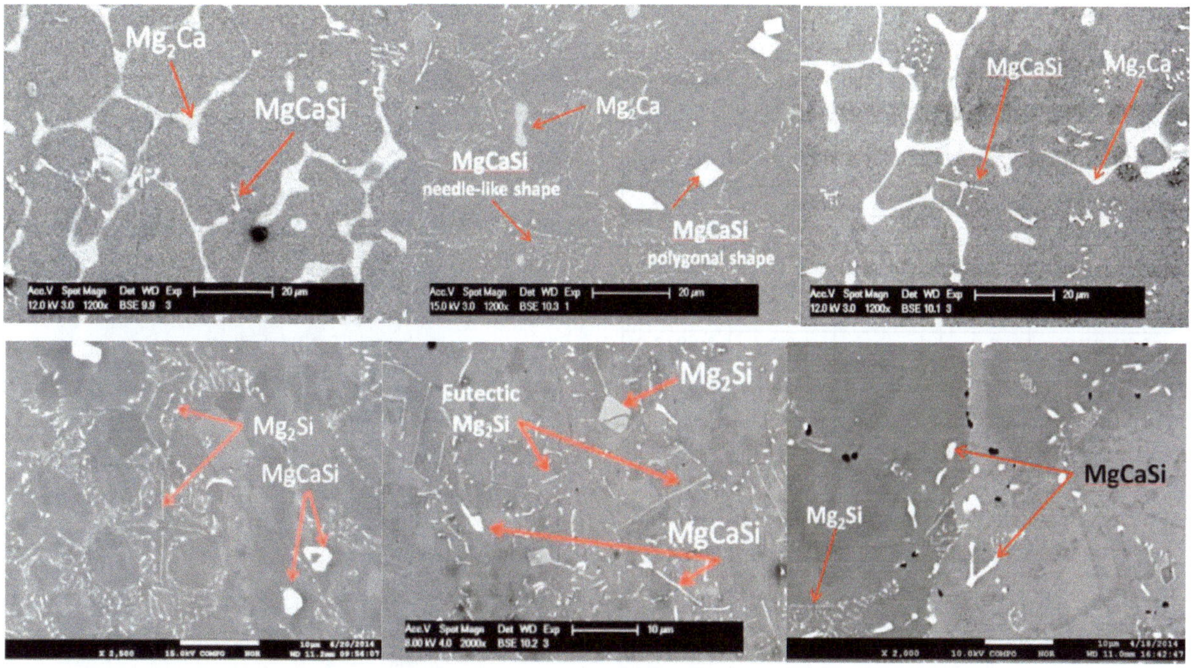

Figure 2. Identification of the phases in samples a) A.Mg1.7Ca0.1Si, b) B.Mg1.43Ca0.16Si, c) C.Mg1.34Ca0.41Si, d) D. Mg0.5Ca0.55Si, e) E. Mg0.24Ca0.31Si and f) F. Mg0.06Ca0.1Si.

A clear difference in the size and the shape of the precipitates in the Mg-Ca-Si alloys can be observed from the micrographs shown in figure 2. The common intermetallic phase that is present in all the samples is MgCaSi; it appears as large precipitates with a polygonal shape in B, C and D alloys. This observation can be linked to the solidification process shown in figures 3 and 4. In the alloys A, B, C and D the first phase forming is MgCaSi followed by hcp-Mg. In alloy A, the temperature gap between the formation of these phases is merely 2 °C, whereas in alloy systems B, C and D this temperature gap is substantially larger, i.e. 15, 59 and 19 degrees, allowing the primary phase to grow extensively. For the alloys E and F, the Mg starts solidifying before MgCaSi does, producing smaller particles of the ternary compound.

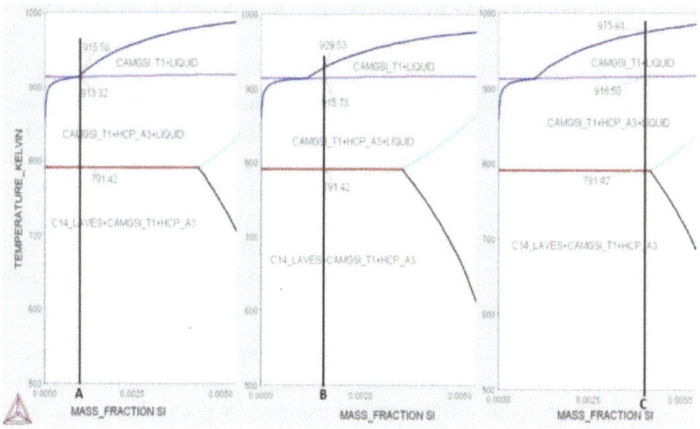

Figure 3. Calculated vertical sections in the Mg-Ca-Si system, at A) 1.7wt.%Ca 0.1wt.%Si, B) 1.43wt.%Ca 0.16wt.%Si and C) 1.34wt.%Ca 0.41wt.%Si.

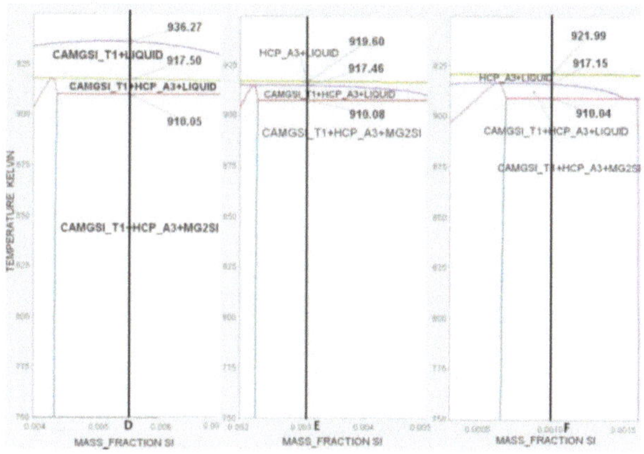

Figure 4. Calculated vertical sections in the Mg-Ca-Si system, at D) 0.5wt.%Ca 0.55wt.%Si, E) 0.24wt.%Ca 0.31wt.%Si and F) 0.06wt.%Ca 0.1wt.%Si

For the Mg-Si-Sr produced alloys (table II), $Mg_{17}Sr_2$ and Mg_2Si are the only binary phases observed that fit suitably well with the calculations carried out (figure 5).

Table I. MgCaSi system. Mass percentages* of the intermetallic phases present in the alloys related to the hardness.

Alloy	%Mg_2Si	%Mg_2Ca	%MgCaSi	HV0.25
A. Mg1,7Ca0,1Si		4.3	0.5	55.0±4.1
B. Mg1,43Ca0,16Si		2.9	0.8	52.7±1.2
C. Mg1,34Ca0,41Si		1.9	5.4	46.8±3.8
D. Mg0,5Ca0,55Si	0.6		0.5	46.3±7.8
E. Mg0,24Ca0,31Si	1.9		0.4	44.1±4.1
F. Mg0,06Ca0,1Si	0.9		0.2	39.8±3.2
Pure Mg				37.0±2.2

*The phase percentage was estimated applying the Rietveld method on the XRD diffraction patterns

Table II. MgSiSr system. Mass percentages* of the intermetallic phases present in the alloys related to the hardness.

Alloy	% Mg_2Si	% $Mg_{17}Sr_2$	HV0.25
Mg0.47Si 0.26Sr	Yes		40.6±6.7
Mg0.38Si 0.51Sr		Yes	46.6±5.8
Mg0.88Ca0.26Si0.51Sr			47.8±4.6
Pure Mg			37.0±2.2

* The phase percentage was not estimated since the necessary thermodynamic data were not available for all the phases.

The Mg-Si-Sr database is only based on the binary systems as can be deduced from figure 5 where only binary compounds are present. Therefore, the MgSrSi and $MgSrSi_2$ ternary compounds could not be predicted.

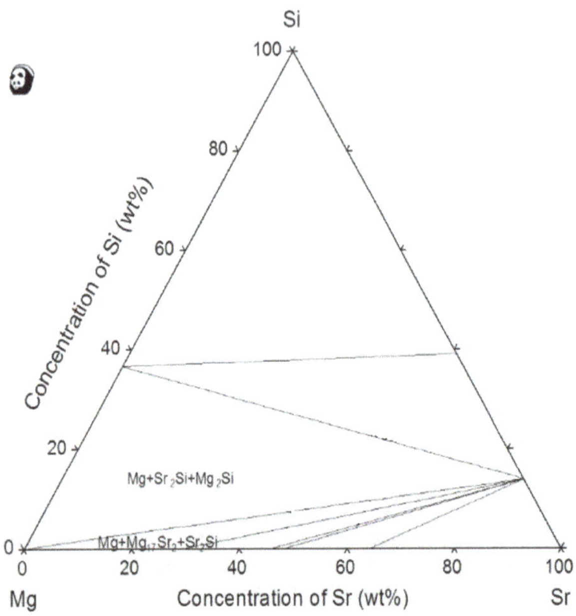

Figure 5. Isothermal sections at room temperature for the Mg-Si-Sr system calculated with PanMagnesium database in Pandat 8.1

Notwithstanding, some ternary compounds were formed during solidification and experimentally detected, as confirmed by the microstructures shown in figure 6. Even if they were not predicted by the isothermal section they could have been easily expected due to the similarities in crystalline structure type, valence and co-ordination number between Ca and Sr [17-19].

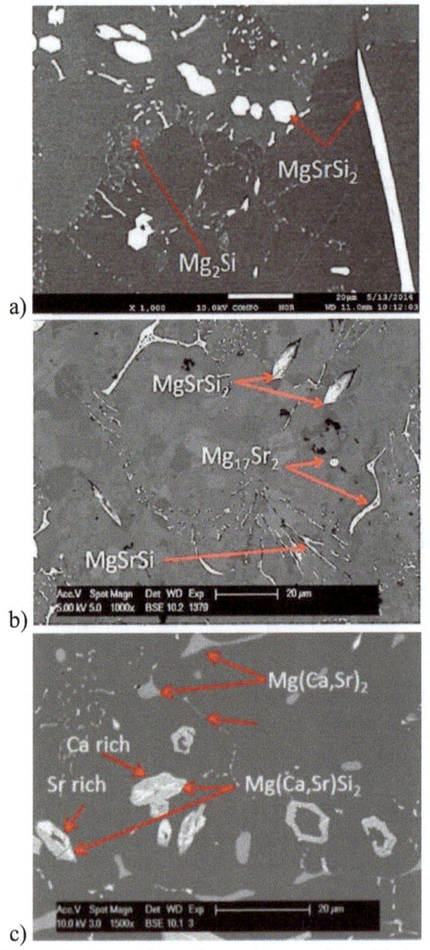

Figure 6. Identification of the phases in samples with Sr content a) Mg0.47Si0.26Sr b) Mg0.38Si0.51Sr c) Mg0.88Ca0.26Si0.51Sr

The phase content could be thus explained by the fact that Ca and Sr form intermetallic phases with the same crystalline structure. Both MgCaSi and MgSrSi phases belong to the same crystallographic space group, i.e. space group no 62, abbreviated as the Pnma (COD ID 4000953) [20]. A similar assumption has also been applied to interpret the results for the quaternary Mg-Ca-Si-Sr system, where the same phase shows distinguishable Sr and Ca rich regions (Image c from figure 6)

In all cases in the Mg-Ca-Si system, the phase content was predicted assuming the Scheil condition and subsequently measured with both X-ray diffraction and using an image analysis based method. The Rietveld method was applied on the XRD diffraction patterns to calculate the phase content taking into account the JCPDS cards numbers 00-035-0821 (Mg), 00-035-0773 (Mg_2Si), 01-073-5122 (Mg_2Ca) and 01-089-1917 (MgCaSi). Image-J was the program used to estimate the phase content from SEM pictures.

The predictions and the two experimental methods were compared and the trends are similar (figure 7). The expected values after simulations were always below the experimentally determined; for the calculations a homogeneous distribution of the precipitates is assumed while in reality this is not true. The content of phases when the Rietveld method was applied was also lower than that calculated from the micrographs. Due to the small quantity of alloying elements, the intensity from the peaks generated by the secondary phases was slightly weak, possibly causing an underestimation of the real phase content.

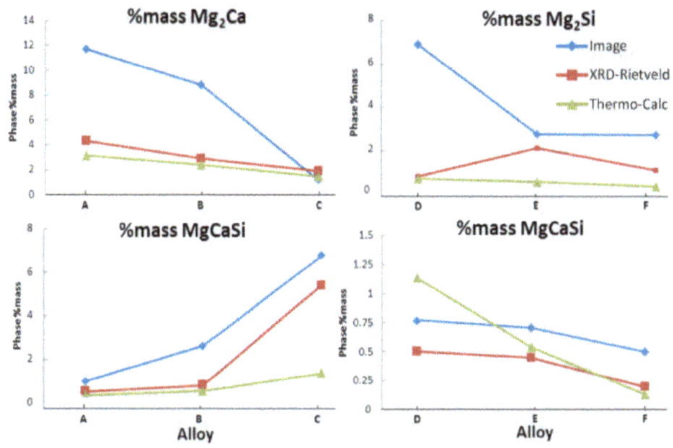

Figure 7. Comparisson betwen the different methods used in this work to estimate the presence of secondary phases (Mg_2Ca, MgCaSi and Mg_2Si) in the alloys.

<u>Mechanical properties:</u>

Table I summarizes the hardness values obtained in Mg-Ca-Si alloys. It can be observed that the alloys in the Mg_2Ca+CaMgSi+Mg region present higher hardness values as compared to those in the Mg_2Si+CaMgSi+Mg region of the Mg-Ca-Si ternary phase diagram (figure 1b). Considering the compression test data (figure 8, table III), the presence of Mg_2Ca can be related with the higher compressive strengths as well as increased ultimate strength and hardness. The MgCaSi phase appears to have the opposite effect with larger amounts resulting in decreased compression strengths and a reduction in the ductility. The influence of Mg_2Si does not appear significant.

Table II reveals that, within the investigated composition ranges, the presence of precipitates in Mg-Si-Sr alloys increases the hardness of the alloys as compared to commercially pure Mg. The separate contribution of both types of precipitates, however, requires further investigation. Considering the compression test results (figure 9, table IV), the secondary Mg_2Si phase appears to both decrease the yield compression strength and the ductility. The highest value of the compression strength was obtained in the alloy containing the $Mg_{17}Sr_2$ phase while the quaternary alloy shows the highest compression at ultimate compressive strength.

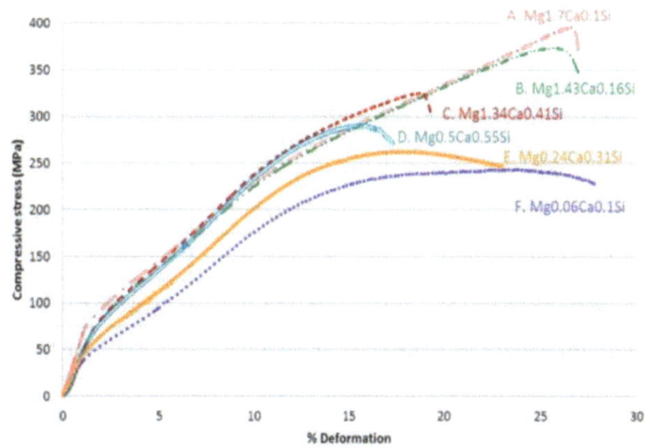

Figure 8. Compressive strength curves (true stress vs true strain) for the different compositions within the Mg-Ca-Si system

Table III. MgCaSi system. Mass percentages of the intermetallics present in the alloys related to compression test values.

ALLOY	%Mg_2Ca	%MgCaSi	Compressive Strength MPa (0.2%strain)	%Compression at Ultimate Comp.Strength
A	4.3	0.5	84.0±0.9	26.6±0.1
B	2.9	0.8	70.9±4.6	23.1±3.2
C	1.9	5.4	74.1±1.4	16.0±1.6

ALLOY	%Mg_2Si	%MgCaSi	Compressive Strength MPa (0.2%strain)	%Compression at Ultimate Comp.Strength
D	0.6	0.5	68.1±1.6	16.0±0.2
E	1.9	0.4	54.0±1.6	17.4±0.4
F	0.9	0.2	43.2±1.0	21.5±2.4

*The phase percentage was applying the Rietveld method on the XRD diffraction patterns.

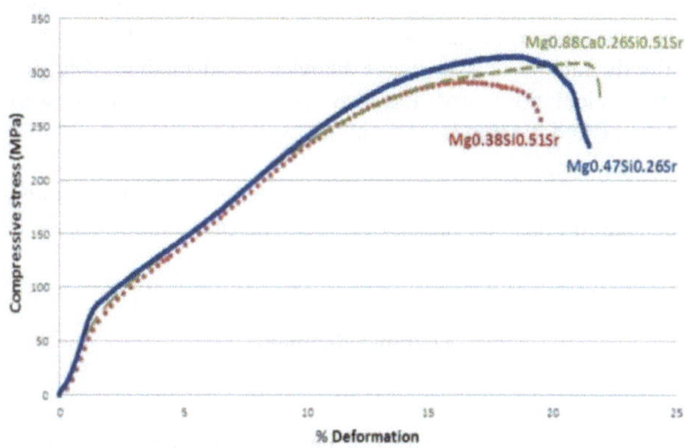

Figure 9. Compressive strength curves (true stress vs true strain) for the different compositions within the Mg-Si-Sr system and the quaternary Mg-Ca-Si-Sr alloy.

Table IV. Compression values of the alloys containing Sr.

ALLOY	Compressive Strength MPa (0.2%strain)	%Compression at Ultimate Comp. Strength
Mg0.47Si0.26Sr	72.5±1.9	16.0±0.5
Mg0.38Si0.51Sr	89.5±3.6	18.6±1.3
Mg0.88Ca0.26Si0.51Sr	74.1±2.8	21.2±1.5

Conclusions

In this study, the influence of Ca, Si and Sr alloying elements on the microstructural development and mechanical properties (Vickers hardness and compressive strengths) of as-cast Mg based alloys was investigated. The phase composition was predicted using thermodynamic databases assuming no diffusion in the solid state (Scheil condition).

The observed phase content of as-cast alloys agrees well with the thermodynamic predictions for Mg-Ca-Si while the Mg-Si-Sr system has to be revised in order to include ternary phase information that is at present not available in the literature.

The microstructure of Mg-Ca-Si alloys is dominated by coarse MgCaSi crystals when MgCaSi phase forms first from the melt. Their presence decreases the compressive strength and reduces the ductility of the alloys. Thermodynamic predictions can be used to select only the compositions where Mg-HCP forms first.

An increase in the Vickers hardness and compression strength was observed with a higher presence of secondary phases and/or alloy elements in the studied alloys. Mg_2Ca significantly increased both the Vickers hardness and the compression strength, while Mg_2Si mainly influences the strength. Ternary MgCaSi reduced the ductility of the alloys.

The predicted phase content assuming Scheil condition was compared with experimental data obtained and quantified using both X-ray diffraction and an image analysis based method. The predicted calculations and the two experimental methods were compared and the trends were similar.

Using Scheil solidification not only the phase content but also the sequence of phase formation can be predicted so that it can be used as a guideline for the alloy production.

Acknowledgements

The authors acknowledge the calculations supplied by C. L. Mendis from Magnesium Innovation Centre, Geesthacht, Germany based on the Pandat software.

The research was supported by the PEOPLE programme (Marie Sklodowska-Curie Action) of the EU FP7 Programme FP7/ 2007-2013/ under REA grant agreement n289163.

References

1. Thermo-Calc software User's guide *"Examples of solidification simulation using TCMG3"* Stockholm, Sweden 2014.
2. H.R. Bakhsheshi-Rad et al. "Mechanical and bio-corrosion properties of quaternary Mg–Ca–Mn–Zn alloys compared with binary Mg–Ca alloys" *Materials and Design* 53 (2014) 283–292.
3. Y.F. Zheng et al. "Biodegradable metals" (2014) *Materials Science and Engineering* R 77:1-34.
4. T.D. Luckey and B. Venugopal, *Metal Toxicity in Mammals* (Plenum Press, New York,1977), 167-168.
5. F. Witte et al. "Degradable biomaterials based on magnesium corrosion" *Current Opinion in Solid State and Materials Science* 12 (2008) 63–72.
6. S.E. Harandi "Effect of forging process on microstructure, mechanical and corrosion properties of biodegradable Mg–1Ca alloy" *Materials and Design* 32 (2011) 2596–2603.
7. S. Sripanyakorn et al. "The silicon content of beer and its bioavailability in healthy volunteers" *British Journal of Nutrition* (2004), 91, 403–409.
8. E.M. Carlisle "Silicon as an essential trace element in animal nutrition" *Silicon Biochemistry Ciba Foundation Symposium,* no. 121(1986), pp. 123–139.
9. E.M. Carlisle "Silicon: a possible factor in bone calcification" *Science.* 1970 Jan 16;167(3916):279-80.
10. N. Patel et al. "A comparative study on the in vivo behavior of hydroxyapatite and silicon substituted hydroxyapatite granules" *J. Mater. Sci. Mater Med.* 2002 Dec;13(12):1199-206.
11. M. Navarro et al. "Review: Biomaterials in orthopaedics" *Journal of the Royal Society Interface* (2008) 5, 1137–1158.
12. S.W. Hwang "A Physically Transient Form of Silicon Electronics" *Science* 28 September 2012, Vol. 337 no. 6102 pp. 1640-1644.
13. G.V. Raynor, *The physical metallurgy of magnesium and its alloy* (Pergamon Press LTD, London 1959).
14. H. E. Friedrich and B. L.Mordike, *Magnesium Technology.* (Springer, Berlin 2006) 122-124.
15. Y .Li et al. *"Mg–Zr–Sr alloys as biodegradable implant materials" Acta Biomaterialia* 8 (2012) 3177–3188.
16. M. Stevenson et al. "The clinical effectiveness and cost-effectiveness of strontium ranelate for the prevention of osteoporotic fragility fractures in postmenopausal women". *Health Technology Assessment* 2007; Vol. 11: No 4.
17. J.W. Bennett et al. *"Orthorhombic ABC semiconductors as antiferroelectrics" Phys. Rev.* Lett. 110, 017603 (2013).
18. J.B. Whalen "Hydrogen Storage and Electronic Characterization of Magnesium-based Intermetallics: Exploratory Flux Synthesis for Advanced Materials" (Ph.D. thesis, Florida State University, 2009) 89-100.
19. http://www.mrl.ucsb.edu/~seshadri/Periodic/
20. Crystallography Open Database http://www.crystallography.net/

Magnesium Technology 2015
Edited by: Michele V. Manuel, Alok Singh, Martyn Alderman, and Neale R. Neelameggham
TMS (The Minerals, Metals & Materials Society), 2015

Investigation of Al Coated Mg for Biomedical Applications

Nabila Elmrabet, Martin Roe, Nigel Neate, David M Grant and Paul D Brown
Division of Materials, Mechanics & Structures, Department of Mechanical, Materials & Manufacturing Engineering, Faculty of Engineering, University of Nottingham, University Park, Nottingham, NG7 2RD, UK

Keywords: Mg biodegradable stents, Magnetron sputtering, Aluminum coatings, Corrosion

Abstract

The corrosion resistant properties of 1-2 μm thick Al coatings deposited by radio frequency magnetron sputtering on polished Mg surfaces, within Ar and Ar/H_2 environments, have been appraised. The coatings were heat-treated at 300°C for 5 h to induce the formation of bioinert Al_2O_3, and samples were corroded within phosphate buffered saline solution at 37°C to mimic the biological environment. Both the as-deposited and heat-treated coatings were found to delay the onset of corrosion, but showed higher initial corrosion rates, once established, as compared with polished Mg surfaces. Slightly improved performance of the coatings was achieved through the addition of H_2 to the system which acted to inhibit Al-Mg alloying and MgO formation. However, localized accelerated corrosion associated with substrate polishing damage emphasized the need for improved process control and coating uniformity.

Introduction

Mg and related alloys are promising candidates for degradable biomaterials combining temporary mechanical properties with biological acceptance within the human body. The target corrosion rate of 0.02 mm/yr for Mg biodegradable stents, set clinically, has yet to be achieved. The corrosion current density i_{cor} and corrosion rate of mechanically polished Mg surfaces exhibit typical values of 0.04 mA/cm^2 and 0.91 mm/yr, respectively. Accordingly, there is interest to develop alumina (Al_2O_3) thin film ceramic diffusion barrier coatings for corrosion and wear resistance [1, 2]. The development of bioinert Al_2O_3 coatings to function within different environments requires appropriate process control [3]. For example, amorphous alumina is most commonly formed [4], whilst the oxidation of Al at temperatures above 400°C can lead to the formation of crystalline structures such as the α-Al_2O_3 series (α, κ, χ-Al_2O_3) or γ-Al_2O_3 series (δ, η, γ-Al_2O_3) [5, 6]. The technique of direct current (DC) reactive sputtering of Al in an oxygen/argon atmosphere can produce stoichiometric Al_2O_3 [7]. However, oxygen is found to interact rapidly with Al forming electrically insulating Al_2O_3 on the target material which tends to charge and discharge, resulting in unwanted arcing during the reactive sputtering process [7]. Hence, radio frequency magnetron sputtering (rf-MS) is becoming a preferred route for the deposition of many thin film ceramics.

In this work, we report on the phase stability and corrosion resistance of pure 1-2 μm thick Al films deposited on Mg substrates using rf-MS, in argon and argon/hydrogen flow conditions, following which the samples were furnace annealed in air at 300°C. Here AlH simply denotes an Al coating deposited in the presence of H_2. Coating structures and near surface chemistries were investigated using XRD, SEM/EDX and XPS. The corrosion and passivation behaviors of uncoated and coated Mg surfaces were examined using potentiodynamic polarization tests in phosphate buffer saline (PBS) solution at 37°C.

Experimental Method

Mg sample discs (20 mm diameter) were prepared from 3 mm thick sheets using electric discharge machining (EDM). The discs were ground sequentially using a series of silicon carbide papers (p240 to p1200) and polished to a mirror finish (R_a = 0.204 μm) using diamond paste. The Mg discs were then cleaned using acetone and dried under flowing air.

Thin film deposition and heat treatment

Both Al and AlH coatings were deposited using a commercial-scale, closed-field, unbalanced magnetron physical vapor deposition (PVD) system (Teer Coatings Ltd. UK), featuring a side-wall mounted circular magnetron and a pure Al target of diameter 57 mm. Polished Mg discs were mounted on a single axis rotating sample holder. A cover glass slip, fixed using Kapton™ high temperature masking tape was used to measure the coating thickness. A base chamber pressure of ~ 10^{-6} torr was established prior to deposition. Prior to film growth, the shuttered Al target was sputtered for 10 min to remove any surface contamination. A controlled flow of Ar (or Ar/H_2) was admitted into the chamber to create the deposition plasma. The coating deposition parameters are summarized in **Table 1**. Growth times of 30 min were used in each case.

Table 1 Deposition parameters for Al and AlH on Mg

Sample	Power/ W	Bias/ V	Base pressure / torr	Gas	Gas flow/ sccm	Chamber pressure / torr	Coating thickness
Al/ Mg	150	40	1.3×10^{-6}	Ar	47.6	1.4×10^{-3}	2.2 μm
AlH/Mg	629	-	1.7×10^{-6}	Ar/ H_2	35.6/ 31.5	1.6×10^{-3}	1.2 μm

Post deposition, thermal oxidation of the coatings was performed using a tube furnace (Lenton Thermal Design, UK) at 300°C for 5 h in air.

Coating corrosion testing and characterization

A potentiostat (ACM Gill-8, UK) was used to appraise the corrosion behavior of the as-prepared and heat-treated Al and AlH coated Mg samples. Corrosion tests were performed using a conventional three-electrode corrosion cell, with the tested metal surface being the working electrode; a 5×10×10 mm Pt sheet the counter electrode; and saturated calomel (Hg_2Cl_2) the reference electrode (Thermo-scientific, UK). 1 L PBS (pH 7.2 ± 0.06) was used as the test solution. The PBS used in this case comprised KCl, NaCl, KH_2PO_4 and Na_2HPO_4 and was free of Ca and Mg for stability. A controlled-temperature water bath was used to establish the test solution at body temperature, 37±1°C. Prior to sample immersion, the O_2 level in the PBS was reduced by bubbling N_2 at a rate of 100 sccm for 30 minutes. A one hour

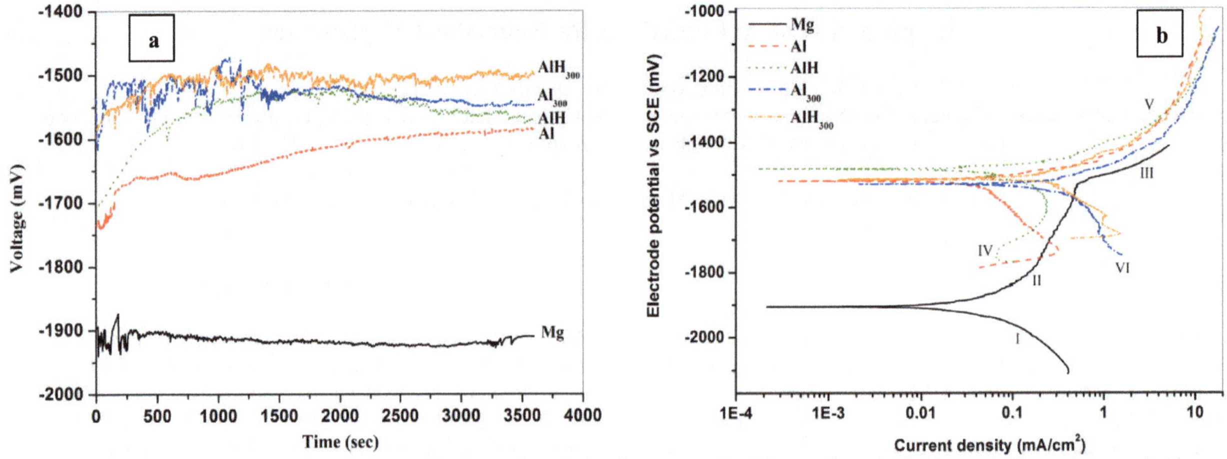

Figure 1 (a) Open circuit potential and (b) potentiodynamic polarization curves for Mg; as-deposited Al and AlH coatings; and Al_{300} and AlH_{300} coatings furnace annealed in air at 300°C for 5 h, immersed in PBS at 37°C.

settlement was used to establish the open circuit potential (OCP, E_{oc}). A potential scan rate of 5 mV min^{-1} was applied, from -200 mV to +500 mV with respect to E_{oc}.

Scanning electron microscopy (Phillips XL-30, 15 kV, spot size 4, working distance 10 mm), equipped with energy dispersive X-ray (EDX) spectroscopy, was used to image sample morphology and appraise the near surface chemistry. X-ray diffractometry (Bruker AXS D8 Advance, Cu K_α (λ= 1.54060 Å), 40 kV, 35 mA) in glancing angle (GA) mode (1°) was used to appraise the microstructure of the as-deposited and heat-treated coatings. Slow 2θ scans were performed from 5 to 80°, step size 0.04°, 32 s / step. Photoelectron spectroscopy (XPS) (VG Scientific ESCALAB MARK II, non-monochromatic Al K_α X-ray source, anode potential 12 kV, chamber vacuum < 7.5×10^{-9} torr) was used to appraise coating surface composition. Survey spectra were acquired for a binding energy range of 0-1100 eV, using pass energy of 50 eV, dwell time 0.2 s and step size 1 eV. High resolution scans for each significant element were acquired using a pass energy of 20 eV, dwell time 0.4 s, and step size 0.4 eV.

Spectra were processed using the CasaXPS program. Charge calibration was performed with respect to the C 1s peak, with peaks referred to a binding energy BE of 285 eV. Peak positions were used to identify chemical bonding states, whilst standard background subtraction and peak fitting procedures were used to quantify coating near surface phase compositions.

Results

Corrosion performance:

Figures 1a,b present typical open circuit potential (OCP) and potentiodynamic polarization curves for polished Mg; as-deposited Al and AlH coatings on Mg; and furnace annealed Al_{300} and AlH_{300} coatings on Mg, immersed in aerated PBS at 37°C. Technical values extracted from these curves are summarized in **Table 2.** The OCP data showed that polished Mg substrates exhibited the lowest rest potential, becoming stable after 20 min, whilst all the coatings as-deposited and after annealing showed more noble behavior, demonstrating increased resistance to the onset of corrosion, with the as-deposited AlH having the most resistance at -1483 mV.

The polarization curves demonstrated significant variation in the behavior of the coatings during corrosion. The polished Mg surface showed normal cathodic hydrogen evolution with the development of a typical Tafel region (I). Fast passivation acted to slow down the generation of Mg^+ ions during the onset of corrosion (II), whilst rapid anodic current generation (III) correlated with the breakdown of the $Mg(OH)_2$ passivation film.

Table 2 Corrosion parameters for as-deposited and heat-treated Al and AlH on Mg

Sample	OCP / mV	E_{cor} / mV	i_{cor} mA/cm^2	CR mm/yr
Mg	-1910	-1907	0.04	0.91
Al	-1584	-1519	0.07	1.60
AlH	-1570	-1483	0.24	5.47
Al_{300}	-1547	-1530	0.29	6.62
AlH_{300}	-1505	-1515	0.25	5.71

The coated Mg surfaces all showed positively offset corrosion potentials as compared with the polished Mg substrates. The as-deposited Al and AlH coated samples both showed anomalous Tafel regions consistent with a mixture of initial corrosion and passivation processes (IV), followed by rapid anodic current generation and extensive corrosion (V). The heat treated AlH_{300} coating similarly showed an anomalous Tafel region and extensive corrosion response; whilst the heat-treated Al_{300} coating showed more uniform behaviour with a conventional Tafel region (VI) prior to rapid corrosion.

Values for corrosion current density (i_{cor}) were obtained from the polarization curves by extrapolating the Tafel regions, whilst corrosion rates (CR) were determined using the standard equation:

$$\mathrm{CR} = 3.27 \times 10^{-3} \frac{i_{cor}}{\rho} EW \qquad (1)$$

with CR in mm/yr, i_{cor} in μA/cm^2, density ρ in g/cm^3 (ρ_{Mg} = 1.74 g/cm^3) and equivalent weight EW_{Mg} = 12.15 [8].

Despite having the lowest E_{cor} value, polished Mg surfaces passivated quickly with an i_{cor} of 0.04±0.01 mA/cm^2 before the onset of corrosion at 0.91 mm/yr (**Table 2**). Intriguingly, the i_{cor}

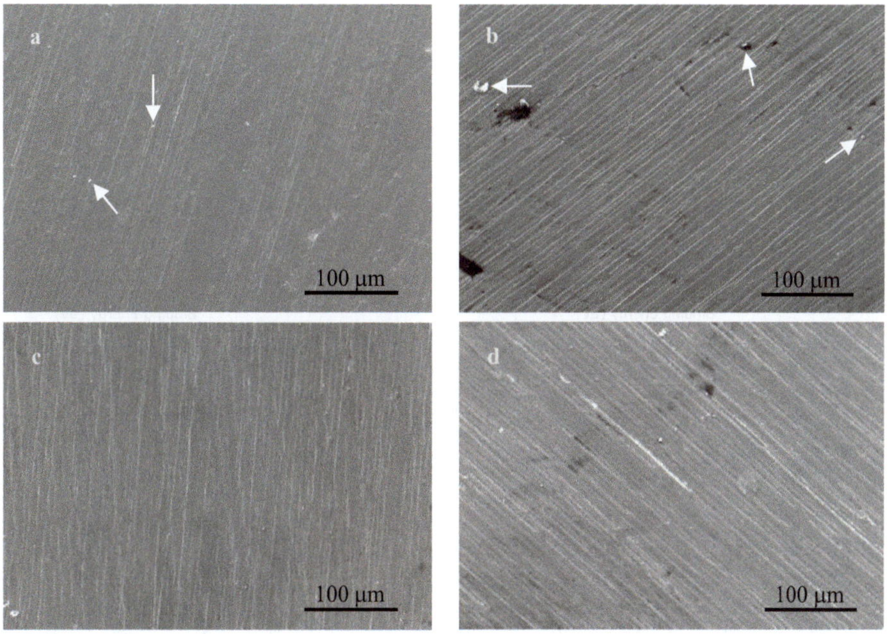

Figure 2 SE images of (a,b) as-deposited Al and AlH coatings, and (c,d) heat-treated Al_{300} and AlH_{300} coatings

values for the as-deposited Al and AlH coatings and heat-treated Al_{300} and AlH_{300} coatings showed improved resistance to the onset of corrosion, but much more rapid initial corrosion rates once established, as compared with polished Mg. The as-deposited Al showed the slowest corrosion rate of 1.60 mm/yr, as compared with the heat-treated Al_{300} coating and the as-deposited and heat-treated AlH_{300} coatings (**Table 2**).

In order to make sense of these findings and to advise on strategies for the design of corrosion resistant coatings in the biological environment, further investigation of the processed coatings was performed using a range of surface characterization techniques. **Figure 2**a,b present secondary electron (SE) images of sputtered Al coating morphologies, as-deposited within Ar and Ar/H_2 environments, respectively. Striations remnant from the Mg polishing process were evident, along with a few embedded SiC particles (**Figure 2**a,b arrowed) as determined by EDX analysis. **Figure 2**c,d present SE images for the Al_{300} and AlH_{300} coatings following heat treatment at 300°C in air. Again the morphologies of the heat-treated samples reflected the striated polishing damage of the Mg substrates.

Figure 3 presents GA-XRD patterns of the as-deposited and heat-treated coatings. In all cases, the patterns indicated the presence of face centered cubic (fcc) Al grains on polycrystalline Mg substrates. The average grain sizes of the as-deposited Al and AlH coatings were 64±1 nm and 94±13 nm, respectively, as determined using the Scherrer equation. It is considered that the intensity variations in these patterns reflects the random orientation of grains within the coating and substrate and does not allow readily for distinction between equiaxed and preferred orientation growth in this instance. The grain sizes appeared to increase slightly with heat treatment for the Al_{300} and AlH_{300} coatings, to 89±1 nm and 136±10 nm respectively. In all these samples, there was no direct evidence for partial hydrogenation of Al or the Mg substrate, but a number of small peaks from the heat treated coatings were consistent with the development of some polycrystalline Al_2O_3, most clearly in the case of the Al_{300} coating.

Figure 4a-h present detailed Al_{2p} and O_{1s} XPS data for the as-deposited and heat-treated coatings, being complementary to the GA-XRD observations. The spectra demonstrated the presence of Al_2O_3 with varying levels of MgO at the near surface for all the coatings. Peaks denoted as Al-Mg were attributed to metallic Al (72.7eV) slightly shifted due to the presence of small quantities of Mg in solid solution. In particular, for the case of the as-deposited coatings, the indications are that the presence of H_2, added to restrict the development of Al-Mg, was associated with an increase in the proportion of Al_2O_3 to MgO (Figures 4a-d). Similarly, the spectra for the heat-treated samples showed enhanced contributions from Al_2O_3 at the sample surfaces, with complete inhibition of MgO for the case of the AlH_{300} coating (**Figure 4**e-h).

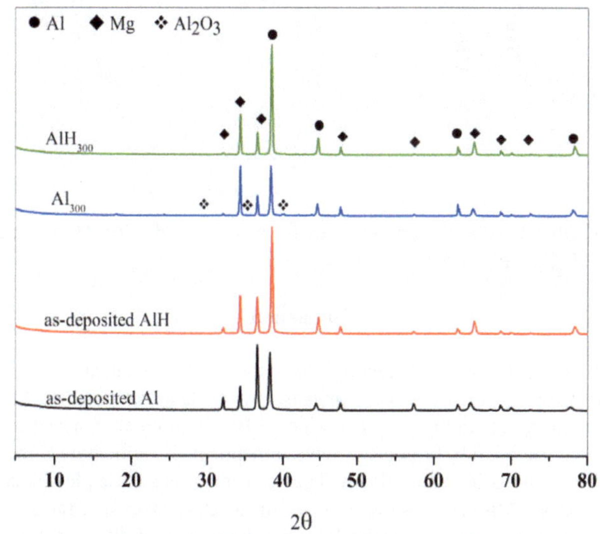

Figure 3 XRD patterns of the as-deposited and heat-treated Al and AlH coatings on Mg.

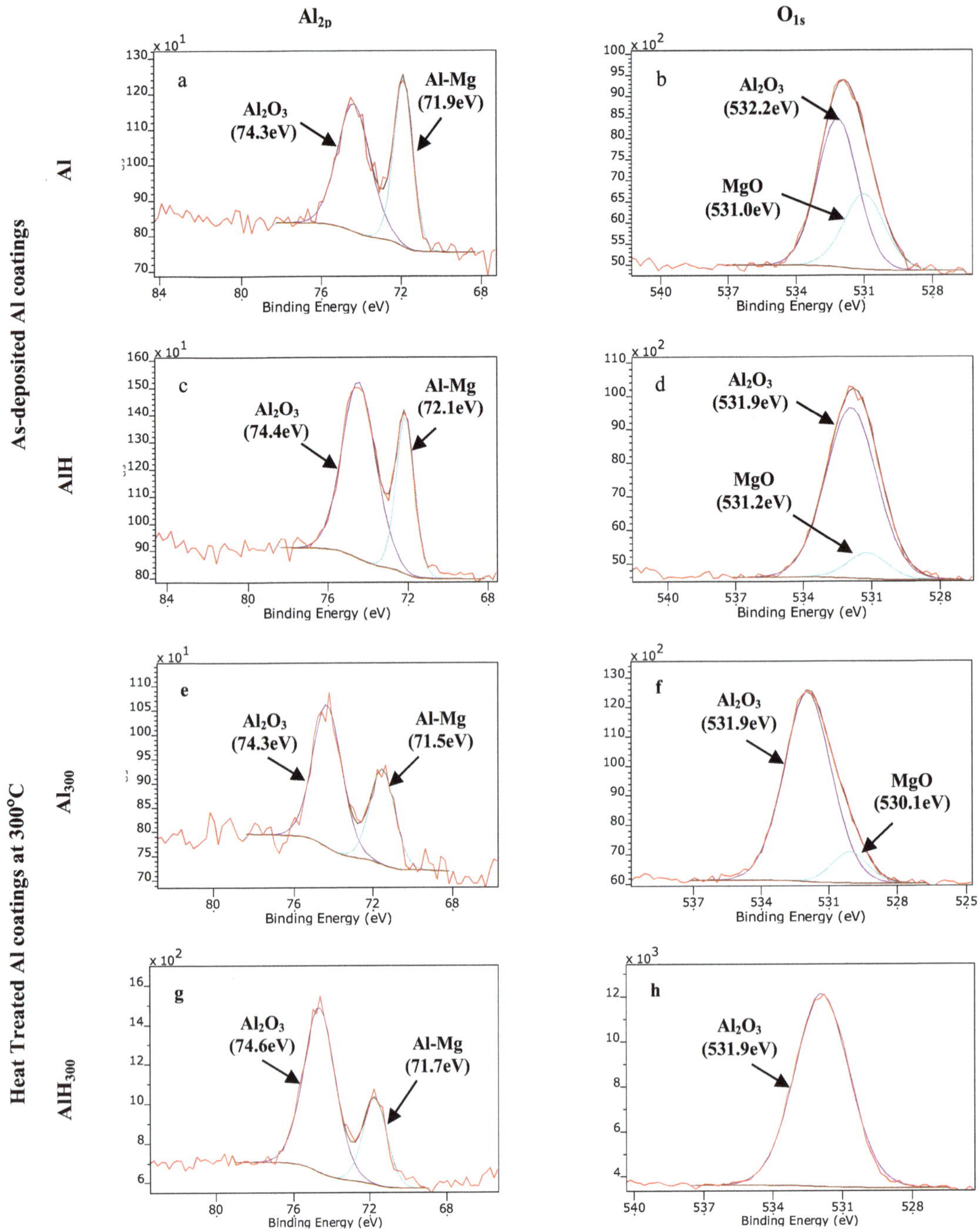

Figure 4 XPS Al_{2p} and O_{1s} data for as-deposited (a,b) Al and (c,d) AlH coatings; and (e,f) Al_{300} and (g,h) AlH_{300} coatings heat-treated at 300°C in air for 5 h.

Discussion

It is recognized that material behaviors vary dramatically when exposed to corrosive environments. It is evident that polished Mg surfaces are highly reactive within PBS, limiting their practical use within the biological environment. It is anticipated that HPO_4^{2-} ions from the PBS will act to form an insoluble phosphate layer on Mg surfaces, along with a protective layer of $Mg(OH)_2$ leading to surface stabilization and inhibition of the corrosion process [9]. However, the presence of Cl^- ions in solution would lead to dissolution of this component of the passivation layer, expressed by:

$$Mg(OH)_2 + 2\,Cl^- \rightleftharpoons MgCl_2 \qquad (2)$$

Further, $Mg(OH)_2$ layers only become thermodynamically stable when the pH increases > 11 (strongly alkaline environment) [10]. This regime could not be attained within PBS which reached a level of pH = 7.39 at the end of the experiment, emphasizing the propensity for breakdown of the passivation layer and hence rapid corrosion of Mg within the biological environment. It is considered also that areas of localized passivation would lead to variations in anodic and cathodic potentials, with variable localized accelerated corrosion rates across these polished surfaces [11].

It is evident that Al-based coatings on Mg are beneficial in terms of delaying the onset of corrosion, but exhibit anomalous behavior with passivation in advance of rapid corrosion. Indeed, the early stages of corrosion of the coated samples, *e.g.* established through pin-holes associated with coating polishing damage, were found to be more rapid than for native Mg surfaces, promoted in part by galvanic action of metal Al/Mg bi-layers, and possibly further promoted by residual strain within these processed coatings. Further, it is considered that embedded artifacts such as SiC polishing particles act as cathodic sites in relation to Mg producing localized galvanic corrosion cells

This emphasizes the need for improved process control to ensure uniformity of thickness and composition of the coatings and associated passivation layers. The challenge is to develop bioinert Al_2O_3 coatings through control of Al deposition and heat treatment, whilst H_2 is added to the sputter system working gas to hinder Al-Mg alloying, by initially forming a magnesium hydride barrier layer on the top surface of the substrate to prevent energetic sputtered Al diffusing into the substrate layer. This layer segregation needs to be demonstrated through future work but it appears from XPS that it enables a more homogenous Al_2O_3 layer to be formed during the annealing in air. For the heat treatment conditions applied here, it is evident that just the near surface regions of the Al_{300} and AlH_{300} coatings were converted to Al_2O_3, indicating that higher processing temperatures or longer annealing times may be required to fully convert all of the Al coatings to bioinert Al_2O_3. It is interesting to note that the thinner 1.2 μm AlH_{300} coatings exhibited slightly better performance than the thicker 2.2 μm Al_{300} in terms of delayed onset of corrosion, suggesting that the addition of H_2 into the system was indeed beneficial through the inhibition of Al-Mg alloying.

Summary

The corrosion response of Mg within PBS has been investigated in order to appraise its suitability for use as a biodegradable stent. The corrosion rate of Mg was found to be too high for clinical applications. Accordingly, rf-MS deposited Al coatings in Ar and Ar/H_2 environments, heat-treated at 300°C for 5 h to induce the formation of bioinert Al_2O_3, have been investigated. Both as-deposited and heat-treated coatings were found to delay the onset of corrosion, but showed higher initial corrosion rates, once established, as compared with polished Mg surfaces. The addition of H_2 to the system was beneficial in terms of inhibiting Al-Mg alloying and MgO formation. However, improved process control is required in view of the effect of localized accelerated corrosion associated with substrate polishing damage.

Acknowledgements

The rf-MS processing was performed in Eminate Ltd., a wholly owned subsidiary company of The University of Nottingham. The authors are very grateful to Dr. Katy Voisey and Dr. Deen Zhang for their contributory discussion and assistance with corrosion tests.

References

1. R. Cremer, *et al.*, "Sputter deposition of crystalline alumina coatings," *Surface & Coatings Technology*, 163 (2003), 157-163.
2. J. Creus, *et al.*, "Reactivity classification in saline solution of magnetron sputtered or EBPVD pure metallic, nitride and Al-based alloy coatings," *Corrosion Science*, 57 (0) (2012), 162-173.
3. K. Koski, J. Holsa and P. Juliet, "Properties of aluminium oxide thin films deposited by reactive magnetron sputtering," *Thin Solid Films*, 339 (1-2) (1999), 240-248.
4. T.-S. Shih and Z.-B. Liu, "Thermally-formed oxide on aluminum and magnesium," *Materials Transactions*, 47 (5) (2006), 1347-1353.
5. S. Zuzjakova, P. Zeman and S. Kos, "Non-isothermal kinetics of phase transformations in magnetron sputtered alumina films with metastable structure," *Thermochimica Acta*, 572 (2013), 85-93.
6. W. Engelhart, *et al.*, "Deposition of alumina thin film by dual magnetron sputtering: Is it gamma-Al2O3?," *Acta Materialia*, 59 (20) (2011), 7757-7767.
7. M.K. Olsson, K. Macak and W. Graf, "Reactive d.c. magnetron sputter deposited Al_2O_3 films: large-area coatings for industrial applications," *Surface & Coatings Technology*, 122 (2-3) (1999), 202-207.
8. A.S.f.T.a.M.-. ASTM, "Standard Practice for Calculation of Corrosion Rates and Related Information from Electrochemical Measurements," *ASTM International*, G102 – 89 (2010), 1-7.
9. Y. Xin, T. Hu and P.K. Chu, "*In vitro* studies of biomedical magnesium alloys in a simulated physiological environment: A review," *Acta Biomaterialia*, 7 (4) (2011), 1452-1459.
10. G.L. Makar and J. Kruger, "Corrosion of magnesium," *International materials reviews*, 38 (3) (1993), 138-153.
11. F. Witte, *et al.*, "*In vitro* and *in vivo* corrosion measurements of magnesium alloys," *Biomaterials*, 27 (7) (2006), 1013-1018.

Magnesium Technology 2015

Functional and Emerging Alloys

Magnesium Technology 2015
Edited by: Michele V. Manuel, Alok Singh, Martyn Alderman, and Neale R. Neelameggham
TMS (The Minerals, Metals & Materials Society), 2015

MICROSTRUCTURE AND MECHANICAL PROPERTIES OF A MAGNESIUM-ALUMINIUM-ERBIUM ALLOY

S. Sankaranarayanan [1], B. M. Ng [1], S. Jayalakshmi [1, 2], M. Ganesh Kumar [1], Q.B. Nguyen [1] and M. Gupta [1]
[1] Department of Mechanical Engineering, National University of Singapore, 9 Engineering Drive 1, Singapore 117576
[2] Department of Mechanical Engineering, Bannari Amman Institute of Technology, Satyamangalam, India 638401

Keywords: Mg-Al-RE alloy, Microstructure, Mechanical Properties

Abstract

Magnesium alloys with their weight saving advantage exhibit unique application potential in the automotive and aerospace industries. Recent years have seen significant progress in the development of rare earth containing Mg alloys, as rare earth addition is considered a promising route to enhance the mechanical characteristics of Mg. In the present study, a new Mg alloy containing 5.16 at. % aluminum and trace (0.05 at. %) erbium was synthesized using the disintegrated melt deposition technique followed by hot extrusion. The microstructural and mechanical properties of the developed Mg-Al-Er alloy were evaluated in comparison to that of pure Mg. Microstructural investigation revealed significant grain refinement and the presence of $Mg_{17}Al_{12}$ intermetallic phases. Evaluation of mechanical properties under indentation loads showed significant improvement in microhardness by +50%. Under tensile loads, the developed Mg-Al-Er alloy exhibited +86%, +115% and +95% enhancement in yield strength, ultimate strength and ductility respectively. Similarly, an enhancement in yield strength by +115%, ultimate strength by +37% and ductility by +25% were observed under compressive loads. The overall effects of Al and Er addition on the mechanical properties of the Mg are discussed using structure-property relationship.

Introduction

The lightest structural metal, magnesium (Mg) has a density ~1.74 g/cm^3, which is only slightly higher than that of the plastics and only 2/3 that of aluminium (Al), and 1/4 that of iron [1]. Besides lightweight, Mg based materials also exhibit good specific mechanical properties, excellent damping capacity, machinability, castability and weldability. However, the inherent limitations of Mg such as low absolute strength, modulus, and ductility limit its extensive utilization. In order to circumvent these limitations, Mg is usually alloyed with other elements such as Al, Zn, Zr, etc. [2]. Despite considerable efforts, the adoption of magnesium alloys in engineering applications remains limited [1]. One important reason for the under-utilization of Mg is that there are limited Mg-alloys for the designers to select for specific applications, and within these limited choices, the most cost-effective magnesium alloys have inadequate properties such as yield strength, creep-resistance, formability, and corrosion resistance.

Recently, alloying of magnesium with rare earth (RE) elements has gained momentum in the context of improved mechanical characteristics due to RE additions [3]. In these Mg-RE alloys, the precipitation of a dense and fine of RE-containing dispersed phases within the grains, and along the grain boundaries generally strengthens the base metal [4, 5]. Further, the presence of RE phases within the α-Mg matrix also favors the formation of protective films to improve the thermal stability and corrosion resistance of the alloy [4]. In spite, RE additions are widely regarded as an expensive choice and the most commonly used RE containing Mg-alloys are limited to AE42 and AE44 which contains about 4 wt. % Al and 2 to 4 wt. % RE respectively [6, 7]. With this background, an attempt is made in this research work, to synthesize and characterize new Mg-Al alloy containing trace amount of RE element, erbium (Er). For this reason, an aluminium-erbium (AlEr) metal alloy powder was incorporated into pure Mg through disintegrated melt deposition (DMD) technique. DMD is a unique processing method that inherits the combined advantages of stir casting and spray processing techniques [8]. Unlike conventional spray processes, DMD method employs higher super heat temperatures and lower impinging velocity which helps in achieving a bulk material with no-over sprayed powders. Further, the DMD process also offers both the features of (i) fine grain structure of spray process and (ii) simplicity/cost effectiveness of conventional stir cast foundry process.

Experimental

<u>Synthesis of Mg-Al-Er alloy</u>

Mg-Al-Er alloy containing 5.16 at% Al and 0.05 at% Er, required for the current study was synthesized using disintegrated melt deposition (DMD) technique. It involves heating of Mg turnings (> 99.9% purity, supplied by ACROS Organics, New Jersey, USA) together with the aluminium-erbium master alloy powder (94.67% Al and 5.33% Er, supplied by Phoenix Scientific Industries, UK) in a graphite crucible to 750 °C in an electrical resistance furnace under inert argon gas protective atmosphere. The superheated melt is then stirred (at 465 rpm for 5 min) and bottom-poured into the steel mold (after disintegration by two jets of argon gas oriented normal to the melt stream). Following deposition, an ingot of 40 mm diameter was obtained. The obtained ingot was then soaked at 400 °C for 1 h and hot extruded at 350 °C into rods of 8 mm diameter. The synthesis of pure Mg (required for the comparison purpose) also involved similar steps except that no Al-Er master alloy powder were added during the melting and casting.

<u>Characterization of Mg-Al-Er alloy</u>

<u>Density and Porosity Measurements:</u> The experimental mass density of the developed Mg-5.16Al-0.05Er alloy was determined using Archimedes' principle which involved weighing the polished samples (~10 mm length) prepared from the extruded rods in air and then in distilled water using an A&D ER-182A electronic balance with an accuracy of ± 0.0001 g. In order to calculate the theoretical density, the relation, ($\rho_{alloy} = \frac{100}{\left(\frac{C_1}{\rho_1}\right)+\left(\frac{C_2}{\rho_2}\right)+\left(\frac{C_3}{\rho_3}\right)}$; wherein 'C' corresponds to the individual elemental concentration in wt.%).

Microstructural Characterization: Microstructural studies were conducted on the polished samples in a direction perpendicular to the extrusion direction in order to determine the average matrix grain size, the presence and distribution of second phases. For this purpose, a small section (8 to 10 mm in length) was cut from the extruded rod and the ends were then ground using 600 and 1200 grit size sand paper in order to remove the large surface scratches and also to produce a flat surface. Once all visible surface scratches and cracks were removed, the sample was polished using a polishing disc with 5 micron diamond slurry, followed by 1 micron diamond slurry and lastly by 0.3 micron alumina slurry. The surface of the polished samples was then etched (to make the grain boundaries visible) with citral (4.2 g citric acid monohydrate in 100 ml of water) and observed under an Olympus optical microscope. Using selected optical micrographs, a total of 120–150 grains were selected to calculate the grain size using the Scion Image analysis software. In order to study the morphology and distribution of secondary phases, the as-polished samples were etched with acetic picral (10 ml acetic acid, 10 ml water, 70ml picral (95% ethanol + 5% 2,4,6-Tri-nitro-phenol)) and observed in a field emission scanning electron microscope (FESEM-S4300, Hitachi Ltd., Tokyo, Japan, SEI mode).

Mechanical Properties: The mechanical properties of the developed Mg-Al-Er alloy were evaluated under indentation, tension and compression loads. Microhardness measurements were carried out on the as-polished samples of developed materials using Matsuzawa MXT 50 automatic digital Microhardness tester in accordance with ASTM standard E3 84-99. Standard ASTM test methods E8/E8M-13a and E9–09 were conducted on the test samples using a fully automated servo-hydraulic mechanical testing machine, to determine the tensile and compressive properties of the developed materials. The fractured surfaces of Mg-materials after tensile and compression tests were studied using the Hitachi S-4300 FESEM.

Results and Discussion

Synthesis

Synthesis of pure Mg and the Mg-5.16Al-0.05Er alloy was successfully carried out by the DMD process followed by hot extrusion. Visual observation of the as-cast ingot and the extruded samples indicates the absence of any macro-defects which clearly designates the suitability of processing parameters used in this study. There was no reaction between the Mg melt slurries with the graphite crucible, which attributes to the inability of Mg to form stable carbides. Also the inert Ar gas atmosphere used during the melting and casting stage has prevented the oxidation of Mg melt.

The results of density measurements conducted on the pure Mg and Mg-Al-Er alloy are tabulated in Table 1. From the tabulated values, it can be seen that near-dense materials have been developed in this study. The results also show an increase in the density due to AlEr addition which can be attributed to the higher density of Al (2.7 g/cm^3) and Er (9 g/cm^3) as compared to Mg (1.738 g/cm^3). The volumetric porosity calculated from the theoretical and experimental density values indicate that the developed Mg-5.16Al-0.05Er alloy exhibit a relatively higher porosity level when compared to pure Mg. However, it is relatively low compared to those materials synthesized using similar technique [8].

Microstructure

The microstructural properties of the developed Mg-Al-Er alloy are studied in terms of: (a) grain morphology, and (b) the presence, distribution, and morphology of the second phase particles. Fig. 1 shows the microstructural characteristics of pure Mg and Mg-5.16Al-0.05Er alloy. It reveals the presence of nearly equiaxed grains in both the cases which indicates the complete recrystallization of Mg-matrix during extrusion. Using Scion Image processing software, the average grain size and aspect ratio of Mg-matrix (Fig. 1(a-b)) were calculated and the results are presented in Table 1. It shows that the addition of AlEr alloy to Mg has refined the average grain size of Mg-matrix [9]. It is known that the addition of Al to Mg would result in the formation of $Mg_{17}Al_{12}$ grain boundary eutectic network as shown in Fig. 1c. This shows that the noticeable grain refinement (~ 85%) observed due to AlEr addition could be attributed to the formation of $Mg_{17}Al_{12}$ intermetallic phase at the grain boundaries near the triple point grain edges and its efficacy in inhibiting the matrix grain growth during recrystallization. The incorporation of Er (in the form of AlEr alloy) is also expected to form Er based intermetallic phases that can influence the dynamic recrystallization effects. However, Er based intermetallic phases were not microscopically detected in the developed Mg-alloy. This could be possibly due to the very low amount of Er (0.05 at %) to effectively form any second phases with Mg or Al.

Mechanical behavior

The results of mechanical property measurements conducted on pure Mg and Mg-5.16Al-0.05Er alloy are summarized in Table 2. It reveals that the developed Mg-5.16Al-0.05Er alloy exhibit superior mechanical properties when compared to pure Mg.

The microhardness measurements shown in Table 2 indicate an increase in the mean hardness value (by 50%) when compared to pure Mg. This increase can be attributed primarily to (i) the increase in the presence of harder intermetallic phases $Mg_{17}Al_{12}$ with an increase in amount of Al, and (ii) the reduction in grain size due to AlEr addition [11].

Table 1 - Results of Density and Microstructural Measurements

Material	Density and Porosity			Microstructure	
	Theoretical Density [g/cc]	Experimental Density [g/cc]	Porosity [%]	Average grain size [μm]	Aspect ratio
Pure Mg	1.7380	1.7366 ± 0.0011	0.26	59 ±16	1.55
Mg-5.16Al-0.05Er	1.7786	1.7731 ± 0.0021	0.96	9 ± 2	1.23

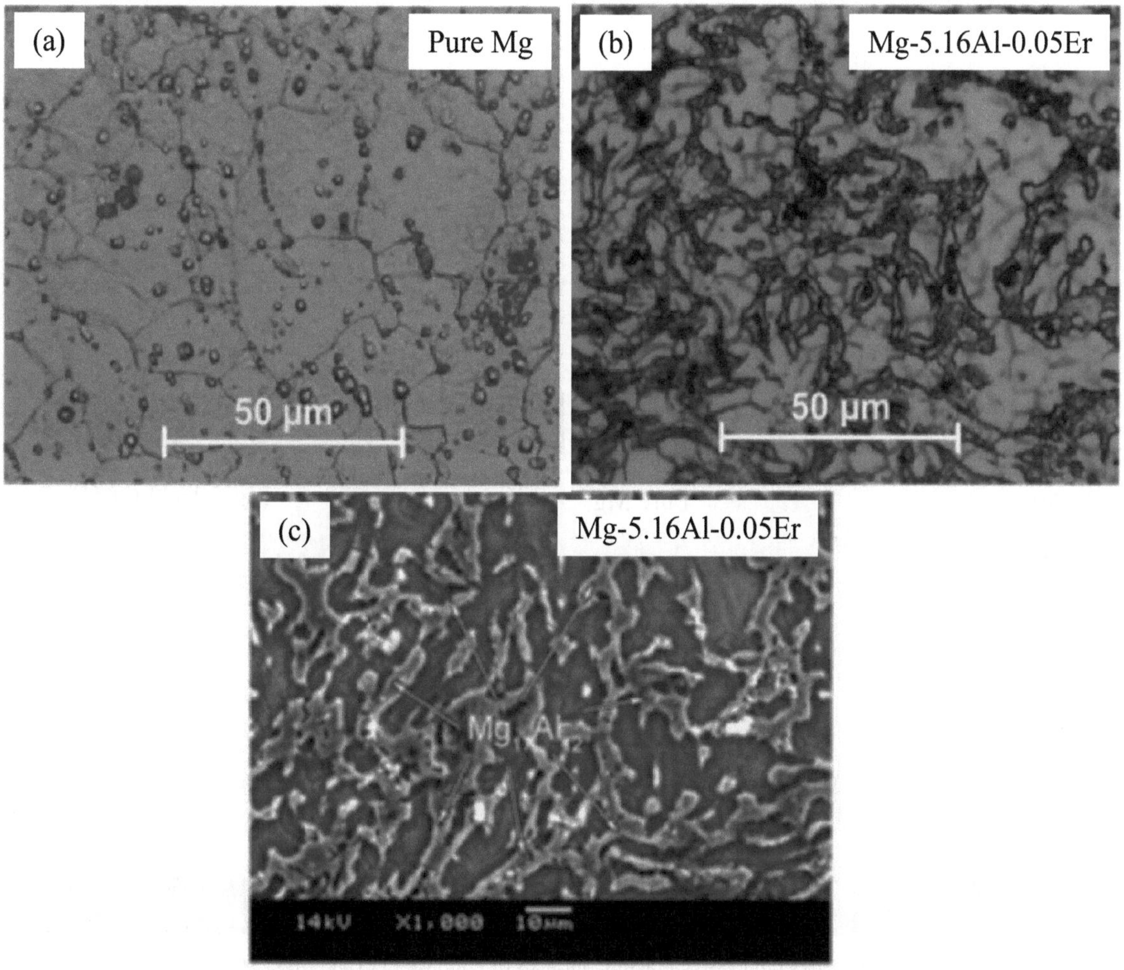

Figure 1 - Microstructures of (a) pure Mg and (b,c) Mg-5.16Al-0.05Er alloy.

The mechanical properties of pure Mg and Mg-5.16Al-0.05Er alloy as listed in Table 2 indicates that the alloying addition of AlEr to Mg has enhanced the tensile and compressive properties of pure Mg. Significant improvement in both yield strength and ultimate strength (under both tensile and compressive loading) was achieved by the addition of Al and Er alloying elements in the form of AlEr alloy powder. Further, it is interesting to note that the developed Mg-5.16Al-0.05Er alloy displayed an increase in tensile and compressive ductility by ~95% and ~25% respectively, when compared to pure Mg. Based on the microstructural analysis; it is evident that the addition of AlEr to Mg has refined the matrix grain size significantly. Hence, in the Mg–Al–Er alloy, the improvement in strength properties under tensile and compressive loads can be attributed to the following strengthening mechanisms:

(i) Strengthening due to the grain size reduction and formation of refined and equiaxed grains [12],
(ii) Strengthening from the reasonably distributed hard grain boundary second phases ($Mg_{17}Al_{12}$) that impedes the stress transfer across the grain boundaries.

The ductility increment due to AlEr addition could be attributed to the activation of non-basal slip systems as suggested by Koike et al. [13] wherein, the non-basal slip activation resulting larger deformation strain (elongation) was observed in the case of fine-grained Mg-alloys.

Table 2 - Results of Mechanical Property Measurements

Materials	Micro-hardness	Tensile Properties			Compressive Properties		
		0.2Offset Yield Strength	Ultimate Tensile Strength	Failure Strain	0.2Offset Yield Strength	Ultimate Tensile Strength	Failure Strain
	[H_v]	[MPa]	[MPa]	[%]	[MPa]	[MPa]	[%]
Pure Mg	69 ± 1	96 ± 8	133 ± 7	8.3 ± 2.9	81 ± 4	368 ± 11	20.1 ± 0.9
Mg-5.16Al-0.05Er	103 ± 7	182 ± 11	285 ± 11	17.5 ± 1.2	175 ± 11	504 ± 8	24.6 ± 0.5

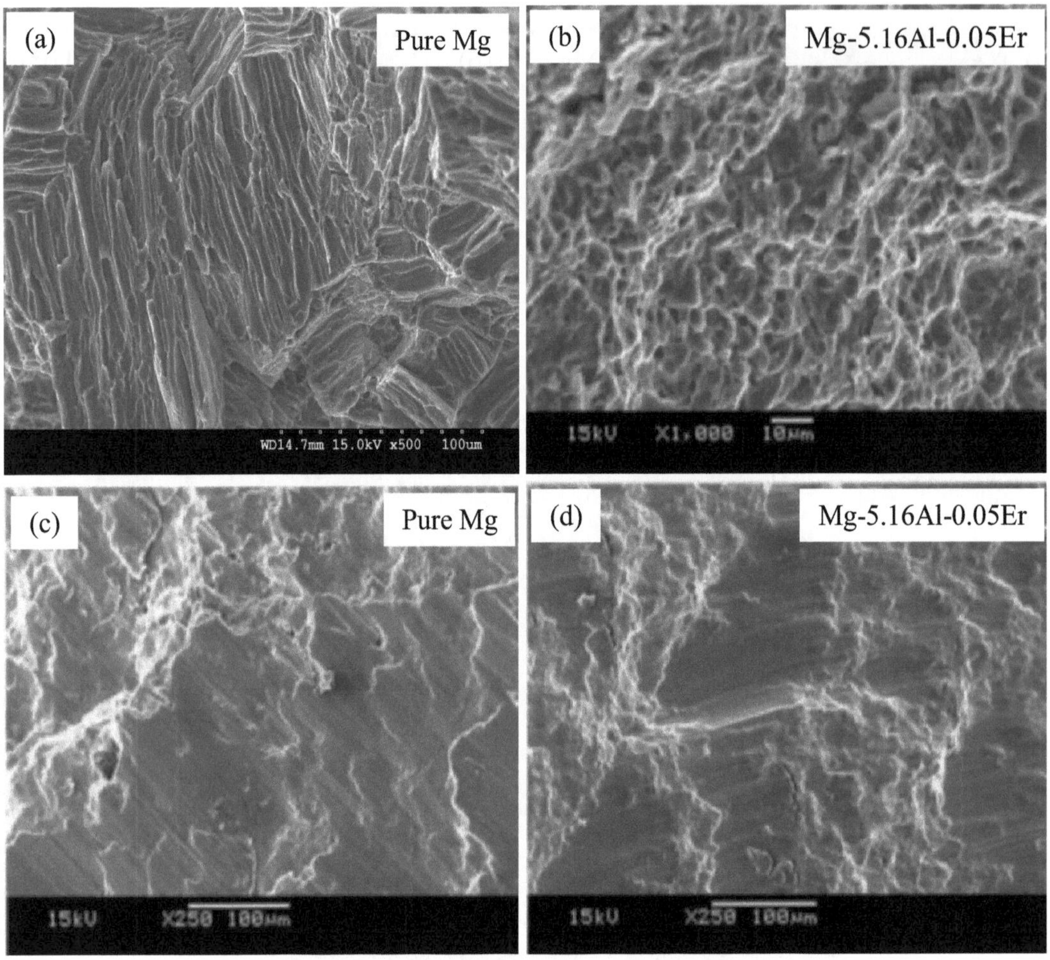

Figure 2 (a-b) Tensile and (c-d) Compressive Fracture Surfaces of Pure Mg (a,c) and Mg-5.16Al-0.05Er alloy (b,d).

The microscopic features of the tension test fractured samples are shown in Fig. 2(a-b). It reveals a typical mixed mode fracture with prominent cleavage features in the case of pure Mg and prominent ductile features in the case of Mg–5.16Al–0.05Er alloy. Typically, the cleavage features observed in magnesium are indicative of the inability to plastic deformation attributed to its HCP crystal structure. The compressive fracture surfaces of both pure Mg and the developed Mg-Al-Er alloy showed shear mode fracture as indicated by the presence of prominent shear bands (Fig. 2(c-d)). However, for Mg–5.16Al–0.05Er, the fracture behavior was observed to be relatively more ductile as indicated by Fig. 2(c-d).

Conclusions

In the present work, a new Mg-alloy containing 5.16 at.% Al and trace (0.05 at.%) Er was produced using the disintegrated melt deposition technique followed by hot extrusion. From the microstructural and mechanical property characterization, the following conclusions can be drawn:

1. The addition of Al and Er in the form of AlEr master alloy powder to Mg resulted in Mg-Al-Er alloy with refined microstructure (~85% grain refinement) containing evenly distributed $Mg_{17}Al_{12}$ intermetallic phases.
2. The developed Mg-5.16Al-0.05Er alloy displayed superior mechanical properties under indentation, tension and compression loads when compared to pure Mg.
3. Microhardness measurements showed ~50% increment in hardness value of Mg due to AlEr addition.
4. Under tensile loads, the developed Mg-Al-Er alloy exhibited +86%, +115% and +95% enhancement in yield strength, ultimate strength and ductility respectively.
5. Under compression loads, an enhancement in yield strength by +115%, ultimate strength by +37% and ductility by +25% were observed.

Acknowledgement

The financial support of Ministry of Education Academic Research Funding (WBS# R-265-000-498-112) is gratefully acknowledged.

References

1. Luo, A. A., Magnesium: current and potential automotive applications. Jom 2002, 54, (2), 42-48.

2. Avedesian, M. M.; Baker, H., ASM specialty handbook: magnesium and magnesium alloys. ASM international 1999, 274.
3. Bohlen, J.; Yi, S.; Letzig, D.; Kainer, K. U., Effect of rare earth elements on the microstructure and texture development in magnesium–manganese alloys during extrusion. Materials Science and Engineering: A 2010, 527, (26), 7092-7098.
4. Bettles, C.; Gibson, M.; Zhu, S.-M., Microstructure and mechanical behaviour of an elevated temperature Mg-rare earth based alloy. Materials Science and Engineering: A 2009, 505, (1), 6-12.
5. Zhu, S.-M.; Gibson, M.; Easton, M.; Nie, J., The relationship between microstructure and creep resistance in die-cast magnesium–rare earth alloys. Scripta Materialia 2010, 63, (7), 698-703.
6. Rokhlin, L., Structure and properties of alloys of the Mg-REM system. Metal science and heat treatment 2006, 48, (11-12), 487-490.
7. Chia, T. L.; Easton, M.; Zhu, S.-M.; Gibson, M.; Birbilis, N.; Nie, J., The effect of alloy composition on the microstructure and tensile properties of binary Mg-rare earth alloys. Intermetallics 2009, 17, (7), 481-490.
8. Gupta, M.; Sharon, N. M. L., Magnesium, magnesium alloys, and magnesium composites. Wiley. com: 2011.
9. Nguyen, Q.; Fan, Y.; Tun, K.; Chan, J.; Kwok, R.; Kuma, J.; Gupta, M., An investigation into the capability of unconventional amount of aluminum and nano-alumina to alter the mechanical response of magnesium. Journal of Materials Science 2012, 47, (1), 234-240.
10. Kiełbus, A.; Rzychoń, T.; Cibis, R., Microstructure of AM50 die casting magnesium alloy. Journal of Achievements in Materials and Manufacturing Engineering 2006, 18, (1-2).
11. Callister, W. D.; Rethwisch, D. G., Materials science and engineering: an introduction. 2007.
12. Dieter, G. E., Mechanical metallurgy. McGraw-Hill New York: 1976; Vol. 3.
13. Koike, J.; Kobayashi, T.; Mukai, T.; Watanabe, H.; Suzuki, M.; Maruyama, K.; Higashi, K., The activity of non-basal slip systems and dynamic recovery at room temperature in fine-grained AZ31B magnesium alloys. Acta Materialia 2003, 51, (7), 2055-2065.

Magnesium Technology 2015
Edited by: Michele V. Manuel, Alok Singh, Martyn Alderman, and Neale R. Neelameggham
TMS (The Minerals, Metals & Materials Society), 2015

MICROSTRUCTURE AND PROPERTIES OF COBALT- AND ZINC-CONTAINING MAGNETIC MAGNESIUM ALLOYS PROCESSED BY HIGH-PRESSURE DIE CASTING

Christian Klose[1], Christian Demminger[1], Hans Jürgen Maier[1]
[1]Institut für Werkstoffkunde (Materials Science), Leibniz Universität Hannover, An der Universität 2, D-30823 Garbsen, Germany

Keywords: magnesium, cobalt, magnetic materials, load-sensitive materials, high-pressure die casting, eddy-current measurements

Abstract

The inherent magnetic properties of lightweight alloys based on magnesium and cobalt offer a novel way in order to measure mechanical loads throughout the entire structural component using the magnetoelastic effect. Because the solubility of cobalt in the magnesium matrix is negligible, the magnetic properties mainly originate from Co-rich precipitates. Thus, the size and distribution of Co-containing phases within the alloy's microstructure wields a major influence on the amplitude of the load-sensitive properties which can be measured by employing the harmonic analysis of eddy-current signals. In this study, Mg-Co-based alloys are produced by several casting methods which allow the application of different cooling rates, e.g. gravity die casting and high-pressure die casting. The differences between the manufactured alloys' micro- and phase structures are compared depending on the applied cooling rate and the superior magnetic and mechanical properties of the high-pressure die cast material are demonstrated.

Introduction

In mechanical engineering, strain gauges are commonly used to determine mechanical loads and stresses during service of a structural component. Still, a limitation of traditional strain sensors is the locally restricted measurement range because only local strains directly at the measuring position are detected. Furthermore, strain gauges are typically applied to the component's surface and thus might be damaged by mechanical impacts. But if the structural component has material-inherent magnetic properties it can serve as a load sensor itself. The external mechanical forces temporarily change the preferred magnetization direction of the ferromagnetic material due to the reversible deformation of the material's crystal lattice [1]. This magnetoelastic effect leads to a macroscopic change of the magnetic properties [2], which can be monitored online by methods of nondestructive component testing, such as the harmonic analysis of eddy-current signals. The application of magnetoelastic force and torque sensors has been reported in literature (e.g. [3, 4]). However, these sensor elements are often made from ferromagnetic materials like metallic glasses which are usually limited to the production of thin strips or wires in order to maintain an amorphous structure.

Magnetic magnesium alloys enable an online-measurement of stresses acting on a structural component due to the changes of the material's magnetic properties under mechanical loading. The sensory properties are caused by alloying the magnesium melt with the ferromagnetic element cobalt, which is not among the usual alloying elements of magnesium alloys. Its high melting point and low solubility in magnesium are major challenges in the production of magnetic magnesium alloys by casting. Furthermore, information about the impact of a cobalt fraction in the range of a few percent on the magnetic and mechanical properties of Mg-Co alloys can be barely found in literature [5, 6]. In addition, the interactions of cobalt and nonmagnetic alloying elements like zinc, which is generally used in order to improve the mechanical strength of magnesium alloys [7], are unknown.

Binary Mg-Co alloys manufactured by gravity die casting (GDC) were investigated earlier by the authors [8]. These alloys clearly exhibited a load-sensitive behavior during the eddy-current measurements, but their mechanical strength was poor, having a yield strength around 30 MPa. The current work emphasizes the improved magnetic and mechanical properties as well as the refined phase structure of a Mg-Co-Zn alloy produced by high-pressure die casting (HP-DC). The overall objective of the project is the development and production of new lightweight materials by means of technically relevant casting processes which can be used in order to functionalize magnesium components through a load-sensor behavior.

Experimental Procedure

In this study, a ternary Mg-Co-Zn alloy was produced using technically pure magnesium (99.95%, Magnesium Elektron) as the base material. The nominal concentrations of the alloying elements were 4 wt.% cobalt (99.9%) and 1 wt.% zinc (99.8%), respectively.

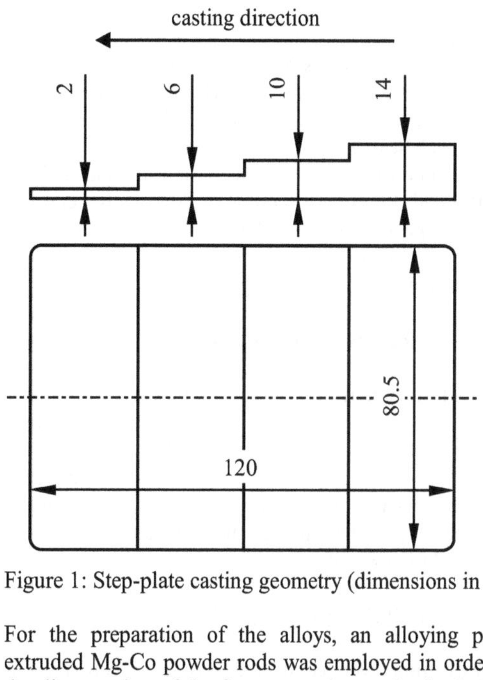

Figure 1: Step-plate casting geometry (dimensions in mm)

For the preparation of the alloys, an alloying process using extruded Mg-Co powder rods was employed in order to facilitate the dispensation of the ferromagnetic powder in the Mg melt and to avoid floating. The rods consisted of a blend of 60 wt.% of pure magnesium and 40 wt.% of pure cobalt powder. The manufacturing process is described in previous publications [8, 9]. During the subsequent casting experiments the magnesium melt was inoculated with the extruded Mg-Co rods.

The magnetic magnesium alloys were manufactured by means of a die casting method using a resistance-heated Nabertherm K4/10 furnace in a shielding-gas atmosphere (N_2 + 0.3% SF_6) and by employing a boron nitride coated, unalloyed steel crucible. The basic material was melted and held at a temperature of 730°C. At the beginning of the mechanical stirring process (45 min. at 300 revolutions/minute), the extruded Mg-Co powder rods were introduced into the melt. For the characterization of the material in GDC condition, the melt was cast into graphite-coated steel molds (test bar geometry ∅ 22 mm x length 250 mm) with horizontal feeders which were preheated to 350°C. Subsequently, threaded cylindrical tensile specimens with a nominal diameter of 6 mm were machined from the GDC test bars.

The HP-DC trials were conducted on a cold chamber die casting machine of the type Frech DAK 350-40. Here, two different die cast tools were used; (1) a special tool for the production of tensile specimens (∅ 6 mm) and (2) a step-plate geometry for the microstructure investigations (Figure 1). In preparation for these experiments, the Mg-Co-Zn alloy was divided into smaller portions matching the shot weight of the casting tools. A standard AZ91D alloy provided by Magnesium Elektron was also cast as a reference in order to compare the mechanical properties. The temperature of the dies was set to 160°C each.

GDC and HP-DC tensile specimens were used as test pieces for the determination of the mechanical strength values in standard tensile tests. Further samples of the Mg-Co-Zn alloy were subjected to cyclic loading tests using stepwise increasing loads from 200 to 1400 N in order to test whether correlations can be established between the forces applied to the specimens and their magnetic properties. During these experiments the magnetic properties were measured by the harmonic analysis of eddy-current signals as described in [8]. For the evaluation of the magnetoelastic behavior, the 3rd harmonic is of great significance.

Sections of the Mg-Co-Zn castings were metallographically prepared by grinding using SiC abrasive papers up to P2500 grit size. Afterwards, the specimens were polished to 1 µm using diamond paste and rinsed with ethanol. The Mg-Co-Zn alloys' microstructures were investigated employing the compositional contrast mode (back-scatter electron imaging; BSE) of a Zeiss LEO 1455VP scanning electron microscope (SEM) and a Zeiss SUPRA 40VP Field Emitter (FE-)SEM. Additionally, energy-dispersive X-ray spectroscopy (EDX) was used in order to determine the local compositions of the different phases in the samples' microstructures.

Results and Discussion

The different mold geometries in conjunction with the technical casting processes used in the present work, i.e. gravity die casting (GDC) and high-pressure die casting (HP-DC), enabled the production of the magnetic magnesium alloy under different cooling conditions. As the results show, cooling wields a major influence on the micro- and phase structure as well as the properties of the Mg-Co-Zn alloy.

Microstructure of the HP-DC Mg-Co-Zn alloy

The SEM investigations were conducted in order to compare the Mg-Co-Zn alloy's micro- and phase structures in samples of the various castings. As previously reported by the authors [8], the element cobalt precipitates in a eutectic structure. Whereas the cobalt-rich phases are distributed unevenly in the magnesium matrix of thick-walled components like the wide area of the step-plate casting (Figure 2 a), a close network of precipitates forms in rapidly cooled castings, e.g. the narrow area of the step-plate (Figure 2 b) or a HP-DC tensile specimen (Figure 3). Concluding from the SEM images, the morphology and density of the phase structures in these last two specimens seem to correspond largely, indicating that the particular cooling rates are similar.

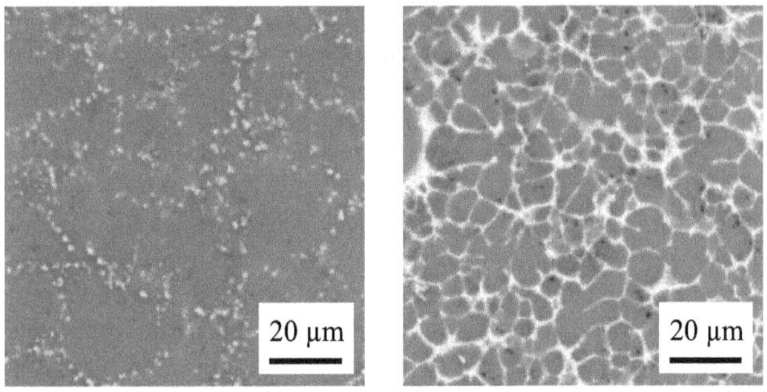

(a) (b)

Figure 2: Comparison of the microstructures in different sections of the step-plate casting (SEM image, BSE mode); (a) wide end – 14 mm, (b) narrow end – 2 mm

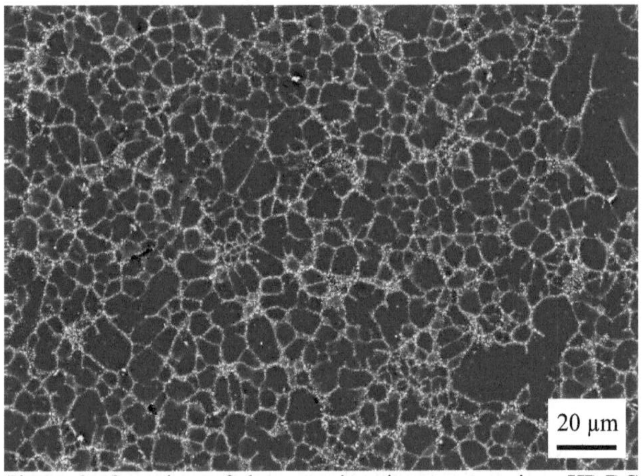

Figure 3: Overview of the eutectic microstructure in a HP-DC tensile specimen (FE-SEM image, BSE mode)

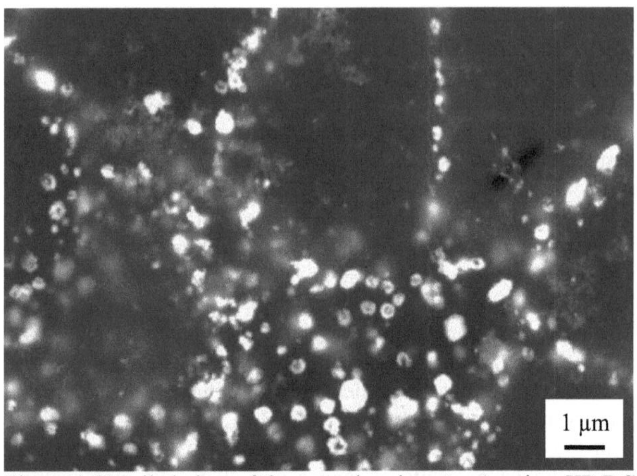

Figure 4: Detailed view of the eutectic microstructure in a HP-DC tensile specimen (FE-SEM image, BSE mode)

In Figure 4, a more detailed view of the precipitations is given by a higher magnification of the SEM image. Here, the refined, almost globular morphology of the individual particles can be seen. In order to determine the composition of these phases, several EDX linescans were performed on the sample intersecting both the magnesium matrix and the brighter areas on the surface (Figure 5). The results of the measurements showed that the amount of cobalt was considerably higher in the precipitates than in the matrix; few also contained zinc slightly above the average concentration. Comparisons with literature concerning the microstructure and properties of Mg-Co based alloys are difficult because cobalt is not often used as an alloying element in magnesium; nevertheless the formation of a cobalt-rich eutectic phase structure is also reported in older publications. In [6] it is noted that, due to the tendency of agglomeration, a completely eutectic microstructure can only be achieved by rapid cooling. This is confirmed in the present study (cf. Figure 3).

Magnetic Properties of GDC and HP-DC Specimens

With the aid of the manufactured tensile specimens, the Mg-Co-Zn alloy's load-sensor abilities were investigated depending on the applied casting process by means of cyclic loading experiments. The results are shown in Figure 6. The magnetic properties of all tested samples, which were measured by an eddy-current sensor, were sufficient to distinguish the values of the 3rd harmonic between the loaded and the unloaded condition. Moreover, an obvious dependency exists between the applied mechanical loads and the measured amplitude values. Following the test sequence from left to right, every value is the average result of three consecutive measurements at the same test force. While a force of 200 N leads to a slightly greater amplitude than in the initial state, the highest load in this experiment (1,400 N) causes the 3rd harmonic to increase significantly. In case of the GDC specimens (Figure 6 a), there is a gradual increase of the amplitudes in the unloaded condition until they reach a constant value after the application of the maximum load. The load-sensitive behavior though is continued throughout the whole measurement. This effect is caused by beginning local plastic deformation and it is thus related to the relatively low yield strength (cf. Table 1).

Compared with the GDC condition the HP-DC specimens generally exhibit a much higher amplitude of the 3rd harmonic (Figure 6 b) which means the magnetic properties are much more pronounced in this material. This is probably due to the refined and dense phase structure in the rapidly cooled castings, as shown in Figure 4. Basically, higher amplitudes are favorable because the useful signal is differentiated easier from the background noise during technical measurements. In addition, the HP-DC alloy shows an almost ideal linear correlation between the measured amplitude and the actual load on the tensile specimen. Apart from the greater absolute values, the difference between the amplitudes under a load of 1,400 N and 50 N seems to be virtually independent from the cooling conditions. In the experiments in this study, the difference was 0.49 mV for the samples in the GDC condition and 0.45 mV for the HP-DC specimens. It is concluded that the sensitivities for the measurement of mechanical loads are nearly the same for both materials.

Strength of the HP-DC Mg-Co-Zn alloy compared with AZ91

The mechanical strength and ductility was determined by means of standard tensile tests. In order to classify the mechanical properties of the experimental Mg-Co-Zn alloy, it was possible to produce corresponding tensile specimens out of AZ91 using identical HP-DC parameters. The results demonstrate that, compared to the binary Mg-Co alloys which were investigated in a previous work [8], the yield strength is clearly improved by the addition of Zn.

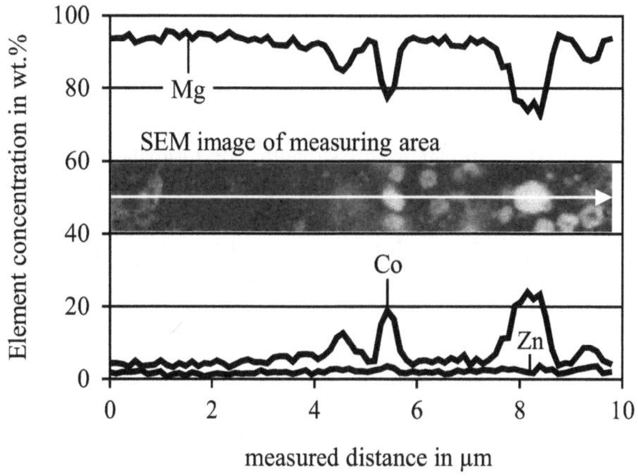

Figure 5: Concentration of alloying elements in a HP-DC tensile specimen as recorded by EDX linescan in the FE-SEM

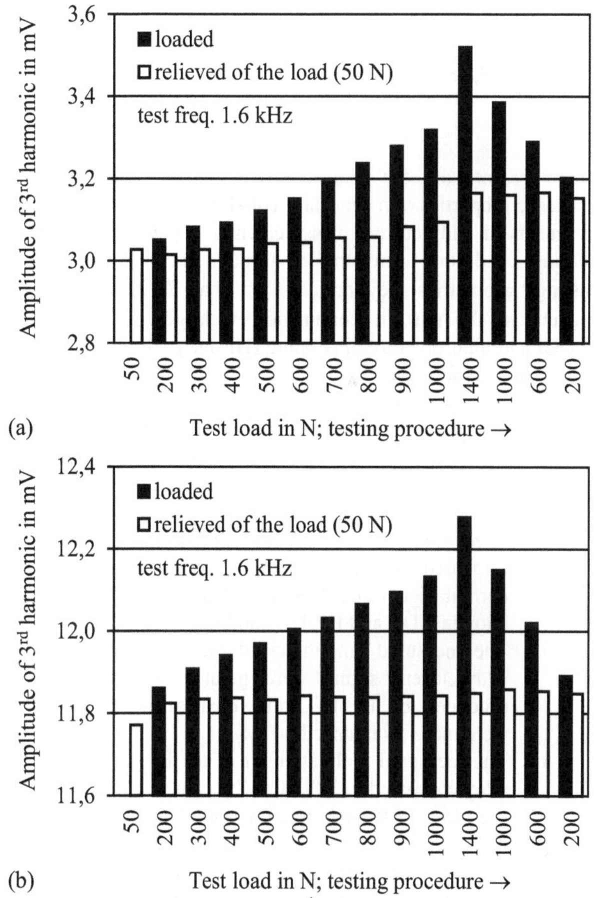

Figure 6: Amplitudes of the 3rd harmonic measured in cyclic loading tests of the Mg-Co-Zn alloy; (a) gravity die cast condition, (b) high-pressure die cast condition

According to the Hall-Petch relationship, further increases in both strength and ductility are achieved through a refined microstructure which can be produced by, for instance, HP-DC.

Here, the yield strength of the Mg-Co-Zn alloy is 84% higher than in GDC condition. Compared with the commercial alloy AZ91, the HP-DC magnetic magnesium alloy shows a lower but acceptable yield strength. However, the tensile strength of the experimental alloy is almost identical to AZ91 and the elongation at fracture is better (Table 1).

Table 1: Mechanical properties (mean values) of the experimental Mg-Co-Zn alloy and the reference alloy AZ91 (YS: yield strength; UTS: ultimate tensile strength; UE: ultimate elongation)

Alloy	Mg-Co-Zn		AZ91
Condition	GDC	HP-DC	HP-DC
YS in MPa	70.3	129.6	152.8
UTS in MPa	135.7	187.1	187.2
UE in %	3.5	5.9	1.5

It should be noted that the ultimate tensile strength and the ultimate elongation of the AZ91 samples tested in this study are considerably lower than the standard values issued by the alloy's producer [10]. This can be attributed to the specific cooling conditions during the casting experiments which are probably not comparable to other HP-DC experiments with different casting geometries. Further investigations should clarify if tensile specimens produced with the special die casting tool used in this work show the expected mechanical properties depending on the casting parameters.

Conclusions

In the present work, the influence of the casting conditions on the magnetic properties of magnesium-cobalt based alloys was shown by means of eddy-current measurements. In particular, different cooling conditions were applied through the use of different casting geometries (cylindrical test bar ∅ 22 mm, tensile specimens ∅ 6 mm) in conjunction with gravity die casting (GDC) and high-pressure die casting (HP-DC). The results can be summarized as follows:

1. Structure and distribution of Co-rich magnetic phases in the alloys varied depending on the cooling conditions. The microstructure in the tensile specimens produced by HP-DC was entirely eutectic with a dense network of Co (and partially Zn) containing precipitations.
2. A relationship between the applied test load and the measured amplitude of the 3rd harmonic of the eddy-current signal could be established for the Mg-Co-Zn alloy in GDC and HP-DC condition.
3. The measured amplitudes of the 3rd harmonic of the eddy-current signal were greater in the specimens produced by HP-DC, thus their magnetic properties were more pronounced.
4. The measured yield strength of the HP-DC Mg-Co-Zn alloy compared to HP-DC AZ91 was slightly lower, but the ultimate tensile strength was almost identical. The elongation to fracture of the magnetic magnesium alloy was higher than for AZ91.

Magnetic magnesium alloys were already applied in the production of load-sensitive lightweight components in the wheel suspension of a formula-student race car [5, 8]. The load-sensitive magnesium components, such as wheel carriers and pushrods, were equipped with eddy-current sensors and could be used successfully in order to measure the applied loads qualitatively during test drives. Further investigations will be done in order to establish a quantitative correlation between the measured eddy-current signal and the loads during test drives. Although the initial series of components was manufactured by sand casting, this study suggests that better magnetic and mechanical properties are achieved through the use of HP-DC components.

Acknowledgments

This research is sponsored by the German Research Foundation (DFG) within the subproject E2 "Magnetic Magnesium Alloys" of the Collaborative Research Center 653 "Gentelligent Components in their Lifecycle". The eddy current measurements were executed in cooperation with subproject S3 "Gentelligent Part Identification and Integrity Assessment".

References

1. R.C. O'Handley, *Modern Magnetic Materials. Principles and Applications* (New York, NY: Wiley-VCH, 2000), 218-221.
2. A. Bieńkowski, J. Kulikowski, "The magneto-elastic Villari effect in ferrites", *J. Magn. Magn. Mater.* 19 (1980), 120-122.
3. A. Bieńkowski, R. Szewczyk, J. Salach, "Industrial Application of Magnetoelastic Force and Torque Sensors" *Acta Phys. Pol. A* (118) (2010), 1008-1009.
4. N.B. Ekreem, A.G. Olabi, T. Prescott, A. Rafferty, M.S.J. Hashmi, "An overview of magnetostriction, its use and methods to measure these properties", *J. Mater. Process. Tech.* (191) (2007) 96-101.
5. C. Klose, *Development of Magnetic Magnesium Alloys with Load-Sensitive Properties* (Hannover, Germany: PZH-Verlag, 2013), IV.
6. A.A. Nayeb-Hashemi, J.B. Clark, "The Co-Mg (Cobalt-Magnesium) system", *Bulletin of Alloy Phase Diagrams* 8 (4) (1984), 352-355.
7. Z. Yang, J.P. Li, J.X. Zhang, G.W. Lorimer, J. Robson, "Review on Research and Development of Magnesium Alloys", *Acta Metallurgica Sinica (English Letters)* 21 (5) (2008), 313-328.
8. C. Klose, C. Demminger, G. Mroz, W. Reimche, F.-W. Bach, H.J. Maier, K. Kerber, "Influence of Cobalt on the Properties of Load-Sensitive Magnesium Alloys", *Sensors* 13 (1) (2013), 106-118.
9. C. Klose, G. Mroz, G.L. Angrisani, K. Kerber, W. Reimche, F.-W. Bach, "Casting Process and Comparison of the Properties of Adapted Load-Sensitive Magnesium Alloys", *Production Engineering Research & Development* 7 (1) (2013), 35-41.
10. Magnesium Elektron Datasheet 440, "Magnesium Casting Alloys", Magnesium Elektron, Manchester, UK (2006).

Magnesium Technology 2015
Edited by: Michele V. Manuel, Alok Singh, Martyn Alderman, and Neale R. Neelameggham
TMS (The Minerals, Metals & Materials Society), 2015

EFFECTS OF La ADDITION ON THERMAL CONDUCTIVITY AND MECHANICAL PROPERTIES OF Mg-4Zn-0.5CaO ALLOYS

Gun-Young Oh[1,2], Young-Ok Yoon[1], Shae K. Kim[1], Hyun Kyu Lim[1] and Young-Jig Kim[2]
[1] Rare metal R&BD group, Korea Institute of Industrial Technology, Incheon, Korea
[2] Dept. Adv. Mater. Eng., Sungkyunkwan Univ., Suwon 440-746, Korea

Keywords: Mg-Zn-CaO, Thermal conductivity, Fluidity, Microstructure, Mechanical properties

Abstract

The thermal conductivities of quaternary Mg-Zn-CaO-La alloys have been investigated by evaluating the effect of La on Mg-4Zn-0.5CaO alloys, with an emphasis to develop a new Mg alloy without compromising thermal conductivity, process-ability and mechanical property. The thermal conductivity of these alloys had been determined by tests for thermal properties, such as specific heat and diffusivity, from room temperature to 200 °C. The microstructures of specimens were observed by OM and SEM, and the phase analysis was performed by XRD, EDS. The fluidity was also investigated by using a spiral fluidity mold for improved process-ability during actual die casting. To evaluate the mechanical properties of tested alloys, tensile tests were carried out at room temperatures.

As a result, although the La addition decreased the thermal conductivity of Mg-Zn-CaO alloy, with increasing La contents, thermal conductivities of La added alloys maintained and the yield strength of room temperature increased. In addition, the average spiral flow lengths of Mg-Zn-CaO-xLa alloys were almost same levels with Mg-Zn-CaO alloy.

Introduction

Mg alloys exhibit promising combined properties including low density, high specific strength, good process-ability, and adequate ductility. Because of these attractive properties, Mg alloys are considered to be the strong potential for weight saving in automotive and aerospace components [1-3]. However, the researches on Mg alloys have been so far mainly concentrated on the improvement of mechanical properties for conventional alloy systems such as AZ91, AM60, and AZ31 alloys [4]. In spite of increasing the usages of Mg alloys as heat sinks for LED lightening system and as brackets for smart phones with large LCD window, there is only limited information concerning about the thermal properties, especially thermal conductivity [5-6]. Thermal conductivity has an important function in the performance of alloys and is essential for the selection of alloys for certain applications [7]. High thermal conductivity ensures uniform temperature distribution which reduces thermally induced stresses and thus prolongs the service life of the alloy [8].

On the other hand, ECO-Mg technique has been developed. The concept of ECO-Mg technique is to incorporate simply CaO into Mg alloys and generate the reaction of CaO on the surface of molten Mg alloys. The surface oxide film of molten Mg alloy changes from porous to dense and complex films. Therefore, the alloys with ECO-Mg technique exhibit the improved oxidation and ignition resistances, safety of parts, castability (i.e. increase of fluidity and reduction of hot tearing), and mechanical properties, mainly due to the improvement of melt cleanliness [9-11].

According to our preliminary test, the Mg-4Zn-0.5CaO alloy has exhibited the relatively high thermal conductivity in Mg-Zn-CaO alloy system, however the strength of this alloy was lower than that of conventional AZ91 alloys. Therefore, in this study, to improve the mechanical properties of the Mg-4Zn-0.5CaO alloy without the decrease in thermal conductivity, La was chosen as a strengthening element and the effect of La addition on thermal and mechanical properties were investigated.

Experimental procedure

As-cast ingots were prepared by melting commercially pure Mg, Zn and La with Mg-5CaO master alloy in an electric resistance furnace under SF_6+CO_2 atmosphere, followed by pouring the melt into a steel mold. The chemical compositions of the as-cast alloys were analyzed by ICP-mass spectrometer and the results were listed in Table 1. The molten Mg-4Zn-0.5-xLa alloys were poured into an automatic spiral mold by three times at the same condition. The thermal conductivities of the alloys ware measured in the temperature range from 25 °C to 200 °C with NETZSCH model LFA447 Flash Analyzer. The size of specimen was Φ 12.7 x 2 mm. The thermal conductivity K (W/mK) was calculated from the diffusivity and specific heat using the following relation:

$$K = \alpha \cdot \rho \cdot Cp \quad (1)$$

where α is the thermal diffusivity; ρ is the density; and Cp is the specific heat capacity. The as-cast alloys were investigated by x-ray diffraction (XRD). Intermetallic phase was identified by x-ray diffraction using a Cu-Kα radiation. The XRD patterns were acquired from polished samples at steps of 0.02 deg. 2 θ and 0.6 s/sep exposure, on an automatic diffractometer. The microstructural analysis was carried out using an optical microscopy (OM) and scanning electron microscopy (SEM) equipped with and energy dispersive X-ray spectrometer (EDS) after etching with a solution of 2 vol. % nitric acid + ethyl alcohol. And volume fraction was measured by using image analyzing program. Also, the tensile tests were performed at room temperatures with specimens from ingots with 30 mm of gauge length and 6 mm of diameter.

Table 1. ICP-MS analysis results of as-cast alloys

Alloy	Chemical composition (wt. %)			
	Zn	Ca*	La	Mg
Mg-4Zn-0.5CaO	4.1	0.557		Bal.
Mg-4Zn-0.5CaO-0.24La	4.19	0.519	0.24	Bal.
Mg-4Zn-0.5CaO-0.78La	4.13	0.578	0.78	Bal.
Mg-4Zn-0.5CaO-0.96La	4.13	0.555	0.96	Bal.

*Ca2Mg6Zn3, generated from CaO reduction, can only be analyzed by Ca amount by ICP.

Result and discussion

Fig. 1 shows XRD results of the as-cast alloys. The main peaks for the Mg-4Zn-0.5CaO and Mg-4Zn-0.5CaO-0.24La alloys were corresponding to α-Mg and $Ca_2Mg_6Zn_3$ phases. However, in Mg-4Zn-0.5CaO-0.78La and Mg-4Zn-0.5CaO-0.96La alloys, the $(Mg,Zn)_{11}La$ phase formed with α-Mg and $Ca_2Mg_6Zn_3$ phases.

These phases seem to be formed by excess La.

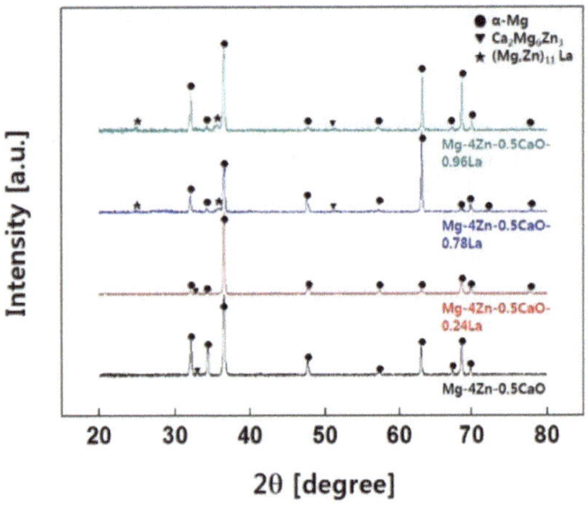

Fig. 1 X-ray diffraction spectra of as-cast alloys.

Table 2. Chemical compositions of area A, B, and C analyzed by EDS in Fig. 3 (in at. %)

Area	Mg	Zn	Ca	La
A	73.2~75.04	17.4~17.7	7.5~9.1	
B	76.1	17.56	1	0.533
C	78.6~82.9	11.9~14.9		5.2~6.4

Table 3. Solidification interval of tested alloys calculated by JMatPro 7.0.

Alloy	Liquidus	Solidus	Interval
Mg-4Zn-0.5CaO	635.28	328.48	306.8
Mg-4Zn-0.5CaO-0.24La	635	325	310
Mg-4Zn-0.5CaO-0.78La	633.75	325	308.75
Mg-4Zn-0.5CaO-0.96La	633.39	325	308.39

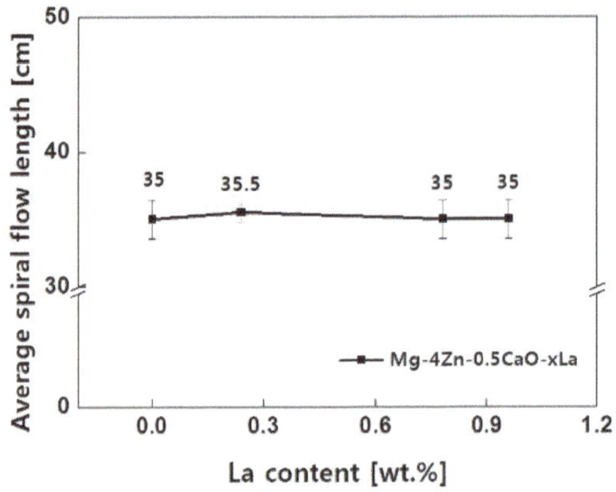

Fig. 4 Average spiral flow lengths of as-cast alloys.

Fig. 2 shows the microstructures of as-cast alloys. The microstructures of alloys were characterized by primary α-Mg dendrites and second phases at grain boundary and interdendritic regions. The volume fraction of secondary phases increased with increasing amounts of La element.

The backscattered electron (BSE) images of the as-cast alloys were shows in Fig. 3. All alloys were consisted of α-Mg and the secondary solidified phases. The EDS results of α-matrix and the secondary phases were shown in Table 2. $Ca_2Mg_6Zn_3$ phase was observed in all alloys. Also, in case of the excess La added Mg-4Zn-0.5CaO-xLa (x=0.78 and 0.98) alloys, the formation of $(Mg,Zn)_{11}La$ phase was confirmed.

Fig. 4 shows the average spiral flow lengths of as-cast alloys. The spiral flow lengths of Mg-4Zn-0.5CaO, Mg-4Zn-0.5CaO-0.24La, Mg-4Zn-0.5CaO-0.78La, and Mg-4Zn-0.5CaO-0.96La alloys were 35, 35.5, 35, and 35 cm, respectively. Regardless of the La contents in Mg-4Zn-0.5CaO alloys, the average spiral flow lengths were similar to the value of Mg-4Zn-0.5CaO alloy, because the La addition in Mg-4Zn-0.5CaO alloys did not change the solidification interval, as shown in Table 3. Spiral fluidity is inversely proportional to solidification interval [12].

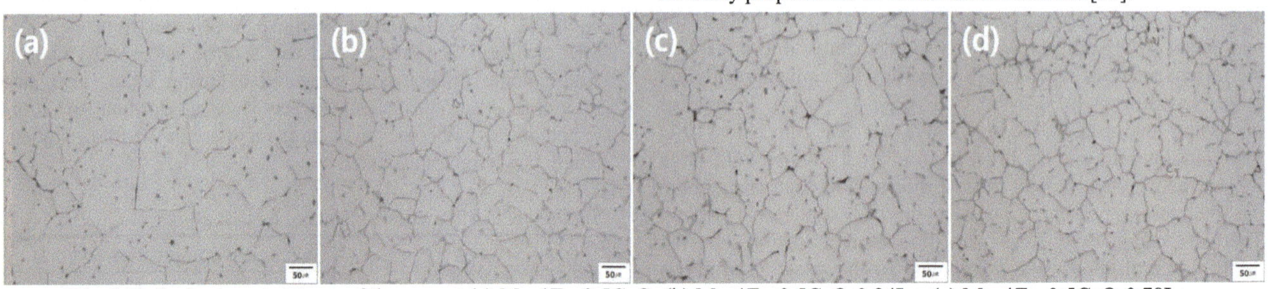

Fig. 2 Optical microstructures of the as-cast (a) Mg-4Zn-0.5CaO, (b) Mg-4Zn-0.5CaO-0.24La, (c) Mg-4Zn-0.5CaO-0.78La, and (d) Mg-4Zn-0.5CaO-0.96La alloys.

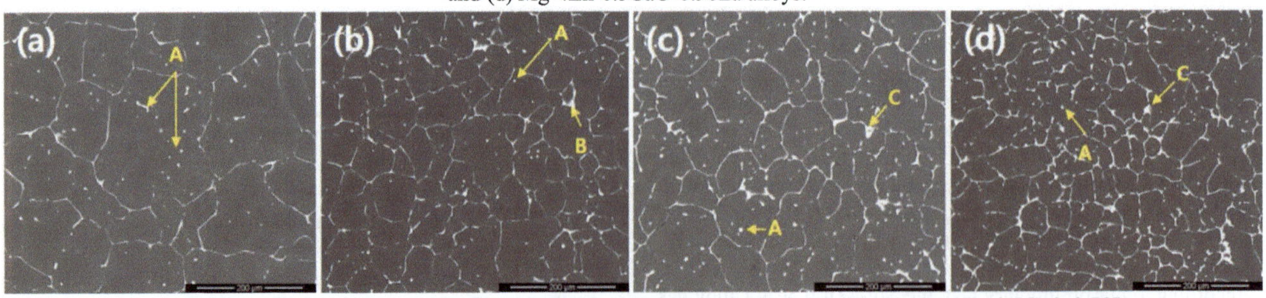

Fig. 3 BSE images of the as-cast (a) Mg-4Zn-0.5CaO, (b) Mg-4Zn-0.5CaO-0.24La, (c) Mg-4Zn-0.5CaO-0.78La, and (d) Mg-4Zn-0.5CaO-0.96La alloys.

alloy.

Fig. 6 and Table 4 show the average contents of alloying elements in α-Mg matrix from EDS line analysis results of as-cast alloys. EDS line analysis showed the similar tendency comparing with thermal conductivity. The amount of solute La atoms in α-Mg matrix maintained with increasing La content in as-cast alloys.

Thermal conductivity is usually degraded by the scattering of carriers in the crystal lattice [13]. In alloys, additional lattice distortions by alloying elements cause similar disturbances; when foreign atoms are added, they cause disturbance in the periodicity of the lattice, and electrons are deflected in the absence of thermal agitation [14]. Different alloying elements do not have equivalent effects. Addition of materials of the same atomic size leads to small differences, whereas addition of materials with different atomic sizes leads to large variations [7].

In Mg-Zn-CaO-xLa alloys, the atomic radii of Mg and La are 1.6 and 1.87 Å, respectively. Also, Mg and La have the same crystal structure with a close-packed hexagonal lattice. Therefore, in Mg-Zn-CaO-xLa alloy, the addition of La atoms to the Mg matrix will have smaller effects on thermal conductivity.

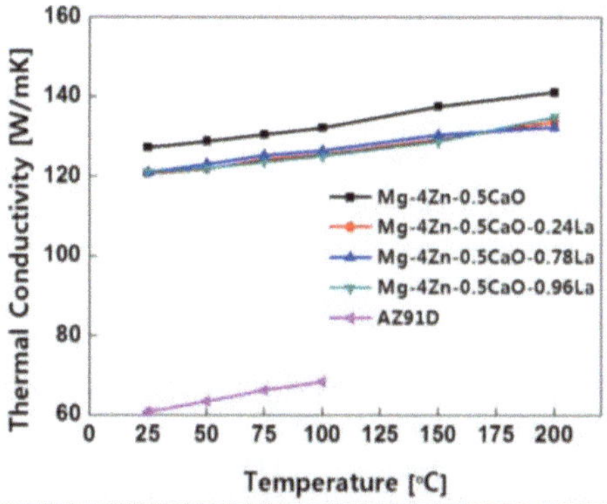

Fig. 5 Temperature dependence of thermal conductivity of as-cast alloys.

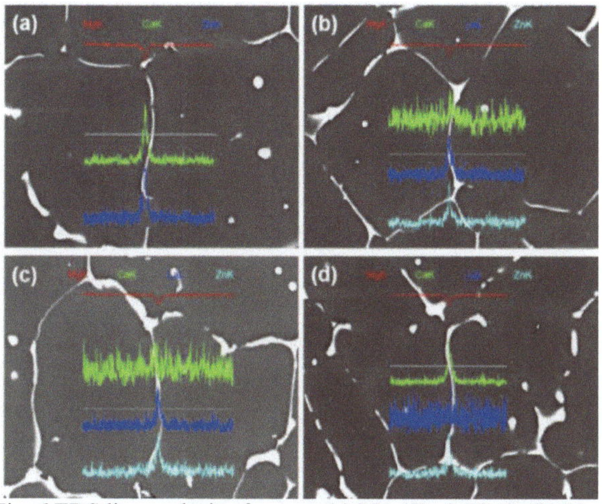

Fig. 6 EDS line analysis of as-cast (a) Mg-4Zn-0.5CaO, (b) Mg-4Zn-0.5CaO-0.24La, (c) Mg-4Zn-0.5CaO-0.78La, and (d) Mg-4Zn-0.5CaO-0.96La alloys.

Table 4. Average contents of alloying elements in α-Mg matrix from EDS line analysis results (Fig. 6) of as-cast alloys.

Alloy	Elements (at. %)			
	Mg	Zn	Ca	La
Mg-4Zn-0.5CaO	98	1.51	0.42	
Mg-4Zn-0.5CaO-0.24La	98.1	1.15	0.36	0.31
Mg-4Zn-0.5CaO-0.78La	98.2	1.044	0.38	0.31
Mg-4Zn-0.5CaO-0.96La	98.3	0.94	0.38	0.31

Fig. 5 shows the temperature dependence of the thermal conductivities for all as-cast alloys under investigation. The thermal conductivities of as-cast alloys were calculated from the thermal diffusivity, specific heat capacity and density of the material with equation (1). The thermal conductivities of the as-cast alloys increased with increasing temperature. At room temperature the thermal conductivities of Mg-4Zn-0.5CaO, Mg-Zn-0.5CaO-0.24La, Mg-4Zn-0.5CaO-0.78La, and Mg-4Zn-0.5CaO-0.96La alloys were 127.24, 120.806, 121.047, and 121.08 W/mK, respectively. Although the La addition decreased the thermal conductivity of Mg-4Zn-0.5CaO alloy, thermal conductivities of the La added alloys maintained with increasing La contents. Compared with that of AZ91D alloy, the thermal conductivity of tested alloys were higher than that of AZ91D

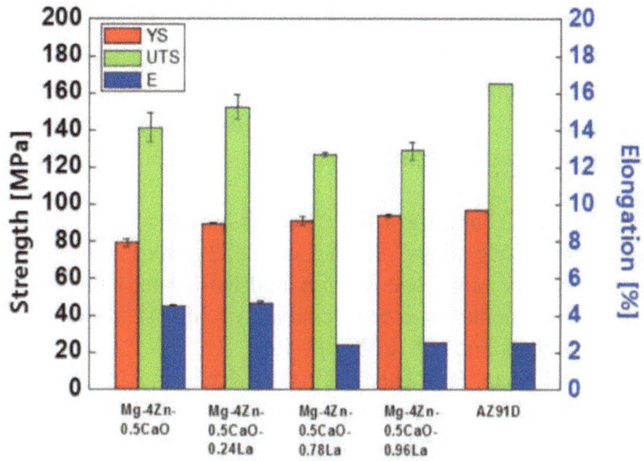

Fig. 7 Room Temperature mechanical properties of as-cast alloys

The tensile properties of as-cast alloys were shown in Fig. 7. The tensile yield strengths of Mg-4Zn-0.5CaO, Mg-4Zn-0.5CaO-0.24La, Mg-4Zn-0.5CaO-0.78La, and Mg-4Zn-0.5CaO-0.96La alloys were 79.15, 86.66, 91.26, and 93.79 MPa, respectively. By adding La in Mg-4Zn-0.5CaO, the tensile yield strength was improved with increasing CaO contents, due to the increased volume fraction of second phases. The ultimate tensile strengths of Mg-4Zn-0.5CaO, Mg-4Zn-0.5CaO-0.24La, Mg-4Zn-0.5CaO-0.78La, and Mg-4Zn-0.5CaO-0.98La alloys were 141.52, 152.54, 126.91, and 128.9 MPa, and the elongation values of Mg-4Zn-0.5CaO, Mg-4Zn-0.5CaO-0.24La, Mg-4Zn-0.5CaO-0.78La, and Mg-4Zn-0.5CaO-0.98La alloys were 4.5, 4.7, 2.4, and 2.5 %, respectively. The ultimate tensile strength and elongation of Mg-4Zn-0.5CaO-0.24La alloy were higher than those of Mg-4Zn-0.5CaO, Mg-4Zn-0.5CaO-0.78La, and Mg-4Zn-0.5CaO-0.96La alloys. In the case of Mg-4Zn-0.5CaO-0.78 and Mg-4Zn-0.5CaO-0.96 alloys, because the formation of brittleness $(Mg,Zn)_{11}La$ phase was increased by increasing La contents, the ultimate tensile strength and elongation were decreased. Compared with that of AZ91D alloy, the tensile yield strengths and ultimate tensile strengths of as-cast alloys were lower than that of AZ91D alloy.

Conclusion

This paper discussed the thermal conductivity, fluidity, and mechanical property of the Mg-4Zn-0.5CaO-xLa alloys. The conclusions are as follows:

(1) In spite of decreasing thermal conductivity of Mg-4Zn-0.5CaO alloy by the La addition, thermal conductivities of La added alloys were still higher than that of commercial AZ91 alloy. In addition, thermal conductivities maintained with increasing La contents.
(2) The tensile yield strength was improved with increasing La contents, because of increment of volume fraction of the second phases. Also, the ultimate tensile strength and elongation of Mg-4Zn-0.5CaO-0.24La alloy were higher than those of Mg-4Zn-0.5CaO, Mg-4Zn-0.5CaO-0.78La and Mg-4Zn-0.5CaO-0.96 alloys.
(3) Regardless of the La contents in Mg-4Zn-0.5CaO alloys, the average spiral flow lengths were similar value of Mg-4Zn-0.5CaO alloy.

As a conclusion, the optimum La content was 0.24 wt.% in the Mg-4Zn-0.5CaO-xLa alloy.

References

[1] B. L. Mordike and T. Ebert: "Magnesium properties application potential", Mater. Sci. Eng., A 302 (2001) 37-45.
[2] H. Friedrich and S. Schumann: "Research for a new age of magnesium in the automotive industry", J. Mater. Proc. Tech., 117 (2001) 276-281.
[3] W. Ha, Young-jig Kim: "Effects of cover gases on melt protection of Mg alloys", J. All. Com., 422 (2006) 208-213.
[4] Hyun-Kyu Lim, Shae K. Kim, and Do-Hyang Kim: "Effect of alloying elements on creep behavior of Mg-Al alloys", J. Kor. Foundrymen's Soc., 32 (2012) 44-49
[5] Jiawei Yuan, Kui Zhang, Ting Li, Xinggang Li, Yongjun Li, Minglong Ma, Ping Luo, Guangqiu Luo, Yonghui Hao: "Anisotropy of thermal conductivity and mechanical properties in Mg-5Zn-1Mn alloy", Materials and Design, 40 (2012) 257–261.
[6] Dimitar Georgiev Todorov, Lazar Georgiev Kapisazov: "LED thermal management", ELECTRONICS'2008, (2008) 139-144.
[7] A. Rudajevova, P. Lukacˇ, Materials Science and Engineering A 397 (2005) 16–21.
[8] A. Rudajevova, F. von Buch, B.L. Mordike, Journal of Alloys and Compounds 292 (1999) 27–30.
[9] Shae K. Kim, Jin-Kyu Lee: "Behavior of alkaline earth metal oxides in magnesium alloys", 135th TMS 2006" (2006) 517-521.
[10] Jin-Kyu Lee, Shae K. Kim, Hyung-Ho Jo: "Development of CaO added wrought Mg alloys for cleaner production", 135th TMS 2006, (2006) 185-189.
[11] Shae K. Kim, Young-Ok Yoon, Jin-Kyu Lee, Hyung-Ho Jo: "Green Manufacturing for Magnesium Alloys" WFC 2006, (2006) 34/1-34/9.
[12] Merton C. Flemings, "Solidification processing", (1974) 219-224)
[13] Y. Terada, K. Ohkubo, T. Mohri, T. Suzuki, Journal of Alloys and Compounds 285 (1999) 233–237.
[14] M.J. Peet, H.S. Hasan, H.K.D.H. Bhadeshia, International Journal of Heat and Mass Transfer 54 (2011) 2602–2608.

Magnesium Technology 2015

STRIP CASTING OF LIGHT METALS

Modeling and Properties

Papers in this section have been presented in the symposium Strip Casting of Light Metals at the 2015 TMS Annual Meeting, organized by K.F. Karhausen, W. Sillekens, M. Dundar, J. Bohlen, and D. Letzig. This section collects the peer-reviewed magnesium-related contributions to this symposium and was edited by Dr. J. Bohlen and Dr. D. Letzig (Chief Scientist and Department Head at the Magnesium Innovation Center of the Helmholtz Research Center Geesthacht, Germany, respectively), and Dr. W. Sillekens (Project Manager at the Research and Technology Center of the European Space Agency in Noordwijk, Netherlands).

Magnesium Technology 2015
Edited by: Michele V. Manuel, Alok Singh, Martyn Alderman, and Neale R. Neelameggham
TMS (The Minerals, Metals & Materials Society), 2015

TWIN ROLL CASTING OF MAGNESIUM STRIP AT CANMET MATERIALS - MODELING AND EXPERIMENTS

A. Javaid[1], Jeremy Hanke[2], C. H. Simha[1], M. S. Kozdras[1]
[1]CanmetMATERIALS Natural Resources Canada
183 Longwood Road South, Hamilton, ON, Canada, L8P 0A5
[2]CD-adapco, 21800 Haggerty Road, Suite 300, Northville, MI, 48167, United States

Keywords: twin roll casting (TRC), modeling, magnesium strip, solidification

Abstract

At CanmetMATERIALS, a pilot-scale Twin Roll Caster (TRC) facility was used to produce AZ31 magnesium strip and this paper describes the experiments and modeling effort. In the experiments, the effects of cast-strip thickness and roll speed on the surface quality and solidification microstructure of the strip was investigated. The TRC is instrumented with process control sensors and the thermocouple recordings are particularly used to validate the finite volume modeling. The finite-volume-based STAR CCM+ software was used to build a three-dimensional model of the TRC; the computations use implicit time integration. A marked feature of the computations is the inclusion of surface tension of melt, the interaction of the melt with the air, and rotating rolls with varying speed. This paper will also discuss some effort to minimize lateral freezing fronts in the edge of the strip by optimizing the nozzle shape and geometry.

Introduction

Twin roll continuous casting, also known as twin roll strip casting, combines casting and rolling into a single step operation, thus reducing the manufacturing cycle time, energy, pollutant emissions and final cost compared to traditional sheet production using a direct chill (DC) ingot casting process. The development of low-cost magnesium strip and sheet is an enabler to downstream processing for vehicle light-weighting applications for hood, trunk and door inner panels, and seat components among other applications.

Furthermore, the near-rapid solidification rates (10^2 to 10^3 °C/s) during strip casting, leads to more homogeneous microstructures with fine grains, reduced chemical segregation, increased solid solubility, and enhanced precipitate nucleation within the matrix [1,2]. Although the twin roll casting process has been successfully used for aluminum strip production for decades, its application for magnesium strip production has proven difficult due to inherent characteristics of magnesium alloys, such as their high reactivity to oxygen, low specific heat and large freezing ranges which can cause formation of casting defects such as segregation, hot cracking and oxide inclusions.

There is a general consensus that high solidification rates, experienced during twin roll casting, have positive effects on the strip microstructure and resulting properties. According to Ref. [2,3], the near-rapid solidification requires the design of new alloys that take advantage of the increased amount of solute in the α-Mg solid solution, thus resulting in enhanced nucleation of precipitates within the matrix and their dense distribution. The features of microstructure modifications, related to rapid solidification, positively affect the strip formability at room temperature. Enhanced formability is an important enabler for the adoption of the sheet material for automotive body parts.

At CanmetMATETRIALS, a pilot-scale TRC facility was used to produce AZ31 magnesium strip. In this article, some of the experimental results are presented. During these runs, magnesium appeared to solidify upstream of the roll gap leaving scalped edges in the strip and inevitably damaging the ceramic delivery nozzle. In order to overcome this casting phenomenon, a computational strategy to gain insight into the performance of the molten metal delivery system of the TRC was adopted, and this effort is described herein.

Experimental Procedure

A model of the pilot-scale twin roll caster used at CanmetMATERIALS is presented in Figure 1. A model of the ceramic nozzle (or tip) used for feeding the molten metal to the casting rolls is shown in Figure 2. The nozzle is instrumented with thermocouples at the inlet and outlet. The TRC is instrumented with optical sensors and thermocouples for monitoring and temperature control in the furnace, head-box, nozzle inlet/outlet, roll surface, cooling water and sheet exit. The TRC is also equipped with sensors for roll separating force, liquid level in the head-box and rolling speed. Basic geometry of the caster and typical operating conditions are shown in Table 1. This TRC can produce 250mm wide magnesium strip using 355mm diameter water-cooled casting rolls that are 610mm wide. It can be seen that the caster is sized to manufacture sheet large enough for both material property measurement at various angles to the rolling direction, as well as assessment in metal forming processes. The casting line also includes a 400kg magnesium melting furnace and automated melt delivery system. The TRC has a fully integrated HMI (Human Machine Interface) system that collects all the process information/data during strip casting. Commercial AZ31B (Al=3%, Zn=1%) ingots were melted in a 400kg resistance furnace under the protective gas mixture of SF_6 and N_2. The molten metal temperature in the furnace was maintained at 705°C and the metal was transferred to the head-box by a metal delivery

Roll Diameter	355 mm	Furnace Temperature	705°C
Roll Width	610 mm	Head-Box Temperature	700°C
Strip Width	250 mm	Initial Roll Temperature	70°C
Furnace Capacity	400 kg	Nozzle Preheat Temperature	>400°C
Strip Thickness	4-6mm	Strip Velocity	2-4 m/min

Table 1: Basic geometry and typical operating conditions.

system. Then the melt was introduced into the twin roll gap through the ceramic nozzle to cast AZ31 strip. It should be noted that temperature control of the ceramic nozzle is very important in that it must be preheated above 400°C for at least 2 hours before introducing the molten metal. Since the commissioning of TRC at Hamilton in 2010-11, sixteen experiments has been conducted to evaluate the effect of various process parameters (such as cast strip thickness, casting roll speed and temperature) on the quality of the strip produced. The recurrent solidification in the ceramic nozzle caused many experiments to be terminated early. TRC design modifications were recently done to improve the ceramic nozzle positional stability and the computational work was initiated to understand and remedy the premature solidification in the ceramic nozzle. Once complete, trials will be conducted to develop the process window and material behavior for AZ31 alloy that will be followed by ZEK100 alloy known to have improved room temperature formability due to the presence of rare earths.

TRC data analysis and the observation of the fractured ceramic nozzle indicated freezing of the molten metal at the edges of the ceramic nozzle resulting in intermittent obstruction in the metal flow leading to scalping of the edges of the strip. Figure 3 presents the histories of temperature during the strip casting process from nozzle thermocouples (TC1 and TC3) located at the edges of ceramic nozzle as well as strip exit temperature and the strip velocity. The thermocouple location in the ceramic nozzle assembly is shown in Figure 2. There is fluctuation in the temperature of thermocouple TC3 which drops below solidus temperature and then jumps back to liquidus temperature due to intermittent obstruction in the metal flow leading to the fracture of ceramic nozzle. Presently, nozzle design (shape and geometry) optimization is being carried out using computer modeling to solve this problem.

For a good quality magnesium strip, solidification of the molten metal starts at the point of first metal-roll contact and is completed

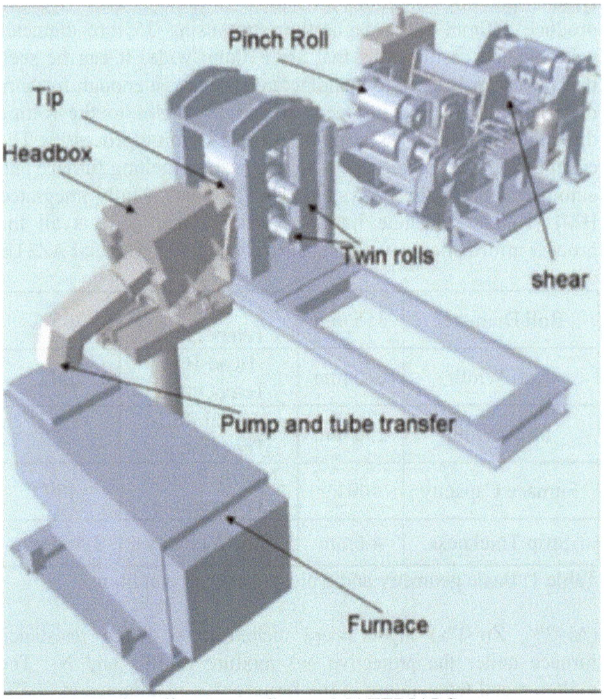

Figure 1: Twin roll caster at CanmetMATERIALS.

before the kissing point (point of least roll separation) of the two rollers. A good quality (defect-free) 5mm thick AZ31 strip produced at the casting speed of 2 to 2.5m/min is shown in Figure 4a. Surface defects appear in the cast strip, when the solidification is not complete before the kissing point, as a result of high casting speed and/or high melt temperature. An example of the surface defects due to hot cracking and excessive oxidation in a 5mm thick AZ31 strip is shown in Figure 4b.

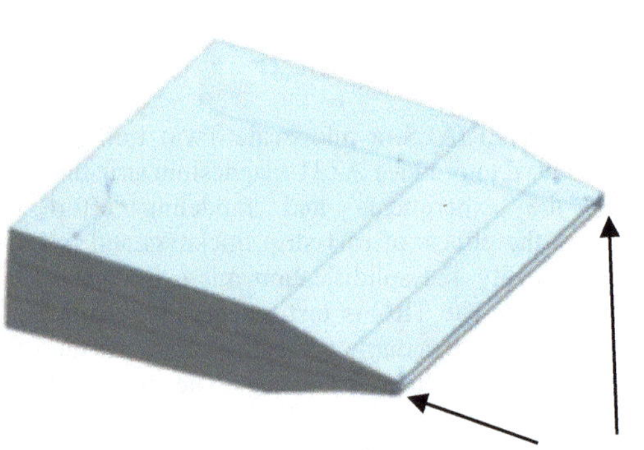

Thermocouples at both edges of ceramic nozzle - TC1, TC3

Figure 2: Ceramic nozzle (or tip), 250-mm wide, for the delivery of molten metal to casting rolls.

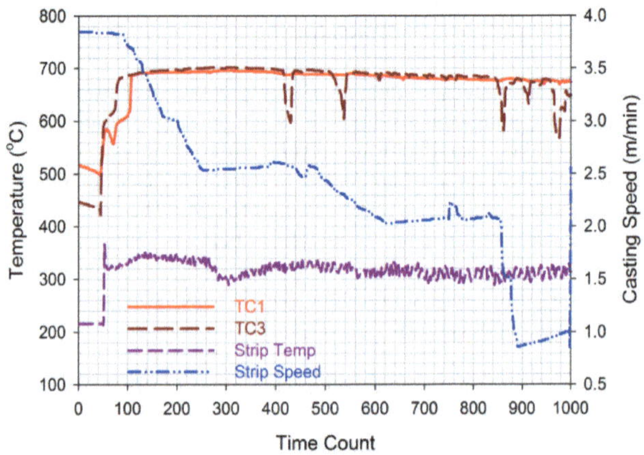

Figure 3: Plots from nozzle thermocouples and strip-velocity sensors.

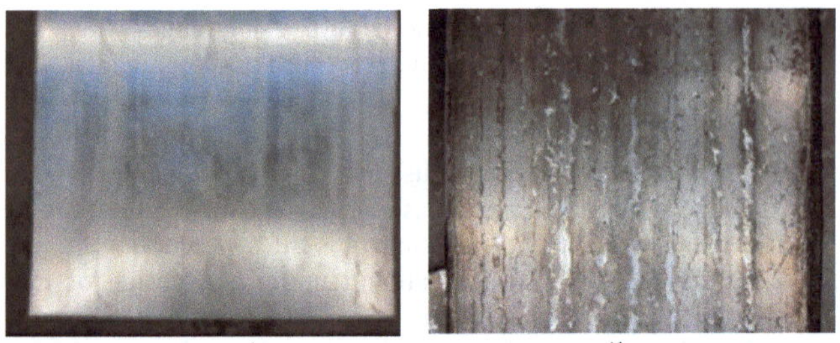

4a 4b

Figure 4a and 4b. A good quality (defect-free) 5mm thick AZ31 magnesium strip (4a) and surface defects due to hot cracking and excessive oxidation in a 5mm thick AZ31 strip (4b)

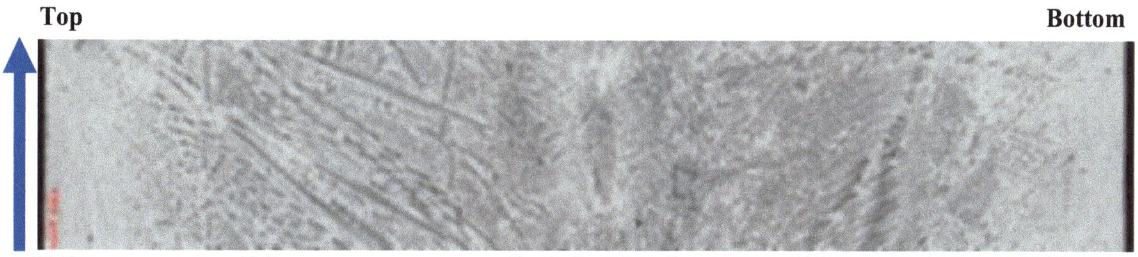

Figure 5. TRC as-cast microstructure observed through 6mm thick AZ31 strip at the transverse direction.

Limited mechanical property characterization has been done to date since steady state conditions were not achieved. Analysis will continue when future runs can be performed. The microstructures of the as-cast strip have been characterized using optical and scanning electron microscopy. Fine dendritic microstructure (chill zone) is observed near the surface of the strip due to the very fast cooling rate by the direct contact with the roll surface. Columnar dendrites can be observed beneath the chill zone through to the center. Centerline segregation is present in the central area, where equiaxed dendrites and coarse eutectic structure are frequently observed. The average value of secondary dendrite arm spacing, in 6 mm thick strip, is about 7 micron. After annealing at 330°C for 2 hours, coarse eutectic structures were still observed. On the other hand, when the magnesium alloy strip was annealed at 400°C for 1 hour, coarse particles were mostly dissolved into the magnesium matrix with only a small amount of particle traces left. It is evident that a heat treatment for 1 hour at 400°C is better than 2 hours at 330°C to reduce heavy centerline segregation [4]. The average grain size varies from 4 to 11 micron from the surface to the strip center. Table 2 compares the tensile properties in longitudinal direction of AZ31 sheet produced by the TRC with commercial sheet produced by direct chill ingot casting process [4]. Yield and tensile strength of AZ31 sheet produced in the TRC was comparable to commercial AZ31 sheet while elongation of the TRC produced sheet was superior to commercial sheet.

Alloy	YS (MPa)	UTS (MPa)	El (%)
AZ31-TRC Sheet Hot rolled and annealed	197	272	21
AZ31B-H24 [5]	220	290	15

Table 2: Comparison of the tensile properties of AZ31 sheet produced by TRC and direct chill ingot casting process.

Other process parameters can have a significant effect on the quality of the strip When the caster is running too hot and too fast, solidification is not completed before the kissing point and the magnesium strip surface defects appear such as burnout (excessive surface oxidation), surface cracks (solidification cracking or hot tearing) and oxide inclusions. Inadequate melt supply, due to the flow characteristics in the nozzle, was hypothesized to lead to lateral freezing of the strip edges. In order to minimize expensive experimental effort, a modeling program was initiated. A preliminary report of the modeling activities is given in the following sections.

Modeling

Background and Modeling Details

Both three- and two-dimensional models were constructed in the finite-volume-based computational fluid dynamics package STAR CCM+ [6]. The package also has a casting module, STAR Cast, which contains material data for a wide variety of casting materials and enables efficient setup of casting simulations.

The two-dimensional computations took about 3 hours of compute time for 1 second of physical time on 4 CPUs. For 6 seconds of physical time, about 24 hours were required.

The two dimensional analyses were used primarily to provide a fast-turnaround simulation to determine the boundary condition, material property and mesh requirements for the three-dimensional case. The three-dimensional computations used 740,000 cells in the mesh and took 6.75 hours of compute time on 32 CPUs to model 1 second of physical time. The three-dimensional simulation modeled 20 seconds of physical time, for a total run-time of 6 days. Figure 6 presents the three-dimensional mesh. Only one-half of the model was constructed due to symmetry. Anisotropic refinements were used to capture gradients efficiently, and a fine prism mesh was used at the solid-fluid interface. Rolls were modelled as rigid bodies with a tangential velocity boundary condition at the fluid interface, and the velocity being governed by the rotational speed; the tangential velocity was varied from 3 m/min to 2 m/min at 13.7 s of physical time.

The initial temperature of the air and the ceramic nozzle were taken to be 400°C. Properties of the AZ31 magnesium alloy were taken from reference [7] and a dynamic viscosity of 0.001 Pa-s or 100 cP was assumed for the melt. The ceramic of the solid nozzle was modelled using the thermal and mechanical properties of ceramic material CALCAST CC450. Surface tension of the melt was also modelled.

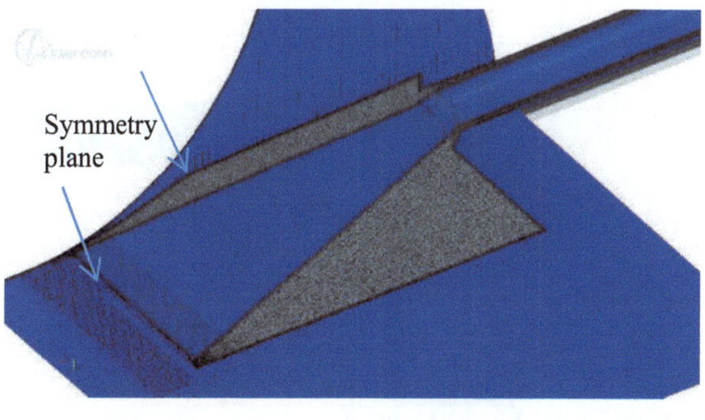

Figure 6: Polyhedral mesh in solid nozzle, hexahedral trim mesh to model the fluid. One-half of model is shown, symmetry is assumed.

The time step in the computation was based on an automatic algorithm, which is governed by the physics of the air-melt interface. Pressure at the inlet of the nozzle was applied based on the pressure head measured during the experiment.

Computational Results

Results from the three-dimensional computations at approximately 9.8 s and 13.9 seconds are shown in Figure 7 and Figure 8, respectively. The speed of the rolls was slowed at ~13.8 seconds in the simulation to show the effect of changing roll speed. These figures show the temperature field of the solidified melt near the nozzle tip, looking from the symmetry plane of the model.

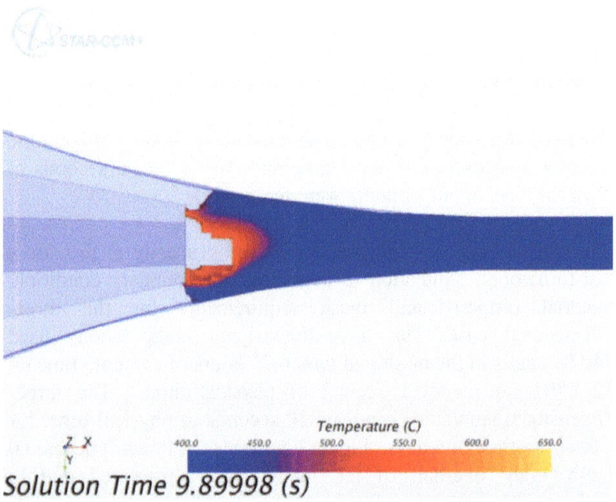

Figure 7: Temperature field from three-dimensional computation at 9.9 seconds.

Two features are marked in the results. Firstly, there is significant flow upstream after the filling of the nozzle. After the rolls are slowed at 13.8 seconds, there is solidification of the melt around the horizontal edges of the nozzle. This behavior could support the drops in temperature measured by the thermocouple TC3 shown in Figure 3.

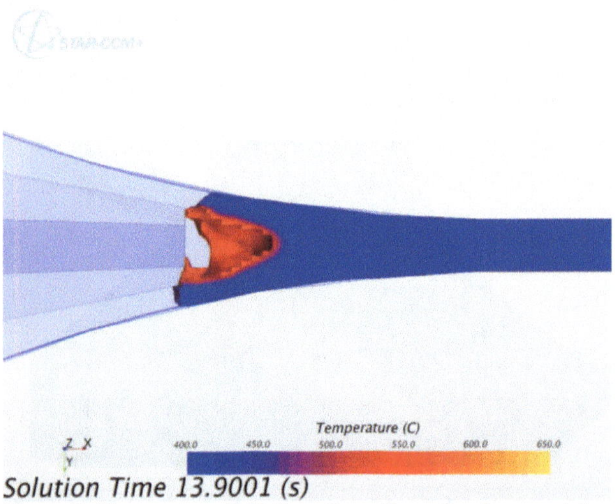

Figure 8: Temperature field from three-dimensional computation at 13.9 seconds.

Furthermore, there are air pockets that travel along the sides of the nozzle to the nozzle-box interface and these pockets persist, Figure 9.

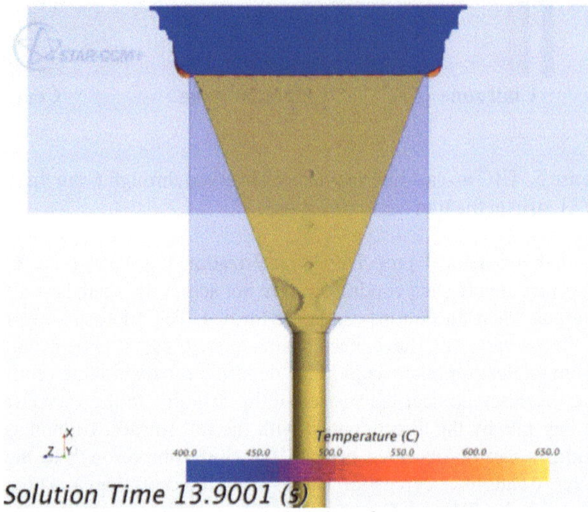

Figure 9: Air pocket at nozzle-box interface.

Concluding Remarks

In an endeavour to create highly formable, low cost magnesium sheet for automotive light-weighting, CANMETMaterials has been performing research on twin roll cast AZ31 magnesium alloy. However, solidification effects in the environ of the ceramic nozzle and rollers are known to adversely affect strip quality. To gain a better understanding of the impact of the nozzle design and operating parameters, a three-dimensional model of the TRC was constructed and multi-phase computations were carried out. Key insights, to date, are that there is significant recirculation of the flow in the nozzle which could lead to insufficient melt supply to the gap between the rolls and contribute to the observed defects. Future modeling work will focus on incorporation of the thermocouples in the model, which we suspect is perturbing the flow field and causing instabilities in the flow, and further validation of the model by comparing with the measured temperature histories. Additionally, optimization of the nozzle design using a parametrized model is planned.

References

[1] F. Czerwinski: *Magnesium Alloys - Design, Processing and Properties*, InTech, January 14, 2011, pp. 299.

[2] H. Chen et al, *Materials Science and Engineering A*, Vol. 492, 2008, pp. 317-326.

[3] A. Hadadzadeh, M.A. Wells and E. Essadiqi, *Magnesium Technology 2012*, pp.141-145.

[4] M. Aljarrah, et al, *Materials Science Forum*, Volume 690, 2011, Pages 331-334

[5] *Magnesium and Magnesium Alloys*, ASM Specialty Handbook, ASM International, September 1999.

[6] STAR-CCM+ (2014). Release version 9.02, CD-adapco, www.cd-adapco.com.

[7] A. Hadadzadeh, M. A. Wells, *International Journal of Cast Metals Research*, Vol. 26, Number 4, Aug 2013, pp. 228-238

Magnesium Technology 2015
Edited by: Michele V. Manuel, Alok Singh, Martyn Alderman, and Neale R. Neelameggham
TMS (The Minerals, Metals & Materials Society), 2015

MICROSTRUCTURE EVOLUTION OF DIFFERENT MAGNESIUM ALLOYS DURING TWIN ROLL CASTING

Gerrit Kurz[1], Joachim Wendt[1], Jan Bohlen[1], Dietmar Letzig[1]
[1]Helmholtz-Zentrum Geesthacht – Magnesium Innovation Centre; Max-Planck-Straße 1, 21502 Geesthacht, GERMANY

Keywords: magnesium, aluminum free, twin roll casting, magnesium strip

Abstract

Twin-roll casting (TRC) is an economic production process for the generation of feedstock materials that can subsequently be rolled to sheets. The melt flows from a tundish by gravity through the tip into the gap of the rolls of the twin roll caster. The metal leaving the tip solidifies on the rolls into a strip that is further deformed by the rolls. Unfortunately, commercial magnesium sheets do not provide the required properties due to limitations like e.g., low tensile strength and low formability. In order to improve the properties of magnesium sheets there are two possibilities. One possibility is to optimize the process parameters of the twin roll casting process and the subsequent rolling process. The other possibility is to develop new magnesium alloys and process them to sheet metal. This paper reports on first results of twin roll casting experiments on new developed aluminum containing and aluminum-free magnesium alloys. The influence of alloy composition on the microstructure of the strip is presented and discussed with respect to arising texture.

Introduction

Currently magnesium alloys are used for different applications in the transportation industry where cast magnesium alloys dominate the market. Although cast alloys predominate over wrought products such as extrusions, forgings, sheets and plates, the latter are also being used in a variety of different applications. Recently, the growing interest in the automotive industry for looking at potential applications for magnesium turned back towards wrought alloys. Typically, applications of magnesium sheets are sought in automotive (Figure 1) or aeronautics industry.

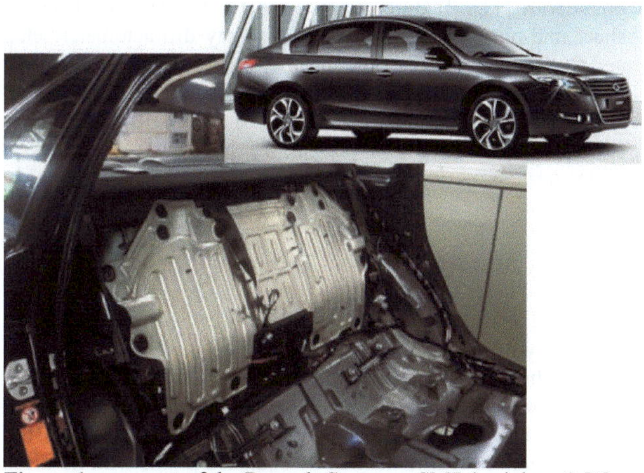

Figure 1: rear seat of the Renault-Samsung SM7 (serial part) [1]

However, the mechanical properties of die cast materials often do not meet essential requirements with regard to endurance, strength, ductility, etc. A promising alternative for thin, large area parts, such as automotive body components is to utilize sheet material. Sheet metal formed parts are characterized by superior mechanical properties and high quality surfaces without pores in comparison to die cast components. Twin-roll casting is an economic production process for the generation of fine-grained feedstock materials that can subsequently be warm rolled to thin sheets. This production process for thin strips combines solidification and rolling into one single production step and so saves a number of rolling and annealing passes in comparison to the conventional rolling process. Twin roll casting technology is already well established in for example the aluminum industry, but applications to magnesium alloys are in their infancy at the present time [2]. There are worldwide a small number of industrial or laboratory scale twin roll casters installed at universities, companies and research facilities. Initial results from these activities on conventional wrought and cast alloys have shown promising sheet properties [3-7]. Substitution of conventional sheet materials such as steel or aluminum by magnesium sheets could lead to significant weight savings. The presented paper describes a new approach allowing to tackle the above drawbacks by applying the twin roll casting (TRC) [8] to newly developed wrought magnesium alloys.

Twin Roll Casting Equipment at HZG

The development of wrought magnesium alloys and their introduction into industrial, structural applications are the main goal of the activities at the Magnesium Innovation Centre MagIC of the Helmholtz-Centre Geesthacht (HZG). The current focus of the research work is on alloy design and the development of processing technologies for semi-finished magnesium products. In the particular case of sheet materials, it has been recognized that the feedstock for the warm-rolling process needs to be in the form of thin bands as they are produced via twin roll casting, if thin magnesium sheets are to become competitive industrial products. For this purpose HZG has installed the twin roll caster shown in Figure 2 and 3.

The twin roll casting line consists of a furnace line (Striko-Westofen) and a twin roll caster (Novelis, Jumbo 3CM). The twin roll caster is designed to produce thin strips of magnesium with a maximum width of 650 mm and a thickness in the range 4 -12 mm at a maximum rolling speed of 6 m/min. The furnace line is configured to allow manual loading of raw materials in ingots and the possibility to vary the alloy composition, i.e. cast strips can be made from different alloys. The furnace line includes an ingot pre-heater to remove humidity from the ingots and accelerate fusion in the melting furnace. After pre-heating, the feedstock is transferred to the melting furnace. A second transfer device conveys the molten metal to the cleaning furnace. From the cleaning furnace the melt is transferred to the head-box which is connected to the tip. The liquid metal flows by gravity through the tip into the gap of the rolls of the twin roll caster. The metal exiting the tip solidifies on the rolls into a strip that is further deformed by the rolls.

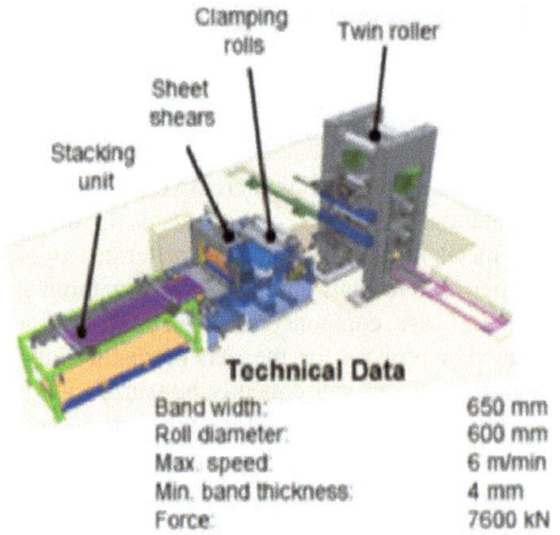

Figure 2: Twin roll caster at HZG and technical data

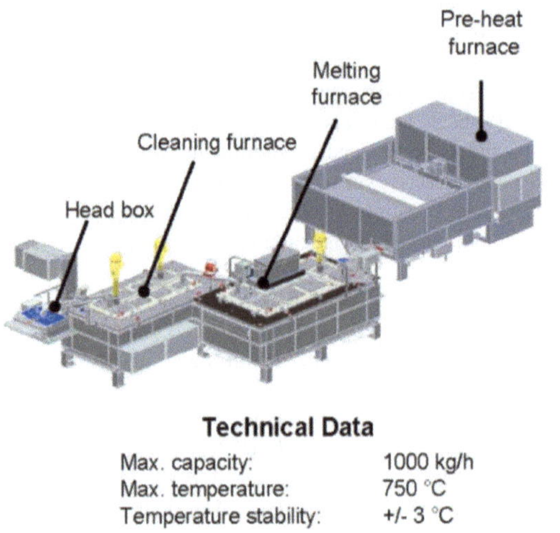

Figure 3: Furnace line of the twin roll caster at HZG and technical data

Twin Roll Casting Trials

In a first step strips were produced from twin roll cast AZ31 strip material, which serve as reference material for further developments. In order to improve the formability of magnesium sheets, the following magnesium alloys were developed and twin roll cast:

- AZ31 (reference)
- Mg-Zn-Zr-Y-RE (alloy1)
- Mg-Zr-Zn-Ca-Y-RE (alloy2)
- Mg-Al-Zn-Mn-Ca–Sr-RE (alloy3)

The exact alloy compositions are confidential. The alloy1 and alloy2 were aluminum-free Zink based alloys and the alloy3 was an modified AZ alloy. In alloy1 the Zn was the main alloying element. Whereas Zn was used to confer improved strength properties, the RE-element additions enable the development of a weak non-basal texture of the rolled sheets, which is favourable for improving formability [9]. The addition of Zr should work as a grain refiner.

For enhanced hardness and strength of the material in alloy2 the amount of RE was significantly increased, in order to harden the material by age hardening. Because Zn has got the tendency to react with the Yttrium and rare earth elements to coarse intermetallic phases the amount of Zn was significantly reduced in alloy2.

As an alternative to the two aluminum-free alloys the aluminum containing alloy3 with a low RE content was developed. In order to improve the strength of the material the amount of Al is decreased in comparison to the reference alloy AZ31. RE in aluminum based magnesium sheets weakens the texture intensity what leads to a better formability of the finish magnesium sheets. [10]. The alloying elements Ca and Sr should also improve the formability and mechanical properties.

By means of the TRC-process this alloy was cast into strips of 5.6 mm thickness and 350 mm width. The process parameters see in the table below.

Table I: Parameters of the twin roll cast process

	AZ31	Alloy1	Alloy2	Alloy3
Casting temperature	650 °C	700°C	700°C	680°C
Rolling speed	1.8 m/min	2.0 m/min	2.0 m/min	2.0 m/min
Final gauge	5.6 mm	5.6 mm	5.6 mm	5.6 mm
Width	350 mm	350 mm	350 mm	350 mm

After twin roll casting, the microstructure of the strips was analyzed using optical microscopy. Standard metallographic sample preparation techniques were employed and an etchant based on picric acid was used to reveal grains and grain boundaries [11]. Texture measurements were performed on the sheet mid-planes using a Panalytical X-ray diffractometer setup and CuKα radiation. The pole figures were measured up to a tilt of 70° which allowed recalculation of full pole figures based on an MTEX software routine [12] The (1000) and (1010) pole figures are used in this work to present the texture of the strips at midplane.

Microstructure and Texture of the Strips

Figures 4 to 7 depict the microstructures of samples extracted from strips obtained from twin roll casting trials of the different alloys. The microstructure of the alloy AZ31 consists of dendritic and equiaxed grains, which are elongated and tilted in the rolling direction in the near-surface regions of the strip (Figure 4). The center region is clearly marked by equiaxed grains with no tilt in rolling direction. An average grain size of 170 µm in the center section was determined by means of the line intersection method. This microstructure is typical for deformed AZ31 strip material. The (0001) pole figure shown in Figure 6 is characterized by higher maximum intensities (I_{max} = 3.7) than in the other strips and a distinct basal character of the texture. This is a different texture compared to typical textures of conventionally rolled magnesium sheets of this alloy where a broader angular distribution is typically found towards the rolling direction, if any. However, Haßlinger et al. [13] were able to show that such textures as observed in this work during twin roll casting can also occur during the course of rolling if the degree of deformation is still low.

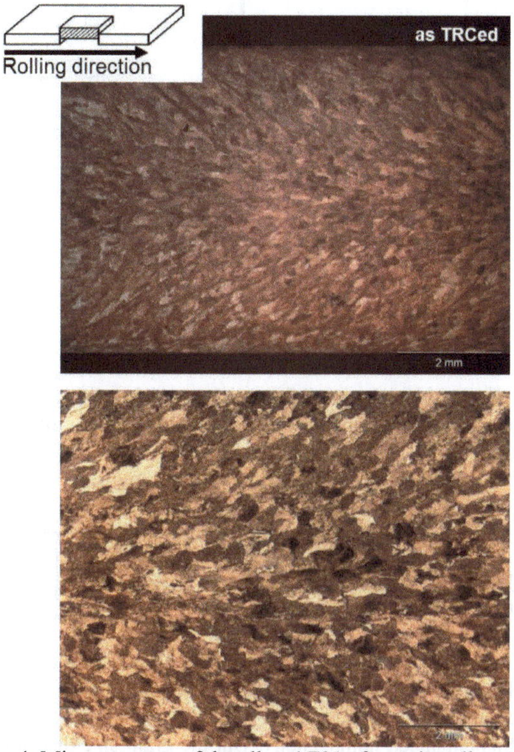

Figure 4: Microstructure of the alloy AZ31 after twin roll casting

In comparison to the AZ31 the twin roll cast strip of the alloy1 shows a homogeneous microstructure with slightly squeezed equiaxed grains with a significantly decreased grain size (average grain size of 26 µm) (Figure 5). The grain refining effect of the Zr leads to very fine grained microstructure with a very homogeneous grain size distribution. A very weak centreline segregation was revealed as well as numerous precipitates especially located at grain boundaries. EDX analyses showed, that the precipitates contain a large content of Zn and a less content of Y and RE. Contrary to the microstructure of the alloy AZ31, the fine grained microstructure of the alloy1 strip is favourable for good rollability. Figure 8 also shows recalculated (0001) pole figure of the strip. A tendency of the development towards a "basal" texture is found, however with a low intensity of I_{max} = 2.0, what is also favorable for good formability.

Figure 6 displays the microstructure of the strip of alloy2. The strip of alloy2 shows also a homogeneous microstructure with slightly deformed equiaxed grains and a similar average grain size like alloy2. The centreline segregation was also not very distinct like in alloy1. Also numerous precipitates were obseved especially located at grain boundaries but also located in the matrix.

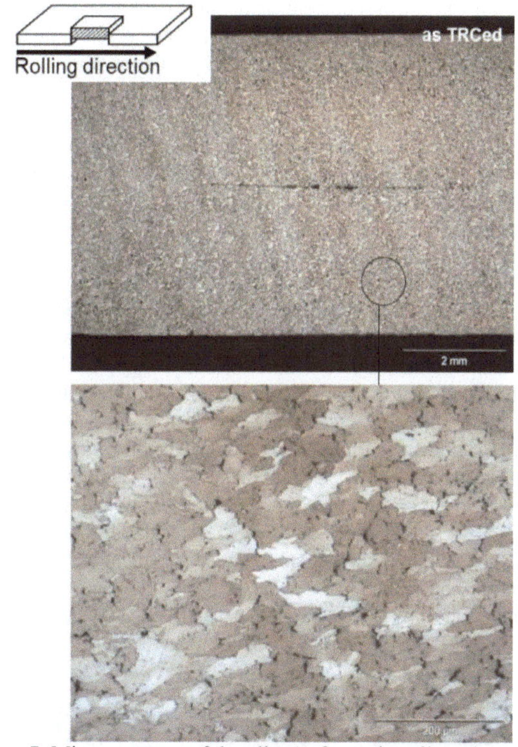

Figure 5: Microstructure of the alloy1 after twin roll casting.

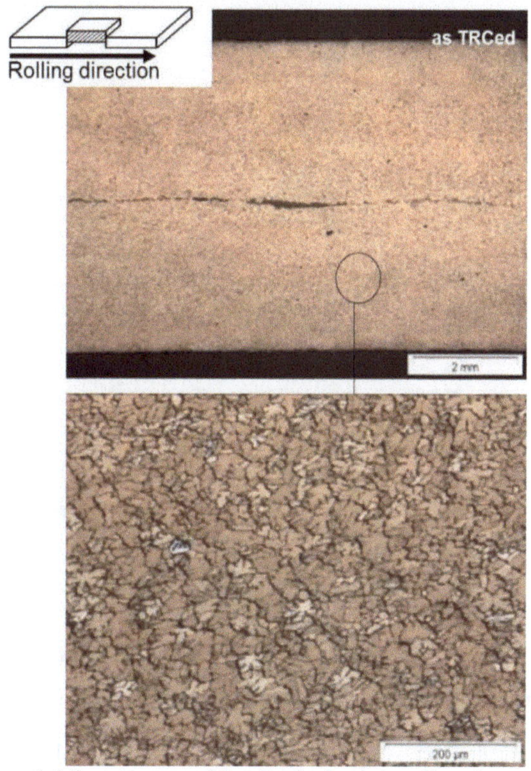

Figure 6: Microstructure of the alloy2 after twin roll casting.

Further investigations by a scanning electron micrograph SEM were carried out, in order to characterize the precipitations. In these investigations two different types of precipitates were found in the intermetallic phase the alloy2. In the segregation zone in the center of the strip coherent lamellar precipitates as well as precipitates at the grain boundaries, that enclose the grains, were observed. The precipitates at the grain boundaries were also found in the regions near the strip surface. The lamellar precipitates contained next to a high content of Mg and RE low contents of Zn, Y, Zr and Ca. In comparison to the lamellar precipitates the precipitates at the grain boundaries contain less contents of RE but similar contents of Zn and Ca. The content of Y and Zr is higher in the precipitates at the grain boundaries. In the matrix are very small contents of solute Zn and RE.

The texture of alloy2 is comparable to the texture to the texture of the alloy1. The very fine and homogeneous microstructure in combination with the low texture intensity, $I_{max} = 1.9$ leads to the conclusion that alloy2 is also favorable for good formability.

Figure 7 illustrates the microstructure of the alloy3. A clear dendritic structure growing from the upper and bottom surfaces of the strip towards the center, with a region containing equiaxed grains at the center can be observed. A remarkable feature is the strong segregation band containing a large amount of impurities at the centre-line of the strip. The large number of dendritic grains in the upper and lower regions of the strip is an indicator for a low degree of deformation during the rolling operation.

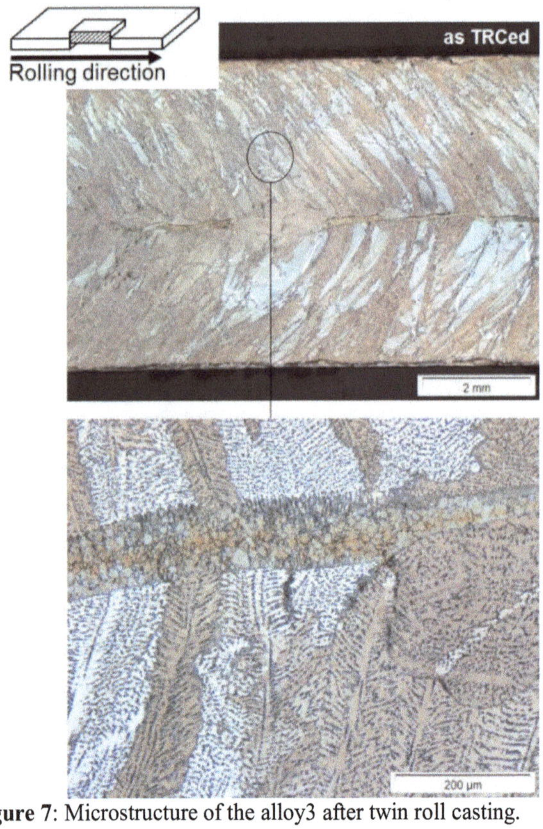

Figure 7: Microstructure of the alloy3 after twin roll casting.

Precipitates were found as interdendritic particles, which often contained Y and Mn, as well as coherent eutectic precipitates. The coherent particles were mainly $Mg_{17}Al_{12}$ precipitates. Also small areas of Ca and Zn containing particles were found.

Figure 8 also shows recalculated (0001) pole figures of the strip which both exhibit low maximum intensities (I_{max} = 2,8). A very weak alignment of basal planes parallel to the strip plane corresponds with a low degree of deformation.

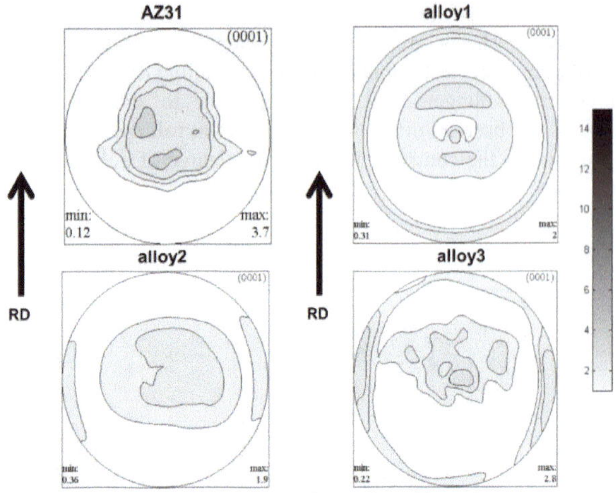

Figure 8: Textures of all alloys after twin roll casting.

Heat Treatment of the Strips

In order to solve the precipitates in the matrix and improve the microstructure the strip material were heat treated as followed:

Table II: Parameters of the heat treatment

	AZ31	Alloy 1	Alloy 2	Alloy 3
Temperature	450 °C	400°C	525°C	450°C
Duration	min	16h	12h	16h

Figure 9 displays the microstructure of the heat treated AZ31. It can be observed that the material still contains both features of directionally solidified grains and recrystallized grains but exhibits enhanced homogeneity. But the grain size distribution is very inhomogeneous and the segregation zone in the center region is still clearly visible. This indicates that the deformation in the strip was low.

Figure 9: Microstructure of the alloy AZ31 after heat treatment

In Figure 10 the microstructure of the homogenized alloy1 can be seen. The homogenization treatment did not affect the grain sizes but allows resolving the precipitates at the grain boundaries. The particles are now spread over the whole microstructure. Contrary to coarse grained AZ31, the fine grained microstructure of the strip of the alloy1 is favorable for good rollability.

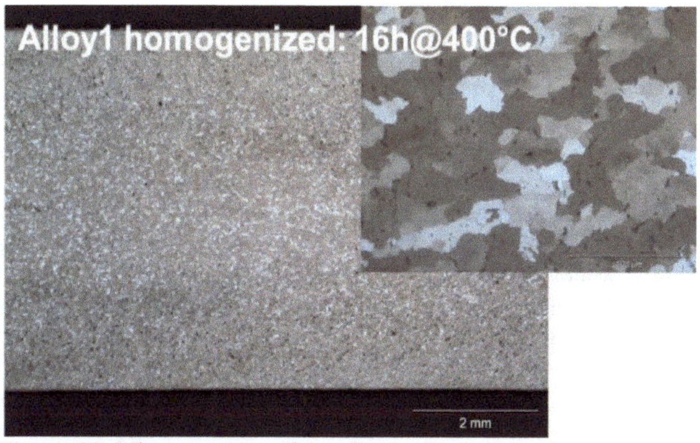

Figure 10: Microstructure of the alloy1 after heat treatment

Figure 11: Microstructure of the alloy2 after heat treatment

Figure 12: Microstructure of the alloy3 after heat treatment

Figure 11 displays the microstructure of the alloy2 after homogenization at 525 °C. The microstructure of the alloy2 shows in general fine grains with a homogenous distribution at the upper and lower surface region (Figure 11). In contrast to this in the center region a significant grain growth can be observed. This grain growth in the center region indicates that the deformation in the strip only near the surface took place. The lamellar precipitates in the center region are disappeared and a band of particles are left. Between the larger particles there are often very small cubic or rod shaped particles. The coherent precipitates at the grain boundaries are widely destroyed or in solution. Only small particles are left. Near the strip surface are only very small precipitates at the grain boundaries visible. In contrast to this in the Matrix new fine, specular phases are occurred.

In Figure 12 the microstructure of the homogenized strip of alloy3 is shown. The columnar structure is still present but the material is partly recrystallized. After the heat treatment the eutectic particles disappeared widely. But during the homogenization in some areas on the strip surface oxidation took place. This microstructure after the heat treatment is an indicator that the material was low deformed in the twin roll casting process.

Conclusions

The results show that alloy1 and alloy2 are two promising new alloys for the twin roll casting process. The alloy3 offers potential for optimization. As the casting parameters are very comparable in all cases and the resulting textures exhibit features of deformed microstructures after melt solidification, it is noteworthy, that the alloy compositions lead to distinct variations in the resulting microstructures. Whereas AZ31 exhibits a classical multi-feature microstructure with columnar grains grown from the surface due to directional solidification, alloy1 and alloy 2 do not exhibit such a feature, but are rather homogeneous and fine-grained in their microstructures. The content of Zr in these alloys as well as the absence of Al allows to hypothezise that heterogeneous grain nucleation during solidification is distinctly enhanced in these alloys.

This is not the case in alloy3, an aluminium containing alloy including a certain amount of rare earth elements. Again, columnar solidification appears to be of higher significance with almost no visible central area of globular grain structure. In correlation to a companion work [14] it is suggested that alloy 3 exhibits slower solidification rates by maintaining columnary grown grains. The central area textures exhibit components with basal planes aligned in the sheet plane, indicating an impact of resulting deformation due to rolling. However, all alloys with the exception of AZ31 also exhibit other texture components where basal planes are tendencially aligned parallel to ND. Obviously, the interaction between solidification and deformation is more complex at this point, indicating an influence of the different, finer grained and more homogeneous microstructure. In alloy 3 it is assumed that the latter component may be a remainder of directional solidification as proposed in [14] for alloy AZ31.

The selected annealing procedures are appropriate to improve the homogeneity of the precipitation structure and may enhance recrystallization slightly. However, the main features of the microstructures persist during annealing.

Former results indicated that the processing parameters in twin roll casting have also a significant influence on the microstructure and texture of the strips. The in this paper presented results show that the alloy composition has got also a significant influence of the strip properties. So the complexity of the twin roll casting and rolling process in combination of the possibilities of the alloy composition offers numerous chances to tune the properties of magnesium sheets. Further rolling trials to final gauge sheet will have to demonstrate in how far the differences in the alloy composition have an impact on the performance of the resulting sheets. Due to the fact that magnesium is about 35 % lighter magnesium sheet out of twin roll cast strip offers a very high potential for the substitution of mild steel and aluminum alloys for automotive components.

Acknowledgements

This research was performed as a collaboration project within the European EUREKA R&D Platform (EUREKA project No 5874). The work was financially supported by the Federal Ministry of Economics and Technology on the basis of a decision by the German Bundestag within the framework of the Central Innovation Programme SME (ZIM) and by the Office of the Chief Scientist (OCS) of the Ministry of Industry, Trade and Labour of the Israeli Government. This research was also supported by funding from the German Federal Ministry of Education.

References

[1] http://www.autoherald.co.kr/news/articlePrint.html?idxno=12185.

[2] F. Basson, D. Letzig, "Aluminium twin roll casting transfers benefits to magnesium", *Aluminium International Today*, November/December 2010, 19-21.

[3] D.H. St John, "Overview of current international magnesium research and recent CAST CRC developments", *J. Advanced Materials Research*, 29/30, 2007, 3-8.

[4] S.S. Parks, G.T. Bae, J.G. Lee, D.H. Kang, K.S. Shin, N.J. Kim, "Microstructure and mechanical properties of twin-roll strip cast Mg alloys", *Materials Science Forum*, 539/543, 2007, 119-126.

[5] R Kawalla, M Oswald, C. Schmidt, M. Ullmann, H.P. Vogt, N.D. Cuong, "Development of a strip-rolling technology for Mg alloys based on the twin-roll-casting process", *Magnesium Technology 2008*, New Orleans, Louisiana, TMS, 2008, 177–182.

[6] H. Watari, T. Haga, R. Paisarn, N. Koga, K. Davey, "Mechanical properties and metallurgical qualities of sheets manufactured by twin-roll casting", *Key Engineering Materials*, 345/346, 2007, 165–168.

[7] M. Aljarrah, E. Essadiqi, D.H. Kang, I.H. Jung, "Solidification microstructure and mechanical properties of hot rolled and annealed Mg sheet produced through twin roll casting route", *Light Metals Technology V*, 5th International Conference on Light Metals Technology, 19-22 July 2011, Lueneburg, Germany, 331-334.

[8] G. Kurz, J. Bohlen, D. Letzig, K.U. Kainer, "Influence of Process parameters on Twin Roll Cast Strip of the Alloy AZ31", *Materials Science Forum Vol. 765 (2013)*, 205-209. Doi: 10.4028/www.scientific.net/MSF.765.205.

[9] K. Hantzsche, J. Bohlen, J. Wendt, K.U. Kainer, S.B. Yi, D. Letzig, "Effect of rare earth additions on microstructure and texture development of magnesium alloy sheets", View Point Paper, *Scripta Materialia 63 (2010)*, 725–730.

[10] J. Bohlen, M.R. Nürnberg, J.W. Senn, D. Letzig, S.R. Agnew, "The texture and anisotropy of magnesium-zinc-rare earth alloy sheets", *Acta Materialia 55 (2007)*, 2101-2112.

[11] V. Kree, J. Bohlen, D. Letzig, K.U. Kainer, "Practical Metallography 5", 2004, 233-246.

[12] F. Bachmann, R. Hielscher, H. Schaeben, "Texture Analysis with MTEX - Free and Open Source Software Toolbox", *Solid State Phenomena (2010)*, 160, 63-68.

[13] U. Haßlinger, L. Fuskova, Ch. Hartig, R. Günther, D. Letzig, K.U. Kainer, R. Bormann, "Effects of Ceramic Inoculants and Rare Earth Phases on Microstructure and related Mechanical Properties of Magnesium alloys", *8th International Conference on Magnesium Alloys and their Applications, 2009*, edited by K.U. Kainer, 1260-1267.

[14] J. Bohlen, S. Yi, J. Victoria-Hernández, N. Schell, B. Schwebke, H.-G. Brokmeier, G. Kurz, D. Letzig, "The Microstructure and Texture Development during Twin Roll Casting and Rolling of Magnesium Alloy AZ31", (this volume).

THE MICROSTRUCTURE AND TEXTURE DEVELOPMENT DURING TWIN ROLL CASTING AND ROLLING OF MAGNESIUM ALLOY AZ31

Jan Bohlen[1], Sangbong Yi[1], José Victoria-Hernández[1], Norbert Schell[1], Bernd Schwebke[1,2], Heinz-Günter Brokmeier[1,2], Gerrit Kurz[1], Dietmar Letzig[1]

[1]Magnesium Innovation Centre (MagIC) and High Energy Materials Science (HEMS) at PETRA III, Helmholtz-Zentrum Geesthacht, Max-Plack-Str. 1, D21502 Geesthacht, Germany

[2]Institute of Materials Science and Engineering, TU Clausthal, Agricolastr. 6, 38678, Clausthal-Zellerfeld, Germany

Twin roll casting, rolling, magnesium alloy AZ31, microstructure, texture, synchrotron diffraction

Abstract

Twin-roll casting of magnesium alloys is seen as a promising processing route to enable a further development of sheet properties. In this regard the influence of melt temperature and rolling speed on the microstructure and texture development of magnesium strips is investigated for alloy AZ31. Microstructure and through-thickness texture measurements are used to reveal effects of grain growth during solidification and deformation of the strip. For this purpose synchrotron radiation at the PETRA III storage ring of DESY in Germany has been used. Twin-roll casting with varied processing parameters leads to microstructures ranging from directionally solidified grains to deformed microstructures with a profound fraction of recrystallized grains. Specifically, sheets rolled from different strips exhibit an alignment of their Mg basal planes in the sheet plane with certain significance depending on grain size and grain orientation effects.

Introduction

Magnesium and especially magnesium alloys find widespread interest as being the lightest available metallic construction materials. It is recognized that the utilization of semi-finished products such as sheets is a prerequisite to expanding technical applications [1]. One limitation is the crystallization of magnesium with a hexagonal close packed lattice structure. It is principally accepted that the comparably low formability of sheets is based on the resulting limitation in the number of active deformation mechanisms, at least in comparison to cubic metals.

Such low formability [2, 3] of magnesium alloys also has a significant impact on the process ability during massive deformation. Even at high temperatures, it leads to complex rolling plans with a high number of rolling passes at comparably low degree of deformation per pass. Although the feasibility of sheet rolling has been shown for a wider selection of magnesium alloys [4, 5], the processing remains complex, time consuming and therefore expensive. It has therefore been recognized that the processing route of magnesium sheets requires an improvement of the efficiency for the production of sheets.

In the specific case of magnesium alloys twin roll casting is a processing route which is seen as an important approach to enable such an improvement [6]. It has been shown that twin roll casting is an efficient process for the generation of fine-grained strip as feedstock for sheet rolling. It combines casting and rolling in one processing step and therefore improves the processing efficiency of magnesium sheets [7].

In this process, liquid metal of the respective alloy is pumped from a furnace or oven into a tundish by also using protective gas coverage to avoid burning. The melt then flows into a flat pipe (tip), leading it into the roll gap of a pair of counter rotating rolls. The metal solidifies upon contact with the initially cooled rolls and is thus rolled to a strip with a typical thickness of a few millimeters. As a result, microstructure development features of both processes are likely to occur, including directional solidification, globular solidification, deformation and recrystallization. Such different mechanisms have a profound influence on the resulting microstructure and texture of twin roll cast strips and may be significantly influenced by a variation of process parameters such as melt temperature and rolling speed.

Twin-roll cast strips are then used as feedstock to conventionally roll sheets to final gauge in only a few rolling passes. The reduction in the number of processing steps to final gauge leads to shorter production times and a decrease in production costs. The microstructure development during rolling is then also influenced by the condition of the feedstock used, i.e. the microstructure and texture of the strips.

The objective of this work is to investigate the influence of process parameters like melt temperature and rolling speed on the microstructure and texture development during twin roll casting of magnesium strips. Through-thickness texture analysis of twin roll cast strips of alloy AZ31 is used to reveal effects of grain growth during solidification as well as of the degree of deformation of solidified material. Exemplarily, strips with different microstructures and textures are used for rolling them to final gauge. The influence of strip microstructures and textures is then investigated with respect to their development during rolling. For the purpose of local texture measurements along the thickness direction of the strips and sheets synchrotron radiation has been used. A synchrotron radiation beam with well-defined size and position along the sample thickness allows a fast and full detection of diffraction patterns to be carried out without the need of sample surface preparation.

Experimental

Commercial ingots of magnesium alloy AZ31 (Mg + 3.0 wt.% Al + 1.0 wt.% Zn + 0.3 wt.% Mn) were used for twin roll casting experiments. A twin roll casting line at the Magnesium Innovation Centre MagIC at the Helmholtz-Zentrum Geesthacht is used consisting of a furnace line (Striko-Westofen) and a twin roll caster (Novelis, Jumbo 3CM). Details can be found elsewhere [8]. Experiments were performed with the melt temperature varied between 650 °C and 715 °C (furnace temperature) and the rolling speed varied between 1.4 m/min and 3.7 m/min. In these experiments the strip width was 350 mm.

A 50 ton laboratory rolling stand with cold rolls of 400 mm diameter was used to carry out rolling experiments on two exemplary strips. The rolling temperature was 400°C which was received after heating the samples for 30 min prior to rolling. A rolling procedure with increasing degree of deformation per pass was applied, starting with 2 passes of 0.1 followed by 4 passes of 0.2. Thus, the rolling procedure consisted of 6 passes with a total degree of deformation of 1.0. The rolled samples were reheated for 20 min between the rolling passes. The rolling speed was constant at 16 m/min. Samples were cut from the rolled strip after each rolling pass. This allowed the microstructure and texture development to be tracked during the rolling procedure.

The microstructure of the strips was analysed using optical microscopy. Typical metallographic sample preparation procedures were used including an etchant based on picric acid [9] in order to reveal grains and grain boundaries on longitudinal sections of the twin roll cast strips and the rolled sheets. Texture measurements were carried out using hard x-ray diffraction at 84 keV at a synchrotron radiation beamline HEMS-P07B (PETRA III storage ring of DESY in Germany) with dimensions 0.7 mm x 0.5 mm. The samples were placed into the beam and the height was varied so that measurements were subsequently carried out at 3 - 5 positions across the thickness in order to investigate the texture gradient. Sample rotation was 180° with 3° steps on the sample at each position. Debye-Scherrer patterns were collected using a Perkin Elmer XRD1622 detector. The {10.0}, (00.2), {10.1}, {10.2}, {11.0} and {10.3} reflections were analysed after background correction and pole figure data were calculated using data specific computer codes [e.g. 10, 11].

Twin roll casting of magnesium alloy AZ31

The microstructure and texture development during twin roll casting of magnesium alloy AZ31 are investigated with respect to the influence of melt temperature and rolling speed. Fig. 1 collects the microstructure and texture results by relating the local texture at different through-thickness positions to the respective local microstructure. The melt temperature was highest in the context of this study at 710 – 715°C. Fig. 1 a-d show results for decreasing rolling speed from 3.7 mm/min in Fig. 1a to 1.4 mm/min in Fig. 1d. The microstructure consists of different fractions including grains of columnar type close to the surface of the sheet growing into the strip centre. Very close to the surface of the strips a layer of fine grains is found. The mid plane area of the strips typically exhibits globular grains interpreted as a result of non-directional solidification. In this area a profound fraction of particles, partly very pronounced, consisting of impurities or concentrated alloying elements can also be seen. Decreasing the rolling speed at this high melt temperature correlated with a decreasing amount of such particles or segregation bands.

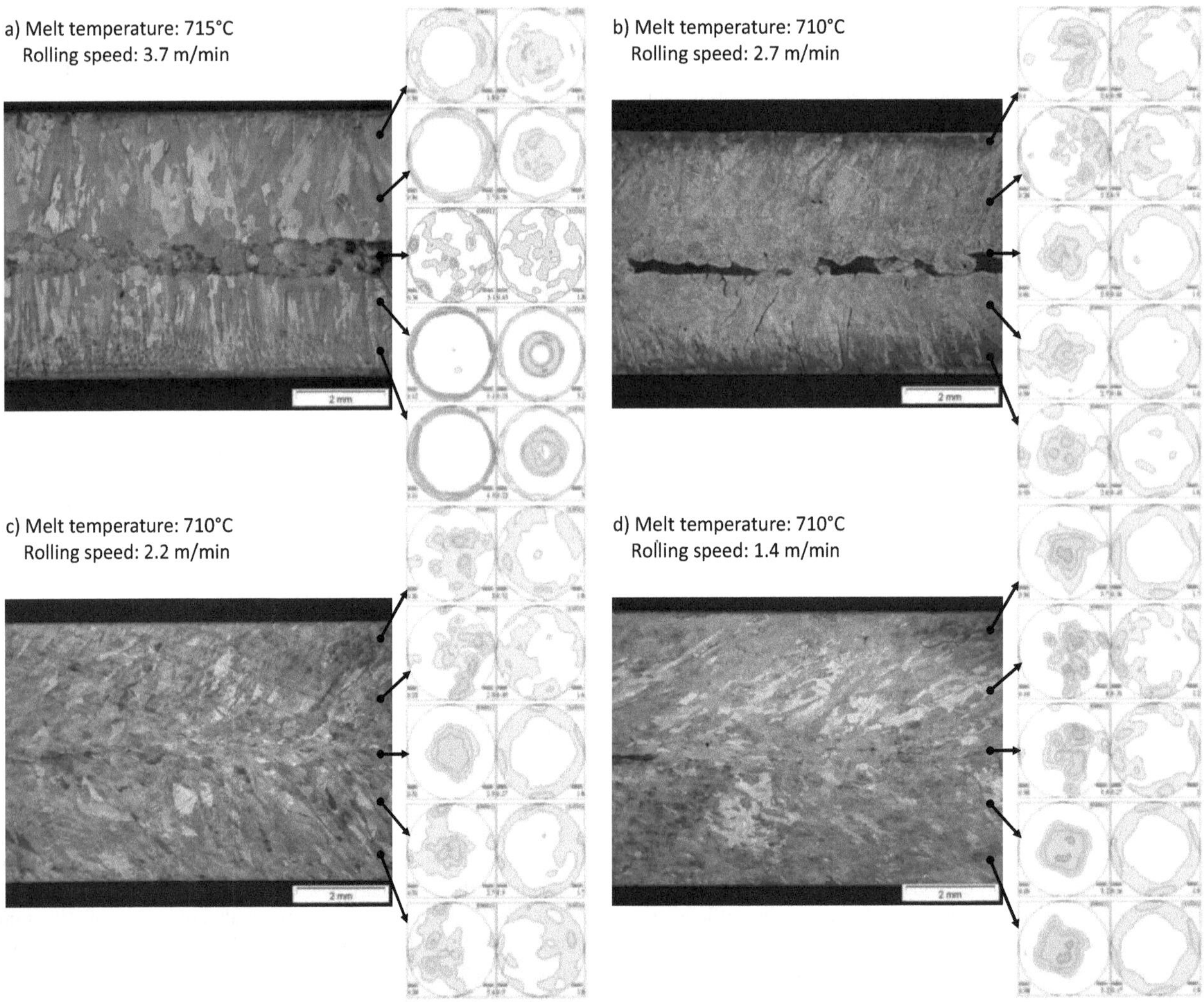

Fig. 1: Longitudinal sections of twin roll cast strips of magnesium alloy AZ31 cast at 710-715°C with varied speed; a) 3.7 m/min, b) 2.7 m/min, c) 2.2 m/min, d) 1.4 m/min (twin roll casting direction horizontal); pole figures: RD vertical, TD horizontal, ND normal, levels: 1.0, 1.5, 2.0, 3.0, 5.0, ... multiples of random distribution

Each micrograph in Fig. 1 is correlated to pole figures measured at the respective local area which is indicated by an arrow. In all cases a pronounced gradient of the orientation distribution vs. the sheet thickness is revealed. In Fig. 1a an almost random texture which no preferential alignment of basal planes is found at mid-plane which correlates well to the globular grain structure. Moving to the surface reveals pronounced textures with basal planes aligned parallel to the normal direction (ND) of the strip and prismatic planes aligned parallel to the sheet plane. This texture is directly correlated to columnar grains that have their long axis parallel to the normal direction. Such a texture is very unusual for sheet materials of magnesium alloys. A preferential crystallographic growth direction of the grains can explain this result, which has also been suggested in earlier work [12]. With decreasing rolling speed in Figs. 1b-d a tendency to develop a preferential alignment of basal planes parallel to the sheet plane and prismatic planes aligned parallel to ND is found, specifically at mid-plane. Closer to the surface this tendency also seems to increase. The former alignment of basal planes parallel to ND in Fig 1a is changed in form of a tilt of such planes towards the sheet plane, i.e. basal planes are re-oriented to align into the sheet plane. Fig. 2 shows the microstructure and texture results at low rolling speed of 2.0 m/min or 1.8 m/min for varied melt temperature, decreasing from 710°C to 650°C. Changes in the microstructure are not very pronounced any more but best described as a further homogenization of the grain structure. Specifically in Fig. 2d at the lowest melt temperature of 650°C a microstructure with elongated grains in the mid-plane area and tilted columnar grains in the outer areas is found. Furthermore, the significance of the center-line segregation decreases. Still, the formation of a basal texture at mid-plane is visible. In the intermediate area of columnar grains a tendency to align into the sheet plane increases with decreasing melt temperature which is comparable to the finding in Fig. 1 with decreasing the rolling speed. At the lowest melt temperature and lowest rolling speed in Fig. 2d a weak alignment of basal planes into the sheet plane and prismatic planes aligned parallel to ND is found throughout the thickness of the strip, leaving it in the most homogeneous condition of the strips of this study.

In Fig. 3 details of the microstructure are collected for the intermediate region of directionally solidified columnar grains. The highest rolling speed in Fig. 3a leads to comparably few particles distributed in the microstructure whereas decreasing the speed in Fig. 3b and 3c shows significant fractions of particles appearing finer with decreasing rolling speed.

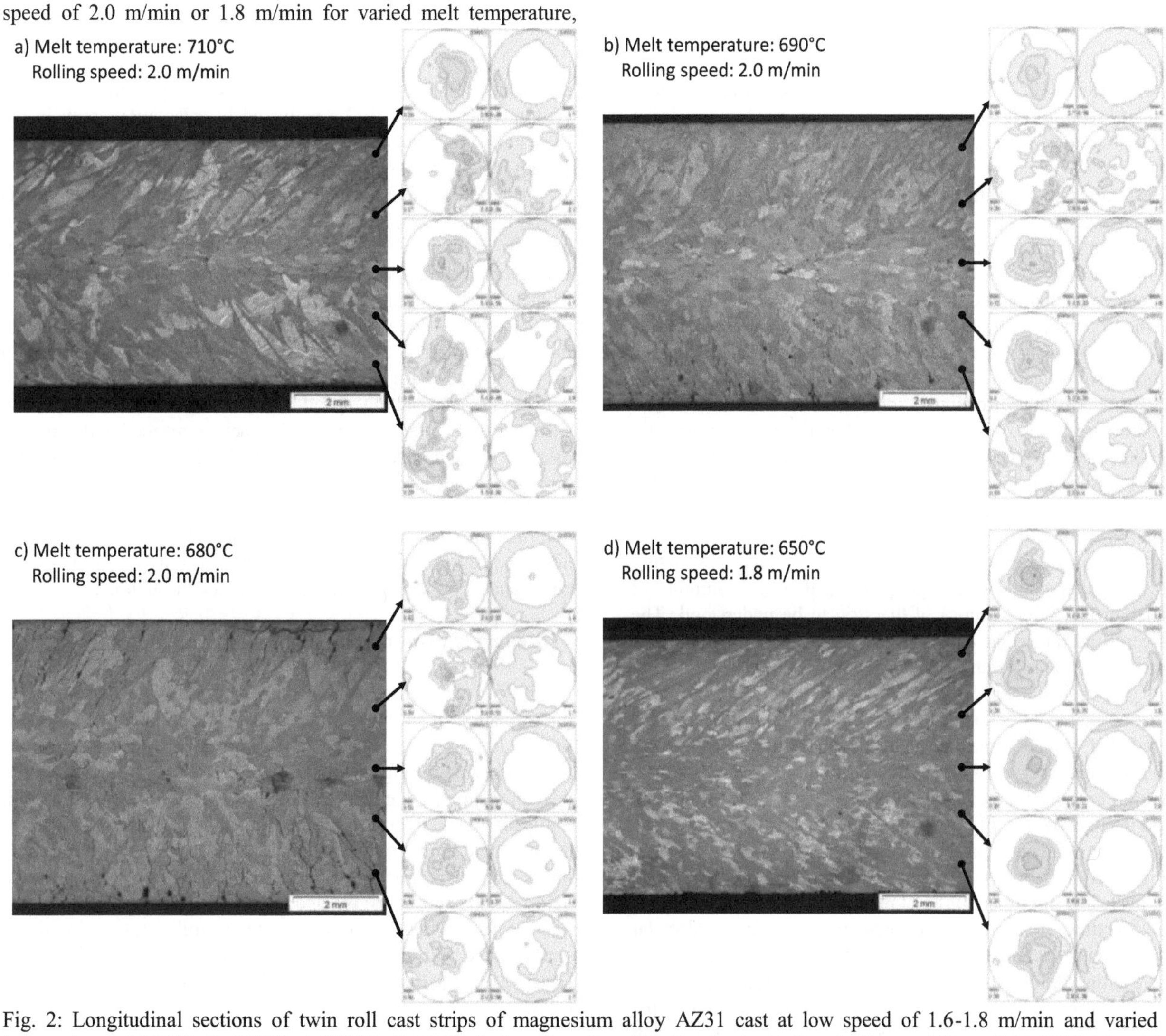

Fig. 2: Longitudinal sections of twin roll cast strips of magnesium alloy AZ31 cast at low speed of 1.6-1.8 m/min and varied melt temperature; a) 710°C, b) 690°C, c) 680°C, d) 650°C (twin roll casting direction horizontal); pole figures: RD vertical, TD horizontal, ND normal, levels: 1.0, 1.5, 2.0, 3.0, 5.0, … multiples of random distribution

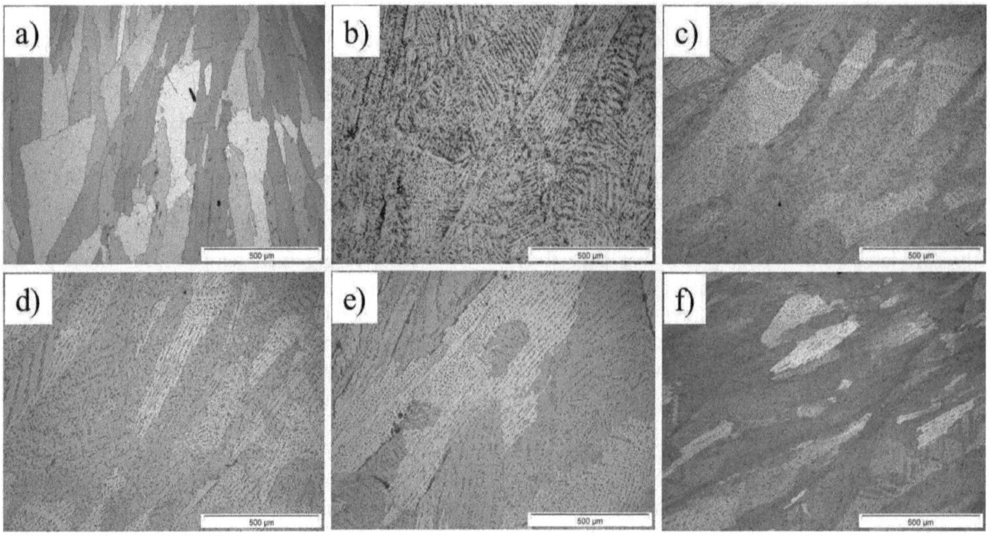

Fig. 3: Detailed longitudinal section like in Figs. 1 and 2 of region between mid-plane and surface at various melt temperature and rolling speed; a) 715°C and 3.7 m/min, b) 710°C and 2.7 m/min, c) 710°C and 2.0 m/min, d) 690°C and 2.0 m/min, e) 680°C and 2.0 m/min, f) 650°C and 1.8 m/min (twin roll casting direction horizontal)

A similar effect with decreasing melt temperature in Fig. 3c - 3e is not clearly seen. In Fig. 3f at lowest speed and melt temperature a fine distribution of particles is revealed.

The resulting microstructures and texture of the presented strips can be understood as a variation of the significance of solidification microstructures and deformation due to hot rolling. Both mechanisms are applied in-situ in this process. Solidification is directly enhanced on the surface of the flowing melt after the tip exit when contact to the cold rolls is apparent. Solidified grains then grow into the strip towards the center establishing a directional solidification. A zone in the center of the strip appears as a result of non-directional solidification with a globular microstructure. If the rolling speed is higher, the contact time after exiting the tip up to the minimum roll gap is low, shifting a point of full solidification towards the rollers. This is consistent with a preferential microstructure resulting from solidification rather than deformation due to rolling as shown in Fig. 1a. Furthermore, segregation of elements during solidification is likely, allowing very significant center-line segregation and a low amount of particles in the outer sections of the strip to be understood. The texture results underline that deformation is not very impacting the process at these parameters as the strong textures with basal planes perpendicular to ND are seen as a result of a preferential orientation due to directional solidification. If the rolling speed is lowered, the contact time between the melt and the rolls after exiting the tip increases. A broad distribution of a high amount of particles, in some cases in dendritic structures, results. In comparison between Figs. 3a and 3b an influence of the solidification kinetics may be hypothesized. With decreasing casting speed a distinct refinement of such particles is concurrent to an increasing alignment of basal planes into the sheet plane, i.e. a basal texture, formed at mid-plane in a region of globular solidification as a result of massive deformation during rolling. Such resulting deformation of directionally solidified grains then causes a re-orientation, e.g. due to basal slip, which explains the ongoing re-orientation of basal planes into the sheet plane with decreasing rolling speed. In some cases in Fig. 3 also twins are visible which can easily be activated in orientations seen in the large long grains rather than in those grains where basal planes are already aligned into the sheet plane [13]. Such an impact of activated deformation mechanism is more likely if solidification is faster, i.e. the solidification point is closer to the exit of the tip or more far away from the rolling point. Like shown in Fig. 2 for an exemplary rolling speed the decrease of the melt temperature by tendency has a similar effect like the decrease of the rolling speed but this effect is not as pronounced.

In summary, all above mentioned features of the developing microstructures are visible despite changes in the melt temperature and rolling speed. Controlling the parameters in a way that early solidification of the melt is reached and a rolling pass is applied leads to enhanced distribution of precipitates throughout the microstructure by minimizing the formation of a center-line segregation zone. Furthermore, the resulting deformation step due to rolling leads to through thickness homogenization of the texture by forming a well-known alignment of basal planes in the sheet plane throughout.

Rolling of twin roll cast strips of magnesium alloy AZ31

It is known that rolling magnesium alloy AZ31 to sheets leads to the formation of strong textures that favour an alignment of basal planes in the sheet plane [4]. Using twin roll cast strip as feedstock for rolling changes the microstructure and texture development for a number of reasons. First, the shape of strip with a thickness of only a few millimetres reduces the total degree of deformation required for rolling to final gauge. This possibly limits the microstructure development as a result of activated deformation mechanisms and recrystallization. Second, the microstructure and texture of twin roll cast strip differ significantly from cast slabs received e.g. as a result of direct-chill casting, an established process for feedstock preparation for rolling.

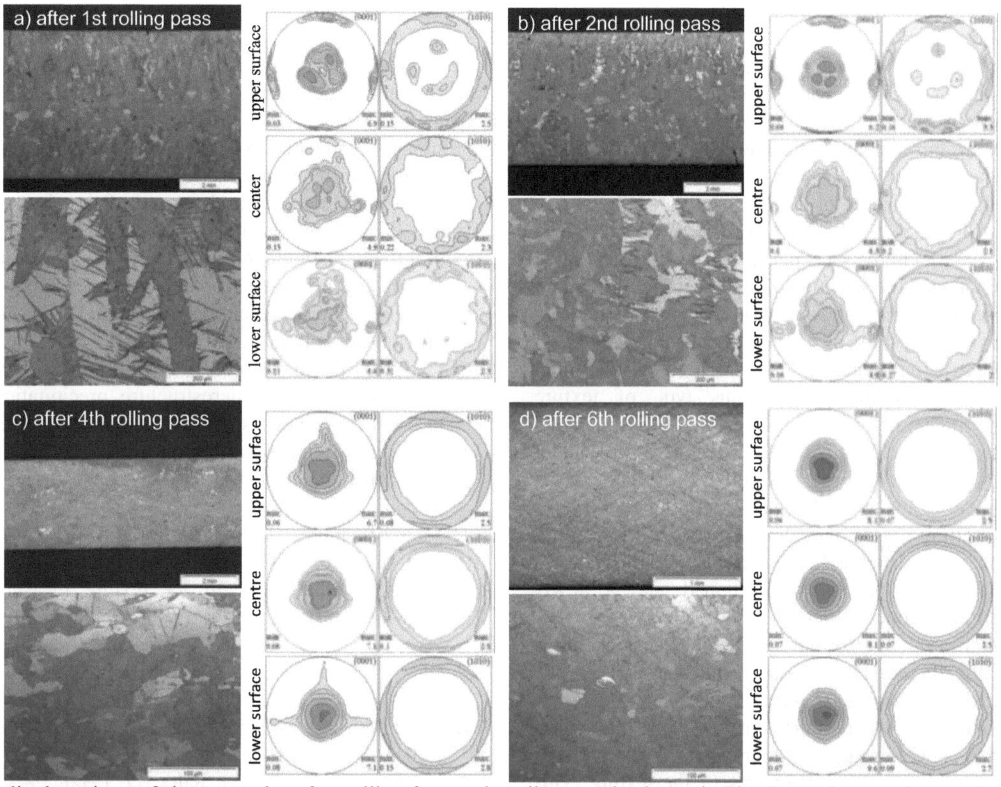

Fig. 4: Longitudinal sections of sheet samples after rolling from twin roll cast strip shown in Fig. 1a: each two micrographs with different magnifications and pole figures from the center, the upper and lower part of the sheet are shown; a) 1st rolling pass, b) 2nd rolling pass, c) 4th rolling pass, d) 6th rolling pass; micrographs: RD horizontal, pole figures: RD vertical, TD horizontal, ND normal, levels: 1.0, 1.5, 2.0, 3.0, 5.0, … multiples of random distribution

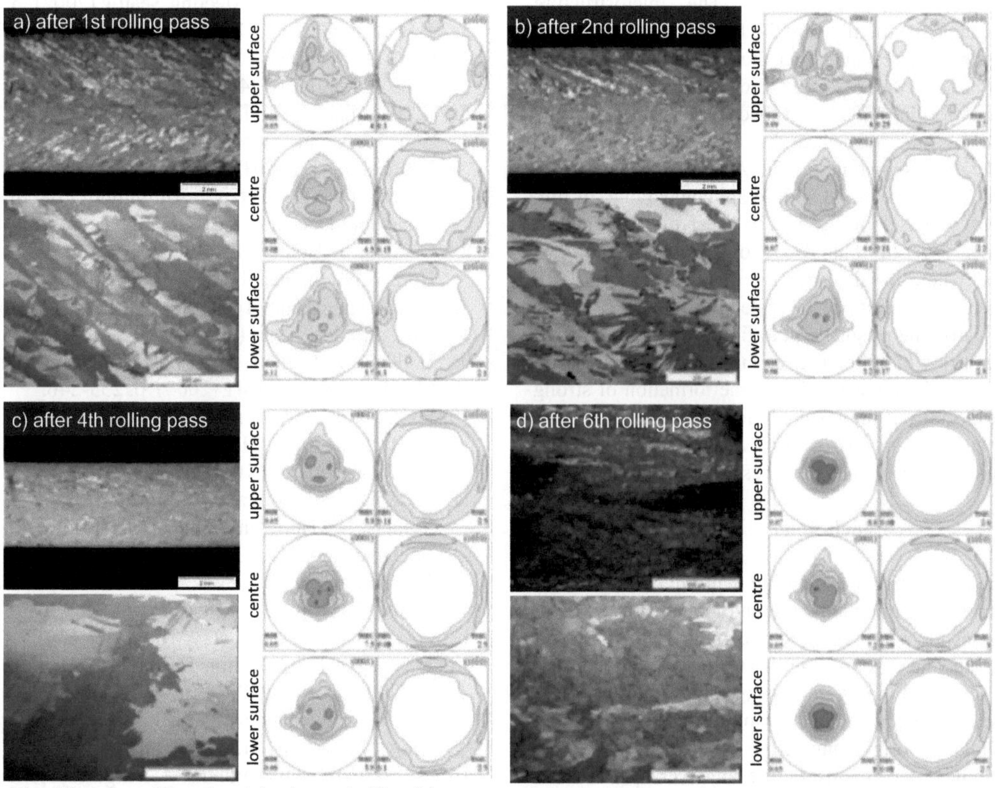

Fig. 5: Same as Fig. 4 but for rolling the strip shown in Fig. 2d

Two strips are selected for rolling experiments, one shown in Fig. 1a with large columnar grains and a pronounced texture not aligning basal planes into the rolling plane. Results of the rolling experiment are shown in Fig. 4. The second strip used for rolling is the one shown in Fig. 2d with enhanced homogeneity and a visible basal texture throughout the thickness of the strip. Rolling results are collected in Fig. 5.

After rolling with a low degree of deformation of 0.1 the strong alignment of basal planes parallel to ND (Fig.1a) as well as of prismatic planes parallel to the sheet plane is almost fully changed in Fig. 4a. A pronounced reorientation of columnar grains is found concurrent to the occurrence of twins. The originally randomly textured section in the center also develops a basal texture. If basal slip and extension twinning [14] are seen as the main deformation mechanisms that control this type of texture development, the impact of twinning is high which is concluded from the almost sudden change of the texture. It is noted that twinning is very favourable for grains oriented with the basal planes perpendicular to the applied tensile loading direction [13]. In the following rolling passes (including intermediate annealing) the alignment of basal planes is homogenized through-thickness and the originally deformed grains sectioned into a newly developed recrystallized microstructure.

The starting condition of the second strip (shown in Fig. 2d) is different as it already reveals a basal texture throughout the thickness. After the first rolling pass in Fig. 5a also twins are directly visible. However, whereas the center of the sheet persists with a basal texture, slight changes are found in the sections closer to the surfaces, revealing off-basal features of the texture. This tendency persists throughout the rolling experiment and is visible even after the 4th rolling pass. Certainly, the amount of twins is not as pronounced as shown in Fig. 4a. Significant changes of the texture are not predominant, also leading to the development of a basal texture after 6 rolling passes. The microstructure development is changed, still allowing the features of the twin roll cast microstructure to be revealed after the second rolling pass in Fig. 5b, which is not the case any more in the counterpart condition in Fig. 4b. In Figs. 4d it becomes visible that the development of the microstructure due to recrystallization is slower than in the sheet shown in Fig. 5d. Although the intensities in the basal pole figure in Fig. 4d and 5d are comparable, in Fig. 5d the alignment of basal planes in the sheet plane is not as pronounced as in Fig. 4d, but also shows a tendency for a transverse spread where basal planes are tilted towards the TD. Such textures have sometimes been revealed in rolled AZ31 strips [15, 16]. Specifically, the prismatic pole figures reveal a delayed formation of the strong basal textures of the sheet. In summary, a strong influence of twinning can enhance the formation of strong basal textures, especially if the texture before rolling differs from such a basal type. Vice versa, strips with weak basal texture may delay the formation of strong basal textures during rolling.

Conclusions

Melt temperature and rolling speed during twin roll casting have a distinct influence on the microstructure and texture development of strips for magnesium alloy AZ31. Reducing the rolling speed is consistent with an increase of the heat transfer, having the material earlier solidified during processing. Thus, deformation is applied during contact with the rolls up to the point of minimum distance. Precipitates originating from the melt are finer distributed throughout the microstructure at lower speed and a center-line segregation of alloying elements or impurities is reduced. Deformed columnar grains due to directional solidification as well as globular solidified grains in the center develop basal textures, i.e the basal planes of the lattice structure are preferentially aligned in the sheet plane. Reducing the melt temperature may have a similar effect which is not that pronounced compared to the impact of the rolling speed.

Rolling of differently textured strips of AZ31 result in the formation of a typical basal texture of the sheets rolled to final gauge. This development is more pronounced for strips that exhibit fractions of the microstructure with non-basal orientations. Twinning appears to be an important mechanism for enhancing the development of the texture, which is not that likely if the strip already reveals a basal texture. In such a case a full development of a basal texture may not result due to rolling to final gauge.

Homogenization of the microstructure, precipitation structure and texture of twin roll cast strips are enhanced at lower rolling speed and lower melt temperature. Strips with weak basal texture tend to develop sheet textures that can differ from a strong basal type.

The authors appreciate the contribution of the twin roll casting team at MagIC of HZG/Germany and the team for establishing the synchrotron radiation beam for texture measurements at the PETRA III storage ring of DESY in Hamburg/Germany.

References

1. H. Friedrich, S. Schumann, J. Mat. Process. Techn. 117 (2001), p. 276-281.
2. L. Stutz, J. Bohlen, G. Kurz, D. Letzig, K.U. Kainer, Key Eng. Mater. 473 (2011), p. 335 – 342.
3. C.E. Dreyer, W.V. Chiu, R.H. Wagoner, S.R. Agnew, J. Mat. Process. Techn. 210 (2010), p. 37-47.
4. S.R. Agnew, in: H.I. Kaplan (ed.) „Magnesium Technology 2002“, TMS Annual Meeting Seattle (2002), p. 169 – 174.
5. J. Bohlen, G. Kurz, S. Yi, D. Letzig, in C. Bettles, M. Barnett (eds.) “Advances in wrought magnesium alloys” Woodhead Publ. Ltd. 2012, p. 346 – 375.
6. E. Essadiqi, I.-H. Jung, in: C. Bettles, M. Barnett (eds.) “Advances in wrought magnesium alloys” Woodhead Publications Ltd. 2012, p. 272 – 303.
7. F. Basson, D. Letzig, Aluminium International Today, November/December 2010, p. 19-21.
8. G. Kurz, J. Bohlen, L. Stutz, D. Letzig, K.U. Kainer, in: N. Hort, S.N. Mathaudhu, N.R. Neelameggham, M. Alderman (eds.) „Magnesium Technology 2013“, TMS Annual Meeting, San Antonio / Texas (2013), p. 365 – 370.
9. V. Kree, J. Bohlen, D. Letzig and K.U. Kainer: Praktische Metallographie Vol. 41 (2004), p. 233-246.
10. Fit2D: http://www.esrf.eu/computing/scientific/FIT2D/
11. F. Bachmann, R. Hielscher and H. Schaeben, Solid State Phenomena 160 (2010) p. 63-68.
12. M. Salgado-Ordorica, W. Punessen, S.Yi, J. Bohlen, K.U. Kainer, N. Hort, in: W.H. Sillekens, S.R. Agnew, N.R. Neelameggham, S.N. Mathaudhu (eds.) “Magnesium Technology 2011”, TMS Annual Meeting, San Diego, CA (USA), 2011, p. 113- 118.
13. P. G. Patridge, Metall. Rev. 12 (1967), p. 169-194.
14. A. Styczynski, C. Hartig, J. Bohlen, D. Letzig, Scripta Mater. 50 (2004), p. 943-947.
15. J. Victoria-Hernandez, S. Yi, J. Bohlen, G. Kurz, D. Letzig, J. Alloys Compds 616 (2014), p. 189 – 197.
16. M. Masoumi, , F. Zarandi, M. Pekguleryuz, Mat. Sci. Eng. A 528 (2011), p. 1268–1279.

Magnesium Technology 2015

STRIP CASTING OF LIGHT METALS

Process Technology

Papers in this section have been presented in the symposium Strip Casting of Light Metals at the 2015 TMS Annual Meeting, organized by K.F. Karhausen, W. Sillekens, M. Dundar, J. Bohlen, and D. Letzig. This section collects the peer-reviewed magnesium-related contributions to this symposium and was edited by Dr. J. Bohlen and Dr. D. Letzig (Chief Scientist and Department Head at the Magnesium Innovation Center of the Helmholtz Research Center Geesthacht, Germany, respectively), and Dr. W. Sillekens (Project Manager at the Research and Technology Center of the European Space Agency in Noordwijk, Netherlands).

Magnesium Technology 2015
Edited by: Michele V. Manuel, Alok Singh, Martyn Alderman, and Neale R. Neelameggham
TMS (The Minerals, Metals & Materials Society), 2015

MAGNESIUM TWIN-ROLL CASTING BENEFITS FROM ALUMINIUM HERITAGE

Dr Frederic Basson[1]
[1]Novelis PAE, Centr'Alp B.P. 24, F-38341 Voreppe Cedex, France

Keywords: continuous casting, metal feeding, tip technology, heat extraction, roll gap control, shell coating

Abstract

The history of twin-roll casting for Aluminium started in the 1950's, with the first Pechiney trials in 1956, a first patent filed in 1958 and the first industrial lines installed at the beginning of the 1960's. 60 years of technology development later, Aluminium twin-roll casting accounts for 20 to 30% of the world flat rolled production and is recognized as a low-cost flexible short route to produce a wide range of alloys with cast width up to 2300 mm.

In comparison, Magnesium twin-roll casting is a young process but the same promising future can be foreseen for it, as it can largely benefit from the technology developed for Aluminium.

From Pechiney to Novelis in the Aditya Birla Group, Novelis PAE has been a major actor in developing twin-roll continuous strip-casting technology for more than five decades, for Aluminium (more than 110 casters installed), and also for Magnesium (3 casters installed).

Twin-roll casting process principles

Twin-roll casting is the only process that combines both solidification and rolling in a single step. The liquid metal is fed through a nozzle (the "tip") into the bite of a 2-high mill where the rolls are designed to fill both a heat exchanger and a rolling function (see Figure 1). The rolls are made of two parts : a core and a shell shrink-fitted on the core. Water circulates between the core and the shell to cool down the shell, which is the part in contact with the metal. The metal exiting the tip solidifies on the rolls into a strip that is further work-hardened by the rolls. To prevent the metal from sticking to the rolls a release agent (most of the time a suspension of water with graphite) is applied to the rolls. This release agent acts both as a lubricant and as a thermal barrier that controls the heat exchange between the rolls and the metal. Metal feeding and roll lubrication are key elements in the control of the strip quality.

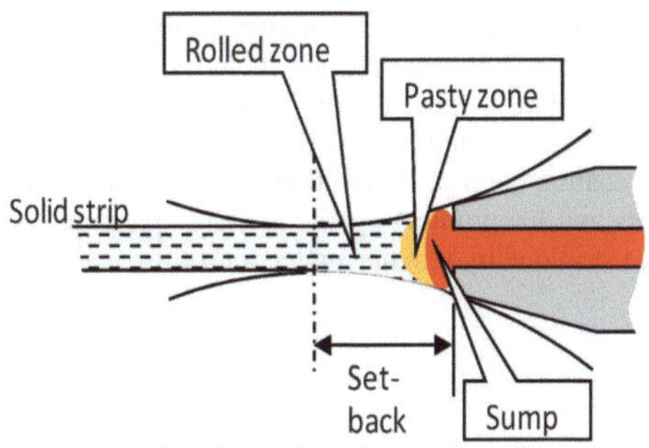

Figure 1. Schematics of the twin-roll casting principle.

The typical cast gauge obtained with the twin-roll casting process is similar to the hot mill exit gauge. Therefore, compared to the conventional DC casting and hot rolling route, the twin-roll casting route is more flexible and economical, avoiding expensive and time-consuming heat treatment steps and rolling passes. A key difference between twin-roll casting and DC casting is the solidification rate of the metal : while in DC casting it is limited to 1 to 50°C/s, it reaches 500 to 1000°C/s on a Jumbo 3CM® thin gauge continuous caster. This gives very fine structures (dendrite arm spacing, or D.A.S., about 5 µm) with small and finely distributed intermetallics, extended solid solution and fine grain size. Final product properties can benefit from that different metallurgical structure if the downstream processing route is adequately adapted.

A little bit of history

From Pechiney to Novelis, and today with Hindalco in the Aditya Birla Group, Novelis PAE has been a major player in developing twin-roll continuous strip-casting technology for more than five decades, mainly for Aluminium (more than 110 casters designed and installed), but more recently also for Magnesium (3 casters designed and installed).

The history of Aluminium twin-roll casting started in the 1950's, with the first Pechiney trials in 1956, a first patent filed in 1958 and the first industrial lines installed at the beginning of the 1960's. The size of the casting rolls (roll diameter and width) tends to determine the overall capacity of the equipment. In the 60's it was limited to 620 mm in diameter and dedicated to relatively narrow cast widths (below 1500 mm) and thick cast gauges (8 to 12 mm). In the late 70's and early 80's, the Jumbo 3C® 960 and 840 (for 960 mm and 840 mm roll diameters) were introduced and the cast width was increased up to 2000 mm, while the cast gauge was typically around 6 to 8 mm. With the 90's came the Jumbo 3CM® generation of machines with rolls of 1150 mm diameter, cast width up to 2300 mm and cast gauges as low as 3 or even 2 mm (see Table I).

Table I. The generations of Novelis PAE Aluminium twin-roll casters

Caster model	3C Caster	Jumbo 3C Caster	Jumbo 3CM Caster
Years	1960-1975	1975...	1995...
Roll diameters (mm)	620	620 to 1150	620 to 1150
Maximum strip width (mm)	1600	1500 to 2300	1500 to 2300
Strip thickness (mm)	5 to 12	5 to 12	2 to 10
Coil weight (t)	10	10 to 24	10 to 24
Productivity range at 1000-mm width (t/h)	0.8 to 1.2	1.3 to 1.6	1.6 to 1.9
Annual capacity (Kt/year)	up to 14	up to 24	up to 27

Today, after 60 years of technology development, Aluminium twin-roll casting accounts for 20 to 30% of the world flat rolled production and is recognized as a low-cost flexible short route to produce a wide range of alloys with cast width up to 2300 mm.

Compared to Aluminium, twin-roll casting of Magnesium is a quite young process with only a few casters in operation, most of them for research and development work [1, 2]. However, the same promising future can be foreseen for this process with little doubt, as it can largely benefit from the technology developed for Aluminium.

Physical and thermal properties

In order to define the proper technology for twin-roll casting of Magnesium and how much can be derived from Aluminium twin-roll casting, a comparison of the physical and thermal properties of both metals shall be done. Looking at the main twin-roll casting process parameters for a reference as-cast strip gauge of 5 mm, it is experienced that the lower density of Magnesium is compensated by a higher casting speed (see Table II). The productivity (expressed in tons of metal cast per hour and per meter of cast width), or the mass metal flow rate through the caster, is indeed very similar for Aluminium and Magnesium. It is therefore reasonable to compare physical and thermal properties per mass unit.

Table II. Comparison of Aluminium and Magnesium main twin-roll casting parameters

Typical TRC process parameters	Aluminium	Magnesium
Reference as-cast gauge	5 mm	5 mm
Typical casting speed	1.6 m/min	2.5 m/min
Density – solid at 25°C	2.70	1.74
Productivity	**1.30 t/h/m of width**	**1.31 t/h/m of width**

The main physical and thermal properties involved in the twin-roll casting process for Aluminium and Magnesium are compared in Table III. Three main conclusions can be derived from this comparison :

- The operating temperature range is almost identical for both metals. Thus, from a thermal point of view, the materials and equipment used for twin-roll casting of Aluminium will be suitable for Magnesium
- Taking the specific heat in the liquid state, the heat of fusion and the specific heat in the solid state into account, it can be calculated that the energy to be extracted in the caster roll gap by the solidification and cooling system is very similar for both metals, in the range of 600 to 650 $kJ.kg^{-1}$. Therefore, the heat extraction technology developed for twin-roll casting of Aluminium will be adequate for Magnesium
- Comparing the surface tension of both metals, it is clear that liquid Magnesium has a higher tendency to leak than liquid Aluminium. As a direct consequence, any sealing in the metal feeding equipment will be more challenging. Moreover, the meniscus that creates between the lips of the tip and the surface of the caster rolls will be less stable with liquid Magnesium, making tip geometry consistency and tip positioning accuracy and stability of paramount importance for twin-roll casting of Magnesium.

Table III. Comparison of Aluminium and Magnesium physical and thermal properties

Physical & thermal properties	Aluminium	Magnesium
Density – solid at 25°C	2.70	1.74
Density – liquid at melting point	2.32	1.60
Melting point	660°C	650°C 630°C for AZ31 595°C for AZ91
Specific heat – solid at 25°C	897 $J.kg^{-1}.K^{-1}$	1025 $J.kg^{-1}.K^{-1}$
Specific heat – liquid at melting point	1025 $J.kg^{-1}.K^{-1}$	1323 $J.kg^{-1}.K^{-1}$
Heat of fusion	400 $kJ.kg^{-1}$	368 $kJ.kg^{-1}$
Surface tension	865 to 1100 $mJ.m^{-2}$	560 $mJ.m^{-2}$

Twin-roll casting technology

Figure 2 shows what would be the typical cross section of a Magnesium twin-roll casting line designed for continuous operation.

Several main functional units can be identified following the metal flow (from left to right on Figure 2) :

- The ingot preheating and loading unit : this is the preparation stage of the incoming raw material to be charged into the melting furnace
- The melting furnace, to melt the incoming raw material and the alloying elements. Continuous mixing ensures a homogeneous alloy composition
- The holding furnace, continuously fed from the melting furnace. It is the molten metal reservoir feeding the tundish or headbox
- The twin-roll caster stand, concentrating the heart of the solidification and rolling process. It converts the incoming molten metal flow into a solid strip
- The strip operation units : pinch rolls to apply tension to the strip, X-ray gauge to measure the strip gauge and cross profile, edge cutter to remove the strip edge heterogeneities (edge cracks and segregations), shear to cut samples or change coil
- The coiler unit, possibly equipped with a wrapper.

The heart of the process is of course localized in the twin-roll caster stand, where three main technology areas shall be considered : liquid metal feeding (which includes the nozzle or tip), solidification and heat extraction in the caster roll gap (which includes roll gap control and roll cooling system), and shell lubrication and coating control. In addition, an in-line monitoring of the strip cross and longitudinal profile is highly recommended to check that the strip quality remains under control.

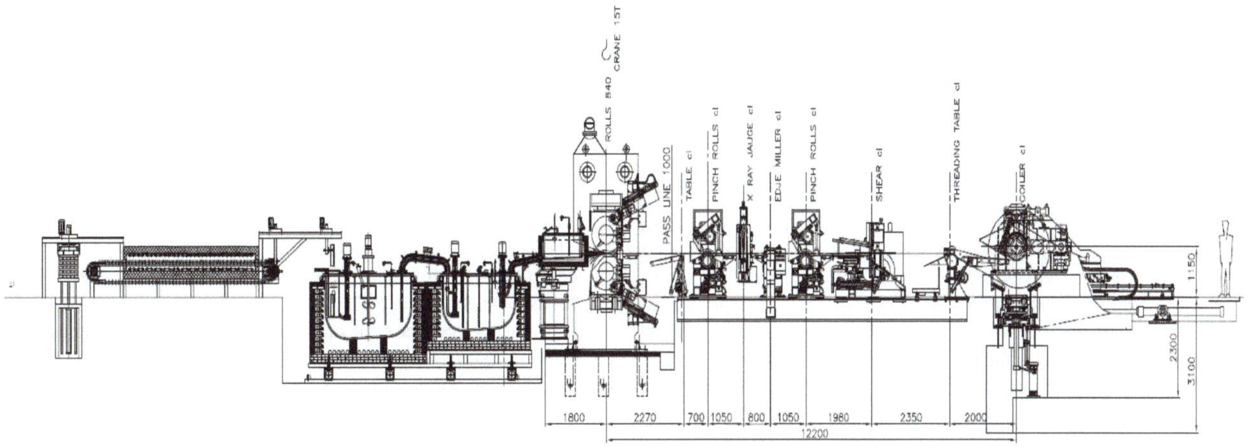

Figure 2. Typical cross section of a Magnesium twin-roll casting line.

Liquid metal feeding

An adequate metal feeding on a twin-roll continuous caster is obtained when the metal temperature field at the outlet of the tip leads to a regular solidification front [3, 4, 5, 6, 7, 8, 9, 10, 11]. This is a condition to obtain a proper strip cross profile and a uniform metal microstructure. The tip (or nozzle) is then the key element of the process to inject molten Magnesium into the roll gap (see Figure 3).

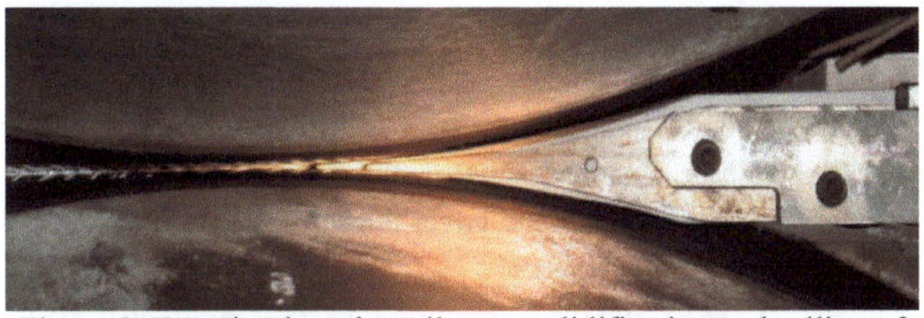

Figure 3. Zooming into the roll gap : solidification and rolling of the cast strip at tip outlet.

Two alternative tip technologies have been developed for twin-roll casting of Magnesium :

- The heated steel tip, where the tip top and bottom plates and lateral spacers are made of high-grade Nickel-free heat-resistant steel. Grooves are machined in the outer face of the plates to insert electrical heating elements that are used for tip preheating and to keep the tip plates at a temperature close to the liquid metal temperature during casting. This technology is relatively expensive and time-consuming for maintenance, but it has the significant advantage of being reusable
- The ceramic tip, based on the design developed for twin-roll casting of Aluminium, where heat losses through the tip are limited by using highly insulating materials. However, as liquid Magnesium strongly reacts with the silica contained in most refractory materials (either as a base component, or as a hardener), the inner face of the tip plates shall be wear-protected and the key is the selection of the right material/coating combination.

When using a ceramic tip, the risk is that the metal freezes in the tip during the priming phase at start-up. A tip preheating device is then required for safe operation. The heating is done by a hot air blower located on the exit side of the caster stand, mounted on a movable frame, the hot air flowing within pipes that come right at the tip outlet. This way, preheating can be done when the tip is already set in the start-up position and goes on until the very last minute before priming. Residual moisture possibly present in the refractory tip boards is then removed and the freezing risk at caster start-up is reduced.

Whatever the selected tip technology (heated steel or ceramic), an accurate positioning of the tip with respect to the rolls is necessary for a regular solidification front across the cast width. As for twin-roll casting of Aluminium, the tip table shall allow for every movement that may be necessary to obtain a perfect adjustment of the tip and a constant position of the tip once it has been set. When the tip is mounted on the tip support before start-up, the operator has to adjust the tip setback (distance from the tip outlet to the vertical plane that contains the roll axes) and the gap between the tip lips and the rolls (the roll gap). As the setback depends in particular on the roll diameter and on the cast gauge, the tip table allows for adjustment of the tip in the longitudinal and vertical directions. It also allows fine-tuning of the tip horizontality. It is very important to maintain a proper and constant gap between the tip lips and the shell surface as it affects the meniscus stability. A contact between the tip and the shell surface can be very detrimental to the integrity of the strip, not to mention the tip itself. A too-large gap can also lead to bleed-out or surface defects on the strip.

Solidification and heat extraction

As explained previously, the heat extraction needs are similar for Aluminium and Magnesium. Consequently, the same technology can be applied for roll gap control and roll water cooling system.

An accurate control of the roll gap and applied separating force is necessary to ensure a constant heat transfer coefficient (positively correlated to the applied force) and tight strip geometrical tolerances, especially regarding longitudinal gauge variations and strip cross profile. This is achieved with hydraulic screwdown cylinders located at the top of the top roll, on each side of the caster. Each screwdown cylinder is equipped with a position sensor, allowing a measurement of the position of each side of the top roll with an accuracy of +/- 1 μm, and a pressure sensor, allowing a measurement of the separating force applied on each side of the caster with an accuracy of +/- 0.5 ton.

This double set of sensors is at the heart of the highly reactive hydraulic roll gap control, allowing two working modes :

- The position mode, where the caster Programmable Logic Controller (PLC) regulates the position of each screwdown cylinder, is used for accurate gap setting during caster preparation before start-up, and during start-up to avoid any bleed-out when the first liquid metal comes into the roll bite (no gap opening)
- The pressure mode, where the PLC regulates the separating force applied through each screwdown cylinder, is used as the standard production mode as it allows an accurate regulation of the separating force, resulting in a constant flexion of the caster rolls, and thus in a strip cross profile kept within tight tolerances
- The switch from the position mode to the pressure mode after start-up is triggered automatically when the separating force reaches a preset level, which depends mainly on the casting width and on the alloy. This automatic switch allows a smooth transition between the start-up transitory phase and the stable casting regime.

As in twin-roll casting the rolls behave primarily as a heat exchanger, Novelis PAE roll design criterion is to have a uniform cooling efficiency across the width and over the circumference as well as a maximum heat extraction power in order to maximize the casting rate without impacting the strip quality negatively.

This approach led to design the cooling channels of the roll core with a checkered pattern of circumferential and transverse grooves at the core-shell interface (see Figure 4). This way, the surface temperature differences over the circumference and across the width of the shells have been reduced to a maximum of 2 to 3°C, providing a homogeneous heat extraction in the solidification and rolling zone, thus reducing longitudinal gauge variations of the cast strip to a minimum [12].

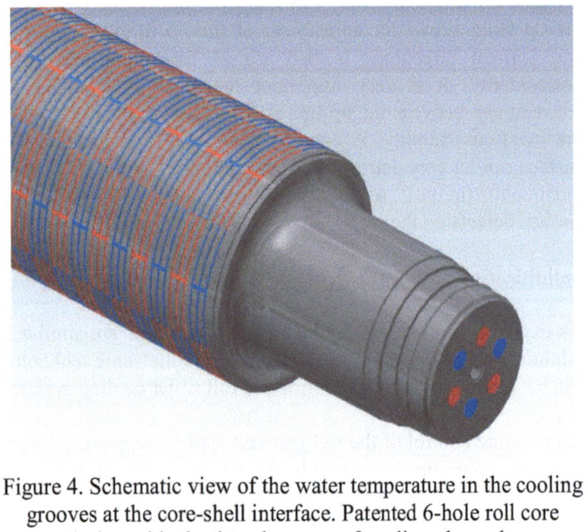

Figure 4. Schematic view of the water temperature in the cooling grooves at the core-shell interface. Patented 6-hole roll core design with checkered pattern of cooling channels.

Shell lubrication and coating control

At the roll/metal interface, the roll coating plays a key role in the quality and geometry of the as-cast strip. The main parameters that govern the roll coating are the composition and consistency of the oxide layer at the roll/strip interface, the amount of release agent that is deposited on the shells, and how the release agent is deposited on the shells.

Therefore, when developing the shell lubrication and coating control system for twin-roll casting of Aluminium, Novelis PAE focused on achieving :

- A good control of the quantity of release agent applied on the rolls, using an accurate lubricant flow rate control device based on peristaltic pump technology (one per caster roll)
- A uniform deposit of the lubricant on the shells. The spraying guns are mounted on a traveling support system and connected to a specific brushless drive to allow smooth and fast motion of the gun support arm across the roll width. The traveling speed can either be set manually but is preferably automatically controlled by the caster PLC, based on a setpoint which is calculated according to the cast width and roll rotation speed. In automatic mode it optimizes the deposit of lubricant on the rolls
- A high output of the guns, i.e. more release agent on the rolls and less in the atmosphere and especially less on the strip. The spray guns use low air pressure to break up the jet of release agent and to get a flat spray pattern. The low-pressure spray leads to less rebound of the release agent droplets on the rolls. These droplets are also much finer and cover the rolls more efficiently and more homogeneously
- A good stability over time of the spray quality, by feeding the spray guns from a mixing station where the mother suspension of release agent is diluted in water and continuously mixed. There is a second mixing vessel to insure continuity of supply of release agent.

Figure 5. Roll lubricant spraying system.

The above technology features can be applied directly for twin-roll casting of Magnesium. However, the addition of the following specific devices is beneficial :

- Roll scrapers, made of adjustable bronze blades positioned very close to the roll surface on the caster exit side, in order to remove any piece of metal that may be stuck on the shell surface at start-up. This device reduces the risk of tip damage during the start-up transitory phase and increases operator safety by limiting the needs to access the caster stand exit side
- Roll brushing device to further improve coating homogeneity, each roll being equipped with a rotating brush positioned on the caster exit side after the spray guns. Roll brushes have a smoothening effect on the shell coating, thus improving the heat exchange homogeneity and reducing the likeliness of strip surface segregations.

In-line monitoring of the strip cross and longitudinal profile

The technological features described here above certainly help in reducing the chance of uncontrolled parameter drift or change, thus keeping the strip quality and geometry within the required specifications and tolerances. Yet, if there has been a change in the casting parameters such as shell lubrication or metal level, then it may be detected only at the first rolling pass. In order to avoid that, a continuous monitoring of the strip gauge is highly recommended. Consequently, most new twin-roll continuous casters are now equipped with a traveling X-ray gauge installed on the exit side of the caster stand.

Figure 6 shows a typical supervision computer display of an X-ray gauge color map of the strip gauge for a complete coil. It is then very easy for the operator to judge if the strip being cast meets the requirements or not on cross profile and longitudinal gauge variations. Any drift is also immediately detected to permit a quick corrective action.

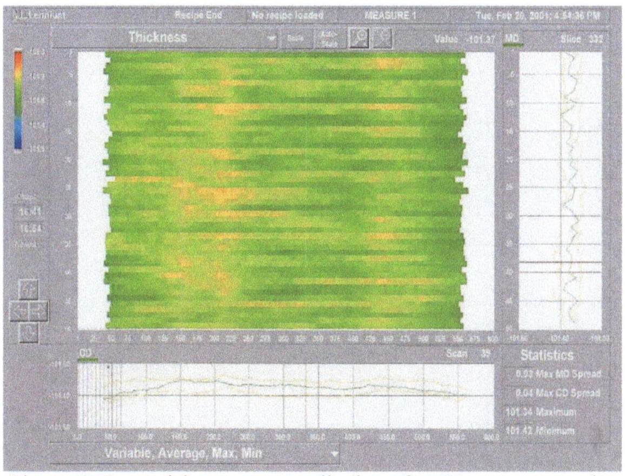

Figure 6. Supervision computer display of an X-ray gauge color map of the strip gauge.

The benefits of an X-ray gauge in-line monitoring could be summarized as follows:

- Constant monitoring of the strip gauge and cross profile, allowing early detection of parameter drift
- Better control on the as-cast quality, resulting in an increase of the casthouse recovery (less scrap generated)
- Improvement of the understanding of the process by the operators. New operators therefore need less training time and follow a faster learning curve.

Conclusion

It has been shown (as well as operationally demonstrated) that twin-roll casting of Magnesium can largely benefit from the technology developed for Aluminium during the last 60 years. Liquid metal feeding and tip are key elements and required specific developments, especially regarding heat losses compensation, compatibility with liquid Magnesium, and positioning. Shell lubrication and roll coating control equipment have been adapted from Aluminium twin-roll casting standards. Finally, heat extraction needs being similar for Aluminium and Magnesium, the same technology can be applied for roll gap control and roll cooling system.

References

1. H. Pircher, M. Weber, R. Kawalla, "Magnesium sheet materials for auto body construction", *Proceedings of the International Conference on New Developments in Sheet Metal Forming Technology*, June 4-5, 2002, Stuttgart.

2. Lothar Löchte, Håkon Westengen, John RØdseth, "An Efficient Route to Magnesium Alloy Sheet: Twin-Roll Casting and Hot Rolling", *Magnesium Technology 2005*, TMS (The Minerals, Metals & Materials Society), 2005.

3. P-Y. Menet and L. Calaque, "Process and device for controlling the thickness profile of a thin metal strip obtained by continuous casting between mobile molds", French patent FR 2 775 916, international patent WO 99/00527.

4. R. Cayol, "Latest Technical Development in Thin Gauge Continuous casting of Aluminium Alloys", *Alumitech '97: 2nd International Aluminum Conference and Trade Exposition* (The Aluminium Association, 1997), 471-476.

5. R. Cayol, F. Basson, K. Maiwald, and P-Y. Menet, "JUMBO 3CM® Thin Strip Casting: An alternative for a wide range of products", *Light Metals 1999* (Canadian Institute of Mining, Metallurgy and Petroleum, Aug. 1999), 265-275.

6. F. Basson, M. Jamet, K. Maiwald, P-Y. Menet, and J-C. Terrier, "JUMBO 3CM® Thin Strip Caster - Recent Developments", *6th Australian Asian Pacific Conference*, July 26-29. 1999, Sydney, Australia.

7. F. Basson, P-Y. Menet, K. Maiwald, and M. Bosch, "Strip Casting Technology... A Key to Product Quality", *Aluminum 2002, Proceedings of the TMS 2002 Annual Meeting : Automotive Alloys and Aluminum Sheet and Plate Rolling and Finishing Technology Symposia,* ed. by S.K. Das and M.H. Skillingberg, TMS (The Minerals, Metals & Materials Society, 2002), 143-150.

8. M. Bosch, P-Y. Menet, F. Basson, and P. Charlier, "Continuous Casting Technology – A Driver of Evolution in the Aluminium Transformation", *Proceedings of the International Conference on*

Aluminium, INCAL '03, April 23-25. 2003, New Delhi, India, Eds. S. Subramanian and D.H. Sastry (2003), 2, 445-454.

9. B. Engl, "Potential uses and new production technology for Magnesium sheet", *Steel Grips 1*, 2003, No. 6, 413-418.

10. B. Engl, "Future aspects of Magnesium sheet materials using a new production technology", *Proceedings of the 62nd Annual World Magnesium Conference*, May 22-24, 2005, Berlin.

11. B. Engl, "A New Technology for Magnesium Sheet Production", *Light Metal Age*, October 2005, 14-19.

12. "Process and device for correcting the out-of-round of rolls for the continuous casting of metal strip", US patent 5 642 772.

Magnesium Technology 2015
Edited by: Michele V. Manuel, Alok Singh, Martyn Alderman, and Neale R. Neelameggham
TMS (The Minerals, Metals & Materials Society), 2015

INNOVATIONS IN THE PROCESS TECHNOLOGY FOR MANUFACTURING MAGNESIUM ALLOY SHEET

Enrico Romano[1], Roberto Passoni[1], Chris Romanowski[2]

[1] FATA Hunter; S.S. 24 km 12; Pianezza (TO), 10044, ITALIA
[2] FATA Hunter Inc.; 1040 Iowa Avenue, Suite 100; Riverside, California, 92507, USA

Keywords: Magnesium Sheet, Twin Roll Casting, Reversing Rolling Mill

Abstract

Personal electronics and lightweight vehicles have driven demand for competitively priced lightweight materials. In recent years magnesium die castings have been used, but further weight reductions require wrought magnesium sheet. Until recently, magnesium sheet was made by hot rolling small ingots, handling/reheating constraints limiting the coil size. This made the sheet prohibitively expensive for consumer applications. Also inherent processing variations limited compatibility with mass production. Recent advances in twin roll casting allowed magnesium alloys to be cast directly into large coils, but the technology to economically roll these coils was unavailable. Working with ORNL (Oak Ridge National Laboratories) and Magnesium Elektron, FATA Hunter recently manufactured a pilot mill which demonstrated an industrial process to economically roll these coils. This paper describes how these casting and rolling technologies can be combined in a twin roll caster based 30,000 tpa plant that can economically produce magnesium sheet for automotive and other consumer applications.

Historical Background

Magnesium alloys have a high potential for use as structural materials due to their excellent properties such as light weight, high specific strength and superior damping capacity. Until the 1990s magnesium was mainly used for aerospace applications (Figure 1) due to the cost of magnesium sheet, the public perception and the lack of any significant technical or economic driver.

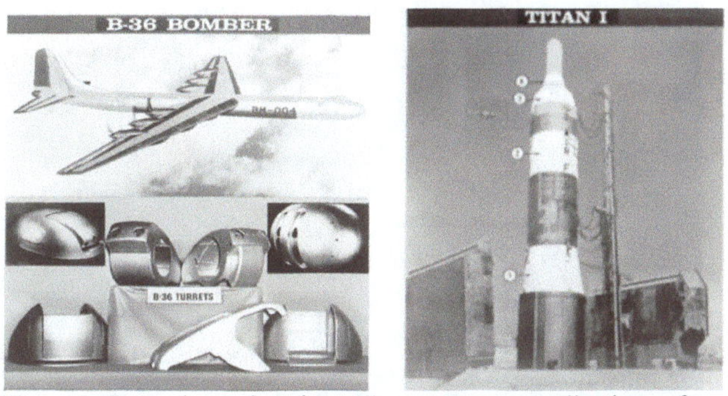

Figure 1. Examples of "vintage" aerospace applications for Magnesium.

The high production cost of magnesium sheet was related to the production methods that were used for hot and warm rolling the magnesium ingot into sheet (Figure 2).

Figure 2. Magnesium Rolling at Dow and at Alcoa in the 1950s.

In the conventional process, Mg alloys are DC cast in the form of slabs that are about 300 mm thick. The slabs are then homogenized at about 400 °C prior to the hot and warm rolling processes. Magnesium is a metal with a Hexagonal Close Packed (HCP) crystal structure that has very limited plasticity at room temperature. This requires frequent reheating in off-line ovens to maintain the temperature between 250°C and 450°C. Below this temperature, the metal has a tendency to crack during rolling.

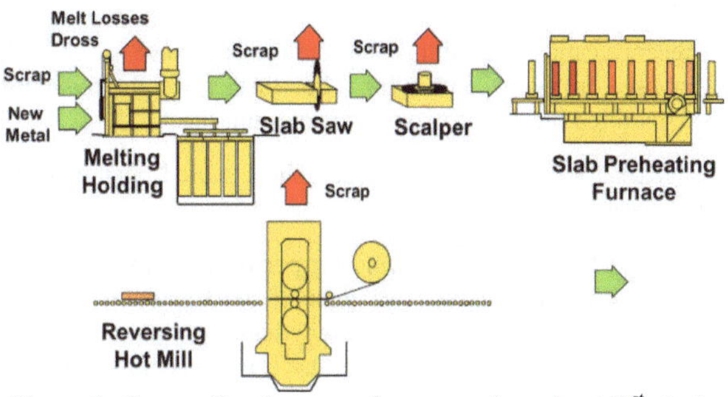

Figure 3. Conventional process for magnesium sheet (1st step).

During the hot rolling process a reversing hot mill (Figure 3) is used to roll the slabs in a series of passes with about 10 to 20% reduction per pass to a "reroll" coil that can be transferred to a warm rolling mill (Figure 4). The warm mill further rolls the reroll coil to a thickness that that can be less than 1 mm. Due to the low heat capacity of magnesium and the heat losses on the hot mill run-out tables, the maximum slab size that could be rolled is limited and traditionally resulted in either very small coils, or rolling the magnesium on a virtually a sheet-by-sheet basis. This is a very labor and energy intensive, inefficient method of production and contributed to the traditionally high cost of magnesium sheet.

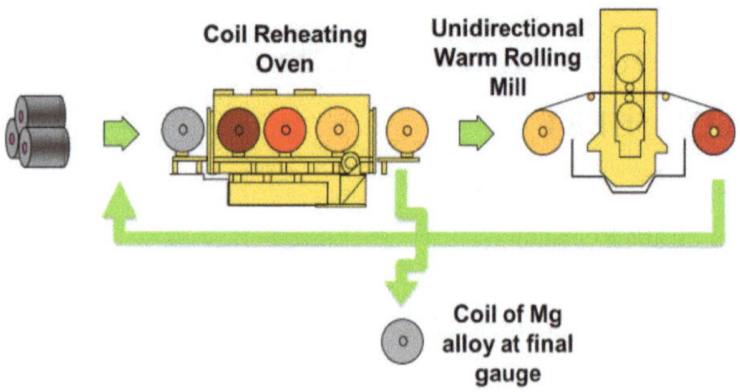

Figure 4. Conventional process for magnesium sheet (2nd step).

In 1982, working in cooperation with Dow Chemical, FATA Hunter demonstrated the feasibility of twin roll casting magnesium on a laboratory scale (Figure 5). The tests were technically successful, but in 1982 Mg ingot was expensive and there was no widespread demand for magnesium sheet, thus the market was insufficient to justify the investment and the technology was put "on-hold".

Figure 5. World's first twin roll cast coil of AZ31B Mg alloy.

In recent years, magnesium alloy die castings have successfully been used in many applications, but further weight reductions have required the use of wrought magnesium sheet [1,2].

The market demand for magnesium sheet started changing after 2000 because of:

- The growth in demand for personal electronics (laptop computer cases, cell phones, MP3's, etc.). Some of these items were thicker than desired due to the minimum wall thickness that can be achieved with die casting, that is around 1~1.5 mm for conventional die casting and 0.5 mm for thixo-cast. Thin wall die castings also had the disadvantage that they had inherently large runner and gate losses and high finishing costs, but the wrought magnesium sheet produced by the conventional method was still too expensive for these applications.
- Energy costs, government mandates, and concerns over CO_2 emissions placed emphasis on light weight car production. Magnesium has been used in automobiles for many years as die castings (e.g. Volkswagen and Porsche crankcases, European car structural components, instrument panel support beams) but conventionally produced wrought magnesium sheet was too expensive for automotive applications. The work in magnesium auto sheet was originally driven by European car manufacturers, but US and Asian producers have gained a progressively more significant role in recent years.
- Magnesium metal prices decreased as Chinese production increased.

FATA Hunter Experience In Mg Casting

Based on FATA Hunter's participation in the 1981 casting trials, in 2004 FATA Hunter cooperated with RIST (Research Institute of Industrial Science and Technology, an affiliate of POSCO) in South Korea to build a pilot casting line for magnesium alloys (Figure 6 and Figure 7).

Figure 6. FATA Hunter pilot Mg caster during pre-assembly at workshop.

The pilot caster was used to prove the technology of Twin Roll Casting Mg and the commercial viability of process. It showed the possibility of casting cast quality magnesium alloy sheet at a very affordable price. The pilot machine was then converted for full-time production of narrow sheet for the personal electronics market, with the furnace configuration progressively modified to meet production demands.

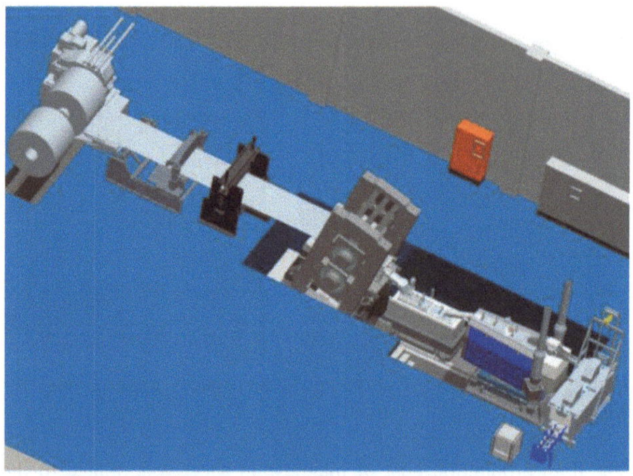

Figure 7. 600 mm wide FATA Hunter Mg strip caster.

Based on the success of the pilot caster, RIST (POSCO) invested in a 2m wide full-scale production caster (exact details subject of Confidentiality Agreement), with the target of supplying wide magnesium alloy sheet for the transportation market. The production machine utilized operating knowhow developed on the 600 mm wide pilot caster.

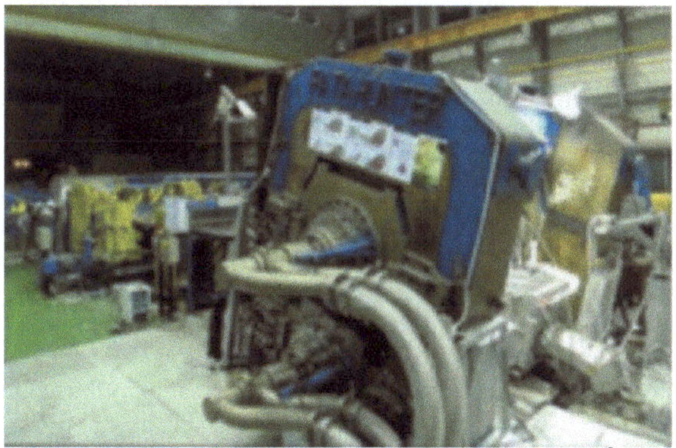

Figure 8. 2,000 mm wide FATA Hunter SuperCaster *Plus*® at RIST (POSCO) in South Korea [3].

The wide caster entered production very quickly and full size coils were produced less than one month after start-up. Wide magnesium (AZ31B Mg alloy) coils of about 16 tons (Fig. 8,9) with 2,000 mm width, 2,500 mm diameter and 4-8 mm thickness were successfully produced by careful control and monitoring of the key operating parameters (casting speed, melt temperature, roll separation force, roll gap and setback distance, etc.).

Figure 9. As-cast AZ31 Mg alloy coil [4].

The molten metal transfer system is a critical safety part of FATA Hunter's magnesium casting technology; it consists of the liquid metal transfer and control equipment to transfer metal from the magnesium furnace to the casting machine. All transfer components were fabricated from nickel free stainless steel and were electrically heated, insulated and housed in stainless steel jackets. A FATA Hunter Optiflow® ceramic fiber nozzle is used to deliver the molten magnesium into the nip between the caster rolls. Any area where liquid metal may potentially be exposed to the atmosphere is protected by a cover gas.

FATA Hunter Experience In Mg Rolling

The above advances in twin rolling casting (TRC) have allowed magnesium alloys to be directly cast into large coils that are in the range of 4 mm to 7 mm thick [3,4,5], but for twin roll casting to bring the price of wrought magnesium sheet to a level where it can economically be used for mass-market transportation applications a low cost way of rolling the TRC Mg coils must be developed.

In 2011 FATA Hunter designed, developed, and manufactured a reversing rolling mill, which demonstrated the feasibility of an industrial rolling process that not only economically reduces cast coils to the gauges required by consumer products, but also can modify the microstructure to improve the formability of the rolled sheet while maintaining a quality surface that requires minimal treatment after rolling. This new technology was implemented on a pilot rolling mill (Figures 10 and 11) built and tested under subcontract to the US Oak Ridge National Laboratories (ORNL) working in conjunction with Magnesium Elektron North America.

Figure 10. ORNL pilot rolling mill during pre-delivery testing in California.

Pre-delivery testing took place at a facility in Southern California, where the mill successfully rolled magnesium strip. The mill can accept material up to 0.5 inches thick by 10 inches wide and can isothermally process magnesium with symmetric or asymmetric roll speed control. The mill is configured for roller table feed with the option to upgrade to a coil-to-coil arrangement in the future.

Figure 11. Pilot magnesium rolling mill located at the Oak Ridge National Laboratories (photo courtesy of UT-Battelle ORNL).

Mg Pilot Mill Description

The main innovative features of this pilot mill were:

- Internally electrically heated and thermostatically controlled tool steel work rolls with a maximum working temperature of 300°C.
- Electric convective heating of roller table enclosures with a maximum working temperature of 530°C.

- Independent work roll drive system where the work rolls had independent high torque drives capable of producing a maximum asymmetry during rolling of 3:1.
- Pressure transducers on the hydraulic gap cylinders and torque sensors on each spindle to accurately measure the dynamic rolling forces.

The preliminary results from this mill showed that when AZ31B magnesium alloy was rolled with highly asymmetric roll speeds the strong basal texture normally found in Mg alloys is reduced and the modified microstructure improves low temperature formability, while maintaining good surface quality [6]. This mill received a "R&D 100 Award" in 2012 (Figure 12).

Figure 12. 2012 R&D 100 award.

Mg Complete Plant Description

Based on the magnesium casting experience with the pilot and full-size production casters and on the results from the pilot magnesium rolling mill, FATA Hunter has developed and designed a complete plant that is able to produce 30,000 tpa of low cost, good quality magnesium strip.

The plant incorporates all the equipment required to convert from magnesium ingot into coils of magnesium alloy at final gauge that have been leveled, cleaned and given a temporary coating to prevent corrosion during in-plant storage and handling. The plant is composed by the following major pieces of equipment:

- (6) Melting Furnaces complete with associated ingot destackers and drying ovens
- (3) Holding Furnaces
- (3) Twin Roll Casters
- (1) Coil Reheating oven
- (1) Rolling Mill
- (1) Finishing Line

If the plant is located near a smelter, then the melting furnaces can be substituted by liquid metal transfer systems.

Melting and Holding Furnaces

The magnesium ingots are melted in melting furnaces that are suitable for supplying a constant flow of clean and uniform temperature metal to the casting lines. The molten magnesium is transferred from the melting furnaces to the holding furnaces by means of electrically heated siphon tubes. The holding furnaces consist of an inner single chamber crucible that is externally electrically heated and supported in an outer insulted shell. The main function of the holding furnace is to stabilize the temperature and each holding furnace is equipped with a pump that is driven by a variable speed motor that adjusts its speed in response to the output from the caster headbox level sensor. This pump transfers the molten magnesium, at constant pressure and temperature, into the entry of a transfer tube that joins the holding furnace to the casting line. The transfer tube assembly is an electrically heated and insulated, nickel-free stainless steel transfer tube and it conveys the metal to an electrically heated and insulated headbox assembly.

Figure 14. Transfer tube and headbox configuration.

When the correct level and temperature are reached in the headbox, the molten magnesium is allowed to flow into a planar ceramic nozzle (the "tip") located between the rolls. The molten magnesium must be shielded from the atmosphere by a protective gas. Traditionally this gas was a mixture of dry air (or nitrogen) and Sulphur Hexafluoride (SF_6). Although SF_6 is highly effective, it is also one of the most potent and long-lasting greenhouse gases; for this reason the equipment is also designed to operate with alternate gases such as such as HFC-134a, sulphur dioxide (SO_2) etc.

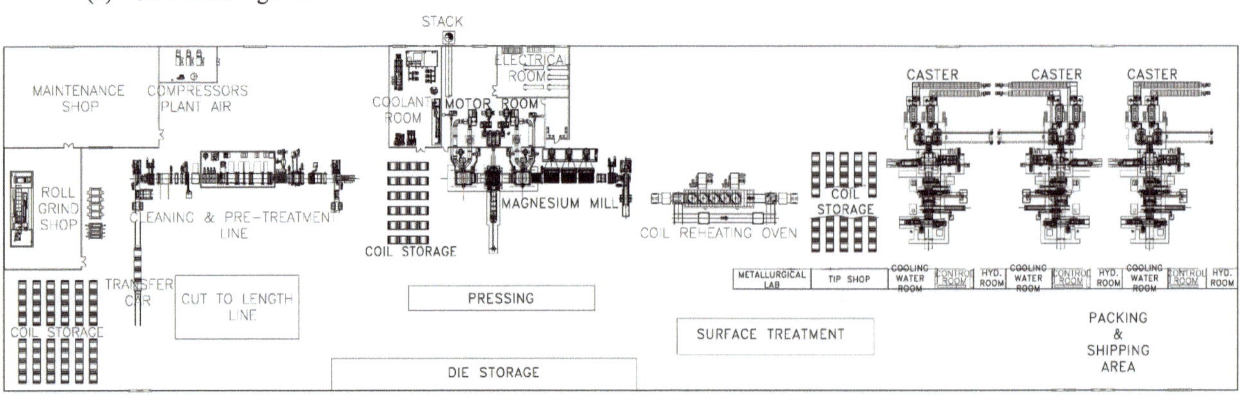

Figure 13. Plant lay-out of a Mg sheet plant.

Twin Roll Casters

The FATA Hunter SuperCaster® is a 2,000 mm wide twin roll casting machine that produces 2,500 mm outer diameter coils (16.6 tons) to minimize downstream head and tail losses and improve the overall plant efficiency. The casting line consists of the following main components and features:

- One piece heavy-duty housings provide the rigidity and high "Mill Modulus" necessary for good gauge response by the hydraulic gap AGC system.
- 1,220mm diameter caster rolls to minimize roll deflection and provide profile stability. Consistent profile control is important for most sheet applications and must be controlled for automotive applications.
- Rolls with an interlaced 60° cooling pattern that combines maximum neck strength and uniform neck modulus with minimal thermal roll eccentricity and the resultant cyclic gauge variations that can cause problems during rolling.
- 3,400 tons of rolling force to cast a wide variety of magnesium alloys and gauges.
- Each caster roll is driven by a separate variable frequency 110kW AC motor and epicyclic gearbox.
- Variable frequency 20kW AC motor driven winder with 600mm diameter expanding mandrel and automatically pre-positioned gripper slot for easy starting of new coils.
- Edge trimmer: Most magnesium alloys tend to form significant edge cracks during the twin roll casting process. These cracks would grow during subsequent processing and thus are normally removed before the rolling operation.

Coil Reheating Oven

Although the FATA Hunter rolling mill can reheat a coil from room temperature to rolling temperature, it is a slow process and the mill is designed to work with pre-heated coils. The coil reheating oven pre-heats the coils for delivery to the rolling mill. The coil reheat oven is a "pusher" style furnace where loading a cold coil at the input end causes a hot coil to exit from the output end. Each oven has the capacity to accommodate six as-cast 2,500mm OD x 2,000mm wide coils. Each oven consists of the following major components and features:

- One entry coil upender: the entry upending station includes a fixed coil rest for delivery of a eye horizontal coil by crane or fork truck from the storage area or directly from a caster.
- One pallet pusher.
- One reheat oven: the unit consists of an insulated "tunnel" through which the coils are pushed while heating and two attached insulated circulation fan/burner boxes. The circulation fans circulate hot air over the coils through ductwork and nozzles directed at the coils. The burners fire through "radiant" tubes in the circulation air stream separating the burner flame from the oven atmosphere. Coil processing time in the oven can be 10 to 18 hours depending on the anneal/homogenization/temperature requirement before rolling. When the coil reaches the exit of the oven it has been brought to a temperature between 300°C and 400°C
- One pallet extractor.
- One exit coil upender.
- One set of pallet conveyors to move the pallets from the exit of the oven to the entry after the coils are removed by the exit coil upender.

Rolling Mill

After being heated to the rolling temperature in the coil reheating oven, the magnesium coils are transferred to the FATA Hunter rolling mill where they are subjected to a series of warm rolling passes to reduce their thickness and refine their microstructure to meet the final product requirements.

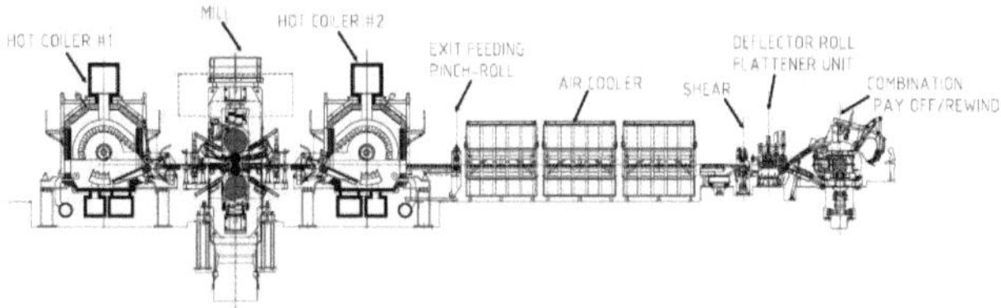

Figure 15. Magnesium rolling mill with a combination payoff and rewind reel.

Although many configurations of the basic warm rolling mill design are possible, the arrangement described and shown in Figure 15 utilizes a single-stand 4-Hi reversing mill with a hot coiler on each side of the stand and a separate combination Payoff/Rewind reel located on the right side of the mill. This configuration incorporates the following features and operating methodology:

- The combination Payoff and Rewind reel works in conjunction with a coil car to load the pre-heated coils onto the mill and then remove the coils after they have been rolled to final gauge. A pre-heated coil at as-cast gauge is loaded onto this reel by the coil car. This reel unwinds the coil into a pinch roll, shear and flattener assembly that conditions and straightens the leading edge of the strip. This strip is then fed through the (inactive) air cooler, through the right hand hot coiler enclosure, through the mill stand and into the left hand hot coiler.
- The left hand hot coiler grips the leading edge of the strip and tightly winds it. Both the left hand hot coiler and right hot coiler are surrounded by insulated enclosures that heat the coil using forced air convection. With this convective heating (versus radiant) no part of the coil can ever be heated above the air temperature, thereby precluding any possibility of ignition.
- Once the coil has been wound onto the left hand coiler and the temperature stabilized at the exact rolling temperature, the coil is fed back into the mill stand. The mill stand consists of two 60 tonne mill housings that contain the back up and work rolls and the high speed servo-hydraulically position and pressure controlled load cylinders that can exert more than 1500 tonnes of rolling force. As with the pilot mill, each of the work rolls incorporates a resistance heater located in the central axis. The work rolls are supported by water cooled chocks that fit within E-blocks on the mill housings for positive and negative roll bending and this, in combination with a full-face segmented induction heating system, provide roll profile control to maintain the flatness of the rolled strip.
- Each of the work rolls is driven by an individual 1000kW variable frequency AC motor and two speed reducer. These individual drives provide a maximum rolling speed of 300m/min and allow the same degree of asymmetric rolling as the pilot mill.
- Thickness gauges and high temperature shape rolls are located either side of the mill stand to monitor the thickness and flatness of the rolled strip. These work in combination with the mill automation system and the mill actuators to ensure uniform rolled thickness and good flatness.

- Upon exiting the mill stand the strip is wound onto the right hand hot coiler which applies the proper tension for tight wrapping while the left hand hot coiler provides the necessary back tension to the rolled strip. At the end of the pass the temperature is allowed to equilibrate and the strip is then passed back through the mill to the left hand hot coiler during the next rolling pass. This reversing rolling process continues until the strip reaches the desired final thickness. The speed, temperature and reduction of each pass are controlled by the mill automation system such that the coil reaches its final gauge and temperature on the left hand hot coiler.
- The strip is then fed through the open bite of the mill to the cooler where it is rapidly cooled to prevent grain growth and the development of any undesirable changes in texture thereby maintaining the enhanced physical properties of the fine grain sheet produced by the mill.
- The cooled strip is then wound onto the combination payoff/rewind reel from which it is removed by the coil car and moved to a storage area for final cooling to room temperature.

A more detailed description of the mill operation has been published by Light Metal Age[7].

Finishing Line

The as-rolled magnesium sheet must then be cleaned and chemically treated to prepare it for forming and subsequent painting. The proposed configuration of this line can be seen in the flow chart of Figure 16.

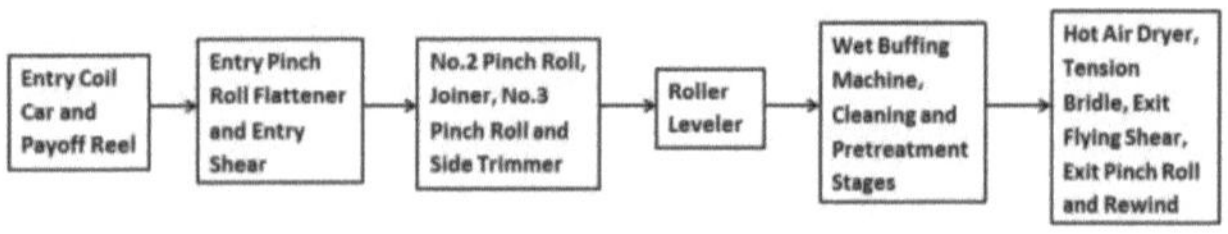

Figure 16. Flow chart of finishing line.

This line incorporates an edge trimmer and a leveling section to remove any edge cracking and shape errors after rolling. The main components of the line are:

- Side Trimmer: a twin rotary knife side trimmer removes edge cracks and defects from the incoming strip.
- Roller Leveler: the roller leveler flexes the strip back and forth over a series of rollers thereby equalizing the residual stresses within longitudinal strip fibers. This flexing process corrects any strip shape issue that may be present from previous processes.
- Wet Buffing Machine: it consists of two (2) top side and two (2) bottom side rotating brushes. This machine removes oxidation from the strip surface that may be present from previous processes and exposure to the atmosphere. At each brush, a back-up roll is fitted to support the strip during brushing, producing a stable and evenly cleaned strip surface. A recirculated cold water spray lubricates and cools the brushes to assure long brush life and remove any magnesium fines that may be generated.
- Cleaning Stages: strip cleaning is necessary to wash remaining particles, oxides, and other debris from the strip surface. It includes a four (4) stage cleaning section for efficient cleaning with minimal fluid consumption and waste discharge.
- Pretreatment Stages: a single spray coating stage is followed by a single final rinse stage. The spray solution for each stage is recirculated from a solution reservoir by a centrifugal pump via multiple spray bars mounted above and below the strip. Included in each recirculation loop is a heat exchanger to heat the solution to the optimal temperature for coating adhesion. A pair of squeegees rolls between the pretreatment and the rinse stages minimizes cross stage contamination.
- Hot Air Dryer: the hot air dryer evaporates the remaining moisture layer left on the strip after it has passed through the final squeegee rolls.

Conclusions

The commercialization of magnesium twin roll casting in 2004 significantly reduced the cost of reroll coil and resulted in the commonplace use of wrought magnesium sheet in personal electronics. However, the utilization of magnesium sheet in transportation applications has been limited by the high cost of warm rolling and the properties that can be obtained from low cost magnesium alloys. Recent developments have shown on a pilot scale that isothermal asymmetric rolling of twin roll cast coil can economically produce high quality wrought magnesium sheet with a formability and price that would make it technically and commercially viable for automotive applications.

This paper shows how the rolling technology developed on the pilot rolling mill can be implemented on an industrial scale to build a 30,000 tpa twin roll caster based magnesium sheet plant capable of reliably supplying high quality magnesium sheet to an automotive panel plant.

References

1. Peter Juchmann and Sebastien Wolff, "Mobile with Magnesium Sheet" (Paper presented at the 65th Annual World Magnesium Conference, Warsaw, Poland, 18-20 May, 2008), 75-84.

2. Faramarz Zarandi and Stephen Yue, *Magnesium Sheet; Challenges and Opportunities*, *Magnesium Alloys - Design, Processing and Properties* (14 January, 2011, INTECH Open Access Publisher).

3. Jae Joong Kim et al., "Wide Strip Casting Technology of Mg Alloys" (Paper presented at the Korea-Germany Workshop on Magnesium, Irsee, Germany, 6-9 February, 2012).

4. W.J. Park et al., "Wide Strip Casting Technology of Magnesium Alloys" (Paper presented at the 140th TMS Annual Meeting , San Diego, California, 27 February – 3 March, 2011), 143-146 (Magnesium Technology 2011).

5. I.H. Jung et al., "Mg Coil Production via Strip Casting and Coil Rolling Technologies" (Paper presented at the 136th TMS Annual Meeting, Orlando, Florida, 25 February – 1 March, 2007), 85-88 (Magnesium Technology 2007).

6. G. Muralidharan et al., "Shear Rolling of Magnesium Sheet for Automotive, Defense and Energy Applications" (ORNL Report TM-2012/541, Oak Ridge National Laboratory, 2014).

7. Roberto Passoni et al., "New Reversing Rolling Mill Can Make Magnesium Sheet Commercially Viable," Light Metal Age, 2 (2013), 20-22.

AUTHOR INDEX
Magnesium Technology 2015

L

M

N

O

P

Q

R

S

T

U

V

W

X

Y

Z

SUBJECT INDEX
Magnesium Technology 2015

M

N

O

P

Q

R

S

T

V

W

Y

Z

MIX
Papier aus verantwortungsvollen Quellen
Paper from responsible sources
FSC® C105338

If you have any concerns about our products,
you can contact us on
ProductSafety@springernature.com

In case Publisher is established outside the EU,
the EU authorized representative is:
Springer Nature Customer Service Center GmbH
Europaplatz 3, 69115 Heidelberg, Germany

Printed by Libri Plureos GmbH
in Hamburg, Germany